The chemistry of
organic silicon compounds
Volume 2

THE CHEMISTRY OF FUNCTIONAL GROUPS

*A series of advanced treatises under the general editorship of
Professors Saul Patai and Zvi Rappoport*

The chemistry of alkenes (2 volumes)
The chemistry of the carbonyl group (2 volumes)
The chemistry of the ether linkage
The chemistry of the amino group
The chemistry of the nitro and nitroso groups (2 parts)
The chemistry of carboxylic acids and esters
The chemistry of the carbon–nitrogen double bond
The chemistry of amides
The chemistry of the cyano group
The chemistry of the hydroxyl group (2 parts)
The chemistry of the azido group
The chemistry of acyl halides
The chemistry of the carbon–halogen bond (2 parts)
The chemistry of the quinonoid compounds (2 volumes, 4 parts)
The chemistry of the thiol group (2 parts)
The chemistry of the hydrazo, azo and azoxy groups (2 volumes, 3 parts)
The chemistry of amidines and imidates (2 volumes)
The chemistry of cyanates and their thio derivatives (2 parts)
The chemistry of diazonium and diazo groups (2 parts)
The chemistry of the carbon–carbon triple bond (2 parts)
The chemistry of ketenes, allenes and related compounds (2 parts)
The chemistry of the sulphonium group (2 parts)
Supplement A: The chemistry of double-bonded functional groups (3 volumes, 6 parts)
Supplement B: The chemistry of acid derivatives (2 volumes, 4 parts)
Supplement C: The chemistry of triple-bonded functional groups (2 volumes, 3 parts)
Supplement D: The chemistry of halides, pseudo-halides and azides (2 volumes, 4 parts)
Supplement E: The chemistry of ethers, crown ethers, hydroxyl groups
and their sulphur analogues (2 volumes, 3 parts)
Supplement F: The chemistry of amino, nitroso and nitro compounds and their derivatives
(2 volumes, 4 parts)
The chemistry of the metal–carbon bond (5 volumes)
The chemistry of peroxides
The chemistry of organic selenium and tellurium compounds (2 volumes)
The chemistry of the cyclopropyl group (2 volumes, 3 parts)
The chemistry of sulphones and sulphoxides
The chemistry of organic silicon compounds (2 volumes, 5 parts)
The chemistry of enones (2 parts)
The chemistry of sulphinic acids, esters and their derivatives
The chemistry of sulphenic acids and their derivatives
The chemistry of enols
The chemistry of organophosphorus compounds (4 volumes)
The chemistry of sulphonic acids, esters and their derivatives
The chemistry of alkanes and cycloalkanes
Supplement S: The chemistry of sulphur-containing functional groups
The chemistry of organic arsenic, antimony and bismuth compounds
The chemistry of enamines (2 parts)
The chemistry of organic germanium, tin and lead compounds

UPDATES

The chemistry of α-haloketones, α-haloaldehydes and α-haloimines
Nitrones, nitronates and nitroxides
Crown ethers and analogs
Cyclopropane derived reactive intermediates
Synthesis of carboxylic acids, esters and their derivatives
The silicon–heteroatom bond
Synthesis of lactones and lactams
Syntheses of sulphones, sulphoxides and cyclic sulphides
Patai's 1992 guide to the chemistry of functional groups — *Saul Patai*

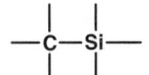

The chemistry of
organic silicon compounds

Volume 2

Part 2

Edited by

ZVI RAPPOPORT

The Hebrew University, Jerusalem

and

YITZHAK APELOIG

Technion–Israel Institute of Technology, Haifa

1998

JOHN WILEY & SONS
CHICHESTER–NEW YORK–WEINHEIM–BRISBANE–SINGAPORE–TORONTO

An Interscience® Publication

Copyright © 1998 John Wiley & Sons Ltd,
Baffins Lane, Chichester,
West Sussex PO19 1UD, England

National   01243 779777
International (+44) 1243 779777
e-mail (for orders and customer service enquiries): cs-books@wiley.co.uk
Visit our Home Page on http://www.wiley.co.uk
or http://www.wiley.com

All Rights Reserved. No part of this publication may be reproduced, stored in a retrieval system, or transmitted, in any form or by any means, electronic, mechanical, photocopying, recording, scanning or otherwise, except under the terms of the Copyright Designs and Patents Act 1988 or under the terms of a licence issued by the Copyright Licensing Agency, 90 Tottenham Court Road, London W1P 9HE, UK, without the permission in writing of the Publisher

Other Wiley Editorial Offices

John Wiley & Sons, Inc., 605 Third Avenue,
New York, NY 10158-0012, USA

WILEY-VCH Verlag GmbH, Pappelallee 3,
D-69469 Weinheim, Germany

Jacaranda Wiley Ltd, 33 Park Road, Milton,
Queensland 4064, Australia

John Wiley & Sons (Asia) Pte Ltd, Clementi Loop #02-01,
Jin Xing Distripark, Singapore 129809

John Wiley & Sons (Canada) Ltd, 22 Worcester Road,
Rexdale, Ontario M9W 1L1, Canada

British Library Cataloguing in Publication Data

A catalogue record for this book is available from the British Library

ISBN 0 471 96757 2

Typeset in 9/10pt Times by Laser Words, Madras, India
Printed and bound in Great Britain by Biddles Ltd, Guildford, Surrey
This book is printed on acid-free paper responsibly manufactured from sustainable forestry, in which at least two trees are planted for each one used for paper production.

Dedicated to

Robert West

a pioneer in silicon chemistry
and a dear friend

Contributing authors

Wataru Ando — Department of Chemistry, University of Tsukuba, Tsukuba, Ibaraki 305, Japan

Yitzhak Apeloig — Department of Chemistry and the Lise-Meitner Minerva Center for Computational Quantum Chemistry, Technion–Israel Institute of Technology, Haifa 32000, Israel

D. A. ('Fred') Armitage — Department of Chemistry, King's College London, Strand, London, WC2R 2LS, UK

Norbert Auner — Fachinstitut für Anorganische und Allgemeine Chemie, Humboldt-Universität zu Berlin, Hessische Str. 1–2, D-10115 Berlin, Germany

David Avnir — Institute of Chemistry, The Hebrew University of Jerusalem, Jerusalem 91904, Israel

Alan R. Bassindale — Department of Chemistry, The Open University, Milton Keynes, MK7 6AA, UK

Rosa Becerra — Instituto de Quimica Fisica 'Rocasolano', C/Serrano 119, 28006 Madrid, Spain

Johannes Belzner — Institut für Organische Chemie der Georg-August-Universität Göttingen, Tammannstrasse 2, D-37077 Göttingen, Germany

M. B. Boisen Jr — Department of Materials Science and Engineering, Virginia Tech, Blacksburg, VA 24061, USA

Mark Botoshansky — Department of Chemistry, Technion–Israel Institute of Technology, Haifa 32000, Israel

A. G. Brook — Lash Miller Chemical Laboratories, University of Toronto, Toronto, Ontario M5S 3H6, Canada

C. Chatgilialoglu — I. Co. C. E. A., Consiglio Nazionale delle Ricerche, Via P. Gobetti 101, 40129 Bologna, Italy

Buh-Luen Cheng — Department of Chemistry, Tsing Hua University, Hsinchu, Taiwan 30043, Republic of China

Nami Choi — Department of Chemistry, University of Tsukuba, Tsukuba, Ibaraki 305, Japan

Ernest W. Colvin — Department of Chemistry, University of Glasgow, Glasgow, G12 8QQ, UK

Contributing authors

Uwe Dehnert	Institut für Organische Chemie der Georg-August-Universität Göttingen, Tammannstrasse 2, D-37077 Göttingen, Germany
Robert Drake	Dow Corning Ltd, Cardiff Road, Barry, South Glamorgan, CF63 2YL, UK
Jacques Dubac	Hétérochimie Fondamentale et Appliqué, ESA-CNRS 5069, Université Paul-Sabatier, 118 route de Narbonne, 31062 Toulouse Cedex, France
Moris S. Eisen	Department of Chemistry, Technion–Israel Institute of Technology, Kiryat Hatechnion, Haifa 32000, Israel
C. Ferreri	Departimento di Chimica Organica e Biologica, Università di Napoli 'Federico II', Via Mezzocannone 16, 80134 Napoli, Italy
Toshio Fuchigami	Department of Electrochemistry, Tokyo Institute of Technology, 4259 Nagatsuta, Midori-ku, Yokohama 226, Japan
Peter P. Gaspar	Department of Chemistry, Washington University, St Louis, Missouri 63130-4899, USA
Christian Guérin	Chimie Moléculaire et Organisation du Solide, UMR-CNRS 5637, Université Montpellier II, Place E. Bataillon, 34095 Montpellier Cedex 5, France
G. V. Gibbs	Department of Materials Science and Engineering, Virginia Tech, Blacksburg, VA 24061, USA
T. Gimisis	I. Co. C. E. A., Consiglio Nazionale delle Ricerche, Via P. Gobetti 101, 40129 Bologna, Italy
Simon J. Glynn	Department of Chemistry, The Open University, Milton Keynes, MK7 6AA, UK
Norman Goldberg	Technische Universität Braunschweig, Institut für Organische Chemie, Hagenring 30, D-38106 Braunschweig, Germany
Edwin Hengge	(Deceased)
Reuben Jih-Ru Hwu	Department of Chemistry, Tsing Hua University, Hsinchu, Taiwan 30043, Republic of China
Jörg Jung	Institut für Organische Chemie der Justus-Liebig Universität Giessen, Heinrich-Buff-Ring 58, D-35392 Giessen, Germany
Peter Jutzi	Faculty of Chemistry, University of Bielefeld, Universitätsstr. 25, D-33615 Bielefeld, Germany
Yoshio Kabe	Department of Chemistry, University of Tsukuba, Tsukuba, Ibaraki 305, Japan
Menahem Kaftory	Department of Chemistry, Technion–Israel Institute of Technology, Haifa 32000, Israel
Inna Kalikhman	Department of Chemistry, Ben-Gurion University of the Negev, Beer Sheva 84105, Israel
Moshe Kapon	Department of Chemistry, Technion-Israel Institute of Technology, Haifa 32000, Israel

Contributing authors

Miriam Karni	Department of Chemistry and the Lise-Meitner Minerva Center for Computational Quantum Chemistry, Technion–Israel Institute of Technology, Haifa 32000, Israel
Mitsuo Kira	Department of Chemistry, Graduate School of Science, Tohoku University, Aoba-ku, Sendai 980-77, Japan
Sukhbinder S. Klair	Department of Chemistry, Loughborough University, Loughborough, Leicestershire, LE11 3TU, UK
Lisa C. Klein	Ceramics Department, Rutgers–The State University of New Jersey, Piscataway, New Jersey 08855-0909, USA
Daniel Kost	Department of Chemistry, Ben-Gurion University of the Negev, Beer Sheva 84105, Israel
Takahiro Kusukawa	Department of Chemistry, University of Tsukuba, Tsukuba, Ibaraki 305, Japan
R. M. Laine	Department of Chemistry, University of Michigan, Ann Arbor, Michigan 48109-2136, USA
David Levy	Instituto de Ciencia de Materiales de Madrid, C.S.I.C., Cantoblanco, 28049 Madrid, Spain
Larry N. Lewis	GE Corporate Research and Development Center, Schenectady, NY 12309, USA
Zhaoyang Li	Department of Chemistry, State University of New York at Stony Brook, Stony Brook, New York 11794-3400, USA
Paul D. Lickiss	Department of Chemistry, Imperial College of Science, Technology and Medicine, London, SW7 2AY, UK
Shiuh-Tzung Liu	Department of Chemistry, National Taiwan University, Taipei, Taiwan 106
Tien-Yau Luh	Department of Chemistry, National Taiwan University, Taipei, Taiwan 106
Gerhard Maas	Abteilung Organische Chemie I, Universität Ulm, Albert-Einstein-Allee 11, D-89081 Ulm, Germany
Iain MacKinnon	Dow Corning Ltd, Cardiff Road, Barry, South Glamorgan, CF63 2YL, UK
Svetlana Kirpichenko	Irkutsk Institute of Chemistry, Siberian Branch of the Russian Academy of Sciences, 1 Favorsky St, 664033 Irkutsk, Russia
Christoph Maerker	Laboratoire de Chimie Biophysique, Institut Le Bel, Université Louis Pasteur, 4 rue Blaise Pascal, F-67000 Strasbourg, France
Günther Maier	Institut für Organische Chemie der Justus-Liebig Universität Giessen, Heinrich-Buff-Ring 58, D-35392 Giessen, Germany
Michael J. McKenzie	Department of Chemistry, Loughborough University, Loughborough, Leicestershire, LE11 3TU, UK
Andreas Meudt	Institut für Organische Chemie der Justus-Liebig Universität Giessen, Heinrich-Buff-Ring 58, D-35392 Giessen, Germany

Philippe Meunier	Synthèse et Electrosynthèse Organométalliques, UMR-CNRS 5632, Université de Bourgogne, 6 Boulevard Gabriel, 21004 Dijon Cedex, France
Takashi Miyazawa	Photodynamics Research Center, The Institute of Physical and Chemical Research, 19-1399, Koeji, Nagamachi, Aoba-ku, Sendai 980, Japan
Thomas Müller	Fachinstitut für Anorganische und Allgemeine Chemie, Humboldt-Universität zu Berlin, Hessische Str. 1–2, D-10115 Berlin, Germany
Shigeru Nagase	Department of Chemistry, Faculty of Science, Tokyo Metropolitan University, Hachioji, Tokyo 192-03, Japan
Iwao Ojima	Department of Chemistry, State University of New York at Stony Brook, Stony Brook, New York 11794-3400, USA
Renji Okazaki	Department of Chemistry, School of Science, The University of Tokyo, Bunkyo-ku, Tokyo 113, Japan
Harald Pacl	Institut für Organische Chemie der Justus-Liebig Universität Giessen, Heinrich-Buff-Ring 58, D-35392 Giessen, Germany
Philip C. Bulman Page	Department of Chemistry, Loughborough University, Loughborough, Leicestershire, LE11 3TU, UK
Vadim Pestunovich	Irkutsk Institute of Chemistry, Siberian Branch of the Russian Academy of Sciences, 1 Favorsky St, 664033 Irkutsk, Russia
Stephen Rosenthal	Department of Chemistry, Loughborough University, Loughborough, Leicestershire, LE11 3TU, UK
Hideki Sakurai	Department of Industrial Chemistry, Faculty of Science and Technology, Science University of Tokyo, Yamazaki 2641, Noda, Chiba 278, Japan
Paul von Ragué Schleyer	Center for Computational Quantum Chemistry, The University of Georgia, Athens, Georgia 30602, USA
Ulrich Schubert	Institute for Inorganic Chemistry, The Technical University of Vienna, A-1060 Vienna, Austria
Helmut Schwarz	Institut für Organische Chemie der Technischen Universität Berlin, Straße des 17 Juni 135, D-10623 Berlin, Germany
Akira Sekiguchi	Department of Chemistry, Graduate School of Science, University of Tsukuba, Tsukuba, Ibaraki 305, Japan
A. Sellinger	Sandia National Laboratory, Advanced Materials Laboratory, 1001 University Blvd, University of New Mexico, Albuquerque, New Mexico 87106, USA
Hans-Ullrich Siehl	Abteilung für Organische Chemie I der Universität Ulm, D-86069 Ulm, Germany
Harald Stüger	Institut für Anorganische Chemie, Erzherzog-Johann-Universität Graz, Stremayrgasse 16, A-8010 Graz, Austria
Reinhold Tacke	Institut für Anorganische Chemie, Universität Würzburg, Am Hubland, D-97074 Würzburg, Germany

Contributing authors

Toshio Takayama	Department of Applied Chemistry, Faculty of Engineering, Kanagawa University, 3-27-1 Rokkakubashi, Yokohama, Japan 221
Yoshito Takeuchi	Department of Chemistry, Faculty of Science, Kanagawa University, 2946 Tsuchiya, Hiratsuka, Japan 259-12
Peter G. Taylor	Department of Chemistry, The Open University, Milton Keynes, MK7 6AA, UK
Richard Taylor	Dow Corning Ltd, Cardiff Road, Barry, South Glamorgan, CF63 2YL, UK
Norihiro Tokitoh	Department of Chemistry, School of Science, The University of Tokyo, Bunkyo-ku, Tokyo 113, Japan
Shwu-Chen Tsay	Department of Chemistry, Tsing Hua University, Hsinchu, Taiwan 30043, Republic of China
Mikhail Voronkov	Irkutsk Institute of Chemistry, Siberian Branch of the Russian Academy of Sciences, 1 Favorsky St, 664033 Irkutsk, Russia
Stephan A. Wagner	Institut für Anorganische Chemie, Universität Würzburg, Am Hubland, D-97074 Würzburg, Germany
Robin Walsh	The Department of Chemistry, The University of Reading, P O Box 224, Whiteknights, Reading, RG6 6AD, UK
Robert West	Department of Chemistry, University of Wisconsin at Madison, Madison, Wisconsin 53706, USA
Anna B. Wojcik	Ceramics Department, Rutgers–The State University of New Jersey, Piscataway, New Jersey 08855-0909, USA
Jiawang Zhu	Department of Chemistry, State University of New York at Stony Brook, Stony Brook, New York 11794-3400, USA
Wolfgang Ziche	Fachinstitut für Anorganische und Allgemeine Chemie, Humboldt-Universität zu Berlin, Hessische Str. 1–2, D-10115 Berlin, Germany

Foreword

The preceding volume in 'The Chemistry of Functional Groups' series, *The chemistry of organic silicon compounds* (S. Patai and Z. Rappoport, Eds), appeared a decade ago and was followed in 1991 by an update volume, *The silicon–heteroatom bond*. Since then the chemistry of organic silicon compounds has continued its rapid growth, with many important contributions in the synthesis of new and novel types of compounds, in industrial applications, in theory and in understanding the chemical bonds of silicon, as well as in many other directions. The extremely rapid growth of the field and the continued fascination with the chemistry of this unique element, a higher congener of carbon — yet so dramatically different in its chemistry — convinced us that a new authoritative book in the field is highly desired.

Many of the recent developments, as well as topics not covered in the previous volume are reviewed in the present volume, which is the largest in 'The Chemistry of Functional Groups' series. The 43 chapters, written by leading silicon chemists from 12 countries, deal with a wide variety of topics in organosilicon chemistry, including theoretical aspects of several classes of compounds, their structural and spectral properties, their thermochemistry, photochemistry and electrochemistry and the effect of silicon as a substituent. Several chapters review the chemistry of various classes of reactive intermediates, such as silicenium ions, silyl anions, silylenes, and of hypervalent silicon compounds. Multiple-bonded silicon compounds, which have attracted much interest and activity over the last decade, are reviewed in three chapters: one on silicon–carbon and silicon–nitrogen multiple bonds, one on silicon–silicon multiple bonds and one on silicon–hereroatom multiple bonds. Other chapters review the synthesis of several classes of organosilicon compounds and their applications as synthons in organic synthesis. Several chapters deal with practical and industrial aspects of silicon chemistry in which important advances have recently been made, such as silicon polymers, silicon-containing ceramic precursors and the rapidly growing field of organosilica sol–gel chemistry.

The literature covered in the book is mostly up to mid-1997.

Several of the originally planned chapters, on comparison of silicon compounds with their higher group 14 congeners, interplay between theory and experiment in organisilicon chemistry, silyl radicals, recent advances in the chemistry of silicon–phosphorous,–arsenic,–antimony and –bismuth compounds, and the chemistry of polysilanes, regrettably did not materialize. We hope to include these important chapters in a future complementary volume. The current pace of research in silicon chemistry will certainly soon require the publication of an additional updated volume.

We are grateful to the authors for the immense effort they have invested in the 43 chapters and we hope that this book will serve as a major reference in the field of silicon chemistry for years to come.

We will be grateful to readers who will draw our attention to mistakes and who will point out to us topics which should be included in a future volume of this series.

Jerusalem and Haifa
March, 1998

ZVI RAPPOPORT
YITZHAK APELOIG

The Chemistry of Functional Groups
Preface to the series

The series 'The Chemistry of Functional Groups' was originally planned to cover in each volume all aspects of the chemistry of one of the important functional groups in organic chemistry. The emphasis is laid on the preparation, properties and reactions of the functional group treated and on the effects which it exerts both in the immediate vicinity of the group in question and in the whole molecule.

A voluntary restriction on the treatment of the various functional groups in these volumes is that material included in easily and generally available secondary or tertiary sources, such as Chemical Reviews, Quarterly Reviews, Organic Reactions, various 'Advances' and 'Progress' series and in textbooks (i.e. in books which are usually found in the chemical libraries of most universities and research institutes), should not, as a rule, be repeated in detail, unless it is necessary for the balanced treatment of the topic. Therefore each of the authors is asked not to give an encyclopaedic coverage of his subject, but to concentrate on the most important recent developments and mainly on material that has not been adequately covered by reviews or other secondary sources by the time of writing of the chapter, and to address himself to a reader who is assumed to be at a fairly advanced postgraduate level.

It is realized that no plan can be devised for a volume that would give a complete coverage of the field with no overlap between chapters, while at the same time preserving the readability of the text. The Editors set themselves the goal of attaining reasonable coverage with moderate overlap, with a minimum of cross-references between the chapters. In this manner, sufficient freedom is given to the authors to produce readable quasi-monographic chapters.

The general plan of each volume includes the following main sections:

(a) An introductory chapter deals with the general and theoretical aspects of the group.

(b) Chapters discuss the characterization and characteristics of the functional groups, i.e. qualitative and quantitative methods of determination including chemical and physical methods, MS, UV, IR, NMR, ESR and PES — as well as activating and directive effects exerted by the group, and its basicity, acidity and complex-forming ability.

(c) One or more chapters deal with the formation of the functional group in question, either from other groups already present in the molecule or by introducing the new group directly or indirectly. This is usually followed by a description of the synthetic uses of the group, including its reactions, transformations and rearrangements.

(d) Additional chapters deal with special topics such as electrochemistry, photochemistry, radiation chemistry, thermochemistry, syntheses and uses of isotopically labelled compounds, as well as with biochemistry, pharmacology and toxicology. Whenever applicable, unique chapters relevant only to single functional groups are also included (e.g. 'Polyethers', 'Tetraaminoethylenes' or 'Siloxanes').

This plan entails that the breadth, depth and thought-provoking nature of each chapter will differ with the views and inclinations of the authors and the presentation will necessarily be somewhat uneven. Moreover, a serious problem is caused by authors who deliver their manuscript late or not at all. In order to overcome this problem at least to some extent, some volumes may be published without giving consideration to the originally planned logical order of the chapters.

Since the beginning of the Series in 1964, two main developments have occurred. The first of these is the publication of supplementary volumes which contain material relating to several kindred functional groups (Supplements A, B, C, D, E, F and S). The second ramification is the publication of a series of 'Updates', which contain in each volume selected and related chapters, reprinted in the original form in which they were published, together with an extensive updating of the subjects, if possible, by the authors of the original chapters. A complete list of all above mentioned volumes published to date will be found on the page opposite the inner title page of this book. Unfortunately, the publication of the 'Updates' has been discontinued for economic reasons.

Advice or criticism regarding the plan and execution of this series will be welcomed by the Editors.

The publication of this series would never have been started, let alone continued, without the support of many persons in Israel and overseas, including colleagues, friends and family. The efficient and patient co-operation of staff-members of the publisher also rendered us invaluable aid. Our sincere thanks are due to all of them.

The Hebrew University
Jerusalem, Israel

SAUL PATAI
ZVI RAPPOPORT

Contents

1. Theoretical aspects and quantum mechanical calculations of silaaromatic compounds ... 1
   Yitzhak Apeloig and Miriam Karni

2. A molecular modeling of the bonded interactions of crystalline silica ... 103
   G. V. Gibbs and M. B. Boisen

3. Polyhedral silicon compounds ... 119
   Akira Sekiguchi and Shigeru Nagase

4. Thermochemistry ... 153
   Rosa Becerra and Robin Walsh

5. The structural chemistry of organosilicon compounds ... 181
   Menahem Kaftory, Moshe Kapon and Mark Botoshansky

6. ^{29}Si NMR spectroscopy of organosilicon compounds ... 267
   Yoshito Takeuchi and Toshio Takayama

7. Activating and directive effects of silicon ... 355
   Alan R. Bassindale, Simon J. Glynn and Peter G. Taylor

8. Steric effects of silyl groups ... 431
   Jih Ru Hwu, Shwu-Chen Tsay and Buh-Luen Cheng

9. Reaction mechanisms of nucleophilic attack at silicon ... 495
   Alan R. Bassindale, Simon J. Glynn and Peter G. Taylor

10. Silicenium ions: Quantum chemical computations ... 513
    Christoph Maerker and Paul von Ragué Schleyer

11. Silicenium ions — experimental aspects ... 557
    Paul D. Lickiss

12. Silyl-substituted carbocations ... 595
    Hans-Ullrich Siehl and Thomas Müller

13. Silicon-substituted carbenes ... 703
    Gerhard Maas

14	Alkaline and alkaline earth silyl compounds—preparation and structure **Johannes Belzner and Uwe Dehnert**	779
15	Mechanism and structures in alcohol addition reactions of disilenes and silenes **Hideki Sakurai**	827
16	Silicon-carbon and silicon-nitrogen multiply bonded compounds **Thomas Müller, Wolfgang Ziche and Norbert Auner**	857
17	Recent advances in the chemistry of silicon-heteroatom multiple bonds **Norihiro Tokitoh and Renji Okazaki**	1063
18	Gas-phase ion chemistry of silicon-containing molecules **Norman Goldberg and Helmut Schwarz**	1105
19	Matrix isolation studies of silicon compounds **Günther Maier, Andreas Meudt, Jörg Jung and Harald Pacl**	1143
20	Electrochemistry of organosilicon compounds **Toshio Fuchigami**	1187
21	The photochemistry of organosilicon compounds **A. G. Brook**	1233
22	Mechanistic aspects of the photochemistry of organosilicon compounds **Mitsuo Kira and Takashi Miyazawa**	1311
23	Hypervalent silicon compounds **Daniel Kost and Inna Kalikhman**	1339
24	Silatranes and their tricyclic analogs **Vadim Pestunovich, Svetlana Kirpichenko and Mikhail Voronkov**	1447
25	Tris(trimethylsilyl)silane in organic synthesis **C. Chatgilialoglu, C. Ferreri and T. Gimisis**	1539
26	Recent advances in the direct process **Larry N. Lewis**	1581
27	Acyl silanes **Philip C. Bulman Page, Michael J. McKenzie, Sukhbinder S. Klair and Stephen Rosenthal**	1599
28	Recent synthetic applications of organosilicon reagents **Ernest W. Colvin**	1667
29	Recent advances in the hydrosilylation and related reactions **Iwao Ojima, Zhaoyang Li and Jiawang Zhu**	1687

30	Synthetic applications of allylsilanes and vinylsilanes **Tien-Yau Luh and Shiuh-Tzung Liu**	1793
31	Chemistry of compounds with silicon-sulphur, silicon-selenium and silicon-tellurium bonds **D. A. ('Fred') Armitage**	1869
32	Cyclic polychalcogenide compounds with silicon **Nami Choi and Wataru Ando**	1895
33	Organosilicon derivatives of fullerenes **Wataru Ando and Takahiro Kusukawa**	1929
34	Group 14 metalloles, ionic species and coordination compounds **Jacques Dubac, Christian Guérin and Philippe Meunier**	1961
35	Transition-metal silyl complexes **Moris S. Eisen**	2037
36	Cyclopentadienyl silicon compounds **Peter Jutzi**	2129
37	Recent advances in the chemistry of cyclopolysilanes **Edwin Hengge and Harald Stüger**	2177
38	Recent advances in the chemistry of siloxane polymers and copolymers **Robert Drake, Iain MacKinnon and Richard Taylor**	2217
39	Si-containing ceramic precursors **R. M. Laine and A. Sellinger**	2245
40	Organo-silica sol–gel materials **David Avnir, Lisa C. Klein, David Levy, Ulrich Schubert and Anna B. Wojcik**	2317
41	Chirality in bioorganosilicon chemistry **Reinhold Tacke and Stephan A. Wagner**	2363
42	Highly reactive small-ring monosilacycles and medium-ring oligosilacycles **Wataru Ando and Yoshio Kabe**	2401
43	Silylenes **Peter P. Gaspar and Robert West**	2463
Author index		2569
Subject index		2721

List of abbreviations used

Ac	acetyl (MeCO)
acac	acetylacetone
Ad	adamantyl
AIBN	azoisobutyronitrile
Alk	alkyl
All	allyl
An	anisyl
Ar	aryl
Bn	benzyl
Bz	benzoyl (C_6H_5CO)
Bu	butyl (also t-Bu or Bu^t)
CD	circular dichroism
CI	chemical ionization
CIDNP	chemically induced dynamic nuclear polarization
CNDO	complete neglect of differential overlap
Cp	η^5-cyclopentadienyl
Cp*	η^5-pentamethylcyclopentadienyl
DABCO	1,4-diazabicyclo[2.2.2]octane
DBN	1,5-diazabicyclo[4.3.0]non-5-ene
DBU	1,8-diazabicyclo[5.4.0]undec-7-ene
DIBAH	diisobutylaluminium hydride
DME	1,2-dimethoxyethane
DMF	N,N-dimethylformamide
DMSO	dimethyl sulphoxide
ee	enantiomeric excess
EI	electron impact
ESCA	electron spectroscopy for chemical analysis
ESR	electron spin resonance
Et	ethyl
eV	electron volt

List of abbreviations used

Fc	ferrocenyl
FD	field desorption
FI	field ionization
FT	Fourier transform
Fu	furyl(OC_4H_3)
GLC	gas liquid chromatography
Hex	hexyl(C_6H_{13})
c-Hex	cyclohexyl(C_6H_{11})
HMPA	hexamethylphosphortriamide
HOMO	highest occupied molecular orbital
HPLC	high performance liquid chromatography
i-	iso
Ip	ionization potential
IR	infrared
ICR	ion cyclotron resonance
LAH	lithium aluminium hydride
LCAO	linear combination of atomic orbitals
LDA	lithium diisopropylamide
LUMO	lowest unoccupied molecular orbital
M	metal
M	parent molecule
MCPBA	m-chloroperbenzoic acid
Me	methyl
MNDO	modified neglect of diatomic overlap
MS	mass spectrum
n	normal
Naph	naphthyl
NBS	N-bromosuccinimide
NCS	N-chlorosuccinimide
NMR	nuclear magnetic resonance
Pc	phthalocyanine
Pen	pentyl(C_5H_{11})
Pip	piperidyl($C_5H_{10}N$)
Ph	phenyl
ppm	parts per million
Pr	propyl (also i-Pr or Pr^i)
PTC	phase transfer catalysis or phase transfer conditions
Pyr	pyridyl (C_5H_4N)

R	any radical
RT	room temperature
s-	secondary
SET	single electron transfer
SOMO	singly occupied molecular orbital
t-	tertiary
TCNE	tetracyanoethylene
TFA	trifluoroacetic acid
THF	tetrahydrofuran
Thi	thienyl(SC_4H_3)
TLC	thin layer chromatography
TMEDA	tetramethylethylene diamine
TMS	trimethylsilyl or tetramethylsilane
Tol	tolyl(MeC_6H_4)
Tos or Ts	tosyl(p-toluenesulphonyl)
Trityl	triphenylmethyl(Ph_3C)
Xyl	xylyl($Me_2C_6H_3$)

In addition, entries in the 'List of Radical Names' in *IUPAC Nomenclature of Organic Chemistry*, 1979 Edition, Pergamon Press, Oxford, 1979, p. 305–322, will also be used in their unabbreviated forms, both in the text and in formulae instead of explicitly drawn structures.

CHAPTER **16**

Silicon–carbon and silicon–nitrogen multiply bonded compounds

THOMAS MÜLLER[†], WOLFGANG ZICHE and NORBERT AUNER[†]

Fachinstitut für Anorganische und Allgemeine Chemie der Humboldt Universität Berlin, D 10115 Berlin, FRG
Fax: +49-30-20936966; e-mail: h0443afs@joker.rz.hu-berlin.de, wolfgang=ziche@chemie.hu-berlin.de, norbert=auner@chemie.hu-berlin.de

I. SILENES	859
A. Synthesis	860
1. Cycloreversion reactions	860
2. Salt elimination	876
3. Donor cleavage	879
4. Isomerization of acylpolysilanes	880
5. Silenes by the sila-Peterson reaction	884
a. Reaction of silyllithium reagents with ketones or aldehydes	885
b. Reaction of polysilylacylsilanes with organometallic reagents	888
c. Deprotonation of polysilylcarbinols	889
d. Reaction of 2-siloxysilenes with organometallic reagents	890
6. Photolysis of disilanes	891
7. Silenes by rearrangement of silylenes and carbenes	900
8. Other group 14 carbon double bonded species	909
B. Reactivity	910
1. Rearrangements	910
a. Silene-to-silylene interconversions	910
b. Silene-to-silene isomerization	911
c. Intramolecular insertion reactions	914

[†] Present address: Institut fur Anorganische Chemie der J.W.-Goethe Universität Marie-Curie St. 11, D-60438 Frankfurt am Main, Germany

The chemistry of organic silicon compounds, Vol. 2
Edited by Z. Rappoport and Y. Apeloig © 1998 John Wiley & Sons Ltd

2. Dimerizations	916
a. Head-to-head dimerization	919
b. Head-to-tail dimerization	931
3. Nucleophilic additions	932
a. Reaction with alcohols	932
b. Reaction with alkoxysilanes	937
c. Reaction with organometallic reagents	937
4. Cycloaddition reactions	940
a. Wiberg-type silenes	940
i. $Me_2Si=C(SiMe_3)_2$	940
α. [2 + 2] Cycloaddition reactions	940
β. [2 + 4] Cycloaddition and ene reactions	941
ii. [2 + 3] Cycloaddition reactions	942
iii. $Ph_2Si=C(SiMe_3)_2$	943
iv. $t\text{-}Bu_2Si=C(SiMe_3)_2$, $Me_2Si=C(SiMe_3)[SiMe(t\text{-}Bu)_2]$ (**104a**) and **104a** ·donor	943
b. Neopentylsilenes: $R_2Si=C(R')-CH_2Bu\text{-}t$	945
i. Diorgano-substituted neopentylsilenes	945
ii. π-Donor-substituted neopentylsilenes	945
c. Brook-type silenes	949
i. [2 + 1] Cycloaddition reactions	949
ii. Cycloadditions with alkenes, dienes and alkynes	953
d. Apeloig–Ishikawa–Oehme-type silenes	957
e. Cycloaddition reactions with carbonyl compounds and derivatives	958
f. Miscellaneous silenes	970
i. [2 + 2] Cycloaddition reactions	970
ii. [2 + 3] Cycloaddition reactions	972
iii. [2 + 4] Cycloaddition reactions	973
5. Ene reactions	974
6. Oxidations	977
7. Miscellaneous	980
C. Structure and Spectroscopic Properties of Silenes	981
1. Structural studies	981
2. ^{29}Si and ^{13}C NMR spectroscopic data	985
a. ^{29}Si NMR spectroscopy	985
b. ^{13}C NMR spectroscopy	991
3. Infrared spectroscopic data	991
4. UV spectroscopic data	996
5. Miscellaneous	996
II. SILAALLENES	997
A. Theoretical Studies	997
B. Synthesis	998
C. Reactivity	1000
1. 1-Silaallenes	1000
2. 2-Silaallenes	1005
III. SiC TRIPLY BONDED COMPOUNDS: THEORETICAL RESULTS	1008
IV. SILICON–NITROGEN MULTIPLY BONDED COMPOUNDS	1010
A. Introduction	1010
B. Synthesis	1010
1. Si=N and Si≡N systems	1010

a. Salt elimination	1010
b. Cycloreversion reactions	1012
c. Photolysis and pyrolysis of silyl azides	1017
d. Retro-ene reactions	1022
e. Dehydrochlorination	1025
f. Silylene addition to azides	1025
g. Reaction of SiH_4 and N_2	1026
h. Ionic Si=N compounds by gas phase reactions	1026
i. Silicon-containing anions	1026
α. [HSiNH]⁻: an anion related to silanimine $H_2Si=NH$	1027
β. Silaformamide ion [HSi(O)NH]⁻	1028
ii. Silicon-containing cations	1028
2. N=Si=N systems	1029
a. Silanediimines	1029
b. Silaamidides [(RN)₂SiR′]⁻	1030
3. Si=N metal systems	1032
C. Reactivity	1033
1. Donor addition	1033
2. Insertion reactions	1034
3. Ene reactions	1035
4. Cycloaddition reactions	1035
a. [2 + 2] Cycloadditions	1035
b. [2 + 3] Cycloadditions	1038
c. [2 + 4] Cycloadditions	1040
5. Metal silanimine complexes	1041
D. X-ray Structures	1042
E. Spectroscopy	1043
1. NMR spectroscopy	1043
2. IR and UV-Vis spectroscopy	1044
3. Miscellaneous	1046
F. Theoretical Studies	1046
G. GeN, SiP and SiAs Multiply Bonded Species and Related Compounds	1051
V. ACKNOWLEDGEMENTS	1053
VI. REFERENCES	1054

I. SILENES

Three decades after the publication of landmark papers by Gusel'nikov and Flowers[1,2] reporting evidence for a silicon–carbon doubly bonded species, the status of silenes has been changed from that of a laboratory curiosity to a not uncommon chemical compound. An impressive amount of information concerning silenes and the behaviour of the Si=C double bond has been discovered and published. While the early days of silene chemistry were marked by gas-phase studies, in the last fifteen years the emphasis of silene chemistry has moved from gas-phase investigations to studies of the reactivity and the physical properties of silenes in the condensed phase. Clearly, the isolation and structural characterization of stable silenes by Brook and coworkers[3] and subsequently by Wiberg and coworkers[4] are the highlights in the short history of silene chemistry, which are adequately followed by the recent synthesis of an air-stable 1-silaallene by West and coworkers[5] and the isolation of the first room-temperature stable silaaromatic compound, the 2-silanaphthalene, by Tokitoh, Okazaki and coworkers[6]. Reactivity studies of silenes and the isolation and characterization of novel organosilicon compounds arising from silene chemistry have been the focus of research in the late eighties and early nineties. Although silenes will, due to their inherent kinetic instability, never play such a dominant role in organosilicon chemistry as alkenes do in organic chemistry, there will be an increasing number of novel compounds and new materials arising from silene chemistry. Physical organic chemistry of silicon, which was restricted for a long time to gas-phase chemistry, has now been developed as an important tool, suggesting detailed mechanisms for the reactivity of silenes also in the condensed phase. Probably no other field of chemistry has been influenced so much by theory as the chemistry of multiple bonded silicon. Silene chemistry has been especially inspired from its infancy until now by the fruitful interplay between theory and experiment.

All these developments of the last thirty years have been well summarized in several accounts. The reviews by Raabe and Michl[7,8] cover most of the work in the silene area till 1988 and the review by Brook and Brook[9] extends this to 1994. The theoretical implications on silene chemistry are reviewed in several articles by Gordon[10a], Grev[10b], Apeloig[11] and up to 1996 by Karni and Apeloig[12]. The photochemistry of organosilanes has received considerable attention in recent years and has also been well reviewed. *Organosilane photochemistry* by Steinmetz is a review that covers the literature from the mid-1980s till 1994[13]. The photolysis of silylene precursors is treated in a review by Gaspar and coworkers[14].

A. Synthesis

1. Cycloreversion reactions

Most of the pioneering and fundamental work of the last thirty years in silene chemistry has been conducted in the gas phase under thermolytic or photolytic conditions. The results of the last ten years in this area of silene chemistry are summarized together with related earlier reports in the next sections of the chapter. For more complete information, which includes the literature before 1987, the interested reader is referred in addition to the previous reviews by Raabe and Michl[7,8] and to references given therein.

The classical pyrolytic source of silenes is the [2 + 2] cycloreversion of silacyclobutanes performed at 700–1000 K at low pressure. Due to the high temperature employed, generally only silenes with small substituents (H, Me, vinyl) can be generated. The classical first evidence for the formation of a transient silene came from the thermolysis of dimethylsilacyclobutane **1**, which resulted in fragmentation into 1,1-dimethylsilene **2** and ethene (equation 1)[1,2]. This reaction proceeds for 1,1-disubstituted silacyclobutanes via an initial homolytic cleavage of a C−C bond followed by extrusion of ethene[15,16]. This mechanism is also favored by recent high level ab initio calculations[16b].

$$\text{Me}_2\text{Si}\square \xrightarrow{\Delta} \text{Me}_2\text{Si}=\text{CH}_2 + \text{H}_2\text{C}=\text{CH}_2 \quad (1)$$

(1) **(2)**

This fragmentation mode is not altered for silacyclobutanes bearing a vinyl group at the silicon[17], as the same Arrhenius parameters are found for the decomposition of **1** and of 1-methyl-1-vinylsilacyclobutane **3** ($\log A = 15.64 \text{ s}^{-1}$, $E_A = 62.6 \text{ kcal mol}^{-1}$), in sharp contrast to the pyrolysis of cyclobutanes where a vinyl group accelerates the pyrolysis by a factor of nearly 600[18]. 2-Silabuta-1,3-diene **4** was produced in a laser-photosensitized (SF$_6$) decomposition (LPD) of 1-methyl-1-vinylsilacyclobutane **3**

16. Silicon–carbon and silicon–nitrogen multiply bonded compounds 861

(equation 2)[19] which, under these conditions, gave polymeric poly(silaisoprene) lacking any carbon–carbon double bonds. The measured Arrhenius parameters ($\log A$ (s^{-1}) = 16.0, $E_A = 59.7$ kcal mol^{-1} for the decomposition reaction are very close to those obtained in the low pressure pyrolysis[18] of **3**, suggesting that the rate-determining step of LPD is the same as that of the pyrolysis, which is assumed to be the cleavage of a C–C bond[19].

$$\text{(3)} \xrightarrow{\text{LPD}} \text{(4)} \tag{2}$$

Conlin and coworkers have prepared (E)- and (Z)-1,1,2,3-tetramethylsilacyclobutanes **5** and have studied the mechanism of their thermal decomposition in order to gain insight into the stereochemistry of the thermal decomposition of silacyclobutanes[20]. The occurrence of transient 1,4-biradicals like **6** in [2 + 2] fragmentations is accompanied by a loss of the reactant stereochemistry. This can be rationalized by rotational processes in the diradical **6** (**6a** → **6b**) which compete effectively with the β-scission steps yielding the silene **2** and E/Z 2-butene **7** (equation 3).

$$\text{Z-(5)} \quad \text{E-(5)}$$
$$\downarrow \Delta \qquad \downarrow \Delta$$
$$\text{(6a)} \rightleftharpoons \text{(6b)} \tag{3}$$
$$\downarrow \qquad \downarrow$$
$$\text{(2)} \quad \text{Z-(7)} \qquad \text{(2)} \quad \text{E-(7)}$$

TABLE 1. Kinetic data for the thermal decomposition of silacyclobutanes

Compound	log A (s^{-1})	E_A (kcal mol^{-1})	$10^4 k$ (s^{-1})	k_{rel}	Reference
(1) Me$_2$Si-cyclobutane	15.64	62.6	0.20	1	2, 21
(12) Me$_2$Si-cyclobutane, 2-Me	15.45	60.6	0.53	2.7	20
(13) Me$_2$Si-cyclobutane, 3-Me	16.39	63.3	0.62	3.1	16
Z-(5) Me$_2$Si-cyclobutane, 2,3-diMe (Z)			7.0	35	20
E-(5) Me$_2$Si-cyclobutane, 2,3-diMe (E)			1.97	10	20
(14) MeHSi-cyclobutane	14.9	59.1			22
(3) Me(vinyl)Si-cyclobutane	15.64	62.6			18
	16.0	59.7			19

Separate pyrolysis of (E)-5 and (Z)-5 led to the same products 2, 7–11 in equation 4 in slightly different ratios: the major products were trimethylsilene 8 and propene 9, indicating preferred cleavage of the most highly substituted C—C bond[20]. Comparison (see Table 1) with kinetic data for the pyrolysis of other small silacyclobutanes like 1 or 12–13 reveals an activating effect of the methyl substituents in 5. The stereospecificity of the 2-butene formation is high in both cases: the fragmentation exhibits a fivefold preference for retention in the 2-butene fragment, indicating that the Si—C bond breaking in 6 is fast compared with bond rotation, if a diradical intermediate is assumed[20]. The results have been compared with pyrolysis data of substituted cyclobutanes. Silacyclobutanes were found to react faster than cyclobutanes and the activating effect of the methyl groups on the decompostion rate is larger in the former. Thus, ring splitting across the more substituted bond is more dominant in the silacyclobutane than in the cyclobutane series[20].

The very similar Arrhenius parameters measured for methylsilacyclobutane **14** suggest that **14** also decomposes by a multistep biradical mechanism similar to **1** and not by a unimolecular reaction as was originally proposed[22].

In contrast, in the excited state the primary cleavage mechanism in silacyclobutanes like **5** involves the breaking of a silicon–carbon bond[23]. The initially formed silyl radicals **15** and **16** are stabilized by an intramolecular disproportionation reaction giving the silenes **17** and **18** and the homoallylsilane **19**. **17** and **18** were identified by their trapping products (**20**, **21**) with methanol (equation 5)[23]. From pyrolysis of Z-**5** a different set of products from 1,4-diradical disproportionation is obtained, which can be attributed to predominant cleavage of the carbon–carbon bond[23].

The laser flash photolysis of gaseous silacyclobutanes **22**[24] and **23**[25] and 1,3-disilacyclobutane **24** produced the transient silenes **25** (from **22** and **24**) and **26** (from **23**) as the major primary product. The silenes **25** and **26** were identified by their UV spectra with $\lambda_{max} \approx 260$ nm. Rate constants for the decay processes of the transient silenes were also measured.

Grobe, Auner and coworkers studied the thermal decomposition of **22**, **1** and methylsilacyclobutane **14** under low pressure flow pyrolysis conditions[26]. They characterized the transient silenes **2**, **25** and **26** by mass spectrometric methods and by low temperature NMR spectroscopy of the adducts **27**–**29** of the silenes with hexadeuteriomethyl ether (equation 6)[27].

For the generation of the parent silene $H_2Si=CH_2$ **25** the vacuum flash photolysis of 5,6-bis(trifluoromethyl)-2-silabicyclo[2.2.2]octa-5,7-diene **30** was shown to be the most favourable approach. IR and UV spectra of matrix isolated **25** have been reported[28,29] and the PES spectra[30] of **25** have also been obtained in the gas phase using **30** as precursor. Only recently Bogey, Bürger and coworkers obtained by pyrolysis of **30** a millimeter microwave spectrum from **25** in the gas phase, thereby characterising for the first time the transient **25** by high resolution spectroscopy[31–33]. Other precursors like silacyclobutane **22** or 1,3-disilacyclobutane **24** were also employed (equation 7), but the required temperatures for the generation of **25** from both molecules were much higher[31]. The structural details of **25** resulting from this study will be discussed in Section I.C.1.

Leigh and coworkers used laser flash photolysis to generate transient 1,1-diphenylsilene **31** from silacyclobutane **32** and measured its UV absorption[34,35]. The method was used to determine Arrhenius parameters for the addition of nucleophiles such as alcohols to **31**[25]

16. Silicon–carbon and silicon–nitrogen multiply bonded compounds

In the absence of a trapping agent the head-to-tail dimer **33** was isolated (equation 8).

Pyrolysis of the bis-silacyclobutane **34** yields under chemical vapour deposition conditions polymeric material **37** (equation 9)[38]. **37** is the mixed product of the polydimerization of the intermediate bis-silene **35** and of the ring-opening polymerization of the benzosilacyclobutane **36**. **36** is formed by a sigmatropic 1,3-hydrogen shift from an aryl carbon to the sp^2 silene silicon atom in **35**, that results in a 1,4-diradical, which closes to **36**. This surprising rearrangement could be verified for the thermolysis of phenylsilacyclobutane **38**, which gives as one of the main products the benzo-annelated silacyclobutene **39** (equation 10)[38].

$$ (10) $$

Conlin and Bobbitt thermolyzed 3-vinyl-1-silacyclobutane **40** in order to liberate the silene **41**. In addition to the expected silene trapping products like **42**, several products **44–46** resulting from the biradical **43** were found (equation 11)[37].

$$ (11) $$

While several highly substituted 1,2-disilacyclobutanes are known to revert under very mild conditions to silenes[38,39], it is generally believed that 1,3-disilacyclobutanes need more drastic conditions to undergo the cycloreversion yielding silenes[40]. Kinetic data for the pyrolyses of several 1,3-disilacyclobutanes (**24**, **47**, **48**) have been reported by Davidson and coworkers and are summarized in Table 2[40]. Silene formation was inferred from detection of trapping products with TMSOMe and HCl. It was found that methyl substitution at silicon slows down the pyrolysis rate. The initial process for the decomposition of

TABLE 2. Kinetic data for the thermal decomposition of 1.3-disilacyclobutanes.[40].

Compound	log A (s^{-1})	E_A (kcal mol^{-1})	K_{rel}
H₂Si–SiH₂ ring (24)	13.3	55.0	1220
MeHSi–SiHMe ring (47)	13.5	61.0	50
Me₂Si–SiMe₂ ring (48)	14.4	70.8	1

hydrogen substituted 1,3-disilacyclobutanes was shown to be the 1,2-hydrogen shift from silicon to carbon which results in a complex decomposition mechanism. Both modes of Si—C bond breaking processes, endo- and exocyclic, are shown to occur in the pyrolysis of the tetramethyl derivative **48**[40].

$$H_2Si\diamond SiH_2 \quad MeHSi\diamond SiHMe \quad Me_2Si\diamond SiMe_2$$

(**24**)   (**47**)   (**48**)

The cleavage of 1,1,3,3-tetraphenyl-2,4-dineopentyl-1,3-disilacyclobutane E/Z-**49** probably gives a 1,4-biradical species **50** in the first step, which may fragment to give silene **51** or reclose to the ring compound[41]. A competing reaction is the disproportionation of the diradical to give **52**. The silene **51** may be trapped to give the species **53** (equation 12).

When one of the phenyl groups is replaced by a vinyl group, a silaallylic diradical intermediate is formed, which produces a six-membered ring silene[42].

Okazaki and coworkers showed that photolysis of benzosilacyclobutenes **54** results in the formation of *ortho*-silaquinonoid compounds **55** (equation 13)[43].

Similarily, the intriguing *ortho* quinodisilane **56** was suggested as the key intermediate in the thermal reaction of benzodisilacyclobutene **57**, which gives a mixture of trisilacyclopentene **58** and the dibenzo-1,4-disilacyclohexadiene **59**[44]. Earlier reports of the formation of the dibenzotetrasilacycloocta-4,7-diene **60**[45,46] were based on wrong structural assignments[44]. The formation of product **58** can be explained in terms of a [4 + 2] dimerization of **56**, followed by elimination of benzosilacyclopropene **61** which would undergo dimerization to **59**[44]. Further evidence for the intermediate **56** was provided by the isolation of its trapping product with *t*-butanol **62** and with other trapping reagents (equation 14)[45,46].

Interestingly, **56** cannot be obtained photochemically from the benzodisilacyclobutene **57**; instead silastyrene **64** is formed as an intermediate[46]. **64** is produced by homolytic bond scission of the Si—Si bond in **57** followed by intramolecular disproportionation of the resulting biradical **65**. **64** can be intercepted by deuteriated *t*-butanol giving the silanol ether **66** in which the deuterium is incorporated in one ethyl group next to the *t*-butoxy group. In the absence of a scavenger reagent the disilane **67** is formed in 64% yield. The reaction proceeds in a somewhat unusual head-to-head coupling to the carbon centered biradical **68**, which abstracts hydrogens to give product **67** as shown (equation 15)[46] (for an alternative mechanism see Reference 9).

(12)

R=Me, Et
E=H, SiMe$_3$, SiH(OEt)$_2$

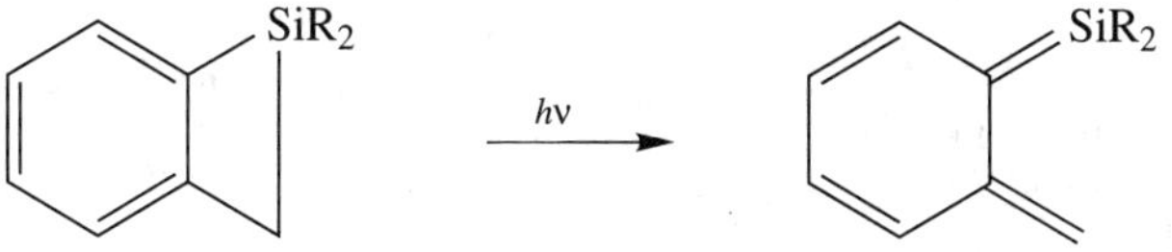

16. Silicon–carbon and silicon–nitrogen multiply bonded compounds 869

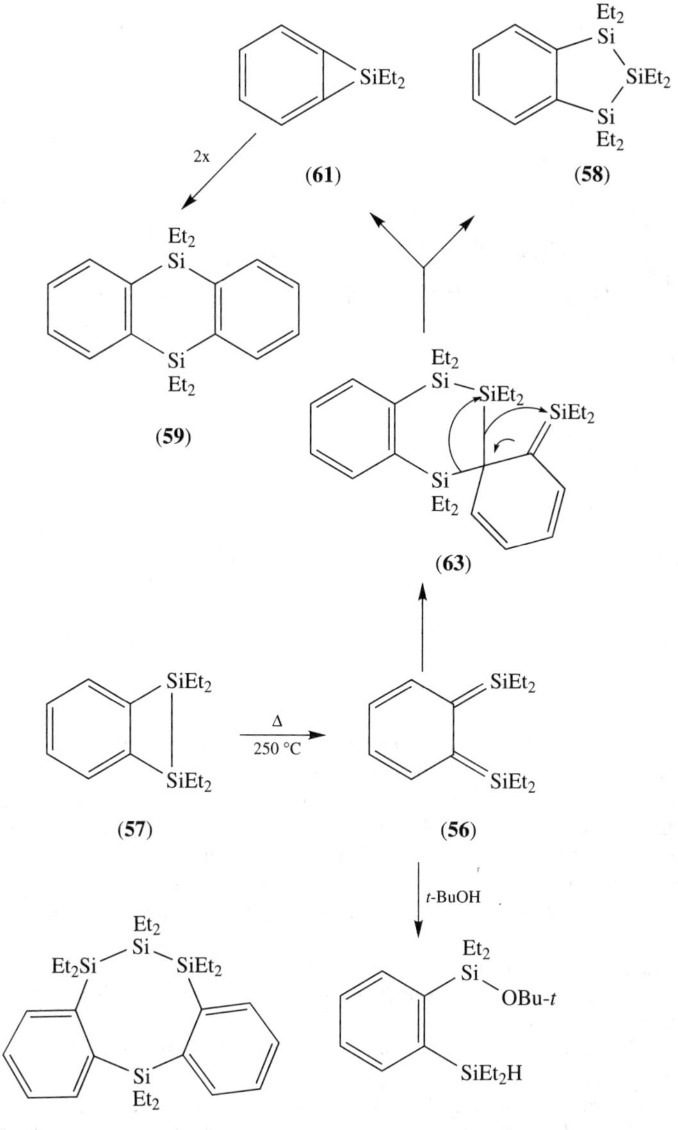

(13)

(14)

(15)

The ring opening of **57** can be achieved thermolytically at milder conditions using catalytic amounts of Ni[0][47], Pd[0][44] or Pt[0][48] catalysts. Under these conditions the reaction is believed to proceed via a metal[0] bissilene complex **69**. Depending on the metal used, different products **60**, **70** and **71** have been isolated from the thermolysis in benzene (equation 16). Catalyzed ring-opening reactions of **57** in the presence of carbonyl compounds[49], alkenes[50], dienes[50] and acetylenes[51] have also been reported.

Photolysis of 2,3-disiloxetanes **72** liberates the corresponding silene **73** and silanone **74** in a retro-[2 + 2] reaction[52]. The transient species are trapped by ethanol to give **75** and **76**. In the absence of traps the silene undergoes a 1,3-hydrogen shift to give **77**. Ene reaction of the initially formed intermediates, or cleavage of the 1,3-disiloxetane **78** yields the silyl ether **79** (equation 17).

The irradiation of a 1,2-disila-3-thietane gives products from a [2 + 2] cycloreversion, which may be trapped by ethanol[53].

Conlin and Namavari pyrolyzed dimethyl-1-silacyclobut-2-ene **80** in order to obtain 1,1-dimethylsilabutadiene **81**. In the presence of excess ethene two silacyclohexenes, **82** and **83**, were obtained in nearly quantitative yield (equation 18)[54].

16. Silicon–carbon and silicon–nitrogen multiply bonded compounds 871

(17)

Steinmetz and coworkers carried out mechanistic studies on the far-UV photochemical ring opening of 1-silacyclobut-2-ene **80**. The intermediates were trapped by alcohols to give **84–87** and by methoxytrimethylsilane to give **88** and **89**[55]. The main reaction is the formation of 1-silabuta-1,3-diene **81**, while the formation of silene **2**, probably via the carbene **90**, is a minor reaction (equation 19). The mechanism suggested was supported by deuterium labelling studies and *ab initio* calculations.

The main reaction occuring in the gas-phase pyrolysis of diallyldimethylsilane **91** is the retro-ene elimination of propene with formation of 1-silabuta-1,3-diene **81** followed by a ring closure reaction to yield the 1-silacyclobut-2-ene **80** (equation 20)[56,57]. In the presence of methanol the adduct of **81** was observed[57]. The Arrhenius parameters for the retro-ene fragmentation of **91** ($\log A$ (s^{-1}) = 11.2, E_A = 47.6 kcal mol^{-1})[57] are very similar to those for the pyrolysis of the carbon analogue hepta-1,6-diene[58].

16. Silicon–carbon and silicon–nitrogen multiply bonded compounds 873

The 'Wiberg'-type silenes like **92**, available through salt elimination reactions from **93**, react with nonenolisable aldehydes, ketones and the corresponding imino derivatives to give in a first step donor adducts **94**[59], which are then transformed to the [2 + 2] and [2+4] cycloadducts **95** and **96**, respectively (equation 21)[60−62]. These cycloadducts may liberate the silene **92** upon heating and it can be trapped by suitable reagents.

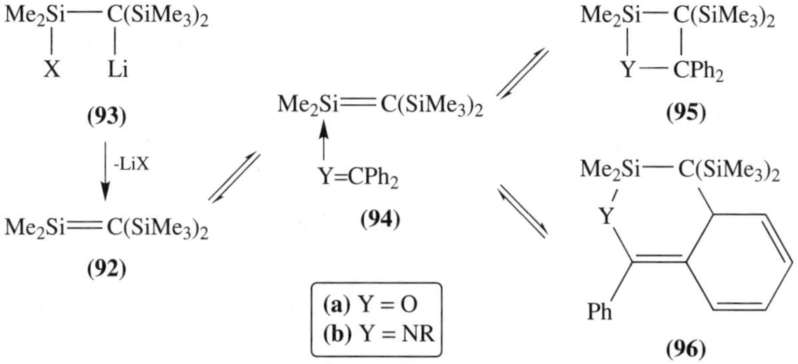

(a) Y = O
(b) Y = NR

(21)

Silene $Ph_2Si=C(SiMe_3)_2$ **97** reacts with benzophenone and gives the [2 + 4] cycloproduct **98**, probably via the donor adduct **99**. **98** is, however, not a silene precursor but decomposes to the alkene **100** and diphenylsilanone **101** (equation 22)[63,64].

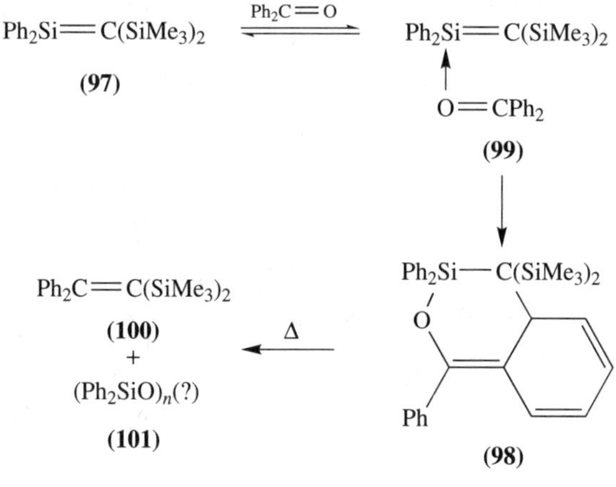

(22)

The primary product from the reaction of **97** with $Me_3Si-N=CPh_2$ is the [2+4] adduct **102**, which, above 80 °C, equilibrates with the [2+2] adduct **103**. **102** as well as **103** serve as silene sources when heated (equation 23)[63,64]. Trapping reactions and rearrangements of **97** are discussed in more detail below.

The attempted synthesis of **104** from its LiF adduct by salt elimination leads exclusively to the silene **104a**[65,66]. **104a** can be reacted with benzophenone to give the [2 + 4] and [2 + 2] cycloadducts **105** and **106**[67]. The [2 + 4] cycloadduct of silene **104** cannot be obtained directly. The adducts **105** and **106**, however, rearrange to the thermodynamically more stable **107**, probably via the donor adducts **108**, **109** and the free silenes **104** and **104a** (equation 24).

Jones and Bates used the combination of t-BuLi/vinylchlorosilane in the presence of anthracene to synthesize the E and Z isomers of the [4 + 2] cycloadducts **110** of the transient neopentylsilene **111** (equation 25)[68]. The thermal retro-Diels–Alder reaction proceeds stereospecifically. Photochemical reactions of similar compounds have been carried out[69].

$$Ph_2Si=C(SiMe_3)_2 \quad (23)$$

$$Me_2Si=C(SiMe_3)(SiMePh_2) \quad (97a) \longrightarrow \text{rearrangements, dimerization}$$

(equation 24)

The photochemical reactions of 2-silabicyclo[2.2.2]octanes **110** and **112** have been investigated[69]. Photolysis of **112** in the presence of methanol gives silyl methyl ether **113** as the main product, while **114** is only a minor byproduct (equation 26). This suggests that silene formation is a minor process and the main reaction proceeds via diradical intermediate similar to **115**. In agreement, interconvertion of E/Z **110**, probably via diradical **115**, is found during photolysis along with extensive polymerization (equation 27).

$$\text{E-(110)} \rightleftharpoons \text{(115)} \rightleftharpoons \text{Z-(110)} \qquad (27)$$

More reaction pathways are opened up when one of the substituents at silicon is a vinyl group as in **116** (equation 28)[70]. The intermediate silaallylic radical **117** (alternatively, the authors suggest an ionic mechanism) can be trapped directly by methanol to give the silyl ether **118**. Alternatively, **117** closes to the silene **119**, which was identified by its reaction product with methanol, **120**. A third reaction channel is the elimination of a silene, which again was identified by its trapping product **121** with methanol.

2. Salt elimination

The most straightforward synthesis of unsaturated Si=C compounds is the formation of the double bond by 1,2-elimination of a salt. This method has been widely used by N. Wiberg's and N. Auner's groups in recent years to produce a variety of different silenes.

The synthesis of the 'Wiberg'-type silenes is primarily achieved through the metalation of halogenotrisilylmethanes **122** with subsequent mild thermal salt elimination from the metalated species **92·LiX**, **97·LiX** and **104·LiX**[71-78]. The silenes formed, **92**, **97** and **104**, rearrange to the silenes **92a**, **97a** and **104a** or are trapped by suitable reagents (equation 29). In the case of R = t-Bu (**104a**) the silene is metastable[65,66] and its structure could be determined by X-ray diffraction[4].

The further increase in steric bulk is achieved in the potential silene precursors $Me_2Si(X)-C(Y)(SiMe_3)(SiBu-t_3)$[77]. For X = F and Y = Na the trisilylmethane **123** transforms into a silene/THF adduct **124·THF** in the presence of Me_3SiCl (equation 30). Removal of THF destabilizes the silene **124**, which can be either trapped or rearranged by a 1,3 H-shift to give **125**.

Recently, Wiberg and coworkers have answered the question whether the formation reaction for silenes of the type **126** can be analogously extended for silene types **127** and **128**[78].

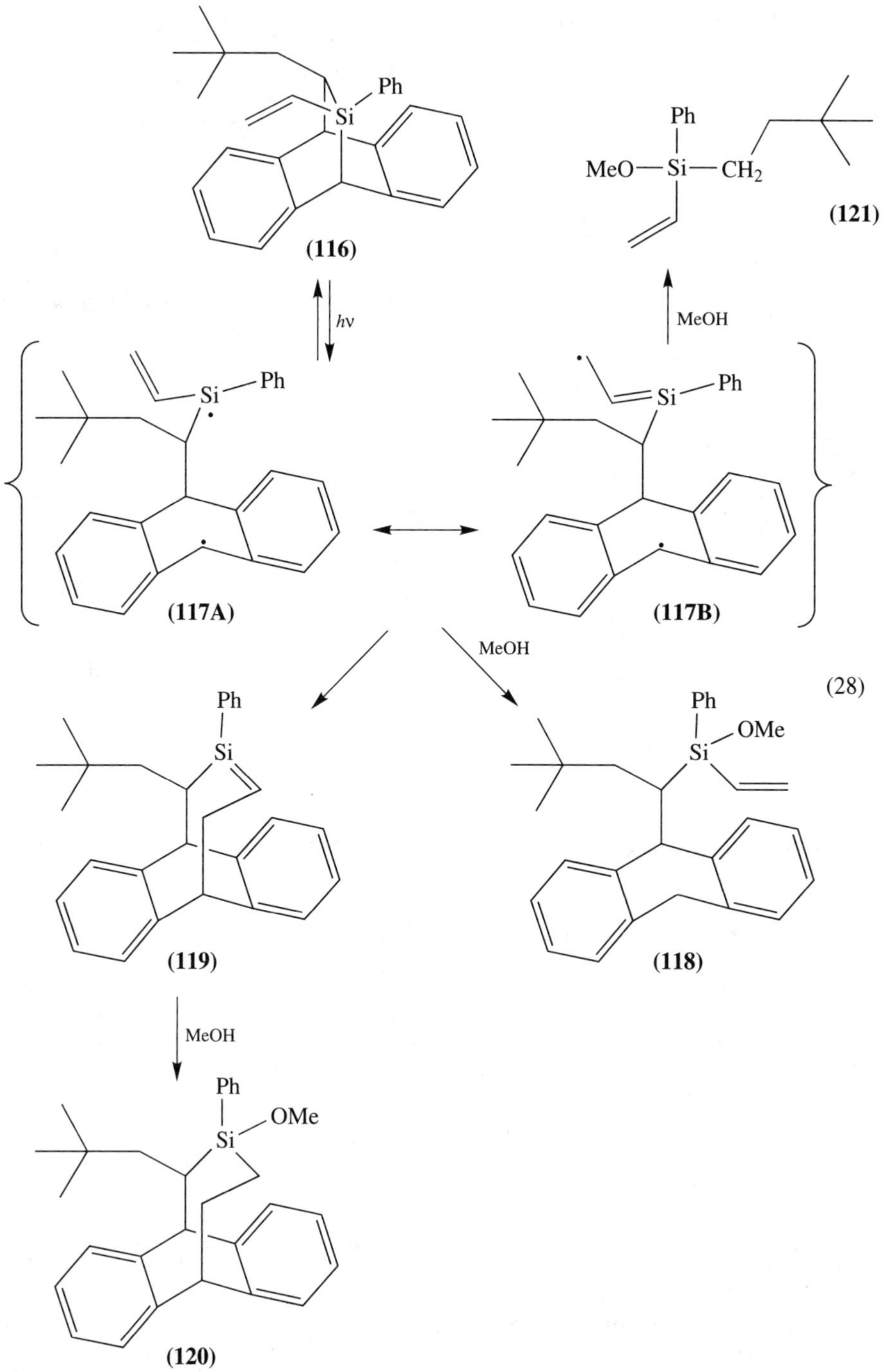

16. Silicon–carbon and silicon–nitrogen multiply bonded compounds 877

(116)

(121)

(117A) ⟷ (117B)

(119)

(118)

(120)

(28)

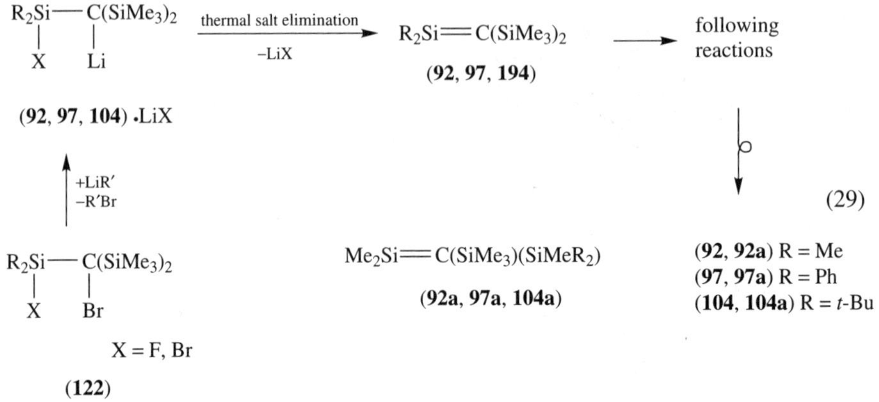

$R_2Si(X)-C(SiMe_3)_2Li$ $\xrightarrow[-LiX]{\text{thermal salt elimination}}$ $R_2Si=C(SiMe_3)_2$ ⟶ following reactions

(92, 97, 104)·LiX   (92, 97, 194)

(29)

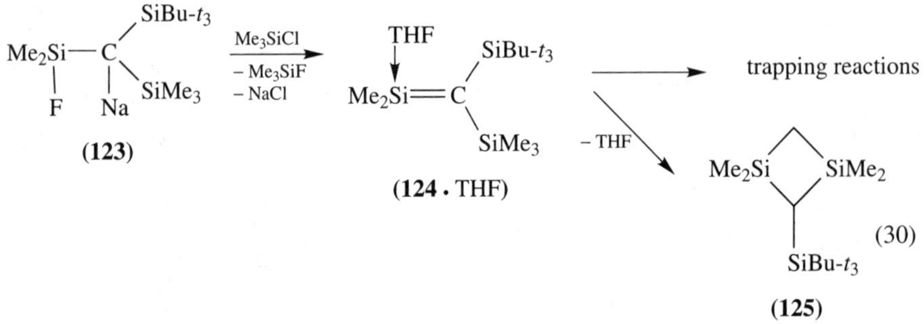

$\uparrow$ +LiR′ / −R′Br

$R_2Si(X)-C(SiMe_3)_2Br$ $Me_2Si=C(SiMe_3)(SiMeR_2)$ (92, 92a) R = Me
(97, 97a) R = Ph
(92a, 97a, 104a)   (104, 104a) R = t-Bu

X = F, Br

(122)

$Me_2Si(F)-C(Na)(SiBu\text{-}t_3)(SiMe_3)$ $\xrightarrow[-Me_3SiF, -NaCl]{Me_3SiCl}$ $Me_2Si=C(SiBu\text{-}t_3)(SiMe_3)$ ·THF $\xrightarrow{\text{THF}}$ trapping reactions

(123)   (124·THF)

$\xrightarrow{-THF}$

$Me_2Si\underset{SiBu\text{-}t_3}{\overset{}{\square}}SiMe_2$

(30)

(125)

$R_2Si=C(SiR_3)_2$ $R_2Si=C(H)(SiR_3)$ $R_2Si=CH_2$ $Me_2Si=CH(SiBu\text{-}t_3)$

(126)   (127)   (128)   (129)

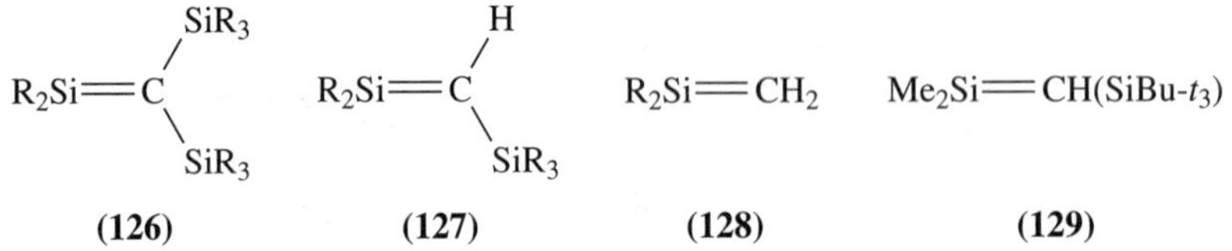

This question can be answered positively for the sterically shielded silenes **129** and **130**. They not only form dimers, but can be trapped by dienes. In the case of **131**, no dimers or trapping products with dienes are obtained. The formation of the products from the reaction $t\text{-}Bu_2Si(Br)-C(Br)H_2 + LiR$, however, is tentatively explained by the authors by the participation of a silene along the reaction pathway.

$t\text{-}BuMeSi=CH(SiBu\text{-}t_3)$ $t\text{-}Bu_2Si=CH_2$

(130)   (131)

The formation of the α-metalated species in the case of the 'Auner/Jones Type' silenes (neopentylsilenes) **132** (which is deduced by formation of dimers **133** or by trapping) is achieved through addition of t-butyllithium to a vinylchlorosilane in an inert solvent[79,80]. The initially formed lithiated intermediate eliminates LiCl at ca 0 °C (equation 31), the exact temperature depends on the substituents R. The multitude of transient silenes that are available through this method will be discussed in Section I.B.4.b on the reactivity of neopentylsilenes. The basic reactions of $Cl_3Si-CH=CH_2$ with t-BuLi have

16. Silicon–carbon and silicon–nitrogen multiply bonded compounds

been investigated[81].

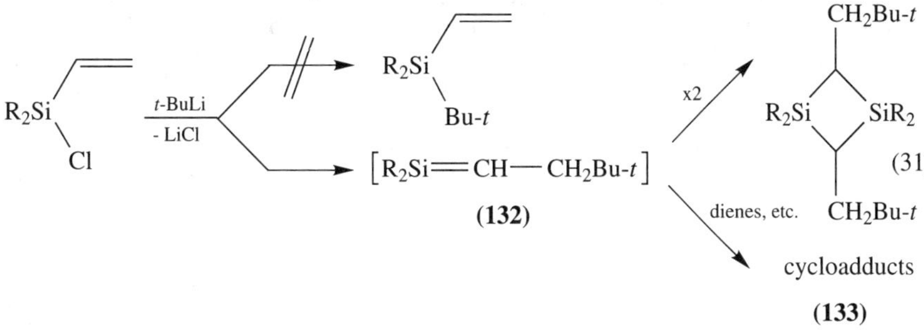

$$[R_2Si=CH-CH_2Bu\text{-}t]$$
(132)

(31)

(133) cycloadducts

A stable neopentylsilene, 1,1-dimesityl-2-neopentyl-1-silene **134**, was synthesized by the Couret group from fluorodimesitylvinylsilane and t-BuLi[82]. It was characterized by NMR spectroscopy.

$$Mes_2Si=CHCH_2C(CH_3)_3$$
(134)

3. Donor cleavage

The silicon atom in Wiberg's silenes is a Lewis acidic center and can be coordinated with donors. These donors may be halide ions [like, e.g., in **104**·LiX(12-C-4)[83]], ethers or nitrogen bases[84]. Coordination with the donor stabilizes the silenes and, when the basicity suffices ($Et_2O < Br^-$, THF, NMe_3, F^-), the adducts serve as stock compounds from which the silenes may be liberated (equation 32). **92** decomposes at $-100\,°C$, but its trimethylamine adduct is stable at room temperature[85]. Noticeable amounts of **97** are available from **97**·LiBr already at $-78\,°C$, whereas from **97**·LiF the temperature has to be raised to $+30\,°C$. Structures of such silene adducts will be discussed in Section I.C.1. The silene adduct **124**·THF has been discussed in Section I.A.2.

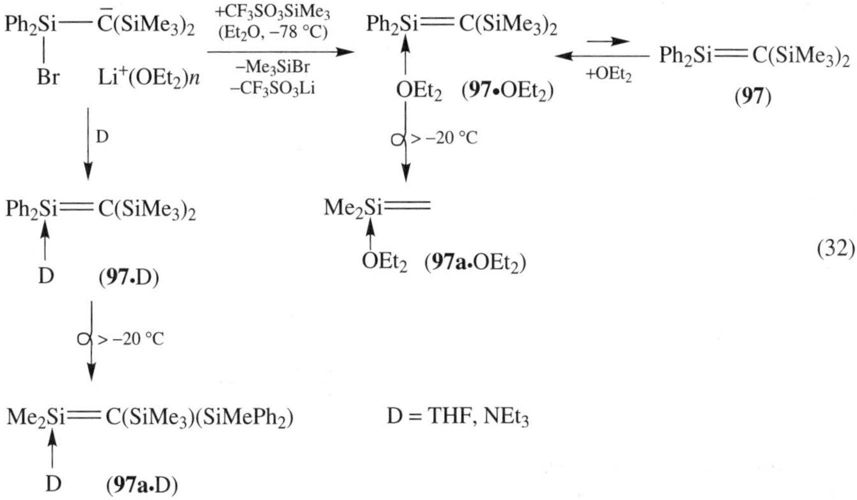

(32)

D = THF, NEt_3

4. Isomerization of acylpolysilanes

The facile photochemical sigmatropic 1,3-trimethylsilyl shift in polysilylacylsilanes from silicon to oxygen (equation 33) was utilized historically to prepare the first relatively stable silenes[3,86,87]. Silenes prepared by isomerization of acylpolysilanes bear, due to the synthetic approach, a trimethylsiloxy group at the sp^2-hybridized carbon and relatively stable silenes of this type have in addition also at least one trimethylsilyl group at the silicon. These substituents strongly influence the physical properties and the chemical behaviour of these silenes. This is noticeable in many reactions in which these 'Brook'-type silenes behave differently from simple silenes or silenes of the 'Wiberg' type.

$$(Me_3Si)_3Si-C{\overset{O}{\underset{R^1}{\diagup\!\!\!\diagdown}}} \underset{\Delta}{\overset{h\nu}{\rightleftharpoons}} (Me_3Si)_2Si=C{\overset{OSiMe_3}{\underset{R^1}{\diagup\!\!\!\diagdown}}} \overset{\Delta}{\rightleftharpoons} \begin{array}{c} Me_3SiO \\ (Me_3Si)_2Si-R^1 \\ | \\ (Me_3Si)_2Si-R^1 \\ Me_3SiO \end{array}$$

(33)

TABLE 3. Silenes synthesized by photolysis or thermolysis of acyl di- or polysilanes

$$Me_3SiR^1R^2Si-C{\overset{O}{\underset{R^3}{\diagup\!\!\!\diagdown}}} \overset{h\nu}{\longrightarrow} R^1R^2Si=C{\overset{OSiMe_3}{\underset{R^3}{\diagup\!\!\!\diagdown}}}$$

R^1	R^2	R^3	Silene characterized by	References
Me$_3$Si	Me$_3$Si	Me	dimerization	86, 88
Me$_3$Si	Me$_3$Si	Et	dimerization	89
Me$_3$Si	Me$_3$Si	i-Pr	dimerization	89
Me$_3$Si	Me$_3$Si	CH$_2$Ph	dimerization	90
Me$_3$Si	Me$_3$Si	t-Bu	dimerization, NMR	3, 38, 86
Me$_3$Si	Me$_3$Si	CEt$_3$	dimerization, NMR	3, 90
Me$_3$Si	Me$_3$Si	bco^a	dimerization, NMR	90
Me$_3$Si	Me$_3$Si	1-Mecyhex^b	dimerization, NMR	91
Me$_3$Si	Me$_3$Si	1-Ad^c	NMR, X-ray	3
Me$_3$Si	Me$_3$Si	Ph	dimerization	86
Me$_3$Si	Me$_3$Si	C$_6$H$_4$Y or C$_6$H$_3$Y$_2^d$	dimerization	92
Me$_3$Si	Me$_3$Si	CF$_3$	dimerization	92
Me$_3$Si	Me$_3$Si	Mes	NMR	91
Me$_3$Si	Me	1-Ad^c	dimerization	90
Me$_3$Si	t-Bu	1-Ad^c	NMR	90
Me$_3$Si	Ph	1-Ad^c	dimerization, UV	90
Me$_3$Si	Ph	CEt$_3$	dimerization, NMR	90
Me$_3$Si	Ph	t-Bu	dimerization, NMR	90
Me$_3$Si	Ph	Mes	trapping	90
Me$_3$Si	Mes	1-Ad^c	trapping, NMR	90
Me$_3$Si	Tip^e	1-Ad^c	trapping, NMR	93
Ph	Ph	1-Ad^c	trapping	94
Mes	Mes	1-Ad^c	trapping	94

^{a}bco: bicyclooctyl.
^{b}1-Mecyhex: 1-methylcyclohexyl.
^{c}1-Ad: 1-adamantyl.
^{d}Y = p-MeO, o-MeO, p-t-Bu, 3,5-dimethyl.
^{e}Tip 2,4,6-triisopropylphenyl.

Depending on the bulk of the group R^1 the produced silenes are reactive intermediates (i.e. $R^1 = Me^{86,88}$, Et^{89}, i-Pr^{89}, CH_2Ph^{89}), or they are in a temperature-sensitive equilibrium with their head-to-head dimers (e.g. $R^1 = t$-Bu)^{3,38,86}. When $R^1 =$ 1-adamantyl^{3,87}, no dimer was formed; rather the pure silene was isolated and its crystal structure was obtained.

Subsequently, this methodology was further widely exploited by Brook and coworkers and siloxysilenes with different substitution pattern have been prepared (see Table 3 for a list of silenes prepared along this route). Many of the siloxysilenes listed in Table 3 have been characterized by NMR spectroscopy. The identity of nearly all silenes is confirmed by the isolation of their dimerization products, the 1,2-disilacyclobutanes or by isolation of their trapping products with alcohols, dienes or acetylenes. Photolysis of acylsilanes of the type $(Me_3Si)_2R^2SiC(O)R^3$ gives rise to a mixture of E and Z isomers^{93,95}. It has been reported that t-Bu(Me$_3$Si)Si=C(OSiMe$_3$)Ad-1 **135** is formed as a single geometric isomer on the basis of NMR evidence^{90}. Trapping experiments with acetylenes do, however, reveal that this silene must have also been formed as a pair of geometric isomers^{93}. The geometrical isomers are conformationally stable even at higher temperatures, thus the (E/Z) isomers of Mes(Me$_3$Si)Si=C(OSiMe$_3$)Ad-1 **136** do not interconvert even upon heating to 100 °C, when they slowly decompose^{95}.

Siloxysilenes Ar$_2$Si=C(OSiMe$_3$)Ad-1 arising from photolysis of acyldisilanes are transient species and can only be identified by their trapping product or by isolation of consecutive isomerization products^{94}.

Silenes of the family Me$_3$SiR^{1}Si=C(OSiMe$_3$)Ad-1 **137** undergo a complex silene-to-silene photoisomerization reaction^{90,94,96}. When silenes **137** are generated by photolysis of acylsilanes **138**, the isomeric silenes **139** and **140** are formed in a subsequent reaction. The reaction was followed by UV and ^{1}H NMR spectroscopy. The disappearance of **138** cleanly follows first-order kinetics and the overall kinetics were consistent with the transformation **138** → **137** → **139**. **137** as well as **139** were characterized by NMR spectroscopy and, in addition, the structure of **137** was established by trapping with methanol. The identity of **139** and **140** was confirmed by the isolation of their head-to-tail dimers from which crystals, suitable for X-ray analyses, were isolated (equation 34)^{90}.

The complex reaction sequence shown in equation 34 might provide some rationalization. The formation of the silylcarbene **141** is suggested, based on experimental results from related reactions^{94}, but there is no evidence for the formation of **141** nor for a silylene intermediate. Thus, the transformation **137** → **142** might proceed via a dyotropic rearrangement as well. The facile 1,3-methyl shift in 2-trimethylsilylsilenes which interconverts **142** → **139** is well known from 'Wiberg'-type silenes^{64,97}. **139** ($R^1 = t$-Bu) is stable in solution at room temperature over days and isomerizes only slowly to **140** ($R^1 = t$-Bu) which rapidly dimerizes giving a 1,3-disilacyclobutane^{90}.

A similar but thermal silene-to-silene rearrangement is reported for the sterically highly crowded silene **143**. Upon prolonged heating to 120° E/Z-**143** isomerizes cleanly to a single isomer of the new silene **144**, which was identified by NMR spectroscopy^{93}. The silenes **145** and **146** were not detected, although they are regarded as intermediates (equation 35).

The thermolysis of neat acylpolysilanes usually gives mixtures of compounds. However, thermolysis in the presence of a scavenger such as an alcohol or an acetylene is much cleaner. Brook and coworkers showed that thermolysis of pivaloyltris(trimethylsilyl)silane **147** in the presence of 1-phenylpropyne yields the 2-silacyclobutene **148** in 72% yield, indicating the thermolytic generation of **149**^{86}. (Note, however, that originally a different regiochemistry for the cycloaddition product **148**^{86}

was suggested[95]; equation 36.)

16. Silicon–carbon and silicon–nitrogen multiply bonded compounds

Recently, Ishikawa and coworkers again adopted this approach[98,99]. Silenes were formed from various substituted tris(trimethylsilyl)acylsilanes by heating them to 140 °C in benzene and they could be trapped by suitable scavenger reagents. For example, **150**, generated by thermolysis of **151**, reacts with acetone yielding the siloxetane **152** which is, however, not stable at the applied conditions (equation 37). In the presence of a Ni catalyst the thermolysis of **151** gives rise to products which can be rationalized by the intermediacy of Ni/silene complexes[98]. In the absence of any trapping agent the starting acylpolysilane was in some cases recovered unchanged[98,99].

$$(Me_3Si)_3Si-C(=O)(Ad\text{-}1) \underset{24\ h,\ 140\ °C}{\overset{\Delta}{\rightleftharpoons}} (Me_3Si)_2Si=C(OSiMe_3)(Ad\text{-}1) \quad (37)$$

(**151**) → (**150**) →[Me$_2$CO] **152**: 4-membered ring with O, (Me$_3$Si)$_2$Si, C bearing Me, Me, Ad-1, OSiMe$_3$

An intermediate siladiene **153** is formed during the thermolysis of an α,β-unsaturated acylsilane **154**. Addition of (Me$_3$Si)$_3$Si• and subsequent hydrogen abstraction led to the isolation of the enol ether **155** formed, presumably, via radical **156** (equation 38). The identity of **155** was confirmed by an X-ray crystal structure[100].

(**154**) (Me$_3$Si)$_3$Si—C(=O)(CH=CMe$_2$) →[130 °C, 30 min]

(**153**) (Me$_3$Si)$_2$Si=C(OSiMe$_3$)(HC=CMe$_2$) + (Me$_3$Si)$_3$Si• → −Me$_2$C=CHCO•

(**156**) (Me$_3$Si)$_2$Si•—C(OSiMe$_3$)=CH—CMe$_2$Si(SiMe$_3$)$_3$ → (**155**) (Me$_3$Si)$_2$SiH—C(OSiMe$_3$)=CH—CMe$_2$Si(SiMe$_3$)$_3$ \quad (38)

5. Silenes by the sila-Peterson reaction

A relatively new synthetic approach to silenes was established independently in the laboratories of Oehme[101–110], Apeloig[39,111] and Ishikawa[112,113]. The key-step is a base-initiated 1,2-elimination of silanolate from α-hydroxydisilanes **157** and formation of silenes **158** analogous to the original Peterson olefination reaction (equation 39).

$$R^1 = R_3Si, Ar$$
$$R^2 = Aryl, Alkyl, OSiMe_3$$
$$R^4 = Aryl, Alkyl$$
$$M = Li, MgX$$

$$R^1; R^2 = R_3Si$$
$$R^3; R^4 = Aryl, alkyl, R_3Si, R_3Ge$$
$$R^5 = Me, Ph, H$$
$$M = Li, MgX$$

(39)

All silenes formed by the sila-Peterson elimination have a common, novel substitution pattern, bearing two silyl substituents at the tricoordinated silicon centre and a wide variety of substituents including hydrogen, alkyl, aryl and vinyl groups, at the doubly bonded carbon atom. This novel substitution pattern has consequences related to the physical properties and the chemical behaviour of the 'Apeloig–Ishikawa–Oehme' type of silenes. Almost all silenes formed via the sila-Peterson elimination reaction are only transient species. They can either be trapped by nucleophilic reagents or undergo cycloadditions with various dienes. In the absence of scavenger reagents they dimerize in a head-to-head fashion yielding 1,2-disilacyclobutanes or linear polysilanes.

Three different routes to the key compounds for the sila-Peterson elimination, the α-alkoxydisilanes **157**, are described in the literature, namely: A, reaction of silyllithium reagents with ketones or aldehydes; B, addition of carbon nucleophiles to acylsilanes; C, deprotonation of the polysilylcarbinols. In addition, method D, which already starts with the reaction of 2-siloxysilenes with organometallic reagents, leads to the same products. The silenes of the 'Apeloig–Ishikawa–Oehme' type synthesized so far are summarized in Table 4.

TABLE 4. Silenes $R^1R^2Si=CR^3R^4$ synthesized by the sila-Peterson reaction and related reactions

R^1	R^2	R^3	R^4	Method of preparation[a]	References
Me$_3$Si	Me$_3$Si	cypen[b]		A	114, 115
t-BuMe$_2$Si	t-BuMe$_2$Si	cypen[b]		A	114
Me$_3$Si	Me$_3$Si	2-Ad[c]		A	39, 111
t-BuMe$_2$Si	Me$_3$Si	2-Ad[c]		A	111
t-BuMe$_2$Si	t-BuMe$_2$Si	2-Ad[c]		A	111
Me$_3$Si	Me$_3$Si	Me	Me	A,B	107, 112
Me$_3$Si	Me$_3$Si	Me	Ph	B	112
Me$_3$Si	Me$_3$Si	Ph	Ar[d]	B	113
Me$_3$Si	Me$_3$Si	H	Sup[e]	A	109
Me$_3$Si	Me$_3$Si	H	Tip[f]	A,C	105
Me$_3$Si	Me$_3$Si	H	Ar[g]	C	g
Me$_3$Si	Me$_3$Si	H	t-Bu	C	104, 107
Me$_3$Si	Me$_3$Si	H	CH=CMe$_2$	C	106
Me$_3$Si	Me$_3$Si	H	CH=CHPh	C	106
Me$_3$Si	Me$_3$Si	cypheptene[h]		A	115
Me$_3$Si	Me$_3$Si	Me	p-DMAP[i]	C	110
Me$_3$Si	Me	Me$_3$Si	t-Bu	D	116
Me$_3$Si	Me	Me$_3$Si	Ph	D	116
Me$_3$Si	Me	Me$_3$Si	1-Ad[j]	D	116
Me$_3$Si	Et	Me$_3$Si	1-Ad[j]	D	116
Me$_3$Si	Me$_3$Si	Me$_3$Si	1-Ad[j]	D	116
Me$_3$Si	Ph	Me$_3$Si	1-Ad[j]	D	116
Me$_3$Si	PhCH$_2$	Me$_3$Si	1-Ad[j]	D	116
Ph	Me	Me$_3$Si	1-Ad[j]	D	116
Ph	Et	Me$_3$Si	1-Ad[j]	D	116

[a] See equation 39.
[b] cypen: 2,3 di-t-butylcyclopropenylidene.
[c] 2-Ad: 2-adamantylidene.
[d] Ar: Ph, o-tolyl, p-xylyl.
[e] Sup: 2,4,6-tri-t-butylphenyl.
[f] Tip: 2,4,6-triisopropylphenyl.
[g] Ar: 1,4-di-t-butylphenyl[109], 2,4,6-trimethylphenyl[107,108,117], 2,5-diisopropylphenyl[118], o-dimethylaminophenyl[110].
[h] cyheptene: [a,e]-dibenzocyclohepta-2,4,6-trienylidene.
[i] DMAP: dimethylaminophenyl.
[j] 1-Ad: 1-adamantyl.

a. Reaction of silyllithium reagents with ketones or aldehydes. Oehme and his group studied the reaction of (Me$_3$Si)$_3$SiLi **159** with simple aliphatic ketones like acetone in THF (equation 40)[101–104]. They found that the reaction of acetone with an excess of **159** leads to α-alkoxypolysilane **160**, which spontaneously eliminates lithium silanolate. The intermediate silene **161** is immediately trapped by the excess of the silyllithium reagent forming polysilyl carbanion **162**. **162** undergoes a 1,3-silyl shift to the polysilaanion **163** and therefore, after hydrolytic work-up, products deriving from **163** were isolated[103].

$$(40)$$

The use of equimolar amounts of a ketone in hydrocarbons prevents the addition of the silyl anion to the silene. Apeloig, Bravo-Zhivotovskii and coworkers found that tris(trimethylsilyl)silyllithium **159** gives in the reaction with 2-adamantanone the unusual head-to-head dimer **164** in 85% yield, suggesting the intermediate formation of silene **165** (equation 41)[39]. More bulky polysilyl anions react with 2-adamantanone to form at room temperature indefinitely stable silenes. Thus, (*t*-BuMe$_2$Si)(Me$_3$Si)$_2$SiLi **166** reacts with 2-adamantanone to give the silene (*t*-BuMe$_2$Si)(Me$_3$Si)Si=Ad-2 (2-Ad = 2-adamantylidene) **167** (equation 42), whose crystal structure was obtained (see Section I.C.1 for details)[111]. A second indefinitely stable silene bearing two (*t*-BuMe$_2$Si) groups at silicon was also obtained[111].

$$(41)$$

The sila-Peterson olefination reaction was also used to synthesize derivatives of 4-silatriafulvene **168**[114,119]. While the reaction of **159** with cyclopropenone **169** without any trapping reagents gives a complex mixture of products including several dimers of the transient silatriafulvene **170**, in the presence of 2,3-dimethylbuta-1,3-diene or anthracene **171** and **172**, respectively, are formed (equation 43). Since the anthracene adduct **172** undergoes a facile *retro* cycloaddition upon heating to 200 °C, **172** is a convenient storage compound of 4-silatriafulvene **170**[119]. The use of more bulky silyl anions like **173** allows the isolation of a stable silatriafulvene **174** as yellow crystals, the identity of which could be proven by NMR spectroscopy[114].

(t-BuMe₂Si)₃Si⁻ (173)

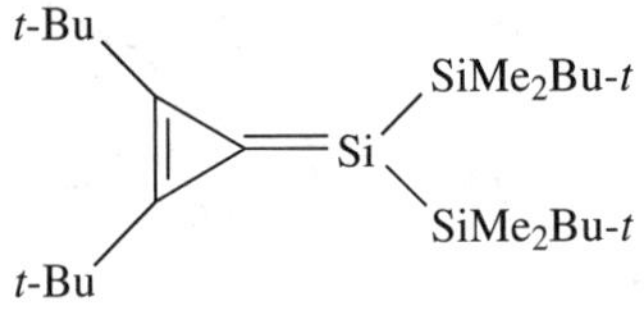

(174)

b. Reaction of polysilylacylsilanes with organometallic reagents. Ishikawa and coworkers studied the addition of organolithium reagents to polysilylacylsilanes such as (Me₃Si)₃SiCOR (R = Me, Ph) **175**[112,113]. In the reaction of **175a**, the initial adduct **176** eliminates spontaneously Me₃SiO⁻ and the intermediate silene **177** could be either trapped or dimerized, yielding head-to-head dimers **178** (equation 44). The scope of the organometallic reagent has been limited to MeLi[112] and ArLi[113].

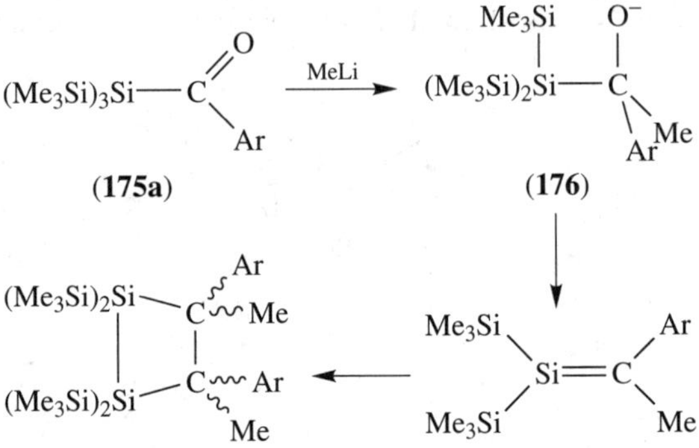

(44)

In contrast, silyl- or germyllithium reagents attack one of the trimethylsilyl groups attached to the central silicon atom rather than the carbonyl group of the acylsilanes **179**[120–122]. Subsequent elimination of disilane or germylsilane, respectively, results in the formation of lithium silenolate anions **180–183** (equation 45), which were characterized by NMR spectroscopy (see Section I.C.2)[120–122] and by trapping experiments[120–123].

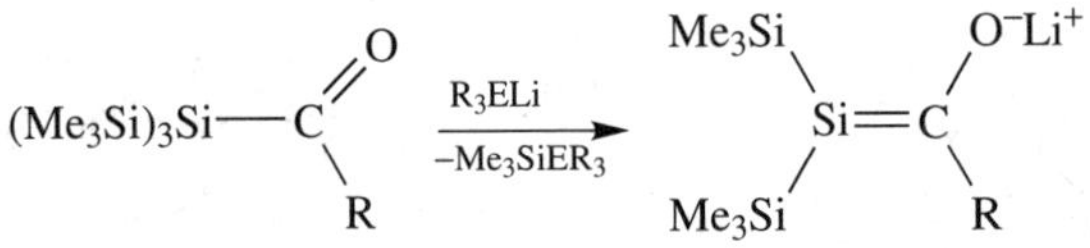

(45)

(180) R = *t*-Bu
(181) R = 1-Ad
(182) R = Mes
(183) R = *o*-Tol

E = Ge, Si

182 is thermally stable, while **180** and **181** could be analysed only by low temperature NMR spectroscopy[121]. Anions **182** and **183** can be silylated at the oxygen atom yielding the 'Brook-type' silenes **184** and **185** respectively, which could be trapped with 2,3-dimethylbuta-1,3-diene and thus identified. In addition, **184** could be characterized by NMR spectroscopy at −80 °C[121]. Interestingly, in the silylation reaction of the alkyl substituted silenolates **180** and **181** a different regiochemistry was observed: **180** and **181**

give acylpolysilanes **186** and **187** upon quenching with Et$_3$SiCl (equation 46)[121].

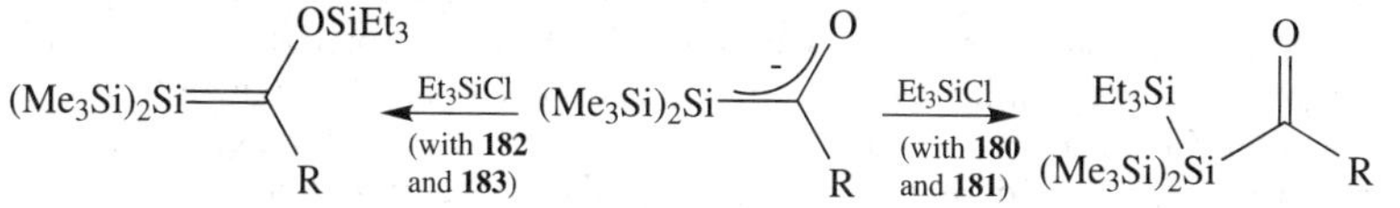

$$(Me_3Si)_2Si=C(OSiEt_3)(R) \underset{\text{(with 182 and 183)}}{\overset{Et_3SiCl}{\longleftarrow}} (Me_3Si)_2Si-C(O^-)(R) \overset{Et_3SiCl}{\underset{\text{(with 180 and 181)}}{\longrightarrow}} Et_3Si-(Me_3Si)_2Si-C(=O)R \quad (46)$$

(**184**) R = Mes
(**185**) R = o-Tol

180–183

(**186**) R = t-Bu
(**187**) R = 1-Ad

c. Deprotonation of polysilylcarbinols. The reaction of tris(trimethylsilyl)silylmagnesium bromide **188** with carbonyl compounds offers easy and versatile access to isolable polysilylcarbinols **189**[106–110,117]. Polysilylcarbinols **189** can be subsequently deprotonated with various bases. Thus, addition of MeLi or RMgX to **189** in ether results in the elimination of silanolate and gives the transient silenes **190** (equation 47), which undergo dimerization reactions yielding various dimers depending on the substituents on carbon.

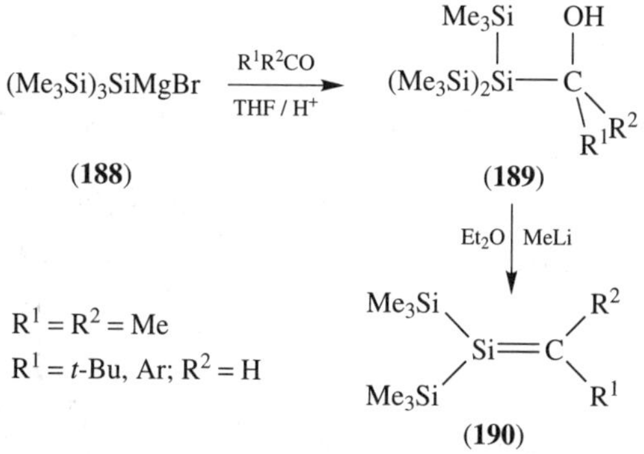

$$(Me_3Si)_3SiMgBr \xrightarrow[THF/H^+]{R^1R^2CO} (Me_3Si)_2Si(Me_3Si)-C(OH)(R^1)(R^2)$$

(**188**)   (**189**)

$\downarrow$ Et$_2$O | MeLi   (47)

R^1 = R^2 = Me
R^1 = t-Bu, Ar; R^2 = H

$$(Me_3Si)(Me_3Si)Si=C(R^2)(R^1)$$

(**190**)

The final product of the deprotonation depends strongly on the deprotonation reagent and/or the reaction conditions. Thus, in THF and using MeLi (or NaH) as base, an anionotropic 1,3-Si,O-trimethylsilyl migration occurs in the alkoxymethylsilane **191** with formation of the silyl anion **192** instead of elimination of silanolate. Therefore, after hydrolytic work-up only trimethylsiloxy(bis(trimethylsilyl)silyl)alkanes **193** were obtained (equation 48).[108,117]

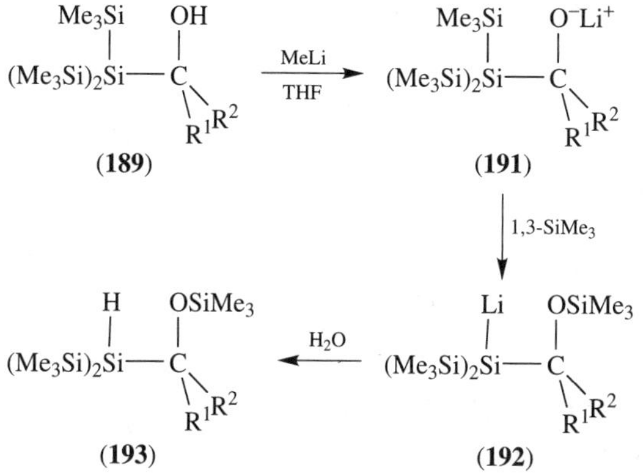

$$(Me_3Si)_2Si(Me_3Si)-C(OH)(R^1)(R^2) \xrightarrow[THF]{MeLi} (Me_3Si)_2Si(Me_3Si)-C(O^-Li^+)(R^1)(R^2)$$

(**189**)   (**191**)

R^1 = R^2 = Me
R^1 = t-Bu, Ar; R^2 = H   (48)

$\downarrow$ 1,3-SiMe$_3$

$$(Me_3Si)_2Si(H)-C(OSiMe_3)(R^1)(R^2) \xleftarrow{H_2O} (Me_3Si)_2Si(Li)-C(OSiMe_3)(R^1)(R^2)$$

(**193**)   (**192**)

d. Reaction of 2-siloxysilenes with organometallic reagents. 'Brook-type' silenes **194**, generated by photolysis of acylsilanes **195**, react with Grignard or organolithium reagents giving new intermediate silenes **196** with a novel substitution pattern[116]. The key step in this reaction sequence is similar to the sila-Peterson reaction. The organometallic reagent is added to the Si=C bond forming the carbanion **197** which rearranges to the alkoxide **198**. In a Peterson-type elimination of silanolate the silene **196** is generated. The excess of the organometallic compound or the silanolate adds readily to the Si=C bond. Thus, after hydrolytic work-up products deriving from the carbanions **199** and **200** have been isolated (equation 49)[116].

$R^1 = Me_3Si, Ph; R^2 = t\text{-Bu}, 1\text{-Ad}, Ph;$
$R^3 = Me, Et, OSiMe_3; M = MgBr, Li$

6. Photolysis of disilanes

The generation of silenes, e.g. **201** from aryl- or vinyl-substituted disilanes like **202**, is possible by irradiation, which may lead to a 1,3-silyl migration to the π-system (equation 50)[124,125].

$$\text{(202)} \xrightarrow{h\nu} \text{(201)} \qquad (50)$$

A number of disilanes have been investigated for their suitability as potential silene precursors. The formation of silenes by the 1,3-silyl migration pathway competes with dehydrosilylation, which also gives silenes, and with homolytic Si—Si bond cleavage to give silyl radicals.

In this connection Ishikawa and coworkers studied the photodegradation of poly(disilanylene)phenylenes **203**[126], and found that irradiation under the same conditions as in the photolysis of the aryldisilanes results in the formation of another type of nonrearranged silene **204** produced together with silane **205** from homolytic scission of a silicon—silicon bond, followed by disproportionation of the resulting silyl radicals **206** to **204** and **205** (equation 51).

(51)

The gradual increase in chain length from monomeric 1,4-disilanylbenzenes to poly(disilanylene)phenylenes was investigated for a series of oligomers **207**[127]. For $n = 2$

and 3 the main route under photolysis conditions is to form rearranged silenes, which can, however, only be trapped in low yields. The homolytic scission of the Si—Si bond is a minor route. For the oligomer with $n = 4$, appreciable amounts of products arising from homolytic scission to give silyl radicals are obtained.

Me$_3$Si—(Me$_2$Si—C$_6$H$_4$—SiMe$_2$)$_n$—SiMe$_3$

$n = 2, 3, 4$ **(207)**

In contrast, the study of isomeric bis(disilanyl)benzenes **208–210** showed that only products **211–213** from ene reaction of isobutene with rearranged silenes **214–216** were formed[128]. The silepines **217** and **218** result from a reaction of the initially formed silenes with isobutene (equation 52–54).

(208) → (hν) → **(214)** → (ene, isobutene) → **(211)** (52)

(214) → (hν) → valence isomerized products

(208): ortho-bis(SiMe$_2$SiMe$_3$)benzene
(214): rearranged silene with SiMe$_2$SiMe$_3$, =SiMe$_2$, SiMe$_3$
(211): benzene with SiMe$_2$SiMe$_3$, SiMe$_2$CH$_2$CHMe$_2$, SiMe$_3$

The irradiation of systems having two different π-electron systems[129] shows, in the case of 1,4-bis(1-phenyltetramethyldisilanyl)benzene, the exclusive migration to the phenylene ring but not to the phenyl ring. When the phenyl groups are in 2-position of the disilanyl moiety, a small fraction of silene formed from migration to this aryl ring is obtained[130].

Disilanyl substituted naphthalenes exhibit unusual photochemical reactivity[131]. 1,4-Bis(pentamethyldisilanyl)naphthalene **219** yields compound **220** in both the absence and presence of methanol, possibly via a biradical **221**. Noteworthy is the 1,8-silyl migration from position 1 to 8 of the naphthalene ring. In the presence of methanol, compound **222** is formed via an initial silene **223**, which then rearranges via **224** (equation 55). In a homogenous solution of methanol/benzene (1 : 1.5) only **225** is formed, probably by direct reaction of the photoexcited disilane with methanol, before migration of a trimethylsilyl

16. Silicon–carbon and silicon–nitrogen multiply bonded compounds

group to the naphthalene ring (equation 56).

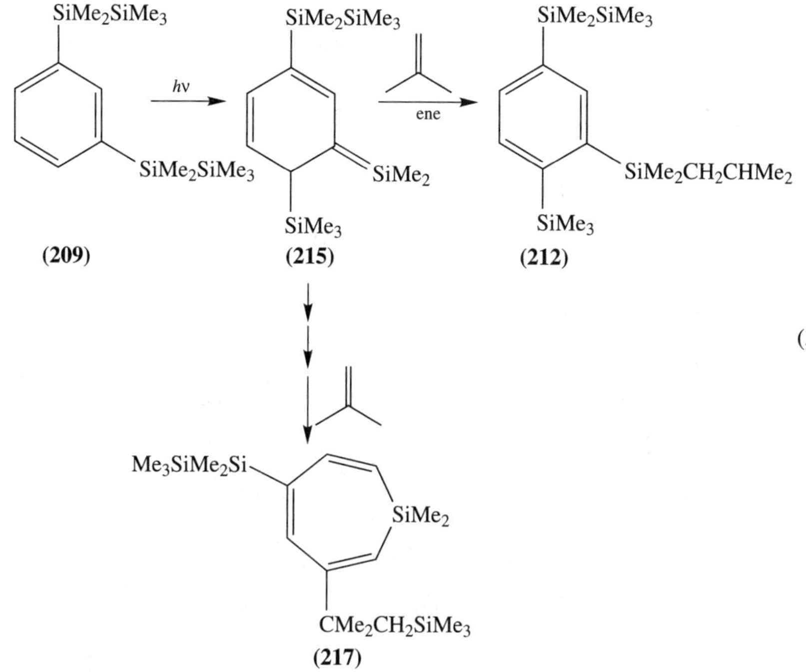

(53)

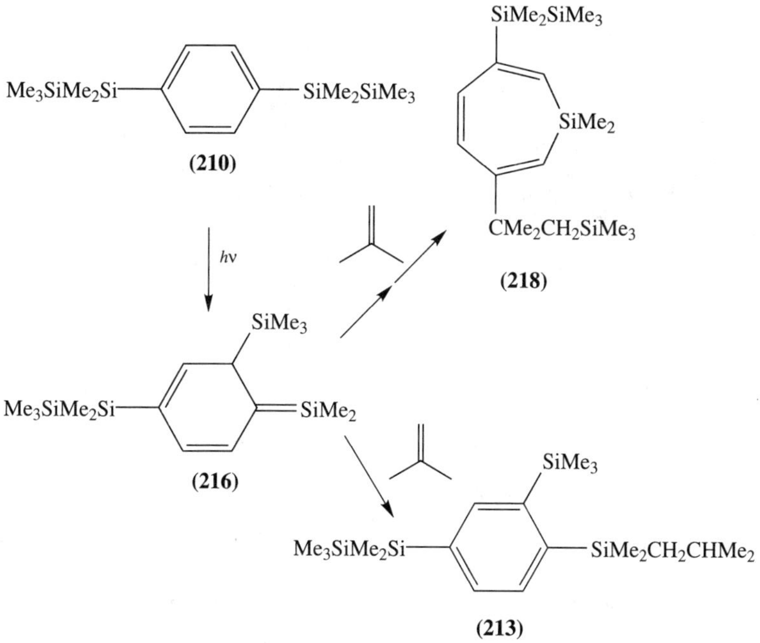

(54)

(55)

(56)

16. Silicon–carbon and silicon–nitrogen multiply bonded compounds

For 1,5-bis(pentamethyldisilanyl)naphthalene a 1,3-silyl migration does not occur. The products formed are analogous to **220** and **225**. 1-(Pentamethyldisilanyl)naphthalene undergoes silyl migration from position 1 to 8. 2-(Pentamethyldisilanyl)naphthalene **226** gives a silene **227** by a silyl shift from position 2 to 1. It can be trapped in an ene reaction with isobutene to yield **228** (equation 57).

(57)

2,6-Bis(pentamethyldisilanyl)naphthalene **229** undergoes 1,3-silyl shift to the 1-position to give the rearranged silene **230**[132]. A siloxetane **231** is discussed as an intermediate in the reaction with acetone. It loses silanone to give **232**, and then restores aromaticity by a 1,3-silyl migration to give the naphthalene **233**. Alternatively, **231** may expand the ring and, along with a 1,2-silyl shift, produce siloxacyclopentanes **234**, which then rearrange photolytically to the cyclic allenes **235**. An ene by-product **236** is also formed in the reaction with acetone (equation 58). 2,7-Bis(pentamethyldisilanyl)naphthalene reacts analogously.

Ishikawa and coworkers investigated the relative ease of migration of a Me₃Si group to vinyl and phenyl groups in precursor compounds containing both groups[133]. In the case of the irradiation of compound **237** in the presence of methanol, silene **238** was found to be the major reactive intermediate, while with acetone and 2,3-dimethylbuta-1,3-diene the number and nature of different ene products can only be explained by the existence of silatriene **239** (equation 59).

Substitution of one of the methyl groups on silicon by an ethyl group gave similar results. The authors state that these findings indicate the existence of an equilibrium between the precursor and the silenes or between the silenes.

The regiochemistry of such reactions was further investigated for the dihydropyranyl-substituted phenyldisilane **240**[134]. It was found that two types of silene intermediates, **241** and **242**, are formed by 1,3-silyl shifts to the dihydropyranyl and the phenyl group, respectively (equation 60), and that the distribution of trapping products from these silenes depends on the trapping agent used.

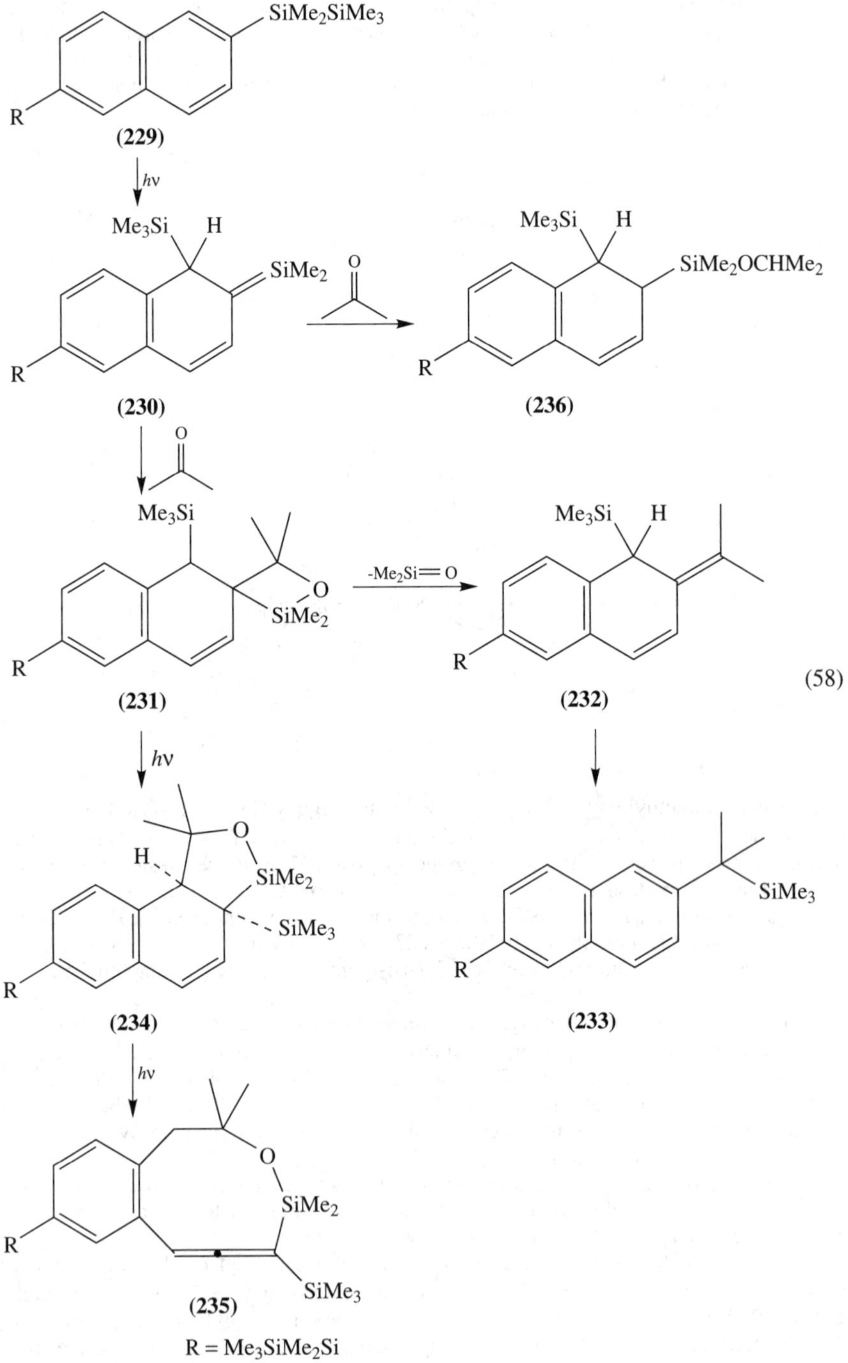
SiMe2SiMe3
R
(229)
hν
Me3Si
H
SiMe2
O
SiMe2OCHMe2
(230)
(236)
Me3Si
O
SiMe2
-Me2Si═O
(231)
(232)
hν
H
SiMe2
SiMe3
SiMe3
(234)
(233)
hν
O
SiMe2
SiMe3
(235)
R = Me3SiMe2Si

(58)

[Structures 237, 238, 239 with equation label (59)]

[Structures 240, 241, 242 with equation label (60)]

Sluggett and Leigh also used these reactions to obtain reactive intermediates, and studied these intermediates by nanosecond laser flash photolysis. Absorption spectra of the transient species were recorded, their decay rates were determined and kinetic studies were conducted. The photolysis of $Ph_3Si-SiMePh_2$ gives a low yield of the silatriene **243**[135], whose absorption spectrum could be recorded and found to be comparable to that of the well known silatriene obtained from $PhMe_2Si-SiMe_3$. Other species ($Ph_2Si=CH_2$, Ph_3SiH, $Ph_3Si^•$ and $MePh_2Si^•$) are also formed in the photolysis reaction. The relative yields depend on the solvent[136,137]. Silene species are formed from the lowest excited singlet state and are the major products in non-polar solvents, while silyl radicals stem from the lowest excited triplet state and are preferentially formed in polar solvents.

[Structure **(243)**]

A photolytic study with trifluoromethyl-substituted phenyldisilanes in the presence of alcohols revealed the nature of the excited states responsible for the photoreactions: the 1,3-silyl migration occurs from an aromatic $\pi\pi^*$ (locally excited) state, while the direct alcoholysis of the aryldisilane takes place from the $\sigma\pi^*$ orthogonal intramolecular charge transfer state[138].

The product types vary as a function of the degree and type of alkyl/aryl substitution at silicon. Thus Ph$_2$t-BuSi—SiPh$_2$(t-Bu) only yields silyl radicals and no Si=C species when irradiated[139].

In the case of the disilane **244** 1,3-silyl migration to give silene **245** competes with dehydrosilylation to give silene **246**[140] (equation 61).

$$\text{(246)} \xleftarrow{-\text{Me}_3\text{SiH}} \text{(244)} \longrightarrow \text{(245)} \quad (61)$$

With the functional groups at different silicon atoms as in **247** only the migration to the vinyl group occurs. A competitive reaction is the cleavage to yield the silyl radicals **248** and **249**.

(247)   PhMe$_2$Si• (248)   (249)

The photolysis of the rigid 1-silabicyclo[2.2.1]heptene **250**, which may be regarded as a cyclic vinylsilane, induces a 1,3-carbon shift to produce a cyclic silene intermediate **251**, which is trapped by alcohols to give **252** (equation 62)[141,142].

$$\text{(250)} \xrightarrow[214\text{ nm}]{h\nu} \text{(251)} \xrightarrow{\text{ROH}} \text{(252)} \quad (62)$$

Sakurai, Kira and coworkers synthesized silene **253** from the cyclic divinyldisilane precursor **254** (equation 63)[143].

$$\text{(254)} \xrightarrow{h\nu} \text{(253)} \quad (63)$$

The corresponding trisilane **255**, however, gives the expected silylene **256** (trapped as **257**) when photolysed, as the irradiation of trisilanes having a chromophore is a good method for the synthesis of silylenes[14]. Silene **258** is not obtained (equation 64).

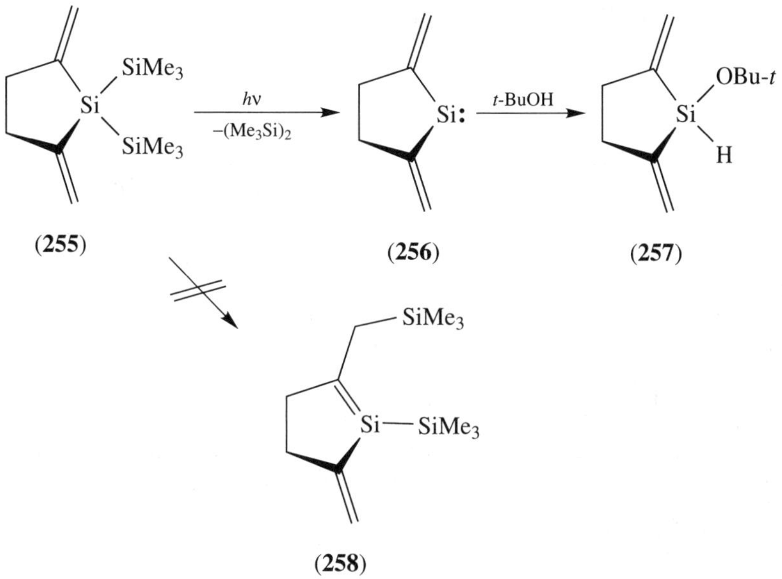

(64)

The 1,3-silyl shift in aryl disilanes is suppressed when the aromatic ring is *ortho*-substituted[144]. An attempted silylene synthesis from 1,3-dimesitylhexamethyltrisilane **259**, however, led to low yields of silylene trapping products (*ca* 30% generation of Me$_2$Si:). The major pathway is the homolytic cleavage of the trisilane, followed by disproportionation of the radicals **260** and **261** to the silene **262** and the disilane **263** (equation 65).

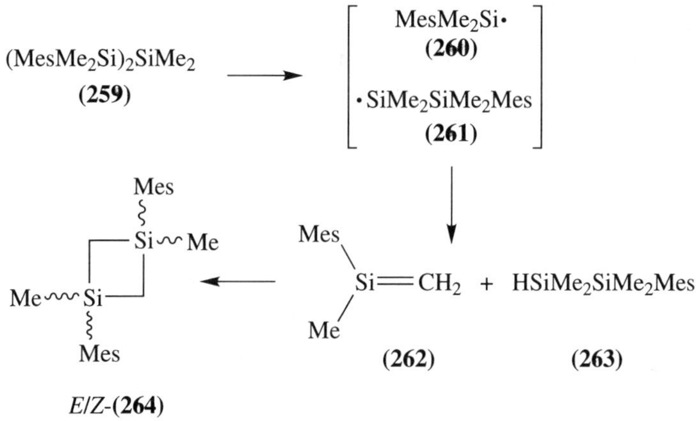

(65)

E/Z-(**264**)

The generation of these reactive species was supported by trapping experiments and the dimerization of the silene to E/Z isomeric 1,3-disilacyclobutanes **264**. The coproduct, 1,2-dimesityltetramethyldisilane **265**, of the Me$_2$Si: extrusion is also photolabile. Silene

262 is also formed as intermediate in this case (equation 66)[145].

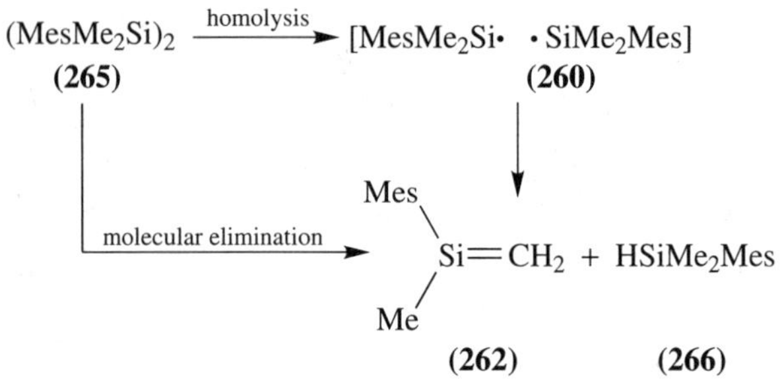

$$(MesMe_2Si)_2 \xrightarrow{homolysis} [MesMe_2Si\cdot \;\; \cdot SiMe_2Mes]$$

(265) → (260) → Mes\Si=CH$_2$ + HSiMe$_2$Mes via molecular elimination (262) (266) (66)

Similarly, the attempted synthesis of diadamantylsilylene by extrusion from a trisilane **267** did not give the wanted product[145]. The predominant photoreactions are silicon-silicon bond homolysis to give the radicals **268** and **269** (equation 67). Disproportionation of **268** and **269** results in the formation of silene **271** and silane **272**. The silene is identified by isolation of its head-to-tail dimer **273**. In the presence of scavenger reagents like 2,3-dimethylbuta-1,3-diene radical trapping products like **270** could be detected in low yields. Secondary photoprocesses involving the disilane **272** take place. Formation of silyl radicals **269** and **274** with subsequent disproportionation of the radicals explain the formation of diadamantylsilane **275**.

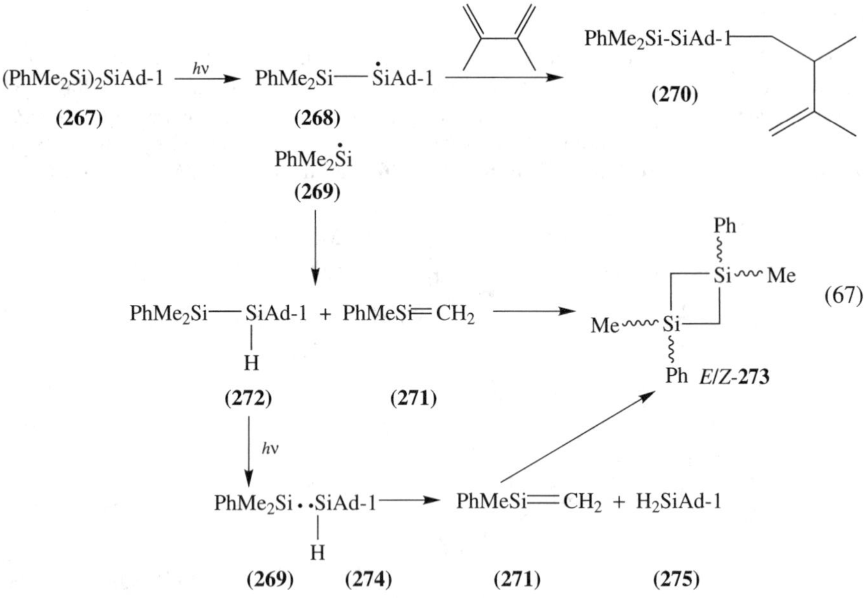

$$(PhMe_2Si)_2SiAd\text{-}1 \xrightarrow{h\nu} PhMe_2Si\text{---}\dot{S}iAd\text{-}1 \longrightarrow PhMe_2Si\text{-}SiAd\text{-}1$$

(267) (268) (270)

PhMe$_2\dot{S}i$ (269)

PhMe$_2$Si—SiAd-1 + PhMeSi=CH$_2$ → E/Z-273
        |
        H
    (272)   (271)

$\downarrow h\nu$

PhMe$_2$Si··SiAd-1 → PhMeSi=CH$_2$ + H$_2$SiAd-1
         |
         H
(269) (274) (271) (275)

(67)

7. Silenes by rearrangement of silylenes and carbenes

In contrast to ethene which is separated by a high potential barrier from the isomeric methylcarbene, silene **25** can undergo a 1,2 shift to either methylsilylene **276** or, less favourably, to silylmethylene **277** (equation 68). The thermochemistry and the kinetics of these reactions have been a point of major disparity between theory and experiment[11,146] and much work has been done by theoreticians and experimentalists to settle this discrepancy. The recent developments have been summarized in great detail[11,12] and we will refer here only to the latest results.

$$\underset{(276)}{\overset{H}{\underset{H}{\diagdown}}\overset{\overset{..}{Si}}{\underset{H}{\diagup}}\overset{H}{\underset{H}{\diagdown}}} \rightleftharpoons \underset{(25)}{\overset{H}{\underset{H}{\diagdown}}Si=C\overset{H}{\underset{H}{\diagup}}} \rightleftharpoons \underset{(277)}{\overset{H}{\underset{H}{\diagdown}}\overset{Si}{\underset{H}{\diagup}}\overset{\overset{..}{C}}{\underset{H}{\diagdown}}} \quad (68)$$

According to the best available calculations, silylcarbene **277** is higher in energy than silene **25** by 47.4 kcal mol^{-1} (MP4SDTQ/6-31G*//3-21G*)[147] while the best experimental value derived from calculated heats of formation is 52 kcal mol^{-1}[148]. In contrast, the silylene—silene rearrangement **25** → **276** is nearly thermoneutral, with the silene slightly more stable. A difference of *ca* 5 kcal mol^{-1} can be derived from the recent experimental heat of formations of **276** (48.3 kcal mol^{-1})[149] and of **25** (43 ± 3 kcal mol^{-1})[150], in agreement with high level calculations by Grev and coworkers which predict an energy difference of 4 kcal mol^{-1}[151].

The barrier for the 1,2-hydrogen shift was calculated by Kudo and Nagase to be 42.2 kcal mol^{-1} (MP3/6-31G*//6-31G*)[152]. This is considerably higher than the experimental values reported by Conlin and Kwak who reported an activation energy $E_A = 30.4 \pm 0.7$ kcal mol^{-1} ($\Delta H^{\ddagger} = 28.9 \pm 0.7$ kcal mol^{-1}) and a preexponential factor $\log A = 9.6 \pm 0.2$ s^{-1} ($\Delta S^{\ddagger} = -18.5 \pm 0.9$ cal deg^{-1} mol^{-1}) for the reaction **25** → **276**[22]. Higher quality calculations are clearly desirable to resolve this disagreement between theory and experiment[146].

Both silene isomers **278** and **279** are ideal precursors for the generation of silenes **284**, since their interconversion to **284** is spontaneous (in the case of **278**) or can be easily induced by irradiation (in the case of **279**). There are numerous well-established methods to prepare transient silylenes **279**. Three important examples are shown in equation 69, namely the photolytic generation from a trisilane **280**[153], thermolytic or photolytic decomposition of cyclic silanes **281**[14,154,155] and degradation of diazidosilanes **282**[153,156]. The photolysis of the diazido silane **282** is an especially clean reaction which has been used in several spectroscopic studies[157]. The photolysis of α-diazo compounds **283** is the only frequently used reaction path to silenes **284** via a carbene–silene rearrangement[8].

Maier and coworkers have shown that it is possible to induce by irridation using the appropriate absorption band, a 1,2-hydrogen shift in silenes and in silylenes[29,158]. Thus, it is possible to 'switch' photochemically between silylenes and silenes. Michl, West and coworkers have used this approach to isomerize dimethylsilylene **285** and 1-methylsilene **26** several times (equation 70). Due to the clean formation of **285** from the diazido precursor **286**[153] it was possible to measure the IR transition moment directions for both **26** and **285**[156].

Nefedov, Michl and coworkers reported the successful spectroscopic characterization of the 1-silacyclopenta-2,4-diene **287** and its photochemical transformation into the isomeric species **288**–**290** (equation 71)[159]. Gaspar and coworkers[160] had suggested that silole **287** might be formed by a rearrangement of the cyclic silylene **288** via the 1-silacyclopenta-1,3-diene **289**. This proposal was confirmed in an elegant UV-visible, IR-matrix isolation study[159]. The silylene **288** was generated by vacuum pyrolysis either of the spiro compound **291** or the diazide **292** or by UV-irradiation of matrix isolated **292**. The complex interrelationship between the silylene **288**, the isomeric silabutadienes **289**, **290** and the silole **287** has been established through matrix isolation, spectroscopic characterization and photoisomerization of the individual compounds. A 1,2-H shift transforms silylene **288** to silabutadiene **289**. Subsequent 1,3- or 1,5-H shifts interconvert the silabutadienes **289–290** and silole **287**. The final product after warmup of the argon matrix is **293**, the [4 + 2] dimer of **287** (equation 71).

(71)

A photochemical approach via a silylene-to-silene rearrangement was followed by Fink and coworkers in their synthesis of silacyclobutadienes in a 3-methylpentane matrix at low temperatures[161]. Irradiation of the cyclopropenyltrisilane **294** gives the relatively stable cyclopropenylsilylene **295**. **295** can be efficiently converted to silacyclobutadiene **296** by irradiation into the visible absorption band of the silylene (equation 72)[161,162].

The structure of the trapping product with TMSOMe has been established by X-ray crystallography, thereby confirming the identity of **296**[161]. Further adducts with amines and alcohols have been isolated[161,163]. The [2 + 2] cycloadduct **297** of **296** with trimethylsilylacetylene (equation 73) is the first room-temperature stable Dewar silabenzene[163].

$$(72)$$

$$(73)$$

Silenes are formed by rearrangement of silylcarbenes. If polysilylated diazomethanes **298–300** are employed, a selective migration of a silyl group to the carbene centre occurs and silenes **301**, **92** and **302** are formed (equations 74–76)[164]. The outcome of trapping reactions is independent of the mode of silene generation: photochemical and pyrolytic methods give the same results.

$$(Me_3Si)SiMe_2\!-\!\underset{\underset{(298)}{N_2}}{\overset{\|}{C}}\!-\!H \longrightarrow \underset{(301)}{Me_2Si\!=\!CH(SiMe_3)} \tag{74}$$

$$(Me_3Si)SiMe_2\!-\!\underset{\underset{(299)}{N_2}}{\overset{\|}{C}}\!-\!SiMe_2 \longrightarrow \underset{(92)}{Me_2Si\!=\!C(SiMe_3)_2} \tag{75}$$

$$(Me_3Si)_3Si\!-\!\underset{\underset{(300)}{N_2}}{\overset{\|}{C}}\!-\!SiMe_3 \longrightarrow \underset{(302)}{(Me_3Si)_2Si\!=\!C(SiMe_3)_2} \tag{76}$$

In the case of the disilanyldiazoacetate **303**, photolysis initially gives the carbene **304** that rearranges to the silaacrylate **305**[165], which then isomerizes to the bissilylketene **306** (equation 77). The trapping reaction with alcohols gives the products **307** and **308**. The latter is obviously formed via ion pair intermediates like **309**. The UV and IR spectra of irradiated matrices at 10 K have been measured. The band at 1670 cm^{-1} is tentatively assigned to the $\nu_{C=O}$ of **305**; λ_{max} is at 288 nm.

The integration of two carbene precursor groups, like in bis(1-diazo-2-oxoalkyl)-disilane **310**, leads to interesting products upon irradiation[166,167]. Intermediate carbenes (**313** and **314**) formed through nitrogen loss and silenes (**315** and **316**) formed through 1,2-silyl migration to the carbene centre are discussed. The formation of the eight-membered ring **312** (R = 1-adamantyl) is not explained. Product **311** is obtained, probably via the intermediates **317**–**320** as shown in equation 78.

The photochemical decomposition of bis(silyldiazomethyl)tetrasilane **320** produces one silene group to give **321** followed by intramolecular [2 + 3] silene–diazo cycloaddition via **322** to give the bicyclic compound **323** as final product, while thermal decomposition gives a bis-silene **324** which then undergoes head-to-tail dimerization[168,169] to

325. Interestingly, compound **323** is one of the most stable siliranes having methyl substituents at silicon. However, under prolonged exposure to air **323** decomposes to give **326** (equation 79).

For a central hexamethyltrisilanyl unit in the precursor **327** a formal head-to-tail dimer **329** and a head-to-head dimer **330** are formed[169] under photolysis conditions. Thermal reaction gave only the bicyclic dimer **329** (equation 80).

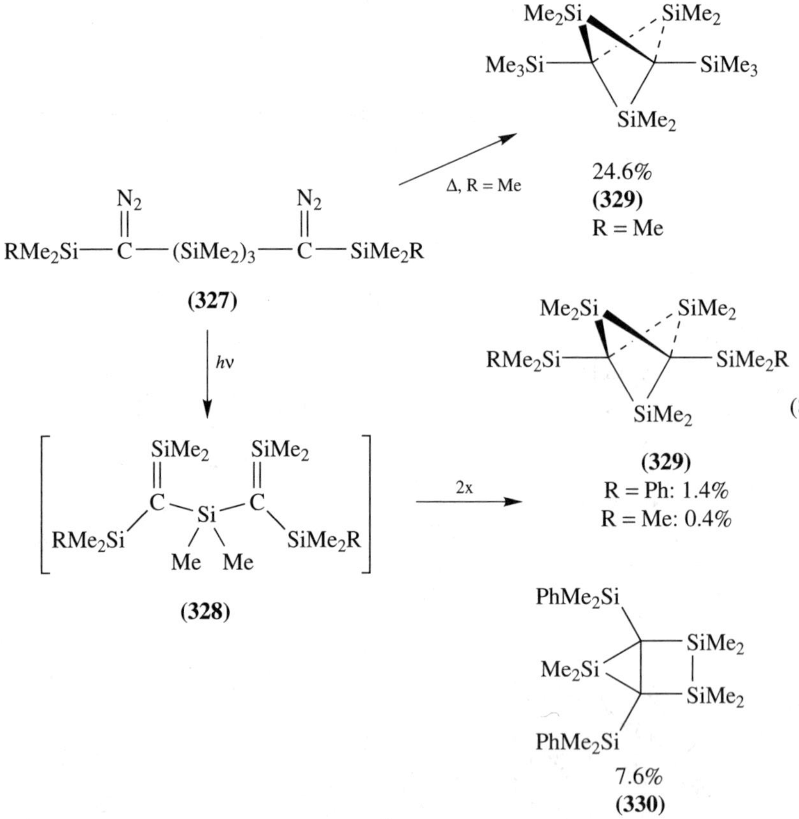

(80)

When the 1-diazo-2-silyl moiety is incorporated into cycles then endocyclic silicon–carbon double bonds should be formed. Most interesting in this connection is the decomposition of 1-diazo-2-sila-3,5-cyclohexadiene **331**, because the initial Si=C product should be silabenzene **332**[170]. The outcome of the nitrogen elimination depends on the conditions used. The products isolated are **333** (equation 81), **335** via bicyclic **334** (equation 82) and **338**, formed via silafulvenes and **336** (equation 83).

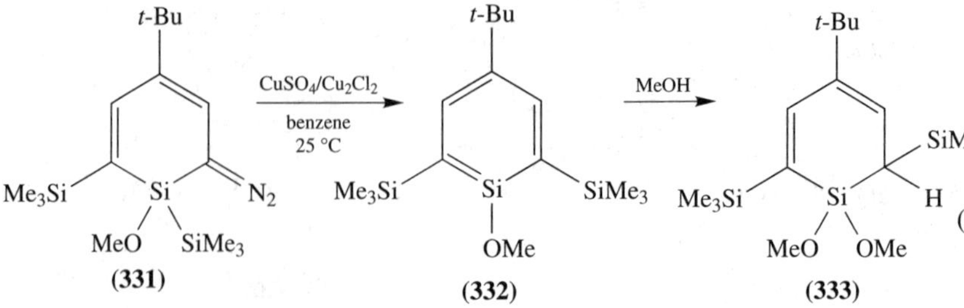

(81)

8. Other group 14 carbon double bonded species

Wiberg's group has prepared other species analogous to the silenes **92**, **97** and **104**, with tin or germanium instead of silicon. With tin and germanium the generation and reactivity of $Me_2E=C(SiMe_3)_2$ has been investigated (E = Sn^{171}, E = Ge^{172}). With germanium a stable species was available: $Me_2Ge=C(GeMe_3)(SiMe$-t-$Bu_2)$ is formed from the germene source t-Bu_2SiF-$CLi(GeMe_3)_2 \cdot 2$ THF[173,174].

The group of Satgé has presented evidence for the formation of the chloro- or fluorogermenes $Me_5C_5(X)Ge=CR_2$, with CR_2 = fluorenylidene[175].

B. Reactivity

1. Rearrangements

a. Silene-to-silylene interconversions. An interesting isomerization reaction is reported for the reactive 4-silatriafulvene **170**. When **170** is generated by thermolysis of **172** in the presence of *t*-BuOH the silacyclobutenes **339** and **340** are formed quantitatively in a relative ratio of 3 : 1[119]. This result indicates a rearrangement of **170** to the silacyclobutadiene **341**, which is faster than the addition of *t*-BuOH to the Si=C double in **170**. This is especially interesting in the context that thermolysis of **172** in the presence of 2,3-dimethylbuta-1,3-diene yields **171** in quantitative yield (equation 84). Thus, **170** reacts with dienes faster than with *t*-BuOH[119]. This unusual behaviour of **170** is illustrated by comparison with the silene Me$_2$Si=C(SiMe$_3$)$_2$ **92**, which reacts with *t*-BuOH nearly 2000 times faster than with 2,3-dimethylbuta-1,3-diene[62,176].

The low reactivity of **170** toward alcohols would be explained by the reduced polarity of the Si=C double bond due to the contribution of the resonance structure **168B**, in addition to the silyl substituent effects. In agreement with experiment a theoretical study for the parent species **168** predicts that water addition to **168** is thwarted by an unusual high barrier (20.0 kcal mol^{-1} at MP2/6-311++G**)[119,177]. Furthermore, the calculations suggest that the transformation **170** → **341** proceeds in two steps via the cyclopropenylsilylene **342** (equation 85). The highest barrier for the isomerization of the parent **168** → **343** is calculated to be 17 kcal mol^{-1}. Thus, the isomerization can effectively compete with the water addition[177].

(85)

b. Silene-to-silene isomerization. The rearrangement process of the 'Wiberg' silenes **92**, **97** and **104** (formed, e.g., by equation 86) by methyl migration gives the silenes **92**, **97a** and **104a** as has already been mentioned in Section I.A.1. Equation 87 demonstrates this degenerate rearrangement.

$$(86)$$

$$(87)$$

The use of deuteriated **92** allows the observation of this process. So the deuteriated 1-aza-2-silacyclobutane **95b$_1$**-d$_6$ [from (D$_3$C)$_2$SiF—CLi(SiMe$_3$)$_2$ and Ph$_2$C=N—SiMe$_3$] releases (D$_3$C)$_2$Si=C(SiMe$_3$)$_2$ **92**-d$_6$ in a reversible [2 + 2] cycloreversion in minor amounts when slightly heated. Fortunately, the methyl migration is faster than intermolecular dimerization and a mixture of 1-aza-2-silacyclobutanes **95b$_1$**-d$_6$–**95b$_4$**-d$_6$ is formed by [2 + 2] cycloaddition reaction with imine.

h = CH$_3$
d = CD$_3$

When deuteriated **92** is generated from irreversibly decomposing precursors, just the head-to-tail dimer [(D$_3$C)$_2$Si—C(SiMe$_3$)$_2$]$_2$ is found[97]. An even faster reaction than the methyl migration is the reaction of the silene with silyl azides. Just one of the possible deuteriated dihydrotetrazoles **344**-d_6 is formed (equation 88).

$$\text{(95b}_1\text{-}d_6) \xrightarrow[{-(h_3Si)_2C=N=N}]{\substack{+2\,t\text{-Bu}_3SiN_3 \\ -Ph_2C=NSih_3}} \text{(344-}d_6\text{)} \quad (88)$$

h = CH_3
d = CD_3

Silene **124** having the very bulky tri-*tert*-butylsilyl group at the linking carbon atom undergoes hydrogen migration to give a disilacyclobutane, as has been mentioned above[77].

The tendency to isomerization by methyl migration is higher in $Ph_2Si=C(SiMe_3)_2$ (**97**), because the less polar and acidic **97a** is formed[64].

A rearrangement quite similar to the above-mentioned methyl migrations is found for the initially formed dichloroneopentylsilene **345**, which possesses a *tert*-butoxy group. The higher electrophilicity of this silene's silicon atom in comparison with that of the resulting dimethylsilene **346** leads to a 1,3 migration of the *tert*-butoxy group[178,179]. The identity of **346** was established by isolation of the dimer **347** and the trapping product with quadricyclane **348** (equation 89).

The thermal 1,3 shift of a trimethylsilyl group from oxygen to silicon in 2-siloxysilenes (equation 90) is the reversal of their photochemical formation from acylpolysilanes. The reaction occurs already at room temperature[38,180].

'Brook-type' silenes with less than two trimethylsilyl groups at the silicon tend to be photolabile[90,96]. Thus silenes $Me_3SiR^1Si=C(OSiMe_3)$1-Ad **137** undergo a complex photoisomerization to novel silenes (Me$_3$SiO)MeSi=C(1-Ad)(SiMe$_2$R^1) **140** (equation 91). This has been summarized in detail together with the related thermal isomerization[93] in Section I.A.4.

$$\begin{array}{c} Me_3Si \\ \diagdown \\ Si=C \\ \diagup\diagdown \\ Me_3Si R^1 \end{array} \begin{array}{c} OSiMe_3 \end{array} \longrightarrow \begin{array}{c} Me_3Si \\ | \\ Me_3Si-Si-C \\ | \diagdown \\ Me_3Si R^1 \end{array} \begin{array}{c} O \end{array} \qquad (90)$$

$$\begin{array}{c} R^1 \\ \diagdown \\ Si=C \\ \diagup \diagdown \\ Me_3Si Ad\text{-}1 \\ (\mathbf{137}) \end{array} \begin{array}{c} OSiMe_3 \end{array} \xrightarrow{h\nu} \begin{array}{c} Me_3SiO Ad\text{-}1 \\ \diagdown \diagup \\ Si=C \\ \diagup \diagdown \\ Me SiMe_2R^1 \\ (\mathbf{140}) \end{array} \qquad (91)$$

c. Intramolecular insertion reactions. In a number of rearrangements of silenes the Si=C bond has been observed to insert into a C—H bond of an *ortho* methyl substituent of an adjacent mesityl group. For example, Brook and coworkers found that upon irradiation of the silene **349** the benzocyclobutane **350** is produced. Formally, this can be regarded as a [2π + 2σ] cycloaddition; however, a 1,5-hydrogen migration with a consecutive electrocyclic ring closure via **351** can also account for the observed product (equation 92)[90,181]

The transient silene **352** undergoes, under the photolytic conditions applied for its generation from **353**, a fast subsequent isomerization yielding the silaindane **354**. The silylcarbene **355** was suggested as short-lived intermediate (equation 93)[94]. Similar formal insertions of the Si=C bond into the C—H bond of an *ortho*-methyl group have been also observed for other 1-mesityl silenes[95] or 1-aminomesityl silenes[182].

(353) → hν → (352) → (355) → (354)   (93)

In an attempt to synthesize a kinetically stabilized silene, Oehme and coworkers used the sterically highly demanding 2,4,6-tri-*t*-butylphenyl (supermesityl = sup) substituent as a protecting group. Although the Si=C double bond in **356** is extremely sterically shielded, the silene is not stable under the applied conditions. The steric congestion of the molecule obviously prevents a dimerization, but the highly reactive silene group is inserted into the C–H bond of an *ortho-t*-Bu group of the supermesityl substituent and the bicyclic product **357** is obtained (equation 94)[199].

(356) → (357)   Sup = 2,4,6-*t*-BuC$_6$H$_2$ (94)

Generally, the isomerization/dimerization behaviour of Ph$_2$Si=C(SiMe$_3$)$_2$ (**97**) is more complex than that of Me$_2$Si=C(SiMe$_3$)$_2$ (**92**)[64]. Silene **97**, formed from **97·**LiBr, isomerizes by fast methyl migration to Me$_2$Si=C(SiMe$_3$)(SiMePh$_2$) **97a**, which undergoes a slower phenyl migration to PhMeSi=C(SiMe$_3$)(SiMe$_2$Ph) **97b**, and finally a fast methyl migration gives Me$_2$Si=C(SiMe$_2$Ph)$_2$ **97c**. Simultaneously **97a** and **97c** dimerize and also irreversibly isomerize into disilaindanes **358a–b** (equation 95). The latter reaction is quantitative when the dimers of **97a** and **97c** are thermolysed.

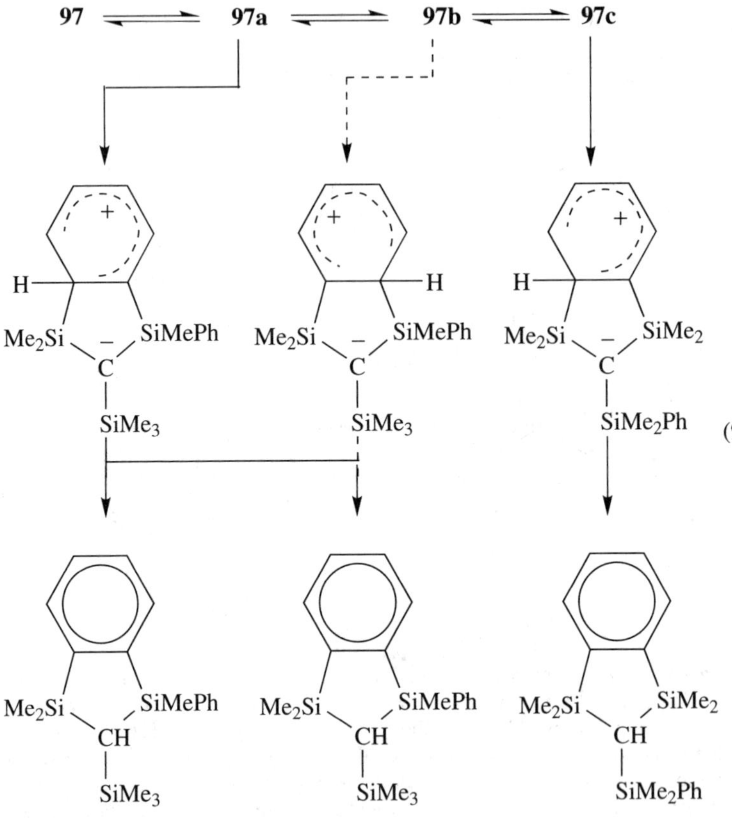

$$97 \rightleftarrows 97a \rightleftarrows 97b \rightleftarrows 97c$$

(95)

(358a)   (358a)   (358b)

The silene **124** is probably formed as its THF adduct and can be trapped by, e.g., 1,3-dimethyl 2,3-butadiene to give a [4+2] cycloadduct. The attempt to liberate the silene **124** from its donor adduct results in the formation of a disilacyclobutane **125**. This is ascribed to the prolonged life-time of the intermediate **359** formed by the methyl migration in the silene (equation 96), which allows for a hydrogen migration to take place.

Photolytically generated 1-silabuta-1,3-dienes undergo a thermal reverse reaction to 2-silacyclobutenes. Thus 2-phenylsilacyclobut-2-ene **360** is easily opened to the 2-phenylsilabuta-1,3-diene **361** by irradiation in 3-methylpentane matrix at 77 K or by flash photolysis at ambient temperature (equation 97)[183]. The rate for the thermal reverse reaction was measured at room temperature and the activation energy for the 1-siladiene ring closure was estimated to be 9.4 kcal mol^{-1} [183].

2. Dimerizations

The dimerization of silenes is probably *the* prominent type of reaction of silenes since the structures of most of the relatively stable and transient silenes have been established by isolation and identification of their dimers. The formation of disilacyclobutanes has

16. Silicon–carbon and silicon–nitrogen multiply bonded compounds 917

now gained acceptance as evidence for the intermediate formation of a transient silene.

[Scheme showing equation (96): Me$_2$Si(F)—C(Na)(SiBu-t_3)(SiMe$_3$) → (with Me$_3$SiCl, −Me$_3$SiF, −NaCl, THF) → Me$_2$Si=C(SiBu-t_3)(SiMe$_3$) (124·THF), then −THF leads to cyclic structure (125) with Me$_2$Si—SiMe$_2$ ring bearing SiBu-t_3; H migration connects to silene (124).]

[Scheme: (124) Me$_2$Si=C(Me)(SiMe$_2$)(SiBu-t_3) ⇌ (359) [bracketed cyclic intermediate] ≠ (124)]

[Scheme (97): (360) Ph/SiMe$_2$ cyclobutene →$^{h\nu}$ (361) Ph/SiMe$_2$ open form →$^{\Delta}$ (360)]

Simple silenes readily couple to yield the head-to-tail dimers, 1,3-disilacyclobutanes[1,2,15,157]. The dimerization is extremely facile and silenes bearing only small alkyl groups dimerize in an argon matrix even at 40 K, i.e., the dimerization proceeds at a diffusion controlled rate. Bulky substituents slow down the dimerization rate and allow the isolation of stable silenes. The head-to-tail dimerization (equation 98) is the predominant dimerization path for silenes, including those of the 'Auner–Jones'[79,80] and 'Wiberg' type[72,73,78].

[Equation (98): head-to-head C—Si / C—Si ring ← C=Si → head-to-tail Si—C / C—Si ring]

head-to-head   head-to-tail

In contrast, 1,1-silylsilenes[39,104,107,108] and 2-siloxysilenes[3,86,90,180] i.e. silenes of the 'Apeloig–Ishikawa–Oehme' and 'Brook' type as well as some 1-silaallenes[184,185], dimerize in a head-to-head mode yielding 1,2-disilacyclobutanes (equation 98).

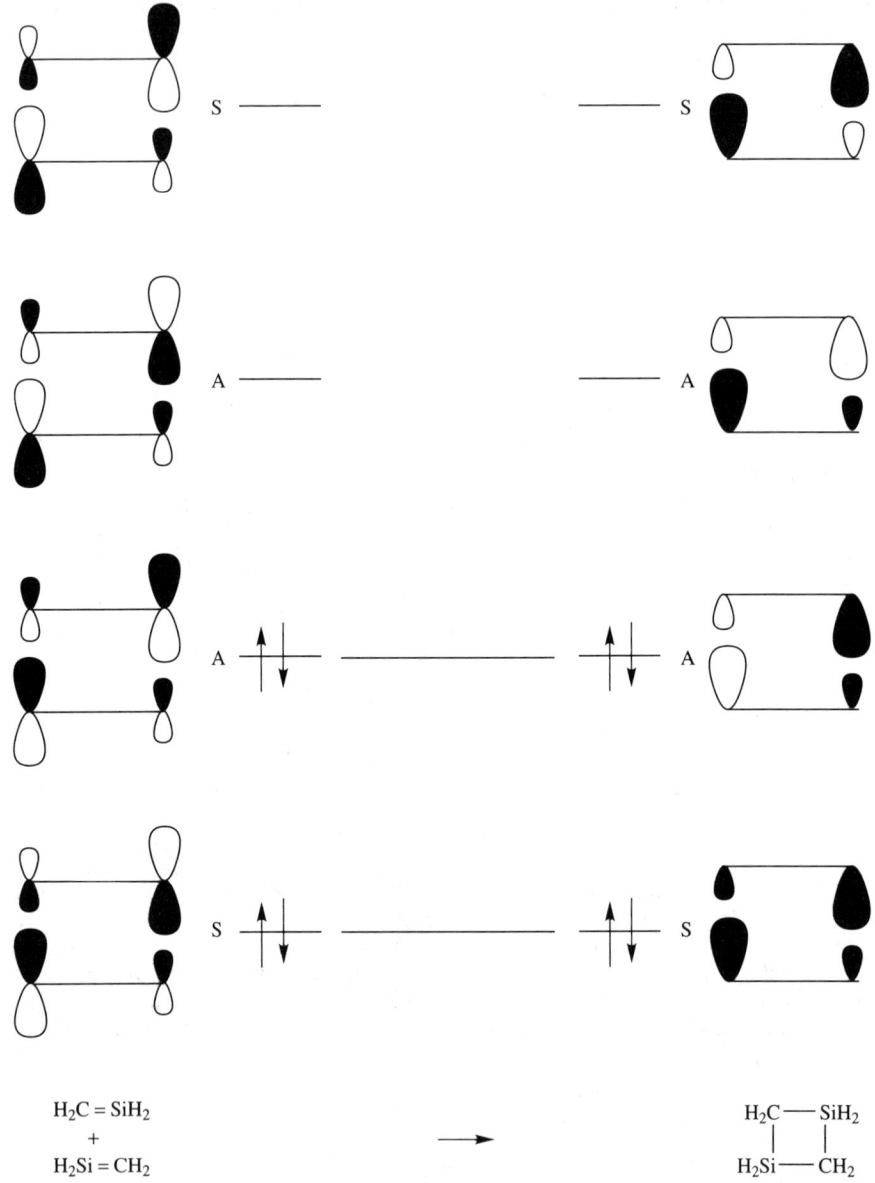

FIGURE 1. Correlation diagram for the head-to-tail dimerization reaction allowing for reduction in orbital symmetry. S and A designate a molecular orbital as symmetric or antisymmetric with respect to the C_2 axis perpendicular to the plane of the page. Reprinted with permission from Ref. 186. Copyright (1992) American Chemical Society

The principal questions arising from the facile occurrence of a formally forbidden concerted process like the dimerization of silenes attracted considerable interest from theoreticians. The theoretical work done is well summarized in several recent reviews[11,12] and we will only discuss briefly the very recent theoretical results on the silene dimerization reaction.

All recent *ab initio* studies on the head-to-tail dimerization reaction for the parent silene $H_2Si=CH_2$ predict a very exothermic reaction with a low barrier, despite the fact that the reaction is formally forbidden[12]. It is believed that the strong polarization of the Si=C double bond leads to a relaxation of the Woodward–Hoffmann rules. A detailed analysis of the symmetry of the head-to-tail silene dimerization reveals, however, that a concerted $[2\pi_s + 2\pi_s]$ reaction in the appropriate point group is not forbidden by symmetry (see Figure 1)[186].

Thus, Seidl, Grev and Schaefer found a barrier for a concerted fully synchronous $[2\pi_s + 2\pi_s]$ reaction path of 5.2 kcal mol^{-1} (at CCSD/DZ+d+ZPE)[186]. At the same level of theory the head-to-tail dimerization is exothermic by 79.1 kcal mol^{-1}. The transition state for this reaction appears very early on the reaction coordinate, i.e., the silene geometry being nearly conserved, as expected for a highly exothermic reaction[186]. The calculated reaction path is shown in Figure 2a.

These conclusions of Schaefer and coworkers were severly challenged by Bernardi, Robb, Olivucci, and coworkers who concluded, based on multiconfigurational calculations at the CASSCF level, that in the course of the head-to-tail silene dimerization a conical intersection occurs[187,188]. That is, the reaction path does not proceed on a single energy surface but the formerly higher $(\pi-\pi^*)$ doubly excited state intercepts with the ground state potential energy surface while the reaction proceeds. Thus, Bernardi and coworkers favoured a stepwise mechanism with a biradical intermediate, lying 18.9 kcal mol^{-1} lower in energy than the two separated silenes with a barrier for its formation of 5.3 kcal mol^{-1} (at CASSCF/DZ + d) (see Figure 2b). At this level of theory the transition state for the concerted reaction channel is 12.6 kcal mol^{-1} higher in energy than the reactants[187,188].

The head-to-head dimerization of $H_2C=SiH_2$ was also studied by both groups[186,188]. Schaefer and coworkers found the 1,2-disilacyclobutane to be less stable than the head-to-tail dimer, 1,3-disilacyclobutane by 19.8 kcal mol^{-1} (at SCF/DZ+d)[186]. Both theoretical studies predict for the head-to-head dimerization a stepwise reaction involving a carbon-centred biradical which is formed after the initial Si—Si bond formation. Bernardi and coworkers located a barrier of merely 2.5 kcal mol^{-1} for Si—Si bond formation to create the biradical[188]. Note that this barrier for the head-to-head dimerization is distinctively smaller than the barrier for the head-to-tail process (5.3 kcal mol^{-1}). This theoretical finding is totally at odds with the experimental experience that simple silenes including $H_2Si=CH_2$ dimerize to give the head-to-tail dimer[8]. Apparently, even more sophisticated computational methods are required to reproduce the correct course of these reactions.

a. Head-to-head dimerization. Head-to-head dimerization of silenes is the typical dimerization mode for silenes of the 'Brook' type, 'Apeloig–Ishikawa–Oehme' type and of 1-silaallenes. The reduced polarity of the Si=C double bond in these families of silenes seems to favour the head-to-head dimerization mechanism and to raise the barrier for the dimerization reaction.

Relatively little is known experimentally about the mechanism of the reaction. A widely accepted mechanism originally suggested by Brook and coworkers starts with the formation of a Si—Si bond, giving a carbon centred 1,4-biradical[86]. This 1,4-biradical then combines in a second step to the 1,2-disilacyclobutane. This mechanism is favoured by the calculations and is also corroborated by experiments: the relatively stable silene **149**

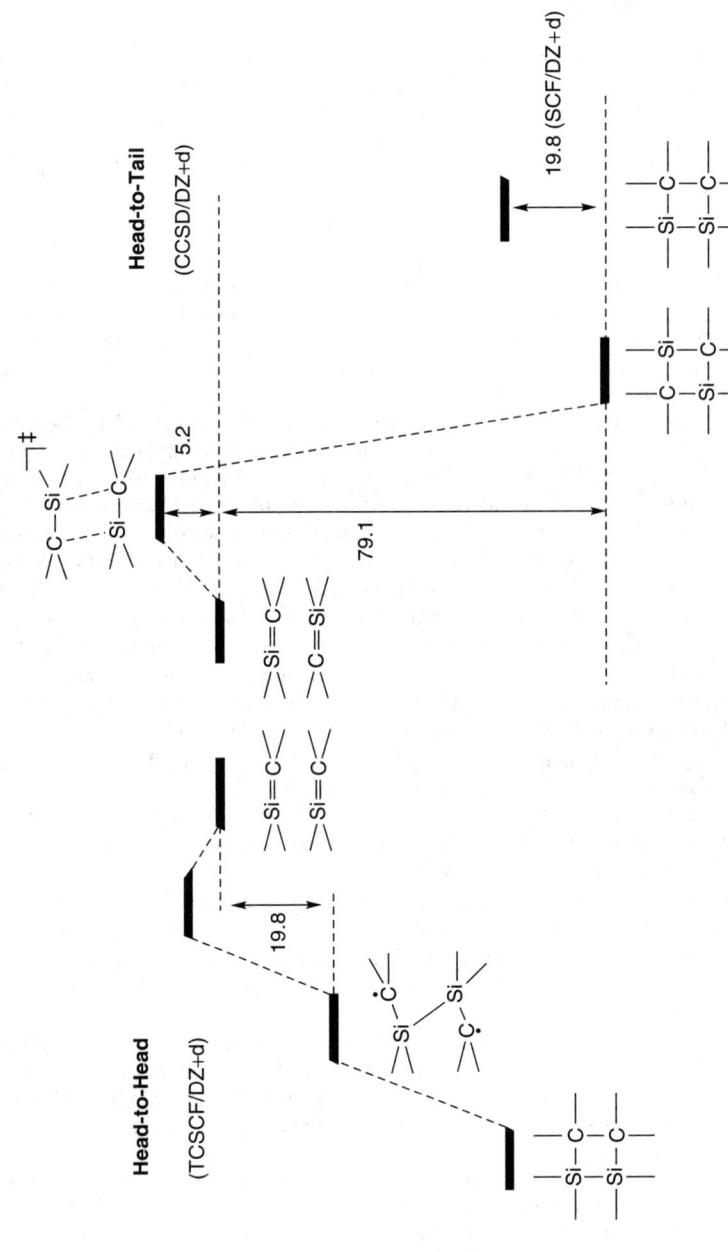

FIGURE 2a. Calculated reaction path for the head-to-head and head-to-tail dimerization of silene (at TCSCF/DZ + d or CCSD/DZ + d, respectively, relative energies in kcal mol^{-1})[186]

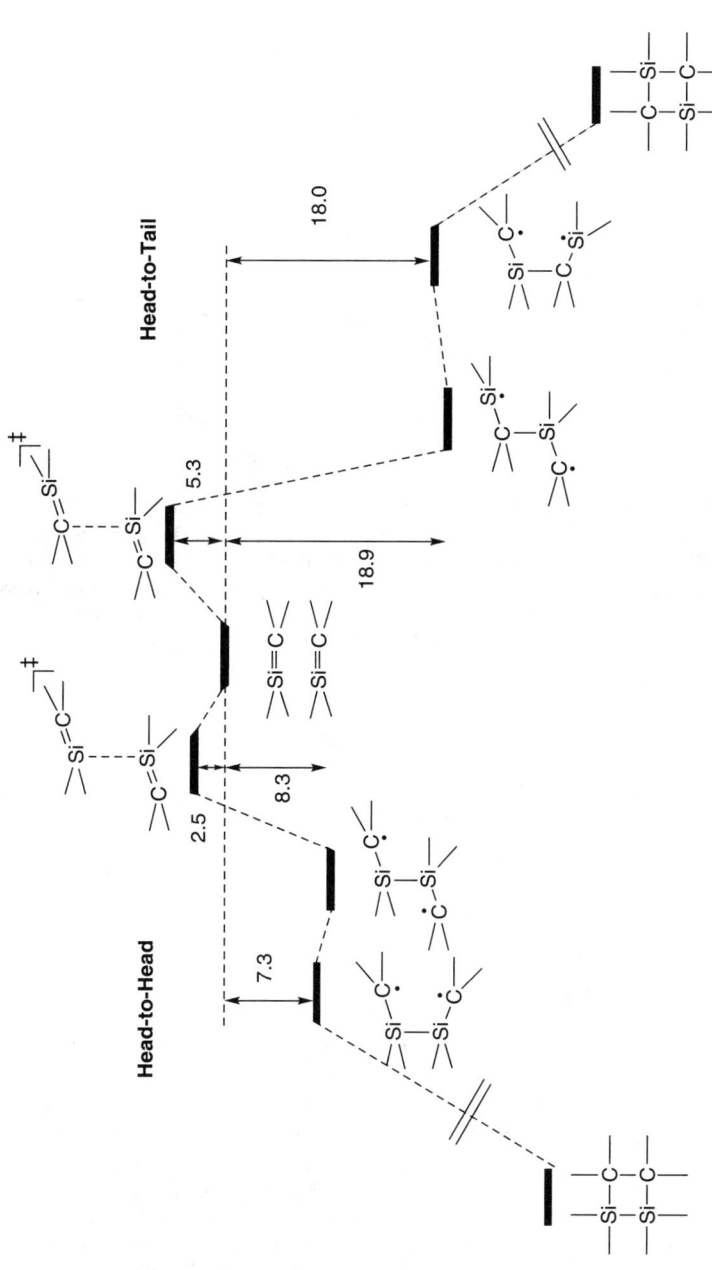

FIGURE 2b. Calculated reaction path for the head-to-head and head-to-tail dimerization of silene: biradical path (relative energies in kcal mol^{-1} at CASSCF/DZ + d)[187,188]

is in a temperature-sensitive equilibrium with its head-to-head dimer **362**, a solid whose crystal structure has been obtained (equation 99).

$$(Me_3Si)_3Si-C(=O)(Bu\text{-}t) \xrightleftharpoons[\Delta]{h\nu} (Me_3Si)_2Si=C(OSiMe_3)(Bu\text{-}t)$$

(**147**)   (**149**)

⇅   (99)

$(Me_3Si)_2Si$—C(Bu-t)(OSiMe$_3$)
|
$(Me_3Si)_2Si$—C(Bu-t)(OSiMe$_3$)

(**362**)

The very long endocyclic C—C bond in **362** (1.66 Å) indicates already relatively facile homolytic C—C bond cleavage[38]. The solution of the dimer and the irradiated pivaloylacylsilane **147** gave rise to a strong broad ESR signal without a fine structure. Attempts to intercept a putative biradical like **363** failed, but they do not rule out the possible existence as an intermediate in low concentrations. A biradical species like **363** would also account for dimerization products isolated from other silenes (see below)[119].

$$R^2_2Si-\overset{\bullet}{C}(R^1)(R^3)-\ldots-R^2_2Si-\overset{\bullet}{C}(R^1)(R^3)$$

(**363**)

Clear evidence for radicals as intermediates in a somewhat unusual head-to-head dimerization of matrix isolated silacyclobutadiene **296** was given by Fink and coworkers[189]. Upon annealing of the 3-methylpentane matrix the unusual head-to-head dimer **364** is formed in 95% yield[190]. When the reaction was followed by ESR spectroscopy two distinctly different triplet biradicals have been identified[189]. This finding discards the originally suggested mechanism for the formation of **364**[190] and corroborates the reaction mode outlined in equation 100[163,189]. After initial formation of a Si—Si bond, the allylic biradical **365** undergoes an internal hydrogen atom abstraction to afford the new biradical species **366**. Ring closure of this ultimate biradical gives the observed dimer **364**.

Silenes, bearing an allylic hydrogen, frequently give 'linear' non-cyclic dimers, in which two silene molecules form a Si—Si bond[39,86,88,89,107,112]. This linear head-to-head dimer can be formed by intramolecular disproportionation of the initially formed biradical (path B in equation 101). Of course, an 'ene'-reaction (path A in equation 101)

16. Silicon–carbon and silicon–nitrogen multiply bonded compounds   923

between two silenes would also account for the formation of the linear dimer. This dimerization mode is dominant for silenes of the type $(Me_3Si)_2Si=C(OSiMe_3)R^1$ and $(Me_3Si)_2Si=CR^2R^1$ (R^1 = alkyl). In some cases, such as for $(Me_3Si)_2Si=C(OSiMe_3)R^1$ (R^1 = i-Pr, CH_2Ph)[89] or $(Me_3Si)_2Si=CPhMe$[112], both the cyclic and linear dimer were formed in appreciable amounts. Table 5 gives more detailed data on the products formed in the head-to-head dimerization of silenes.

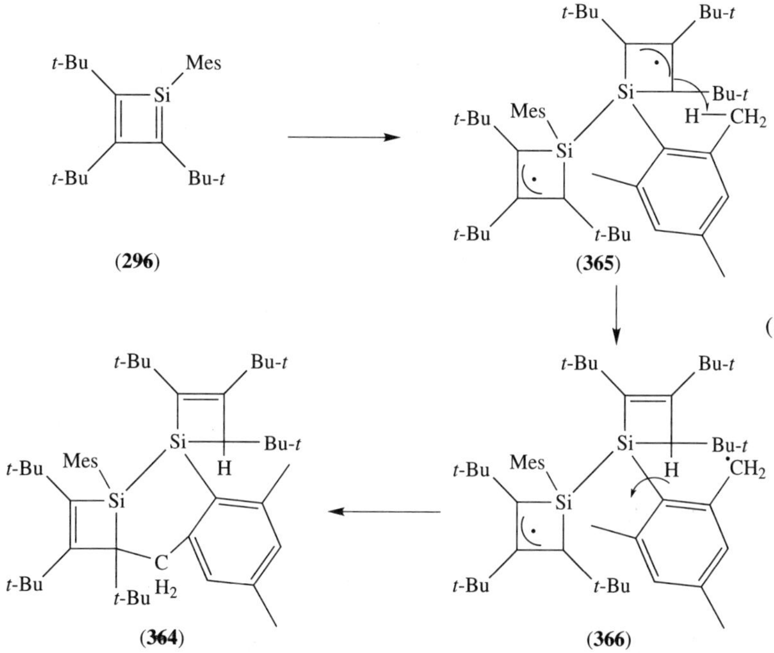

(100)

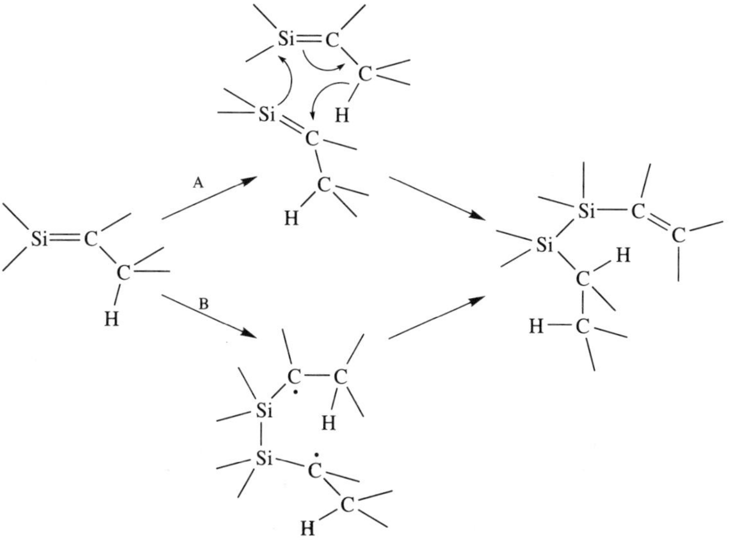

(101)

TABLE 5. Formation of different head-to-head dimers from silenes and silaallenes

Silene	Linear dimer	1,2-Sila-cyclobutane	Tetrahydro-2,3-disilanaphthalene	Remarks	References
$(Me_3Si)_2Si=C(OSiMe_3)R^1$					
$R^1 = Me$	exclusively				86,88
$R^1 = Et$	exclusively				89
$R^1 = i\text{-}Pr$	24%	76%			89
$R^1 = CH_2Ph$	20%	80%			89
$R^1 = t\text{-}Bu$		exclusively		see Ref. for details	86
$R^1 = Ph$		exclusively		equilibrium with silene	89
$R^1 = bco^a$		exclusively		equilibrium with silene	90
$R^1 = CF_3$		exclusively			92
$R^1 = Ar$		exclusively			92
$(Me_3Si)R^2Si=C(OSiMe_3)R^1$					
$R^1 = 1\text{-}Ad^b,\ R^2 = Me$		exclusively			90
$R^1 = t\text{-}Bu,\ R^2 = Ph$		exclusively			90
$R^1 = 1\text{-}Ad^b,\ R^2 = Ph$		exclusively			90
$Ph_2Si=C(OSiMe_3)1\text{-}Ad^b$		exclusively		equilibrium with silene	95
$(Me_3Si)_2Si=CMe_2$	exclusively				112
$(Me_3Si)_2Si=CPhMe$	67%	33%			112
$(Me_3Si)_2Si=cyhex^c$	exclusively				39

$(Me_3Si)_2Si=2-Ad^d$		exclusively		39
$(Me_2EtSi)(Me_3Si)Si=2-Ad^d$		exclusively	reverts at 70°C	191
$Ph_2Si=C=C(SiMe_3)_2$		exclusively		129, 192
$(Me_3Si)_2Si=CR^1R^2$				
$R^1 = p$-diethylaminophenyl, $R^2 =$ Me	80%	20%		110
$R^1 = 2,4$-di-t-Bu-phenyl, $R^2 =$ H		88%	12%	109
$R^1 =$ mesityl, $R^2 =$ H		exclusively	exclusively	107, 117
		82%	18%	108
$R^1 = t$-Bu, $R^2 =$ H		exclusively		107
$R^1 =$ Me, $R^2 =$ Me	exclusively		MeLi, Et_2O, $-78°C$	104, 107
			$PhMgBr$, THF, rt	
$R^1 = 2,5$-di-isopropylphenyl-, $R^2 =$ H		46%	54%	118
$R^1 =$ HC=CMe_2, $R^2 =$ H		exclusively		106
$R^1 =$ HC=CHPh, $R^2 =$ H		exclusively		106
$R^1 =$ Tip^e, $R^2 =$ H		exclusively		105

[a] bco = bicyclooctyl.
[b] 1-Ad = 1-adamantyl.
[c] 2-cyhex = cyclohexenylidene.
[d] 2-Ad = 2-Adamantylidene
[e] Tip = 2,4,6-triisopropylphenyl.

Interestingly, dimerization of thermolytically generated **8** gives two products. The regular head-to-tail dimer **367** is formed preferentially, but minor amounts of the linear head-to-head dimer **368** could also be detected (equation 102)[20].

$$\text{(8)} \quad \xrightarrow{2x,\ 398.2\,°C} \quad \text{(367) major} \quad + \quad \text{(368) minor} \quad (102)$$

Conlin and coworkers have studied the kinetics of the dimerization of the silene **369** by laser flash photolysis. **369** gives in a clean reaction exclusively the linear dimer **370** (equation 103)[88]. Second-order kinetics were observed with a bimolecular rate constant of $k = 1.3 \times 10^7\ \text{M}^{-1}\,\text{s}^{-1}$. The temperature dependence of the rate constant k gave a preexponential factor $\log H(\text{A/s}^{-1})$ of 7 ± 1 and an activation energy $E_A = 0.2 \pm 0.1\ \text{kcal mol}^{-1}$. These activation parameters suggest a highly ordered transition state requiring only small enthalpic changes for reaction[88]. This low A value and the small E_A are more in accord with a concerted 'ene'-reaction than with an intermediate biradical. However, more experimental data are needed in order to draw more definite conclusions.

$$\text{(369)} \quad \longrightarrow \quad \text{(370)} \quad (103)$$

The surprising product distribution of the dimerization of some silenes having a bulky aryl group at the doubly bonded carbon studied by Oehme and his group[107–109,117,118] might be rationalized by the occurrence of biradicaloid intermediates. Oehme and coworkers found that, for example, mesitylsilene **371** dimerizes in a kinetically controlled reaction to a formal head-to-head [4 + 2] cycloadduct, the tetrahydro-2,3-disilanaphthalene **372**[117]. **372** gradually decomposes to the thermodynamically more stable 1,2-disilacyclobutane **373** (equation 104)[108]. For other aryl groups [R = bis(2,4-di-*t*-Bu)phenyl[109], R = bis(2,5-di-*i*-Pr)phenyl[107,118]] both types of dimers have been isolated directly from the reaction mixture. In all cases the 2,3-disilanaphthalene could be converted by heating to the thermodynamically more stable 1,2-disilacyclobutane. An intermediate carbon-centred, 1,4-benzyl-type biradical **374** can account for both types of dimers[109]. 1,4-Recombination of **374** affords the 1,2-disilacyclobutane **373**.

16. Silicon–carbon and silicon–nitrogen multiply bonded compounds

Alternatively, the reaction may proceed via 1,6-ring closure giving the isolated 2,3-disilanaphthalene **372**[109]. It should be mentioned, however, that under the reaction conditions it is possible that silyl lithium compound **375** formed from the precursor of the silene **376** (obtained from metalation of **377**) by a 1,3-trimethylsilyl shift adds to the intermediate silene **371** and that subsequent ring closure with elimination of lithium silanolate gives rise to the observed products **372** and **373** (equation 105).

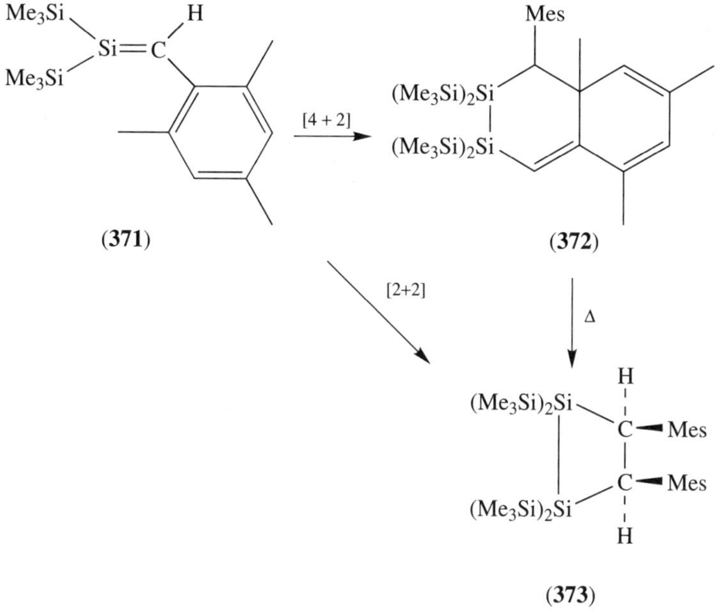

(104)

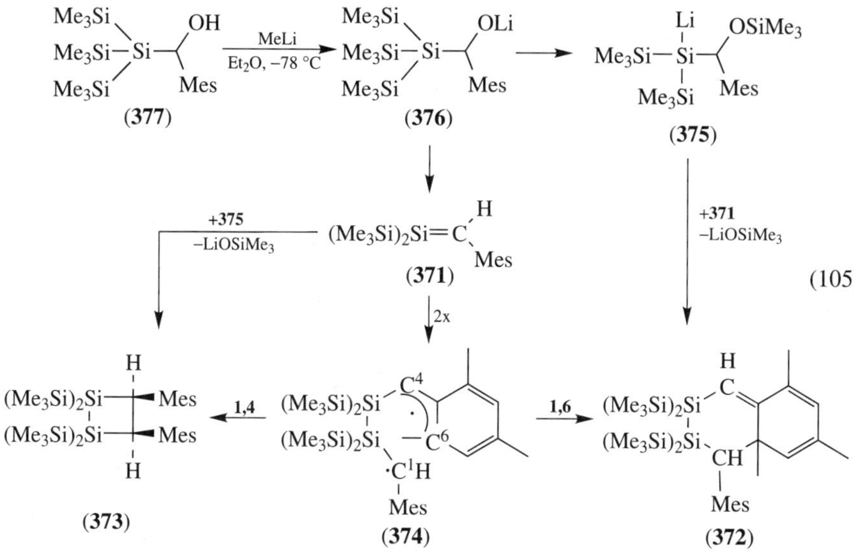

(105)

Head-to-head dimers with bulky substituents at the C—C bond tend to be thermolabile and they can be used as convenient sources for relatively stable or transient silenes. Examples are the already-mentioned dimer of the relatively stable silene **149**[86] and the 1,2-disilacyclobutane **164**, which liberates the adamantylidene silene **165** smoothly upon heating to 70 °C in benzene (equation 106)[39].

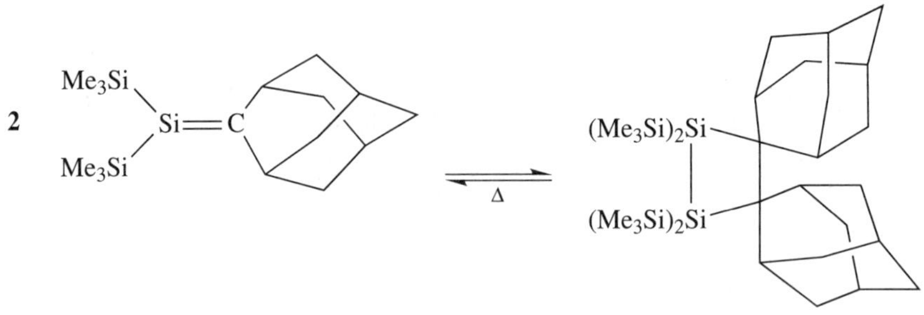

(106)

In contrast, the cleavage of the 1,2-disilacyclobutane **378** needs more drastic conditions (250 °C) and yields the thermodynamically more stable linear dimer **379**. It was suggested that this reaction proceeds via the 1,4 biradical **380** (equation 107)[112].

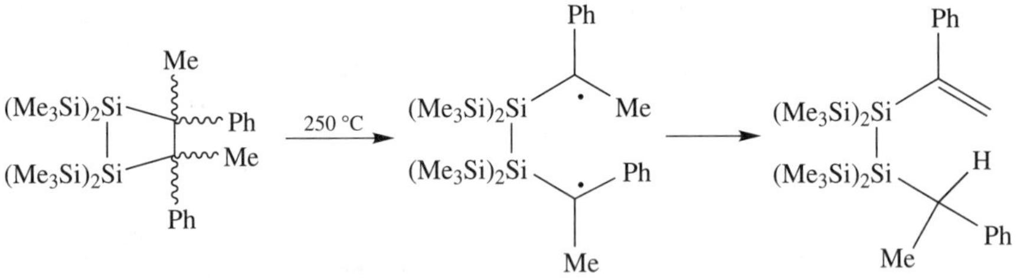

(107)

Obviously, the interplay between subtle steric and electronic effects determines the final product of the dimerization of 'Apeloig–Ishikawa–Oehme' silenes. Thus, the *p*-dimethylaminophenyl-substituted silene **381** gives the expected dimerization products, the 1,2-disilacyclobutane **382** and the linear dimer **383**, in a relative ratio of 1 : 4 (equation 108)[110]. In sharp contrast the very similar *o*-dimethylaminophenyl-substituted silene **384** gives in 63% yield a 1 : 2 mixture of the *E/Z*-1,3-disilacyclobutane **385** (equation 109)[110]. It was suggested that this peculiar regiospecific outcome of the dimerization of **384** results from a directive *intermolecular* donor–acceptor interaction between the *ortho*-dimethylamino group and the Si=C double bond[110]. Clearly, more work has to be done in order to understand the small effects which determine the dimerization behaviour of 'Apeloig–Ishikawa–Oehme' silenes[110].

The dimerization of silatrienes **216** obtained from photolysis of disilanylbenzenes **210** was studied by Ishikawa and coworkers[193,194]. A head-to-head dimerization takes place under participation of the vinylogous double bonds to give the isomers **386a** and **386b**.

A ring-opened isomer **387** is obtained through thermolysis of **386b** (equation 110).

(**381**)

(**382**)   +   (**383**)   (108)

(**384**)   →   (**385**)   (109)

(110)

b. *Head-to-tail dimerization.* The head-to-tail dimerization is found for 'Wiberg'-type[78] and neopentyl silenes[79,80].

Heating the azasiletane **95b** in the absence of trapping reagents for **92** gives **388**, the head-to-tail dimer of **92** (equation 111)[72,73]. This dimer is even formed at $-100\,°C$ when **92** is synthesized via salt elimination of $Me_2SiF-CLi(SiMe_3)_2$.

(95b) → $-Me_3SiN=CPh_2$ → (92) → (388)   (111)

The mode of dimerization of silaacrylates **389** depends on the steric bulk of the substituents[195,196]. Large substituents R prevent formation of dimers **390** and lead to intramolecular [2 + 2] cycloaddition to give 1-oxa-2-silacyclobut-3-enes **391** (equation 112). Smaller substituents R allow head-to-tail dimerization to the eight-membered ring compound **390**.

(389); R = 1-Ad, *t*-Bu → (391); R = *i*-Pr, Me → (390)   (112)

For a central hexamethyltrisilanyl unit in the bisdiazomethyloligosilane **327** the silenes resulting from photolytic decomposition give a formal head-to-tail dimer **329** and a head-to-head dimer **330** (equation 80)[169].

3. Nucleophilic additions

Nucleophiles easily attack the Si=C double bond at the silicon. Complexes between diethyl ether and the parent silene $H_2Si=CH_2$ have been detected by low temperature NMR[27] and stable silenes of the 'Wiberg'-type form adducts with THF, triethylamine, pyridine and even with the fluoride anion[83,84,197]. These complexes have been structurally characterized by X-ray structure analysis and details are presented in Section I.C.1. Similarly, the initial step of the alcohol addition to silenes is the formation of a short-lived complex between the silene and the attacking alcohol[198], as is evident from the measured negative activation energy in the reaction of simple silenes with alcohols[35,34,199].

a. Reaction with alcohols. In the presence of trapping reagents of the general type A−H (O−H: water, alcohols, carbonic acids; S−H: thiols; N−H: amines, imines) addition across the Si=C bond takes place. The general reaction is shown for a 'Wiberg'-type silene in equation 113[62]

$$Me_2Si=C(SiMe_3)_2 \xrightarrow{A-H} \underset{A\quad H}{Me_2Si-C(SiMe_3)_2} \qquad (113)$$
$$(92)$$

The addition reaction of alcohols to silenes is a strictly regiospecific process; the OR group in the product is always attached to the silicon. The reaction is of special importance since silenes are easily trapped by reagents like methanol and *t*-butanol, and isolation of the adducts is normally taken as evidence for an intermediate silene. Furthermore, the alcohol addition to silenes is the prototypical reaction for the 1,2-addition of polar bonds across the Si=C double bond. A complete account on the alcohol addition to silenes, including the most recent mechanistic implications, is given in the chapter by Sakurai. We will therefore give only a brief survey of the reactivity of several families of silenes towards alcohols.

The addition of methanol to Brook-type silenes at room temperature using an excess of methanol was found to be a non-stereospecific process, as is evident from its stereochemical outcome. Methanolysis of a 1 : 4 mixture of *E/Z*-**392** gives a mixture of two diastereomeric methanol adducts *SS*- and *RS*-**393** in a relative ratio of 2.5 : 1 (equation 114)[95].

(114)

Fink and coworkers showed, however, that the ethanol addition to the sterically highly hindered silacyclobutadiene **296** is a stereospecific *syn* addition process[163]. When **296** is photolytically generated from the silylene **295** in the presence of ethanol, only the *syn* adduct *E*-**394** is formed (equation 115). Upon further photolysis the silacyclobutene **394** undergoes a photoisomerization process involving photolytic ring opening of *E*-**394** to give silabutadiene *EZ*-**395**, followed by a thermal ring closure to *Z*-**394** with the opposite stereochemistry[163]; a second cycle equilibrates the isomers *E*/*Z*-**394** via silabutadiene *ZZ*-**395** (equation 116).

In agreement, the thermolytically generated silacyclobutadiene **341** reacts with *t*-butanol stereospecifically, yielding the regioisomers **339** and **340** (equation 84)[119].

The stereospecificity of methanol addition to neopentylsilenes has been investigated by Jones and Bates[68]. The mild thermal retro-Diels–Alder reaction (at *ca* 200 °C) of *E* and *Z* anthracene [4 + 2] cycloadducts **110** liberates stereospecifically the corresponding silenes **111**, which are trapped by methanol. The ratio of the diastereomeric products **396a/396b** coincides with the *E/Z* ratio of the precursors **110** (equation 117). In photochemical reactions of similar silene precursors, alcohols were used also to probe the decomposition mechanism[69].

An alternative mechanism to explain the stereospecificity has been suggested in the review by Brook and Brook[9].

The most widely accepted and applicable mechanism for the alcohol addition was suggested by Kira, Maruyama and Sakurai[198]. They found that the ratio of the *syn* adduct **397** to the *anti* adduct **398** obtained by the reaction of the photolytically generated silene **253** is dependent on the alcohol concentration and on its acidity. Thus, the **397/398** ratio increased in the order MeOH < *n*-PrOH < *i*-PrOH ≪ *t*-BuOH. *t*-Butanol gave only *syn* adduct **397**[198]. At high concentrations of the alcohol the *anti* isomer **398** is preferentially formed. This is compatible with an initial formation of a complex **399** between the silene and the alcohol. Intramolecular proton migration in **399** (route b) competes with the intermolecular proton transfer from a second alcohol to **399** (route a). These two processes give the *syn* and *anti* isomers, **397** and **398**, respectively (equation 118). Higher concentrations of alcohol favour process (a) and higher acidity of the protonated alcohol and lower acidity of the alcohol facilitate the intramolecular proton transfer (b)[198]. Further detailed mechanistic studies were conducted by Leigh and coworkers[25,34,200] and these

16. Silicon–carbon and silicon–nitrogen multiply bonded compounds

are summarized in the chapter by Sakurai.

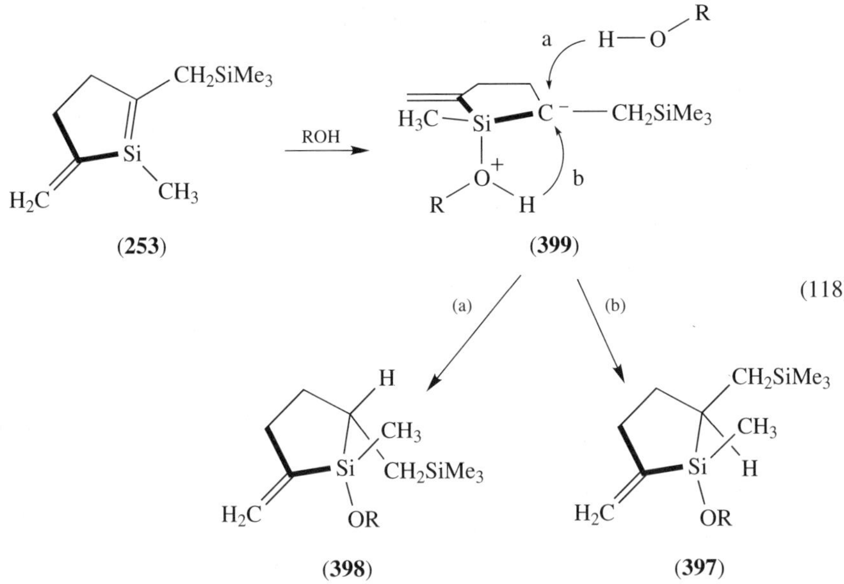

(118)

The competitive migration to either a vinyl or phenyl group has been investigated by Ishikawa and coworkers[133]. While with traps other than methanol there is proof for the intermediacy of both silenes **238** and **239**, formed from **237** (of equation 53), just one product, i.e., **400**, is found for the reaction with the alcohol (equation 119).

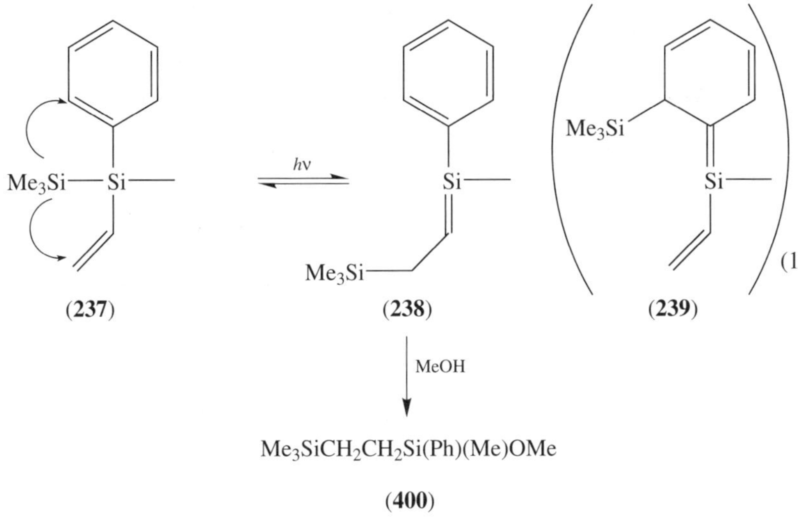

Me₃SiCH₂CH₂Si(Ph)(Me)OMe

(**400**)

With the dihydropyranyl-substituted phenyldisilane **240** Ishikawa and coworkers found the formation of both possible types of silenes **241** and **242**; they were formed and trapped by methanol giving adducts **401–403** (equation 120)[134].

b. *Reaction with alkoxysilanes*. Polar single bond species, e.g., A—EMe$_3$ are also suitable for trapping silenes as long as one of the atoms has a free electron pair (E = Si, A = OR, NR$_2$, N=CPh$_2$; E = Ge, Sn, A = Cl). The general scheme is shown for the 'Wiberg'-type silene **92** (equation 121)[62].

$$\text{Me}_2\text{Si}=\text{C(SiMe}_3)_2 \xrightarrow{\text{A}-\text{EMe}_3} \underset{\underset{\text{A}}{|}}{\text{Me}_2\text{Si}}-\underset{\underset{\text{EMe}_3}{|}}{\text{C(SiMe}_3)_2}$$
(**92**)

(121)

Alkoxysilanes are frequently used as scavenger reagents for silenes. They add regiospecifically to the Si=C bond. In the case of silacyclobutadiene **296** the reaction was shown to be also stereospecific. Thus Z-**404** is the sole product of the addition of trimethylsilyl methyl ether (TMSOMe) to **296** (equation 122)[163]. Characteristically 'Brook'-type silenes do not react with alkoxysilanes like TMSOMe.

(122)

c. *Reaction with organometallic reagents*. In the course of the sila-Peterson reaction a nucleophilic addition of an organometallic reagent to the incipient silenes is often observed. Organolithium reagents[63,103,104,107,113,117,201], silyllithium reagents[103], lithium trimethylsilanolate[103,113,116], and Grignard reagents[103,105,116] add regiospecifically across the Si=C double bond. In each case the anionic part of the reagent adds to the silicon, yielding a carbanion which can be protonated, deuteriated by hydrolysis, or may rearrange prior to hydrolysis (equation 123). The addition of organometallic reagents to 'Brook-type' silenes is complicated by subsequent reactions leading to new silenes[116] (see Section I.A.5.d). Addition of silyl anions gives polysila-anions as final products which result from a silyl shift in the intermediate polysilyl carbanion[103] (see Section I.A.5.a for an example). More examples are summarized in Table 6.

(123)

Addition reactions of lithium organyls have been investigated for **97**[63,201]. The reaction is not a nucleophilic substitution of Br$^-$ by R$^-$ in **97**·LiBr, but rather a two-step mechanism with an initial dissociation via **97** (equation 124). This was proven by the addition of the very efficient silene trap *t*-Bu$_2$MeSiN$_3$, that competes with the lithium organyl for the silene.

TABLE 6. Reactions of silenes with organometallic reagents

Silene	Organometallic reagent	Product after hydrolysis	Remarks	References
(Me$_3$Si)$_2$Si=CPh$_2$	PhLi	(Me$_3$Si)$_2$PhSi—CHPh$_2$		113
(Me$_3$Si)$_2$Si=CPh$_2$	Me$_3$SiOLi	(Me$_3$Si)$_2$(Me$_3$SiO)Si—CHPh$_2$		113
Me$_3$SiMeSi=CSiMe$_3$Ad-1	MeMgBr	Me$_3$SiMe$_2$Si—CH(SiMe$_3$)Ad-1		103
(t-Bu)$_2$Si=CHCH$_2$Bu-t	t-BuLi	(t-Bu)$_3$Si—CH=CHBu-t	product after LiH elimination	199
(Me$_3$Si)$_2$Si=CHBu-t	PhLi	(Me$_3$Si)$_2$PhSi—CH$_2$Bu-t		104, 200
(Me$_3$Si)$_2$Si=CHMes	PhLi	(Me$_3$Si)$_2$PhSi—CH$_2$Mes		117, 200
(Me$_3$Si)$_2$Si=CHMes	MeLi	(Me$_3$Si)$_2$MeSi—CH$_2$Mes		107, 200
(Me$_3$Si)$_2$Si=CHTip	(Me$_3$Si)$_3$SiMgBr	(Me$_3$Si)$_3$Si(Me$_3$Si)$_2$Si—CH$_2$Tip		105
(Me$_3$Si)$_2$Si=CMe$_2$	(Me$_3$Si)$_3$SiLi	H(Me$_3$Si)$_2$Si(Me$_3$Si)$_2$Si—CMe$_2$SiMe$_3$	product after 1,3-silyl shift	103
(Me$_3$Si)$_2$Si=C(OSiMe$_3$)Ad-1	Me$_3$SiOLi	Me$_3$Si(Me$_3$SiO)$_2$Si—C(1-Ad)(SiMe$_3$)H	formation of a new silene	116
(Me$_3$Si)$_2$Si=C(OSiMe$_3$)1	MeMgBr	Me$_3$SiMe$_2$Si—C(1-Ad)(SiMe$_3$)H	formation of a new silene	116
Ph$_2$Si=C(SiMe$_3$)$_2$	PhLi	Ph$_3$Si—C(SiMe$_3$)$_2$H		63, 201
Ph$_2$Si=C(SiMe$_3$)$_2$	t-BuLi	HPh$_2$Si—C(SiMe$_3$)$_2$H	elimination of isobutene	63, 201

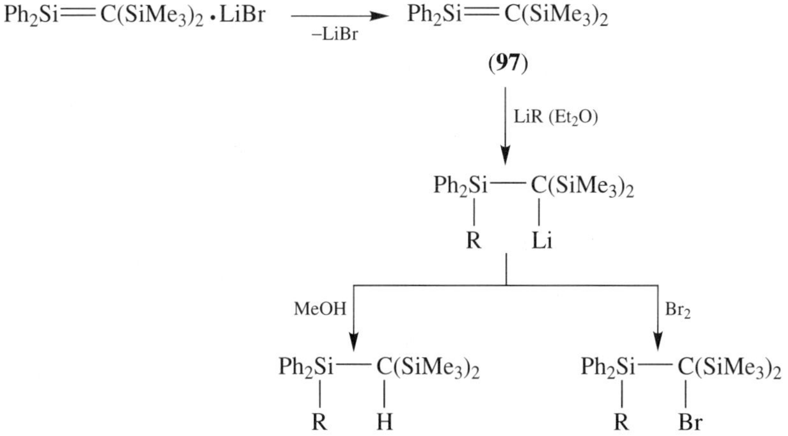

$Ph_2Si=C(SiMe_3)_2 \cdot LiBr \xrightarrow{-LiBr} Ph_2Si=C(SiMe_3)_2$
(97)

$\downarrow$ LiR (Et$_2$O)

```
        Ph₂Si——C(SiMe₃)₂
          |      |
          R      Li
```

MeOH ↙   ↘ Br$_2$

```
Ph₂Si——C(SiMe₃)₂       Ph₂Si——C(SiMe₃)₂
  |      |               |      |
  R      H               R      Br
```
(124)

The stable adamantylsilene **150** was found to react with the phosphorous ylide $Ph_3P^+CH_2^-$ to form the zwitterion **405** which cyclizes to the silacyclopropane **406**, as shown by NMR spectroscopy and by derivatization with methanol or acetophenone (equation 125)[202].

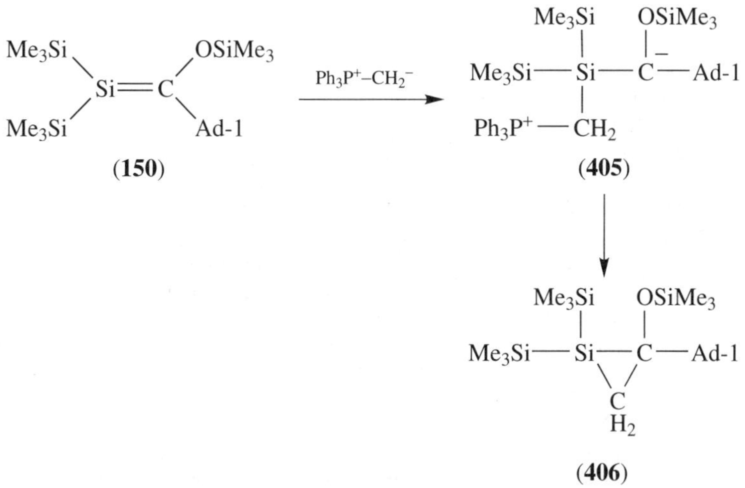

```
  Me₃Si     OSiMe₃                    Me₃Si    OSiMe₃
      \    /                            |       |  ⁻
       Si=C              Ph₃P⁺–CH₂⁻    Me₃Si—Si——C—Ad-1
      /    \         ─────────────▶           |
  Me₃Si     Ad-1                        Ph₃P⁺—CH₂
    (150)                                  (405)
```

$\downarrow$

```
                      Me₃Si     OSiMe₃
                        |         |
                  Me₃Si—Si————————C—Ad-1
                          \      /
                           C
                           H₂
                         (406)
```
(125)

An interesting addition reaction to silenes was recently described by Oehme and coworkers. They found that at high concentrations of LiBr the dimerization of the transient silene **371** yields 33% of the head-to-head dimer **373** and the head-to-tail dimer **407** in a 1 : 5.6 relative ratio[108]. The formation of the unexpected dimer **407** was rationalized by the addition of LiBr to the Si=C bond and intermolecular cyclization of the α-lithiosilyl bromide **408** or its reaction with the transient silene **371** with subsequent cyclization to the 1,3-disilacyclobutane **407** (equation 126)[108].

$$\begin{array}{c}
\text{Me}_3\text{Si} \\
\phantom{\text{Me}_3\text{Si}}\diagdown \\
\phantom{\text{Me}_3\text{Si}\diagdown}\text{Si}=\text{C} \\
\phantom{\text{Me}_3\text{Si}}\diagup \phantom{\text{Si}=\text{C}}\diagdown \\
\text{Me}_3\text{Si} \phantom{\diagup\text{Si}=\text{C}\diagdown} \text{Mes}
\end{array} \quad \xrightarrow{2x} \quad \text{(371)} \quad \xrightarrow[-2\text{LiBr}]{+408} \quad \text{(373)}$$

(126)

Scheme showing interconversion between (371), (373), (407), (408) with +LiBr / −LiBr and 2x / −2LiBr steps.

4. Cycloaddition reactions

In this section we will summarize bimolecular reactions of silenes with alkenes, alkynes and dienes which might be regarded nominally as cycloaddition reactions.

It is well accepted that silenes undergo concerted cycloadditions of various types[8]. Most silenes readily form $[4\pi + 2\pi]$ cycloadducts with dienes, although products arising from the 'ene' reaction are often present and may occasionally even be the major product. In contrast to carbon chemistry $[2\pi + 2\pi]$ cycloadditions compete with the above processes. Silenes react even with alkenes, giving products arising from a formal $[2\pi + 2\pi]$ cycloaddition, provided no allylic hydrogen is present. The presence of an allylic hydrogen results normally in the products of an 'ene' reaction. The [4+2] and 'ene' reactions of silenes are believed to be concerted processes, but the [2 + 2] addition reaction is often considered to be a stepwise process involving 1,4-biradicals or 1,4-zwitterionic species. It should be pointed out, however, that true mechanistic studies, both experimental and theoretical, on the cycloadditions of silenes are very rare[54,88], and most of the mechanistic conclusions are based on product studies.

It is evident that the relative proportions of products from [2 + 2], [4 + 2] and 'ene' reactions depend strongly on the polarity of the Si=C bond, on the polarization of the frontier orbitals and on steric factors both in the silenes and in the reactant. Therefore, we will first discuss cycloaddition reactions of silenes without separating them into subclasses like [2+2] or [2+4] cycloadditions, but we will concentrate on the cycloaddition behaviour of several classes of silenes towards alkenes, alkynes and dienes. We will then discuss some special cycloaddition reactions in subsequent sub-sections.

a. Wiberg-type silenes. i. $Me_2Si=C(SiMe_3)_2$ 92. One of the greatest assets of Wiberg's silenes is that they may be stored as cycloadducts of, e.g., $Ph_2C=N-SiMe_3$, from which they are simply regenerated by heating (*vide supra*). The liberated silenes can then be studied in detail. $Me_2Si=C(SiMe_3)_2$ (**92**) has been investigated most closely[72−76,83,84,97] and the [2 + 2] imine adduct **95b** has been mostly used as its source.

α. *[2+2] Cycloaddition reactions.* The polar double bond of silene **92** can react in a [2 + 2] cycloaddition reaction with systems of the general formula a=b[62]. The cycloadducts **95a** and **95b** from $Ph_2C=Y$ (Y = O, N—SiMe$_3$) have already been mentioned in Section I.A.1. Their formation is reversible, and the concomitant formation of [2 + 4] cycloadducts is described below. With C=C bonds, [2 + 2] adducts **409** are formed when an electron-donating substituent is bound to the π-system as in vinyl ethers or *cis*-1-propenylethene (equation 127)

$$92 \xrightarrow{H_2C=CHR} \underset{(409)}{\overset{Me_2Si—C(SiMe_3)_2}{\underset{R}{\square}}} \quad R = OMe, \quad \overset{CH_3}{\underset{H}{>}}\overset{}{\underset{H}{<}} \quad (127)$$

The formation of [2 + 2] cycloaddition products obviously occurs in a stepwise fashion via an adduct of the Lewis acid **92** and the Lewis base a=b. Kinetic data, investigation of isotope effects and the isolation of **92**·donor adducts support this assumption.

β. [2 + 4] Cycloaddition and ene reactions. Dienes >C=C—C=C< such as buta-1,3-diene, isoprene, 2,3-dimethylbuta-1,3-diene, *trans*-piperylene, cyclopentadiene or anthracene react with **92** in Diels–Alder fashion to give [2 + 4] cycloadducts **410** (equation 128)[62]. Ene products **411** are formed additionally when the relative reaction rates for the [2 + 4] cycloaddition reaction and the ene reaction are comparable (e.g. for isoprene and 2,3-dimethyl-1,3-butadiene); Alkenes with allylic hydrogen (propene, 2-butene, isobutene) give ene products see equation 129.

$$92 + \bigcirc\!\!\!\!\!\!\!\diagdown \longrightarrow \underset{(410)}{\overset{Me_2Si—C(SiMe_3)_2}{\bigcirc\!\!\!\!\!\!\!\diagdown}} \quad (128)$$

$$92 + \overset{}{\underset{}{>}}\!\!\!C\!\overset{H}{\underset{}{<}} \longrightarrow \underset{(411)}{\overset{Me_2Si—C(SiMe_3)_2}{\underset{}{\bigcirc}\!\!\!\!\!\diagdown\! H}} \quad (129)$$

A one-step mechanism is proposed for the formation of the [2 + 4] cycloadducts and the ene products[62], since the sila-Diels–Alder and sila-ene reactions are regio- and stereo-selective. The stereospecifity of Diels–Alder reactions of **92** with hexadienes has been investigated[203], as well as the regio- and stereoselectivity and the relative reaction rates of ene reactions[176] and [2 + 4] cycloadditions[204]. A selection of relative reaction rates of **92** with trapping reagents is given in Table 7.

With heterodienes, like $Ph_2C=Y$ (Y = O, N—SiMe$_3$), [2 + 2] as well as [2 + 4] cycloadducts **95** and **96**, respectively, are obtained from the reaction with **92**.

When $Me_2SiX—CLi(SiMe_3)_2$ [X = F, Br, $(PhO)_2PO_2$] is decomposed in the presence of benzophenone, only the [2 + 4] cycloadduct **96a** is formed[60]. Similarly to **96b**, **96a** can be rearranged thermally to the [2 + 2] cycloadduct **95a** in quantitative yield; **95a** is the first example of an isolated oxasilacyclobutane[59]. The rearrangement occurs through the intermediate formation of **92**. This is supported by thermolysis reactions of **96a/95a** in the presence of trapping reagents, since the trapping products are essentially the same as those obtained from other sources of **92**.

TABLE 7. Relative rates of Me$_2$Si=C(SiMe$_3$)$_2$ (**92**) with trapping reagents[a]

Trapping reagent	Relative rate	Reaction type
i-PrNH$_2$	23000	addition
MeOH	21000	addition
i-PrOH	10000	addition
Ph$_2$C=O	3000	[2 + 2]
t-BuN$_3$	230	[2 + 3]
CpH[b]	2.9	[2 + 4]
Propene	2.9	ene
Buta-1,3-diene	1	[2 + 4]

[a] For a comprehensive compilation see References 62 and 176.
[b] Cyclopentadiene.

ii. [2 + 3] Cycloaddition reactions. Azides R—N$_3$ (R = Me$_3$C, Me$_3$Si, t-Bu$_2$MeSi, t-Bu$_3$Si) react with **92** (obtained from **95b** in Et$_2$O at 100 °C) to give [2+3] cycloadducts **412a**[62], which are, however, not stable under the reaction conditions. They decompose to the diazomethanes **413a** by rearrangement, or to (Me$_3$Si)$_2$C=N=N and the silanimines **414a**. The latter can be trapped, e.g., by excess azide to give the addition product **415a** and/or the dihydrotetrazoles **344a** (see also Section III) (equation 130).

(130)

16. Silicon–carbon and silicon–nitrogen multiply bonded compounds   943

An analogous reaction is observed with N_2O: the [2 + 3] cycloadduct **412b** rearranges to **413b**, or, with elimination of $(Me_3Si)_2CN_2$, to the silanone **414b**. In the presence of Me_3SiCl, the insertion product of **414b**, $Me_2SiCl-OSiMe_3$ is obtained. No [2 + 3] adduct is formed with $(Me_3Si)_2CN_2$.

iii. $Ph_2Si=C(SiMe_3)_2$ 97. The replacement of the two methyl groups at the silene silicon atom in **92** by two phenyl groups to give silene **97** is not accompanied by a profound change in silene reactivity[63,64]. The slight differences can be attributed to an electronically enhanced double bond polarity (i.e. a greater Lewis acidity of the silene silicon atom) and to a sterically greater shielding of the double bond. The differences in reactivity between **92** and **97** are therefore due to (a) more stable donor adducts of **97** due to the greater Lewis acidity, and (b) changes in the relative reactivity of the trapping reagents.

iv. $t\text{-}Bu_2Si=C(SiMe_3)_2$ (104), $Me_2Si=C(SiMe_3)[SiMe(t\text{-}Bu)_2]$ (104a) and 104a·donor. By LiF elimination from $t\text{-}Bu_2SiF-CLi(SiMe_3)_2$ only **104a** instead of **104** is isolated[67]. With benzophenone, silene **104a** (from **104a·THF**) forms a [2 + 4] cycloadduct **416** and a [2 + 2] cycloadduct **417** at $-78\,°C$. With increasing temperature **416** transforms to **417**, and **417** more slowly transforms to the [2 + 4] cycloadduct **418** of **104** with benzophenone. This means that an equilibrium exists between **104** and **104a** (equation 131). **418**, whose X-ray structure has been determined, serves as a source for **104**, which is trapped by acetone (to give an ene product) and by benzaldehyde (to form a [2 + 2] cycloadduct, unstable against cycloreversion into $PhCH=C(SiMe_3)_2$ and $t\text{-}Bu_2SiO$ containing substances). Less reactive traps, like trimethylsilyl azide, allow for rearrangement of **104** (again formed from **418**) to **104a**, which then gives an unstable [2 + 3] cycloadduct **420**, which transforms into **421** and **422** (equation 132).

(131)

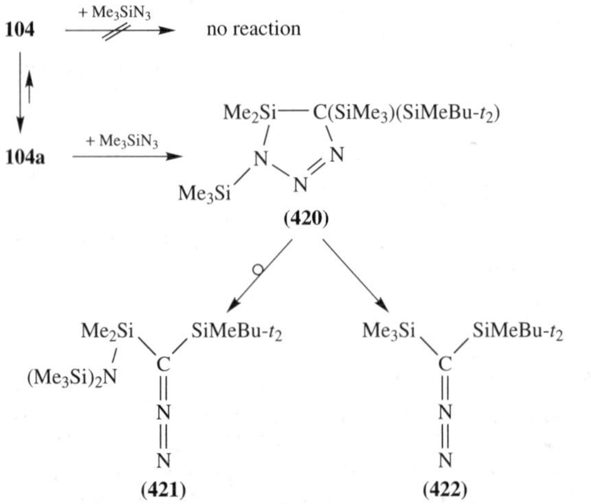

$$104 \xrightarrow{+ \text{Me}_3\text{SiN}_3} \text{no reaction}$$

$$104 \updownarrow$$

$$104\mathbf{a} \xrightarrow{+ \text{Me}_3\text{SiN}_3} (420)$$

(132)

(420) → (421) and (422)

104a and its donor adducts **104a·donor** react in essentially the same manner; this implies that the reactions are preceded by the dissociation of the donor to give the free silene **104a**[83]. Most investigations have been conducted with the THF adduct[197]. The reactivity of **104a** is summarized in equation 133. Ene reactions with, e.g., propene, [2+4] reactions with butadiene, and [2 + 3] cycloadditions with silyl azides are principally the same as for other 'Wiberg-type' silenes.

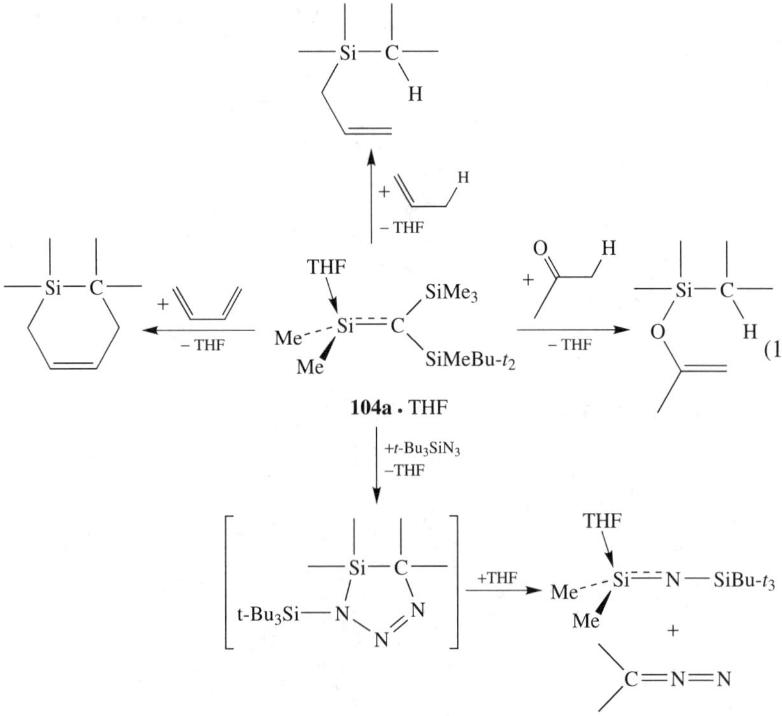

(133)

TABLE 8. Products from selected cycloaddition reactions of $R_2Si=CHCH_2Bu$-t **132** (R = t-Bu, Me, Ph)

R	Diene[a]	Dimer	[2 + 2]	[4 + 2]	Ene	Others	Reference
t-Bu	DMB			+			81
	Cp			+			81
Me	BD	+	+[b]	+		+[c]	79
	DMB	+		+		+[c]	79
	Cp	+		+		+[c]	79
	anthracene	+		+		+[c]	79
	CHD	+	+	+			205
Ph	MBD		+	+	+		206
	DB			+	+		206
	Cp			+			206
	CHD	+	+				206
	NBD	+				+[d]	206
	anthracene	+		+			206

[a]DMB : 2,3-dimethylbuta-1,3-diene; MBD : 2-methylbuta-1,3-diene; BD : buta-1,3-diene; Cp : cyclopentadiene; CHD : cyclohexadiene; NBD : norbornadiene.
[b]Two regioisomers.
[c]Coupling product from lithiated species to chlorosilane.
[d][2 + 2 + 2] Cycloadducts.

b. Neopentylsilenes: $R_2Si=C(R')CH_2Bu$-t—132. i. Diorgano-substituted neopentylsilenes. The products formed in the system vinylchlorosilane/t-butyllithium/trap depend strongly on the substituents at the silicon atom. Diorgano-substituted neopentylsilenes tend to give predominantly 'normal', i.e. [4 + 2] cycloadducts, with diene systems (see Table 8). The cycloaddition reactions are mostly regiospecific, but give stereoisomers as far as the orientation of the neopentyl group is concerned.

For R_2 = cyclobutyl the results are not very much different from those for R = alkyl[207]. The substitution of one of the alkyl groups by Cp(CO)$_2$Fe leads to the formation of the dimer only when the remaining alkyl group is methyl. The proposed intermediate silene may be trapped by dienes like DMB or Cp[208].

ii. π-Donor-substituted neopentylsilenes. The more interesting neopentylsilenes are those that have electronegative π-donor substituents like alkoxy, siloxy and, especially, chlorine. They have a quite pronounced tendency towards [2 + 2] cycloaddition and react even with reagents of only low reactivity, e.g. alkynes[209,210] (see Table 9). Thus, dichloroneopentylsilene **423** reacts with cycloheptatriene in [2 + 2] and [6 + 2] cycloaddition fashion[211]. No [4 + 2] cycloadducts are formed. The [2 + 2] cycloadducts **424** rearrange thermally into the [6+2] adduct **425** probably via a zwitterionic intermediate (equation 134).

The exceptional reactivity is explained by the authors as due to the enhanced polarity of the Si=C double bond. For a discussion of the effects see the references in the tables. In some cases, and preferably when R = Cl, the initially formed cycloadducts like *E*-**426** or **429** undergo ring enlargements to other silacycles like **427** or **430**. Interestingly,

TABLE 9. Products of cycloaddition reactions of $R_2Si=CHCH_2Bu$-t (R = Cl, OBu-t, OSiMe$_3$)

R	Diene[a]	Dimer	[2 + 2]	[4 + 2]	Ene	Others	References
Cl	BD		+	+			212
	MBD		+				212
	DMB		+	+			212
	Cp			+			81
	Cp*			+			81
	furan	+		+		+[b]	213
	2-methylfuran	+		+		+[b]	213
	2,5-dimethylfuran	+		+		+[b]	213
	CHD		+	+			205, 214
	CHpD		+	+	+		211
	CHpT		+			+[c]	211
	NB		+				215
	NBD		+		+	+[d]	215
	naphthalene			+			214
	anthracene			+			81
	aldehydes					+[e]	216
	imines		+				217
	alkynes		+				209, 210
	vinyl ethers		+				218
	pentafulvenes		+	+	+		219, 220
	terpenes		+	+	+		221
	styrene		+	+			222
t-BuO	BD		+				223
OSiMe$_3$	BD		+				224
	DMB		+				224
	CHD		+	+			224
	CHpD		+	+			224
	NB		+		+		224
	NBD		+		+	+[d]	224
	QUC		+				224
	anthracene			+			224
	styrene		+				224

[a] See footnote a in Table 8; Cp* : pentamethylcyclopentadiene; CHpD : cycloheptadiene; CHpT : cycloheptatriene; NB : norbornene; QUC : quadicyclane.
[b] Oxasilaheptene from rearrangement of the [4 + 2] adduct; see scheme above.
[c] [6 + 2] Cycloadduct.
[d] [2 + 2 + 2] Cycloadducts.
[e] Rearranged non-cyclic product.

16. Silicon–carbon and silicon–nitrogen multiply bonded compounds 947

silacyclobutane Z-**426** does not give silacyclohexenes like **427**, but it isomerizes in a retro-ene reaction to the acyclic chlorosilane **428** (equations 135 and 136).

Cl_2Si=$CHCH_2t$-Bu + [furan with R, R'] → [bicyclic adduct (429)] (136)

(423)

(429) →(170 °C)→ (430)

Nitrogen ligands can replace oxygen substituents bound to silicon and several aminocyclic precursors have been synthesized and their reaction towards t-BuLi has been investigated. The reactivity of the aminoneopentylsilenes is close to that of the well investigated dichloroneopentylsilene[225–228]. The substitution of just one chlorine atom by an amino group does not change much the resulting silene's reactivity[229]. Precursors for neopentylsilenes having one alkyl or aryl group and one amino group at silicon hardly yield cycloadducts in the presence of trapping agents but rather undergo dimerization upon reaction with t-BuLi.[229]

The substitution of one of the chlorine atoms by a vinyl group reduces the strong tendency to [2 + 2] cycloaddition reactions[225,230–232]. With the additional possibility of the formation of a second Si=C bond from dichlorodivinylsilane/2t-BuLi the intermediacy

TABLE 10. Products from cycloaddition reactions of Cl_2Si=CR^1CH_2Bu-t (R^1 = $SiMe_3$, Ph)

R^1	Diene[a]	[2 + 2]	[4 + 2]	Ene	Others	Reference[1]
$SiMe_3$	BD	+				233
	DMB		+	+		
	Cp*		+			
	CHD		+			
	NBD		+		+[b], +[c]	
	QUC				+	
	anthracene		+			
Ph	BD		+	+		234
	MBD		+	+		
	DMB		+	+		
	CHD	+	+	+		
	NBE	+		+		
	NBD	+		+	+	
	QUC				+	
	styrene	+	+		+[d]	

[a] See footnotes a in Tables 8 and 9.
[b] [2 + 2 + 2] Cycloadducts.
[c] Coupling product.
[d] Rearomatized [4 + 2] cycloadduct.

of a 2-silaallene is conceivable[232]. An additional substitution of the vinyl group at the α-carbon atom is possible (see Table 10 and equation 137 for an example of cycloadducts with R′ = SiMe$_3$)[233].

(137)

The rearrangement of the silene **345** to **346** has been described in Section I.B.1.b. **346** undergoes head-to-tail dimerization, [2+4] and [2+2+2] cycloaddition reactions[178,179].

c. *Brook-type silenes. i. [2+1] Cycloaddition reactions.* The formal addition of silylenes to Si=C double bonds has been reported by Brook and Wessely[181]. They found that the reaction of silene **349** with hexamethylsilirane **431** in the dark yields the disilirane **432** which was identified by NMR spectroscopic methods as a main product. **432** was found to be thermally unstable. It isomerizes quantitatively into the disilaindane derivative **433** during 48 h. This reaction is photoreversible: upon photolysis of **433** the disilirane **432** is cleanly formed (equation 138; note that in the original paper a different assignment of the regiochemistry of the disilane unit was suggested[181,235]).

$$(138)$$

The products arising from the reaction of **431** with the alkyl-substituted silenes **149** and **150** suggest that the reaction occurs by a radical pathway, initiated by a homolytic Si−C bond cleavage of **431** and subsequent Si−Si bond formation giving the biradical **434**. Intramolecular disproportionation of **434** gives **435**, while **436** and **437** are the results of ring closure reactions without or with expulsion of tetramethylethene, respectively (equation 139)[181].

(**149**) $R^1 = t$-Bu
(**150**) $R^1 = 1$-Ad

$$(139)$$

Alkylisocyanides do react readily with Brook-type silenes $(Me_3Si)_2Si=C(OSiMe_3)R^1$ giving silaaziridines **438** as isolated products[236,237]. The initially formed silacyclopropanimines **439** are very unstable. Only one could be detected in the reaction of **150** with *t*-butylisocyanide as a transient species by low temperature ^1H and ^{13}C NMR at $-70\,°C$. At $-40\,°C$, **439** ($R^1 = 1$-Ad) undergoes a rapid thermal rearrangement to the isomeric **438** ($R^1 = 1$-Ad)[236]. When one of the groups R^1 or R^2 is an aryl group, the first observed isolated products are 1-sila-3-azacyclobutanes **440** resulting from the dark reaction between **438** and a second molecule of isocyanide (equation 140)[237].

$$
\begin{array}{c}
Me_3Si \\
\diagdown \\
Si=C \\
\diagup \diagdown \\
Me_3Si R^1
\end{array}
\xrightarrow{R^2NC}
(Me_3Si)_2Si-\overset{\overset{\displaystyle R^1}{\mid}}{\underset{\underset{\displaystyle C}{\parallel}}{C}}-OSiMe_3
$$

(150) R^1 = 1-Ad
(149) R^1 = *t*-Bu
(349) R^1 = Mes

(439) ... R^2

(140)

$$(Me_3Si)_2Si \cdots \text{(ring structure)} \cdots OSiMe_3$$

(440) ← R^2NC, R^1 or R^2 = aryl ← (438) ← $-40\,°C$ ← (439)

Ab initio calculations by Nguyen and coworkers for the principal reaction $H_2Si=CH_2$ and HNC at the QCISDT/6-31G**//MP2(fc)/6-31G** + ZPE level of theory corroborate the experimental findings[238a]. The initial reaction between the HNC and the Si=C double bond giving the siliranimine **441** is a concerted but asynchronous [2 + 1] process in which the carbon lone pair of HNC attacks first the Si. The reaction is exothermic by *ca* 16 kcal mol^{-1} and has a barrier of 11.1 kcal mol^{-1} (see Figure 3). It was suggested that the electron-releasing substituents OSiMe$_3$ and Me$_3$Si in the experimentally investigated species will decrease the barrier for the cycloaddition[238a]. According to the calculation the isomerization of **441** to the silaaziridine **442**, which is 6.3 kcal mol^{-1} more stable, occurs in two distinct steps involving the four-membered cyclic carbene intermediate **443**[238a]. This is in contrast to the original suggestion by Brook[238b] who favoured a zwitterionic intermediate **444**. The carbene **443** is of nearly similar stability to **441** and its formation is the rate-determining step of the entire transformation **441** → **442**. Although **443** is a formal [2 + 2] cycloadduct of HNC + $H_2Si=CH_2$, the possibility of its direct formation was excluded based on experimental evidence.

Relatively stable adducts of sulphur and selenium can be prepared by direct addition to silenes, provided that the latter bear appropriate substituents. Thus, silene **392** reacts with sulphur or selenium at room temperature giving the silathiirane **445** or silaselenirane **446** in 70% or 60% yield, respectively (equation 141). In contrast, the reaction of other silenes of the $(Me_3Si)_2Si=C(OSiMe_3)R^1$ family give only complex mixtures with sulphur[239].

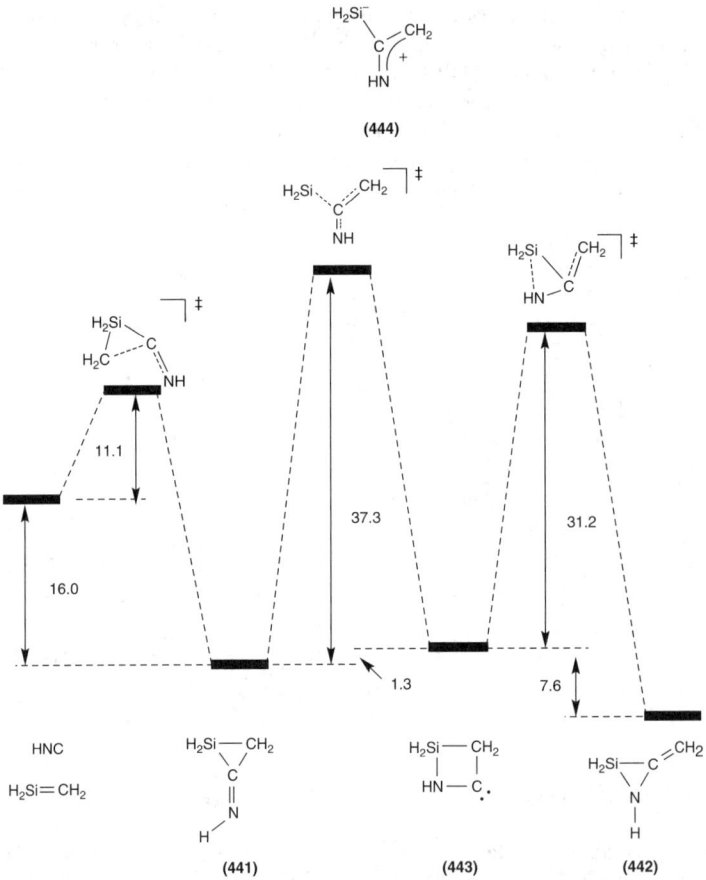

FIGURE 3. Schematic potential energy profile for the cycloaddition of $H_2Si=CH_2$ and HNC and the rearrangement to a silaziridine (calculated at QCISD(T)/6-31G**//MP2(fc)/6-31**)[238a]

(141)

ii. Cycloadditions with alkenes, dienes and alkynes. The cycloaddition behaviour of some typical Brook-type silenes with alkenes, dienes and acetylenes is summarized in Table 11 and will be discussed only briefly. Experimentally, most of the reactions have been conducted in two different ways: (1) The diene was cophotolysed ($\lambda > 340$ nm) with the acylsilane precursor of the 2-siloxysilene and the reaction took place immediately during the photolysis. (2) The silene was preformed by photolysing the precursor acylsilane and only then is the reactant added. Thus, this is a 'dark' reaction and no photochemistry can interfere. Both experimental methodologies give in nearly all cases exactly the same results with respect to formed products and product distribution. Frequently, different reaction rates for the dark and for the photolytic reaction have been observed.

The 'Brook-type' silenes $(Me_3Si)_2Si=C(OSiMe_3)R^1$ give with dienes having an allylic hydrogen normally a mixture of the Diels–Alder cycloadduct 1-silacyclohex-3-ene and the product of the intermolecular 'ene' reaction[240]. Thus, **149** or **150** give with 2,3-dimethylbuta-1,3-diene 60 : 40 mixtures of **447** and **448** (equation 142)[240]. The amount of 'ene' product **448** is larger than that found for sterically less hindered silenes, i.e. the 'Wiberg'-type silene $Me_2Si=C(SiMe_3)_2$ **92** gave only 20% of 'ene' product[176]. Steric factors alone cannot account for the observed product distributions. For example, the mesitylsilene **349** gives exclusively the [4 + 2] cycloadduct **447** (R = Mes), although **349** is not less crowded than **149** (see Table 11)[240].

$$Me_3Si\diagdown_{Si=C}\diagup^{OSiMe_3}_{\diagdown R}$$
$$Me_3Si\diagup$$

(**150**) R = 1-Ad
(**149**) R = *t*-Bu

(142)

(**447**)   (**448**)

In the reaction of 'Brook-type' silenes with buta-1,3-diene the products **449** arising from a [2 + 2] cyclization are predominant (see Table 11). Silacyclohexenes like **450**, resulting from formal [4 + 2] cycloadditions are formed only as byproducts[240]. Furthermore, with the silenes E/Z-$(Me_3Si)MesSi=C(OSiMe_3)Ad$-1 **136** the [2 + 2] adduct is formed even exclusively as a mixture of 4 diastereomers. Usually, only the regioisomers **449** having the vinyl group attached to the C^3 atom of the ring are formed (equation 143). There is no experimental evidence for the other regioisomer **451**[240].

Brook and coworkers attempted to establish a mechanism for these formally forbidden [2 + 2] cycloadditions[240]. Attempts to intercept a possible biradical intermediate **452** in the reaction of **150** with buta-1,3-diene with tributyltin hydride failed. In addition, when

TABLE 11. Product distribution (given as % of characterized product) from the cycloaddition reactions of 2-siloxysilenes with dienes, alkenes and alkynes.

Silene[a]	Reactant	[2+4]	[2+2]	Ene	Reference
(Me$_3$Si)$_2$Si=C(OSiMe$_3$)R[a]	buta-1,3-diene	20	80 (2 stereoisomers, 1:1)		240
(Me$_3$Si)$_2$Si=C(OSiMe$_3$)Mes	buta-1,3-diene	25	75 (2 stereoisomers, 4:1)		240
E/Z-(Me$_3$Si)MesSi=C(OSiMe$_3$)Ad-1	buta-1,3-diene		100 (4 stereoisomers)		95
(Me$_3$Si)$_2$Si=C(OSiMe$_3$)Ad-1	isoprene	88 (2 regioisomers 3:1)		11	240
(Me$_3$Si)$_2$Si=C(OSiMe$_3$)R[a]	2,3-dimethylbuta-1,3-diene	60		40	240
(Me$_3$Si)$_2$Si=C(OSiMe$_3$)Mes	2,3-dimethylbuta-1,3-diene	100			240
E/Z-(Me$_3$Si)MesSi=C(OSiMe$_3$)Ad-1	2,3-dimethylbuta-1,3-diene		no reaction		95
(Me$_3$Si)$_2$Si=C(OSiMe$_3$)R[a]	cyclopentadiene	100 (1 stereoisomer)			240
(Me$_3$Si)$_2$Si=C(OSiMe$_3$)R[a]	cyclohexadiene	80 (2 stereoisomers)	20 (2 stereoisomers)		240
(Me$_3$Si)$_2$Si=C(OSiMe$_3$)Mes	cyclohexadiene	no reaction			240
(Me$_3$Si)$_2$Si=C(OSiMe$_3$)Ad-1	1-octene			100	240
(Me$_3$Si)$_2$Si=C(OSiMe$_3$)Mes	1-octene			no reaction	240
(Me$_3$Si)$_2$Si=C(OSiMe$_3$) Ad-1	styrene		100 (2 stereoisomers of 1 regioisomer)		240
(Me$_3$Si)$_2$Si=C(OSiMe$_3$)Mes	styrene		100		240
(Me$_3$Si)$_2$Si=C(OSiMe$_3$)R[a]	PhMeC=CH$_2$			100	240
(Me$_3$Si)$_2$Si=C(OSiMe$_3$)Mes	PhMeC=CH$_2$			100	240
(Me$_3$Si)$_2$Si=C(OSiMe$_3$) Bu-t	PhC≡CMe		100 (one regioisomer)		95
E/Z-(Me$_3$Si)MesSi=C(OSiMe$_3$)Ad-1[a]	PhC≡CH		100 (one regioisomer)		95
E/Z-(Me$_3$Si)MesSi=C(OSiMe$_3$)Ad-1[a]	Me$_3$SiC≡CH		100 (one regioisomer)		95

[a]R : t-Bu.
[b]1-Ad : 1-adamantyl.

the reaction was performed in different solvents (pentane, benzene and diethyl ether) no significant change in the product distribution was observed. This led Brook and coworkers to discard also the zwitterionic species **453** as a possible intermediate. They concluded that the reaction of 2-siloxysilenes like **150** with buta-1,3-diene does not involve a two-step process but is consistent with a concerted [2 + 2] process[240].

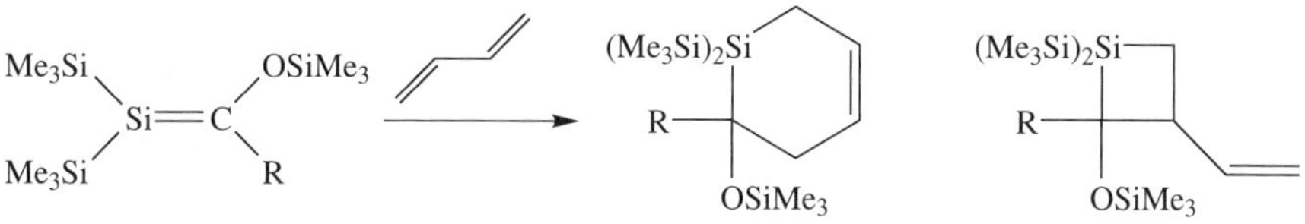

(450) **(449)**

(143)

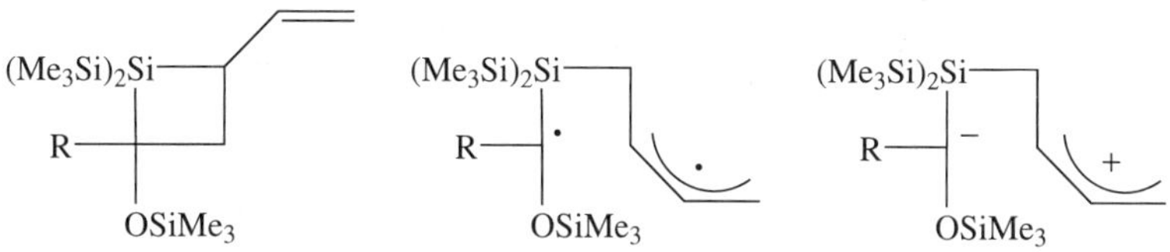

(451) **(452)** **(453)**

When simple alkenes were employed as reaction partners for silenes of the type $(Me_3Si)_2Si{=}C(OSiMe_3)R^1$, silacyclobutanes were obtained, provided that no allylic hydrogen is present in the alkene. In the reaction with alkenes with allylic hydrogens the 'ene' reaction becomes predominant (see Table 11). Thus, while the reaction with styrene exclusively gives the four-membered ring compound **454**, with 1-methylstyrene the 'ene' products **455** were obtained (equation 144). Similarly, from the reaction of **150** with 1-octene only the 'ene' product **456** was isolated (equation 145).

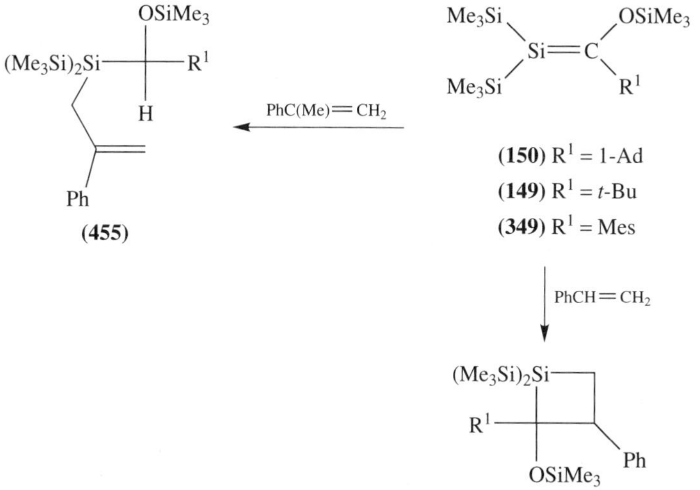

(455)

(150) R^1 = 1-Ad
(149) R^1 = *t*-Bu
(349) R^1 = Mes

(144)

(454)

Alkynes also react with sterically, highly hindered Brook-type silenes giving 1-silacyclobut-2-enes in high yields[95]. This reaction is strictly regiospecific: thus from E/Z-**136** only a E/Z mixture of **457** is formed, whereas the regioisomer **458** was not detected (equation 146)[95].

Similarly, it was shown by a crystal structure analysis that the adduct of silene **149** with phenylpropyne has the constitution **148** (equation 147)[95] in contrast to earlier reports which favoured the reversed regiochemistry[86].

16. Silicon–carbon and silicon–nitrogen multiply bonded compounds

d. Apeloig–Ishikawa–Oehme-type silenes. Compared with the well established cycloaddition chemistry of the 'Brook-type silenes, relatively little is known about the behaviour of 'Apeloig–Ishikawa–Oehme'-type silenes in these reactions. Characteristically these silenes undergo [4 + 2] cycloadditions with buta-1,3-dienes yielding silacyclohexenes; e.g. silene **459** gives with 2,3-dimethylbuta-1,3-diene the silacyclohexene **460** in 60% isolated yield (equation 148)[104].

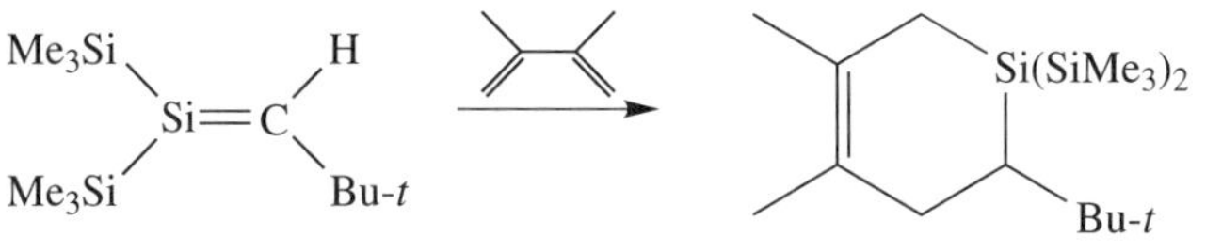

(148)

(**459**)   (**460**)

There are, however, reports in which the formation of a [2 + 2] cycloadduct was also observed. Thus, the stable silene **167** reacts with 1-methoxybuta-1,3-diene yielding regiospecifically only one [4 + 2] regioisomer **461**[111]. The product **462** arising from a formal [2 + 2] cycloaddition is formed, however, in appreciable amounts (20–30% yield) (equation 149)[111].

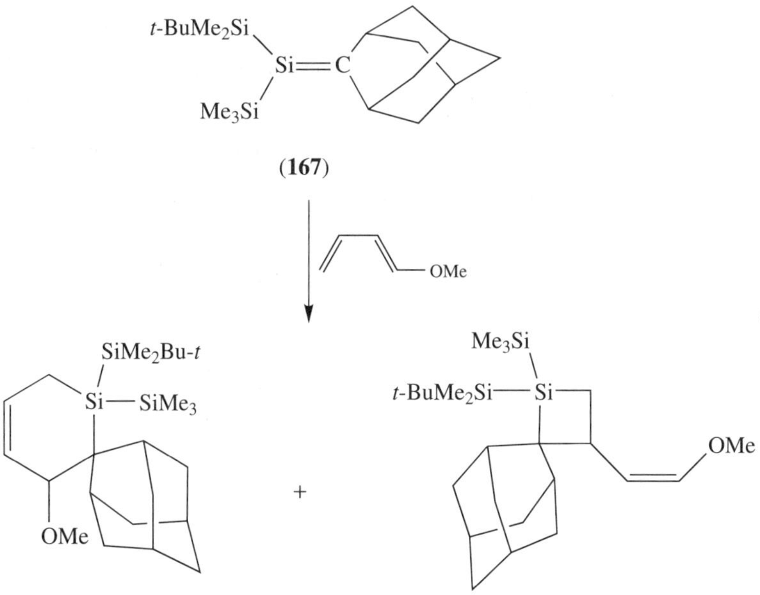

(**167**)

(149)

(**461**)   (**462**)

The [4 + 2] reaction with dienes has been frequently used to establish the intermediacy of a silene in the sila-Peterson olefination reaction[39,107,112,119]. Thus, further examples for cycloadditions can be found in Section I.A.5.

Silenolates like **180** do react with 2,3-dimethylbuta-1,3-diene in a formal [4 + 2] fashion and, when the alkoxide **463** is trapped by triethylsilyl chloride, the 1-silacyclohex-3-ene

464 is isolated (equation 150)[120].

$$(180) \quad \longrightarrow \quad (463)$$

$$\downarrow \text{Et}_3\text{SiCl} \qquad (150)$$

$$(464)$$

A very interesting reaction of a disilene with a silenolate forming a stable trisilacyclobutane was reported by Bravo-Zhivotovskii, Apeloig and coworkers[241]. The decomposition of **182**, prepared by the reaction of Brook's ketone $(\text{Me}_3\text{Si})_3\text{SiC(O)Ad-1}$ with triethylgermyllithium, yields in a complex reaction sequence the trisilacyclobutane **465** in 54% isolated yield (equation 151). The key step in this reaction is believed to be the formal cycloaddition between the transient tetra(trimethylsilyl)disilene **466** and **182** giving the trisilacyclobutane **467** (equation 152).

$$(182) \xrightarrow{48\text{ h, RT}} (465) \qquad (151)$$

$$(182) + (466) \longrightarrow (467) \qquad (152)$$

e. Cycloaddition reactions with carbonyl compounds and derivatives. The reaction of transient silenes with carbonyl compounds leads normally to an olefin and silanone oligomers[8]. It is usually assumed that a 2-siloxetane **468** is formed first and then fragments at the applied reaction conditions (equation 153). As pointed out by Raabe and Michl, a concerted 2-siloxetane formation is thermally not forbidden, when lone pair electrons of the carbonyl oxygen are involved; however, a two-step process is equally possible[8]. Streitwieser and Bachrach studied the reaction using *ab initio* methods[242]. They found that the overall reaction between $H_2Si=CH_2$ and H_2CO to produce $H_2C=CH_2$ and H_2SiO is exothermic by *ca* 30 kcal mol^{-1} (HF/3-21G*) and that the intermediate 2-siloxetane **468** is more stable than the products by at least 50 kcal mol^{-1}, thus indicating that siloxetanes similar to **468** are stable toward unimolecular decomposition and should therefore be isolable compounds[242].

$$H_2Si=CH_2 \atop O=CH_2 \quad \longrightarrow \quad \substack{H_2Si-CH_2 \\ | \quad\quad | \\ O-CH_2} \quad \longrightarrow \quad \substack{H_2Si \\ \| \\ O} + \substack{CH_2 \\ \| \\ CH_2} \quad (153)$$

(**468**)

Reaction of simple aromatic aldehydes with 'Brook' - type silenes $(Me_3Si)_2Si=C(OSiMe_3)R^1$ gives exclusively a *cis/trans* mixture of the 2-siloxetanes **469**, the nominal [2 + 2] cycloadducts, which can be identified by NMR spectroscopy (equation 154)[235].

$$(Me_3Si)_2Si=C\genfrac{}{}{0pt}{}{OSiMe_3}{R^1} \quad \xrightarrow{R^2CHO} \quad \substack{(Me_3Si)_2Si-O \\ | \quad\quad\quad | \\ R^1-C-C-R^2 \\ | \quad\quad | \\ Me_3SiO \quad H} \quad (154)$$

(**469**)

In contrast, the reactions with aromatic ketones are rather complex and hardly predictable. In some cases, only the [2 + 2] cycloadducts were formed, but they isomerized in the dark to the bicyclic [4 + 2] isomers. Thus, mesitylsilene **349** gives with benzophenone the 2-siloxetane **470**, which undergoes spontaneously a ring expansion reaction at room temperature in the dark to give a 1.8 : 1 mixture of **471** and unchanged **470** (equation 155)[235].

$$(Me_3Si)_2Si=C\genfrac{}{}{0pt}{}{OSiMe_3}{Mes} \quad \xrightarrow{Ph_2CO} \quad \substack{(Me_3Si)_2Si-O \\ | \quad\quad\quad | \\ Mes-C-C-Ph \\ | \quad\quad | \\ Me_3SiO \quad Ph}$$

(**349**)   (**470**)

$h\nu$ or dark ↑↓ dark   (155)

(**471**)

When **349** is generated thermally at 140 °C from mesitoyl-tris(trimethylsilyl)silane, the silyl enol ether **472** was isolated as the sole product[99]. **472** results from isomerization of the siloxetane **470** (equation 156)[99].

$$(156)$$

In contrast, the initial product of the reaction of phenylsilene **473** with aromatic ketones is the formal [4 + 2] cycloadduct **474** in which the silene serves as the 4π component and no evidence could be found that a siloxetane **475** is formed initially. On photolysis and also in the dark, although much more slowly, **474** isomerizes to the siloxetane **475**, indicating that **475** is the thermodynamically more stable isomer. This isomerization is not reversible[235]. When bis(p-tolyl) ketone was used in the reaction, the isomerization cascade did not end at the stage of the siloxetane but a second isomerization step giving **476** took place (equation 157). **476** is the formal [4 + 2] cycloadduct with the ketone serving as the diene and the silene as the dienophile[235].

$$(157)$$

Reaction of phenylsilene **473** with benzophenone in boiling dioxane gives directly the silyl enol ether **477**, the thermal decomposition product of the initially formed oxasilirane *E*/*Z*-**478** (equation 158)[243].

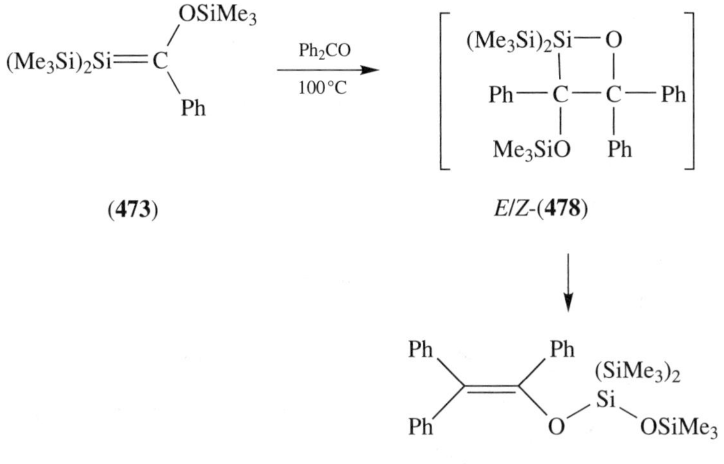

(**473**) *E*/*Z*-(**478**) (**477**)

(158)

The same product was obtained in 52% yield along with considerable amounts of 1,2,2-triphenyl-1-(trimethylsilyl)ethene (22%) when the benzoyl-tris(trimethylsilyl)silane was thermolysed in the presence of benzophenone[99].

The [4π(ketone)–2π (silene)] reaction mode is dominant for the alkyl substituted silenes **149** and **150**: The initially formed product is **479**, which isomerizes in the dark in a non-reversible reaction to the siloxetane **480** (equation 159)[235].

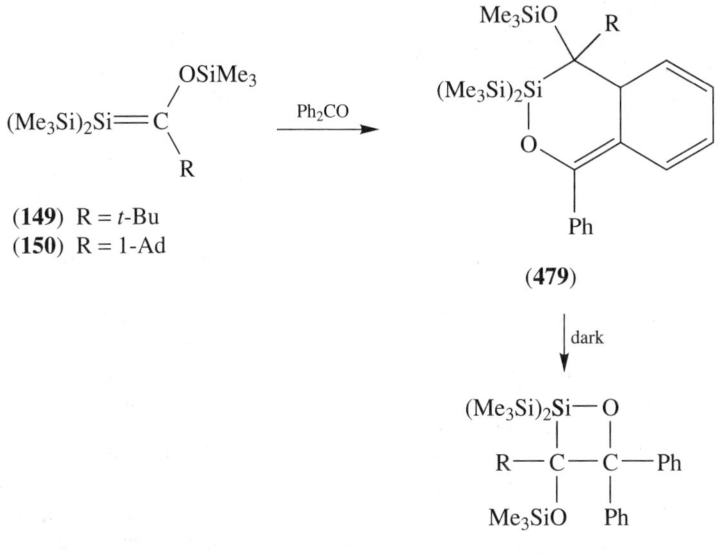

(**149**) R = *t*-Bu
(**150**) R = 1-Ad

(**479**) (**480**)

(159)

Under thermolytic conditions the cycloheptatriene **481** and the silyl enol ether **482** were obtained in 46% and 24% yield, respectively, along with 29% of the unreacted starting compound, the acylpolysilane which is the precursor for the reactive silene (equation 160)[99].

$$(Me_3Si)_2Si=C\begin{matrix}OSiMe_3\\R\end{matrix}$$

(**149**) R = *t*-Bu
(**150**) R = 1-Ad

[Scheme showing conversion: **479** → **480** via 140°C, 24h, Ph$_2$CO; **479** → **481**; **480** → **482**; equation (160)]

The reaction of thermolytically generated 'Brook'-type silenes with acetone is believed to give initially the siloxetane **483**, which undergoes diverse isomerizations or fragmentations depending on the substituent R, yielding products **484–491** (equation 161)[99].

The complexity of the reactions is even increased when α,β-unsaturated carbonyl compounds have been employed as reactants. Silenes of the family $(Me_3Si)_2Si=C(OSiMe_3)R^1$ react with propenal **492** ($R^2 = R^3 = H$) in a [4 + 2] manner giving surprisingly both possible regioisomers, the 1-sila-3-oxacyclohex-4-ene **493** as the major product and the 1-sila-2-oxocyclohex-3-ene isomer **494**. The siloxetane **495** was also formed in minor amounts[244]. Aldehydes **492** bearing a β-substituent (R^2 and/or $R^3 \neq H$) did not give the regioisomer **493** upon reaction with silenes of the type $(Me_3Si)_2Si=C(OSiMe_3)R^1$ but yield a mixture of the Si—O bonded isomer **494** and, as the minor component, **495**[244]. The reaction between the mesitylsilene **349** and cinnamaldehyde, however, gives exclusively the siloxetane **495** (R^1 = Mes, R^2 = Ph, R^3 = H) (equation 162)[244].

(162)

α,β-Unsaturated ketones behave similarly to aldehydes in their reaction with $(Me_3Si)_2Si=C(OSiMe_3)R$, i.e. in the reaction with methyl vinyl ketone, a ketone lacking a β-substituent, the 1-sila-3-oxocyclohex-4-enes are formed as major products[244]. The reaction of silenes (Me$_3$Si)$_2$Si=C(OSiMe$_3$)R with β-substituted α,β-unsaturated ketones give exclusively the 1-sila-2-oxocyclohex-3-enes **496** (equation 163). In contrast to the reaction with aldehydes, in both cases no formation of a siloxetane similar to **495** was observed[244].

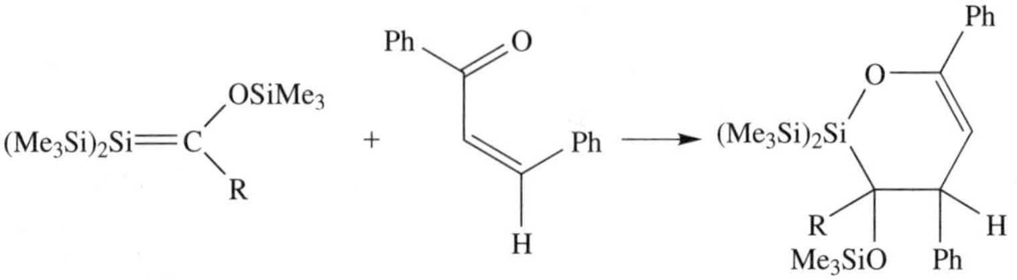

(163)

(**496**)

In contrast to all investigated α,β-unsaturated ketones, cyclopentadienones like **497** react with **149** and **473** exclusively to give the spiro compounds **498** i.e. the nominal [2 + 2] cycloadduct, in 61 and 63% yield (equation 164)[243].

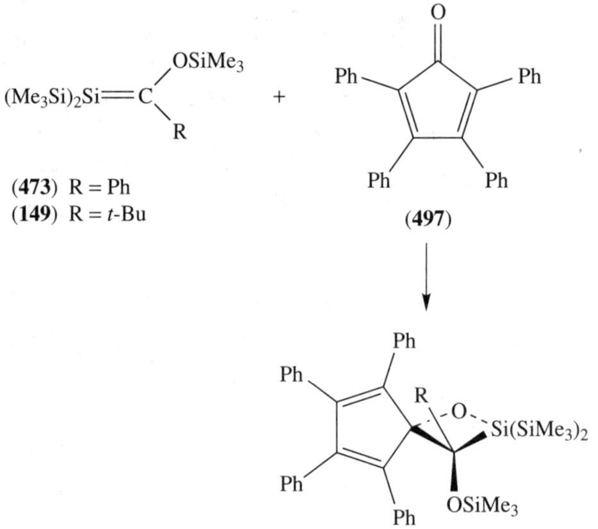

(**473**) R = Ph
(**149**) R = t-Bu

(**497**)

(164)

(**498**)

Brook-type silenes are unreactive toward simple esters such as ethyl acetate or methyl benzoate[245]. α,β-Unsaturated esters, however, readily undergo [2+4] cycloaddition reactions with **149** or **150**. In these reactions no evidence was found for a product arising from a [2 + 2] cycloaddition, neither involving the C=C bond nor the C=O double bond[245]. Anomalous behaviour was observed in the regiochemistry of the [2+4] cyclizations. With acrylate esters only one type of regioisomer, the 1-sila-3-oxacyclohex-4-enes **499** having a new Si—C bond, were formed as the sole product (equation 165). In contrast, 1-sila-2-oxacyclohex-3-enes **500** having a new Si—O bond were the products of the reaction of **149** or **150** with ethyl cinnamate (equation 166). When diethyl maleate and diethyl fumarate were employed, diastereomeric mixtures of 1-sila-3-oxacyclohex-4-enes similiar to **499** were obtained[245].

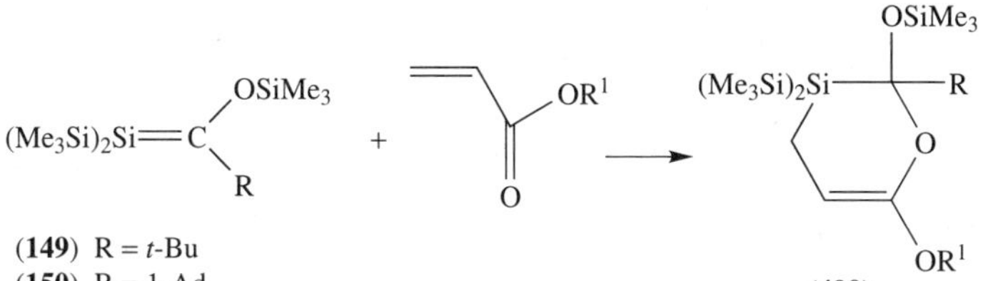

$$(Me_3Si)_2Si=C\begin{smallmatrix}OSiMe_3\\R\end{smallmatrix} + \begin{smallmatrix}\\OR^1\\O\end{smallmatrix} \longrightarrow \begin{smallmatrix}OSiMe_3\\(Me_3Si)_2Si\!-\!\!\!-\!R\\O\\OR^1\end{smallmatrix}$$ (165)

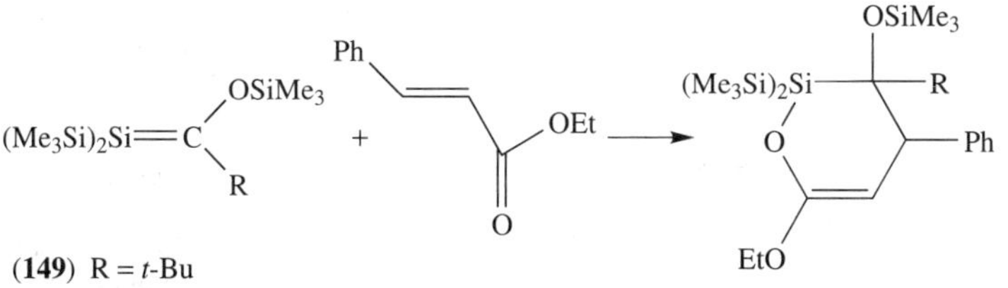

(**149**) R = *t*-Bu
(**150**) R = 1-Ad
(**499**)

$$(Me_3Si)_2Si=C\begin{smallmatrix}OSiMe_3\\R\end{smallmatrix} + \begin{smallmatrix}Ph\\\\OEt\\O\end{smallmatrix} \longrightarrow \begin{smallmatrix}OSiMe_3\\(Me_3Si)_2Si\!-\!\!\!-\!R\\O\!\!-\!Ph\\EtO\end{smallmatrix}$$ (166)

(**149**) R = *t*-Bu
(**150**) R = 1-Ad
(**500**)

When crotonate esters were employed the regioisomers **501** were mainly formed, accompanied by a significant amount of non-cyclic compounds **502** formally arising from insertion of the Si=C group in an allylic C—H bond (equation 167). The reaction of silenes **149** and **150** with crotonate esters are the only examples known so far where the product distribution is different when the reaction is conducted under photolytic conditions or in the absence of light. Thus, in the dark reaction of **150** with methyl crotonate the relative ratio **501** (R^1 = 1-Ad, R^2 = Me): **502** (R^1 = 1-Ad, R^2 = Me) = 3.75 : 1 was obtained, while under photolytic conditions only minor amounts of **501** were detected[245]. The nominal insertion product **503** is also the major product in the reaction of methacrylate esters with **149** and **150** (equation 168). The minor product in this reaction is the 1-sila-3-oxacyclohex-4-ene **504** (relative ratio **503** : **504** = 5.7 : 1)[245].

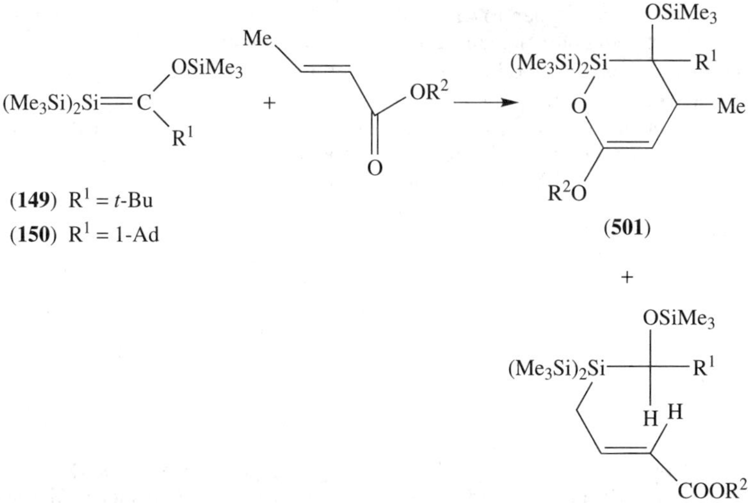

$$(Me_3Si)_2Si=C\begin{smallmatrix}OSiMe_3\\R^1\end{smallmatrix} + \begin{smallmatrix}Me\\\\OR^2\\O\end{smallmatrix} \longrightarrow \begin{smallmatrix}OSiMe_3\\(Me_3Si)_2Si\!-\!\!\!-\!R^1\\O\!\!-\!Me\\R^2O\end{smallmatrix}$$

(**149**) R^1 = *t*-Bu
(**150**) R^1 = 1-Ad
(**501**)

+ (167)

$$\begin{smallmatrix}OSiMe_3\\(Me_3Si)_2Si\!-\!\!\!-\!R^1\\H\,H\\\\COOR^2\end{smallmatrix}$$

(**502**)

Evidently, the reactivity of Brook-type silenes with α,β-unsaturated carbonyl compounds[243–245] is often very sensitive to factors other than polarities of the silene double bond and of the α,β-unsaturated system. Brook and coworkers pointed out that frontier orbital control of the [4 + 2] cyclization reactions might account for the predominant formation of 1-sila-3-oxacyclohex-4-enes like **505** in the reaction of acrylate esters or propenal with silenes $(Me_3Si)_3Si=C(OSiMe_3)R$ (equation 169)[244]. Thus, the HOMO of the silene $(H_3Si)_2Si=C(OSiH_3)R$ with its higher orbital coefficient located on silicon as calculated by Apeloig and Karni[246] will interact with the LUMO of the unsubstituted α,β-unsaturated carbonyl compound, where the higher orbital coefficient of the frontier orbital appears to be located on the terminal carbon atom of the C=C bond. The formation of 1-sila-2-oxacyclohex-3-ene like **506** from crotonaldehyde, crotonate esters or from ethyl cinnamate (equation 169) is in accord with a cyclization reaction dominated by the high affinity of oxygen towards silicon and is consistent with the polarity of the reagents, as most commonly prevails in [2+4] cycloadditions. Steric effects introduced by the β-substituents may also be important in determining the preferred regiochemistry[244]. The occurrence of significant amounts of byproducts in some reactions[245] suggest, however, that other non-concerted reaction routes with different intermediates may be important. More synthetic and theoretical work is clearly needed to understand in more detail the mechanism of this formal cycloaddition of silenes.

In view of the evident reactivity of 'Brook'-type silenes toward carbonyl compounds and the fact that they are generated photochemically from acylpolysilanes, the question arises whether the silenes do react with their precursors. The very reactive silene **507**

16. Silicon–carbon and silicon–nitrogen multiply bonded compounds 967

reacts in the absence of suitable trapping reagents with the acyldisilane **508** leading to the siloxetane **509**[94]. **509** is thermally unstable and rearranges to the siloxyalkene **510** (equation 170)[94].

$$\text{Me}_3\text{Si}-\underset{\underset{\text{Ph}}{|}}{\overset{\overset{\text{Ph}}{|}}{\text{Si}}}-\text{C}\overset{\diagup\text{O}}{\underset{\diagdown\text{Ad-1}}{}} \quad + \quad \text{Ph}_2\text{Si}=\text{C}\overset{\diagup\text{OSiMe}_3}{\underset{\diagdown\text{Ad-1}}{}} \quad \longrightarrow \quad \text{1-Ad}-\underset{\underset{\text{Me}_3\text{SiO}}{|}}{\overset{\overset{\text{Ph}_2\text{Si}-\text{O}}{|}}{\text{C}}}-\underset{\underset{\text{SiPh}_2\text{SiMe}_3}{|}}{\overset{\overset{|}{|}}{\text{C}}}-\text{Ad-1}$$

(**507**)   (**508**)   (**509**)

$$\underset{\text{Me}_3\text{SiOSiPh}_2\text{O}}{\overset{\text{1-Ad}}{\diagdown}}\text{C}=\text{C}\underset{\text{Ad-1}}{\overset{\text{SiPh}_2\text{SiMe}_3}{\diagup}}$$

(**510**)

(170)

The reaction shown in equation 170 is, however, the only known example where a silene reacts with its acylpolysilane precursor. Whereas silenes of the type $(\text{Me}_3\text{Si})_2\text{Si}=\text{C}(\text{OSiMe}_3)\text{R}^1$ do not react with their polysilane precursors, they readily react with less sterically congested acylsilanes[247]. Thus, mesitylsilene **349** gives with benzoyl trimethylsilane a mixture of the diastereomeric siloxetanes E/Z-**511** in 95% yield (equation 171). Typically, the adamantylsilene **150** gives with benzoyl trimethylsilane the nominal [4+2] cycloadduct **512** in 89% yield (equation 172). The silyl enol ether **513** is the result of an 'ene'-type reaction which occurs between **150** and acetyl triphenylsilane, a carbonyl compound having an activated hydrogen (equation 173)[247].

$$(\text{Me}_3\text{Si})_2\text{Si}=\text{C}\overset{\diagup\text{OSiMe}_3}{\underset{\diagdown\text{Mes}}{}} \quad \xrightarrow{\text{Me}_3\text{SiCOPh}} \quad \underset{\underset{\text{Me}_3\text{SiO}}{|}}{\overset{\overset{(\text{Me}_3\text{Si})_2\text{Si}-\text{O}}{|}}{\text{Mes}-\text{C}}}-\underset{\underset{\text{Ph}}{|}}{\overset{\overset{|}{|}}{\text{C}}}-\text{SiMe}_3 \quad (171)$$

(**349**)   E/Z-(**511**)

$$(\text{Me}_3\text{Si})_2\text{Si}=\text{C}\overset{\diagup\text{OSiMe}_3}{\underset{\diagdown\text{Ad-1}}{}} \quad \xrightarrow{\text{Me}_3\text{SiCOPh}} \quad \text{(512)} \quad (172)$$

(**150**)   (**512**)

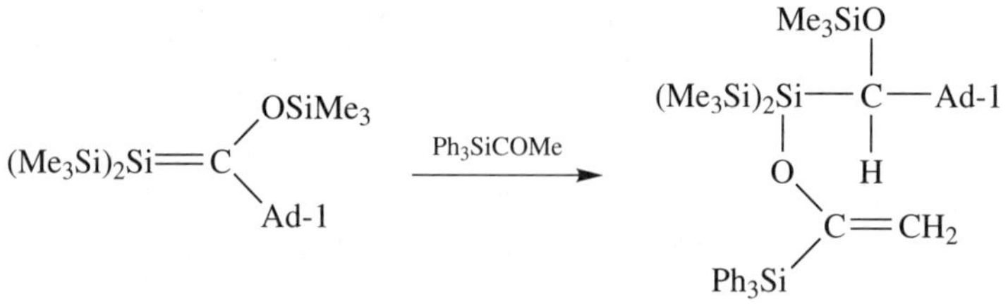

$$(Me_3Si)_2Si=C\genfrac{}{}{0pt}{}{OSiMe_3}{Ad\text{-}1} \xrightarrow{Ph_3SiCOMe} \begin{array}{c} Me_3SiO \\ | \\ (Me_3Si)_2Si-C-Ad\text{-}1 \\ || \\ OH \\ \diagdown/ \\ C=CH_2 \\ / \\ Ph_3Si \end{array} \quad (173)$$

(150) \hspace{4cm} (513)

'Brook'-type silenes have been found to undergo [4+2] and [2+2] cycloadditions with imines $R_2C=NR$ to give silatetrahydroisoquinolines or silaazetidines, respectively[248]. If formed, the [4 + 2] adducts generally rearrange slowly in the dark or more rapidly when photolysed to the thermodynamically more stable [2 + 2] isomers.

Thus, the alkyl-substituted silenes of the family $(Me_3Si)_2Si=C(OSiMe_3)R$ give with triphenylimine the unstable **514** which is converted completely, faster upon photolysis or more slowly in the dark, into the silaazetidine **515** (equation 174). For the adamantylsilene **150** the complete conversion from the acylpolysilane to **515** (R = 1-Ad) requires 5 days and proceeds in an overall yield of 94%[248]. The mesitylsilene **349** forms no [4 + 2] cycloadduct, and the only product of the reaction with triphenylimine detected after 24 h is the silaazetidine **515** (R = Mes)[248]. The imine component also influences the product distribution of the reaction. For example, no [4+2] cycloadducts are formed in the reaction of silenes $(Me_3Si)_2Si=C(OSiMe_3)R$ with N-fluorenylidineaniline and only silaazetidines have been detected[248].

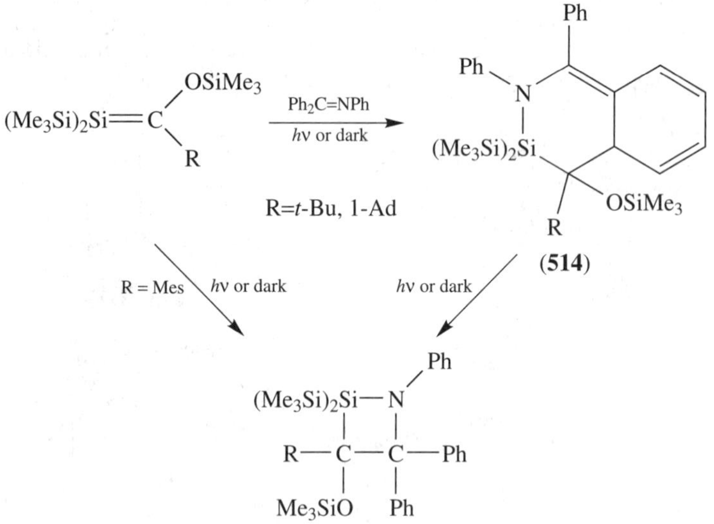

(174)

(515)

Most of the silaazetidines are remarkably stable and prolonged heating over days at 70 °C is required to reconvert them to precursors acylpolysilane and imine (equation 175)[248]. This is in contrast to the behaviour of silaazetidines deriving from

'Wiberg'-type silenes which can be easily cleaved thermolytically to the silene and the imine[60–64].

$$
\begin{array}{c}
(515) \quad \underset{70\,°C}{\rightleftarrows} \quad (Me_3Si)_2Si{=}C(OSiMe_3)(R) \;+\; Ph_2C{=}NPh \\
\Big\updownarrow 70\,°C \\
(Me_3Si)_3Si\text{—}C(=O)R \;+\; Ph_2C{=}NPh
\end{array}
\qquad (175)
$$

An exception is the silaazetidine **516** prepared from **349** and *N*-fluorenylidineaniline, which decomposes slowly already at room temperature, yielding the head-to-head dimer **517** of the intermediate silanimine **518** and siloxyalkene **519** (equation 176)[248].

(516) → (518) + (519) ; 2× (518) → (517)   (176)

Silaacrylate **305** undergoes reactions with ketones[165]. The initial coordination of the ketones is reminiscent of the donor adducts to silenes. The products **520–522** are formed by an ene or a formal [2 + 2] cycloaddition reaction, depending on the substituents on the ketones (equation 177).

Leigh could prove the formation of siloxetanes **525** from carbonyl compounds and silatrienes **524** obtained through the photolysis of aryldisilanes **523** (equation 178)[136,137,140,249]. They are thermally and hydrolytically unstable but an unambiguous assignment was possible through NMR spectroscopy. The yield of siloxetanes

525 varies with the degree of aryl substitution at the silene's **524** silicon atom.

A siloxetane **526** as an intermediate from a [2 + 2] cycloaddition of silene **241** with acetone has been formulated by Ishikawa[134]. It extrudes a silanone equivalent to give the vinyl ether **527**. The second regioisomeric silene **242** generated together with **241** by photolysis of **240** undergoes an ene reaction instead (equation 179)[134].

f. Miscellaneous silenes. i. [2 + 2] Cycloaddition reactions. Conlin and Bobbitt photolysed the symmetric 1,2-divinyldisilane **528** and obtained silene **42** which was trapped by buta-1,3-diene to form the E/Z isomeric [2 + 2] cycloadducts **40**, along with minor amounts of the [4 + 2] cycloadduct (equation 180)[37].

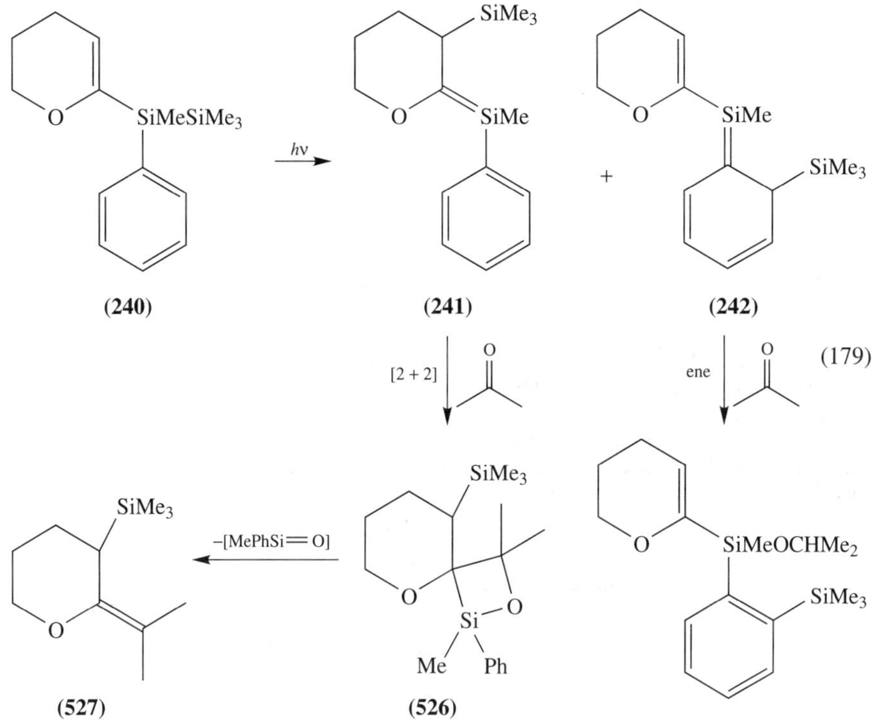

(179)

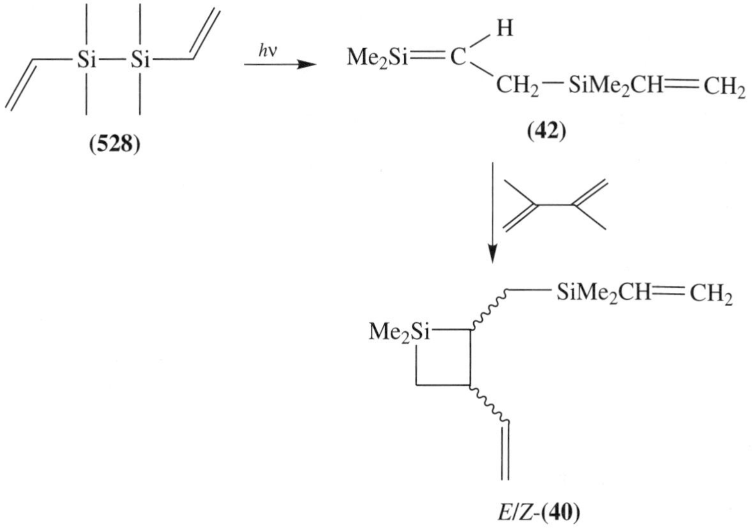

(180)

The silene **42** itself is liberated from the vinylsilacyclobutanes **40** and undergoes, in competition with other reactions, an intramolecular [2 + 2] addition giving **44** (equation 181).

The pyrolysis of 1,1-dimethyl-1-silacyclobut-2-ene **80** gives 1,1-dimethylsilabutadiene **81**[54,250]. The consecutive reactions of **81** yield several compounds **529–533** (with the yields given at 363 °C after 180 min of pyrolysis) (equation 182). While **529** is formed by the dimerization of **81**, the authors suggest a pathway to the formation of the formal dimers **530** and **531** that includes cycloaddition reactions with the starting compound **80**. A Diels-Alder-type reaction gives regiospecifically **530** and the [2+2] cycloreaction gives initially the bicyclic compound **534**, which undergoes subsequent ring expansions to **535** and to **531** (equation 183). In a secondary fragmentation reaction **535** decomposes into dimethylsilene **2** and **532**. The occurrence of small amounts of **533** can be rationalized by a [4 + 2] cycloaddition between **81** and **2**.

ii. [2 +3] Cycloaddition reactions. The photochemical decomposition of bis(silyldiazomethyltetrasilane **321** produces one silene group in **322** followed by [2 + 3] silene–diazo cycloaddition to give **323** and finally **324** (equation 184)[168,165].

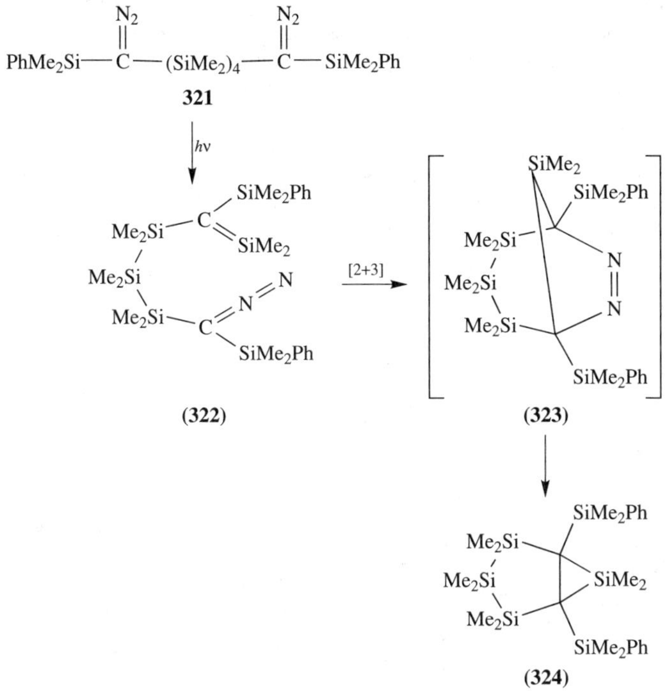

16. Silicon–carbon and silicon–nitrogen multiply bonded compounds

(184)

iii. [2 + 4] Cycloaddition reactions. Conlin and Namavari have demonstrated that 1-silabuta-1,3-dienes may undergo a stereospecific [4 + 2] and non-stereospecific [2 + 2] cycloaddition with alkenes[54]. Isomerization of 1,1-dimethylsilacyclobutene **80** in the presence of 20-fold excess of ethylene at 350 °C leads cleanly to a mixture of the isomeric silacyclohexenes **82** and **83** in nearly quantitative yields. Formation of the silacyclohex-2-ene **82** might be anticipated from a Diels–Alder-type reaction between ethylene and sila-1,3-diene **81** (equation 185).

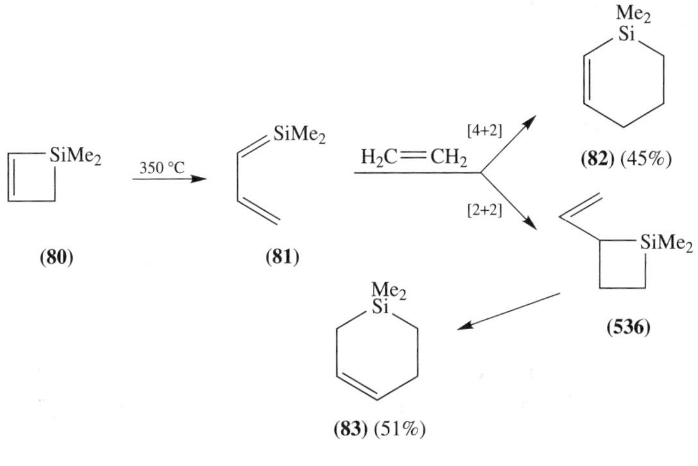

(185)

The observation of the silacyclohex-3-ene **83** was rationalized by a [2+2] cycloaddition of ethylene to give unusual 2-vinylcyclobutane **536**. A facile ring expansion via a 1,3-silyl shift to the terminal methylene group yields **83**. Surprisingly, the relative product ratio **82/83** suggests that $\Delta G^{\ddagger}$ for the forbidden [2 + 2] reaction is slightly smaller than for the stereospecific allowed Diels–Alder path[54].

The [4 + 2] reaction was shown to be stereospecific by the formation of only one geometric silacyclohex-2-ene isomer (**537** or **538**) in the reaction of **81** with Z- or E-but-2-ene, respectively, whereas the distribution of products deriving from the [2 + 2] cycloaddition (**539**–**541**, equation 186) suggests a non-stereospecific course of the reaction[54].

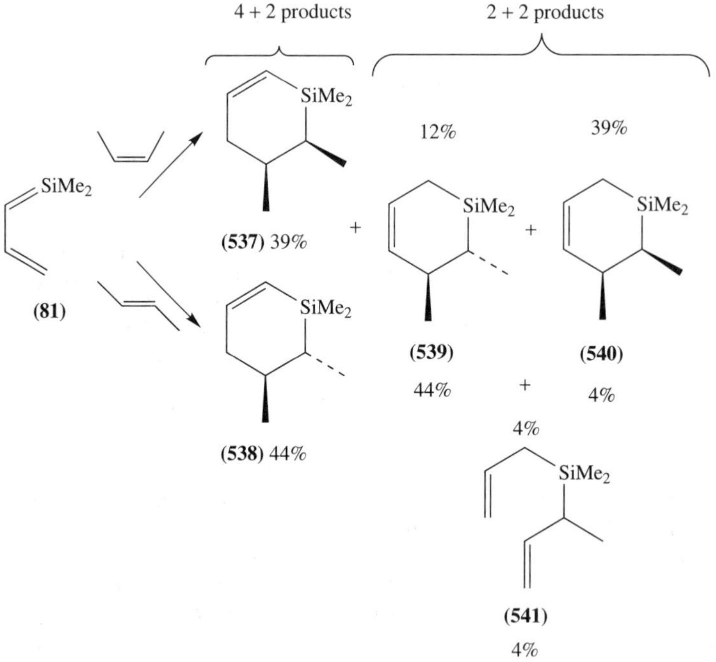

(186)

The *ortho* quinodisilane **57** reacts with acetylenes exclusively in a [4 + 2] fashion. The only product isolated in 89% yield from the reaction with phenylacetylene is **542** (equation 187)[46,45].

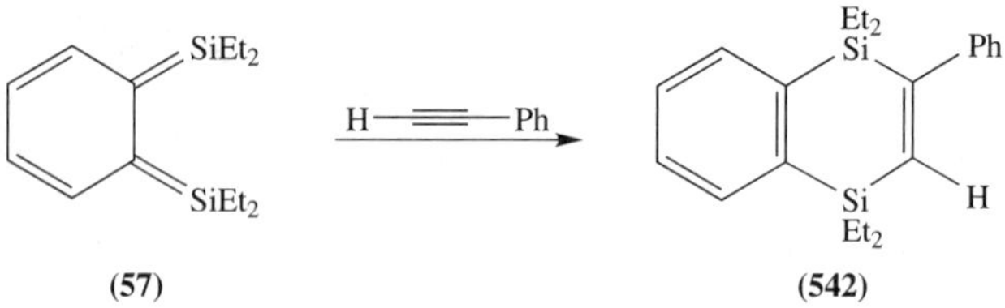

(187)

5. Ene reactions

The competitive silyl migration to either a vinyl or phenyl group in aryl- and vinyl-disilanes has been investigated by Ishikawa and coworkers[133]. With 2,3-dimethylbuta-1,3-diene, two modes of ene reaction are possible for **238** to give **543** and **544**. Another ene product **547** arises from silene **239**. Both silenes react with acetone to give the respective ene products **545** and **546** (equation 188).

(188)

With dihydropyranyl-substituted phenyldisilane **240** Ishikawa and coworkers found the formation of both possible types of silenes **241** and **242**[134]. Both were trapped by acetone, but **242**, however, only through an ene reaction to give **549**. With isobutene only the silene **242** gives an ene product **548** (equation 189).

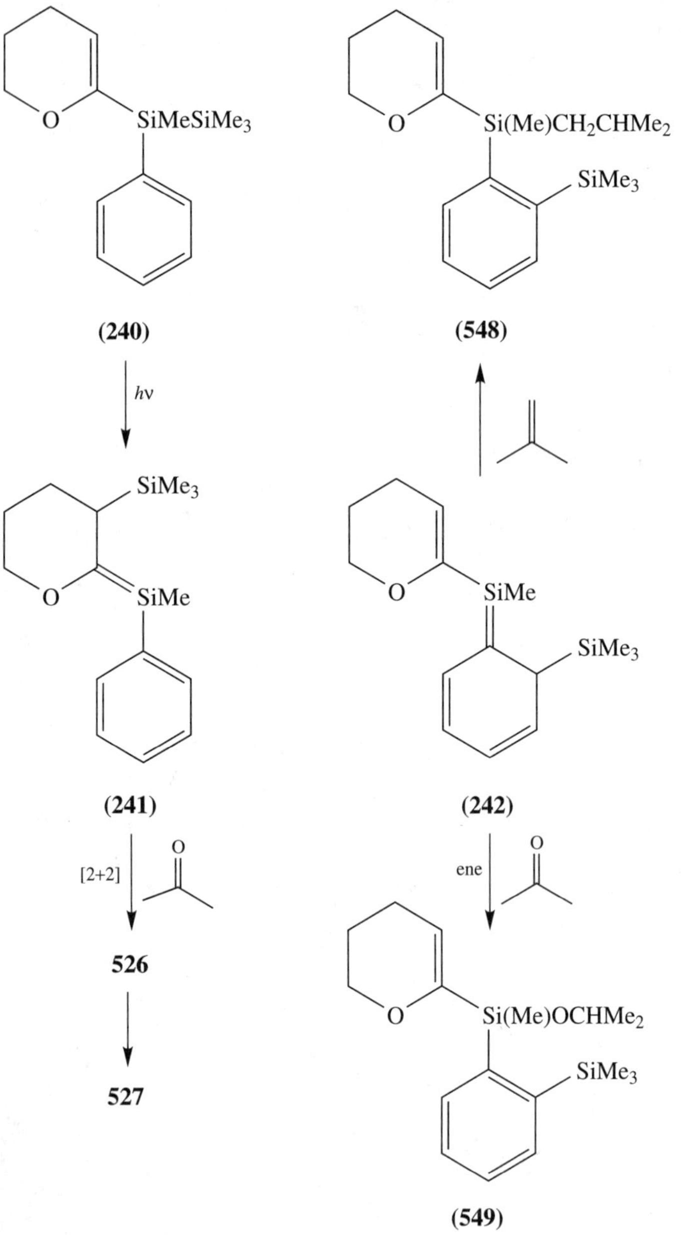

(240) (548)

(241) (242) (189)

526

527 (549)

The stereochemistry of the ene reaction with carbonyl compounds has been investigated by Ishikawa and coworkers with meso-(**550**) and rac-Ph(Et)MeSi−SiMe(Et)Ph as silene precursors[251]. With acetone and acetaldehyde the reactions were found to be highly diastereoselective when run in ether or toluene. In electron-donating solvents like THF or acetonitrile the diastereoselectivity decreased, and this was ascibed to a stepwise reaction.

With benzophenone the reactions are stereoselective. Equation 190 outlines the reaction mechanism for the case of acetone ($R^1 = R^2 = Me$). The disilanes rearrange via a concerted suprafacial 1,3-silyl shift under the photochemical conditions to produce the silenes diastereospecifically.

(190)

The ene reactions of the silatriene from $Ph_3Si-SiMe_3$ **552** with acetone and 2,3-dimethylbuta-1,3-diene have been investigated by Leigh and Sluggett[140]. In addition, a [2 + 2] cycloadduct is formed in the reaction with acetone.

6. Oxidations

All known silenes react violently with molecular oxygen with the exception of the stable 1-silaallenes which are air-stable. Along with carbonyl compounds the products of the oxidation of silenes are cyclic siloxanes from the corresponding silanone, depending on the reaction conditions[3,8,28]. For example, when the stable adamantylsilene **150** is exposed to dilute dry oxygen, the major products of the reaction are the cyclic trimer **553** and the trimethylsilyl ester **554**[3]. As an intermediate the disila-2,3-dioxetane **555** was suggested (equation 191)[13].

$$\underset{(150)}{\overset{Me_3Si}{\underset{Me_3Si}{>}}Si=C\overset{OSiMe_3}{\underset{Ad-1}{<}}} \xrightarrow{O_2} \left[\underset{(555)}{Me_3Si-\underset{\underset{O-O}{|}}{\overset{\overset{Me_3Si}{|}}{Si}}-\overset{OSiMe_3}{\underset{Ad-1}{|}}} \right]$$

(191)

$$\underset{(554)}{Me_3SiO\overset{O}{\underset{}{\overset{\|}{C}}}Ad-1} + [(Me_3Si)_2Si=O]$$

$$\downarrow$$

$$\underset{(553)}{-C-((Me_3Si)_2Si-O-)_3}$$

Only a recent matrix isolation study by Sander and coworkers gives further evidence for these type of intermediates[252,253]. They followed the reaction of small silenes with molecular oxygen in an argon matrix at 10–25 K with IR spectroscopy. Their results for the oxidation of $Me_2Si=CHMe$ **8** suggests the mechanism shown in equation 192. Initially the triplet biradical **T556** is formed. Intersystem crossing of **T556** to **S556** and fast exothermic ring closure gives the siladioxetane **557**. This opens easily, giving a complex **558** between dimethylsilanone and acetaldehyde. Finally, hydrogen transfer and silicon oxygen bond formation gives the product **559**, which could be identified by IR spectroscopy. The intermediacy of the biradical could be established by the spectroscopic identification of the hydroperoxide **560** (equation 192). A complex between methylsilanone and formaldehyde similar to **558** has been detected in the oxidation reaction of methylsilene. Although the existence of dioxetane intermediate **557** could not be proven directly, the identified oxidation product **559** gives strong evidence that siladioxetane **557** is an intermediate in the oxidation of silenes. Further support comes from the spectroscopic identification of complexes similar to **558** in the related oxidation of methylsilene[253]. Whether the excess energy produced in the exothermic formation of **557** leads to the fragmentation or if siladioxetanes like **557** are intrinsically unstable could, however, not be answered[253].

Some evidence for the intermediacy of a siladioxetane was given by Brook and coworkers, who detected in the careful oxidation of E/Z-**392** at $-70\,°C$ a new species whose ^{29}Si NMR spectrum is consistent with a structure like **561**[95]. The originally proposed heterolytic ring opening to the peroxocarbene **562** and the consecutive reactions via the radicals **563** and **565**, followed by recombination to **564** which is the only product (equation 193)[95], is however not consistent with the results of the matrix isolation study by Sander and coworkers[252,253]. We suggest therefore that also in the oxidation of **392** a complex **566**, between a silanone and an ester, is formed as an intermediate which, after initial nucleophilic attack by the oxygen on the silicon and subsequent trimethylsilyl shift, yields **564** (equation 194).

(192)

(193)

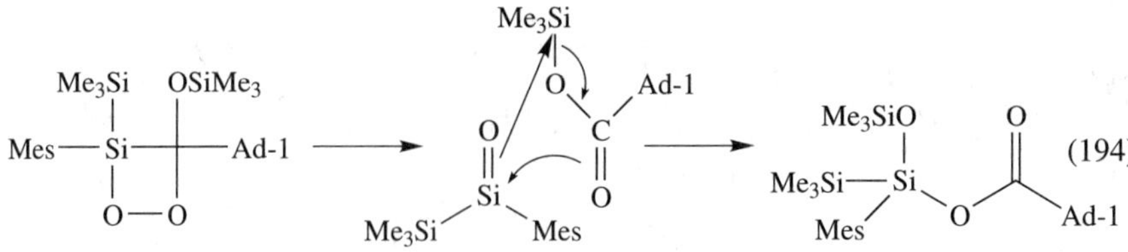

$$\underset{(561)}{\text{Mes}-\overset{\overset{\displaystyle\text{Me}_3\text{Si}}{|}}{\underset{\underset{\displaystyle\text{O}-\text{O}}{|}}{\text{Si}}}-\text{Ad-1}} \longrightarrow \underset{(566)}{\text{[intermediate]}} \longrightarrow \underset{(564)}{\text{Me}_3\text{SiO}-\underset{\underset{\displaystyle\text{Mes}}{|}}{\overset{\overset{\displaystyle\text{OSiMe}_3}{|}}{\text{Si}}}-\text{O}-\overset{\displaystyle\text{O}}{\underset{}{\text{C}}}-\text{Ad-1}} \qquad (194)$$

7. Miscellaneous

Different reaction modes between free **104a** and **104a·**donor adducts are found in two cases:

(a) The adduct **104a·**pyridine does not decompose in the regular way to give **104a** and pyridine, but isomerizes well below 0 °C by formation of the insertion product **567** into an *ortho* CH bond of pyridine (equation 195).

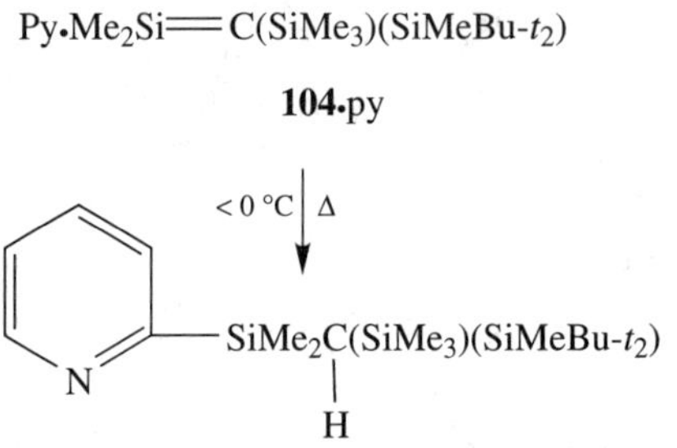

$$\text{Py·Me}_2\text{Si}=\text{C}(\text{SiMe}_3)(\text{SiMeBu-}t_2)$$

104·py

$$\downarrow <0\,°C \quad \Delta$$

2-[SiMe$_2$C(SiMe$_3$)(SiMeBu-t_2)(H)]-pyridine

(567) (195)

(b) The bromination of **104a** leads to the expected dibromo compound **568**, whereas with **104a·**THF, **568** and **569** — with incorporated THF — are formed in a 1 : 1 ratio (equation 196).

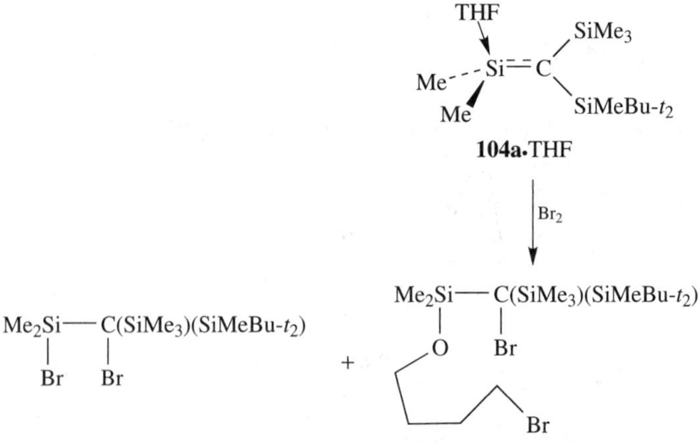

104a·THF: THF→SiMe$_2$=C(SiMe$_3$)(SiMeBu-t_2)

$$\downarrow \text{Br}_2$$

(568) Me$_2$Si(Br)—C(Br)(SiMe$_3$)(SiMeBu-t_2) + **(569)** Me$_2$Si(—O—(CH$_2$)$_4$—Br)—C(Br)(SiMe$_3$)(SiMeBu-t_2)

(196)

16. Silicon–carbon and silicon–nitrogen multiply bonded compounds 981

C. Structure and Spectroscopic Properties of Silenes

1. Structural studies

In the early days of silene chemistry most of the structural data for simple silenes came from IR, UV[1,2,7,8,29,157,158,254–256] and microwave studies[257,258]. By 1997 only three X-ray studies of stable silenes had been published[3,4,111] and much of the information about structure and bonding in silenes came from *ab initio* molecular orbital calculations. These data are well summarized in several review articles by Brook and Brook[9] and by Michl and Raabe[7,8]. The important role which *ab initio* calculations play in this field is outlined in the comprehensive review by Apeloig[11] and by Karni and Apeloig[12]. We will therefore emphasize in this paragraph the new important findings and we will outline the general trends of structural silene chemistry. Some structural data and original references are summarized in Table 12.

Only very recently a millimeter microwave study by Bogey, Bürger and coworkers established finally the ground state geometry of the parent silene **25**[31–33]. **25** was generated

TABLE 12. Structural parameter of silenes and related compounds (bond length in Å)

Compound	r (Si=C)	Method	Remarks	References
Silenes				
$H_2Si=CH_2$ **25**	1.704	millimeter microwave		31–33
$Me_2Si=CH_2$ **2**	1.692	microwave		254, 257
$Me_2Si=C(SiMe_3)(SiMeBu\text{-}t_2)$ **104a**	1.702	X-ray	twisted by 1.6° $\Sigma SiCSi = 360°^f$	4
$(Me_3Si)_2Si=C(OSiMe_3)Ad\text{-}1$ **150**[a]	1.764	X-ray	twisted by 16°	3
$(Me_3Si)_2Si=Ad\text{-}2$ **167**[b]	1.741	X-ray	twisted by 4.6°	111
1-Silaallenes				
1-AdSupSi=C=fluorene-2 **571**[a,c,d]	1.704	X-ray	*trans* bent	5
$Tip_2Si=C=C(Ph)Bu\text{-}t$ **572**[e]	1.693	X-ray	Σ: 357.2^f	259
Silaaromatics				
$(PhC)_4Si^{2-}$ **574a**	1.850	X-ray	dilithio compound	260
$(MeC)_4Si^{2-}$ **575**	1.830	X-ray	dipotassium compound	261, 262
Silene donor adducts				
$Me_2Si=C(SiMe_3)(SiMe Bu\text{-}t_2)\cdot THF$ **104a·THF**	1.747	X-ray	$r(D \rightarrow Si) = 1.878^g$ $\Sigma SiCSi = 348.7°^f$	197
$Me_2Si=C(SiMe_3)(SiMe Bu\text{-}t_2)\cdot F^-$ **104·F$^-$**	1.777	X-ray	$r(D \rightarrow Si) = 1.647^g$ $\Sigma SiCSi = 341.7°^g$	83
$Me_2Si=C(SiMe_2Ph)_2\cdot NEtMe_2$ **97·NEtMe$_2$**	1.761	X-ray	$r(D \rightarrow Si) = 1.988^g$	84

[a] 1-Ad = 1-adamantyl.
[b] 2-Ad = 2-adamantylidene.
[c] Sup = 2,4,6-*t*-butylphenyl.
[d] fluoren = 1,3,6,8-tetraisopropyl-2,7-dimethoxyfluorenylidene.
[e] Tip = 2,4,6-tri-isopropylphenyl.
[f] pyramidalization.
[g] D = donor.

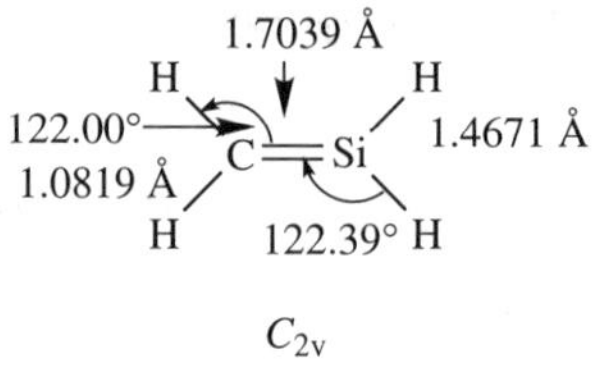

```
                    1.7039 Å
            H          ↓       H
    122.00°   ╲                ╱  1.4671 Å
    1.0819 Å ╱   C═══Si  ╲
            H       122.39°   H

                    $C_{2v}$
```

FIGURE 4. Experimental equilibrium structure of silene **25**, deduced from the rotational constants of six different isotopomers (for error margins see Table 13)[31–33].

by pyrolysis of 5,6-bis(trifluoromethyl)-2-silabicyclo[2.2.2]octa-5,7-diene **30** at 600 °C and at 1000 °C from silacyclobutane **22** and in traces also from 1,3-disilacyclobutane **24** in a flow of argon. The search for the millimeter wave transitions in the 180–473 GHz frequency range was guided by high level *ab initio* calculations of the rotational constants up to the CCSD(T) level of theory by Breidung, Thiel and coworkers[31]. The perfect agreement between the experimentally observed rotational constants A_0, B_0 and C_0 and the values predicted by the calculations (see Table 13) confirms beyond doubt that the rotational spectrum of $H_2C=SiH_2$ was observed[31]. From the rotational constants of six isotopomers of **25**, the experimental equilibrium structure shown in Figure 4 was deduced[32,33]. The geometric parameters are consistent with the results of *ab initio* calculations summarized in Table 13.

The Si=C bond length of 1.704 Å for **25** is in accord with a microwave study of $Me_2Si=CH_2$ **2** by Gutowsky and coworkers[257,258]. They found for **2** a Si=C bond length of 1.692 Å, in agreement with all *ab initio* calculations[246] but in contrast with a previous electron diffraction study which predicted an erroneously long Si=C bond length of 1.83 Å[263].

TABLE 13. Molecular constants of $H_2Si=CH_2$[31–33]

Parameter[a]	Ab initio					Experimental
	SCF[b]	MP2[b]	CCSD[b]	CCSD(T)[b]	CCSD(T)[c]	
r_e(C=Si)	1.6922	1.7085	1.7084	1.7167	1.7043	1.7039(18)[d]
r_e(CH)	1.0739	1.0777	1.0781	1.0799	1.0824	1.0819(12)[d]
r_e(SiH)	1.4684	1.4680	1.4713	1.4728	1.4670	1.4671(9)[d]
α_e(HCSi)	122.32	121.84	122.12	122.04	122.06	122.00(4)[d]
α_e(HSiC)	123.01	122.6	122.6	122.50	122.49	122.39(3)[d]
A_0	106554	105200	105119	104557	104945.0	104716.6
B_0	14944	14719	14700	14583	14772.6	14786.7
C_0	13082	12890	12875	12776	12927.9	12936.3
μ_e	1.114	0.867		0.810		0.700

[a]Equilibrium distances r_e in Å and angles α_e in deg; ground state rotational constants (A_0, B_0, C_0) in MHz; dipole moment μ_e in D.
[b]TZ2Pf basis set was used.
[c]cc-pVQZ(C,Si)/cc-pVTZ(H) basis set was used.
[d]The quoted accuracy corresponds to 15 standard deviations.

Three X-ray structures of stable silenes were published by 1997. All three silenes differ markedly in their substitution pattern and we will discuss them below in more detail. Historically, the first silene which was subjected to an X-ray analysis was the 'Brook'-type silene $(Me_3Si)_2Si=C(OSiMe_3)Ad-1$ **150**[3]. **150** had a relatively long Si=C bond length of 1.764 Å and its double bond is twisted by 16°. The increased bond length in **150** could be satisfactorily explained by Apeloig and Karni[246] by calculating the substituent effects on the Si=C bond length. The combined effect of two silyl substituents at silicon, and especially of the one siloxy substituent on carbon, lengthens the Si=C bond by 0.070 Å. Taking this substituent effect into account, the relatively long Si=C bond length is fully consistent with the Si=C bond length in the unsubstituted **25**[246]. This excellent experimental — theoretical agreement implies that in **150** the elongation of the Si=C bond due to the steric repulsion between the bulky substituents is small[246]. In contrast, the 16° twist in **150** is explained by steric interaction between the bulky groups at the ends of the double bond. Small twist angles in silenes have only minimal effects on the bond strengths[10]. Steric congestion in silenes leads to a twisting of the Si=C bond rather than to a bond elongation.

A much less perturbed Si=C bond is found in the Wiberg silene $Me_2Si=C(SiMe_3)(SiMe(Bu-t)_2)$ **104a**[4]. **104a** was shown to be nearly planar (twist angle of 1.8°), and the length of the Si=C bond was 1.702 Å, very near to the bond length of **25** and in perfect agreement with calculated substituent effects.

(150)

(104a)

(162)

Only very recently Apeloig, Bravo-Zhivotovskii and coworkers published the X-ray structure of the indefinitely stable silene **162**, synthesized in a sila-Peterson reaction[111]. Due to the synthetic approach to it, **162** has a novel substitution pattern, and it bears two silyl groups at silicon and two alkyl substituents on carbon. **162** exhibits an essentially planar arrangement around the C=Si bond with a small twist angle of 4.6° around the Si=C bond. The Si=C bond length in **162** is 1.741 Å, intermediate between that in **150** and **104a**. Similar to **150** the elongation of the Si=C bond in **162** is due to electronic effects of the substituents and not to their steric bulk. Thus, the calculated (HF/6-31G*) $r(Si=C)$ in **162** of 1.734 Å is only 0.005 Å longer than for the strain-free $Me_2C=Si(SiH_3)_2$ **570** (1.729 Å at HF/6-31G*)[111].

The Si=C bond lengths in the recently synthesized novel stable 1-silaallenes **571**[5] and **572**[259] are relatively short (1.704 Å and 1.693 Å in **571** and **572**, respectively). This might be expected for a Si=C double bond formed from a sp-hybridized carbon and a sp^2-hybridized silicon. Both 1-silaallenes are slightly bent at the central carbon [α(SiC^{1}C^2): 173.5° and 172.0° in **571** and **572**, respectively][5,259] and **571** is *trans* bent about the Si=C^1 bond in agreement with theoretical predictions[264].

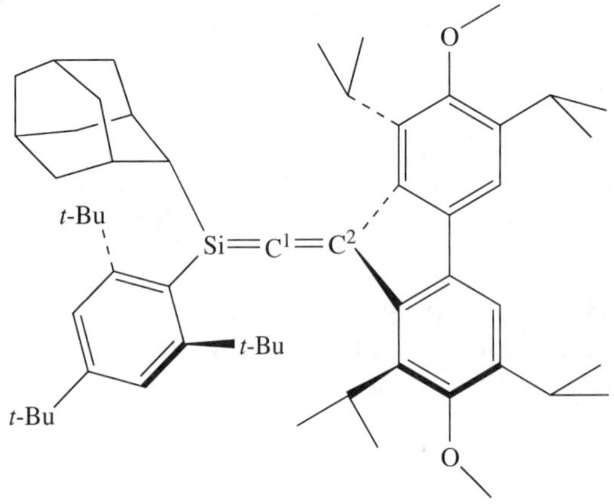

(571)

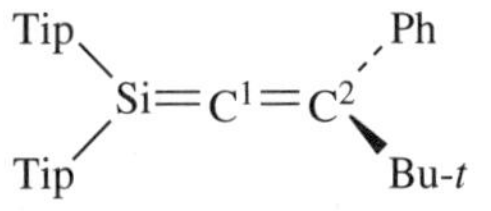

(572)

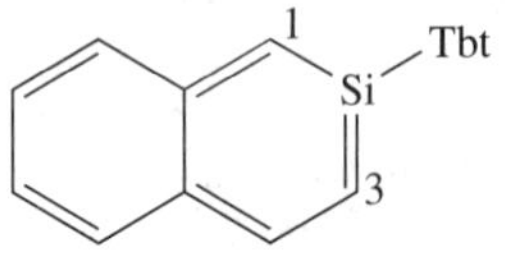

(573)

Tbt = 2,4,6-[(Me$_3$Si)$_2$CH]$_3$C$_6$H$_2$

Since 1994 several silaaromatic compounds have been synthesized[265] and structurally characterized[260–262]. The crystal structure of the remarkably stable 2-silanaphthalene **573** reveals planar environment around silicon, suggesting a delocalization of π-electrons in the 2-silanaphthalene ring system of **573**[6]. Further structural details were not reported due to severe disordering of the 2-silanaphthalene ring in the crystal. The crystal structures of the two silole dianions **574a**[260] and **575**[261] have been reported by the groups of West and Don Tilley. In both compounds the five-membered ring is planar and the inner-cyclic C–C bond lengths are nearly equalized, indicating the occurrence of cyclic conjugation in both compounds. The SiC bond lengths in **574a** and **575** are in the usual range for SiC single bonds. Calculations reveal, however, that silyl dianions are expected to have very long C–Si bonds, i.e. MeHSi^{2-}: 2.078 at MP2(fc)/6-31+G*)[266]. Thus the comparatively short Si–C$^\alpha$ distance in dianions **574a** and **575** indicates some degree of π-bonding. This view is supported by further theoretical analysis[266,267].

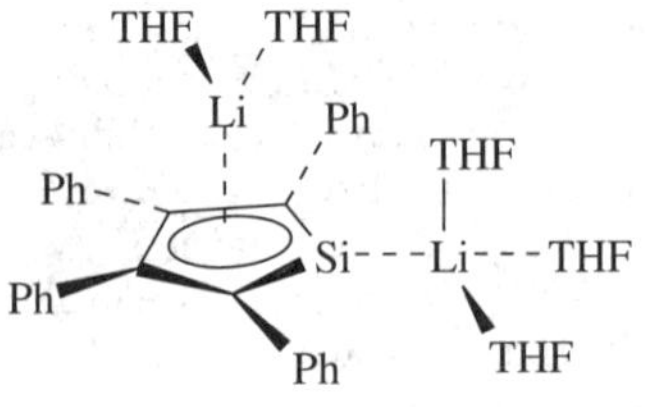

(574a)

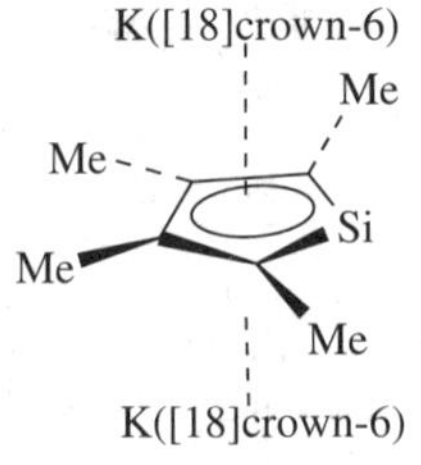

(575)

From the accumulated experimental and theoretical structural data one can draw the following general picture: The $R_2Si=CR_2$ unit is essentially planar with a Si=C bond length of about 1.70 Å. Electronic substituent effects do influence strongly the measured Si=C bond length. It is noteworthy that the substitution pattern is also important for $r(Si=C)$. Thus, for $Me_2C=Si(SiH_3)_2$ **576** a much longer $r(Si=C)$ is calculated than for the isomeric $Me_2Si=C(SiH_3)_2$ **577** [1.754 Å compared to 1.722 Å for **576** and **577** at MP2(fc)/6-31G*, respectively][111]. This theoretically predicted behaviour is supported by the structural data for **104a**[4] and **162**[111] (see Table 13). Steric interactions between the substituents at the Si=C bond do not contribute significantly to the $Si=C^1$ bond length but do enforce small twisting of the double bond. The $Si=C^1$ bond in 1-silaallenes is somewhat distorted towards a *trans* bent arrangement.

2. ^{29}Si and ^{13}C NMR spectroscopic data

a. ^{29}Si NMR spectroscopy. The chemical shifts of sp^2-hybridized silicon in silenes are found significantly downfield from the position normally observed for tetracoordinated, sp^3-hybridized silicon atoms. This is reminiscent of the carbon analogs alkanes and alkenes. Although the experimental data are limited to a small number of silenes with only small variances in the substitution pattern, the data summarized in Table 14 allows one to already draw some conclusions. The ^{29}Si chemical shift of silenes depends markedly on the substituents at silicon. Thus, silenes bearing two silyl substituents, i.e. silenes from the Brook family **149, 150, 578–580**[3,38,90,91,93,95] or those synthesized in the group of Apeloig[111] **162** and **585**, absorb in the relatively small region from 41 to 54 ppm (see Table 14). At the other extreme, two methyl groups at the silicon, like in Wiberg's silene **104a**[268] or in $Me_2Si=C(SiMe(t-Bu)OSiMe_3)Ad-1$ **583**[90], lead to an extremely deshielded ^{29}Si resonance ($\delta^{29}Si$ = 144 ppm and 126.5 ppm in **104a** and **583**, respectively). This reflects already the different polarization of the Si=C bond in those silenes, which is also revealed in their different chemical behaviour. When in Brook-type silenes one trimethylsilyl group is replaced by an aryl group, little effect on the ^{29}Si chemical shift is observed (0.3 ppm for the mesityl substituent in E-**392**[95] and 7.3 ppm for the phenyl group in **581**[90], while the t-butyl group in **582** leads to a significant downfield shift by 32.3 ppm compared with **150**[91]. The neopentyl silene **134** recently isolated by Delpon-Lacaze and Couret with two aryl groups at the trigonal silicon resonates at 77.6 ppm[269].

A comparison with the substituent effects on the chemical shift in disilenes might be of interest here: The sp^2-hybridized silicon in tetramesityldisilene **590** has a ^{29}Si chemical shift ($\delta^{29}Si$ = 63.3 ppm)[269,270] similar to that of the neopentylsilene **134**. In contrast with the relatively small shielding effect of two silyl groups on $\delta^{29}Si$ in silenes ($\delta^{29}Si$ = 49.7 ppm for **585**), silyl substitution in the disilenes leads to a markedly downfield shift. Thus, for tetra(triisopropylsilyl)disilene **591** $\delta^{29}Si$ = 164 ppm is found[270,254]. Apparently, substituent effects on the ^{29}Si chemical shift found in silenes do not simply parallel those in disilenes.

$$R_2Si=SiR_2$$

(**590**) R = Mes

(**591**) R = $(i\text{-Pr})_3Si$

Coupling constants across the Si=C bond are measured for some stable Brook-type silenes[29]. All $^1J_{Si=C}$ coupling constants cluster around 84 Hz, which is appreciably larger than $^1J_{SiC}$ between sp^3-hybridized silicon and attached methyl carbons (47–48 Hz), consistent with a double bond between carbon and silicon[91].

TABLE 14. δ^{29}Si and δ^{13}C chemical shifts (ppm) of silenes and related compounds

Compound	δ^{29}Si	δ^{13}C	Remarks	References
(Me$_3$Si)$_2$Si=C(OSiMe$_3$)R				
R = t-Bu (**149**)	41.5	212.7	$^1J_{C=Si}$ = 83.5 Hz	38, 91
R = 1-Ad(**150**)[a]	41.4	214.2	$^1J_{C=Si}$ = 84.4 Hz	3, 91
R = CEt$_3$ (**578**)	54.3	207.3	$^1J_{C=Si}$ = 83.9 Hz	3, 91
R = (1-Me)cyhex (**579**)[b]	43.5	212.9	$^1J_{C=Si}$ = 85.0 Hz	90, 91
R = bco (**580**)[c]	42.4	212.7		90
(Me$_3$Si)$_2$Si=C(OSiMe$_3$)Mes (**349**)	37.8	197.7		181
(Me$_3$Si)$_2$Si=C(OSiEt$_3$)Mes (**184**)	34.3	197.7		121
(Me$_3$Si)(Ph)Si=C(OSiMe$_3$)CEt$_3$ (**581**)	61.6	191.2		90
(Me$_3$Si)(t-Bu)Si=C(OSiMe$_3$)Ad-1 (**582**)[a]	73.7	195.6		90
E-(Me$_3$Si)(Mes)Si=C(OSiMe$_3$)Ad-1 (E-**392**)[a]	41.8	195.8		95
Z-(Me$_3$Si)(Mes)Si=C(OSiMe$_3$)Ad-1 (Z-**392**)[a]	44.0	191.6		95
E-(Me$_3$Si)TipSi=C(OSiMe$_3$)Ad-1 (E-**143**)[a,d]	40.3	195.9		93
Z-(Me$_3$Si)TipSi=C(OSiMe$_3$)Ad-1 (Z-**143**)[a,d]	43.3	191.0		93
Me$_2$Si=C(SiMe(t-Bu)OSiMe$_3$)Ad-1 (**583**)[a]	126.5	118.1		90
Me$_2$Si=C(SiMePhOSiMe$_3$)CEt$_3$ (**584**)	k	110.6		90
Me$_2$Si=C(SiMe$_3$)SiMe(t-Bu)$_2$ (**104a**)	144.2	77.2		268
TipMeSi=C(SiMe$_2$OSiMe$_3$)Ad-1 (**144**)[a,d]	108.1	127.3		93

Compound				
Mes$_2$Si=CHCH$_2$Bu-t (134)	77.6	110.4	δ^1H(=CH) = 5.53	82
(Me$_3$Si)(t-BuMe$_2$Si)Si=Ad-2 (162)^e	51.7	196.8		111
(t-BuMe$_2$Si)$_2$Si=Ad-2 (585)^e	49.7	198.2		111
(t-BuMe$_2$Si)$_2$Si=Cypen (174)^f	−71.0	157,160		114
Silaaromatics				
2-Tbt-Silanaphthalene^g (573)	87.4	116.0(C1), 122.6(C3)	$^1J_{C1=Si}$ = 92 Hz; $^1J_{C3=Si}$ = 76 Hz	6
1,4-di-t-Bu-2,6-bis(Me$_3$Si)-1-silabenzene (586)	26.8	126–127		274
(PhC)$_4$Si^{2-} (574b)	68.5	151.2^h	dilithio compound in solution	265
(PhC)$_4$Si^{2-} (574a)	87.3		dilithio compound in the solid state	260
1-Silaallenes				
1-AdSupSi=C=fluoren^{a,i,j} (571)	48.4	225.7		5
Tip$_2$Si=C=CPhBu-t (572)^d	13.1	213.6		259
t-BuSupSi=C=CPhBu-t (587a)^i	55.1^l	216.3	$^1J_{C=Si}$ = 142.4 Hz	259
t-BuSupSi=C=CPh$_2$ (587b)^d	58.7	227.9		259
t-BuSupSi=C=fluoren(a) (588)^{d,l,m}	48.0			271
t-Bu$_2$Si=C=fluoren(b) (589)^n	44.0			271

(continued overleaf)

TABLE 14. (continued)

Compound	δ^{29}Si	δ^{13}C	Remarks	References
Silenolate anions				
(Me$_3$Si)$_3$Si=C(O$^-$)R				
R = Mes (**182**)	−59.9	262.7		121
R = 1-Ad (**181**)^a	−70.5	274.1		121
R = *t*-Bu (**180**)	−70.3	274.3		121
Silene donor adducts				
Me$_3$N/Me$_2$Si=C(SiMe$_3$)$_2$ (**92•NMe$_3$**)	36.9	56.5		83
Me$_3$N/Me$_2$Si=C(SiMe$_3$)SiMe(Bu-*t*)$_2$ **104a•NMe$_3$**	34.7	^{k}		83
THF//Me$_2$Si=C(SiMe$_3$)SiMe-Bu-*t*)$_2$ (**104a•THF**)	52.4	^{k}		83
(D$_3$C)$_2$O//H$_2$Si=CH$_2$ (**27**)	−25.2	10.8		27
(D$_3$C)$_2$O//MeHSi=CH$_2$ (**28**)	−1.8	^{k}		27
(D$_3$C)$_2$O//Me$_2$Si=CH$_2$ (**29**)	16.8	^{k}		27

^{a}1-Ad = 1-adamantyl.
^{b}(1-Me)cyhex = 1-methylcyclohexyl.
^{c}bco = bicyclooctyl.
^{d}Tip = 2,4,6-trisopropylphenyl.
^{e}2-Ad = 2-adamantylidene.
^{f}Cypen = 2,3-di-*t*-Bu-cyclopropenylidene.
^{g}Tbt = 2,4,6-tris[bis(trimethylsilyl)methyl]phenyl.
^{h}assignment according to Ref 267.
^{i}Sup = 2,4,6-*t*-butylphenyl.
^{j}fluoren = 1,3,6,8-tetraisopropyl-2,7-dimethoxyfluorenylidene.
^{k}Not assigned.
$^l\delta^{29}$Si = 52.8 ppm reported in Ref. 271.
^{m}fluoren(a) = octaisopropylfluorenylidene.
^{n}fluoren(b) = fluorenylidene.

The silicon atom in the recently synthesized 4-silatriafulvene **174** is strongly shielded and therefore has a remarkably high field shifted ^{29}Si resonance (δ^{29}Si = −71 ppm)[114]. The cyclopropenyl carbons, however, were observed at relatively low field, suggesting substantial contribution from the canonical structure **174B** (equation 197)[114].

The trigonal silicon in the 1-silaallenes **571**[5], **587a-b**[259], **588** and **589**[271] gives rise to a signal in the region between 44.0–58.7 ppm. The unexpected high field shift of δ^{29}Si in **572**[259] might be reasonably explained by the substituent effect of the second aryl group which has in silenes a large shielding influence, i.e. $\Delta\delta \approx 32$ ppm for **582** compared with *E*-**392** (see Table 14). For **587a** an extremly large $^1J_{C=Si} = 142$ Hz is reported.

	R^1/R^2	R^3/R^4/R^5		R^1/R^2	R^3/R^4
(**571**)	Sup/1-Ad	*i*-Pr/OMe/H	(**572**)	Tip/Tip	Ph/*t*-Bu
(**588**)	Sup/*t*-Bu	Me/Me/Me	(**587a**)	Sup/*t*-Bu	Ph/*t*-Bu
(**589**)	*t*-Bu/*t*-Bu	H/H/H	(**587b**)	Sup/*t*-Bu	Ph/Ph

The dicoordinated silicon in the aromatic silole dianion **574** resonates at an unexpectedly low field for a silyl dianion (δ^{29}Si = 68.5 ppm)[265]. The different chemical shifts in the solid state (δ^{29}Si = 87.3 ppm)[260] and in solution suggest different structures. While in the solid state a η^1,η^5 structure **574a** is proven by an X-ray structure, in solution a highly symmetric η^5,η^5 structure **574b** seems to dominate[260]. According to *ab initio* calculations both isomers show indications of cyclic conjugation[260].

The first neutral silaaromatic compound stable at room temperature was recently synthesized by Okazaki and coworkers[6,272]. The 2-silanaphthalene **573** is characterized by

                              t-Bu
                               |
         Li
         | Ph
   Ph-- | ,'
      \ | /
       \ |/
   Ph /‾‾\ Si   Me₃Si   Si   SiMe₃
         | |
         Li   Ph   t-Bu

       (574b)   (586)

δ^{29}Si $= 87.4$ ppm, well in the region for trigonal sp^2-hybridized silicon. The coupling constants $^1J_{C=Si}$ to the neighbouring carbons C^1 and C^3 of 92 Hz and 76 Hz, respectively, are similar to those reported for the stable 'Brook' silenes. Schlosser and Märkl[274] reported that the ^{29}Si chemical shift of the highly substituted silabenzene **586** at $-100\,°$C occurred at 26.8 ppm. It is not clear, however, if the relative high field shift of the ^{29}Si signal is characteristic for the silabenzene or is caused by complexation with the solvent THF.

        Me₃Si   O
             \   //
              Si === C
             /   \
        Me₃Si   R

        (180) R = t-Bu
        (181) R = 1-Ad
        (182) R = Mes

The silenolate anion **182** is characterized by a very high field shifted signal in the ^{29}Si NMR (δ^{29}Si $= -59.9$ ppm). It is shifted to a lower field by 15.4 ppm relative to that of the acylsilane (Me₃Si)₃SiC(O)Mes. Due to hindered rotation around the SiC(O) bond the two trimethylsilyl groups at the silicon in **182** are magnetically not equivalent and give rise to different signals in the NMR spectrum, which coalesce at 25 °C[121]. The barrier for rotation around the SiC(O) bond was calculated from these data to be 14.3 kcal mol^{-1}, considerably smaller than that for Si=C double bonds[121] (43 ± 6 kcal mol^{-1} suggested by Jones and Lee)[273]. **181** and **180** have magnetically equivalent trimethylsilyl groups even at $-80\,°$C, indicating free rotation around the SiC(O) bond[121]. [Note that for **181** and **180** quite different chemical shifts are reported by Ishikawa and coworkers[21] (δ^{29}Si(**181**) $= -70.5$ ppm) and by Bravo-Zhivotovskii, Voronkov and coworkers (δ^{29}Si(**181**) $= 64.9$ ppm)[122]; we will refer here only to results of the first group.

Auner, Grobe and coworkers could detect the complexes **27–29** of simple silenes with perdeuterio dimethyl ether by low temperature NMR spectroscopy[27]. The NMR spectra of the complex **27** recorded at $-140\,°$C showed a triplet at -25.2 ppm in the ^{29}Si NMR spectra and a triplet at 10.8 ppm in the ^{13}C NMR spectra. The spectral data are consistent with the formation of a complex between silene **25** and dimethyl ether yielding the betain-like complex **27**. High level *ab initio* calculations of structure and NMR chemical shift for **27–29** strongly corroborate the experimental findings and establish the importance of both canonicals structures **27A** ↔ **27B**[27].

$$\begin{array}{c}
\text{CD}_3 \\
| \\
\text{H} \quad \quad \text{O} \quad \text{CD}_3 \\
\text{C}=\text{Si}-\text{R}^2 \\
\text{H} \quad \quad \text{R}^1
\end{array}$$

(**27**) $R^1 = R^2 = H$
(**28**) $R^1 = H, R^2 = Me$
(**29**) $R^1 = R^2 = Me$

(**27A**) ↔ (**27B**)

The complexation of silene **104a** with THF or trimethylamine leads to a shielding of the silicon by 92 ppm or 110 ppm, respectively, indicating the strong interaction between the Si and the donor molecule[4].

b. *^{13}C NMR spectroscopy*. The observed range for the ^{13}C resonance for stable silenes is very large (77–214 ppm), much larger than for most alkenes (95–155 ppm). For the group of silenes (Me$_3$Si)$_2$=C(OSiMe$_3$)R **149, 150, 578–580**, the sp^2-hybridized carbon atom is observed in the range of 207–214 ppm. The carbon in silenolates absorbs at even lower field (262.7–274.1 ppm). Replacing one trimethylsilyl group at the silicon in silenes by an aryl or alkyl group shifts the ^{13}C resonance to higher field by 15–20 ppm. The trimethylsiloxy group in **149, 150, 578–580** is responsible for a deshielding of the carbon by approximatively 10–15 ppm compared with an alkyl group. Thus, the silenes **162** and **585**, having only alkyl substituents at the carbon, resonate at 197–198 ppm. The tertiary vinylic carbon in the neopentylsilene **134** is found at 110.4 ppm. The shielding influence of two directly attached trimethylsilyl groups is responsible for the relatively high field resonance of the doubly bonded carbon at 77.2 ppm in Wiberg's silene **104a**, although the different polarity of the Si=C bond might also contribute to the shielding of the carbon atom. The ^{13}C signal of the central carbon in 1-silaallenes (δ^{13}C = 213.6–227.9 ppm) can be found at the lowfield end of the usual range for the ^{13}C resonance of the central allenic carbon (δ^{13}C = 195–215 ppm).

3. Infrared spectroscopic data

Much IR data has been collected for simple silenes in matrix isolation work and these data are summarized in several previous reviews[7,8].

Simple silenes have a Si=C stretching frequency around 1000 cm^{-1} (see Table 15 for a few examples). Thus, for silene **25** a weak band is found at 985 cm^{-1} which is shifted by methyl substitution to higher frequencies. Me$_2$Si=CH$_2$ **2** has a $\nu_{Si=C}$ of 1003.5 cm^{-1}.

TABLE 15. UV [λ (nm) (ϵ_0)], IR [$\nu_{(Si=C)}$ (cm^{-1})] and PE [IP (eV)] data of silenes and related compounds

Compound	$\lambda_{max}(\epsilon_0)$	$\nu_{(Si=C)}$	$\nu_{(Si=H)}$	IP	Remarks	References
Silenes						
H$_2$Si=CH$_2$ (25)	258	985	2239,2219	8.85		28–30
MeHSi=CH$_2$ (26)	260	989	2187			156, 158
Me(HO)Si=CH$_2$ (592)		899			Si–O 777.5, 781	275
Me$_2$Si=CH$_2$ (2)	244	1001;1003.5		7.71(ad),7.98(v)		254–256, 276
						256
Me$_2$Si=CHMe (8)	255	978				157, 164
Me$_2$Si=CHSiMe$_3$ (301)	265					164
Me$_2$Si=CMeSiMe$_3$ (593)	274					164
Me$_2$Si=C(SiMe$_3$)$_2$ (92)	278					164
Me$_2$Si=CPhCO$_2$Me (594)	280					18, 165
Me$_2$Si=CPhCO$_2$Et (595)	288					165, 277
Me$_2$Si=C(SiMe$_3$)CO Ad-1 (596)	284					165, 277
Me$_2$Si=C(OSiMeBu-*t*OSiMe$_3$) Ad-1 (583)[a]	290					90
Ph$_2$Si=CH$_2$ (31)	325					35b
(Me$_3$Si)$_2$Si=C(OSiMe$_3$)R						
R = Me (369)	330 (6500)					88
R = 1-Ad (150)[a]	340 (7400)			7.7		3, 91
R = CEt$_3$ (578)	342 (7060)					3, 91
R = *t*-Bu (149)	339 (5200)					3, 91
(Me$_3$Si)(*t*-BuMe$_2$Si)Si=Ad-2 (162)[b]	322 (6300)					111
(Me$_3$Si)$_3$*t*-BuSi=C(OSiMe$_3$)Ad-1 (582)[a]	340					90
Silaallenes						

Me$_2$Si=C=C(SiMe$_3$)$_2$ (**597**)	275, 325			278
1-Ad(Sup)Si=C=fluoren^{a,c,d} (**571**)	267, 276, 297, 318, 334			5
Silabutadienes				
Me$_2$Si=CPh–CH=CH$_2$ (**361**)	338			183
(**290**) R = H R = Me	296 312	929 917	2210 2202	159 159
(**289**) R = H R = Me	270 274	936 933	2216 2212	159 159
(**296**)	278		further λ_{max} at 328, 400	
Silatrienes				

(*continued overleaf*)

TABLE 15. (continued)

Compound	$\lambda_{max}(\epsilon_0)$	$\nu_{(Si=C)}$	$\nu_{(Si=H)}$	IP	Remarks	References
R^{1}R^{2}Si=... Me$_3$Si H						
R^1=R^2=Me (201)	425					137
R^2=Ph, R^2=Me (524A)	460					137
R^1=R^2=Ph (524B)						
Ph$_2$Si=... MePh$_2$Si H (598)	490					137
Silaaromatics						
2-Tbt-silanaphthalene (573)^e	267(20000),312(7000) 387(3000)				1368 cm^{-1} (Raman (C—C))	6
Silabenzene (599)	212,272,298,305,313,321				for IR see Ref.	279,281
1,4-Disilabenzene (600)	275,396,385,396,408					282
4-Silatoluene (601)	301,307,314,322					283

^{a}1-Ad = 1-adamantyl.
^{b}2-Ad = 2-adamantylidene.
^{c}Sup = 2,4,6-tri-*t*-butylphenyl.
^{d}fluoren = 1,3,6,8-tetraisopropyl-2,7-dimethoxyfluorenylidene.
^{e}Tbt = 2,4,6-tris[bis(trimethylsilyl) methyl]phenyl.

16. Silicon–carbon and silicon–nitrogen multiply bonded compounds 995

From that value a force constant of $k = 5.6$ mdyn Å^{-1} for the Si=C double bond is deduced[255]. This frequency is clearly higher than the usual range for Si–C stretch vibrations but substantially less than for C=C stretches, both because Si is heavier than C and because the Si=C bond is weaker than the C=C bond. More suitable for the experimental characterization is the vinylic Si–H stretch vibration which gives rise to a medium band at 2239 cm^{-1} (**25**) or 2187 cm^{-1} (**2**)[29], hypsochromically shifted by around 100 cm^{-1} relative to the Si–H stretch in simple silanes. A detailed analysis of the vibrational spectra of matrix-isolated MeHSi=CH$_2$ **26** using polarized IR spectroscopy established IR transition moment directions relative to the $\pi\pi^*$-transition moment (Si–C axis) in **26**[156]. These data provide detailed information about the vibrational modes and about the structure of **26**[156]. The bathochromic shift of the Si=C stretch in the isomeric 1,3-silabuta-1,3-dienes **289** and **290** by around 70 cm^{-1} compared with the Si=C stretch in simple silenes (Table 15), was interpreted as an indication of Si=C–C=C and C=Si–C=C π-conjugation[159].

(289)   (290)

R = H, Me

Stable Brook silenes like (Me$_3$Si)$_2$Si=C(OSiMe$_3$)Ad-1 **150** are characterized by several bands in the region 1300–930 cm^{-1}, but it is not known with certainty which bond is associated with the Si=C bond[3]. The same is true for the heavily substituted 1-silaallene **571**. Several strong bands have been detected by IR spectroscopy but no assignment was given[5]. Two silabenzenes **599**[279–281] and **600**[282] and silatoluene **601**[283] have been investigated in the matrix at low temperatures and the IR spectra are reported without any assignment. For the 2-silanaphthalene **573**, an absorption at 1368 cm^{-1} for the C–C bond stretch is found in the Raman spectrum, which corresponds to the maximum absorption at 1382 cm^{-1} for naphthalene[6].

(599)   (600)   (601)

A matrix reaction of trimethylsilane with oxygen atoms (from ozone) at 14–17 K gives a product that was tentatively identified as the silenol H$_2$C=Si(OH)Me **592** by IR spectroscopy[275]. The finding was supported by isotopic labeling with ^{18}O and D.

4. UV spectroscopic data

UV spectroscopic data have been summarized in detail in earlier reviews and we will outline only the fundamental UV characteristics of silenes and will concentrate on new achievements of interest.

Small silenes absorb in the region 245–260 nm, a region which is assigned to the $\pi\pi^*$ transition[89]. For example, the parent silene **25** is characterized by a UV transition at 258 nm[28,29]. Me$_2$Si=CH$_2$ **2** has a λ_{max} at 244 nm and further substitution at the vinylic carbon gives a bathochromic shift of the $\pi\pi^*$ band (see Table 15). The highly substituted stable silenes of the 'Brook' family have UV absorbtions around 340 nm and extinction coefficients ϵ_0 between 5200 and 7400[3,91]. Similarly, the stable 'Apeloig' silene **162** has a UV maximum at 322 nm (ϵ_0 = 6300)[111]. This indicates a relatively large bathochromic shift induced by the two silyl groups at the sp^2-hybridized silicon. Interestingly, silyl substitution at the carbon atom induces a smaller bathochromic shift, thus for Me$_2$Si=C(SiMe$_3$)$_2$ **92**, λ_{max} = 278 nm[164]. Conjugation with aryl or C=O groups gives expectedly a bathochromic shift. Thus the UV band is red shifted in **595** to λ_{max} = 288 nm[165].

The UV absorption bands of the siladienes **289** and **290** attributed to $\pi\pi^*$ transition on the basis of multireference CI calculations are considerably red shifted in comparison with the absorptions of the parent silene **25** (Table 15) and suggest that π-conjugation is significant in these compounds[159]. The large difference between the 1-silabuta-1,3-diene **290** and the 2-silabuta-1,3-diene **289** was rationalized by the opposite influences of the small SiC resonance integral on the HOMO–LUMO gap in **290** (decreasing HOMO–LUMO gap) and **289** (increasing HOMO–LUMO gap). Phenyl substitution at the silabuta-1,3-diene moiety increases the bathochromic shift further: **361** absorbs at 338 nm (Table 15)[183]. Transient silacyclobuta-1,3-diene **296** is characterized by an UV absorption at 278 nm[161]. Further red-shifted to 425–490 nm is the $\pi\pi^*$ transition in a series of 1-silatrienes **201**, **524** and **598** observed by Leigh and coworkers[137].

The transient 1-silaallene, Me$_2$Si=C=C(SiMe$_3$)$_2$ **597**, absorbs at 275 nm and 325 nm[278]. The ca 30 nm red shift of the spectral maximum for **597** relative to that of Me$_2$Si=CH$_2$ was ascribed to hyperconjugative interactions between the β-trimethylsilyl groups and the Si=C bond[278]. The UV spectrum of the stable 1-silaallene **571** was also reported, but no assignment of the bands was given (Table 15)[5].

For several silaaromatics (**573, 599–601**) UV spectra are reported (Table 15). The compounds were found to behave like pertubed benzenes (or naphthalenes).

5. Miscellaneous

Only a few photoelectron spectra of silenes have been measured. Silene **25** has a first vertical ionization energy of ca. 8.85 eV, in agreement with the theoretical value of 8.95 eV[30]. The assignment is further supported by the observed fine structure of the first band which is similar to that found for ethene. The first ionization potential of the stable 'Brook' silene **150** is 7.7 eV[3], similar to what is found for Me$_2$Si=CH$_2$ **2** (7.98 eV)[276].

Silene **2**, 1-fluorosilene **602** and 1,1-difluorosilene **603** have been examined by FT ion cyclotron resonance spectroscopy in order to estimate the π-bond strength. Bracketing studies with various bases (B) and fluoride acceptors (A) gave values for the proton and fluoride affinity of the silenes[284]. The π-bond energy was calculated by thermochemical cycles and found to increase with fluorine substitution. The data are given in Table 16.

16. Silicon–carbon and silicon–nitrogen multiply bonded compounds

TABLE 16. Thermochemical data for substituted silenes[284]

Silene	Proton affinity (kcal mol^{-1})	Fluoride affinity (kcal mol^{-1})	π-Bond energy (kcal mol^{-1})
Me$_2$Si=CH$_2$, **2**	228±2	37±2	39±6
MeFSi=CH$_2$, **602**	187±2	40±2	45±5
F$_2$Si=CH$_2$, **603**	178±2	46±2	50±5

The proton and fluoride silene adducts were produced according to equations 198–201.

$$Si(CH_3)_4 + e^- \longrightarrow Si(CH_3)_3^+ + CH_3 + 2e^-$$

$$\begin{array}{c}H_3C\\ \diagdown\\ \quad\quad Si{-}CH_3 + B \longrightarrow BH^+ + Me_2Si{=}CH_2 \quad\quad \text{proton affinity}\\ \diagup\\ H_3C\end{array} \quad (198)$$

(**2**)

$$NF_3 + e^- \longrightarrow F^- + NF_2 \quad (199)$$

$$SiMe_4 + F^- \longrightarrow [FSiMe_4]^{-*} \longrightarrow FSiMe_2CH_2^- + CH_4 \quad \text{fluoride affinity} \quad (200)$$

$$FSiMe_2CH_2^- + A \longrightarrow AF^- + Me_2Si{=}CH_2 \quad (201)$$

II. SILAALLENES

A. Theoretical Studies

The isomeric 1- and 2-silaallenes have been the subject of numerous computational studies. According to calculations the 1-silaallene **604** is more stable than the isomeric **605** by 14.7 kcal mol^{-1} (at MP2/6-31G**//MP2/6-31G**)[285a], but both are relatively high lying minima on the C$_2$H$_4$Si potential energy surface. Thus, **604** is less stable than the syn/anti- isomeric silylenes **606** and **607** by 5.7 and 5 kcal mol^{-1}, respectively. Note that according to high level *ab initio* calculations silene H$_2$Si=CH$_2$ **25** is more stable than methylsilylene H$_3$CSiH **277** by *ca* 4 kcal mol^{-1}[151].

$$\begin{array}{cccc} H_2C{=}C{=}SiH_2 & H_2C{=}Si{=}CH_2 & H_2C{=}CH\overset{\displaystyle Si\diagup}{\diagdown H} & H_2C{=}CH\overset{\displaystyle \diagdown Si{:}}{\diagup}^H \\ \mathbf{(604)} & \mathbf{(605)} & \mathbf{(606)} & \mathbf{(607)} \end{array}$$

In contrast to carbon chemistry, the three-membered cyclic C$_2$H$_4$Si isomers are relatively stable. Thus, silacyclopropylidene **608** and 2-silacyclopropene **609** are more stable than **604** by 10.9 and 13.8 kcal mol^{-1}, respectively[285b], while cyclopropene **610** is by 22.4 kcal mol^{-1} less stable than allene **611** and more stable by 62.0 kcal mol^{-1} than singlet cyclopropylidene **612**[286].

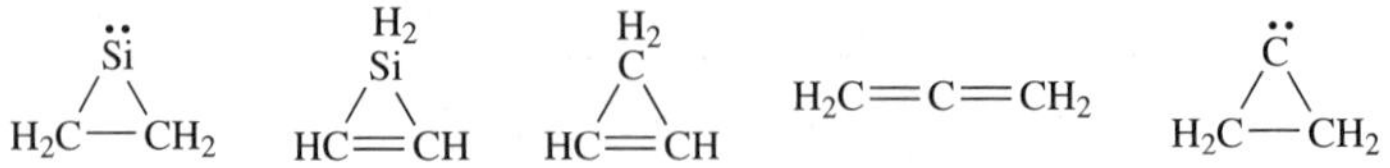

(608) (609) (610) (611) (612)

B. Synthesis

It was suggested that 1-silaallene [1-D] **604** is a short-lived intermediate in the isomerization of the isotopomeric vinylsilylenes [1-D] **606** and [2-D] **606** (equation 202) which have been generated in the high vaccum flash pyrolysis of 1,1,1-trimethyl-2-vinyldisilane[285b].

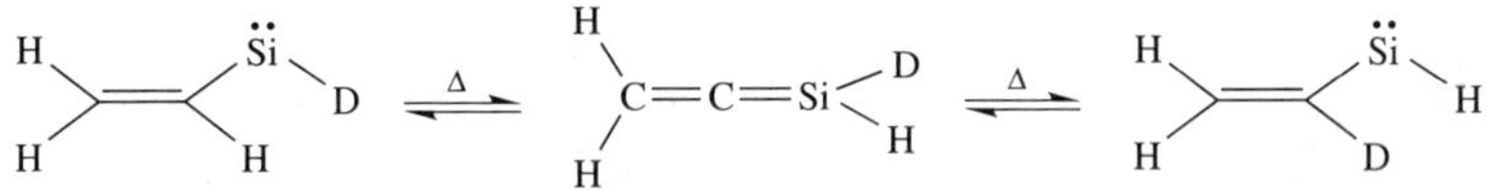

[1-D] **(606)** [1-D] **(604)** [2-D] **(606)**

(202)

1-Silaallenes **613** have been proposed by Ishikawa, Kumada and coworkers as transient reactive intermediates in the photolytic or thermolytic degradation of alkynyldisilanes **614** as minor byproducts, the main product being silacyclopropenes **615**[184,192,287] (equation 203).

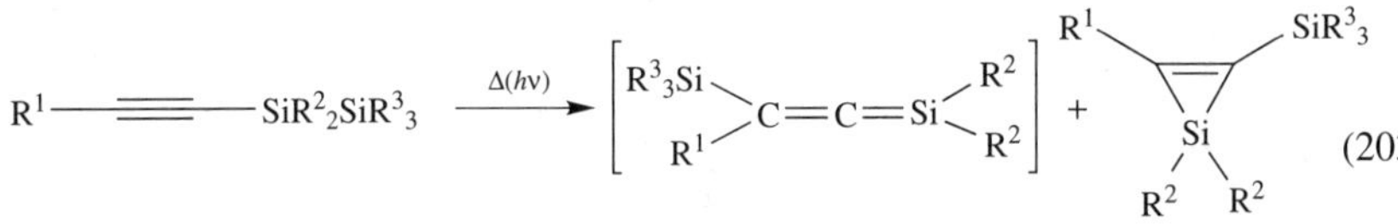

(203)

(614) **(613)** **(615)**

In the presence of scavenger reagents, consecutive products of 1-silaallenes **613** have been characterized[287,288]. Optimization of the substitution pattern in the precursor ethynylsilane increases the yield of the transient 1-silaallene. Also, silacyclopropenes themselves may serve as starting materials for 1-silaallenes[289,290]. Photolysis or thermolysis of 2-silacyclopropene **616** gives, after trapping of the intermediate silaallene **617** with methanol, the vinylsilane *Z*-**618** in 21% yield[291] (equation 204).

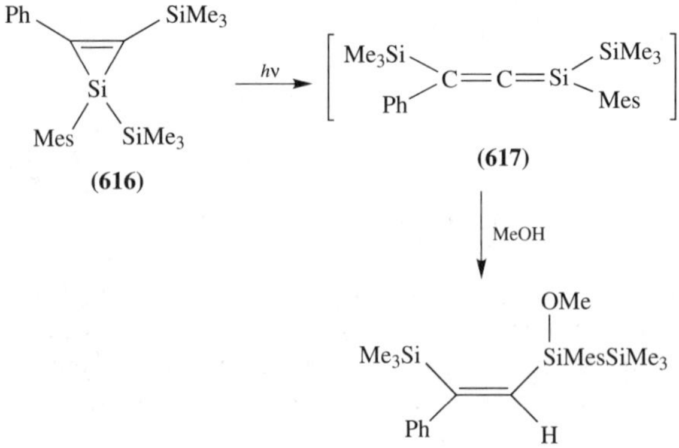

(204)

Z-(618)

Following this reaction sequence established by Kumada and Ishikawa, Leigh and coworkers detected the transient 1-silaallene **597** in a laser flash photolysis of the ethynyldisilane **619**. **597** was identified by its characteristic UV absorptions at 275 nm and 325 nm[278] and by the trapping reaction with methanol. It is formed, however, in a mixture with silacyclopropene **620**, dimethylsilylene **621** and acetylene **622**. Based on the quantitative analysis of the products of methanolysis (**623–626**) a chemical yield of **597** of 12–15% was deduced[278] (equation 205).

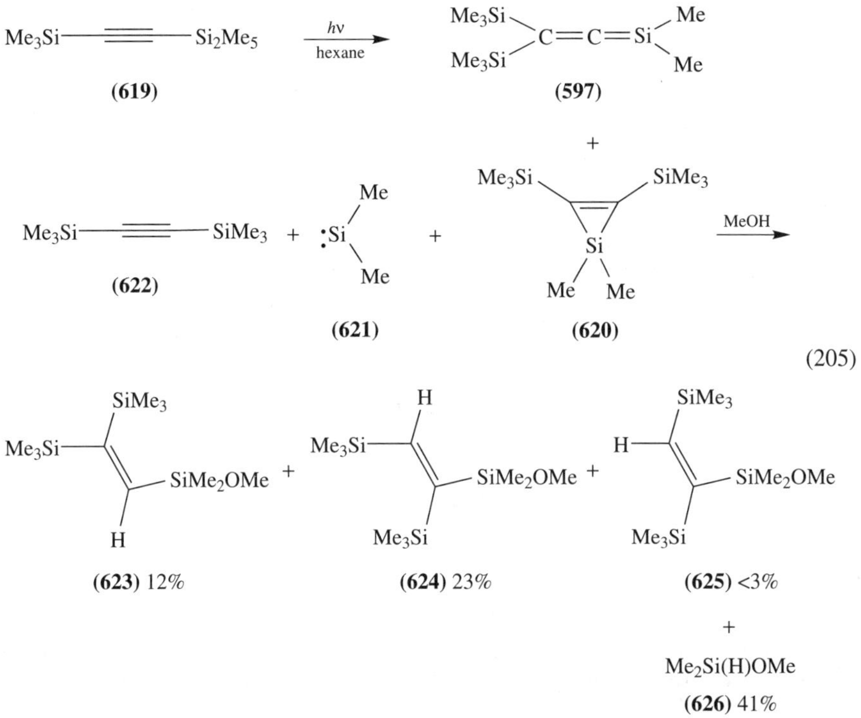

$$Me_3Si-\!\!\equiv\!\!-Si_2Me_5 \quad \xrightarrow[\text{hexane}]{h\nu} \quad \underset{Me_3Si}{\overset{Me_3Si}{>}}C\!=\!C\!=\!Si\underset{Me}{\overset{Me}{<}}$$

(**619**) → (**597**)

+

$$Me_3Si-\!\!\equiv\!\!-SiMe_3 \quad + \quad :Si\underset{Me}{\overset{Me}{<}} \quad + \quad \underset{Me_3Si}{\overset{Me_3SiSiMe_3}{\triangle\!\!\!\!\!\underset{Si}{}}} \quad \xrightarrow{MeOH}$$

(**622**)   (**621**)   (**620**)

(205)

$$\underset{H}{\overset{SiMe_3}{Me_3Si\!-\!\!=\!\!-SiMe_2OMe}} \quad + \quad \underset{Me_3Si}{\overset{H}{Me_3Si\!-\!\!=\!\!-SiMe_2OMe}} \quad + \quad \underset{Me_3Si}{\overset{SiMe_3}{H\!-\!\!=\!\!-SiMe_2OMe}}$$

(**623**) 12%   (**624**) 23%   (**625**) <3%

+

Me$_2$Si(H)OMe

(**626**) 41%

The first synthesis of stable 1-silaallenes follows closely an established route developed in carbon chemistry[292]. An overall nucleophilic 1,3 substitution at propargylic halides gives allenes in good yields (equations 206 and 207). In organosilicon chemistry, special attention must be paid to avoid direct substitution at the silicon which is the preferred site for nucleophilic attack.

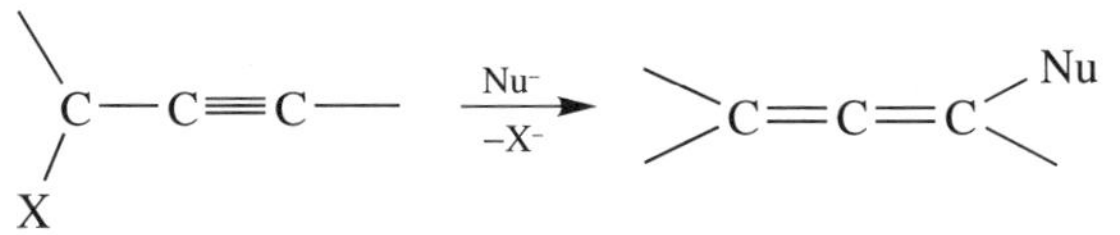

$$\underset{X}{\overset{\diagdown}{\underset{\diagup}{C}}}-C\!\equiv\!C-\quad\xrightarrow[-X^-]{Nu^-}\quad \underset{}{\overset{\diagdown}{\underset{\diagup}{>}}}C\!=\!C\!=\!C\underset{}{\overset{Nu}{<}} \qquad (206)$$

$$\underset{X}{\overset{\diagdown}{\underset{\diagup}{Si}}}-C\!\equiv\!C-\quad\xrightarrow[-X^-]{Nu^-}\quad \underset{}{\overset{\diagdown}{\underset{\diagup}{>}}}Si\!=\!C\!=\!C\underset{}{\overset{Nu}{<}} \qquad (207)$$

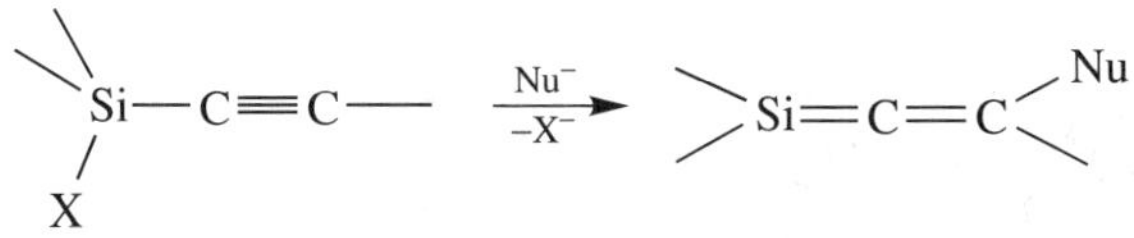

Two different approaches to solve this problem have been developed by the group of West[5,259,271]. In the steric method the substituents R^1 and R^2 at the silicon block the attack of the organometallic reagent by virtue of their bulkiness, thus attack occurs at the β-carbon atom[271]. Three 1-silaallenes **572**, **587a** and **587b**, which are stable at room temperature, have been prepared from ethynylfluorsilanes **627** and **628** according to equation 208 and were identified by NMR spectroscopy[259]. From **572** a crystal structure was also obtained (see Section I.C.1). The organolithium compounds **629–631** can be detected by low temperature NMR and **629** is even stable around 0 °C and was clearly identified by trapping reactions and by X-ray crystal diffraction of its TMEDA complex[259].

$R^1/R^2/R^3$

(627) Tip/Tip/Ph
(628) Sup/t-Bu/Ph

$R^1/R^2/R^3/R^4$

(629) Tip/Tip/Ph/t-Bu
(630) Sup/t-Bu/Ph/t-Bu
(631) Sup/t-Bu/Ph/Ph

(208)

$R^1/R^2/R^3/R^4$

(572) Tip/Tip/Ph/t-Bu
(587a) Sup/t-Bu/Ph/t-Bu
(587b) Sup/t-Bu/Ph/Ph

In the intramolecular method the organometallic reagent is attached to the alkyne in such a way that it cannot reach the silicon atom intramolecularly, yet it can easily reach the β-carbon atom[271]. The synthesis of 1-silaallenes **571**, **588**, **589** by the dehalogenative intramolecular carbometalation–elimination (DICE) reaction is shown in equation 209. In the reaction of 2-bromo-2'[(fluorosilyl)]ethynyl]biphenyls **632–634** with 2 eq of t-butyllithium at 0 °C, the 1-silallenes **571**, **588**, **589** were obtained in nearly quantitative yields[5,192]. Due to extreme steric congestions, **571** is stable at air and moisture and can be dissolved in boiling ethanol without observable transformation[5].

Of the 2-silaallenes no sufficiently stable species have yet been isolated, so that no spectroscopic verification is available up to date. The synthesis of possible 2-silaallenes is therefore treated along with the description of their reactivity in Section II.C below.

C. Reactivity

1. 1-Silaallenes

The polarity of the Si=C bond in 1-silaallenes is reduced compared with that in silenes. Thus, for the parent 1-silaallene $H_2Si=C=CH_2$ **604** the calculated Mullikan atomic charges on Si and on the central C are +0.17 and -0.10[293], respectively, compared to 0.46 and -0.65 in $H_2Si=CH_2$ **25**. This is expected to reduce the reactivity of 1-silaallenes

16. Silicon–carbon and silicon–nitrogen multiply bonded compounds 1001

compared with silenes[246]. The theoretical prediction is strongly supported by kinetic data collected by Leigh and coworkers[278]. 1-Silaallene **597** exhibits a reactivity characteristic of the Si=C bond but the absolute rate constants for reaction of **597** towards characteristic silene trapping reagents like MeOH are significantly lower than those determined for other transient silenes. For example, $Ph_2Si=CH_2$ **31** reacts with MeOH more than 1000 times faster than **597** ($k_{MeOH} = 1.9\pm0.2\times10^9$ $M^{-1}s^{-1}$ for **31** $K_{MeOH} = 1.7\pm0.1\times10^6$ $M^{-1}s^{-1}$ for **597**[278]) (For the mechanistic details of the alcohol addition to silenes, see the chapter by Sakurai.) Also, the reactivity towards other silene scavenger reagents like HOAc, acetone, O_2 or octa-1,3-diene is strongly reduced[278]. Silylenes have also been employed as trapping reagents. Thus, the transient silylene Mes_2Si: **635** reacts with **636** giving the intriguing bissilacyclopropylidene **637** in 29% yield[185,288] (equation 210).

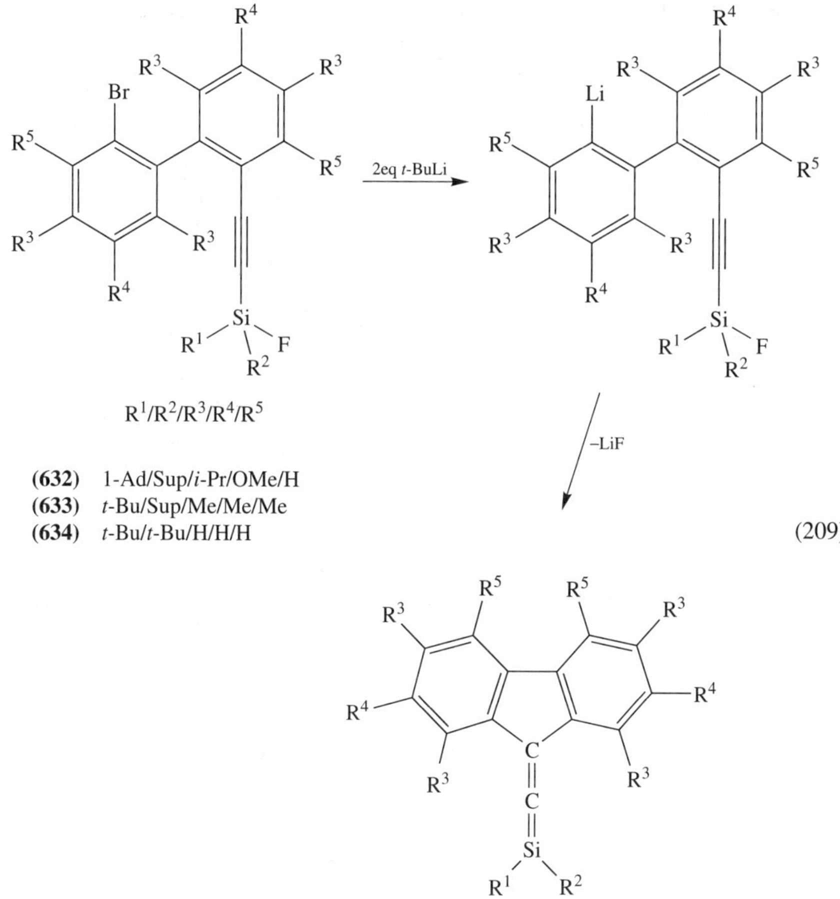

$R^1/R^2/R^3/R^4/R^5$

(**632**) 1-Ad/Sup/i-Pr/OMe/H
(**633**) t-Bu/Sup/Me/Me/Me
(**634**) t-Bu/t-Bu/H/H/H

(209)

$R^1/R^2/R^3/R^4/R^5$

(**571**) 1-Ad/Sup/i-Pr/OMe/H
(**588**) t-Bu/Sup/Me/Me/Me
(**589**) t-Bu/t-Bu/H/H/H

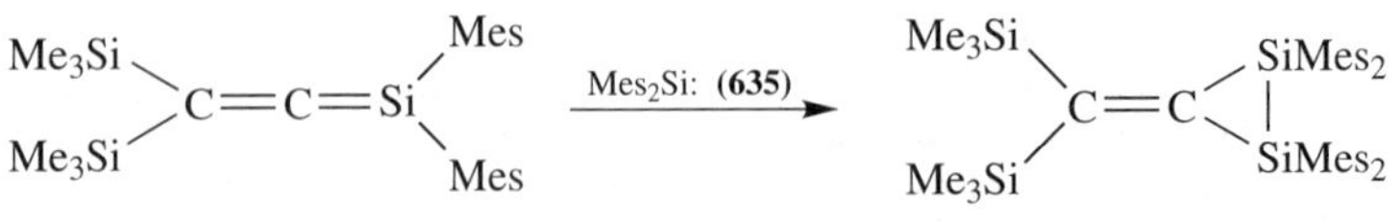

$$\underset{(636)}{\underset{Me_3Si}{Me_3Si}}C=C=Si\underset{Mes}{\overset{Mes}{\diagup}} \xrightarrow{Mes_2Si: (635)} \underset{Me_3Si}{\underset{(637)}{Me_3Si}}C=C\underset{SiMes_2}{\overset{SiMes_2}{\diagdown|\diagup}} \quad (210)$$

In the absence of trapping reagents the transient 1-silapropadiene **638** dimerizes in contrast to usual simple silenes in a head-to-head fashion forming the 1,2-disilacyclobutane **639** in 18% yield[184,192]. This parallels the dimerization behaviour of 'Brook'-type and 'Apeloig–Ishikawa–Oehme'-type silenes in which the inherent polarity of the Si=C double bond is reduced by the substituents (see Section I.B.2.a). In contrast, the silene $Ph_2Si=C(SiMe_3)_2$ **97** cleanly gives the head-to-tail dimer[63,64]. **639** undergoes under diverse reaction conditions a further rearrangement, yielding the cummulenic 1,2-disilacyclobutane **640** (equation 211)[184,192]. 1-Silabutatrienes like $(Me_3Si)_2C=C=C=Si(o\text{-tol})_2$, however, dimerize in a head-to-tail fashion, like most simple silenes[192].

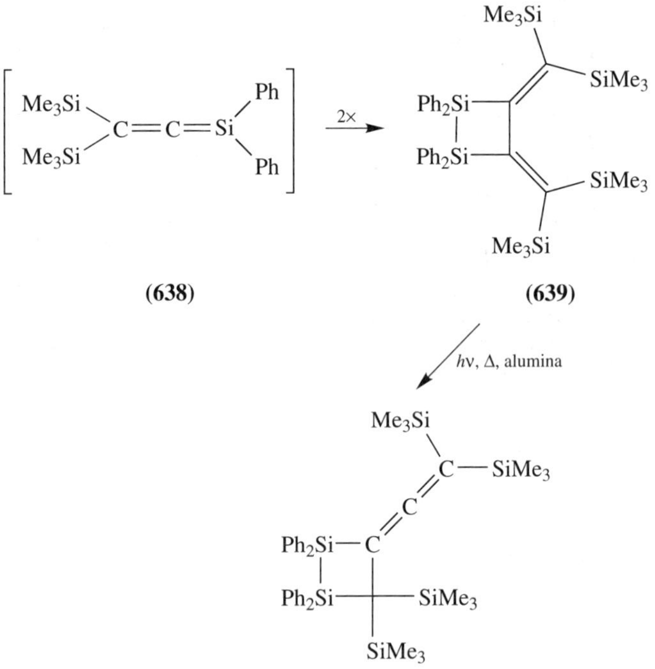

$$\left[\underset{Me_3Si}{\underset{Me_3Si}{\diagdown}}C=C=Si\underset{Ph}{\overset{Ph}{\diagup}}\right] \xrightarrow{2\times} \underset{(639)}{\text{structure}} \quad (211)$$

(638)   (639)

↓ $h\nu$, Δ, alumina

(640)

The smaller polarity of the Si=C bond in 1-silaallenes probably contributes also to the remarkable kinetic stability of the stable silaallenes prepared by West and coworkers, but obviously steric factors play a dominant role in this case. Some product studies have been undertaken with the stable silaallene **572** and they are summarized in equation 212[259].

572 reacts instantaneously with water and methanol to yield hydroxyvinylsilane **641** and methoxyvinylsilane **642**, respectively. In each case the sterically less hindered Z-isomer was formed preferentially ($E:Z = 1:11$ and $1:15$ for **641** and **642**, respectively) (equation 212). The addition of alcohols proceeds probably by initial protonation of the central carbon atom; the incipient silicenium ion is trapped by the nucleophile or, alternatively, can be trapped intramolecularly. Thus, acidic ethanolysis of **587a** or **588** yields **645** or **646** (equations 213 and 214) in subsequent reactions of the transient silylium ions with the primary C−H bond of an adjoining *t*-Bu group[259].

(212)

572 gives with benzophenone, in a formal [2 + 2] cycloaddition, a mixture of the stereoisomeric 1,2-oxasiletane E/Z-**644** ($E:Z = 6:1$) (equation 212). When **572** is heated in C_6D_6, an insertion of the Si=C double bond into one of the adjacent tertiary aliphatic C−H bonds of the isopropyl groups occurs yielding a mixture of the isomeric benzosilacyclobutenes E/Z-**643** (equation 212). Similarly, the benzosilacyclopentene **645**, the formal insertion product of Si=C in the primary C–H bond of the *ortho* t-butyl group of the supermesityl (Sup) substituent, is isolated when **587a** is heated in ether (equation 213)[259]. Related insertions of Si=X bonds in the *ortho* position of adjacent mesityl groups are well documented[90,181,259].

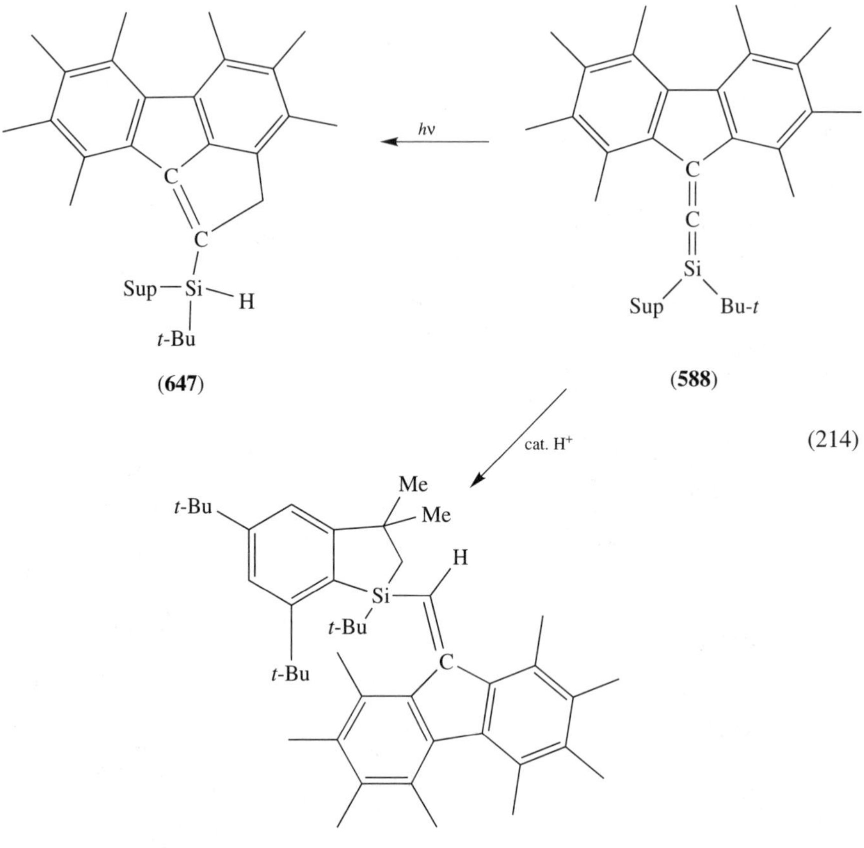

(**647**)

(**588**)

(214)

(**646**)

Interestingly, photolysis of fluorenylidenesilene **588** is reported to give insertion of the Si=C bond in one of the *ortho* methyl groups of the fluorenylidene substituent yielding the polycyclic compound **647** (equation 214). This product has the opposite regiochemistry in respect to the formal addition of a C–H bond across the Si=C double bond of the 1-silaallene moiety than that found in the protonation reaction[271].

A different decomposition channel is utilized by the silyl-substituted 1-silaallene **617**[289,290]. In the absence of trapping reagents the transient **617** formed in the thermolysis of **646** at 280 °C undergoes a 1,2-trimethylsilyl shift giving the silylene **648** and finally the 3,5-disilacyclopentene **649** in 25% yield. Alternatively **648** can also be formed from the silacyclopropene **616**. The silaindene **650**, the formal insertion product of the Si=C bond into the *ortho* C–H of the phenyl ring, is isolated in 18% yield[289]. The formation of **650** was rationalized by the intermediacy of the biradical **651** (equation 215)[289,290].

(215)

2. 2-Silaallenes

The first report about the possible formation of a short-lived 2-silaallene dated back to 1978. Bertrand and coworkers found as products from the pyrolysis of the silaspirocycle **652** in the presence of benzaldehyde, silica and styrene **653** as major products[294]. They suggested that 2-silaallene **605** is a possible intermediate in this reaction which can account for the formation of the observed products via **654** (equation 216)[296,294].

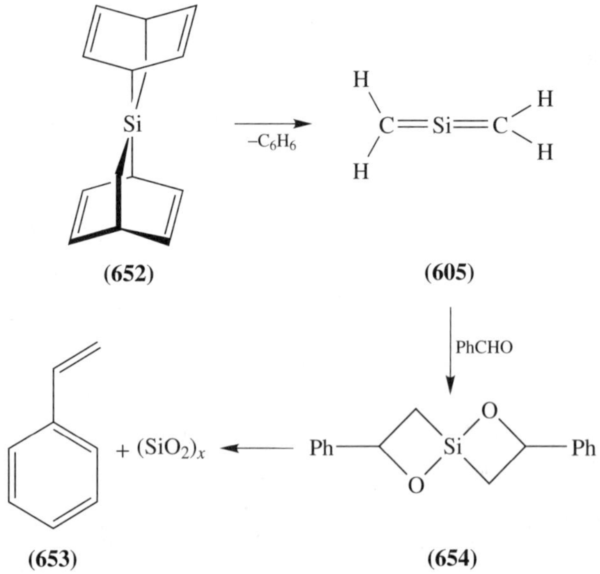

(216)

Silaspirocycles are the ideal precursors for 2-silaallenes. Thus, Pola and coworkers used 4-silaspiro[3,3]heptane **655** in a continuous-wave laser-photosensitized decomposition reaction in an attempt to generate 2-silaallene as an intermediate[295]. In the absence of scavenger reagents a transparent material is formed, which results from polymerization of the highly reactive intermediates **605, 656, 657** and **658**, which are produced either directly from the precursor **655** or from the silene **659**. (equation 217). ESCA, Auger and FT-IR analysis of the SiC$_2$ polymer give indirect indications for transient species like **605** or its isomers. In the presence of alcohols as trapping reagents dimethyldialkoxysilanes **660** are formed, which might result from the double addition of ROH to **605** and/or **657**, **660** (equation 218)[295].

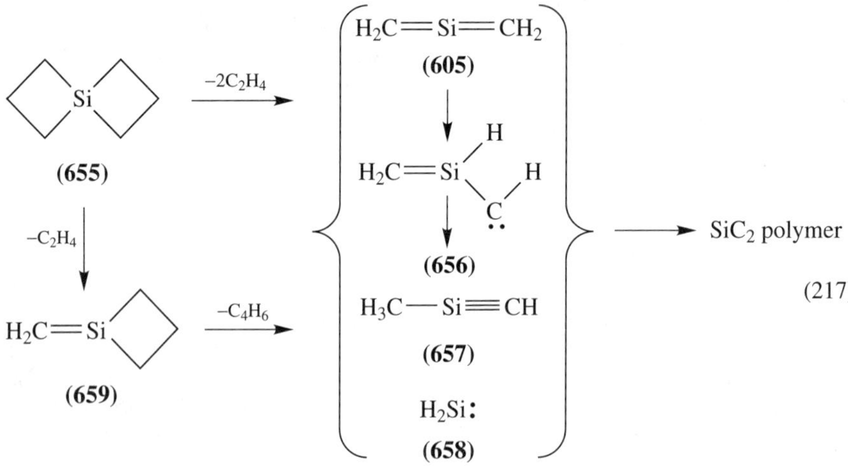

(217)

16. Silicon−carbon and silicon−nitrogen multiply bonded compounds 1007

$$H_2C=Si=CH_2 \xrightarrow{ROH} (H_3C)_2Si(OR)_2 \xleftarrow{ROH} H_3C—Si\equiv CH \quad (218)$$
$$\quad (605) \qquad\qquad (660) \qquad\qquad (657)$$

A different approach to 2-silallenes was utilized by Auner and coworkers[232]. They found that in the reaction of dichlorodivinylsilane and 2 equivalents *t*-BuLi, highly reactive intermediates are formed, which could be trapped with suitable scavenger reagents like anthracene, TMSOMe and norbornadiene (equation 219).

(219)

For example, in the presence of norbornadiene the spirocyclic compound **661** is formed. According to the authors this provides evidence for the intermediate formation of bisneopentyl-2-silaallene **662**. The stepwise formation of **661** via the 3-silabuta-1,3-diene **663**, vinylsilane **664** and silene **665** was excluded based on the different observed reactivities of dichlorodivinylsilane and **664** versus *t*-BuLi/norbornadiene. While dichlorodivinylsilane reacts immediately with 2 equivalents of *t*-BuLi in the presence of norbornadiene yielding **661**, the stepwise synthesized vinylsilane **664** reacts only sluggishly with 1 equivalent of *t*-BuLi in the presence of norbornadiene (equation 219)[232]. The experiments conducted by Auner and coworkers, however, do not exclude the possiblity that organolithium species like **666** and **667** give rise to the observed products. Nevertheless, the synthesized spirocycles like **661** are tailor-made molecules for the pyrolytic or photolytic generation of 2-silaallenes[232].

(**666**)   (**667**)

III. SiC TRIPLY BONDED COMPOUNDS: THEORETICAL RESULTS

Triple bonds to silicon still have the touch of unreachableness. At the time this review is being completed (summer 1997) no clear experimental evidence for a species containing a Si≡C bond is available. Some rather indirect evidence for the intermediacy of MeSi≡CH and XSi≡CH (X = F, Cl) in laser-induced decompositions of substituted silacyclobutanes is given by Pola and coworkers[295,296a]. Positively identified by spectroscopic methods were, however, several compounds with triple bonds between nitrogen and silicon (see Section IV.B) and dimethyldisilyne MeSi≡SiMe has been suggested as a transient intermediate[286b,c]. At the moment, the only reliable information about molecules with Si≡C bond comes from theory, which allows one to study their fundamental properties and reactions. However, theory also has faced considerable difficulties in describing triple bonds to silicon and the calculational results have been converging only in the last years to attain a uniform picture. These developments are well summarized in the reviews by Apeloig[11] and Karni and Apeloig[12] and we will give here only the most interesting and recent theoretical results for silynes and related compounds.

H—C≡Si—H   H\C≡Si/H   H\C=Si:/H   :C=Si\H/H

(**668a**)   (**668b**)   (**668c**)   (**668d**)

All recent high level calculations predict that a classical linear silyne **668a**, the analogue to acetylene, is not a minimum on the potential energy surface (PES)[12,297]. The *trans*-bent isomer **668b** is, however, a local minimum, but it is thermodynamically and kinetically very unstable. The barrier for the isomerization to the global minimum, i.e, the silavinylidene **668c** which is by 34.1 kcal mol^{-1} more stable than **668b**, is very low. The most sophisticated calculations at the CCSD(T)/TZ2p(fd)//CCSD(T)/TZ2p+ZPE level predict a barrier of 5.1 kcal mol^{-1}[297b]. **668c** has been recently identified in a glow-discharge plasma of a gaseous mixture of SiH$_4$ and CO and rotational spectral lines of H$_2$C=Si: have been observed[298a], confirming the calculated structures. In addition, vibrational frequencies of **668c** and its deuterio analogue D$_2$C=Si: have been measured in a photoelectron study of the anion SiCH$_2^-$[298b]. The fourth isomer, **668d,** is a high-lying minimum *ca* 87–89 kcal mol^{-1} higher in energy than **668c**[12,297]. The PES of H$_2$SiC, calculated at the CCSD(T)/TZ2p(fd)//CCSD(T)/TZ2p+ZPE level of theory, is shown in Figure 5[297b]. A theoretical treatment of hydrogen tunneling across the small barrier of isomerization of **668b** to **668c** suggests a lifetime for **668b** and its deuterio analogue DSi≡CH of about 10^{-8} and 5 × 10^{-6} s, respectively[297a]. Thus, detection of **668b** or of DSi≡CH with a nanosecond resolution spectrometer might be feasible[297a].

The calculated Si≡C bond lengths in *trans*-bent silynes are distinctively shorter than Si=C bond lengths in silenes (i.e. 1.632 Å and 1.714 Å in HSi≡CH and in H$_2$Si=CH$_2$, respectively, at MP2/6-31G**)[297c]. A theoretical analysis in the framework of natural bond theory predicts for **668b** and substituted derivatives bond orders for the SiC bond which are intermediate between double and triple bonds with considerable ionic contributions[297b,c].

Apeloig and Karni investigated computationally the influence of substitution on the relative energies of the different silyne isomers **668a-d** and the barriers separating them[297c]. They found that fluorine and oxygen substitution at silicon stabilizes the *trans*-bent structure RSi≡CH relative to the silavinylidene isomer RHC=Si:. Thus, FSi≡CH is even more stable by 6.4 kcal mol^{-1} than HFC=Si:. These large substituent effects are best understood in terms of R—Si versus R—C bond energies. Thus, the larger Si—F bond energy compared with the C—F bond energy overrides the intrinsic preference of **668c** over **668b**. Furthermore, the barriers for the interconversion of both isomers are increased by substitution at silicon (from 9 kcal mol^{-1} for R = H to 17.5–25.0 kcal mol^{-1} for

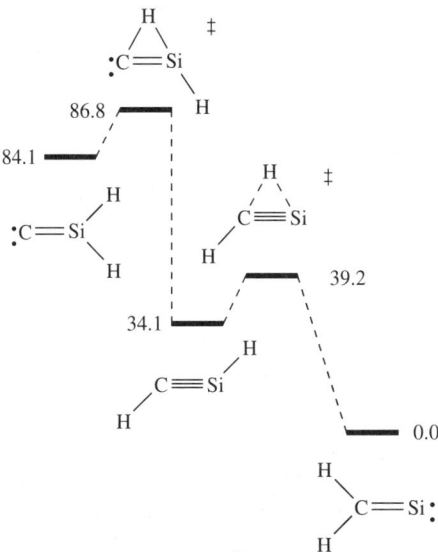

FIGURE 5. Calculated PES (relative energies kcal mol^{-1}) of H$_2$SiC at the CCSD(T)/TZ2p(fd)//CCSD(T)/ TZ2p+ZPE level of theory[297b]

R = F, Me, OH at MP4/6-31G**//MP2/6-31G**+ZPE), while the activation energies for the 1,2-hydrogen shift from RHSi=C:, forming the *trans*-bent isomer RSiCH, remain very small for different substituents (0.6–2.1 kcal mol^{-1} for R = Me, F, OH). According to Apeloig and Karni, a possible strategy for the synthesis of *trans*-bent silynes $R^2Si\equiv CR^1$ is to generate substituted 2-silavinylidenes $R^1R^2Si=C$:, preferently from suitable substituted silenes. An immediate 1,2-shift of the group R^1 might yield the *trans*-bent silyne $R^2Si\equiv CR^1$, which is kinetically stable towards unimolecular isomerization to the silavinylidenes $R^1R^2C=Si$:[297c].

$$\underset{R^2}{\overset{R^1}{\diagdown}}Si=C\underset{Y}{\overset{X}{\diagup}} \xrightarrow{-XY} \underset{R^2}{\overset{R^1}{\diagdown}}Si=C: \longrightarrow \underset{}{\overset{R^2}{\diagdown}}Si\equiv C\underset{R^1}{\diagup} \quad \begin{array}{l} R^1 = \text{alky, aryl} \\ R^2 = \text{F, OR} \end{array}$$

IV. SILICON–NITROGEN MULTIPLY BONDED COMPOUNDS

A. Introduction

This section deals with the literature since 1987 on silicon-nitrogen multiply bonded species as a sequel to Raabe and Michl's chapter in Vol 1 of *The Chemistry of Organic Silicon Compounds*[8]. A comprehensive treatment of silicon–nitrogen compounds has since appeared in 1989 in the *Gmelin Handbook of Inorganic Chemistry* covering the literature until the end of 1987[298]. A review on iminosilanes and related compounds was published by Hemme and Klingebiel in 1995[299].

The nomenclature of Si=N-systems used in the literature is somewhat confusing, especially for unusual systems. The nomenclature used in this review corresponds with the most commonly used description.

$R_2Si=NR$: iminosilane, silaketimine, silanimine
SiNR: iminosilylene, iminosilicon, silaisonitrile, silane isonitrile, isosilacyanide, silaisocyanide
RSiN: silanitrile, silane nitrile
RN=Si=NR: silanediimine
RN=Si=NR$^-$: silaamidide

B. Synthesis

This section comprises synthetic approaches to SiN multiply bonded systems. Whenever neccessary reactions of Si=N species have been included.

1. Si=N and Si≡N-systems

a. Salt elimination. Salt elimination is a well investigated route for the synthesis of unsaturated Group 14 species. Their stability depends on the size of the substituents at the Si–E skeleton and the subsequent replacement of small by bulky substituents eventually leads to stable sterically shielded Si=E species. This approach has been successfully exploited by Wiberg and coworkers in the field of silene chemistry (see Section I) and was later extended to silanimine synthesis. Instead of, or in addition to, the introduction of bulky substituents, stabilization of the Si=N moiety is also achieved by donor coordination to the Lewis acidic silicon atom. In this way Wiberg's group isolated adducts of Me$_2$Si=NSi*t*-Bu$_3$ **672** by salt elimination from N-metalated aminohalogenosilanes **670** (formed from **669**) in the presence of donors (Et$_2$O, THF, NR$_3$, X$^-$)[4]

$$\begin{array}{ccccc}
\text{Me}_2\text{Si}-\text{NSiBu-}t_3 & & \text{Me}_2\text{Si}-\text{NSiBu-}t_3 & & \text{Me}_2\text{Si}-\text{NSiBu-}t_3 \\
|\quad\quad\quad | & \xrightarrow{\text{solvent D}} & |\quad\quad\quad | & \xrightarrow{+\text{RLi, }-\text{LiX}} & |\quad\quad\quad | \\
\text{X}\quad\text{H} & +\text{RLi, }-\text{RH} & \text{X}--\text{LiD}_n & & \text{R}\quad\text{Li} \\
\textbf{(669)} & & \textbf{(670)} & & \textbf{(673)}
\end{array}$$

$$\Big\updownarrow \text{solvent D} \qquad\qquad \Big\downarrow {+\text{MeOH} \atop -\text{LiOMe}} \quad (220)$$

$$\begin{array}{ccccc}
\text{Me}_2\text{Si}=\text{NSiBu-}t_3 & & \text{Me}_2\text{Si}=\text{NSiBu-}t_3 & & \text{Me}_2\text{Si}-\text{NSiBu-}t_3 \\
\uparrow & \xrightleftharpoons{-\text{LiX}} & \uparrow\quad\quad & & |\quad\quad\quad | \\
\text{D} & & \text{X}\quad\text{LiD}_4^+ & & \text{R}\quad\text{H} \\
\textbf{(672)} & & \textbf{(671)} & & \textbf{(674)}
\end{array}$$

X = F$^-$, Cl$^-$

The LiX elimination equilibrium depends on the kind of solvent D used and it may be shifted via the ion pair **671** to the silanimine ·D (**672**) side by the addition of F$_3$CSO$_3$SiMe$_3$ (which results in the formation of the weak base LiOSO$_2$CF$_3$ and removal of LiCl), of 12-crown-4 [causing formation of insoluble Li([12]-C-4)Cl] or of solvents in which LiCl is badly soluble. A side reaction may take place under unfavourable conditions (slow metalation of **669**, fast X/R exchange with **670** to give **673**); the rate ratio **669**→**670**/**670**→**673** is a function of the solvent. An X-ray structure of the adduct Me$_2$Si=NSiBu-t_3·THF has been published[301,302]. The further substitution of methyl groups by *tert*-butyl groups to give the sterically overloaded bis(silyl)amine t-Bu$_2$Si(Cl)−NHSiBu-t_3 was not possible but the reverse reaction by addition of HX to the required silanimine leads to this amine. An elegant detour devised by the Wiberg group finally led to the stable silanimine **678** by the reaction of NaSiBu-t_3 with the azide **675**, probably via the triazenide **676** and the α-sodio chlorosilane **677** (equation 221)[303]

$$\begin{array}{ccc}
t\text{-Bu}_2\text{Si}-\text{N}=\text{N}=\text{NH} & & t\text{-Bu}_2\text{Si}-\text{N}=\text{N}-\text{N}-\text{SiBu-}t_3 \\
|\quad\quad\quad\quad\quad\quad & \xrightarrow{+\text{NaSiBu-}t_3} & |\quad\quad\quad\quad\quad\quad\quad\quad\quad\quad\quad | \\
\text{Cl} & & \text{Cl} \quad\quad\quad\quad\quad\quad\quad\quad \text{Na} \\
\textbf{(675)} & & \textbf{(676)}
\end{array}$$

$$\Big\downarrow -\text{N}_2 \quad (221)$$

$$\begin{array}{ccc}
t\text{-Bu}_2\text{Si}=\text{N}-\text{SiBu-}t_3 & \xleftarrow{-\text{NaCl}} & t\text{-Bu}_2\text{Si}-\text{N}-\text{SiBu-}t_3 \\
 & & |\quad\quad | \\
 & & \text{Cl}\quad\text{Na} \\
\textbf{(678)} & & \textbf{(677)}
\end{array}$$

The X-ray structure of **678** has been determined (see below). The addition of donors leads, as expected, to adducts.

The synthesis of silanimines by salt elimination has been independently developed in the laboratories of Klingebiel[304−307]. An example of the synthesis of a stable donor adduct (**681**) is shown in equation 222 and equation 223 illustrates the general dimerization reaction of iminosilanes.

$$t\text{-}Bu_2SiF_2 + LiNHSiMeBu\text{-}t_2 \xrightarrow{-LiF} \underset{(679)}{t\text{-}Bu_2Si(F)-N(H)(SiMeBu\text{-}t_2)} \xrightarrow{BuLi} \underset{(680a)}{t\text{-}Bu_2Si(F)-N(SiMeBu\text{-}t_2)-Li(THF)_2}$$

$$\downarrow THF \qquad\qquad THF \Big| \begin{array}{c} +Me_3SiCl \\ -Me_3SiF \end{array} \quad (222)$$

$$\underset{(681)}{t\text{-}Bu_2Si=N(SiMeBu\text{-}t_2)} \xleftarrow{-LiCl} \underset{(680b)}{t\text{-}Bu_2Si(Cl)-N(SiMeBu\text{-}t_2)Li(THF)_n}$$

Aminofluorosilanes **679** are formed in the reaction of difluorosilanes with lithiated amines. **679** can then be metalated by RLi (R = Me, Bu) to give the corresponding lithium salts (e.g. **680a**). The structure of salts **680** depends on the basicity of the nitrogen atom and on the solvent. Representative structure types (**680a–c**) will be discussed in the subsection dedicated to structural features. The thermal elimination of LiF leads to silanimine dimers (e.g. **682** from **680c**), if sterically possible (equation 223), or to rearrangements.

$$(THF)_3LiF\text{-}SiR_2=N\text{-}R' \xrightarrow[-THF]{\Delta, -LiF} \underset{(682)}{R_2Si(N R')_2 SiR_2} \quad (223)$$

(**680c**) → (**682**)

The mode of reaction of the salts **680** with Me$_3$SiCl depends on the bulk of the substituents. With less sterically demanding residues a trimethylsilyl group is introduced at the nitrogen atom, whereas F/Cl exchange takes place for larger substituents, provided that a Lewis base (e.g. THF) is present. In comparison with LiF the cleavage of LiCl is more facile. Thus, when compounds like **683c** are heated to 80 °C in a vacuum, THF is cleaved and the silanimine **684** (formed via **683b**) begins to sublime (equation 224)[306]. The donor adduct **681**, obtained in a similar fashion, can be distilled without decomposition.

b. Cycloreversion reactions. While the salt elimination method is only applicable without problems to systems having bulky substituents, the thermal cycloreversion reactions of suitable precursors are possible for a wide variety of substituents[308]. First results have been obtained by Wiberg and coworkers in the synthesis of the thermally labile silanimine **414a** from either the triazole **412a** or from the 2,3-diazasiletane **685** (equation 225)[309].

The synthesis of siladihydrotriazoles **412a** is easily achieved by the [2+3] cycloaddition reaction of azides RN$_3$ with the silenes Me$_2$Si=C(SiMe$_3$)$_2$ **92** (equation 226).

The thus formed heterocycles **412a** decompose or isomerize thermally; the required reaction temperature depends on the substituents. The isomerization leads to diazomethane derivatives **413a**, whereas the decomposition by [2 + 3] cycloreversion reaction gives bis(trimethylsilyl)diazomethane and short-lived silanimines **414a**, which dimerize in most cases. The ratio isomerization/cycloreversion depends on R, the solvent and the temperature and more cycloreversion is observed at higher temperature. An unfavourable side reaction is the insertion of Me$_2$Si=NR into Si–N bonds (formation of **415a** and **686**; see below).

16. Silicon–carbon and silicon–nitrogen multiply bonded compounds

The formation of silanimines can further be proven by trapping reactions. One such reaction is the [2 + 3] cycloaddition with azides to give siladihydrotetrazoles **344a** (equation 227).

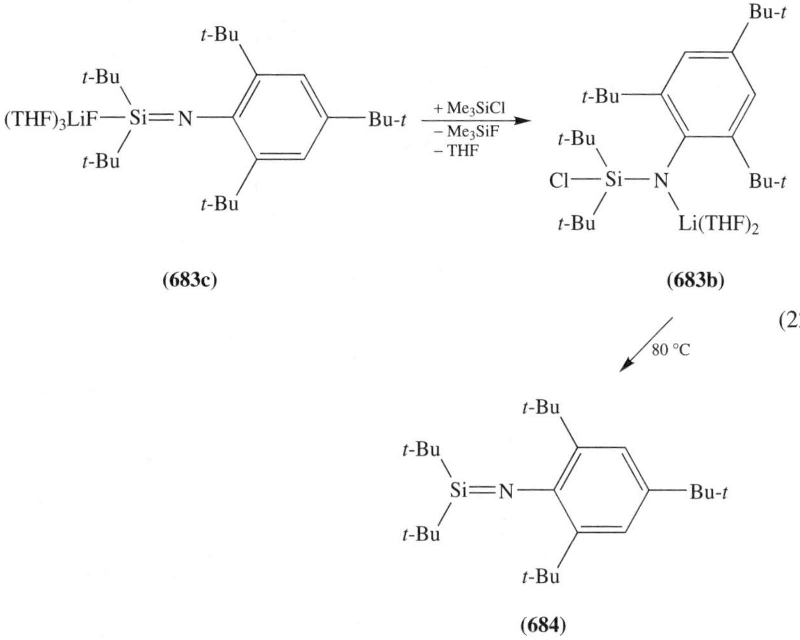

(224)

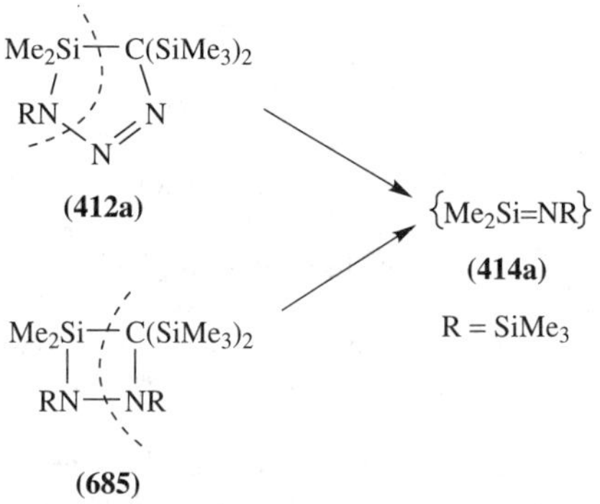

(225)

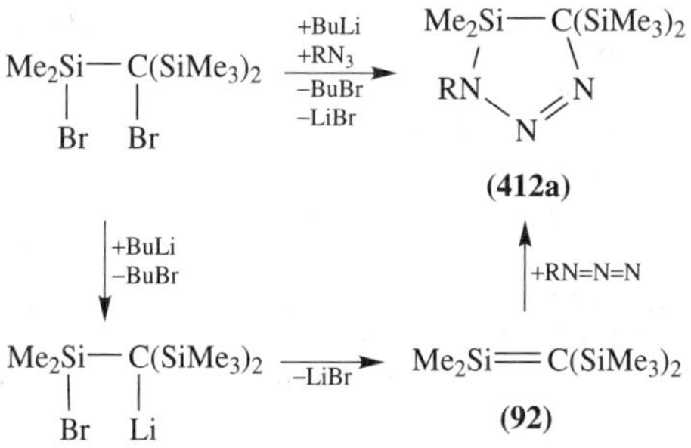

(226)

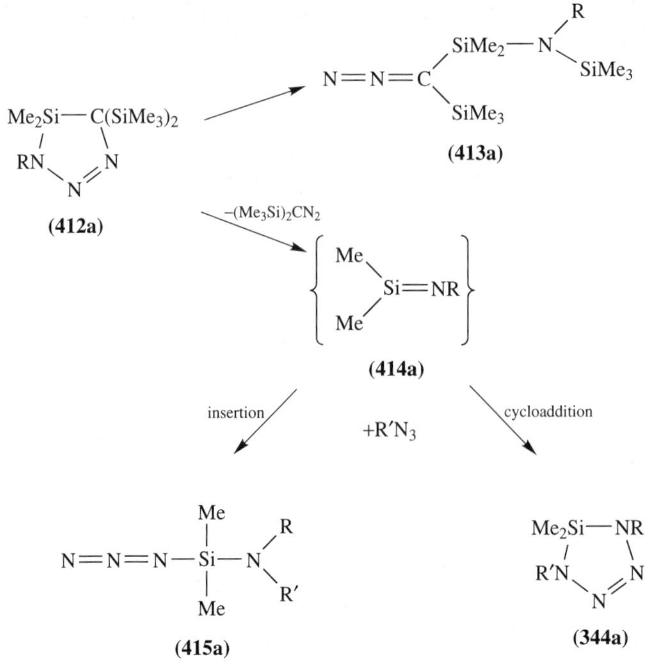

<!-- Scheme (227) -->

(227)

<!-- Scheme (686) -->

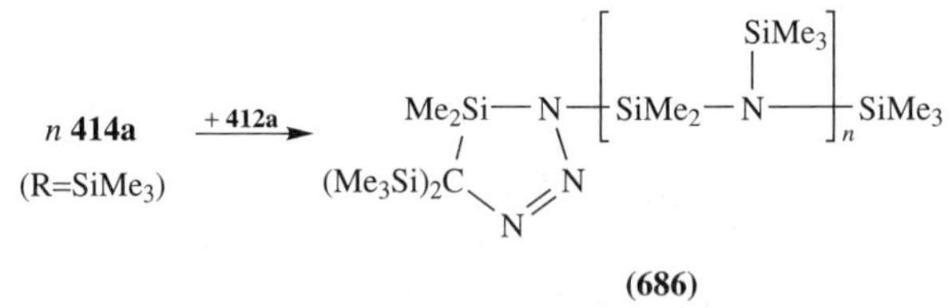

The yield of the [2 + 3] cycloadducts lies between 15 and 100%. The siladihydrotetrazoles **344a** are thermally quite stable and decompose in solution at temperatures higher than 130 °C in the reversal of their formation reaction to give silanimines and azides. The Si=N species dimerize to **687** or may be trapped by, e.g., acetone to give the ene product **688** (equation 228)[310].

The cleavage of unsymmetric siladihydrotetrazoles may lead to two different silanimines. The ease of formation increases in the order $Me_2Si=NSiMe_3$ < $Me_2Si=NSiMe_2Bu\text{-}t$ < $Me_2Si=NSiMeBu\text{-}t_2$ < $Me_2Si=NSiBu\text{-}t_3$, which might reflect the increasing stability of the silanimines.

The [2 + 2] cycloreversion of 1-aza-2-silacyclobutanes (silaazetidines) is a possible mode of silanimine preparation (equation 229).

While route B has been observed and utilized by Wiberg to synthesize silenes (see Section I), route A has been found to occur when less bulky substituents are used. Tamao and coworkers prepared the heterocycles (e.g. **689**) by intramolecular hydrosilylation of silylated allylamines and decomposed them in toluene solution at 204 °C[311]. From the thermolysis of **689** they obtained the corresponding *trans* alkene and the dimerization product **691**. In the presence of Me_3SiOEt the insertion product **690** was formed at the expense of the dimer **691** (equation 230).

16. Silicon–carbon and silicon–nitrogen multiply bonded compounds 1015

The [2+2] cycloreversion reaction proceeds stereospecifically with retention at carbon atoms, its rate obeys first-order kinetics and it depends on the stereochemistry and number of substituents at the ring carbon atoms. It is therefore concluded that the reaction mechanism is a concerted $[2_s + 2_a]$ cycloreversion.

Auner and coworkers synthesized dichlorosilaazetidines **692** by the reaction of $H_2C=CH-SiCl_3/t$-BuLi with imines[217,312]. The reaction possibly proceeds through the silene $Cl_2Si=CHCH_2Bu$-t **423** (equation 231). The attempted liberation of this silene from the silaazetidine analogous to one of Wiberg's silene syntheses failed, and instead reaction via pathway A of equation 229 took place. The silanimine **693a** obviously polymerizes, but can be trapped by $MeOSiMe_3$ and $Ph_2C=NBu$-t to give **694** and **695**, respectively (equation 232). In contrast to Tamao's results both stereoisomers of the alkene **696** were found (equation 233). This is explained by a stepwise reaction.

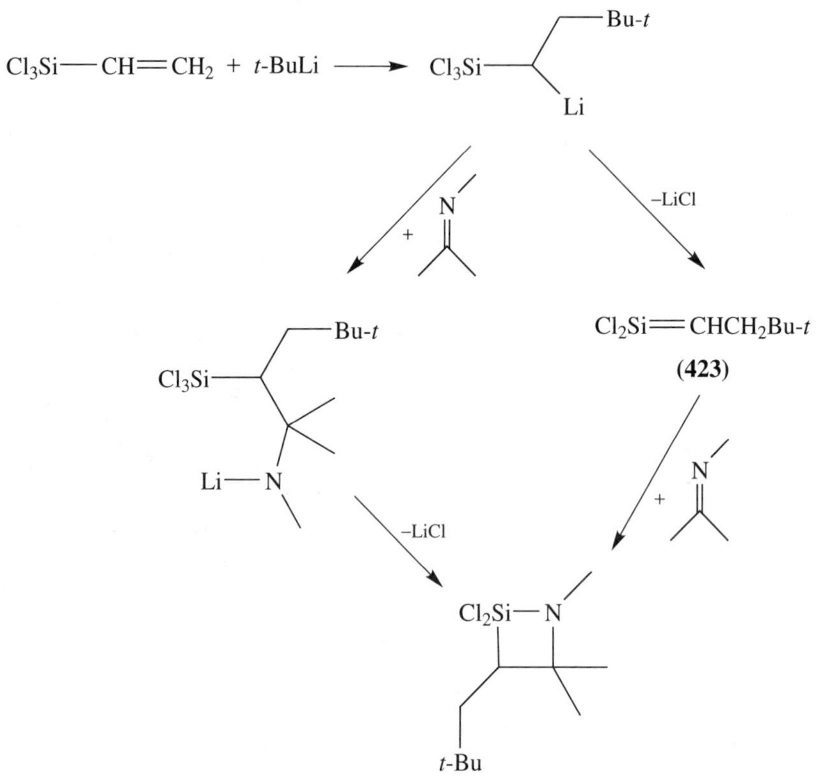

$$Cl_3Si-CH=CH_2 + t\text{-BuLi} \longrightarrow Cl_3Si-\overset{Bu\text{-}t}{\underset{Li}{\bigvee}}$$

(692)

(231)

The cyclization of allyl silyl amine **697** by hydrosilylation led to silaazetidine **698**, which was subjected to flash vacuum thermolysis at 700–900 °C at 10^{-4} hPa[313]. The silanimines **699** and **700** themselves were too reactive to be observed by high resolution mass spectrometry of the reaction mixture, but their cyclic dimers, the cyclodisilazane **701** and **702** and a trapping product with t-BuOH **703**, were definitely confirmed (equation 234; see also the section on retro-ene reactions).

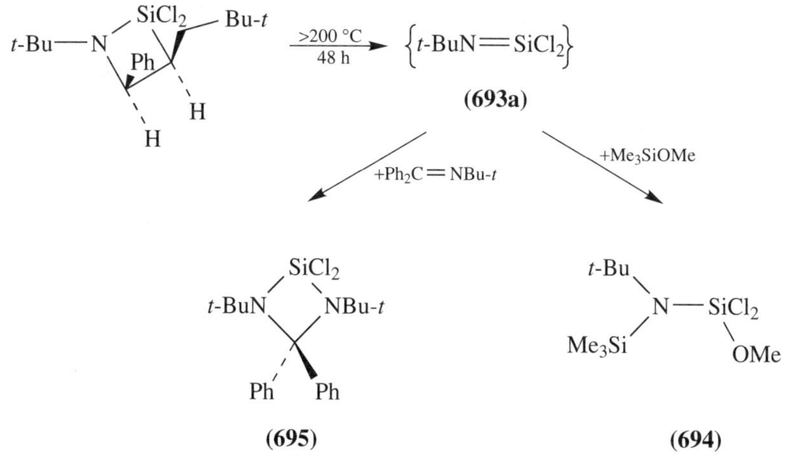

$$t\text{-Bu}-\underset{\underset{H}{|}}{\overset{\overset{SiCl_2}{|}}{N}}\overset{Bu\text{-}t}{\underset{H}{\diagdown\!\!\!\!\diagup}} \xrightarrow[48\ h]{>200\ °C} \{t\text{-BuN}=SiCl_2\}$$

(693a)

$\xrightarrow{+Ph_2C=NBu\text{-}t}$ on the left; $\xrightarrow{+Me_3SiOMe}$ on the right

$$t\text{-BuN}\underset{\underset{Ph}{}\underset{Ph}{}}{\overset{SiCl_2}{\diagup\!\!\!\!\diagdown}}NBu\text{-}t \qquad\qquad \underset{Me_3Si}{\overset{t\text{-Bu}}{\diagdown}}N-\underset{OMe}{\overset{}{SiCl_2}}$$

(695)   (694)

(232)

$$i\text{-Pr}-\underset{\underset{H}{|}}{\overset{\overset{SiCl_2}{|}}{N}}\overset{Bu\text{-}t}{\underset{H}{\diagdown\!\!\!\!\diagup}} \xrightarrow[48\ h]{>200\ °C} \{i\text{-PrN}=SiCl_2\}$$

(693b)

+

Z-(696)   Ph—CH=CH—Bu-t   40%

E-(696)   Ph—CH=CH—Bu-t   60%

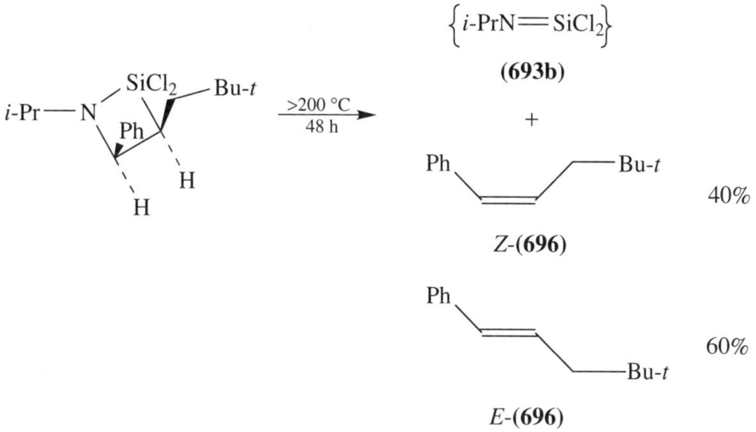

(233)

c. Photolysis and pyrolysis of silyl azides. For the study of highly reactive silicon species in the gas phase or at low temperatures in solution or matrix, azides are most suitable precursors, as the by-product of photolysis or pyrolysis is just molecular nitrogen that does not impede the spectroscopic characterization.

The first report of a silicon–nitrogen multiple bond molecule obtained from an azide is comparatively old: already in 1966 Cradock and Ogilvie isolated silaisocyanide **704** at 4 K (Ar matrix) after photolysis of H_3SiN_3[314].

The flash pyrolysis of trimethylsilyl azide **705** (1100 K, 0.01 mbar) was monitored by photoelectron spectroscopy, a method already successfully employed in the detection of phenylsilaisocyanide, PhNSi[315]. Accompanied by theoretical studies Guimon and Pfister-Guillouzo were able to assign the PE bands to two products: dimethylsilylimine **706** and silaisocyanide **704** (equation 235)[316].

$$(CH_3)_3SiN_3 \xrightarrow[1100 \text{ K}, 10^{-2} \text{ mbar}]{} (CH_3)_2HSiN=CH_2 + N_2 + Si\equiv NH \quad (235)$$
$$(705) \qquad\qquad\qquad (706) \qquad\qquad (704)$$

West and coworkers photolysed a series of hindered azidosilanes **707–710** in a 3-methylpentane (3-MP) glass at 77 K and in solution at low temperatures[317]. The products were analysed by UV-Vis, GC/MS and ^{1}H NMR spectroscopy.

Trimesitylsilanimine **711** is identified by its UV absorption at 296 and 444 nm and by trapping experiments with alcohols. **711** is stable in solution up to ca −125 °C, as could be shown by the disappearence of its characteristic UV bands above this temperature. The major product of the photolysis, regardless of temperature, is, however, the C−H

16. Silicon–carbon and silicon–nitrogen multiply bonded compounds   1019

insertion compound **712** (55% yield) (equation 236).

$$\text{Mes}_3\text{SiN}_3 \quad \underset{\underset{\text{N}_3}{|}}{\text{Mes}_2\text{Si}} \!\!-\!\! \text{SiPh}_2\text{Bu-}t \quad \underset{\underset{\text{N}_3}{|}}{\text{Me}_2\text{Si}} \!\!-\!\! \text{SiMes}_2\text{Bu-}t \quad \underset{\underset{\text{N}_3}{|}}{i\text{-Pr}_2\text{Si}} \!\!-\!\! \text{SiMes}_2\text{Bu-}t$$

(**707**)   (**708**)   (**709**)   (**710**)

Mes = Mesityl

$$\text{Mes}_3\text{SiN}_3 \xrightarrow[\text{3-MP, 77 K}]{h\nu,\ 254\ \text{nm}} \text{Mes}_2\text{Si}\!=\!\text{NMes}\ +\ \text{(712)}$$

(**707**)   (**711**)

$$\downarrow \text{ROH}$$

$$\underset{\underset{\text{RO}}{|}}{\text{Mes}_2\text{Si}}\!\!-\!\!\underset{\underset{\text{H}}{|}}{\text{NMes}}$$

(236)

where **712** is a 2,2-dimesityl-4,7-dimethyl-2,3-dihydro-1H-benzo[b]silole-type structure with Mes$_2$Si–NH.

The photolysis of the three other azidosilanes **708**–**710** in 3-MP glass at 77 K produced no new UV bands, nor was a trapping product obtained in the presence of alcohols. In solution, however, the silanimines **711**–**713** were produced by photolysis and could be intercepted with alcohols (equation 237).

$$\underset{\underset{\text{N}_3}{|}}{\text{R}_2\text{Si}}\!\!-\!\!\text{SiR}'_2t\text{-Bu} \xrightarrow[\text{ROH}]{h\nu} \left[\text{R}_2\text{Si}\!=\!\text{NSiR}'_2t\text{-Bu}\right] \xrightarrow{\text{ROH}} \underset{\underset{\text{RO}}{|}}{\text{R}_2\text{Si}}\!\!-\!\!\underset{\underset{\text{H}}{|}}{\text{NSiR}'_2t\text{-Bu}}$$

(237)

(**708**) R = Mes; R' = Ph   (**711**) R = Mes; R' = Ph
(**709**) R = Me$_3$; R' = Mes   (**712**) R = Me$_3$; R' = Mes
(**710**) R = i-Pr; R' = Mes   (**713**) R = i-Pr; R' = Mes

The authors propose that N-silanimines result from a migration of a silyl group with simultaneous loss of N$_2$ in the excited state of the azidosilane. This is in accordance with evidence published earlier by Kyba and Abramovich[318,319] that the photochemical decomposition of alkyl azides does not proceed via a nitrene intermediate (see also Reference 320).

An elegant study to probe the effects of π-bonding on the properties of the Si=N bond has been carried out with bridgehead silanimines **714**–**716**, in which the double bond is twisted[321].

The photolyses of the azides **717** and **718** and the trapping of the silanimines **714**–**716** was repeated according to an earlier study[322]. The irradiation (248 nm) of an argon matrix of **717** at 10 K led to a new absorption band at 397 nm and a new set of IR peaks. The experimental data are in good agreement with *ab initio* calculations. Trapping experiments and isotope labeling support the evidence of silanimine **714** as the primary photoproduct (equation 238).

In the case of azide **718** the matrix containing the primary photoproducts, silanimines **715** and **716**, was also irradiated shortly at 514.5 nm; this obviously destroyed the more stable isomer **715** (equation 239).

The authors stated that their results support the common belief[320] that the loss of N_2 is concerted with the rearrangement of the silyl group and that a singlet nitrene does not appear as a distinct intermediate. As far as the geometry of the silanimines is concerned the calculations (which predicted the IR spectra quite well) give optimized Si=N bond lengths which are *ca* 5 pm longer than in planar $Me_2Si=NMe$. This implies that the Si=N bond is weakened by twisting. Other geometric factors (e.g. pyramidalization) also play a role. The observed trends leave no doubt about the importance of π bonding in the electronic structure of the Si=N double bond, but they are not easily predictable.

The first spectroscopic observation of the radical SiN, a silicon compound that can be formulated as triply bonded in one of its formal valence-bond structures (·Si≡N:) dates back to 1913[323–325]! Despite this early appearance it was not until recently that enhanced effort was applied to the synthesis of triply bonded silicon species. In this connection a group of authors photolysed and pyrolysed the geminal triazide, **719**, in the attempt to isolate a silicon analogue of a nitrile or an isocyanide.[326]

The pyrolysis at 900 °C and in a stream of argon gave the silaisocyanide **721** (also as the ^{15}N-labeled species) which could be trapped in a matrix (12 K), and characterized unequivocally by its UV and IR spectra and GC-MS of the trapping product with *t*-BuOH. The spectra correspond well to calculated values. The assignment of the structure of the pyrolysis product agrees with the prior attribution by photoelectron spectroscopy by Bock and Dammel[315]. Photolyses of matrices containing **719** also give **721**; it was ruled out that spectral features of minor impurities could be attributed to the silanitrile **720**. Chemical trapping experiments with *t*-BuOH, however, gave evidence for the intermediacy of a species still containing the C−Si−N skeleton: two volatile trapping products (**722** and **723**) were detected by GC-MS analysis (equation 240). The authors state that, although **722** could be formed by the addition of two equivalents of *t*-BuOH to **720**, this is not ultimate proof for the existence of the silanitrile. The structure of the silaisocyanide **721** as determined by *ab initio* methods (see Section IV.F) exhibits a linear C−N−Si group and a Si=N bond length of 1.587 Å, which coincides well with the measured length of the Si=N bond in hindered, stable silanimines (1.568 Å)[327]. The energy difference between **720** and **721** is 55 kcal mol^{-1} in favour of the silaisocyanide.

$$PhSi(N_3)_3 \xrightarrow[-4N_2]{h\nu \text{ or } \Delta} PhSi\equiv N \xrightarrow{h\nu \text{ or } \Delta} PhN=Si:$$
$$(719) \qquad\qquad (720) \qquad\qquad (721)$$

$$\downarrow t\text{-BuOH} \qquad\qquad \downarrow t\text{-BuOH} \qquad (240)$$

$$PhSi(NH_2)(OBu\text{-}t)_2 \qquad PhNHSiH(OBu\text{-}t)_2$$
$$(722) \qquad\qquad (723)$$

This fact did not deter Maier and coworkers in the quest for the still 'missing' silanitriles, $R-Si\equiv N$[328–330].

After the elimination of four equivalents of nitrogen from methyltriazidosilane (**724**), one nitrogen should remain on the silicon atom, i.e. a silanitrile or silaisocyanide could be formed. The photochemistry of **724** is relatively complex. The final product of 254 nm irradiation is silyl isocyanide **725**, which is in photoequilibrium with silyl cyanide **726**. It is assumed that the silanitrile **727** is an intermediate, which isomerizes by methyl migration to the silaisocyanide **728** (identified by IR). **728** then rearranges by three-fold 1,3-hydrogen shift to **725**. The latter compound is obtained in traces upon irradiation of the silanimine **729**, which is available from methylazidosilane **730**, along with aminosilylene

731, which itself is in equilibrium with 2-aminosilene **732** (equation 241).

$$H_3C-Si(N_3)_3 \xrightarrow[-4N_2]{h\nu\ 254} H_3C-Si\equiv N$$
$$(724) \qquad\qquad (727)$$

$$\downarrow h\nu\ 254$$

$$:Si=N-CH_3$$
$$(728)$$

$$\downarrow h\nu\ 254$$

$$H_3Si-C\equiv N \xrightleftharpoons[h\nu\ 254]{h\nu\ 254} H_3Si-N=C:$$
$$(726) \qquad\qquad (725)$$

$$\uparrow h\nu\ 254 \qquad\qquad (241)$$

$$H_3C-\underset{\underset{H}{|}}{\overset{\overset{H}{|}}{Si}}-N_3 \xrightarrow{h\nu\ 254} \underset{H_3C}{\overset{H}{\diagdown}}Si=N\overset{H}{\diagup}$$
$$(730) \qquad\qquad (729)$$

$$\downarrow h\nu\ 254$$

$$\underset{H_2C}{\overset{H}{\diagdown}}Si-NH_2 \xrightleftharpoons[h\nu\ 254]{h\nu\ >310} \underset{H_3C}{\overset{\cdot\cdot}{\diagdown}}\overset{H}{\underset{H}{\diagup}}Si-N\overset{H}{\diagdown}$$
$$(732) \qquad\qquad (731)$$

The ideal starting compound for the matrix studies of the parent silanitrile, $H-Si\equiv N$, is silyl azide **734**. By judicious choice of irradiation wavelengths the preparation and interconversion of a series of H,Si,N containing species **735–737**, characterized by UV and IR spectra, was possible. It was found that silanitrile **733** can indeed be obtained, at first in a mixture with the hydrogen associate **733·**H_2. Tempering the matrix to 30 K leaves only **733**. The interconversion of **737** and **733** is also possible (equation 242).

In addition, these experiments were the first proof for the simplest of all silanimines, $H_2Si=NH$ (**735**). The results were interpreted with the help of corresponding calculations and supported by deuterium labeling experiments. The bond orders, as determined from the force constants (experimental IR spectra), are 2.0 for $HSi\equiv N$ **733** and 2.3 for $HN=Si$ **737**.

d. Retro-ene reactions. The retro-ene reaction of allyldimethylsilyl ether, $H_2C=CHCH_2O-SiMe_2H$, has been successfully used to prepare dimethylsilanone, $Me_2Si=O$[331]. This led a group of French researchers to anticipate the silanamine **738** as a precursor of the silanimine **739**[332]. In fact, the flash vacuum thermolysis (FVT, 900 °C, 10^{-4} hPa) of **738** mainly gave the expected products, namely propene and silanimine **739**. The formation of **739** was unambiguously shown by trapping it with *tert*-butanol to give **740** (equation 243). Direct evidence of the presence of **739** has been obtained by coupling the FVT oven with a high-resolution mass spectrometer (HRMS).

$$\text{(242)}$$

$$\text{(243)}$$

Other precursors like **741** and **742** were pyrolysed under similar conditions, but here the energetically favoured, competitive retro-ene process leading to the silyl imines **743** was predominant (equation 244).

$$\text{(244)}$$

Similarly to the preparation of **739**, substituted silanimine **744** was prepared and characterized along with the silyl substituted analogue **745**[313]. HRMS showed **745** and the dimer **702** of **744**, which is expected to polymerize rapidly in the condensed phase. (equation 245).

The detection of the *t*-BuOH trapping product **703** and of the dimer **746** gives further evidence for the intermediacy of **744** and **745**, respectively. Silanimine **735** is an intermediate in the synthesis of silaisocyanide **737** by retro-ene reaction from **747** under FVT conditions (equation 246)[333]. **737** was characterized by millimeter wave spectroscopy.

16. Silicon–carbon and silicon–nitrogen multiply bonded compounds 1025

e. Dehydrochlorination. The dehydrochlorination of volatile halogenated silanimines by non-volatile bases in a vacuum gas solid reaction (VGSR)[332] is reminiscent of the salt elimination procedure used by Wiberg and Klingebiel for the synthesis of silanimines.

The compounds resulting from the reaction of **748** were characterized by HRMS directly coupled to the reactor. The stable products **750** and **751** were analysed by GC and ^{1}H NMR spectroscopy. The formation of the cyclodisilazane **750** is explained by dimerization of the unstable silanimine **749** only in the cold trap, as the reaction is carried out under high dilution conditions (equation 247). It was also shown that the hydrogen chloride elimination did not occur in the ion source of the mass spectrometer.

$$\begin{array}{c} \text{Pr-}i \\ \text{Me}_2\text{Si}-\text{N} \\ | \quad | \\ \text{Cl} \quad \text{H} \\ \textbf{(748)} \end{array}$$

VGSR, 60 °C, 10^{-4} h Pa, (Ph$_2$MeSi)$_2$NK → Me$_2$Si=NPr-i **(749)**

VGSR, 60 °C, 10^{-3} h Pa, (Me$_3$Si)$_2$NK →

$$\begin{array}{c} i\text{-PrN}-\text{SiMe}_2 \\ | \quad\quad | \\ \text{Me}_2\text{Si}-\text{NPr-}i \\ \textbf{(750)} \end{array} \quad + \quad \begin{array}{c} \quad\quad\text{NHPr-}i \\ \text{Me}_2\text{Si} \\ | \\ \text{N} \\ \text{Me}_3\text{Si}\quad\text{SiMe}_3 \\ \textbf{(751)} \end{array}$$

(247)

f. Silylene addition to azides. Weidenbruch and coworkers reacted silylenes, *t*-Bu$_2$Si: and Mes$_2$Si:, with sterically hindered *t*-Bu$_3$SiN$_3$ **752** in order to obtain either [2 + 1] or [3 + 1] cycloadducts[334]. With *t*-Bu$_2$Si: they obtained products (for **678** see above) already known from another synthesis (**753** by prolonged heating of **678**, or double ene reaction of **678** with isobutene) by Wiberg (equation 248)[303].

$$t\text{-Bu}_3\text{Si}-\text{N}=\text{N}=\text{N} + t\text{-Bu}_2\text{Si}: \xrightarrow{h\nu} \left\{ \begin{array}{c} \quad\quad\quad\text{N} \\ t\text{-Bu}_3\text{Si}-\text{N}\quad\text{N} \\ \quad\quad\quad\text{Si} \\ t\text{-Bu}\quad\text{Bu-}t \end{array} \right\}$$

(752)

$$\begin{array}{c} t\text{-Bu}_3\text{Si}-\text{N}-\text{SiBu-}t_2 \\ \quad\quad\quad\text{H}\quad\quad\backslash \\ \quad\quad\quad\quad\quad\quad\text{CH}_2 \\ \quad\quad\quad\quad\quad\quad | \\ \quad\quad\quad\quad\text{C}=\text{CH}_2 \\ \quad\quad\quad\quad\quad\quad | \\ \quad\quad\quad\text{H}\quad\quad/\text{CH}_2 \\ t\text{-Bu}_3\text{Si}-\text{N}-\text{SiBu-}t_2 \\ \textbf{(753)} \end{array} \xleftarrow{h\nu,\ 0°C} t\text{-Bu}_3\text{Si}-\text{N}=\text{SiBu-}t_2 \quad \xleftarrow{-\text{N}_2}$$

(678)

(248)

Under photolysis conditions silanimine **678** was always obtained along with its consecutive product **753**. With Mes_2Si**:** another silanimine **(754)** was formed in the reaction with the azide, but in this case it rearranged to give a benzosilacyclobutene **(755)**, that was characterized by X-ray diffraction (equation 249).

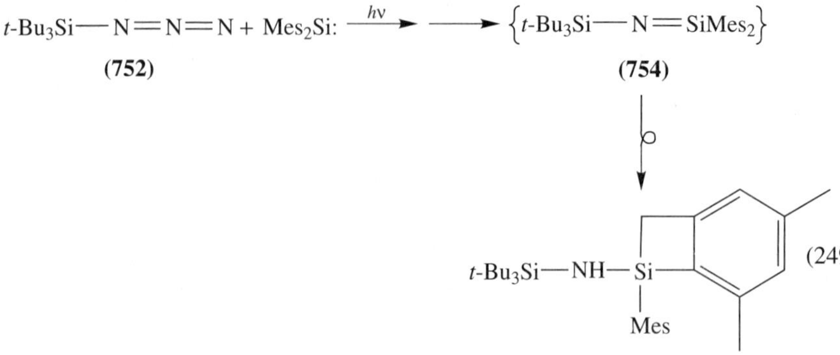
t-Bu3Si—N=N=N + Mes2Si:
hν
{t-Bu3Si—N=SiMes2}
t-Bu3Si—NH—Si
Mes

(752) **(754)** **(755)**

(249)

The stable silylene **756** was reacted with triphenylmethyl azide to give a silanimine **(757)** (equation 250)[335].

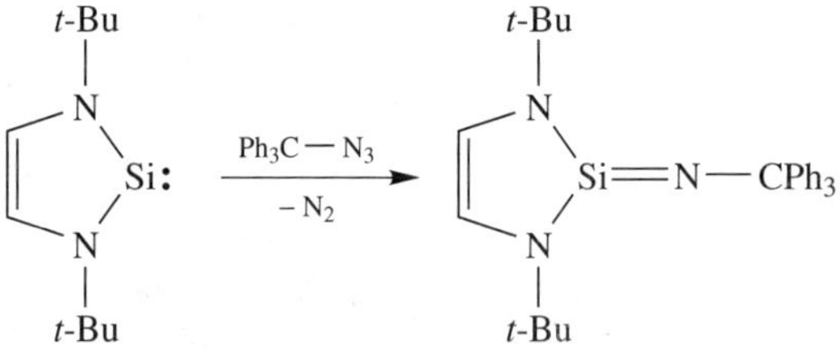
t-Bu
N
Si:
N
t-Bu
Ph3C—N3
− N2
t-Bu
N
Si=N—CPh3
N
t-Bu

(756) **(757)**

(250)

g. Reaction of SiH_4 and N_2. A radio frequency excited discharge of a mixture of silane and nitrogen was monitored by Fourier transform spectroscopy. A band at 3584 cm^{-1} was assigned to the fundamental ν_1 band (NH stretch) of silaisocyanide, HNSi **737**[336]. This was the first observation of this molecule in the gas phase by high resolution infrared vibration rotation spectra. The experimental data fit the calculated values assuming a linear molecule.

h. Ionic Si=N compounds by gas phase reactions. i. Silicon-containing anions. Damrauer and coworkers have perfected the preparation of simple silicon-containing anions in the gas phase whose conjugate acids are highly reactive, low valent neutral silicon compounds[337,338]. These studies have the advantage that many anions are readily prepared in the gas phase, even though their conjugate acids would be expected to be exceptionally reactive in the gas and condensed phase. The reaction chemistry of anions is an indirect probe of the properties of their corresponding acids.

The technique used is the Flowing Afterglow Selected Ion Flow Tube (FA-SIFT), a short description of which is given elsewhere[337,338]; it allows the preparation of ions in a first flow tube, their separation from complex reaction mixtures with a quadrupole, the

16. Silicon–carbon and silicon–nitrogen multiply bonded compounds

study of their reactivity in a second flow tube and the detection of the reaction products. It is demonstrated for [HSiNH]$^-$ in equation 251.

$$NH_3 \xrightarrow{EI} H_2N^-$$
$$C_6H_5SiH_3 + H_2N^- \longrightarrow H_3SiNH^- + C_6H_6 \quad \text{flow tube 1}$$
$$H_3SiNH^- \xrightarrow{CID} [HSiNH]^- + H_2 \quad \text{flow tube 2}$$

or (251)

$$C_6H_5SiH_3 + H_2N^- \longrightarrow [HSiNH]^- + H_2 + C_6H_6 \quad \text{flow tube 1}$$

EI = electron impact
CID = collisionally induced dissociation

α. *[HSiNH]$^-$: an anion related to silanimine* $H_2Si=NH^{337}$. Experimental verification of the connectivity for [HSiNH]$^-$ rests on labeling and H/D exchange reactions (equation 252).

$$D_3SiNH^- \xrightarrow{CID} [DSiNH]^- \text{ only product}$$
$$HSiNH^- \xrightarrow[\text{or EtOD}]{CD_3OD} [HSiND]^- \quad (252)$$
$$[DSiNH^-] \xrightarrow{EtOH} \text{no exchange}$$
$$\xrightarrow{EtOD} [DSiND]^-$$

The reactivity (see below) and the greater electronegativity of nitrogen with respect to silicon suggests that the charge will be concentrated on the nitrogen of HSiNH$^-$. The gas phase anion proton affinity has been determined by bracketing techniques and $\Delta G^0_{acid} = 352 \pm 3$ kcal mol^{-1} is found as the acidity of HSiNH$_2$. A variety of reactions with small neutral molecules, their reaction products, rate coefficients, efficiencies and branching ratios are reported. Equation 253 illustrates most of the important features of the reaction of HSiNH$^-$ **758** with COS.

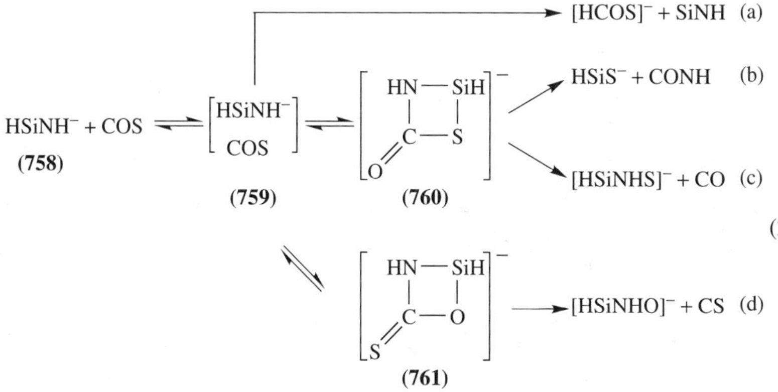

$$HSiNH^- + COS \rightleftharpoons \begin{bmatrix} HSiNH^- \\ COS \end{bmatrix} \rightleftharpoons \begin{bmatrix} HN-SiH \\ | \quad | \\ C-S \\ \parallel \\ O \end{bmatrix}^-$$

(758) (759) (760)

$$\longrightarrow [HCOS]^- + SiNH \quad (a)$$
$$\longrightarrow HSiS^- + CONH \quad (b)$$
$$\longrightarrow [HSiNHS]^- + CO \quad (c)$$

(253)

$$\begin{bmatrix} HN-SiH \\ | \quad | \\ C-O \\ \parallel \\ S \end{bmatrix}^- \longrightarrow [HSiNHO]^- + CS \quad (d)$$

(761)

1028 Thomas Müller, Wolfgang Ziche and Norbert Auner

At first an ion–molecule complex **759** is formed from which either a hydride is transferred to COS (path a) or addition to COS takes place (paths b–d). Intermediate **760** can react by ring cleavage leading to HSiS$^-$ and CONH (path b), or by extrusion leading to [HSiNHS]$^-$ and CO (path c). Intermediate **761** can also undergo extrusion (path d).

Other neutral reaction partners used were CO_2, CS_2, O_2, C_6F_6 and alcohols. Computational studies of the [H_2,Si,N]$^-$ system show the *cis* and *trans* isomers [HSiNH]$^-$ to be more stable than either [H_2SiN]$^-$ (by 24 kcal mol^{-1}), or [SiNH$_2$]$^-$ (by 24 kcal mol^{-1}).

β. *Silaformamide ion* [HSi(O)NH]$^{-338}$. The reactions which can lead to the silaformamide ion are given in equation 254.

$$NH_3 \xrightarrow{EI} H_2N^-$$

$$C_6H_5SiH_3 + H_2N^- \longrightarrow H_2NSiH_2^- + C_6H_6$$

$$H_2NSiH_2^- + H_2O \longrightarrow H_2NSiH_2O^- + H_2$$

$$H_2NSiH_2O^- \xrightarrow{CID} [H_2, Si, N, O]^- + H_2$$

flow tube 1 (254)

The connectivity of the silaformamide ion [HSi(O)NH]$^-$ **762** was established by deuteriation studies using both ND$_2^-$ and D$_2$O. The reaction chemistry with neutral reagents such as CO_2, CS_2, COS, fluoro and aliphatic alcohols has been studied. An example is shown in equation 255, which displays the ambident reactivity versus COS.

(255)

Ab initio calculations probed the stability of **762**: it was found to be less stable than the silylene isomer [Si(O)NH$_2$]$^-$ **763** by 1.32–7.5 kcal mol^{-1}, depending on the method. Calculation of the charge density shows nearly equal charges on oxygen and nitrogen in the anion **762** and suggests an explanation for its ambident behaviour. The gas phase acidity of the conjugate acid of **762** has been difficult to determine experimentally (ca 350–355 kcal mol^{-1}) but a value of 345 kcal mol^{-1} is given by calculation. Silaformamide HSi(O)NH$_2$ **764** itself is calculated to be the most stable [H$_3$SiON] isomer.

ii. *Silicon containing cations*. Bohme and coworkers have reacted ground state (^{2}P) Si$^+$ ions, obtained by electron impact from tetramethylsilane, and have studied their reactivity in ca 100 ion/molecule reactions using the selected ion flow tube (SIFT) technique[339,340].

16. Silicon–carbon and silicon–nitrogen multiply bonded compounds   1029

Reaction with ammonia initiates the sequence of equation 256 which, according to theoretical studies, selectively establishes hydrogen silaisocyanide, H—N=Si **737**[341].

$$Si^+ + NH_3 \longrightarrow SiNH_2^+ + H\cdot$$

$$SiNH_2^+ + NH_3 \longrightarrow HNSi + NH_4^+ \quad (256)$$
(765)

Calculated and experimentally determined proton affinities for HNSi (**737**) confirm its formation. The ion $SiNH_2^+$ **765** is best described as :Si^+-NH_2. It inserts into C—C and C—S bonds of $(H_3C)_2CO$ and $(H_3C)_2S$ or, in the case of $(H_3C)_2S$, in a side reaction a hydride H^- is transferred to form the aminosilylene :Si(H)NH$_2$ **736**.

Schwarz and coworkers have used the technique of neutralization–reionization mass spectrometry (NRMS) to structurally characterize numerous elusive silicon-containing molecules of interstellar interest[342]. The identification of HNSi was supported by *ab initio* calculations. The radical ion [HNSi]$^{\bullet+}$ was produced from N_2 and SiH_3I in the chemical ionization source of the mass spectrometer.

2. N=Si=N systems

a. Silanediimines. The first silanediimine was isolated in matrix by West and coworkers[343]. Photolysis of **766** in the presence of Me$_3$SiOMe in solution at room temperature gives **768** and further photolysis gives **770**. Both products probably result from addition of the alkoxysilane to intermediate silanimines **767** and **769** formed by 1,2-trimethylsilyl migration (equation 257).

$$(Me_3Si)_2Si(N_3)_2 \xrightarrow{h\nu\ 254\ nm} [Me_3SiSi(N_3)\!=\!\!=\!NSiMe_3]$$
(766)   **(767)**

$\Big\downarrow$ Me$_3$SiOMe

$$[(Me_3Si)_2NSi(OMe)\!=\!\!=\!NSiMe_3] \xleftarrow{h\nu\ 254\ nm} Me_3SiSi(OMe)(N_3)N(SiMe_3)_2 \quad (257)$$
(769)   **(768)**

$\Big\downarrow$ Me$_3$SiOMe

$$[(Me_3Si)_2N]_2Si(OMe)_2$$
(770)

Irradiation of **766** in glassy 3-methylpentane at 77 K gave two new UV bands at 274 and 324 nm. The two species responsible for this spectrum were identified as **767** and **771**. The following experimental facts support the assignment: (i) the 274 nm band grows more rapidly, (ii) bleaching at >270 nm reduces the 274 nm band with a concomitant increase in the 324 nm band, (iii) trapping experiments give **768** and **770** (equation 258). The ratio depends on the irradiation conditions.

$$(Me_3Si)_2Si(N_3)_2 \xrightarrow[77\ K]{h\nu\ 254\ nm} Me_3SiSi(N_3)=NSiMe_3 \xrightarrow{(Me_3SiOMe)} (768)$$
(766)   (767)

$\lambda_{max}= 274$ nm

$\downarrow h\nu > 270$ nm   (258)

$$Me_3SiN=Si=NSiMe_3 \xrightarrow{Me_3SiOMe} [(Me_3Si)_2N]\,Si(OMe)_2$$
(771)   (770)

$\lambda_{max}= 324$ nm

The photolysis of $Ph_2Si(N_3)_2$ **772** in the presence of *tert*-butanol has been reported to give one product from the migration of both phenyl groups to a nitrogen and the addition of 2 equivalents of alcohol. This was interpreted as evidence for the intermediacy of N,N'-diphenylsilandiimine PhN=Si=NPh **773**[344]. The above results, however, suggest that **773** is not formed, but that sequential migration–addition steps take place.

The irradiation of a similar diazidosilane (**774**) gives as the major product the silylene t-$Bu_2Si\!:$, which undergoes a subsequent photochemical C–H insertion to give the stable silacyclopropane **775**[345]. Evidence for the intermediacy of t-Bu_2SiN_2 **776** as a photochemical precursor to **775** (<5%) is presented and supported by calculations. Another minor product of the photolysis of **774** is N,N'-di-*tert*-butylsilanediimine (**777**, ca 10%), which exhibits UV bands at 240 and 385 nm; a definite IR spectrum was impossible to obtain (equation 259). Chemical trapping, isotope labeling (^{15}N) and calculations support the spectroscopic evidence.

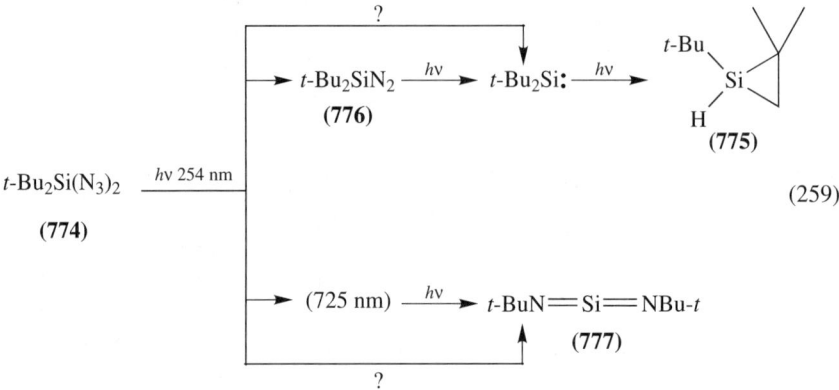

(259)

b. *Silaamidides* $[(RN)_2SiR']$. The first silaamidide ion was isolated by Underiner and West[346]. The reaction of the silanediamine **778** with two equivalents of *t*-BuLi gives the silaamidide **779**. The ether-free salt **780** can be obtained by heating **779** in vacuo. The lithium cation may be complexed by 12-crown-4 to give **781** (equation 260). The NMR spectroscopic data are consistent with symmetric species. The ^{29}Si NMR chemical shifts reflect the extent of ion pair separation, e.g. $\delta^{29}Si$ is +7 ppm for $[SupN)_2SiPh]Li$

780 and −36.3 ppm for [(SupN)$_2$SiPh][Li([12]-C-4)$_2$] **781**. An intense IR peak in the range of 1303–1284 cm^{-1} is assigned to ν_{as} of the [N−Si−N]$^-$ system and, as expected, these bands are at lower energy than for the Si=N stretching absorption. The crystal structure of [SupN)$_2$SiBu-t] [Li([12]-C-4)THF] (see Table 17) shows that there is indeed no interaction between the ions. The geometry around the tricoordinated silicon atom is planar. The Si−N bond lengths, 1.594 and 1.626 Å, are intermediate between single (ca 1.75 Å) and double bond lengths (ca 1.57 Å), but closer to a double bond, and are thus consistent with a considerable Si−N π bonding in the allylic system.

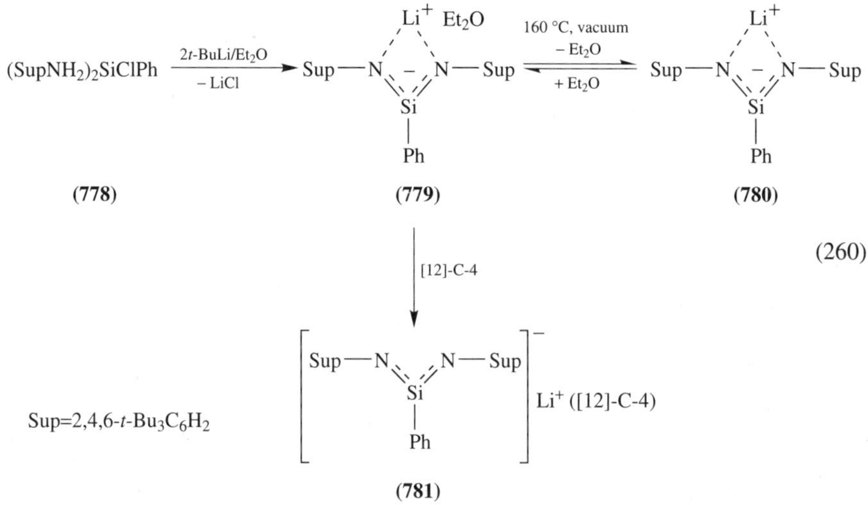

(260)

Sup = 2,4,6-t-Bu$_3$C$_6$H$_2$

Salts **782** and **783** having chloro or triflate substituents at silicon could also be prepared by partial metal–halogen exchange[347]. The elimination of MX (X = Cl, OTf; M = Li, K) from these compounds to give silanediimines did not occur, but may lead to substitution products like **784** (equation 261). Addition reactions and a [2 + 2] cycloaddition are reported (equation 262 and 263).

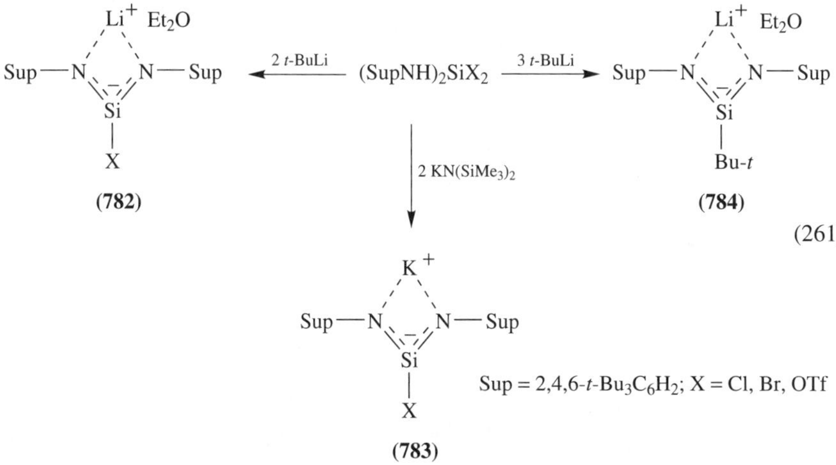

(261)

Sup = 2,4,6-t-Bu$_3$C$_6$H$_2$; X = Cl, Br, OTf

$$(SupNH)_2SiR(NHR') \xleftarrow[-LiNHR']{R'NH_2} \text{[Li}^+ \text{Et}_2\text{O complex, Sup—N—N—Sup, Si, R]} \xrightarrow[-LiOR']{R'OH} (SupNH)_2SiR(OR') \quad (262)$$

with PPh$_3$·HBr → (SupNH)$_2$SiRBr

and 1. n-BuLi, 2. H$_2$O → (SupNH)$_2$SiR(Bu-n)

$$\text{[Li}^+ \text{Et}_2\text{O complex, Sup—N—N—Sup, Si, t-Bu]} \xrightarrow[-LiOH]{\substack{1.\ PhCHO \\ 2.\ H_2O}} \text{SupNH—Si(t-Bu)(NSup)—O—CHPh} \quad (263)$$

(784)

3. Si=N metal systems

The stabilization of small reactive species is mostly always possible by the coordination to transition metals. For silicon such complexes are known for silylenes, silenes and disilenes[348,349]. The first example of a metal silanimine complex was reported by Berry and coworkers in 1991[350]. The synthetic strategy employed involves formation of the unsaturated fragment by β-hydrogen abstraction and loss of alkane. Alkylation of the amido complex **785** with LiCH$_2$SiMe$_3$ gives **786**, which is observed as unstable intermediate by ^{1}H NMR. It decomposes in solution, probably via the reactive 16-electron intermediate **787**. In the presence of PMe$_3$ as a trapping ligand the silanimine complex **788** can be isolated (equation 264).

$$[Cp_2Zr(H)(Cl)]_n \xrightarrow[-LiCl]{LiN(t\text{-}Bu)SiMe_2H \cdot THF} Cp_2Zr(H)(N(t\text{-}Bu)SiMe_2H) \xrightarrow[-CH_4]{MeI} Cp_2Zr(I)(N(t\text{-}Bu)SiMe_2H) \quad \textbf{(785)}$$

$$\textbf{(786)}\ Cp_2Zr(CH_2SiMe_3)(N(t\text{-}Bu)SiMe_2H) \xleftarrow[-LiI]{LiCH_2SiMe_3}$$

$$\textbf{(787)}\ \{Cp_2Zr(SiMe_2)(N\text{-}Bu\text{-}t)\} \xleftarrow{-SiMe_4}$$

$$\xrightarrow{PMe_3} Cp_2Zr(PMe_3)(SiMe_2=N\text{-}t\text{-}Bu) \quad \textbf{(788)}$$

$$(264)$$

In analogy to metal alkene complexes, two extreme resonance forms (metallacycle, sp^3 Si; π-donor complex, sp^2 Si) can be formulated. X-ray structural analysis of **788** gives a Si—N bond length of 1.687 Å, which is within the normal range for Si—N single bonds (1.64–1.80 Å); Wiberg's silanimine t-Bu$_2$Si=NSiBu-t_3 (1.568 Å) and a donor adduct THF·Me$_2$Si=NSiBu-t_3 (ca 1.58 Å) show significantly shorter bonds[301,303]. Along with the short Zr—Si distance (2.654 Å) this implies that the best description of the complex is that of a metallacycle.

C. Reactivity

The reactivity of SiN multiply bonded species has in part been discussed in Section IV. B, dedicated to synthesis, because the existence of reactive and short-lived species is often proven by trapping experiments. This part of the review therefore focusses on more comprehensive reactivity studies of Si=N systems.

1. Donor addition

The addition of donors to the Lewis acidic sp^2-hybridized silicon in Si=N systems is a general method to stabilize the multiple bond, even when non-bulky substituents are used. In most cases the donor adduct reacts just like the free Si=N compound, obviously by initial dissociation of the donor molecule. In the absence of trapping agents and when the steric bulk allows it, silanimines dimerize[300,308,311], e.g. **672** gives cyclodisilazane **789** (equation 265).

$$
\begin{array}{c}
\text{D} \\
\downarrow \\
\text{Me}_2\text{Si}\!=\!\!=\!\text{NSiBu-}t_3 \\
\textbf{(672)} \\
\text{(D=Et}_2\text{O, THF, NR}_3\text{)}
\end{array}
\quad \xrightarrow{\Delta,\ \text{vacuum},\ -\text{D}} \quad
\begin{array}{c}
\text{SiBu-}t_3 \\
| \\
\text{N} \\
\text{Me}_2\text{Si} \diagup \diagdown \text{SiMe}_2 \\
\diagdown \diagup \\
\text{N} \\
| \\
\text{SiBu-}t_3 \\
\textbf{(789)}
\end{array}
\qquad (265)
$$

The proximity of the donor to the reactive species does, however, lead to the reaction of the Si=N bond with the coordinated molecule. Wiberg has encountered several cases of such behaviour. When **672**·Et$_2$O is heated to 60 °C in diethyl ether, the dimer **789** is only formed in 10% yield; the other 90% are compound **790** (equation 266)[300]. Its formation may be understood as β-elimination reaction analogous to that occurring in oxonium salts, and subsequent insertion of the alcohol. No dimer is obtained when **672** (D = Me$_2$EtN) is heated: the products **791** and **793** are formed in the sense of a Stevens migration from the silanimine donor adducts **672** and **792**, respectively. In excess Me$_2$EtN the displacement of the amine donor (D = Me$_2$EtN) from **672** by **791** to give **792** does not occur (equation 267)[300].

$$
\begin{array}{c}
\text{EtO}\!-\!\text{C}_2\text{H}_4\!-\!\text{H} \\
| \\
\text{Me}_2\text{Si}\!=\!\!=\!\text{NSiBu-}t_3 \\
\textbf{(672)}\ \text{D=Et}_2\text{O}
\end{array}
\quad \xrightarrow[-\text{C}_2\text{H}_4]{\Delta} \quad
\begin{array}{c}
\text{EtO} \quad \text{H} \\
|\quad\quad| \\
\text{Me}_2\text{Si}\!-\!\text{NSiBu-}t_3 \\
\textbf{(790)}
\end{array}
\qquad (266)
$$

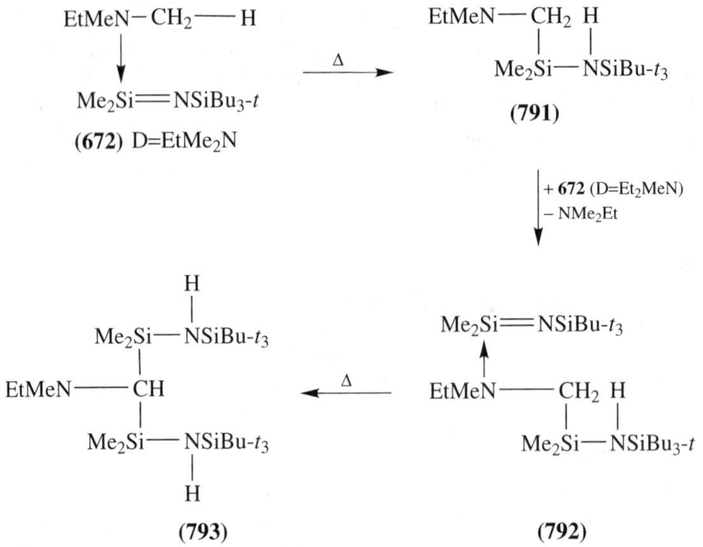

$$
\begin{array}{c}
\text{EtMeN—CH}_2\text{—H} \\
\downarrow \\
\text{Me}_2\text{Si}=\text{NSiBu}_3\text{-}t \\
\text{(672) D=EtMe}_2\text{N}
\end{array}
\xrightarrow{\Delta}
\begin{array}{c}
\text{EtMeN—CH}_2\ \text{H} \\
\ \ \ \ \ \ \ \ \ \ \ \ \ \ \ |\ \ \ \ \ \ | \\
\text{Me}_2\text{Si—NSiBu-}t_3 \\
\text{(791)}
\end{array}
$$

$$\downarrow +\textbf{672}\ (D=Et_2MeN)\ -NMe_2Et$$

(793) ← Δ ← (792)

```
        H
        |
 Me2Si—NSiBu-t3                    Me2Si=NSiBu-t3
        |                                ↑
EtMeN——CH               EtMeN——CH2 H
        |                           |    |
 Me2Si—NSiBu-t3                   Me2Si—NSiBu3-t
        |
        H
      (793)                          (792)
```

(267)

The expected exchange of donors does not take place with compound **678·THF** and NMe$_2$Et; instead the THF ring is cleaved, ethylene is eliminated and compound **794** is formed (equation 268). The reaction of **678·THF** with N$_2$O, in which the THF ligand also participates, will be described below.

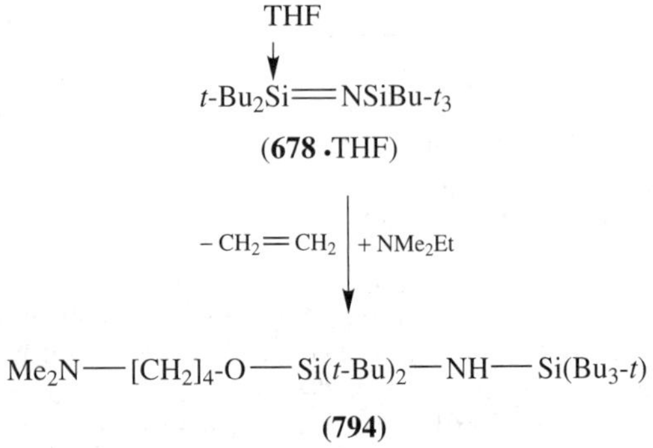

$$
\begin{array}{c}
\text{THF} \\
\downarrow \\
t\text{-Bu}_2\text{Si}=\text{NSiBu-}t_3 \\
\textbf{(678 ·THF)}
\end{array}
$$

$$\xrightarrow[\ -\text{CH}_2=\text{CH}_2\ ]{+\text{NMe}_2\text{Et}}$$

$$\text{Me}_2\text{N—[CH}_2\text{]}_4\text{-O—Si(}t\text{-Bu)}_2\text{—NH—Si(Bu}_3\text{-}t\text{)}$$
(794)

(268)

2. Insertion reactions

Insertion of the Si=N bond into polar bonds is the most used reaction for the characterization of very reactive and transient silanimines. Especially, the reaction with alcohols is often the first reaction to be carried out with silicon–nitrogen multiple bond systems. Other reagents used for insertion reactions are amines, water and alkoxysilanes (equation 269)[300,303,306,311,351–353]. The insertion into E—X bonds (E = Si, Ge, Sn; X = Cl, OR, NR$_2$, N$_3$) has been shown earlier in this review for the insertion into the Si—N bond of silyl azides[310,351]. A reaction reported is the insertion of **678** into the C—H bond of benzene at elevated temperatures[303].

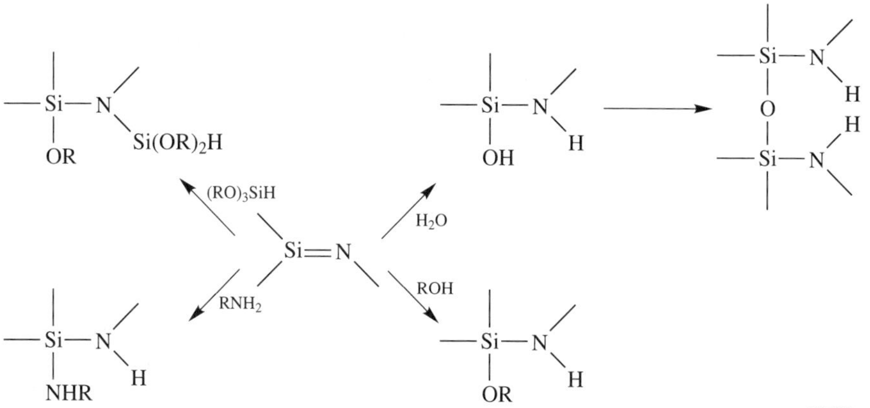

(269)

3. Ene reactions

In contrast to the silenes, silanimines undergo ene reactions (equation 270) rather than [2 + 4] cycloaddition reactions with diene reagents. $Me_2Si=NSiBu-t_3 \cdot D$ (**672·D**)[300] and $Me_2Si=NR$ [R = $SiMe_nBu-t_{3-n}$, $SiPh_3$, $EMe_2N(SiMe_3)_2$ with E = Si, Ge][351] react with propene, isobutene, 2,3-dimethylbuta-1,3-diene and cyclopentadiene; $t\text{-}Bu_2Si=NSiBu-t_3$ (**678**), its donor adducts, and $Me_2Si=NR$ react also with acetone[303,351].

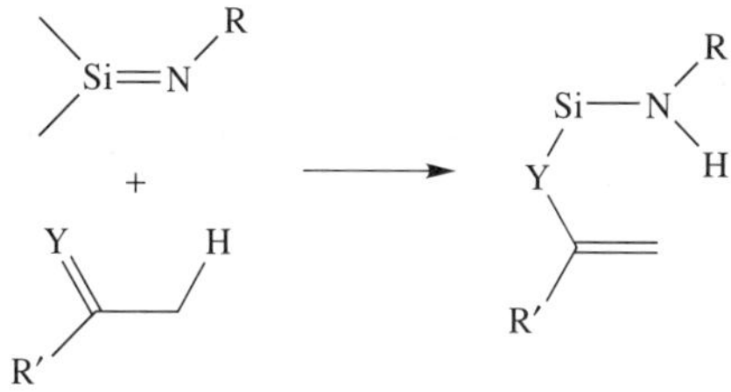

(270)

(Y=CH_2, O; R′=H, Me, CH_2=CMe; R = $SiMe_n t\text{-}Bu_{3-n}$, $SiPh_3$, $EMe_2N(SiMe_3)_2$, E = Si, Ge)

4. Cycloaddition reactions

a. [2 + 2] Cycloadditions. The dimerization reaction as a special case of the [2 + 2] cycloaddition reaction has already been mentioned above. It was mentioned that the dimerization occurs when the steric bulk allows the reaction. In the case of **795** a direct dimerization is thus not possible. The system seeks to gain stability in another way. The authors suggest an interaction between the Lewis acidic silicon atom in **795** and the oxygen atom of the siloxy group of a second molecule of **795**, which leads to an isomerized silanimine **796** that then dimerizes to give the head-to-tail dimer **797** (equation 271)[354].

$$\text{(271)}$$

(**795**) → (**796**) → (**797**)

Silanimine **672·D** reacts with benzophenone to give a cycloadduct **798**, which cannot be isolated, and eliminates the benzophenoneimine $Ph_2C=NSiBu-t_3$, giving the cyclus **799** with another equivalent of **672·D** (equation 272)[300].

$$\text{(272)}$$

(**672·D**) → (**798**) → (**799**)

678 just gives an adduct with benzophenone, while with the less bulky benzaldehyde a [2 + 2] cycloadduct **800** is obtained (equation 273)[59].

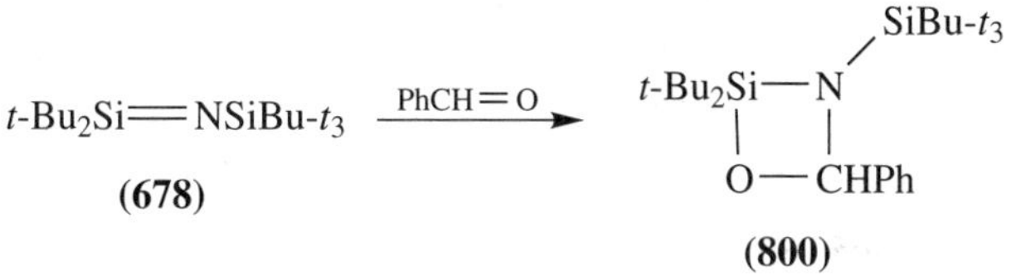

$$t\text{-Bu}_2\text{Si}=\!\!=\!\!\text{NSiBu-}t_3 \xrightarrow{\text{PhCH}=\text{O}} \begin{array}{c} \text{SiBu-}t_3 \\ t\text{-Bu}_2\text{Si}-\text{N} \\ | \quad\quad | \\ \text{O}-\text{CHPh} \end{array}$$

(**678**)   (**800**)

(273)

Lithiated fluorosilanes such as **801** and **802** may be considered as fluoride donor adducts to silanimines and behave as the free silanimines (e.g. **803**) in cycloaddition reactions with C=O compounds. The stability of the four-membered cycloadducts obviously depends on the substituents (equation 274)[355].

$$(\text{THF})_3\text{LiF}-\underset{R}{\overset{R}{\underset{|}{\overset{|}{\text{Si}}}}}=N-R' \quad\quad\quad \underset{i\text{-Pr}}{\overset{i\text{-Pr}}{\diagdown}}\!\!\text{Si}=N-R'$$

(**801**) R = t-Bu, R' = 2,6-(i-Pr)$_2$C$_6$H$_3$ (**803**) R' = Sup
(**802**) R = i-Pr, R' = Sup

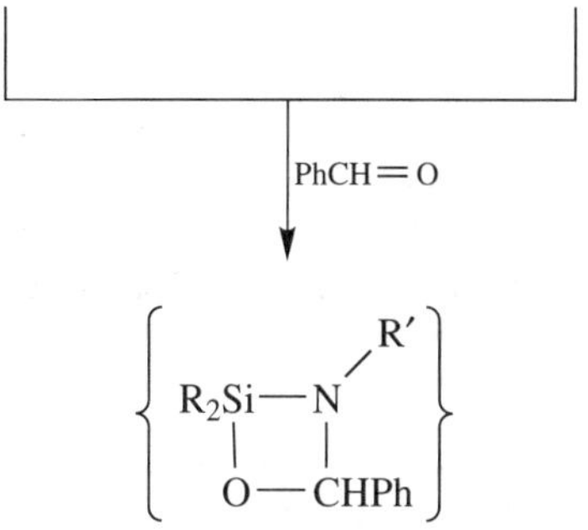

$$\downarrow \text{PhCH}=\text{O}$$

$$\left\{ \begin{array}{c} R' \\ R_2\text{Si}-\text{N} \\ | \quad\quad | \\ \text{O}-\text{CHPh} \end{array} \right\}$$

(**804**) R = i-Pr, t-Bu; R' = Sup, 2,6-(i-Pr)$_2$C$_6$H$_3$

$$\downarrow$$

R'N=CHPh + 1/3 (R$_2$SiO)$_3$

(274)

The choice of other differently substituted silanimine precursors like **805** and **806** allows the isolation of stable cycloadducts, e.g. **807** and **808** (equation 275). While **808** slowly decomposes at room temperature (equation 276), **807** can even be distilled at 110 °C/10^{-2} mbar.

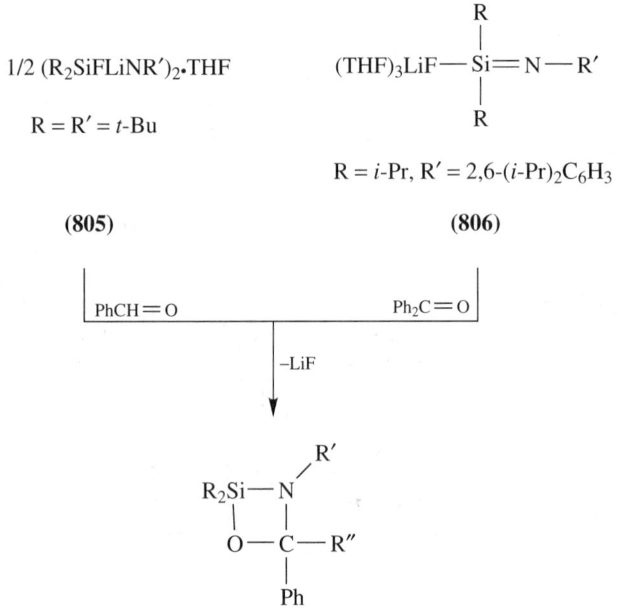

$$\frac{1}{2}\,(R_2SiFLiNR')_2 \cdot THF \qquad (THF)_3LiF—Si(R)(R)=N—R'$$

R = R' = t-Bu

R = i-Pr, R' = 2,6-(i-Pr)$_2$C$_6$H$_3$

(805)   (806)

PhCH=O   Ph$_2$C=O   (275)

−LiF

$$R_2Si—N(R')—O—C(Ph)(R'')$$

(807) R = R' = t-Bu, R'' = H
(808) R = i-Pr, R' = 2,6-(i-Pr)$_2$C$_6$H$_3$, R'' = Ph

$$808 \xrightarrow{25\,°C} R'N{=}C(Ph)R'' + 1/3\,(R_2SiO)_3 \qquad (276)$$

The lithiated aminofluorosilane [t-Bu$_2$Si−N−Si(OSiMe$_3$)Pr-i$_2$·LiF]$_2$ **809** behaves in the same way, and its cycloadduct with benzaldehyde decomposes into the respective imine and cyclotrisiloxane above 105 °C[354].

Silanimine **678** reacts with methyl vinyl ether in a clean reaction to give the [2 + 2] cycloadduct **810** (equation 277)[303].

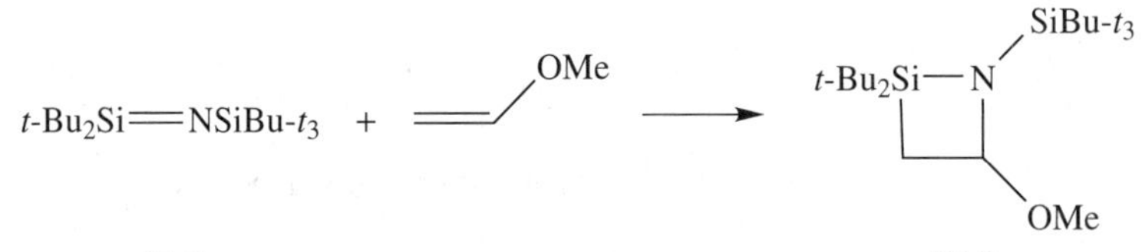

$$t\text{-Bu}_2Si{=}NSiBu\text{-}t_3 \;+\; {=}\!\!/\!OMe \;\longrightarrow\; \text{(810)} \qquad (277)$$

(678)   (810)

b. *[2 + 3] Cycloadditions*. The silanimine **414a** reacts with azides R'N$_3$ (R' = t-Bu, silyl) to give siladihydrotetrazoles **344a**[300,310,351]. As a byproduct the insertion product **415a** is also formed (equation 278).

Silanimines **678**[303], t-Bu$_2$Si=NSiBu-t$_2$Ph (**811**) and t-Bu$_2$Si=NSiBu-t$_2$Me·THF[353] (**812**·THF) also react with silyl azides in the same way as shown in equation 278. The reaction of **678** with N$_2$O obviously gives the [2 + 3] cycloadduct **813** as the initial product, but **813** then decomposes, probably via the silanone t-Bu$_2$Si=O, to give **814** as main product (equation 279)[356].

16. Silicon–carbon and silicon–nitrogen multiply bonded compounds

$$
\begin{array}{c}
\text{Me}_2\text{Si}\!-\!\text{C(SiMe}_3)_2 \\
|\backslash \\
\text{RN}\text{N} \\
\backslash\diagup \\
\text{N}
\end{array}
$$

(412a)

$\Big\downarrow -(\text{Me}_3\text{Si})_2\text{CN}_2$

$$
\left\{
\begin{array}{c}
\text{Me} \\
\diagdown \\
\text{Si}\!=\!\text{NR} \\
\diagup \\
\text{Me}
\end{array}
\right\}
\qquad (278)
$$

(414a)

insertion ↙ +R′N$_3$ ↘ cycloaddition

$$
\begin{array}{c}
\text{Me} \\
|\text{R} \\
\diagdown\diagup \\
\text{N}_3\!-\!\text{Si}\!-\!\text{N} \\
|\diagdown \\
\text{Me}\text{R}'
\end{array}
\qquad
\begin{array}{c}
\text{Me}_2\text{Si}\!-\!\text{NR} \\
|\backslash \\
\text{R}'\text{N}\text{N} \\
\backslash\diagup \\
\text{N}
\end{array}
$$

(415a) **(344a)**

$$
t\text{-Bu}_2\text{Si}\!=\!\text{N}\!-\!\text{SiBu-}t_3 \xrightarrow{\text{N}_2\text{O}} \left\{
\begin{array}{c}
\text{O}\!-\!\text{N} \\
t\text{-Bu}_2\text{Si}\diagup\parallel \\
\diagdown\text{N}\!-\!\text{N} \\
t\text{-Bu}_3\text{Si}
\end{array}
\right\}
$$

(678) **(813)**

$$(279)$$

$$
\begin{array}{c}
t\text{-Bu}\text{Bu-}t \\
\diagdown\diagup \\
\text{Si} \\
\text{HN}\diagup\diagdown \\
t\text{-Bu}\diagdown \\
\text{Si} \\
t\text{-Bu}\diagup \\
\text{O}\!-\!\text{Si} \\
|\diagdown\text{Bu-}t \\
t\text{-Bu}
\end{array}
\qquad \xleftarrow[\substack{+\,\mathbf{678}\\-t\text{-Bu}_3\text{SiN}_3}]{}
$$

(814)

The authors suggest that the formation of **814** probably proceeds via the silanone $t\text{-Bu}_2\text{Si}\!=\!\text{O}$ and an insertion into one C—H bond of a *tert*-butyl group.

A different reaction occurs when the donor adduct **678**·THF is used for the reaction with N_2O. In this reaction the THF ring is incorporated into the product **815** (equation 280)[303].

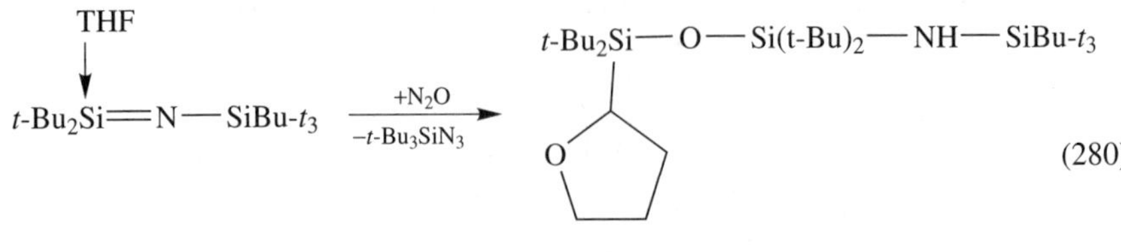

$$t\text{-}Bu_2Si{=}N{-}SiBu\text{-}t_3 \cdot THF \xrightarrow[-t\text{-}Bu_3SiN_3]{+N_2O} \mathbf{815} \quad (280)$$

(**678**)·THF (**815**)

c. [2+4] Cycloadditions. With the Wiberg silanimines [2+4] cycloaddition reactions do not occur, rather ene reaction takes place. There is one example by Klingebiel and coworkers who obtained [2+2] and [2+4] cycloadducts **816** and **817** in a 1 : 3 ratio respectively, in the reaction of $t\text{-}Bu_2Si{=}N(2,6\text{-}C_6H_3i\text{-}Pr_2)$*LiF(THF)$_3$ (**801**·LiF(THF)$_3$) with 2-methyl-2-propen-1-al (equation 281)[355]. The [2 + 2] adduct decomposes as described above.

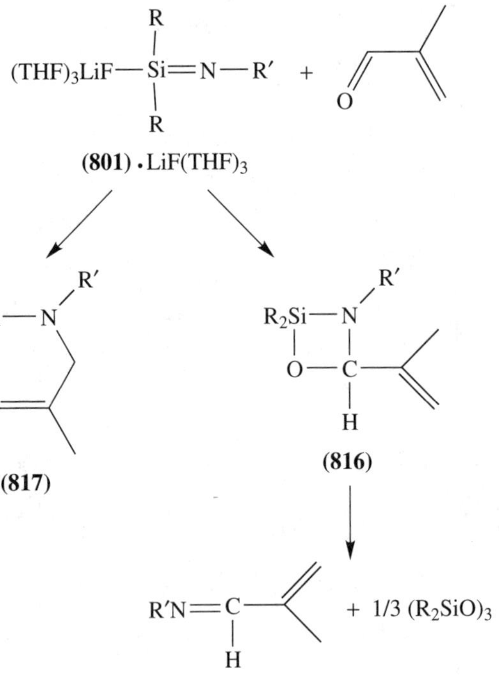

$$(THF)_3LiF{-}\underset{R}{\overset{R}{Si}}{=}N{-}R' + \text{(2-methyl-2-propen-1-al)} \quad (281)$$

(**801**)·LiF(THF)$_3$

→ **817** → **816** → $R'N{=}CH{-}C(CH_3){=}CH_2$ + 1/3 $(R_2SiO)_3$

$R = t\text{-}Bu, R' = 2,6\text{-}(i\text{-}Pr)_2C_6H_3$

5. Metal silanimine complexes

Berry describes a variety of reactions of the silanimine complex **788** to give **818–821**[348–350]. The phosphine ligand is quite labile and can be replaced by CO yielding the carbonyl complex **818** (equation 282).

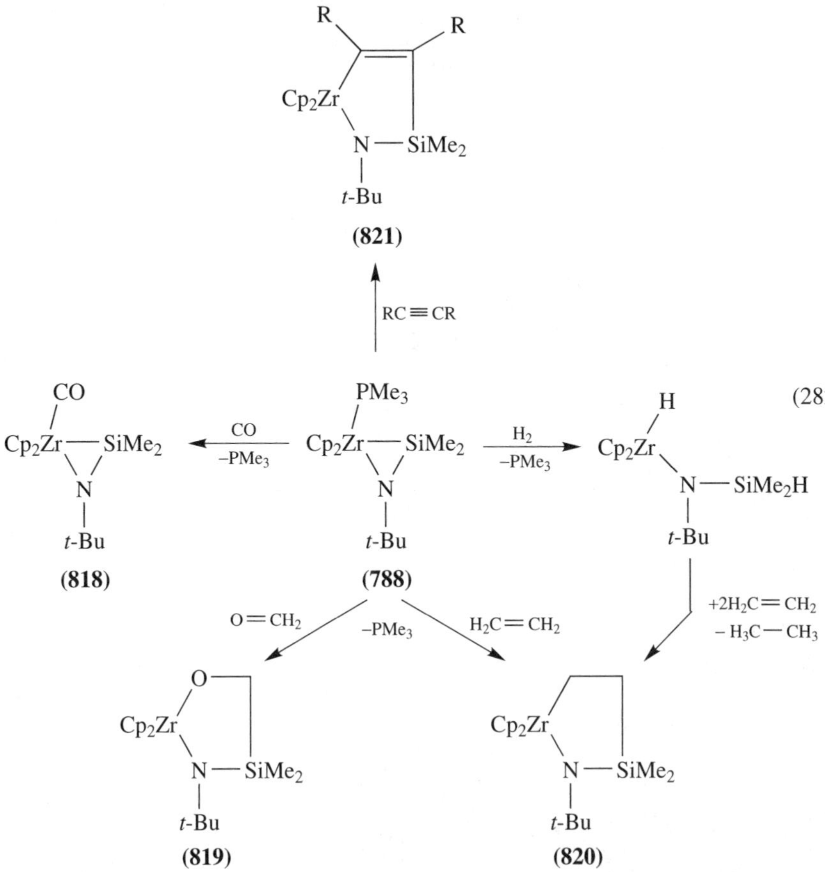

16. Silicon–carbon and silicon–nitrogen multiply bonded compounds 1041

(282)

The carbonyl adduct **818** might be expected to react with PMe_3 in the reversal of its formation reaction, but the product isolated after prolonged reaction at 50 °C is the five-membered metallacycle **822** (equation 283).

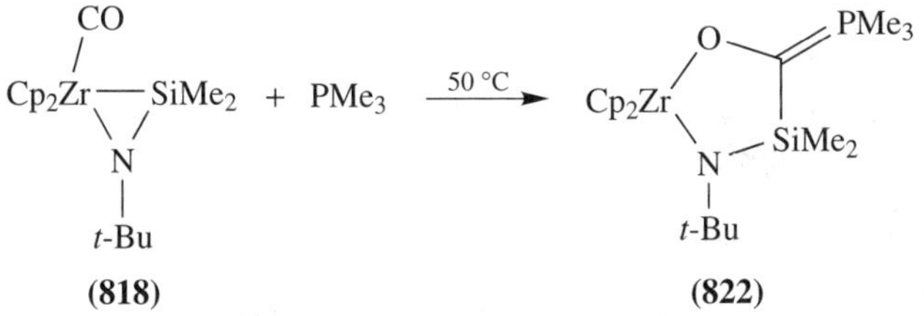

(283)

In particular, the insertion of C=X compounds giving metallacycles like **819**, **820** and **813** suggested that molecules with cumulenic bonds might insert in a similar fashion. CO_2, however, leads to a net oxygen atom insertion into the Zr–Si bond and the dimeric complex **823** is formed. The initial intermediate cyclo-$Cp_2Zr[O(C=O)SiMe_2NBu\text{-}t]$ **824**, formed by insertion of CO_2 into the Zr–Si bond, has been observed by low temperature NMR studies (equation 284).

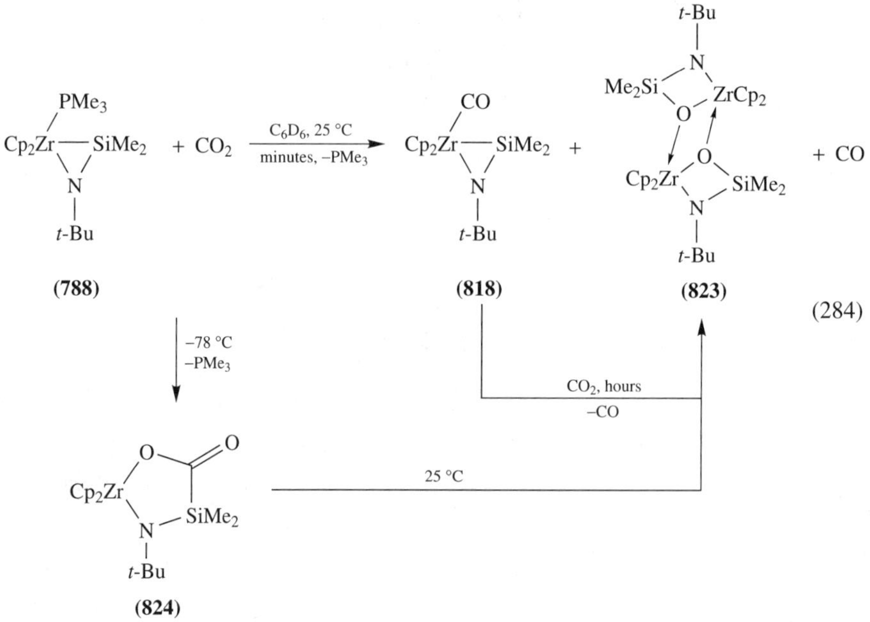

[Scheme for equation (284) showing reaction of 788 with CO₂ to give 818 and 823, with 824 formed at −78 °C]

(284)

The reaction with CS$_2$ led to a monomeric sulphur insertion product **825** (equation 285). With COS, a mixture of the oxygen and sulphur insertion products (**823** and **825**) and the carbonyl adduct **818** are obtained.

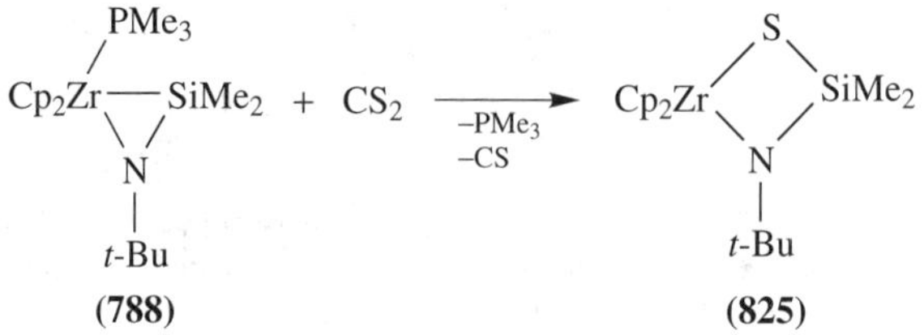

[Scheme for equation (285): 788 + CS₂ → 825 with loss of PMe₃ and CS]

(285)

D. X-ray Structures

A comparison of the calculated Si=N bond length with that measured by single crystal X-ray diffraction was possible for just one compound, Wiberg's silanimine t-Bu$_2$Si=N−SiBu-t_3[327]. The Si=N bond length was determined to be 1.568 Å (Table 17) and, despite the bulky groups in this molecule, this value corresponds well with the calculated length for H$_2$Si=NMe (1.569 Å, Table 21)[316].

The coordination of donors lengthens the Si=N bond and leads to pyramidalization at the silicon centre, as can be seen from the deviation of the sum of angles around the silicon atom from 360° (see examples in Table 17). Coordination of silanimines to transition metal centres also increases the silicon–nitrogen bond length to 1.66–1.69 Å, which corresponds to a Si−N single bond. It has already been mentioned above that these metal compounds are best described as metallacycles[349,350].

16. Silicon–carbon and silicon–nitrogen multiply bonded compounds

TABLE 17. Structural data of silanimines and related compounds (bond distances r in Å, angles α in deg)

Compound	r(Si=N)	α(Si=N–R)	r(Si→ D)	$\alpha(\angle$Si$)^b$	References
t-Bu$_2$Si=NSiBu-t_3	1.568	177.8		359.9	327
Me$_2$Si=NSiBu-t_3·THF^a	1.588	161.5	1.888	349.5	327
	1.57	161.0	1.866	349.2	
t-Bu$_2$Si=NSiBu-t_3· Ph$_2$C=O	1.601	169.3	1.927	350.3	327
t-Bu$_2$Si=NSiBu-t_2Me·THF	1.596	174.3	1.902	347.7	306
(2,4,6-t-Bu$_3$C$_6$H$_2$)N=SiPr-i_2·FLi(THF)$_3$	1.619	160.6			307, 357
(2,4,6-t-Bu$_3$C$_6$H$_2$)N=SiBu-tMe·FLi(THF)$_3$	1.606	161.3	1.676	342.3	358
(t-Bu$_2$PhSi)N=SiBu-t_2· FLi(THF)$_3$	1.608^c	176.3	1.692		357, 307
[(t-Bu$_2$PhSi)N=SiBu-t_2F]$^-$[Li(TMEDA)$_2$]$^{+a}$	1.616^d	169.7	1.654		357, 307
	1.618^e	176.2	1.651		
[(t-Bu$_2$MeSi)N=SiBu$_2$FLi]$_2$	1.647^f	148.9	1.699	349.5	307
[(t-Bu$_2$SiF)N]$^-$[Li(THF)$_2$]$^+$	1.636	176.7	1.659		359
			1.653		
[(t-Bu$_2$SiF)N]$^-$[Li([12]-C-4)$_2$]$^+$	1.606	162.6	1.644		359
	1.630		1.637		
[t-Bu$_2$Si=NSi-i-Pr$_2$(OSiMe$_3$)FLi]$_2$	1.636	155.6	1.680		354
[Ph$_3$Si–N–SiPh$_3$]$^-$ [Li([12]-C-4)$_2$]$^+$	1.633	154.9			360
[(2,4,6-t-Bu$_3$C$_6$H$_2$N)$_2$SiBu-t]$^-$[Li([12]-C-4)THF]$^+$	1.594	161.9		360	347
	1.626	136.5			
Cp$_2$Zr(η^2-t-BuN=SiMe$_2$)PMe$_3$	1.687	133.8	2.654^g	336.5	350
Cp$_2$Zr(η^2-t-BuN=SiMe$_2$)CO	1.661	131.1	2.706	348.6	349

^{a}Two molecules in the unit cell.
^{b} Sum of the three angles around the silicon atom, excluding the D→ Si bond.
$^c d$[(t-Bu$_2$PhSi)–N] 1.652 Å.
$^d d$[(t-Bu$_2$PhSi)–N] 1.645 Å.
$^e d$[(t-Bu$_2$PhSi)–N] 1.640 Å.
$^f d$[(t-Bu$_2$MeSi)–N] 1.694 Å.
^{g}'D' = Zr.

E. Spectroscopy

1. NMR spectroscopy

The NMR spectroscopic chemical shifts of the silicon atoms in multiply bonded compounds serve as the best evidence for the existence of sp^2-hybridized silicon species. Data for known silanimine and related compounds are summarized in Table 18. The ^{29}Si resonances of the stable silanimines are significantly shifted to lower field (60–80 ppm) as compared to singly bonded Si–N compounds[303,307]. The ^{14}N chemical shift of the nitrogen atom in the stable silanimine t-Bu$_2$Si=N–SiBu-t_3 confirms the existence of the Si=N bond with a shift of -230 ppm[303]. The coordination of donors lessens the double bond character and the ^{29}Si and ^{14}N resonances are found at higher field; e.g. for t-Bu$_2$Si=N–SiBu-t_3·THF, δ^{29}Si $= 1.0$, δ^{14}N $= -330$ ppm[303]. Fluoride as donor gives rise to quite large Si–F coupling constants and thus the 'fluoride adducts' are best

TABLE 18. NMR spectroscopic data (δ in ppm) of silanimines and related compounds

Compound	δ^{29}Si	δ^{n}E	References
t-Bu$_2$Si^a=NSi^{b}Bu-t_3	a: 78; b: −7.7	^{14}N: −230	303
t-Bu$_2$Si^a=NSi^{b}Bu-t_3·Ph$_2$CO	a: 54.2; b: −9.6		303
t-Bu$_2$Si^a=NSi^{b}Bu-t_3·THF	a: 1.0; b: −14.7	^{14}N: −330	303
t-Bu$_2$Si^a=NSi^{b}Bu-t_3·MeTHF	a: 3.60; b: −14.57	^{14}N: −329.08	306
t-Bu$_2$Si^a=NSi^{b}Bu-t_3·NMe$_2$Et	a: 18.1; b: −13.6		303
Me$_2$Si^a=NSi^{b}Bu-t_3·THF	a: −1.51; b: −11.1	^{14}N: −330	300
Me$_2$Si^a=NSi^{b}Bu-t_3·Et$_2$O	a: −1.51; b: −11.1		300
Me$_2$Si^a=NSi^{b}Bu-t_3·NEt$_3$	a: −11.2; b: −11.1		300
Me$_2$Si^a=NSi^{b}Bu-t_3·NMe$_2$Et	a: −8.85; b: −10.25		300
t-Bu$_2$Si^{a}F−N−Si^{b}Bu-t_2Me ·Li (TMEDA)	a: −2.27; b: −10.19	^{19}F: 11.29	307
i-Pr$_2$Si^a=N−Si^bi-Pr$_2$(OSiMe$_3$) ·FLi(THF)$_3$	a: −5.39; b: −24.28	^{19}F: 7.22 ^1J(F,Si) = 261 Hz	354
t-Bu$_2$Si^{a}F=N−Si^{b}Bu-t_2Me ·Li([12]-C-4)	a: −19.69; b: −17.83	^{19}F: 14.78	307
i-Pr$_2$Si=N(2,4,6-t-Bu$_3$C$_6$H$_2$)	60.3		307
t-Bu$_2$Si=N(2,4,6-t-Bu$_3$C$_6$H$_2$)	63.1		307, 305^a
t-Bu$_2$Si=N−SiPhBu-t_2	80.4		307
[(t-Bu$_2$SiF)$_2$N]$^−$[Li(THF)$_2$]$^+$	−12.3	^{19}F: 11.7 ^1J(F,Si) = 272.5 Hz	359
[(t-Bu$_2$SiF)$_2$N]$^−$[Li(TMEDA)]$^+$	−12.0	^{19}F: 13.1 ^1J(F,Si) = 271.4 Hz	359
[(t-Bu$_2$Bu$_2$SiF)$_2$N]$^−$[Li(12-C-4)$_2$]$^+$ ^{c}		^{19}F: −22.5 ^1J(F,Si) = 295.2 Hz	359
[(2,4,6-t-Bu$_3$C$_6$H$_2$)N]$_2$SiPh$^−$ Li(OEt)$^+$	3.2		346
[(2,4,6-t-Bu$_3$C$_6$H$_2$)N]$_2$SiPh$^−$Li$^+$	7.0		346
[(2,4,6-t-Bu$_3$C$_6$H$_2$)N]$_2$SiPh$^−$ Li([15]-C-5)$^+$	−37.5		346
[(2,4,6-t-Bu$_3$C$_6$H$_2$)N]$_2$SiCl$^−$ K$^+$	−54.2		346^b
Cp$_2$Zr(η^2-Me$_2$Si=NBu-t)(CO)	−69.9		349

^{a} For further data on lithiated aminofluorosilanes see Reference 305.
^{b} For further data on silaamidides see Reference 346.
^{c} $\delta(^{29}$Si) not reported.

described as covalently bonded fluorosilanes[359]. The negative charge in silaamidides leads to strong shielding of the silicon atom, which resonates at −54.2 ppm[346].

2. IR and UV-Vis spectroscopy

IR and UV spectroscopy are the primary analytical tools for the identification of highly reactive species trapped in low temperature matrices. Along with calculations the assignment of vibrations and electronic transitions to various species is possible. Additional verification for IR assignments is obtained through isotopic labeling. The stretching vibration of the Si=N bond is found to be 1326 cm^{-1} for the stable compound t-Bu$_2$Si=N−Sit-Bu$_3$ (see Table 19)[303]. The silanimine H$_2$Si=NH, without bulky substituents, exhibits a $\nu_{\text{Si}=\text{N}}$ at 1097 cm^{-1}[326]. The Si=N bond is significantly weakened

TABLE 19. IR and UV spectroscopic data of silanimines and related compounds

Compound	IR [cm^{-1}]	UV-vis (nm)	References
t-Bu$_2$Si=N–SiBu-t_3	$\nu_{Si=N}$ 1326		303
[(2,4,6-t-Bu$_3$C$_6$H$_2$)N]$_2$SiR$^-$	ν_{as} (N–Si–N)$^-$ 1303–1284		346, 347
(Me$_3$Si)(N$_3$)Si=N–SiMe$_3$		274	343
Me$_3$Si–N=Si=N–SiMe$_3$		324	
t-Bu$_2$Si=N$_2$	2150 (tentative)	300 (tentative)	345
t-Bu$_2$Si=N=^{15}N/t-Bu$_2$Si=^{15}N=N	2110		
t-BuN=Si=NBu-t		240, 385 (5 : 1 ratio)	
Mes$_2$Si=N–Mes		296, 444	317
Me$_2$Si=N–SiMes$_2$Bu-t			
i-Pr$_2$Si=N–SiMes$_2$Bu-t			
PhNSi	ν_{as} (CNSi) 1530 ip ring stretch	ca 315	325
Ph^{15}NSi	1514		
(Si=N bicyclic structure)	$\nu_{Si=N}$ 1088; ^{15}N 1077	397	321
(Si=N bicyclic structure)	$\nu_{Si=N}$ 1014; ^{15}N 998	557	321
(Si=N bicyclic structure)	$\nu_{Si=N}$ 1050; ^{15}N 1042	406	321
HNSi	$\nu_{Si=N}$ 1202	250	328
HNSi·H$_2$	$\nu_{Si=N}$ 1200; ν_{HH} 4178		
H$_2$Si=NH	$\nu_{Si=N}$ 1097	240	
D$_2$Si=ND	$\nu_{Si=N}$ 1063		
HSi≡N	$\nu_{Si≡N}$ 1163	238, 258, 266, 350	
HSi≡N·H$_2$	$\nu_{Si≡N}$ 1161; ν_{HH} 4164		
DSi≡N·D$_2$	$\nu_{Si≡N}$ 1145		
Si≡N·D$_2$	$\nu_{Si≡N}$ 1144; ν_{DD} 3010, 3005		
HNSi	IR emission: fundamental ν_1 vibration-rotation band (NH stretch) 3584		336
Cp$_2$Zr(η^2-Me$_2$Si=NBu-t)(CO)	ν_{co} 1797		349
Cp$_2$Zr(η^2-Me$_2$Si=NBu-t)(CO)	ν_{co} 1756		

TABLE 20. Miscellaneous spectroscopic data of silaisocyanide

Compound	Method/Result	Reference
HNSi	mm wave spectroscopy/J: 6 → 7 rotational transition 266234.888 MHz	333
HNSi	HeI photoelectron spectroscopy/ionization potentials: 9.9, 9.9, 10.7 eV	316

by twisting. Thus, the incorporation into bicyclic systems gives rise to Si=N stretching vibrations at 1014–1088 cm^{-1}[321].

The parent silaisocyanide HNSi has a $\nu_{Si=N}$ of 1202 cm^{-1}[328], while the phenyl derivative PhNSi vibrates at the higher wavenumber of 1530 cm^{-1}[325]. The silanitrile HSiN has a stretching vibration at 1163 cm^{-1}, which indicates that the silicon–nitrogen bond is weaker in this isomer[328].

3. Miscellaneous

The theoretical analysis of the gas phase thermal fragmentation (flash pyrolysis with subsequent photoelectron spectroscopy) of trimethylsilyl azide combined with calculations of the ionization potentials has led to the conclusion that HNSi (calc. 10.18, 10.18, 10.71 eV; exptl. 9.9, 9.9, 10.7 eV) is formed[316].

The detection and identification of molecules in interstellar space is possible by millimeter wave spectroscopy. The independent synthesis and detection of such reactive species, e.g. by flash vaccuum thermolysis and mm wave spectroscopy, provides proof for their cosmochemical existence. The detection of the J: 6 → 7 rotational transition in the decomposition products of t-Bu$_2$HSi–NH(CH$_2$–C≡CH) indicated the formation of HNSi (Table 20)[333].

F. Theoretical Studies

The interpretation and verification of analytical data, obtained especially for highly reactive species in the gas phase or matrix, is often possible only through the concomitant use of quantum chemical calculations. The listing of experimental data as above would be incomplete without at least mentioning the wealth of information available through theoretical methods. Thus interesting geometric and spectroscopic features have been included in the Table 21 which is sorted with respect to the systems studied. The geometries are always optimized at the indicated level. Details are given in the corresponding references.

In connection with the experimental data on silanimines (structures and NMR), it can generally be said that (a) donor addition lengthens the Si=N bond, (b) the bond order of silicon–nitrogen 'triply' bonded species is ca 2 and (c) the flexibility of the angle Si=N–R is greatest when R = silyl. Silyl-substituted O and N atoms (e.g. in disiloxanes and disilazanes) are known for the easy variability of the Si–E–Si angle. Computations preceding 1987 (not included in the table) on H$_2$Si=NH by Truong and Gordon found a planar geometry with d(SiN) = 157.6 pm and a Si=N–H angle of 125.2°[373]. The $syn/anti$ isomerization through inversion requires only 5.6 kcal mol^{-1}; the rotation around the Si–N bond needs 37.9 kcal mol^{-1}. The Si=N bond in the linear transition state is even shorter with 153.2 pm. Schleyer and Stout found similar results for H$_2$Si=N–SiH$_3$: d(SiN) = 154.9 pm, ∠Si=N–Si = 175.6°[374]. The angle flexibility at the silicon atom is also found for siladiimine, HN=Si=NH, where the conformational potential energy surface is very flat[377].

TABLE 21. Calculated data for various silicon-nitrogen species

Compound(s)/System(s)	Method(s)/Basis set(s)	Calculated properties	Reference
Silanimines			
Me$_2$Si=NMe	RHF/6-31G*	r(SiN) 1.551 pm	321
Me$_2$Si=NMe (twisted to fit the 3 compds below)	CIS/DZ+d	UV-Vis	
(bicyclic Si=N structure)	RHF/6-31G*	r(SiN) 1.592 Å	
(bicyclic Si=N structure)	RHF/6-31G*	r(SiN) 1.596 Å	
(bicyclic Si=N structure)	RHF/6-31G*	r(SiN) 1.600 Å	
H$_2$Si=NH	IGLOBII/HF/6-31G* MP2/DZ+d	δ^{29}Si: 52 IR	361
H$_2$Si=NH	RHF/3-21G*	r(SiN) 1.580 Å	316
H$_2$Si=NMe		r(SiN) 1.569 Å	
MeHSi=NH		r(SiN) 1.579 Å	
H$_3$Si–N$_3$	RHF/3-21G*	reaction pathways, ionization potentials	
H$_3$Si–N$_3$ and possible pyrolysis products	RHF/4-21G*		
[H$_3$, Si, N, O]	MP4/6-31G*//RHF/6-31G*; MP4/6-311G*//MP2/6-31G*	internal rotations MP2/6-31G*; H$_2$SiN–OH d(SiN) 1.655 Å (HO)HSi=NH r(SiN) 1.595 Å	362

(continued overleaf)

TABLE 21. (continued)

Compound(s)/System(s)	Method(s)/Basis set(s)	Calculated properties	Reference
H_3SiN: → $H_2Si=NH$	(U)HF/3-21G(*); (U)MP4/6-31G++G*//HF/3-21G*; RHF-CI	Curtius-type rearrangement reaction	320
[Si,N,H,F] system	BAC-MP4/6-31G**/(U)HF/6-31G*	thermochemistry, bond dissociation enthalpies	363
$H_nX=YH_m-ZH_2$ (X,Y = C,N,Si,P; Z = B,N)		rotational isomers; conjugation of double bonds	364
Si=N=Si systems			
$(H_3Si)_2N^-$	RHF/6-31+G*	$r(SiN)$ 1.624 Å; $\alpha SiNSi$ 180° +25 kcal mol^{-1} for α SiNSi 120°	307 359
N=Si=N systems and isomers			
MeN=Si=NMe	MNDO INDO/S	$r(SiN)$ 1.51 Å, α NSiN 172°, α CSiN 153°, IR UV-Vis	345
$Me_2Si=N_2$		$r(SiN)$ 1.64 Å, α CSiC 128°, IR UV-Vis	
cyclo-$(MeN)_2Si$: cyclo-(MeSi−NMe=N)		relative energies, UV-Vis	
MeN=N−MeSi:			
HN=Si=NH	RHF/6-31G*; MP4/6-31G**//MP2/6-31G*	RHF: $r(SiN)$ 1.516 Å, α NSiN 180° MP2: $r(SiN)$ 1.597 Å, α NSiN 156.9° conformational PES^a very flat	365
HN=Si=CH_2	RHF/6-31G*	$r(SiN)$ 1.549 Å, α CSiN 172.3°	

Silanitrile and silaisocyanides

Compound	Method	Results/Notes	Ref
PhSi≡N/PhN=Si:	INDO/S; MP2/6-31G*	UV-Vis; IR	325
[HNSi]•+ ($^1\Sigma^+$)	G1	PES^a	342
[HNSi]•+ ($^2\Sigma^+$) ions			
[HNSi], [DNSi]	MP2/6-31G*; MP2/D95**	IR	328
HNSi / DNSi	DFT method	PES^a, vibrational-rotational energies transition moments	366
HNSi		microwave transitions	336
PhSiN	AM1; RHF/6-31G or 3-21G*; MP2/6-31G//RHF/6-31G or 3-21G*	3-21G*: $r(\mathrm{SiN})$ 1.537 Å $r(\mathrm{SiN})$ 1.534 Å	367
PhNSi			
MeSiN	RHF/6-31G**; MPn/6-31G**//RHF/6-31G**	6-31G**: $r(\mathrm{SiN})$ 1.527 Å $r(\mathrm{SiN})$ 1.530 Å	
MeNSi			
RNSiN/RSiN	MP4/6-311+G**//MP2(fu)/6-31G*; also in part: G2; QCISD(T)/6-311G*//QCISD/6-31G**; CASSCF/6-31G*; QCISD(T)/6-311G(2df,p)//QCISD/6-31G**	$\Delta E = [E(\mathrm{MeSiN}) - E(\mathrm{MeNSi})] = 48.3$ kcal mol^{-1} relative stabilities, energy barriers ΔE positive (22–76 kcal mol^{-1}) $\Rightarrow$ RNSi more stable for R = H, Li, BeH, BH$_2$, CH$_3$, SiH$_3$, PH$_2$, SH $\Delta E = 10.5$ kcal mol^{-1} for R = NH$_2$; -0.5 kcal mol^{-1} for R = Cl ΔE negative $\Rightarrow$ RSiN more stable for R = OH (6.5 kcal mol^{-1}); R = F (26.6 kcal mol^{-1}) Barrier heights: 22.7 kcal mol^{-1} for R = OH 29.4 kcal mol^{-1} for R = F	368
R = F, Cl			

(continued overleaf)

TABLE 21. (continued)

Compound	Method	Results/Comments	Ref.
HNSi	HF, MPn/various basis sets	interstellar production of HNSi from H$_2$NSi$^+$; HNSi proton affinity; IR	369
Metal compound			
Cp$_2$Zr(η^2-Me$_2$Si=NBu-t)(CO)	ZINDO	Zr–Si bond order 0.34	349
Ionic and radical systems			
[HNSi]$^{•+}$ ($^2\Sigma^+$) ions	G1	PES^a	342
[HNSi]$^+$ ($^1\Sigma^+$)			
[HSi(O)NH]$^-$, [Si(O)NH$_2$]$^-$, and 8 other related anionic isomer pairs	MP2/6-31+G**; MP4/6-311++G**	charge density; gas phase acidity	338
[H$_2$,Si,N]$^-$	MP2/6-31+G**	$trans$-HSiNH$^-$ r(SiN) 165.0 Å cis-HSiNH$^-$ 164.6° [H$_2$SiN]$^-$ 1.581 Å [SiNH$_2$]$^-$ 1.764 Å IR	336
H$_2$SiN$^+$	MPn (n = 2,3,4), CCSDT, CASSCF, CI, MCSCF/6-311++G**	r(SiN) 1.772–1.794 Å; rotational constants IR; electronic absorption	370
[H$_5$,Si,N]$^{•+}$	(U)HF/6-31G**; (U)MP4/6-31+G(2df,p)//MP2/6-31G**	spin density; unimolecular rearrangements	371
[SiNNH]$^•$ and isomers	ROHF, UHF, MCSCF/6-31G*	rotational constants	372

^{a}PES = potential energy surface.

G. GeN, SiP and SiAs Multiply Bonded Species and Related Compounds

Germanimines, $R_2Ge=NR'$ have been studied in parallel to silanimines by Wiberg's group. Their formation is achieved by salt elimination via germaethenes, which are trapped with azides to yield the [3+2] cycloadducts, the germadihydrotriazoles[308]. Germanimines may be thermally liberated from the latter and their reactivity can then be studied[351]. Another storage form are the germadihydrotetrazoles, obtained from germanimines and azides[310].

Ando and coworkers used a different approach to bis(germanimines): the reaction of bis[bis(trimethylsilyl)methyl]germylene with diazidosilanes gave in the first step N-(azidosilyl) germanimines, which reacted with a second equivalent of germylene to give bis(germanimines)[375]. The X-ray structures of two such compounds were determined r(Ge=N): 1.704 Å.

The chemistry of Si=P and Si=As compounds has been well reviewed by Bickelhaupt[378,379] and by Driess[376,377]. A short overview is, however, given.

A transient phosphasilene **827**, which was generated from 1,2-phosphasiletane **826**, was first reported in 1979 (equation 286)[380]. Just recently, this cycloreversion reaction was used to produce the phosphasilene in the gas phase and to prove its existence by high resolution mass spectrometry and photoelectron spectroscopy[381]. Higher temperatures are needed for the decomposition of the diphosphadisiletane **828** to generate **829**. The first ionization potentials are listed (equation 287).

$$\underset{(826)}{\begin{array}{c}\text{SiMe}_2\\|\\\text{P}\\|\\\text{Ph}\end{array}} \xrightarrow[-C_2H_4]{80°C} \underset{(827)}{Me_2Si=P\backslash_{Ph}} \quad \text{Ip: 8.3, 9.3, 9.8 eV} \quad (286)$$

$$\underset{(828)}{\begin{array}{c}t\text{-Bu}\\ \backslash\\ P-SiMe_2\\ |\quad\quad |\\ Me_2Si-P\\ \quad\quad\backslash\\ \quad\quad Bu\text{-}t\end{array}} \xrightleftharpoons{500°C} \underset{(829)}{Me_2Si=P\backslash_{Bu\text{-}t}} \quad \text{Ip: 7.0, 8.2 eV} \quad (287)$$

The most common route to phosphasilenes is by salt elimination from lithium(fluorosilyl)phosphanylides. Examples of transient to moderately stable species are t-Bu$_2$Si=PBu-t **830**, which could be observed by ^{31}P NMR spectroscopy[382], and t-Bu$_2$Si=P(SiBu-t_3) **831** whose existence could only be proved by trapping reactions[383].

The first synthesis of a crystalline phosphasilene **832** was achieved by a two-step reaction of t-Bu$_3$SiCl with four equivalents of SupPHLi (Sup = 2,4,6,-t-Bu$_3$C$_6$H$_2$) and Ph$_2$PCl[384,385] via the diphosphasila-allyl salt **833**[386], a remarkable conjugated sila-π system (equation 288). The Si=P bond length in **832** is 2.094 Å and the silicon atom is pyramidalized ($\Sigma <= 356.7°$). The $^1J_{P=Si}$ of **832** is remarkably high with 203 Hz.

$$t\text{-BuSiCl}_3 \xrightarrow[-3\text{ LiCl, }-2\text{ H}_2\text{PSup}]{4\text{ LiPHSup}} \underset{(833)}{\begin{array}{c}t\text{-Bu}\\|\\\text{Si}\\\text{Sup}-P\diagdown\quad\diagup P-\text{Sup}\\\quad\quad Li^+\end{array}} \xrightarrow[-2\text{ LiCl}]{Ph_2PCl} \underset{(832)}{\begin{array}{c}t\text{-Bu}\quad\quad\text{Sup}\\\quad\backslash\quad\quad\diagup\\\quad\text{Si}=P\\\diagup\quad\quad\backslash\\\text{Sup}\quad\quad PPh_2\end{array}} \quad (288)$$

Bickelhaupt published an improved route for the synthesis of phosphasilenes **836** in a one-pot reaction via the dilithiated phosphine **835** obtained from the primary phosphine **834** (equation 289). The drawback is that the substituent at phosphorous may not be smaller than the supermesityl (Sup) group[387].

$$R^1PH_2 \xrightarrow[-2\ C_4H_{10}]{2\ n\text{-BuLi}} R^1PLi_2 \xrightarrow[-2\ LiCl]{R^2R^3SiCl_2} R^1P{=}SiR^2R^3 \qquad (289)$$
$$(834) \qquad\qquad (835) \qquad\qquad (836)$$

The concept of the stabilization of unsaturated silicon centres by coordination has been applied in the synthesis of phosphasilene **837**[388].

(837)

Phosphasilene **832** is rather labile in solution, and thus derivatives with silyl groups at phosphorous were synthesized and proved to be more stable[389,390]. The reaction of the silyl phosphines **838** with *n*-BuLi gave the lithiated species **839**, which eliminated LiF upon heating in solution to 60–80 °C and thus liberated the phosphasilenes **840** (equation 290).

(838) → (839) → (840) (290)

R = Tip, *t*-Bu
R′ = silyl, germyl
Tip = (2,4,6-*i*-Pr$_3$C$_6$H$_2$)

For **840** with R = *t*-Bu and R′ = Si(Pr-i)$_3$ an X-ray structure could be obtained[381]. The Si=P bond length is 2.062 Å and the silicon atom has a trigonal planar geometry. The NMR chemical shifts of the phosphorous nuclei in **840** are found in the region of −30 to 28 ppm; the ^{29}Si resonances are mainly around 170 ppm. The $^1J_{Si=P}$ is found to be 150 to 160 Hz[389,390].

The attempt to stabilize transients in the formation reaction of phosphasilenes by transition metal centres directly bound to the precursors was not successful[392]. The metal substituent was either cleaved or behaved like a 'normal' substituent, as is shown for the synthesis of phosphasilene **842** from **841** (equation 291). The NMR parameters for **842** are as expected: $\delta^{29}\text{Si} = 201$ ppm, $^1J_{\text{P=Si}} = 163$ Hz; $\delta^{31}\text{P} = 57$ ppm.

$$\begin{array}{c}\text{Fp}\diagdown\quad\diagup\text{F}\\ \text{P}-\text{Si}-\text{Tip}\\ \text{Li}\diagup\quad\diagdown\text{Tip}\\ \textbf{(841)}\end{array}\quad\xrightarrow[-\text{LiF}]{\Delta}\quad\begin{array}{c}\text{Fp}\diagdown\\ \text{P}=\text{SiTip}_2\\ \\ \textbf{(842)}\end{array}\qquad(291)$$

Fp = Cp(CO)$_2$Fe

The synthesis of arsasilenes followed the same route as that employed for phosphasilenes[390,391,393]. Heating solutions of the precursors **843** to 60 °C gave the arsasilenes **844** (equation 292).

$$\begin{array}{c}\text{R}\diagdown\quad\diagup\text{Li}\\ \text{F}-\text{Si}-\text{As}\\ \text{Tip}\diagup\quad\diagdown\text{R}'\\ \textbf{(843)}\end{array}\quad\xrightarrow[-\text{LiF}]{\Delta}\quad\begin{array}{c}\text{R}\diagdown\quad\\ \text{Si}=\text{As}\\ \text{Tip}\diagup\quad\diagdown\text{R}'\\ \textbf{(844)}\end{array}\qquad(292)$$

R = Tip, *t*-Bu
R' = silyl

The ^{29}Si NMR shift is found in the region from 180 to 190 ppm, with the exception of 228.8 ppm for R = *t*-Bu, R' = Si(*i*-Pr)$_3$. The X-ray structure of the latter derivative was determined and the Si=As bond length was found to be 2.164 Å (Si–As = 2.363 Å). The silicon atom is trigonal planar.

Phospha- and arsasilenes undergo $[2+n]$ cycloadditions ($n = 1$–4), thermal decomposition, and reaction with P$_4$, sulphur and tellurium. In order to avoid excessive redundancy the interested reader should consult the two reviews by Driess on this subject[376,377]. A number of theoretical treatments on unsaturated silicon phosphorous compounds, apart from those cited in the above-mentioned reviews, has been published[394–398].

V. ACKNOWLEDGEMENTS

The authors express their gratitude to Professors M. Kira, R. West, R. Okazaki and to Dr M. Karni, who have kindly provided us with manuscripts prior to publication. We thank also Mr D. Parker from the Open University in Milton Keynes for reading Section IV and Mrs U. Kätel for assistance. The authors are particularly grateful to Prof. Zvi Rappoport. The Hebrew University of Jerusalem, for carefully reading the manuscript.

N.A. is especially indebted to his previous and present coworkers, who contributed to the research in the Berlin and Munich groups in the field of silene chemistry: M. Backer, R. Gleixner, B. Goetze, M. Grasmann, Dr C.-R. Heikenwälder, Dr B. Herrschaft, Dr M. Kersten, Dr M. Päch, Dr E. Penzenstadler, Dr R. Probst, Dr C. Seidenschwarz, U. Steinberger, Dr C. Wagner, Dr A. Weingartner, Dr A. Wolff and Dr W. Ziche. T. M. thanks the Fonds der Chemischen Industrie for a Liebig Scholarship.

VI. REFERENCES

1. L. E. Gusel'nikov and M. C. Flowers, *J. Chem. Soc., Chem. Commun.*, 864 (1967).
2. M. C. Flowers and L. E. Gusel'nikov, *J. Chem. Soc. (B)*, 419 (1968).
3. A. G. Brook, S. C. Nyburg, F. Abdesaken, B. Gutekunst, G. Gutekunst, R. K. M. R. Kallury, Y. C. Poon, Y.-M. Chang and W. Wong-Ng, *J. Am. Chem. Soc.*, **104**, 5667 (1982).
4. N. Wiberg, G. Wagner, J. Riede and G. Müller, *Organometallics*, **6**, 32 (1987).
5. G. E. Miracle, J. L. Ball, D. R. Powell and R. West, *J. Am. Chem. Soc.*, **115**, 11598 (1993).
6. N. Tokitoh, K. Wakita, R. Okazaki, S. Nagase, P. v. R. Schleyer and H. Jiao, *J. Am. Chem. Soc.*, **119**, 6951 (1997).
7. J. Michl and G. Raabe, *Chem. Rev.*, **85**, 419 (1985).
8. G. Raabe and J. Michl, in *The Chemistry of Organic Silicon Compounds*, Vol. 1 (Eds. S. Patai and Z. Rappoport), Wiley, Chichester, 1989.
9. G. A. Brook and M. A. Brook, *Adv. Organomet. Chem.*, **39**, 71 (1995).
10. (a) K. K. Baldridge, J. A. Boatz, S. Koseki and M. S. Gordon, *Ann. Rev. Phys. Chem.*, **38**, 211 (1987).
    (b) R. S. Grev, *Adv. Organomet. Chem.*, **33**, 125 (1991).
11. Y. Apeloig, in *The Chemistry of Organic Silicon Compounds* (Eds. S. Patai and Z. Rappoport), Wiley, Chichester, 1989.
12. M. Karni and Y. Apeloig, *Chem. Rev.*, to appear.
13. M. Steinmetz, *Chem. Rev.*, **95**, 1527 (1995).
14. P. P. Gaspar, D. Holten, S. Konieczny and J. Y. Corey, *Acc. Chem. Res.*, **20**, 329 (1987).
15. T. J. Barton, G. Marquardt and J. A. Kilgour, *J. Organomet. Chem.*, **85**, 317 (1975).
16. (a) N. S. Nametkin, N. N. Dolgopolov and L. E. Gusel'nikov, *J. Organomet. Chem.*, **169**, 165 (1979).
    (b) M. S. Gordon, T. J. Barton and H. Nakano, *J. Am. Chem. Soc.*, **119**, 11966 (1997).
17. J. Grobe and N. Auner, *J. Organomet. Chem.*, **197**, 13 (1980).
18. A. M. Fenton, P. Jackson, F. T. Lawrence and I. M. T. Davidson, *J. Chem. Soc., Chem. Commun.*, 806 (1982).
19. J. Pola, E. A. Volnina and L. E. Gusel'nikov, *J. Organomet. Chem.*, **391**, 275 (1990).
20. R. T. Conlin, M. Namavari, J. S. Chickos and R. Walsh, *Organometallics*, **8**, 168 (1989).
21. I. M. T. Davidson, R. Laupert, P. Potzinger and S. Basu, *Ber. Bunsenges. Phys. Chem.*, **83**, 1282 (1979).
22. R. T. Conlin and Y.-W. Kwak, *J. Am. Chem. Soc.*, **108**, 834 (1986).
23. M. Steinmetz and H. Bai, *Organometallics*, **8**, 1112 (1989).
24. A. Kumar, R. K. Vatsa, R. D. Saini, J. P. Mittal, J. Pola and S. Dhanya, *J. Chem. Soc., Faraday Trans.*, **92**, 179 (1996).
25. R. K. Vatsa, A. Kumar, P. D. Naik, H. P. Upadhyaya, U. B. Pavanaja, R. D. Saini, J. P. Mittal and J. Pola, *Chem. Phys. Lett.*, **255**, 129 (1996).
26. J. Grobe and N. Auner, *Z. Anorg. Allg. Chem.*, **459**, 15 (1979).
27. (a) N. Auner, Abstracts of the XI-th International Symposium on Organosilicon Chemistry, Montpellier, France (1996) OB20.
    (b) N. Auner, J. Grobe, H. Rathmann and T. Müller, to appear.
28. G. Mihm, H. P. Reisenauer and G. Maier, *Angew. Chem., Int. Ed. Engl.*, **20**, 597 (1981).
29. G. Mihm, H. P. Reisenauer and G. Maier, *Chem. Ber.*, **117**, 2351 (1984).
30. H. Bock, B. Soulouki, G. Maier, G. Mihm and P. Rosmus, *Angew. Chem., Int. Ed. Engl.*, **20**, 598 (1981).
31. S. Bailleux, M. Bogey, J. Breidung, H. Bürger, R. Fajgar, Y. Liu, J. Pola, M. Senzlober and W. Thiel, *Angew. Chem., Int. Ed. Engl.*, **35**, 2513 (1996).
32. M. Bogey, H. Bürger, R. Fajgar, Y. Liu, J. Pola, M. Senzlober and S. Bailleux, Abstracts of the XI-th International Symposium on Organosilicon Chemistry, Montpellier, France (1996) PA13-14.
33. S. Bailleux, M. Bogey, J. Demaison, H. Bürger, M. Senzlober, J. Breidung, W. Thiel, R. Fajgar and J. Pola, *J. Chem. Phys.*, **106**, 10016 (1997).
34. W. Leigh, C. Bradaric, C. Kerst and J.-A. Banisch, *Organometallics*, **15**, 2246 (1996).
35. (a) W. J. Leigh, C. Bradaric and G. W. Sluggett, *J. Am. Chem. Soc.*, **115**, 5332 (1993).
    (b) C. J. Bradaric and W. J. Leigh, *J. Am. Chem. Soc.*, **118**, 8971 (1996).
36. V. Volkova, L. Gusel'nikov, E. Volnina, and E. Buravtseva, *Organometallics*, **13**, 4661 (1994).
37. R. T. Conlin and K. L. Bobbitt, *Organometallics*, **6**, 1406 (1987).

38. A. G. Brook, S. C. Nyburg, W. F. Reynolds, Y. C. Poon, Y.-M. Chang and J.-S. Lee. *J. Am. Chem. Soc.*, **101**, 6750 (1979).
39. D. Bravo-Zhivotovskii, V. Braude, A. Stanger, M. Kapon and Y. Apeloig, *Organometallics*, **11**, 2326 (1992).
40. N. Auner, I. M. T. Davidson, S. Ijadi-Maghsoodi and F. T. Lawrence, *Organometallics*, **5**, 431 (1986).
41. I. Jung, D. Pae, B. Yoo, M. Lee and P. Jones, *Organometallics*, **8**, 2017 (1989).
42. B. Yoo, M. Lee and I. Jung, *Organometallics*, **11**, 1626 (1992).
43. K.-T. Kang, N. Inamoto and R. Okazaki, *Tetrahedron Lett.*, **22**, 235 (1981).
44. A. Naka, M. Hayashi, S. Okazaki and M. Ishikawa, *Organometallics*, **13**, 4994 (1994).
45. M. Ishikawa, H. Sakamoto and T. Tabuchi, *Organometallics*, **10**, 3173 (1991).
46. M. Ishikawa, A. Naka and S. Okazaki, in *Progress in Organosilicon Chemistry* (Eds. B. Marciniec and J. Chojnowski), Gordon and Breach, Basel, 1995, p. 309.
47. M. Ishikawa, S. Okazaki, A. Naka and H. Sakamoto, *Organometallics*, **11**, 4135 (1992).
48. M. Ishikawa, A. Naka and J. Ohshita, *Organometallics*, **12**, 4987 (1993).
49. M. Ishikawa, A. Naka, S. Okazaki and H. Sakamoto, *Organometallics*, **12**, 87 (1993).
50. M. Ishikawa, S. Okazaki, A. Naka, A. Tachibana, S. Kawauchi and T. Yamabe, *Organometallics*, **14**, 114 (1995).
51. A. Naka, M. Hayashi, S. Okazaki, A. Kunai and M. Ishikawa, *Organometallics*, **15**, 1101 (1996).
52. A. Fanta, D. DeYoung, J. Belzner and R. West, *Organometallics*, **10**, 346 (1991).
53. K. Kabeta, D. Powell, J. Hanson and R. West, *Organometallics*, **10**, 827 (1991).
54. R. Conlin and M. Namavari, *J. Am. Chem. Soc.*, **110**, 3689 (1988).
55. M. Steinmetz, B. Udayakumar and M. Gordon, *Organometallics*, **8**, 530 (1989).
56. L. K. Revelle and E. Block, *J. Am. Chem. Soc.*, **100**, 1630 (1978).
57. N. Auner, I. M. T. Davidson and S. Ijadi-Maghsoodi, *Organometallics*, **4**, 2210 (1985).
58. K. W. Egger and P. Vitins, *J. Am. Chem. Soc.*, **96**, 2714 (1974).
59. N. Wiberg, K. Schurz, G. Müller and J. Riede, *Angew. Chem., Int. Ed. Engl.*, **27**, 935 (1988).
60. N. Wiberg, G. Preiner, K. Schurz and G. Fischer, *Z. Naturforsch.*, **43b**, 1468 (1988).
61. N. Wiberg, G. Preiner, G. Wagner, H. Köpf and G. Fischer, *Z. Naturforsch.*, **42b**, 1055 (1987).
62. N. Wiberg, G. Preiner, G. Wagner and H. Köpf, *Z. Naturforsch.*, **42b**, 1062 (1987).
63. N. Wiberg and M. Link, *Chem. Ber.*, **128**, 1231 (1995).
64. N. Wiberg and M. Link, *Chem. Ber.*, **128**, 1241 (1995).
65. N. Wiberg and G. Wagner, *Chem. Ber.*, **119**, 1455 (1986).
66. N. Wiberg and G. Wagner, *Chem. Ber.*, **119**, 1467 (1986).
67. N. Wiberg, H.-S. Hwang-Park, H.-W. Lerner and S. Dick, *Chem. Ber.*, **129**, 471 (1996).
68. P. R. Jones and T. Bates, *J. Am. Chem. Soc.*, **109**, 913 (1987).
69. B. R. Yoo, M. E. Lee and I. N. Jung, *J. Organomet. Chem.*, **410**, 33 (1991).
70. I. Jung, B. Yoo, M. Lee and P. Jones, *Organometallics*, **10**, 2529 (1991).
71. N. Wiberg, in *Organosilicon Chemistry II—From Molecules to Materials* (Eds. N. Auner and J. Weis), VCH Verlag, Weinheim, **1996**, p. 367.
72. N. Wiberg, G. Preiner, O. Schieda and G. Fischer, *Chem. Ber.*, **114**, 3505 (1981).
73. N. Wiberg, G. Preiner, O. Schieda and G. Fischer, *Chem. Ber.*, **114**, 3518 (1981).
74. N. Wiberg, G. Preiner, O. Schieda and G. Fischer, *Chem. Ber.*, **122**, 409 (1989).
75. N. Wiberg, G. Fischer and K. Schurz, *Chem. Ber.*, **120**, 1605 (1987).
76. N. Wiberg, K. Schurz and G. Fischer, *Chem. Ber.*, **119**, 3498 (1986).
77. N. Wiberg, T. Passler and K. Polborn, *J. Organomet. Chem.*, **531**, 47 (1997).
78. N. Wiberg, C. Finger, T. Passler, S. Wagner and K. Polborn, *Z. Naturforsch.*, **51b**, 1744 (1996).
79. P. R. Jones, T. F. O. Lim and R. A. Pierce, *J. Am. Chem. Soc.*, **102**, 4970 (1980).
80. N. Auner, in *Organosilicon Chemistry—From Molecules to Materials* (Eds. N. Auner and J. Weis), VCH Verlag, Weinheim, 1994, p. 103.
81. N. Auner, *Z. Anorg. Allg. Chem.*, **558**, 55 (1988).
82. G. Delpon-Lacaze and C. Couret, *J. Organomet. Chem.*, **480**, C14 (1994).
83. N. Wiberg, G. Wagner, G. Reber, J. Riede and G. Müller, *Organometallics*, **6**, 35 (1987).
84. N. Wiberg, K.-S. Joo and K. Polborn, *Chem. Ber.*, **126**, 67 (1993).
85. N. Wiberg and H. Köpf, *J. Organomet. Chem.*, **315**, 9 (1986).
86. A. G. Brook, J. W. Harris, J. Lennon and M. El Sheikh, *J. Am. Chem. Soc.*, **101**, 83 (1979).
87. A. G. Brook, F. Abdesaken, B. Gutekunst, G. Gutekunst and R. K. M. R. Kallury, *J. Chem. Soc. Chem. Commun.*, 191 (1981).

88. R. T. Conlin, P. F. McGarry, J. C. Scaiano and S. Zhang, *Organometallics*, **11**, 2317 (1992).
89. K. M. Baines and A. G. Brook, *Organometallics*, **6**, 692 (1987).
90. K. M. Baines, A. G. Brook, R. R. Ford, P. D. Lickiss, A. K. Saxena, W. J. Chatterton, J. F. Sawyer and B. A. Behnam, *Organometallics*, **8**, 693 (1989).
91. A. G. Brook, F. Abdesaken, G. Gutekunst and N. Plavac, *Organometallics*, **1**, 994 (1982).
92. A. G. Brook, R. K. M. R. Kallury and Y. C. Poon, *Organometallics*, **1**, 987 (1982).
93. P. Lassacher, A. G. Brook and A. J. Lough, *Organometallics*, **14**, 4359 (1995).
94. A. G. Brook, A. Baumegger and A. J. Lough, *Organometallics*, **11**, 310 (1992).
95. A. G. Brook, A. Baumegger and A. J. Lough, *Organometallics*, **11**, 3088 (1992).
96. A. G. Brook, K. D. Safa, P. D. Lickiss and K. M. Baines, *J. Am. Chem. Soc.*, **107**, 4338 (1985).
97. N. Wiberg and H. Köpf, *Chem. Ber.*, **120**, 653 (1987).
98. J. Ohshita, H. Hasebe, Y. Masaoka and M. Ishikawa, *Organometallics*, **13**, 1064 (1994).
99. M. Ishikawa, S. Matsui, A. Naka and J. Ohshita, *Organometallics*, **15**, 3836 (1996).
100. A. G. Brook, A. Ionkin and A. J. Lough, *Organometallics*, **15**, 1275 (1996).
101. H. Oehme and R. Wustrack, *Z. Anorg. Allg. Chem.*, **552**, 215 (1987).
102. R. Wustrack and H. Oehme, *J. Organomet. Chem.*, **352**, 95 (191988).
103. H. Oehme, R. Wustrack, A. Heine, G. M. Sheldrick and D. Stalke, *J. Organomet. Chem.*, **452**, 33 (1993).
104. C. Krempner and H. Oehme, *J. Organomet. Chem.*, **464**, C7 (1994).
105. F. Luderer, H. Reinke and H. Oehme, *J. Organomet. Chem.*, **510**, 181 (1996).
106. C. Wendler and H. Oehme, *Z. Anorg. Allg. Chem.*, **622**, 801 (1996).
107. C. Krempner, H. Reinke and H. Oehme, *Chem. Ber.*, **128**, 143 (1995).
108. C. Krempner, H. Reinke and H. Oehme, *Chem. Ber.*, **128**, 1083 (1995).
109. F. Luderer, H. Reinke and H. Oehme, *Chem. Ber.*, **129**, 15 (1996).
110. C. Krempner, D. Hoffmann, H. Oehme and R. Kempe, *Organometallics*, **16**, 1828 (1997).
111. Y. Apeloig, M. Bendikov, M. Yuzefovich, M. Nakash, D. Bravo-Zhivotovskii, D. Bläser and R. Boese, *J. Am. Chem. Soc.*, **118**, 12228 (1996).
112. J. Ohshita, Y. Masaoka and M. Ishikawa, *Organometallics*, **10**, 3775 (1991).
113. J. Ohshita, Y. Masaoka, M. Ishikawa and T. Takeuchi, *Organometallics*, **12**, 876 (1993).
114. J. Ogasawara, K. Sakamoto and M. Kira, *30th Organo-silicon Symposium*, London, Ontario, Canada, May 1997, Abstr. P53.
115. D. Hoffmann, H. Reinke and H. Oehme, *J. Organomet. Chem.* **526**, 185 (1996).
116. A. G. Brook, P. Chiu, J. McClenaghnan and A. J. Lough, *Organometallics*, **10**, 3292 (1991).
117. C. Krempner, H. Reinke and H. Oehme, *Angew. Chem., Int. Ed. Engl.*, **33**, 1615 (1994).
118. D. Hoffmann, H. Reinke and H. Oehme, *Z. Naturforsch.*, **51b** 371 (1996).
119. K. Sakamoto. J. Ogasawara, H. Sakurai and M. Kira, *J. Am Chem. Soc.*, **119**, 3405 (1997).
120. J. Ohshita, Y. Masaoka, S. Masaoka, M. Ishikawa, A. Tachibana, T. Yano and T. Yamabe, *J. Organomet. Chem.*, **473**, 15 (1994).
121. J. Ohshita, S. Masaoka, Y. Masaoka, H. Hasebe, M. Ishikawa, A. Tachinaba, T. Yano and T. Yamabe, *Organometallics*, **15**, 3136 (1996).
122. I. S. Biltueva, D. A. Bravo-Zhivotovskii, I. D. Kalikhman, V. Y. Vitkovskii, S. G. Shevchenko, N. S. Vyazankin and M. G. Voronkov, *J. Organomet. Chem.*, **368**, 163 (1989).
123. J. Ohshita, S. Masaoka, Y. Morimoto and M. Ishikawa, *Organometallics*, **16**, 910 (1997).
124. M. Ishikawa, T. Fuchikami, T. Sugaya and M. Kumada, *J. Am. Chem. Soc.*, **97**, 5923 (1975).
125. M. Ishikawa and M. Kumada, *Adv. Organomet. Chem.*, **19**, 51 (1981).
126. K. Nate, M. Ishikawa, H. Ni, H. Watanabe and Y. Saheki, *Organometallics*, **6**, 1673 (1987).
127. M. Ishikawa, K. Watanabe, H. Sakamoto and A. Kunai, *J. Organomet. Chem.*, **455**, 61 (1993).
128. M. Ishikawa, H. Sakamoto and F. Kanetani, *Organometallics*, **8**, 2767 (1989).
129. M. Ishikawa, Y. Nishimura and H. Sakamoto, *Organometallics*, **10**, 2701 (1991).
130. M. Ishikawa, K. Watanabe, H. Sakamoto and A. Kunai, *J. Organomet. Chem.*, **435**, 249 (1992).
131. J. Ohshita, H. Ohsaki, M. Ishikawa, A. Tachibana, Y. Kurosaki and T. Yamabe, *Organometallics*, **10**, 880 (1991).
132. J. Ohshita, H. Ohsaki, M. Ishikawa, A. Tachibana, Y. Kurosaki, T. Yamabe, T. Tsukihara, K. Kiso, and Y. Takahashi, *Organometallics*, **10**, 2685 (1991).
133. M. Ishikawa, Y. Nishimura and H. Sakamoto, *Organometallics*, **10**, 2701 (1991).
134. K. Takaki, H. Sakamoto, Y. Nishimura, Y. Sugihara and M. Ishikawa, *Organometallics*, **10**, 888 (1991).
135. G. W. Sluggett and W. J. Leigh, *J. Am. Chem. Soc.*, **114**, 1195 (1992).

136. G. W. Sluggett, and W. J. Leigh, *Organometallics*, **13**, 1005 (1994).
137. W. J. Leigh and G. W. Sluggett, *Organometallics*, **13**, 269 (1994).
138. M. Kira, T. Miyazawa, H. Sugiyama, M. Yamaguchi and H. Sakurai, *J. Am. Chem. Soc.*, **115**, 3116 (1993).
139. G. W. Sluggett and W. J. Leigh, *Organometallics*, **11**, 3731 (1992).
140. W. J. Leigh and G. W. Sluggett, *J. Am. Chem. Soc.*, **115**, 7531 (1993).
141. M. Steinmetz and C. Yu. *J. Org. Chem.*, **57**, 3107 (1992).
142. M. Steinmetz and Q. Chen, *J. Chem. Soc., Chem. Commun.*, 133 (1995).
143. M. Kira, T. Maruyama and H. Sakurai, *J. Am. Chem. Soc.*, **113**, 3986 (1991).
144. H. Sakurai, H. Sugiyama and M. Kira, *J. Phys. Chem.*, **94**, 1837 (1990).
145. (a) J. Braddock-Wilking, M. Y. Chiang and P. P. Gaspar, *Organometallics*, **12**, 197 (1993).
    (b) D. Pae, M. Xiao, M. Chiang and P. P. Gaspar, *J. Am. Chem. Soc.*, **113**, 1281 (1991).
146. H. F. Schaefer III, *Acc. Chem. Res.*, **15**, 283 (1982).
147. B. T. Luke, J. A. Pople, M.-B. Krogh-Jespersen, Y. Apeloig, M. Karni, J. Chandrasekhar and P. v. R. Schleyer, *J. Am. Chem. Soc.*, **108**, 270 (1986).
148. C. F. Melius and M. D. Allendorf, *J. Phys. Chem.*, **96**, 428 (1992).
149. H. M. Frey, B. P. Mason, R. Walsh and R. Becerra, *J. Chem. Soc., Faraday Trans.*, **89**, 411 (1993).
150. S. K. Shin, K. K. Irikura, J. L. Beauchamp and W. A. Goddard, *J. Am. Chem. Soc.*, **110**, 24 (1988).
151. R. S. Grev, G. E. Scuseria, A. C. Scheiner, H. F. Schaefer III and M. S. Gordon, *J. Am. Chem. Soc.*, **110**, 7337 (1988).
152. T. Kudo and S. Nagase, *J. Chem. Soc., Chem. Commun.*, 141 (1984).
153. H. Vancik, G. Raabe, M. J. Michalczyk, R. West and J. Michl *J. Am. Chem. Soc.*, **107**, 4097 (1985).
154. C. A. Arrington, K. A. Klingensmith, R. West and J. Michl, *J. Am. Chem. Soc.*, **106**, 525 (1984).
155. T. J. Drahnak, J. Michl and R. West, *J. Am. Chem. Soc.*, **103**, 1845 (1981).
156. G. Raabe, H. Vancik, R. West and J. Michl, *J. Am. Chem. Soc.*, **108**, 671 (1986).
157. C.-C. Chang, J. Kolc, M. E. Jung, J. A. Lowe and O. L. Chapman, *J. Am. Chem. Soc.*, **98**, 7846 (1976).
158. H.-P. Reisenauer, G. Mihm and G. Maier, *Angew. Chem., Int. Ed. Engl.*, **21**, 854 (1982).
159. V. N. Khabashesku, V. Balaji, S. E. Boganov, O. M. Nefedov and J. Michl, *J. Am. Chem. Soc.*, **116**, 320 (1994).
160. P. P. Gaspar, R.-J. Hwang and W. C. Eckelman, *J. Chem. Soc., Chem. Commun.*, 242 (1974).
161. M. J. Fink, D. B. Puranik and M. P. Johnson, *J. Am. Chem. Soc.*, **110**, 1315 (1988).
162. D. B. Puranik and M. J. Fink, *J. Am. Chem. Soc.*, **111**, 5951 (1989).
163. M. J. Fink, in *Frontiers of Organosilicon Chemistry* (Eds. A. R. Bassindale and P. P. Gaspar), Royal Society of Chemistry, Cambridge, 1991, p. 285.
164. A. Sekiguchi and W. Ando, *Organometallics*, **6**, 1857 (1987).
165. A. Sekiguchi, T. Sato and W. Ando, *Organometallics*, **6**, 2337 (1987).
166. G. Maas and A. Fronda, *J. Organomet. Chem.*, **398**, 229 (1990).
167. M. Fronda and G. Maas, *Angew. Chem., Int. Ed. Engl.*, **28**, 1663 (1989).
168. W. Ando, M. Sugiyama, T. Suzuki, C. Kato, Y. Arakawa and Y. Kabe, *J. Organomet. Chem.*, **499**, 99 (1995).
169. W. Ando, H. Yoshida. K. Kurishima and M. Sugiyama, *J. Am. Chem. Soc.*, **113**, 7790 (1991).
170. G. Märkl and W. Schlosser, *Tetrahedron Lett.*, **29**, 467 (1988).
171. N. Wiberg and S. Vasisht, *Angew. Chem., Int. Ed. Engl.*, **30**, 93 (1991).
172. N. Wiberg and S. Wagner, *Z. Naturforsch.*, **51b**, 838 (1996).
173. N. Wiberg and H. S. Hwangpark, *J. Organomet. Chem.*, **519**, 107 (1996).
174. N. Wiberg, H. S. Hwangpark, P. Mikulcik and G. Müller, *J. Organomet. Chem.*, **511**, 239 (1996).
175. M. Chaubon, J. Escudié, H. Ranaivonjatovo and J. Satgé, *J. Chem. Soc., Dalton Trans.*, **6**, 893 (1996).
176. N. Wiberg and S. Wagner, *Z. Naturforsch.*, **51b**, 629 (1996).
177. M. Takahashi, J. Ogasawara, K. Sakamoto, M. Kira and T. Veszpremi, submitted.
178. W. Ziche, N. Auner and P. Kiprof, *J. Am. Chem. Soc.*, **114**, 4910 (1992).
179. W. Ziche, N. Auner and J. Behm, *Organometallics*, **11**, 3805 (1992).
180. K. M. Baines and A. G. Brook, *Adv. Organomet. Chem.*, **25**, 1 (1986).
181. A. G. Brook and H.-J. Wessely, *Organometallics*, **4**, 1487 (1985).

182. U. Klingebiel, S. Pohlmann and L. Skoda, *J. Organomet. Chem.*, **291**, 277 (1985).
183. R. T. Zhang, S. Conlin, M. Namavari, K. L. Bobbitt and M. J. Fink, *Organometallics*, **8**, 571 (1989).
184. M. Ishikawa, D. Kovar, T. Fuchikami, K. Nishimura, M. Kumada, T. Higuchi and S. Miyamoto, *J. Am. Chem. Soc.*, **103**, 2324 (1981).
185. M. Ishikawa and S. Matsuzawa, *J. Chem. Soc., Chem. Commun.*, 588 (1985).
186. E. T. Seidl, R. S. Grev and H. F. Schaefer III, *J. Am. Chem. Soc.*, **114**, 3643 (1992).
187. F. Bernardi, A. Bottoni, M. Olivucci, M. A. Robb and A. Venturino, *J. Am. Chem. Soc.*, **115**, 3322 (1993).
188. F. Bernardi, A. Bottoni, M. Olivucci, A. Venturini and M. A. Robb, *J. Chem. Soc. Faraday Trans.*, **90**, 1617 (1994).
189. W. A. Howard, G. L. McPherson, M. J. Fink, and J. R. Gee, *J. Am. Chem. Soc.*, **113**, 5461 (1991).
190. D. B. Puranik, M. P. Johnson and M. J. Fink, *J. Chem. Soc., Chem. Commun.*, 706 (1989).
191. D. Bravo-Zhivotovskii, personal communication to T. M., 1995.
192. M. Nishimura, K. Ishikawa, H. Ochial and M. Kum, *J. Organomet. Chem.*, **236**, 7 (1982).
193. M. Ishikawa, M. Kikuchi, H. Watanabe, H. Sakamoto and A. Kunai, *J. Organomet. Chem.*, **443**, C3 (1993).
194. M. Ishikawa, M. Kikuchi, A. Kunai, T. Takeuchi, T. Tsukihara and M. Kido, *Organometallics*, **12**, 3474 (1993).
195. G. Maas, K. Schneider and W. Ando, *J. Chem. Soc., Chem. Commun.*, 72 (1988).
196. K. Schneider, B. Daucher, A. Fronda and G. Maas, *Chem. Ber.*, **123**, 589 (1990).
197. N. Wiberg, G. Wagner, G. Müller and J. Riede, *J. Organomet. Chem.*, **271**, 381 (1984).
198. W. J. Leigh and G. W. Sluggett, *J. Am. Chem. Soc.*, **116**, 10468 (1994).
199. N. Auner, *Z. Anorg. Allg. Chem.*, **558**, 87 (1988).
200. C. Krempner, Ph. D. Thesis, University of Rostock, Rostock, FRG, 1996.
201. N. Wiberg, M. Link and G. Fischer, *Chem. Ber.*, **122**, 409 (1989).
202. A. MacMillan and A. G. Brook, *J. Organomet. Chem.*, **341**, C9 (1988).
203. N. Wiberg, G. Fischer and S. Wagner, *Chem. Ber.*, **124**, 769 (1991).
204. N. Wiberg, S. Wagner and G. Fischer, *Chem. Ber.*, **124**, 1981 (1991).
205. N. Auner, C. Seidenschwarz and N. Sewald, *Organometallics*, **11**, 1137 (1992).
206. N. Auner, W. Ziche and E. Herdtweck, *J. Organomet. Chem.*, **426**, 1 (1992).
207. N. Auner, *J. Organomet. Chem.*, **336**, 83 (1987).
208. N. Auner, J. Grobe, T. Schäfer, B. Krebs and M. Dartmann, *J. Organomet. Chem.*, **363**, 7 (1989).
209. N. Auner, C. Seidenschwarz and E. Herdtweck, *Angew. Chem., Int. Ed. Engl.*, **30**, 1151 (1991).
210. N. Auner, C.-R. Heikenwälder and C. Wagner, *Organometallics*, **12**, 4135 (1993).
211. W. Ziche, C. Seidenschwarz, N. Auner, E. Herdtweck and N. Sewald, *Angew. Chem., Int. Ed. Engl.*, **33**, 77 (1994).
212. N. Sewald, W. Ziche, A. Wolff and N. Auner, *Organometallics*, **12**, 4123 (1993).
213. N. Auner and A. Wolff, *Chem. Ber.*, **126**, 575 (1993).
214. N. Auner, C. Seidenschwarz, N. Sewald and E. Herdtweck, *Angew. Chem., Int. Ed. Engl.*, **30**, 444 (1991).
215. N. Auner and A. Wolff, *Organometallics*, submitted.
216. N. Auner and C. Seidenschwarz, *Z. Naturforsch.*, **45b**, 909 (1990).
217. N. Auner, A. W. Weingartner and G. Bertrand, *Chem. Ber.*, **126**, 581 (1993).
218. N. Auner and C.-R. Heikenwälder, *Organometallics*, submitted.
219. C.-R. Heikenwälder and N. Auner, in *Organosilicon Chemistry II — From Molecules to Materials* (Eds. N. Auner and J. Weis), VCH Verlag, Weinheim, 1996, p. 399.
220. N. Auner and C.-R. Heikenwälder, *Z. Naturforsch.*, **52b**, 500 (1997).
221. C.-R. Heikenwälder and N. Auner, in *Organosilicon Chemistry III — From Molecules to Materials* (Eds. N. Auner and J. Weis), VCH Verlag, Weinheim, 1997, p. 101.
222. W. Ziche, N. Sewald, C. Seidenschwarz, E. Herdtweck, V. Popkova and N. Auner, *Organometallics*, submitted.
223. J. Grobe, H. Schröder and N. Auner, *Z. Naturforsch.*, **45b**, 785 (1990).
224. N. Auner, C.-R. Heikenwälder and W. Ziche, *Chem. Ber.*, **126**, 2177 (1993).
225. N. Auner and E. Penzenstadler, *Z. Naturforsch.*, **47b**, 217 (1992).
226. N. Auner and E. Penzenstadler, *Z. Naturforsch.*, **47b**, 795 (1992).
227. N. Auner and E. Penzenstadler, *Z. Naturforsch.*, **47b**, 805 (1992).

228. N. Auner, E. Penzenstadler and E. Herdtweck, *Z. Naturforsch.*, **47b**, 1377 (1992).
229. N. Auner, A. W. Weingartner and E. Herdtweck, *Z. Naturforsch.*, **48b**, 318 (1993).
230. N. Auner, *J. Organomet. Chem.*, **377**, 175 (1989).
231. N. Auner and R. Gleixner, *J. Organometal. Chem.*, **393**, 33 (1990).
232. B. Goetze, B. Herrschaft and N. Auner, *Chem. Eur. J.*, **3**, 948 (1997).
233. W. Ziche, N. Auner and J. Behm, *Organometallics*, **11**, 2494 (1992).
234. N. Auner, C. Wagner and W. Ziche, *Z. Naturforsch.*, **49b**, 831 (1994).
235. A. G. Brook, W. J. Chatterton, J. F. Sawyer, D. W. Hughes and K. Vorspohl, *Organometallics*, **6**, 1246 (1987).
236. A. G. Brook, Y. Kun Kong, A. K. Saxena and J. F. Sawyer, *Organometallics*, **7**, 2245 (1988).
237. A. G. Brook, A. K. Saxena and J. F. Sawyer, *Organometallics*, **8**, 850 (1989).
238. (a) M. T. Nguyen, H. Vansweevelt, A. De Neef and L. G. Vanquickborne, *J. Org. Chem.*, **59**, 8015 (1994).
    (b) A. G. Brook, in *Heteroatom Chemistry* (Ed. E. Block), VCH verlag, Weinheim, **1990**, p. 105.
239. A. G. Brook, R. Kumarathasan and A. J. Lough, *Organometallics*, **13**, 424 (1994).
240. A. G. Brook, K. Vorspohl, R. R. Ford, M. Hesse and W. J. Chatterton, *Organometallics*, **6**, 2128 (1987).
241. D. Bravo-Zhivotovskii, Y. Apeloig, Y. Ovchinnikov, V. Igonin and Y. T. Struchkov, *J. Organomet. Chem.*, **446**, 123 (1993).
242. S. M. Bachrach and A. Streitwieser, *J. Am. Chem. Soc.*, **107**, 1186 (1985).
243. G. Märkl and M. Horn, *Tetrahedron, Lett.*, **14**, 1477 (1983).
244. A. G. Brook, S. S. Hu, W. J. Chatterton, and A. J. Lough, *Organometallics*, **10**, 2752 (1991).
245. A. G. Brook, S. S. Hu, A. K. Saxena and A. J. Lough, *Organometallics*, **10**, 2758 (1991).
246. (a) Y. Apeloig, and M. Karni, *J. Am. Chem. Soc.*, **106**, 6676 (1984).
    (b) V. Braude, Ph. D. Thesis, Technion, Haifa, Israel, 1993.
247. A. G. Brook, R. Kumarathasan and W. Chatterton, *Organometallics*, **12**, 4085 (1993).
248. A. G. Brook, W. J. Chatterton and R. Kumarathasan, *Organometallics*, **12**, 3666 (1993).
249. N. Toltl and W. Leigh, *Organometallics*, **15**, 2554 (1996).
250. R. Conlin and M. Namavari, *J. Organomet. Chem.*, **376**, 259 (1989).
251. J. Ohshita, H. Niwa and M. Ishikawa, *Organometallics*, **15**, 4632 (1996).
252. W. Sander and M. Trommer, *Chem. Ber.*, **125**, 2813 (1992).
253. M. Trommer, W. Sander and A. Patyk, *J. Am. Chem. Soc.*, **115**, 1175 (1993).
254. V. V. Volkova, V. G. Avakyan, N. S. Nametkin and L. E. Gusel'nikov, *J. Organomet. Chem.*, **201**, 137 (1980).
255. O. M. Nefedov, A. K. Maltsev, V. N. Khabashesku and V. A. Korolev, *J. Organomet. Chem.*, **201**, 123 (1980).
256. G. Mihm, H. P. Reisenauer, D. Littmann and G. Maier, *Chem. Ber.*, **117**, 2369 (1984).
257. J. Chen, P. J. Hajduk, J. D. Keen, C. Chuang, T. Emilsson and H. S. Gutowsky, *J. Am. Chem. Soc.*, **113**, 4747 (1991).
258. J. Chen, P. J. Hajduk, J. D. Keen, T. Emilsson and H. S. Gutowsky, *J. Am. Chem. Soc.*, **111**, 1901 (1989).
259. M. Trommer, G. E. Miracle, D. R. Powell and R. West, *Organometallics*, in press (1997).
260. H. Sohn, U. Bankwitz, J. Calabrese, Y. Apeloig, T. Müller and R. West, *J. Am. Chem. Soc.*, **117**, 11608 (1995).
261. T. D. Tilley, G. P. A. Yap, A. L. Rheingold and W. P. Freeman, *Angew. Chem., Int. Ed. Engl.*, **35**, 976 (1996).
262. T. D. Tilley, L. M. Liable-Sands, A. L. Rheingold and W. P. Freeman, *J. Am. Chem. Soc.*, **118**, 10457 (1996).
263. R. Gutowsky, L. K. Montgomery and P. G. Mahaffy, *J. Am. Chem. Soc.*, **102**, 2854 (1980).
264. J.-P. Malrieu and G. Trinquier, *J. Am. Chem. Soc.*, **109**, 5303 (1987).
265. P. Boudjouk, S. Castellino and J-H. Hong, *Organometallics*, **13**, 3387 (1994).
266. R. West, H. Sohn, Y. Apeloig and T. Müller, in *Organosilicon Chemistry III—From Molecules to Materials* (Eds. N. Auner and J. Weis) VCH Verlag, Weinheim, 1997, p. 144.
267. P. v. R. Schleyer, F. Hampel and B. Goldfuss, *Organometallics*, **15**, 1755 (1996).
268. N. Wiberg, G. Wagner and G. Müller, *Angew. Chem., Int. Ed. Engl.*, **24**, 229 (1985).
269. R. Okazaki and R. West, *Adv. Organomet. Chem.*, **39**, 232 (1995).
270. J. D. Cavalieri, J. J. Buffy, C. Fry, K. W. Zilm, J. C. Duchamp, M. Kira, T. Iwamoto, T. Müller, Y. Apeloig and R. West, *J. Am. Chem. Soc.*, **119**, 4972 (1997).

271. G. E. Miracle, J. L. Ball, S. R. Bielmeier, D. R. Powell and R. West, in *Progress in Organosilicon Chemistry* (Eds. B. Mariciniec and J. Chojnowski) Gordon and Breach Basel, 1995, pp. 83–100.
272. K. Wakitu, N. Tokitoh, R. Okazaki and S. Nagase, 30th Organosilicon Symposium, London, Ontario. Canada, May 1997, Abstr. P-52.
273. P. R. Jones and M. E. Lee, *J. Organomet. Chem.*, **271**, 299 (1984).
274. W. Schlosser and G. Märkl, *Angew. Chem., Int. Ed. Engl.*, **27**, 963 (1988).
275. R. Withnall and L. Andrews, *J. Phys. Chem.*, **92**, 594 (1988).
276. G. D. Josland, R. A. Lewis, A. Morris and J. M. Dyke, *J. Phys. Chem.*, **86**, 2913 (1982).
277. W. Ando and A. Sekiguchi, *Chem. Lett.*, 2025 (1986).
278. C. Kerst, C. W. Rogers, R. Ruffolo and W. Leigh, *J. Am. Chem. Soc.*, **119**, 466 (1997).
279. G. Mihm, H. P. Reisenauer and G. Maier, *Angew. Chem., Int. Ed. Engl.*, **19**, 52 (1980).
280. G. Mihm, H. P. Reisenauer and G. Maier, *Chem. Ber.*, **115**, 801 (1982).
281. G. Mihm, R. O. W. Baumgärtner, H. P. Reisenauer and G. Maier, *Chem. Ber.*, **117**, 2337 (1984).
282. K. Schöttler, H. P. Reisenauer and G. Maier, *Tetrahedron Lett.*, **26**, 4079 (1985).
283. O. L. Chapman, G. T. Burns, T. J. Barton and C. L. Kreil, *J. Am. Chem. Soc.*, **102**, 841 (1980).
284. C. Allison and T. McMahon, *J. Am. Chem. Soc.*, **112**, 1672 (1990).
285. (a) T. Müller and W. Ziche, unpublished results.
    (b) G. Maier, H. Pacl and H. P. Reisenauer, *Angew. Chem., Int. Ed. Engl.*, **34**, 1439 (1995).
286. J. Pacansky, N. Honjou and M. Yoshimine, *J. Am. Chem. Soc.*, **111**, 4198 (1989).
287. M. Ishikawa, H. Sugisawa, T. Fuchikami, M. Kumada, T. Yamabe, H. Kawakami, K. Fukui, Y. Ueki and H. Shizuka, *J. Am. Chem. Soc.*, **104**, 2872 (1982).
288. M. Ishikawa, H. Sugisawa, M. Kumada, T. Higichi, K. Matsui, K. Hirotsu and J. Iyoda, *Organometallics*, **2**, 174 (1983).
289. M. Ishikawa, T. Horio, Y. Yuzuriha and A. Kunai, *J. Organomet, Chem.*, **402**, C20 (1991).
290. M. Ishikawa, T. Horio, Y. Yuzuriha, A. Kunai, T. Tsukihara and H. Naitou, *Organometallics*, **11**, 597 (1992).
291. J. Ohshita, Y. Isomure and M. Ishikawa, *Organometallics*, **8**, 2050 (1989).
292. S. Landor, *The Chemistry of Allenes*, Academic Press, London, 1982.
293. K. Krogh-Jespersen, *J. Comput. Chem.*, **3**, 571 (1982).
294. G. Manuel, P. Mazerolles and G. Bertrand, *Tetrahedron*, **34**, 1951 (1978).
295. M. Urbanova, E. A. Volnina, L. E. Gusel'nikov, Z. Bastl and J. Pola, *J. Organomet. Chem.*, **509**, 73 (1996).
296. (a) M. Jakoubková, R. Fajgar, J. Tláskal and J. Pola, *J. Organomet. Chem.*, **466**, 29 (1994) and references cited there.
    (b) A. Sekiguchi, S. S. Ziegler, R. West and J. Michl, *J. Am. Chem. Soc.*, **108**, 4241 (1986).
    (c) A. Sekiguchi, G. R. Gillette and R. West, *Organometallics*, **7**, 1226 (1988).
297. (a) M. T. Nguyen, D. Sengupta and L. G. Vanquickenborne, *Chem. Phys. Lett.*, **244**, 83 (1995).
    (b) R. Stegmann and G. Frenking, *J. Comput. Chem.*, **17**, 781 (1996).
    (c) Y. Apeloig and M. Karni, *Organometallics*, **16**, 310 (1997).
298. (a) M. Izuha, S. Yamamoto and S. Saito, *J. Chem. Phys.*, **105**, 4923 (1996).
    (b) D. G. Leopold and A. A. Bengali, unpublished results, cited in C. D. Sherill and H. F. Schaefer III, *J. Phys. Chem.*, **99**, 1949 (1995).
    For mass spectroscopic evidence for $H_2C=Si\colon$ see
    (c) R. Srinivas, D. Sülzle and H. Schwarz *J. Am. Chem. Soc.*, **113**, 52 (1991).
    (d) R. Damrauer, C. H. DePuy, S. E. Barlow and S. Gronert, *J. Am. Chem. Soc.*, **110**, 2005 (1988).
299. (a) *Silicon Nitrogen Compounds in Gmelin Handbook of Inorganic Chemistry* (Ed. F. Schröder) 8th ed., Springer Verlag, Berlin, 1989.
    (b) I. Hemme and U. Klingebiel, *Adv. Organomet. Chem.*, **39**, 159 (1995).
300. N. Wiberg and K. Schurz, *J. Organomet. Chem.*, **341**, 145 (1988).
301. N. Wiberg, K. Schurz, G. Reber and G. Müller, *J. Chem. Soc., Chem. Commun.*, 591 (1986).
302. N. Wiberg, K. Schurz and G. Fischer, *Angew. Chem., Int. Ed. Engl.*, **24**, 1053 (1985).
303. N. Wiberg and K. Schurz, *Chem. Ber.*, **121**, 581 (1988).
304. M. Hesse and U. Klingebiel, *Angew. Chem., Int. Ed. Engl.*, **25**, 649 (1986).
305. D. Stalke, N. Keweloh, U. Klingebiel, M. Noltemeyer and G. M. Sheldrick, *Z. Naturforsch.*, **42b**, 1237 (1987).
306. S. Walter, U. Klingebiel and D. Schmidt-Bäse, *J. Organomet. Chem.*, **412**, 319 (1991).

307. D. Grosskopf, L. Marcus, U. Klingebiel and M. Noltemeyer, *Phosphorus, Sulfur, Silicon Relat. Elem.*, **97**, 113 (1994).
308. N. Wiberg, P. Karampatses and C.-K. Kim, *Chem. Ber.*, **120**, 1203 (1987).
309. N. Wiberg and G. Preiner, *Angew. Chem., Int. Ed. Engl.*, **17**, 362 (1978).
310. N. Wiberg, P. Karampatses and Ch.-K. Kim, *Chem. Ber.*, **120**, 1213 (1987).
311. K. Tamao, Y. Nakagawa and Y. Ito, *J. Am. Chem. Soc.*, **114**, 218 (1992).
312. A. Weingartner, W. Ziche and N. Auner, in *Organosilicon Chemistry: From Molecules to Materials* (Eds. N. Auner and J. Weis), VCH, Weinheim, 1994, p. 115.
313. M. Letulle, A. Systermans, J.-L. Ripoll and P. Guenot, *J. Organomet. Chem.*, **484**, 89 (1994).
314. S. Cradock and J. F. Ogilive, *J. Chem. Soc., Chem. Commun.*, 364 (1966).
315. H. Bock and R. Dammel, *Angew. Chem., Int. Ed. Engl.*, **24**, 111 (1985).
316. C. Guimon and G. Pfister-Guillouzo, *Organometallics*, **6**, 1387 (1987).
317. S. S. Zigler, L. M. Johnson and R. West, *J. Organomet. Chem.*, **341**, 187 (1988).
318. R. Abramovich and E. Kyba, *J. Am. Chem. Soc.*, **93**, 1537 (1971).
319. E. Kyba and R. Abramovich, *J. Am. Chem. Soc.*, **102**, 735 (1980).
320. M. T. Nguyen, M. Faul and N. J. Fitzpatrick, *J. Chem. Soc., Perkin Trans. 2*, 1289 (1987).
321. J. G. Radziszewski, P. Kaszynski, D. Littmann, V. Balaji, B. A. Hess and J. Michl, *J. Am. Chem. Soc.*, **115**, 8401 (1993).
322. M. Elseikh and L. Sommer, *J. Organomet. Chem.*, **186**, 301 (1980).
323. W. Jevons, *Proc. R. Soc. London, Ser. A*, **89**, 187 (1913).
324. R. S. Mulliken, *Phys. Rev.*, **26**, 319 (1925).
325. S. Saito, Y. Endo and E. Hirota, *J. Chem. Phys.*, **78**, 6447 (1983).
326. J. G. Radziszewski, D. Littmann, V. Balaji, L. Fabry, G. Gross and J. Michl, *Organometallics*, **12**, 4816 (1993).
327. G. Reber, J. Riede, N. Wiberg, K. Schurz and G. Müller, *Z. Naturforsch.*, **44**, 786 (1989).
328. G. Maier and J. Glatthaar, *Angew. Chem., Int. Ed. Engl.*, **33**, 473 (1994).
329. G. Maier, J. Glatthaar and H. Reisenauer, *Chem. Ber.*, **122**, 2403 (1989).
330. G. Maier and J. Glatthaar, in *Organosilicon Chemistry: From Molecules to Materials* (Eds. N. Auner and J. Weis), VCH, Weinheim, 1994, p. 131.
331. T. J. Barton and S. Bain, *Organometallics*, **7**, 528 (1988).
332. J.-M. Denis, P. Guenot, M. Letulle, B. Pellerin and J.-L. Ripoll, *Chem. Ber.*, **125**, 1397 (1992).
333. A. Chive, V. Lefevre, A. Systermans, J.-L. Ripoll, M. Bogey and A. Walters, *Phosphorus, Sulfur, Silicon*, **91**, 281 (1994).
334. M. Weidenbruch, B. Brand-Roth, S. Pohl and W. Saak, *J. Organomet. Chem.*, **379**, 217 (1989).
335. M. Denk and R. West *Pure Appl Chem.* **68**, 785 (1996).
336. M. Elhanine, R. Farrenq and G. Guelachvili, *J. Chem. Phys.*, **94**, 2529 (1991).
337. R. Damrauer, M. Krempp and R. A. J. O'Hair, *J. Am. Chem. Soc.*, **115**, 1998 (1993).
338. J. A. Hankin, M. Krempp and R. Damrauer, *Organometallics*, **14**, 2652 (1995).
339. D. K. Bohme, *Int. J. Mass Spectrom. Ion Processes*, **100**, 719 (1990).
340. D. K. Bohme, *Chem. Rev.*, **92**, 1487 (1992).
341. S. Wlodek, C. F. Rodriguez, M. H. Lien, A. C. Hopkinson and D. K. Bohme, *Chem. Phys. Lett.*, **143**, 385 (1988).
342. N. Goldberg, M. Iraqi, J. Hrušúk and H. Schwarz, *Int. J. Mass Spectrom. Ion Processes*, **125**, 267 (1993).
343. S. S. Zigler, K. M. Welsh and R. West, *J. Am. Chem. Soc.*, **109**, 4392 (1987).
344. W. Ando, H. Tsumaki and M. Ikeno, *J. Chem. Soc., Chem. Commun.*, 597 (1981).
345. K. M. Welsh, J. Michl and R. West, *J. Am. Chem. Soc.*, **110**, 6689 (1988).
346. G. E. Underiner and R. West. *Angew. Chem., Int. Ed. Engl.*, **29**, 529 (1990).
347. G. E. Underiner, R. Tan, D. Powell and R. West, *J. Am. Chem. Soc.*, **113**, 8437 (1991).
348. L. J. Procopio, P. J. Carroll and D. H. Berry, *Organometallics*, **12**, 3087 (1993).
349. L. J. Procopio, P. J. Carroll and D. H. Berry, *Polyhedron*, **14**, 45 (1995).
350. L. J. Procopio, P. J. Carroll and D. H. Berry, *J. Am. Chem. Soc.*, **113**, 1870 (1991).
351. N. Wiberg, G. Preiner, P. Karampatses, C.-K. Kim and K. Schurz, *Chem. Ber.*, **120**, 1357 (1987).
352. N. Wiberg, E. Kühnel, K. Schurz, H. Borrmann and A. Simon, *Z. Naturforsch.*, **43**, 1075 (1988).
353. D. Grosskopf, U. Klingebiel, T. Belgardt and M. Noltemeyer, *Phosphorus, Sulfur, Silicon*, **91**, 241 (1994).
354. S. Walter, U. Klingebiel and M. Noltemeyer, *Chem. Ber.*, **125**, 783 (1992).
355. S. Vollbrecht, U. Klingebiel and D. Schmidt-Bäse, *Z. Naturforsch.*, **46**, 709 (1991).

356. N. Wiberg, G. Preiner and K. Schurz, *Chem. Ber.*, **121**, 1407 (1988).
357. R. Boese and U. Klingebiel, *J. Organomet. Chem.*, **315**, C17 (1986).
358. D. Stalke, U. Piper, S. Vollbrecht and U. Klingebiel, *Z. Naturforsch.*, **45b**, 1513 (1990).
359. U. Pieper, S. Walter, U. Klingebiel and D. Stalke, *Angew. Chem., Int. Ed. Engl.*, **29**, 209 (1990).
360. H. Chen, R. A. Bartlett, H. V. R. Dias, M. M. Olmstead and P. P. Power, *J. Am. Chem. Soc.*, **111**, 4338 (1989).
361. M. Driess and R. Janoschek, *J. Mol. Structure (Theochem)*, **119**, 129 (1994).
362. P. Marshall, *Chem. Phys. Lett.*, **201**, 493 (1993).
363. C. F. Melius and P. Ho, *J. Phys. Chem.*, **95**, 1410 (1991).
364. A. Korkin, *Int. J. Quantum Chem.*, **38**, 245 (1990).
365. M. S. Gordon, M. W. Schmidt and S. Koseki, *Inorg. Chem.*, **28**, 2161 (1989).
366. D. P. Chong, D. Papousek, Y.-T. Chen and P. Jensen, *J. Chem. Phys.*, **98**, 1352 (1992).
367. M. S. El-Shall, *Chem. Phys. Lett.*, **159**, 21 (1989).
368. Y. Apeloig and K. Albrecht, *J. Am. Chem. Soc.*, **117**, 7263 (1995).
369. J. R. Flores and J. Largo-Cabrerizo, *J. Mol. Structure (Theochem)*, **52**, 17 (1989).
370. O. Parisel, M. Hanus and Y. Ellinger, *J. Chem. Phys.*, **104**, 1979 (1996).
371. M. Sana, M. Decrem, G. Levroy, M. T. Nguyen and L. G. Vanquickenborne, *J. Chem. Soc., Faraday Trans.*, **90**, 3505 (1994).
372. K. Fan and S. Iwata, *Chem. Phys. Lett.*, **195**, 475 (1992).
373. T. N. Truong and M. S. Gordon, *J. Am. Chem. Soc.*, **108**, 1775 (1986).
374. P. v. R. Schleyer and P. D. Stout, *J. Chem. Soc., Chem. Commun.*, 1373 (1986).
375. W. Ando, T. Ohtaki and Y. Kabe, *Organometallics*, **13**, 434 (1994).
376. M. Driess, *Coord. Chem. Rev.*, **145**, 1 (1995).
377. M. Driess, *Adv. Organomet Chem.*, **39**, 183 (1995).
378. C. Smit and F. Bickelhaupt, *Phosphorus, Sulfur, Silicon*, **30**, 357 (1987).
379. Y. v. d. Winkel, H. Bastiaans and F. Bickelhaupt, *Phosphorus, Sulfur, Silicon*, **49–50**, 333 (1990).
380. C. Couret, J. Escudié, J. Satgé, J. Andriamizaka and B. Saint-Roch *J. Organomet. Chem.*, **182**, 9 (1979).
381. V. Lefevre, J. Ripoll, Y. Dat, S. Joanteguy, V. Metail, A. Chrostowskasenio and G. Pfisterguillouzo, *Organometallics*, **16**, 1635 (1997).
382. U. Klingebiel, R. Boese, D. Bläser, and M. Andrianarison, *Z. Naturforsch.*, **44b**, 265 (1989).
383. N. Wiberg and H. Schuster, *Chem. Ber.*, **124**, 93 (1991).
384. H. Bender, E. Klein, E. Niecke, M. Nieger and H. Ranaivonjatovo, in *Organosilicon Chemistry From Molecules to Materials* (Eds. N. Auner and J. Weis), VCH, Weinheim, 1994, p. 143.
385. H. Bender, E. Niecke and M. Nieger, *J. Am. Chem. Soc.*, **115**, 3314 (1993).
386. E. Niecke, E. Klein and M. Nieger, *Angew. Chem., Int. Ed. Engl.* **28**, 751 (1989).
387. Y. v. d. Winkel, H. Bastiaans and F. Bickelhaupt, *J. Organomet. Chem.*, **405**, 183 (1991).
388. R. Corriu, G. Lanneau and C. Priou, *Angew. Chem., Int. Ed. Engl.*, **30**, 1130 (1991).
389. M. Driess, *Angew. Chem., Int. Ed. Engl.*, **30**, 1022 (1991).
390. M. Driess, H. Pritzkow, S. Rell and U. Winkler, *Organometallics*, **15**, 1845 (1996).
391. M. Driess, S. Rell and H. Pritzkow, *J. Chem. Soc., Chem. Commun.*, 253 (1995).
392. M. Driess, H. Pritzkow and U. Winkler, *J. Organomet. Chem.*, **529**, 313 (1997).
393. M. Driess and H. Pritzkow, *Angew. Chem., Int. Ed. Engl.*, **31**, 316 (1992).
394. A. Baboul and H. Schlegel, *J. Am. Chem. Soc.*, **118**, 8444 (1996).
395. M. Nguyen, A. Vankeer and L. Vanquickenborne, *J. Organomet. Chem.*, **529**, 3 (1997).
396. W. Schoeller and T. Busch, *Chem. Ber.*, **125**, 1319 (1992).
397. W. Schoeller and T. Busch, *J. Chem. Soc., Chem. Commun.*, 234 (1989).
398. W. Schoeller, J. Strutwolf, U. Tubbesing and C. Begemann, *J. Phys. Chem.*, **99**, 2329 (1995).

CHAPTER **17**

Recent advances in the chemistry of silicon–heteroatom multiple bonds

NORIHIRO TOKITOH and RENJI OKAZAKI

Department of Chemistry, Graduate School of Science, The University of Tokyo, 7-3-1 Hongo, Bunkyo-ku, Tokyo 113, Japan
Fax: +81-3-5800-6899

I. INTRODUCTION	1064
A. Multiple Bonds to Silicon	1064
B. Theoretical Calculations	1065
1. Silanones and silanethiones	1065
2. Other double bonds to silicon	1067
3. Miscellaneous	1067
II. DOUBLE BONDS BETWEEN SILICON AND GROUP 16 ELEMENTS	1068
A. Introduction	1068
B. Silicon–Oxygen Double Bonds (Silanones)	1068
1. Formation from hydrosilanes	1068
2. Formation from cyclic silyl ethers	1069
3. Formation from allyloxysilane derivatives	1071
4. Formation from silanol derivatives	1074
5. Formation from silylenes	1075
6. Formation from silenes and silicon–heteroatom double-bond compounds	1080
C. Silicon–Sulfur Double Bonds (Silanethiones)	1083
1. Formation from hydrosilanes	1084
2. Formation from diaminosilanes	1086
3. Formation from silylenes	1086
4. Formation from silicon-containing cyclic polysulfides	1090
5. Formation from a disilene	1094
6. Formation from thioketenes	1095
7. Matrix isolation of SiS_2, Cl(H)Si=S and $Cl_2Si=S$	1096

The chemistry of organic silicon compounds, Vol. 2
Edited by Z. Rappoport and Y. Apeloig © 1998 John Wiley & Sons Ltd

D. Silicon–Selenium and Silicon–Tellurium Double Bonds (Silaneselones and Silanetellones) 1097
  1. Transient silaneselones. 1097
  2. Silaneselone derived from silicocene. 1099
  3. Thermodynamically stabilized silaneselone. 1099
  4. Sterically protected silaneselone stable in solution. 1099
III. OUTLOOK FOR THE CHEMISTRY OF SILICON–HETEROATOM MULTIPLE BONDS 1100
IV. REFERENCES 1101

I. INTRODUCTION

A. Multiple Bonds to Silicon

For the past decades, it was considered that double-bond compounds of heavier main-group elements would be neither stable nor synthetically accessible because of their weak $p\pi$–$p\pi$ bonding (the so-called 'double-bond rule'). However, since the isolation of the first stable doubly-bonded compounds containing heavier group 14 or 15 elements such as silene (Si=C)[1], disilene (Si=Si)[2] and diphosphene (P=P)[3], remarkable progress has been made in the chemistry of unsaturated compounds of these elements, especially in the field of group 14 metals[4]. Thus far, as for the chemistry of multiply-bonded compounds to silicon, the syntheses and characterization of stable iminosilanes (Si=N)[5], phosphasilenes (Si=P)[6] and their heavier congeners[7] have been reported in addition to those of a variety of stable silenes and disilenes[8]. In most cases, these novel low-coordinate silicon compounds were synthesized and isolated by taking advantage of kinetic stabilization with bulky substituents, i.e. steric protection. With this stabilization method the highly reactive molecules can be isolated with only small electronic perturbation, and hence one can reveal the intrinsic nature of the newly obtained multiple bonds which will be of great importance for the investigation of their non-stabilized, transient analogues.

On the other hand, the chemistry of doubly-bonded compounds between silicon and group 16 elements (chalcogen atoms) has been less explored due to their much higher reactivity than that of the above-mentioned silicon-containing multiple bonds to group 14 and 15 elements. This results from the fact that bulky substituents for steric protection can be introduced only on the silicon atom and hence their oligomerization cannot be efficiently suppressed. Although thermodynamically stabilized silanethione (Si=S) and silaneselone (Si=Se)[9] have been reported, they are considerably perturbed by intramolecular coordination of a nitrogen substituent in the neighborhood of the double bond. Only recently was the first kinetically stabilized silicon–chalcogen doubly-bonded compound i.e. a stable silanethione with a new steric protection system, been successfully isolated[10].

Since several excellent reviews on the chemistry of multiple bonds to silicon [see the chapter written by G. Raabe and J. Michl in a previous volume of this series (1989)[4e]] and also the reviews in *Advances in Organometallic Chemistry*, Vol. 39 (1995) (e.g. 'The Chemistry of Silenes' by A. G. Brook and M. A. Brook[8a], 'Iminosilanes and Related Compounds—Synthesis and Reactions' by I. Hemme and U. Klingebiel[5d], 'Silicon–Phosphorus and Silicon–Arsenic Multiple Bonds' by M. Driess[7c] and 'Chemistry of Stable Disilenes' by R. Okazaki and R. West[8b]) have appeared in recent years and as some parts of this chapter will be discussed in detail in related chapters of this book (e.g. 'Silenes and Iminosilanes' by N. Auner and coworkers[11] and 'Recent Advances in the Chemistry of Disilenes' by H. Sakurai[12]), we will concentrate our attention in this chapter on the chemistry of silicon–chalcogen doubly-bonded compounds.

B. Theoretical Calculations

1. Silanones and silanethiones

In 1986, Nagase and coworkers reported a theoretical study of silanethione ($H_2Si=S$, **1**) in the ground, excited and protonated states together with a comparison with silanone ($H_2Si=O$, **2**) and the parent carbonyl system ($H_2C=O$, **3**) using *ab initio* calculations including polarization functions and electron correlation[13]. Scheme 1 shows unimolecular reactions pertinent to the stability of $H_2Si=S$ and Figure 1 shows the equilibrium structures on the ground singlet potential energy surface of H_2SiS species and a related silanethiol (H_3SiSH) obtained at the HF/6-31G* level of theory. Similar theoretical calculations for species on the $H_2S=O$ ground state potential surface are shown in Figure 2[14].

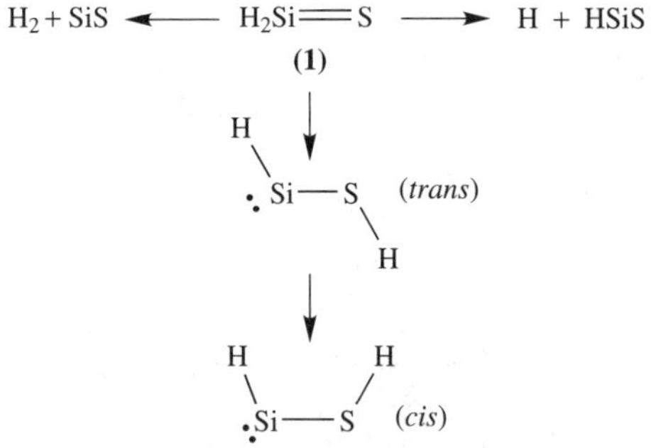

$$H_2 + SiS \longleftarrow H_2Si=S \longrightarrow H + HSiS$$
$$(1)$$

$$\overset{H}{\underset{H}{\cdot Si-S}} \quad (trans)$$

$$\overset{H\ \ \ \ H}{\underset{}{\cdot Si-S}} \quad (cis)$$

SCHEME 1

The equilibrium structure of $H_2Si=S$ is calculated to be planar with C_{2v} symmetry, as in the case of $H_2C=O$ and $H_2Si=O$. The Si—S bond length in **1** (1.936 Å) is by 0.438 and 0.752 Å longer than the Si—O and C—O bond lengths in **2** and **3**, respectively,

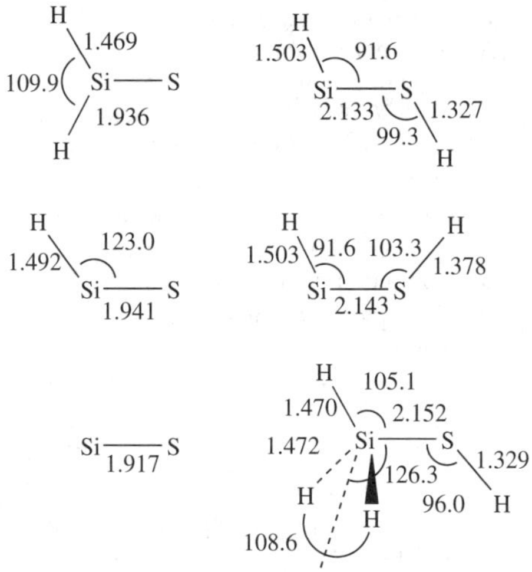

FIGURE 1. Equilibrium structures (in Å and deg) at the HF/6-31G* level

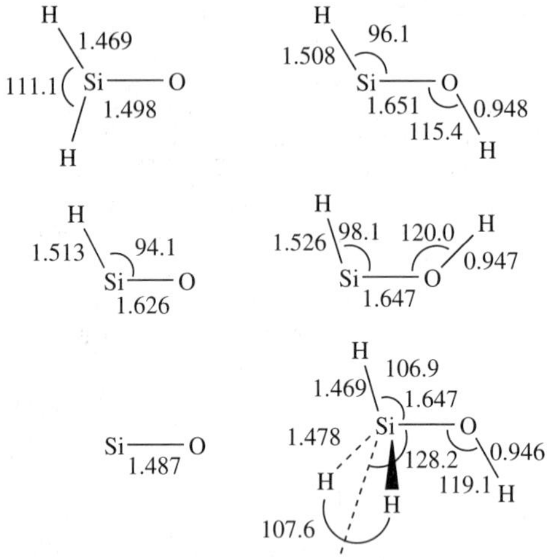

FIGURE 2. Equilibrium structures (in Å and deg) at the HF/6-31G* level

but it is by 0.216 Å shorter than the Si−S single bond length in H_3Si-SH (2.152 Å), indicating that a real π-bond exists between the Si and S atoms. The bond length shortening of 10% from H_3Si-SH to $H_2Si=S$ is comparable to that of 9% from H_3Si-OH to $H_2Si=O$, but it is smaller than the 15% shortening found on going from H_3C-OH to $H_2C=O$. A smaller shortening of silicon-containing π-bonds is found in the ethene analogues: 14% ($H_2C=CH_2$), 10% ($H_2Si=CH_2$) and 9% ($H_2Si=SiH_2$). According to the calculations, $H_2Si=S$ is found to be kinetically stable toward unimolecular dissociations such as $H_2Si=S \rightarrow H_2 + SiS$, $H_2Si=S \rightarrow H + HSiS$ and isomerization, i.e. $H_2Si=S \rightarrow H(HS)Si:$, as in the cases of **2** and **3**. For example, the barrier for the 1,2-H shift in $H_2Si=S$ to $H(HS)Si:$ is 54.8 kcal mol^{-1}, suggesting that **1** is kinetically stable to unimolecular isomerization and is expected to be isolable. Furthermore, it is found that $H_2Si=S$ is theremodynamically more stable than $H_2Si=O$. In an attempt to assess the strength of the silicon−sulfur double bond, they compared the calculated hydrogenation energies of **1, 2** and **3** at the MP3/6-31G**//6-31G* level. Using these calculated hydrogenation energies, the π-bond energies $E_\pi(Si=S)$, $E_\pi(Si=O)$ and $E_\pi(C=O)$ are estimated to be 42, 33 and 63 kcal mol^{-1}, respectively. The calculated vibrational frequencies of $H_2Si=X$ (**1** and **2**) at the HF/6-31G* level were calculated to be 682 and 1203 cm^{-1} for the SiS and SiO stretching vibrations of **1** and **2**, respectively. The calculated proton affinities of **1, 2** and **3** were also reported and they are given in Table 1. The zero-point energy corrected MP3/6-31G**//6-31G* proton-affinity value of 174.7 kcal mol^{-1} for $H_2C=O$ agrees well with the experimental value of 171.7 kcal mol^{-1}.

The proton affinities increase in the order of **3** (174.7 kcal mol^{-1}) <**1** (190.5 kcal mol^{-1}) <**2** (208.3 kcal mol^{-1}). This was explained in terms of the predominance of the electrostatic over charge transfer interactions, thus the charge separations in the double bonds increase in the order of: $H_2C^{+0.2}-O^{-0.4}$ < $H_2Si^{+0.7}-S^{-0.4}$ < $H_2Si^{+1.0}-O^{-0.7}$, while the energy of the frontier n orbital levels rise in the order **3** (−11.8 eV) ≈ **2** (−11.9 eV) <**1** (−9.8 eV)[13]. The most important result of this theoretical study was the conclusion that silicon is significantly less reluctant to form double bonds with sulfur than with oxygen. Thus according to these calculations

17. Recent advances in the chemistry of silicon–heteroatom multiple bonds 1067

TABLE 1. Calculated proton affinities (kcal mol^{-1}) of H$_2$Si=S (**1**), H$_2$Si=O (**2**) and H$_2$C=O (**3**) at various theoretical levels[a]

Level of theory	H$_2$Si=S (**1**)	H$_2$Si=O (**2**)	H$_2$C=O (**3**)
HF/6-31G*	192.6	215.6	182.0
HF/6-31G**	195.4	220.6	186.6
MP2/6-31G**	193.1	209.4	180.3
MP3/6-31G**	196.1	215.7	183.2
MP3/6-31G** + ZPC[b]	190.5	208.3	174.7

[a] All calculations used the HF/6-31G* optimized structures.
[b] Zero-point energy corrected calculations.

silanethiones are expected to be more stable and less reactive than the corresponding silanones[13]. The authors concluded that the major obstacle to the successful isolation of silanethiones is their relatively high reactivity.

2. Other double bonds to silicon

The π-bond strengths of double bonds to silicon were studied theoretically by Gordon and coworkers (in 1987)[15] and by Schleyer and coworkers (in 1988)[16]. In the former paper, all possible π-bonds between the elements C, N, O, Si, P and S were considered and the π-bond strengths were estimated by calculating the *cis–trans* rotation barriers (where possible) and by hydrogenation energies. The ability of these elements to form strong π-bonds was found to be in the order: O > N $\approx$ C $\gg$ S > P > Si. They reported the computed bond lengths and vibrational stretching frequencies for singly- and doubly-bonded compounds of all the possible combinations of these elements[15]. Schleyer and coworkers evaluated the π-bond energies (E_π) for the doubly-bonded systems H$_2$Y=XH$_n$ (Y = C, Si; X = B, C, N, O, Al, Si, P, S) employing the MP4SDTQ/6-31G*//6-31G* + ZPE level of theory. The difference between the energy of two single bonds, X–Y, and the corresponding double bond, X=Y, was calculated by means of isodesmic equations. E_π is given by subtraction of this difference from the dissociation energies of the single bonded X=Y system. Si=X bonds were found to have significantly lower E_π energies than the corresponding C=X bonds. For C=X and Si=X, the π-bond energies correlate with the electronegativities of X with different lines for second- and third-row substituents. Families of linear correlations are also observed between E_π and the Y=X bond lengths. Schleyer and coworkers criticized alternative procedures for estimating π-bond energies (such as rotation barriers and diradical components)[16].

3. Miscellaneous

The vertical electronic spectrum of silanethione (H$_2$Si=S; **1**) was investigated by using a multi-reference CI method and an expanded AO basis set including 4s and 4p Rydberg functions[17]. The equilibrium geometry of the X$^1A_1$ ($\sigma^2\pi^2n^2$) ground state was determined at the calculated MP2 level. The excited states studied include all those generated through single excitations from the σ, π and n MOs into the vertical π^*, σ^* and 4s, 4p Rydberg species. The lowest-lying states are $^3(n\pi^*)$ and $^3(\pi\pi^*)$, in the range from 2.72 to 4.55 eV above the ground state. Four allowed transitions lie in the energy range from 4.55 to 6.15 eV, namely $^1A_1(\pi\pi^*)$, $^1B_2(n\sigma^*)$, $^2B_1(\sigma\pi^*)$ and $^1B_2(n4s)$. The 0–0 origin for the strong absorption X$^1A_1 \rightarrow$ $^1A_1(\pi\pi^*)$ is expected near 3.0 eV. Excitations into σ^* lie below transitions into the first 4s Rydberg species. The adiabatic ionization potentials (in eV) are 9.08 (X^2B_2, $n \to \infty$) and 9.84 ($^2B_1$, $n \to \infty$). The ground state of H_2SiS is highly polar, with a dipole moment of about 3.18 D ($H_2Si^+S^-$). A comparison of the vertical spectrum of H_2SiS with those of H_2CO, H_2CS and H_2SiO indicates a general stabilization of the valence electronic transitions relative to those of Rydberg character as C is replaced by Si.

II. DOUBLE BONDS BETWEEN SILICON AND GROUP 16 ELEMENTS

A. Introduction

Great advances have been made in recent years in studies of multiple bonding to silicon, and by 1990 we witnessed a wealth of excellent work on the isolation of stable compounds having Si double bonds where X belongs to group 14 or 15, i.e. silenes ($R_2Si=CR_2$)[1,8a,11], disilenes ($R_2Si=SiR_2$)[2,8b,12], silanimines ($R_2Si=NR$)[5,11] and phosphasilenes ($R_2Si=PR$)[6]. The kinetic stability of all these compounds depended on the presence of four or at least three bulky substituents [e.g. mesityl (Mes = 2,4,6-trimethylphenyl), 2,4,6-triisopropylphenyl (Tip), 2,4,6-tri-*t*-butylphenyl, *t*-butyl, adamantyl, trialkylsilyl, etc.). In contrast to these doubly-bonded species between silicon and group 14 or 15 elements which can be sterically protected by bulky substituents from both ends of the double bond, no kinetically stabilized silanones and silanethiones have been isolated, except in rigid matrices, until the recent successful isolation of a stable silanethione Tbt(Tip)Si=S {**4**; Tbt = 2,4,6-tris[bis(trimethylsilyl)methyl]phenyl}[10]. Since only the silicon atom can be sterically protected by substituents in the cases of silanones and silanethiones, the conventional steric protection groups such as those mentioned above are not large enough to prevent their oligomerization.

Although stable silanones are still unknown at this time, there have recently been several interesting reports on the chemistry of doubly-bonded compounds between silicon and chalcogen atoms. In the following sections, the new findings since 1987 on the generation method and some properties of these silicon–chalcogen doubly-bonded compounds are summarized separately for each chalcogen atom, together with some important reports which were not discussed in the previous reviews[4a,b,c].

B. Silicon–Oxygen Double Bonds (Silanones)

Evidence for the formation of silanones depends on two types of experiments. In the first, silanones are generated as unstable intermediates in reactions, and their formation is inferred from the isolation of trapping products with suitable substrates. The second approach is based on their generation in a low temperature matrix and their characterization by infrared spectroscopy which reveals $\nu_{(Si=O)}$ at *ca* 1200 cm^{-1}. These two groups of experiments are described below.

1. Formation from hydrosilanes

When a mixture of methylsilane (or dimethylsilane) deposited with ozone in an argon matrix at 17 K was photolyzed, the corresponding silanone, MeHSi=O (or Me$_2$Si=O), was generated and characterized by their infrared spectra. In the case of the parent silane, H$_3$SiH, a similar photolytic reaction with ozone in an Ar matrix led to the identification of SiO, H$_2$Si=O (**2**), (HO)HSi=O (silanoic acid) and (HO)$_2$Si=O (silicic acid)[18].

In 1989, Corriu and workers reported the thermal decomposition (85 °C) of the silylformate **6**, having a remote stabilizing amino group, which was prepared by the insertion reaction of the corresponding pentacoordinated functional silane **5** with CO$_2$,

17. Recent advances in the chemistry of silicon–heteroatom multiple bonds   1069

into formaldehyde and the transient silanone **7** (Scheme 2)[9]. Silanone **7** was found to be trapped with $(Me_2SiO)_3 (D_3)$ to give the eight-membered ring compound **8**, while it underwent ready trimerization leading to the formation of **9** in the absence of a trapping reagent.

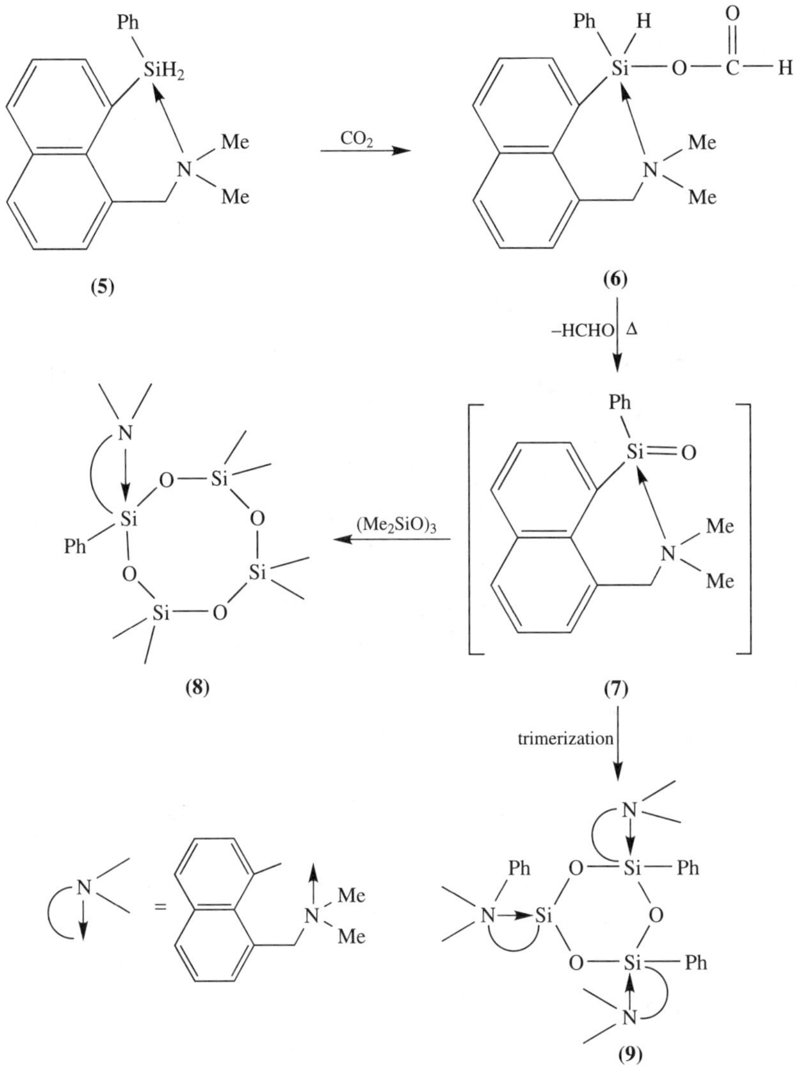

SCHEME 2

2. Formation from cyclic silyl ethers

The silanones $(CH_3)_2Si=O$ (**10**) and $(CD_3)_2Si=O$ (**11**) have been generated by vacuum pyrolysis of the corresponding 6-oxa-3-silabicyclo[3.1.0]hexanes **12** (Scheme 3)[19–21].

The Diels–Alder adduct **13** of silapyranes with maleic anhydride was also used as an alternative precursor for **10** (Scheme 3)[19]. In the pyrolysis of **12**, a 2-siloxetane intermediate **14** was proposed as a transient precursor leading to dimethylsilanone **10** (or **11**)[19–21].

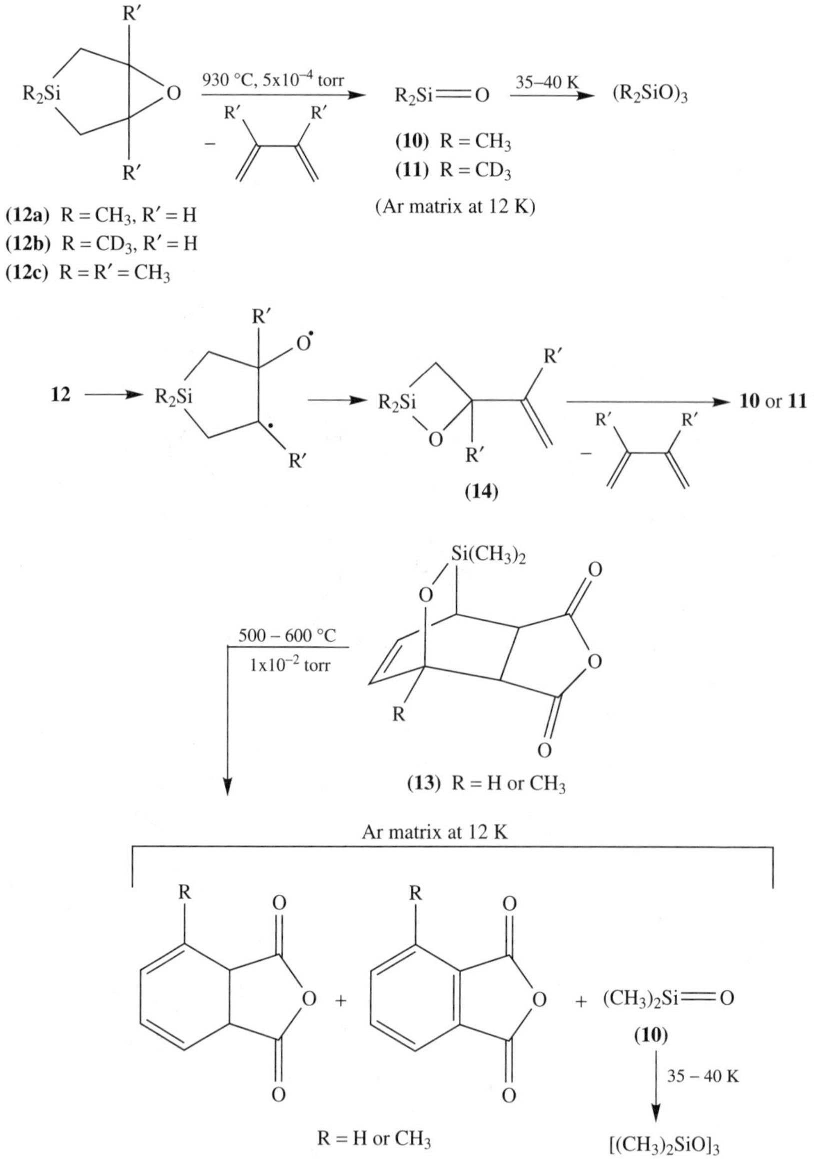

(**12a**) R = CH$_3$, R′ = H
(**12b**) R = CD$_3$, R′ = H
(**12c**) R = R′ = CH$_3$

(**10**) R = CH$_3$
(**11**) R = CD$_3$

(Ar matrix at 12 K)

(**13**) R = H or CH$_3$

R = H or CH$_3$

SCHEME 3

Silanones **10** and **11** have also been trapped in argon matrices at 12 K and they were studied by IR spectroscopy[19]. Using the dependence of the spectra on temperature and pressure in the pyrolysis zone or in warming-up experiments (to 35–40 K) the following vibrational bands of silanones have been revealed: 1244, 1240, 1210, 822, 798, 770, 657 cm^{-1} for **10**; 1215, 1032, 1007, 995, 712, 685, 674 cm^{-1} for **11**. The limits of thermal (*ca* 850 °C) and kinetic (5×10^{-4} torr) stability of dimethylsilanone were determined. By comparison of the measured and calculated frequencies the band at 1210 cm^{-1} in **10** (1215 cm^{-1} in **11**) was assigned to the Si=O stretching vibration. The calculated force constant (8.32 mdyn/Å) and bond order (1.45) of the Si=O bond have been presented as evidence for a significant double-bond character in dimethylsilanone.

The tricyclic silyl ether **13** (R = H) also undergoes photochemical fragmentation into dimethylsilanone **10** as shown in Scheme 4[22]. Of the two possible fragmentation mechanisms, the decomposition via a bicyclic intermediate **15** (path B) was preferred by the authors, based on the observed reaction products.

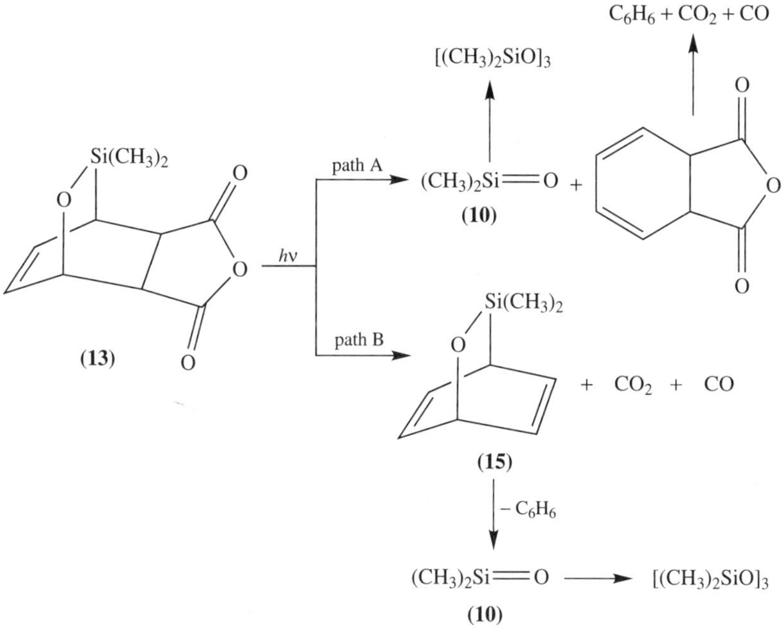

SCHEME 4

3. Formation from allyloxysilane derivatives

It is well known that thermal decomposition of allyl-substituted silanes proceeds by retro-ene reaction with formation of transient species having a Si=C bond, such as silabenzene, silatoluene and dimethylsilaethylene[4b,e]. The kinetic data on the gas-phase pyrolysis of a similar allyloxysilane derivative, (1,1-dimethylallyloxy)dimethylsilane (**16**), and the results on thermolysis of allyloxydimethylsilane (**17**) in a flow system both indicate the participation of an intermediate silanone, $(CH_3)_2Si=O$ (**10**), as shown in Scheme 5[23].

Allyl(allyloxy)dimethylsilane (**18**) was also proposed by Barton and coworkers as being the source of **10** in gas-phase pyrolysis[24]. It has been suggested that the generation of **10** by pyrolysis of **18** results from consecutive loss of two allyl radicals. 1,5-Hexadiene and cyclosiloxanes **19** (D$_3$) and **20** (D$_4$) were the final pyrolysis products of **18** (Scheme 6).

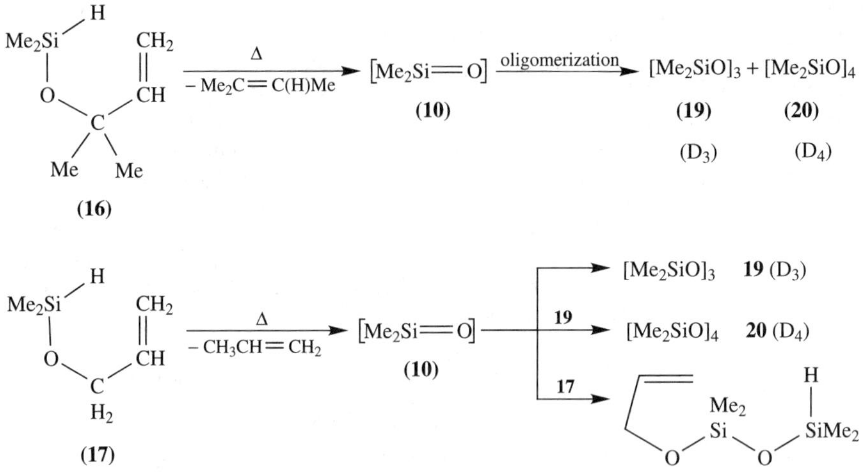

SCHEME 5

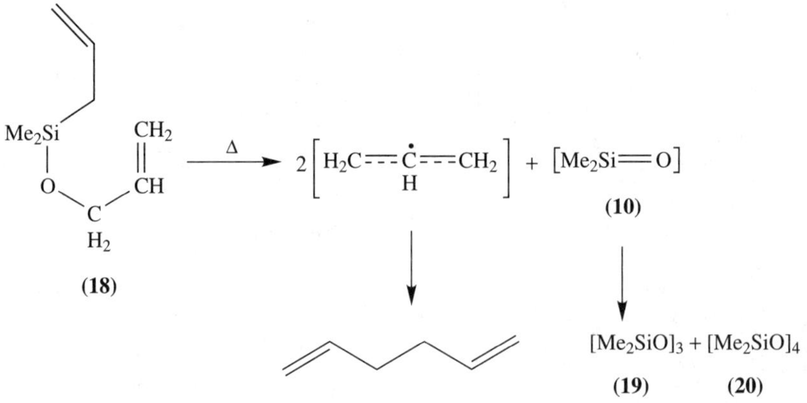

SCHEME 6

In 1989, Nefedov and coworkers have reinvestigated the thermolysis of the abovementioned allyloxysilane derivatives **16–18** and of 2,2,6-trimethyl-2-silapyrane (**21**) using vacuum pyrolysis and matrix isolation techniques[23]. IR spectroscopic studies on the products isolated in the matrices enabled them to probe directly the intermediacy of **10** in these reactions and to discuss its thermal stability. Only in the case of allyloxydimethylsilane (**17**) did they find direct spectroscopic evidence for the formation of **10** by observation of its most intense band at 798 cm^{-1} in the matrix IR spectrum of the pyrolysis products. In all other cases silanone **10** was not detected and it was assumed that it is thermally unstable, undergoing fragmentation into SiO and CH$_3$ radicals as shown in Schemes 7, 8 and 9 (the species actually observed in the matrix are indicated). In this paper, Nefedov and coworkers have reaffirmed the thermal and kinetic stability of dimethylsilanone **10** in the gas phase, which they had previously described[19].

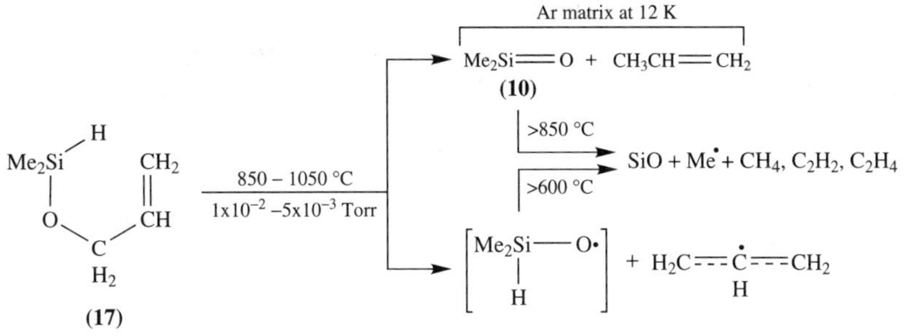

17. Recent advances in the chemistry of silicon–heteroatom multiple bonds

SCHEME 7

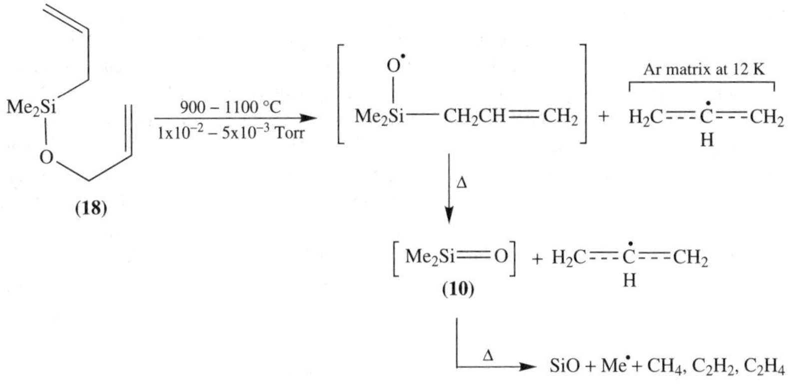

SCHEME 8

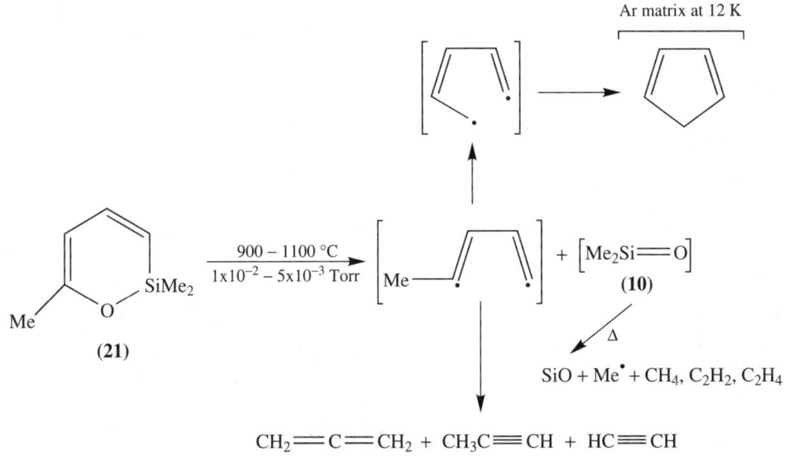

SCHEME 9

4. Formation from silanol derivatives

Eaborn and Stanczyk deduced from kinetic data that under basic conditions silanone **10** can be formed by loss of a proton and a carbanion R⁻ from the silanols **22** (R = m-ClC$_6$H$_4$CH$_2$ or PhC≡C) (Scheme 10)[25].

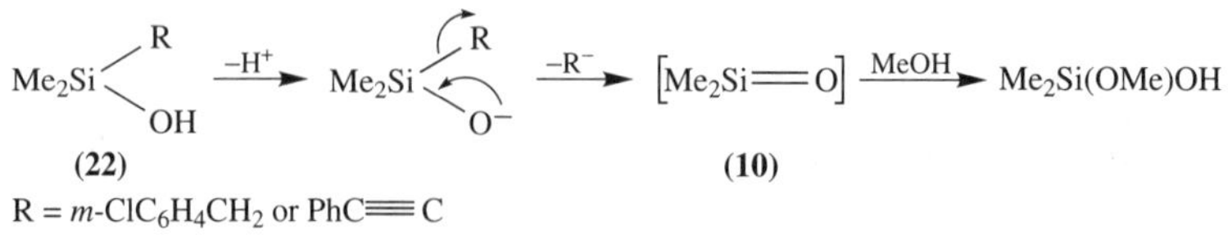

$$\text{Me}_2\text{Si}\begin{array}{c}R\\OH\end{array} \xrightarrow{-H^+} \text{Me}_2\text{Si}\begin{array}{c}R\\O^-\end{array} \xrightarrow{-R^-} [\text{Me}_2\text{Si}=O] \xrightarrow{\text{MeOH}} \text{Me}_2\text{Si}(\text{OMe})\text{OH}$$

(22)   (10)

R = m-ClC$_6$H$_4$CH$_2$ or PhC≡C

SCHEME 10

Although attempted dehydration of di-t-butylsilanediol into di-t-butylsilanone gave instead 1,1,3,3-tetra-t-butyl-1,3-dihydroxydisiloxane, a similar approach using di-t-butylsilanol (**23**) was found to be useful to generate an anion radical of silanone. Thus, while irradiation of a mixture of **23** and di-t-butyl peroxide in t-butyl alcohol resulted in the observation of an ESR spectrum of the corresponding di-t-butylhydroxysilyl radical (**24**), the photolysis of a solution of **23** in di-t-butyl peroxide and t-butyl alcohol containing also potassium t-butoxide (in excess with respect to the silanol) gave an ESR spectrum assignable to the di-t-butylsilanone radical anion (**25**), which is probably formed principally via the silanolate anion **26** (Scheme 11)[26]. The magnitude of the hyperfine coupling to ^{29}Si in the ESR spectrum of **25** shows that the radical anion is pyramidal at silicon. This approach seems to be particularly promising for the preparation of anion radicals of silanones since silanols are more acidic than the corresponding alcohols (e.g. Me$_3$SiOH compared with Me$_3$COH or Ph$_3$SiOH compared with Ph$_3$COH)[27] and alkoxy radicals abstract hydrogen from a Si—H bond more readily than from a C—H bond[28].

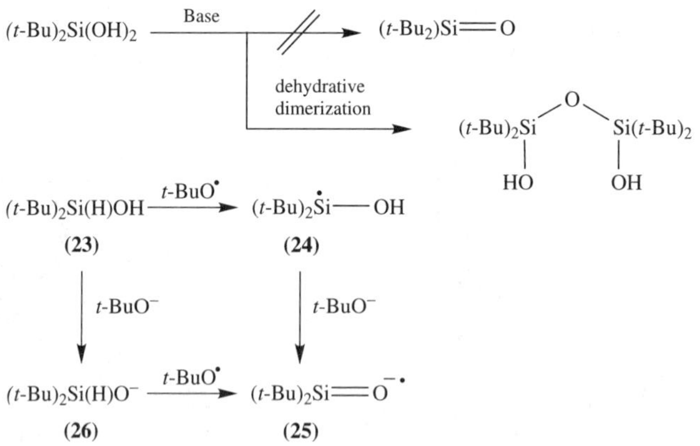

SCHEME 11

Most recently, Milstein and coworkers reported a new type of homogeneous catalytic reaction that generates silanones from secondary silanols under extremely mild conditions. The platinum complex (dmpe)Pt(Me)OTf) (**27**) (dmpe = Me$_2$PCH$_2$CH$_2$PMe$_2$, OTf = OSO$_2$CF$_3$ was treated with an equimolar amount of the silanol, (i-Pr)$_2$SiH(OH) (**28**), in acetone to yield a new dimeric hydrido-bridged complex **29** and the trimer **30** of diisopropylsilanone, **31** (Scheme 12)[29].

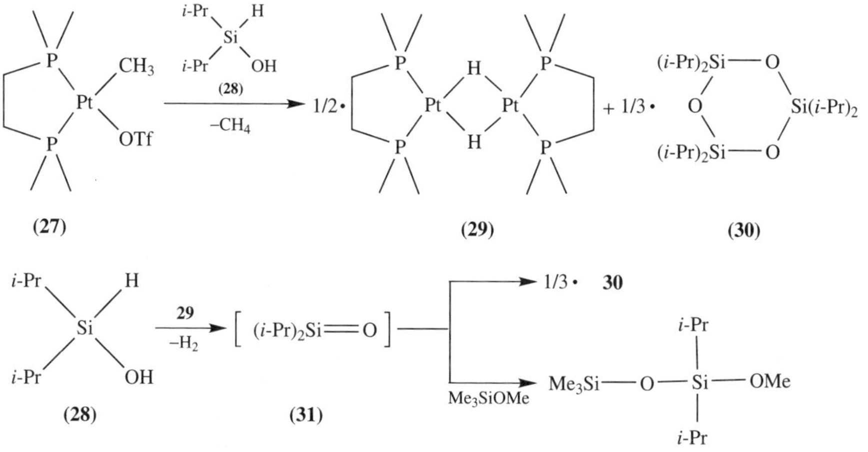

SCHEME 12

The generation of methane in the reaction was evidenced by the ^{1}H NMR spectrum of the reaction mixture. It was also shown that the newly obtained complex **29** reacts catalytically with silanol **28** to give the trimer **30** (presumably from trimerization of **31**) with the evolution of hydrogen gas. In the presence of Me$_3$SiOMe the same reaction resulted in the formation of an insertion product of the intermediate silanone **31** as shown in the lower part of Scheme 12. The proposed catalytic cycle for the dehydrogenation of **28** with **29** is shown in Scheme 13. It should be noted, however, that spectroscopic evidence for the proposed silanones was not presented.

Although no intermediates except for **29** are detected by NMR with **28** as a substrate, the use of bulkier di-t-butylsilanol **32** as a starting material allowed the authors to isolate and fully characterize by spectroscopy and X-ray crystallography the corresponding platinum–silanol complex **33**, which strongly suggested the intermediacy of **33** in this catalytic reaction as shown in Scheme 13. This reaction is formally analogous to (although mechanistically different from) the well-known catalytic dehydrogenation of alcohols[30] that leads to ketones.

5. Formation from silylenes

The oxidation of silylenes (silandiyls) by oxygen transfer reagents such as dimethyl sulfoxide[31], tertiary amine N-oxides[31] or epoxides[32] in the presence of trapping agents usually yields products, whose formation may be explained in terms of the capture of initially generated silanones. In those cases oligomerization products are usually obtained in the absence of trapping reagents, although in low yields. As mentioned in the previous reviews, unstable silanones are frequently postulated as intermediates in these reactions, but the involvement of free silanones in these oxidation reactions is still questionable[4e]. On the other hand, when silylenes are treated with oxidants such as N$_2$O[33] or O$_2$[34] in argon matrices, the formation of silanones as the oxidation products of silylenes was unambiguously demonstrated by IR spectroscopy.

Most recently, Belzner and coworkers reported a new type of oxygen transfer reaction from isocyanates to bis[2-(dimethylaminomethyl)phenyl]silylene (**35**)[35], which was thermally generated from the corresponding cyclotrisilane **34**[36], and they were able to obtain some convincing results that support the involvement of silanone **36** as an intermediate[37].

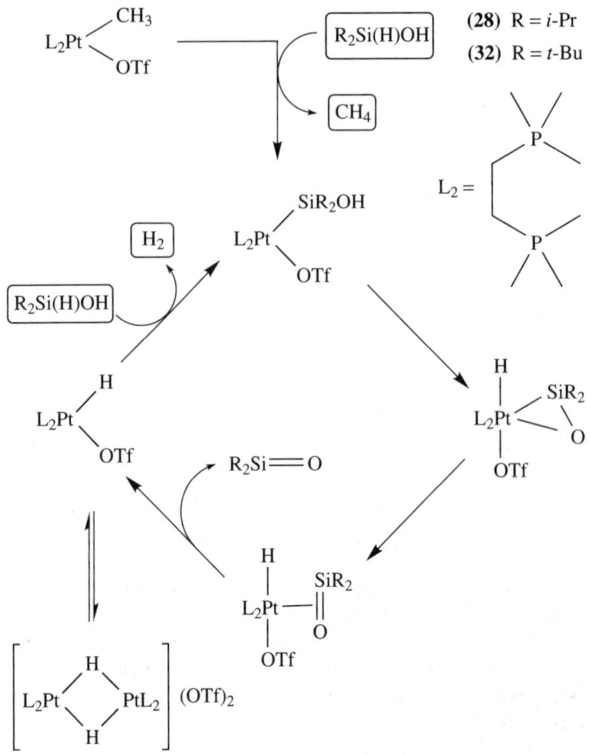

SCHEME 13

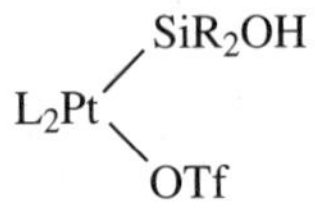

(33) R = t-Bu

When a toluene solution of a mixture of cyclotrisilane **34** and cyclohexyl isocyanate (or t-butyl isocyante) was heated at 70 °C, cyclic di- and trisiloxanes **37** and **38**, i.e. the cyclic dimer and trimer of the silanone **36**, were obtained together with the corresponding isonitrile RN=C. The formation of **37** as well as **38** was completely suppressed in the presence of hexamethylcyclotrisilane (**19**; D_3); instead, quantitative conversion of **35** into **39**, the formal insertion product of the silanone **36** into the Si—O bond of D_3, occurred (Scheme 14). Since neither cyclodisiloxane **37** nor cyclotrisiloxane **38** reacted with D_3 under the reaction conditions, the possibility that **37** or **38** is the precursor of **39** was ruled out. Whereas the oxidation of **35** with cyclohexyl and t-butyl isocyanates proceeded with exclusive formation of **37** and **38** (as the silicon-containing compounds) the reaction of **35** with phenyl isocyanate resulted in the formation of **37** in low yield. Furthermore, in this case the presence of D_3 did not totally suppress the formation of **37**. According to the authors, these results indicate that the oxidation of **35** with cyclohexyl and t-butyl isocyanates appears to use other reaction channels than that with phenyl isocyanate.

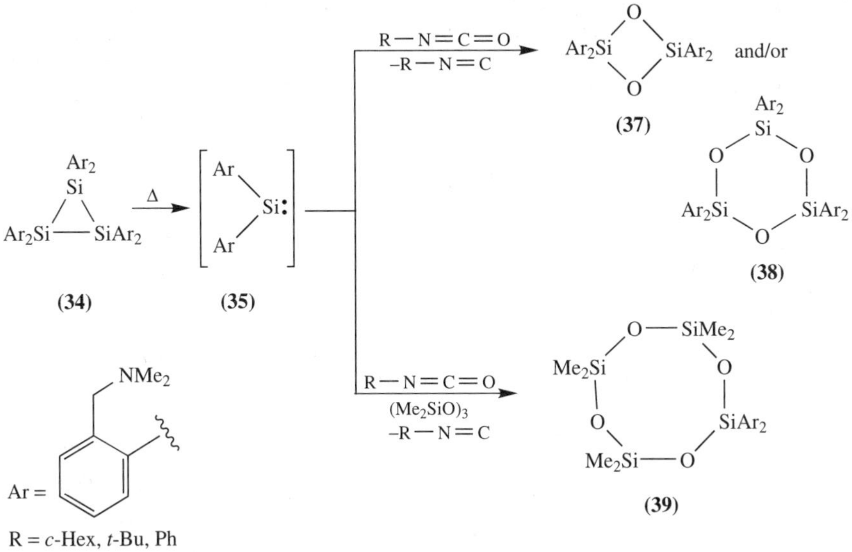

17. Recent advances in the chemistry of silicon–heteroatom multiple bonds

SCHEME 14

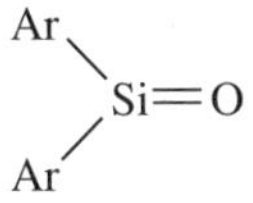

Ar = 2-(dimethylaminomethyl)phenyl (o-CH₂NMe₂-C₆H₄)

R = c-Hex, t-Bu, Ph

Structures: (34) Ar₂Si–SiAr₂–SiAr₂ cyclotrisilane; (35) Ar₂Si: silylene; (36) Ar₂Si=O silanone; (37) four-membered (Ar₂SiO)₂; (38) six-membered (Ar₂SiO)₃; (39) mixed cyclic siloxane with Me₂Si and Ar₂Si units.

The reaction of **34** with isocyanates thus takes a different course from the photolysis of hexa-*t*-butylcyclotrisilane in the presence of tri-*t*-butylsilyl isocyanate, which was reported by Weidenbruch and coworkers to yield the five-membered ring product **40** (Scheme 15)[38].

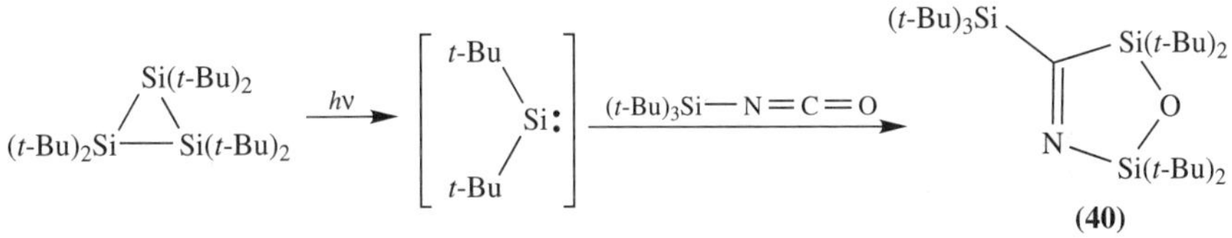

SCHEME 15

The formation of the cyclic siloxanes **37** and **38** from cyclotrisilane **34** and isocyanates is most likely explained by the oligomerization of an intermediate silanone **36**. This is further supported by the fact that the product ratio in this reaction is considerably affected by the concentration of the starting material. A high concentration of **36** favors trimerization to **38**, whereas in dilute solutions the dimerization product **37** is formed preferentially. The latter reaction is surprising in view of the fact that only one silanone bearing a very bulky 2,4,6-triisopropylphenyl substituent was reported to undergo dimerization[39]; less hindered silanones are known to form cyclic trimers and tetramers[4e,40]. The reason for this different behavior might be due to the fact that silanone **36** is electronically stabilized by intramolecular coordination with the nitrogen functionalities, which was confirmed by the X-ray structural analysis of the cyclic dimer **37**. The authors proposed various pathways to silanone **36** as a crucial intermediate in the oxidation of silylene **35** by isocyanates as shown in Scheme 16.

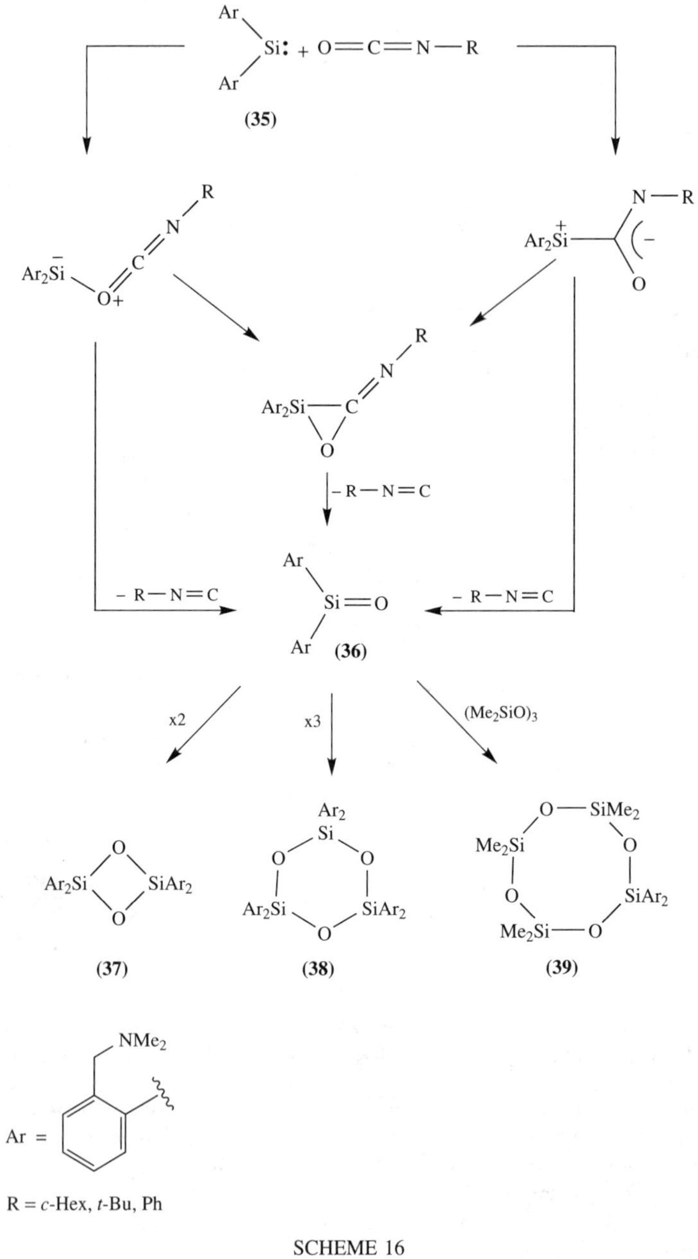

SCHEME 16

17. Recent advances in the chemistry of silicon–heteroatom multiple bonds 1079

Oxygen transfer reactions were also examined by Jutzi and coworkers[41] for the nucleophilic decamethylsilicocene **41**. Silicocene **41** reacts with carbon dioxide under mild conditions to give two types of products, **46** and **47**, depending on the solvent used, as shown in Scheme 17.

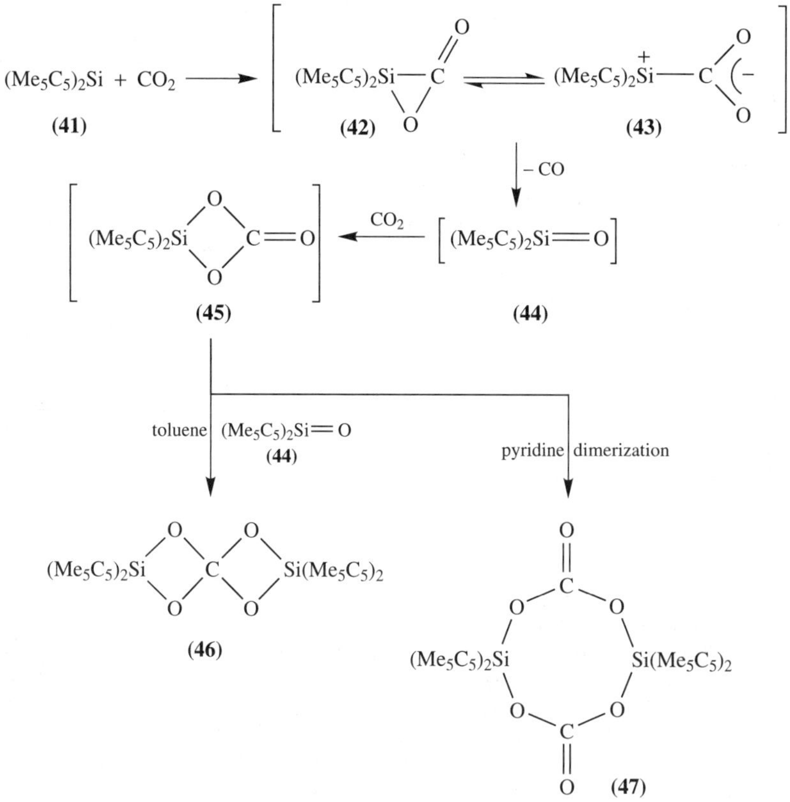

SCHEME 17

In these reactions the formal oxidation state of the silicon atom changes from $+2$ in **41** to $+4$ in the final products, and hapticity of the pentamethylcyclopentadienyl ligands changes from η^5 to η^1. The authors propose that the first intermediate is a highly reactive [2 + 1] cycloaddition product **42** or its ring-opened isomer **43**, which easily loses carbon monoxide to give the silanone intermediate **44**. The silanone **44** is not stable under the reaction conditions and is transformed by excess CO_2 to the [2+2] cycloaddition product **45**. In toluene as solvent, **45** reacts with the silanone **44** present in the reaction mixture to give the final product **46** in a further [2+2] cycloaddition step. In pyridine as solvent, the intermediate silanone **44** is deactivated by coordination to the solvent, and hence **45** does not react with **44** but forms the dimerization product after ring-opening of one of the Si–O bonds. The intermediacy of the silanone **44** was further supported by trapping reactions. In the presence of t-butyl methyl ketone or acetone, the addition products **48** or **49** were formed as expected if the silanone **44** undergoes an ene-type reaction[42] (Scheme 18).

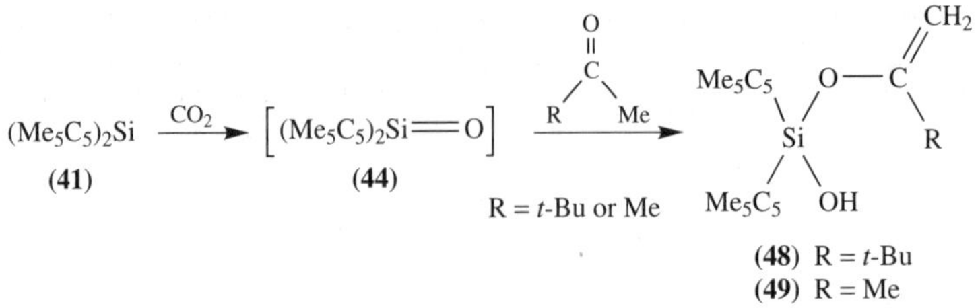

SCHEME 18

6. Formation from silenes and silicon–heteroatom double-bond compounds

Although silenes were long believed to undergo [2 + 2] cycloaddition reaction with carbonyl compounds to form the four-membered 2-siloxetane rings, only the products of retro-cleavage of the siloxetane ring, i.e. an alkene and oligomers of a silanone $R_2Si=O$ (presumed to be formed from individual silanone product molecules), were observed in most cases (Scheme 19)[4e,43].

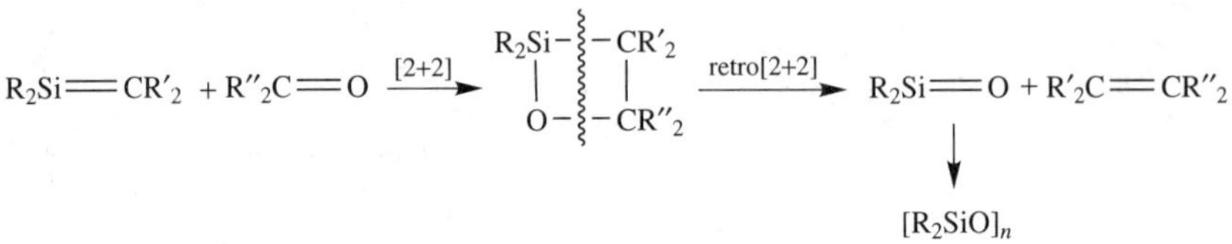

SCHEME 19

Bachrach and Streitwieser studied theoretically at the SCF level the cycloaddition of formaldehyde to the parent silanone ($H_2Si=O$)[44]. They found that the [2+2] cycloaddition reaction to form siloxetane is exothermic by about 80 kcal mol^{-1}, but that the further fragmentation of the siloxetane to ethylene and silanone is endothermic by ca 50 kcal mol^{-1}, an energy which is too large to be consistent with a spontaneous unimolecular retro-[2+2] cycloreversion process. Therefore, a free silanone is probably not involved as an intermediate in the thermal decomposition of a siloxetane, and bimolecular reactions between two siloxetanes are an energetically more feasible way to account for the experimentally obtained alkenes and silanone oligomers. This theoretical suggestion was experimentally supported by Brook and coworkers by a kinetic investigation of the thermolysis of a stable siloxetane obtained from $(Me_3Si)_2Si=C(OSiMe_3)(Ad-1)$ and benzophenone[45].

The reaction of a silene with molecular oxygen followed by the retro-[2+2] cylcoreversion with oxygen–oxygen bond cleavage seems to be another promising way to generate a silanone. The pyrolysis of 1,1-dimethylsilacylcobutane (50), which would lead to the formation of the unstable 1,1-dimethylsilene (51), gave in the presence of oxygen several oligomers of dimethylsilanone 10 among the products, indicating according to the authors the intermediacy of 10, which was produced by a retro-[2 + 2] reaction of the dioxasiletane 52 (Scheme 20)[46]. The reaction of the stable silene $(Me_3Si)_2Si=C(OSiMe_3)(Ad-1)$ (53) with oxygen was also examined and it was found to give a carboxylic acid silyl ester (55) and a cyclic trimer (56) of the silanone $(Me_3Si)_2Si=O$ (54) (Scheme 21)[47]. The mechanism of these reactions has not been fully investigated.

Driess and Pritzkow have recently succeeded in the synthesis and isolation of a stable silicon–arsenic double-bond compound (57), the first stable arsasilene (arsanilidenesilane)[7a]. The arsasilene 57 was found to undergo a ready [2 + 2],

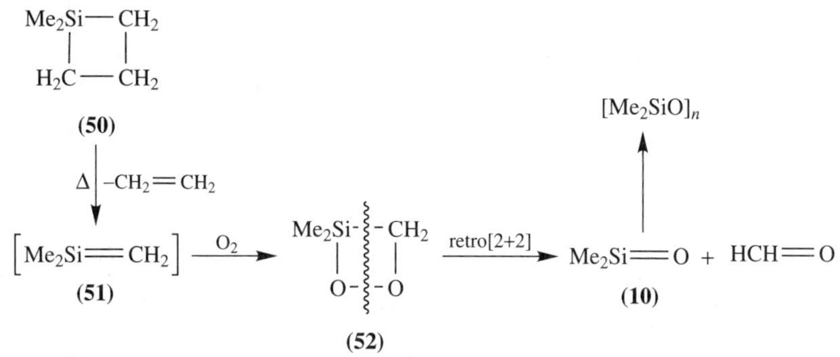

17. Recent advances in the chemistry of silicon–heteroatom multiple bonds

SCHEME 20

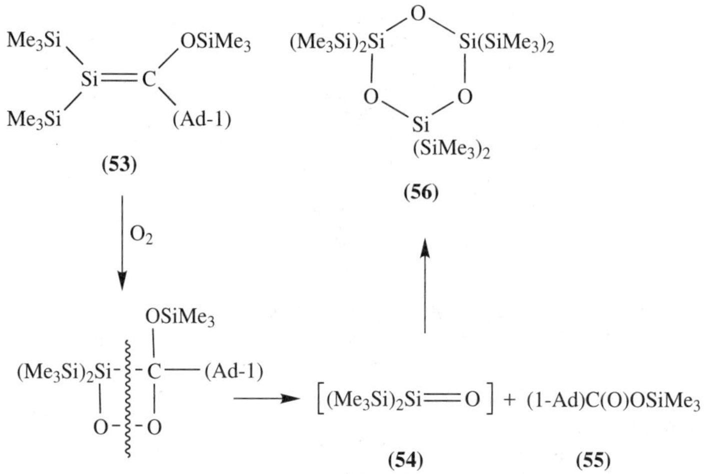

SCHEME 21

cycloaddition reaction with benzophenone to give the four-membered ring product **58** as a stable crystalline compound. As in the case of above-mentioned siloxetanes, the arsenic-containing siloxetane **58** was found to decompose thermally under drastic conditions (160 °C, 10 h) to give the cyclic dimer (**62**) of a silanone, Tip$_2$Si=O (**59**), together with the arsaalkenes **60** and **61** (Scheme 22)[7a].

As previously mentioned (see Section II.B.5), the formation of the dioxadisiletane **62** should be noted as the first example of a dimerization of a silanone, in contrast to the common trimerization or tetramerization of less hindered silanones. This metathetical reaction of arsasilene **57** with benzophenone represents a new approach to arsaalkenes via arsasilenes by means of a pseudo-Wittig reaction. More recently, Driess and coworkers reported the similar sila-Wittig-type reaction via the phosphorus analogue **64** of the above-mentioned arsenic-containing siloxetane **58**[48]. Thus, the [2 + 2] cycloadduct (**64**) of phosphasilene (**63**) with benzophenone afforded, upon heating at 160 °C for 2 days, the dioxadisiletane **62** along with two phosphaalkenes **65** and **66**, suggesting the intermediacy of silanone **59** (Scheme 22)[48].

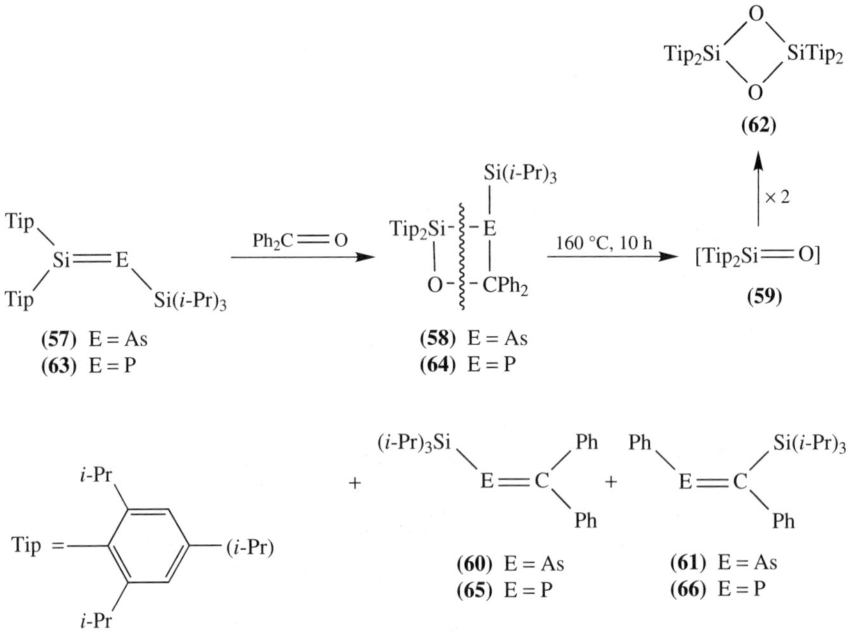

SCHEME 22

Kudo and Nagase have reported a theoretical *ab initio* investigation at both HF/3-21G and HF/6-31G* on the potential energy surface for the dimerization of H$_2$Si=O (**2**), revealing that this reaction proceeds with no barrier to yield the cyclic dimer (H$_2$SiO)$_2$ by a stepwise formation of two new bonds[49]. They have also discussed the structures and stability of the cyclic trimer (H$_2$SiO)$_3$ and tetramer (H$_2$SiO)$_4$, and calculated that the disproportionation for the reaction 3(H$_2$SiO)$_2$ → 2(H$_2$SiO)$_3$ is exothermic by −110 kcal mol^{-1} at HF/6-31G*//3-21G. This very high negative value clearly favors the trimer over the dimer. Furthermore, the insertion of H$_2$Si=O into the Si−O bond of the dimer (H$_2$SiO)$_2$, i.e. (H$_2$SiO)$_2$ + H$_2$Si=O → (H$_2$SiO)$_3$, is calculated to be exothermic by 117.3 kcal mol^{-1} (HF/6-31G*//3-21G). These high exothermicities suggest that the ring expansion proceeds rapidly with no significant barrier. Similar calculations revealed that the energy released upon going from 4(H$_2$SiO)$_3$ to 3(H$_2$SiO)$_4$ is 74.2 kcal mol^{-1} (HF/3-21G level). This energy is by 88.1 kcal mol^{-1} smaller than that for 3(H$_2$SiO)$_2$ → 2(H$_2$SiO)$_3$, suggesting that the formation of the tetramer is less favorable than the formation of the trimer. However, tetramerization is likely to occur readily, as expected from the fact that the reactions (H$_2$SiO)$_3$ + H$_2$Si=O → (H$_2$SiO)$_4$ and (H$_2$SiO)$_2$ + (H$_2$SiO)$_2$ → (H$_2$SiO)$_4$ are calculated to be exothermic by 120.8 and 132.9 kcal mol^{-1}, respectively, at the HF/3-21G level. To the extent that the difference in the exothermicities is meaningful, the tetramer may be produced more favorably by the reaction of two dimers than by the insertion of H$_2$Si=O into the Si−O bond of the trimer[49]. The electronically stabilized silanone **7** (see also Section II.B.1) was postulated as an intermediate in the oxidation by air of the corresponding silanethione **67** (the synthesis of which will be described in the following section) (Scheme 23)[9]. The fast oxidation of **67** resulted in the formation of the same cyclic trimer **9** as previously mentioned.

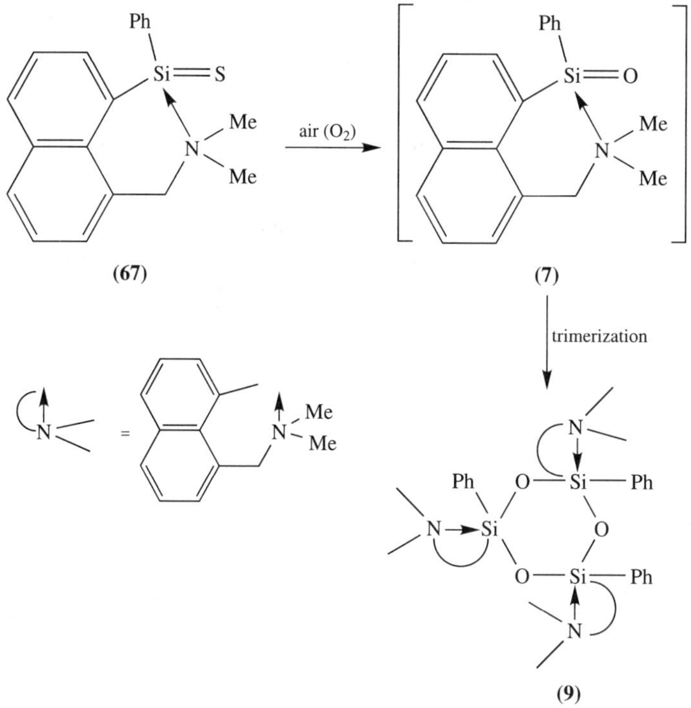

SCHEME 23

C. Silicon−Sulfur Double Bonds (Silanethiones)

Among the silicon−chalcogen double-bond compounds, the silicon−sulfur doubly-bonded compounds (silanethiones) are considered to be easier to synthesize, since it has been predicted by the theoretical calculations that a silicon−sulfur double bond is thermodynamically and kinetically more stable than a silicon−oxygen double bond (silanone)[13,14]. According to the calculations, the lower polarization of Si=S compared to Si=O should lead to a lower reactivity of Si=S. In addition, $H_2Si=S$ (1) is calculated to be by 8.9 kcal mol^{-1} more stable than its divalent isomer, $H(HS)Si:$, whereas $H_2Si=O$ (2) is by 2.4 kcal mol^{-1} less stable than $H(HO)Si:$.

As described in the preceding reviews on this field, most of the early work on silicon−sulfur doubly-bonded compounds was restricted to simple dialkylsilanethiones, which are all transient in solution or in the gas phase[4]. However, in contrast to the successful matrix isolation and spectroscopic identification of dimethylsilanone 10[23], no spectroscopic detection of transient dialkylsilanethiones in matrices has been reported up to now, although the matrix isolation of $Cl_2Si=S$[50] and $Cl(H)Si=S$[51], the silicon analogues of thiophosgene and thioformyl chloride, has been reported.

In recent years, however, impressive progress has been made in the field of silicon− sulfur double-bond chemistry; the first examples of kinetically stabilized and electronically stabilized silanethiones were successfully synthesized and fully characterized by spectroscopic and X-ray crystallographic data[9,10]. These results together with the theoretical studies have revealed the intrinsic nature of this unique double bond to silicon.

The following sections review recent progress in the chemistry of silanethiones.

1. Formation from hydrosilanes

In 1980, Weidenbruch and coworkers reported the formation of 1,1,3,3-tetraalkyl-1,3,2,4-disiladithietanes **70**, i.e. the formal head-to-tail dimers of dialkylsilanethiones **69**, in the copyrolysis of the corresponding 1,1,2,2-tetraalkyldisilanes **68** with elemental sulfur (Scheme 24)[52]. Minor isolated products in these reactions were 1,1,3-trialkyl-1,3,2,4-disiladithietanes **71**, the formation of which was interpreted in terms of the elimination of R_3SiH from the substrates leading to the generation of R(H)Si**:** under the reaction conditions used (Scheme 24). The reaction of R(H)Si**:** with sulfur, followed by the [2 + 2]cycloaddition of the resulting R(H)Si=S **72** with the major intermediate R_2Si=S **69** gives rise to the unusual minor reaction products **71** (Scheme 24).

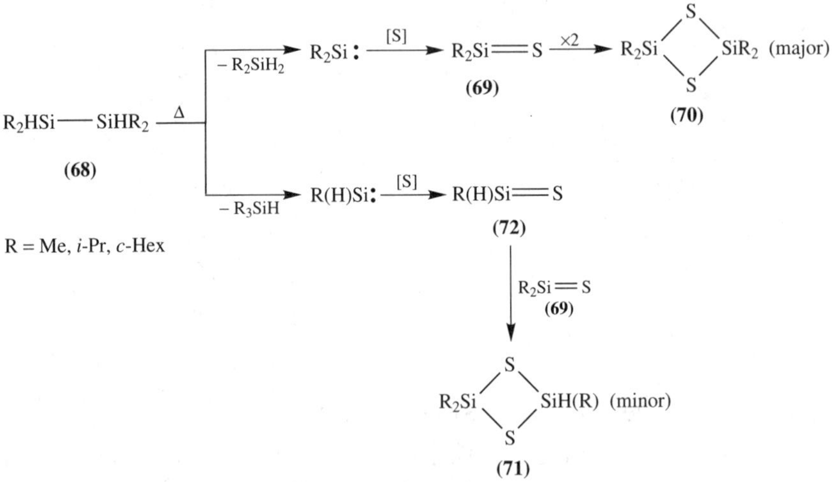

SCHEME 24

In 1989, Corriu and coworkers reported the first synthesis of an isolable silanethione **67** (although internally coordinated) by the reactions of the pentacoordinated functionalized silane **5** (see also Section II.B.1) with carbon disulfide or elemental sulfur (Scheme 25)[9].

In contrast to the fast trimerization of the silanone **7** having the same ligands, silanethione **67** was found to be relatively long-lived in solution ($t_{1/2}$: 3d in $CDCl_3$ at 25 °C) as a monomeric species, though extreme precautions must be taken to avoid exposure to minute amounts of air. Silanethione **67** shows a ^{29}Si chemical shift at $\delta + 22.3$ ppm, suggesting a strong intramolecular coordination with the nitrogen atom. The coordinated silanethione **67** showed unexpectedly low reactivity toward electrophiles and nucleophiles. Thus phosphanes, phosphites, ketones, epoxides, methyl iodide and hydrogen chloride are unreactive, as are alkoxysilanes, siloxanes and hydrosilanes. Only when **67** was treated with a large excess of methanol was the corresponding dimethoxysilane **73** produced (Scheme 26). Hydrolysis and oxidation lead to the silanone trimer **9**.

Crystallographic structural analysis of the coordinated silanethione was established with the more bulky silanethione **74** having a 1-naphthyl group replacing the phenyl group in **67**[9]. Figure 3 shows a schematic drawing of the molecular structure of **74** with some selected bond lengths.

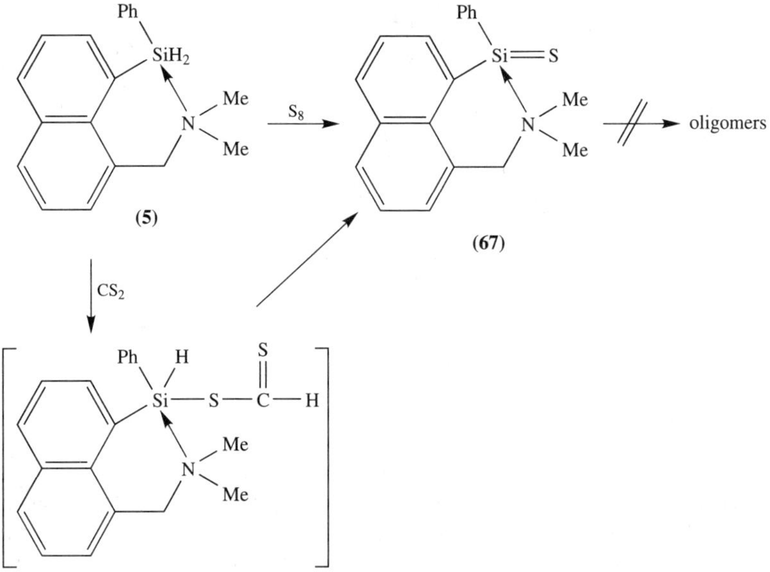

17. Recent advances in the chemistry of silicon–heteroatom multiple bonds 1085

SCHEME 25

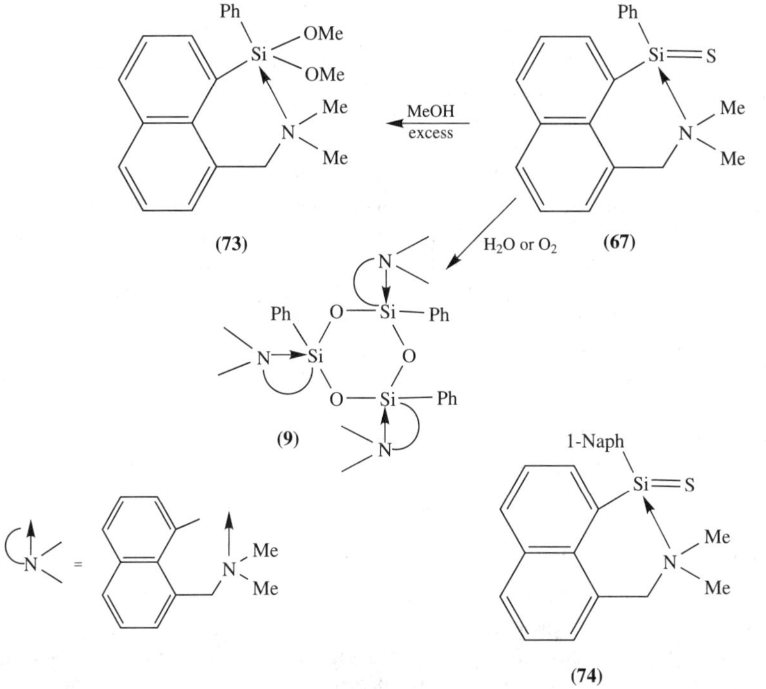

SCHEME 26

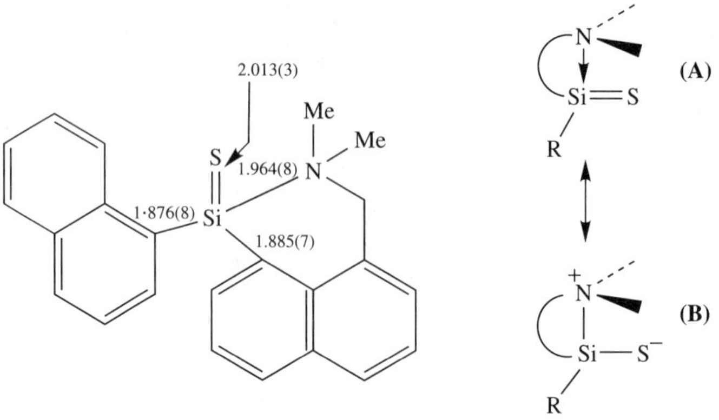

FIGURE 3. Molecular structure of silanethione **74** with selected bond lengths (Å)[9]

Although the Si−S bond (2.013 Å) in **74** is shorter than a typical Si−S single bond (2.13−2.16 Å)[53] suggesting its double-bond character to some extent, it is still 0.07 Å longer than the calculated value for the parent silanethione $H_2Si=S$ (**1**) (Section I.B.1). The Si−N distance (1.964 Å), which is only slightly longer than a Si−N σ bond (1.79 Å), supports a very strong coordination of the nitrogen atom of the dimethylaminomethyl group to the central silicon atom, which in turn makes the silathiocarbonyl unit of **74** considerably deviating from the ideal trigonal planar geometry around silicon; i.e. the sum of the angles around the central silicon atom is 344.9°. The authors concluded that the resonance betaine structure **B** (see Figure 3) contributes strongly to the electronic structure of the internally coordinated silanethiones **67** and **74**[9].

2. Formation from diaminosilanes

Corriu and coworkers also described an alternative synthetic method for internally coordinated silanethiones starting from pentacoordinated diaminosilanes. As shown in Scheme 27, the pentacoordinated diaminosilanes **75** are allowed to react with sulfur-containing heterocumulenes such as carbon disulfide or phenyl isothiocyanate to give the corresponding insertion products **76**, which undergoes thermal decomposition to produce the corresponding silanethiones **67**, **77** and **78**[54].

The silanethiones generated by these reactions were not isolated but their characteristic ^{29}Si NMR chemical shifts were reported to be +22.3, +34.2 and +41.1 ppm (CDCl$_3$) for **67**, **77** and **78**, respectively. The ^{29}Si chemical shift values of **77** and **78** are similar to those observed for **67** and **74** and therefore are indicative of their intramolecularly coordinated betaine structure.

3. Formation from silylenes

In analogy to silanone **36**, the reaction of the silylene, bis[2-(dimethylaminomethyl)phenyl]silanediyl (**35**), with phenyl isothiocyanate was examined (see Section II.B.5)[37]. In this reaction the expected silanethione **79** was obtained as a single product (Scheme 28), even in the presence of D$_3$ no insertion product of **79** into a Si−O bond of D$_3$ was observed, as shown by mass spectrometry and elemental analysis.

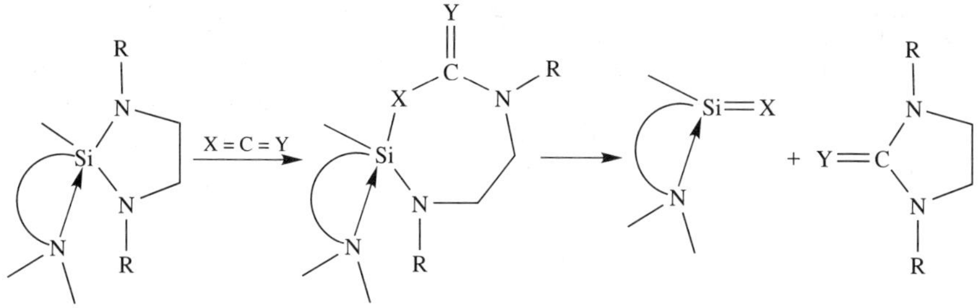

17. Recent advances in the chemistry of silicon–heteroatom multiple bonds 1087

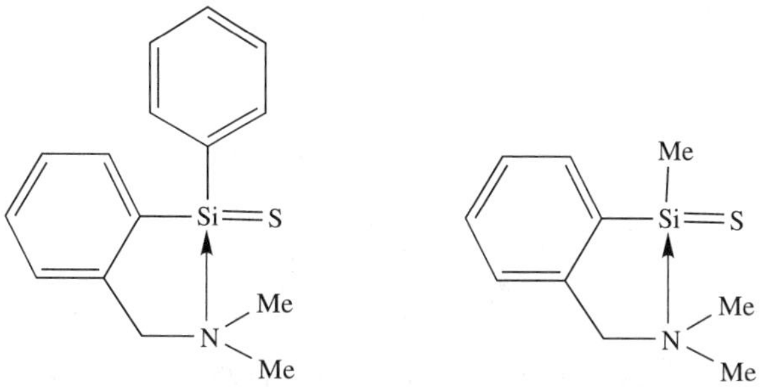

SCHEME 27

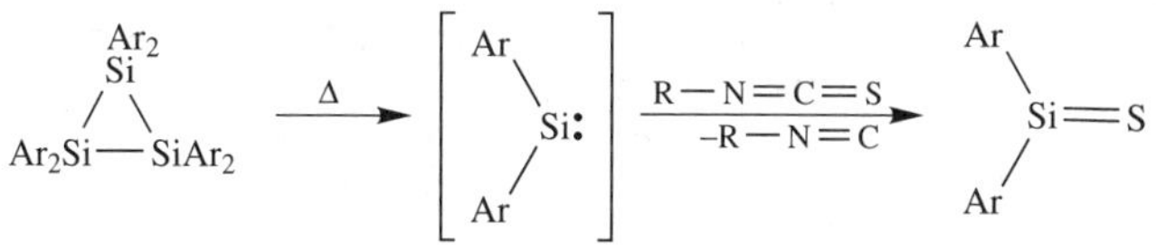

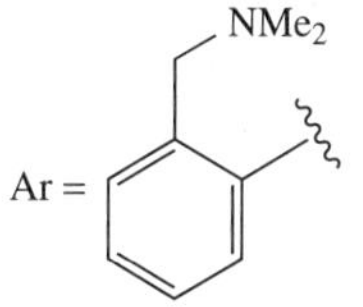

SCHEME 28

Silanethione **79** showed a characteristic ^{29}Si NMR chemical shift at $\delta -21.0$, which is apparently upfield shifted from that reported recently for the noncoordinated silanethione **4** ($\delta +166.56$ ppm), and is also at a significantly higher field than those of the tetracoordinated silanethiones **67**, **77** and **78** mentioned above. Accordingly, the silicon center of **79** is assumed to be pentacoordinated due to its intramolecular interaction with the two available terminal amino groups attached to the aryl substituents. This type of twofold coordination of the nucleophilic side arm of the 2-(dimethylaminomethyl)phenyl substituent to a coordinatively unsaturated silicon center has been shown to be also effective in thermodynamically stabilizing silicenium ions[55]. Although it was reported that silanethione **79** can be isolated as a white powder [mp 215 °C (dec)], there was no description of its reactivity or crystallographic structural analysis.

In analogy to the reaction of decamethylsilicocene **41** with carbon dioxide leading to the generation of intermediary silanone **44**, silicocene **41** was allowed to react with carbon oxysulfide under very mild conditions (−78 °C/toluene) to give the corresponding 1,3,2,4-dithiadisiletane derivative **81**, a formal head-to-tail [2 + 2]cycloaddition reaction product of the initially formed silanethione **80** (Scheme 29)[41b].

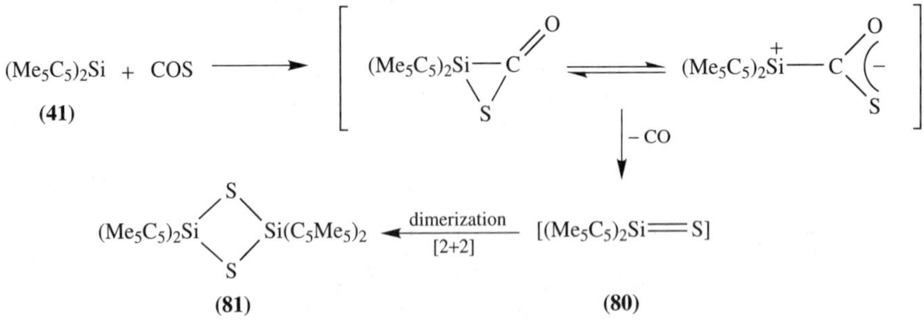

SCHEME 29

Silanethione **80** was also postulated as an intermediate in the reaction of silicocene **41** with isothiocyanates (room temperature/toluene/16 h for methyl isothiocyanate and 65 °C/toluene/5 h for phenyl isothiocyanate), which resulted in the formation of the corresponding intermolecular [2 + 2] cycloadducts (**82** and **83**) of **80** with the isothiocyanates (Scheme 30)[41a,b]. Under even more drastic conditions (100 °C/toluene/20 h), the reaction of **41** with phenyl isothiocyanate gave the five-membered heterocycles **84**, which is most likely produced by a second attack of the silicocene **41** on the initially formed [2 + 2] cycloadduct **83** (Scheme 30).

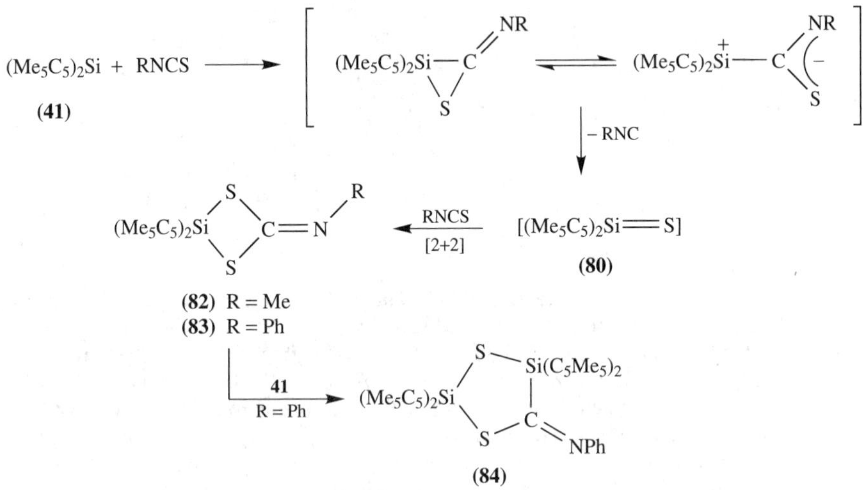

SCHEME 30

17. Recent advances in the chemistry of silicon–heteroatom multiple bonds   1089

In contrast, the reaction of silicocene **41** with carbon disulfide did not give the reaction products derived from the expected silanethione **80**; instead a surprising multistep reaction product **89** was isolated (Scheme 31)[41b]. In the preliminary communication[41a], the authors misinterpreted the structure of this reaction product as being a simple dimer **90** of the initial intermediate, thiasiliranethione **85**, but the X-ray crystallographic analysis later revealed that **89** is the actual product, as shown in Scheme 31[41b].

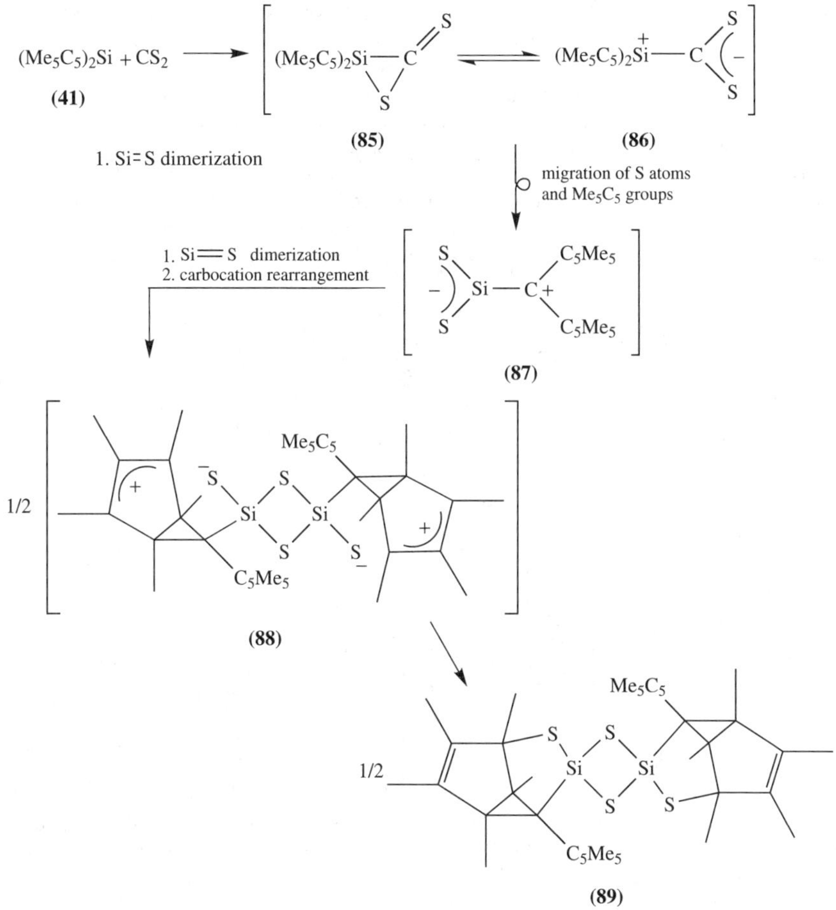

SCHEME 31

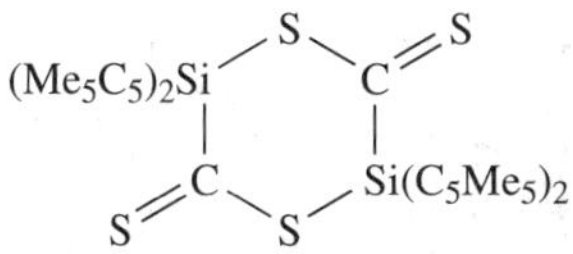

4. Formation from silicon-containing cyclic polysulfides

In order to synthesize a stable silanethione, it is important to choose a proper precursor and a suitable methodology, since it can be anticipated that the desired silanethione might be extremely reactive toward atmospheric oxygen and moisture to allow its purification by chromatography. Recently, the first kinetically stabilized silanethione **4** has been successfully synthesized and isolated by taking advantage of a new and efficient steric protection group, 2,4,6-tris[bis(trimethylsilyl)methyl]phenyl[10] (denoted as Tbt group in this chapter). The strategy for synthesizing this stable silanethione **4** is based on the simple desulfurization by a phosphine reagent of the corresponding silicon-containing cyclic polysulfides. Thus, a hexane solution of Tbt- and Tip-substituted tetrathiasilolane **91**[56], which was prepared by the reaction of the dibromosilane **92** with lithium naphthalenide followed by treatment with elemental sulfur, was refluxed in the presence of 3 molar equivalents of triphenylphosphine to produce quantitatively the silanethione **4** together with triphenylphosphine sulfide (Scheme 32). Since triphenylphosphine sulfide is almost insoluble in hexane, silanethione **4** was easily separated from the reaction mixture by filtration of the precipitated phosphine sulfide and was purified by recrystallization from hexane under argon atmosphere in a glovebox, yielding a yellow crystalline compound.

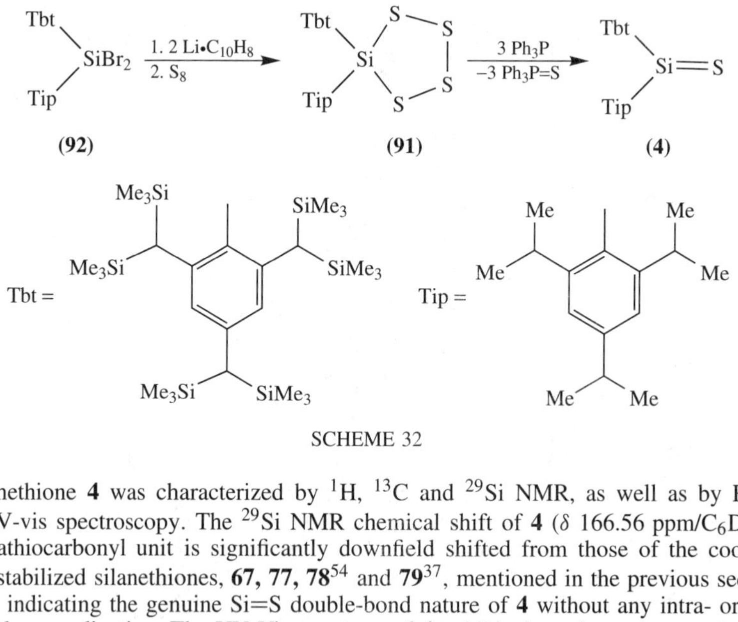

SCHEME 32

Silanethione **4** was characterized by ^{1}H, ^{13}C and ^{29}Si NMR, as well as by Raman and UV-vis spectroscopy. The ^{29}Si NMR chemical shift of **4** (δ 166.56 ppm/C$_6$D$_6$) for the silathiocarbonyl unit is significantly downfield shifted from those of the coordinatively stabilized silanethiones, **67, 77, 78**[54] and **79**[37], mentioned in the previous sections, clearly indicating the genuine Si=S double-bond nature of **4** without any intra- or intermolecular coordination. The UV-Vis spectrum of **4** exhibited an absorption maximum at 396 nm which was assigned to the n–π* transition. The Raman spectrum of **4** in the solid state showed an absorption at 724 cm^{-1} attributable to the Si=S stretching; density functional calculations for H$_2$Si=S (**1**) at the B3LYP/TZ (d,Z) level led to a Si=S stretching frequency of 723 cm^{-1} (see also Section I.B.1). The molecular structure of **4** was successfully solved by X-ray crystallographic analysis and the ORTEP and schematic drawings of the molecular structure of **4** are shown in Figure 4.

The silathiocarbonyl unit of **4** has a completely trigonal planar geometry, the sum of the bond angles around the central silicon atom being 359.9°. The dihedral angles between the trigonal plane and the two aryl planes are 41.8° for the Tbt ring and 67.8°

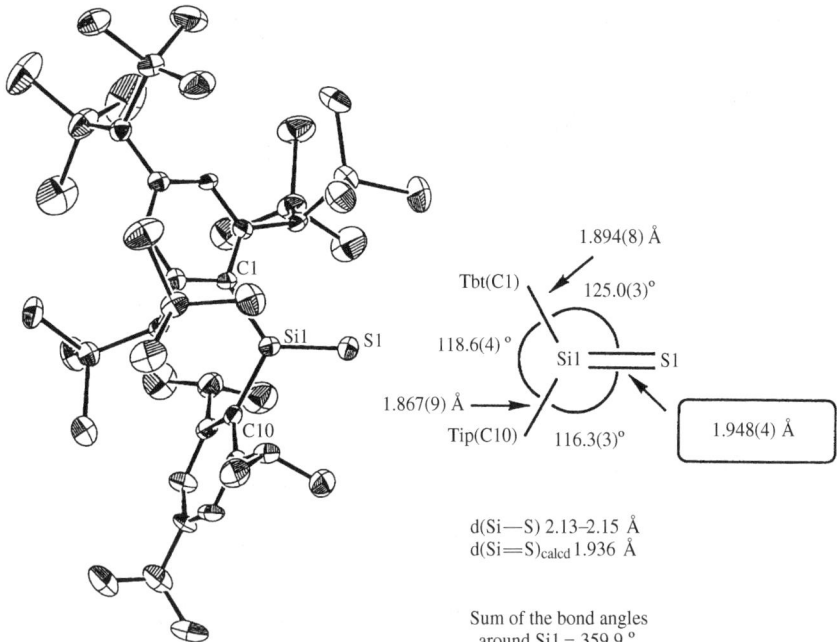

FIGURE 4. Molecular structure of silanethione **4** with selected bond lengths (Å) and angles (deg)[10]. Reprinted with permission from Ref. 10. Copyright (1994) American Chemical Society

for the Tip ring. The silicon–sulfur double-bond length is 1.948(4) Å, which is about 0.2 Å shorter than typical Si–S single-bond lengths[53] (*ca* 9% shortening) and significantly shorter than that reported for Corriu's silanethione **74** [2.013(3) Å][9]. These results support unambiguously a double-bond character between silicon and sulfur (a 'pure' silanethione nature) in compound **4**.

Though silanethione **4** is thermally very stable (up to its melting point of 185–189 °C), it has high chemical reactivity toward various reagents (Scheme 33). Methanol reacts instantaneously with **4** at room temperature to afford the expected addition product **93** (53%). Reactions of **4** with phenyl isothiocyanate and mesitonitrile oxide at room temperature resulted in the rapid formation of the [2 + 2] and [2 + 3] cycloadducts **94** (63%) and **95** (54%), respectively[10]. In contrast, 2,3-dimethyl-1,3-butadiene was very reluctant to react with **4**. The yellow color due to **4** did not disappear even upon heating a hexane solution of **4** and the diene in a sealed tube up to 150 °C. The reaction finally proceeded at 180 °C to give after 3 h the expected [2 + 4] cycloadduct **96** in 74% yield[10,57]. The formation of the [2 + 4] cycloadduct **96** from **4** demonstrates that the silanethione has a considerable extent of ene character like its carbon analogs such as thioketones and thioaldehydes, which are known to have high reactivities in Diels–Alder reactions[58]. It is surprising that silanethione **4** shows these high reactivities in spite of the extremely severe steric congestion around the Si=S group.

The synthesis of **4** makes it possible to compare the electronic spectra (n → π^*) of a series of R^1R^2M=S (M = C[59], Si[10], Ge[60], Sn[61]) compounds. In Table 2 are listed the observed spectra of these compounds along with the calculated spectra (at the CIS/DZ+d level) for the corresponding parent molecules H_2M=S (M = C, Si, Ge, Sn). The data in

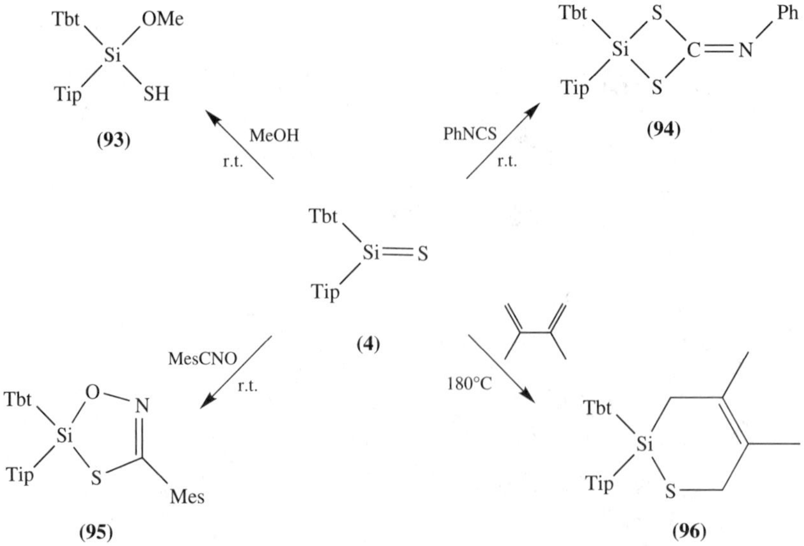

SCHEME 33

Table 2 show interesting change in the observed λ_{max} as a function of M; λ_{max} is significantly blue-shifted on going from thioaldehyde **97** to silanethione **4**, whereas λ_{max} values for **4**, germanethione **98** and stannanethione **99** are red-shifted when the atomic number of the group 14 element is increased. This trend is also found in the calculated values for H$_2$M=S (M = C, Si, Ge, Sn). Since the calculated $\Delta\epsilon_{n\pi^*}$ values increase continuously from H$_2$Sn=S to H$_2$C=S, the long-wavelength absorption for H$_2$C=S (and hence for **97**) most likely results from a large repulsion integral ($J_{n\pi^*}$) for the carbon–sulfur double bond (as in the case of H$_2$C=O vs H$_2$Si=O[62]), which causes λ_{max} to be longer than that expected from the HOMO–LUMO gap[10].

The less hindered tetrathiasilolane, Tbt(Mes)SiS$_4$ (**100**), was also desulfurized with triphenylphosphine (3 equiv.) in hexane at $-78\,^\circ$C to give a yellow-colored solution[63],

TABLE 2. Electronic spectra ($n \rightarrow \pi^*$) of doubly-bonded compounds between group 14 elements and sulfur

Observed[a]		Calculated[e]		
compound	λ_{max} (nm)	compound	λ_{max} (nm)	$\Delta\epsilon_{n\pi^*}$ (eV)[f]
Tbt(H)C=S (**97**)	587[b]	H$_2$C=S	458	10.81
Tbt(Tip)Si=S (**4**)	396	H$_2$Si=S	345	10.39
Tbt(Tip)Ge=S (**98**)	450[c]	H$_2$Ge=S	363	9.97
Tbt(Tip)Sn=S (**99**)	473[d]	H$_2$Sn=S	380	9.30

[a] In hexane.
[b] Reference 59.
[c] Reference 60.
[d] Reference 61.
[e] At CIS/DZ + d.
[f] $\epsilon_{LUMO(\pi^*)} - \epsilon_{HOMO(n)}$.

suggesting in analogy with the desulfurization of **91** the generation of the corresponding silanethione **101** (Scheme 34). The yellow color of the solution disappeared on warming the solution to room temperature, resulting in the exclusive formation of cis-1,3,2,4-dithiadisiletane **102**, the stereospecific [2 + 2] cycloaddition product of silanethione **101**. Although the silanethione **101** could not be isolated and it rapidly dimerized at room temperature to dithiadisiletane **102**, the formation of the silanethione **101** is strongly supported by the isolation of the expected trapping products with various trapping reagents such as methanol, phenyl isothiocyanate and mesitonitrile oxide, i.e. **103, 104** and **105**, respectively, as shown in Scheme 34.

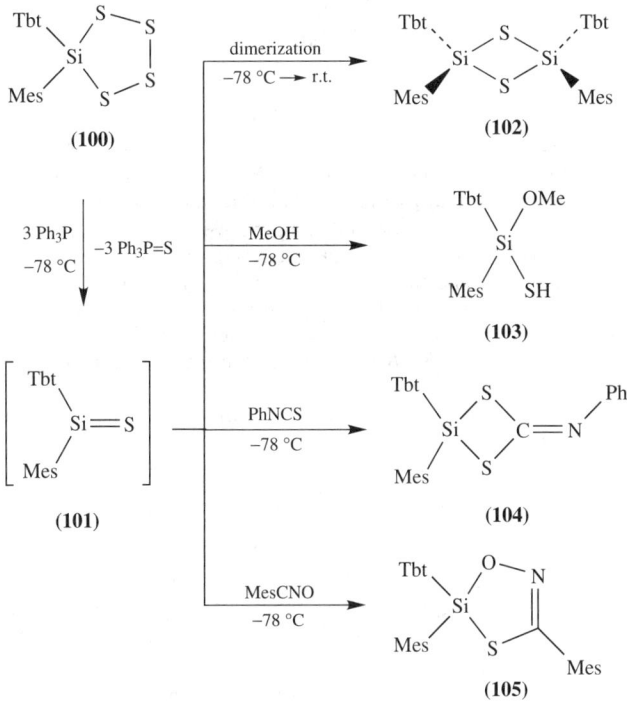

SCHEME 34

The synthetic strategy for the preparation of the stable silanethione **4** based on the desulfurization of the corresponding overcrowded tetrathiasilolane described above was found to be applicable to the syntheses of the other species having double bonds between group 14 and group 16 elements, such as thioaldehydes[59], selenoaldehydes[64], gemanethiones[60], germaneselones[65], stannanethiones[61], stannaneselones[66] and plumbanethiones[67] as shown in Scheme 35. The only exceptions are the germanetetellones[68] which were synthesized by the reaction or a germylene with elemental tellurium. All the germanium–chalcogen double-bond compounds were isolated as stable crystalline compounds and were fully characterized by spectroscopic data and by X-ray structural analysis. All of them showed a substantial bond shortening of the Ge=X double bond compared to a Ge–X single bond by ca 8–9% and a completely planar trigonal geometry, as in the case of the silanethione **4**.

5. Formation from a disilene

As a new approach to a silanethione, West and coworkers reported the photolysis of a sterically crowded 1,2,3-thiadisiletane derivative **106** derived from the [2 + 2] cycloaddition reaction of tetramesityldisilene with thiobenzophenone[69]. Crystal structural analysis revealed that **106** has abnormally long Si−Si (2.443 Å) and Si−S (2.177 Å) distances and a highly distorted four-membered ring with a dihedral angle of 45.6° between the Si(1)−Si(2)−C and Si(1)−S−C planes. As shown in Scheme 36, photolysis of this strained silacyclic compound **106** in the presence of ethanol afforded two products, dimesityl(diphenylmethyl)ethoxysilane (**107**) and dimesitylethoxysilanethiol (**108**). In the absence of ethanol, photolysis of **106** resulted in the formation of a silene, 1,1-dimesityl-2,2-diphenylsilene (**109**), together with an oligomer of dimesitylsilanethione **110** (**111**; for this compound the authors proposed a dimer or a trimer structure).

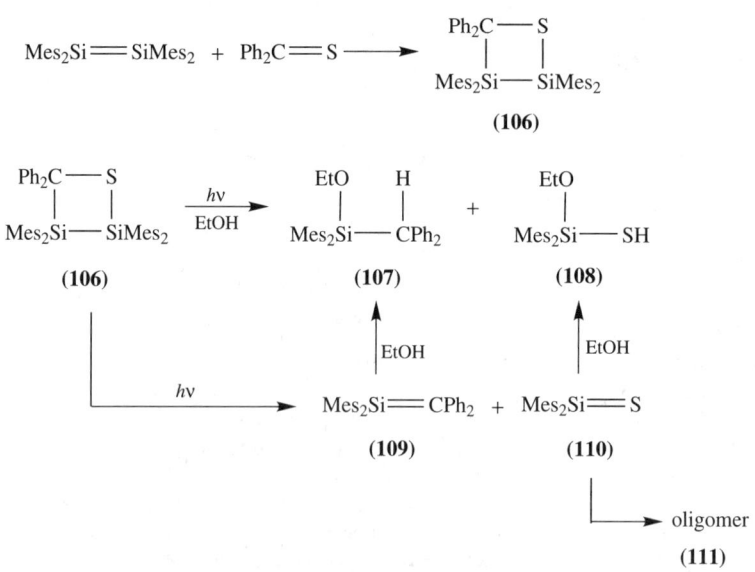

SCHEME 36

17. Recent advances in the chemistry of silicon–heteroatom multiple bonds 1095

These results are most likely interpreted in terms of the photochemical cycloreversion of the thiadisiletane **106** leading to the formation of the silene **109** and silanethione **110**, though both of them were not isolated.

6. Formation from thioketenes

As in the case of extrusion of dimethylsilanone, $Me_2Si=O$ (**10**), in the thermolysis of certain silaketenes[70], a similar type of silanethione ($Me_2Si=S$; **112**) extrusion was postulated in the flash vacuum pyrolysis of bis(trimethylsilyl)thioketene (**113**) and (dimethylsilyl)(trimethylsilyl)thioketene (**121**) as shown in Schemes 37 and 38[71]. In both cases, the formation of all the reaction products (compounds **115–120** for the pyrolysis of **113** shown in Scheme 37 and compounds **118, 120, 123** and **124** for the pyrolysis of **121** shown in Scheme 38) can be mechanistically rationalized by processes each initiated by isomerization of the starting thioketenes via a 1,2-shift of a trimethylsilyl group to the corresponding α-thioketocarbenes **114** and **122**. Under the pyrolytic reaction conditions used (700 or 768 °C) the intermediate silanethione **112** underwent ready oligomerization to give its dimer **120** and/or trimer **124**.

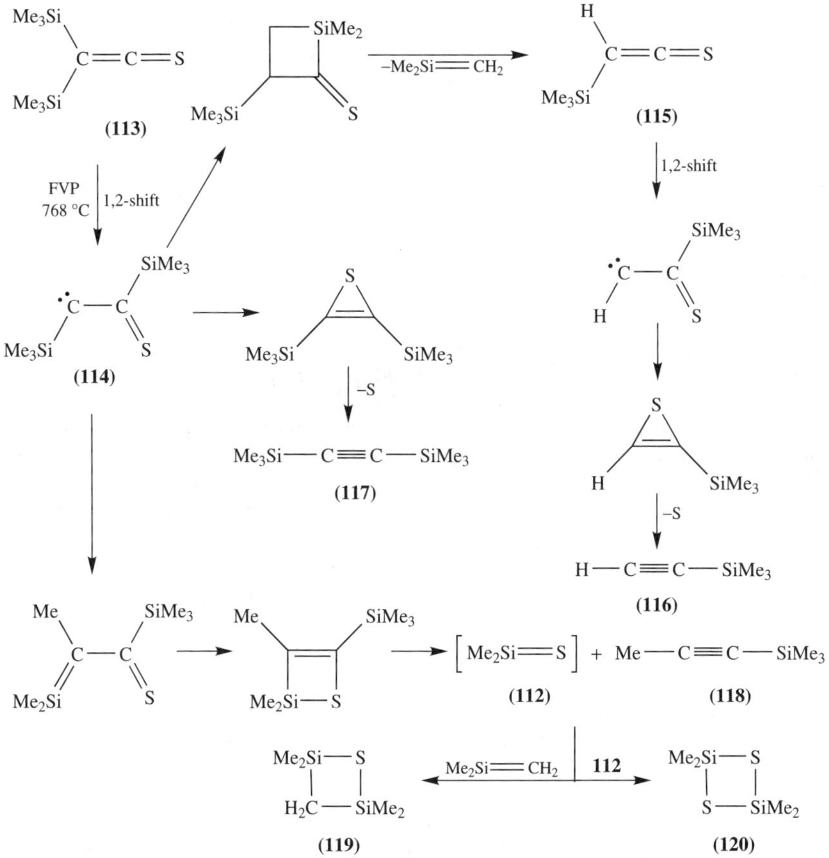

SCHEME 37

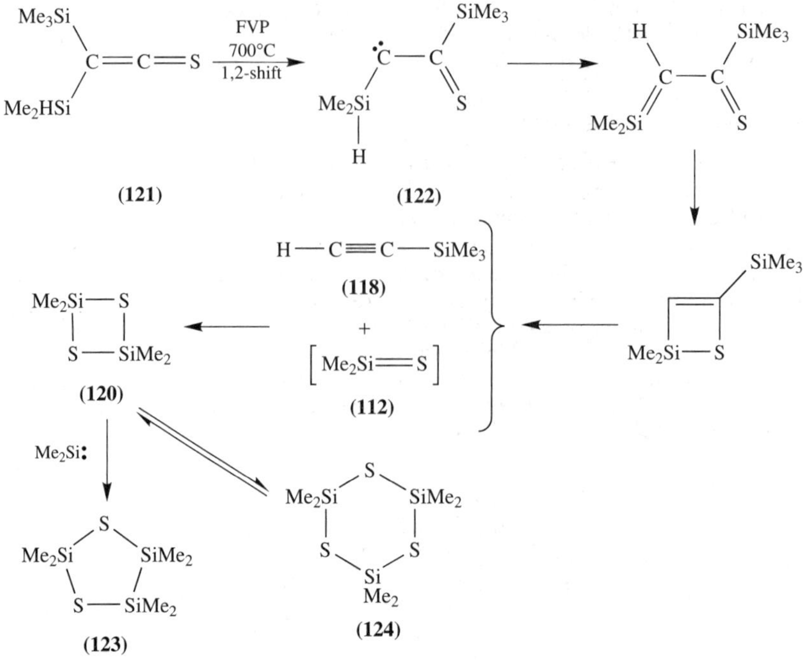

SCHEME 38

7. Matrix isolation of SiS$_2$, Cl(H)Si=S and Cl$_2$Si=S

SiS$_2$ (**125**)[72], Cl(H)Si=S (**126**)[51] and Cl$_2$Si=S (**127**)[50] were prepared and isolated in low temperature matrices. The properties of these species are of interest as the silicon analogues of carbon disulfide (CS$_2$), thioformyl chloride [S=C(H)Cl] and thiophosgene (S=CCl$_2$), which are well-known thiocarbonyl compounds.

Molecular SiS$_2$ (**125**) was generated in a solid argon matrix by a reaction of SiS with S atoms. The antisymmetric stretching vibration ν_{as}(SiS) is observed at 918 cm^{-1}. Bonding and structure properties (force constants from experimentally observed frequencies and results from *ab initio* SCF calculations) of SiS$_2$ were compared with those of similar molecules: CO, CS, CO$_2$, COS, CS$_2$, SiO, SiS, SiO$_2$ and SiOS[72].

Molecular Cl(H)Si=S (**126**) was also formed in an argon matrix in a photochemically induced reaction of SiS with HCl. From the isotopic splittings (H/D and ^{35}Cl/^{37}Cl) of the IR absorptions the C_s structure of the species with silicon as the central atom is deduced. By a normal coordinate analysis a value of 4.83 mdyn Å^{-1} is obtained for the SiS force constant, a value which was confirmed by *ab initio* SCF calculations of the IR spectrum[51].

Under similar reaction conditions, Cl$_2$Si=S (**127**) was formed in a matrix reaction between SiS and Cl$_2$. The formation of **127** was also concluded from some isotopic shifts in the IR spectra. The force constant of the Si–S bond in **127** has a value of 4.9 mdyn Å^{-1}[50].

Schnöckel and coworkers have also calculated at SCF level, the molecular structures and charge distributions for H$_2$S=S (**1**), Cl(H)Si=S (**126**) and Cl$_2$Si=S (**127**) as well as those of Cl$_2$Si=O (**128**), the oxygen analogue of **127**, and the results are presented in Figure 5.

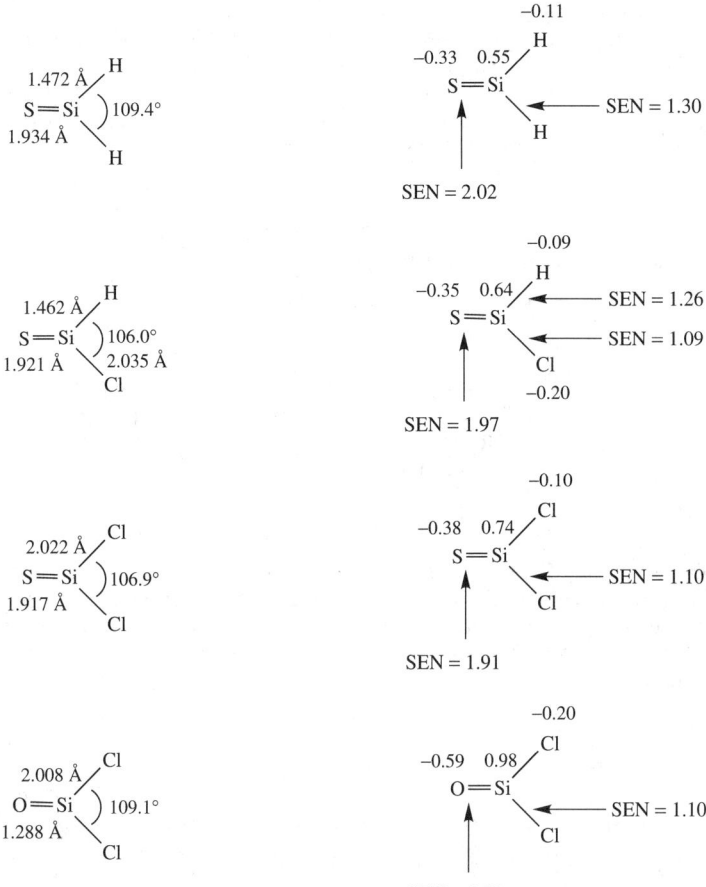

FIGURE 5. Molecular structures and charge distributions calculated at the SCF level for $H_2Si=S$ (**1**), ClHSi=S (**126**), $Cl_2Si=S$ (**127**) and $Cl_2Si=O$ (**128**). SEN = shared electron numbers. See: C. Ehahardt and R. Ahlrichs, *Theor. Chim. Acta*, **68**, 231 (1985).

D. Silicon–Selenium and Silicon–Tellurium Double Bonds (Silaneselones and Silanetellones)

As can be seen in the previous sections of this chapter, remarkable progress has been made in the chemistry of silanones and silanethiones, but still very little is known on the chemistry of their heavier chalcogen analogues, i.e. silaneselones and silanetellones. Thus, up to now, no report has appeared on the preparation or the spectroscopic detection of a silanetellone, and only limited information is available on silaneselones.

1. Transient silaneselones

For the generation of silaneselones, Boudjouk and Thompson have described the photochemical and thermal fragmentations of cyclosilaselenanes[73]. Thus, several cyclodisiladiselenanes and cyclotrisilatriselenanes, $(R_2SiSe)_n$ (**129–134**; $n = 2, 3$), were generated from Na_2Se and the corresponding R_2SiCl_2 (Scheme 39)[74]. The properties of the cyclosilaselenanes are found to depend on the groups attached to silicon. Thus, while alkyl-substituted systems [e.g. **129** and **130** (R = Me) and **131** and **132** (R = Et) are thermally unstable air-sensitive yellow oils, the silyl-substituted one **134** (R = $SiMe_3$) was isolated as green crystals stable to air for several days.

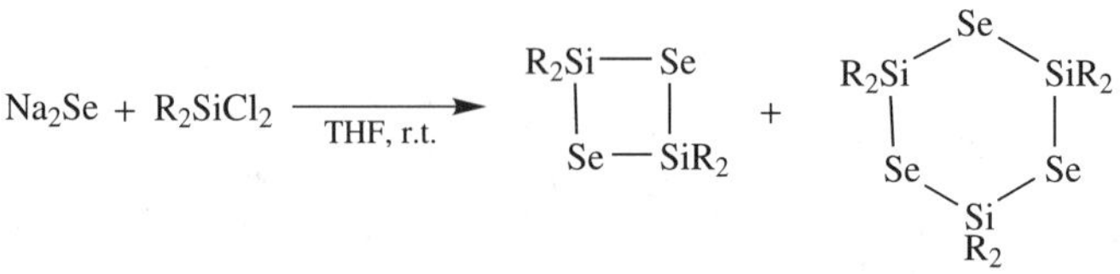

R	4-membered	6-membered
R = Me	(**129**) 6%	(**130**) 35%
R = Et	(**131**) 40%	(**132**) 30%
R = Ph	–	(**133**) 40%
R = $SiMe_3$	(**134**) 35%	–

SCHEME 39

The above-mentioned cyclosilaselenanes generate the corresponding silaneselones, $R_2Si=Se$ (**135**), both thermally and photochemically when R = Me, Et and $n = 3$ but only thermally when R = Me, Et and $n = 2$ (Scheme 40). The silaneselones **135** are easily trapped with hexamethylcyclotrisiloxane (D_3) to give the corresponding insertion product **136**.

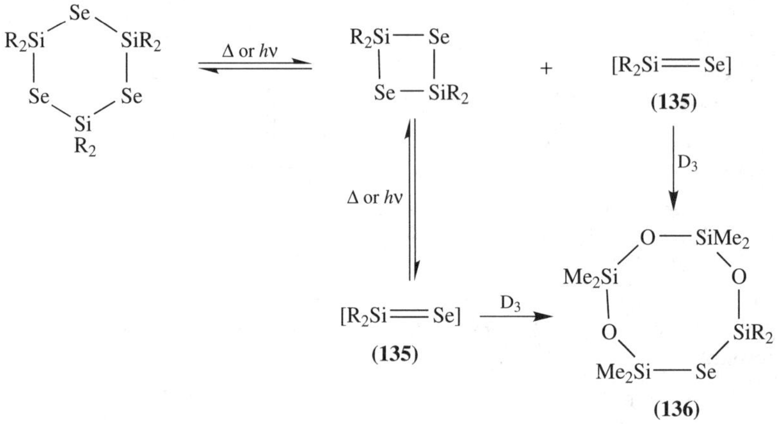

SCHEME 40

In the case of R = Ph only the six-membered ring product **133** was available as the precursor, and upon thermolysis it gave a modest yield of an insertion reaction product of $Ph_2Si=Se$ to D_3, but a complex mixture resulted upon photolysis. On the other hand, the trimethylsilyl substituted cyclodisiladiselenane **134** was found to be stable toward ring fragmentation either on thermolysis (250 °C for 5 days in a sealed tube) or on photolysis[73b].

2. Silaneselone derived from silicocene

In 1989 Jutzi and coworkers reported the reaction of decamethylsilicocene **41** with tri-n-butylphosphine selenide in benzene at room temperature leading to almost quantitative formation of 1,3,2,4-diselenadisiletane derivative **138**, a head-to-tail [2 + 2] cycloaddition reaction product of the initially formed silaneselone **137**[75]. The intermediacy of silaneselone **137** was supported by the fact that the reaction in the presence of 2,3-dimethyl-1,3-butadiene resulted in the formation of the corresponding [2+4] cycloaddition reaction product **139** (Scheme 41). As in the cases of silanone **44** and silanethione **80**, the ligands on silicon undergo a haptotropic rearrangement from η^5-C$_5$Me$_5$ in **41** to η^1-C$_5$Me$_5$ in **138** or **139**. Apparently, silaneselone **137** is not kinetically stable enough to be isolated under normal conditions.

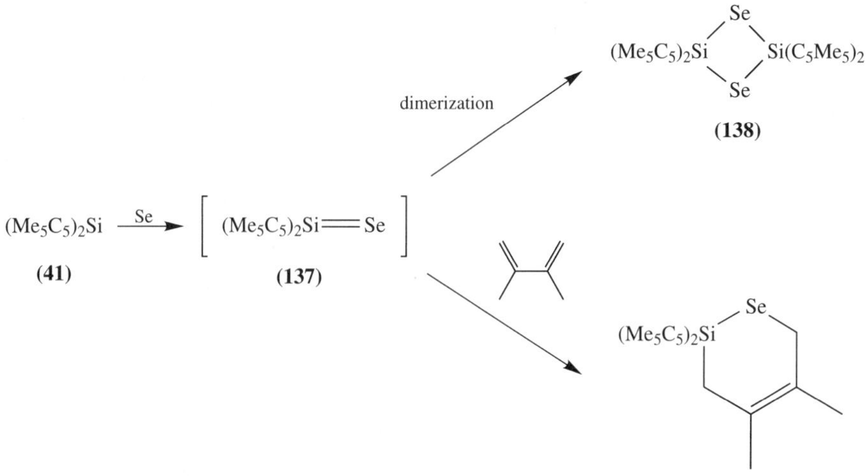

SCHEME 41

3. Thermodynamically stabilized silaneselone

Corriu and coworkers found that the reaction of the pentacoordinated functionalized silane **6** (see also Sections II.B.1 and II.C.1) with elemental selenium leads to a stable silaneselone **140**, which is stabilized by intramolecular coordination of the nitrogen-containing substituent to the doubly-bonded silicon (Scheme 42)[9]. Although the structure of **140** was supported by ^{29}Si and ^{13}C NMR and MS spectra, including a downfield ^{29}Si chemical shift ($\delta = +29.4$) and a high coupling constant with ^{77}Se ($J_{SeSi} = 257$ Hz), neither the crystallographical structure analysis nor the reactivity of this isolable Si=Se compound has been reported.

4. Sterically protected silaneselone stable in solution

In analogy to the case of stable silanethione **4**, kinetic stabilization of a silaneselone by the large Tbt group was examined (see Section II.C.3). Since the synthesis and isolation of the cyclic polyselenides Tbt(R)SiSe$_4$ (R = Mes or Tip) were unsuccessful, probably due to their lower stabilities than those of the corresponding tetrathiasilolanes such as **91**

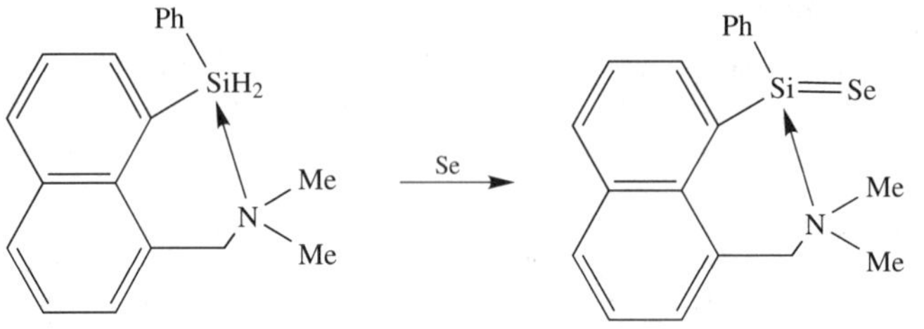

SCHEME 42

and **100**, the direct selenation of the sterically hindered diarylsilylene Tbt(Dip)Si : (**142**; Dip = 2,6-diisopropylphenyl) was examined. Thus, the bicyclic silirane derivative (**141**) bearing Tbt and Dip groups on the silicon atom was treated with elemental selenium to afford a red solution, suggesting the formation of the corresponding silaneselone **143** (Scheme 43)[76].

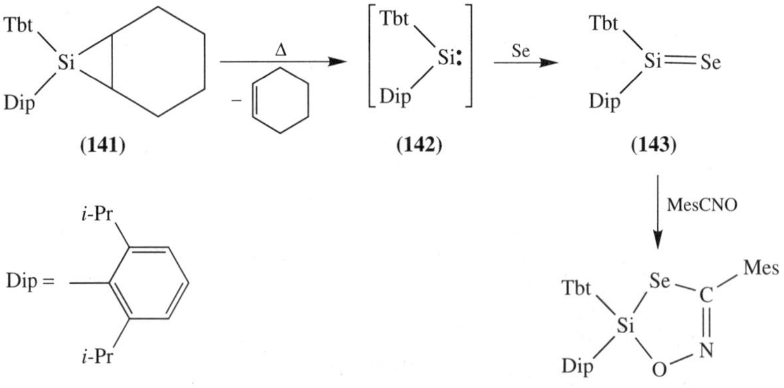

SCHEME 43

The formation of silaneselone **143** was supported by the trapping reaction with mesitonitrile oxide leading to the expected corresponding cycloadduct **144** as well as by the observation of a remarkable downfield ^{29}Si chemical shift (δ_{Si} = 174 ppm) indicative of the Si=Se double bond of **143**.

III. OUTLOOK FOR THE CHEMISTRY OF SILICON–HETEROATOM MULTIPLE BONDS

As mentioned in this chapter, in recent years much progress has been made in the chemistry of silicon–chalcogen multiple bonds. For silicon–sulfur doubly-bonded compounds, we have now several isolated examples, both kinetically stabilized and thermodynamically stabilized. Furthermore, there have been reports of the synthesis and characterization of stable compounds with silicon–nitrogen double bonds (i.e. silanimines or iminosilanes) as well as their heavier group 15 element analogues such as phosphasilenes and arsasilenes.

17. Recent advances in the chemistry of silicon–heteroatom multiple bonds 1101

In view of the recent investigations it seems that kinetic stabilization with bulky substituent(s) (steric protection) of silicon–heteroatom double bonds is superior to their thermodynamic stabilization [e.g. by mesomeric effects or intramolecular coordination of heteroatom-containing substituent(s) or metal complexation to the silicon–heteroatom π-bond]. Although except for the Si=S double bond (silanethiones) other silicon–chalcogen atom doubly-bonded species still remain a challenging and fascinating target for isolation and characterization, the remarkable progress in molecular design and steric protection will make it possible in the future, so we believe, to reveal the interesting bond character of these novel silicon-containing π-bond systems.

IV. REFERENCES

1. A. G. Brook, F. Abdesaken, B. Gutekunst, G. Gutekunst and R. K. Kallury, *J. Chem. Soc., Chem. Commun.*, 191 (1981).
2. R. West, M. J. Fink and J. Michl, *Science*, **214**, 1343 (1981).
3. M. Yoshifuji, I. Shima, N. Inamoto, K. Hirotsu and T. Higuchi, *J. Am. Chem. Soc.*, **103**, 4587 (1981).
4. (a) L. E. Gusel'nikov and N. S. Nametkin, *Chem. Rev.*, **79**, 529 (1979).
   (b) G. Raabe and J. Michl, *Chem. Rev.*, **85**, 419 (1985).
   (c) A. G. Brook and K. M. Bains, *Adv. Organomet. Chem.*, **25**, 1 (1986).
   (d) R. West, *Angew. Chem., Int. Ed. Engl.*, **26**, 1201 (1987).
   (e) G. Raabe and J. Michl, in *The Chemistry of Organic Silicon Compounds, Part 2* (Eds. S. Patai and Z. Rappoport), Wiley, New York, 1989, p. 1015.
   (f) J. Barrau, J. Escudié and J. Satgé, *Chem. Rev.*, **90**, 283 (1990).
   (g) T. Tsumuraya, S. A. Batcheller and S. Masamune, *Angew. Chem., Int. Ed. Engl.*, **30**, 902 (1991).
5. (a) N. Wiberg, K. Schurz and G. Fischer, *Angew. Chem., Int. Ed. Engl.*, **24**, 1053 (1985).
   (b) N. Wiberg, K. Schurz, G. Reber and G. Müller, *J. Chem. Soc., Chem. Commun.*, 591 (1986).
   (c) M. Hesse and U. Klingebiel, *Angew. Chem., Int. Ed. Engl.*, **25**, 649 (1986).
   (d) For a review on the synthesis and reactions of iminosilanes and related compounds see I. Hemme and U. Klingebiel, *Adv. Organomet. Chem.*, **39**, 159 (1996).
6. (a) C. N. Smit and F. Bickelhaupt, *Organometallics*, **6**, 1156 (1987).
   (b) Y. van den Winkel, H. M. M. Bastiaans and F. Bickelhaupt, *J. Organomet. Chem.*, **405**, 183 (1991).
   (c) M. Driess, *Angew. Chem., Int. Ed. Engl.*, **30**, 102 (1991).
7. (a) M. Driess and H. Pritzkow, *Angew. Chem., Int. Ed. Engl.*, **31**, 316 (1992).
   (b) For a review on stable doubly-bonded compounds of germanium and tin see K. M. Baines and W. G. Stibbs, *Adv. Organomet. Chem.*, **39**, 275 (1996).
   (c) For a review on silicon–phosphorus and silicon–arsenic multiple bonds see M. Driess, *Adv. Organomet. Chem.*, **39**, 193 (1996).
8. For recent reviews see
   (a) A. G. Brook and M. A. Brook (The Chemistry of Silenes), *Adv. Organomet. Chem.*, **39**, 71 (1996).
   (b) R. Okazaki and R. West (Chemistry of Stable Disilenes), *Adv. Organomet. Chem.*, **39**, 232 (1996).
9. P. Arya, J. Boyer, F. Carré, R. Corriu, G. Lanneau, J. Lapasset, M. Perrot and C. Priou, *Angew. Chem., Int. Ed. Engl.*, **28**, 1016 (1989).
10. H. Suzuki, N. Tokitoh, S. Nagase and R. Okazaki, *J. Am. Chem. Soc.*, **116**, 11578 (1994).
11. N. Auner Chapter 16 in this volume.
12. H. Sakurai Chapter 15 in this volume.
13. T. Kudo and S. Nagase, *Organometallics*, **5**, 1207 (1986).
14. T. Kudo and S. Nagase, *J. Phys. Chem.*, **88**, 2833 (1984).
15. M. W. Schmidt, P. N. Truong and M. S. Gordon, *J. Am. Chem. Soc.*, **109**, 5217 (1987).
16. P. v. R. Schleyer and D. Kost, *J. Am. Chem. Soc.*, **110**, 2105 (1988).
17. P. J. Bruna and F. Grein, *Chem. Phys.*, **165**, 265 (1992).
18. R. Withnall and L. Andrews, *J. Am. Chem. Soc.*, **107**, 2567 (1985).

19. V. N. Khabashesku, Z. A. Kerzina, E. G. Baskir, A. K. Maltsev and O. M. Nefedov, *J. Organomet. Chem.*, **347**, 277 (1988).
20. G. Manuel, G. Bertrand, W. P. Weber and S. A. Kazoura, *Organometallics*, **3**, 1340 (1984).
21. I. M. T. Davidson, A. Fenton, G. Manuel and G. Bertrand, *Organometallics*, **4**, 1324 (1985).
22. G. Hussmann, W. D. Wulff and T. J. Barton, *J. Am. Chem. Soc.*, **105**, 1263 (1983).
23. V. N. Khabashesku, Z. A. Kerzina, E. G. Baskir, A. K. Maltsev and O. M. Nefedov, *J. Organomet. Chem.*, **364**, 301 (1989).
24. T. J. Barton, Contribution to International Symposium of Orgaonsilicon Reactive Intermediates, Sendai, Japan, September 1984.
25. C. Eaborn and W. A. Stanczyk, *J. Chem. Soc., Perkin Trans. 2*, 2099 (1984).
26. A. G. Davies and A. G. Neville, *J. Organomet. Chem.*, **436**, 255 (1992).
27. R. West and R. H. Baney, *J. Am. Chem. Soc.*, **81**, 6145 (1959).
28. C. Chatgilialoglu, K. U. Ingold, J. C. Scaiano and H. Woynar, *J. Am. Chem. Soc.*, **103**, 3231 (1981).
29. R. Goikhman, M. Aizenberg, L. J. W. Shimon and D. Milstein, *J. Am. Chem. Soc.*, **118**, 10894 (1996).
30. T. Matsubara and Y. Saito, *J. Mol. Catal.*, **92**, 1 (1994) and references cited therein.
31. H. S. D. Soysa, H. Okinoshima and W. P. Weber, *J. Organomet. Chem.*, **133**, C-17 (1977).
32. (a) W. F. Gore and T. J. Barton, *J. Organomet. Chem.*, **199**, 33 (1980).
    (b) D. Tzeng and W. P. Weber, *J. Am. Chem. Soc.*, **102**, 1451 (1980).
    (c) W. Ando, M. Ikeno and Y. Hamada, *J. Chem. Soc., Chem. Commun.*, 621 (1981).
33. C. A. Arrington, R. West and J. Michl, *J. Am. Chem. Soc.*, **105**, 6176 (1983).
34. A. Patyk, W. Sander, J. Gauss and D. Cremer, *Angew. Chem., Int. Ed. Engl.*, **28**, 898 (1989).
35. (a) J. Belzner and H. Ihmels, *Tetrahedron Lett.*, **34**, 6541 (1993).
    (b) J. Belzner, H. Ihmels, B. O. Kneisel, R. O. Gould and R. Herbst-Irmer, *Organometallics*, **14**, 305 (1995).
    (c) R. Corriu, G. Laneau, C. Priou, F. Soulairol, N. Auner, R. Probst, R. Conlin and C. Tan, *J. Organomet. Chem.*, **466**, 55 (1994).
36. J. Belzner, *J. Organomet. Chem.*, **430**, C51 (1992).
37. J. Belzner, H. Ihmels, B. O. Kneisel and R. Herbst-Irmer, *Chem. Ber.*, **129**, 125 (1996).
38. M. Weidenbruch, B. Flintjer, S. Pohl and W. Saak, *Angew. Chem. Int. Ed. Engl.*, **28**, 95 (1989).
39. M. Driess and H. Prizkow, *Angew. Chem., Int. Ed. Engl.*, **31**, 316 (1992).
40. N. Wiberg, G. Preiner and K. Schurz, *Chem. Ber.*, **121**, 1407 (1988).
41. (a) P. Jutzi and A. Möhrke, *Angew. Chem. Int. Ed. Engl.*, **28**, 762 (1989).
    (b) P. Jutzi, D. Eikenberg, A. Möhrke, B. Neumann and H.-G. Stammler, *Organometallics*, **15**, 753 (1996).
42. N. Wiberg, G. Preiner and G. Wagner, *Z. Naturforsch. B*, **B426**, 1062 (1987).
43. (a) C. M. Golino, R. D. Bush, P. On and L. H. Sommer, *J. Am. Chem. Soc.*, **97**, 1957 (1975).
    (b) W. Ando, A. Sekiguchi and T. Migita, *J. Am. Chem. Soc.*, **97**, 7159 (1975).
    (c) W. Ando, M. Ikeno and A. Sekiguchi, *J. Am. Chem. Soc.*, **99**, 6447 (1977).
44. S. Bachrach and A. Streitwieser, Jr., *J. Am. Chem. Soc.*, **107**, 1186 (1985).
45. A. G. Brook, W. J. Chatterton, J. F. Sawyer, D. W. Hughes and K. Vorspohl, *Organometallics*, **6**, 1246 (1987).
46. I. M. T. Davidson, C. E. Dean and F. T. Lawrence, *J. Chem. Soc., Chem. Commun.*, 52 (1981).
47. A. G. Brook, S. C. Niburg, F. Abdesaken, B. Gutekunst, G. Gutekunst, R. K. M. R. Kallury, Y. C. Poon, J.-M. Chang and W. Wong-Ng, *J. Am. Chem. Soc.*, **104**, 5667 (1982).
48. M. Driess, H. Pritzkow, S. Rell and U. Winkler, *Organometallics*, **15**, 1845 (1996).
49. T. Kudo and S. Nagase, *J. Am. Chem. Soc.*, **107**, 2589 (1985).
50. H. Schnöckel, H. J. Göcke and R. Köppe, *Z. Anorg. Allg. Chem.*, **578**, 159 (1989).
51. R. Köppe and H. Schnöckel, *Z. Anorg. Allg. Chem.*, **607**, 41 (1992).
52. M. Weidenbruch, A. Schäfer and R. Rankers, *J. Organomet. Chem.*, **195**, 171 (1980).
53. W. S. Sheldrick, in *The Chemistry of Organic Silicon Compounds, Part 1*, (Eds. S. Patai and Z. Rappoport), Wiley, New York, 1989, pp. 227–304. See also R. K. Sibao, N. L. Keder and H. Eckert, *Inorg. Chem.*, **29**, 4163 (1990).
54. R. J. P. Corriu, G. F. Laneau and V. D. Mehta, *J. Organomet. Chem.*, **419**, 9 (1991).
55. J. Beltzner, D. Schär, B. O. Kneisel and R. Herbst-Irmer, *Organometallics*, **14**, 1840 (1995).
56. N. Tokitoh, H. Suzuki, T. Matsumoto, Y. Matsuhashi, R. Okazaki and M. Goto, *J. Am. Chem. Soc.*, **113**, 7047 (1991).

17. Recent advances in the chemistry of silicon–heteroatom multiple bonds 1103

57. N. Tokitoh, H. Suzuki and R. Okazaki, *Xth International Symposium on Organosilicon Chemistry*, Poznan, Poland, Abstract O-62, 1993, p. 110.
58. (a) W. J. Middleton, *J. Org. Chem.*, **30**, 1390 (1965).
    (b) A. Schönberg and B. König, *Chem. Ber.*, **101**, 725 (1968).
    (c) Y. Ohnishi, Y. Akasaki and A. Ohno, *Bull. Chem. Soc. Jpn.*, **46**, 3307 (1973).
    (d) J. E. Baldwin and R. C. G. Lopez, *Tetrahedron*, **39**, 1487 (1983).
    (e) G. W. Kirby, A. W. Lochead and G. N. Sheldrake, *J. Chem. Soc., Chem. Commun.*, 922 (1984).
    (f) G. A. Krafft and P. T. Meinke, *Tetrahedron Lett.*, **26**, 1947 (1985).
    (g) E. Vedejs, T. H. Eberlein, D. J. Mazur, C. K. McClure, D. A. Penny, R. Ruggeri, E. Schwartz, J. S. Stults, D. L. Varie, R. G. Wilde and S. Wittenberger, *J. Org. Chem.*, **51**, 1556 (1986).
    (h) M. Segi, T. Nakajima, S. Suga, S. Murai, A. Ogawa and N. Sonoda, *J. Am. Chem. Soc.*, **110**, 1976 (1988).
59. (a) N. Tokitoh, N. Takeda and R. Okazkai, *J. Am. Chem. Soc.*, **116**, 7907 (1994).
    (b) N. Takeda, N. Tokitoh and R. Okazaki, *Chem. Eur. J.*, **3**, 62 (1997).
60. N. Tokitoh, T. Matsumoto, K. Manmaru and R. Okazaki, *J. Am. Chem. Soc.*, **115**, 8855 (1993).
61. N. Tokitoh, M. Saito and R. Okazaki, *J. Am. Chem. Soc.*, **115**, 2065 (1993).
62. T. Kudo and S. Nagase, *Chem. Phys. Lett.*, **128**, 507 (1986).
63. H. Suzuki, Ph. D. Thesis, The University of Tokyo (1994); H. Suzuki, N. Tokitoh, R. Okazaki, M. Goto and S. Nagase, submitted to *J. Am. Chem. Soc.*
64. N. Takeda, N. Tokitoh and R. Okazaki, *Angew. Chem., Int. Ed. Engl.*, **35**, 660 (1996).
65. T. Matsumoto, N. Tokitoh and R. Okazaki, *Angew. Chem., Int. Ed. Engl.*, **33**, 2316 (1994).
66. (a) Y. Matsuhashi, N. Tokitoh and R. Okazaki, *Organometallics*, **12**, 2573 (1993).
    (b) M. Saito, N. Tokitoh and R. Okazaki, *J. Organomet. Chem.*, **499**, 43 (1995).
67. N. Tokitoh and R. Okazaki, *Main Group Chemistry News*, **3**, 4 (1995).
68. (a) N. Tokitoh, T. Matsumoto and R. Okazaki, *The IVth International Conference on Heteroatom Chemistry*, Seoul, Korea, Abstract OB-11, 1995, p. 54.
    (b) N. Tokitoh, T. Matsumoto and R. Okazaki, *The VIIIth International Conference on the Organometallic Chemistry of Germanium, Tin and Lead*, Sendai, Japan, Abstract O 10, 1995, p. 36.
69. K. Kabeta, D. R. Powel, J. Hanson and R. West, *Organometallics*, **10**, 827 (1991).
70. T. J. Barton and B. L. Groh, *J. Am. Chem. Soc.*, **107**, 7221 (1985).
71. T. J. Barton and G. C. Paul, *J. Am. Chem. Soc.*, **109**, 5292 (1987).
72. H. Schnöckel and R. Köppe, *J. Am. Chem. Soc.*, **111**, 4583 (1989).
73. (a) D. P. Thompson and P. Boudjouk, *J. Chem. Soc., Chem. Commun.*, 1466 (1987).
    (b) P. Boudjouk, S. R. Bahr and D. P. Thompson, *Organometallics*, **10**, 778 (1991).
74. D. P. Thompson and P. Boudjouk, *J. Org. Chem.*, **53**, 2109 (1988).
75. P. Jutzi, A. Möhrke, A. Müller and H. Bögge, *Angew. Chem., Int. Ed. Engl.*, **28**, 1518 (1989).
76. T. Sadahiro, N. Tokitoh and R. Okazaki, unpublished results.

CHAPTER 18

Gas-phase ion chemistry of silicon-containing molecules[†]

NORMAN GOLDBERG[‡] and HELMUT SCHWARZ

Institut für Organische Chemie der Technischen Universität Berlin, Straße des 17. Juni 135, D-10623 Berlin, Germany

I. INTRODUCTION	1106
II. THERMOCHEMISTRY	1106
A. Cationic Silicon Hydrides	1106
B. Substituted Silicenium Ions	1108
C. Anionic Silicon Species	1109
III. ION–MOLECULE REACTIONS OF SILICON-CONTAINING MOLECULES	1109
A. Reactions of Transition-metal Ions with Silicon-containing Molecules	1110
B. Use of Silicon Compounds as Precursors for the Generation of Elusive Carbanions	1115
C. Reactions of Atomic Silicon Cations with Neutral Molecules	1117
D. Reactions of Silicon-containing Ions with Neutral Molecules	1118
IV. GENERATION AND CHARACTERIZATION OF SMALL MULTIPLY BONDED SILICON-CONTAINING IONS AND THEIR NEUTRAL COUNTERPARTS	1122
A. Small SiC_xH_y Cations with Si—C Multiple Bonds and Their Neutral Counterparts	1123
B. Nitrogen- and Oxygen-containing Silicon Ions and Their Neutral Counterparts	1125
V. REARRANGEMENTS OF ORGANOSILICON IONS	1130
A. Cationic Rearrangements	1130
B. Anionic Rearrangements	1133

[†] This article is dedicated to Saul Patai.
[‡] *Present address*: Cornell University, Department of Chemistry, Baker Laboratory, Ithaca, NY 14853-1301, USA.

The chemistry of organic silicon compounds, Vol. 2
Edited by Z. Rappoport and Y. Apeloig © 1998 John Wiley & Sons Ltd

VI. DOUBLY CHARGED IONS	1134
VII. ACKNOWLEDGMENTS	1135
VIII. REFERENCES	1136

I. INTRODUCTION

The chemistry of silicon compounds has experienced tremendous research activities over the last decades, and researchers originally driven to find analogies between carbon and its higher homologue have had to realize that in many cases the differences between these elements far outweigh the similarities. The diversity of silicon chemistry and depth of penetration into other research areas can easily be seen by taking a look at the contributions to this compendium. One of the research fields that has contributed a great deal to our current understanding of the chemistry of this main-group element is that of mass spectrometry. Not only do some of the many mass-spectrometric techniques allow us to obtain thermochemical data on small charged and neutral silicon-containing species, but we can also obtain insight into intrinsic properties of this intriguing element. The absence of numerous perturbing effects which are common in liquids or solids is one of the main advantages of gas-phase studies.

The aim of this review is to give the reader an overview of the progress to date in the field. A detailed account by Schwarz[1a] on the 'Positive and Negative Ion Chemistry of Silicon in the Gas Phase' appeared in this series in 1989 as well as an exhaustive review by Bock and Solouki on 'Organosilicon Radical Cations'[1b], and the present manuscript is intended to cover the literature that has appeared since the publication of this work until early 1996. A brief discussion of thermochemistry will be given in Section II. Section III gives a detailed account of the advances made in the exploration of ion–molecule reactions of atomic silicon-containing ions. Here in particular the reactions of neutral silicon-containing molecules with main-group and transition-metal ions, as well as the reactions of atomic silicon cations with neutral molecules will be reported. Section IV covers the field of small multiply bonded silicon species — ionic as well as neutral, as explored by mass-spectrometric techniques. The employment of methods such as neutralization–reionization mass spectrometry (NRMS) has allowed for the generation and detection of neutral highly reactive species by mass spectrometric techniques, and a number of such studies have been carried out. Section V will report on current advances in the investigations of rearrangements of silicon-containing ions as explored by mass spectrometric techniques. The generation and investigation of doubly charged cationic and anionic silicon ions will be mentioned in Section VI.

II. THERMOCHEMISTRY

There have been numerous experimental as well as theoretical studies dealing with the structural and thermochemical properties of cationic silicon hydrides, $Si_n H_m^{+\bullet}$, and a detailed discussion of these species would certainly exceed the limited space available. We will therefore confine ourselves to the discussion of only a few exemplary cases. For further information on Si-ion thermochemistry the reader is referred to several reviews on the experimental[2–4] as well as computational[5] determination of thermodynamic properties of silicon-containing ions.

A. Cationic Silicon Hydrides

The properties of ionized monosilane, $SiH_4^{+\bullet}$, have been reevaluated by Berkowitz and coworkers[6]. In agreement with earlier studies[2,7] it was found that ionized monosilane,

$SiH_4^{+\bullet}$, is a very unstable species. Generation of $SiH_4^{+\bullet}$ was accomplished by Photoionization Mass Spectrometry (PIMS) and only observed within a very narrow energy range. The original discrepancy between earlier Photoelectron Spectroscopic (PES) measurements which had obtained a value of 11.60 eV for the adiabatic ionization energy and the more recent PIMS measurements (appearance potential of $SiH_4^{+\bullet}$ = 11.00 ± 0.02 eV) could be explained by theoretical calculations which reveal that the $SiH_4^{+\bullet}$ cation undergoes a strong Jahn-Teller distortion[8-10]. The ground state of $SiH_4^{+\bullet}$ possesses a D_{4h} symmetry, **1**, whereas the tetrahedral $SiH_4^{+\bullet}$ ion **2** lies 7.1 kcal mol^{-1} higher in energy[8].

(1) **(2)**

Thus the overlap between the vibrational wave functions of the tetrahedral **2** and Jahn-Teller distorted **1** is very small (weak Franck–Condon factors), and the small onset signal in the original PES measurement was simply overlooked in the significantly less sensitive PES experiments (as compared with PIMS).

Further support for the instability of $SiH_4^{+\bullet}$ comes from studies of the charge transfer reactions of a variety of small ions with SiH_4[11]. Although almost resonant charge transfer was achieved in the reaction of SiH_4 with $Xe^{+\bullet}$ (ΔIE = 0.08 ± 0.21 eV), no $SiH_4^{+\bullet}$ cations were observed in these guided-ion beam studies.

SiH_3^+ ions are easily formed in the reactions of noble gas cations and other small ions with SiH_4 via dissociative charge transfer (equation 1)[11].

$$X^{+\bullet} + SiH_4 \longrightarrow SiH_3^+ + X + H^{\bullet} \quad (X = He, Ne, Ar) \qquad (1)$$

The appearance potential of SiH_3^+ formed from SiH_4 has been measured at 12.086 eV via PIMS[6].

The silylene cation, $SiH_2^{+\bullet}$ has been studied in a number of experimental setups[12,13]. As indicated by MO calculations, the species possesses a C_{2v} structure[8-10,14-16]. The heat of formation of $SiH_2^{+\bullet}$ has been determined as 12.06 eV[11]. This ion is one of the main products found upon ionization of SiH_4. Besides generation via electron ionization (EI), SiH^+ can be generated in the dissociative charge transfer reaction of SiH_4 with small ions[11].

(3)

An interesting study on the hypervalent silanium ions SiH_7^+ has been carried out by Cao and coworkers[17]. These authors measured the infrared spectrum of **3** by vibrational predissociation spectroscopy. The ions were generated from a SiH_4/H_2 mixture in a high-pressure glow discharge source. A structure corresponding to a $H_2-SiH_3^+-H_2$

complex was assigned to this species. These findings are rather interesting in view of the fact that the hypervalent CH_7^+ congener is believed to possess a structure corresponding to a CH_5^+ complexed with a H_2 spectator molecule.

B. Substituted Silicenium Ions

While silicenium ions are still a rather elusive species in the solid state or in solution[18], their generation in the gas phase represents no major difficulties and their properties have been studied in great detail. In fact, the wealth of data that has been accumulated does not allow for a full treatment of all the results in this short section. Rather, we will confine ourselves to the discussion of a number of the more prominent examples here. Further details will also be discussed in Sections III and IV.

The methyl-substituted silicenium ions $SiMe_3^+$, $SiMe_2H^+$ and $SiMeH_2^+$ have been investigated by a number of groups. A particularly interesting study has been carried out by Shin and Beauchamp[19]. In this work the authors measured the hydride affinities of the Me_xSiH_{3-x} cations and compared these values with those of the carbon analogues. The results clearly indicate that the corresponding silyl-substituted ions are significantly more stable than the carbon species (when H^- is used as a reference base). The heats of formation for the silicenium ions $SiMeH_2^+$, $SiMe_2H^+$ and $SiMe_3^+$ (calculated from these hydride affinities) were determined to be 204, 172 and 147 kcal mol^{-1}, respectively, as compared to values of 215, 192 and 166 kcal mol^{-1} for $CMeH_2^+$, CMe_2H^+ and CMe_3^+. The apparent smaller stabilization of silicenium ions by successive methyl substitution can be explained by a smaller hyperconjugative interaction between the carbon-hydrogen σ bonds in the Si—C species due to the longer Si—C bond distance and the larger size of the empty Si 3p orbital, as compared with the C—C bond distance and the C 2p orbital. The relative stabilities of the MR_3^+ ions were shown to be strongly dependent on the reference anion that is used. When F^- or Cl^- are employed the $SiMe_3^+$ ion is significantly less stable than the CMe_3^+ cation. The higher affinity of silicenium ions for bases such as halogens or OH^- readily explains the fact that such species are not observed in the condensed phase[20].

The thermal-induced dissociation of tetraethylsilyl cations, $SiEt_4^{+\bullet}$, has been reported by Lin and Dunbar[21]. This technique which makes use of the blackbody background radiation field for the dissociation of a weakly bound cluster[22] allows for the study of weakly bound clusters with dissociation energies in the range of 0.5 to 1 eV. Thus, by measuring the rate constants for the loss of an ethyl radical from $SiEt_4^{+\bullet}$, a heat of formation of 131.7 ± 2 kcal mol^{-1} was derived for the $SiEt_3^+$ ion.

The thermochemistry of small SiX_n^+ cations, where X = F and Cl and $n = 1-4$, has been evaluated by Armentrout and coworkers[23-29]. ΔH_f° of SiF_3^+ was found to be -29.3 kcal mol^{-1}, while the value for the chlorine-analogue $SiCl_3^+$ is 99.8 kcal mol^{-1}. This difference reflects the exceptional thermodynamic stability of the Si—F bond. The significant differences in measured bond dissociation energies (BDE) when going from SiF_4^+ to SiF_3^{+}[27] (0.84 ± 0.16 eV and 6.29 ± 0.10 eV, respectively) can be explained by an enhancement of the s-orbital character in the lone-pair orbitals[30].

The resonance stabilization of the sila-allyl ion, $H_2C=CHSiH_2^+$, as compared to its other homologues, $H_2C=CHXH_2^+$ (C, Ge, Sn, Pb), has been studied theoretically by Gobbi and Frenking[31]. On the basis of *ab initio* calculations (MP2 level of theory), they predict that the resonance interaction decreases from 37.8 kcal mol^{-1} to 14.1 kcal mol^{-1} when going from the allyl cation to the sila-allyl species. The authors conclude in this study that in contrast to the allyl cations, the heavier analogues experience a significantly reduced stabilization by π-conjugative interactions.

C. Anionic Silicon Species

The thermochemistry of silicon-containing anions has very recently been compiled in an excellent review by Damrauer and Hankin[4], as well as in an earlier work by Damrauer[3]. In these reviews the authors give a detailed introduction into the experimental techniques as well as the cycles used to obtain thermodynamic data from negative-ion gas-phase chemistry. We will therefore confine ourselves here to the discussion of a few exemplary cases, and for a more detailed overview the reader is referred to the above-mentioned publications and the literature cited therein.

The formation and thermochemical properties of pentacoordinate silicon hydride anions, $SiH_{5-n}X_n^-$, have been reported by Squires and coworkers[32,33]. These ions can be generated via the addition of nucleophilic anions to silanes or, in the case of $SiH_5^{-[34,35]}$, via hydride transfer from alkyl silicon hydride anions to SiH_4 in flowing afterglow experiments. It is interesting to note that Squires and coworkers observed hydride–deuteride exchange in the reaction of n-BuSiH$_4^-$ and SiD_4 of all four hydrogen atoms in the hydride anion. The hydride affinity ordering[36] of various alkylsilanes was determined from bracketing experiments, and was found to decrease with increasing alkyl substitution. However, the differences between the various alkylsilanes were found to be smaller than 2 kcal mol^{-1}. The absolute hydride affinity for SiH_4 was determined to lie between 19 and 20 kcal mol^{-1}.

The fact that α-silyl substitution leads to a significant stabilization of carbanionic species is well-known and has been exploited in synthetic chemistry. On the other hand, silyl anions themselves are in general much more stable than their carbon analogues. The stabilization of carbanions by silyl substituents in the α position has been measured by Brauman and coworkers[37]. The anions were generated via nucleophilic displacement reactions (equation 2) of a silyl group with F$^{-[38]}$ (see also Section III.B).

$$F^- + (Me_3Si)_3SiH \longrightarrow (Me_3Si)_2SiH^- + Me_3SiF \qquad (2)$$

The electron affinities of a number of α-silyl substituted silyl and carbon radicals were determined in photodetachment experiments and confirmed by data obtained from *ab initio* calculations. The authors conclude in this study that the stabilization a carbanion experiences through α-silyl substitution is approximately 14–20 kcal mol^{-1} per silyl group; that of a silyl anion is approximately 6–14 kcal mol^{-1}. The larger stabilization in the carbanionic systems is readily explained by stronger hyperconjugation of the anionic carbon center with the silyl groups as compared to that of the silyl anion with a silyl group.

Interestingly, theoretical studies[39] indicate that substituted silane radical anions $RSiH_3^{-\bullet}$ are stabilized more by second-row substituents than by first-row substituents. This increased stabilization was explained by the difference in the diffuseness of first- and second-row elements and thus a better overlap of the larger second-row orbitals with the anionic center.

III. ION–MOLECULE REACTIONS OF SILICON-CONTAINING MOLECULES

The field of ion–molecule reactions has profited immensely from the introduction of new techniques in mass spectrometry such as Fourier Transform Ion-Cyclotron Resonance Mass Spectrometry (FT-ICR) and Selected-Ion Flow Tube methods (SIFT). Several accounts of the progress in the field of ion–molecule reactions involving silicon-containing molecules have been published. For review articles the reader is referred elsewhere[1,3,4,40–51] A detailed compilation of the kinetic data for bimolecular ion-molecule reactions of positive silicon ions has been published by Anicich[52].

A. Reactions of Transition-metal Ions with Silicon-containing Molecules

As already mentioned, due to progress on the experimental side the last decade has seen tremendous research efforts in the area of ion–molecule reactions, and especially, the reactions of metal cations with neutral organic molecules have received a great deal of interest[53]. In particular, the possibilities of studying biomimetic effects and obtaining data about the intrinsic properties of unligated ('bare') metal ions in mass spectrometric experiments have prompted intense research activities.

Although not as numerous and detailed as the studies of organic molecules with metal cations, there have been a number of experiments reported which deal with the reactions of silanes and metal cations. Kickel and Armentrout have studied the reactions of *all* first-row d metals with monosilane, SiH_4, and the data gathered in these guided ion beam studies have been used to evaluate the thermochemistry of the metal–silicon bond in $M-SiH_x^+$ ions[54-57]. The authors analyzed the kinetic energy dependence found for the reactions of the metal cations with SiH_4 to obtain the bond energies for the metal–silicon bond in the $M-SiH_x^+$ ions ($x = 0-3$) (Table 1). The nature of the metal–silicon double bonds in these cationic $M=SiH_2^+$ complexes has been analyzed by *ab initio* (FORS-MCSCF) calculations[58,59].

Geribaldi and coworkers have examined the oligomerization of SiH_4 with the rare earth ions Sc^+, Y^+ and Lu^{+}[60,61a]. In these FT-ICR studies, clustering of up to seven silane molecules around a metal ion was observed. In the additions of silane molecules to the metal center, the excess energy of the initial complex was found to be dissipated via sequential H_2 loss, thus giving rise to $MSi_nH_{2n}^+$ ions among various other products (equation 3).

$$M^+ \text{ (M = Sc, Y, Lu)} + n SiH_4 \longrightarrow M(SiH_2)_n^+ + nH_2 \qquad (3)$$

The reactions of the cations Fe^+, Co^+ and Ni^+ with cyclic silanes have been investigated in two separate ICR studies[62,63]. Bjarnason and Arnason reacted 1,3,5-trisilacyclohexane with bare and Cp-ligated metal ions. The authors observed the dehydrogenation of up to six hydrogen atoms from the trisilacyclohexane (equation 4) which led them to postulate a benzenoid structure **4** for the new metal-ligated $Si_3C_3H_6$ ligand. Further support for this structural assignment came from labeling studies and collisional induced dissociation (CID) experiments. Upon collisional activation the $(Cp)M-Si_3C_3H_6^+$ ions were found to lose an intact $Si_3C_3H_6$ unit as a main dissociation product, thus regenerating the $(Cp)M^+$ cations.

$$(Cp)M^+ + c\text{-}C_3Si_3H_{12} \longrightarrow (Cp)M-Si_3C_3H_6^+ + 3H_2 \quad (M = Fe, Co, Ni) \qquad (4)$$

Similar experiments have been carried out with the monosilyl substituted cyclohexane, C_5SiH_{12}[63]. Reactions of the silane with Fe^+, Co^+ and Ni^+ in this case, however, did not lead to dehydrogenation, but instead to mainly ethene elimination. Labeling studies indicated a reaction mechanism as depicted in equation 5. Initially, insertion of the metal

TABLE 1. Experimental bond energies (D) at 0 K in eV[54-57]

D(eV)	Sc	Ti	V	Cr	Mn	Fe	Co	Ni	Cu	Zn
$D(M-Si^+)$	2.51	2.54	2.37	2.10	>0.87	2.87	3.03	3.38	2.65	2.84
$D(M-SiH^+)$	2.33	2.30	2.09	1.02	—	2.63	2.29	2.39	2.55	3.37
$D(M-SiH_2^+)$	2.17	2.17	2.02	0.99	0.43	1.88	2.66	2.66	2.39	1.65
$D(M-SiH_3^+)$	1.76	1.69	1.54	0.78	—	1.90	1.96	1.91	1.00	3.11

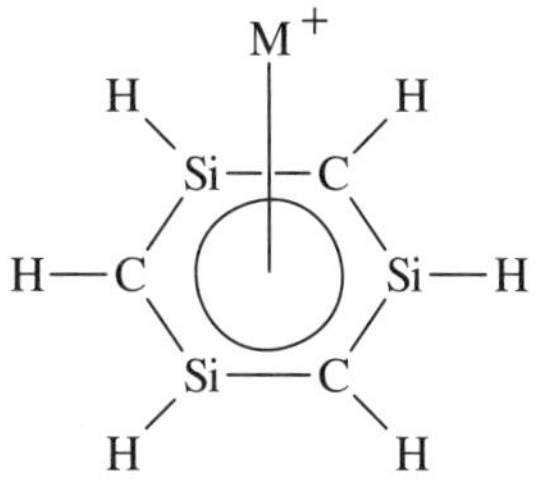

(4)

ion into the weak Si—C bond is believed to take place forming the metallacycle **5**, which subsequently can eliminate an ethene molecule.

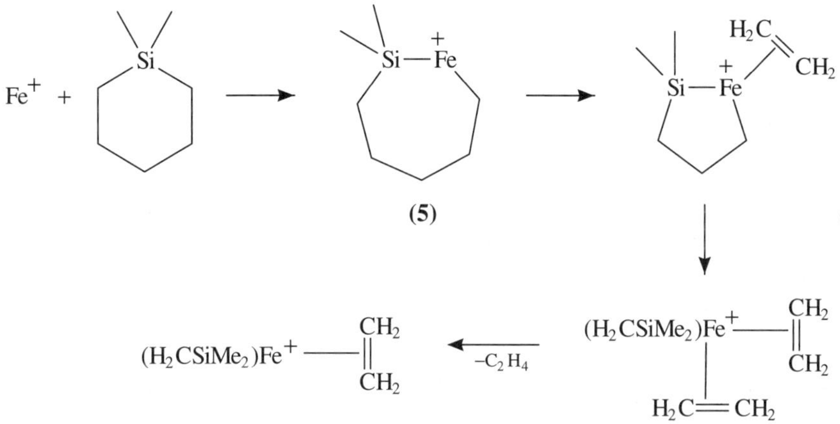

(5)

The enhanced tendency of the monosilacyclohexane to expel an ethene molecule and the fact that no ring cleavage was observed in the case of the trisilacyclohexane was tentatively explained by the lack of a stable leaving group in the latter case. In fact, the much smaller thermodynamic stability of the $H_2Si=CH_2$ leaving group seems to strengthen this argument.

Jacobson and coworkers[64,65] have recently investigated the reactions of Fe^+ with a variety of organosilanes. They reported compelling evidence for the formation of cationic iron–silylene, FeSiRR′, as well as iron–silene complexes, Fe—RR′Si=CR″R‴ (R″ = alkyl, H), which can be produced independently by reacting Fe^+ with different organosilicon molecules. Thus, reaction of Fe^+ with **6** can give rise to two different insertion products **7** and **8**, which form the metal–silene complex **10** via formation of the π-bonded species **9** and subsequent loss of an ethene molecule (equation 6). Whether the metal atom inserts into the thermodynamically weak Si—C bond or into the strained C1—C2 bond of the ring could not be distinguished on the basis of isotopic labeling studies; however, on the grounds that the silicon–carbon bond is considerably weaker than the C—C bond, insertion into a C—C bond seems less likely. Reaction of bare Fe^+ with dimethylsilane and ethenylsilanes (equation 7), on the other hand, was found to result in the formation of the isomeric iron–silylene complexes. The proposed mechanism for this reaction involves insertion into the Si—H bond giving **11**, as shown in path 7a, or into the vinylic C—Si bond giving **12** (path 7b). Subsequent formation of the silylene complex **13** can then proceed via ethene elimination following either an α- or β-H shift.

$$(6)$$

$$(7)$$

Data obtained from collision-induced dissociation experiments did not allow for a distinction of the isomeric metal–silene and –silylene species; however, structure-specific ion–molecule reactions of the complexes with labeled ethene were used to clearly differentiate between the metal silene and the silylene. In this intriguing study, Jacobson and coworkers also bracketed the bond dissociation energies of the isomeric ions.

They found that the dissociation energy for the silylene Fe—Si(Me)H$^+$ lies between 56 and 78 kcal mol^{-1}, and that of the silene isomer Fe—H$_2$Si=CH$_2{}^+$ between 55 and 70 kcal mol^{-1}. The similarity of the two energy ranges once again demonstrates the relative lability of the metal–silylene bond and suggests a potential stabilization of molecules which contain silicon, π-bonded to transition metals.

Replacement of CO in FeCO$^+$ by small silanes was observed to lead to strongly bound σ-complexes (equation 8)[66]. On the basis of simple displacement reactions it was concluded that the silane–metal ions actually correspond to the non-inserted species **14** (for $x=0$) rather than the inserted isomer **15**. The binding energies of the metal–silane bond were observed to increase in the order of Me$_2$SiH$_2$ > Me$_3$SiH > MeSiH$_3$ > Me$_4$Si > SiH$_4$. The actual energies were found to lie in the range 39.9 ± 1.4 kcal mol^{-1} for D^o Fe$^+$–SiMe$_2$H$_2$ and 31.3 ± 1.8 kcal mol^{-1} for D^o [Fe$^+$–SiH$_4$]—values almost twice as high as those reported for simple alkanes (e.g. D^o[Fe–propane] = 19 ± 2 kcal mol^{-1})[67]. The interesting trends in bond strength upon methyl substitution were tentatively explained by an interplay between polarizability of the silanes and σ-donation/σ^*-back donation features in these adducts.

$$\text{Fe(CO)}^+ + \text{H}_{4-x}\text{SiMe}_x \ (x = 0\text{--}4) \longrightarrow \text{FeH}_{4-x}\text{SiMe}_x{}^+ + \text{CO} \tag{8}$$

(14) **(15)**

An interesting example for a methide transfer was observed in the reaction of Fe$^+$ and Co$^+$ cations with bis(trimethylsilyl)amine **(16)** and bis(trimethylsilyl)methane **(17)**[68]. The complexes of these molecules with the metal cations unimolecularly lose a neutral MCH$_3$ species (equation 9). Such processes had hitherto only been observed for the reactions of atomic Cu$^+$ with organic molecules. In these cases the formation of the strongly bound CuCH$_3$ species[69] (formally a d^{10} compound) compensates for the endothermicity of the carbenium ion formation. The observation of methide transfer from a trimethylsilyl group thus reflects the enhanced intrinsic stability of silicenium ions as compared with their carbon analogues (see Section II).

$$\text{X(SiMe}_3)_2 + \text{M}^+ \longrightarrow \text{M}\text{—Me} + \text{Me}_3\text{SiXSiMe}_2^+ \tag{9}$$

(16) X = NH
(17) X = CH$_2$

The reactions of Fe$^+$ cations with a number of bis(trimethylsilyl) substituted amines and alkanes have been studied by Karrass and Schwarz[70]. Specific labeling studies indicate that the unimolecular decomposition of the ion–molecule complex proceeds via two possible intermediates, i.e. **18** or **19**. Both intermediates, however, require a β-methyl shift to be operative, in order to explain the experimentally observed loss of a neutral ethane molecule, which was found to occur from **20** (equation 10).

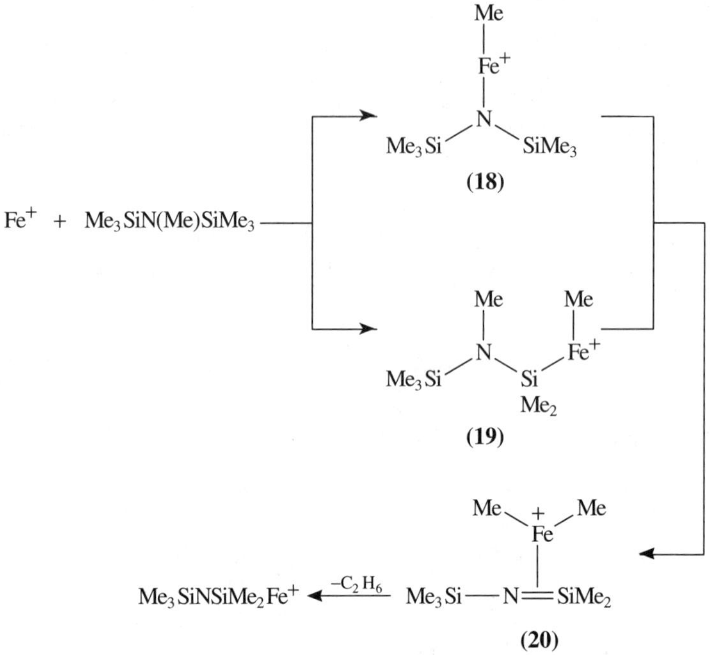

$$Fe^+ + Me_3SiN(Me)SiMe_3 \longrightarrow \begin{bmatrix} \text{(18)} \\ \text{(19)} \end{bmatrix} \longrightarrow$$

(18): $Me_3Si-N(Fe^+Me)-SiMe_3$

(19): $Me_3Si-N(Me)-Si(Me)(Fe^+Me)$

$Me_3SiNSiMe_2Fe^+ \xleftarrow{-C_2H_6} Me_3Si-N=SiMe_2$ with $Fe^+(Me)(Me)$ above N

(20)

(10)

The mechanism of the C—H and C—C bond activation of bare Fe^+ with *n*-heptyltrimethylsilane has been elucidated with the help of extensive labeling studies[71]. The system was found to display a rather rich chemistry. Loss of neutral tetramethylsilane from the ion–molecule complex (equation 11) was explained by an initial insertion of the metal ion into the C1—C2 bond to form **21**, and a subsequent β-H shift giving rise to the iron–hydride complex **22**. This ion can then lose a tetramethylsilane molecule via reductive elimination.

The observed competitive loss of ethene in this system, together with labeling studies, point to an extensive skeletal rearrangement of the heptyl chain prior to the dissociation.

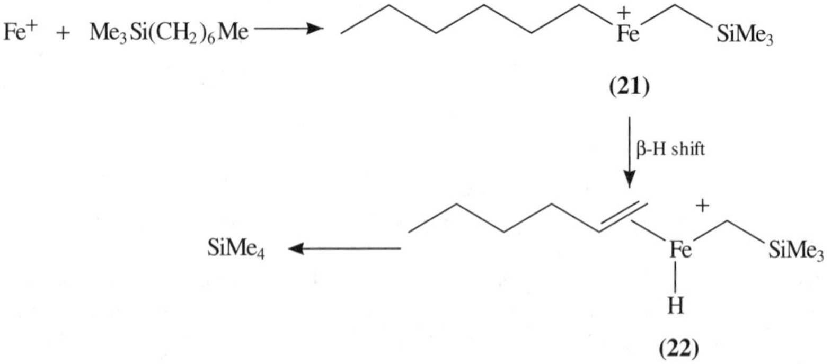

$Fe^+ + Me_3Si(CH_2)_6Me \longrightarrow$ [pentyl–Fe^+–CH$_2$–SiMe$_3$]

(21)

↓ β-H shift

$SiMe_4 \longleftarrow$ [butyl–CH=CH–Fe^+(H)–CH$_2$–SiMe$_3$]

(22)

(11)

Interestingly, the closely related but now bifunctional system of the Fe^+ complex with 7-trimethylsilyl heptanenitrile displays a significantly different reactivity[72a]. No C—C bond activation was observed in the unimolecular decay of this ion–molecule complex, and the exclusive loss of an H_2 molecule was explained by the operation of a metal-ion mediated cooperative effect of the two functional groups. A suggested reaction mechanism involves an 'anchoring' of the metal cation in an end-on mode to the nitrogen atom of the nitrile group[53] and the loss of H_2 via an initial formation of a site-specific, silyl directed C—H bond activation (cf **23a**). A β-H shift can then give rise to the iron–dihydride complex **23b** and subsequent formation of H_2 via reductive elimination (equation 12). The fact that the hydrogen atoms were found to stem exclusively from the C6 and C7 positions of the methylene chain further supported this reaction mechanism as the postulated intermediate **23a** should gain some stabilization through the silicon atom in the β-position.

$$Fe^+ + Me_3Si(CH_2)_6CN \longrightarrow \text{(23a)} \longrightarrow \text{(23b)} \longrightarrow H_2 \quad (12)$$

Recently the unprecedented example of stereoselective C—Si bond activation in ω-silyl-substituted alkane nitriles by 'bare' Co^+ cations has been reported by Hornung and coworkers[72b]. Very little is known of the gas-phase reactions of anionic metal complexes with silanes. In fact there seems to be only one such study which has been carried out by McDonald and coworkers[73]. In this work the reaction of the metal–carbonyl anions $Fe(CO)_n^-$ ($n = 2, 3$) and $Mn(CO)_n^-$ ($n = 3, 4$) with trimethylsilane and SiH_4 have been examined. The reactions of $Fe(CO)_3^-$ and $Mn(CO)_4^-$ anions exclusively formed the corresponding adduct ions via an oxidative insertion into the Si—H bonds of the silanes. The 13- and 14-electron ions $Fe(CO)_2^-$ and $Mn(CO)_3^-$ were observed to form dehydrogenation products $(CO)_xM(\eta^2-CH_2=SiMe_2)^-$ besides simple adduct formation with trimethylsilane. The reaction of these metal carbonyl anions with SiH_4 afforded the dehydrogenation products $(CO)_2Fe(H)(SiH)^-$ and $(CO)_3Mn(H)(SiH)^-$.

B. Use of Silicon Compounds as Precursors for the Generation of Elusive Carbanions

The utility of the 'DePuy fluoride-induced desilylation reaction'[38] for the regioselective generation of gas-phase carbanions from trimethylsilyl compounds (equation 13) has been exploited by Squires and coworkers in a number of very elegant flowing-afterglow studies[74–78]. The driving force for this reaction is the formation of the extremely strong bond between silicon and fluorine, and Squires and colleagues have further pioneered this technique for the generation of a number of distonic anions. In these studies they used a second fluoride-induced desilylation reaction between the trimethylsilyl-containing anions and F_2 to generate biradical anions. This subsequent reaction allowed the generation of such elusive species as the trimethylenemethane anion **25**[76], via formation of anion **24b** (equation 14) from **24a**. The photoelectron spectrum of **25** generated via desilylation has thus recently been obtained and the electron affinity, as well as the singlet–triplet splitting of the neutral trimethylenemethane ($^3A'_2 - {}^1A_1 = 16.1$ kcal mol^{-1}), have been determined from these data[79].

$$R-SiMe_3 + F^- \longrightarrow R^- + FSiMe_3 \qquad (13)$$

(14)

There is still a significant interest in the generation of such anions as a straightforward synthesis of these ions would allow the measurement of fundamental properties such as the acid–base character of neutral diradicals[80], their electron affinities and, most importantly, the singlet–triplet gaps of the diradical species[81,82]. Thus Squires and coworkers employed the desilylation reaction for the regioselective generation of chlorophenyl anions from the *meta*- and *para*-trimethylsilyl-substituted chlorobenzenes (**26** and **27**) (equation 15). Subsequent collision-induced dissociation of anions **28** and **29** gave rise to the *meta* (**30**) and *para* isomers (**31**)[76] of the elusive *ortho*-benzyne (**32**)[83]. Upon measuring the threshold for the dissociation reactions, the heats of formation for the neutral cyclic C_6H_4 isomers could thus be experimentally determined for the first time ($\Delta H°_{298} = 122.0 \pm 3.1$; 137.3 ± 3.3 and 106.6 ± 3.0 kcal mol^{-1} for **30**, **31** and **32**, respectively)[78].

The desilylation procedure has furthermore been employed in the targeted synthesis of a number of enolate anions. In these intriguing flowing-afterglow studies the enolate anions were generated regioselectively by desilylation of the corresponding trimethylsilyl enol ethers with fluoride anions (equation 16)[74,75]. The rate coefficients measured for methanol-catalyzed tautomerization of the corresponding enolate anions were employed to derive the equilibrium ratios of the corresponding tautomers.

18. Gas-phase ion chemistry of silicon-containing molecules

[Scheme showing structures (26), (28), (30), (27), (29), (31), (32) with reactions +F⁻/−FSiMe₃ and CID/−Cl⁻]

(15)

$$RR'C=CR''O-SiMe_3 + F^- \longrightarrow RR'C=CR''O^- + Me_3SiF$$

$$R = R' = R'' = \text{alkyl} \qquad (16)$$

C. Reactions of Atomic Silicon Cations with Neutral Molecules

The amount of literature which has accumulated over the last few years concerning the subject of ion–molecule reactions of silicon cations with neutral molecules is tremendous. In a review of the various reactions of $Si^{+\bullet}$ with neutral molecules published in 1990, Böhme[44] lists almost 100 reactions studied by selected-ion flow tube (SIFT) techniques alone[84,85]. In the meantime, this list will have grown to even larger dimensions and we can therefore only confine ourselves to the description of a few exemplary cases[45,50].

One of the reactions which has received special attention is the oxidation of $Si^{+\bullet}$ by oxygen-containing molecules. Thus the reaction of $Si^{+\bullet}$ in its ^2P ground state[86] with N_2O has been shown to give rise to $SiO^{+\bullet}$ which, in subsequent reactions with N_2O, has been reported to lead to the formation of the highly oxidized $SiO_4^{+\bullet}$ cation (equation 17)[87,88].

$$Si^{+\bullet}(^2P) \xrightarrow[]{N_2O N_2} SiO^{+\bullet} \xrightarrow[]{N_2O N_2} SiO_2^{+\bullet} \xrightarrow[]{N_2O N_2} SiO_3^{+\bullet} \xrightarrow[N_2]{N_2O} SiO_4^{+\bullet} \qquad (17)$$

A subsequent ICR study of the reaction of Si$^{+\bullet}$ with N$_2$O has been carried out by Stöckigt and coworkers[89] in order to examine the structures of these polyoxides. Although the formation of a SiO$_4$$^{+\bullet}$ cation under ICR conditions was not observed, the SiO$_3$$^{+\bullet}$ ion was obtained in high abundance. Ligand displacement reactions and *ab initio* calculations (MP4/SDTQ) were used to elucidate the structure of this intriguing oxide. The studies suggest that **33** possesses a D_{3h} symmetrical $^4A_1'$ ground state. However, it should be noted that the [Si,O$_3$]$^{+\bullet}$ potential energy surface is probably rather flat and conversion to several other minima such as **34** and **35** is easily achieved. These quartet states which are ion–dipole complexes of SiO$^{+\bullet}$ and O$_2$ were calculated to be only 5.0 and 2.0 kcal mol^{-1} less stable than **33**.

(33)   (34)   (35)

The reactions of Si$^{+\bullet}$ with small silanes have been examined in a number of experimental and theoretical studies. One of the reasons for the particular interest in these systems is due to a desire to understand plasma processes which are employed for the growth of silicon surfaces. Armentrout and coworkers studied the reactions of Si$^{+\bullet}$ with small silanes and hydrocarbons[24,90–92], and the data obtained have been used to evaluate the thermochemistry of a number of small organosilicon species (see also Section II). Thus the reactions of Si$^{+\bullet}$ (^2P) with methane[90] and ethane[91] have been investigated by guided ion beam techniques. The initial reaction step was found to correspond to an insertion into the covalent bonds of the hydrocarbon molecule. In the reaction of Si$^{+\bullet}$ with ethane, insertion into the C–C bond is exothermic by 65 kcal mol^{-1}. Thus, most products in this reaction evolve through an insertion into the C–C bond. However, a number of low-energy products that have been observed in this reaction, such as SiH$^+$ and SiH$_2$$^{+\bullet}$, cannot evolve directly via this reaction mechanism, and it was proposed that a relatively facile interconversion of silyl cations via a silacyclopropanium ion may take place.

There have been a number of theoretical studies describing the thermochemistry and structural properties of SiC$_n$H$_m$$^+$ cations. Ketvirtis and coworkers[93–95] investigated several small organosilicon cations by means of *ab initio* molecular orbital calculations, and found that, in general, the isomers which contain Si atoms that have no hydrogens bonded to them are considerably more stable than other isomers. This certainly is a consequence of the significantly weaker Si–H bond strength as compared to that of C–H bonds.

The association reactions of Si$^{+\bullet}$ (^2P) with acetylene and benzene have been measured by Glosik and coworkers[96], who found that the rate coefficients for these reactions have strong negative-temperature dependencies. These observations were rationalized in terms of a negative entropy change in the reactions.

D. Reactions of Silicon-containing Ions with Neutral Molecules

So-called clustering reactions of SiH$_n$$^+$ ions ($n = 0$–3) with silanes and other small molecules have attracted considerable interest. In particular, the reaction sequences which might lead to the formation of large-sized clusters have received attention, as these processes are thought to be involved in undesirable dust-formation which occurs during silicon-film depositions from silane plasmas and vapors[97–99]. Thus, Mandich and Reents have thoroughly studied the ion–molecule reactions of SiD_n^+ ions ($n = 0-3$) and Si_n^+ clusters ($n = 2-7$) with $SiD_4{}^{100-105}$. These experimental studies were published together with companion papers by Raghavachari[106–109] in which quantum mechanical calculations (MP4/6-31G** level of theory) on the mechanisms of the clustering reactions were reported. The primary reaction pathway for the reactions of the SiD_n^+ ions was found to correspond to an insertion of the ion into a Si–D bond of SiD_4 and a subsequent D_2 loss from the ion–molecule complex (equation 18).

$$SiD_n^+ + SiD_4 \longrightarrow Si_2D_{n+2}^+ + D_2 \qquad (18)$$

The clustering reactions of the SiD_n^+ ions with SiD_4 were observed to stop at rather small cluster sizes, all leading to energetic bottlenecks. Addition of three SiD_2 units to SiD^+ gave rise to a $Si_4D_7^+$ ion[103] which, according to the theoretical calculations[108], corresponds to the highly stable silacyclobutyl cation. Further reactions of this ion were found to be endothermic.

The interesting nonclassical C_{2v}-symmeterical structure **36** was predicted to be the ground state structure for the $S_3H_7^+$ cation (and its deuterated analogue) which was calculated to correspond to the ground state of the potential energy surface[107]. This ion was obtained upon sequential SiH_2 addition to SiH_3^+ and its high stability was used to explain the inertness of $Si_3H_7^+$ in an SiH_4 atmosphere[102]. The termination of the clustering reactions of SiH_3^+ with SiH_4 was suggested to be due to the formation of this isomer. Thus it was concluded that clustering of silane in ion–molecule reactions reaches unreactive structures which contain less than six silicon atoms.

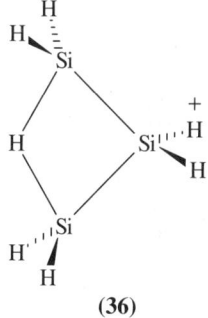

(**36**)

The clustering reactions of SiD_n^+ ($n = 0-3$) and $Si_2D_n^+$ ($n = 0-6$) cations with deuterated disilane, Si_2D_6, have been measured in a FTMS study[110]. The dominant pathway for these reactions was found to correspond to silylene transfer and SiD_4 elimination. The overall reactivity of disilane compared to monosilane was found to be higher, and this was explained by the fact that the silicon–silicon bond in disilane is considerably weaker (76 kcal mol^{-1}) than the Si–H bond of SiH_4 (88 kcal mol^{-1})[111]. Thus the insertion of $Si^{+\bullet}$ into the Si–Si bond was calculated to be 17 kcal mol^{-1} more favorable than $Si^{+\bullet}$ insertion into the Si–H bond of $SiH_4{}^{106,112}$.

Mandich and Reents[104] succeeded in observing the first reaction sequence which in fact leads to large-sized clusters in silane plasmas. They introduced a small amount of water along with the silane atmosphere and obtained clusters with masses of 650 amu and larger. Interestingly, the SiD^+ cation was found to correspond to the only SiD_x^+ cation which leads to the formation of a number of highly specific species. Among other clusters, the perdeuterated $Si_3D_7^+$ ion (**36**-D_7) is observed which then gives rise to the larger clusters upon reaction with D_2O and subsequent addition of a silylene unit from SiD_4 (equation 19).

$$Si_3D_7^+ + D_2O \longrightarrow Si_3D_7O^+ + D_2$$
$$Si_3D_7O^+ + nSiD_4 \longrightarrow Si_{3+n}D_{7+2n}O^+ + nD_2 \qquad (19)$$

A number of other groups have investigated the clustering reactions of small cation silicon species with silanes and other small molecules. The ion–molecule reactions occurring between SiH_x^+ ($x = 0-3$) cations and neutral ammonia, as well as the reactions between NH_x^+ cations and SiH_4 were studied by FTMS[113,114]. The main channel for the reaction between SiH_x^+ ions and NH_3 was found to correspond to the elimination–addition reaction, well-known for silanes (equation 20), which formally corresponds to the transfer of a nitrene-unit (NH)[113].

$$SiH_x^+ + NH_3 \longrightarrow SiH_{x+1}N^+ + H_2 \qquad (20)$$

Gal and coworkers[114] observed the formation of ionic clusters with up to five silicon and nitrogen atoms upon ionization of silane/ammonia mixtures. Again, the silylene and nitrene transfers via addition–elimination reactions were found to be the most important pathways leading to the formation of larger clusters (equation 21). However, it was concluded in these studies that chain propagation leading to the formation of large-sized clusters was not possible in these systems.

$$Si_xN_yH_n^+ + SiH_4 \longrightarrow Si_{x+1}N_yH_{n+2} + H_2$$
$$Si_xN_yH_n^+ + NH_3 \longrightarrow Si_xN_{y+1}H_{n+1} + H_2 \qquad (21)$$

The ion–molecule reactions that occur in mixtures of SiH_4 and PH_3 have recently been described[115]. As in the aforementioned cases, the main reaction channels for the formation of larger-sized clusters from SiH_n^+ cations ($n = 0-3$) and PH_3, and from the reactions of PH_x^+ cations ($n = 0-3$) with SiH_4, correspond to the addition of the neutral molecule and dissipation of the excess energy via loss of a hydrogen molecule. It was concluded in this study that the nucleation of mixed silicon–phosphorous ions was favored upon ionization of mixtures that contain a large excess of SiH_4.

Experiments similar to those described above have been performed by Operti and coworkers[116–118] who investigated the gas-phase ion–molecule reactions taking place in silane/germane mixtures and their methyl derivatives. For the GeH_4/SiH_4 mixtures, the formation of mixed germanium and silicon-containing ions $GeSiH_n^+$ ($n = 2-5$) was observed. Ionization of SiH_3Me/GeH_4 mixtures was found to give rise to a number of ions containing the C, Si and Ge atoms.

Lim and Lampe[119] have studied the ion–molecule reactions in mixtures of SiH_4 and CO. During the course of these studies the authors observed that the intermediate $CSiH^+$ ion readily protonates CO and proposed that this reaction (equation 22) might be a possible source for the neutral SiC molecule in interstellar space.

$$CSiH^+ + CO \longrightarrow HCO^+ + SiC \qquad (22)$$

The structure of ionic silicon clusters, the mixed silicon–carbon species and their reactions with neutral molecules have been investigated by a number of experimental[151,120–133] and theoretical groups[134–137]. Ion–molecule reactions of the silicon clusters have also been thoroughly studied by Jarrold and coworkers[51]. These authors reported for the reactions of larger silicon clusters with neutral molecules a simple association reaction (chemisorption); e.g. Si_n^+ reacts with C_2H_4 according to equation 23.

$$Si_n^+ + nC_2H_4 \longrightarrow Si_n(C_2H_4)_n^+ \qquad (23)$$

Magic numbers for the stability of the clusters, and large variations in reactivity with respect to the number of silicon atoms per ionic cluster unit have not been observed in these experiments. The only cluster ion that displays some unique behavior was found to be $Si_{13}^{+\bullet}$. This cluster has been shown to be particularly unreactive in its ion–molecule reactions with O_2, H_2O and C_2H_4. As initially suggested in theoretical studies[138,139], the low reactivity of $Si_{13}^{+\bullet}$ was believed to be related to its proposed icosahedral structures (13, 19 and 23 atoms are magic numbers for an icosahedral packing sequence). However, recent Carr–Parrinello calculations by Andreoni and coworkers[140] show that the icosahedral structure corresponds to a high-energy isomer for the $Si_{13}^{+\bullet}$ cation.

The chemistry of mixed carbon- and silicon-containing cluster ions has been studied by several groups[128–133,141]. Parent reported the rate constants for the reactions of SiC_n^+ ($n = 2$–8) and $Si_2C_n^+$ ($n \leqslant 6$) with acetylene[130]. In this study an interesting exchange of carbon atoms upon reaction of ^{13}C-labeled $Si_2{}^{13}C_n^+$ clusters was observed (equation 24).

$$Si_2{}^{13}C_n^+ + {}^{12}C_2H_2 \longrightarrow Si_2{}^{13}C_{n-1}{}^{12}C^+ + {}^{13}C{}^{12}CH_2$$
$$\longrightarrow Si_2{}^{13}C_{n-2}{}^{12}C_2^+ + {}^{13}C_2H_2 \qquad (24)$$

Interestingly, this carbon exchange was not observed in the reactions of acetylene with cluster ions which contain only one silicon atom. On the basis of the exchange reactions and CID experiments, it was concluded that the mixed clusters possess linear chain-like geometries. These findings were later confirmed by theoretical calculations of the [Si_2, C_2]$^+$ surface [RHF/CCSD(T) and UB3LYP-level of theory][137] which established that the linear Si=C=C=Si$^{+\bullet}$ cation does indeed correspond to the energetically most favorable structure. Very recently Negishi and coworkers[133] measured the reactions of cationic as well as anionic silicon–carbon clusters, $SiC_n^{+/-}$ ($n = 1$–6) with O_2. For all the cations, only the O atom adducts were observed in these experiments. The anion SiC_2^- was found to form the oxygen adduct, SiC_2O^-, whereas the larger SiC_n^- species did not react with O_2.

New reactions of neutral molecules with the trimethylsilyl ion, Me_3Si^+, have been studied by several groups. The Me_3Si^+ ion has been demonstrated to readily undergo radiative association reactions[48] (equation 25) with a number of bases[48,142,143]. In these reactions the initial complexation energy gained and stored in the ion–molecule complex is dissipated by emission of photons. It was concluded that the efficiency of these association reactions increased with increasing bond dissociation energy and higher number of degrees of freedom in the adducts.

$$SiMe_3^+ + B \longrightarrow [BSiMe_3^+]^* \longrightarrow BSiMe_3^+ + h\nu \qquad (25)$$

The gas-phase affinities of primary, secondary and tertiary alkyl amines towards Me_3Si^+ were determined in a high-pressure mass spectrometer. The Me_3Si^+ affinities were found to increase linearly with the proton affinities of the amines. In these measurements trialkyl amines were found to undergo only very slow associations with Me_3Si^+ ions, which was explained by steric effects. The proton affinities of trimethylsilylamines which have been measured in these studies demonstrated again that the ability of an α-silicon to stabilize a positive charge on nitrogen is somewhat smaller than that at a carbon atom.

The aromatic silylation of benzene and substituted benzenes by Me_3Si^+ have been studied by several groups. In a high-pressure mass-spectrometric study, Stone and Stone[144] investigated the nature of the bonding in the adduct ions. They determined that the entropy value for the association of Me_3Si^+ with toluene (33.6±2.0 cal K^{-1} mol^{-1}) is significantly smaller than that of Me_3C^+ (54.6 ± 0.8 cal K^{-1} mol^{-1}). On the basis of these values the authors argue that the Me_3Si^+ ion in its toluene adduct must have considerable rotational freedom and is likely to correspond to a loose complex rather than to an arenium ion, or the energy difference between these two structures must be very small.

The ion–molecule reactions of Me_3Si^+ have been employed to distinguish between the *cis*- and *trans*- isomers of 1,2-cyclopentanediol[145]. The formation of a $[Me_3SiOH_2]^+$ cation (equation 26, top) was found to be indicative for a *cis*-structure of the reacted diol **37**. Thus the formation of the $[Me_3SiOH_2]^+$ cation from the adduct **38** proceeds without a barrier, whereas the production of this ion from the *trans*-diol **39** via adduct **40** (equation 26, bottom) was shown to be an endothermic process which possesses a translational energy onset. These studies are the first to use the reactivity of the Me_3Si^+ ion as a probe of the stereochemistry of the reacted neutrals in a mass spectrometer.

(26)

IV. GENERATION AND CHARACTERIZATION OF SMALL MULTIPLY BONDED SILICON-CONTAINING IONS AND THEIR NEUTRAL COUNTERPARTS

Over the past years a number of small silicon-containing ions have been structurally characterized by collisional-activation mass spectrometry. These ions have been used as precursors for the generation of their neutral counterparts via Neutralization–Reionization Mass Spectrometry (NRMS)[146–152]. To a large extent the interest in small elusive silicon-containing molecules is due to the role which these species have been postulated to play as intermediates and building blocks in the genesis of interstellar matter[46,85]. In particular, the different conceivable pathways which lead from atomic silicon cations to the formation of neutral silicon-containing molecules have been studied intensely[43–46,153]. A second driving force for the elucidation of the gas-phase chemistry of small silicon-containing ions and neutrals is the increasing importance of silicon-etching technologies[154]. These processes are known to lead to a large variety of ionic as well as neutral molecules and a better understanding of their chemistry is highly desirable.

A. Small SiC$_x$H$_y$ Cations with Si—C Multiple Bonds and Their Neutral Counterparts

There has been particular interest in the chemistry of small organosilicon ions and their neutral counterparts for a number of reasons: (i) Small SiC$_x$H$_y$ molecules are well suited to draw analogies between the structural chemistry of silicon and carbon — or to state differences. (ii) The chemistry of small silicon compounds is viewed as fundamental in astrophysics and astrochemistry, and a large number of cationic and neutral SiR$_x$ molecules have been detected in interstellar and circumstellar matter.

The SiCH$_x^+$ ($x = 1-3$) (**41–43**) cations have been generated upon electron ionization (EI) of H$_3$SiCH$_2$Cl[155]. Collisional activation mass spectrometry (CA-MS) confirms that the hydrogen atoms reside on the carbon atom rather than on silicon in these ions — findings which are in keeping with the results from *ab initio* calculations[156].

$$HC\equiv Si^+ \qquad H_2C\equiv Si^{+\bullet} \qquad H_3C-Si^+$$
$$(41) \qquad\qquad (42) \qquad\qquad (43)$$

Subsequent NR experiments[155] proved for the first time that the corresponding neutral counterparts of cations **41–43** are viable species in the gas phase.

EI of Si(CH$_3$)$_4$ yields ions of [H$_4$,C,Si]$^{+\bullet}$ and [H$_6$,C$_2$,Si]$^{+\bullet}$ compositions[157]. These ions have been previously observed in ion–molecule reactions of Si$^{+\bullet}$ with alkanes[90,91,158]. CA-mass spectrometry was used to establish that these ions correspond to the silylenes **44** and **45**.

$$H\diagdown Si^{+\bullet} \diagup CH_3 \qquad\qquad H_3C \diagdown Si^{+\bullet} \diagup CH_3$$
$$(44) \qquad\qquad\qquad (45)$$

These findings are in agreement with the results of earlier theoretical results[158] which had predicted the cationic silylene (**44**) to correspond to the global minimum on the [H$_4$,C,Si]$^{+\bullet}$ surface. MP4 6-31G* calculations predicted the silaethene isomer H$_2$CSiH$_2^{+\bullet}$ to lie only 6.7 kcal mol^{-1} above the silylene, yet separated by a substantial barrier of 42.6 kcal mol^{-1}. The successful generation and characterization of the corresponding neutral silylenes HSiCH$_3$ and H$_3$CSiCH$_3$ by NRMS was in accordance with the results of earlier matrix-isolation experiments[159–163] which had shown that the silylene isomers of these molecules are energetically favored over the silaethene species.

Schwarz and coworkers[164,165] have been successful in characterizing several cationic silicon-acetylides Si(C≡C)$_n$H$^+$ ($n = 1-3$) (**46–48**) and their neutral counterparts via NRMS. Common to these and related species is the absence of (weak) Si—H bonds.

$$^+Si-C\equiv C-H \qquad\qquad ^+Si-C\equiv C-C\equiv C-H$$
$$(46) \qquad\qquad\qquad (47)$$

$$^+Si-C\equiv C-C\equiv C-C\equiv C-H$$
$$(48)$$

The SiC$_2$H$^+$ cations were generated by electron impact ionization of (H$_3$C)$_3$SiC≡CH. In agreement with *ab initio* data[166] which predict the Si-protonated C$_2$SiH$^+$ to be

94 kcal mol^{-1} less stable than the corresponding C-protonated molecule, a connectivity corresponding to a SiC$_2$H$^+$ ion was established via CA experiments. The neutral SiC$_2$H$^•$ radical was successfully generated upon neutralization of the cationic precursor (**46**). As suggested by theory[167] the linear acetylene, SiC$_2$H$^•$ is approximately 4.5 kcal mol^{-1} more stable than the cyclic isomer and 70 kcal mol^{-1} energetically favored over the silylene radical.

The higher acetylene radicals SiC$_4$H$^•$ and SiC$_6$H$^•$ have received some experimental interest as these acetylenes are believed to play an important role in the interstellar genesis of silicon carbides (equation 27)[168,169].

$$Si^{+•} + HCCH \xrightarrow{-H^•} SiC_2H^{+•} \xrightarrow{+e^-, -H^•} SiCC$$

$$\downarrow +HCCH, -H_2$$

$$SiC_4H^{+•} \xrightarrow{+e^-, -H^•} SiC_4 \qquad (27)$$

$$\downarrow +HCCH, -H_2$$

$$SiC_6H^{+•} \xrightarrow{+e^-, -H^•} SiC_6$$

The cationic SiC$_4$H$^+$ and SiC$_6$H$^+$ molecules have been generated upon electron bombardment of phenylsilane. CA experiments indicate that, as in the case of the smaller homologue **46**, the hydrogen atom resides on the terminal carbon atom in these species. The successful neutralization of these ionic species by electron capture is believed to have some interesting implications for the proposed interstellar genesis of silicon–carbide species SiC$_n$ as the experiments indicate that these acetylene molecules are stable species (at least on the μsec time-scale); of course, due to the different energetics neutralization of SiC$_x$H$^+$ under NR conditions does not necessarily lead to detachment of a hydrogen atom.

Another example which clearly demonstrates the advantageous interplay between different experimental techniques and state-of-the-art *ab initio* methods is that of the ionic and neutral 3-silacyclopropenylidene (**49**) and its isomer **50–53**[170].

(**49**) (**50**) (**51**) (**52**) (**53**)

In the ion-source EI spectrum of ClSi(CH$_3$)$_3$, ions of the formal composition [H$_2$,C$_2$,Si]$^{+•}$ have been observed[171]. Collisional activation experiments performed with these ions indicated that they possess a cyclic structure **49**[170]. The successful neutralization of this ion demonstrates nicely the stability of the corresponding cyclic silylene, and NRMS experiments could in fact distinguish this isomer from other conceivable species, such as **52** or **53**. These results were supported by *ab initio* MO calculations[172,173] which predicted the singlet isomer **51** to correspond to the energetically most stable species on the [H_2,C_2,Si] hypersurface. Surprising experimental results on this system have been reported recently by Maier and coworkers[174,175] who also found the cyclic silacyclopropyne (**53**) to be a stable species under matrix isolation conditions[176,177].

The reactivity exhibited by silicon cations π-bonded to arenes such as benzene (**54**) and naphthalene[178] (**55**) has been demonstrated to be significantly different from that of atomic silicon ions[153,178]. In this context the parent system $Si-C_6H_6^{+\bullet}$ generated by co-ionization of $SiMe_4$ and benzene, has been studied by CA and NRMS methods[179]. The π-complex, **54** is readily distinguishable from the inserted σ-bonded silylene $PhSiH^{+\bullet}$ (**56**) by mass spectrometric means. Theoretical results[179] (MP2/6-31G**) reveal the cationic $Si-C_6H_6^{+\bullet}$ to prefer a C_s structure rather than the silicon being bonded on the center of the C_6 axis. This latter structure was found to correspond to a transition state. The isomeric phenylsilylene cation (**56**) was found to lie 5.5 kcal mol^{-1} higher in energy, thus indicating that in fact the ionic chemistry of silicon π-bonded to polyaromatic hydrocarbons (PAHs) represents an important alternative to conventional ion–molecule chemistry of atomic $Si^{+\bullet}$. A third isomer — the inserted seven-membered ring structure **57** — was located 30.8 kcal mol^{-1} above the π-bonded species **54**.

The ion $\cdot CH_2OSi^+$ is an interesting species in its own right as it provides the first direct proof for the existence of a distonic ion[180,181] which incorporates a silicon atom[182]. It can be generated via electron-impact ionization of tetramethoxysilane[183]. The genesis of this distonic ion presumably takes place via a sequence of unimolecular hydrogen and formaldehyde losses — a metastable behavior that has been observed previously in methoxy-substituted silane cations[184]. Neutralization of $\cdot CH_2OSi^+$ leads to the stable $\cdot CH_2OSi\cdot$ diradical which, in accordance with *ab initio* calculations (MP2/6-31G**), corresponds to a local minimum on the [H_2,C,O,Si] hypersurface[184].

B. Nitrogen- and Oxygen-containing Silicon Ions and Their Neutral Counterparts

Several new oxygen- and nitrogen-containing silicon ions have been studied (Table 2) and the stability of their neutral counterparts has been probed by NRMS experiments[148,151,152]. In view of the emerging insight into the remarkable differences between the π-bonding ability of carbon and silicon, the structural and energetic properties of small unsaturated silicon-containing ions has become the subject of considerable scrutiny. A further understanding of the chemistry of small oxygen-containing silicon radicals and molecules of the type H_xSiO_y is also desirable as these species are known to be relevant to the combustion chemistry taking place in silane–oxygen flames[185–188].

TABLE 2. Oxygen- and nitrogen-containing neutral molecules with silicon as studied by neutralization–reionization mass spectrometry

$Si_xO_yH_z$	$HSiO^{35}$, $HOSi^{193}$, H_2SiO^{194}, $HSiOH^{194}$, H_2SiOH^{194}, $H_3SiO_4^{194}$, Si_2O^{195}, c-$Si_2O_2^{196}$
$Si_xN_yH_z$	$HNSi^{197}$, H_2NSi^{198}, Si_2N^{195}, Si_3N^{199}

For recent theoretical studies of such species, see elsewhere[189–191]. It was found to be particularly useful to corroborate these experimental findings by theoretical studies dealing with the structures and thermochemistry of the observed molecules. Table 2 gives an overview of the species studied so far.

The careful reader will encounter a structural pattern exhibited by all these low-valent silicon species. Whereas the preferred bonding of silicon to other main group elements in these ions and neutral molecules can no longer be deduced on the grounds of simple 'Lewis-structure type reasoning'—as it is so successfully employed for the lower homologue carbon—the higher homologue silicon prefers to form cyclic or bridged 'nonclassical' structures[192]. This is presumably a consequence of the far more diffuse p orbitals of this element as compared with carbon, which not only leads to significantly weaker π-type interactions but also to much less 'directionality' in its bonding. Thus the successful thinking in terms of sp hybridizations for carbon molecules cannot be stringently applied to silicon!

Protonation of silicon monoxide can in principle give rise to two different isomers: the O-protonated $SiOH^+$ and the silicon-protonated form $HSiO^+$. $SiOH^+$ is formed upon reaction of water with silicon cations[200] and is believed to play a key role in the depletion mechanism of atomic silicon in the earth's ionosphere[50]. Theory (MP4) predicts the two cationic isomers to be separated by a substantial barrier of 91.5 kcal mol^{-1} (with respect to the formation of $HSiO^+$ from $SiOH^+$)[193]. The neutral counterpart $SiOH^{\bullet}$ has been generated in a NR experiment from the cationic precursor (equation 28)[193,201] (accessible via electron impact ionization of tetramethoxysilane). Neutral $HSiO^{\bullet}$—in line with theoretical predictions[202,203]—was successfully generated via Charge-Reversal NRMS (CR-NRMS) of anionic $HSiO^{-\,204}$ (equation 29).

$$SiOH^+ \longrightarrow SiOH^{\bullet} \longrightarrow SiOH^+ \quad (28)$$

$$HSiO^- \longrightarrow HSiO^{\bullet} \longrightarrow HSiO^+ \quad (29)$$

According to the computational studies, a species with $HSiO^{\bullet}$ connectivity forms the global minimum on both the neutral and cationic [H,O,Si] potential energy surfaces. The barrier for the 1,2-hydrogen shift generating $HSiO^{\bullet}$ from $HOSi^{\bullet}$ was estimated to be 35.9 kcal mol^{-1} above $HOSi^{\bullet}$; the latter represents the global minimum, lying 6.4 kcal mol^{-1} below $HSiO^{\bullet}$. This is an ordering of stability similar to that found on the cationic surface where $HOSi^+$ corresponds to the global minimum[204,205], and these findings once again underline the significant differences between carbon and its higher homologue silicon: on the neutral as well as ionic [H,C,O] surface[205] the order of stabilities is reversed, compared to that of [H,O,Si].

Among the other silicon-containing molecules and ions of the composition [Si,O,H$_x$], Srinivas and coworkers have characterized the ions $HSiOH^{+\bullet}$, $H_2SiO^{+\bullet}$, H_2SiOH^+ and H_3SiO^+ as well as their neutral counterparts in collision experiments[194]. The neutral and cationic H_xSiO species ($x = 2,3$) were generated by neutralization and charge-reversal experiments which employed the anionic counterparts as precursor molecules. The $H_2SiO^{-\bullet}$ and H_3SiO^- anions[206] can be readily obtained in the reaction of $O^{-\bullet}$ with phenylsilane according to equations 30 and 31.

$$O^{-\bullet} + C_6H_5SiH_3 \longrightarrow H_2SiO^{-\bullet} + C_6H_6 \quad (30)$$

$$O^{-\bullet}(OH^-) + C_6H_5SiH_3 \longrightarrow H_3SiO^- + C_6H_5 \cdot (C_6H_6) \quad (31)$$

The cations $HSiOH^{+\bullet}$ and H_2SiOH^+ were generated by electron ionization of tetramethoxysilane. As demonstrated by these NRMS experiments, all four species HSiOH, H_2SiO, $H_2SiOH^{\bullet}$ and $H_3SiO^{\bullet}$ are accessible, stable molecules in the gas phase (for matrix isolation studies on silanone, H_2SiO, and hydroxysilylene, HSiOH, see elsewhere)[207-209]. These findings are in line with earlier computational studies on the neutral and cationic [H_n,O,Si] potential energy surfaces ($n = 2,3$) which had predicted high barriers for the interconversion of these isomers[209-212].

Two different silicon–oxygen cluster ions $Si_xO_y^{+\bullet}$ have been characterized by Schwarz and coworkers in the gas phase[195,196]. These clusters may be viewed as the simplest prototypes for the interaction of oxygen at silicon surfaces[210] and thus their structures are of particular interest. The ion $Si_2O^{+\bullet}$ can be generated via electron ionization of disilyl ether, $H_3SiOSiH_3$. The $Si_2O^{+\bullet}$ ion has been successfully neutralized and its dissociation pattern was found to agree well with the theoretically predicted energetics for Si_2O by Boldyrev and Simons[211]. According to the theoretical calculations[211,212], the global minimum structure for Si_2O corresponds to a C_{2v} symmetric species **58** with a $^1A_1$ ground state, with the neutral isomer **59** being 27 kcal mol^{-1} less stable than **58**. While an unambiguous differentiation between a cyclic and an open structure in these collision experiments was not possible, recourse to the *ab initio* data very much suggests that neutral Si_2O as well as its cationic counterpart possess a cyclic structure[213].

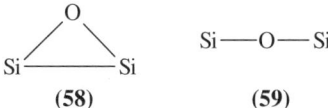

(58)   **(59)**

The $Si_2O_2^{+\bullet}$ cation has been generated upon ionizing a mixture of disilyl ether and N_2O under chemical ionization conditions[196]. The genesis of this interesting ion most likely proceeds via loss of the hydrogen atoms from $H_3SiOSiH_3$ upon electron impact, thus giving rise to $Si_2O^{+\bullet}$ which is subsequently oxidized in an ion–molecule reaction with N_2O (equation 32).

$$H_3SiOSiH_3 + e^- \longrightarrow Si_2O^{+\bullet} + 3H_2 + 2e^-$$
$$Si_2O^{+\bullet} + N_2O \longrightarrow Si_2O_2^{+\bullet} + N_2 \quad (32)$$

These results from collisional activation experiments have been augmented by *ab initio* calculations [MP2/6-311 G(df) CCSD(T) level] on the neutral and cationic [Si_2,O_2] surface[196,214]. The theoretical results point to the existence of three low-lying cationic isomers **60**, **61** and **62**.

Isomer **60** was calculated to be 27.5 kcal mol^{-1} and 20.9 kcal mol^{-1} more stable than **61** and **62**, respectively. The relative abundances of fragment ions in the CA mass spectra reflected the predicted energetics for the dissociation processes. While CA mass spectrometry does not allow for a distinction of cyclic versus noncyclic isomers, the fact that the CA mass spectrum contained an $Si_2^{+\bullet}$ and $SiO_2^{+\bullet}$ ion leads to the exclusion of isomers **61** and **62**. On the neutral [Si_2, O_2] surface[215] only two energetically low-lying

(60)   (61)   (62)

isomers (**63** and **64**) were located by the *ab initio* calculations. Isomer **63** possesses a D_{2h} symmetry and bond lengths rather similar to **60**, while **64** was found to constitute a monocyclic ring with an externally bound oxygen atom. In line with these calculations, neutralization of the $[Si_2,O_2]^{+\bullet}$ ion was found to lead to a strong signal for the survivor ions, indicating rather favorable Franck–Condon factors[216–218] for the electron-transfer processes — a finding that is expected to be valid for geometrically similar species. Thus the combined experimental as well as theoretical work, revealing that neutral Si_2O_2 forms a cyclic structure, once again point to the differences between carbon and silicon chemistry. While Si_2O_2 can be readily generated from its ionic precursor, the C_2O_2 molecule which has been theoretically predicted to correspond to a linear structure, O=C=C=O, still remains elusive despite numerous experimental attempts[219].

(63)   (64)

Among multiple-bond-containing silicon compounds, neutral HNSi holds a special place. The [H,N,Si] potential energy surface and several possible N- and Si-substituted isomers have formed the subject of several experimental[220–225] and theoretical investigations[226–228]. In a recent experimental and theoretical study, the ionic as well as the neutral [H,N,Si] surfaces were explored[197]. Ionization of a mixture of iodosilane and NH_3 gives rise to $[H,N,Si]^{+\bullet}$ cations, which according to CA experiments exist in an N-protonated (**65**) rather than the isomeric Si-protonated form (**66**).

(65)   (66)

Ab initio MO calculations (GAUSSIAN-1) predict isomer **65** to correspond to the global minimum on the $[H,N,Si]^{+\bullet}$ surface; isomer **66** was calculated to be 42.5 kcal mol^{-1} less stable. The barrier for the 1,2-hydrogen shift was calculated to be 8 kcal mol^{-1} above **66**. Neutral HNSi (**67**) has been successfully generated from ionic **65** and, according to the calculations, exists in a singlet state ($^1\Sigma^+$). Isomer **68** also prefers a singlet electronic ground state (65 kcal mol^{-1} higher than **67**) with a barrier for formation of **68** from **67** of 78 kcal mol^{-1}[224].

Again it is interesting to note the apparent differences between carbon and its higher homologue: HCN corresponds to the global minimum on the neutral [H,C,N] surface[229];

Si—N—H   H—Si—N

(67)   (68)

thus, the order of stability is reversed upon silicon substitution. In this context Apeloig and Albrecht[228] have carried out instructive studies on the R—SiN ⇌ SiN—R isomerization and the dependence of the stabilities of the two possible isomers upon the substituent R. They found that substitution of R by a more electronegative ligand such as F or OH leads to a preference for the silanitrile structure over the N-substituted isomer. An experimental verification of these intriguing findings still remains to be done.

The N-protonated form of HNSi, the H_2NSi^+ cation **69** and its possible isomers have been studied extensively by theoretical methods[230–234]. Böhme[47] suggested that the ion **69**, which is formed upon reaction of silicon cations with ammonia and subsequent dissociation of **69** upon electron capture (equation 33)[235], is of prime importance for the formation of Si—N bonded species such as **67** in the chemistry of interstellar matter.

$$Si^{+\bullet}(^2P) + NH_3 \longrightarrow SiNH_2^+ + H^\bullet$$
$$SiNH_2^+ + e^- \longrightarrow SiNH + H^\bullet$$
(33)

Thus, $SiNH_2^+$ has been proposed to give rise to neutral HNSi (**67**) upon neutralization via dissociative recombination. Electron bombardment of trisilylamine (or a mixture of iodosilane and ammonia) leads to the formation of $[Si,N,H_2]^+$ cations, which according to collision experiments can be assigned a H_2NSi^+ structure (**69**)[198]. Thus, the preferred site of protonation of the silaisonitrile molecule HNSi is the nitrogen atom. Theoretical calculations on the ionic $[H_2,N,Si]^+$ surface and all possible dissociation channels carried out in the same study indicate that the $^1A_1$ state of **69** corresponds to the most stable isomer. The isomers **70** ($^1A'$) and **71** ($^3A_2$) are predicted to lie respectively 50.3 and 116.7 kcal mol^{-1} higher in energy than **69**.

```
      H                H               H
     /                  \               \   +
 +Si—N              Si—N             Si—N
     \              /  +              /  \
      H            H                 H    H
     (69)          (70)              (71)
```

Neutralization of **69** was found to give rise to a stable $H_2NSi^\bullet$ radical **72**. The other possible isomers **73** and **74** were computationally estimated to be 21.8 and 45.8 kcal mol^{-1} less stable than **72**. However, significant barriers for the possible interconversion of the neutral isomers, found in these calculations, indicate that in principle isomers **73** and **74** should also be viable species.

```
   •  H                H                H
     /                  \ •              \   •
 Si—N              Si—N             Si—N
     \              /                  /  \
      H            H                 H    H
     (72)          (73)              (74)
```

The Si_2N^+ cluster molecule has been generated upon ionization of trisilylamine[195]. Theory [QCISD(T)/6-311+G*][199] predicts a linear $D_{\infty h}$ structure for the cationic as well as the neutral NSiN$^\bullet$ radical (**75**). The linear species was found to correspond to the global minimum on both the neutral and cationic potential energy surfaces and, in line with the theoretical predictions, NSiN• has been identified in NRMS experiments[195]. According to the theoretical and recent spectroscopic measurements[35] NSiN• possesses a linear structure (**75**) and a $^2\Pi_g$ ground electronic state.

$$Si—\overset{\bullet}{N}—Si \qquad (75)$$

(**76**) shows a C_{2v} symmetric Si$_3$N$^+$ cluster; (**77**) shows the analogous neutral Si$_3$N• radical.

Electron ionization of trisilylamine was also found to give rise to the ionic cluster molecule Si$_3$N$^+$ [199]. As indicated by *ab initio* calculations[198] [MP2(full)/6-311 +G*] on the ionic [N, Si$_3$]$^+$ surface, the ground state structure of the Si$_3$N$^+$ cluster corresponds to a C_{2v} symmetric species (**76**) with the nitrogen atom being bound to all three silicon atoms. The calculated dissociation energies for **76** were found to agree well with the relative abundances obtained in CA experiments. Neutralization of the ionic precursor **76** demonstrated the viability of the Si$_3$N• radical in the gas phase. For this neutral species, the computational results predict a ground state $^2A_2$ structure (**77**) possessing C_{2v} symmetry with a geometry that is very similar to that of cation **76**.

It can be concluded from these studies on small silicon clusters which are bound to main-group elements, that the tendency of silicon to form significantly stronger σ-bonds with first-row main-group elements leads to the formation of cyclic 'nonclassical' structures, unknown in carbon chemistry. For carbon, the ability to form strong π-bonding interactions and the more pronounced directionality of its σ-bonding orbitals do not allow for an observation of such cyclic clusters as ground state species; it preferably forms linear clusters.

V. REARRANGEMENTS OF ORGANOSILICON IONS

A. Cationic Rearrangements

The unimolecular dissociation of Me$_3$Si$^+$ (**78**) has been studied experimentally[236]. As indicated by CAD studies and isotopic exchange reactions, ion **78** readily interconverts upon excitation to isomer **79** (equation 34). The reaction mechanism for this interesting interconversion has been proposed to proceed via a concerted 1,2-hydrogen/1,2-methyl migration. This dyotropic rearrangement[237] corresponds to a thermally allowed [$\sigma_s^2+\sigma_s^2$] process[238]. In a similar fashion the ethene elimination from **78** was proposed to proceed via a dyotropic multicenter reaction mechanism (equation 35).

$$Me_3Si^+ \longrightarrow \left[\begin{array}{c} H \\ H_2C \cdots Si—Me \\ Me \end{array} \right]^{\ddagger} \longrightarrow Et(Me)SiH^+ \quad (34)$$
(**78**) → (**79**)

$$Me_3Si^+ \longrightarrow \left[\begin{array}{c} Me \\ H—Si—H \\ CH_2 \;\; CH_2 \end{array} \right]^{\ddagger} \longrightarrow MeSiH_2^+ + C_2H_4 \quad (35)$$
(**78**)

Excited ions of **78** were found to exchange an ethene molecule for the isotopomer C_2D_4, which lends further support to this hypothesis.

Bakhtiar and coworkers also studied the isomerization of $SiMe_2R^+$ (**80**) to $SiHEtR^+$ (**81**) upon collisional activation[239]. As already described for the trimethylsilyl cation, this isomerization seems to proceed via a concerted $[\sigma_s^2 + \sigma_s^2]$ process (equation 36).

$$RMe_2Si^+ \rightleftharpoons \left[R-\overset{+}{Si} \underset{Me}{\overset{H}{\cdots}} CH_2 \right]^{\ddagger} \rightleftharpoons \overset{H}{\underset{R}{\diagdown}} \overset{+}{Si} - Et \quad (36)$$

(**80**)   (**81**)

The rearrangements preceding the dissociation of α-silyl substituted carbenium ions have been investigated earlier by Drewello and coworkers[240]. In this study, extensive isotopic labeling has been successfully employed to demonstrate that the metastable MeC^+HSiMe_3 ions (**82**) rapidly undergo 1,2-methyl migrations prior to dissociation to C_2H_4 and $SiMe_3^+$ with an almost complete exchange of the four methyl groups (equation 37). These findings are in line with earlier *ab initio* molecular orbital calculations[236] which predict that the ion $Me_2CH-Si^+Me_2$ (**83**) is approximately 28 kcal mol^{-1} more stable than the corresponding carbenium ion **82**. Thus the 1,2-methyl migration giving rise to **83** is a highly exothermic process. For the slow 1,2-hydrogen shift a kinetic isotope effect k_H/k_D of 2.5 was observed. In this study also a similarly rapid exchange of methyl groups was reported for the trimethylsilyl isopropyl carbenium ion, $Me_3SiC^+Me_2$.

$$\underset{(82)}{\overset{Me}{\underset{H}{\diagup}}\overset{+}{C}-\underset{Me}{\overset{Me}{\diagdown}}Si-Me} \underset{fast}{\rightleftharpoons} \underset{(83)}{Me-\underset{H}{\overset{Me}{\diagup}}C-\underset{Me}{\overset{+}{\diagdown}}\overset{Me}{\diagup}Si-Me}$$

$$\Big\downarrow \text{slow} \qquad\qquad\qquad\qquad\qquad\qquad (37)$$

$$\underset{H\ H}{\overset{H_2\overset{+}{C}}{\diagdown}}C-\underset{Me}{\overset{Me}{\diagdown}}\overset{Me}{\diagup}Si-Me \xrightarrow{fast} SiMe_3^+ + C_2H_4$$

The migratory aptitude of hydrogen, alkyl and phenyl groups in the silicon analogue of the Wagner–Meerwein rearrangement (migration from a silicon center to a carbenium ion) has been studied by Bakhtiar and coworkers[241]. The corresponding α-silyl-substituted carbenium ion **84** was generated by chloride elimination from the α-chloroalkylsilane and the nascent ions were found to undergo rearrangement to the thermodynamically favored silicenium ions upon 1,2-migration from silicon to carbon. The rearrangement of the ion $PhMeSiHCH_2^+$ allowed for a direct comparison of the competitive migrations of a phenyl, methyl and hydrogen (equation 38). The authors found that 94% of the obtained silicenium ions resulted from phenyl migration, the other 6% corresponded to a hydrogen migration and no methyl migration was observed.

$$
\text{(84)} \xrightarrow{94\%} \text{PhCH}_2\text{Si}^+\text{HMe}
$$
$$
\xrightarrow{6\%} \text{PhSi}^+\text{Me}_2
$$
$$
\not\rightarrow \text{PhSi}^+\text{HEt}
$$
(38)

As an explanation for the preferred aryl migration, delocalization of the charge into the ring in the bridged transition state (leading to a phenonium ion) was invoked. It was concluded that the corresponding α-silylcarbenium ions as well as the silicenium ions may be important intermediates in the solvolytic rearrangements of α-functionalized silanes. These findings have been questioned by a recent theoretical study of Cho[242], who argues that the α-silylcarbenium ions do not correspond to energy minima on the PES and that the rate-determining step for the migration is really the departure of the Cl^- ion. On the basis of MP2/6-31G* calculations, the Cl^- dissociation and silicenium ion formation is predicted to correspond to a single process and thus Cho concludes that the differences in dissociation energies of the chloroalkylsilanes are responsible for the migratory tendencies. Further experimental as well as theoretical studies are needed to settle this point.

The phenylsilyl cation $C_6H_5SiH_2^+$ (85) and its interconversion to other possible isomers such as the silacyloheptatrienyl cation (86) have attracted considerable experimental as well as theoretical interest in view of the possible similarities or differences of the well-known benzyl and cycloheptatrienyl cations[243] and their silicon analogues. Thus, Beauchamp and coworkers[244,245] have found that electron ionization of phenylsilane gives rise to both isomers 85 and 86, which do not interconvert at room temperature and can be readily distinguished by their ion–molecule reactions.

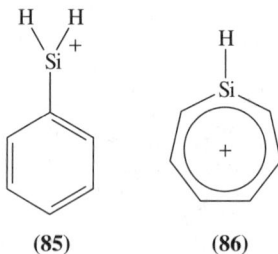

(85)   (86)

Recent theoretical results by Nicolaides and Radom[246], who employed isodesmic and isogyric reactions for an accurate determination of the heats of formation of both ions at the G2 level of theory, suggest that the silabenzyl cation 85 is more stable than the silatropylium ion 86 by approximately 40 kcal mol^{-1}. For the corresponding $[C_7,H_7]^+$ surface, they found a reverse ordering, here the tropylium ion is 29 kcal mol^{-1} more stable than the benzyl cation. These findings presumably reflect the well-documented inability of silicon to form π-bonds with significant strength.

18. Gas-phase ion chemistry of silicon-containing molecules 1133

Tajima, Okada and coworkers[247] reported the unimolecular metastable decompositions of alkoxysilanes in a number of recent studies[184,248,249]. These authors found significant differences in the ion fragmentation characteristics between the silanes and their carbon analogues. Thus, for example, the principal fragmentation process of ionized diethoxydimethylsilane was found to correspond to a consecutive loss of ethylene and aldehyde molecules from the silicenium ion formed by loss of an ethoxy radical. In contrast, the carbon analogue acetone diethyl acetal does not exhibit a significant loss of aldehyde molecules in its metastable ion mass spectrum.

B. Anionic Rearrangements

Bowie's group has investigated a number of rearrangements of sila anions. The authors reported the unimolecular chemistry of deprotonated α-alkoxyvinyltrimethylsilane[250]. Interestingly, these ions were found to lose a ketene molecule upon excitation via collisional activation. This unexpected behavior points to the fact that ions such as **87** rearrange via a methyl migration (equation 39) for which two different pathways are conceivable. The methyl either migrates to the silicon of an incipient dimethylsilene ($Me_2Si=CH_2$) (39a), or to the methylene carbon (39b). Charge reversal spectra indicate that the anions formed in this rearrangement possess a structure **88** and thus pathway 39b seems to be favored.

$$[(H_2CCO)Me_2EtSi]^- \xrightarrow{H_2CCO} Me_2EtSi^-$$

$$[(H_2CCO)Me_3SiCH_2]^- \xrightarrow{H_2CCO} Me_3SiCH_2^-$$

(39)

(87)

$$\begin{array}{c} H_3C \\ H_3C \\ H_3C \end{array} Si-\bar{C}H_2$$

(88)

The rearrangements preceding collisional excitation of trimethylsilyl-substituted alkoxides, ketones or carboxylic acids carrying anions have been investigated[251]. Most of the products obtained correspond to a species in which a Si—O bond has been formed. Thus, for example, all species $Me_3Si(CH_2)_nO^-$ ions ($n = 2-5$), e.g. **89** where $n = 3$, were found to react via an intramolecular S_N2 reaction (equation 40) by an attack of the O^- at the silicon center and a subsequent formation of a fivefold coordinated intermediate (**90**), which then reacts further to give the final products such as Me_3SiO^- and **91**.

Another interesting case of migration of a methoxide to a neutral silicon center has been reported by Krempp and Damrauer[252]. These authors reported that silanamide anions, H_3SiNH^- (**92**), react with methyl formate to form the neutral methyl silyl ether and the formamide anion (equation 41). The thermodynamic driving force for this reaction as well as the above-mentioned rearrangements are most likely the formation of the strong Si—O bond in the silyl ether.

$$Me_3Si\text{(89)} \longrightarrow Me_3\bar{S}i\text{(90)} \longrightarrow Me_3SiO^- + C_3H_6 \tag{40}$$

$$\text{(90)} \xrightarrow{-CH_4} \text{(91)}$$

$$H_3SiNH^- + HCO_2CH_3 \longrightarrow \begin{bmatrix} H_3Si-NH \\ H_3C-O \quad H \end{bmatrix}^- \tag{41}$$

$$\downarrow$$

$$H_3SiOCH_3 + HNC(H)O^-$$

O'Hair and coworkers[253] have investigated the gas-phase chemistry of siloxide and silamide ions in flowing afterglow experiments. The reaction of siloxide anions, H_3SiO^- (as well as substituted analogues), with CS_2 was found to lead to the exclusive formation of COS and the corresponding H_3SiS^- anion (equation 42). As suggested by *ab initio* calculations carried out in the same study, the reaction quite likely proceeds via a fivefold coordinated cyclic transition state (93).

$$H_3SiO^- + CS_2 \longrightarrow \begin{bmatrix} S=C\underset{S}{\overset{O}{\diagup}}SiH_3 \end{bmatrix}^{\ddagger} \longrightarrow H_3SiS^- + COS \tag{42}$$

(93)

VI. DOUBLY CHARGED IONS

Although a large number of doubly charged cations can be observed in a typical mass spectrum of organo-silicon compounds[254–262], there have been relatively few recent studies dealing with these species. In general it must be noted that silicon-containing dications are far more stable than their carbon analogues due to the more electropositive character of silicon.

Schwarz and coworkers[258] have been successful in generating the diatomic dication SiF^{2+} by charge-stripping mass spectrometry (equation 43)[259–262].

$$SiF^+ [E_{kin} = 8\,keV] + O_2 \longrightarrow SiF^{2+} + O_2^* + e^- \tag{43}$$

The theoretically predicted existence of SiF^{2+}[263] could be confirmed in these experiments. In addition, the vertical ionization energy ($MF^+ \rightarrow MF^{2+} + e^-$) of SiF^+ was calculated using single- and multi-reference configuration interaction methods. The experimental value of 21.7 eV mirrored the theoretically obtained values. An explanation for the stability of this diatomic dication may be seen in the extraordinary strength of the covalent silicon–fluoride bond (see Section III.C).

A number of small dications such as Si_2O^{2+} [195], Si_2N^{2+} [195], Si_3N^{2+} [199] have been observed in the CA mass spectra of their corresponding monocations. However, the ionization energies of the monocations have not been determined and no theoretical results are available on these interesting species as yet.

There are still very few examples of small stable doubly charged negative ions that have been characterized in the gas phase[264-266]. The main reason for the instability of dianions is certainly the strong electron–electron repulsion, causing species which possess two spatially close additional electrons to autodetach the second electron rather than form a stable system. In view of this, it is interesting to find that Sommerfeld, Scheller and Cederbaum[267] have recently proposed a number of small $Si_mO_n^{2-}$ clusters which should be stable with respect to electron detachment. In their theoretical study (DZ/SDCI level of theory) the authors predict that the smallest system which should be stable with respect to vertical electron loss (and hence to possess a finite lifetime) is the D_{2d} symmetrical cluster **94**, the silicon analogue of the oxalate dianion, $O_2CCO_2^{2-}$.

(94)

Interestingly, the authors find that the silicon-containing dianions are more stable with respect to electron detachment and dissociation than their corresponding carbon analogues. The predominantly covalent and stronger bonding in the silicon–oxygen clusters has been invoked as an explanation for this behavior. In a continuation of their search for small stable dianionic systems, these authors theoretically investigated the structures and stabilities of several $Si_2O_5^{2-}$ isomers[268]. They predict that the dianions **95–97** could be stable species with respect to dissociation into SiO_3^- and SiO_2^-. With respect to possible electron autodetachment only **97** is found to lie in an unstable energy regime, and **95** as well as **96** might be further candidates for stable dianionic systems. The experimental verification of this theoretical work should be a challenging task.

(95) (96) (97)

VII. ACKNOWLEDGMENTS

N.G. warmly thanks Professor Roald Hoffmann, Cornell University, for his kind hospitality and support whilst he worked on this review. He also thanks Alison for her invaluable help. The authors are grateful to many colleagues, in particular Professors P. B. Armentrout, D. K. Böhme, J. H. Bowie, R. Damrauer, D. B. Jacobson, R. Squires and F. Turecek, for providing them with re- and preprints of current results obtained in their laboratories. Financial support by the Deutsche Forschungsgemeinschaft (DFG) (research grant to N.G.) and the Fonds der Chemischen Industrie is gratefully acknowledged.

VIII. REFERENCES

1. (a) H. Schwarz, in *The Chemistry of Organic Silicon Compounds* (Eds. S. Patai and Z. Rappoport), Chap. 7, Wiley, Chichester, 1989, pp. 445–510; (b) H. Bock and B. Solouki, *Chem. Rev.*, **95**, 1161 (1995).
2. J. Berkowitz, *Acc. Chem. Res.*, **22**, 413 (1989).
3. R. Damrauer, in *Advances in Silicon Chemistry* (Ed. G. L. Larson), Vol. 2, JAI Press, Greenwich, Conn., 1993, p. 91.
4. R. Damrauer and J. A. Hankin, *Chem. Rev.*, **95**, 1137 (1995).
5. M. S. Gordon, J. S. Francisco and H. B. Schlegel, in *Advances in Silicon Chemistry* (Ed. G. L. Larson), Vol. 2, JAI Press, Greenwich, Conn., 1993, p. 137.
6. J. Berkowitz, J. P. Greene, H. Cho and B. Ruscic, *J. Chem. Phys.*, **86**, 1235 (1987).
7. K. Börlin, T. Heinis and M. Jungen, *J. Chem. Phys.*, **103**, 93 (1985).
8. J. A. Pople and L. A. Curtiss, *J. Phys. Chem.*, **91**, 155 (1987).
9. J. A. Pople, B. T. Luke, M. J. Frisch and J. S. Binkley, *J. Phys. Chem.*, **89**, 2198 (1985).
10. J. A. Pople and L. A. Curtiss, *J. Phys. Chem.*, **91**, 3637 (1987).
11. E. R. Fisher and P. B. Armentrout, *J. Chem. Phys.*, **93**, 4858 (1990).
12. D. I. Hall, A. P. Levick, P. J. Sarre, C. J. Whitham, A. A. Alijah and G. Duxbury, *J. Chem. Soc., Faraday Trans. 2*, **89**, 177 (1993).
13. B. H. Boo and P. B. Armentrout, *J. Am. Chem. Soc.*, **109**, 3549 (1987).
14. P. J. Bruna, G. Hirsch, R. J. Buenker and S. D. Peyerimhoff, in *Molecular Ions* (Eds. J. Berkowitz and K.-O. Groeneveld), Chap. 4, Plenum Press, New York, 1993, p. 309.
15. S. P. Mort, N. A. Jennings, G. G. Balint-Kurti and D. M. Hirst, *J. Chem. Phys.*, **101**, 10576 (1994).
16. C. Bauer, D. M. Hirst, D. I. Hall, P. J. Sarre and P. Rosmus, *J. Chem. Soc., Faraday Trans. 2*, **90**, 517 (1994).
17. Y. Cao, J.-H. Choi, B.-M. Haas, M. S. Johnson and M. Okumura, *J. Phys. Chem.*, **97**, 5215 (1993).
18. J. B. Lambert, L. Kania and S. Zhang, *Chem. Rev.*, **95**, 1191 (1995).
19. S. K. Shin and J. L. Beauchamp, *J. Am. Chem. Soc.*, **111**, 900 (1989).
20. G. A. Olah and Y. K. Mo, *J. Am. Chem. Soc.*, **93**, 4942 (1971).
21. C.-Y. Lin and R. C. Dunbar, *J. Phys. Chem.*, **100**, 655 (1996).
22. D. S. Tonner, D. Thölmann and T. B. McMahon, *Chem. Phys. Lett.*, **233**, 324 (1995).
23. M. E. Weber and P. B. Armentrout, *J. Chem. Phys.*, **90**, 2213 (1989).
24. M. E. Weber and P. B. Armentrout, *J. Phys. Chem.*, **93**, 1596 (1989).
25. E. R. Fisher and P. B. Armentrout, *Chem. Phys. Lett.*, **179**, 435 (1991).
26. E. R. Fisher and P. B. Armentrout, *J. Phys. Chem.*, **95**, 4765 (1991).
27. E. R. Fisher, B. L. Kickel and P. B. Armentrout, *J. Phys. Chem.*, **97**, 10204 (1993).
28. B. L. Kickel, E. R. Fisher and P. B. Armentrout, *J. Phys. Chem.*, **97**, 10198 (1993).
29. B. L. Kickel, J. B. Griffin and P. B. Armentrout, *Z. Phys. D*, **24**, 101 (1992).
30. J. Walsh, *Acc. Chem. Res.*, **14**, 246 (1981).
31. A. Gobbi and G. Frenking, *J. Am. Chem. Soc.*, **116**, 9287 (1994).
32. D. J. Hajdasz, Y. Ho and R. R. Squires, *J. Am. Chem. Soc.*, **116**, 10751 (1994).
33. Pentacoordinated anionic silicon adducts had earlier been demonstrated to be intermediates in anionic gas-phase reactions: G. Angelini, C. E. Johnson and J. I. Brauman, *Int. J. Mass Spectrom. Ion Proc.*, **109**, 1 (1991).
34. For theoretical calculations on SiH_5^-, see: A. E. Reed and P. v. R. Schleyer, *Chem. Phys. Lett.*, **133**, 533 (1987).
35. A theoretical study on the pseudorotation in $SiH_{5-n}X_n^-$ (X = F, Cl) anions has been reported: T. L. Windus, M. S. Gordon, L. P. Davis and L. W. Burggraf, *J. Am. Chem. Soc.*, **116**, 3568 (1994).
36. R. R. Squires, in *Structure/Reactivity and Thermochemistry of Ions* (Eds. P. Ausloos and S. G. Lias), Reidel, Dordrecht, 1987, p. 177.
37. E. A. Brinkman, S. Berger and J. I. Brauman, *J. Am. Chem. Soc.*, **116**, 8304 (1994).

38. C. H. DePuy, V. M. Bierbaum, L. A. Flippin, J. J. Grabowski, G. K. King, R. J. Schmitt and S. A. Sullivan, *J. Am. Chem. Soc.*, **101**, 6443 (1979).
39. R. Yoshimura and T. Tada, *J. Phys. Chem.*, **97**, 845 (1993).
40. J. H. Bowie, *Mass Spectrom. Rev.*, **3**, 1 (1984).
41. A. Oppenstein and F. W. Lampe, *Rev. Chem. Intermed.*, **6**, 275 (1986).
42. C. H. DePuy, R. Damrauer, J. H. Bowie and J. C. Sheldon, *Acc. Chem. Res.*, **20**, 127 (1987).
43. D. K. Böhme, S. Wlodek and A. Fox, in *Rate Coefficients in Astrochemistry* (Eds. T. J. Millar and D. A. Williams), Kluwer Academic Publisher, Amsterdam, 1988, p. 193.
44. For an excellent earlier review on the chemistry of atomic silicon ions, see: D. K. Böhme, *Int. J. Mass Spectrom. Ion Proc.*, **100**, 719 (1990).
45. D. K. Böhme, in *Chemistry and Spectroscopy of Interstellar Molecules* (Eds. D. K. Böhme, E. Herbst, N. Kaifu and S. Saito), Chap. 3, University of Tokyo Press, Tokyo, 1992, p. 155.
46. D. K. Böhme, *Chem. Rev.*, **92**, 1487 (1992).
47. D. K. Böhme, *Adv. Gas-Phase Ion Chem.*, **1**, 225 (1992).
48. R. C. Dunbar, in *Unimolecular and Bimolecular Ion–Molecule Reaction Dynamics* (Eds. C.-Y. Ng, T. Baer and I. Powis), Wiley, Chichester, 1994.
49. J. M. Farrar and J. Saunders (Eds.), *Techniques for the Study of Ion–Molecule Reactions*, Vol. 20 in Techniques of Chemistry, Wiley, New York, 1988.
50. For a review of the chemistry of silicon in interstellar clouds, see: E. Herbst, T. J. Millar, S. Wlodek and D. K. Böhme, *Astron. Astrophys.*, **222**, 205 (1989).
51. For an excellent review on the structures and properties of Si-cluster ions, see: M. Jarrold, in *Cluster Ions* (Eds. C.-Y. Ng, T. Baer and I. Powis), Chap. 3, Wiley, Chichester, 1993, p. 165.
52. V. G. Anicich, *J. Phys. Chem. Ref. Data*, **22**, 1993 (1993).
53. For details on the reaction of transition-metal cations with organic molecules, see: K. Eller and H. Schwarz, *Chem. Rev.*, **91**, 1121 (1991).
54. B. L. Kickel and P. B. Armentrout, *J. Am. Chem. Soc.*, **116**, 10742 (1994).
55. B. L. Kickel and P. B. Armentrout, *J. Phys. Chem.*, **99**, 2024 (1995).
56. B. L. Kickel and P. B. Armentrout, *J. Am. Chem. Soc.*, **117**, 764 (1995).
57. B. L. Kickel and P. B. Armentrout, *J. Am. Chem. Soc.*, **117**, 4057 (1995).
58. T. R. Cundari and M. S. Gordon, *J. Phys. Chem.*, **96**, 631 (1992).
59. For a recent analysis of the nature of the metal–silicon bond in neutral transition-metal complexes, see: H. Jacobsen and T. Ziegler, *Inorg. Chem.*, **35**, 775 (1996).
60. M. Decouzon, J.-F. Gal, S. Geribaldi, M. Rouillard and J.-M. Sturla, *Rapid Commun. Mass Spectrom*, **3**, 298 (1989).
61. (a) M. Azzaro, S. Breton, M. Decouzon and S. Geribaldi, *Rapid Commun. Mass Spectrom*, **6**, 306 (1992), (b) For the reactions of atomic W^+ with SiH_4, see: A. Ferhati, T. B. MacMahon and G. Ohannesian *J. Am. Chem. Soc.*, **188**, 5997 (1996).
62. A. Bjarnason and I. Arnason, *Angew. Chem., Int. Ed. Engl.*, **31**, 12 (1992).
63. A. Bjarnason, I. Arnason and D. P. Ridge, *Org. Mass Spectrom.*, **28**, 989 (1993).
64. D. B. Jacobson and R. Bakhtiar, *J. Am. Chem. Soc.*, **115**, 10830 (1993).
65. (a) R. Bakhtiar, C. M. Holznagel and D. B. Jacobson, *J. Am. Chem. Soc.*, **115**, 345 (1993), (b) For a theoretical study of the interaction of Fe^+ with silene, see: J. Moc and M. S. Gordon, *J. Am. Chem. Soc.*, **16**, 27 (1997).
66. R. Bakhtiar and D. B. Jacobson, *Organometallics*, **12**, 2876 (1993).
67. R. H. Schultz and P. B. Armentrout, *J. Am. Chem. Soc.*, **113**, 729 (1991).
68. S. Karrass and H. Schwarz, *Int. J. Mass Spectrom. Ion Proc.*, **98**, R1 (1990).
69. For the observation of methide transfer in the reactions of Cu^+ with alcohols, see: D. A. Weil and C. L. Wilkins, *J. Am. Chem. Soc.*, **107**, 7316 (1985).
70. S. Karrass and H. Schwarz, *Organometallics*, **9**, 2409 (1990).
71. (a) A. Hässelbarth, T. Prüsse and H. Schwarz, *Chem. Ber.*, **123**, 213 (1990), (b) G. Hornung, D. Schröder and H. Schwarz, *J. Am. Chem. Soc.*, **119**, 2273 (1997).
72. A. Hässelbarth, T. Prüsse and H. Schwarz, *Chem. Ber.*, **123**, 209 (1990).
73. R. N. McDonald, M. T. Jones and A. K. Chowdhury, *Organometallics*, **11**, 356 (1992).
74. M. D. Brickhouse, L. J. Chyall, L. S. Sunderlin and R. R. Squires, *Rapid Commun. Mass Spectrom*, **7**, 383 (1993).
75. L. J. Chyall, M. D. Brickhouse, M. E. Schnute and R. R. Squires, *J. Am. Chem. Soc.*, **116**, 8681 (1994).
76. P. G. Wenthold, J. Hu and R. R. Squires, *J. Am. Chem. Soc.*, **116**, 6961 (1994).

77. P. G. Wenthold and R. R. Squires, *J. Am. Chem. Soc.*, **116**, 11890 (1994).
78. P. G. Wenthold and R. R. Squires, *J. Am. Chem. Soc.*, **116**, 6401 (1994).
79. P. G. Wenthold, J. Hu, R. R. Squires and W. C. Lineberger, *J. Am. Chem. Soc.*, **118**, 475 (1996).
80. Y. Guo and J. J. Grabowski, *J. Am. Chem. Soc.*, **113**, 5923 (1991).
81. D. G. Leopold, A. E. S. Miller and W. C. Lineberger, *J. Am. Chem. Soc.*, **108**, 1379 (1986).
82. K. M. Ervin and W. C. Lineberger, in *Advances in Gas Phase Ion Chemistry* (Eds. N. G. Adams and L. M. Babcock), Vol. 1, JAI Press, Greenwich, CT, 1993.
83. J. Lee and J. J. Grabowski, *Chem. Rev.*, **92**, 1611 (1992).
84. A. B. Raksit and D. K. Böhme, *Int. J. Mass Spectrom. Ion Phys.*, **55**, 69 (1983).
85. D. Smith, *Chem. Rev.*, **92**, 1473 (1992).
86. S. Wlodek, A. Fox and D. K. Böhme, *J. Am. Chem. Soc.*, **109**, 6663 (1987).
87. W. R. Creasy, A. O'Keefe and J. R. McDonald, *J. Phys. Chem.*, **91**, 2848 (1987).
88. S. Wlodek and D. K. Böhme, *J. Chem. Soc., Faraday Trans. 2*, **85**, 1643 (1989).
89. D. Stöckigt, N. Goldberg, J. Hrušák, D. Sülzle and H. Schwarz, *J. Am. Chem. Soc.*, **116**, 8300 (1994).
90. B. H. Boo, J. L. Elkind and P. B. Armentrout, *J. Am. Chem. Soc.*, **112**, 2083 (1990).
91. B. H. Boo and P. B. Armentrout, *J. Am. Chem. Soc.*, **113**, 6401 (1991).
92. B. L. Kickel, E. R. Fisher and P. B. Armentrout, *J. Phys. Chem.*, **96**, 2603 (1992).
93. A. E. Ketvirtis, D. K. Böhme and A. C. Hopkinson, *J. Phys. Chem.*, **98**, 13225 (1994).
94. A. E. Ketvirtis, D. K. Böhme and A. C. Hopkinson, *J. Phys. Chem.*, **99**, 16121 (1995).
95. A. E. Ketvirtis, D. K. Böhme and A. C. Hopkinson, *Organometallics*, **14**, 347 (1995).
96. J. Glosik, P. Zakouril, V. Skalsky and W. Lindinger, *Int. J. Mass Spectrom. Ion Proc.*, **149/150**, 499 (1995).
97. S. M. Sze, (Ed.), *VLSI Technology*, McGraw-Hill, New York, 1983.
98. M. J. Kushner, *J. Appl. Phys.*, **63**, 2532 (1991).
99. L. Boufendi, A. Bouchoule, A. Plain, J. P. Blondeau and C. Laure, *Appl. Phys. Lett.*, **60**, 169 (1992).
100. M. L. Mandich, W. D. Reents and M. F. Jarrold, *J. Chem. Phys.*, **88**, 1703 (1988).
101. M. L. Mandich and W. D. Reents, *J. Chem. Phys.*, **90**, 3121 (1989).
102. M. L. Mandich, W. D. Reents and K. D. Kolenbrander, *J. Chem. Phys.*, **92**, 437 (1990).
103. M. L. Mandich and W. D. Reents, *J. Chem. Phys.*, **95**, 7360 (1991).
104. M. L. Mandich and W. D. Reents, *J. Chem. Phys.*, **96**, 4233 (1992).
105. W. D. Reents and M. L. Mandich, *J. Chem. Phys.*, **96**, 4429 (1992).
106. K. Raghavachari, *J. Chem. Phys.*, **88**, 1688 (1988).
107. K. Raghavachari, *J. Chem. Phys.*, **92**, 452 (1990).
108. K. Raghavachari, *J. Chem. Phys.*, **95**, 7373 (1991).
109. K. Raghavachari, *J. Chem. Phys.*, **96**, 4440 (1992).
110. W. D. Reents, M. L. Mandich and C. R. C. Wang, *J. Chem. Phys.*, **97**, 7226 (1992).
111. L. A. Curtiss, K. Raghavachari, P. W. Deutsch and J. A. Pople, *J. Chem. Phys.*, **95**, 2433 (1991).
112. M. A. Al-Laham and K. Raghavachari, *J. Chem. Phys.*, **95**, 2560 (1991).
113. I. Haller, *J. Phys. Chem.*, **94**, 4135 (1990).
114. J.-F. Gal, P. Grover, P.-C. Maria, L. Operti, R. Rabezzana, G.-A. Vaglio and P. Volpe, *J. Phys. Chem.*, **98**, 11978 (1994).
115. P. Antoniotti, L. Operti, R. Rabezzana, G. V. Vaglio, P. Volpe, J.-F. Gal, R. Grover and P.-C. Maria, *J. Phys. Chem.*, **100**, 155 (1996).
116. L. Operti, M. Splendore, G. A. Vaglio and P. Volpe, *Organometallics*, **12**, 4516 (1993).
117. L. Operti, M. Splendore, G. A. Vaglio and P. Volpe, *Organometallics*, **12**, 4509 (1993).
118. L. Operti, M. Splendore, G. A. Vaglio and P. Volpe, *Spectrochim. Acta*, **49A**, 1213 (1993).
119. K. P. Lim and F. W. Lampe, *Int. J. Mass Spectrom. Ion Proc.*, **101**, 245 (1990).
120. U. Ray and M. F. Jarrold, *J. Chem. Phys.*, **93**, 5709 (1990).
121. M. F. Jarrold, J. E. Bower and K. M. Creegan, *J. Chem. Phys.*, **90**, 3615 (1989).
122. K. M. Creegan and M. F. Jarrold, *J. Am. Chem. Soc.*, **112**, 3768 (1990).
123. S. Maruyama, L. R. Anderson and R. E. Smalley, *J. Chem. Phys.*, **93**, 5349 (1990).
124. M. F. Jarrold and V. A. Constant, *Phys. Rev. Lett.*, **67**, 2994 (1991).
125. J. M. Alford, R. T. Laaksonen and R. E. Smalley, *J. Chem. Phys.*, **94**, 2618 (1991).
126. M. F. Jarrold and E. C. Honea, *J. Am. Chem. Soc.*, **114**, 459 (1992).
127. M. F. Jarrold and J. E. Bower, *J. Chem. Phys.*, **96**, 9180 (1992).
128. D. Consalvo, A. Mele, D. Stranges, A. Giardini-Guidoni and R. Teghil, *Int. J. Mass Spectrom. Ion Proc.*, **91**, 319 (1989).

129. R. J. Tench, M. Balooch, L. Bernadez, M. J. Allen and W. J. Siekhaus, *J. Vac. Sci. Technol. B*, **9**, 820 (1991).
130. D. C. Parent, *Int. J. Mass Spectrom. Ion Proc.*, **116**, 257 (1992).
131. P. F. Greenwood, G. D. Willett and M. A. Wilson, *Org. Mass Spectrom.*, **28**, 831 (1993).
132. A. Nakajima, T. Takuwa, N. Nakao, M. Gomei, R. Kishi, S. Iwata and K. Kaya, *J. Chem. Phys.*, **103**, 2050 (1995).
133. Y. Negishi, A. Kimura, N. Kobayashi, H. Shiromura, Y. Achiba and N. Watanabe, *J. Chem. Phys.*, **103**, 9963 (1995).
134. K. Raghavachari, *Z. Phys. D*, **12**, 61 (1989).
135. K. Raghavachari and C. M. Rohlfing, *J. Chem. Phys.*, **89**, 2219 (1989).
136. C. H. Patterson and R. P. Messmer, *Phys. Rev. B*, **42**, 9241 (1990).
137. I. S. Ignatyev and H. F. Schaefer, III, *J. Chem. Phys.*, **103**, 7025 (1995).
138. J. R. Chelikowsky and J. C. Phillips, *Phys. Rev. Lett.*, **63**, 1653 (1989).
139. J. R. Chelikowsky and J. C. Phillips, *Phys. Rev. B*, **41**, 5735 (1990).
140. U. Rothlisberger, W. Andreoni and P. Giannozi, *J. Chem. Phys.*, **96**, 1248 (1992).
141. D. C. Parent, *Int. J. Mass Spectrom. Ion Proc.*, **138**, 307 (1994).
142. Y. Lin, D. P. Ridge and B. Munson, *Org. Mass Spectrom.*, **26**, 550 (1991).
143. R. C. Dunbar, in *Radiative Association Workshop in Fundamentals of Gas-Phase Ion Chemistry*, (Eds. K. R. Jennings), *NATO ASI Series, Ser. C*, Kluwer, Dordrecht, 1991.
144. J. M. Stone and J. A. Stone, *Int. J. Mass Spectrom. Ion Proc.*, **109**, 247 (1991).
145. W. J. Meyerhoffer and M. M. Bursey, *Org. Mass Spectrom.*, **24**, 246 (1989).
146. J. K. Terlouw and H. Schwarz, *Angew. Chem., Int. Ed. Engl.*, **26**, 805 (1987).
147. F. W. McLafferty, *Science*, **247**, 925 (1990).
148. F. W. McLafferty, *Int. J. Mass Spectrom. Ion Proc.*, **118/119**, 221 (1992).
149. J. L. Holmes, *Mass Spectrom. Rev.*, **11**, 53 (1989).
150. F. Turecek, *Org. Mass Spectrom.*, **27**, 1087 (1992).
151. N. Goldberg and H. Schwarz, *Acc. Chem. Res.*, **27**, 347 (1994).
152. D. V. Zagorevskii and J. L. Holmes, *Mass Spectrom. Rev.*, **13**, 133 (1994).
153. D. K. Böhme, S. Wlodek and H. Wincel, *Astrophys. J.*, **342**, L91 (1989).
154. M. J. J. Kushner, *J. Appl. Phys.*, **63**, 2532 (1988).
155. R. Srinivas, D. Sülzle and H. Schwarz, *J. Am. Chem. Soc.*, **113**, 52 (1991).
156. B. T. Luke, J. A. Pople, M. Krogh-Jespersen, Y. Apeloig, M. Karni, J. Chandrashekhar and P. v. R. Schleyer, *J. Am. Chem. Soc.*, **108**, 270 (1986).
157. R. Srinivas, D. K. Böhme and H. Schwarz, *J. Phys. Chem.*, **97**, 13643 (1993).
158. S. Wlodek, A. Fox and D. K. Böhme, *J. Am. Chem. Soc.*, **113**, 4461 (1991).
159. P. Rosmus, H. Bock, B. Solouki, G. Maier and G. Rihm, *Angew. Chem., Int. Ed. Engl.*, **20**, 598 (1981).
160. G. Maier, G. Rihm and H. P. Reisenauer, *Chem. Ber.*, **117**, 2351 (1984).
161. G. Maier, G. Rihm, H. P. Reisenauer and D. Littmann, *Chem. Ber.*, **117**, 2369 (1984).
162. N. Auner and J. Grobe, *Z. anorg. allg. Chem.*, **459**, 15 (1979).
163. J. E. Baggott, M. A. Blitz, H. M. Frey, P. D. Lightfoot and R. Walsh, *Chem. Phys. Lett.*, **135**, 39 (1987).
164. R. Srinivas, D. Sülzle and H. Schwarz, *Chem. Phys. Lett.*, **175**, 575 (1990).
165. M. Iraqi and H. Schwarz, *Chem. Phys. Lett.*, **205**, 183 (1993).
166. J. R. Flores, A. Largo-Cabrerizo and J. Largo-Cabrerizo, *J. Mol. Struct. THEOCHEM*, **148**, 33 (1986).
167. J. Largo-Cabrerizo and J. R. Flores, *Chem. Phys. Lett.*, **90**, 147 (1988).
168. A. E. Glassgold and G. A. Manson, *The Chemistry and Spectroscopy of Interstellar Molecules*, Univ. of Tokyo Press, Tokyo, 1992, p 261.
169. K. P. Lim and F. W. Lampe, *Int. J. Mass Spectrom. Ion Proc.*, **101**, 245 (1991).
170. R. Srinivas, D. Sülzle, T. Weiske and H. Schwarz, *Int. J. Mass Spectrom. Ion Proc.*, **107**, 369 (1991).
171. W. R. Creasy and S. W. McElvany, *Surface Sci.*, **201**, 59 (1988).
172. G. Frenking, R. B. Remington and H. Schaefer III, *J. Am. Chem. Soc.*, **108**, 2169 (1986).
173. M.-D. Su, R. D. Amos and N. C. Handy, *J. Am. Chem. Soc.*, **112**, 1499 (1990).
174. G. Maier, H. P. Reisenauer and H. Pacl, *Angew. Chem., Int. Ed. Engl.*, **33**, 1248 (1994).
175. G. Maier, H. Pacl, H. P. Reisenauer, A. Meudt and R. Janoschek, *J. Am. Chem. Soc.*, **117**, 12712 (1995).

176. W. Sander, *Angew. Chem., Int. Ed. Engl.*, **33**, 1455 (1994).
177. An account on matrix-isolation spectroscopy of low-coordinated silicon atoms has recently been published in: V. K. Korolev, in *Advances in Physical Organic Chemistry* (Ed. D. Bethell), Vol. 30, Academic Press, London, 1995, p. 1.
178. D. K. Böhme, S. Wlodek and H. Wincel, *J. Am. Chem. Soc.*, **113**, 6396 (1991).
179. R. Srinivas, J. Hrušák, D. Sülzle, D. K. Böhme and H. Schwarz, *J. Am. Chem. Soc.*, **114**, 2803 (1992).
180. Radical ions are termed 'distonic' if they possess spatially separated charge and radical centers. For a definition and an excellent review, see: B. F. Yates, W. J. Bouma and L. Radom, *J. Am. Chem. Soc.*, **106**, 5805 (1984).
181. S. Hammerum, *Mass Spectrom. Rev.*, **7**, 123 (1988).
182. For a theoretical work on the stability of silicon-based distonic ions, see: K. Pius and J. Chandrasekhar, *Int. J. Mass Spectrom. Ion Proc.*, **87**, R15 (1989).
183. R. Srinivas, D. K. Böhme, J. Hrušák, D. Schröder and H. Schwarz, *J. Am. Chem. Soc.*, **114**, 1939 (1992).
184. S. Tajima, H. Ida, S. Tobita, F. Okada, E. Tabei and S. Mori, *Org. Mass Spectrom.*, **25**, 441 (1990).
185. O. Horie, R. Taege, B. Reimann, N. L. Arthur and P. Potzinger, *J. Phys. Chem.*, **95**, 4393 (1991).
186. M. Koshi, A. Myoshi and H. Matsui, *J. Phys. Chem.*, **95**, 9869 (1991).
187. S. Koda, *Prog. Energy Combust. Sci.*, **18**, 513 (1992).
188. L. Ding and P. Marshall, *J. Chem. Phys.*, **98**, 8545 (1993).
189. M. R. Zacharia and W. Tsang, *J. Phys. Chem.*, **99**, 5308 (1995).
190. C. L. Darling and H. B. Schlegel, *J. Phys. Chem.*, **97**, 8207 (1993).
191. D. J. Lucas, L. A. Curtiss and J. A. Pople, *J. Chem. Phys.*, **99**, 6697 (1993).
192. For a discussion and references of 'Polymorphism in the Heavier Analogues of the Ethyl Cation', see, for example: G. Trinquier, *J. Am. Chem. Soc.*, **114**, 6807 (1992).
193. R. Srinivas, D. Sülzle, W. Koch, C. H. DePuy and H. Schwarz, *J. Am. Chem. Soc.*, **113**, 5970 (1991).
194. R. Srinivas, D. K. Böhme, D. Sülzle and H. Schwarz, *J. Phys. Chem.*, **95**, 9836 (1991).
195. M. Iraqi, N. Goldberg and H. Schwarz, *J. Phys. Chem.*, **97**, 11371 (1993).
196. N. Goldberg, M. Iraqi, W. Koch and H. Schwarz, *Chem. Phys. Lett.*, **225**, 404 (1994).
197. N. Goldberg, M. Iraqi, J. Hrušák and H. Schwarz, *Int. J. Mass Spectrom. Ion Proc.*, **125**, 267 (1993).
198. N. Goldberg, J. Hrušák, M. Iraqi and H. Schwarz, *J. Phys. Chem.*, **97**, 10687 (1993).
199. N. Goldberg, M. Iraqi and H. Schwarz, *J. Chem. Phys.*, **101**, 2871 (1994).
200. S. Wlodek, D. K. Böhme and E. Herbst, *Mon. Not. Astron. Soc.*, **242**, 674 (1990).
201. For earlier reports on the matrix isolation of this species, see: R. J. VanZee, P. F. Ferrante and W. Weltner, *J. Chem. Phys.*, **83**, 6181 (1985).
202. G. Frenking and H. F. Schaefer, III, *J. Chem. Phys.*, **82**, 4584 (1985).
203. Y. Xie and H. F. Schaefer, III, *J. Chem. Phys.*, **93**, 1196 (1990).
204. S. Gronert, R. A. O'Hair, S. Prodnuk, D. Sülzle, R. Damrauer and C. H. DePuy, *J. Am. Chem. Soc.*, **112**, 997 (1990).
205. A. Fox, S. Wlodek, A. C. Hopkinson, M. H. Lien, M. Sylvain, C. Rodriguez and D. K. Böhme, *J. Phys. Chem.*, **93**, 1549 (1989) and references cited therein.
206. For a computational study of the silanol anion, H_3SiO^-, see: J. N. Nicholas and M. Feyereisen, *J. Chem. Phys.*, **103**, 8031 (1995).
207. R. Withnall and L. Andrews, *J. Phys. Chem.*, **89**, 3261 (1985).
208. R. J. Glinski, J. L. Gole and D. A. Dixon, *J. Am. Chem. Soc.*, **107**, 5891 (1985).
209. Z. K. Ismail, R. H. Hauge, L. Fredin, J. W. Kauffmann and J. L. Margrave, *J. Chem. Phys.*, **77**, 1617 (1982).
210. C.-M. Chiang, B. R. Zegarski and L. H. Dubois, *J. Phys. Chem.*, **97**, 6984 (1993).
211. A. I. Boldyrev and J. Simons, *J. Phys. Chem.*, **97**, 5875 (1993).
212. For an earlier theoretical study of Si_2O, see: R. L. DeKock, B. F. Yates and H. F. Schaefer III, *Inorg. Chem.*, **28**, 1680 (1989).
213. For earlier results from matrix spectroscopic studies which assigned a linear SiSiO structure to the triplet state of this molecule, see: R. J. VanZee, R. F. Ferrante and W. Weltner, *Chem. Phys. Lett.*, **139**, 426 (1987).
214. For previous calculations on Si_2O_2, see: L. C. Snyder and K. Raghavachari, *J. Chem. Phys.*, **80**, 5076 (1984).

215. For earlier results from matrix isolation experiments, see: H. Schnöckel, T. Mehner, H. S. Plitt and S. Schnuck, *J. Am. Chem. Soc.*, **111**, 4578 (1989) and references cited therein.
216. P. Fournier, J. Appell, F. C. Fehsenfeld and J. Durup, *J. Phys. B*, **5**, L58 (1972).
217. F. C. Fehsenfeld, J. Appell, P. Fournier and J. Durup, *J. Phys. B*, **6**, L268 (1973).
218. J. C. Lorquet, B. Leyh-Nihaut and F. W. McLafferty, *Int. J. Mass Spectrom. Ion Proc.*, **100**, 465 (1990).
219. For attempts to generate this long-sought-after triplet species and reasons why this molecule has still not been unequivocally identified, see: D. Sülzle, T. Weiske and H. Schwarz, *Int. J. Mass Spectrom. Ion Proc.*, **125**, 75 (1993).
220. The phenyl-substituted isomer PhNSi has been characterized by photoelectron spectroscopy some time ago: H. Bock and R. Dammel, *Angew. Chem.*, **97**, 128 (1985).
221. J. F. Ogilvie and S. Cradock, *J. Chem. Soc., Chem. Commun.*, 364 (1966).
222. M. Bogey, C. Demuynck, J. L. Destombes and A. Walters, *Astron. Astrophys.*, **244**, L47 (1991).
223. M. Elhanine, R. Farrenq and G. Guelachvili, *J. Chem. Phys.*, **94**, 2529 (1991).
224. For a recent report on the characterization of HSiN via photoisomerization of HNSi, see: G. Maier and J. Glatthaar, *Angew. Chem., Int. Ed. Engl.*, **33**, 473 (1994).
225. J. G. Radziszewski, D. Littmann, V. Balaji, L. Fabry, G. Gross and J. Michl, *Organometallics*, **12**, 4816 (1993).
226. R. Preuss, R. J. Buenker and S. D. Peyerimhoff, *J. Mol. Struct. THEOCHEM*, **49**, 171 (1978).
227. M. S. El-Shall, *Chem. Phys. Lett.*, **159**, 21 (1989).
228. Y. Apeloig and K. Albrecht, *J. Am. Chem. Soc.*, **117**, 7263 (1995).
229. For experimental work on R−CN and CN−R compounds, see: C. Rüchardt, M. Meier, K. Haaf, J. Pakusch and W. K. L. Wo, *Angew. Chem., Int. Ed. Engl.*, **103**, 907 (1991) and references cited therein.
230. J. R. Flores and J. Largo-Cabrerizo, *Chem. Phys. Lett.*, **142**, 159 (1987).
231. J. R. Flores and J. Largo-Cabrerizo, *J. Mol. Struct. THEOCHEM*, **183**, 17 (1989).
232. J. R. Flores, F. G. Crespo and J. Largo-Cabrerizo, *Chem. Phys. Lett.*, **147**, 84 (1988).
233. C. F. Melius and P. Ho, *J. Phys. Chem.*, **95**, 1410 (1991).
234. O. Parisel, M. Hanus and Y. Ellinger, *J. Chem. Phys.*, **100**, 2926 (1996).
235. For a recent theoretical investigation of this ion–molecule reaction, see: J. R. Flores, P. Redondo and S. Azpeleta, *Chem. Phys. Lett.*, **240**, 193 (1995).
236. Y. Apeloig, M. Karni, A. Stanger, H. Schwarz and T. Drewello, *J. Chem. Soc., Chem. Commun.*, 989 (1987).
237. M. T. Reetz, *Adv. Organomet. Chem.*, **16**, 33 (1977).
238. R. B. Woodward and R. Hoffmann, *The Conservation of Orbital Symmetry*, Academic Press, New York, 1970.
239. R. Bakhtiar, C. M. Holznagel and D. B. Jacobson, *Organometallics*, **12**, 621 (1993).
240. T. Drewello, P. C. Burgers, W. Zummack, Y. Apeloig and H. Schwarz, *Organometallics*, **9**, 1161 (1990).
241. R. Bakhtiar, C. M. Holznagel and D. B. Jacobson, *J. Am. Chem. Soc.*, **114**, 3227 (1992).
242. S. G. Cho, *J. Organomet. Chem.*, **510**, 25 (1996).
243. For a review on this interesting topic of gas-phase ion chemistry, see: C. Lifshitz, *Acc. Chem. Res.*, **27**, 138 (1994).
244. S. Murthy, Y. Nagano and J. L. Beauchamp, *J. Am. Chem. Soc.*, **114**, 3573 (1992).
245. Y. Nagano, S. Murthy and J. L. Beauchamp, *J. Am. Chem. Soc.*, **115**, 10805 (1993).
246. A. Nicolaides and L. Radom, *J. Am. Chem. Soc.*, **116**, 9769 (1994).
247. E. Tabei, S. Mori, F. Okada, S. Tajima, K. Ogino, H. Tanabe and S. Tobita, *Org. Mass Spectrom.*, **28**, 412 (1993).
248. S. Tobita, S. Tajima and F. Okada, *Org. Mass Spectrom.*, **24**, 373 (1989).
249. S. Tobita, S. Tajima, F. Okada, S. Mori, E. Tabei and M. Umemura, *Org. Mass Spectrom.*, **25**, 39 (1990).
250. P. C. H. Eichinger and J. H. Bowie, *Rapid Commun. Mass Spectrom.*, **5**, 629 (1991).
251. K. M. Downard, J. H. Bowie and R. N. Hayes, *Aust. J. Chem.*, **43**, 511 (1990).
252. M. Krempp and R. Damrauer, *Organometallics*, **14**, 170 (1995).
253. R. A. J. O'Hair, J. C. Sheldon, J. H. Bowie, R. Damrauer and C. H. DePuy, *Aust. J. Chem.*, **42**, 489 (1989).
254. H. Schwarz, *Pure Appl. Chem.*, **61**, 685 (1989).
255. K. Lammertsma, P. v. R. Schleyer and H. Schwarz, *Angew. Chem., Int. Ed. Engl.*, **28**, 1321 (1989).

256. L. M. Roth and B. S. Freiser, *Mass Spectrom. Rev.*, **10**, 303 (1991).
257. D. Matur, *Phys. Rep.*, **225**, 193 (1993).
258. C. Heinemann, D. Schröder and H. Schwarz, *J. Phys. Chem.*, **99**, 16195 (1995).
259. T. Ast, C. J. Proctor, C.J. Porter and J. H. Beynon, *Int. J. Mass Spectrom. Ion Phys.*, **40**, 111 (1981).
260. C. J. Procter, C. J. Porter, T. Ast and J. H. Beynon, *Int. J. Mass Spectrom. Ion Phys.*, **41**, 251 (1982).
261. C. J. Porter, C. J. Procter, T. Ast and J. H. Beynon, *Int. J. Mass Spectrom. Ion Phys.*, **41**, 265 (1982).
262. M. Rabrenovic, T. Ast and J. H. Beynon, *Int. J. Mass Spectrom. Ion Proc.*, **61**, 31 (1984).
263. M. Kolbuszewski and J. S. Wright, *J. Phys. Chem.*, **99**, 16196 (1995).
264. K. Leiter, W. Ritter, A. Stamatovic and T. D. Mark, *Int. J. Mass Spectrom. Ion Proc.*, **68**, 341 (1986).
265. W. P. M. Maas and N. M. M. Nibbering, *Int. J. Mass Spectrom. Ion Proc.*, **88**, 257 (1989).
266. S. N. Schauer, P. Williams and R. N. Compton, *Phys. Rev. Lett.*, **65**, 625 (1990).
267. T. Sommerfeld, M. K. Scheller and L. S. Cederbaum, *J. Chem. Phys.*, **103**, 1057 (1995).
268. T. Sommerfeld, M. K. Scheller and L. S. Cederbaum, *J. Chem. Phys.*, **104**, 1464 (1996).

CHAPTER 19

Matrix isolation studies of silicon compounds

GÜNTHER MAIER, ANDREAS MEUDT, JÖRG JUNG and HARALD PACL

Institut für Organische Chemie der Justus-Liebig-Universität Giessen, Heinrich-Buff-Ring 58, D-35392 Giessen, Federal Republic of Germany

I. INTRODUCTION	1144
A. Motivation	1144
B. Coverage	1145
II. SILICON–CARBON MULTIPLE BONDS	1145
A. Silenes	1145
B. Silynes	1148
C. Silaaromatics	1148
D. Silaantiaromatics	1152
III. SILICON–SILICON MULTIPLE BONDS	1156
IV. SILICON–NITROGEN MULTIPLE BONDS	1158
A. Silanimines, Silanitriles and Silaisonitriles	1158
B. Silanediimines and Diazosilanes	1160
V. SILICON–OXYGEN MULTIPLE BONDS	1161
A. Silanones, Silacarbonates and Silacarboxylic Esters	1161
B. Silicon Oxides and Related Compounds	1162
VI. SILICON–SULFUR MULTIPLE BONDS	1162
VII. STRAINED SILACYCLES	1163
A. Silacyclopropyne and Other C_2H_2Si Isomers	1163
B. Silacyclopropenes	1165
VIII. SILYLENES	1166
A. Unsubstituted Silylene	1167
B. Carbon Substituted Silylenes	1167
1. Silylenes by photoisomerization of silenes	1167
a. Methylsilylene	1167
b. Dimethylsilylene	1167
2. Silylenes by photolysis of oligosilanes	1168
3. Silylenes by photolysis of diazides	1170
4. Silylenes by trimethylsilane extrusion	1170

The chemistry of organic silicon compounds, Vol. 2
Edited by Z. Rappoport and Y. Apeloig © 1998 John Wiley & Sons Ltd

a. C_3H_4Si isomers	1171
b. $C_2H_4Si_2$ isomers	1172
c. C_4H_2Si isomers	1172
5. Silylenes by cocondensation reactions	1175
a. C_2H_4Si isomers	1175
b. Silicon–carbon clusters	1175
C. Silicon Substituted Silylenes	1176
1. Silylenes with one silyl substituent	1176
2. Silylenes with two silyl substituents	1177
D. Nitrogen Substituted Silylenes	1178
1. Silylenes by photolysis of oligosilanes	1178
2. Silylenes by photolysis of diazides	1178
3. Silylenes by cocondensation reactions	1179
a. CHNSi isomers	1179
b. SiN_2 isomers	1179
E. Oxygen Substituted Silylenes	1179
1. Silylenes by photolysis of oligosilanes	1179
2. Silylenes by pyrolysis of suitable precursors	1180
3. Silylenes by cocondensation reactions	1180
F. Halogen Substituted Silylenes	1180
IX. REFERENCES	1181

I. INTRODUCTION

A. Motivation

Though numerous molecules involving tetravalent silicon atoms singly bonded to first- and second-row elements are known, the chemistry of compounds containing the element in 'unusual' low-valent bonding situations is a relatively young area of research. Some exotic representations like the very first compound of multiply bonded silicon, the SiN radical, were identified as early as 1925[1]. Nevertheless, countless investigations indicating the nonaccessibility of π-systems[2] of silicon and other third period elements resulted in the formulation of the classical double-bond rule, which predicted high reactivity for such compounds[1]. After some reports[3,4] on 'successful' syntheses, all of which failed to stand up to scrutiny, the existence of transient species with silicon–silicon multiple bond systems was first demonstrated convincingly as late as 1969 by the pyrolysis of 1,2-disila-4-cyclohexenes yielding disilenes, whose intermediate existence was demonstrated indirectly[5]. In the same decade, 1,1-dimethylsilaethene as the first silene was generated by pyrolysis of the appropriate silacyclobutane[6–8].

Novel highly reactive substances are often first proposed as transient intermediates, then later isolated in noble gas or hydrocarbon matrices at very low temperatures[9] and still later obtained as stable species in the form of highly substituted derivatives. The second step in this classic historical pattern, namely the matrix isolation of reactive intermediates, is mostly concentrated on the study of the prototype systems unadulterated by substituent effects. This area has become a field of especially fruitful interaction between experiment and theory, which allows the unequivocal identification of the often long sought-after substances. A good example for the development described above is the fate of silylenes. Soon following the proposal of these species as fleeting intermediates, some organic derivatives could be matrix-isolated starting in 1979 by photolytic decomposition of suitable precursors[10]. These investigations finally smoothened the way for the isolation of a destillable silylene which survives even heating in toluene solution to 150 °C for many months[11].

The matrix technique allowed even the isolation of compounds with silicon–nitrogen triple bonds such as HSiN and HNSi[12].

These isolations of highly reactive prototype systems are important for our understanding of chemical bonding.

B. Coverage

This survey covers the matrix isolation and spectroscopic analysis of a variety of molecules containing doubly and triply bonded silicon, strained silacycles and silylenes. In most cases, the matrix-isolated substances and their photoproducts were identified by IR and UV spectroscopy and the comparison of experimental with calculated spectra using quantum mechanical methods. The explosive growth of this field of organic chemistry is well documented by numerous reviews, for example the excellent article of Raabe and Michl[1] on 'Multiple Bonds to Silicon' and the review about 'Silylenes' by Gaspar[13] which serve as a basis for this review. Publications which appeared prior to the two cited reviews are only mentioned here if necessary in the context of the discussion. Our literature research relying on computerized search services in *Chemical Abstracts* extends to the end of 1995. Maybe we have failed to reach completeness in finding all relevant articles covering the present subject. If this is the case we apologize to both the readers and the authors.

It was impossible to avoid a certain amount of overlap between the sections of this review, especially because of the easy conversion of substituted silylenes into species with multiply bonded silicon and vice versa. But we tried to reduce duplications of this type as far as possible.

II. SILICON–CARBON MULTIPLE BONDS

A. Silenes

The chemistry of silenes presents an especially well-investigated area of organosilicon chemistry. Despite their high reactivity and tendency to dimerize, more than 150 different silenes were the subject of a large number of experiments[14]. The isolation of a crystalline silene — indefinitely stable at room temperature in the absence of air — represented a breakthrough in the chemistry of silicon π-systems[15]. In the meantime, a two-digit number of such stable silenes is known[1], crowned by the recent isolation of the first stable 1-silaallene[16].

The parent silaethylene **2** could be matrix-isolated by our group in 1981[17] by pyrolysis of the bicyclic system **1** (equation 1). Later, a new entry into this system was found by photolysis of methyldiazidosilane **3**[18] (equation 1). This simplest representative **2** shows already most aspects of the chemistry of the whole class of silenes, e.g. the photoisomerizability into the corresponding silylene (see Section VIII).

$$\mathbf{1} \xrightarrow{\Delta} H_2Si{=}CH_2 \xleftarrow{254\ nm} H_3C{-}SiH(N_3)_2 \quad (1)$$

(1)   (2)   (3)

TABLE 1. Substituent effects on the UV maxima of silenes

Silene	λ_{max} (nm)	Literature
$H_2Si=CH_2$	258	20
$D_2Si=CH_2$	259	20
$Cl_2Si=CH_2$	246	20
$(Me_3Si)_2Si=C(OSiMe_3)$ (R)	ca 340	21
R = t-Bu, CEt$_3$, 1-Ad		22
$Me_2Si=CHMe$	255	19
$Me_2Si=CH(SiMe_3)$	265	19
$Me_2Si=CMe(SiMe_3)$	274	19
$Me_2Si=C(SiMe_3)_2$	278	19
$Me_2Si=CPh(CO_2Me)$	280	19
$Me_2Si=C(SiMe_3)(CO\ 1\text{-}Ad)$	284	19
$Me_2Si=C(SiMe_3)(CO_2Et)$	293	19

In the last decade, the interest in this area focussed on the spectroscopic properties of silenes and their intermolecular reactions in matrices.

Before that time, little was known about their UV spectra. Ando and coworkers extended our knowledge about this facet of organosilicon chemistry by photolysis of silyldiazomethanes **4** in 3-methylpentane at 77 K yielding the expected silenes **5**[19] (equation 2). The measured UV spectra together with previous results are summarized in Table 1.

$$R_3Si-C\begin{matrix}R'\\ \\N_2\end{matrix} \xrightarrow[-N_2]{h\nu} R_2Si=CRR' \qquad (2)$$

(4) (5)

The UV absorptions of the tri- or tetrasubstituted silenes are red-shifted compared to the parent silene **2** ($\lambda = 258$ nm) but occur at shorter wavelengths than in the highly substituted silenes of Brook ($\lambda = ca$ 340 nm). The introduction of trimethylsilyl groups on the carbon atoms results in slight red-shifts just as the conjugation of the silene system with carbonyl groups.

Sander and coworkers studied intermolecular reactions of methylsilenes in argon matrices[23]. 1,1-Dimethylsilene (**7**) was generated by photolysis of precursor **6** in quantitative yield (equation 3). Warming to 35 K resulted in dimerization of silene **7** to the corresponding disilacyclobutane **8**. This kind of silene dimerization leading to the formation of 1,3-disilacyclobutane was originally one of the structural proofs for the parent silene **2**[20].

$$Me_2HSi-C\begin{matrix}H\\ \\N_2\end{matrix} \xrightarrow[-N_2]{\lambda>305\text{ nm}} Me_2Si=CH_2 \xrightarrow{35\text{ K}} \begin{matrix}Me_2Si\!-\!\!\!\!-\!\!\!\!-\\ \ \ \ \ \ \ \ \ \ \ \ \ \ \ \ \ |\\ \ \ \ \ \ \ \ \ \ \ \ \ \ \ \ SiMe_2\end{matrix} \qquad (3)$$

(6) (7) (8)

Photolysis of **6** in a formaldehyde-doped argon matrix (0.6%) yielded also silene **7** as the major product but also minor quantities of other products. Warming of the matrix to 30–35 K allowed the direct spectroscopic observation of the reaction between **7** and formaldehyde. The intensity of the IR absorptions of these compounds disappeared to a large extent, and in addition to **8** a new product, which could be identified as the siloxycarbene **10** was observed. This compound was stable under the reaction conditions (equation 4). Siloxycarbenes were frequently postulated[24–26] as fleeting intermediates in the photolysis of acylsilanes, but up to now only in one other case observed spectroscopically[27].

$$Me_2Si=CH_2 \quad \xrightarrow[30 \text{ K}]{H_2CO} \quad \left[\begin{array}{c} H \\ \diagdown \\ O=C-H \\ \vdots \quad \vdots \\ Me_2Si=CH_2 \end{array} \right] \quad \longrightarrow \quad \begin{array}{c} Me_2Si-O-CH_2-H \\ | \\ CH_3 \end{array}$$

(7) (9) (10)

(4)

$\lambda > 570$ nm

$$\begin{array}{c} O \\ \| \\ Me_2Si-C-H \\ | \\ CH_3 \end{array}$$

(11)

Intermediate **10** turned out to be easily photoisomerizable yielding the stable acylsilane **11**. According to CCSD(T)/6-31G(d,p)//MP2/6-31G(d,p) calculations[23] carbene **10** originates from the initial complex **9**.

2-Silapropene **12** reacted much faster under analogous conditions with formaldehyde giving the expected 1,2-siloxetane, whereas 2-methyl-2-sila-2-butene **13** proved to be unreactive. Obviously the reactivity of silenes against H_2CO decreases with increasing degree of methyl substitution. Another interesting aspect of this reaction is the high kinetic isotope effect which totally inhibits the reaction even of silapropene **12** with D_2CO.

$$\begin{array}{cc} H_3C \diagdown \quad \diagup \quad & H_3C \diagdown \quad \diagup H \\ Si=CH_2 & Si=C \\ H \diagup & H_3C \diagup \quad \diagdown CH_3 \end{array}$$

(12) (13)

The same authors investigated the photochemical and thermal oxidation of 2-silapropene **12**, silaisobutene **7** and 1,1,2-trimethylsilene **13** in O_2-doped argon matrices (equation 5)[28]. These silenes are easily photooxidized in matrices containing more than 1% O_2, but only trimethylsilene **13** exhibits thermal reactivity toward oxygen at temperatures as low as 20–40 K.

The photochemical reactivity goes up with increasing number of methyl groups at the double bond and decreasing ionization potential[28]. Key intermediates in both the photochemical and the thermal oxidation of silenes **12**, **7** and **13** are the siladioxetanes **14**. These species are labile even in low-temperature matrices and could not be identified spectroscopically. Evidence for their formation comes from the observed oxidation products such as complexes **15** between silanones and formaldehyde and formylsilanols **16**.

The complexes **15** were identified by comparison of *ab initio* calculated and observed IR spectra[28]. They show the expected red-shifts of $\nu_{C=O}$ and $\nu_{Si=O}$.

$$\begin{array}{c} R \\ \diagdown \\ H_3C \end{array} Si = \begin{array}{c} H \\ \diagup \\ H \end{array} \xrightarrow{O_2} \left[\begin{array}{c} O-O \\ | \quad | \\ H_3C \diagup Si \\ | \\ R \end{array} \right] \longrightarrow \begin{array}{c} \cdots CH_2 \\ O \quad \| \\ \| \quad \cdots O \\ Si \\ \diagup \quad \diagdown \\ H_3C \quad R \end{array} \tag{5}$$

(7) R = CH₃
(12) R = H

(14) (15)

$\lambda = 480$ nm

$$\begin{array}{c} OH \\ | \\ H_3C-Si-CHO \\ | \\ R \end{array}$$

(16)

An additional oxidation pathway is observed for **13** with a methyl group at the C-atom of the silene moiety. Here, the primary adduct of **13** and $^3O_2$ under photolytic conditions, i.e. triplet diradical **17**, can either ring-close to give the corresponding dioxetane (the dioxetane is not stable under the photolytic conditions required to induce the photooxidation, and therefore cannot be observed) leading to products derived from it, or produce dimethylvinylsilyl hydroperoxide (**18**) via H-abstraction from a methyl group (equation 6).

$$Me_2Si=CHMe \xrightarrow[25\,K]{^3O_2} Me_2Si-\overset{\bullet}{C}HMe \longrightarrow Me_2Si\diagdown \hspace{-1em}\overset{OOH}{} \tag{6}$$

(13) (17) (18)

B. Silynes

No experimental evidence concerning compounds with silicon–carbon triple bonds is known, so that experimental efforts should be concentrated on the search for these fascinating species. 2-Silaallenes and 2-silaketenes, which also would contain a sp-hybridized silicon atom, could not be isolated up to now and remain a challenge of organosilicon chemistry (for 1-silaallene and for silylene–CO complexes cf Section VIII).

C. Silaaromatics

Theoretical studies have shown that the resonance energy of silabenzene should be 2/3 that of benzene[29]. Even though other isomers of C_5H_6Si such as Dewar silabenzene may be competitive in stability, silabenzene appears to have all the attributes expected for an analogue of benzene. Nevertheless, silaaromatics were for a long time thought to be elusive molecules despite the relative stability of similar heteroatomic derivatives of benzene such as phosphabenzene, arsabenzene and stibabenzene[30].

The first unambiguous evidence for the existence of a substituted silabenzene was reported by Barton and coworkers in 1978[31,32]. The pyrolysis of precursor **19** in the presence of perfluoro-2-butyne yielded adduct **21** and copyrolysis of **19** and MeOH(D) gave methoxysilane **22**[33], strongly suggesting the intermediate existence of silatoluene **20** (equation 7). In the absence of trapping agents, vacuum flow pyrolysis of silacyclohexadiene **19** resulted in the Diels–Alder dimer **23** of silatoluene **20** rather than the predicted[34] [2 + 2] dimer. The deposition of the products of neat pyrolysis of **19** together with argon on a spectroscopic window at low temperatures allowed the direct observation of silatoluene **20** whose UV spectrum (λ_{max} = 310 nm) showed the expected[29,34] bathochromic shift compared to the spectrum of benzene. The position of the band and its well-resolved vibrational structure are in good agreement with the prediction that silabenzene should be an aromatic system. Silatoluene **20** was stable to UV irradiation but vanished upon warming of the matrix producing a polymer, which was not characterized[33].

The parent silabenzene **24** was first matrix-isolated by our group in 1980[35] by pyrolysis of precursors **25** and **26**, which yield the expected silabenzene by retro-ene fragmentation. Later, it could be shown that in analogy to carbon chemistry the hydrogen elimination from silacyclohexadiene **27** also gives the silaaromatic **24**[36]. This reaction is allowed by the Woodward–Hoffmann rules. In accordance with the Woodward–Hoffmann rules, it could be demonstrated that silabenzene **24** is not accessible by pyrolysis of the conjugated silacyclohexadiene **28** (equation 8).

Silabenzene **24** reveals a characteristic Si—H stretching vibration at 2217 cm^{-1}, as expected for hydrogen attached to a sp^2-hybridized silicon atom. Compound **24** shows a typical benzene-type UV spectrum with absorptions at λ = 217, 272 and 320 nm, which fit into the series of the already known donor-substituted heterobenzenes[30]. An additional structural proof was the partially reversible photochemical conversion of **24** into Dewar silabenzene **29** whose Si–H stretching vibration is shifted to 2142 cm^{-1}, as expected for a hydrogen connected to a sp^3-hybridized silicon atom. Silabenzene **24** survived warming of the matrix to a temperature of 37 K, which allows intermolecular reactions. At still higher temperatures (80 K) the absorptions of **24** disappeared completely.

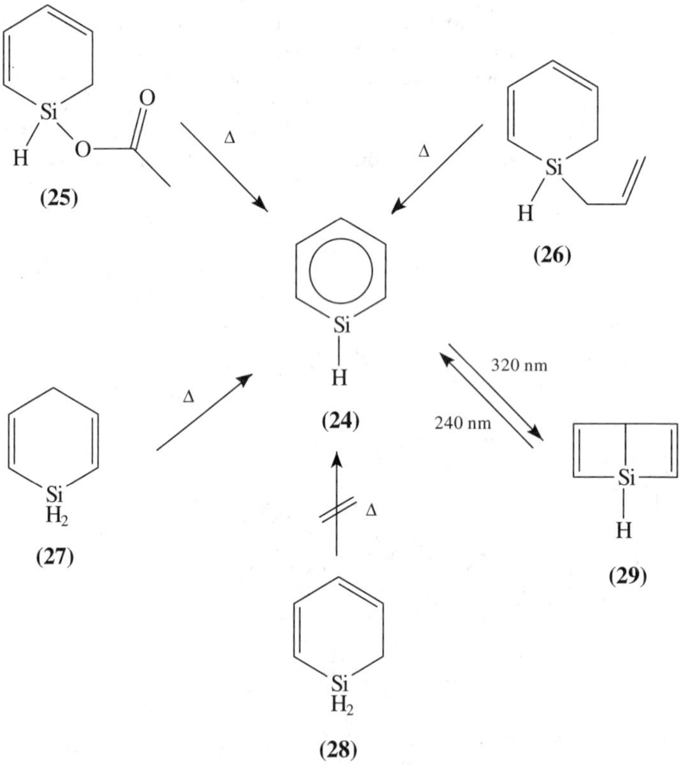

(8)

The properties of silabenzene **24** resemble strongly those of silatoluene **20** described above not only in the UV spectra, but also in the measured PE spectra[37–40]. In conclusion, one has to note that both species are best considered as symmetry-distorted delocalized 6π-electron systems in conformity with the above-mentioned calculations[40].

After the successful preparation of silabenzene **24** by hydrogen elimination from **27**, it was tempting to try to generate 1,4-disilabenzene **31** by pyrolysis of the easily accessible disilacyclohexadiene **30**[41] (equation 9).

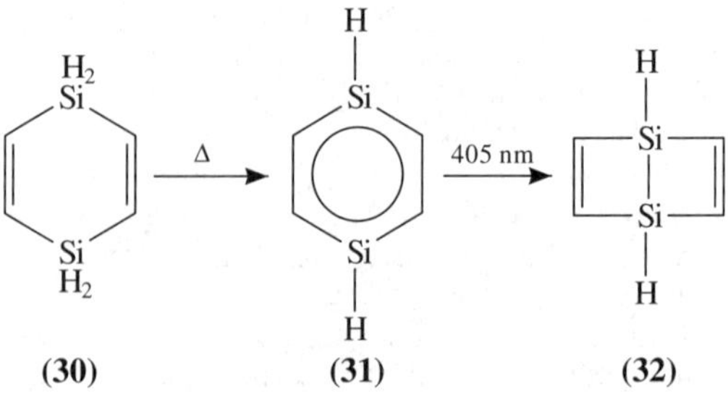

(9)

The dehydrogenation of **30** turned out to be more difficult than the analogous conversion of **27** to **24**. The IR spectrum of the matrix-isolated products showed only one absorption at 1273 cm^{-1}, which disappeared upon irradiation with $\lambda = 405$ nm probably yielding Dewar disilabenzene **32**. The UV spectrum proved to be of higher diagnostic value. 1,4-Disilabenzene **31** shows a typical heterobenzene electronic spectrum with absorptions at $\lambda = 408$, 340 and 275 nm displaying another bathochromic shift compared to silabenzene **24**.

No experimental data are available for derivatives of 1,2- or 1,3-disilabenzenes[42]. The same is true for the even more fascinating hexasilabenzene[43].

According to theoretical investigations[44,45] the reason for the high reactivity of silabenzenes is the polar SiC double bond. Silabenzenes bearing large substituents should be kinetically more stable than the parent system and could allow possibly the synthesis of isolable derivatives. As outlined in the other sections of this review, the strategy of kinetic stabilization by introducing bulky substituents led to a variety of silicon-element π-bond systems stable up to room temperature. Another way to stabilize the heterobenzene should be the benzoannelation of the central ring. This idea was the motivation for the preparation of 9-silaanthracenes by pyrolytic[46] and photolytic[47] methods. Older procedures, such as in the elimination of a hydrogen halide from suitable dihydro precursors, gave only dimeric or polymeric material[48]. Similarly, a thermolytic approach which had been developed for phosphorus and arsenic rings was unsuccessful when applied to silicon-containing educt molecules[49]. A breakthrough was achieved by pyrolysis of **33** (R = Ph), which gives the substituted 9-silaanthracene **34** (R = Ph) by elimination of benzene[46] (equation 10).

Evidence for the formation of **34** (R = Ph) was provided by neutralization reionization mass spectrometry and more directly by the matrix isolation and spectroscopic investigations on **34** (R = Ph) in an argon matrix at 12 K. The UV spectrum of **34** (R = Ph) exhibits characteristic bands at $\lambda = 364$, 386, 404, 420, 440, 470 and 502 nm, resembling those of the electronic spectrum of anthracene, but with the expected bathochromic shifts. If one irradiates into the maximum at $\lambda = 502$ nm, all bands shown in the spectrum disappear completely within 5 minutes. The vanishing of these characteristic bands can again be explained by the photoisomerization of silaanthracene **34** (R = Ph) to the corresponding Dewar valence isomer.

In the same way pyrolysis of unsubstituted dihydrosilaanthracene **33** (R = H) formed the parent silaanthracene **34** (R = H), which likewise could be photoisomerized into the corresponding Dewar silaanthracenes[46].

Silaanthracenes turned out to be also available by the 254 nm photolysis of the appropriate dihydrosilaanthracenes in rigid hydrocarbon glass at 77 K[47].

In summary, the benzoannelation of the central silabenzene ring is not enough to stabilize compounds of this type to a high extent. They still proved to be very thermolabile. Even phenyl substitution as in **34** (R = Ph) does not furnish sufficient steric protection for isolation outside of an argon matrix.

The sterically overcrowded silabenzenes **36** and **38** were generated by pyrolysis of the appropriate precursors **35** and **37** (equations 11a and 11b, respectively) and isolated in an argon matrix[50].

$$
\begin{array}{c}
(35) \xrightarrow[-Me_3SiOMe]{\Delta} (36)
\end{array}
\qquad (11a)
$$

$$
\begin{array}{c}
(37) \xrightarrow[-Me_3SiOMe]{\Delta} (38)
\end{array}
\qquad (11b)
$$

The silaaromatics **36** and **38** reveal in the observed UV spectra an additional bathochromic shift compared to silabenzene **24**. Both, **36** and **38**, disappear in an irreversible process upon irradiation with $\lambda = 290-420$ nm, probably due to the formation of the corresponding Dewar silabenzenes. The kinetically stabilized silaaromatics **36** and **38** turned out to be stable up to 90 K even without an argon cage, but decomposed unspecifically at higher temperatures.

In 1988, Märkl and Schlosser[51] reported the synthesis of the substituted silabenzene **40** by irradiation of the diazo compound **39**, which was found to be stable in solution up to 170 K (equation 12).

$$
(39) \xrightarrow[-N_2]{h\nu} (40) \qquad (12)
$$

Nevertheless, the synthesis of the first silabenzene stable up to room temperature remains a challenging problem.

D. Silaantiaromatics

Whereas the isomeric hydrocarbon skeletons cyclobutadiene and tetrahedrane have been the object of many investigations[52], little is known about the analogous silacyclobutadienes and silatetrahedranes. *Ab initio* calculations carried out by Gordon and coworkers[53,54] show that the most stable species among the cyclic C_3SiH_4 species should be 2-methylsilacyclopropenylidene, followed by other silylenic species which might be suitable precursors for unsubstituted silacyclobutadiene. Silacyclobutadiene is almost 60 kcal mol^{-1} less stable than methylsilacyclopropenylidene as a consequence of its calculated large antiaromatic destabilization energy. Silatetrahedrane lies 32 kcal mol^{-1} higher in energy than silacyclobutadiene[53,54].

Attempts to generate silacyclobutadienes through the thermal rearrangements of cyclopropenyl silylenes in solution proved to be unsuccessful[55], in contrast to the known similar rearrangements of cyclopropenyl carbenes[52,56,57] and nitrenes[58].

Our group tried to observe the formation of silacyclobutadienes via a photochemical pathway[59-61]. Photolysis of cyclopropenyl silanes **41 a–l** in hydrocarbon matrices at 77 K could produce the corresponding silylenes **42**, which should be suitable silacyclobutadiene precursors (equation 13). In only two cases (**41a,d**), the photolysis produced the expected silylenes **42**, but it was not possible to initiate their photochemical conversion into the desired silacyclobutadienes **43**.

	R^1	R^2	R^3
a	SiMe$_3$	SiMe$_3$	SiMe$_3$
b	Ph	SiMe$_3$	SiMe$_3$
c	*t*-Bu	SiMe$_3$	SiMe$_3$
d	SiMe$_3$	SiMe$_3$	SiPhMe$_2$
e	*t*-Bu	SiMe$_3$	SiPhMe$_2$
f	SiMe$_3$	SiPhMe$_2$	SiPhMe$_2$
g	*t*-Bu	SiPhMe$_2$	SiPhMe$_2$
h	*t*-Bu	I	I
i	SiMe$_3$	N$_3$	N$_3$
j	Ph	N$_3$	N$_3$
k	Mes	N$_3$	N$_3$
l	*t*-Bu	N$_3$	N$_3$

(13)

Irradiation of compound **41e** in solution yields a yellow species to which we tentatively assign structure **45** (equation 14). The radical character of **45** is supported by EPR spectroscopy and trapping experiments. The formation of cyclobutenyl radical **45** can be explained by homolytic cleavage of the Si—SiMe$_2$Ph bond in **41e** and rearrangement of the resulting primary product **44**.

(41e) → (λ = 254 nm, −•SiMe$_2$Ph) → **(44)** → **(45)** (14)

In this context it is astonishing that Fink and Purani succeeded (1987) in observing a photochemical rearrangement of a cyclopropenyl silylene of the same type to give the corresponding silacyclobutadiene[55]. The 254-nm photolysis of cyclopropenyltrisilane **46** (Ar = Mes) in a 3-methylpentane glass yielded the yellow mesityl(1,2,3-tri-*tert*-butylcyclopropenyl)silylene (**47**, Ar = Mes), which could be trapped in the presence of ethanol to give the corresponding ethoxysilane **48** (equation 15).

(46) Ar = Mes, Tip → (254 nm) → **(47)** Ar = Mes, Tip → (EtOH, Ar = Mes) → **(48)**; or (hν) → **(49)** (15)

Irradiation of silylene **47** (Ar = Mes) with visible light resulted in the vanishing of the silylene absorption bands and the formation of a new species, which could be identified as the first example of a silacyclobutadiene **49** (Ar = Mes). Trapping of **49** (Ar = Mes) with methoxytrimethylsilane or ethanol yielded the expected products **50** and **51** in a completely stereospecific *syn* addition (equation 16). This result is in good agreement with reports by Jones and coworkers about *syn* additions of these reagents to simpler acyclic silenes[62,63] (equation 16).

Subsequent warming of matrix-isolated **49** (Ar = Mes) to room temperature results in the formation of a dimer[64,65]. As could be shown by EPR spectrometry and product analysis, triplet diradical **52** (Ar = Mes) seems to be the first product which can be

19. Matrix isolation studies of silicon compounds

detected during annealing of the matrix[65] (equation 17).

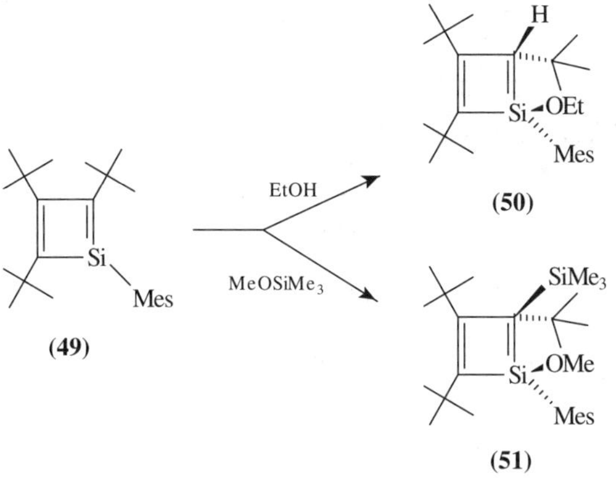

(16)

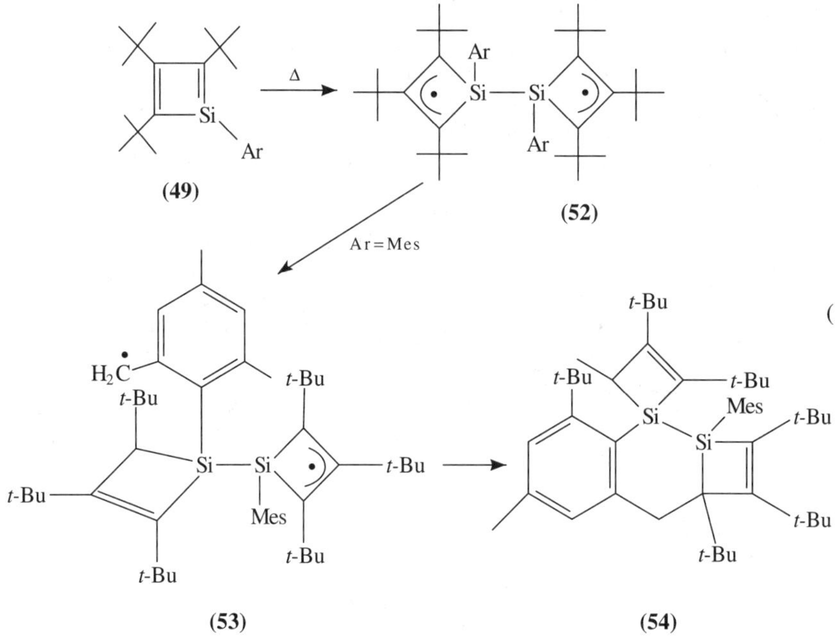

(17)

The dimerization of **49** (Ar = Mes) yields compound **54**, whose X-ray crystal structure could be determined[64], via diradicals **52** and **53**. This reaction may be considered as a special case of silene dimerization. Although the typical course of silene dimerization leads to head-to-tail adducts (cf Section II.A), highly substituted analogues often give the head-to-head isomer. Brook and Baines have proposed that many of these dimerizations also proceed via initial formation of a Si—Si bond to give 1,4-diradicals which, in some cases, isomerize by intramolecular hydrogen abstractions[66,67].

Diradical **52** (Ar = Mes) and the corresponding 2,4,6-triisopropylphenyl substituted **52** (Ar = Tip), which is available in the same manner starting from trisilane **46** (Ar = Tip), were the first directly observed diradicals of this type.

III. SILICON–SILICON MULTIPLE BONDS

If the field of silicon–silicon multiple bond systems is compared with corresponding organic molecules, it becomes clear that there are still large gaps. For example, even the parent disilene $H_2Si=SiH_2$ could not be isolated up to now, in spite of the fact that its intermediate existence was strongly supported by copyrolysis of trisilane **55** and 1,3-butadiene as early as 1975[68] in which the trapping products **56**, **57** and **58** of SiH_2, disilene and its isomer silylsilylene, respectively, were found (equation 18).

$$Si_3H_8 \xrightarrow{\Delta} \text{(55)} \quad \rightarrow \quad \text{(56)} + \text{(57)} + \text{(58)} \tag{18}$$

(55)   (56)   (57)   (58)

Our group has been interested for quite some time in the matrix isolation of disilene, the prototype molecule for Si,Si π-systems. We have synthesized a great variety of precursors and used different methods of energy transfer, but none of the precursors or methods turned out to be suitable[69–73]. We think that only new, unconventional methods can bring the breakthrough in the isolation of this long sought-after molecule. Our own attempts are briefly summarized below.

Disilene and its isomer silylsilylene were neither available by standard vacuum flash pyrolysis of precursors **59–63**, nor by the more elaborate method of pulsed flash pyrolysis of **60–63**, a pulsed discharge in mixtures of argon and mono- and disilane[74] or by the matrix photolysis of educts **59–66** using various light sources (Hg lamps, excimer laser)[69,70,72], the microwave discharge in disilane **66** or the cocondensation of silicon atoms with SiH_4.

(59) $Me_3Si-Si_2H_5$ (60) $HI_2Si-SiH_3$ (61)

$IH_2Si-SiH_2I$ (62) $PhH_2Si-SiH_3$ (63)

$PhMe_2Si-Si_2H_5$ (64) $H_3Si-SiH(SiMe_3)_2$ (65) $H_3Si-SiH_3$ (66)

Silyldisilene, which could be photoisomerizable into the likewise unknown cyclotrisilane, has also not yet been observed[69].

The only study in the field of matrix-isolated compounds containing silicon–silicon multiple bonds in the last decade exhibiting new aspects concerns the photolysis of cyclotetrasilanes, which shows a remarkable dependence of the reaction course on molecular structure[75]. It is known that the photolysis of peralkylcyclotetrasilanes leads to ring contraction[76–80]. However, steady-state photolysis of planar persilylcyclotetrasilanes **67** in hydrocarbon glass at 77 K gave rise to the corresponding disilenes **68**, identified by their UV spectra[75]. On the contrary, peralkylcyclotetrasilanes **69** having bent structures were photostable under these conditions, but reacted upon laser flash photolysis in solution at higher temperatures under ring contraction yielding the corresponding silylenes and cyclotrisilanes (equation 19)[75].

$$\begin{array}{c} R_2Si - SiR_2 \\ | \quad\quad | \\ R_2Si - SiR_2 \end{array} \xrightarrow{h\nu} R_2Si = SiR_2$$

(**67**)   (**68**)

R = SiMe$_3$, SiEtMe$_2$

(19)

$$\begin{array}{c} R^1R^2Si - SiR^1R^2 \\ | \quad\quad\quad | \\ R^1R^2Si - SiR^1R^2 \end{array}$$

(**69**)

R^1 = R^2 = i-Pr; R^1 = t-Bu, R^2 = Me

Tetrasilyldisilenes of type **68** have been the focus of much attention in the last years from an experimental[81,82] and theoretical[83,84] point of view. According to the latter, silyl-substituted disilenes should have significantly increased bond dissociation energies and are therefore interesting synthetic targets. Furthermore, silyl substituents are predicted to favor planar arrangement around the silicon–silicon double bond, a prediction which so far has not been supported experimentally[81,82]. It is tempting to try the photoisomerization of such persilylated disilenes to the corresponding silylenes by 1,2-silyl migration because the resulting highly substituted disilylsilylenes could be the first silylenes with a triplet ground state[85].

In analogy to the above-mentioned parent disilene, no compounds containing silicon–silicon triple bonds have so far been found. Even their intermediacy could not be proved unequivocally. Two isomers of Si$_2$H$_2$, namely **70** and **71**, were detected in a low pressure, low power plasma in a mixture of SiH$_4$ and argon by the measurement of their millimeter and submillimeter wave rotational spectrum[86,87]. Both compounds have no acetylene-type silicon–silicon triple bonds, in agreement with the predictions of theoretical studies[86,87].

$$\begin{array}{cc} \underset{\text{Si}-\text{Si}}{\overset{H\quad\quad H}{\diagdown\;\diagup}} & \underset{\text{Si}-\text{Si}}{\overset{H\quad H}{\diagdown\diagup}} \\ (\mathbf{70}) & (\mathbf{71}) \end{array}$$

Recently, we were able to matrix-isolate a species which we also tentatively assume to be the energetically lowest-lying isomer on the H_2Si_2 energy hypersurface, namely compound **71**. The matrix-isolated products of pulsed flash pyrolysis of 1,1,1-trimethyltetrasilane **72**, originally believed to be a suitable cyclotrisilane precursor, reveal the absorptions of trimethylsilane and monosilane together with a band at 1093 cm^{-1}, which by comparison with the calculated IR spectrum (BLYP/6-31G*: strongest absorption at 1081 cm^{-1}; all other vibrations were calculated with much weaker intensities) can tentatively be assigned to the butterfly molecule **71** (equation 20)[69].

$$Me_3Si-SiH_2-SiH_2-SiH_3 \xrightarrow[-SiH_4]{\Delta, -Me_3SiH} \mathbf{71} \longleftarrow \ddot{Si}: + H_2 \quad (20)$$

(**72**)   (**71**)

The same compound **71** was also formed together with monosilane in the evaporation and matrix-isolation of pure silicon (equation 20; for the origin of hydrogen cf Section VIII)[73]. This observation can be taken as a further support for structure **71**. By pyrolysis of a partially deuteriated tetrasilane precursor, HDSi$_2$ and D$_2$Si$_2$ were matrix-isolated showing the expected isotopic shifts (HDSi$_2$: 1061 cm^{-1} {BLYP/6-31G*: 1044}; D$_2$Si$_2$: 785 {784})[69].

Despite these successes, a compound containing a silicon–silicon triple bond is still lacking.

IV. SILICON–NITROGEN MULTIPLE BONDS

A. Silanimines, Silanitriles and Silaisonitriles

Compounds containing silicon–nitrogen multiple bonds have been investigated intensively in recent years[1]. The first silanimine, that is stable at room temperature, was synthesized in 1986[88]. Other sterically hindered silanimines were observed in matrices and glasses after irradiation of the corresponding azidosilane precursors[89–91].

In accordance with these results, Maier and Glatthaar showed that azidomethylsilane (**73**) and its deuteriated analogue azido(dideuterio)methylsilane [D$_2$]-**73** eliminate nitrogen upon irradiation with $\lambda = 254$ nm in an argon matrix[18]. The first observable product after nitrogen elimination is silanimine **74** which, based on the comparison of the experimental and calculated IR wavenumbers, should have the (E) configuration (equation 21). During further irradiation a second H-shift occurs yielding aminosilylene **75** (UV: $\lambda_{max} = 330$ nm). Additional isomerization can be achieved after the exposition of **75** to longer wavelengths. The resulting aminosilene **76** (UV: $\lambda_{max} = 256$ nm) reacts upon irradiation with shorter wavelengths to regenerate **75**.

(**73**) → 254 nm → (**74**) → 254 nm → (**75**) ⇌ (**76**)   (21)

19. Matrix isolation studies of silicon compounds

Triazidophenylsilane is a good photochemical precursor for the preparation of phenylsilaisonitrile and presumably phenylsilanitrile[92,93]. Analogously triazidomethylsilane (**77**) should produce the methylated series upon irradiation. But the photochemistry of matrix-isolated **77** in argon turned out to be complex. The final product of the photofragmentation of **77** is isonitrile **80** (equation 22)[18]. **80** was identified by its IR spectrum. The same is true for methylsilaisonitrile (**79**), which is a migration product of the assumed silanitrile intermediate **78**, that should be initially formed from **77**. A threefold 1,3 H-shift in **79** gives **80**, a molecule which also could be generated by photoisomerization of the long known[94] silylnitrile (**81**). This was shown by irradiation of independently synthesized unmarked and perdeuteriated silylnitrile (**81**) in an argon matrix with the same wavelength[18]. In both cases the corresponding silylisonitrile **80** was formed. Furthermore, the irradiation of silanimine **74** ($\lambda = 254$ nm) under the same conditions yields traces of **80**, which are formed in addition to the main product **75** (equation 21)[18].

$$\underset{(\mathbf{77})}{\begin{array}{c}N_3\diagdown\quad\diagup N_3\\ Si\\ H_3C\diagup\quad\diagdown N_3\end{array}} \xrightarrow{254\text{ nm}} \underset{(\mathbf{78})}{\left(H_3C\text{---}Si\equiv N\right)} \xrightarrow{254\text{ nm}} \underset{(\mathbf{79})}{:Si\equiv N\text{---}CH_3}$$

$$\Big\downarrow 254\text{ nm} \qquad (22)$$

$$\underset{(\mathbf{81})}{H_3Si\text{---}C\equiv N} \xrightarrow{254\text{ nm}} \underset{(\mathbf{80})}{H_3Si\text{---}N\!=\!C:}$$

The photochemical behavior of the parent azidosilane (**82**) is also rather complex. Early investigations showed that **82** is converted into silaisonitrile (**87**) upon irradiation in an argon matrix[95]. Surprisingly, a dehydrogenation takes place and no H_3SiN isomer could be observed. Later reinvestigations of **82** by Maier, Glatthaar and Reisenauer led to new important findings[96]. Depending on the wavelength of the irradiating light four different products **83**, **85**, **86** and **87** are formed (equation 23). Firstly, irradiation of **82** with $\lambda = 254$ nm yields aminosilylene (**85**) (UV: $\lambda_{max} = 330$ nm; SiN stretching vibration: 866 cm^{-1}). Secondly, if **82** is irradiated with 222 nm light of a $KrCl_2$ excimer laser, the main product is the well-known[95] silaisonitrile (**87**)[12]. In contrast to earlier investigations[95] silanimine (**86**) (UV: $\lambda_{max} = 240$ nm; SiN stretching vibration: 1097 cm^{-1}) could be identified as an intermediate of the reaction $\mathbf{82} \rightarrow \mathbf{87}$[12]. Lastly, if the 193 nm emission of an ArF excimer laser is used, an IR spectrum can be registered which indicates that in this case the newly formed compound is silanitrile (**83**) (UV: $\lambda_{max} = 350$ nm; SiN stretching vibration: 1162 cm^{-1})[12]. Silanitrile (**83**) is a species with a formal silicon–nitrogen triple bond. An additional structural proof is the fact that **83** looses its hydrogen atom upon prolonged irradiation and the long known[97] SiN molecule **84** remains as the final product.

In addition to these observations some more interconversions are worth noting, which for reasons of clearness were not included in equation 23. Aminosilylene (**85**) can be photoisomerized into silanimine (**86**) and vice versa[12]. The transition state for such an isomerization has been calculated[98]. The same is true for the dehydrogenated species **83** and **87**[12]. Irradiation of **87** with $\lambda = 254$ nm results in the formation of **85** after hydrogen recapture[96]. Thus hydrogen does not escape the argon cage except that the matrix has been tempered. Warm-up results in diffusion of the reagents. Moreover, there is also a direct spectroscopic proof that both **87** and **83** can associate with the eliminated hydrogen molecule[12]. The complexes with the 'disturbed' hydrogen molecule were identified by their infrared features at around 4170 cm^{-1} (HH stretching vibration, forbidden for undisturbed hydrogen).

$$\underset{(82)}{\overset{H}{\underset{H}{\diagdown}}\text{Si}\overset{H}{\underset{N_3}{\diagup}}} \quad \overset{h\nu}{\searrow} \quad \underset{(83)}{H-\text{Si}\equiv N} \quad \xrightarrow{193 \text{ nm}} \quad \underset{(84)}{\cdot\text{Si}\equiv N} \qquad (23)$$

$$+$$

$$\underset{(85)}{\overset{\cdot\cdot}{H}\diagdown\text{Si}-NH_2} + \underset{(86)}{\overset{H}{\underset{H}{\diagdown}}\text{Si}=N\overset{H}{\diagup}} + \underset{(87)}{:\text{Si}=N-H}$$

Due to the observed isotopic shifts the matrix experiments with deuteriated azidosilane [D$_3$]-**82**[12] support the described results. According to calculations (cf reference 6 in Reference 12), it is not surprising that the energy-rich silylnitrene, H$_3$SiN, was not observed upon the photolysis of **82**.

B. Silanediimines and Diazosilanes

Silanediimines are compounds which contain a digonal, sp-hybridized silicon atom. In 1987 *N,N*-bis(trimethylsilyl)silanediimine, the first silanediimine, was observed in a low temperature glass[1,99].

By using the same method, namely the photolysis of the sterically hindered diazidosilane **88**, a second silanediimine with a new substitution pattern [di-*tert*-butyl instead of bis(trimethylsilyl) substituents] was likewise accessible[100]. In solid argon as well as in glassy 3-methylpentane diazidosilane **88** is partially converted into silanediimine **89** (λ_{max} = 385 nm) upon irradiation (equation 24). Diimine **89** is the minor product of the photolysis of **88** and was identified by UV spectroscopy and results of chemical trapping experiments. The major product of the irradiation of **88** is di-*tert*-butylsilylene (**91**), which undergoes a subsequent photochemical insertion to give silacyclopropane **92**. Silylene **91** is formed in at least two steps. The authors observed a small amount of a photochemical precursor to **91** with a UV band at λ_{max} = 300 nm and an IR band at 2150 cm^{-1}. Furthermore, ^{15}N-labelling proved that this species still contains nitrogen. The unknown substance was tentatively assigned as di-*tert*-butyldiazosilane (**90**). If this assignment is correct, **90** is the first example of a diazosilane (for N$_2$Si cf Section VIII.D.3.b).

$$\underset{(88)}{t\text{-Bu}_2\text{Si}(N_3)_2} \xrightarrow{254 \text{ nm}} \underset{(89)}{t\text{-BuN}=\text{Si}=\text{N}t\text{-Bu}}$$

$$\Bigg\downarrow 254 \text{ nm} \qquad\qquad\qquad\qquad\qquad\qquad\qquad (24)$$

$$\underset{(90)}{t\text{-Bu}_2\text{SiN}_2} \xrightarrow{254 \text{ nm}} \underset{(91)}{t\text{-Bu}_2\text{Si}:} \xrightarrow{500 \text{ nm}} \underset{(92)}{\overset{t\text{-Bu}}{\underset{H}{\diagdown}}\text{Si}\triangleleft}$$

V. SILICON–OXYGEN MULTIPLE BONDS
A. Silanones, Silacarbonates and Silacarboxylic Esters

Up to 1986 several silanones have been matrix-isolated and their IR spectra were recorded[1]. Most of these silanones are accessible by intermolecular reactions in the matrix which take place under specific conditions, that allow the diffusion of the reagents.

As could be shown by Whitnall and Andrews, dimethylsilanone (**97**) is formed when dimethylsilane (**93**) and ozone (**94**) are codeposited and irradiated in solid argon at 14–17 K (equation 25)[101]. Beside **97**, methylsilanone[102] is generated as well.

The use of methylsilane in this procedure results in the formation of methylsilanone and the parent silaformaldehyde H_2SiO[101,103]. The relative yields of the products show that H_2 elimination (Si–H instead of Si–C bond cleavage) is favored in both cases. Notwithstanding the behavior of methylsilane and dimethylsilane, a trimethylsilane/ozone mixture presumably gives $H_2C=Si(OH)CH_3$, the enol isomer of **97** (Si–C cleavage), when irradiated in the region 290 nm $< \lambda <$ 1000 nm[101]. However, full mercury arc irradiation of the trimethylsilane/ozone mixture produces the keto isomer, silanone **97**[101].

Dimethylsilanone (**97**) is also formed in the oxygenation of **95** with oxygen atoms generated by a microwave discharge of an Ar/O_2 stream and subsequent matrix isolation of the reaction products[101].

Pyrolytic fragmentations of other suitable precursors also lead to dimethylsilanone (**97**). For example, 6-oxa-3-silabicyclo[3.1.0]hexane **96** is reported to be split into **97** and 1,3-butadiene upon pyrolysis[104]. Other possible routes to **97** consist in the pyrolysis of (allyloxy)dimethylsilane (**99**)[105] or Diels–Alder adduct **98**[104]. Matrix-isolated **97** gives its trimer hexamethylcyclotrisiloxane when the matrix is warmed up to 35–40 K. The SiO stretching vibration of **97** was found at 1210 cm^{-1}. This frequency fits the calculated force constant and bond order and has to be considered as evidence for significant double bonding in **97**[104]. Octamethylcyclotetrasiloxane, allyl(allyloxy)dimethylsilane and 2,2,6-trimethyl-2-silapyrane failed as precursors for **97** and only the SiO molecule (cf Section V.B) and CH_3 radicals were found on the matrix holder[105].

Using the corresponding derivatives of epoxide **96**, D_6-dimethylsilanone ([D_6]-**97**)[104], diphenylsilanone and silacarbonic acid dimethyl ester[106] were matrix-isolated and studied by IR spectroscopy.

Examples of matrix-isolated silacarboxylic esters are scarce. Only the argon matrix photolysis of dimethoxysilylene (**100**) (cf Section VIII.E.2) results in the formation of infrared bands and a UV maximum ($\lambda_{max} = 232$ nm) which are consistent with structure **101** (equation 26)[107,108].

$$H_3CO-\ddot{Si}-OCH_3 \xrightarrow{254 \text{ nm}} H_3C-\underset{OCH_3}{\overset{O}{\underset{\|}{Si}}} \quad (26)$$

(**100**) (**101**)

B. Silicon Oxides and Related Compounds

In principle this section can be divided into two parts. The first one deals with the matrix isolation of various silicon oxides whereas reactions of such oxides with metal atoms are presented in the second part.

Cocondensation in solid argon of SiO and oxygen atoms, generated by a microwave discharge, yields SiO_2[109]. SiO, $(SiO)_2$ and $(SiO)_3$ are well-known species[110,111] and can be obtained when molecular oxygen is passed over heated silicon (at about 1500 K)[112].

More recently, Schnöckel and coworkers reexamined the structure of dimeric SiO[113]. Using the method described above $(SiO)_2$ was generated and condensed together with an excess of argon on a liquid-helium-cooled copper surface. The IR spectra of the matrix-isolated species and their ^{29}Si and ^{18}O isotopomers are in good agreement with a planar cyclic structure of D_{2h} symmetry. The trimer $(SiO)_3$ has also a planar cyclic structure (D_{3h} symmetry)[113,114].

The reaction M+SiO → MSiO can be carried out with various metals. For that purpose the metal atoms are cocondensed with monomeric SiO onto an argon matrix holder. The identification of the reaction products is based on both their IR spectra and isotopic shifts.

In the case of sodium or potassium atoms[115] the SiO stretching vibration of the reaction product is detected at around 1020 cm^{-1} (uncoordinated SiO: 1226 cm^{-1}). Together with the observed isotopic splittings, the formation of an ionic species M$^+$(SiO)$^-$ seems reasonable, this species is very likely to have a strongly bent structure[114].

Less pronounced charge transfer to SiO and side-on coordination takes place in AgSiO (SiO stretching vibration in solid argon: 1163 cm^{-1})[116]. Ag$_2$SiO is probably a side product formed in the cocondensation of silver atoms with monomeric SiO[116]. The structure of AgSiO has been confirmed by hydrocarbon matrix electron magnetic resonance studies[117]. In addition to AgSiO the authors also observed AgSi$_2$O$_2$, AgSi$_3$O$_3$ and AgSi$_n$O$_n$ [derived from polymeric $(SiO)_n$] after the reaction of silver atoms with SiO in an adamantane matrix at 77 K.

In contrast to the results described above, experiments with palladium atoms and SiO lead to a different behavior. It is clear that PdSiO is formed, but compared with monomeric SiO the corresponding stretching vibration of PdSiO is shifted to higher wavenumbers (1246 cm^{-1} in solid argon)[118]. With the aid of a normal coordinate analysis involving different isotopomers, a linear structure of PdSiO is deduced. Bonding in PdSiO is similar to that in typical transition metal carbonyl complexes.

VI. SILICON–SULFUR MULTIPLE BONDS

Following the 'classical' double-bond rule, multiple bonds between silicon and sulfur, both elements of the third period, should be very difficult to obtain. Except for SiS[119,120],

SiOS[121] and a sterically hindered silanethione[122] no compounds with silicon–sulfur multiple bonds have been isolated[1].

Recently, Schnöckel and Köppe generated SiS_2 in solid argon[120]. The authors cocondensed SiS with COS and, during irradiation of this mixture, the IR bands of SiS_2 appeared. In consideration of the formed CO it was concluded that sulfur atoms have been produced from COS which subsequently react with the SiS molecule to give SiS_2. Another possible but unsatisfying access (the observed spectra were not of comparable quality with those discussed above) to molecular SiS_2 consists in the vaporization of solid SiS_2 and matrix isolation of the products[120].

$S=SiCl_2$ was found in a photochemically induced argon matrix reaction between monomeric SiS and Cl_2[123]. With COS cocondensed $SiCl_2$ yields $S=SiCl_2$ after irradiation as well[123]. Later, $S=Si(H)Cl$ was accessible via the photoreaction of SiS and HCl under argon matrix conditions[124]. The use of DCl instead of HCl confirmed the structural assignment for this species (C_s symmetry). $S=SiCl_2$ and $S=Si(H)Cl$ have nearly the same values of the SiS force constants as the SiS multiple bonds in SiS and SiS_2[124].

AgSiS can be prepared by reaction of silver atoms with molecular SiS. Studies of AgSiS in both hydrocarbon matrices (electron paramagnetic resonance study)[125] and solid argon[126] demand a triangular structure with a Ag–S bond (similar to the structure of AgSiO; cf Section V.B).

VII. STRAINED SILACYCLES

A. Silacyclopropyne and Other C_2H_2Si Isomers

Maier and coworkers achieved the isolation of silacyclopropyne (**106**), a C_2H_2Si isomer, by irradiation (λ = 254 nm) of the matrix-isolated C_2H_2Si isomers 1-silacyclopropenylidene (**104**) or ethynylsilylene (**103**)[127–129]. The best access to the C_2H_2Si potential energy surface is offered by pulsed flash pyrolysis of 2-ethynyl-1,1,1-trimethyldisilane (**102**). Using this procedure apart from trimethylsilane only 1-silacyclopropenylidene (**104**), which is the most stable C_2H_2Si isomer[130–133] and possibly plays a role in the chemistry of interstellar clouds (like the cyclic C_2Si molecule **107**), could be isolated in an argon matrix and was identified by its IR spectrum. Trapping experiments make it likely that ethynylsilylene (**103**) is formed from **102** in the first step, but under the conditions of the pulsed flash pyrolysis a rearrangement immediately takes place resulting in the formation of 1-silacyclopropenylidene (**104**), which after leaving the hot zone can be isolated in solid argon (equation 27)[128]. In contrast, the high-vacuum flash pyrolysis of **102** followed by the direct condensation of the reaction products onto a spectroscopic window at 10 K remained unsuccessful; no C_2H_2Si isomer was obtained[127–129].

A particle whose connectivities are indicative of structure **104** has earlier been detected by neutralization–reionization mass spectrometry[134]. Furthermore, 1-silacyclopropenylidene (**104**) was recently identified by microwave spectroscopy[135]. 1-Silacyclopropenylidene (**104**) (λ_{max} = 286 nm[136]; a delocalized three-center π-bond orbital[128]) generated from **102** and matrix-isolated is transformed into ethynylsilylene

(**103**) upon irradiation with $\lambda = 313$ nm (equation 28)[127-129]. 2-Ethynyl-1,1,3,3-tetramethyl-1, 3-diphenyltrisilane, too, can be used as a photochemical precursor for **103** and other C_2H_2Si isomers (irradiation with $\lambda = 254$ nm). But the IR spectra measured in these experiments were not of comparable quality to those obtained after pyrolysis of 2-ethynyl-1,1,1-trimethyldisilane (**102**)[136].

$$\underset{(\mathbf{104})}{\overset{H}{\underset{H}{>}}C=C\overset{\cdot\cdot}{\underset{H}{<}}} \underset{500\ nm}{\overset{254\ nm,\ 313\ nm}{\rightleftharpoons}} \underset{(\mathbf{103})}{H-C\equiv C-\overset{\cdot\cdot}{Si}\overset{\cdot}{\underset{H}{}}} \underset{340\ nm}{\overset{500\ nm}{\rightleftharpoons}} \underset{(\mathbf{105})}{\overset{H}{\underset{H}{>}}C=C=Si:}$$

$$\Big\downarrow_{395\ nm}^{254\ nm} \quad \Big\downarrow^{254\ nm}$$

$$\underset{(\mathbf{106})}{\overset{H\ \ H}{\underset{C\equiv C}{\overset{Si}{\triangle}}}} + \underset{(\mathbf{107})}{C_2Si} + \underset{(\mathbf{107})\cdot H_2}{C_2Si\cdot H_2} \qquad \underset{(\mathbf{108})}{\overset{H\ \ H}{\underset{C---C}{\overset{Si}{\triangle}}}}$$

(28)

Ethynylsilylene (**103**) ($\lambda_{max} = 500$ nm[136]; calculated: $\lambda_{max} = 520$ nm[137]) isomerizes into 1-silacyclopropenylidene (**104**) and vinylidenesilylene (**105**) when irradiated with $\lambda = 500$ nm[127-129]. Vinylidenesilylene (**105**) shows two weak infrared bands and a UV absorption with vibrational fine structure ($\lambda = 340$, 325 and 310 nm[136]). Irradiation into this absorption leads back to ethynylsilylene (**103**) (equation 28)[127-129].

Irradiation of **104** with $\lambda = 254$ nm (as with $\lambda = 313$ nm) yields **103** as the first photoproduct. However, upon longer irradiation, silacyclopropyne (**106**) as well as the well-known[138-142] cyclic C_2Si molecule **107** (cf reference 128 for the electronic structure of **107**) and its adduct, $C_2Si\cdot H_2$ (**107**·H_2), can be detected[127-129]. Additional experiments with the corresponding ^{13}C and D isotopomers are in good agreement with the assigned structures. The structural assignment of silacyclopropyne (**106**), which is the first example of a 'formal' cyclopropyne[143], is based on comparison of the experimental and calculated IR spectroscopic data. The CC stretching vibration of **106** was observed at 1769.8 cm^{-1}. Experiments with the corresponding ^{13}C and D isotopomers again support the cyclopropyne structure[128]. Upon long-wavelength irradiation ($\lambda > 395$ nm) **106** and **107**·H_2 are retransformed into **104**. Therefore these species have to be isomers.

In order to understand the unique bonding situation in silacyclopropyne (**106**) detailed *ab initio* calculations have been carried out[128]. It was found that the Lewis structure **108**, where the dashed line indicates an electron pair which occupies a CC nonbonding orbital, is the best possible compromise to describe the electronic structure of **106** correctly. Silacyclopropyne (**106**) is best considered as a closed-shell singlet diradical but with only one 'nonbonding' orbital. The energies (AU) and symmetries of the HOMO and LUMO describing the unique bonding situation in **106** are shown in Figure 1. The most striking feature of the canonical orbitals is the fact that the CC-σ' orbital (A_1) is the HOMO and lies higher in energy by 0.01 AU than the CC-π orbital. The occupation numbers of HOMO and LUMO were found to be $(A_1)^{1.913} (B_2)^{0.095}$ [CASSCF (8.8)/6-31G(d,p)].

Quite recently, Maier and coworkers discovered another possible route to the C_2H_2Si potential energy surface[144]. Cocondensation of silicon atoms and acetylene onto an argon

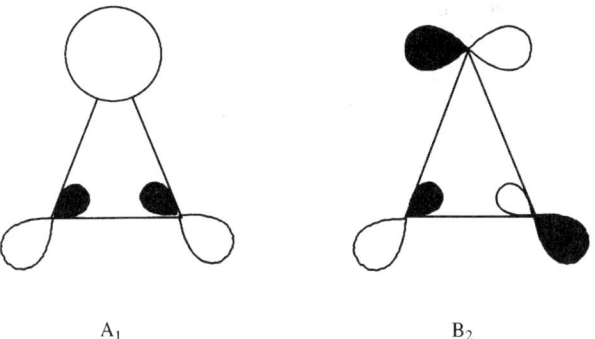

FIGURE 1. Schematic representation of HOMO and LUMO of **106** (C_{2v}) discussed in the text. Symmetries and orbital energies (AU): A_1 (CC-σ', HOMO) -0.389, B_2 (CC-σ'^*, LUMO) 0.005.[MP2/6-311G(d,p)]

matrix holder results in the formation of 1-silacyclopropenylidene (**104**) as well (cf Section VIII.B.5) and again all the photoreactions shown in equation 28 can be carried out.

B. Silacyclopropenes

Tetramethylsilacyclopropene, the first silacyclopropene derivative, was prepared in 1976[145]. With the help of the matrix-isolation techniques, attempts were made to synthesize the parent silacyclopropene (**114**). Experiments with this goal culminated in the isolation of 1,1-dimethylsilacyclopropene (**110**) in solid argon[146]. Sander and coworkers showed that **110** is accessible by photolysis of the corresponding bis(diazomethyl)silane (**109**) (equation 29). After subsequent irradiation with shorter wavelengths, isomerization into the photostable ethynyldimethylsilane (**111**) takes place.

$$\underset{(109)}{\underset{N_2HC}{\overset{H_3C}{\diagdown}}\underset{CHN_2}{\overset{CH_3}{\diagup}}Si} \xrightarrow{405 \text{ nm}} \underset{(110)}{\underset{H}{\overset{H_3C \quad CH_3}{\diagdown \diagup}}\underset{H}{\overset{Si}{\diagup\diagdown}}\underset{H}{\overset{}{C=C}}} \xrightarrow{>254 \text{ nm}} \underset{(111)}{H-C\equiv C-SiH(CH_3)_2} \quad (29)$$

Later on, Maier and coworkers succeeded in the synthesis of the unsubstituted silacyclopropene (**114**)[147]. Once more (cf Section VII.A) a precursor molecule was used which fragments into trimethylsilane and the corresponding silylene. Trapping experiments show that 1,1,1-trimethyl-2-vinyldisilane (**112**) (upon pyrolysis) gives vinylsilylene (**113**) in the first step (for **113** also cf Section VIII.B). However, only small amounts of this species were isolated on the matrix holder[136,144] because an isomerization into both silacyclopropene (**114**), which according to calculations contains a localized π-bond orbital[128], and ethynylsilane (**115**)[148] occurs (equation 30). The latter two products were identified with the aid of their infrared spectra (CC stretching vibration of **114**: 1467.2 cm^{-1}). The same experiment carried out with 2,2-dideuterio-1,1,1-trimethyl-2-vinyldisilane ([D$_2$]-**112**) gives evidence for either a 1-silaallene (**161**) or a 1-silacyclopropylidene (**160**) participation (compare equations 44 and 45) in the discussed process **113** → **114** + **115**[136,147]. In accordance with the photochemical behavior of **110**, parent silacyclopropene (**114**) is transformed into the photostable ethynylsilane (**115**) upon irradiation with λ = 254 nm under argon matrix conditions.

$$
\underset{(112)}{\begin{array}{c}H\\H\end{array}\!\!>\!C\!\!=\!\!C\!\!<\!\!\begin{array}{c}Si(H)(SiMe_3)\\H\end{array}} \xrightarrow[-HSiMe_3]{650\,°C} \underset{(113)}{\begin{array}{c}H\\H\end{array}\!\!>\!C\!\!=\!\!C\!\!<\!\!\begin{array}{c}\ddot{S}iH\\H\end{array}}
$$

$\Big\downarrow$ 650 °C

(30)

silacyclopropane (114) + H—C≡C—SiH₃ (115)

254 nm ⇌

Attempts to generate the parent silacyclopropane from 2-ethyl-1,1,1-trimethyldisilane in an analogous manner failed[69] (for an example of a matrix-isolated silacyclopropane cf **92**, Section IV.B).

VIII. SILYLENES

Whereas the analogous carbenes easily isomerize wherever possible to compounds containing doubly bonded carbon atoms even under the conditions of matrix isolation, silylenes are almost as stable as the corresponding substances with doubly bonded silicon atoms. For example, methyl- and silylsilylene lie just 4 and 8 kcal mol^{-1} above silaethene and disilene, whereas the difference between ethene and methylcarbene is as high as 70 kcal mol^{-1} [149–151]. As a consequence, silylenes are often key intermediates on the way to other highly reactive silicon compounds discussed above.

Furthermore, all known silylenes are singlets in their ground states. According to calculations, the singlet ground state of the parent compound :SiH$_2$ lies 21 kcal mol^{-1} lower than the first triplet state[152]. In the carbon analogue the situation is reversed. The first singlet state is calculated to lie 9 kcal mol^{-1} above its triplet ground state[152,153], in agreement with experimental data[154,155].

Several factors were mentioned as explanation for this important feature[156]. In addition to pairing energy and electrostatic effects, the small extent of s,p mixing in silicon seems to play an important role[156]. The last factor is manifested not only in the large stabilization of the singlet state but also in the bond angle of :SiH$_2$, calculated to be as small as 93.4°[157]. This is a consequence of the relatively large difference of the sizes of the 3s and 3p orbitals compared with the difference between 2s and 2p[158].

According to *ab initio* calculations[85], to have triplet ground state silylenes it is necessary to enlarge the RSiR′ angle of substituted silylenes to more than 141° in the case of carbon substituted silylenes and to more than 115° for silyl substituted silylenes. However, even bis(1-adamantyl)silylene with its sterically very demanding substituents led only to trapping products of the singlet species[159]. Hence the generation of a triplet ground state silylene remains one of the most challenging goals in silylene chemistry.

A. Unsubstituted Silylene

The matrix isolation and spectroscopic examination of :SiH$_2$, the parent compound of the entire class, was reported twice. In 1970, Milligan and Jacox[160] photolyzed mono- and disilane and found new absorptions that they assigned to :SiH$_2$. But as mentioned by Margrave and coworkers[161], this assignment is erroneous and the observed bands are due to the formation of the ·SiH$_3$ radical. The same authors were able to generate silylene by the cocondensation reaction of silicon atoms with molecular hydrogen. Caused by the uptake of another hydrogen molecule, silane was obtained as a by-product (equation 31).

$$:\!\ddot{S}i \xrightarrow{H_2} :SiH_2 \xrightarrow{H_2} SiH_4 \tag{31}$$

The observed IR spectrum agrees well with that expected for :SiH$_2$. Four bands were registered, three of them fundamentals and one an overtone. Furthermore, the assignment is supported by deuterium labelling.

On the other hand, up to now we were not able to generate :SiH$_2$ from a 'chemical' precursor. Despite the fact that Ring and coworkers could obtain trapping products of :SiH$_2$ (among others, see equation 18) in the copyrolysis of trisilane and 1,3-butadiene[68], we failed in matrix isolation of silylene generated by pyrolysis of di- or trisilane[69]. Obviously, silylenes bearing hydrogen atoms at the low-valent silicon center are too reactive to pass a pyrolysis tube.

B. Carbon Substituted Silylenes

1. Silylenes by photoisomerization of silenes

a. Methylsilylene. The photochemical interconversion of silylenes and silenes is an important link between these two classes of compounds. It was first established by Maier and coworkers[162], who irradiated the parent silene with light at the wavelength of $\lambda = 254$ nm and obtained methylsilylene (equation 32). The reaction is reversible by using 320 nm light.

$$H_2Si=CH_2 \underset{>320\,nm}{\overset{254\,nm}{\rightleftarrows}} \underset{H}{\overset{\cdot\cdot}{Si}}\!-\!CH_3 \tag{32}$$

(30)   (116)

As discussed later, the location of the UV-maximum of silylene **116** strongly depends on the matrix material. So irradiation of **30** in solid argon results in the growing of an absorption with $\lambda_{max} = 480$ nm, whereas $\lambda_{max} = 330$ nm is found if nitrogen is used instead[162]. This large hypsochromic shift of 150 nm is caused by complexation of methylsilylene with nitrogen[73b].

b. Dimethylsilylene. Although long assumed to be intermediates in several reactions, no carbon substituted silylene was directly observed for many years. In 1979 Michl, West and Drahnak[10] detected a broad UV absorption band ($\lambda_{max} = 453$ nm) after photolysis of dodecamethylcyclohexasilane (**117**) in 3-methylpentane. This band was assigned to dimethylsilylene (**118**). Many different approaches to this intermediate, either photochemically or pyrolytically[163], were examined in the following years. They are shown in equation 33 in order to give a survey on the methods known to generate silylenes.

$$\text{(33)}$$

As in the case of methysilylene (**116**), dimethylsilylene (**118**), is accessible by irradiation of 2-silapropene (**12**) at λ = 254 nm. If light of a longer wavelength is used, the reaction is reversible again[162], which serves as an important structure proof for both species.

Dimethylsilylene (**118**) is one of the best examined silylenes. Not only IR and UV spectra are known[163,164], but also bimolecular reactions of the matrix-isolated compound, leading to new, highly reactive species. Arrington and coworkers[165] as well as West and Pearsall[166] reacted matrix-isolated dimethylsilylene (**118**) with carbon monoxide (equation 34). The resulting adduct (**119**) absorbs at about 340 nm, thus showing a blue-shift of more than 100 nm compared with the free silylene (**118**). According to *ab initio* calculations[167], the structure of **119** is most likely pyramidal. The same result was obtained for the parent compound H_2SiCO. The planar silaketene-like isomer is the transition state between two pyramidal structures lying 18 kcal mol^{-1} above the pyramidal species[167].

The reaction of **118** with oxygen in an O_2 doped argon matrix yielded dioxasilirane **120**, identified by its vibrational spectrum[108] (equation 34).

$$\text{(34)}$$

2. Silylenes by photolysis of oligosilanes

As mentioned above, the photolysis of cyclopolysilanes was the earliest method for the generation of silylenes in matrices. Indeed, such extrusions from oligosilanes were used for the generation of most silylenes known so far[10,137,163,168−175] (equation 35).

$$(RR'Si)_n \xrightarrow[\lambda = 254 \text{ nm}]{h\nu} RR'Si: + (RR'Si)_{n-1}$$

$$RR'Si\begin{subarray}{l}\diagup SiMe_3 \\ \diagdown SiMe_3\end{subarray} \xrightarrow[\lambda = 254 \text{ nm}]{h\nu} RR'Si: + Si_2Me_6$$

(35)

The effect of substituents on the properties and structure of silylenes was analyzed in detail by Apeloig and Karni on the basis of *ab initio* calculations[137].

In principle, two kinds of substituent effects can be distinguished. The first effect is steric; widening of the RSiR' angle results in a bathochromic shift. This becomes evident if the phenyl substituents in diphenylsilylene are gradually replaced by more bulky mesityl groups. Whereas Ph_2Si: absorbs at 495 nm, MesPhSi: (Mes = 2,4,6-trimethylphenyl) has its UV maximum at 530 nm and Mes_2Si: at 577 nm[169]. These results agree nicely with theoretical predictions[85,176].

The second type of effects are electronic ones, which can further be divided into inductive and conjugative effects.

Inductive Effects: Electronegative substituents increase the n(Si) → 3p(Si) transition energy and thus induce a blue-shift. The opposite is true for electropositive substituents.

Conjugative Effects: Ligands acting as n-donors generally cause blue-shifts[107,137,171]. However, the n-donor does not have to be a real substituent at the subvalent silicon atom. Lewis bases which are matrix-isolated along with the silylene also induce blue-shifts by the formation of adducts. This was shown for phosphines, sulfides, amines, alcohols, ethers[177] and carbon monoxide[166]. As mentioned above, in the case of methylsilylene even the use of N_2 as the matrix material instead of argon led to a blue-shift of 150 nm[162].

A second possible conjugative interaction is that with a π-system. In contrast to n-donors, π-donors result in red-shifts, which is best understood qualitatively as a consequence of the presence of π^*-orbitals in these substituents[137]. As is the case in the examples above, also for π-donors the experimental data fit well with the theoretically predicted values[137]. So species **121**, **122** and **123**, accessible by photolysis of the appropriate trisilanes[174], absorb at 436, 475 and 505 nm.

(121)   (122)   (123)

Frequently, UV absorptions of the respective silylenes are not the only observed absorptions from 3-methylpentane matrix experiments. Since the corresponding disilenes are often formed upon annealing of the matrix, their spectra can be registered as well[169].

As in the case of dimethylsilylene (**118**), for the dimesityl compound (**124**) not only the spectral features are known but also its reaction with molecular oxygen[178] (equation 36). Surprisingly this reaction, performed by photochemical generation of **124** in an oxygen matrix, yielded silanone-O-oxide **125** and not the dioxasilirane as with **118**.

$$\text{Mes}_2\text{Si:} \xrightarrow{O_2} \text{Mes}_2\overset{+}{\text{Si}}=\overset{\bar{O}}{O} \tag{36}$$

(**124**)   (**125**)

3. Silylenes by photolysis of diazides

A third photochemical access to silylenes, beside the isomerization of silenes and the photolysis of tri- or cyclopolysilanes, is the irradiation of geminal diazidosilanes, which under nitrogen loss often gives the corresponding divalent silicon species. However, silanimines are frequent by-products[164] or in some cases even the only products (Section IV).

A silylene which can be generated in this way is, beside the methyl (Section II.A, equation 1) and dimethyl species, bis(t-butyl)silylene (**91**)[100]. Its chemistry was already discussed in Section IV.A (equation 24). A second silylene, indirectly accessible from the corresponding diazide **126**, is the silacyclopentadiene isomer **130**[179] (equation 37). Remarkably, in the photolysis of **126** as well as in the the pyrolysis of spirosilane **128** (a second access), silacyclopentadiene **127** is the first spectroscopically detectable substance. Upon irradiation with broad-band UV light (260–390 nm) silylene **130** is formed along with traces of its silene isomers **129** and **131**. The latter were formed in larger amounts upon irradiation with light of an appropriate wavelength. For the compounds **127, 129, 130** and **131** IR as well as UV spectra were recorded. The structural assignment of silole (**127**) and the silylenic compound **130** is unambiguous, whereas the distinction between the isomeric **129** and **131** is based solely on mechanistic arguments. A tetramethylated analogue of **128** was also examined, leading to the 3,4-dimethylsilole in the primary step and showing the same behavior as the parent compound[179].

4. Silylenes by trimethylsilane extrusion

A relatively long known access to silylenes is the thermolysis of disilanes[13], used for preparative purposes[180] as well as for matrix-isolation studies of silylenes[107]. In our group this method was recently used in the generation of C_2H_2Si isomers. Trimethylsilane was extruded from 2-ethynyl-1,1,1-trimethyldisilane (**102**) pyrolytically, and silacyclopropenylidene (**104**) was the first species detectable after subsequent trapping in solid argon, as discussed in detail in Section VII.A (equations 27 and 28).

a. C_3H_4Si isomers. In the meantime this successful access to silylenes was extended to other hypersurfaces. The formal addition of a methylene unit to **102** leads to compounds that should yield C_3H_4Si species upon pyrolysis. Some of these intermediates are remarkable, e.g. the parent silacyclobutadiene (**132**), for which only mass spectroscopic indications are available[181], and even more for the unsubstituted silatetrahedrane (**133**).

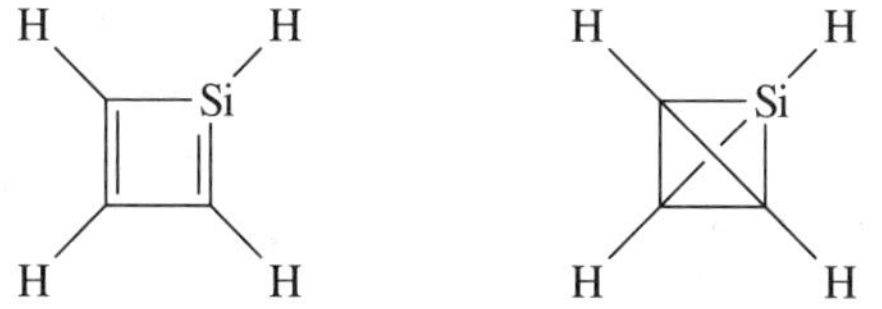

(**132**)   (**133**)

It was not expected that one could trap one of these target molecules directly from the pyrolysis of one of the appropriate precursors. But it was hoped to succeed in generating one of these compounds by photochemical conversion of the silylenes expected as primary products upon pyrolysis. So all stable open-chain precursors that should lead to C_3H_4Si isomers upon pyrolytic extrusion of trimethylsilane were prepared and examined[182,183]. The first precursor, propynyldisilane **134**, behaved completely analogous to the ethynyl compound **102** (equation 38). After pyrolysis, the methyl substituted silacyclopropenylidene **135** was matrix-isolated, representing the global minimum of the C_3H_4Si hypersurface, as is the case for the unsubstituted analogue (equations 27 and 28). Irradiation with light of an appropriate wavelength yielded propynylsilylene (**136**) and ethynylmethylsilylene (**138**) lying 14 and 16 kcal mol^{-1} (BLYP/6-31G*) above the global minimum, respectively. The given energy values derive from calculations at the BLYP/6-31G* level of theory and are corrected by zero-point vibrational energies[182]. Silylene **138** is also pyrolytically accessible from precursor **137**. An additional band was registered in this experiment that was later recognized to be the most intense absorption of silacyclobutenylidene **139**.

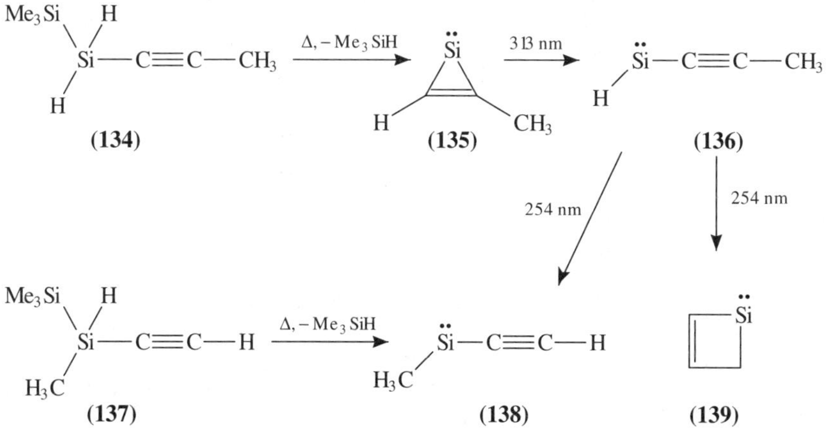

(38)

The two remaining stable open-chain isomers of **134** and **137**, propargyldisilane **140** and allenyldisilane **142**, yielded upon photolysis the cyclic silylene **139** as the first detectable product (equation 39). **139** lies only 5 kcal mol^{-1} above the global minimum **135** and was hoped to undergo a 1,2 H shift to silacyclobutadiene (45 kcal mol^{-1} above **135**) upon irradiation. Depending on the wavelength of the light used in the photolysis, silabutadienylidene **143** (13 kcal mol^{-1} above **135**) or allenylsilylene (**141**) (19 kcal mol^{-1} above **135**) was formed, i.e. ring opening occurred in all cases. Further irradiation led to **138**, therefore being an important link between the two described series. In all cases and irrespective of the wavelength of the used light, no silacyclobutadiene or silatetrahedrane could be observed in the photolysis experiments.

(39)

b. $C_2H_4Si_2$ isomers. If one saturated carbon atom in precursors **134**, **137** and **140** is formally substituted by a silicon atom, compounds **144**, **147** and **148** are formed, all being appropriate precursors for $C_2H_4Si_2$ species[182,69] (equation 40). This hypersurface contains, among others, two disilacyclobutadienes which are of even more interest than the monosilicon analogue[184]. As in the examples above, trimethylsilane was extruded upon pyrolysis and in all cases the silyl substituted silacyclopropenylidene **145** was the first species detectable in solid argon. Silylene **145**, the global minimum of the hypersurface according to our calculations at the BLYP/6-31G* level of theory, and the open-chain compound **146**, lying 12 kcal mol^{-1} above **145**, could be interconverted photochemically.

c. C_4H_2Si isomers. A further target in our recent investigations on silylenes was the C_4H_2Si hypersurface[69]. **149** is an appropriate precursor using the reliable extrusion of trimethylsilane (equation 41). The pyrolysis of **149** yielded diethynylsilylene (**150**) as the primary product, being interconvertible with butadiynylsilylene (**153**) upon irradiation with light of appropriate wavelengths. In addition, some bands that do not belong to silylene **150** appeared after the pyrolysis and are assigned to the substituted silacyclopropenylidene **151**, which represents again the global minimum on the C_4H_3Si hypersurface according to our calculations at the BLYP/6-31G* level of theory (corrected by zero-point vibrational energies). Like **150** (13 kcal mol^{-1} above **151**), **151** is photochemically interconvertible with silylene **153** (22 kcal mol^{-1} above **151**). The structural assignments were confirmed by preparing the perdeuteriated and ^{13}C-labelled isotopomeres of these reactive intermediates.

19. Matrix isolation studies of silicon compounds

Another entry to the C_4H_2Si hypersurface was found in the pyrolysis of triethynylsilane (**152**) (equation 41).

$$\text{(40)}$$

$$\text{(41)}$$

If the behavior of the related precursors **102**, **144**, **137** and **149** is compared (equation 42), it is remarkable that the two former yield the silacyclopropenylidenes **104** and **145** as the first detectable species, whereas the fragmentation of the methyl substituted compound **137** stops at the open-chain silylene **138**. In the case of the diethynyl precursor **149** both products (**150** and **151**) are observed.

$$\text{(42)}$$

R = H	(**102**)	[R = H	(**103**)]	R = H	(**104**)
R = SiH$_3$	(**144**)	[R = SiH$_3$	(**146**)]	R = SiH$_3$	(**145**)
R = CH$_3$	(**137**)	R = CH$_3$	(**138**)	[R = CH$_3$	(**135**)]
R = C$_2$H	(**149**)	R = C$_2$H	(**150**)	R = C$_2$H	(**151**)

Since, as could be shown by trapping experiments with 2,3-dimethylbutadiene, the primary products upon trimethylsilane extrusion are always the expected open-chain silylenes, the formation of the cyclic compounds must arise from a rearrangement reaction. Whether really a one-step process with a transition state of type **155** occurs as indicated in equation 43, or whether the reaction pathway from **154** to **156** involves intermediates, cannot be decided here. But obviously, the more facile the migration of the group R from the silicon to the carbon atom, the more cyclic silylene **156** is formed. In addition to the migration tendency of the group R, there is another factor that could determine the product distribution, namely the sensitivity of the initially formed open-chain silylene **154**. So it is known that silylenes with a hydrogen atom at the subvalent silicon center are generally too sensitive to pass through a pyrolysis tube[69].

$$\underset{R}{\overset{\cdot\cdot}{Si}}-C\equiv C-H \overset{\Delta}{\longrightarrow} \left[\underset{R}{\overset{R\diagdown Si}{\underset{C\equiv\equiv C\diagup}{}}}\right]^{\ddagger} \overset{\Delta}{\longrightarrow} \underset{R}{\overset{\cdot\cdot}{\underset{H}{Si}}} \qquad (43)$$

(154)   **(155)**   **(156)**

The observed product distributions reflect the influence of both factors. In the parent compound **102** and in the silyl substituted **104**, the initially formed open-chain silylenes **103** and **146** are relatively sensitive and the migration tendency of the group R is quite large, and therefore the silacyclopropenylidenes **104** and **145** are found. In contrast, the methyl and the ethynyl compound **138** and **150** are stable enough to pass the pyrolysis tube and both can be matrix-isolated upon pyrolysis of the appropriate precursors **137** and **149**. Due to the small migration tendencies of the methyl and the ethynyl group, the rearrangement product **135** is not found at all and **151** is detected only in relatively minor amounts.

Another transformation that should be discussed here briefly is the photochemical conversion of silylenes bearing a hydrogen atom and a substituted ethynyl function at the subvalent silicon center, i.e. **158**. If R is methyl (equation 38) or ethynyl (equation 41), irradiation with 254 nm light results in an exchange of the R group and the hydrogen atom. This can be rationalized in several ways (equation 44). Firstly, a series of two 1,3 shifts with either a silylenic **(159)** or a carbenic **(157)** intermediate is able to explain the formation of **154**. However, silylenes are much more stable than the isomeric carbenes and therefore a 1,3 R shift as the initial step is highly unlikely. A second possible pathway from **158** to **154** involves the cyclic intermediate **156**. At this time, it is not possible to differentiate between these possibilities.

All compounds of compositions C_3H_4Si, $C_2H_4Si_2$ and C_4H_2Si discussed above were identified by comparison of their calculated and experimental vibrational spectra, some by their UV bands as well. These results are new examples for silylene–silylene interconversions, a class of reactions first observed recently in the related C_2H_2Si system (Section VII.B).

5. Silylenes by cocondensation reactions

All methods used for the generation of silylenes discussed so far are based on the fragmentation of 'chemical' precursors yielding the subvalent silicon compounds upon photolysis or pyrolysis. Hitherto, the formation of an organic silylene was never performed otherwise. For 'inorganic' silylenes like $:SiH_2$[161] or silicon–carbon clusters[185–187], the cocondensation of silicon atoms and the respective element is a long known procedure. Recently we were able to extend this method to 'organic' silylenes. For example, the cocondensation of silicon atoms with acetylene yields silacyclopropenylidene[73], a compound well known from the pyrolysis of 1,1,1-trimethyl-2-ethynyldisilane (102) (Section VII.A).

a. C_2H_4Si isomers. In the case of C_2H_4Si isomers the pyrolysis of the 'chemical' precursor 1,1,1-trimethyl-2-vinyldisilane (112) and the cocondensation of silicon atoms with ethylene led to complementary results[144]. Whereas silacyclopropene (114) and ethynylsilane (115) are the first species detectable in solid argon after pyrolysis of disilane 112 (Section VII.B), silacyclopropylidene (160) is the primary product in the cocondensation reaction (equation 45). In the subsequent photoisomerizations of silylene 160 three hydrogen atoms migrate consecutively to the silicon atom, resulting in the formation of ethynylsilane (115) as the final product, which represents the global minimum of this hypersurface. However, the assignment of 1-silaallene (161) has to be made with some caution since the agreement between the observed and calculated vibrational spectra is not as good as for all the other species. Remarkably, ethynylsilane (115) is the only link between the C_2H_4Si species generated in the pyrolysis of 1,1,1-trimethyl-2-vinyldisilane (112) and the reaction between silicon and ethene.

$$:Si: \xrightarrow{C_2H_4} \underset{(160)}{\overset{\ddot{S}i}{\triangle}} \xrightarrow{>395\,nm} \underset{(113)}{H-\overset{\ddot{S}i}{\underset{H}{\diagdown}}\diagup\overset{H}{\underset{H}{\diagdown}}} \xrightarrow{>310\,nm} \underset{(161)}{\overset{H}{\underset{H}{\diagdown}}Si=C=C\overset{H}{\underset{H}{\diagup}}} \quad (45)$$

$$\xrightarrow{254\,nm} \underset{(115)}{H_3Si-C\equiv C-H}$$

b. Silicon–carbon clusters. Further subjects to be discussed in this section are silicon–carbon clusters, since some exhibit silylenic structures. Besides the long known C_2Si[139], three other species consisting only of silicon and carbon were isolated in noble gas matrices. All are accessible by cocondensation of silicon and carbon atoms in suitable relative amounts. Firstly, CSi_2 (164) was examined by Margrave and coworkers[188] who assigned two absorptions (1188.9 and 658.2 cm^{-1}) to the species discussed. According to calculations, CSi_2 should not be a linear (162) but a bent molecule (164) with an angle of about 120° at the central carbon atom[189]. The system was reinvestigated later by Presilla-Márquez and Graham[185] who confirmed the assignment of the stronger absorption (1188.9 cm^{-1}), but showed that the less intense absorption is not due to CSi_2. Instead, they observed two additional weak vibrations at 1354.8 and 839.5 cm^{-1} that belong to this species.

Presilla-Márquez and coworkers also investigated the tetra-atomic silicon–carbon clusters C_2Si_2[186] and CSi_3[187]. According to calculations, a rhombic (167)[186] or a rhomboidal (166)[190] ground state geometry, respectively, is predicted for both. In the case of C_2Si_2 the vibrations of this species (167) are observed in solid argon along with others, that are tentatively assigned to the distorted trapezoidal isomer 165 and the linear compound 163. The latter is predicted to lie 8 kcal mol^{-1} above the ground state (167) whereas 165 has an intermediate position[186]. For CSi_3 only the most stable species (166) was observed in matrix-isolation experiments, and five of the six fundamental modes could be assigned[187]. Interestingly, one of these vibrations, at 658.2 cm^{-1}, was originally assigned to CSi_2 by Margrave and coworkers[188].

:Si═C═Si: :Si═C═C═Si:
(162)   (163)

(164)   (165)   (166)   (167)

C. Silicon Substituted Silylenes

1. Silylenes with one silyl substituent

Silicon substituted silylenes attract the chemists' interest for a special reason. According to *ab initio* calculations[85,137,176], substituents acting as σ-donors should induce a relatively large red-shift of the UV maximum or, in other words, the n(Si) → 3p(Si) transition energy should be relatively small. Therefore these species are potential candidates for the long-sought triplet ground state silylenes, especially if this electronic effect is supported by a steric one. Nevertheless, reports on the matrix isolation of silicon substituted silylenes are comparatively scarce.

West and coworkers reported the generation of several silylenes[169] by photolysis of the corresponding trisilanes, among them (trimethylsilyl)mesitylsilylene (168b). Surprisingly an absorption maximum at 368 nm is reported there for 168b, which would indicate an unexpected blue-shift compared to dimethyl- (453 nm) or dimesitylsilylene (577 nm). This result was reexamined twice in the following years. In 1993 Kira, Maruyama and Sakurai prepared several (trimethylsilyl)- and (trimethylgermyl)arylsilylenes from the corresponding trisilanes in glassy 3-methylpentane[173]. In this paper a maximum at 760 nm is given for compound 168b, being in much better agreement with the calculations[137] than the value reported first. However, this is the highest value feasible with aryl(trimethylsilyl)silylenes (168) since this system can avoid additional sterical pressure by a twist around the Si−C bond, i.e. the coplanarity between the p orbitals of the phenyl ring and the divalent silicon atom will not be maintained. So substitution of the *ortho* hydrogen atoms of the phenyl substituent in 168a (660 nm for the unsubstituted phenyl derivative) leads to a bathochromic shift of 100 nm (760 nm for the mesityl compound), whereas replacement of the mesityl groups in 2,6-position by the more bulky ethyl or isopropyl substituents does not result in a further bathochromic shift, as might be expected. Actually these compounds absorb at shorter wavelengths, namely both (168c and 168d) at 570 nm.

The formal substitution of the silicon atom in the trimethylsilyl group by a germanium atom yields silylenes that absorb approximately at the same wavelengths, but shifted hypsochromically by about 30 nm.

(a) Ar = Ph;
(b) Ar = Mes;

(c) Ar = 2,6-Et₂C₆H₃ (2,6-diethylphenyl);

(d) Ar = 2,4,6-triisopropylphenyl

Me₃Si–Si(Ar): (**168**)

In 1995, Conlin and coworkers also reinvestigated the aryl(trimethylsilyl)silylenes (**168**) in organic glasses[172]. The value reported for λ_{max} of the mesityl compound **168b**, namely 776 nm, is approximately the same as given in the paper by Kira and coworkers. Again, no compound with an even stronger bathochromic shift could be generated. In addition, the absorption at 372 nm appearing together with that at 776 nm upon photolysis of the corresponding trisilane is tentatively assigned to the corresponding disilene, resulting from the dimerization product of **168b**. This result might explain the erroneous early assignment[169] of the 368 nm absorption to the silylene. Trapping experiments with 2,3-dimethylbutadiene and triethylsilane were also performed, confirming the appearance of the assumed silylenes[172].

2. Silylenes with two silyl substituents

As pointed out above, silyl substitution is evidently a promising strategy on the way to triplet ground state silylenes. The next consequent step is the investigation of silylenes carrying two silyl substituents. The first experiments concerning this subject were performed by Apeloig and coworkers[171]. The photolysis of 1,2-disilacyclobutane **169** (equation 46) yielded a deep violet solution with an absorption maximum higher than 750 nm (the calculated value is 886 nm[171]). Trapping products of silylene **170** as well as of disilene **171** and butadiene were isolated, too. It was not reported if the observed silylene is a ground-state singlet or triplet.

$$(Me_3Si)_2Si\text{–}Si(SiMe_3)_2\text{-adamantyl} \xrightarrow[77 \text{ K}, 3\text{-MP}]{254 \text{ nm}} (Me_3Si)_2Si: \rightleftharpoons (Me_3Si)_2Si=Si(SiMe_3)_2$$

(**169**) → (**170**) (**171**) (46)

Even though no triplet ground-state silylene could be isolated so far, silyl or germyl substitution is a promising strategy for this challenging goal. It remains to be seen if additional substitution of the methyl substituents in bis(trimethylsilyl)silylene by more bulky groups will result in the formation of the desired compounds. Nevertheless, the replacement of one methyl group by a *t*-butyl or two by *i*-propyl groups does not seem to be sufficient for the generation of triplet ground-state silylenes, since trapping products of the intermediates generated upon photolysis of the appropriate cyclotrisilanes **172a** and **172b** in solution in the presence of *cis*- or *trans*-butene indicate singlet species[191]. In matrix-isolation experiments the precursors turned out to be photostable[191].

$$R_2Si\begin{array}{c}SiR_2\\|\\SiR_2\end{array}$$

(a) R = Me(*i*-Pr)$_2$Si
(b) R = (*t*-Bu)Me$_2$Si

(**172**)

D. Nitrogen Substituted Silylenes

1. Silylenes by photolysis of oligosilanes

The first investigations dealing with nitrogen substituted silylenes were performed by West and coworkers in glassy 3-methylpentane with the corresponding trisilanes as precursors[169]. A strong hypsochromic shift is predicted for silylenes with substituents acting as n-donors[137,171] and this was actually found in the examples examined. Compared with mesitylmethylsilylene (λ_{max} = 496 nm) the nitrogen substituted species **173** (405 nm) and **174** (404 nm) exhibit a blue-shift of about 100 nm[169].

Mes\Si:/Me$_2$N   Mes\Si:/(Me$_3$Si)$_2$N

(**173**)   (**174**)

2. Silylenes by photolysis of diazides

Maier and coworkers were able to isolate the parent aminosilylene (**85**) in solid argon[96]. As already described in Section IV.A (equation 22), this species, representing the global minimum of the H$_3$NSi-hypersurface, was accessible by irradiation of silylazide and it was characterized by its vibrational and electronic spectrum. The latter shows three absorptions at 208, 220 and 348 nm. The structure was also confirmed by deuterium labelling.

Veith and coworkers examined the photochemistry of the highly substituted diazidosilane **175**[192]. The actual synthetic target was the preparation of a stable silylene like **176**. This silylene is expected to be isolable, since the homologous germanium (**177a**), tin (**177b**) and lead compounds (**177c**) are accessible species and stable at room temperature[193,194]. The photolysis of **175** in benzene yielded only polymeric products in addition to the expected amount of nitrogen. However, if the photolysis of **175** is not carried out in benzene at room temperature but in solid argon, the expected silylene is formed (equation 47), as could be confirmed by comparison of the vibrational spectra with that of the homologues **177a** and **177b**.

(a) M = Ge
(b) M = Sn   (47)
(c) M = Pb

(**175**)   (**176**)   (**177**)

3. Silylenes by cocondensation reactions

a. CHNSi isomers. The cocondensation reaction of silicon atoms and hydrogen cyanide was examined by Maier and coworkers[144]. In contrast to the reaction of ethene or acetylene with silicon atoms, not the π-system serves as a Lewis base, but the nitrogen lone pair, resulting in the formation of **178** as the primary product (equation 48). Irradiation of **178** yields, depending on the wavelength, either the azasilacyclopropenylidene **179** or the isocyanosilylene **181**, the latter being convertible to a species that most likely is the SiNC radical (**180**). The identity of all CHNSi isomers was confirmed by their IR spectra[144].

$$\text{Si:} + \text{HCN} \longrightarrow \text{H}-\overset{..}{\text{C}}=\overset{+}{\text{N}}=\text{Si:} \xrightarrow{>570 \text{ nm}} \underset{\text{H}}{\overset{\overset{..}{\text{Si}}}{\triangle}}_{\text{N}} \tag{48}$$

$$\begin{array}{c} (178) \\ \downarrow 366 \text{ nm} \\ \text{H}\cdot + \cdot\overset{..}{\text{Si}}-\overset{+}{\text{N}}\equiv\overset{-}{\text{C}}\text{:} \underset{254 \text{ nm}}{\overset{366 \text{ nm}}{\rightleftarrows}} \underset{\text{H}}{\overset{..}{\text{Si}}}-\overset{+}{\text{N}}\equiv\overset{-}{\text{C}}\text{:} \\ (180) \quad\quad\quad\quad (181) \end{array}$$

b. SiN$_2$ isomers. The hydrogen cyanide used in the reaction described above might be seen as an acetylene with one CH group replaced by the isoelectronic nitrogen atom. So the next consequent step is the replacement of the remaining CH group, leading to molecular nitrogen. In our cocondensation experiments of silicon atoms and N_2[73a] the already known linear :SiNN (**182**)[195] was the primary product (equation 49). As in the case of hydrogen cyanide, the nitrogen lone pair serves as a Lewis base and not the π-system. Upon irradiation cyclic SiN$_2$ (**183**) was formed. It is necessary to use nitrogen as the matrix material in these experiments (mixtures of argon and nitrogen fail). Again, both SiN$_2$ isomers were identified by their IR spectra[73]. The same sequence **182** → **183** could be observed when either diazidosilane or tetraazidosilane was irradiated in a nitrogen matrix[73b].

$$\text{Si:} + \text{N}_2 \longrightarrow \text{:Si}=\overset{+}{\text{N}}=\overset{-}{\text{N}} \xrightarrow{313 \text{ nm}} \overset{\overset{..}{\text{Si}}}{\underset{\text{N}=\text{N}}{\triangle}} \tag{49}$$

$$(182) \quad\quad\quad (183)$$

In all, the cocondensation with silicon atoms is a useful tool for the generation of silylenes. Sometimes, the same species are obtained as with 'chemical' precursors and sometimes new, otherwise not accessible compounds can be observed. So both methods nicely complement one another.

E. Oxygen Substituted Silylenes

1. Silylenes by photolysis of oligosilanes

The first isolation of an oxygen substituted silylene was reported by West and coworkers[169]. The generation of (*t*-butoxy)mesitylsilylene, (*t*-BuO)MesSi:, upon photolysis of the corresponding trisilane in 3-methylpentane was described[169]. As expected for a n-donor substituted silylene[137], the UV absorption (396 nm) is blue-shifted by more than 100 nm compared with (*t*-butyl)mesitylsilylene (505 nm). In a further study by the same group the UV absorptions of five additional silylenes of the general structures RO(Mes)Si: were reported[168]. Remarkably, the silylenes with bulky substituents at the oxygen atom (R = mesityl and 2,6-diisopropylphenyl) show UV spectra that vary with matrix viscosity. Conformational changes are proposed as explanation.

2. Silylenes by pyrolysis of suitable precursors

Maier and coworkers were able to generate silylenes that carry two oxygen substituents at the divalent silicon atom[107]. The dimethoxy compound **185a** was accessible from two totally different precursors upon pyrolysis, namely the silepine derivative **184a** and the disilane **186a** (equation 50). The structural assignment was confirmed by IR and UV spectra, the latter showing an absorption at 243 nm, which was in good agreement with the expected value[107]. In addition, a weak band was observed at 340 nm that was assigned to methoxysilylene (**187**). Its formation is plausible if an intramolecular CH insertion of the initially formed dimethoxysilylene (**185a**) with subsequent formaldehyde elimination is considered. This interpretation is supported by the appearance of the aldehyde in the matrix. Methoxymethylsilylene (**185b**) is also accessible from the analogous precursors **184b** and **186b**. Again, **185b** was characterized by its vibrational and electronic spectra, the latter showing an absorption at 355 nm. Furthermore, the pyrolysis of $Si_2(Oi\text{-}Pr)_6$ yielded the expected diisopropoxysilylene, showing UV absorption at 247 nm.

3. Silylenes by cocondensation reactions

The parent system, (HO)HSi:, was investigated by Margrave and coworkers[196]. In the cocondensation of silicon atoms and water, a complex was initially formed that spontaneously rearranged to hydroxysilylene. Photolysis led to loss of the hydrogen atoms and thus SiO was formed.

F. Halogen Substituted Silylenes

The members of this 'inorganic' subclass of silylenes were matrix-isolated much earlier than their 'organic' counterparts. Dihalosilylenes are accessible in several ways.

The dichloro compound was first generated upon photolysis of dichlorosilane by Milligan and Jacox[197]. In addition, all dihalosilylenes except of the diiodo compound are accessible from the reaction of the respective silicon tetrahalide with silicon at high temperatures[198,199]. A more recent approach to a halogen substituted silylene is the cocondensation of hydrogen fluoride with silicon atoms, yielding fluorosilylene[200].

In 1989, the reaction of difluoro- and dichlorosilylene (**188**) with molecular oxygen was investigated by Sander and coworkers[201]. The thermolysis of the corresponding hexahalodisilanes was used as access to the corresponding silylenes, which were isolated in pure oxygen or oxygen doped argon matrices. In contrast to dimethylsilylene **118**[108], the reaction of **188** with O_2 occurred only upon irradiation. The cyclic structure of compounds **189** was confirmed by isotope labelling (equation 51).

$$X_2Si: + O_2 \xrightarrow{h\nu} X_2Si\underset{O}{\overset{O}{\diagdown\diagup}}\Big|$$

(**188**) (**189**) (51)

X = F, Cl

IX. REFERENCES

1. G. Raabe and J. Michl, in *The Chemistry of Organic Silicon Compounds* (Eds. S. Patai and Z. Rappoport), Part 2, Wiley, Chichester, 1989.
2. P. Jutzi, *Chem. Unserer Zeit*, **15**, 149 (1981).
3. I. Pierre, *Justus Liebigs Ann. Chem.*, **69**, 73 (1849).
4. W. Schlenk and J. Renning, *Justus Liebigs Ann. Chem.*, **394**, 221(1912).
5. G. J. D. Peddle, D. N. Roark, A. M. Good and S. G. McGeachin, *J. Am. Chem. Soc.*, **91**, 2807 (1969).
6. N. S. Nametkin, V. M. Vdovin, L. E. Gusel'nikov and V. I. Zav'yalov, *Izv. Akad. Nauk SSSR, Ser. Khim.*, 584 (1966) *Chem. Absr.*, **65**, 5478e (1966).
7. L. E. Gusel'nikov and M. C. Flowers, *J. Chem. Soc., Chem. Commun.*, 864 (1967).
8. L. E. Gusel'nikov and M. C. Flowers, *J. Chem. Soc. (B)*, 419 (1968).
9. For a review about this topic, see M. J. Almond and A. J. Downs, in *Advances in Spectroscopy* (Vol. 17): *Spectroscopy of Matrix Isolated Species* (Eds. R. J. H. Clark and R. E. Hester), Wiley, Chichester, 1989.
10. T. J. Drahnak, J. Michl and R. West, *J. Am. Chem. Soc.*, **101**, 5427 (1979).
11. M. Denk, R. Lennon, R. Hayashi, R. West, A. V. Belyakov, H. P. Verne, A. Haaland, M. Wagner and N. Metzler, *J. Am. Chem. Soc.*, **116**, 2691 (1994).
12. G. Maier and J. Glatthaar, *Angew. Chem.*, **106**, 486 (1994); *Angew. Chem., Int. Ed. Engl.*, **33**, 473 (1994).
13. P. P. Gaspar, in *Reactive Intermediates* (Eds. M. Jones Jr. and R. A. Moss), Vol. 3, Wiley, New York, 1985, p. 333.
14. A. G. Brook and K. M. Baines, *Adv. Organomet. Chem.*, **25**, 1 (1986).
15. A. G. Brook, F. Abdesaken, B. Gutekunst, G. Gutekunst and R. K. Kallury, *J. Chem. Soc., Chem. Commun.*, 191 (1981).
16. G. E. Miracle, J. L. Ball, D. R. Powell and R. West, *J. Am. Chem. Soc.*, **115**, 11598 (1993).
17. G. Maier, G. Mihm and H. P. Reisenauer, *Angew. Chem.*, **93**, 615 (1981).
18. G. Maier and J. Glatthaar, in *Organosilicon Chemistry—From Molecules to Materials* (Eds. N. Auner and J. Weis), VCH, Weinheim, 1994, p. 131.
19. A. Sekiguchi and W. Ando, *Chem. Lett.*, 2025 (1986).
20. G. Maier, G. Mihm and H. P. Reisenauer, *Angew. Chem., Int. Ed. Engl.*, **20**, 597 (1981).
21. A. G. Brook, J. W. Harris, J. Lennon and M. El Sheikh, *J. Am. Chem. Soc.*, **101**, 83 (1979).
22. A. G. Brook, S. C. Nyburg, F. Abdesaken, B. Gutekunst, G. Gutekunst, R. K. M. R. Kallury, Y. C. Poon, Y.-M. Chang and W. Wong-Ng, *J. Am. Chem. Soc.*, **104**, 5667 (1982).
23. M. Trommer, W. Sander, C.-H. Ottoson and D. Cremer, *Angew. Chem.*, **107**, 999 (1995).
24. J. M. Duff and A. G. Brook, *Can. J. Chem.*, **51**, 2869 (1973).
25. R. A. Bourque, P. D. Davis and J. C. Dalton, *J. Am. Chem. Soc.*, **103**, 697 (1981).

26. H. M. Perrin, W. R. White and M. S. Platz, *Tetrahedron Lett.*, **32**, 4443 (1991).
27. M. Guth, Doctoral Dissertation, Universität Bochum, 1993.
28. M. Trommer, W. Sander and A. Patyk, *J. Am. Chem. Soc.*, **115**, 11775 (1993).
29. H. B. Schlegel, B. Coleman and M. Jones Jr., *J. Am. Chem. Soc.*, **100**, 6499 (1978).
30. A. J. Ashe III., R. R. Sharp and J. W. Tolan, *J. Am. Chem. Soc.*, **98**, 5451 (1976).
31. T. J. Barton and D. Banasiak, *J. Am. Chem. Soc.*, **99**, 5199 (1977).
32. T. J. Barton and G. T. Burns, *J. Am. Chem. Soc.*, **100**, 5246 (1978).
33. C. L. Kreil, O. L. Chapman, G. T. Burns and T. J. Barton, *J. Am. Chem. Soc.*, **102**, 841 (1980).
34. M. J. S. Dewar, D. H. Lo and C. A. Ramsden, *J. Am. Chem. Soc.*, **97**, 1311 (1975).
35. G. Maier, G. Mihm and H. P. Reisenauer, *Angew. Chem.*, **92**, 58 (1980).
36. G. Maier, G. Mihm and H. P. Reisenauer, *Chem. Ber.*, **115**, 801 (1982).
37. H. Bock, B. Solouki and G. Maier, *J. Organomet. Chem.*, **271**, 145 (1984).
38. G. Maier, G. Mihm, R. O. W. Baumgärtner and H. P. Reisenauer, *Chem. Ber.*, **117**, 2337 (1984).
39. J. N. Murrell, H. W. Kroto and M. F. Guest, *J. Chem. Soc., Chem. Commun.*, 619 (1977).
40. M. S. Gordon and J. A. Pople, *J. Am. Chem. Soc.*, **103**, 2945 (1981).
41. G. Maier, K. Schöttler and H. P. Reisenauer, *Tetrahedron Lett.*, **26**, 4079 (1985).
42. For a theoretical study see: K. K. Baldridge and M. S. Gordon, *J. Organomet. Chem.*, **271**, 369 (1984).
43. For theoretical studies, see A. Sax and R. Janoschek, *Angew. Chem., Int. Ed. Engl.*, **25**, 651 (1986); Z. Slanina, *J. Mol. Struct. (Theochem.)*, **65**, 143 (1990).
44. J. Chandrasekhar and P. v. R. Schleyer, *J. Organomet. Chem.*, **298**, 51 (1985).
45. K. K. Baldridge and M. S. Gordon, *Organometallics*, **7**, 144 (1988).
46. Y. van den Winkel, B. L. M. van Baar, F. Bickelhaupt, W. Kulik, C. Sierakowski and G. Maier, *Chem. Ber.*, **124**, 185 (1991).
47. H. Hiratsuka, M. Tanaka, T. Okutsu, M. Oba and K. Nishiyama, *J. Chem. Soc., Chem. Commun.*, 215 (1995).
48. See, for example, Reference 46 and references cited therein.
49. Y. van den Winkel, O. S. Akkerman and F. Bickelhaupt, *Main Group Metal Chem.*, **11**, 91 (1988).
50. P. Jutzi, M. Meyer, H. P. Reisenauer and G. Maier, *Chem. Ber.*, **122**, 1227 (1989).
51. G. Märkl and W. Schlosser, *Angew. Chem.*, **100**, 1009 (1988).
52. G. Maier, *Angew. Chem., Int. Ed. Engl.*, **27**, 309 (1988).
53. M. S. Gordon, *J. Chem. Soc., Chem. Commun.*, 1131 (1980).
54. G. W. Schriver, M. J. Fink and M. S. Gordon, *Organometallics*, **6**, 1977 (1987).
55. M. J. Fink and D. Puranik, *Organometallics*, **6**, 1809 (1987).
56. T. Bally and S. Masamune, *Tetrahedron*, **36**, 343 (1980).
57. G. Maier, *Pure Appl. Chem.*, **58**, 95 (1986).
58. U.-J. Vogelbacher, M. Regitz and R. Mynott, *Angew. Chem., Int. Ed. Engl.*, **25**, 842 (1986).
59. G. Maier and A. Kratt, unpublished results.
60. P Lingelbach, Doctoral Dissertation, Universität Giessen, Germany, 1989.
61. A. Schick, Doctoral Dissertation, Universität Giessen, Germany, 1992.
62. P. R. Jones, T. F. Bates, A. F. Cowley and A. M. Arif, *J. Am. Chem. Soc.*, **108**, 3123 (1986).
63. P. R. Jones and T. F. Bates, *J. Am. Chem. Soc.*, **109**, 913 (1987).
64. D. B. Puranik, M. P. Johnson and M. J. Fink, *J. Chem. Soc., Chem. Commun.*, 706 (1989).
65. J. R. Gee, W. A. Howard, G. L. McPherson and M. J. Fink, *J. Am. Chem. Soc.*, **113**, 5461 (1991).
66. A. G. Brook and K. M. Baines, *Adv. Organomet. Chem.*, **25**, 1 (1986).
67. K. M. Baines and A. G. Brook, *Organometallics*, **6**, 692 (1987).
68. A. J. Vanderwielen, M. A. Ring and H. E. O'Neal, *J. Am. Chem. Soc.*, **97**, 993 (1975).
69. A. Meudt, Doctoral Dissertation, Universität Giessen, Germany, 1996.
70. H. Büttner, Doctoral Dissertation, Universität Giessen, Germany, 1993.
71. U. Wessolek-Kraus, Doctoral Dissertation, Universität Giessen, Germany, 1988.
72. D. Littmann, Doctoral Dissertation, Universität Giessen, Germany, 1985.
73. (a) G. Maier, H. P. Reisenauer and H. Egenolf, unpublished results.
    (b) G. Maier and J. Glatthaur, unpublished results.
74. A. Thoma, B. E. Wurfel, R. Schlachter, G. M. Lask and V. E. Bondybey, *J. Phys. Chem.*, **96**, 7231 (1992).
75. H. Shizuka, K. Murata, Y. Arai, K. Tonokura, H. Tanaka, H. Matsumoto, Y. Nagai, G. Gillette and R. West, *J. Chem. Soc. Faraday Trans. 1*, **85**, 2369 (1989).

76. B. J. Helmer and R. West, *Organometallics*, **1**, 1458 (1982).
77. C. W. Carlson and R. West, *Organometallics*, **2**, 1792 (1983).
78. H. Watanabe, Y. Kongo and Y. Nagai, *J. Chem. Soc., Chem. Commun.*, 66 (1984).
79. H. Watanabe, T. Okawa, M. Kato and Y. Nagai, *J. Chem. Soc., Chem. Commun.*, 781 (1983)
80. H. Watanabe, Y. Kongo, M. Kato, H. Kuwabara, T. Okawa and Y. Nagai, *Bull. Chem. Soc. Jpn.*, **57**, 3019 (1984).
81. M. Kira, T. Maruyama, C. Kabuto, K. Ebata and H. Sakurai, *Angew. Chem.*, **106**, 1575 (1994).
82. S. Masamune, Y. Eriyama and T. Kawase, *Angew. Chem.*, **99**, 601 (1987); *Angew. Chem., Int. Ed. Engl.*, **26**, 584 (1987).
83. M. Karni and Y. Apeloig, *J. Am. Chem. Soc.*, **112**, 8589 (1990).
84. C. Liang and L. C. Allen, *J. Am. Chem. Soc.*, **112**, 1039 (1990).
85. R. S. Grev, H. F. Schaefer III and P. P. Gaspar, *J. Am. Chem. Soc.*, **113**, 5638 (1991).
86. M. Bogey, H. Bolvin, C. Demuynck and J. L. Destombes, *Phys. Rev. Lett.*, **66**, 413 (1991).
87. M. Cordonnier, M. Bogey, C. Demuynck and J.-L. Destombes, *J. Chem. Phys.*, **97**, 7984 (1992).
88. N. Wiberg, K. Schurz, G. Reber and G. Müller, *J. Chem. Soc., Chem. Commun.*, 591 (1986).
89. A. Sekiguchi, W. Ando and K. Honda, *Chem. Lett.*, 1029 (1986).
90. S. S. Zigler, R. West and J. Michl, *Chem. Lett.*, 1024 (1986).
91. R. West, S. S. Zigler, J. Michl and G. Gross, presented at the 19th Organosilicon Symposium, Louisiana State University, Baton Rouge, LA, April 26-27, 1985.
92. H. Bock and R. Dammel, *Angew. Chem.*, **97**, 128 (1985); *Angew. Chem., Int. Ed. Engl.*, **24**, 111 (1985).
93. G. Gross, J. Michl and R. West, presented at the 19th Organosilicon Symposium, Lousiana State University, Baton Rouge, LA, April 26-27, 1985.
94. A. G. MacDiarmid, *J. Inorg. Nucl. Chem*, **2**, 88 (1956).
95. J. F. Ogilvie and S. Cradock, *J. Chem. Soc., Chem. Commun.*, 364 (1966).
96. G. Maier, J. Glatthaar and H. P. Reisenauer, *Chem. Ber.*, **122**, 2403 (1989).
97. R. S. Mulliken, *Phys. Rev.*, **26**, 319 (1925).
98. N. T. Truong and M. S. Gordon, *J. Am. Chem. Soc.*, **108**, 1775 (1986).
99. S. S. Zigler, K. M. Welsh and R. West, *J. Am. Chem. Soc.*, **109**, 4392 (1987).
100. K. M. Welsh, J. Michl and R. West, *J. Am. Chem. Soc.*, **110**, 6689 (1988).
101. R. Whithnall and L. Andrews, *J. Phys. Chem.*, **92**, 594 (1988).
102. R. Whithnall and L. Andrews, *J. Am. Chem. Soc.*, **108**, 8118 (1986).
103. R. Whithnall and L. Andrews, *J. Phys. Chem.*, **89**, 3261 (1985).
104. V. N. Khabashesku, Z. A. Kerzina, E. G. Baskir, A. K. Maltsev and O. M. Nefedov, *J. Organomet. Chem.*, **347**, 277 (1988).
105. V. N. Khabashesku, Z. A. Kerzina, A. K. Maltsev and O. M. Nefedov, *J. Organomet. Chem.*, **364**, 301 (1989).
106. V. N. Khabashesku, Z. A. Kerzina and O. M. Nefedov, *Izv. Akad. Nauk SSSR, Ser. Khim.*, **9**, 2187 (1988).
107. G. Maier, H. P. Reisenauer, K. Schöttler and U. Wessolek-Kraus, *J. Organomet. Chem.*, **366**, 25 (1989).
108. A. Patyk, W. Sander, J. Gauss and D. Cremer, *Angew. Chem.*, **101**, 920 (1989); *Angew. Chem., Int. Ed. Engl.*, **28**, 898 (1989).
109. H. Schnöckel, *Angew. Chem.*, **90**, 638 (1978); *Angew. Chem., Int. Ed. Engl.*, **17**, 616 (1978).
110. J. S. Anderson and J. S. Ogden, *J. Chem. Phys.*, **51**, 4189 (1969).
111. J. W. Hastie, R. H. Hauge and J. L. Margrave, *Inorg. Chim. Acta*, **3**, 601 (1969).
112. H. Schnöckel, *Z. Anorg. Allg. Chem.*, **460**, 37 (1980).
113. H. Schnöckel, T. Mehner, H. S. Plitt and S. Schunck, *J. Am. Chem. Soc.*, **111**, 4578 (1989).
114. H. Schnöckel and R. Köppe, in *Organosilicon Chemistry—From Molecules to Materials* (Eds. N. Auner and J. Weis), VCH, Weinheim, 1994, p. 147.
115. R. Köppe and H. Schnöckel, *Heteroatom Chem.*, **3**, 329 (1992).
116. T. Mehner, H. Schnöckel, M. J. Almond and A. J. Downs, *J. Chem. Soc., Chem. Commun.*, 117 (1988).
117. J. H. B. Chenier, J. A. Howard, H. A. Joly, B. Mile and P. L. Timms, *J. Chem. Soc., Chem. Commun.*, 581 (1990).
118. T. Mehner, R. Köppe and H. Schnöckel, *Angew. Chem.*, **104**, 653 (1992); *Angew. Chem., Int. Ed. Engl.*, **31**, 638 (1992).
119. R. M. Atkins and P. L. Timms, *Spectrochim. Acta*, **33A**, 853 (1977).

120. H. Schnöckel and R. Köppe, *J. Am. Chem. Soc.*, **111**, 4583 (1989).
121. H. Schnöckel, *Angew. Chem.*, **92**, 310 (1980); *Angew. Chem., Int. Ed. Engl.*, **19**, 323 (1980).
122. H. Suzuki, N. Tokitoh, S. Nagase and R. Okazaki, *J. Am. Chem. Soc.*, **116**, 11578 (1994).
123. H. Schnöckel, H. J. Göcke and R. Köppe, *Z. Anorg. Allg. Chem.*, **578**, 159 (1989).
124. R. Köppe and H. Schnöckel, *Z. Anorg. Allg. Chem.*, **607**, 41 (1992).
125. J. A. Howard, R. Jones, J. S. Tse, M. Tomietto, P. L. Timms and A. J. Seeley, *J. Chem. Phys.*, **96**, 9144 (1992).
126. R. Köppe, H. Schnöckel, F. X. Gadea, J. C. Barthelat and C. Jouany, *Heteroatom Chem.*, **3**, 333 (1992).
127. G. Maier, H. P. Reisenauer and H. Pacl, *Angew. Chem.*, **106**, 1347 (1994); *Angew. Chem., Int. Ed. Engl.*, **33**, 1248 (1994).
128. G. Maier, H. Pacl, H. P. Reisenauer, A. Meudt and R. Janoschek, *J. Am. Chem. Soc.*, **117**, 12712 (1995).
129. G. Maier, H. P. Reisenauer and H. Pacl, in *Organosilicon Chemistry II — From Molecules to Materials* (Eds. N. Auner and J. Weis), VCH, Weinheim, 1996, p. 303.
130. G. Frenking, R. B. Remington and H. F. Schaefer III, *J. Am. Chem. Soc.*, **108**, 2169 (1986).
131. G. Vacek, B. T. Colgrove and H. F. Schaefer III, *J. Am. Chem. Soc.*, **113**, 3192 (1991).
132. D. L. Cooper, *Astrophys. J.*, **354**, 229 (1990).
133. M.-D. Su, R. D. Amos and N. C. Handy, *J. Am. Chem. Soc.*, **112**, 1499 (1990).
134. R. Srinivas, D. Sülzle, T. Weiske and H. Schwarz, *Int. J. Mass Spectrom. Ion Processes*, **107**, 369 (1991).
135. M. Izuha, S. Yamamoto and S. Saito, *Can. J. Phys.*, **72**, 1206 (1994).
136. H. Pacl, Doctoral Dissertation, Universität Giessen, Germany, 1995.
137. Y. Apeloig, M. Karni, R. West and K. Welsh, *J. Am. Chem. Soc.*, **116**, 9719 (1994).
138. B. Kleman, *Astrophys. J.*, **33**, 473 (1956).
139. W. Weltner Jr. and D. McLeod Jr., *J. Chem. Phys.*, **41**, 235 (1964).
140. R. A. Shepherd and W. R. M. Graham, *J. Chem. Phys.*, **88**, 3399 (1988).
141. R. A. Shepherd and W. R. M. Graham, *J. Chem. Phys.*, **82**, 4788 (1985).
142. J. D. Presilla-Márquez, W. R. M. Graham and R. A. Shepherd, *J. Chem. Phys.*, **93**, 5424 (1990).
143. W. Sander, *Angew. Chem.*, **106**, 1522 (1994); *Angew. Chem., Int. Ed. Engl.*, **33**, 1455 (1994).
144. G. Maier, H. P. Reisenauer and H. Egenolf, presented at the III. Münchener Silicontage, Ludwig-Maximilians-Universität, München, April 1–2, 1996; in *Organosilicon chemistry III — From Molecules to Materials* (Eds. N. Auner and J. Weis), VCH, Weinheim, 1997, to appear.
145. R. T. Conlin and P. P. Gaspar, *J. Am. Chem. Soc.*, **98**, 3715 (1976).
146. M. Trommer, W. Sander and C. Marquard, *Angew. Chem.*, **106**, 816 (1994); *Angew. Chem., Int. Ed. Engl.*, **33**, 766 (1994).
147. G. Maier, H. Pacl and H. P. Reisenauer, *Angew. Chem.*, **107**, 1627 (1995); *Angew. Chem., Int. Ed. Engl.*, **34**, 1439 (1995).
148. E. A. V. Ebsworth and S. G. Frankiss, *J. Chem. Soc.*, 661 (1963).
149. B. T. Luke, J. A. Pople, M.-B. Krogh-Jespersen, Y. Apeloig, J. Chandrashekar, M. J. Karni and P. v. R. Schleyer, *J. Am. Chem. Soc.*, **108**, 270 (1986).
150. R. S. Grev, G. E. Scuseria, A. C. Scheiner, H. F. Schaefer III and M. S. Gordon, *J. Am. Chem. Soc.*, **110**, 7337 (1988).
151. J. A. Boatz and M. S. Gordon, *J. Phys. Chem.*, **94**, 7331 (1990).
152. K. Balasubramanian and A. D. McLean, *J. Chem. Phys.*, **85**, 5117 (1986).
153. H.-J. Werner and E.-A. Reinsch, *J. Chem. Phys.*, **76**, 3144 (1982).
154. J. Berkowitz, J. P. Greene, H. Cho and B. Ruscic, *J. Chem. Phys.*, **86**, 1235 (1987).
155. T. J. Sears and P. R. Bunker, *J. Chem. Phys.*, **79**, 5265 (1983).
156. M. Denk, R. West, R. Hayashi, Y. Apeloig, R. Pauncz and M. Karni, in *Organosilicon Chemistry II — From Molecules to Materials* (Eds. N. Auner and J. Weis), VCH, Weinheim, 1996, pp. 251–261.
157. B. T. Luke, J. A. Pople, M.-B. Krogh-Jespersen, Y. Apeloig, J. Chandrashekar and P.v.R. Schleyer, *J. Am. Chem. Soc.*, **108**, 260 (1986).
158. R. Janoschek, *Chem. Unserer Zeit*, **21**, 128 (1988).
159. D. H. Pae, M. Xiao, M. Y. Chiang and P. P. Gaspar, *J. Am. Chem. Soc.*, **113**, 1281 (1991).
160. D. E. Milligan and M. E. Jacox, *J. Chem. Phys.*, **52**, 2594 (1970).
161. L. Fredin, R. H. Hauge, Z. H. Kafafi and J. L. Margrave, *J. Chem. Phys.*, **82**, 3542 (1985).
162. G. Maier, G. Mihm, H. P. Reisenauer and D. Littmann, *Chem. Ber.*, **117**, 2369 (1984).

163. H. Vancik, G. Raabe, M. J. Michalczyk, R. West and J. Michl, *J. Am. Chem. Soc.*, **107**, 4097 (1985).
164. G. Raabe, H. Vancik, R. West and J. Michl, *J. Am. Chem. Soc.*, **108**, 671 (1986).
165. C. A. Arrington, J. T. Petty, S. E. Payne and W. C. K. Haskins, *J. Am. Chem. Soc.*, **110**, 6240 (1988).
166. M.-A. Pearsall and R. West, *J. Am. Chem. Soc.*, **110**, 7228 (1988).
167. T. P. Hamilton and H. F. Schaefer III, *J. Chem. Phys.*, **90**, 1031 (1989).
168. G. R. Gillette, G. Noren and R. West, *Organometallics*, **9**, 2925 (1990).
169. M. J. Michalczyk, M. J. Fink, D. J. De Young, C. W. Carlson, K. M. Welsh, R. West and J. Michl, *Silicon, Germanium, Tin Lead Compd.*, **9**, 75 (1986).
170. A. Sekiguchi, K. Hagiwara and W. Ando, *Chem. Lett.*, 209 (1987).
171. Y. Apeloig, M. Karni and T. Müller, in *Organosilicon Chemistry II—From Molecules to Materials* (Eds. N. Auner and J. Weis), VCH, Weinheim, 1996, pp. 263–288.
172. S. G. Bott, P. Marshall, P. E. Wagenseller, Y. Wang and R. T. Conlin, *J. Organomet. Chem.*, **499**, 11 (1995).
173. M. Kira, T. Maruyama and H. Sakurai, *Chem. Lett.*, 1345 (1993).
174. M. Kira, T. Maruyama and H. Sakurai, *Tetrahedron Lett.*, **33**, 243 (1992).
175. T. J. Drahnak, J. Michl and R. West, *J. Am. Chem. Soc.*, **103**, 1845 (1981).
176. Y. Apeloig and M. Karni, *J. Chem. Soc., Chem. Commun.*, 1048 (1985).
177. G. R. Gilette, G. H. Noren and R. West, *Organometallics*, **8**, 487 (1989).
178. T. Akasaka, S. Nagase, A. Yabe and W. Ando, *J. Am. Chem. Soc.*, **110**, 6270 (1988).
179. V. N. Khabashesku, V. Balaji, S. E. Boganov, O. M. Nefedov and J. Michl, *J. Am. Chem. Soc.*, **116**, 320 (1994)
180. G. Maier, K. Schöttler and H. P. Reisenauer, *Tetrahedron Lett.*, **26**, 4079 (1985).
181. T. M. Gentle and E. L. Muetterties, *J. Am. Chem. Soc.*, **105**, 304 (1983).
182. G. Maier, H. P. Reisenauer, J. Jung, A. Meudt and H. Pacl, presented at the III. Münchener Silicontage, Ludwig-Maximilians-Universität, München, April 1–2, 1996; in *Organosilicon chemistry III—From Molecules to Materials* (Eds. N. Auner and J. Weis), VCH, Weinheim, 1997, to appear.
183. J. Jung, Doctoral Dissertation, Universität Giessen, Germany, 1996.
184. T. A. Holme, M. S. Gordon, S. Yabushita and M. W. Schmidt, *Organomet.*, **3**, 583 (1984).
185. J. D. Presilla-Márquez and W. R. M. Graham, *J. Chem. Phys.*, **95**, 5612 (1991).
186. J. D. Presilla-Márquez, S. C. Gay, C. M. L. Rittby and W. R. M. Graham, *J. Chem. Phys.*, **102**, 6354 (1995).
187. J. D. Presilla-Márquez and W. R. M. Graham, *J. Chem. Phys.*, **96**, 6509 (1992).
188. Z. H. Kafafi, R. H. Hauge, L. Fredin and J. L. Margrave, *J. Phys. Chem.*, **87**, 797 (1983).
189. C. M. L. Rittby, *J. Chem. Phys.*, **95**, 5609 (1991).
190. C. M. L. Rittby, *J. Chem. Phys.*, **96**, 6768, (1992).
191. M. Kira and H. Sakurai, *Kagaku (Kyoto)*, **49**, 876 (1994) *Chem. Abstr.*, **122**, 56065e (1995).
192. M. Veith, E. Werle, R. Lisowsky, R. Köppe and H. Schnöckel, *Chem. Ber.*, **125**, 1375 (1992).
193. M. Veith, M. Grosser and V. Huch, *Z. Anorg. Allg. Chem.*, **513**, 89 (1984).
194. M. Veith, *Angew. Chem.*, **87**, 287 (1975).
195. R. R. Lembke, R. F. Ferrante and W. Weltner Jr., *J. Am. Chem. Soc.*, **99**, 416 (1977).
196. Z. K. Ismail, R. H. Hauge, L. Fredin, J. W. Kauffman and J. L. Margrave, *J. Chem. Phys.*, **77**, 1617 (1982).
197. D. E. Milligan and M. E. Jacox, *J. Chem. Phys*, **49**, 1938 (1968).
198. J. W. Hastie, R. H. Hauge and J. L. Margrave, *J. Am. Chem. Soc.*, **91**, 2536 (1969).
199. G. Maass, R. H. Hauge and J. L. Margrave, *Z. Anorg. Allg. Chem.*, **392**, 295 (1972).
200. Z. K. Ismail, L. Fredin, R. H. Hauge and J. L. Margrave, *J. Chem. Phys.*, **77**, 1626 (1982).
201. A. Patyk, W. Sander, J. Gauss and D. Cremer, *Chem. Ber.*, **123**, 89 (1990).

CHAPTER 20

Electrochemistry of organosilicon compounds

TOSHIO FUCHIGAMI

Department of Electrochemistry, Tokyo Institute of Technology, Midori-ku, Yokohama 226, Japan
Fax: 81-45-921-1089; e-mail: fuchi@echem.titech.ac.jp

I. INTRODUCTION	1188
II. ANODIC OXIDATION OF ORGANOSILICON COMPOUNDS	1188
A. Alkylsilanes	1188
B. Arylsilanes	1189
C. Benzylsilanes, Allylsilanes and Related Compounds	1190
D. Organosilicon Compounds Bearing Heteroatoms	1196
1. Organosilicon compounds bearing sulfur	1197
2. Organosilicon compounds bearing nitrogen	1199
3. Organosilicon compounds bearing phosphorus	1201
4. Organosilicon compounds bearing oxygen	1201
E. Acylsilanes	1203
F. Silyl Enol Ethers and Related Compounds	1204
G. Carboxylic Acid Bearing a β-Trimethylsilyl Group	1206
H. Organosilicon Compounds Having Si−Si Bonds and Related Compounds	1207
I. Hydrosilanes	1208
J. Heterocycles Containing a Si Atom	1210
III. CATHODIC REDUCTION OF ORGANOSILICON COMPOUNDS	1211
A. Arylsilanes	1211
B. Acylsilanes	1212
C. Polysilanes	1212
D. Halosilanes	1214
1. Cathodic reduction of halosilanes	1214
2. Cathodic reduction of halomethylsilanes	1218
3. Utilization as electrophiles	1220
E. Hydrosilanes	1229
IV. ACKNOWLEDGMENT	1229
V. REFERENCES	1229

The chemistry of organic silicon compounds, Vol. 2
Edited by Z. Rappoport and Y. Apeloig © 1998 John Wiley & Sons Ltd

I. INTRODUCTION

Organosilicon compounds are readily available and fairly stable compared with other organometallic compounds. Over the past two decades, organosilicon chemistry has grown rapidly and nowadays organosilicon compounds are widely utilized not only as various functional materials, but also as valuable organic synthetic reagents owing to their unique chemical and physical properties. There has been a remarkable growth in electrochemistry of organosilicon compounds since the last decade. This is mainly due to the following unique electrochemical and chemical properties: (1) A silyl substituent increases the HOMO energy level of π-electron systems and heteroatoms at the β-position and promotes anodic oxidation, (2) a silyl substituent decreases the LUMO energy level of π-electron systems at the α-position and enhances cathodic reduction, (3) a silyl group is an excellent cationic leaving group ('super proton') particularly when cationic intermediates are generated at the β-position and (4) a silyl group stabilizes α-anions and consequently enhances their generation.

This review covers the recent remarkable advances in the electrochemistry of organosilicon compounds, although the emphasis herein is mainly on the electrochemical reactions rather than on electrochemical mechanistic details. Only few reviews dealing with organosilicon electrochemistry have been published so far[1,2].

II. ANODIC OXIDATION OF ORGANOSILICON COMPOUNDS

A. Alkylsilanes

Similarly to the low chemical reactivity of (simple) alkylsilanes devoid of functional groups, the electrochemical reactivity of simple alkylsilanes is quite low. Klingler and Kochi measured the oxidation potentials of tetraalkyl derivatives of group-14-metal compounds by using cyclic voltammetry[3]. These compounds exhibit an irreversible anodic peak in acetonitrile. The oxidation potential (E_p) decreases in the order of Si>Ge>Sn>Pb as illustrated in Table 1. This order is the same as that of the gas-phase ionization potentials (I_p). The absence of steric effects on the correlation of E_p with I_p indicates that the electron transfer should take place by an outer-sphere mechanism. Since tetraalkylsilane has an extremely high oxidation potential (>2.5 V), it is generally difficult to oxidize such alkylsilanes anodically.

However, interestingly, the oxidation potential of tetraalkylsilanes decreases considerably in the presence of fluoride ions[4]. For example, a cathodic shift from 2.6 V to 2.3 V vs Ag/AgNO$_3$ was observed in the oxidation potential of phenyltrimethylsilane in the presence of Et$_4$NF·3HF. This suggests that one-electron transfer from alkylsilanes is assisted by fluoride ions.

The constant potential anodic oxidation of tetraalkylsilanes in the presence of fluoride ions provides the corresponding fluorosilanes derived from cleavage of the C−Si bond[4].

TABLE 1. Oxidation potentials (E_p) and ionization potentials (I_p) of aliphatic group-14-metal compounds

Substrate	E_p (V vs SSCE)[a]	I_p (eV)
Et$_4$Si	2.56	9.78
Et$_4$Ge	2.24	9.41
Et$_4$Sn	1.76	8.93
Et$_4$Pb	1.26	8.13

[a]Saturated NaCl-SCE.

TABLE 2. Anodic oxidation of organosilicon compounds in the presence of fluoride ions

Substrate	Product yield (%)
PhSiMe$_3$	PhSiMe$_2$F (80), CH$_4$, C$_2$H$_6$
PhCH$_2$SiMe$_3$	Me$_3$SiF (50), PhCHO
PhCH$_2$SiEt$_3$	Et$_3$SiF (70), PhCHO

The proposed intermediate is the pentacoordinate silyl radical, which eliminates the most stable alkyl radical to give the corresponding fluorosilane (equation 1 and Table 2).

$$R_3SiR' \xrightarrow[F^-]{-e} R_3\overset{\bullet}{Si}\begin{smallmatrix}F\\ \\R'\end{smallmatrix} \xrightarrow{-R'^\bullet} R_3SiF \qquad (1)$$

The anodic oxidation of silyl-substituted tetrahedranes was studied by cyclic voltammetry[5]. Tri-t-butyl(trimethylsilyl)tetrahedrane is oxidized more easily compared with tetra-t-butyltetrahedrane owing to the σ-donating silyl group (equation 2).

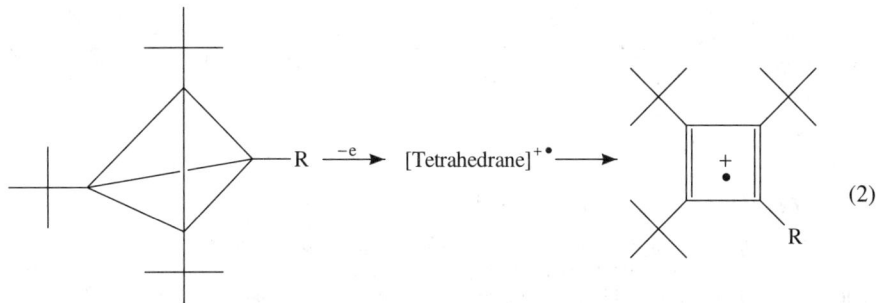

R = t-Bu : E_p^{ox}, +0.50 V vs SCE
R = Me$_3$Si : E_p^{ox}, +0.40 V vs SCE

B. Arylsilanes

Anodic oxidation of arylsilanes is slightly easier than that of alkylsilanes. The oxidation potential (E_p) of tetraphenylsilane was reported to be 2.1 V (vs Ag/Ag$^+$ in 0.01 M LiClO$_4$/MeCN)[6]. The cation radicals of aromatic silanes anodically generated in anhydrous CH$_2$Cl$_2$ in the presence of activated alumina are generally stable. However, in the presence of residual water, a black deposit, mainly consisting of conjugated aromatic polymers, was formed during electrolysis (equation 3)[6].

$$Ph_4Si \xrightarrow{-e} Ph_4Si^{+\bullet} \longrightarrow Ph_3Si^+ + Ph^\bullet \longrightarrow polymer \qquad (3)$$

Anodic oxidation of α-trimethylsilylated thiophene derivatives in nitrobenzene results in desilylation and formation of polythiophenes (equation 4), which have higher conductivity than those prepared from unsilylated thiophenes[7–9]. The silyl group increases the regioselectivity of carbon–carbon bond formation since the silyl group is an excellent cationic leaving group.

$$\text{Me}_3\text{Si}-\text{[thiophene]}-\text{SiMe}_3 \text{ or } \text{[thiophene]}-\overset{\text{Me}}{\underset{\text{Ph}}{\text{Si}}}-\text{[thiophene]} \xrightarrow[\text{Et}_4\text{NPF}_6/\text{nitrobenzene}]{-e} \text{[polythiophene]}_n \quad (4)$$

100–105 S cm^{-1}

In contrast, anodic oxidation of tetra-2-(β'-methylterthienyl)silane provides polythiophene without desilylation[9]. The polymer has a 3D electroactive π-conjugated system (equation 5).

$$\left(\text{[methylterthienyl]}\right)_4\text{Si} \xrightarrow[\text{Bu}_4\text{NPF}_6/\text{CH}_2\text{Cl}_2]{-e} \text{3D electroactive polymer} \quad (5)$$

$E_p^{ox} = 1.20$ V vs SCE

C. Benzylsilanes, Allylsilanes and Related Compounds

Benzylsilanes and allylsilanes are easily oxidized anodically compared with alkylsilanes and arylsilanes. Benzylsilanes exhibit irreversible cyclic voltammetric waves. It is notable that their oxidation potentials (E_p) are markedly less positive than those of the unsilylated parent compounds owing to the $\sigma-\pi$ interaction (Table 3)[10a]. It is interesting that α-trimethylsilylation of xylenes markedly decreases their oxidation potential while additional α'-trimethylsilylation makes a little change (Table 3). It has also been reported that a σ, σ-interacting system (the neighboring C–Si bonds) in addition to a $\sigma-\pi$ interaction caused a significant decrease of the oxidation potentials[10b].

Fuchigami and coworkers and Yoshida and coworkers independently found that anodic oxidation of benzylsilanes in the presence of nucleophiles such as alcohols and carboxylic acids resulted in a selective cleavage of the C–Si bond and the oxygen nucleophiles were introduced exclusively into the benzylic position (equation 6)[11–13]. In the absence of nucleophiles, the benzylsilane itself plays a role of a nucleophile and benzyl(trimethylsilylmethyl)benzene is formed (equation 7)[11,12].

Anodic oxidation of α, α'-bis(trimethylsilyl)xylenes in alcohols provides the corresponding monoalkoxylated products solely. Among α, α'-bis(trimethylsilyl)-o-, -m- and -p-xylenes, the p-xylene derivative gave the best yields of alkoxylated products as shown in equation 8[10].

Allylsilanes are also easily oxidized compared with nonsilylated olefins, as shown in Table 4[13].

20. Electrochemistry of organosilicon compounds 1191

TABLE 3. Oxidation potentials (E_p) of benzylsilanes and their unsilylated parent compounds[7]

Substrate	$E_p{}^a$	Substrate	$E_p{}^a$
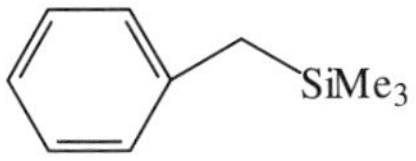 PhCH$_2$SiMe$_3$	1.38	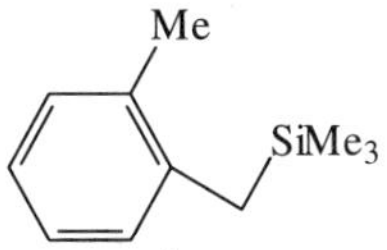 2-MeC$_6$H$_4$CH$_2$SiMe$_3$	1.18
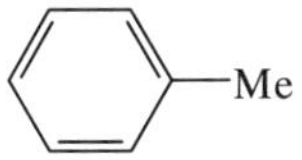 PhMe	1.98	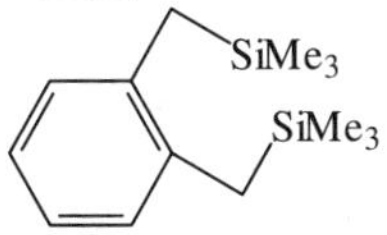 2-(Me$_3$Si)C$_6$H$_4$CH$_2$SiMe$_3$	1.13
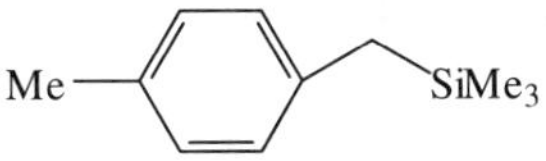 4-MeC$_6$H$_4$CH$_2$SiMe$_3$	1.17	 2,5-(Me$_3$SiCH$_2$)$_2$C$_6$H$_4$ (para)	1.06
 4-MeC$_6$H$_4$Me	1.70	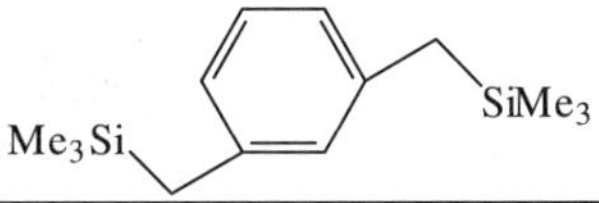 (Me$_3$SiCH$_2$)C$_6$H$_4$(CH$_2$SiMe$_3$) (meta)	1.12

^{a}V vs Ag/AgNO$_3$(sat).

$$X\text{-}C_6H_4\text{-}CH_2SiMe_3 \xrightarrow[\text{ROH}]{-2e} X\text{-}C_6H_4\text{-}CH_2OR$$

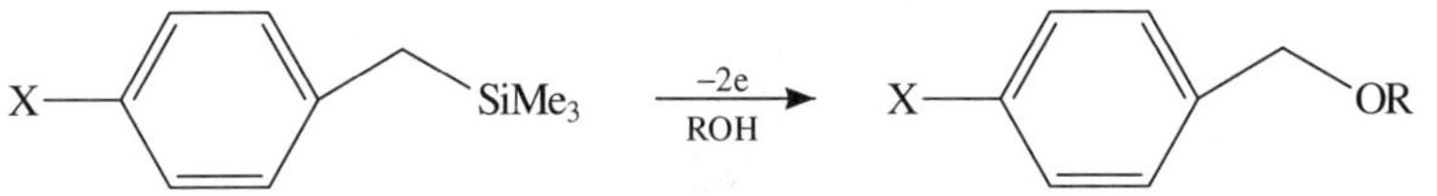

X = H, R = Me (91%)
X = SiMe$_3$, R = Me (100%)
X = Cl, R = Me (72%)
X = H, R = allyl (34%)
X = H, R = CF$_3$CO (24%)   (6)

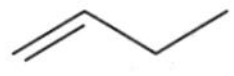

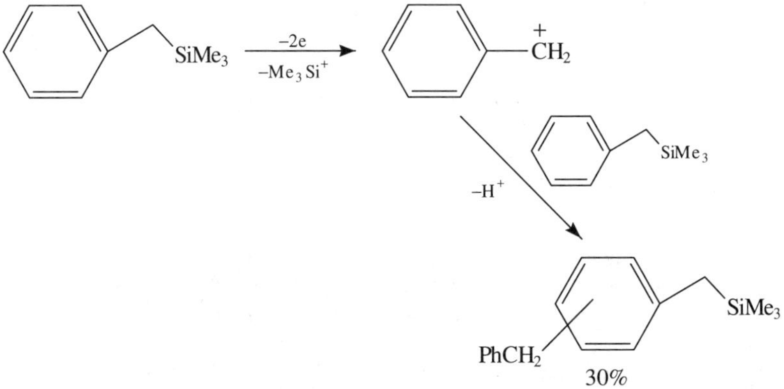

$$PhCH_2SiMe_3 \xrightarrow[-Me_3Si^+]{-2e} PhCH_2^+ \xrightarrow{-H^+} \text{4-PhCH}_2\text{-C}_6H_4\text{-}CH_2SiMe_3 \quad (30\%) \quad (7)$$

TABLE 4. Oxidation potentials (E_p) of allylsilanes and olefins[10]

Substrate	E_p (V vs Ag/AgCl)
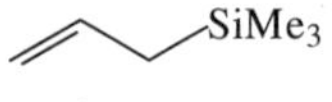 CH₂=CH-CH₂-SiMe₃	1.65
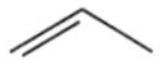 CH₂=CH-CH₂-CH₃	2.0
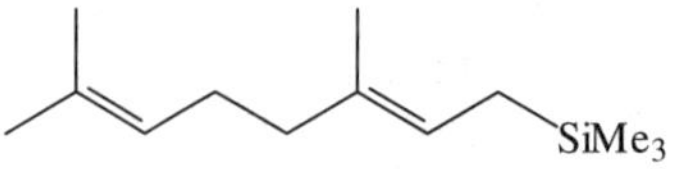 (Me)₂C=CH-CH₂-CH₂-C(Me)=CH-CH₂-SiMe₃	1.30
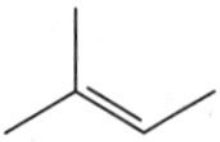 (Me)₂C=CH-CH₃	1.85

Me₃Si-CH₂-C₆H₄-CH₂-SiMe₃ $\xrightarrow[\text{ROH} \atop -\text{Me}_3\text{Si}^+]{-2e}$ Me₃Si-CH₂-C₆H₄-CH₂-OR

R = Me (67%) (8)
R = CF₃CH₂ (50%)
R = CH₂=CH-CH₂- (50%)
R = Cl-CH₂-CH₂-CH₂- (48%)

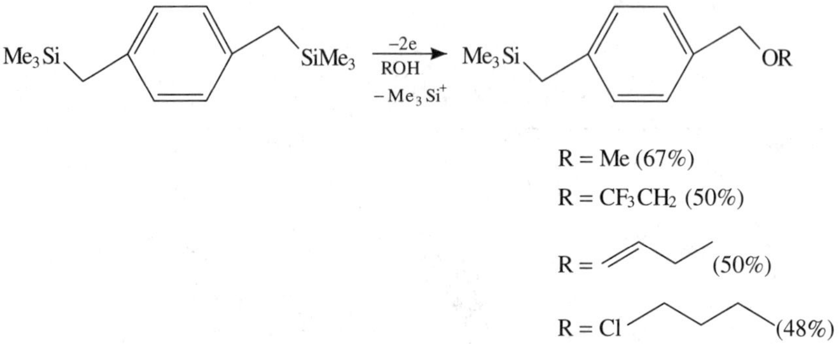

Anodic oxidation of allylsilanes in the presence of nucleophiles results in replacement of a trimethylsilyl group by a nucleophile which is introduced into the allylic carbon. Various oxygen and nitrogen nucleophiles such as alcohols, water, caboxylic acids, p-toluenesulfonic acid, carbamates or a sulfonamide can be employed in this reaction (equations 9–11)[11–13].

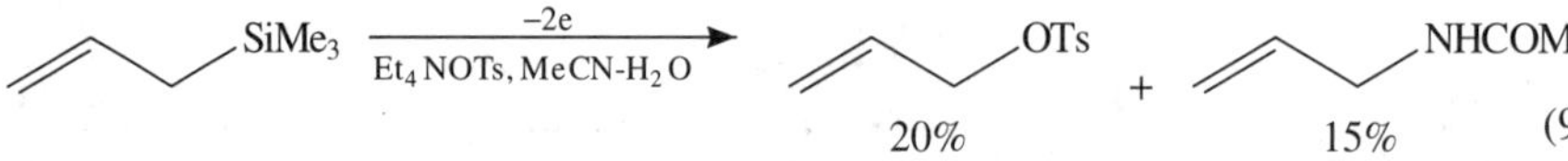

CH₂=CH-CH₂-SiMe₃ $\xrightarrow[\text{Et}_4\text{NOTs, MeCN-H}_2\text{O}]{-2e}$ CH₂=CH-CH₂-OTs + CH₂=CH-CH₂-NHCOMe
                                                                    20%   15%   (9)

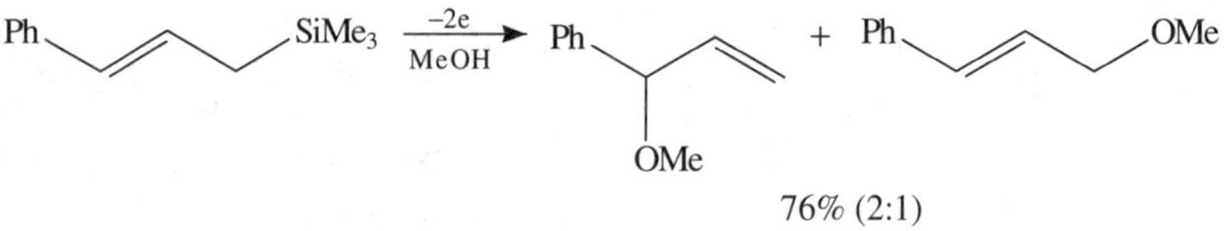

Ph-CH=CH-CH₂-SiMe₃ $\xrightarrow[\text{MeOH}]{-2e}$ Ph-CH(OMe)-CH=CH₂ + Ph-CH=CH-CH₂-OMe
                                                                    76% (2:1)   (10)

It should be noted that in the case of geranyltrimethylsilane, the carbon–carbon double bond of the allylsilane moiety is oxidized selectively to give two regioisomeric products (equation 11)[13]. This is due to the electron-donating effect of the β-silyl group. Since a methoxylated silane was formed in the anodic oxidation of an allylsilane as shown in equation 12, the reaction mechanism can be illustrated as in equation 13[13].

In contrast to the case of allylsilanes, anodic oxidation of disubstituted olefins provides in general four regioisomeric products because all the allylic carbon–hydrogen bonds can be cleaved. In the case of allylsilane, the cleavage of a C—Si bond takes place preferentially since a C−Si bond is more easily cleaved than a C−H one (equation 14), whereas allylsilanes usually act as nucleophiles; in this reaction the allylsilane works as a strong electrophile. Thus, the electrochemical reaction is an excellent method for *'oxidizable umpolung'* of allylsilanes.

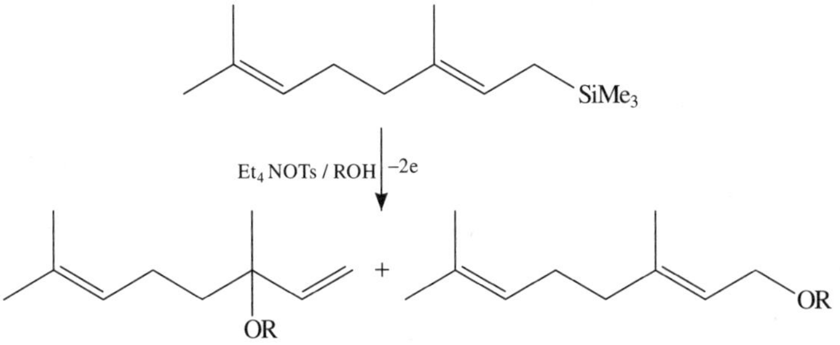

SiMe3
Et4NOTs / ROH
−2e
+
OR
OR

(11)

R = Me, 69% (68 : 32)
R = H, 62% (60 : 40)
R = Ac, 26% (63 : 27)

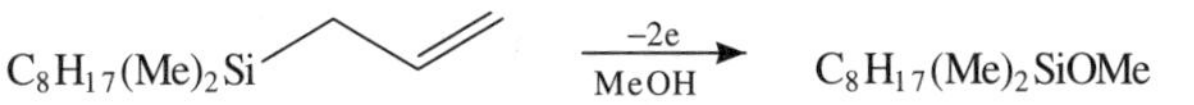

C8H17(Me)2Si
−2e
MeOH
C8H17(Me)2SiOMe

(12)

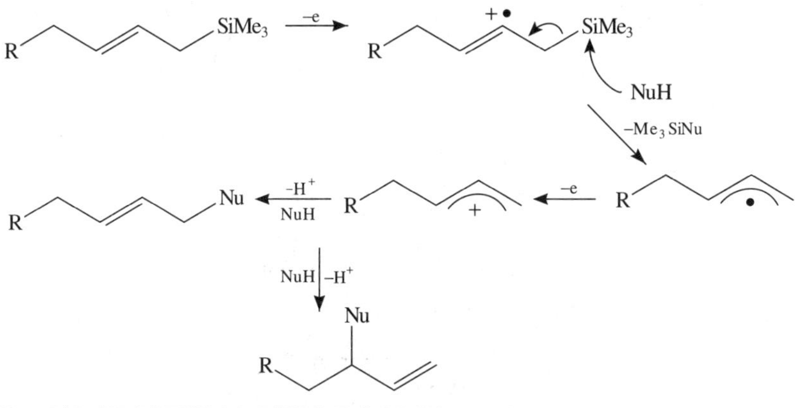

R
SiMe3
−e
+•
NuH
−Me3SiNu
Nu
-H+
NuH
+
NuH
−H+

Nu = R′O, OH, R′NHCO$_2$Me, R′COO, TsO, MeCN (13)

Chemical[14] and photochemical[15] oxidizable umpolung of allylsilanes has been reported and a similar mechanism was proposed.

1-Silyl-1,3-dienes undergo anodic methoxylation in methanol to give 1,4-addition products with an allylsilane structure as intermediates. Therefore, they are further oxidized to give 1,1,4-trimethoxy-2-butene derivatives as the final products. The products are easily hydrolyzed to provide the corresponding γ-methoxy-α, β-unsaturated aldehydes. Since 1-trimethylsilyl-1,3-dienes are readily prepared by the reaction of the anion of 1,3-bis(trimethylsilyl)propene with aldehydes or ketones, 1,3-bis(trimethylsilyl)propene offers α, β-formylvinyl anion equivalent for the reaction with carbonyl compounds (equation 15)[16].

Anodic oxidation of a mixture of allylsilane and cyclic 1,3-diketones in the presence or absence of oxygen provides cyclic peroxides (equation 16) and dihydrofuran (equation 17) derivatives, respectively[17,18]. In these reactions, the cyclic diketones are discharged to generate radical intermediates, which are trapped with allylsilane as shown. In the former case, a catalytic amount of electricity (0.043 F/mol) is enough to complete the reactions[17].

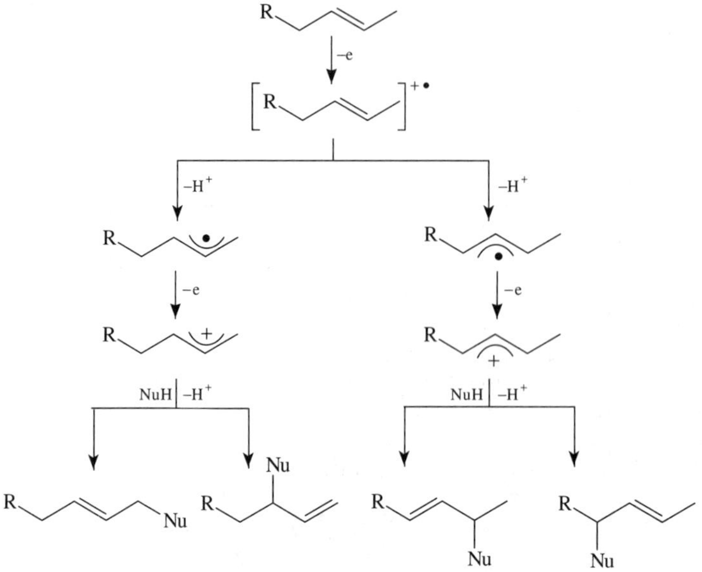

(14)

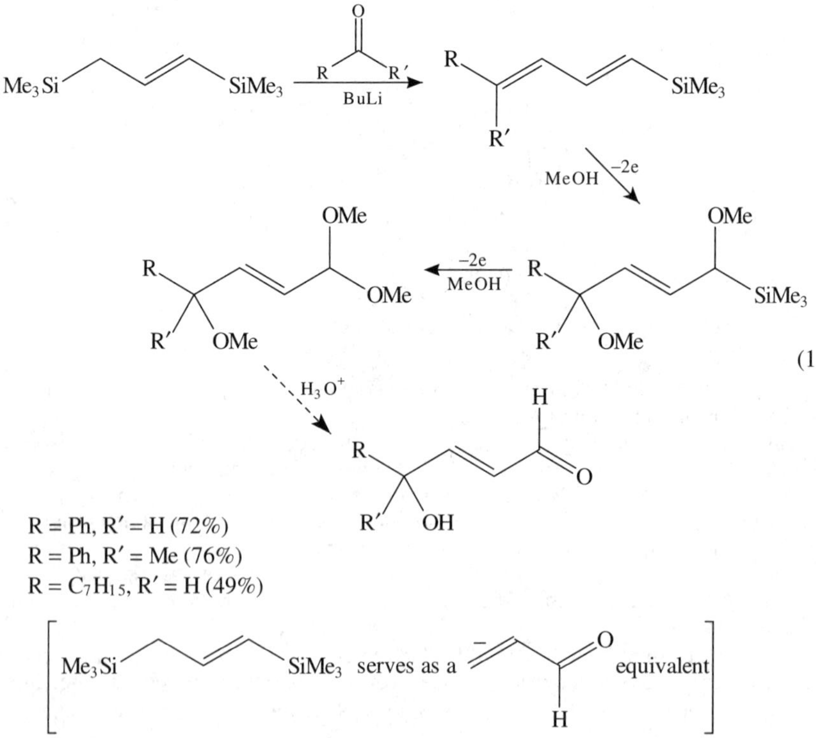

(15)

R = Ph, R' = H (72%)
R = Ph, R' = Me (76%)
R = C$_7$H$_{15}$, R' = H (49%)

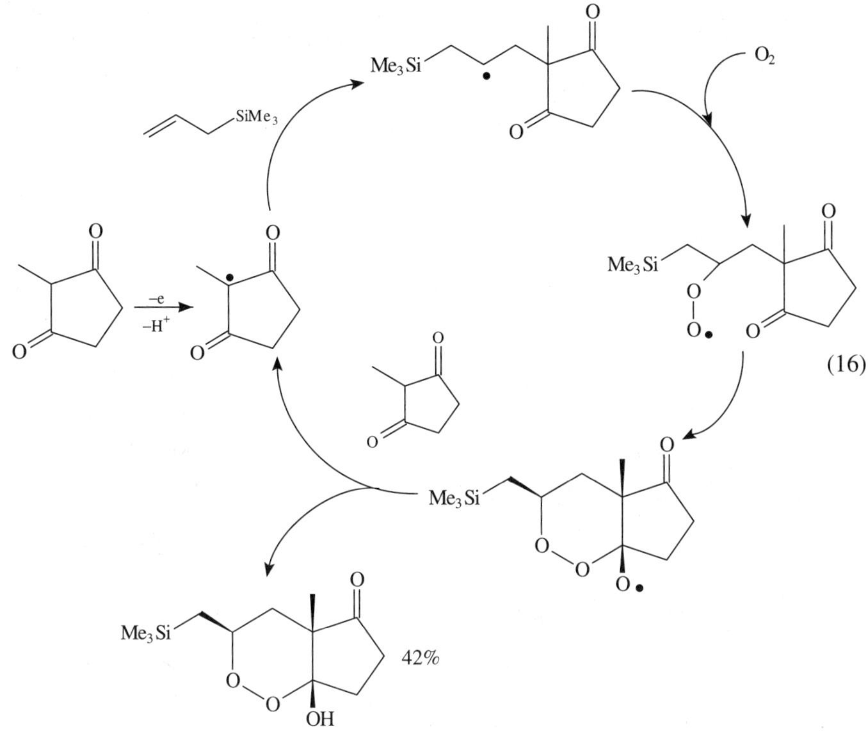

20. Electrochemistry of organosilicon compounds 1195

(16)

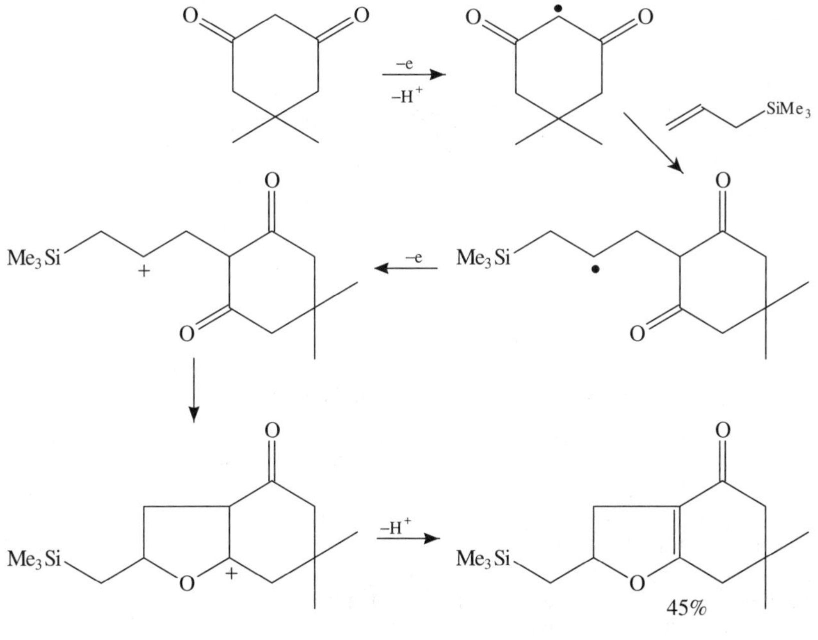

42%

(17)

45%

Intramolecular anodic olefin coupling reactions involving allyl- (equation 18) and vinylsilanes (equation 19) can lead to good yields of quaternary carbons with control of the relative stereochemistry[19,20]. This is the first example of an electrochemical reaction that makes use of a temporary silicon tether.

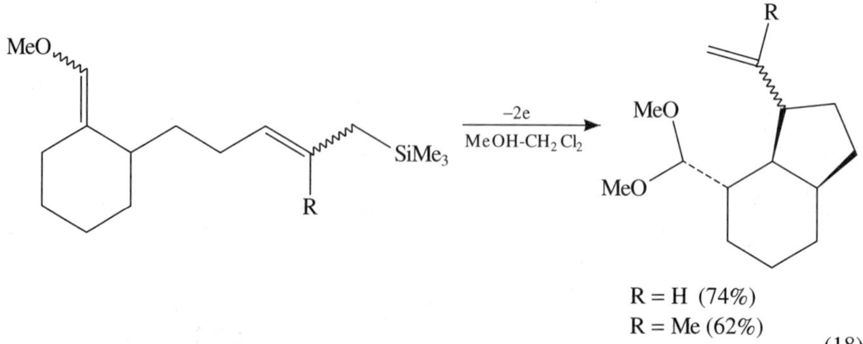

R = H (74%)
R = Me (62%)

(18)

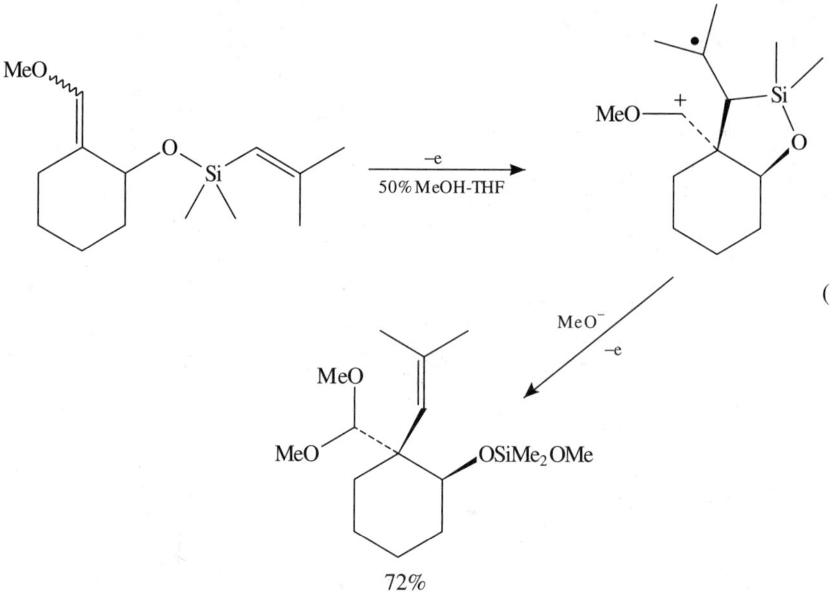

(19)

72%

D. Organosilicon Compounds Bearing Heteroatoms

Anodic oxidation of organosilicon compounds bearing heteroatoms and its synthetic applications have been studied intensively. The oxidation potentials of organosilicon compounds bearing nitrogen, sulfur and phosphorus atoms at the α-, β- and γ-positions were shown in Table 5[12,21–23]. It should be noted that the silyl group α to the heteroatom decreases appreciably the oxidation potential.

Organosilicon compounds bearing heteroatoms generally undergo anodic substitutions with the elimination of a silyl group in a manner similar to that observed for benzylsilanes and allylsilanes as shown in equation 20.

TABLE 5. Oxidation potentials ($E_{p/2}$) of amines, sulfides and phosphines bearing a trimethylsilyl group[12,21-23]

Substrate	$E_{p/2}$ (V vs SCE)	Substrate	$E_{p/2}$ (V vs SCE)
PhNHCH$_2$CMe$_3$	0.60[21]	PhSMe	1.05[b,12]
PhNHCH$_2$SiMe$_3$	0.44[21]	PhSCH$_2$SiMe$_3$	1.15[21] (0.92)[b,12]
PhNH(CH$_2$)$_2$SiMe$_3$	0.60[21]	PhS(CH$_2$)$_2$SiMe$_3$	1.26[21]
Ph(CH$_2$)$_2$NCOOMe | Me	1.95[a,22]	PhS(CH$_2$)$_3$SiMe$_3$	1.28[21]
		PhS(CH$_2$)$_7$Me	1.35[a,23]
Ph(CH$_2$)$_2$NCOOMe ⌐SiMe$_3$	1.45[a,22]	PhCH(SiMe$_3$)(CH$_2$)$_6$Me	1.25[a,23]
Ph$_2$PCH$_2$SiMe$_3$	0.63[21]	MeSCH$_2$Ph	1.25[b,12]
Ph$_2$P(CH$_2$)$_2$SiMe$_3$	0.88[21]	MeSCH(SiMe$_3$)Ph	0.99[b,12]

[a] E_p vs Ag/AgCl.
[b] E_p vs Ag/AgNO$_3$ (sat).

$$RYCH_2SiMe_3 \xrightarrow{-e} RYCH_2SiMe_3^{+\bullet} \xrightarrow[-NuSiMe_3]{Nu^-} RYCH_2^{\bullet} \xrightarrow{-e} RYCH_2^{+}$$
$$RY\!\!=\!\!CH_2^{+} \xrightarrow{Nu^-} RYCH_2Nu \quad (Y = S, N, O) \tag{20}$$

1. Organosilicon compounds bearing sulfur

Anodic oxidation of α-thiomethylsilanes in alcohols provides the α-alkoxylated sulfides selectively[23-25]. In this case, neither formation of sulfoxide nor C—S bond cleavage take place. Generally, anodic alkoxylation of sulfides is quite difficult except for sulfides carrying strong electron-withdrawing groups such as α-cyano[26] and α-(trifluoromethyl)[27,28]. Since the silyl group is a much better cationic leaving group than a proton, the alkoxylation proceeds selectively.

It is notable that allyloxylation can also be performed in relatively good yields (Table 6) although allyl alcohols are easily oxidized anodically[12]. The allyloxylated sulfides thus obtained are easily converted into the corresponding β, γ-unsaturated ketones by a [2,3] Wittig rearrangement using bases as shown in equation 21. Anodic desilylation/carboxylation of α-thiomethylsilanes also takes place similarly as shown in an example in Table 6.

$$MeSCH(Ph)SiMe_3 \xrightarrow[HO\diagdown\!\!\!=]{-2e} \underset{66\%}{MeSCH(Ph)\text{–}O\text{–}CH_2CH\!\!=\!\!CH_2} \xrightarrow[-80\sim20\,°C]{LDA/THF} \underset{74\%}{PhC(O)CH_2CH\!\!=\!\!CH_2} \tag{21}$$

TABLE 6. Anodic oxidation of α-thiomethylsilanes in the presence of alcohols and carboxylic acid

$$R^1SCH(R^2)SiMe_3 \xrightarrow[Et_4NOTs/YOH-MeCN]{-2e} R^1SCH(R^2)OY$$

Substrate		YOH	Product yield (%)
R^1	R^2		
Ph	H	MeOH[a]	66
		EtOH	65
		CH₂=CHCH₂OH (allyl OH)	59
		(CH₃)₂C=CHCH₂OH	58
Me	Ph	CH₂=CHCH₂OH	66
		(CH₃)₂C=CHCH₂OH	77
Ph	H	AcOH	66
Ph	H	MeCH₂OH	61

[a]Without MeCN.

On the other hand, when about 4 F mol^{-1} of electricity is passed in the anodic oxidation of α-thiomethylsilanes, both desilylation and desulfurization take place to give the corresponding acetals in good yields as follows. Since acetals are easily hydrolyzed to aldehydes, such an α-thiomethylsilane is a synthon of a formyl anion (equation 22).

$$R^1\text{CH(SPh)(SiMe}_3) \xrightarrow[R^2OH]{-2e} R^1\text{CH(SPh)(OR}^2) \xrightarrow[R^2OH]{-e} R^1\text{CH(OR}^2)_2$$

$R^1 = \text{CH}_2=\text{CH(CH}_2)_6\text{Me}, R^2 = \text{Me} (64\%)$

$R^1 = C_8H_{17}, R^2 = HOCH_2CH_2\text{-- (from HOCH}_2\text{CH}_2\text{OH)} (60\%)$ (22)

$R^1 = \text{2-Hydroxycyclohexyl}, R^2 = \text{Me} (81\%)$

[RS–CH₂–SiMe₃ is an equivalent synthon to RO–CH₂–OR or ⁻CHO]

Furthermore, sulfides bearing two silyl groups can also be oxidized anodically in methanol to provide the corresponding esters as shown in equation 23[21,22]. Therefore, the α, α-disilylated sulfide provides a synthon of the anion of $^-$C(CO)OMe.

Indirect anodic oxidation of α-phenylthiomethylsilane in an alcohol using Ni^{2+}/Ni^{3+}-cyclam mediator also provides α-alkoxylated sulfides, although the turnover of the mediator is low (equation 24)[29].

It is interesting that anodic oxidation of 2-silyl-1,3-dithianes at a platinum anode provides the corresponding acylsilanes (equation 25)[30]. In this case, only a C—S bond cleavage takes place selectively and the C—Si bond is not cleaved. It is noted that this anodic transformation can be easily applied to the synthesis of α, β-unsaturated acylsilanes. The reactions are affected by the anode material used. A glassy carbon anode gives poor yields of acylsilanes.

$$\text{Me}_3\text{Si}\diagdown\!\!\!\!\diagup\text{SPh} \quad \xrightarrow[\text{2. RX}]{\text{1. BuLi}} \quad \text{R}\diagdown\!\!\!\!\diagup\text{SPh}(\text{SiMe}_3)(\text{SiMe}_3) \quad \xrightarrow[\text{MeOH}]{-e} \quad \text{RCOOMe}$$

$$R = C_{12}H_{25} \ (81\%)$$
$$R = C_8H_{17} \ (72\%) \tag{23}$$

$$\left[\text{Me}_3\text{Si}\diagdown\!\!\!\!\diagup\text{SPh}(\text{SiMe}_3)^-\right] \text{ is a synthon equivalent to } \left[\overset{O}{\underset{-}{\|}}\text{C—OR}\right]$$

$$\text{PhS}\diagdown\!\!\!\!\diagup\text{SiMe}_3 \quad \xrightarrow{\text{ROH}} \quad \text{PhS}\diagdown\!\!\!\!\diagup\text{OR}$$

(cycle: $2\text{Ni}^{3+} \rightleftarrows 2\text{Ni}^{2+}$, $2e$)

$$\text{Ni}^{2+} = [\text{Ni(II)cyclam}](\text{ClO}_4)_2 \tag{24}$$

<chemical structure: 1,3-dithiane with R and SiMe$_3$ substituents> $\xrightarrow[\text{NaClO}_4/\text{H}_2\text{O-MeCN} \\ 4 \text{ F mol}^{-1}]{-e}$ $\text{R}\overset{O}{\underset{\|}{\text{C}}}\text{SiMe}_3$

$$R = \text{Ph} \ (96\%)$$
$$R = \text{Ph—CH=CH—} \ (95\%)$$
$$R = \text{CH}_3(\text{CH}_2)_5\text{CH=CH} \ (84\%) \tag{25}$$

2. Organosilicon compounds bearing nitrogen

Introduction of a silyl group at the position α to the nitrogen decreases the oxidation potentials of organonitrogen compounds by at most 0.3 V (Table 7)[31]. β- and γ-silyl groups do not cause a significant decrease in the oxidation potential. The introduction of a silyl group directly on the amino nitrogen atom does not decrease the oxidation potentials of amines.

α-Silylcarbamates have much lower oxidation potentials of 0.5 V compared with the unsilylated carbamates (Table 5)[22].

TABLE 7. Oxidation potentials (E_p) of aminosilanes

Substrate	E_p (V vs SCE)	Reference
piperidine-NH	0.95	28
piperidine-NSiMe$_3$	1.0	28
piperidine-N-CH$_2$SiMe$_3$	0.65	28
piperidine-N-CH$_2$CH$_2$CH$_2$SiMe$_3$	0.88	28
(HOCH$_2$CH$_2$)$_3$N	0.90	29
(Me$_3$SiOCH$_2$CH$_2$)$_3$N	0.92	29
silatrane R = H	1.70	29
silatrane R = Me	1.43	29
C$_9$H$_{19}$C(=NNHTs)H	1.72^a	30
C$_9$H$_{19}$C(=NNHTs)SiMe$_3$	1.43^a	30

^{a}E_d vs Ag/AgCl (decomposition potential determined by rotating-disk electrode voltammetry).

In sharp contrast to the cases of α-silylamines and α-silylcarbamates, silatranes have higher oxidation potentials than unsilylated triethanolamine and open-chain tris(trimethylsiloxyethyl)amine (Table 7)[32]. Transannular interaction between nitrogen and silicon atoms leads to the strong transmission of electron density from the nitrogen to silicon atom via donor–acceptor bonding. Considering the nitrogen as a site of anodic oxidation, such intramolecular charge transfer should result in higher oxidation potentials of the silatranes compared with those of unsilylated amines and open-chain silylated amines, where donor–acceptor interactions are excluded.

Anodic oxidation of N-benzyl-N-(α-silylmethyl)carbamate provides α-methoxylated product as a single regioisomer (equation 26), while the unsilylated parent carbamate gives a mixture of regioisomeric products (equation 27). Thus, the introduction of a silyl group can control completely the regiochemistry of the anodic methoxylation and can also activate the nitrogen atom toward anodic oxidation.

$$\text{Ph-N(CH}_2\text{SiMe}_3\text{)COOMe} \xrightarrow[\text{MeOH}]{-e} \text{Ph-N(CH(OMe))COOMe} \quad 97\% \tag{26}$$

$$\text{Ph}\diagup\!\!\diagdown\text{NCOOMe} \xrightarrow[\text{MeOH}]{-e} \text{Ph}\diagup\!\!\diagdown\text{NCOOMe} + \underset{\underset{\text{Me}}{|}}{\text{Ph}\diagup\!\!\diagdown\overset{\overset{\text{OMe}}{|}}{\text{C}}\!\diagdown\text{NCOOMe}} \quad (27)$$

$$\underset{\text{Me}}{|} \qquad \underset{\text{MeO}}{|} \qquad \underset{\text{Me}}{|}$$
$$ 65\% 35\%$$

Anodic oxidation of 4-silylazetidin-2-ones in the presence of fluoride ions provides 4-fluoroazetidin-2-ones in high yields[33]. This fluorination is completely regioselective. Even in the case of the *N*-benzyl derivative, a fluorine atom is selectively introduced into the C-4 position of the β-lactam ring (equation 28). In contrast, unsilylated azetidin-2-ones give no fluorinated product.

$$\text{[azetidinone-SiMe}_3\text{]} \xrightarrow[\text{Et}_3\text{N} \cdot 3\text{HF}]{-2e} \text{[azetidinone-F]} \quad (28)$$

Y = H (80%)
Y = Et (88%)
Y = CH(OCH$_2$OMe)Me (75%)

Electrochemical properties of tosylhydrazones of acylsilanes were also investigated. A decrease in the oxidation potential of tosylhydrazones caused by silylation is much smaller than that for carbonyl compounds (see Tables 7 and Section II.E., Table 9). Anodic oxidation of tosylhydrazones of acylsilanes provides the corresponding nitriles with consumption of a catalytic amount of electricity (equation 29)[34].

$$\underset{\text{SiMe}_3}{\overset{\text{NNTs}}{\underset{\|}{\text{R}\diagup\!\!\diagdown}}} \xrightarrow[\substack{\text{CH}_2\text{Cl}_2 \\ (\text{ca } 0.8\,\text{F mol}^{-1})}]{-e} \text{RCN} \quad \text{R} = \text{cyclohexyl (81\%), PhCH}_2\text{CH}_2 \text{ (72\%)} \quad (29)$$

3. Organosilicon compounds bearing phosphorus

Trimethylsilyl phosphites can be anodically oxidized in the presence of arenes to form arylphosphonates (equation 30)[35].

$$(\text{EtO})_2\text{POSiMe}_3 \xrightarrow{-e} (\text{EtO})_2\overset{+\bullet}{\text{POSiMe}_3} \xrightarrow[\text{Ph}\!-\!\text{H}]{-e} (\text{EtO})_2\text{P(O)Ph} \quad (30)$$

4. Organosilicon compounds bearing oxygen

Although anodic oxidation of aliphatic ethers is generally difficult, the ethers bearing a silyl group at a position α to the oxygen atom can be oxidized rather easily. As shown in Table 8, α-silyl substitution causes significant decrease in the oxidation potential of ethers and alcohols[36,37]. The magnitude of the silicon effect is much greater than that for organo nitrogen and sulfur compounds. This is in accord with the better overlap of the nonbonding p orbital of oxygen with the C—Si σ orbital than that of the orbitals of nitrogen and sulfur.

TABLE 8. Oxidation potentials (E_p) of silylated ethers and related compounds

Substrate	E_p (V vs Ag/AgCl)	Substrate	E_p (V vs Ag/AgCl)
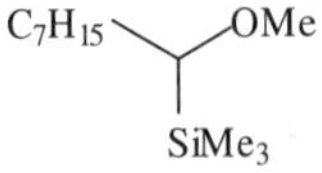 C_7H_{15}–C(SiMe$_3$)(OMe)–	1.72	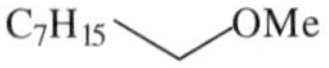 C_7H_{15}–CH(OMe)–	>2.5
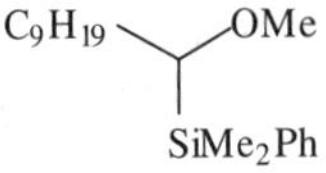 C_9H_{19}–C(SiMe$_2$Ph)(OMe)–	1.60	 C_7H_{15}–CH(SiMe$_2$Ph)–	2.25
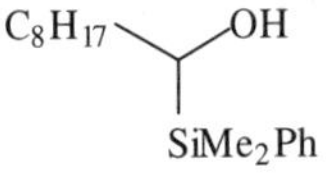 C_8H_{17}–C(SiMe$_2$Ph)(OH)–	1.70	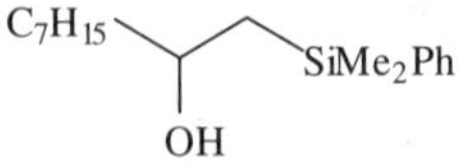 C_7H_{15}–C(SiMe$_2$Ph)(OH)–	2.26

It is quite interesting that such a silicon effect depends strongly on the geometry of the molecule. Yoshida and coworkers found a linear correlation on plotting the oxidation potentials of α-silylated ethers, where the rotation around the C—O bond is restricted, against the HOMO energy–torsion angle (Si—C—O—C) curve obtained by MO calculation[36].

Anodic oxidation of α-silyl ethers in methanol provides acetals. Even a silylated allyl ether gives the corresponding acetals in good yield (equation 31)[2].

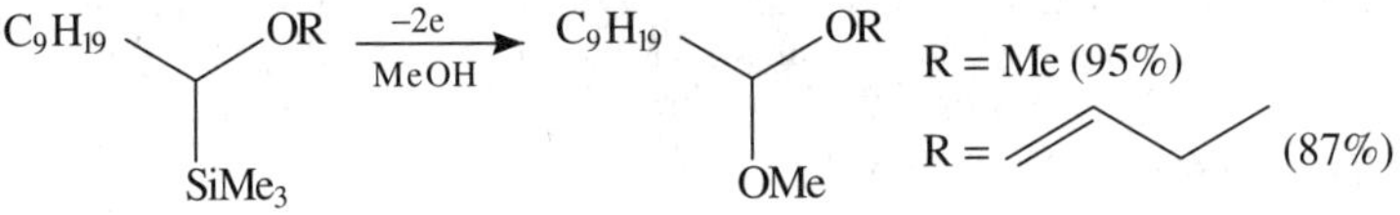

$$C_9H_{19}-C(SiMe_3)(OR)H \xrightarrow[MeOH]{-2e} C_9H_{19}-C(OR)(OMe)H \quad R = Me \ (95\%)$$
$$R = CH_2CH=CHCH_3 \ (87\%) \tag{31}$$

This electrochemical method can be widely applicable as shown in equation 32[25,38]. It is noted that anodic oxidation in dichloromethane containing water provides the aldehyde directly under neutral conditions. Thus, the methoxy(trimethylsilyl)methyl anion is an equivalent to the formyl anion[25,38].

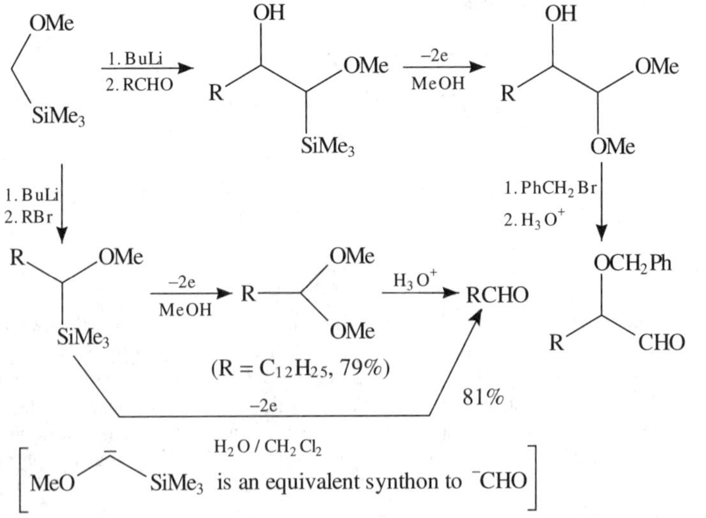

$$\text{MeOCH}_2\text{SiMe}_3 \xrightarrow[\text{2. RCHO}]{\text{1. BuLi}} R-CH(OH)-CH(OMe)(SiMe_3) \xrightarrow[MeOH]{-2e} R-CH(OH)-CH(OMe)_2$$

(with further transformations)

$$R-CH(SiMe_3)(OMe) \xrightarrow[MeOH]{-2e} R-C(OMe)_2H \xrightarrow{H_3O^+} RCHO$$

(R = C$_{12}$H$_{25}$, 79%)

$$\xrightarrow[H_2O/CH_2Cl_2]{-2e} 81\%$$

$$[\text{MeO-CH}^-\text{-SiMe}_3 \text{ is an equivalent synthon to } ^-\text{CHO}]$$

(32)

The anodic oxidation of ethers bearing two α-silyl groups is also useful. As shown in equation 33, the anion of methoxybis(trimethylsilyl)methane can be utilized as a synthon of the anion of the alkoxycarbonyl group.

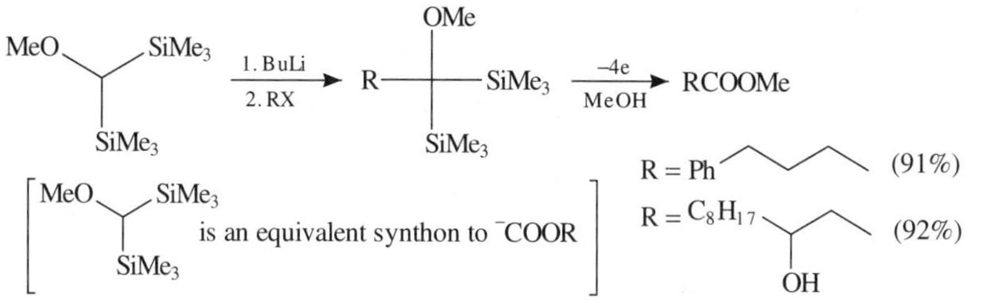

$$\text{MeO-C(SiMe}_3\text{)}_2\text{H} \xrightarrow[\text{2. RX}]{\text{1. BuLi}} \text{R-C(OMe)(SiMe}_3\text{)}_2 \xrightarrow[\text{MeOH}]{-4e} \text{RCOOMe}$$

[MeO-C(SiMe₃)₂H is an equivalent synthon to ⁻COOR]

R = Ph~~~ (91%) (33)

R = C₈H₁₇~~~OH (92%)

This methodology is also applicable to the synthesis of γ-lactones[25] and enantiomers or diastereomers of straight-chain 1,2-polyols[39].

Anodic oxidation of α-silyl ethers in the presence of nitrogen nucleophiles such as sulfonamides and carbamates provides N,O-acatals (equations 34 and 35)[2].

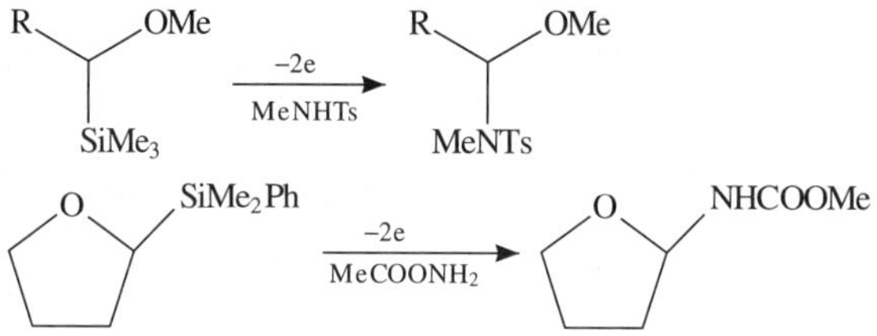

$$\text{R-CH(OMe)(SiMe}_3\text{)} \xrightarrow[\text{MeNHTs}]{-2e} \text{R-CH(OMe)(NMeTs)} \qquad (34)$$

$$\text{(2-SiMe}_2\text{Ph-tetrahydrofuran)} \xrightarrow[\text{MeCOONH}_2]{-2e} \text{(2-NHCOOMe-tetrahydrofuran)} \qquad (35)$$

E. Acylsilanes

Anodic oxidation of aliphatic aldehydes and ketones is generally difficult because their oxidation potentials are very high (>2.5 V). However, silylation at the carbonyl carbon causes a marked decrease in the oxidation potential as shown in Table 9[33,40]. This silicon effect is much smaller in the case of aromatic carbonyl analogues[40]. The silicon effect is attributed to the rise of the HOMO level by the interaction between the C—Si σ orbital and the nonbonding p orbital of the carbonyl oxygen, which in turn favors the electron transfer.

Anodic oxidation of acylsilane in the presence of various nucleophiles provides the corresponding ester, acid and carbamate as shown in equation 36[41].

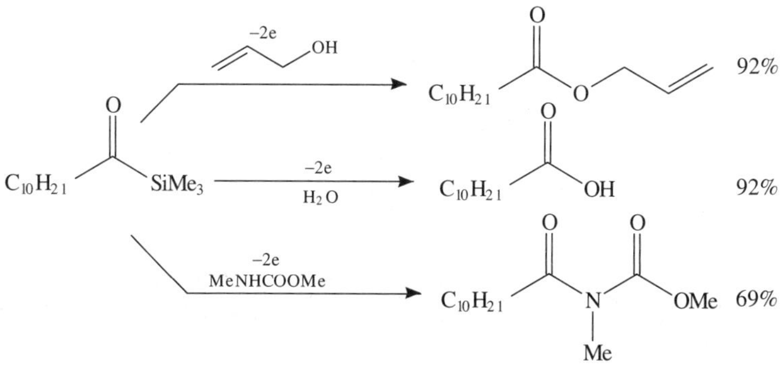

$$\text{C}_{10}\text{H}_{21}\text{C(O)SiMe}_3 \begin{cases} \xrightarrow[\text{CH}_2\text{=CHCH}_2\text{OH}]{-2e} \text{C}_{10}\text{H}_{21}\text{COOCH}_2\text{CH=CH}_2 & 92\% \\ \xrightarrow[\text{H}_2\text{O}]{-2e} \text{C}_{10}\text{H}_{21}\text{COOH} & 92\% \\ \xrightarrow[\text{MeNHCOOMe}]{-2e} \text{C}_{10}\text{H}_{21}\text{C(O)N(Me)COOMe} & 69\% \end{cases} \qquad (36)$$

TABLE 9. Oxidation potentials (E_p) of acylsilanes and related compounds[30,37]

Substrate	E_p (V vs Ag/AgCl)	Substrate	E_p (V vs Ag/AgCl)
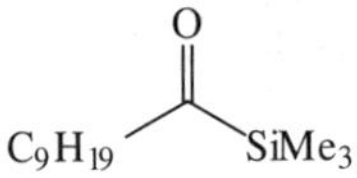 C_9H_{19}–CO–SiMe$_3$	1.45	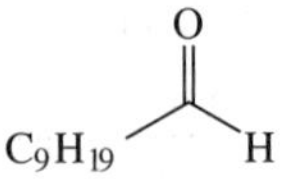 C_9H_{19}–CO–H	>2.5
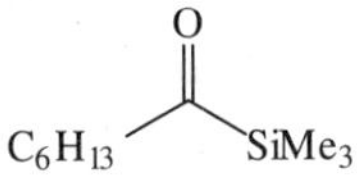 C_6H_{13}–CO–SiMe$_3$	1.70	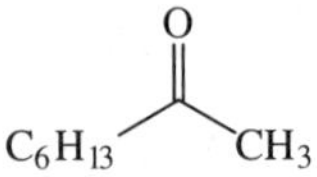 C_6H_{13}–CO–CH$_3$	>2.5
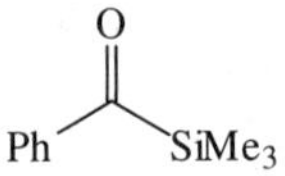 Ph–CO–SiMe$_3$	1.88	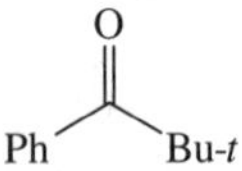 Ph–CO–Bu-t	1.96

F. Silyl Enol Ethers and Related Compounds

Since silyl enol ethers have a silyl group β to the π-system, anodic oxidation of silyl enol ethers takes place easily. In fact, anodic oxidation of silyl enol ethers proceeds smoothly to provide the homo-coupling products, 1,4-diketones (equations 37 and 38)[42]. This dimerization of the initially generated cation radical intermediate is more likely than the reaction of acyl cations formed by two electron oxidation of unreacted silyl enol ethers in these anodic reactions.

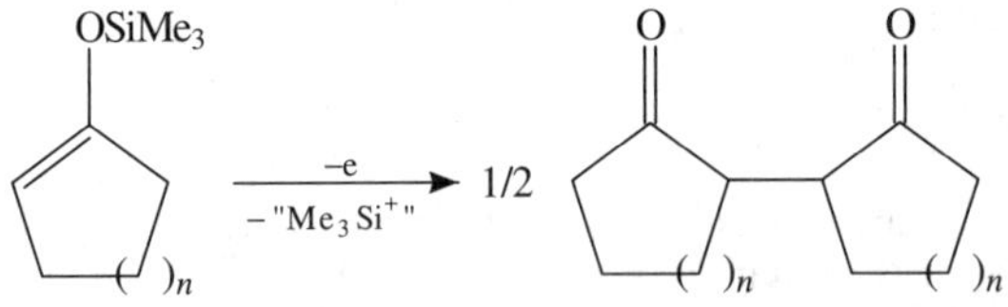

$$n = 1\,(66\%)$$
$$n = 2\,(58\%)$$

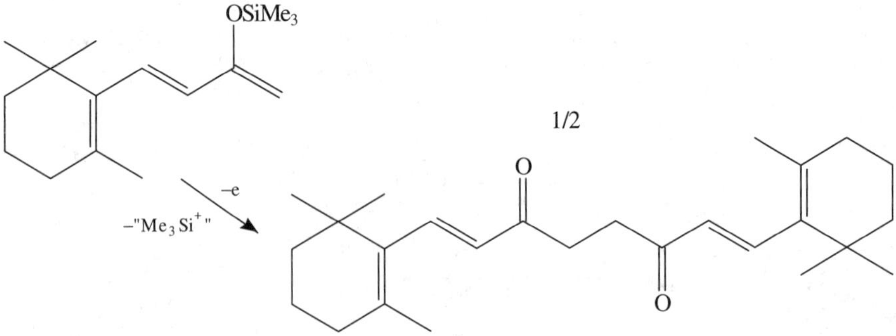

(37)

(38) 64%

TABLE 10. Oxidation potentials (E_p) of trimethylsiloxyarenes

Substrate	E_p (V vs SCE)
Me₃SiO–C₆H₄–OSiMe₃ (1,4-)	1.10
Me₃SiO–C₆H₂(Me)₂–OSiMe₃ (2,5-dimethyl)	1.05
9,10-bis(trimethylsiloxy)anthracene	0.78
MeO–C₆H₄–OSiMe₃	1.14

On the other hand, indirect anodic oxidation of cyclic silyl enol ethers in the presence of iodide ions gives α-iodocyclic ketones (equation 39)[43].

Anodic oxidation of hydroquinone disilyl ethers also takes place easily at around 1 V vs SCE, as shown in Table 10[44]. It was proposed that an initial one-electron oxidation generates a cation radical which decomposes by a Si—O bond cleavage to form quinones (equation 40)[44].

$$\text{R–cyclohexenyl–OSiMe}_3 \xrightarrow[\text{I}^-]{2e, [\text{I}^+]} \text{R–(2-oxocyclohexyl)–I} \quad (39)$$

R = C₅H₁₁ (97%)

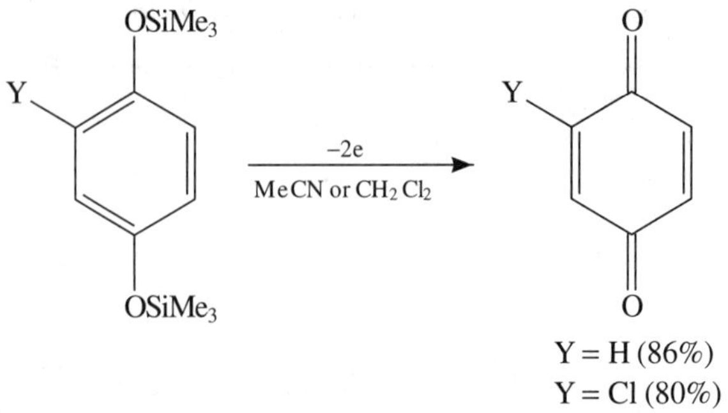

$$\text{(40)}$$

Y = H (86%)
Y = Cl (80%)

It is interesting that α-alkoxycarbonylcycloalkanones are formed from fused silyloxycarbonylcyclopropanes by anodic oxidation in methanol or ethanol in the presence of Fe(NO$_3$)$_3$ (equation 41)[45].

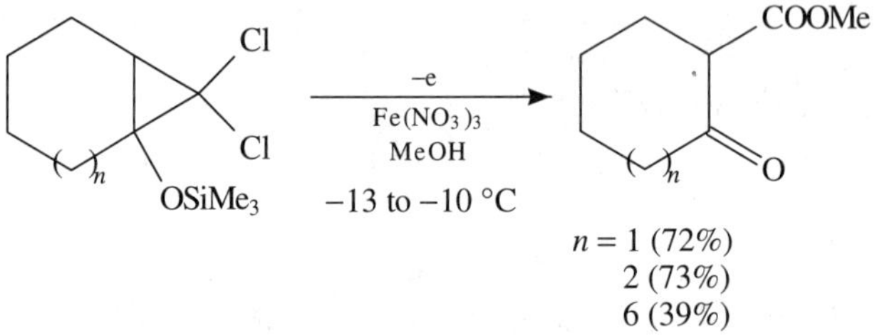

$$\text{(41)}$$

n = 1 (72%)
2 (73%)
6 (39%)

G. Carboxylic Acid Bearing a β-Trimethylsilyl Group

Anodic oxidation of carboxylic acids bearing a trimethylsilyl group on the β-position gives exclusively terminal olefins in rather good yields. The reaction seems to proceed via a carbocation intermediate formed by the oxidative elimination of CO_2 (equation 42)[46].

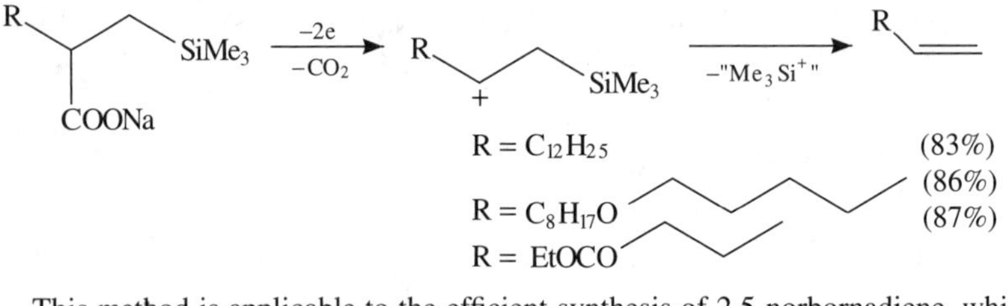

$$\text{(42)}$$

R = C$_{12}$H$_{25}$ (83%)
R = C$_8$H$_{17}$O (86%)
R = EtOCO (87%)

This method is applicable to the efficient synthesis of 2,5-norbornadiene, which can be obtained in low yield by anodic decarboxylation of 5-norbornene-2,3-dicarboxylic acid (equation 43)[47].

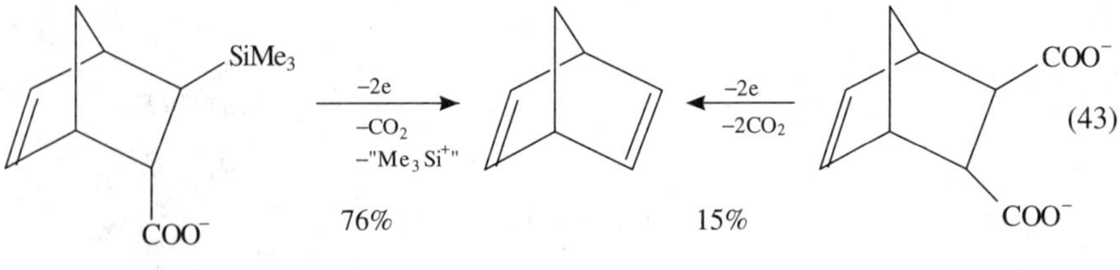

$$\text{(43)}$$

76%   15%

H. Organosilicon Compounds Having Si−Si Bonds and Related Compounds

The reactivity of the metal–metal bond of group-14-dimetals, such as the Si−Si bond, has attracted much interest in comparison with that of a carbon–carbon double bond. Table 11 shows oxidation potentials of group-14-dimetals[48].

The oxidation potential decreases in the order: Si−Si ∼ Si−Ge>Ge−Ge>Si−Sn> Ge−Sn >Sn−Sn in accord with the ionization potential (I_p) of the corresponding dimetal. Anodic generation of silicenium ions from disilanes was also reported. The reduction potentials of silicenium ions were determined by cyclic voltammetry of neutral precursor disilanes[49]. The reduction potential shifted to the negative direction as the center element changed from C to Ge as shown in equation 44.

$$Ph_3E\text{—}EMe_3 \xrightarrow[Bu_4NClO_4/MeCN]{-2e} [Ph_3E^+] + [Me_3E^+]$$

$$E = Si : E_{p/2}^{ox}\ 1.29\ V \quad E_{p/2}^{red}\ [Ph_3Si^+] = -0.3\ V\ vs\ Fc/Fc^+$$

$$E = Ge : E_{p/2}^{ox}\ 1.16\ V \quad E_{p/2}^{red}\ [Ph_3Ge^+] = -0.69\ V\ vs\ Fc/Fc^+$$

$$E_{p/2}^{red}\ [Ph_3C^+] = -0.04\ V\ vs\ Fc/Fc^+$$

(44)

Table 12 shows oxidation potentials ($E_{1/2}$) of a series of permethylpolysilanes determined by a.c. polarography or cyclic voltammetry[50,51]. The $E_{1/2}$ values decrease as the chain length increases. This reflects the HOMO energy levels of these permethylpolysilanes[52].

Cyclic peralkylsilanes exhibit unique behavior which distinguishes these compounds from saturated catentaes of carbon. In some ways, the properties of the cyclosilanes resemble these of poly-unsaturated or aromatic hydrocarbons. As shown in Table 12, five-membered permethyl cyclic silane shows a higher oxidation potential compared with the linear analog[51,53].

TABLE 11. Oxidation potentials of group-14-dimetals

Substrate	E_p (V vs Ag/AgCl)
$Me_3SiSiMe_3$	1.76
$Me_3SiGeMe_3$	1.76
$Me_3GeGeMe_3$	1.70
$Me_3SiSnMe_3$	1.60
$Me_3GeSnMe_3$	1.44
$Me_3SnSnMe_3$	1.28

TABLE 12. Lowest oxidation potentials ($E_{1/2}$) of permethylpolysilanes

Substrate	$E_{1/2}$ (V vs SCE)	Reference
$Me(SiMe_2)_2Me$	1.88^a	50
$Me(SiMe_2)_3Me$	1.52^a	50
$Me(SiMe_2)_4Me$	1.33^a	50
$Me(SiMe_2)_5Me$	1.18^a	50
$Me(SiMe_2)_6Me$	1.08^a	50
$(Me_2Si)_5$	1.35^b	51
$(Ph_2Si)_5$	1.50^b	51

^{a}Determined by a.c. polarography.
^{b}Determined by cyclic voltammetry.

Becker and coworkers have studied intensively anodic behavior of cyclic polysilanes[53–55]. Oxidation potentials of cyclic polysilanes are affected by the counter anions of the supporting electrolytes. For example, in the presence of ClO_4^-, the first oxidation potential (E_p) of $[t\text{-Bu(Me)Si}]_4$ is significantly lower compared with other electrolytes such as BF_4^- and HSO_4^- (1.05 V, 1.15 V, 1.25 V vs Ag/AgCl, respectively)[54].

On controlled potential oxidation, cyclic peralkylsilanes undergo ring opening followed by further Si—Si bond cleavage and reaction with BF_4^- to form α,ω-difluorosilanes as the major products (equation 45)[53].

$$(Pr_2Si)_5 \xrightarrow[\substack{Et_4NBF_4/CH_2Cl_2MeCN \\ +1.25 \text{ V vs Ag/AgCl} \\ \text{pulse electrolysis}}]{-e} F(Pr_2Si)_nF \quad \begin{array}{l} n = 2\ (54\%) \\ 3\ (30\%) \\ 4\ (10\%) \end{array} \quad (45)$$

On the other hand, when this anodic oxidation was carried out in the presence of ClO_4^- or AcO^- salts instead of BF_4^-, it led to both oxygen insertion and ring-opening processes, to form mainly cyclic and linear siloxanes (equation 46). The ratio of these products greatly depends both on the amount of electricity passed and on the presence of oxygen[55].

$$(Pr_2Si)_5 \xrightarrow[\substack{Et_4NClO_4/CH_2Cl_2MeCN \\ \text{under air} \\ 4F\ mol^{-1}}]{-e} \text{[cyclic siloxane 25\%]} + \text{[cyclic siloxane 31\%]}$$

$+(n\text{-Pr})_2[(n\text{-Pr})_2Si]_4O_3$ 18% (46)
+ other types of cyclic siloxanes

The mechanism given in equation 47 has been proposed for this reaction. The initially formed cation radical reacts with molecular oxygen to generate an intermediate, which may couple with a neutral cyclic silane to form species **A**. The intermediate **A** decomposes to the final product **B** and its cation radical **B$^{+\bullet}$**, which could also be generated by direct anodic oxidation of the siloxane **B**. A further oxygen insertion step could take place via intermediate **C$^{+\bullet}$**[54].

Disilenes undergo irreversible anodic oxidation at much less positive potentials, as shown in Table 13[56]. The oxidation potentials for these compounds are similar, indicating that the HOMO of each species lies at approximately the same energy level.

I. Hydrosilanes

The electrochemical properties of hydrosilanes were also examined and the hydrosilanes were found to be electrode active. Dimethylphenylsilane and methyldiphenylsilane show irreversible peaks at 2.2 and 2.1 V vs SCE, respectively, in their cyclic voltammograms measured by a glassy carbon in $LiClO_4/MeCN$[57].

TABLE 13. Oxidation potentials (E_p) of disilenes

$$\begin{array}{c} R^1 \\ \\ R^2 \end{array} Si=Si \begin{array}{c} R^2 \\ \\ R^1 \end{array}$$

Substrate		E_p (V)[a]
R^1	R^2	
Mes[b]	Mes	+0.38
Xyl[c]	Xyl	+0.47
Tbp[d]	Tbp	+0.48
Mes	t-Bu	+0.54
Mes	Ad[e]	+0.36

[a] vs Fc/Fc$^+$.
[b] Mes = 2,4,6-trimethylphenyl.
[c] Xyl = 2,4-dimethylphenyl.
[d] Tbp = 4-tert-butyl-2,6-dimethylphenyl.
[e] Ad = 1-adamantyl.

(47)

Anodic oxidation of dimethylphenylsilane in the presence of $CuCl_2$ or CuCl using platinum electrodes provides chlorodimethylphenylsilane in high yield (>90%), while similar electrolysis in the presence of BF_4^- affords fluorodimethylphenylsilane (equation 48)[57].

TABLE 14. Electrochemical polymerization of hydrosilanes

$$\text{H}-\underset{\underset{R^2}{|}}{\overset{\overset{R^1}{|}}{\text{Si}}}-\text{H} \xrightarrow[\text{Bu}_4\text{NBF}_4/\text{DME}]{\text{electrolysis}} \text{H}\left[\underset{\underset{R^2}{|}}{\overset{\overset{R^1}{|}}{\text{Si}}}\right]_n\text{H}$$

Monomer		Electricity	Yield	mol wt[a]	
R^1	R^2	(F mol^{-1})	(%)	M_w	M_w/M_n
Ph	Me	2.0	60	477	1.05
Ph	H	3.0	32	640	1.42
Hex	H	3.0	70	1240	1.09

[a]GPC vs polystyrene.

When a copper anode is used instead of platinum, the resulting chlorosilane is subsequently reduced to a Si—Si coupling product in a one-pot reaction (equation 49)[57]. Interestingly, when a mixture of hydrosilane and chlorosilane is electrolyzed using a copper anode and a platinum cathode, a Si—Si coupling product is obtained in 64% yield on the basis of the sum of both reagents used. Thus, the paired electrolysis of hydrosilane on the anode and chlorosilane on the cathode proceeds to give disilane (equation 50)[57].

$$\text{Me}_2\text{PhSiH} \xrightarrow[\text{CuCl or CuCl}_2]{-2e,\,-\text{H}^+} \text{Me}_2\text{PhSiCl} \xrightarrow[-\text{Cl}^-\,\text{(halogen exchange)}]{\text{BF}_4^-} \text{Me}_2\text{PhSiF} \quad (48)$$
$$>90\% \qquad\qquad 90\%$$

$$\text{Me}_2\text{PhSiH} \xrightarrow[\text{Cl}^-]{-2e,\,-\text{H}^+} [\text{Me}_2\text{PhSiCl}] \xrightarrow[\text{Cu anode}]{e} 1/2\,\text{Me}_2\text{PhSiSiPhMe}_2 \quad (49)$$

at anode $\quad$ MePh$_2$SiH + Cl$^-$ − 2e ⟶ MePh$_2$SiCl + H$^+$

$\qquad\qquad$ MePh$_2$SiCl + e ⟶ 1/2 (MePh$_2$Si)$_2$ $\qquad$ (50)

at cathode $\quad$ H$^+$ + e ⟶ 1/2 H$_2$

Di- and trihydrosilanes are also active on the anode[58,59]. For example, methylphenylsilane, phenylsilane and hexylsilane have their decomposition oxidation potentials at +0.6–+0.7 V vs Ag/AgCl in TBAF/MeCN[59]. Constant current anodic oxidation of these hydrosilanes at platinum electrodes in an undivided cell provide linear polysilanes as shown in Table 14[59]. In the polymerization, silyl radical cations generated by the one-electron oxidation on the anode might attack another monomer to form a Si—Si bond needed for the propagation while reductive generation of hydrogen gas is occurring on the cathode. Although the molecular weight of the polymer formed is limited, this method does not require any sacrificial electrodes, which are necessary for electroreductive polymerization of chlorosilanes (see Section III.D.1).

J. Heterocycles Containing a Si Atom

Anodic oxidation of phenothiasilane and phenazasiline provides the corresponding neutral and oxidized polymers, respectively (equation 51)[60].

N-Trimethylsilylpyrrole is anodically oxidized at a higher oxidation potential (E_p: +0.55 V vs Fc/Fc$^+$) than simple unsilylated pyrrole (E_p: +0.47 V)[61]. This is due to the steric effect of the electrode. N-Trimethylsilylpyrrole is electrochemically polymerized to an electroconductive polymer (equation 52). The polymer has good film producibility and conductivity as high as that for polypyrrole[61].

$$\text{(51)}$$

$$\text{(52)}$$

III. CATHODIC REDUCTION OF ORGANOSILICON COMPOUNDS

A. Arylsilanes

Owing to a $d\pi-p\pi$ effect due to unoccupied silicon 3d orbitals, arylsilanes are reduced at less negative potentials compared with the corresponding unsilylated aromatic compounds. As shown in Table 15, a trimethylsilyl group decreases the reduction potential $E_{1/2}$ of naphthalenes while a t-butyl group increases $E_{1/2}$. An additional silyl substitution causes further anodic shift of the cathodic potential[62–64]. Although monosilylation of naphthalene causes a decrease in $E_{1/2}$, the addition of a second and third silicon atom to the chain (Me$_5$Si$_2$, Me$_7$Si$_3$) does not change $E_{1/2}$ any more[65].

The reduction potentials of o-, m- and p-nitrophenylsilanes are also less negative by 20–40 mV compared to that of unsilylated nitrobenzene[66].

The polarographic study of 1,4-bis(trimethylsilyl)benzene and its polyphenyl derivatives suggests that the first one-electron reduction wave is reversible[67].

TABLE 15. Reduction potential of silyl-substituted aromatics[62,63]

Substrate	Substituents	$E_{1/2}$ (V vs Ag/AgCl)[a]
Naphthalene	none	−2.58
	1-SiMe$_3$	−2.52
	2-SiMe$_3$	−2.57
	1,4-(SiMe$_3$)$_2$	−2.38
	2,6-(SiMe$_3$)$_2$	−2.44
	1-t-Bu	−2.64
	1,4-(t-Bu)$_2$	−2.67
Biphenyl	none	−2.05[b]
	4-SiMe$_3$	−2.03[b]
	4,4′-(SiMe$_3$)$_2$	−1.94[b]
	4-t-Bu	−2.10[b]

[a]Determined by polarography in Bu$_4$NI/DMF.
[b]vs Ag/AgNO$_3$ (sat. in DMF).

TABLE 16. Reduction potentials of cyclooctatetraene derivatives[a]

Substrate R	$E^1_{1/2}$ (V vs SCE)	$E^2_{1/2}$ (V vs SCE)
H	−1.61	−1.92
SiMe$_3$	−1.64	−1.82
t-Bu	−1.88	−2.00

[a]Determined by linear sweep voltammetry in HMPA.

Cathodic reduction of arylsilanes in methylamine using LiCl as a supporting electrolyte in an undivided cell gives 1,4-cyclohexadiene derivatives. The reaction seems to proceed in a manner similar to the Birch-type reduction. The cathodic reduction in a divided cell provides desilylation products (equation 53)[68].

(53)

Linear sweep voltammetry of silylated cyclooctatetraene shows that silylation decreases the first reduction potential but increases the second one (Table 16). The effect on the second reduction potential seems to be due to the stabilization of the dianion by the $d\pi-p\pi$ electron-withdrawing effect of the silyl group[69].

B. Acylsilanes

Cathodic reduction potentials of acylsilanes have been determined by polarography in Et$_4$NI/DMF[70] or by cyclic voltammetry in Et$_4$NClO$_4$/MeCN using a glassy carbon cathode[40] as shown in Table 17.

Controlled potential cathodic reduction of benzoylsilane provides benzyl as a main product[40]. Two possible mechanisms are illustrated in equation 54.

C. Polysilanes

One characteristic of Si—Si bonds is the low energy level of the LUMO. The LUMO levels also decrease with increasing chain length of polysilanes.

20. Electrochemistry of organosilicon compounds

TABLE 17. Cathodic reduction potentials of acylsilanes and related compounds[40,70]

Substrate	$E_{1/2}$ (V vs Hg pool)	
Ph–C(=O)–SiMe$_3$	−1.30	(−1.98)[a]
Ph–C(=O)–t-Bu	−1.58	
Ph–C(=O)–SiMe$_2$Ph	—	(−1.80)[a]
2-NaPh–C(=O)–SiMe$_3$	−1.17	
2-NaPh–C(=O)–t-Bu	−1.32	

[a] E_p (V vs Ag/AgCl) determined by cyclic voltammetry[40].

(54)

Reduction of cyclic polysilanes is possible. Permethylcyclopolysilanes can be reduced chemically and electrochemically at low temperatures to form anion radicals. Interestingly, permethylcyclopolysilanes, $(Me_2Si)_n$, where $n = 5, 6$ and 7, give only $(Me_2Si)_5{}^-$ upon reduction with alkali metal while $(Me_2Si)_5$ and $(Me_2Si)_6$ give the corresponding anion radicals, respectively, upon electrochemical reduction[71]. Electrochemical reduction of $(Me_2Si)_6$ in HMPA provides $Me_{11}Si_6{}^-$, which is a stable and useful intermediate that can be derivatized with a variety of electrophiles[72].

TABLE 18. Reduction potentials (E_p) of disilenes

$$\begin{array}{c} R^1 \quad\quad R^2 \\ \diagdown \quad / \\ Si=Si \\ / \quad \diagdown \\ R^2 \quad\quad R^1 \end{array}$$

Substrate		E_p (V)^a
R^1	R^2	
Mes^b	Mes	−2.12
Xyl^c	Xyl	−2.03
Tbp^d	Tbp	−2.20
Mes	t-Bu	−2.66
Mes	Ad^e	−2.64

^{a} vs Fc/Fc$^+$.
^{b} Mes = 2,4,6-trimethylphenyl.
^{c} Xyl = 2,4-dimethylphenyl.
^{d} Tbp = 4-*tert*-butyl-2,6-dimethylphenyl.
^{e} Ad = 1-adamantyl.

Cathodic reduction of 1,1,1-trimethyl-2,2,2-triphenyldisilane under constant current in MeCN provides triphenylsilane quantitatively (equation 55). The reaction involves an electrochemically initiated chain reaction[73]. The source of oxygen in $(Me_3Si)_2O$ may be the residual water in acetonitrile.

$$Ph_3Si-SiMe_3 \xrightarrow[MeCN]{e} \underset{quant.}{Ph_3SiH} + (Me_3Si)_2O \qquad (55)$$

Cathodic reduction potentials of disilenes were determined by cyclic voltammetry[56]. As shown in Table 18, tetraaryldisilenes are reduced at less negative potentials than dialkyldiaryl derivatives. This is in sharp contrast to the fact that anodic oxidation potentials are similar for both types of these disilanes (see Table 13).

D. Halosilanes

1. Cathodic reduction of halosilanes

Dessy and coworkers first studied cathodic behavior of halosilanes using polarography, constant potential electrolysis and triangular voltammetry[74]. They confirmed the formation of Ph_3SiH in the reduction of Ph_3SiCl in DME using a mercury cathode at −3.1 V vs AgClO$_4$/Ag (−2.5 V vs SCE). The cathodic reduction of Ph_2SiCl_2 at −1.3 V vs SCE under the same conditions gave Ph_2SiH_2.

On the other hand, Hengge and colleagues first showed the possibility of the Si−Si coupling through cathodic reduction of various mono- and disilanes in DME using a platinum cathode and a mercury or lead anode without control of the applied potential[75,76]. In this case, the mercury and lead anodes work as a 'sacrificial anode'.

Allred and coworkers investigated cathodic reduction of Me_3SiCl at a platinum cathode in MeCN[77]. They pointed out that the Si−Si bond formation was affected by the supporting electrolytes. Namely, Bu$_4$NCl was effective while Bu$_4$NClO$_4$ gave only siloxane.

Corriu and colleagues also re-examined the cathodic reduction of halosilanes in anhydrous DME. They found that Ph_3SiCl and Me_3SiCl gave the corresponding disilanes in reasonable yields (equation 56). In this case, a large amount of siloxanes was also formed[78].

$$R_3SiCl \xrightarrow[\substack{-2.6 \text{ V vs SCE} \\ \text{Hg cathode}}]{e} 1/2 \; R_3Si-SiR_3 \quad (56)$$
$$R = Ph \; (42\%)$$
$$R = Me \; (28\%)$$

Generally, it is quite difficult to determine the precise reduction potentials of halosilanes because halosilanes such as Me_3SiCl are readily hydrolyzed during the measurement. Corriu and coworkers determined the precise reduction potential of Ph_3SiCl in anhydrous 0.1 M TBAP/DME using polarography[79]. Thus, Ph_3SiCl showed a one-electron single irreversible cathodic peak ($E_{1/2}$: -2.4 V vs SCE), which was not affected by addition of phenol[80]. Therefore, generation of Ph_3Si radical species was suggested in this case.

Umezawa, Fuchigami, Nonaka and coworkers found that organodichloromonosilanes such as Ph_2SiCl_2, $PhMeSiCl_2$ and Me_2SiCl_2 could be electroreductively polymerized by using a platinum cathode in Bu_4NBF_4/DME to give the corresponding oxygen-free organopolysilanes in 43, 40 and 26% yields, respectively (equation 57)[81].

$$RR'SiCl_2 \xrightarrow{ne} \left(\!\!\begin{array}{c} R \\ | \\ -Si- \\ | \\ R' \end{array}\!\!\right)_n \quad \begin{array}{l} R = R' = Ph \; (43\%) \\ R = Ph, R' = Me \; (40\%) \\ R = R' = Me \; (26\%) \end{array} \quad (57)$$

They also found that the yield of oxygen-free polysilane increased markedly by electrolysis of a 1 : 1 mixture of Me_2SiCl_2 and Ph_3SiCl[82]. Because Ph_3SiCl has a less negative reduction potential than that of Me_2SiCl_2[81], the bulkier Ph_3SiCl is reduced to the corresponding silyl anion which nucleophilically attacks Me_2SiCl_2 (equation 58, route a). This is desirable and the resulting $[Ph_3Si \, (Me_2Si)_n Cl]$ is more easily reduced than Me_2SiCl_2. This may improve the yield of oxygen-free polysilanes if the polymerization is then initiated by the reduction of Ph_3SiCl at a less negative potential than the reduction potential of Me_2SiCl_2 itself.

$$\begin{array}{c} Ph_3SiCl \xrightarrow{2e^- - Cl^-} Ph_3Si^- \xrightarrow[(b)]{Ph_3SiCl - Cl^-} Ph_3SiSiPh_3 \end{array} \quad (58)$$

(with branches showing routes (a) and (b) leading to $Ph_3Si-(Me_2Si)_n-SiPh_3$ (10%) and $Ph_3Si-(Me_2Si)_n-H$ (68%) via intermediate $Ph_3Si-(Me_2Si)_{(n-1)}-Si^-$ with Me substituents)

Recently, three research groups found almost independently that reactive metal anodes such as Al, Hg, Cu, Ag and Mg electrodes were highly effective for Si—Si bond formation from halosilanes even in an undivided cell.

Umezawa, Nonaka and coworkers demonstrated efficient electroreductive polymerization of Me_2SiCl_2 by using polarity-alternated electrolysis with sacrificial aluminum

TABLE 19. Cathodic Si—Si coupling of chlorosilanes

Chlorosilanes		Ratio B/A	Cathode	Anode	Electrolyte[a]	Product (Yield %)	Reference
A	B						
Me$_3$SiCl			Mg	Mg	A	Me$_3$SiSiMe$_3$ (82)	87
Me$_2$PhSiCl			Mg	Mg	A	Me$_2$PhSiSiPhMe$_2$ (92)	87
Me$_2$PhSiCl			Pt	Hg	B	Me$_2$PhSiSiPhMe$_2$ (84)	85
Me$_2$PhSiCl			Mg	Mg	B	Me$_2$PhSiSiPhMe$_2$ (50)	88
Me$_2$PhSiCl			Mg	Mg	C	Me$_2$PhSiSiPhMe$_2$ (75)	88
MePh$_2$SiCl			Mg	Mg	A	MePh$_2$SiSiPh$_2$Me (77)	87
MePh$_2$SiCl			Cu	Cu	B	MePh$_2$SiSiPh$_2$Me (83)	86
MePh$_2$SiCl			Pt	Hg	B	MePh$_2$SiSiPh$_2$Me (89)	85
MePh$_2$SiCl			Pt	Al	B	MePh$_2$SiSiPh$_2$Me (40)	85
Ph$_3$SiCl			Mg	Mg	A	Ph$_3$SiSiPh$_3$ (85)	87
MePh$_2$SiCl	Me$_3$SiCl	2.2	Pt	Hg	B	MePh$_2$SiSiMe$_3$ (94)	85
MePh$_2$SiCl	Me$_3$SiCl	2.6	Cu	Cu	B	MePh$_2$SiSiMe$_3$ (83)	86
MePhSiCl$_2$	Me$_3$SiCl	10.3	Cu	Cu	B	Me$_3$Si–SiPMe–SiMe$_3$ (61)	86
						Me$_3$Si–(SiPhMe)$_2$–SiMe$_3$ (23)	86
MePhSiCl$_2$	Me$_3$Si–SiMe$_2$Cl	4	Cu	Cu	B	Me$_3$Si–SiMe$_2$–SiMePh–Me$_2$Si–SiMe$_3$ (79)	86

[a] A=LiClO$_4$/THF, B=Bu$_4$NClO$_4$/DME, C=Bu$_4$NClO$_4$/THF.

electrodes (equation 59)[83].

$$\text{Me}_2\text{SiCl}_2 \xrightarrow[\substack{2\text{F mol}^{-1} \\ \text{Al electrodes}}]{+e} \underset{81\%}{\left(\begin{array}{c} \text{Me} \\ | \\ -\text{Si}- \\ | \\ \text{Me} \end{array}\right)_n} \quad \begin{array}{l} M_n = 2.5\text{--}4 \times 10^3 \\ M_w / M_n = 1.58 \end{array} \quad (59)$$

Bordeau and coworkers also reported that cathodic reduction of Me_2SiCl_2 without solvent using a sacrificial aluminum anode and a stainless steel cathode in an undivided cell produces polydimethylsilane with a very high current efficiency[84].

Kunai, Ishikawa and coworkers found that cathodic Si—Si coupling took place efficiently by using a mercury or silver anode and a platinum cathode in DME[85]. They also showed that disilanes, trisilanes, tetrasilanes and pentasilanes were readily obtained in high yields by use of copper nets as the cathode and anode[86].

On the other hand, Shono, Kashimura and coworkers found that alternating the polarity of Mg electrodes in LiClO_4/THF is quite effective for the formation of Si—Si bonds (Table 19)[87,88]. This method was successfully applied to the preparation of polysilanes. Sonication resulted in a marked increase in the yields of polysilanes as shown in Table 20[87,88]. Under higher concentration of dichlorosilane, polysilane of higher molecular weight was obtained.

The electroreduction of dichlorosilane also gave a polysilane in low yield, although its molecular weight is much higher than that of the polymer prepared by the reduction with Na (equation 60)[87]. Under similar conditions, ladder polysilanes were also formed (equation 61)[88].

$$\text{Me}_2\text{ClSiSiClMePh} \xrightarrow[\substack{\text{LiClO}_4 / \text{THF} \\ \text{Mg electrodes}}]{e, 8.6 \text{ F mol}^{-1}} \underset{5\%}{\left(\begin{array}{cc} \text{Me} & \text{Me} \\ | & | \\ -\text{Si}- & \text{Si}- \\ | & | \\ \text{Me} & \text{Ph} \end{array}\right)_n} \quad \begin{array}{l} M_n = 7170 \\ M_w/M_n = 1.8 \end{array} \quad (60)$$

TABLE 20. Cathodic formation of polysilanes[88]

$$\text{PhMeSiCl}_2 \xrightarrow[\substack{\text{LiClO}_4 / \text{THF} \\ \text{Mg electrodes}}]{e} \left(\begin{array}{c} \text{Ph} \\ | \\ -\text{Si}- \\ | \\ \text{Me} \end{array}\right)_n$$

Dichlorosilane Concentration(M)	Electricity (F mol^{-1})	Yield (%)	M_n	M_w/M_n	Mg electrode	
					Sonication	Alternation
0.33	4	43	5200	1.5	O	O
0.33	4	17	3900	1.4	×	O
0.33	4	7	4000	1.4	O	×
0.67	4	79	9900	2.1	O	O
6.3	2.2	43	18000	2.1	O	O
12	0.5	8	31000	1.8	O	O

$$\text{PhSiCl}_3 \xrightarrow[\text{Mg electrodes}]{\underset{\text{LiClO}_4 / \text{THF}}{e, 6 \text{ F mol}^{-1}}} \left(\begin{array}{c} \text{Ph} - \text{Si} - \\ | \\ \text{Ph} - \text{Si} - \\ | \\ \text{Ph} - \text{Si} - \\ | \end{array} \right)_n \quad \begin{array}{l} M_n = 7100 \\ M_w / M_n = 2.2 \end{array} \quad (61)$$

71%

Polycarbosilanes have also attracted much interest as starting material for silicon carbide fiber production. Reactive metal such as Mg[88], Cu[86,89,90] and Al[91] electrodes have been shown to be highly effective for electrochemical synthesis of polycarbosilanes, as shown in Table 21.

The alternating magnesium electrode system is also applicable to the Si–Ge[92] or Si–Sn[93] bond formation (equations 62–64).

$$\text{R}_2\text{PhSiCl} + \text{Me}_3\text{GeCl} \xrightarrow[3\text{F mol}^{-1}]{2e} \text{R}_2\text{PhSiGeMe}_3 + (\text{Me}_3\text{Ge})_2 \quad (62)$$
$$(1 : 3) \qquad \qquad \text{R} = \text{Ph (60\%)} \quad (30\%)$$
$$\text{R} = \text{Me (52\%)}$$

$$\text{PhMeSiCl}_2 + \text{PhMeGeCl} \xrightarrow[3.8 \text{ F mol}^{-1}]{e} \begin{array}{cc} \text{Ph} & \text{Ph} \\ | & | \\ -(\text{Si})_n - (\text{Ge})_n - \\ | & | \\ \text{Me} & \text{Me} \end{array} \quad (63)$$
$$(1 : 1.04) \qquad \qquad 33\% \text{ (Ge content 33\%)}$$
$$M_n = 20600$$

$$\text{Me}_3\text{SiCl} + \text{Me}_3\text{SnCl} \xrightarrow[6\text{F mol}^{-1}]{2e} \text{Me}_3\text{SiSnMe}_3 + (\text{Me}_3\text{Sn})_2 \quad (64)$$
$$(10 : 1) \qquad \qquad 50\% \qquad 41\%$$

Recently, Hengge and coworkers reported that the use of a silicon carbide and a hydrogen electrode instead of sacrificial anodes was effective for the electrochemical coupling of halosilanes without formation of metal chlorides (equation 65)[94].

$$\text{Me}_3\text{SiCl} \xrightarrow[\text{Fe cathode}]{\underset{\text{Et}_4\text{NBF}_4/\text{THF-HMPA (1:1)}}{e}} \text{Me}_3\text{SiSiMe}_3 \quad (65)$$
$$\text{SiC anode (65\%)}$$
$$\text{Hydrogen anode (73\%)}$$

Cathodic formation of disilane from bulky dichlorosilane was reported as shown in equation 66[95]. However, this reaction seems to be difficult to reproduce.

$$\text{Ar}_2\text{SiCl}_2 \xrightarrow[\text{Hg cathode, Ag anode}]{\underset{\text{Bu}_4\text{NClO}_4 / \text{DME}}{e}} \begin{array}{c} \text{Ar} \\ \diagdown \\ \text{Ar} \end{array} \text{Si} = \text{Si} \begin{array}{c} \text{Ar} \\ \diagup \\ \text{Ar} \end{array} \quad (66)$$
$$\text{Ar} = \text{Mesityl (20 \%)}$$

2. Cathodic reduction of halomethylsilanes

Since silicon stabilizes an α-carbanion by d π–p π interaction, silicon should promote electron transfer to a carbon–halogen bond which generates the α-carbanion. In fact,

TABLE 21. Electrosynthesis of polycarbosilanes

Substrate Concentration(M)	Conditions Anode	Conditions Electrolyte	Electricity (F/mol)	Yield (%)	M_n	M_w/M_n	Reference
ClPhMeSi—⟨C₆H₄⟩—SiPhMeCl (0.27)	Mg	LiClO$_4$/THF	8.0	73	9450	1.88	88
ClPhMeSi—⟨C₆H₄⟩—SiPhMeCl (0.24)	Cu	Bu$_4$NBPh$_4$/DME	6.7	27	3900[a]	2.1	90
ClPhMeSi—⟨C₆H₄⟩—SiPhMeCl (0.24)	Cu	Bu$_4$NBPh$_4$/DME	3.1	19	6100[a]	3.4	90
ClMeEtSi—⟨C₆H₄⟩—SiEtMeCl (0.27)	Mg	LiClO$_4$/THF	8.0	50	8570	1.61	88
ClPhMeSiCH$_2$CH$_2$SiPhMeCl (0.14)	Cu	Bu$_4$NBPh$_4$/DME	3.5	13	61000[a]	16	89
ClPhMeSiCH$_2$CH$_2$SiPhMeCl (0.12)	Cu	Bu$_4$NBPh$_4$/DME	12.4	17	120000[a]	8.0	89
ClMe$_2$SiCH$_2$CH$_2$SiMe$_2$Cl (0.18)	Al	Bu$_4$NCl/THF	2.0	68	2500	—	91

[a]M_w.

TABLE 22. Reduction potentials of (iodomethyl)silanes and related compounds[96]

Substrate	$E_{1/2}$ (V vs SCE)
Et_3SiCH_2I	-1.54^a
$CF_3(CH_2)_2SiMe_2CH_2I$	-1.33^a
$PhSiMe_2CH_2I$	-1.43^a
$p\text{-}FC_6H_4SiMe_2CH_2I$	-1.40^a
MeI	-1.63^b

^{a} 0.09 M KCl/75% alcohol.
^{b} 0.05 M Et_4NBr/75% dioxane–water.

(iodomethyl) trialkylsilanes were shown by polarographic study to be reduced at less negative potentials compared with unsilylated simple iodoalkanes, as shown in Table 22[96].

Cathodic reduction of (chloromethyl)dimethylchlorosilane using an aluminum anode provided polycarbosilanes besides a large amount of di- and trisilacyclic compounds (equation 67)[97]. On the other hand, the electrolysis in the presence of Me_2SiCl_2 gave bis(dimethylchlorosilyl)methane, a useful polycarbosilane precursor (equation 68)[97]. Similarly, polycarbosilanes were prepared from dichlorocarbosilanes using sacrificial aluminum electrodes in DME (equation 69)[91].

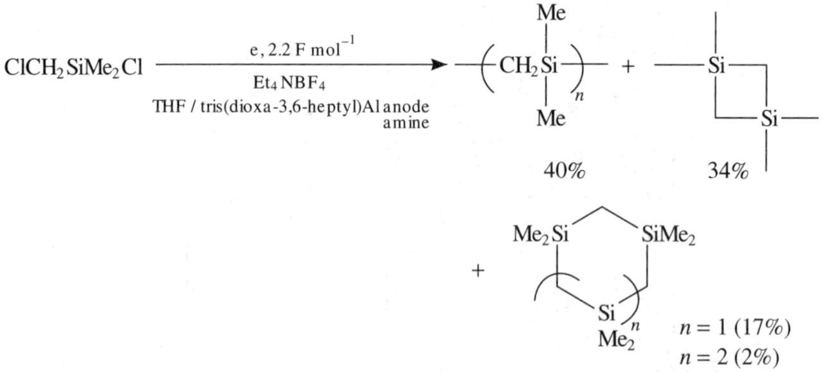

$$ClCH_2SiMe_2Cl \xrightarrow[\substack{Et_4NBF_4 \\ \text{THF / tris(dioxa-3,6-heptyl)Al anode} \\ \text{amine}}]{e, 2.2 \text{ F mol}^{-1}} \left(CH_2\underset{Me}{\overset{Me}{Si}}\right)_n + \text{Si—Si—Si} \quad (67)$$

40% 34%

$$+ \quad Me_2Si\underset{\underset{Me_2}{Si}_n}{\overset{}{\diagup\diagdown}} SiMe_2 \quad n=1\ (17\%)\ n=2\ (2\%)$$

$$ClCH_2SiMe_2Cl + Me_2SiCl_2 \xrightarrow[\text{Al anode}]{e, 2.2 \text{ F mol}^{-1}} ClMe_2SiCH_2SiMe_2Cl \quad (68)$$

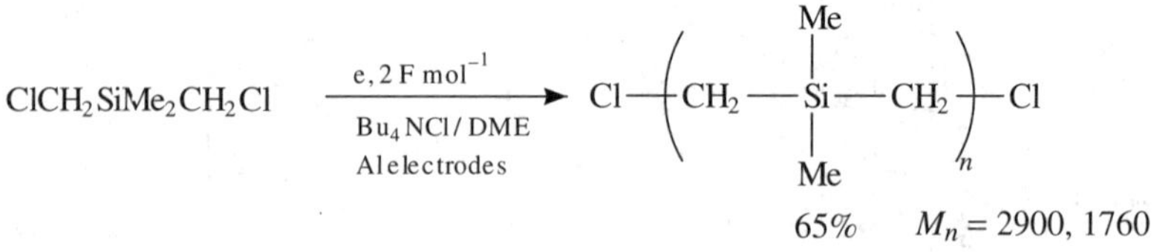

60%

$$ClCH_2SiMe_2CH_2Cl \xrightarrow[\substack{Bu_4NCl/DME \\ \text{Al electrodes}}]{e, 2 \text{ F mol}^{-1}} Cl\left(CH_2-\underset{Me}{\overset{Me}{Si}}-CH_2\right)_n Cl \quad (69)$$

65% $M_n = 2900, 1760$

3. Utilization as electrophiles

Since a carbon–halogen bond is more easily reduced than a silicon–halogen bond, cathodic reduction of organic halides such as allyl, benzyl, aryl and vinyl halides in the presence of a halosilane provides the corresponding organosilicon compounds. This is a convenient method for the introduction of a silyl group into organic molecules and a mechanism involving a cathodically generated carbanion intermediate is suggested.

The cathodic reduction of benzyl chlorides in the presence of a chlorosilane using a divided cell provides the corresponding benzylsilanes while the electrolysis in an undivided cell gives chlorinated benzylsilanes (equations 70–71)[98]. Similarly, vinyl halides provide the corresponding vinylsilanes as stereoisometic mixtures (equation 72)[99,100].

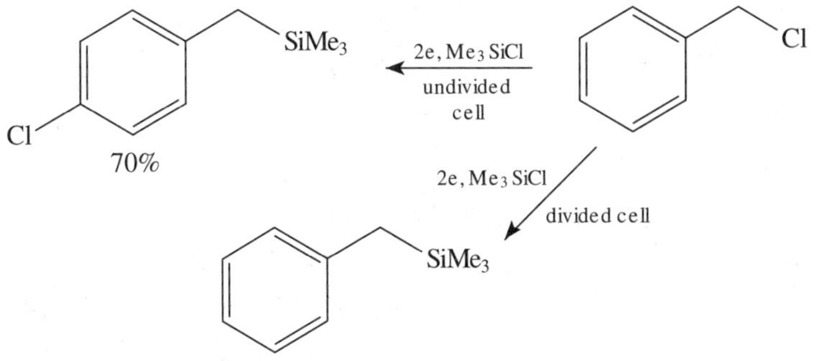

(70)

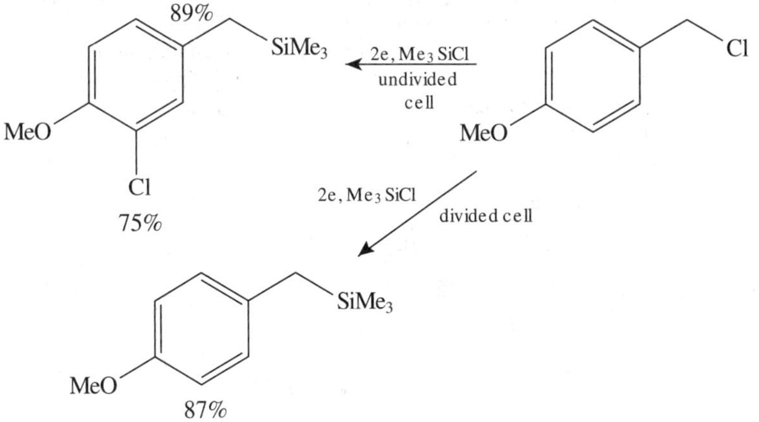

(71)

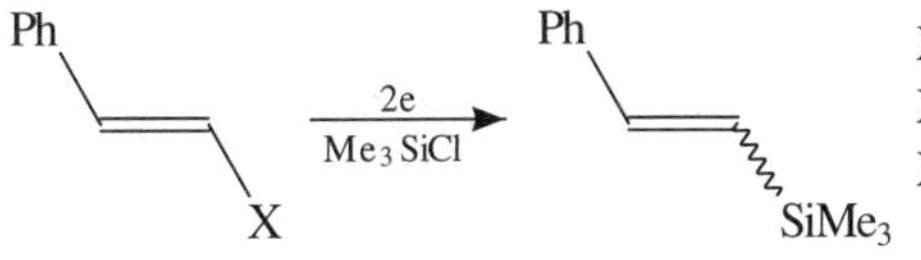

(72)

X = I, 51% (E/Z = 57/43)
X = Br, 23% (E/Z = 68/32)
X = Cl, 11% (E/Z = 78/22)

The regioselectivity of the reaction of allylic halides depends on the nature of the halosilanes: trimethylsilyl and dimethylphenylsilyl groups are introduced at the less substituted end of the allyl group, whereas the dimethylsilyl group is introduced at both ends of the allyl group (equation 73)[99,100]. This method is also highly chemoselective as demonstrated by selective monosilylation of p-bromoiodobenzene and p-bromocinnamyl chloride (equations 74 and 75). Whereas the electroreductive silylation of organic halides is efficient, this method cannot however, be applied to alkyl halides and aryl chlorides. Moreover, as already mentioned, benzene ring chlorination takes place simultaneously in the electrosynthesis of benzylsilanes in an undivided cell (equations 70 and 71).

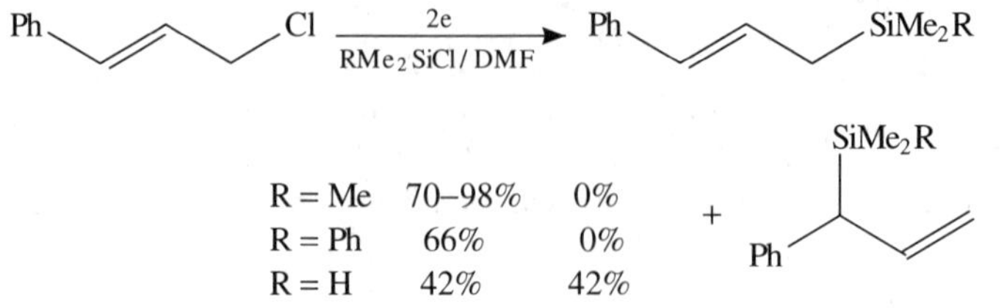

$$Ph\diagdown\!\!\diagup\!\!\diagdown\!\!Cl \xrightarrow[RMe_2SiCl/DMF]{2e} Ph\diagdown\!\!\diagup\!\!\diagdown\!\!SiMe_2R$$

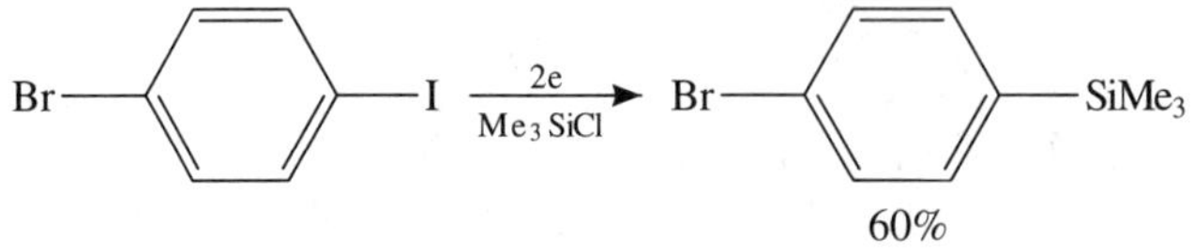

$$+ \quad Ph\diagdown\!\!\overset{SiMe_2R}{\underset{}{\diagup}}\!\!\diagdown \quad (73)$$

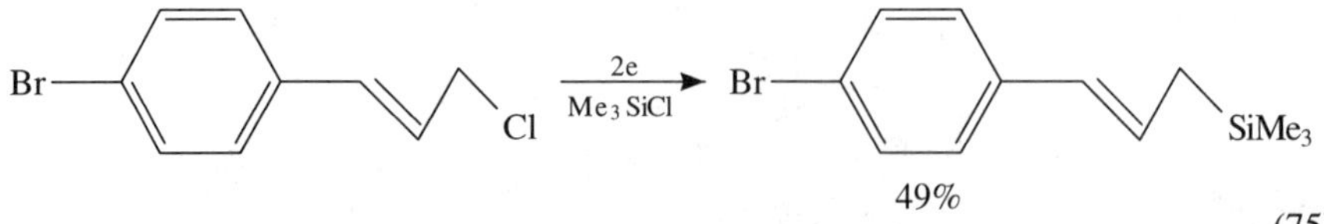

R = Me	70–98%	0%
R = Ph	66%	0%
R = H	42%	42%

$$Br\text{-}C_6H_4\text{-}I \xrightarrow[Me_3SiCl]{2e} Br\text{-}C_6H_4\text{-}SiMe_3 \quad (74)$$

60%

$$Br\text{-}C_6H_4\text{-}CH=CH\text{-}CH_2Cl \xrightarrow[Me_3SiCl]{2e} Br\text{-}C_6H_4\text{-}CH=CH\text{-}CH_2SiMe_3 \quad (75)$$

49%

Reactive metal anodes are quite effective in reactions of aryl chlorides in the presence of a large excess of chlorotrimethylsilane in an undivided cell using a sacrificial aluminum anode in THF/HMPA (4 : 1) which provide the corresponding aryltrimethylsilanes (equations 76 and 77)[101,102]. This method does not require any diaphragms since oxidation of the aluminum anode takes place predominantly as the anodic reaction (equation 76). When excess amount of electricity is passed, *trans*-tris(trimethylsilyl)chlorohexa-1,3-dienes are formed predominantly (equation 78).

at cathode : $Ar-X + 2e \longrightarrow Ar^- + X^-$

$Ar^- + Me_3SiCl \longrightarrow ArSiMe_3 + Cl^-$ (76)

at anode : $2/3\,Al - 2e \longrightarrow 2/3\,Al^{3+}$

$$R\text{-}C_6H_4\text{-}Cl \xrightarrow[\substack{Al\ anode \\ 2.2\ F\ mol^{-1}}]{2e,\ Me_3SiCl} R\text{-}C_6H_4\text{-}SiMe_3 \quad (77)$$

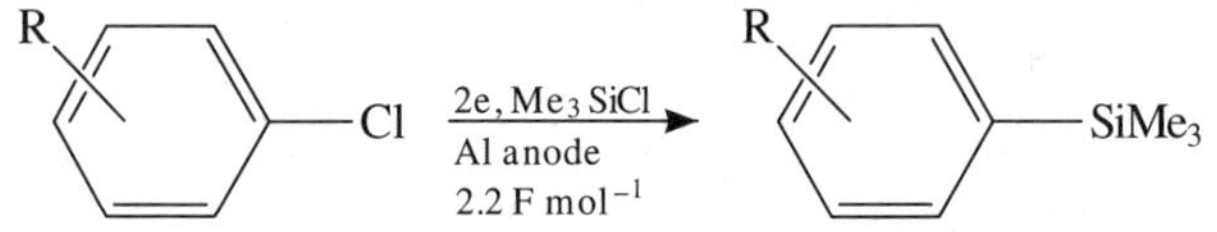

R = H (84%) ; *o*-Me (72%) ; *m*-Me (80%) ; *p*-Me (70%)

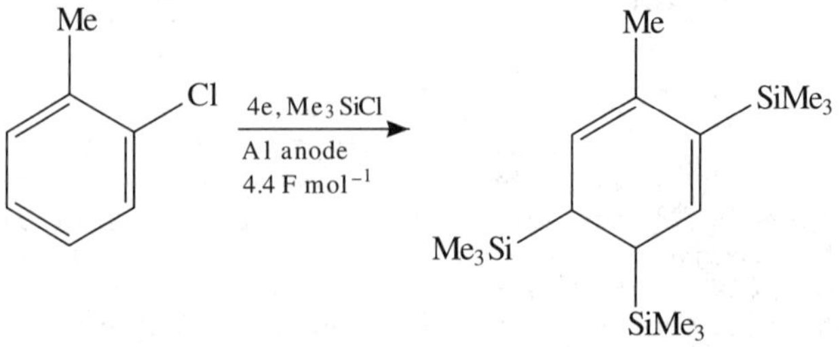

$$\text{o-MeC}_6H_4Cl \xrightarrow[\substack{Al\ anode \\ 4.4\ F\ mol^{-1}}]{4e,\ Me_3SiCl} \text{(bis-silylated chlorohexadiene)} \quad (78)$$

84% (*trans/cis* = 92/8)

Cathodic trimethylsilylation of simple arenes such as benzene and toluene is also possible by this method and gives bis(trimethylsilyl)cyclohexa-1,4-dienes[103]. 1-Methoxy-3,6-bis(trimethylsilyl)cyclohexa-1,4-diene, formed in equation 79, is a useful precursor to ketoprofen.

$$\text{ArR} \xrightarrow[\text{2 F mol}^{-1}]{\text{2e, Me}_3\text{SiCl, Al anode}} \text{product} \quad (79)$$

R = H, 62% (*trans/cis* = 7 : 3)
R = Me, 58% (*trans* only)
R = MeO, 70% (single isomer)

Polysilylation of *o*-dichlorobenzene was also successfully carried out similarly to provide mono-, di-, tetra- and hexasilylated products in fairly good yields, respectively, depending on the amount of electricity passed. This is a useful route to silylated cyclo-C_6 products (equation 80)[104]. Similarly, selective electrochemical mono- and polysilylation of halothiophenes was performed (equation 81)[105].

(80)

2.2 F mol^{-1} → 90%
4.4 F mol^{-1} → 87%
6.6 F mol^{-1} → 89%
8.8 F mol^{-1} → 40%

$$\text{(81)}$$

Selective electrochemical silylation of polychloromethane (equation 82)[106] and benzal chloride (equation 83)[107] is also possible using zinc and magnesium as a sacrificial anode.

$$XCCl_3 + Me_3SiCl \longrightarrow \begin{cases} Me_3SiCCl_3 & (2e, 2.2\ F\ mol^{-1}, Zn\ anode)\quad X = Cl\ (94\%) \\ (Me_3Si)_2CCl_2 & (4e, 2.2\ F\ mol^{-1}, Mg\ anode)\quad X = Cl\ (68\%) \\ Me_3SiCHCl_2 & (2e, 2.2\ F\ mol^{-1}, Zn\ anode)\quad X = H\ (94\%) \end{cases} \quad (82)$$

$$F\text{-}C_6H_4\text{-}CHCl_2 \xrightarrow[\text{Mg anode, DMF}]{2e,\ Me_3SiCl} F\text{-}C_6H_4\text{-}CHClSiMe_3\ (73\%) \xrightarrow[\text{Mg anode, DMF}]{2e,\ Me_3SiCl} F\text{-}C_6H_4\text{-}CH(SiMe_3)_2\ (91\%) \quad (83)$$

(Trifluoromethyl)trimethylsilane is a highly useful trifluoromethylating reagent. Efficient electrochemical trimethylsilylation of bromotrifluoromethane has been developed (equation 84)[108].

$$CF_3Br + Me_3SiCl \xrightarrow[\substack{\text{PhOMe/HMPA} \\ \text{Al anode}}]{2e} F_3CSiMe_3 \quad 73\% \quad (84)$$

Trifluoromethylbenzene is also trimethylsilylated electrochemically to provide α, α-difluoro-α-(trimethylsilyl)toluene (equation 85)[109].

$$PhCF_3 + Me_3SiCl \xrightarrow[\substack{PhOMe/HMPA \\ Al\ anode}]{2e} PhCF_2SiMe_3 \quad 70\% \tag{85}$$

Recently, selective electrochemical trimethylsilylation of tetrachlorocyclopropene has been achieved to provide 1-(trimethylsilyl)trichlorocyclopropene, which has been converted into hexakis(trimethylsilyl)-3,3'-bicyclopropenyl upon successive electrolysis (equation 86)[110–112].

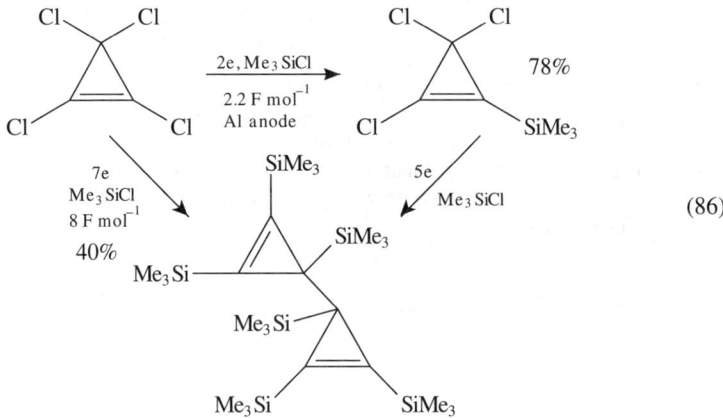

(86)

Electrochemical synthesis of various cyclic alkylsilanes has been performed similarly[113]. It should be noted that 5-silaspiro[4,4]nonane is formed despite the high probability of polymer formation due to the high functionality of the silicon. Such high selectivity in the electrochemical ring closure seems to be due to the orientating effect of an electrode in the course of an irreversible reduction of a carbon–halogen bond in the monosilylated intermediate (equations 87 and 88).

$$Br(CH_2)_n Br + Me_2SiCl_2 \xrightarrow[\substack{THF/DMF \\ Al\ anode}]{2e} (CH_2)_n\ SiMe_2$$

$$\begin{aligned} n &= 2\ (0\%) \\ &= 3\ (18\%) \\ &= 4\ (91\%) \\ &= 5\ (68\%) \\ &= 6\ (57\%) \end{aligned} \tag{87}$$

$$Br(CH_2)_4 Br + SiCl_2 \xrightarrow[Al\ anode]{4e} \text{[5-silaspiro[4,4]nonane]} \quad 24\% \tag{88}$$

Various activated olefins can also be employed instead of organic halide for the formation of a carbon–silicon bond. Thus, cathodic reduction of α,β-unsaturated esters, nitriles and ketones in the presence of chlorotrimethylsilane using a sacrificial magnesium anode provides the corresponding β-trimethylsilyl compounds (equation 89)[114].

$$\underset{EWG}{\underset{|}{Ar}}\overset{R}{\underset{|}{C}}=CH_2 + Me_3SiCl \xrightarrow[\text{DMF}]{2e, \text{Mg anode}} \underset{Me_3Si}{\underset{|}{Ar}}\overset{R}{\underset{|}{C}}-\overset{|}{C}\underset{EWG}{\underset{|}{H}}$$

Ar = Ph, R = H, EWG = CO$_2$Et (66%)
Ar = p-ClC$_6$H$_4$, R = H, EWG = CO$_2$Et (70%)
Ar = Ph, R = EWG = CO$_2$Et (77%)
Ar = Ph, R = H, EWG = CN (70%)
Ar = Ph, R = H, EWG = Ac (68%)

(89)

Similar electroreductive silylation of a series of unsaturated nitrogen compounds such as trimethylsilyl cyanide, cyanamide or isocyanide, bis(trimethylsilyl)carbodiimide and trimethylsilyl isocyanate provide precursors of bis(trimethylsilyl)methylamine, which is useful for organic synthesis (equations 90 and 91)[115]. This electrochemical method is a safer and more economical process compared with the chemical process using an alkali metal such as lithium.

$$Y\text{-}CN + Me_3SiCl \xrightarrow[\text{Al anode}]{e} (Me_3Si)_2CHN\begin{matrix} SiMe_3 \\ \diagdown \\ SiMe_2CH_2R \end{matrix} \quad (90)$$

Y = Me$_3$Si, 78% (R = H/Me$_3$Si = 5/95)
Y = Me$_2$N, 65% (R = H/Me$_3$Si = 25/75)

$$Me_3SiN=C=Y + Me_3SiCl \xrightarrow[\text{Al anode}]{e} (Me_3Si)_2CHN\begin{matrix} SiMe_3 \\ \diagdown \\ SiMe_2CH_2R \end{matrix} \quad (91)$$

Y = O, 56% (R = H/Me$_3$Si = 12/88)
Y = NSiMe$_3$, 75% (R = H/Me$_3$Si = 10/90)

N,N-Disilylenamines of acylsilanes, which are excellent precursors for α-(trimethylsilyl)alkylamine, can be prepared electrochemically using a similar procedure (equation 92)[116].

$$RR'C\begin{matrix} CN \\ \diagup \\ \diagdown \\ OSiMe_3 \end{matrix} + Me_3SiCl \xrightarrow[\substack{4.4\ F\ mol^{-1} \\ \text{Al anode}}]{4e} RR'C=C\begin{matrix} CN \\ \diagup \\ \diagdown \\ N(SiMe_3)_2 \end{matrix} \longrightarrow \underset{Me_3Si}{\overset{RR'}{\diagup\diagdown}}NH_2 \quad (92)$$

R = H, R′ = Et, 75% (E/Z = 88/12)
R = Me, R′ = n-Pr, 82% (E/Z = 58/42)
R = R′ = c-C$_6$H$_{11}$, 75%

Cathodic reduction of acylimidazoles in the presence of chlorotrimethylsilane gives the corresponding acylsilanes in satisfactory yields (equation 93)[117]. Acylsilanes having a functional group such as an alkoxycarbonyl or a chlorine are also synthesized effectively.

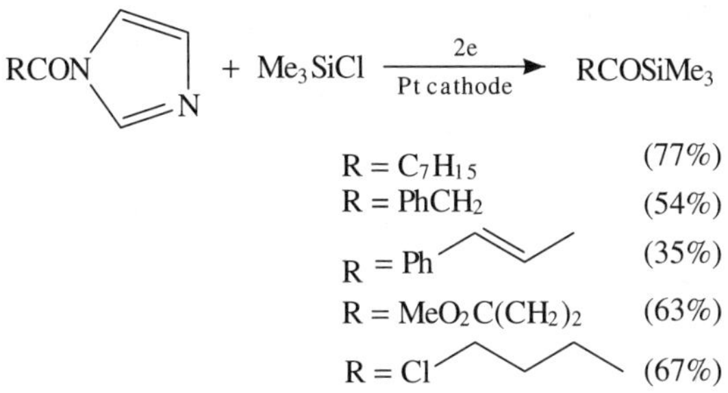

$$RCON\text{(Im)} + Me_3SiCl \xrightarrow[\text{Pt cathode}]{2e} RCOSiMe_3$$

R = C$_7$H$_{15}$ (77%)
R = PhCH$_2$ (54%)
R = Ph–CH=CH– (35%)
R = MeO$_2$C(CH$_2$)$_2$ (63%)
R = Cl(CH$_2$)$_3$– (67%)

(93)

Cathodically induced silylation of unsaturated compounds such as phenylacetylene, styrene and cyclohexene has also been reported (equations 94 and 95)[118,119]. Since chlorotrimethylsilane is reduced at a less negative potential ($E_{1/2}$ = −1.95 V vs Hg pool) compared with phenylacetylene ($E_{1/2}$ = −2.05 V vs Hg pool), cathodic reduction of chlorotrimethylsilane to generate a silyl anion is possible in the presence of phenylacetylene.

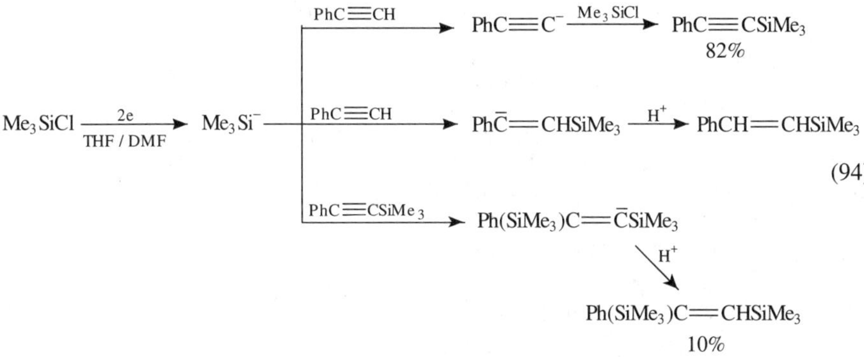

$$Me_3SiCl \xrightarrow[\text{THF / DMF}]{2e} Me_3Si^-$$

PhC≡CH → PhC≡C$^-$ $\xrightarrow{Me_3SiCl}$ PhC≡CSiMe$_3$ 82%

PhC≡CH → PhC̄=CHSiMe$_3$ $\xrightarrow{H^+}$ PhCH=CHSiMe$_3$

PhC≡CSiMe$_3$ → Ph(SiMe$_3$)C=C̄SiMe$_3$ $\xrightarrow{H^+}$ Ph(SiMe$_3$)C=CHSiMe$_3$ 10%

(94)

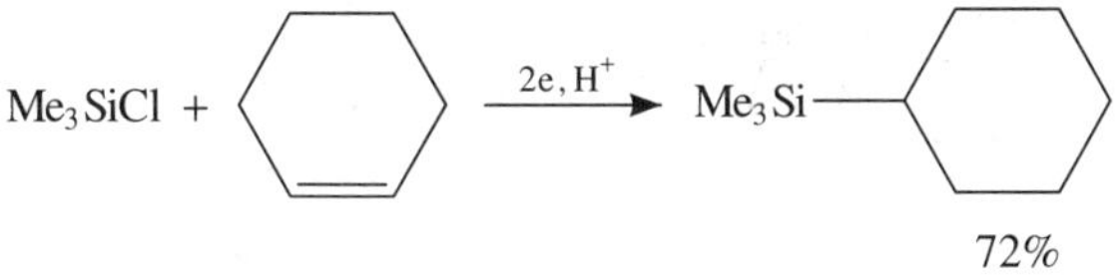

$$Me_3SiCl + \text{cyclohexene} \xrightarrow{2e, H^+} Me_3Si\text{-cyclohexyl}$$

72%

(95)

Cathodic reduction of dichlorosilanes in the presence of 2,3-dimethyl-1,3-butadiene provides cyclic silyl compounds (equation 96)[120].

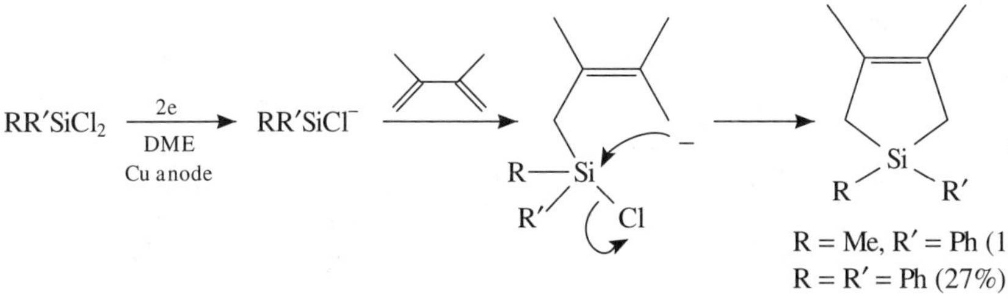

$$RR'SiCl_2 \xrightarrow[\text{Cu anode}]{\substack{2e \\ DME}} RR'SiCl^-$$

R = Me, R' = Ph (10%)
R = R' = Ph (27%)

(96)

It is quite interesting that cathodic intra- and intermolecular coupling of a ketone with a vinylsilane (equations 97 and 98) takes place by using a carbon fiber cathode[121]. The carbon fiber cathode is essential for these reactions.

$$R^1 = Me, R^2 = n\text{-Pr}, R^3 = Me\ (75\%)$$
$$R^1 = Me, R^2 = n\text{-Hex}, R^3 = H\ (72\%)$$

Chlorotrimethylsilane (CTMS) has been shown to promote various cathodic hydrocoupling reactions. Thus, cathodic reduction of a mixture of carbonyl compounds and activated olefins[122] or imines[123] in the presence of CTMS provides carbon–carbon crossed-coupling products efficiently. The cathodic reduction of aromatic esters and an amide to produce aldehydes was also promoted in the presence of CTMS[124].

N-Methyl o-phthalimide and o-phthalic anhydride undergo cathodic trimethylsilylation to give monomeric and dimeric products, respectively (equation 99)[125].

Quite recently, stereoselective electrochemical synthesis of silyl enol ethers using a sacrificial magnesium anode was reported, as shown in equation 100[126].

$$\text{Anode:} \quad Mg \longrightarrow Mg^{2+} + 2e$$

Cathode: 2 pyrrolidinone + 2e $\xrightarrow{-H_2}$ 2 pyrrolidinone anion ⎫ EGB

Ph-CH(R)-C(=O) + Me$_3$SiCl $\xrightarrow{\text{EGB}}$ Ph-C(=CR)-OSiMe$_3$ (100)

R = Ph, 90% (Z only)
R = Me, 100% (Z/E = 95/5)

E. Hydrosilanes

Cathodic reduction of triphenylhydrosilane in Bu$_4$NBF$_4$/MeCN provided triphenylfluorosilane besides a large amount of siloxane. Whereas radical species are likely, the reaction mechanism is still unclear[127].

It is interesting that electrocatalyzed reduction of carbonyl compounds to alcohols takes place by using trialkoxysilanes (equation 101)[128].

$$R^1R^2C=O + HSi(OEt)_3 \xrightarrow[\text{DMF}]{e(cat)} R^1R^2CH\text{-}OSi(OEt)_3 \xrightarrow{H_2O} R^1R^2CH\text{-}OH \quad (101)$$

R^1 = Ph, R^2 = H (80%)
R^1 = Ph, R^2 = t-Bu (69%)

IV. ACKNOWLEDGMENT

The author wishes to thank Professors James Y. Becker of Ben-Gurion University and Tsutomu Nonaka of Tokyo Institute of Technology for encouraging him to write this chapter. Last but not least, he is grateful to Miss Motoko Miyazaki, his graduate student, who assisted in preparing the manuscript.

V. REFERENCES

1. E. M. Geniès and F. El Omar, *Electrochim. Acta*, **28**, 541 (1983).
2. J. Yoshida, in *Topics in Current Chemistry, Vol. 170, Electrochemistry V* (Ed. E. Steckhan), Springer-Verlag, Berlin, 1994, pp. 39–81.
3. R. J. Klingler and J. K. Kochi, *J. Am. Chem. Soc.*, **102**, 4790 (1980).
4. I. Y. Alyev, I. N. Rozhkov and I. L. Knunyants, *Tetrahedron Lett.*, 2469 (1976).
5. B. Hong, M. A. Fox, G. Maier and C. Hermann, *Tetrahedron Lett.*, **37**, 583 (1996).
6. E. M. Geniès and F. El Omar, *Electrochim. Acta*, **28**, 547 (1983).
7. M. Lemaire, W. Büchner, R. Garreau, H. A. Hoa, A. Guy and J. Roncali, *J. Electroanal. Chem.*, **312**, 277 (1991).
8. H. Masuda, Y. Taniki and K. Kaeriyama, *J. Polym. Sci., Part A*, **30**, 1667 (1992).

9. J. Roncali, C. Thobie-Gautier, H. Brisset, J. -F. Favart and A. Guy, *J. Electroanal. Chem.*, **381**, 257 (1995).
10. (a) T. Koizumi, T. Fuchigami and T. Nonaka, *Electrochim. Acta*, **33**, 1635 (1988).
    (b) K. Nishiwaki and J. Yoshida, *Chem. Lett.*, 787 (1996).
11. T. Koizumi, T. Fuchigami and T. Nonaka, *Chem. Express*, **1**, 355 (1986).
12. T. Koizumi, T. Fuchigami and T. Nonaka, *Bull. Chem. Soc. Jpn.*, **62**, 219 (1989).
13. J. Yoshida, T. Murata and S. Isoe, *Tetrahedron Lett.*, **27**, 3373 (1986).
14. (a) M. Ochiai, M. Arimoto and E. Fujita, *Tetrahedron Lett.*, **22**, 4491 (1981).
    (b) T. Fuchigami and K. Yamamoto, *Chem. Lett.*, 937 (1996).
15. K. Ohga and P. S. Mariano, *J. Am. Chem. Soc.*, **104**, 617 (1982).
16. J. Yoshida, T. Murata and S. Isoe, *Tetrahedron Lett.*, **28**, 211 (1987).
17. J. Yoshida, K. Sakaguchi and S. Isoe, *Tetrahedron Lett.*, **28**, 667 (1987).
18. J. Yoshida, K. Sakaguchi and S. Isoe, *Tetrahedron Lett.*, **27**, 6075 (1986).
19. K. D. Moeller, C. M. Hudson and L. V. Tinao-Wooldridge, *J. Org. Chem.*, **58**, 3478 (1993).
20. C. M. Hudson and K. D. Moeller, *J. Am. Chem. Soc.*, **116**, 3347 (1994).
21. B. E. Cooper and W. J. Owen, *J. Organometal. Chem.*, **29**, 33 (1971).
22. J. Yoshida and S. Isoe, *Tetrahedron Lett.*, **28**, 6621 (1987).
23. J. Yoshida and S. Isoe, *Chem. Lett.*, 631 (1987).
24. T. Koizumi, T. Fuchigami and T. Nonaka, *Chem. Lett.*, 1095 (1987).
25. J. Yoshida, S. Matsunaga, T. Murata and S. Isoe, *Tetrahedron*, **47**, 615 (1991).
26. M. Kimura, K. Koie, S. Matsubara, Y. Sawaki and H. Iwamura, *J. Chem. Soc., Chem. Commun.*, 122 (1987).
27. T. Fuchigami, Y. Nakagawa and T. Nonaka, *Tetrahedron Lett.*, **27**, 3869 (1986).
28. T. Fuchigami, K. Yamamoto and Y. Nakagawa, *J. Org. Chem.*, **56**, 137 (1991).
29. T. Takiguchi and T. Nonaka, *Chem. Lett.*, 1217 (1987).
30. K. Suda, J. Watanabe and T. Takanami, *Tetrahedron Lett.*, **33**, 1355 (1992).
31. K. Broka, J. Stradins, I. Sleiksa and E. Lukevics, *Latv. J. Chem.*, **5**, 575 (1992).
32. K. Broka, J. Stradins, V. Glezer, and G. Zelcans and E. Lukevics, *J. Electroanal. Chem.*, **351**, 199 (1993).
33. K. Suda, K. Hotoda, M. Aoyagi and T. Takanami, *J. Chem. Soc., Perkin Trans. 1*, 1327 (1995).
34. J. Yoshida, M. Itoh, S. Matsunaga and S. Isoe, *J. Org. Chem.*, **57**, 4877 (1992).
35. A. S. Romakhin, Yu. A. Babkin, E. V. Nikitin and Yu. M. Kargin, *Zh. Obshch. Khim.*, **58**, 13 (1988); *Chem. Abstr.*, **109**, 100628y (1988).
36. J. Yoshida, T. Maekawa, T. Murata, S. Matsunaga and S. Isoe, *J. Am. Chem. Soc.*, **112**, 1962 (1990).
37. J. Yoshida, H. Tsujishima, K. Nakano and S. Isoe, *Inorg. Chim. Acta*, **220**, 129 (1994).
38. J. Yoshida, S. Matsunaga and S. Isoe, *Tetrahedron Lett.*, **30**, 219 (1989).
39. J. Yoshida, T. Maekawa, Y. Morita and S. Isoe, *J. Org. Chem.*, **57**, 1321 (1992).
40. K. Mochida, S. Okui, K. Ichikawa, O. Kanakubo, T. Tsuchiya and K. Yamamoto, *Chem. Lett.*, 805 (1986).
41. J. Yoshida, S. Matsunaga and S. Isoe, *Tetrahedron Lett.*, **30**, 5293 (1989).
42. H. J. Schäfer, *Angew. Chem., Int. Ed. Engl.*, **20**, 911 (1981).
43. S. Torii, T. Inokuchi, S. Mishima and T. Kobayashi, *J. Org. Chem.*, **45**, 2731 (1980).
44. R. F. Stewart and L. L. Miller, *J. Am. Chem. Soc.*, **102**, 4999 (1980).
45. S. Torii, T. Okamoto and N. Ueno, *J. Chem. Soc., Chem. Commun.*, 293 (1978).
46. T. Shono, H. Ohmizu and N. Kise, *Chem. Lett.*, 1517 (1980).
47. D. Hermeling and H. J. Schäfer, *Angew. Chem., Int. Ed. Engl.*, **23**, 233 (1984).
48. K. Mochida, A. Itani, M. Yokoyama, T. Tsuchiya, S. D. Worley and J. K. Kochi, *Bull. Chem. Soc. Jpn.*, **58**, 2149 (1985).
49. M. Okano and K. Mochida, *Chem. Lett.*, 819 (1991).
50. W. G. Boberski and A. L. Allred, *J. Organometal. Chem.*, **88**, 65 (1975).
51. A. Diaz and R. D. Miller, *J. Electrochem. Soc.*, **132**, 834 (1985).
52. M. Okano and K. Mochida, *Chem. Lett.*, 701 (1990).
53. J. Y. Becker and E. Shakkour, *Tetrahedron Lett.*, **33**, 5633 (1992).
54. Z.-R. Zhang, J. Y. Becker and R. West, *Electrochim. Acta*, in press.
55. J. Y. Becker, M.-Q. Shien and R. West, *Electrochim. Acta*, **40**, 2775 (1995).
56. B. D. Shepherd and R. West, *Chem. Lett.*, 183 (1988).
57. A. Kunai, T. Kawakami, E. Toyoda, T. Sakurai and M. Ishikawa, *Chem. Lett.*, 1945 (1993).

58. Y. Kimata, H. Suzuki, S. Satoh and A. Kuriyama, *Chem. Lett.*, 1163 (1994).
59. Y. Kimata, H. Suzuki, S. Satoh and A. Kuriyama, *Organometallics*, **14**, 2506 (1995).
60. G. Casalbore-Miceli, G. Beggiato, N. Camaioni, L. Favaretto, D. Pietropaolo and G. Poggi, *Ann. Chim.*, **82**, 161 (1992).
61. M. Okano A. Toda and K. Mochida, *Bull. Chem. Soc. Jpn.*, **63**, 1716 (1990).
62. A. G. Evans, B. Jerome and N. H. Rees, *J. Chem. Soc., Perkin Trans. 2*, 447 (1973).
63. M. D. Curtis and A. L. Allred, *J. Am. Chem. Soc.*, **87**, 2554 (1965).
64. F. Correa-Duran, A. L. Allred, D. E. Glover and D. E. Smith, *J. Organometal. Chem.*, **49**, 353 (1973).
65. C. G. Pitt, R. N. Carey and E. C. Toren, Jr., *J. Am. Chem. Soc.*, **94**, 2554 (1972).
66. H. Watanabe, M. Aoki, H. Matsumoto, Y. Nagai and T. Sato, *Bull. Chem. Soc. Jpn.*, **50**, 1019 (1977).
67. A. L. Allred and L. W. Bush, *J. Am. Chem. Soc.*, **90**, 2554 (1968).
68. C. Eaborn, R. A. Jackson and R. Pearce, *J. Chem. Soc., Perkin Trans. 1*, 2055 (1973).
69. L. A. Paquette, C. D. Wright, III, S. G. Traynor, D. L. Taggart and G. D. Ewing, *Tetrahedron*, **32**, 1885 (1976).
70. H. Bock, H. Alt and H. Seidl, *J. Am. Chem. Soc.*, **91**, 355 (1969).
71. E. Carberry, R. West and E. Glass, *J. Am. Chem. Soc.*, **91**, 5446 (1969).
72. A. L. Allred, R. T. Smart and D. A. Van Beek, Jr., *Organometallics*, **11**, 4225 (1992).
73. M. Okano and K. Mochida, *Denki Kagaku*, **61**, 772 (1993); *Chem. Abstr.*, **119**, 203519z (1993).
74. R. E. Dessy, W. Kitching and T. Chivers, *J. Am. Chem. Soc.*, **88**, 453 (1966).
75. E. Hengge and G. Litscher, *Angew. Chem., Int. Ed. Engl.*, **15**, 370 (1976).
76. E. Hengge and H. Firgo, *J. Organometal. Chem.*, **212**, 155 (1981).
77. A. L. Allred, C. Bradley and T. H. Newman, *J. Am. Chem. Soc.*, **100**, 5081 (1978).
78. R. J. P. Corriu, G. Dabosi and M. Martineau, *J. Organometal. Chem.*, **186**, 19 (1980).
79. R. J. P. Corriu, G. Dabosi and M. Martineau, *J. Chem. Soc., Chem. Commun.*, 457 (1979).
80. R. J. P. Corriu, G. Dabosi and M. Martineau, *J. Organometal. Chem.*, **222**, 195 (1981).
81. M. Umezawa, M. Takeda, H. Ichikawa, T. Ishikawa, T. Koizumi, T. Fuchigami and T. Nonaka, *Electrochim. Acta*, **35**, 1867 (1990).
82. M. Umezawa, M. Takeda, H. Ichikawa, T. Ishikawa, T. Koizumi and T. Nonaka, *Electrochim. Acta*, **36**, 621 (1991).
83. M. Umezawa, H. Ichikawa, T. Ishikawa and T. Nonaka, *Denki Kagaku*, **59**, 421 (1991); *Chem. Abstr.*, **115**, 265356v (1991).
84. M. Bordeau, C. Brain, M. -P. Léger-Lambert and J. Dunodués, *J. Chem. Soc., Chem. Commun.*, 1476 (1991).
85. A. Kunai, T. Kawakami, E. Toyoda and M. Ishikawa, *Organometallics*, **10**, 893 (1991).
86. A. Kunai, T. Kawakami, E. Toyoda and M. Ishikawa, *Organometallics*, **10**, 2001 (1991).
87. T. Shono, S. Kashimura, M. Ishifune and R. Nishida, *J. Chem. Soc., Chem. Commun.*, 1160 (1990).
88. T. Shono, *Kino Zairyo*, **12**, 15 (1992); *Chem. Abstr.*, **117**, 7964 (1992).
89. A. Kunai, E. Toyoda, T. Kawakami and M. Ishikawa, *Organometallics*, **11**, 2899 (1992).
90. A. Kunai, E. Toyoda, T. Kawakami and M. Ishikawa, *Electrochim. Acta*, **39**, 2089 (1994).
91. M. Umezawa, M. Kojima, H. Ichikawa, T. Ishikawa and T. Nonaka, *Electrochim. Acta*, **38**, 529 (1993).
92. T. Shono, S. Kashimura and H. Murase, *J. Chem. Soc., Chem. Commun.*, 896 (1992).
93. T. Ishiwata, T. Nonaka, and M. Umezawa, *Chem. Lett.*, 1631 (1994).
94. Ch. Jammegg, S. Graschy and E. Hengge, *Organometallics*, **13**, 2397 (1994).
95. P. Boudjouk, B. -H. Han and K. R. Anderson, *J. Am. Chem. Soc.*, **104**, 4992 (1982).
96. S. G. Mairanovski, V. A. Ponomarenko, N. V. Barashkova, and A. D. Snegova, *Dokl. Akad. Nauk SSSR*, **134**, 387 (1960); *Chem. Abstr.*, **55**, 16221g (1961); S. G. Mairanovski, V. A. Ponomarenko, N. V. Barashkova and M. A. Kadina, *Izv. Akad. Nauk SSSR, Ser. Khim.*, 1951 (1964); *Chem. Abstr.*, **62**, 6378c (1965).
97. M. Bordeau, C. Biran, P. Pons, M. -P. Léger and J. Dunoguès, *J. Organomet. Chem.*, **382**, C21 (1990).
98. T. Shono, Y. Matsumura, S. Katoh and N. Kise, *Chem. Lett.*, 463 (1985).
99. J. Yoshida, K. Muraki, H. Funahashi and N. Kawabata, *J. Organomet. Chem.*, **284**, C33 (1985).
100. J. Yoshida, K. Muraki, H. Funahashi and N. Kawabata, *J. Org. Chem.*, **51**, 3996 (1986).
101. P. Pons, C. Biran, M. Bordeau, J. Dunoguès, S. Sibille and J. Perichon, *J. Organomet. Chem.*, **321**, C27 (1987).

102. M. Bordeau, C. Biran, P. Pons, M. -P. Léger-Lambert and J. Dunoguès, *J. Org. Chem.*, **57**, 4705 (1992).
103. C. Biran, M. Bordeau, F. Serein-Spirau, M. -P. Léger-Lambert and J. Dunoguès, *Synth. Commun.*, **23**, 1727 (1993).
104. D. Deffieux, M. Bordeau, C. Biran and J. Dunoguès, *Organometallics*, **13**, 2415 (1994).
105. D. Deffieux, D. Bonafoux, M. Bordeau, C. Biran and J. Dunoguès, *Organometallics*, **15**, 2041 (1996).
106. P. Pons, C. Biran, M. Bordeau and J. Dunoguès, *J. Organomet. Chem.*, **358**, 31 (1988).
107. A. J. Fry and J. Touster, *J. Org. Chem.*, **54**, 4829 (1989).
108. G. K. S. Prakash, D. Deffieux, A. K. Yudin and G. A. Olah, *Synlett*, 1057 (1994).
109. M. Bordeau, D. Deffieux, M. -P. Léger-Lambert, C. Biran and J. Dunoguès, Fr. Demande FR 2,681,866 (Cl. C07F7/12); *Chem. Abstr.*, **119**, 72833x (1993).
110. G. K. S. Prakash, S. Quaiser, H. A. Buchholz, J. Casanova and G. A. Olah, *Synlett*, 113 (1994).
111. G. K. S. Prakash, H. A. Buchholz, D. Deffieux and G. A. Olah, *Synlett*, 819 (1994).
112. G. K. S. Prakash, H. A. Buchholz, D. Deffieux and G. A. Olah, *J. Org. Chem.*, **59**, 7532 (1994).
113. V. Jouikov and V. Krasnov, *J. Organomet. Chem.*, **498**, 213 (1995).
114. T. Ohno, H. Nakahiro, K. Sanemitsu, T. Hirashima and I. Nishiguchi, *Tetrahedron Lett.*, **33**, 5515 (1992).
115. S. Grelier, T. Constantieux, D. Deffieux, M. Bordeau, J. Dunoguès, J. -P. Picard, C. Palomo and J. M. Aizpurua, *Organometallics*, **13**, 3711 (1994).
116. T. Constantieux and J. -P. Picard, *Organometallics*, **15**, 1604 (1996).
117. N. Kise, H. Kaneko, N. Uemoto and J. Yoshida, *Tetrahedron Lett.*, **36**, 8839 (1995).
118. V. Jouikov and L. Grigorieva, *Electrochim. Acta*, **41**, 469 (1996).
119. V. Jouikov and G. Salaheev, *Electrochim. Acta*, **41**, 2623 (1996).
120. A. Kunai, T. Ueda, E. Toyoda and M. Ishikawa, *Bull. Chem. Soc. Jpn.*, **67**, 287 (1994).
121. T. Shono, in *Organic Synthesis in Japan, Past, Present, and Future* (Ed. R. Noyori), Tokyo Kagaku Dozin, Tokyo, 1992, pp. 389–392.
122. T. Shono, H. Ohmizu, S. Kawakami and H. Sugiyama, *Tetrahedron Lett.*, **21**, 5029 (1980).
123. T. Shono, N. Kise, N. Kunimi and R. Nomura, *Chem. Lett.*, 2191 (1991).
124. P. -R. Goetz-Schatowitz, G. Struth, J. Voss and G. Wiegand, *J. Prakt. Chem.*, **335**, 230 (1993).
125. T. Troll and G. W. Ollmann, *Tetrahedron Lett.*, **22**, 3497 (1981).
126. D. Bonafoux, M. Bordeau, C. Biran and J. Dunoguès, *J. Organomet. Chem.*, **493**, 27 (1995).
127. M. Okano and K. Mochida, *Bull. Chem. Soc. Jpn.*, **64**, 1381 (1991).
128. M. Kimura, H. Yamagishi and Y. Sawaki, *Denki Kagaku*, **62**, 1119 (1994); *Chem. Abstr.*, **122**, 117417v (1995).

CHAPTER **21**

The photochemistry of organosilicon compounds

A. G. BROOK

Lash Miller Chemical Laboratories, University of Toronto, Toronto M5S 3H6, Canada

I. INTRODUCTION	1234
II. COMPOUNDS CONTAINING A SINGLE SILICON ATOM	1236
A. Acyclic Compounds	1236
1. Photochemical rearrangements of organosilicon compounds	1236
2. Laser-induced reactions involving small organosilicon compounds	1236
3. Miscellaneous organosilicon compounds	1237
B. Cyclic Compounds Containing One Silicon Atom	1238
1. Silacyclopropanes and silacyclopropenes	1238
2. Silicon in four-membered rings	1240
3. Other silacycloalkanes	1242
4. Silaaromatics	1246
III. COMPOUNDS CONTAINING TWO SILICON ATOMS DISILANES	1247
A. Introduction	1247
B. Vinyldisilanes	1248
C. Alkynyldisilanes	1249
D. Aryldisilanes	1251
E. Arylvinyldisilanes	1259
F. Transition Metal-substituted Disilanes and Related Compounds	1260
G. Cyclic Disilanes and Related Species	1262
IV. TRI-, TETRA- AND POLYSILANES	1265
A. Introduction	1265
B. Acyclic Polysilanes	1265
C. Cyclic Polysilanes	1267
1. Homocyclic systems	1267
2. Heterocyclic tri-, tetra- and polysilanes	1269

The chemistry of organic silicon compounds, Vol. 2
Edited by Z. Rappoport and Y. Apeloig © 1998 John Wiley & Sons Ltd

V.	ACYLSILANE PHOTOCHEMISTRY	1270
	A. Simple Acylsilanes	1270
	B. Acyldisilanes	1272
	C. Acylpolysilanes	1273
VI.	SILYLDIAZOALKANES AND SILYLCARBENES	1276
VII.	PHOTOCHEMICAL ROUTES TO SILENES, SILYLENES AND DISILENES	1282
VIII.	PHOTOCHEMICAL REARRANGEMENTS OF SILENES, SILYLENES AND DISILENES	1284
	A. Photolysis of Silenes	1284
	B. Photolysis of Silylenes	1286
	C. Photolysis of Disilenes	1288
IX.	OTHER PHOTOLYSES LEADING TO MULTIPLY BONDED SILICON	1289
X.	PHOTOLYSIS OF MISCELLANEOUS COMPOUNDS	1290
	A. Photochemically-induced Electron Transfer Involving Organosilicon Compounds	1290
	B. Miscellaneous Photolyses of Organosilicon Compounds Where Bonds to Si are Not Directly Involved	1294
	C. Cophotolyses of Various Organosilicon Compounds with Fullerenes	1301
XI.	ADDENDUM	1303
XII.	SUMMARY	1305
XIII.	REFERENCES	1305

I. INTRODUCTION

The dramatic increases in the breadth and scope of the field of organosilicon chemistry in recent years have been significantly influenced by the application of photochemical techniques to the studies. Thus the photolyses of organosilicon compounds, particularly in matrices, have allowed the creation, detection, characterization and, in some cases, isolation of unstable reaction intermediates such as silenes, silylenes and disilenes. In some cases photochemical methods have led to the formation of stable, isolable silenes and disilenes, species previously considered to be unobtainable. Increasingly, the applications of photochemistry have become more sophisticated, with greater concern for mechanistic and theoretical details, particularly for polysilane systems, where considerable industrial interest is found, because of the unusual photochemical behavior often displayed, which can be applied to photoresists, semiconductors and related applications. As well, in the last 3–5 years, considerable effort has been expended in studying cophotolyses of mixtures of an organosilicon compound with a co-reactant (usually an unsaturated compound) because of the recognition that the products formed (which frequently lack the original silyl group) are structures of synthetic interest, not readily prepared by other routes.

The present review mainly deals with publications from 1989 to early 1996, with some key earlier references as background. The photochemistry of organosilicon compounds has been reviewed previously, most recently in a detailed article by Steinmetz in 1995[1]. An earlier review was written by Brook in 1989[2], annual reviews have been published[3] and much material including photochemical studies can be found in general reviews of organosilicon chemistry[4–6]. Other specialized reviews will be noted in the appropriate sections of the text.

The chemistry reviewed herein is organized into the following main categories.

1. Compounds containing only one silicon atom (or, in a few cases, where there is more than one silicon atom but where they are separated from each other by carbon atoms and so behave as a single silicon atom) where the photochemistry observed normally involves cleavage (usually homolytically) of a carbon–silicon bond.

2. Compounds with two or more silicon atoms directly attached to one another, subdivided into sections based first on the number of silicon atoms and then on the carbon functionality attached to the silicon atoms. Frequently, but not exclusively, the main photochemical behavior involves homolysis of a silicon–silicon bond yielding silyl radicals, but in some cases silylenes result directly from the photochemistry. The resulting compounds are frequently the products of a molecular rearrangement.

3. Compounds in which one or more silicon atom(s) is (are) directly attached to a carbonyl group (acylsilanes) or to a diazo group (silyldiazoalkanes) where the main chromophore is the carbofunctional group but where the silicon atom, being strongly coupled, becomes involved in the reaction pathway. Radicals, siloxycarbenes or silenes may result from the acylsilanes, and silylcarbenes result from the silyldiazoalkanes, each of which shows interesting and unusual behavior.

4. Routes to the reactive species silenes, silylenes and disilenes, and the photochemical changes which they themselves undergo.

5. Other photochemical studies, including cases where compounds contain one or more silicon atoms but where the major chemical behavior observed may not involve a bond to silicon. Thus silicon may be used as a 'tether' to locate a pair of reacting functional groups in some desirable geometry, or an arylsilyl group may be an 'antenna' for radiation to promote reactivity at another site in the molecule. Also in this category are many cophotolyses involving an organosilicon compound and an unsaturated co-reactant, leading frequently to complex structures of synthetic interest. Thus these sections might have been titled 'New Synthetic Routes Applying the Photochemistry of Organosilicon Compounds'.

Within these main categories a number of subdivisions will generally be found, usually based on the nature of the groups attached to silicon. The emphasis in the subject matter is chiefly in terms of the chemical transformations effected by the radiation, and little emphasis is placed on the photochemical mechanisms involved, which are dealt with in an accompanying chapter of this treatise[7]. Also, the photochemistry of high molecular weight polysilane polymers, or of polymers containing silicon as part of the backbone of the molecules, is not dealt with in this chapter, but some relevant material will be found in other chapters in this volume.

As a background for what is to follow it is useful to make some general comments about the photochemical behavior of organosilicon compounds. The photolysis of simple organosilicon compounds generally leads to homolysis of a bond to silicon leading to silyl and alkyl radicals which subsequently react in conventional ways including hydrogen abstraction, disproportionation, coupling etc. Generally, short-wavelength radiation is required, and the reactions are not very chemically interesting, at least from the synthetic point of view, because of the multiplicity of compounds formed. However, when there are unsaturated groups present in the molecule, or present in the photolysis mixture, interesting transformations may occur, as the following sections will illustrate.

Organosilicon compounds are very susceptible to photochemical rearrangements. This is because di-, tri- and polysilanes absorb at much longer wavelengths than their all-carbon analogs. This is particularly true for compounds in which a chromophore is directly attached to a silicon atom, as in vinylsilanes, acylsilanes and silyldiazoalkanes. Many simple organosilicon compounds also undergo interesting photochemical rearrangements. 1,2-Photochemical rearrangements of a silyl group from carbon to oxygen are well known with acylsilanes, and 1,2-carbon-to-carbon rearrangements are commonly found in ring-opening reactions of silylcyclopropenes and related species as will be seen below. 1,3-Photochemical rearrangements of a silyl group from carbon to carbon are commonly observed with allyl- and propargyl silanes, or with vinyl- or alkynyldisilanes, or from silicon to oxygen with polysilylacylsilanes. The ease of migration of silicon has been attributed in part to the β-silicon effect[8] whereby silicon stabilizes radical or cationic sites in a β-relationship as a result of hyperconjugative interactions, and to the inherent tendency of silicon to migrate, which has been shown to be much greater than that of hydrogen or alkyl groups in many circumstances[9].

II. COMPOUNDS CONTAINING A SINGLE SILICON ATOM

A. Acyclic Compounds

1. Photochemical rearrangements of organosilicon compounds

A few studies involving photochemical rearrangements of relatively simple organosilicon compounds have recently been reported.

Sakurai and coworkers[10] observed several years ago that benzyltrimethylsilane underwent a photochemical carbon-to-carbon 1,3-silyl migration and more recently[11] observed that allylsilanes such as **1** underwent clean 1,3-silyl rearrangements with inversion of configuration at a silicon stereocenter yielding the isomeric compound **2** as shown in equation 1, when photolyzed at 254 nm. The observed stereochemistry is contrary to the predictions of the Woodward–Hoffmann rules. Similar anti-Woodward–Hoffmann behavior had also been observed in the related thermal rearrangements of β-ketosilanes which occurred with retention of configuration at a chiral silicon stereocenter[12]. A related photochemical 1,3-silyl rearrangement of a vinylpolysilane is described below in Section IV.B.

$$\text{Si*R}_3 = \text{SiMePh(1-Np)}$$

(1)

Silyl-substituted cyclopropenes have been observed to rearrange photochemically giving high yields of allenes. A typical example is shown in equation 2 where the silylcyclopropene **3** apparently ring-opened to the carbene **4** which, following 1,2-silyl migration, led to the allene **5**[13].

(2)

2. Laser-induced reactions involving small organosilicon compounds

In recent years the application of laser techniques has become an important tool in the study of the kinetics of relatively simple reactive organosilicon species (silylenes, silenes etc.). Gaspar and coworkers have reviewed the use of laser techniques to study the generation and reactions of silylenes[14].

Pola and coworkers have employed lasers, frequently using sulfur hexafluoride as a sensitizer, to generate silicon-containing reactive intermediates which polymerize under the reaction conditions, yielding high molecular weight materials of potential interest. Thus laser photolysis of 1-methyl-1-vinyl-1-silacyclobutane **6** gave rise to 2-silaisoprene **7** and ethylene following first-order kinetics. The reactions of siladiene **7** with tetrafluoroethylene to give 1,1-difluoro-2-methyl-2-silabutadiene **8**, were investigated, as was the subsequent polymerization of **8**, as shown in equation 3. The reactions of the siladiene **7** with hexafluoroacetone were also studied[15].

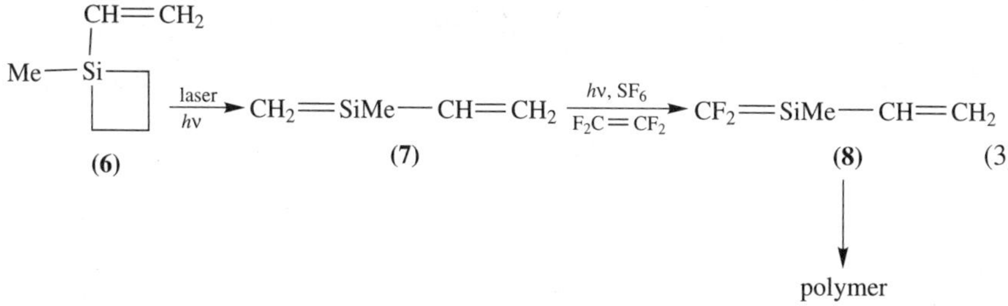

$$\underset{(6)}{Me-Si\underset{|}{\overset{CH=CH_2}{\square}}} \xrightarrow{\text{laser}\atop h\nu} \underset{(7)}{CH_2=SiMe-CH=CH_2} \xrightarrow[F_2C=CF_2]{h\nu,\,SF_6} \underset{(8)}{CF_2=SiMe-CH=CH_2} \quad (3)$$

$$\downarrow$$

polymer

In another study several simple silenes $RR'Si=CH_2$ (R, R' = Me, Vinyl etc.) were formed by laser-powered pyrolysis and were found to form linear polymers, in contrast to the usual behavior of silenes which yield cyclodimers when formed by conventional thermolysis techniques[16]. Reactions of the silenes in the presence of several monomers such as vinyl acetate, allyl methyl ether and methyl acrylate were also studied. Laser-induced decomposition of silacyclobutane and 1,3-disilacyclobutane gave rise to silenes and other oxygen-sensitive deposits[17,18].

3. Miscellaneous organosilicon compounds

The photochemical behavior of numerous organic compounds containing one (or several separated) silicon atoms have been reported in recent years. A few interesting examples are mentioned below.

Pannell and coworkers[18b] prepared the novel compound **9** which contains the four group 14 elements in the order C–Si–Ge–Sn. On photolysis, the products isolated, **10** and **11**, indicate the elimination of dimethylgermylene and elimination of dimethylstannylene, respectively, from the starting material, while **12** indicates radical cleavage of the Ge–Sn bond followed by hydrogen abstraction of the resulting germyl radical, and **13** appears to have been derived from radical dimerization, as shown in Scheme 1. Compound **13** itself suffered further photolysis with loss of dimethylgermylene leading to **14**.

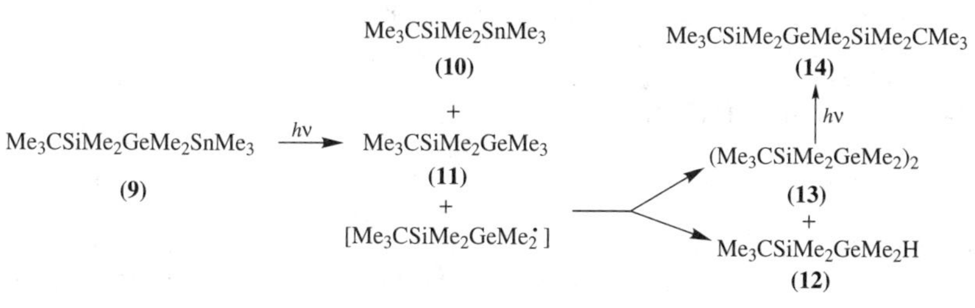

$$\underset{(9)}{Me_3CSiMe_2GeMe_2SnMe_3} \xrightarrow{h\nu} \begin{array}{c} \underset{(10)}{Me_3CSiMe_2SnMe_3} \\ + \\ \underset{(11)}{Me_3CSiMe_2GeMe_3} \\ + \\ [Me_3CSiMe_2GeMe_2^\cdot] \end{array} \begin{array}{c} \underset{(14)}{Me_3CSiMe_2GeMe_2SiMe_2CMe_3} \\ \uparrow h\nu \\ \underset{(13)}{(Me_3CSiMe_2GeMe_2)_2} \\ + \\ \underset{(12)}{Me_3CSiMe_2GeMe_2H} \end{array}$$

SCHEME 1

Hexakis(trimethylsilyl)benzene, **15**, a severely sterically hindered compound which exists in a chair conformation to minimize steric strain, when photolyzed at >300 nm is quantitatively converted to the Dewar benzene isomer **16**[19] or, in part, to the bis(cyclopropenyl) species **17** at 254 nm as shown in Scheme 2[20]. Compound **15** was thermally converted to the bis-allene **18**.

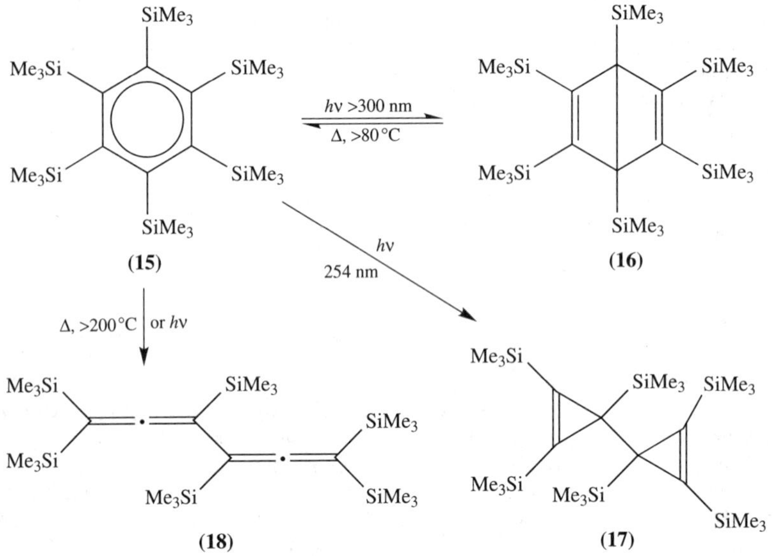

SCHEME 2

B. Cyclic Compounds Containing One Silicon Atom

1. Silacyclopropanes and silacyclopropenes

Silacyclopropanes are commonly formed from the reactions of alkenes with silylenes, which themselves are commonly prepared by photolysis of linear trisilanes (see Section VII). When photolyzed, silacyclopropanes are known to give rise to silylenes or, under other conditions, to undergo 1,3-shifts[2]. Boudjouk and coworkers have prepared a number of silylenes with various substituents on both the carbon and silicon atoms[21], and have shown that significant differences in photochemical behavior are obtained depending on the steric bulk of the substituents. For example, as shown in Scheme 3, the 1,1-di(*t*-butyl)-2,3-dimethylsilirane **19** readily forms di(*tert*-butyl)silylene **20** as does the monomethyl compound **21**, but the unmethylated silirane **22** fails to form the silylene when photolyzed. The differences in behavior were due, it was suggested, to varying degrees of steric strain between the groups on carbon and the groups on silicon[22]. The formation of silylenes from dialkylsiliranes is also discussed in Section VII.

Silacyclopropenes are commonly formed from the addition of a silylene to an alkyne, or in some cases as the result of photolysis of an alkynyldisilane (see Section III.C). Substituted silacyclopropenes have been shown to undergo both 1,2- or 1,3-shifts when photolyzed, yielding silyl-substituted allenes or alkynes, respectively[2]. More complex behavior was observed with methylenesilacyclopropenes such as **23**[23] which ring-opened to a diene, as shown in Scheme 4.

The unusual siladigermacyclopropane **24** on photolysis was observed to decompose to the silagermene **25** and the germylene **26**[24], as shown in equation 4. See Section IV.C.1

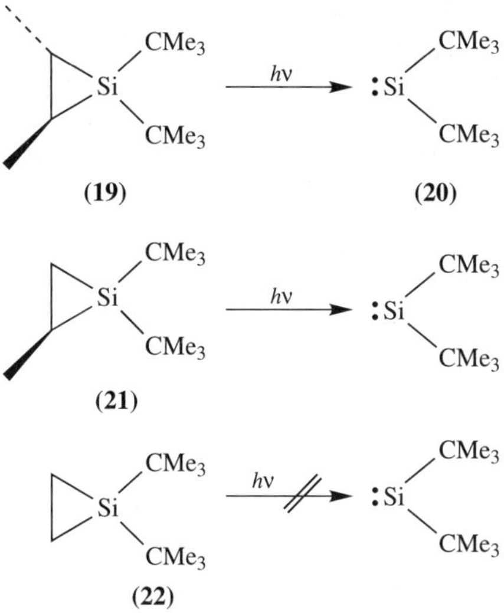

21. The photochemistry of organosilicon compounds

SCHEME 3

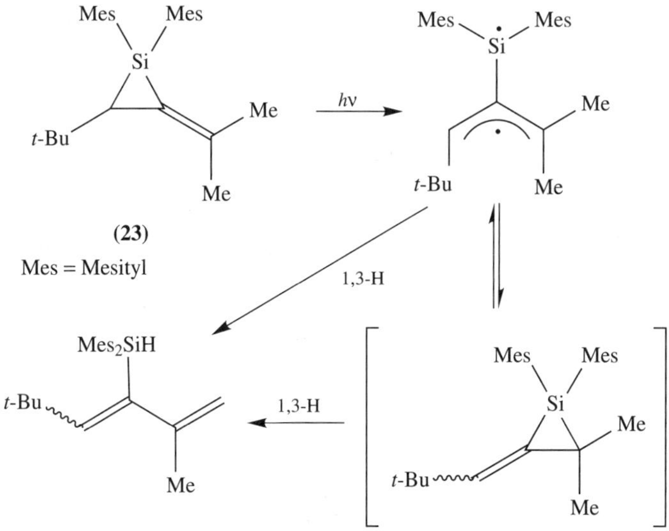

SCHEME 4

for the behavior of the related trisilacyclopropane.

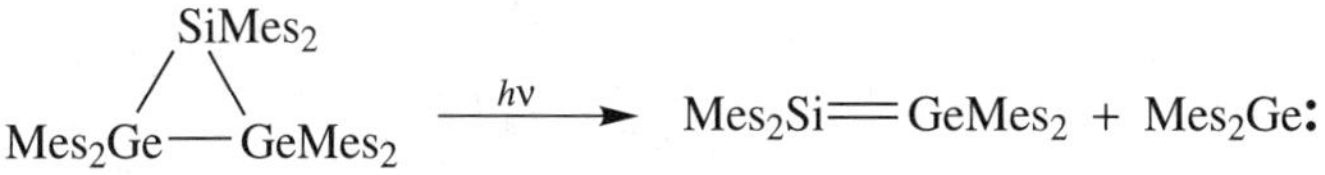

$$Mes_2Ge\text{—}GeMes_2\text{ (with SiMes}_2\text{ bridge)} \xrightarrow{h\nu} Mes_2Si\!\!=\!\!GeMes_2 + Mes_2Ge\!: \qquad (4)$$

(24) (25) (26)

2. Silicon in four-membered rings

A detailed study of the photolysis of the relatively simple silacyclobutane **27** was made by Steinmetz and Bai[25a]. When photolyzed at 185 nm in pentane containing methanol, not only was the initial Z isomer converted to an E,Z mixture (indicative of ring opening and reclosure), but also three ring-opened products, **28**, **29** and **30**, were obtained. Two of these seem clearly to be as the result of methanol addition to the intermediary silenes **31** and **32**, and both the silenes and product **30** were logically explained as arising from intramolecular hydrogen abstraction of diradical intermediates **33** and **34**, arising from the photochemical cleavage of a ring silicon–carbon bond, as shown in Scheme 5. The formation of *cis*- and *trans*-butene and propene in the course of the photolysis were explained as being derived from further cleavage of the intermediary diradicals.

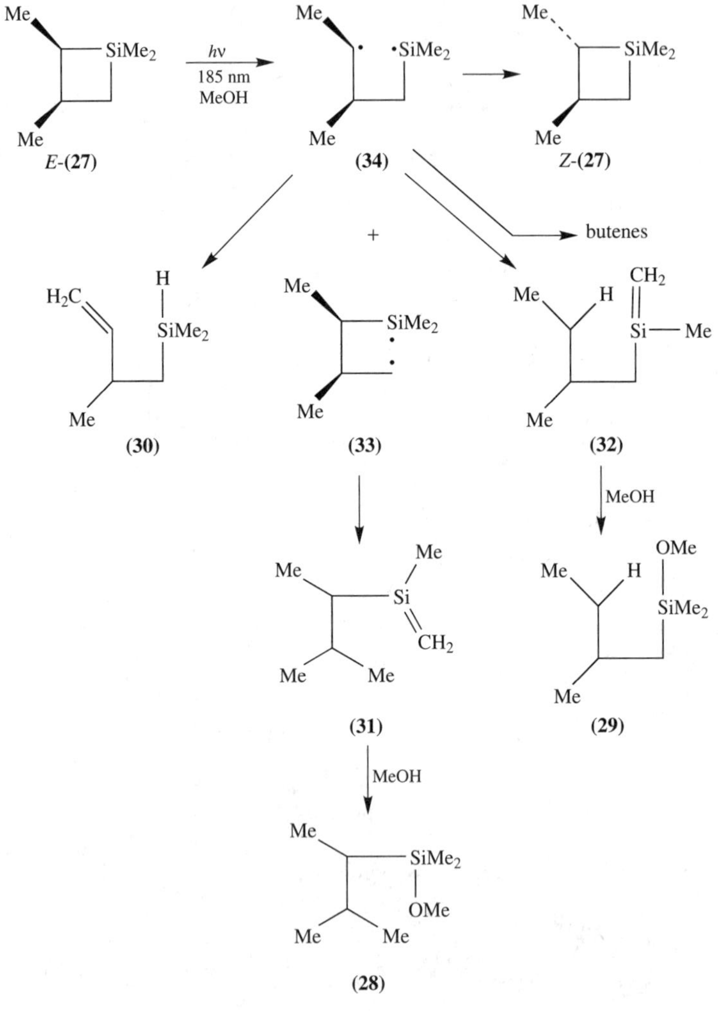

SCHEME 5

The photolysis and flash photolysis of 1,1-diphenyl-1-silacyclobutane has been reinvestigated and much new detail about the reactions of 1,1-diphenylsilene with alcohols and other nucleophiles are reported[25b].

The photolysis and flash photolysis of 1,1-dimethyl-1-sila-2-phenylcyclobut-2-ene, **35**, in the inert solvent cyclohexane have been studied by Conlin, Fink and coworkers[26]. The reaction yielded the siladiene **36**, as shown in equation 5; the kinetics of its reversion to the precursor silacyclobutene **35** were also studied.

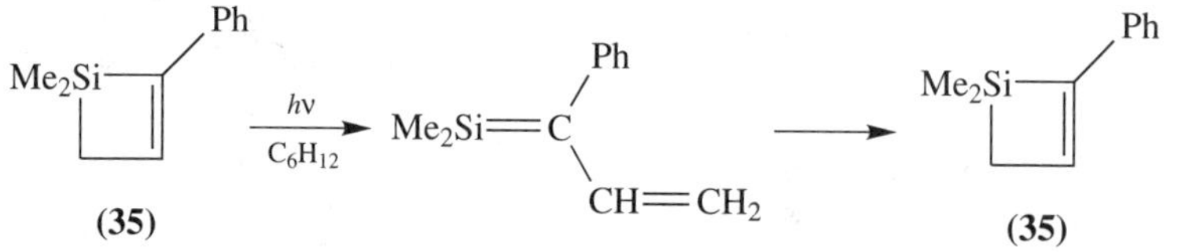

$$(35) \xrightarrow[C_6H_{12}]{h\nu} Me_2Si=C\begin{matrix}Ph\\CH=CH_2\end{matrix} \longrightarrow (35) \qquad (5)$$

Steinmetz and coworkers[27] further investigated the behavior of the related 1,1-dimethyl-1-silacyclobut-2-ene and observed that fragmentation to acetylene and dimethylsilene occurred in addition to electrocyclic ring opening. These two studies are probably the most detailed investigations of siladienes to date.

In a related study Jones and coworkers photolyzed the 1,3-disilacyclobutane **37** and observed that a single diastereomer gave rise to a mixture of products[28]. They suggested that ring cleavage to a diradical **38** occurred which could ring-close to either the starting material or the isomer **39**, or the diradical could dissociate to two molecules of silene **40**, or disproportionate to the alkene **41**, which is photochemically convertible to its isomer **42**, as shown in Scheme 6. These products were proposed based on the structure of their trapping products when the photolysis was carried out in methanol. Attempts to trap the diradical with excess tributyltin hydride failed.

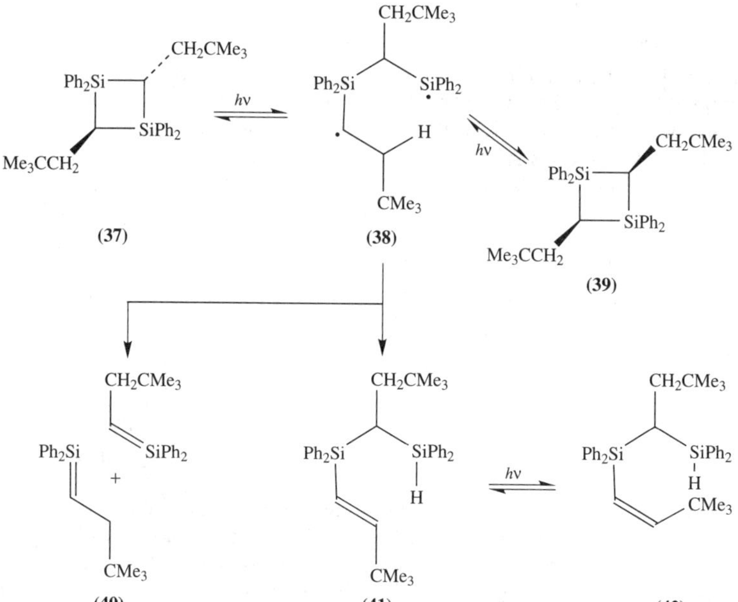

SCHEME 6

Ando and coworkers have recently described the photolysis of the novel 2,4,5,6-tetrathia-1,3-disilabicyclo[2.1.1]pentane, **43**[29]. As shown in Scheme 7, when **43** was photolyzed in the presence of 6 equivalents of trimethylphosphine, sulfur was extruded and the new compound **44** was formed quantitatively. However, when **43** was photolyzed with 10 equivalents of triphenylphosphine, both **44** and the 1,3-disila-2,4-dithiacyclobutane **45** were formed.

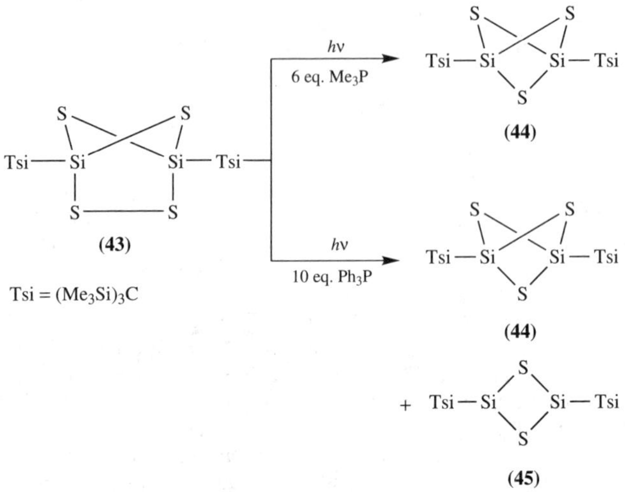

SCHEME 7

3. Other silacycloalkanes

The photolysis at 214 nm of a bridgehead silanorbornene **46** in alcohols has recently been reported by Steinmetz and Chen who proposed that a 1,3-Si to C migration occurred leading to the cyclic silene **47**, which then reacted with the alcohol to give the final product **48**[30]. The reactions are summarized in equation 6.

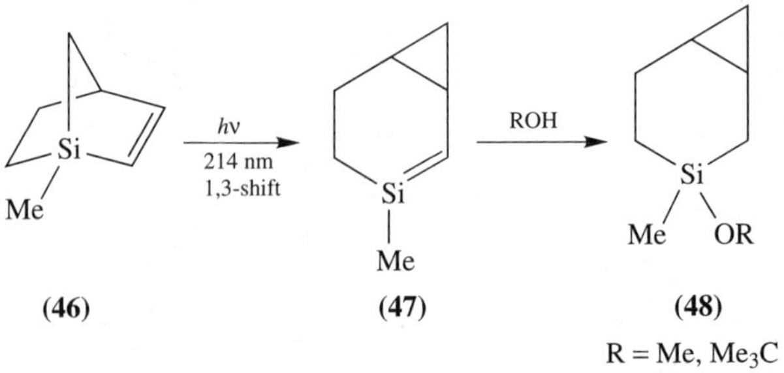

$$R = Me, Me_3C \tag{6}$$

Jones and coworkers observed an interesting rearrangement of a 9,10-bridged dihydroanthracene **49** which on photolysis led, via a diradical **50**, to a mixture of the siladiene **51** and the eight-membered ring silene **52**, as judged by the products of methanol trapping. In addition, the methanol adduct **53** of the diradical was isolated[31]. This is the first example of a silicon–carbon double bond in an eight-membered ring. The reactions are shown in Scheme 8.

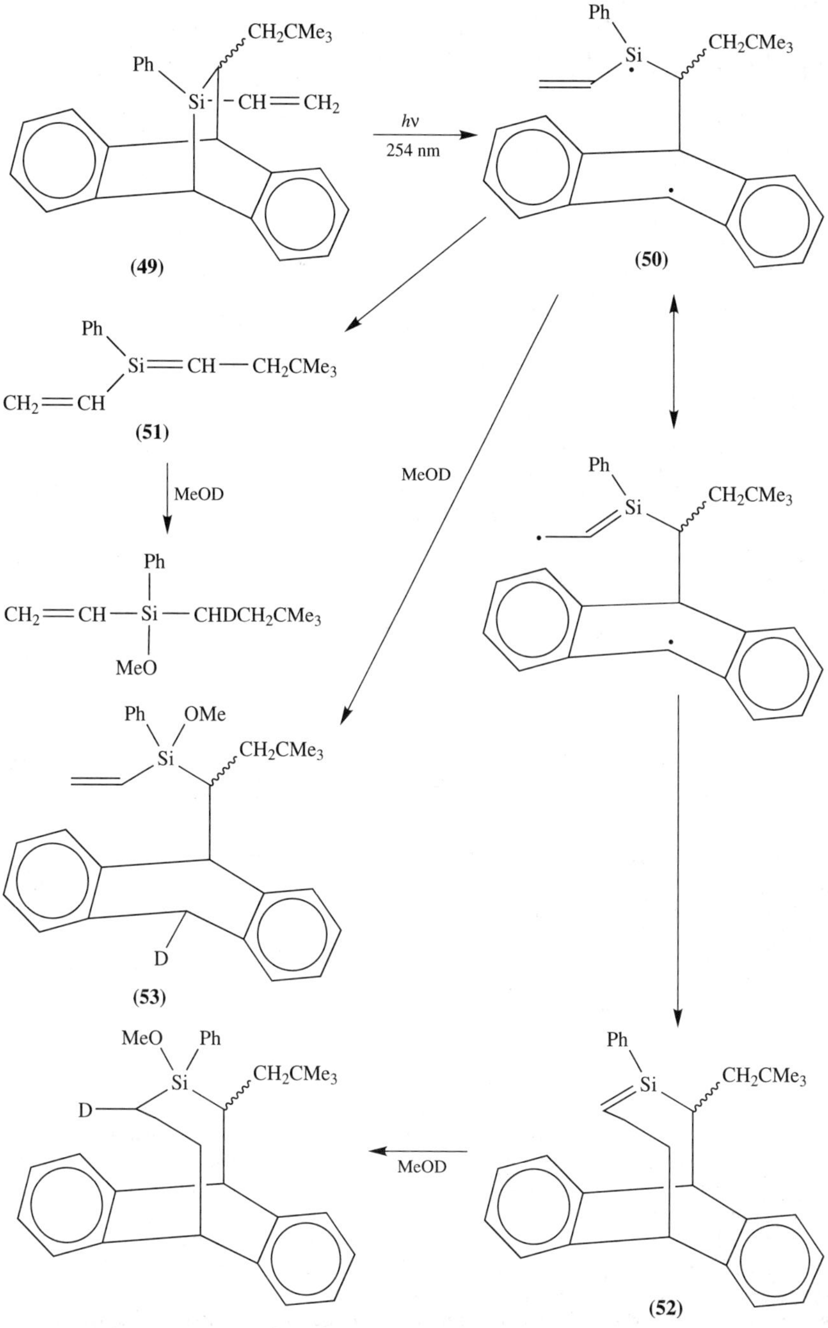

21. The photochemistry of organosilicon compounds

SCHEME 8

The mechanism of photochemical decomposition of the silyl-bridged naphthalene **54** was studied by Nefedov and coworkers using the CIDNP ^{1}H technique[32]. It was suggested that initially an excited diradical **55** was formed, which subsequently decomposed irreversibly to dimethylsilylene and polarized tetraphenylnaphthalene (equation 7). Other studies interpreting the UV absorptions during the photolysis of **54** have been reported[33]. See Section VII for other reactions leading to silylenes.

Compound **54** was also utilized as a photochemical source of dimethylsilylene to prepare the unusually substituted silirene **56**[34] as shown in equation 8.

Several detailed studies by Steinmetz and coworkers[35–37] have concerned the photolysis, in the presence of alcohols or other trapping reagents, of cyclic unsaturated compounds containing one or two silicon atoms in the ring. Thus photolysis of the 1,4-disilacyclohept-2-ene **57** gave products of intramolecular rearrangement, **58**, ring opening, **59**, and evidence, based on the structure of a dimer and of a trapping product, that the initial *cis* isomer, **57**, was converted in part to the *trans* geometric isomer **60**. A β-silylcarbocationic intermediate was proposed, as shown in Scheme 9, to account for the rearrangement products isolated. Formation of the *trans* disilacycloheptene reflects the flexibility and decreased ring strain present when silicon atoms replace carbon in medium-sized rings.

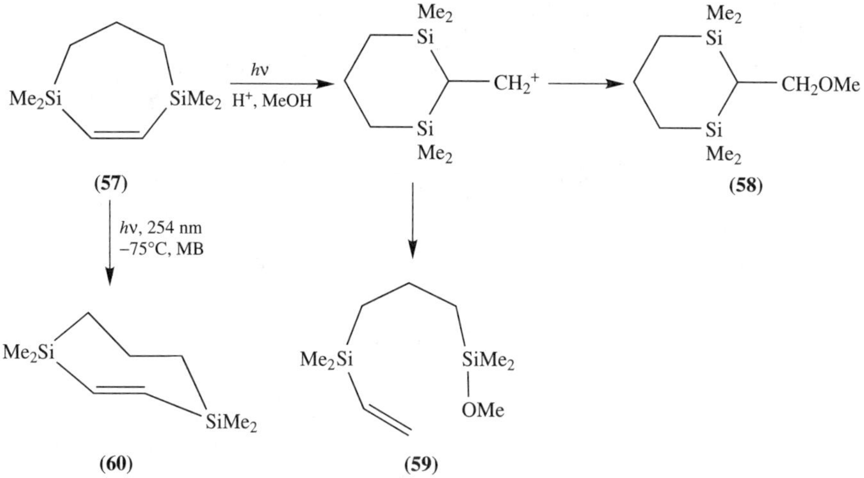

SCHEME 9

The simpler 1-silacyclopent-3-ene **61** gave evidence, based on the reaction products identified, both for silylene extrusion, trapped as **62** by deuteriomethanol, as well as for a 1,3-Si shift to a silirane intermediate **63** which ring-opened during reaction with the alcohol present yielding **64** (Scheme 10). In contrast, the 1-silacyclopent-2-ene **65** on photolysis in alcohols gave products best accounted for by protonation of the silacyclopentene α to silicon to form a β-silyl cation **66**, which then reacted with alcohol either at carbon β to silicon to give **67** or at silicon to give the product **68**. However, photolysis in the presence of Me₃SiOMe or acetone suggested that a silene intermediate **69** was formed as a result of a 1,3-C shift, and trapping led to **70** and **71**, respectively (Scheme 10). This photochemical behavior is very different from that observed for the analogous homocyclic cyclopentenes, demonstrating that the presence of the silicon atom had a significant influence on the course of the reaction.

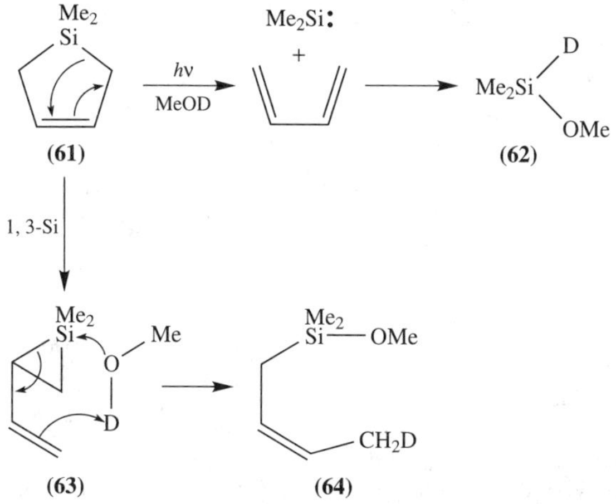

SCHEME 10. (*continued*)

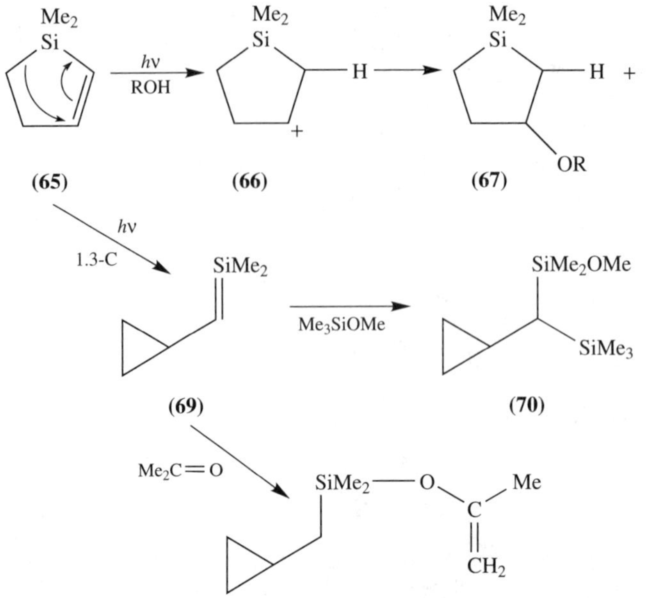

SCHEME 10. (*continued*)

4. Silaaromatics

Photochemical methods have been used in the past to prepare a variety of rather unstable silaaromatic compounds. The photolyses of cyclic unsaturated silyldiazo compounds, which yield silabenzenes and/or silafulvenes, are described in Section VI.

The flash photolysis of several 9,10-dihydrosilaanthracene derivatives **72** in argon matrices led to the detection of 9-silaanthracenes, **73**, characterized by their UV spectra[38] (equation 9).

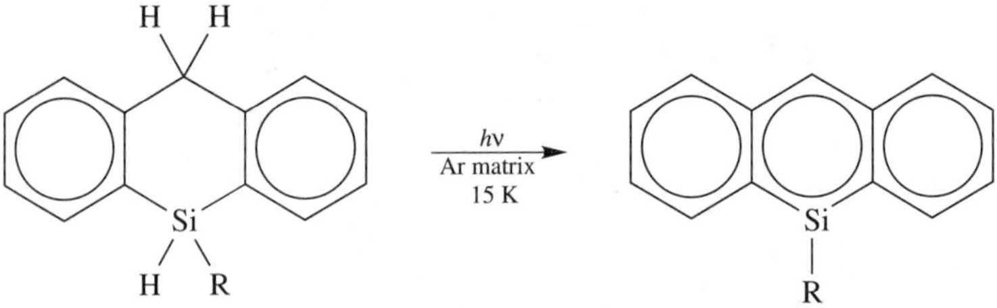

$$R = H, Me, Ph$$

West and coworkers[39] used a 9,10-disilacyclohexadienyl-bridged anthracene **74** to synthesize a permethyl-1,4-disilabenzene, **75**, as shown in equation 10, and to study its reactions.

III. COMPOUNDS CONTAINING TWO SILICON ATOMS. DISILANES

A. Introduction

The photochemistry of disilanes has been investigated for several decades. Photolysis yielding two silyl radicals was recognized early on as one of the common modes of reaction. Studies of Ph$_3$Si–SiPh$_2$Me or its trideuteriomethyl analog **76**, by Boudjouk and coworkers[40], indicated the homolysis of the silicon–silicon bond to a pair of radicals which subsequently disproportionated, yielding the silene **77** (equation 11). This provided some of the early evidence for the existence of the silicon–carbon double bond.

$$\text{Ph}_3\text{Si–SiPh}_2\text{CD}_3 \xrightarrow{h\nu} \text{Ph}_3\text{Si}\cdot + \cdot\text{SiPh}_2\text{CD}_3 \longrightarrow \text{Ph}_3\text{SiD} + \text{Ph}_2\text{Si}=\text{CD}_2 \quad (11)$$
(**76**) \qquad\qquad\qquad\qquad\qquad\qquad\qquad (**77**)

Subsequent studies by Sakurai and coworkers of the photolysis of disilyl-bridged aromatic compound **78**, indicated that two interesting (not necessarily independent) processes were occurring, namely elimination of the disilyl group as a disilene, **79**, another reactive intermediate of silicon chemistry, and secondly, Si–C homolysis to a diradical **80**, which overall underwent a 1,2-silyl shift yielding the isomeric disilyl-bridged compound **81**[41,42] (Scheme 11).

Further investigation of the photolysis of disilanes has been carried out in recent years in which 1,2-silyl and 1,3-silyl shifts were frequently observed. Some of these studies are reported below in sections based on the nature of the group(s) to which the disilyl group is attached.

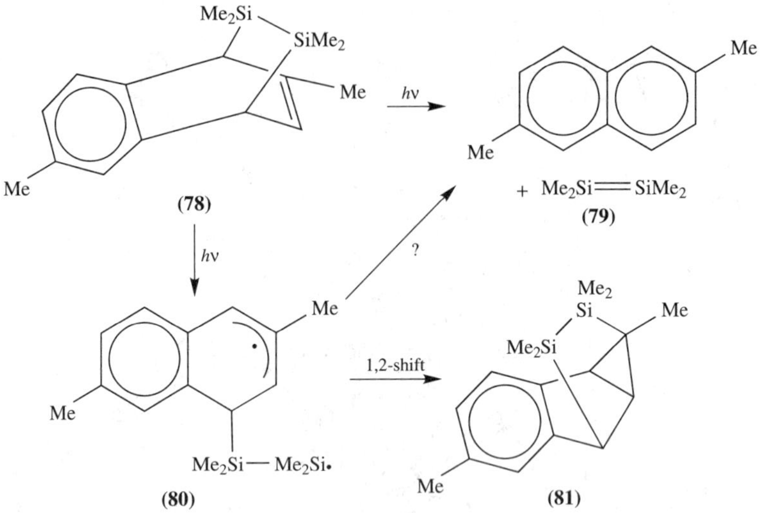

SCHEME 11

B. Vinyldisilanes

Earlier studies of the photolysis of vinyldisilanes **82**[43] clearly established the tendency of 1,3-silyl migrations to occur, leading to silenes **83** (equation 12).

$$CH_2=CHSiR_2SiMe_3 \xrightarrow{h\nu} Me_3SiCH_2CH=SiR_2 \qquad (12)$$
$$\text{(82)} \hspace{4cm} \text{(83)}$$

Sakurai and coworkers[44] took advantage of this rearrangement to prepare from the cyclic divinyldisilane **84** the cyclic silene **85**, in order to investigate the stereochemistry of alcohol addition to a silicon–carbon double bond. Diastereomeric products resulting from both *cis* and *trans* addition to the double bond were formed, whose proportions depended on the concentration of the alcohol in the system. A mechanism accounting for the results proposed that initial coordination of an alcohol molecule to the sp^2-hybridized silicon atom of the silene gave the complex **86**, which either underwent intramolecular proton transfer (path a) yielding the *cis* adduct **87** or else was protonated by a second molecule of alcohol located on the opposite side of the plane of the ring (intermolecular proton transfer, path b) yielding the *trans* product **88** as shown in Scheme 12. The mechanism satisfactorily accounts for several observations on the stereochemistry of alcohol addition to a variety of silenes. A further elaboration is reported in Section III.D.

Conlin and Bobbitt used 1,2-divinyl-tetramethyldisilane **89** to generate a simple silene **90** used to explore competing [2 + 2] and [2 + 4] cycloaddition reactions with butadiene (see Scheme 13), where the *E*-**91** and *Z*-**91** isomers of the [2 + 2] adduct predominated over the [2 + 4] cycloadduct **92**[45].

1,1-Divinyl-tetramethyldisilane similarly rearranged to the analogous silene $CH_2=CHMeSi=CHCH_2SiMe_3$, which showed a UV absorption maximum at 336 nm in 3-MP (3-methylpentane) glass at 77 K[46,47].

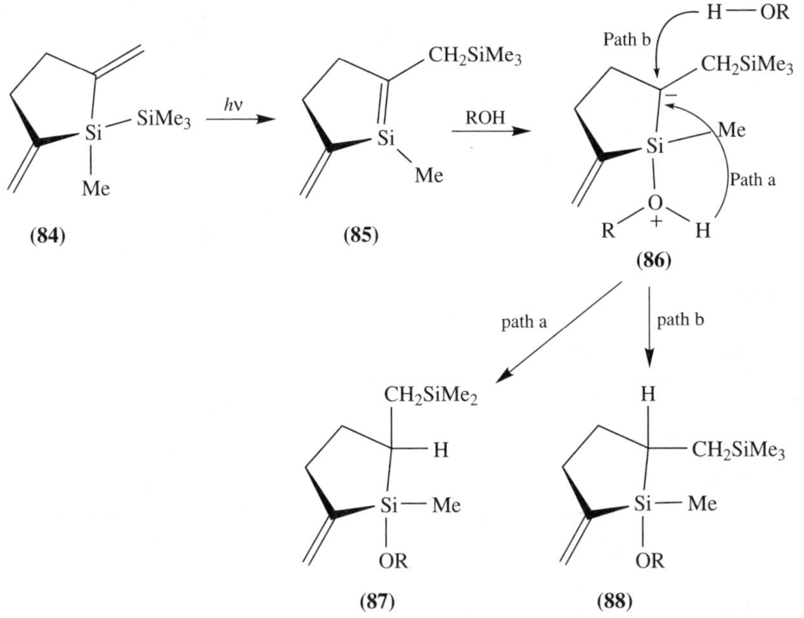

SCHEME 12

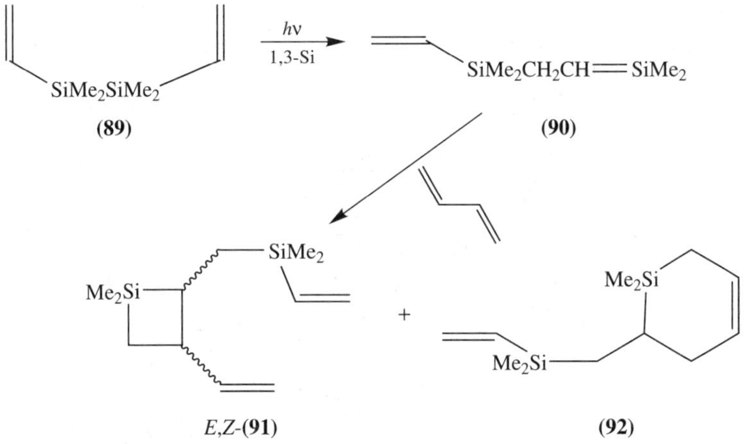

SCHEME 13

C. Alkynyldisilanes

Relatively simple alkynyldisilanes, **93**, have been observed to undergo both 1,2- and 1,3-silyl migrations during photolysis leading to silacyclopropenes **94** and silaallenes **95**, respectively[2] (equation 13).

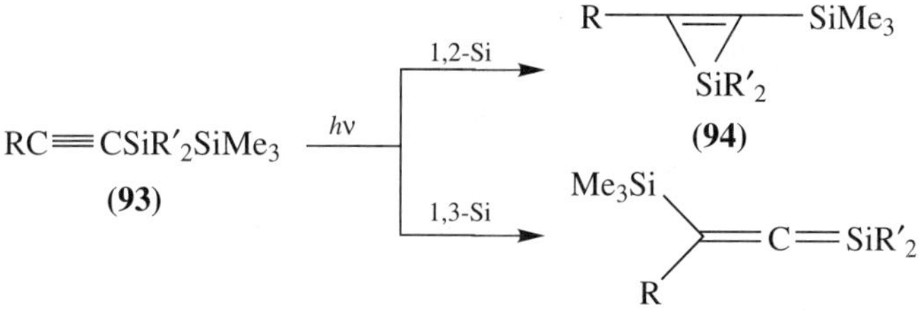

$$RC\equiv CSiR'_2SiMe_3 \xrightarrow{h\nu} \begin{cases} \text{1,2-Si} \to (94) \\ \text{1,3-Si} \to (95) \end{cases} \quad (13)$$

where (94) is the silacyclopropene with R and SiMe₃ on the ring carbons and SiR′₂ as the ring silicon, and (95) is Me₃Si(R)C=C=SiR′₂.

Recently, detailed mechanistic studies have been made of the photolysis of 1-aryl-4-(pentamethyldisilanyl)-1,3-butadiynes **96** in various solvents. Photolyses in the presence of methanol led to the isomeric products **97** and **98**, derived from solvolytic ring opening of the initially formed silacyclopropene, **99**, resulting in turn from 1,2-silyl migration[48], while photolyses in acetone led to the products **100** and **101** arising from two-atom insertion into the three-membered ring[49] (Scheme 14).

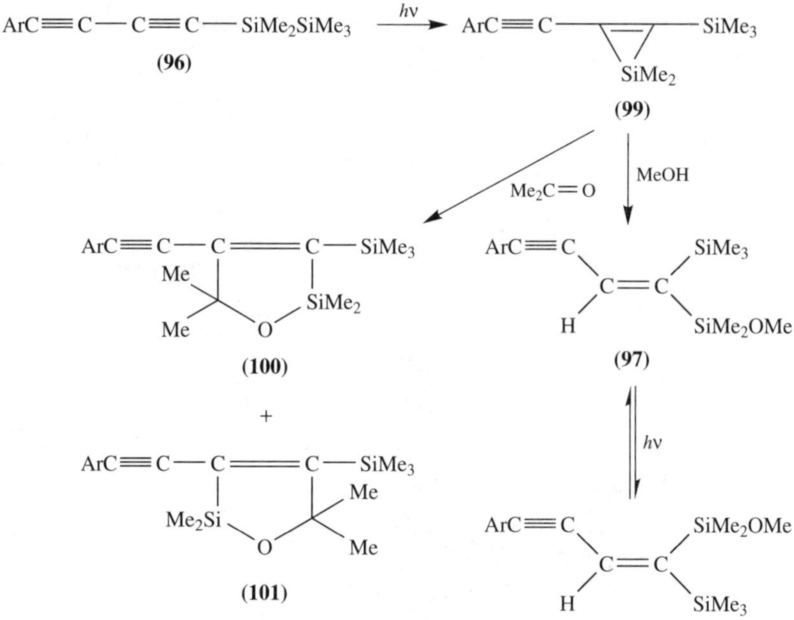

SCHEME 14

The palladium chloride-(PPh₃)₂ catalyzed photoisomerization of a disilylbutadiyne **102** has recently been described. It was shown that isomerization of the diyne initially led to the substituted silacyclopropene **103**, which subsequently dimerized to three isomers of 1,4-disilacyclohexadiene **104**, (equation 14). A mechanism was proposed.

The photochemistry of 1,4-bis(pentamethyldisilanyl)butadiyne in the presence of methanol, acetone and acetaldehyde has also been investigated[50].

$$\text{PhC}\equiv\text{C}-\text{C}\equiv\text{C}-\text{SiMe}_2\text{SiMe}_3 \xrightarrow[\substack{\text{PhH} \\ \text{PdCl}_2\cdot(\text{PPh}_3)_2}]{h\nu} \text{PhC}\equiv\text{C}-\text{C}\!=\!\text{C}-\text{SiMe}_3$$
(102)   Si
                                                            Me$_2$
                                                           (103)

                Me   Me
      Me$_3$Si   \\ /
              Si   C≡CPh   (14)

       PhC≡C   Si   SiMe$_3$
               Me   Me
                  (104)
            three isomers

D. Aryldisilanes

The much studied photochemistry of aryldisilanes carried out in earlier years has been reviewed[51,52]. Cleavage of the silicon–silicon bond of the disilyl moiety is always involved, but various other reactions have been observed depending on the structure of the disilane and the conditions employed. Thus cleavage to a pair of silyl radicals, path a of Scheme 15, is normally observed, and their subsequent disproportionation to a silene and silane, path b, is often observed. There is evidence that the formation of this latter pair of compounds may also occur by a concerted process directly from the photoexcited aryldisilane (path c). Probably the most common photoreaction is a 1,3-silyl shift onto the aromatic ring to form a silatriene, **105**, path d, which may proceed via radical recombination[52]. A very minor process, observed occasionally, is the extrusion of a silylene from the molecule (path e), as shown in Scheme 15.

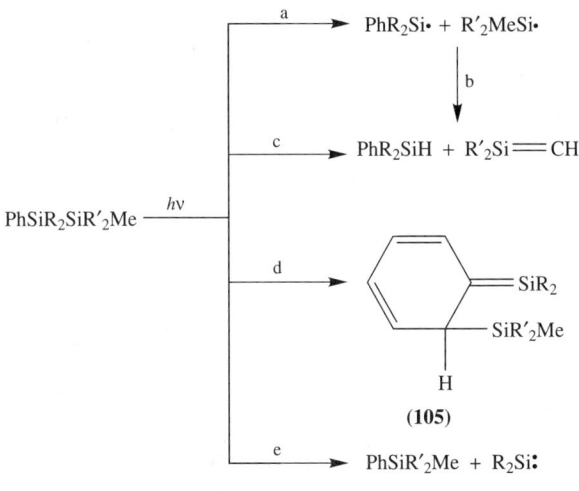

SCHEME 15

Much additional information has been accumulated on these systems in recent years. On the one hand Ishikawa and Sakamoto investigated tolyl-**106** and xylylpentamethyldisilanes and found that the main reaction involved the 1,3-shift of a trimethylsilyl group to the ring to yield the silatriene **107**, provided that an *ortho* position on the aromatic ring was vacant[53]. An example is shown in equation 15, where the silatriene was trapped as the adduct **108** by an ene reaction with isobutene.

The same behavior was observed for bis(pentamethyldisilyl)-substituted aromatics[54] and for more complex systems such as $Me_3Si(SiMe_2C_6H_4\text{-}p\text{-}SiMe_2)_n SiMe_3$ ($n = 2, 3, 4$)[55], as confirmed by the structures of the products obtained by trapping of the intermediate silatrienes with isobutene or other reliable trapping reagents.

With 2-naphthyl systems **109**, 1,3-silyl migration to the 1-position on the ring was observed[56] since trapping by MeOD gave the product **110**, indicating the intermediacy of an arylsiladiene **111** (equation 16).

21. The photochemistry of organosilicon compounds

1-Naphthyldisilanes **112** were shown[57], after reinterpretation of earlier results, to undergo Si—Si bond cleavage with migration of a silyl group to the 8-position of the naphthyl ring to give the bis-silyl adduct **113**, the overall 1,4-rearrangement almost certainly being the result of homolytic cleavage, followed by homolytic aromatic substitution via the postulated intermediate **114**. Most conventional silene traps failed to detect the presence of a silene, but when 1,4-bis(pentamethyldisilyl)naphthalene **115** was photolyzed in MeOD a novel dimethylsilylene bridged species **116** was isolated in about 50% yield (Scheme 16).

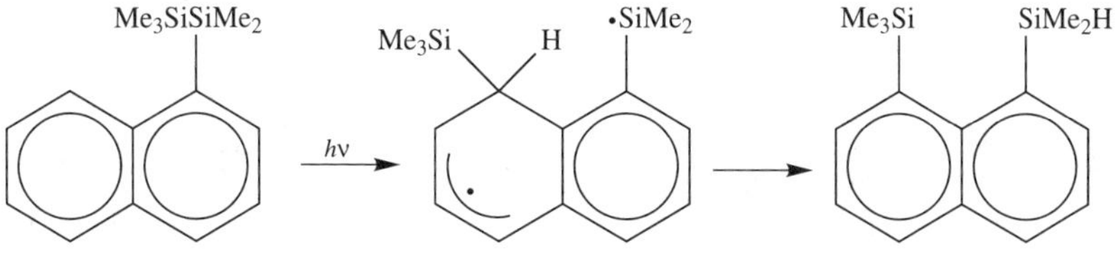

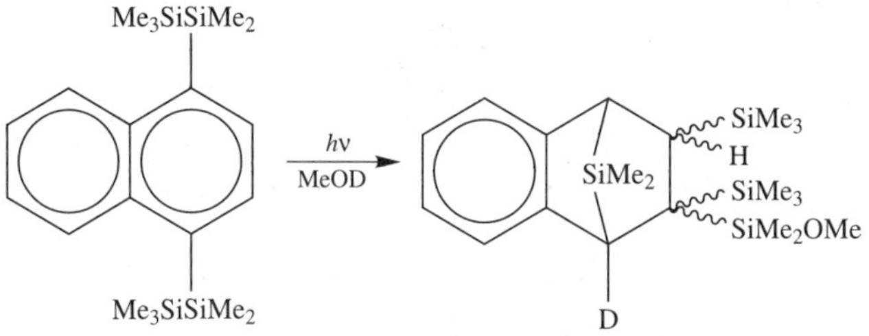

SCHEME 16

Homolytic cleavage of the Si—Si bond, followed by homolytic aromatic substitution, was also invoked to explain the photochemical reactions of 1,1-di(1-naphthyl)- **117**, and 1,2-di(1-naphthyl)-tetramethyldisilane **118**[58] which yielded **119** and **120**, respectively, as the main reaction products (Scheme 17). A minor reaction pathway of the latter disilane involved dimethylsilylene expulsion.

Very recently Sluggett and Leigh have reinvestigated the photochemistry of several aryldisilanes having either methyl or phenyl groups on the disilane side chain. Both direct photolysis as well as laser flash photolysis produced a wealth of new information including fluorescence data for the disilanes, absorption maxima of silene and silyl radical transients, rate constants of quenching processes by a variety of quenching reagents (particularly those used to trap intermediary silatrienes), deuterium isotope effects for the reactions of the silatrienes with water, etc.[59–62]. These studies have contributed to the further understanding of the behavior, the mechanism of formation and the mechanisms of reaction with trapping agents such as alcohols, acetone, dimethylbutadiene and oxygen, of the silatriene intermediates formed in these reactions.

The more recent studies showed that methylpentaphenyldisilane, **121**, on photolysis formed the silatriene **122** in low yield by a 1,3-silyl migration, in addition to forming the silene **123**[59]. Flash photolysis of phenylpentamethyldisilane, **124**, for comparative purposes, formed the silatriene **125**, λ_{max} 420, $\tau = 5$ μs, in isooctane with no evidence for formation of the silene PhMeSi=CH$_2$. The silatrienes were readily quenched either by

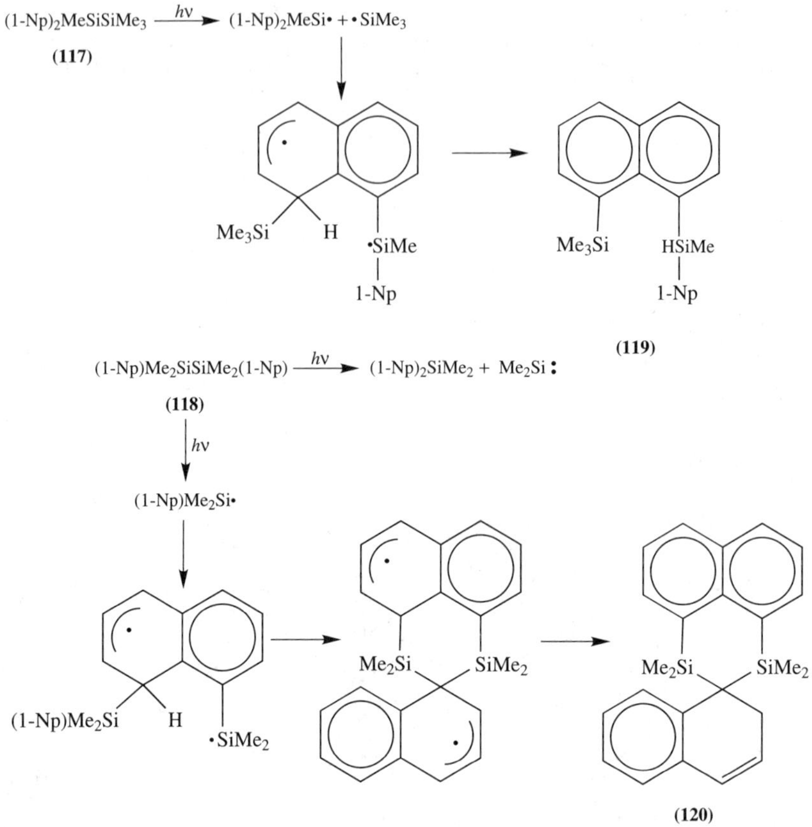

SCHEME 17

reagents which add in a 1,2 manner across the ends of the Si=C bond, or by reagents which react through an ene mechanism as shown in Scheme 18[59]. It was suggested that bulky groups like phenyl, attached to the silicon atoms, tended to hinder silatriene formation, relative to the formation of the simple silene RR'Si=CH$_2$. Consistent with this view was the observation that direct (or laser flash) photolysis of 1,2-di(*t*-butyl)-1,1,2,2-tetraphenyldisilane in chloroform exclusively yielded the silyl radical (*t*-Bu)Ph$_2$Si•, which was captured as silyl chloride in 95% chemical yield with a quantum efficiency of 0.64. The bimolecular rate constants for reactions of the radicals with a variety of reagents were determined[60]. While hexaphenyldisilane is known to mainly yield Ph$_3$Si• radicals on photolysis[63] it was suggested, based on absorption measurements, that the related silatriene was also formed, but only in low yield[60].

Re-examination of the photolyses of the family of aryldisilanes PhRR'Si—SiMe$_3$ (R,R' = Me, Ph) in solution containing acetone was found to give not only the previously observed products believed to be derived from an ene reaction between the silatriene and acetone, **126**[51], but in addition, and not previously observed, the [2+2] siloxetanes, **127**, were detected and isolated, as were the silyl enol ethers **128** derived from the simple silene **129** formed by elimination of trimethylsilane[61], as shown in Scheme 19. The biradical **130** was proposed as an intermediate for some of the reactions involving acetone. Rate

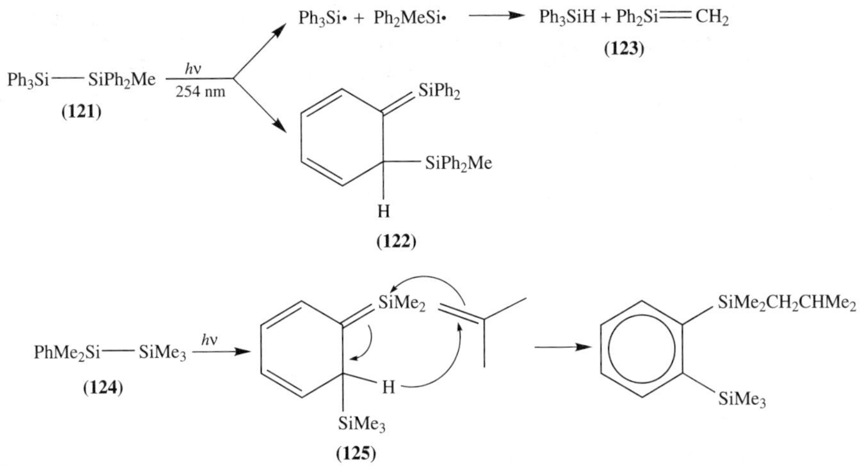

21. The photochemistry of organosilicon compounds

SCHEME 18

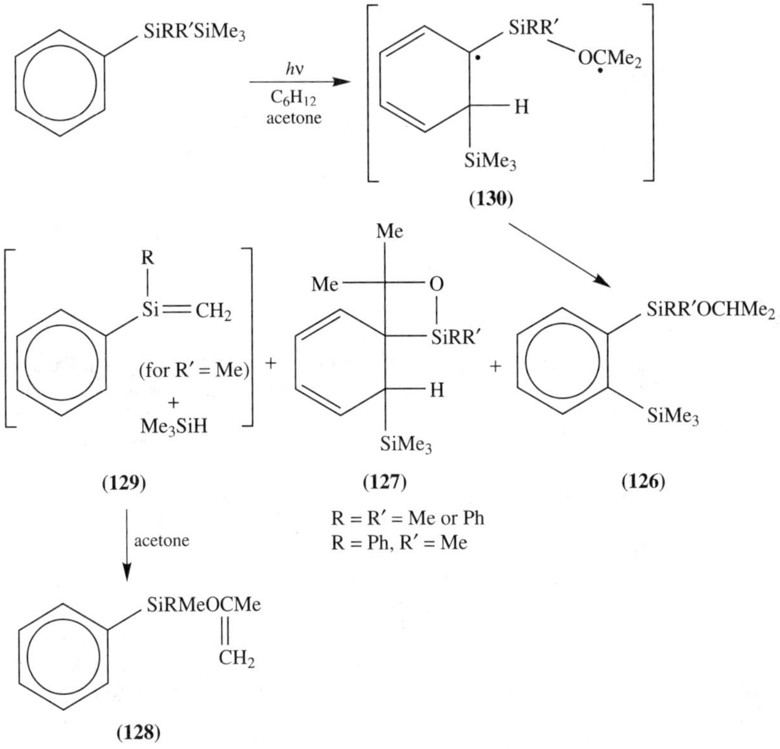

SCHEME 19

constants for the trapping of the silatrienes with several reagents, 2,3-dimethylbutadiene, methoxytrimethylsilane, oxygen and carbon tetrachloride, were also measured.

In earlier studies of the reactions of some of the silatrienes, e.g. **132**, derived from the photolysis of *p*-tolylpentamethyldisilane, **131**, with alcohols, it was reported[51], as shown in equation 17, that only the 1,4- adduct **133** and the 1,6-adduct **134** were formed, and there was no evidence that the 1,2-adduct **135** was formed. The absence of **135** was always difficult to rationalize.

More recent studies by Leigh and Suggett[62] using phenylpentamethyldisilane and other aryldisilanes showed (Scheme 20) that the 1,2-methanol adduct **138** was indeed formed along with the 1,4-adduct **139** and the 1,6-adduct **140**, indicating that the behavior of the silatriene **136** (and its analogs) was consistent with the assigned structure as containing a silicon–carbon double bond. Much rate data were also presented.

Also arising from these important studies was further evidence about the mechanism of the addition of alcohols to silenes, and the silatriene **136** in particular, as illustrated in Scheme 20. In a minor revision of the Sakurai mechanism[44] it was suggested that the reaction was initiated by rapid reversible nucleophilic attack of the alcohol oxygen atom at the sp^2-hybridized silicon atom of the double bond (involving rate constants k_1 and k_{-1}) to form an alcohol–silene complex **137**. This is followed by either intramolecular (intracomplex) proton transfer (rate constant k_2) leading to the 1,2-adduct **138** or by an intermolecular proton transfer from the complex to further molecules of alcohol leading to **138** and its isomers **139** and **140**. The relative rates of these processes were shown to be affected by the nucleophilicities of the alcohol, its concentration and the nature of the groups on the sp^2-hybridized silicon atom of the silene.

Kira and Tikurs[64] observed an interesting variation of the 1,3-silyl rearrangement of aryldisilanes since, during the photolysis of pentafluorinated aryldisilanes such as **141**, in addition to the 1,3-silyl migration to the 2-position of the ring forming the silatriene **142**, a

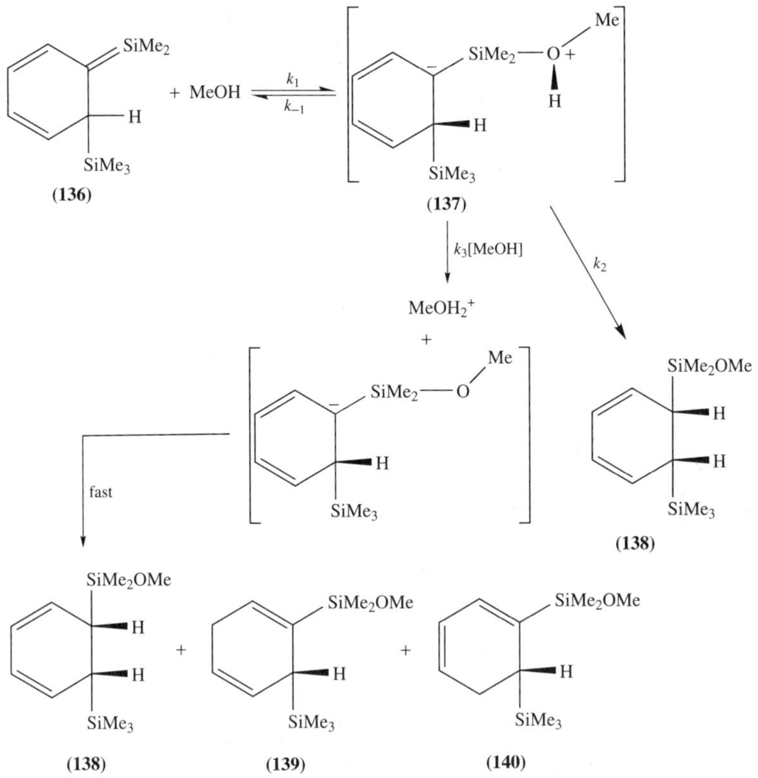

SCHEME 20

reverse 1,3-fluorine migration from the aromatic ring to silicon occurred yielding the product **143** as well as the product of alcohol addition when alcohol was present (Scheme 21). Similar behavior was found with a 2,6-difluorophenyldisilane as was observed with the pentafluoro compound (Scheme 21), but not with a 2,6-dichlorophenyldisilane.

Braddock-Wilking and Gaspar studied the photolysis of mesityl-substituted disilanes such as $Mes_2MeSiSiMeMes_2$, which was found to form $Mes_2MeSi\cdot$ radicals which disproportionated into the silene $Mes_2Si=CH_2$ and Mes_2MeSiH in an inert solvent. There was also evidence that these products were formed in part directly during the photolysis[65].

Mochida and coworkers observed that photolysis of 2-furylpentamethyldisilane also yielded silyl radicals on photolysis, the radicals being trapped as silyl chlorides by reaction with CCl_4[66].

Recently, several studies have been made of the photolysis of disilanes or polysilanes in the presence of an electron-deficient alkene using a photosensitizer (such as phenanthrene) and acetonitrile as solvent. These conditions result in the addition of silyl groups to one end of the alkene double bond and hydrogen to the other end (equation 18) and evidently involve the reaction of the radical anions of the electron-deficient silene with silyl radicals[67] (see also Section VIII.A).

$$PhCH=C(CN)_2 + Me_3SiSiMe_2Ar \xrightarrow[\substack{\text{phenanthrene}\\CH_3CN}]{h\nu} Ph(Me_3Si)CHCH(CN)_2 \\ +Ph(ArMe_2Si)CHCH(CN)_2 \qquad (18)$$

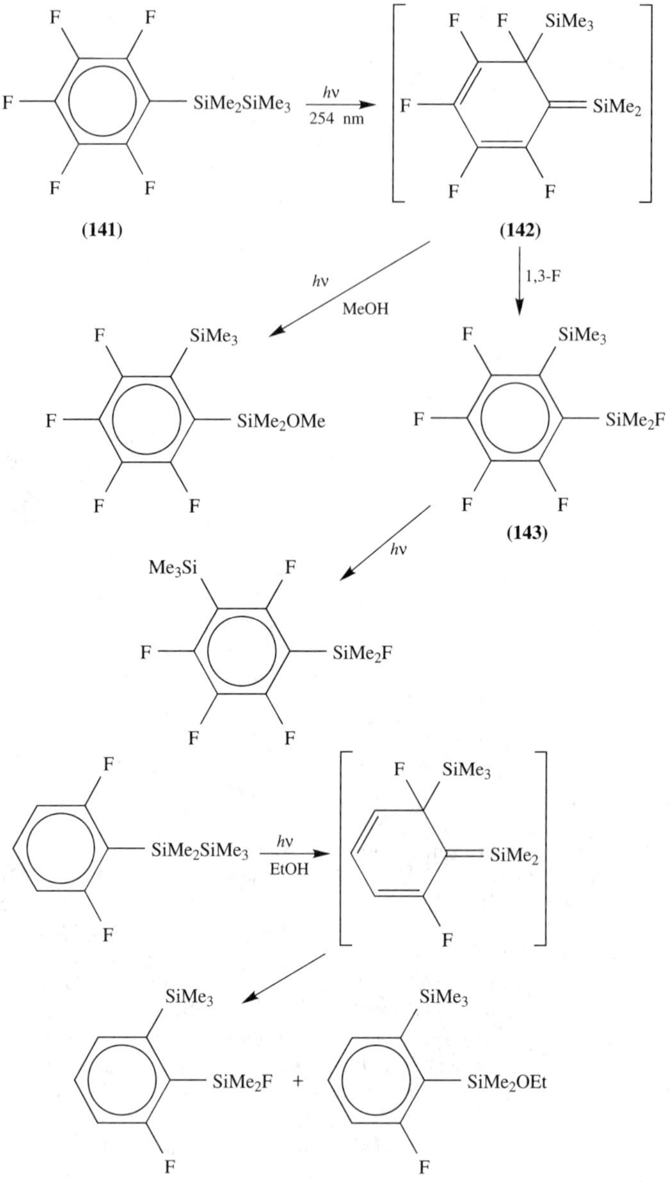

SCHEME 21

The photolysis of the benzodisilacyclobutane **144** was studied[68]. It was postulated that homolysis to the diradical **145** occurred which then disproportionated intramolecularly leading to the silene **146**, that was trapped by *t*-butyl alcohol. An unusual dimer **147** was also formed, postulated by the authors to be derived by head-to-head coupling of the silene to give the diradical **148**, which then abstracted hydrogen. However, intermolecular coupling of two molecules of the diradical to give **149**, followed by hydrogen abstraction, is perhaps a more reasonable explanation, as shown in Scheme 22.

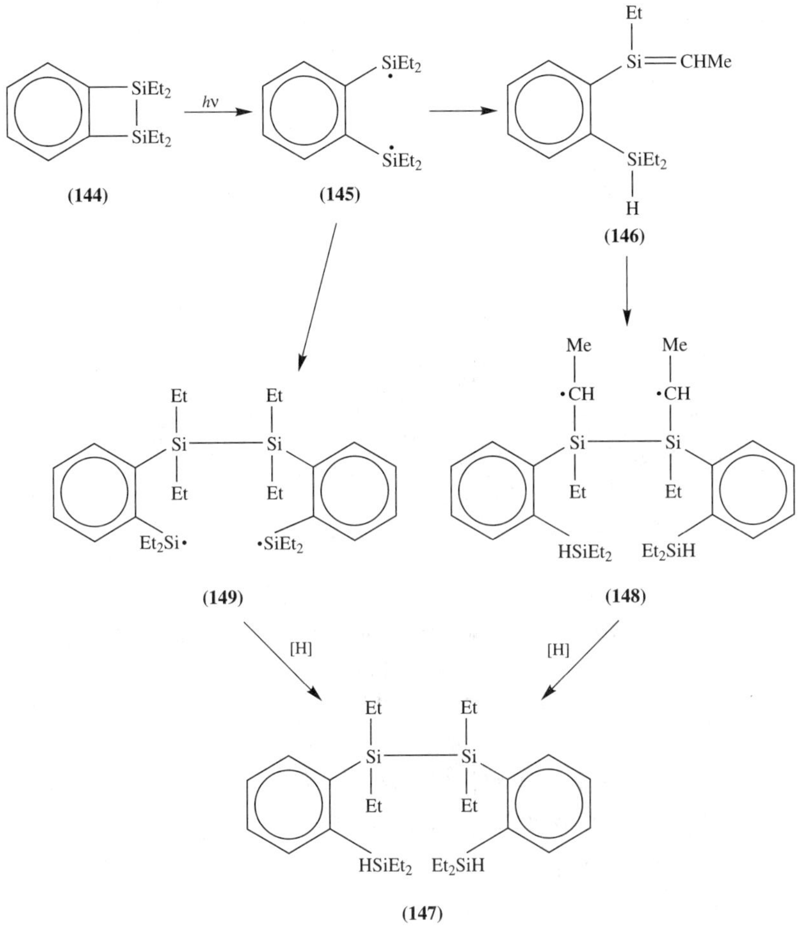

SCHEME 22

E. Arylvinyldisilanes

When a disilane contains both an aryl and a vinyl group as in **150** (or a substituted vinyl such as dihydropyranyl), the possibility exists that during photolysis 1,3-silyl migration to either or both groups may occur. Two studies[69,70] of model compounds clearly established that 1,3-silyl shifts to both the vinyl and phenyl groups generally occurred, which produced isomeric silene intermediates **151** and **152** that could be trapped subsequently by a variety of trapping agents (methanol, acetone, isobutene, 2,3-dimethylbutadiene). An example with acetone is given in Scheme 23. In most cases migration to the vinyl group was favored over migration to the phenyl group (where the usual silatriene was produced), but the extent of migration (based on the yields of trapped products) depended on both the structure of the starting material and the nature of the trapping agent.

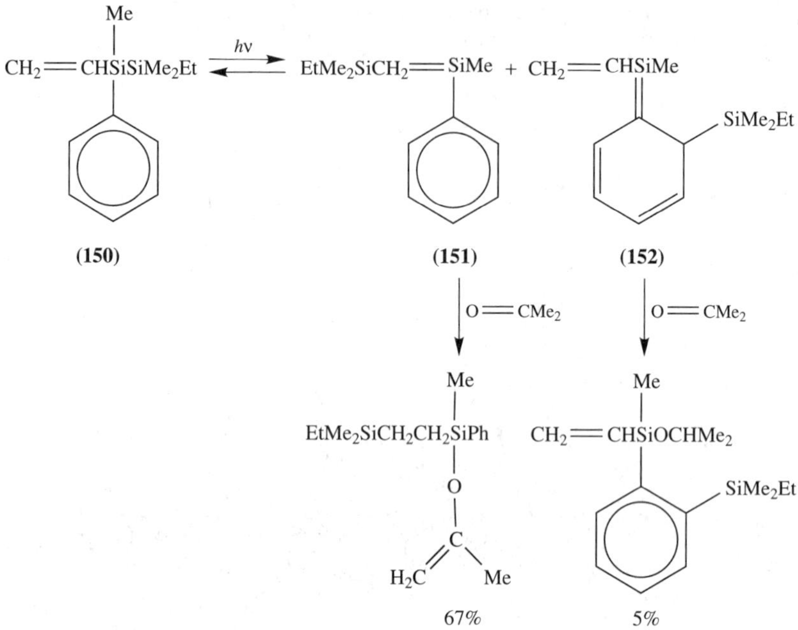

SCHEME 23

F. Transition Metal-substituted Disilanes and Related Compounds

In recent years considerable study has been made of the photochemical behavior of a number of di- or polysilanes which are bonded to a transition metal. Typical compounds are shown below.

$$\eta^5 - [(C_5H_5)Fe(CO)_2](SiR_2)_n SiR_3$$

$$\eta^5 - [(C_5H_5)M(CO)_3](SiR_2)_n SiR_3 \quad M = Cr, Mo, W$$

$$\eta^5 - [(C_9H_7)Fe(CO)_2](SiR_2)_n SiR_3$$

$$\eta^5 - [(C_5R_5)M(CO)_2](SiR_2)_n SiR_3 \quad M = Fe, Ru, Os$$

A review of the chemistry of these species has recently appeared[71]. These readily made compounds undergo interesting photochemistry, which includes silylene expulsion and/or scrambling of R groups within the disilyl group. An example involving $FpSiMe_2SiPh_3$, **153** {$Fp = \eta^5 - [(C_5H_5)Fe(CO)_2]$} is shown in Scheme 24.

As shown, the process is believed to involve reversible elimination of a CO ligand followed by a 1,2-silyl migration to the Fe atom leading to **154**, which can undergo R group scrambling to **155** and **156** due to 1,3-alkyl or 1,3-aryl group migrations from Si to Si. Each of the three metal–silylene complexes can lose a silylene and regain CO, leading to the observed product mixture of **157**, **158** and **159**.

Support for this mechanism has been obtained from isolation of stabilized silylene intermediates[72], isotope labeling experiments[73] and low temperature studies[74]. Electron-donating groups on a phenyl group attached to silicon stabilize the intermediate ($\eta^5 - C_5H_5)FeCO(=SiMeAr)SiMe_3$[75].

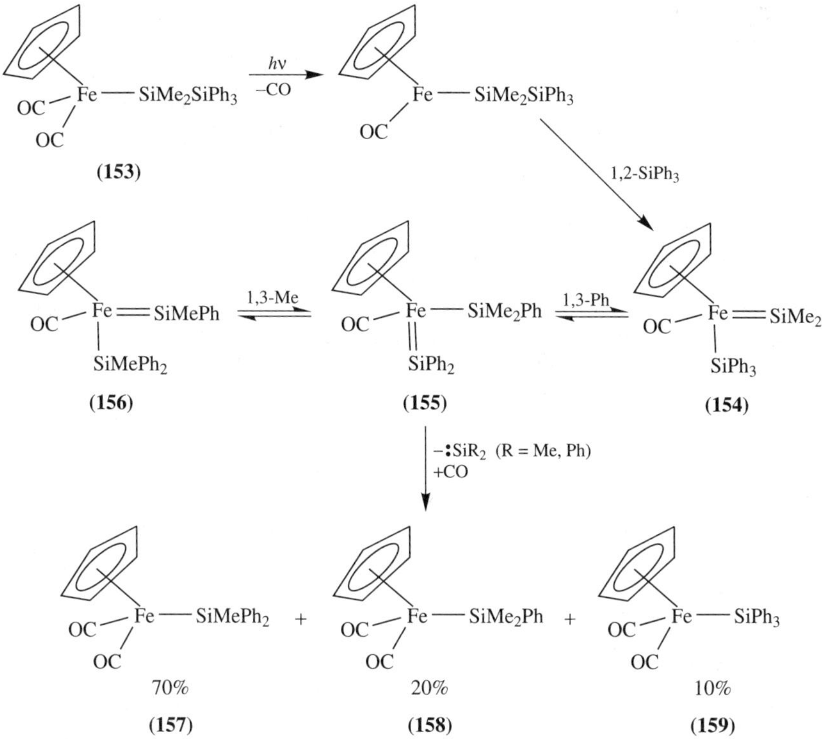

21. The photochemistry of organosilicon compounds

SCHEME 24

With trisilyl systems, e.g. $\eta^5 - [(C_5H_5)Fe(CO)_2](Si_3Me_6R)$, R = Me, Ph, studies have shown that isomerization occurs prior to loss of the silylene $:SiMeR$[76], while tetrasilyl systems primarily isomerize with little loss of silylene groups[77].

In contrast to the above, indenyl compounds such as **160**, with a silyl, disilyl or trisilyl group attached to the metal atom, fail to undergo loss of a silylene on photolysis, and simply rearrange to products such as **161**, as shown by the example given in equation **19**. The locations of aryl groups present in the silyl side chain become scrambled during photolysis[78].

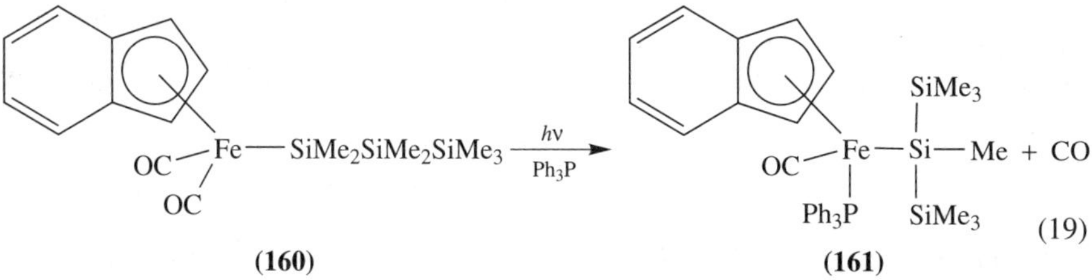

(19)

Whereas photolysis of cyclohexasilanes usually results in silylene elimination and the successive formation of cyclopentasilanes and cyclotetrasilanes[79], attachment of the Fp group to the ring as in **162** results in rearrangement to a silyl-substituted cyclopentasilane, **163**, revealing that the presence of the transition metal alters the chemistry of these systems in a major way (equation 20).

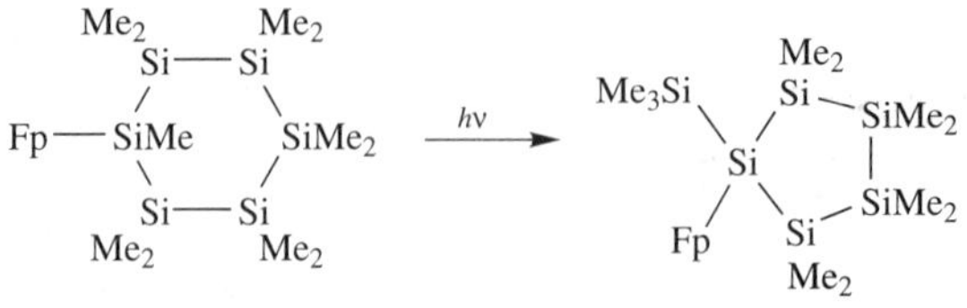

$$
\begin{array}{c}
\text{(162)} \xrightarrow{h\nu} \text{(163)}
\end{array}
\quad (20)
$$

Fp = η^5-[C$_5$H$_5$Fe(CO)$_2$]

When a methylene or dimethylgermyl group intervenes between the transition metal and the silyl group, as in **164**, photolysis results in rearrangement to **165**, which appears, for the methylene case, to involve the intermediacy of a metal–silene complex (Scheme 25).

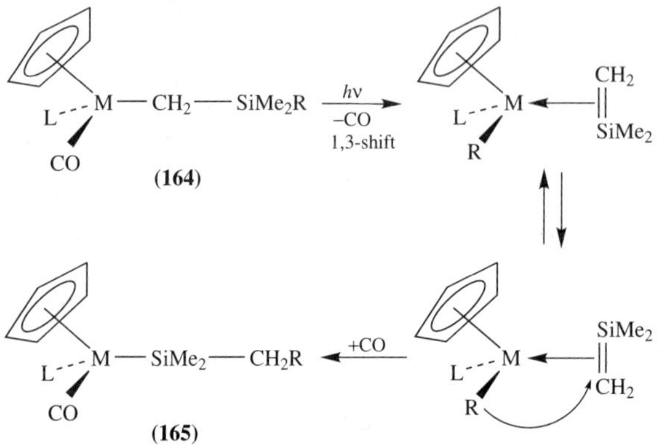

R = Me$_3$Si, Me$_3$SiSiMe$_2$, etc.
M = Fe, W
L = CO or another ligand

SCHEME 25

While the germanium-containing tungsten complex **166** rearranged to **167** on photolysis, the isomeric species **168** directly eliminated the silene Me$_2$Si=CH$_2$ yielding **169** (Scheme 26)[80].

The related system FpSiMe$_2$GeMe$_2$Fp, **170**, and other group 14 analogs have also been studied[81]. The initially formed products **171** involved only bridging GeMeR groups, but on extended photolysis further carbon monoxide was lost leading to isomeric species **172** involving both bridging GeMe$_2$ and SiMe$_2$ groups (Scheme 27).

G. Cyclic Disilanes and Related Species

The photochemistry of a number of cyclic systems containing a two- or three-silicon atom segment has been investigated. Sakurai and coworkers[82] showed that photolysis of the two atom-bridged biphenyl **173** led to formation of the silafluorene **174** and dimethylsilylene. Subsequently it was shown that the three atom-bridged biphenyl **175** led, via the diradical **176**, to the two-atom bridged **173** and dimethylsilylene, which was trapped with 2,3-dimethylbutadiene. An alternate reaction pathway led to silafluorene

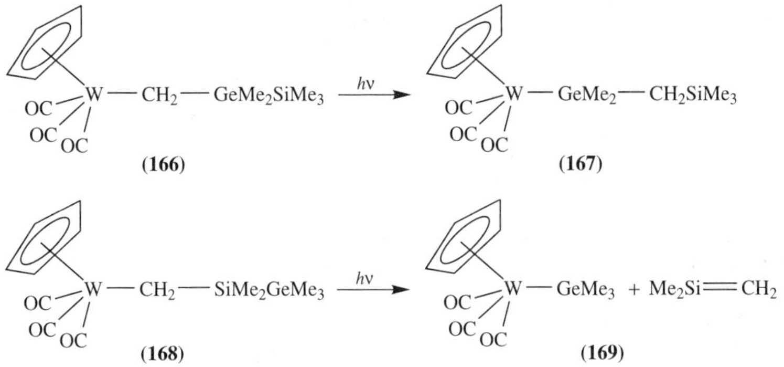

21. The photochemistry of organosilicon compounds

SCHEME 26

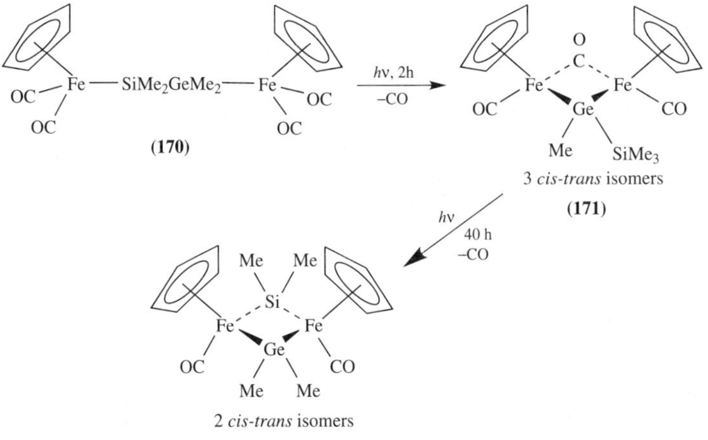

SCHEME 27

174 and tetramethyldisilene, which was also trapped. Convincing evidence that the reaction proceeded by a diradical pathway was obtained through photolysis in the presence of carbon tetrachloride, when the diradical centers were trapped as chlorosilanes (Scheme 28).

The photolysis of silicon-bridged dihydroaromatic compounds has been utilized to prepare reactive organosilicon species such as disilenes. Thus, photolysis of the

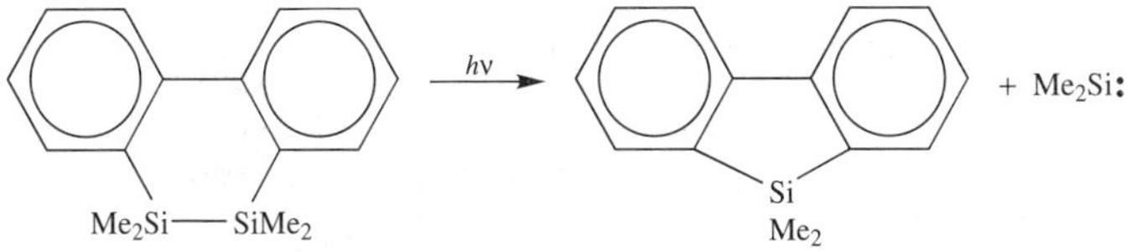

SCHEME 28

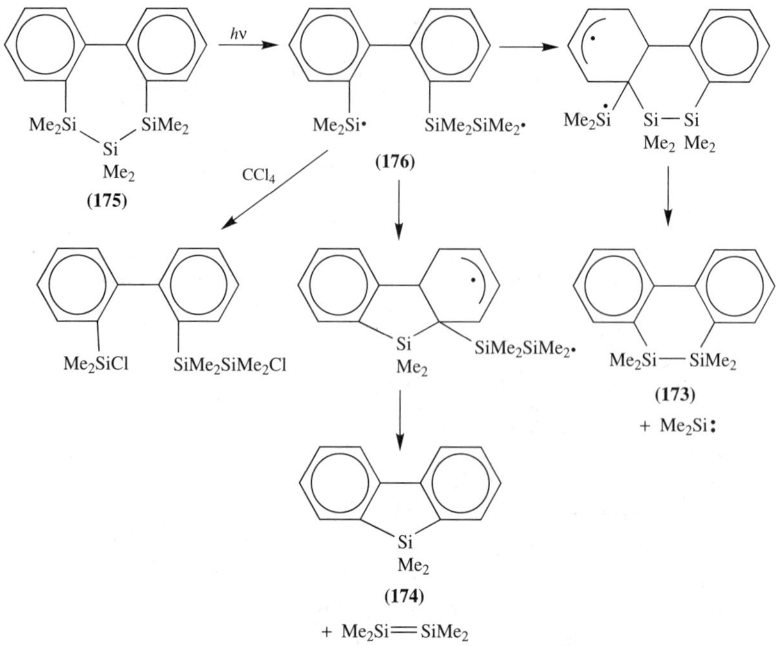

SCHEME 28. (*Continued*)

typical disilyl-bridged compound **177** was shown to yield tetra (*t*-butyl)disilene, **178** (equation 21).

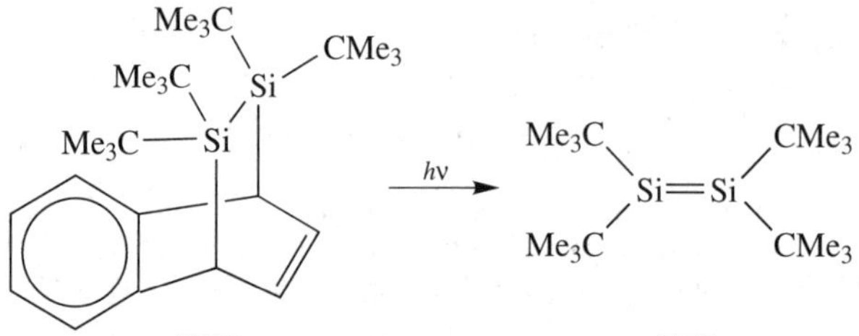

(21)

(**177**)   (**178**)

West and coworkers studied the photolysis of several adducts of disilenes with ketones, i.e. 1,2-disiloxetanes[83]. Based on the products obtained when the photolysis was carried out in ethanol as a trapping agent, it appears that the heterocyclic disiloxetane **179** decomposed to the silanone **180** and the silene **181**, each trapped by ethanol to give the adducts **182** and **183**, respectively (Scheme 29). In the absence of a trapping agent the silene photochemically rearranged to **184**. A related 1,2-disilathietane **185** showed similar behavior (Scheme 29)[84].

Peralkylcyclodisilagermanes and peralkylgermadisilaoxetanes on photolysis at 254 nm were observed to yield predominantly the corresponding disilene and disiloxirane, respectively, in addition to the dialkylgermylene[85].

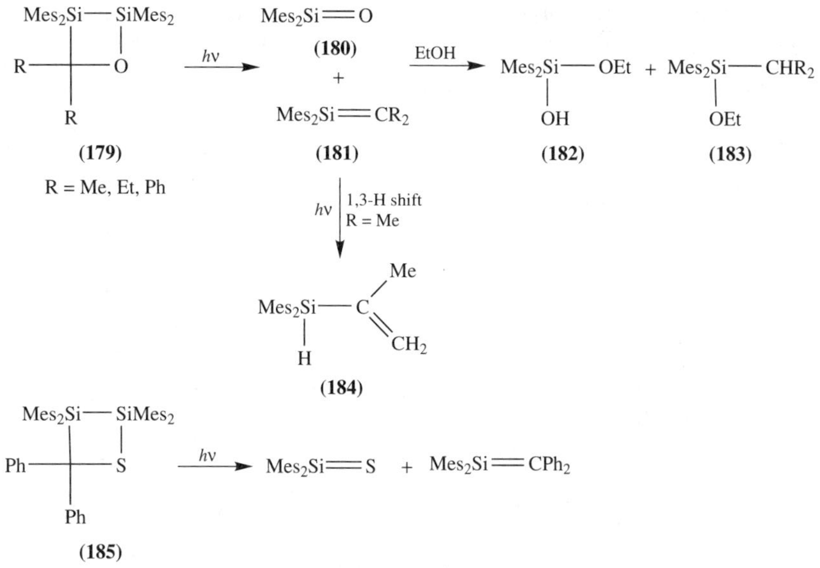

SCHEME 29

IV. TRI-, TETRA- AND POLYSILANES

A. Introduction

As a result of several decades of research it is now known that a polysilane of three or more contiguous silicon atoms is susceptible to reaction by one or more of several pathways when photolyzed, each associated with cleavage of a silicon–silicon bond. The two most common processes observed are the homolysis of a silicon–silicon bond to yield a pair of silyl radicals, and the elimination of a silicon atom from the chain in the form of a silylene. As discussed in Section VII, the use of trisilanes, particularly where the central silicon atom bears aryl groups, has become an important route for the preparation of a wide variety of diarylsilylenes, Ar_2Si:, many of which have been captured in glasses at low temperature, or have been allowed to dimerize to disilenes by warming.

B. Acyclic Polysilanes

Understanding of the behavior of tri-, tetra- or polysilanes during photolysis has been actively pursued in recent years, in part because of the potential application of the polysilanes as photoresists, and for other industrial applications. In addition to the reactions mentioned above, other reactions have now been recognized as occurring in the course of photolysis. These are summarized below.

The photolysis of polysilanes generally results in one or more of the following reaction pathways being followed, as illustrated in Scheme 30, where ch represents a segment of polysilane chain. Most common is the formation of a pair of silyl radicals, path **a**. When there are aryl groups on the silicon radical centers and particularly when a methyl group is one of the R groups on one of the silicon radical centers, it is found that these readily disproportionate to a silane —SiH, and a silene, Si=C, path **b**. However, in some cases it appears that the same products also can arise directly during the photolysis, path **c**. Also, a silylene R_2Si: is commonly extruded from the polysilane chain, path **d**. Finally, it was recently observed that persistent silyl radicals were produced during photolysis, apparently resulting from an internal 1,2-rearrangement of an R group to yield a shortened chain and a silylsilylene, path **e**. As a result of rearrangement of the silylsilylene to a disilene, a known facile process, followed by addition of a silyl radical produced by path **a** at the less hindered end of the silicon–silicon double bond, a highly hindered silyl radical **186** is produced, and recent studies have established the structure of this species[86,87]. Similar processes have subsequently been shown to occur in oligosilanes with as few as three silicon atoms[88].

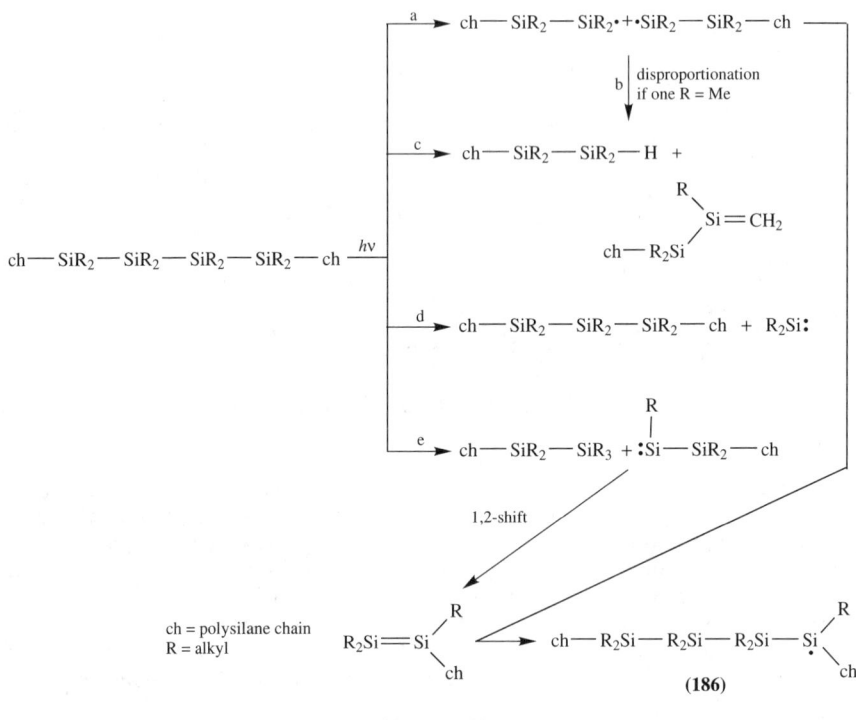

SCHEME 30

As an example of the application of these processes to a simple trisilane, the compound **187** was studied[88] where three of the five processes described above were detected on the basis of product analysis (Scheme 31). In this example no persistent silyl radical was observed because of the lack of steric crowding, but the products observed clearly indicate that processes **a**, **c** and **d** were being followed.

Other recent examples[89–92] of the photolysis of trisilanes to prepare silylenes as precursors to disilenes will be found in Section VII.

Conlin and coworkers photolyzed vinyltris(trimethylsilyl)silane **188** in the presence of a variety of trapping reagents such as butadiene, substituted butadienes or silanes and observed products derived from intermediate silenes **189** (formed by rearrangement) or from silylenes **190** resulting from elimination of hexamethyldisilane[93]. In some cases complex mixtures of products which could have been derived from intermediate silyl radicals were also observed. The reaction products formed from the silene and the silylene in the presence of butadiene, **191** and **192** respectively, are shown in Scheme 32.

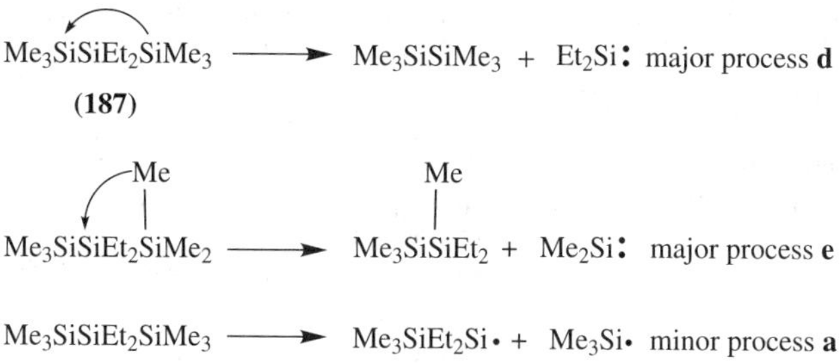

21. The photochemistry of organosilicon compounds

Me$_3$SiSiEt$_2$SiMe$_3$ ⟶ Me$_3$SiSiMe$_3$ + Et$_2$Si: major process **d**
(187)

Me$_3$SiSiEt$_2$SiMe$_2$—Me ⟶ Me$_3$SiSiEt$_2$—Me + Me$_2$Si: major process **e**

Me$_3$SiSiEt$_2$SiMe$_3$ ⟶ Me$_3$SiEt$_2$Si• + Me$_3$Si•   minor process **a**

SCHEME 31

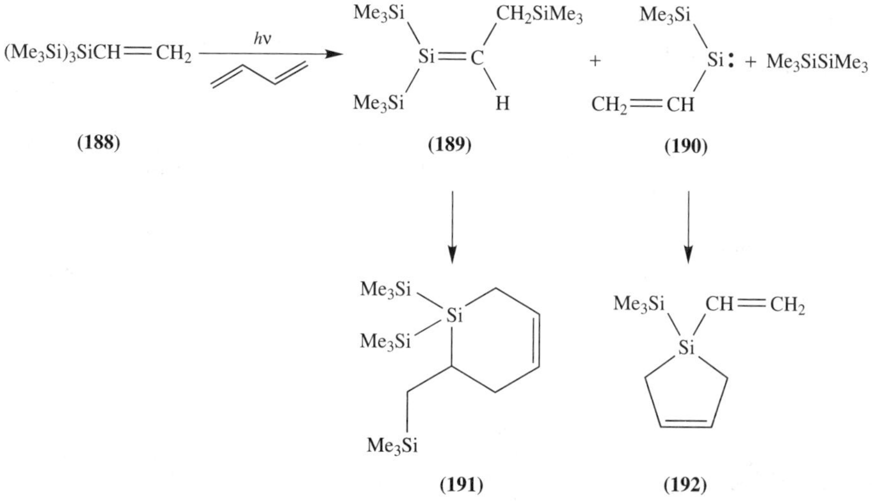

(Me$_3$Si)$_3$SiCH=CH$_2$ $\xrightarrow{h\nu}$

(188)

$\mathrm{\begin{array}{c}Me_3Si\\ \\ Me_3Si\end{array}}$Si=C$\mathrm{\begin{array}{c}CH_2SiMe_3\\ \\ H\end{array}}$ + $\mathrm{\begin{array}{c}Me_3Si\\ \\ CH_2=CH\end{array}}$Si: + Me$_3$SiSiMe$_3$

(189)   (190)

↓   ↓

(191)   (192)

SCHEME 32

C. Cyclic Polysilanes

1. Homocyclic systems

The photolysis of a number of homocyclic silanes with various sized rings has been investigated in recent years: much of this material appears in a recent review by Hengge and Janoschek[94]. It is well known that the photolysis of dodecamethylcyclohexasilane leads to loss of dimethylsilylene, Me$_2$Si:, with the successive formation of the cyclopenta- and cyclotetrasilanes. This reaction constitutes a useful way to generate the silylene for various purposes. Further photolysis generally leads to complete degradation of the ring system.

The photolytic behavior of the cyclotetrasilanes depends on the structure of the molecule and, more particularly, on the shape of the ring. When silyl groups (Me$_3$Si or EtMe$_2$Si) are the substituents on the ring, the ring adopts a planar configuration, and photolysis in methylcyclohexane leads to the silyl-substituted disilenes. Alternatively, it was found that when isopropyl groups occupied all eight sites on the cyclotetrasilane ring, or where one methyl and one *t*-butyl group were attached to each silicon atom of the ring, the cyclotetrasilane ring was bent, and photolysis resulted in silylene expulsion and the formation of a cyclotrisilane. Details of the absorption processes are discussed in the original paper[95].

Photolysis at 300 nm of one of the stereoisomers of methyl(*t*-butyl)cyclotetrasilane (Me(*t*-Bu)Si)$_4$ resulted in the formation of an equilibrium mixture of all four possible stereoisomers indicating cleavage of a ring silicon–silicon bond to a diradical, followed by reclosure. Silylenes, which could be trapped, were also formed during the process[96] and low yields of linear trisilanes were also obtained.

Trisilacyclopropanes are also quite susceptible to photolysis, yielding disilenes R$_2$Si=SiR$_2$ and silylenes R$_2$Si: (equation 22). A review of their behavior has recently appeared[97]. When the substituents on the ring silicon atoms were aromatic (e.g. Ar = 2,6-dimethylphenyl, 2,6-diethylphenyl or mesityl) the silylenes formed on photolysis dimerized, giving nearly quantitative yields of disilenes. On the other hand, when alkyl groups were attached to the ring silicon atoms, it was suggested that the disilenes and silylenes were formed in a cage, and as a result the silylene tended not to dimerize. With suitable added trapping reagents it was common to find products derived from both the expected disilene and the silylene[98].

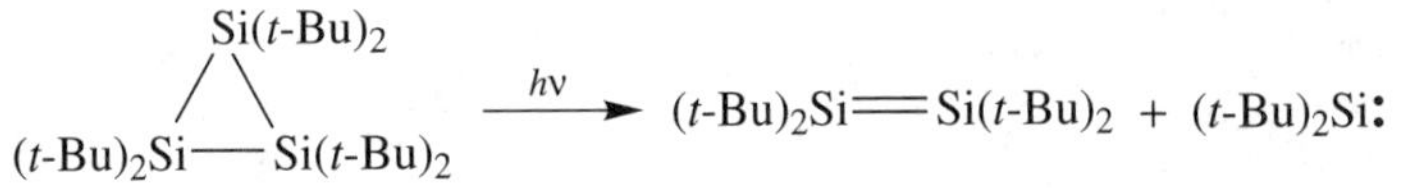

$$(t\text{-Bu})_2\text{Si}\underset{\text{Si}(t\text{-Bu})_2}{\overset{\text{Si}(t\text{-Bu})_2}{\triangle}} \xrightarrow{h\nu} (t\text{-Bu})_2\text{Si}=\text{Si}(t\text{-Bu})_2 + (t\text{-Bu})_2\text{Si}: \qquad (22)$$

In a few special cases, as shown in Scheme 33, both the di(*t*-butyl)silylene **193** and the tetra(*t*-butyl) disilene **194** reacted with the same molecule of trapping agent on photolysis[99].

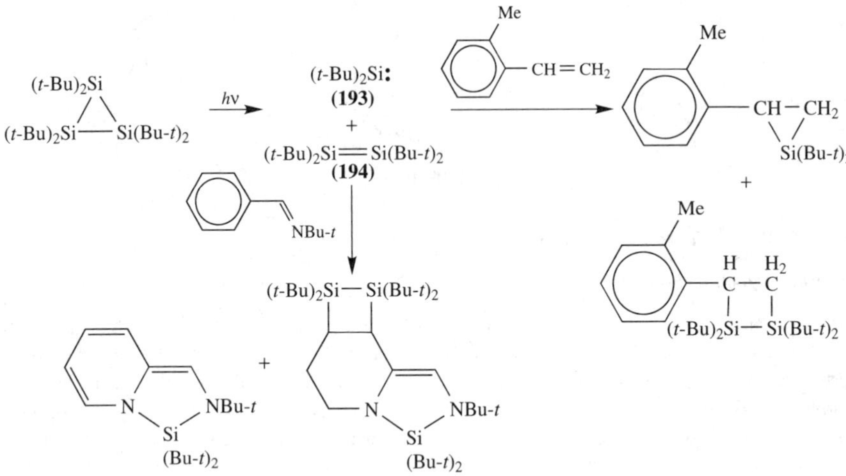

SCHEME 33

Several classes of polycyclic polysilanes, such as the octasilacubanes and the hexasilaprismanes, have been synthesized[100], but little is known about the photochemical behavior of the silacubanes which appear to be stable toward ultraviolet radiation. The silaprismanes, such as **195**, on the other hand, were observed to rearrange to the Dewar benzene isomer **196**, using 340–380 nm radiation[101]. The Dewar benzene reverted to the

21. The photochemistry of organosilicon compounds

silaprismane if irradiated with radiation of wavelength longer than 460 nm (equation 23).

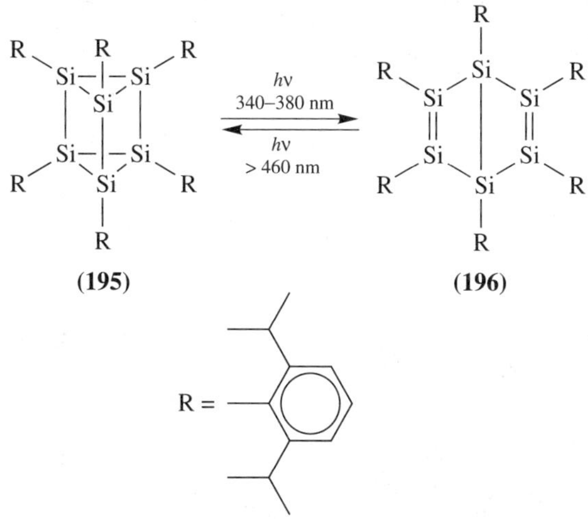

(195) ⇌ (196)   (hv 340–380 nm / hv > 460 nm)

R = 2,6-diisopropylphenyl

(23)

2. Heterocyclic tri-, tetra- and polysilanes

Watanabe and coworkers synthesized several four-membered rings containing three silicon atoms and a heteroatom, either germanium or oxygen, and studied their photochemistry. Thus photolysis of the germatrisilacyclobutane **197** led to the extrusion of the germylene and, it was suggested, the diradical **198**, which cyclized to the trisilacyclopropane **199**. The latter either was trapped with oxygen as a mono- **200** or dioxa compound, **201**, or itself was photolyzed to yield the silylene and the expected disilene[102,103] (Scheme 34).

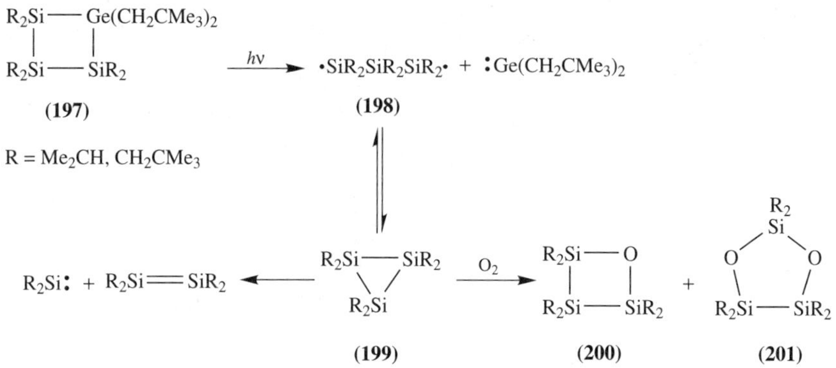

$R = Me_2CH, CH_2CMe_3$

SCHEME 34

The photochemistry of the above oxatrisilacyclobutane **200** (R = CH_2CMe_3) was also investigated. Photolysis resulted in the extrusion of a silylene and the formation of a diradical **202**, which in the presence of cyclohexane afforded the dihydride **203** or in its absence cyclized to the disilaoxirane **204**, as shown in Scheme 35. If the photolysis was done in ethanol, the silylene was trapped, and ring opening of the oxirane occurred at both the Si—O bond and the Si—Si bond, the expected ethanol adducts **205** and **206** being obtained. In addition, the ring of the initial oxatrisilacyclobutane opened in ethanol to give the product **207**. It is of interest to note that there was no evidence that the oxatrisilacyclobutane underwent retro-cleavage to a disilene and silanone, under the photolytic conditions.

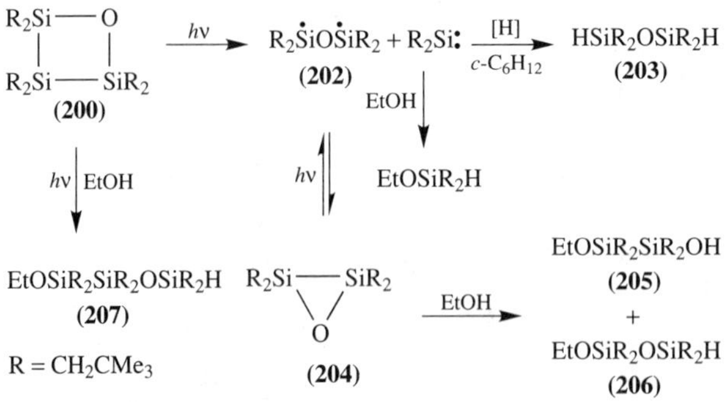

SCHEME 35

The preparations and photolyses of the tetrasilagermacyclopentanes **208** have recently been described[104]. The photolysis of **208** in cyclohexane gave the products **209** and **210**, and in the presence of dimethylbutadiene both the germacyclopentene **211** and the silacyclopentene **212** were formed. Thus it appeared that both a germylene ($Ph_2Ge:$) and a silylene ((i-Pr)$_2$Si: or (t-BuCH$_2$)$_2$Si:)) were extruded during the photolyses, as shown in Scheme 36.

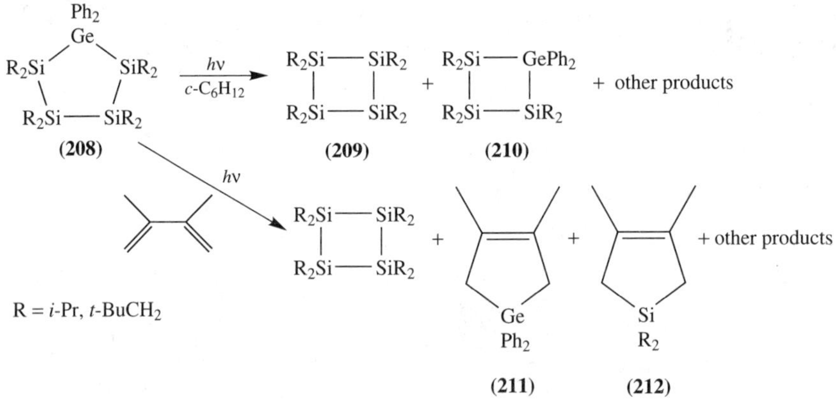

SCHEME 36

V. ACYLSILANE PHOTOCHEMISTRY

A. Simple Acylsilanes

As described in more detail in an earlier review[2] simple acylsilanes, R_3SiCOR', undergo two types of reaction when photolyzed. One involves the rapid reversible 1,2-silyl shift from carbon to oxygen of the carbonyl group leading to a siloxycarbene which can be trapped, particularly by alcohols. A slower competing reaction which becomes more prominent in less polar solvents is homolysis leading to a silyl radical and an acyl radical. In carbon tetrachloride (or other halogenated solvents) the silyl radical is efficiently trapped as the chlorosilane. In both reactions, if the silicon atom is a stereocenter, retention of configuration is largely preserved[105,106]. In hydrocarbon solvents, the silyl radical usually reacts with a hydrocarbon radical derived by decarbonylation of the acyl radical. In some cases disilanes are also produced. These reactions are summarized in Scheme 37.

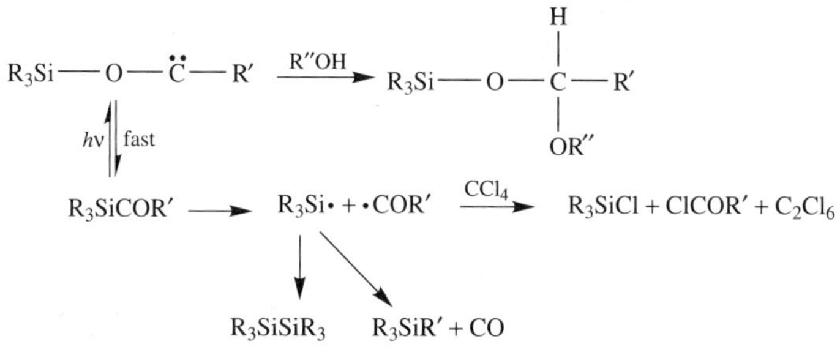

SCHEME 37

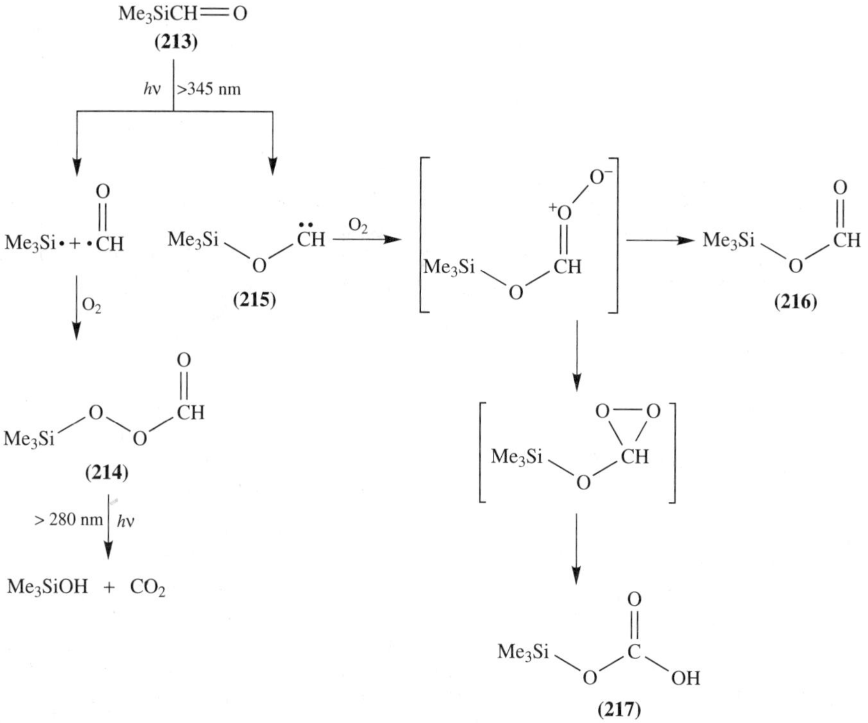

SCHEME 38

The photolyses and photooxidations of formyl-, acetyl-, and benzoyltrimethylsilane in argon or O_2 doped argon matrices have recently been reported[107]. The results are explained in terms of both Norrish type 1 cleavage to silyl and acyl radicals as well as 1,2-silyl shifts resulting in short-lived siloxycarbenes. The major product from photolysis of the formylsilane **213** in the presence of oxygen is trimethylsilyl peroxyformate, **214**, which if photolyzed at >280 nm yields trimethylsilanol and carbon dioxide. The component of the reaction proceeding via the siloxycarbene **215** apparently led to formation of trimethylsilyl formate, **216**, and trimethylsilyl carbonate, **217** (Scheme 38). These proposals were supported by isotopic labeling studies.

Photolysis of the acetylsilane **218** at 10 K in the absence of oxygen was said to yield initially the s-Z vinyloxysilane **219** which slowly isomerized to the s-E conformer **220**, or more rapidly at 45 K as shown in equation 24.

$$Me_3Si-\underset{\underset{(218)}{}}{\overset{\overset{O}{\|}}{C}}-Me \xrightarrow[\substack{280\text{ nm}\\10\text{ K}}]{h\nu} Me_3Si\diagdown_O\diagup\overset{\overset{Me}{|}}{C:} \longrightarrow Me_3Si\diagdown_O\diagup\overset{\overset{CH_2}{\|}}{CH}$$

(219)

$$\downarrow 45\text{ K}$$ (24)

$$Me_3Si\diagdown_O\diagup CH=CH_2$$

(220)

In the presence of oxygen, products analogous to those shown in Scheme 38 were formed.

Photolysis of benzoyltrimethylsilane at 10 K in the absence of oxygen showed no changes in the IR spectrum of the matrix after several hours. In the presence of oxygen a variety of products was identified including trimethylsilyl benzoate and the main product trimethylsilyl perbenzoate.

B. Acyldisilanes

Since the previous review[2] a further study of the photolysis of several acyldisilanes has been reported[108]. Like previously described di- or polysilylacylsilanes, the disilane **221** on photolysis underwent a 1,3-shift of a trimethylsilyl group from silicon to oxygen, yielding the silene **222** which could be trapped as the methanol adduct **223** if the photolysis was done in methanol. In no case was the possible 1,2-shift of the disilyl group from carbon to oxygen to yield a disilyloxycarbene observed, suggesting that the 1,3-shift is a much lower-energy pathway. In inert solvents, there was NMR evidence that the silene underwent reversible head-to-head dimerization to yield **224**. However, an unusual dimeric species, **225**, not previously observed in these photolyses, was also obtained as a major product, apparently derived from the reaction of the silene with the carbonyl group of unphotolyzed acyldisilane, a type of reaction recently shown to be possible (see below). In addition there was some product **226** derived from radical cleavage of the Si—Si bond followed by decarbonylation and recombination (Scheme 39).

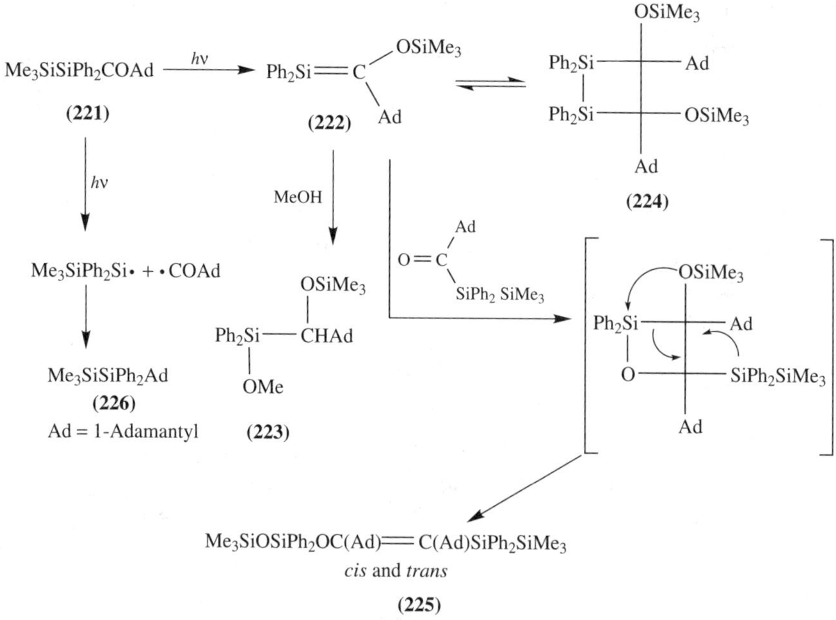

21. The photochemistry of organosilicon compounds

$$Me_3SiSiPh_2COAd \xrightarrow{h\nu} Ph_2Si=C\begin{matrix}OSiMe_3\\ \\Ad\end{matrix} \rightleftharpoons \begin{matrix}&OSiMe_3&\\Ph_2Si&\!\!\!\!-\!\!\!\!&Ad\\|&&|\\Ph_2Si&\!\!\!\!-\!\!\!\!&OSiMe_3\\&Ad&\end{matrix}$$

(221)   (222)   (224)

```
           hv                MeOH                   Ad
           ↓                   ↓                 O=C
                                                    \
  Me3SiPh2Si• + •COAd      OSiMe3                    SiPh2SiMe3
           ↓                 |
                         Ph2Si—CHAd                  [   Ph2Si—OSiMe3  ]
  Me3SiSiPh2Ad              |                        [    |      |     ]
     (226)                 OMe                       [    |      Ad    ]
 Ad = 1-Adamantyl          (223)                     [    O——SiPh2SiMe3]
                                                     [       Ad        ]
```

Me₃SiOSiPh₂OC(Ad)=C(Ad)SiPh₂SiMe₃
cis and trans
(225)

SCHEME 39

The analogous dimesitylacyldisilane **227** showed somewhat different behavior[108]. The silene **228** initially formed by photolysis apparently underwent a subsequent photochemically induced 1,2-trimethylsiloxy shift from carbon to silicon (a previously observed reaction[109]) leading to a silylcarbene **229**, which then inserted into the benzylic C—H bond of the *ortho*-methyl group of one of the mesityl groups to give the silaindane **230**. There were also products **231** and **232** derived from dimerization of each of the radicals produced by Norrish type 1 cleavage of the acyldisilane and, as well, when the photolysis was carried out in methanol, a product **233** derived by methanol trapping of an intermediary disilyloxycarbene **234** was observed (Scheme 40). While there were precedents known for each of these reactions, the mixture of products observed was derived from an unusual combination of these known photochemical processes.

C. Acylpolysilanes

The photolysis of acylpolysilanes has been extensively used as a route for the preparation of silenes, both transient and long-lived. Research to 1989 has been reviewed[2].

When photolyzed in inert solvents, the most common behavior of a polysilylacylsilane such as **235**, with three trimethylsilyl groups on silicon, is for a 1,3-shift of a silyl group from the central silicon atom to the carbonyl oxygen to occur, leading to the formation of a silene **236** bearing a trimethylsiloxy group on the carbon atom of the Si=C bond. (1,2-Shifts of the polysilyl group to oxygen, which would lead to a siloxycarbene, have never been observed in these polysilyl systems.) Silenes such as **236** may be stable if the attached groups are sufficiently bulky, as when groups like adamantyl or mesityl are attached to the carbon atom of the Si=C bond. Otherwise, the silenes dimerize in a head-to-head manner, probably via a diradical **237** leading either to 1,2-disilacyclobutanes,

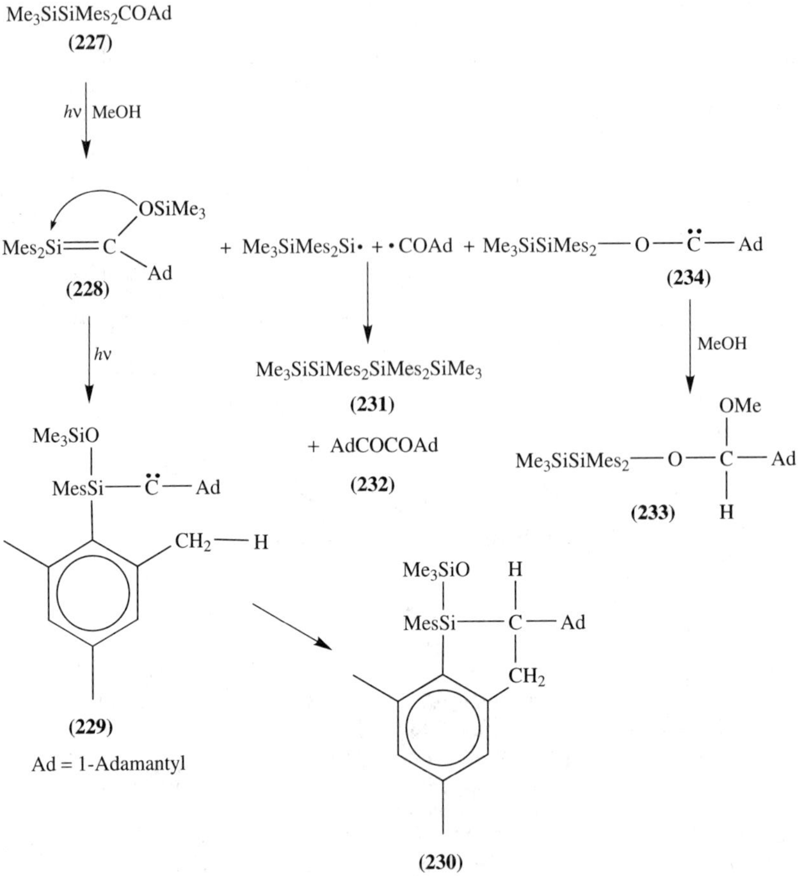

SCHEME 40

238, or to acyclic disilanes such as **239** if R' has appropriately located hydrogen alpha to the original carbonyl–carbon atom. These acyclic dimers could be formed either by the 1,4-diradical **237** undergoing disproportionation, or by an ene reaction between two silene molecules as illustrated in Scheme 41.

Conlin and coworkers studied the silene $(Me_3Si)_2Si=C(OSiMe_3)Me$ using laser flash photolysis, and while much was learned about some details of its reactions, it was not possible to establish unambiguously the mechanism by which dimerization occurred[110].

The reactions of the silenes formed by photolysis of polysilylacylsilanes with a wide variety of reagents, e.g. alkenes[111] alkynes[112], dienes[111], carbonyl compounds[113], α,β-unsaturated carbonyl compounds[114,115], isonitriles[116,117], Grignard reagents[118], imines[119], sulfur[120], selenium[120], acylsilanes[121] and a silylketene[122], have been studied using two methods in most cases. On the one hand, the relatively stable and long-lived silenes were prepared by photolysis and then the reagent was added in the dark. Alternatively, a mixture of the polysilylacylsilane and reagent were co-photolyzed. In all cases the experimental results were virtually identical, indicating that the reaction of the silene with the reagent was a non-photochemical process.

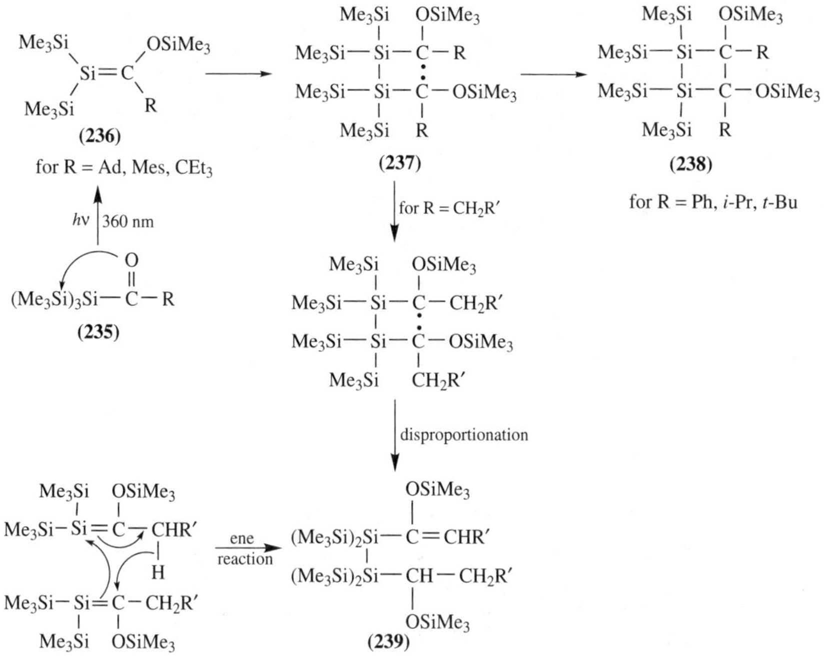

21. The photochemistry of organosilicon compounds

SCHEME 41

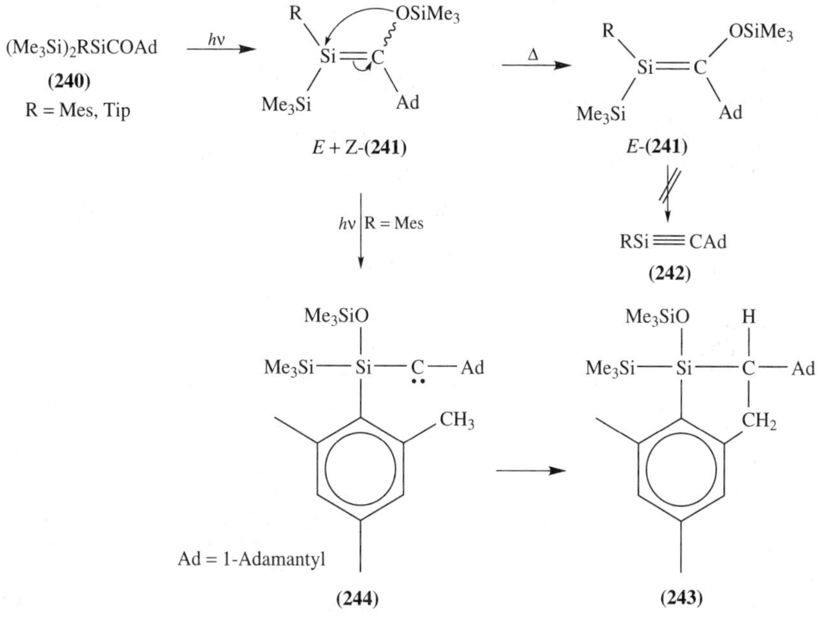

SCHEME 42

Polysilylacylsilanes with the structure $(Me_3Si)_2RSiCOAd$ (R = t-Bu, Ph, Mes, Tip) have also shown interesting behavior. In each case a 1,3-shift of a trimethylsilyl group occurred leading to a silene. When R = Mes[112] or Tip (= 2,4,6-tri-isopropylphenyl)[123] photolysis of **240** led to mixtures of the E- and Z-isomers of the silene **241** (Scheme 42) which were stable at room temperature, although at elevated temperatures they slowly decomposed, leading initially to essentially pure E-**241** without evidence for $E \rightleftharpoons Z$ interconversion, and ultimately to fragmentation, without any evidence that a silyne **242** was formed by loss of hexamethyldisiloxane from the silene. Prolonged photolysis gave rise primarily to silaindanes **243** resulting from a photochemical rearrangement of the silene by migration of the trimethylsiloxy group from carbon to silicon, followed by insertion of the carbene **244** into the benzylic C—H bond of the mesityl group. When R = t-Bu or Ph there was also evidence that the initially formed silene itself underwent photochemical rearrangement (see Section VIII)[109].

Photolysis of α,β-unsaturated acylpolysilanes of the general structure $(Me_3Si)_3$SiCOCR=CR$'_2$ failed to yield significant amounts of the isomeric siladienes, and gave instead products derived from the Norrish type 1 cleavage of the acylsilane into the radicals $(Me_3Si)_3Si\cdot$ and $\cdot COCR=CR'_2$[124].

VI. SILYLDIAZOALKANES AND SILYLCARBENES

The photolysis of silyldiazoalkanes, species which are much stabler than their all-carbon analogs, leads to the formation of silylcarbenes. The subsequent behavior of the silylcarbenes depends greatly on the nature of the groups attached to the diazo-carbon atom, as well as to the silicon atom. Early results have been reviewed[2]. In Scheme 43 several representative reactions are listed.

Me_3SiCHN_2 $\xrightarrow{h\nu}$ $Me_3Si\ddot{C}H$ $\longrightarrow$ $Me_2Si=CHMe$

$Me_3SiCN_2SiMe_3$ $\xrightarrow{h\nu}$ $Me_3Si\ddot{C}SiMe_3$ $\longrightarrow$ $Me_2Si=C(Me)SiMe_3$

$PhMe_2SiCN_2CO_2Me$ $\xrightarrow{h\nu}$ $PhMe_2Si\ddot{C}CO_2Me$ $\longrightarrow$ $Me_2Si=C(Ph)CO_2Me$

$R_3SiCN_2CH_2R'$ $\xrightarrow{h\nu}$ $R_3Si\ddot{C}CH_2R'$ $\longrightarrow$ $R_3SiCH=CHR'$

$Me_3SiSiMe_2CHN_2COOEt$ $\xrightarrow{h\nu}$ $Me_3SiSiMe_2\ddot{C}HCOOEt$ $\longrightarrow$ $Me_2Si=C(SiMe_3)COOEt$

$Me_3SiSiMe_2CN_2COTol$-p $\xrightarrow{h\nu}$ $Me_3SiSiMe_2\ddot{C}COTol$-$p$ $\longrightarrow$ $Me_2Si=C(SiMe_3)COTol$-p

SCHEME 43

The first three examples illustrate that facile migration of Me[125,126] or Ph[125] from silicon to the carbene carbon atom can occur giving rise to silenes, while the fourth example[127] illustrates that hydrogen, if adjacent to the carbene carbon, migrates in preference to a group on silicon, forming an alkene. On the other hand, as illustrated by the last two examples[126,128], migration of a trimethylsilyl group from silicon to the carbene carbon was found to occur easily and cleanly. Many silenes have been prepared by photolysis of the silyldiazoalkanes at low temperature in glasses, allowing the determination of their UV absorption maxima[125,126,129].

More recently, more complex systems have been studied by the groups of Ando[129] and Maas[128,130,131]. As illustrated in Scheme 44, photolysis of the acyldisilyldiazoalkanes **245** yielded successively the disilylcarbene **246**, and then the acylsilene **247** which underwent a variety of rearrangements depending on the R group present. Intramolecular electrocyclic ring closure to give **248** occurred when bulky groups like t-Bu and 1-adamantyl were present, while smaller groups (Me, i-Pr) led to products of intermolecular dimerization, **249**. In addition, 1,3-migration of the R group from C to Si led to ketenes **250**.

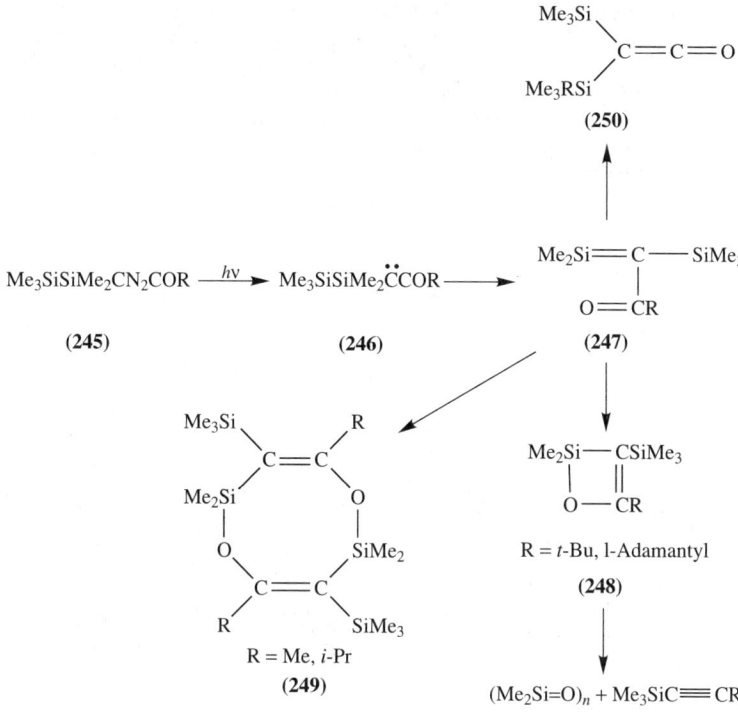

SCHEME 44

The photolysis of dimethylsilyldiazomethane **251** in an argon matrix at 10 K (Scheme 45) was shown by Sander and coworkers to lead to the diazirine **252** and, via the carbene **253** and a 1,2-H shift, to dimethylsilene **254**, used to study the thermal reaction of the silene with formaldehyde[132].

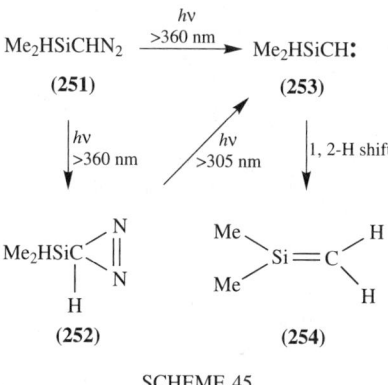

SCHEME 45

Sander and coworkers have recently studied the photolysis of 2,2-dimethyl-2-sila-1,3-bis(diazo)propane **255**[133]. When photolyzed at >305 nm in ethanol or EPA (2 : 5 : 5

EtOH:isopentane:Et$_2$O) glasses at 77 K, four products **256**, **257**, **258** and **259** were obtained in overall >95% yield. The major product **256** was shown by isotopic labeling studies to have been derived from the silirene **260** formed by intramolecular coupling of the bis-carbene **261**, and pathways for the formation of the other products were proposed. The photolysis of the bis-diazo compound **255** was studied in detail at 405 nm and it was found that the diazirine **262** was formed: this on photolysis at 305 nm also gave **260** on the pathway to **256**. The chemistry is shown in Scheme 46.

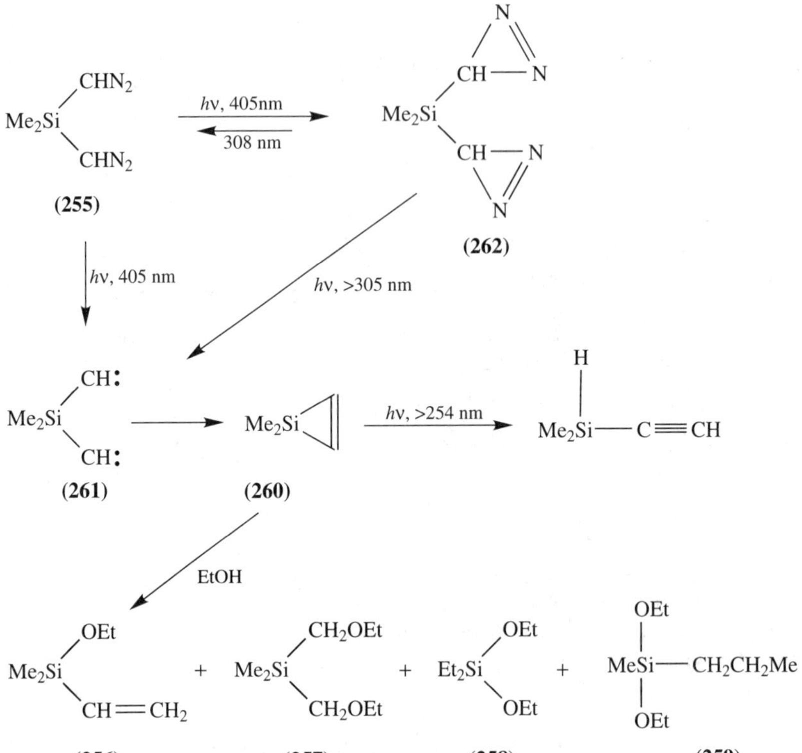

SCHEME 46

The silylcarbene :CHSi(OMe)$_3$ was generated by low-temperature photolysis at 345 nm of the related diazomethane, and was characterized. Photolysis of the silyldiazomethane at 280 nm in the cold led to 1,1-dimethoxy-1-sila-2-oxetane, arising from insertion of the carbene into one of the methoxy groups[134].

Ando and coworkers have described unusual photochemical behavior of a bis-diazo compound **263** which on photolysis gave rise to a bis-silene **264**[135]. In inert solvents this spontaneously underwent two modes of reaction, 'criss-cross' (head-to-tail) addition and 'parallel' (head-to-head) addition, leading to bridged disilacyclobutanes **265** and **266** (Scheme 47).

In a subsequent paper[136] further studies of these and other bis(diazo) compounds having two, three, or four SiMe$_2$ groups between the two diazo groups were described. In contrast to the photolyses in inert solvents, where products of intramolecular [2 + 2]

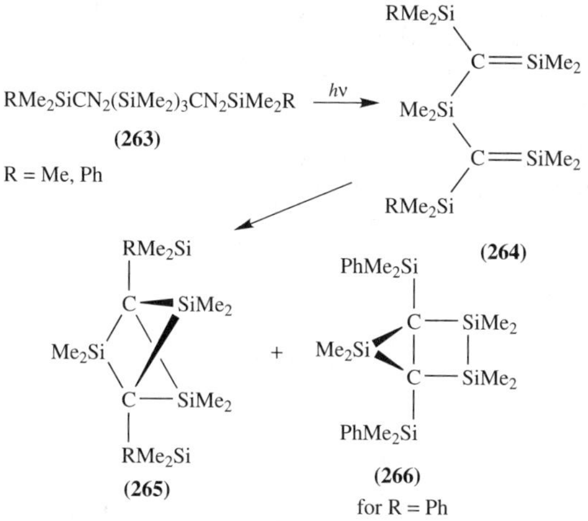

21. The photochemistry of organosilicon compounds

SCHEME 47

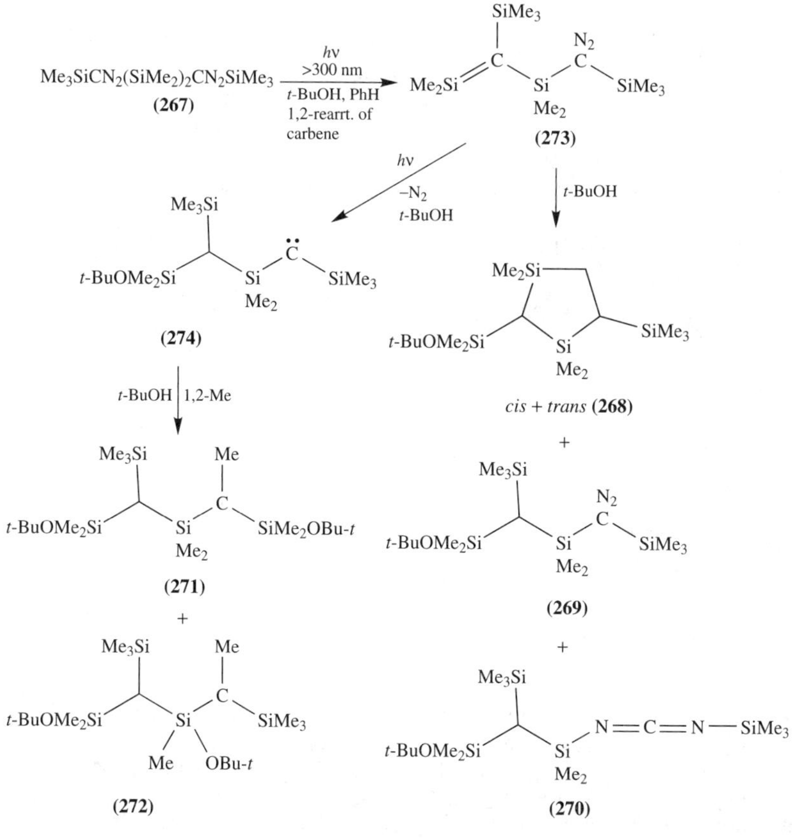

SCHEME 48

cyclization were obtained, when compound **267**, as an example, was photolyzed in *t*-butyl alcohol five structurally different products, **268–272**, derived from several different intermediates, were identified (Scheme 48). Thus the carbene formed from initial photolysis, followed by a 1,2-silyl migration of the internal Me$_2$Si-terminated group, led to the silene **273**, which reacted with one mole of the alcohol in different ways yielding **268**, **269**, and **270**. Alternatively, further photolysis apparently led to a carbene, **274**, which then underwent a 1,2-Me shift prior to reacting with the alcohol, leading to **271**, and **272**.

As another example, photolysis in inert solvents of compound **275**, where four SiMe$_2$ groups bridged the two diazo functions, led to the bis-silene **276**, which reacted by either head-to-head or head-to-tail pathways resulting in the formation of the bicyclic compounds **277** and **278**, respectively. Alternatively, [2 + 3] cycloaddition gave the intermediate **279** which led to the formation of the bicycloadduct **280** (Scheme 49).

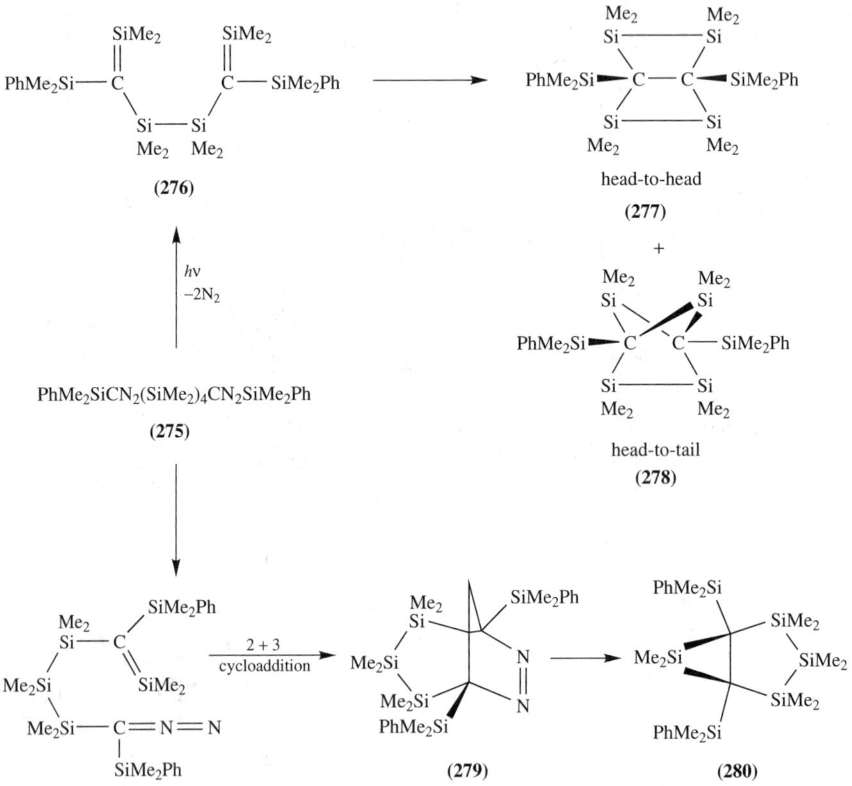

SCHEME 49

The photolyses of several transition metal-substituted silyldiazoalkanes have recently been described[137,138]. As indicated by the examples given in Scheme 50, the silyldiazoalkane **281** initially formed the carbene **282**, which underwent a phosphine 1,2-rearrangement from the transition metal to the carbene–carbon yielding the ylide **283**, which subsequently dimerized to form the product **284**. The related nickel compound also gave a dimer when photolyzed (Scheme 50).

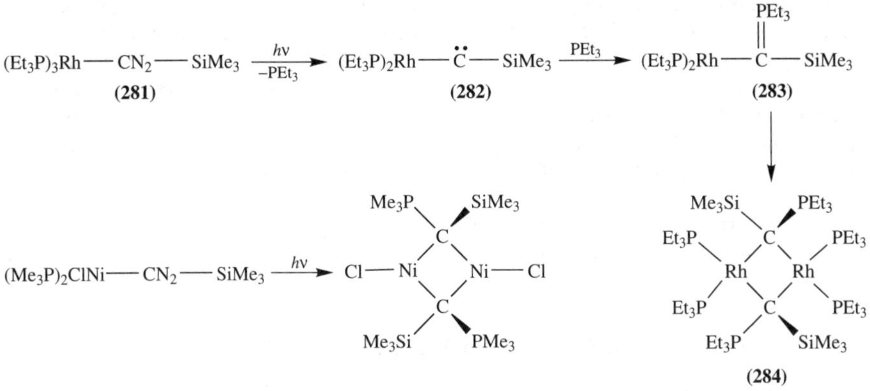

21. The photochemistry of organosilicon compounds

SCHEME 50

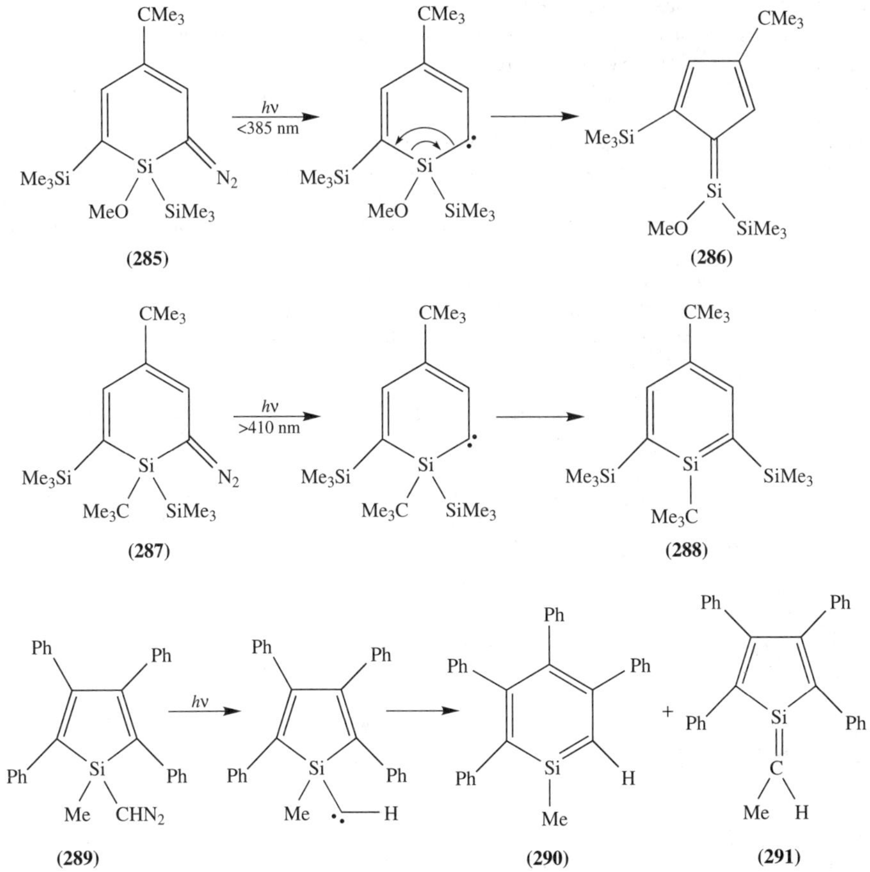

SCHEME 51

The photolysis of unsaturated cyclic silyldiazoalkanes has also been employed as a route to several silaaromatic compounds such as silabenzenes and silafulvenes. Märkl and coworkers showed that the diazo compound **285** was converted on photolysis at <385 nm to the silafulvene **286**[139] and that photolysis of the diazo compound **287** led to the silabenzene **288**[140]. Each of the processes involved a 1,2-shift of a group from silicon to carbon. Earlier, Ando and coworkers[141] had shown that photolysis of the silacyclopentadienyldiazomethane **289** led to the formation of both a silabenzene **290** and a silafulvene **291** (Scheme 51).

These silaaromatics are not stable, although one example described by Märkl and coworkers was sufficiently stable at $-100\,°C$ to be characterized by ^{29}Si NMR spectroscopy[140].

VII. PHOTOCHEMICAL ROUTES TO SILENES, SILYLENES AND DISILENES

Some of the reactions given in the sections above are important routes to reactive organosilicon intermediates such as silenes and silylenes, and because of their tendency to dimerize readily, to disilenes, the latter being formed when matrix-isolated silylenes are warmed up. It appears worthwhile to summarize some of the more useful reactions leading to silenes and silylenes, and their subsequent behaviors. The topics of silylenes[97,142] and disilenes[97,142,143] have recently been reviewed.

Simple silenes are probably most easily and gently generated by the photolysis of silyldiazoalkanes, often done in low-temperature matrices, since the subsequent 1,2-rearrangement that the initially formed silylcarbene undergoes on warming is normally a clean reaction. More stable silenes are best prepared by the photolysis of polysilylacylsilanes as was described in Section V.C, although further rearrangements may occur, as will be described below, depending on the substituents on the Si=C bond, and the length of photolysis. The chemistry of silenes has recently been reviewed[144].

Silylenes are produced by several photochemical routes. Dimethylsilylene, $Me_2Si:$, is produced during the photolysis at 254 nm of dodecamethylcyclohexasilane, $(Me_2Si)_6$, but the cyclopentasilane and lower homologues are by-products of this approach, which

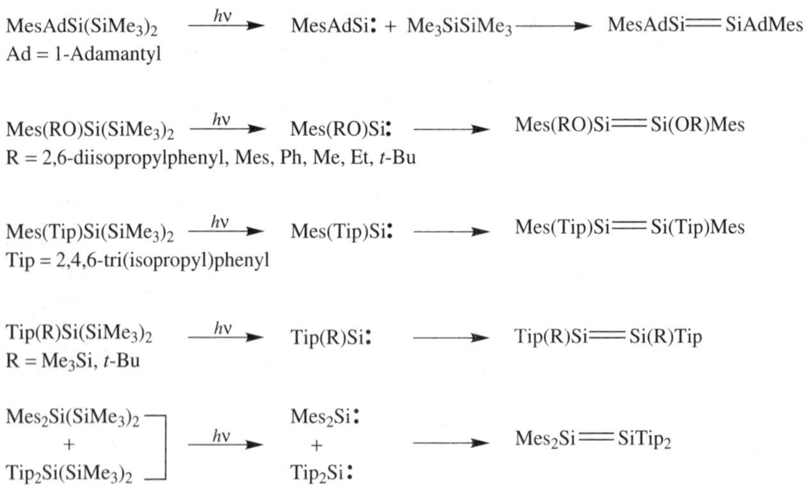

MesAdSi(SiMe₃)₂ —hν→ MesAdSi: + Me₃SiSiMe₃ ——→ MesAdSi=SiAdMes
Ad = 1-Adamantyl

Mes(RO)Si(SiMe₃)₂ —hν→ Mes(RO)Si: ——→ Mes(RO)Si=Si(OR)Mes
R = 2,6-diisopropylphenyl, Mes, Ph, Me, Et, t-Bu

Mes(Tip)Si(SiMe₃)₂ —hν→ Mes(Tip)Si: ——→ Mes(Tip)Si=Si(Tip)Mes
Tip = 2,4,6-tri(isopropyl)phenyl

Tip(R)Si(SiMe₃)₂ —hν→ Tip(R)Si: ——→ Tip(R)Si=Si(R)Tip
R = Me₃Si, t-Bu

Mes₂Si(SiMe₃)₂   Mes₂Si:
     +   —hν→   +   ——→ Mes₂Si=SiTip₂
Tip₂Si(SiMe₃)₂   Tip₂Si:

SCHEME 52

may be inconvenient. Fink and coworkers have recently shown that when $(Me_2Si)_6$ was photolyzed in a molecular beam using a pulsed supersonic jet, a single photolysis photon was capable of producing the silylene $Me_2Si\colon$ in contrast to the behavior of other common silylene precursors such as $PhMe(SiMe_3)_2$ or $PhSi(SiMe_3)_3$, where a single photon removed only a single Me_3Si group[145]. The conventional photolysis of the latter two compounds, and the photolysis of linear trisilanes, particularly those such as $Me_3Si-SiAr_2-SiMe_3$, where the groups on the central silicon atom are aromatic (e.g. Ph, Mes, Tip etc.), is a very convenient source of silylenes for synthetic and other related purposes by trapping in the cold, or for disilene formation by dimerization at higher temperatures. Several examples of disilenes formed from silylenes by this route are shown in Scheme 52[146–148].

When the R groups on the central silicon atom of the trisilanes are aliphatic, e.g. as in $PhMe_2Si-SiR_2-SiMe_2Ph$ [R = t-Bu, 1-Adamantyl (Ad)], photolysis does not give rise to much silylene[149,150]. However, dialkyl-substituted silylenes, e.g. $Ad_2Si\colon$, are readily available by the photolysis of dialkylsiliranes **292**[150,151] or by the photolysis of the reasonably readily available cyclotrisilanes such as **293** (Scheme 53)[97,142]. These photochemically generated singlets add with considerable stereospecificity to alkenes yielding siliranes, the loss of complete stereospecificity being attributed to photochemical ring opening and reclosure of the silirane ring.

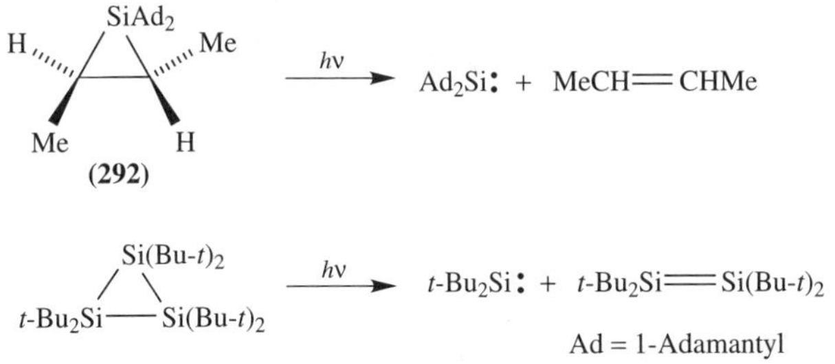

$$\underset{\textbf{(292)}}{\overset{\overset{\displaystyle SiAd_2}{\triangle}}{\underset{Me\ \ \ \ \ H}{H\ \ \ \ \ Me}}} \xrightarrow{h\nu} Ad_2Si\colon + MeCH{=}CHMe$$

$$\underset{\textbf{(293)}}{\overset{\overset{\displaystyle Si(Bu\text{-}t)_2}{\triangle}}{t\text{-}Bu_2Si-Si(Bu\text{-}t)_2}} \xrightarrow{h\nu} t\text{-}Bu_2Si\colon + t\text{-}Bu_2Si{=}Si(Bu\text{-}t)_2$$

Ad = 1-Adamantyl

SCHEME 53

Cyclotrisilanes are easily converted to a mixture of silylene and disilene. Reactions of the silylene $(t$-$Bu)_2Si\colon$ and the disilene t-$Bu_2Si{=}Si(Bu$-$t)_2$ produced by photolysis of the cyclotrisilane **293** with cyclopentadiene, furan, and thiophene have been reported recently[152], as were their reactions with 2,2′-bipyridyl, pyridine-aldimine and ketoimines[153].

Bobbitt and Gaspar have described preparations of $Me_2Si\colon$, $PhMeSi\colon$ and $Ph_2Si\colon$, as well as the kinetics and products formed from their reactions with butadiene[154].

Three unusual and interesting examples of silylene formation from photolyses are shown in Scheme 54. In the first, Fink and coworkers[155] photolyzed the trisilane **294** at 254 nm and produced the relatively stable silylene **295**. This on further photolysis gave rise to the sterically crowded silacyclobutadiene **296** which was trapped with several reagents. In the second example Michl and coworkers[156] photolyzed the matrix-isolated bis-azide **297** to form the cyclic silylene **298**, and this on further photolysis at selected wavelengths, using matrices and low temperatures, isomerized to the silacyclopentadienes **299** and **300** and finally to the 1-sila-2,4-cyclopentadiene **301**. Finally, Sakurai and coworkers[157] were able to convert the trisilane **302** to the cyclic divinylsilylene **303**.

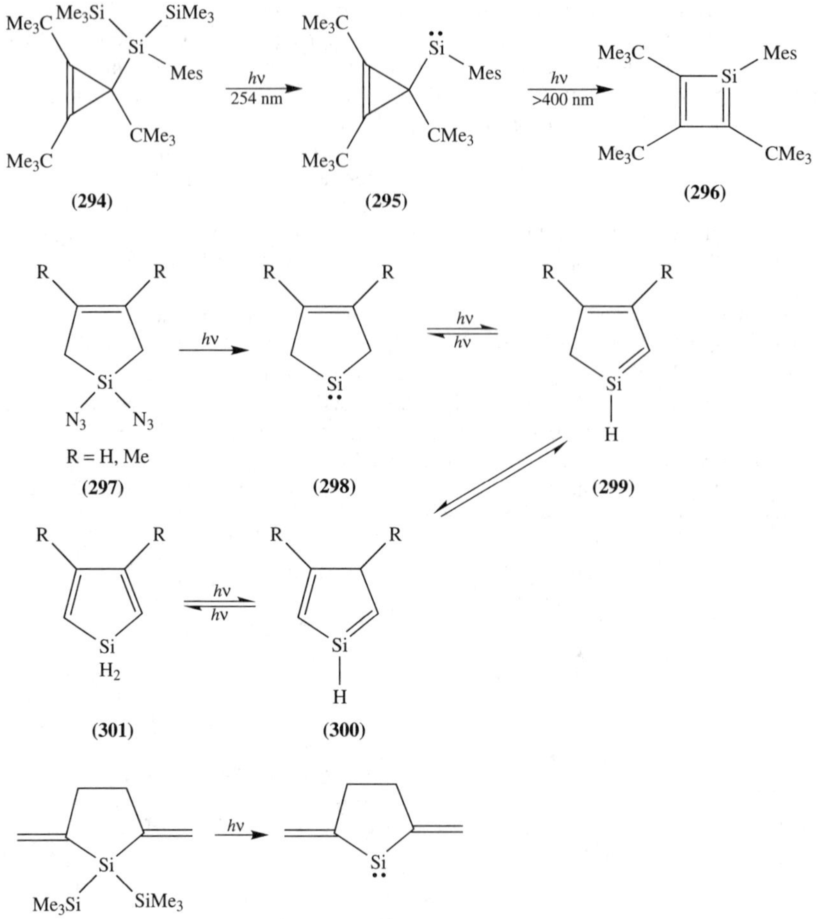

SCHEME 54

VIII. PHOTOCHEMICAL REARRANGEMENTS OF SILENES, SILYLENES AND DISILENES

Knowledge concerning the photochemistry of silenes, silylenes or disilenes is somewhat limited, largely because they are usually reactive transient species difficult to handle. However, a number of interesting and important observations have been made which are reported below.

A. Photolysis of Silenes

The earlier controversy about whether silene–silylene interconversions could occur was settled by studies of West, Michl and coworkers[158]. They generated dimethylsilylene Me$_2$Si: from several precursors in glasses at temperatures from 10 to 77 K and characterized it by its UV and IR spectra. This yellow species was bleached by irradiation with visible light to form colorless methylsilene, MeHSi=CH$_2$, which was also characterized spectroscopically. Its irradiation at 248 nm reconverted it back to dimethylsilylene (equation 25).

$$Me_2Si: \xrightleftharpoons[\substack{h\nu \\ 248 \text{ nm}}]{\substack{h\nu \\ > 400 \text{ nm}}} MeSiH=CH_2 \quad (25)$$

Photochemical rearrangements of more complex silenes have also been observed. Thus the *t*-butylsilene **304**, (generated photochemically from its acylsilane precursor, as shown by trapping reactions), when further photolyzed at wavelengths >360 nm, rearranged via an intermediary silene **305** to an isomeric dimethylsilene **306**, as shown by the X-ray crystal structure of its anticipated head-to-head dimer **307**[109]. It was suggested that **306** was the result of a 1,2-carbon-to-silicon migration of the trimethylsiloxy group and a 1,2-silicon-to-carbon migration of the trimethylsilyl group of **304**, giving **305**, both previously observed rearrangements, followed by a 1,3-silicon-to-silicon methyl shift. Originally it was suggested, since attempts to trap an intermediate silylcarbene **308** failed, that the above transformation might be a dyotropic process, but subsequent studies (see below) indicated that the silylcarbene **308** was probably formed as an intermediate, at least in some cases (Scheme 55). Closely related behavior was also observed for an analog of **304** with Ph instead of *t*-Bu.

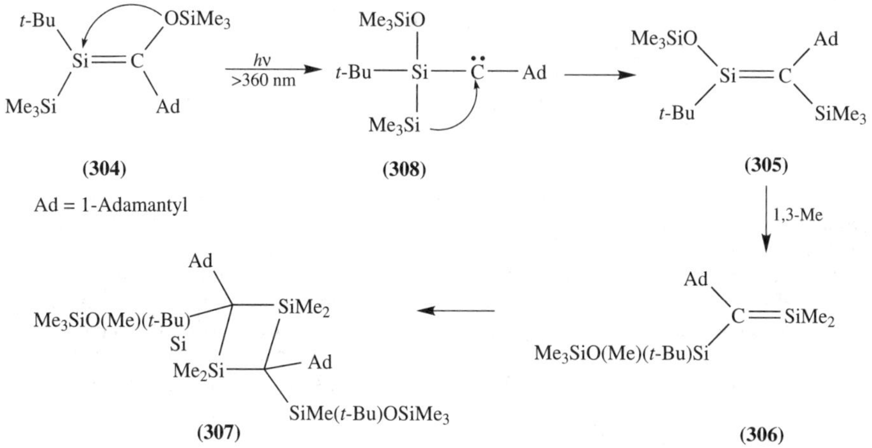

SCHEME 55

When the mesityl analog **309** of **304** was photolyzed with >360 nm radiation, the major product was a diastereomeric pair of silaindanes **310**, due to the two stereocenters present. Their formation was explained as involving the above-described 1,2-trimethylsiloxy shift to give the silylcarbene **311**, which then inserted into the benzylic C—H bond of the methyl group of mesityl (instead of undergoing a 1,2-trimethylsilyl group shift as in Scheme 55); see Scheme 56. Based on this finding it seems reasonable to suggest that the two groups involved in the silene-to-silene **304** → **305** rearrangement rearrange sequentially rather than simultaneously.

In another example West and coworkers photolyzed the stable silaallene **312** which led to the product **313** (Scheme 57). Nominally, this corresponds to an addition of a benzylic C—H bond across the ends of the Si=C bond of the silaallene[159].

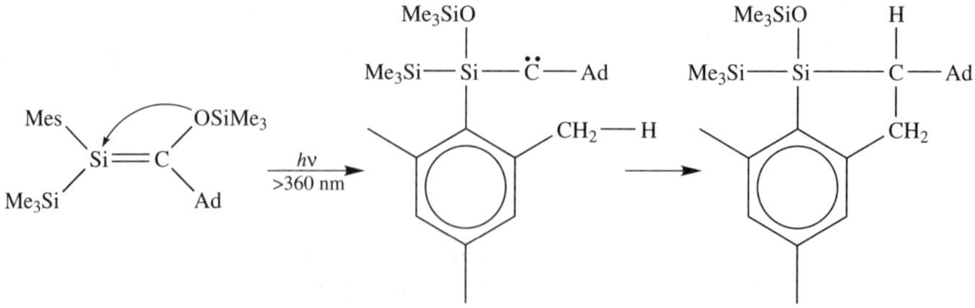

SCHEME 56

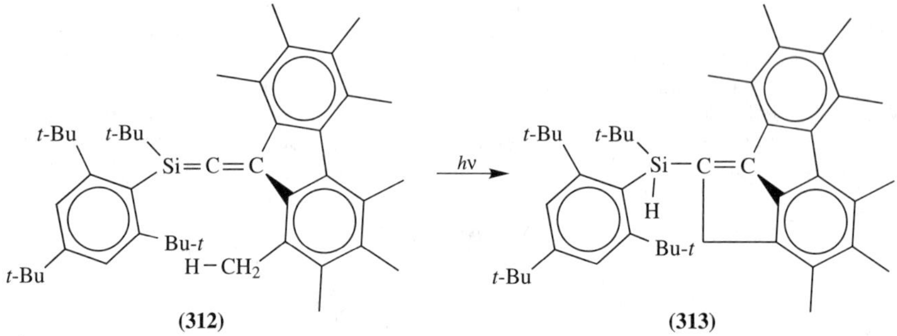

SCHEME 57

A recent study of the photolysis of simple diazoalkanes **314** or diazirines **315**, compounds known to lead to the formation of silenes under inert conditions, led, in oxygen-doped argon matrices, via the silene **316** to the siladioxirane **317**. While previously postulated as an intermediate in silene oxidations, this is important experimental evidence for this intermediate. Continued photolysis of the system led to a compound identified as the silanone–formaldehyde complex **318**, which on further irradiation led to the silanol–aldehyde **319**. The latter compound itself underwent further photochemical oxidation leading to the silanediol **320**[160]. The reactions are summarized in Scheme 58. Detailed infrared studies, including the use of isotopes, and calculations, were used to establish the structures of the compounds.

B. Photolysis of Silylenes

When di(*t*-butyl)silylene **321**, generated in a 3-methylpentane glass at 77 K or in an argon matrix at 10 K by photolysis of a precursor bis-azide, was irradiated with 500-nm light, intramolecular C—H insertion occurred yielding the silacyclopropane **322** (equation 26)[161].

Maier and coworkers have found that pulsed flash pyrolysis of an acetylenic disilane **323** gave rise to the acetylenic silylene **324**, which subsequently rearranged to the cyclic silylene 1-silacyclopropenylidene **325**[162]. Irradiation of this cyclic silylene resulted in its isomerization to the isomeric acetylenic silylene **324**, which itself could be photochemically converted to the allenic silylene **326**. As well, both **324** and **325** were said to isomerize on photolysis to the unusual silacycloalkyne **327**, which was characterized

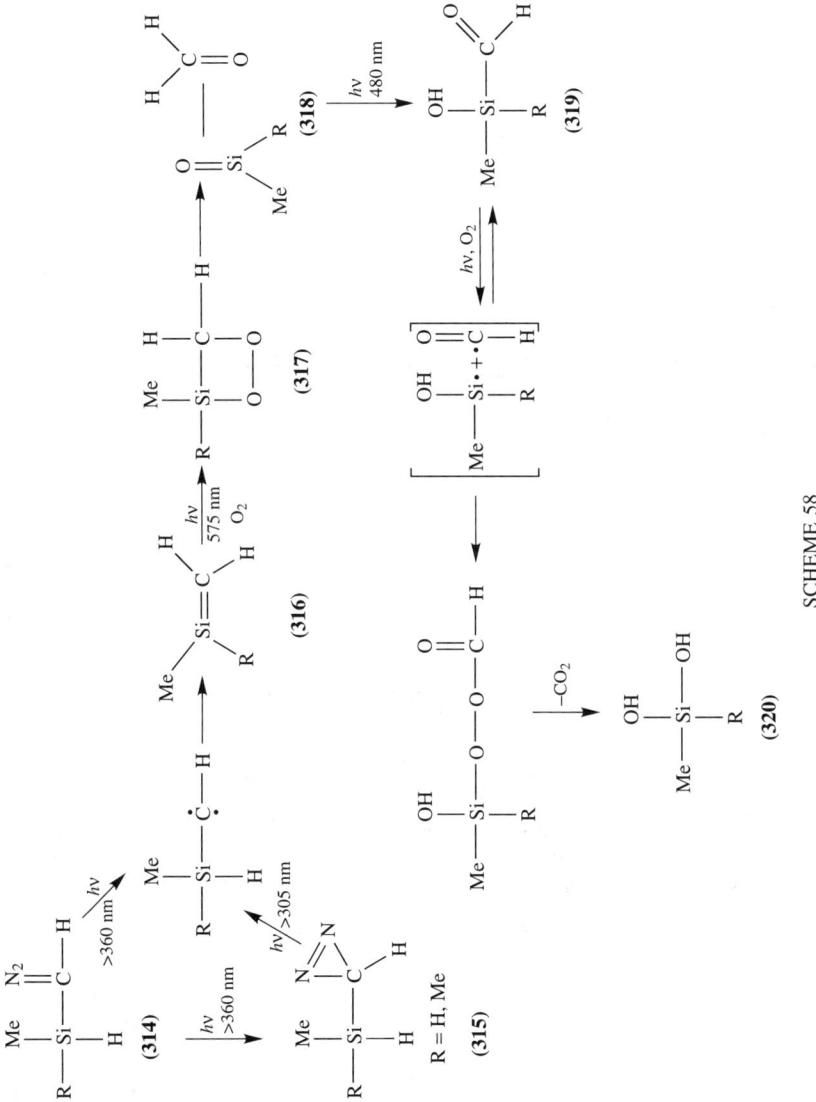

SCHEME 58

spectroscopically (Scheme 59).

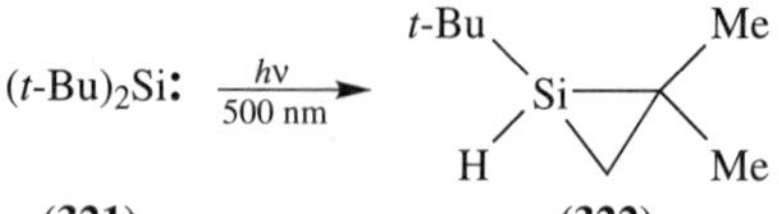

$$(t\text{-Bu})_2\text{Si:} \xrightarrow[500 \text{ nm}]{h\nu} \underset{\textbf{(322)}}{\overset{t\text{-Bu}}{\underset{H}{\diagdown}}\text{Si}\overset{Me}{\underset{Me}{\diagup}}} \quad (26)$$

(321)

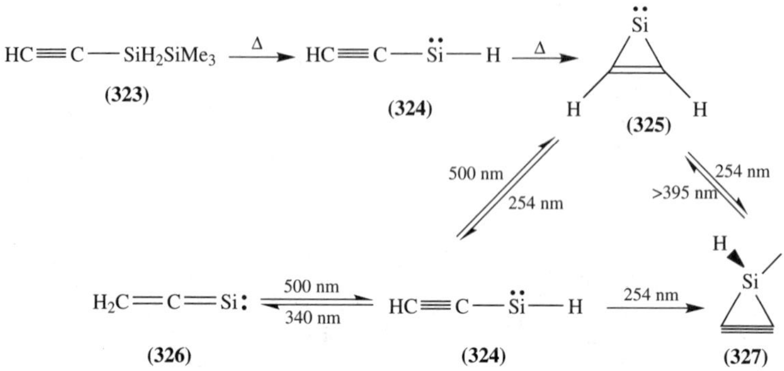

SCHEME 59

C. Photolysis of Disilenes

Among the limited studies of the photochemistry of disilenes, it has been reported that photolysis of tetramesityldisilene, **328**, at 254 nm in pentane gave rise to the dihydrido compound **329** (equation 27), probably as a result of the radical character of the excited state[163].

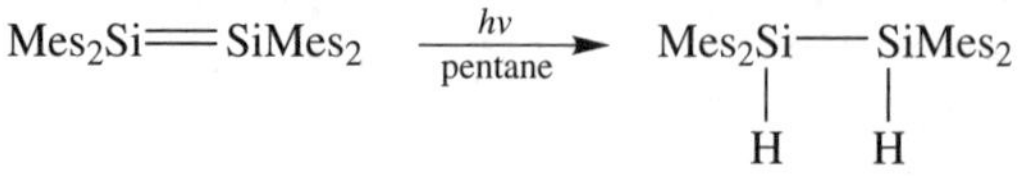

$$\underset{\textbf{(328)}}{\text{Mes}_2\text{Si}=\text{SiMes}_2} \xrightarrow[\text{pentane}]{h\nu} \underset{\textbf{(329)}}{\underset{H\ \ \ H}{\text{Mes}_2\text{Si}-\text{SiMes}_2}} \quad (27)$$

The *trans* isomers of two disilenes were observed to undergo photoisomerization to mixtures enriched in the *cis* isomers when irradiated at either 254 or 350 nm (equation 28), and the kinetics of the thermal *cis–trans* isomerization have been determined[164].

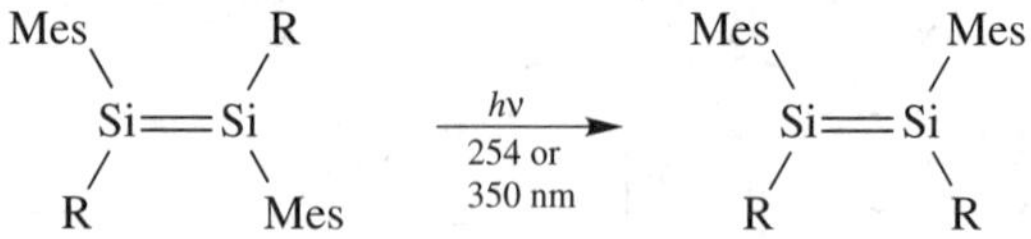

$$\underset{R}{\overset{Mes}{\diagdown}}\text{Si}=\text{Si}\underset{Mes}{\overset{R}{\diagup}} \xrightarrow[254 \text{ or } 350 \text{ nm}]{h\nu} \underset{R}{\overset{Mes}{\diagdown}}\text{Si}=\text{Si}\underset{R}{\overset{Mes}{\diagup}} \quad (28)$$

R = *t*-Bu or N(SiMe$_3$)$_2$

In other studies, the photochemical rearrangements of disilenes with the general structures A$_2$Si=SiB$_2$ ⇌ ABSi=SiAB were observed[165]. This was believed to be a dyotropic process.

There is now significant evidence that highly sterically hindered disilenes readily dissociate to the related silylenes, which then undergo their characteristic reactions[166].

An example where this was effected either thermally or photochemically is given in equation 29[167].

$$\underset{\text{Tbt}}{\overset{\text{Mes}}{\diagdown}}\text{Si}=\text{Si}\underset{\text{Tbt}}{\overset{\text{Mes}}{\diagup}} \underset{hv}{\rightleftarrows} \underset{\text{Tbt}}{\overset{\text{Mes}}{\diagdown}}\text{Si:} \tag{29}$$

Tbt = 2,4,6-tris[bis(trimethylsilyl)methyl]phenyl

IX. OTHER PHOTOLYSES LEADING TO MULTIPLY BONDED SILICON

The photolysis of various cyclic compounds containing silicon atoms in the ring have been found to lead to the formation of transient intermediates containing multiple bonds to silicon. For example, Boudjouk and coworkers observed that photolysis of the trisilatriselenacyclohexane **330** gave rise to the silaselenone **331**, captured as the insertion product into the ring of D$_3$, yielding the product **332**[168] (equation 30).

(330) R = Me, Et $\xrightarrow[254 \text{ nm}]{hv}$ R$_2$Si=Se + (R$_2$Si—Se)$_2$ (331) $\xrightarrow{(D_3)}$ (332) (30)

The photolysis of the phenylsilyltriazide **333** in a matrix indicated, based on UV and infrared absorption spectral data, the formation of compound **334**, the first example of an isolated compound containing a triple bond to silicon[169]. Further irradiation (or thermolysis) of **334** led to an isomer, **335**, believed to be the analog of an isonitrile. Both these compounds were characterized spectroscopically, and were trapped with t-butyl alcohol (equation 31).

PhSi(N$_3$)$_3$ $\xrightarrow{hv}$ PhSi≡N $\xrightarrow{hv}$ PhN=Si:
(333)   (334)   (335)

                ↓ t-BuOH   ↓ t-BuOH   (31)

          PhSi(NH$_2$)(O-Bu-t)$_2$ PhSiNHSiH(O-Bu-t)$_2$

X. PHOTOLYSIS OF MISCELLANEOUS COMPOUNDS

A. Photochemically-induced Electron Transfer Involving Organosilicon Compounds

In the past few years many publications have appeared in which a variety of classes of organosilicon compound have been photolyzed in the presence of aromatic cyano compounds. In many cases the photolyses resulted in substitution of a ring cyano group by the hydrocarbon portion R of the organosilicon compound employed, $RSiMe_3$, and mechanistic evidence[170], which will not be discussed here, suggested that single-electron transfers were being effected, leading to the loss of the silyl group as a cation radical. Many of these reactions could be of synthetic importance to organic chemists, and some examples are shown below.

Nakadaira and coworkers[171] described the reaction of 6-trimethylsilylhex-1-ene with 1,2,4,5-tetracyanobenzene in which both the cyclic adduct **336** and the acyclic product **337** were formed (equation 32), and other workers further investigated the mechanistic details of this and related alkylations[172].

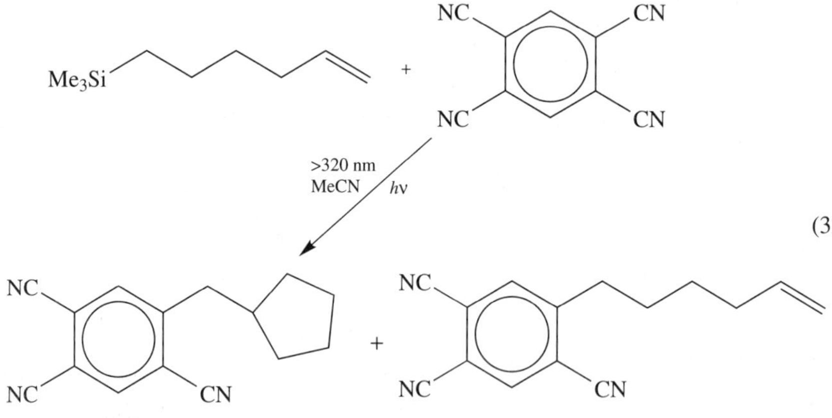

(32)

Closely related studies by Mizuno and coworkers[173] involved the photolysis of dicyanobenzenes with allylsilanes where, as shown by the two examples in equation 33, a cyano group was replaced by an allyl group.

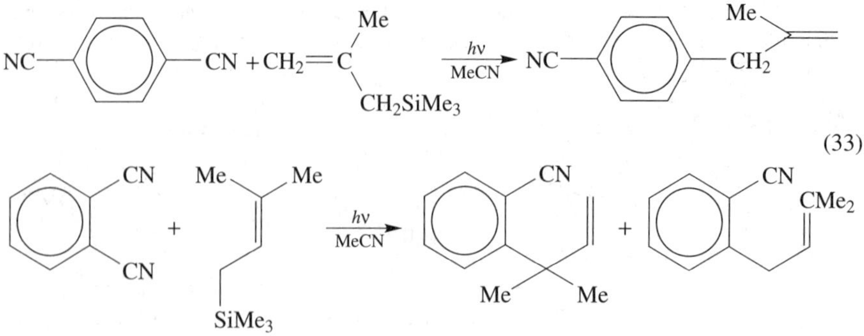

(33)

21. The photochemistry of organosilicon compounds

More recently Mizuno and coworkers studied the photoallylation of various cyanonaphthalenes with allyltrimethylsilane, using added phenanthrene which, in some cases, facilitated the formation of the benzotricyclic compound **338**, as shown in equation 34[174].

$$\text{cyanonaphthalene} + CH_2{=}CHCH_2SiMe_3 \xrightarrow[\text{MeOH}]{\substack{h\nu \\ \text{MeCN}}} \textbf{(338)} \quad (34)$$

A number of aromatic dicarboxylate esters were also found to undergo allylation with loss of the silyl group when photolyzed in the presence of allyltrimethylsilane in acetonitrile-methanol[175]. As illustrated in equation 35 the naphthalene dicarboxylate **339** evidently underwent allylation, but then subsequently underwent photochemical [2 + 2] cycloaddition to give the polycyclic product **340**; in other cases the reaction stopped after allylation.

(35)

(**339**) + allyl-SiMe₃ $\xrightarrow[\text{MeCN}]{\text{MeOH}, h\nu}$ [anion intermediate] $\xrightarrow{H^+}$ [neutral intermediate] $\xrightarrow{h\nu}$ (**340**)

10-Methylacridinium ion also underwent allylation using allyltrimethylsilanes under photochemical conditions (equation 36)[176].

$$(36)$$

It was also shown that under somewhat different experimental conditions benzyl- and related silanes added to the ring of aryl nitriles, as exemplified in equation 37[177].

$$(37)$$

Ar = 4-RC_6H_4
R = Cl, Me, OMe

Modification of these procedures by using only catalytic amounts of the cyano compound dicyanonaphthalene (DCN) allowed the cyclization of silyl enol ethers to bicyclic compounds in good yield (equation 38)[178].

$$(38)$$

21. The photochemistry of organosilicon compounds

Ketene silyl acetals underwent photoaddition involving single electron transfer with electron-deficient alkenes and sensitizers such as phenanthrene (phen)[179]. Two examples of the types of products, which were formed regioselectively, are illustrated in Scheme 60.

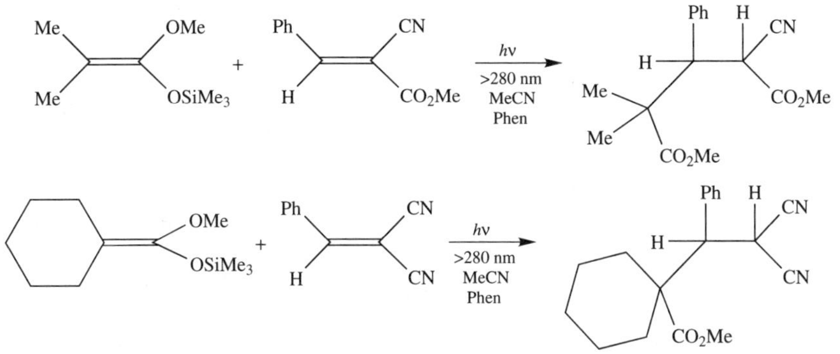

SCHEME 60

Detailed studies of the photochemistry of the (*p*-cyanophenyl)vinyldisilane **341** in methanol gave rise to products with the structures **342** and **343**, believed to be the result of both charge transfer and silyl radical involvement (equation 39)[180].

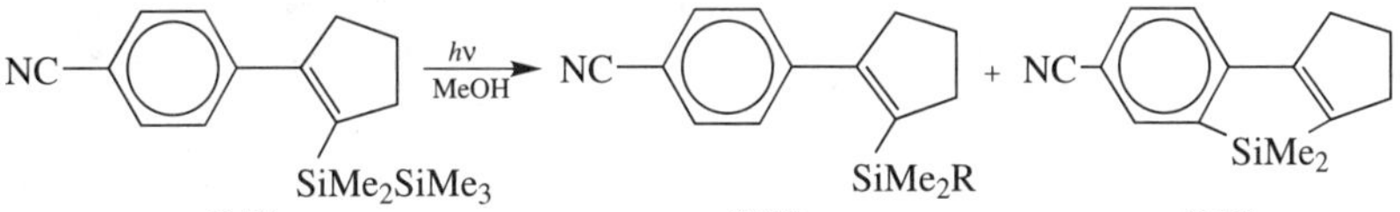

(341) SiMe$_2$SiMe$_3$
(342) SiMe$_2$R
(343) SiMe$_2$
R = H, Me, OMe

(39)

The photolysis of silanorbornadienes such as **344** sensitized by dicyanoanthracene (DCA) induced electron transfer and led to rearrangement affording the products **345**, **346**, and anthracene[181] (equation 40).

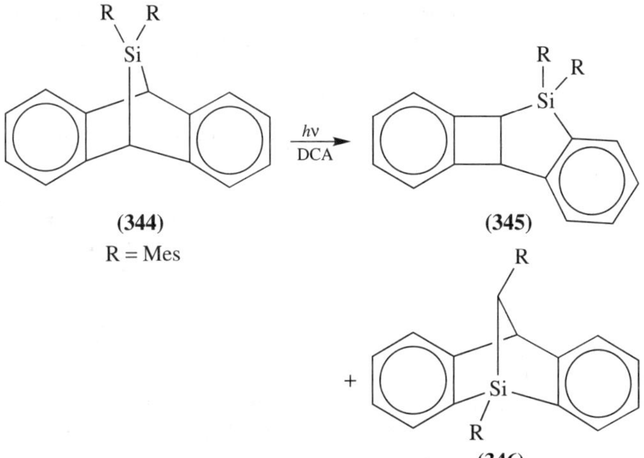

(344) R = Mes
(345)
(346)

(40)

Silyl carbamates have recently been shown to undergo photoinduced electron transfer with substituted alkenes, in the presence of catalytic amounts of dicyanoanthracene and biphenyl (BP), to yield more complex carbamates[182]. Two examples which illustrate the complex structures created in good yield are shown in Scheme 61.

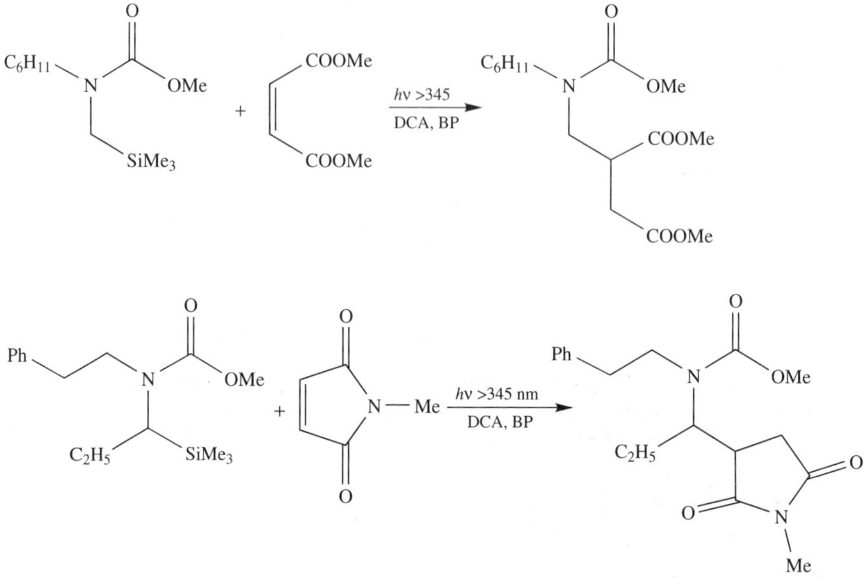

SCHEME 61

The photolysis of N-silylmethylphthalimides, **347**, in the presence of various reagents has been shown to give rise to products effectively derived from azomethine ylides **348**, which subsequently reacted with the reagents present[183]. The reactions are of potential synthetic use. Some examples, which involve a 1,4-silyl migration from carbon to oxygen, followed by reaction with the added reagent, are illustrated in Scheme 62. Other phthalimido compounds were also investigated. Among the mechanisms discussed was the possibility that single electron transfer was involved.

B. Miscellaneous Photolyses of Organosilicon Compounds Where Bonds to Si are Not Directly Involved

A number of photolyses of compounds containing a silicon atom have been studied in which the bonds to the silicon atom are not directly involved in the photolysis, but where the changes effected are potentially of synthetic interest. In many cases the presence of the silyl group may have an influence on the regiochemistry of the reaction, or an activating influence on an adjacent functional group. It is difficult to categorize these reactions, and thus they will simply be listed in an arbitrary order.

The Norrish type 1 cleavage of cyclic ketones has been found to be strongly influenced by the presence of a silyl group β to the carbonyl group, where the acyl radical intermediate was believed to be stabilized by the β effect[184]. For example, as shown in equation 41 with the ketone **349**, C_1-C_2 cleavage led to the diradical **350**, and thence to the products **351**

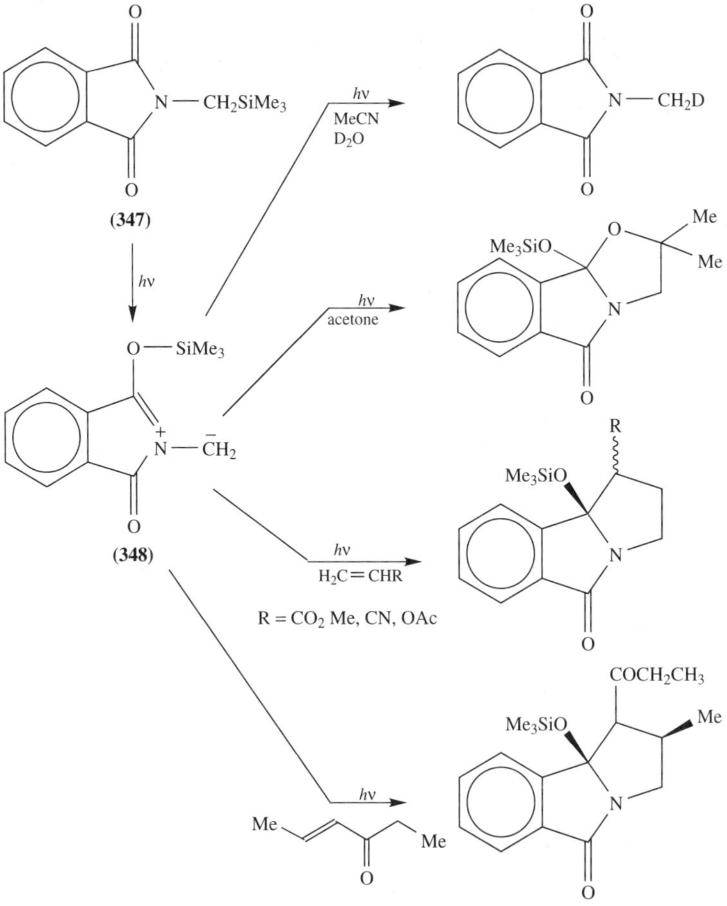

SCHEME 62

and **352**, whereas C_1-C_6 cleavage was known to be the major pathway occurring from the photolysis of the corresponding methylcyclohexanone. Not only was the regiospecificity much higher, but the rates of the reactions were enhanced when the silyl group was present.

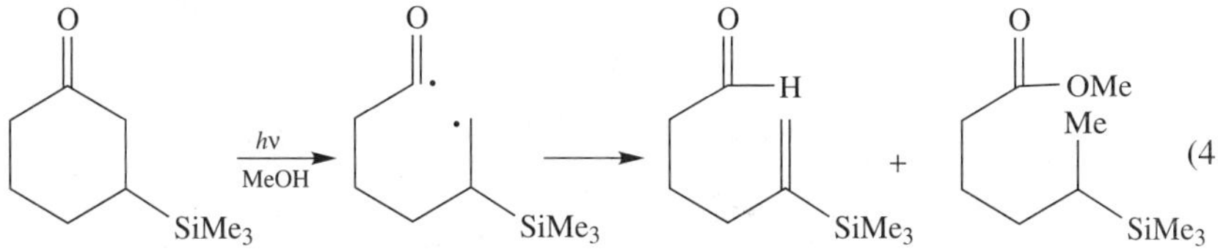

(349) (350) (351) (352) (41)

The sensitized photochemically-induced [2 + 2] cycloadditions of silyl enol ethers with acrylonitrile or methyl acrylate have been found, because of the regiospecificity of the reactions, to be a useful route for the preparation of functionalized eight-membered rings. Thus Suginome and coworkers[185] showed, for example, that the trimethylsilyl enol ether of β-tetralone **353** underwent regiospecific *syn* addition with acrylonitrile to give the tricyclic ring compound **354** which, after removal of the silyl group, could be converted to the benzocyclooctanone **355** (Scheme 63).

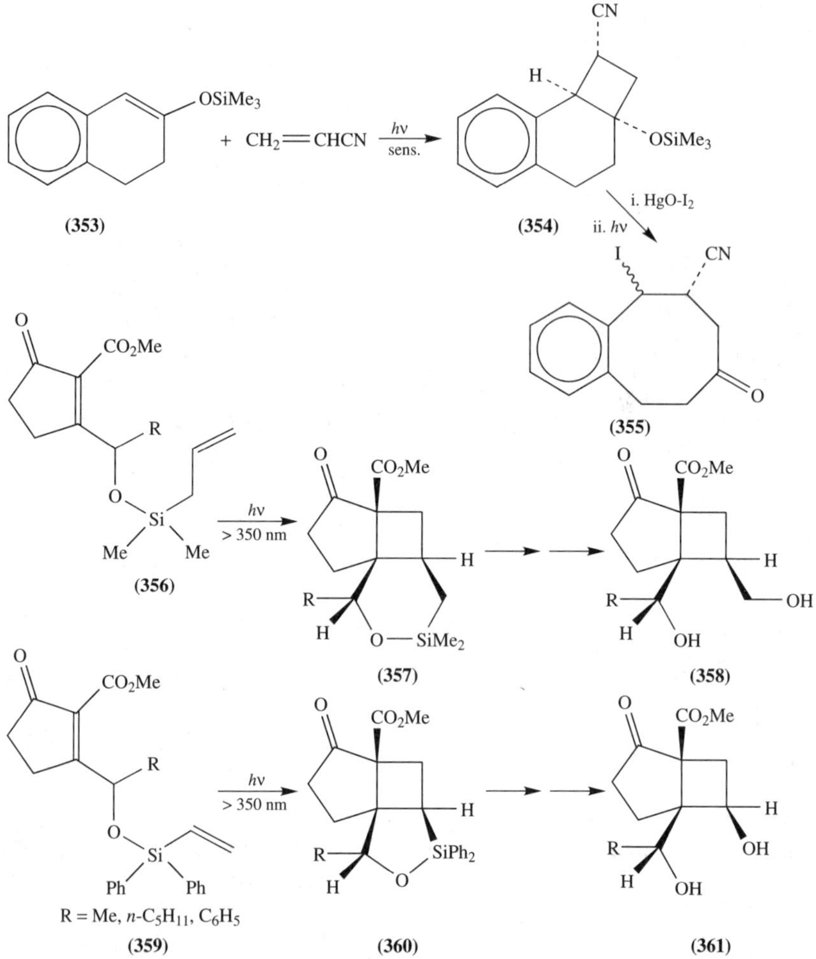

SCHEME 63

Closely related studies by Crimmins and Guise[186], also shown in Scheme 63, described the photochemical ring closure of the allyl silyl ether **356**, leading by virtue of the siloxane tether to the tricyclic species **357**, which after removal of the silyl group gave the bicyclic compound **358**. Also, the vinyl silyl ether **359** photochemically closed to compound **360**, which after removal of the silyl group led to the bicyclic species **361** (Scheme 63). These compounds were prepared with complete regiospecificity and high stereospecificity.

Photocycloadditions of silyl enol ethers **362** with aromatic aldehydes **363** have been used to prepare substituted oxiranes with high diastereoselectivities favoring the isomer

364 relative to other isomers such as **365**[187]. Hydrogenolysis of the ring with loss of the silyl group gave the vicinal diols **366**, useful for further synthetic elaboration (Scheme 64).

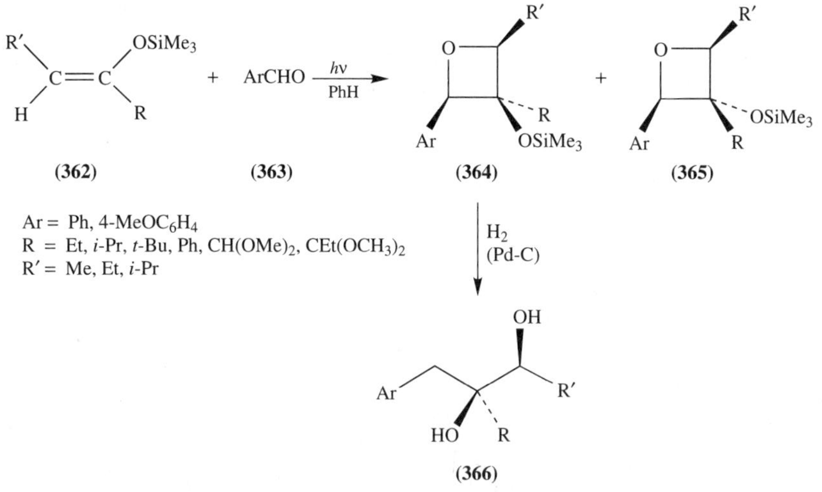

Ar = Ph, 4-MeOC$_6$H$_4$
R = Et, i-Pr, t-Bu, Ph, CH(OMe)$_2$, CEt(OCH$_3$)$_2$
R' = Me, Et, i-Pr

SCHEME 64

Pirrung and Lee[188] have developed protecting groups, (hydroxystyryl)dimethylsilyl and (hydroxystyryl)di-isopropylsilyl, for primary and secondary alcohols which can be removed conveniently by irradiation with 254-nm light in a polar solvent such as acetonitrile. An example, the protected menthol silyl ether **367**, when photolyzed, gave menthol, **368**, and also yielded the heterocycle **369**. In nonpolar solvents, photolysis resulted in a 1,5-shift of hydrogen or deuterium to presumably give **370** followed by a 1,5-shift of the dimethylalkoxysilyl group giving the isomerized product **371** (Scheme 65).

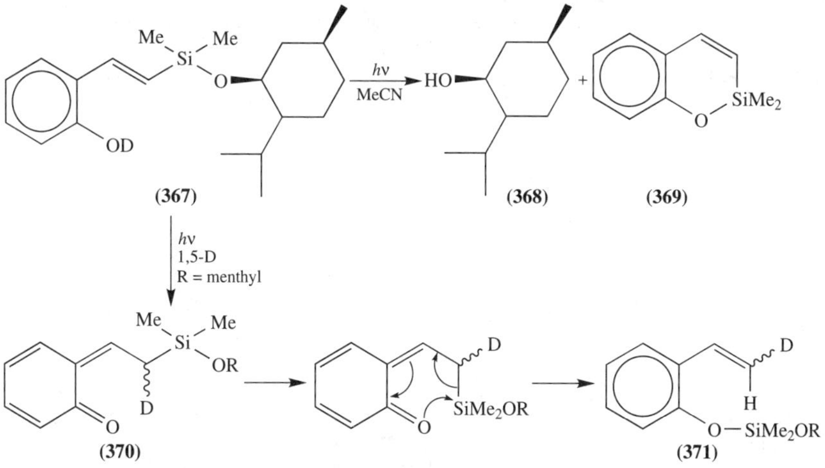

SCHEME 65

In a related vein 'antenna' chromophores have been developed for the self-sensitized photoreduction of ketones using isopropyl alcohol as a mild reducing agent[189]. One of the most effective chromophores for reduction of the model ketone 4-hydroxycyclohexanone was found to be the dimethylphenylsilyl group, as illustrated in equation 42.

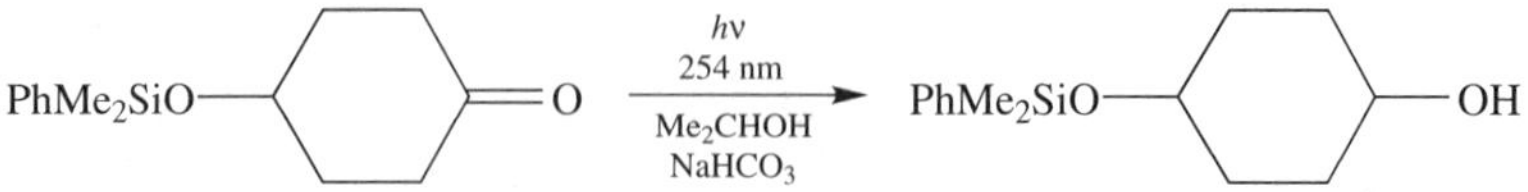
PhMe2SiO
O
hν
254 nm
Me2CHOH
NaHCO3
PhMe2SiO
OH

(42)

The synthesis of heterocycles containing both silicon and sulfur in the ring from linear thiols have recently been described by Kirpichenko and coworkers[190] (Scheme 66).

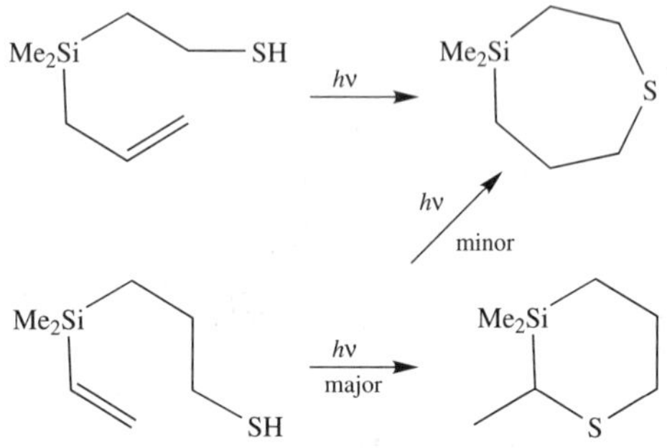
Me2Si
SH
hν
Me2Si
S
hν
minor
Me2Si
hν
major
Me2Si
SH
S

SCHEME 66

Photochemical [2 + 2] ring closures have been used to synthesize cyclophanes and related polycyclic ring systems containing silicon in the ring using triplet sensitizers such as benzophenone or dicyanonaphthalene[191] as shown in Scheme 67. The photochemistry of simpler allylsilanes was also investigated, as shown in the scheme[192]. In these reactions the silyl group and its substituents acted as tethers which controlled the orientation of the two carbon–carbon double bonds involved in the photocyclization.

Similarly, the bis(allyl)disiloxane **372** when photolyzed underwent [2 + 2] cycloaddition to yield the compound **373** as the sole product, the disiloxane linkage acting as a tether to assist in the orientation of the carbon–carbon double bonds involved in the cycloaddition[193]. The product had the all-*trans* structure (equation 43).

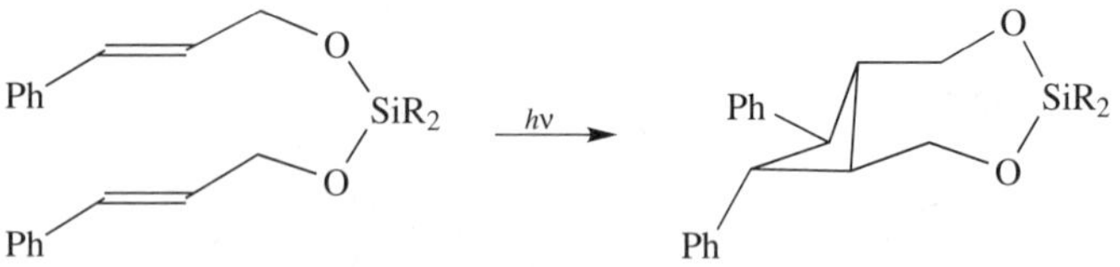
Ph
O
SiR2
O
Ph
hν
Ph
O
SiR2
O
Ph

(372) **(373)** (43)

R = Me, Ph

The photodimerization of a silylalkynylbenzene to cyclooctatetraenes has been reported by West and coworkers[194]. It was proposed that this occurred via intermolecular [2 + 2] dimerization followed by ring opening to the cyclooctatetraene (equation 44).

As illustrated in equation 45, Ando and coworkers photolyzed the cyclic trisilacycloheptene **374** at 254 nm to give the remarkably stable *trans* isomer **375** in modest

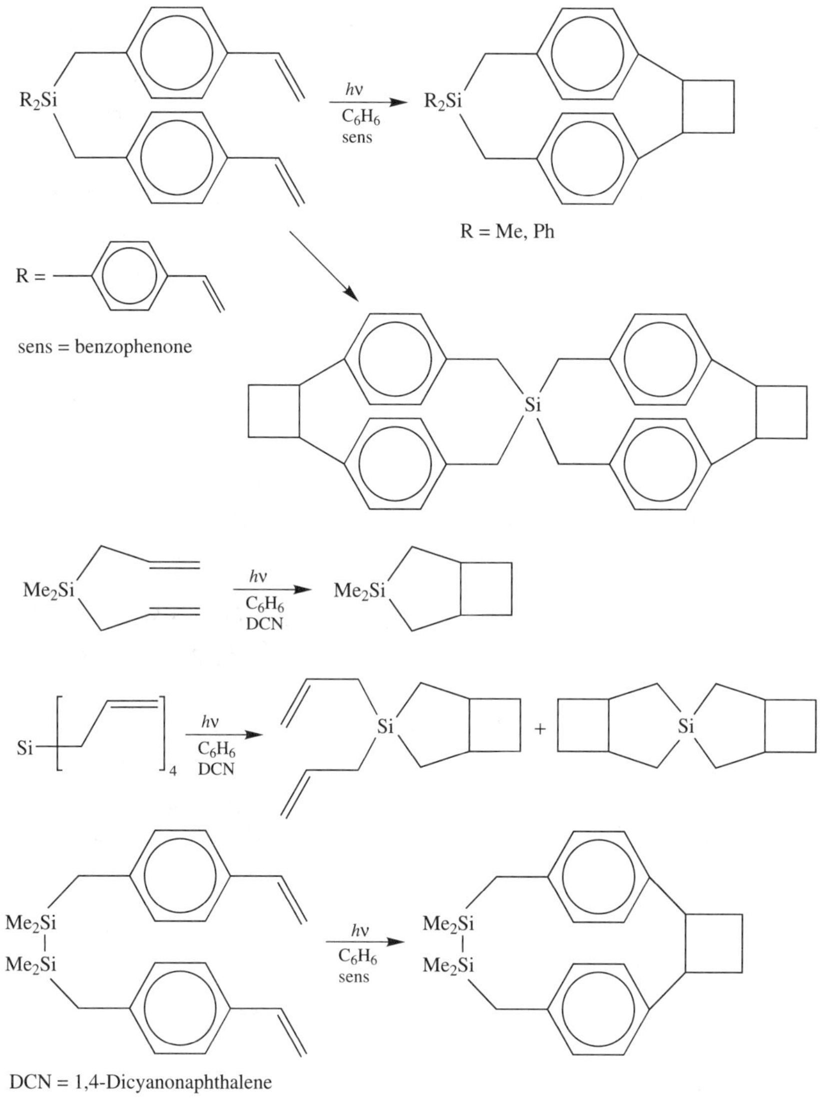

SCHEME 67

yield: the X-ray structure was obtained[195]. This is another case where silicon atoms in a seven-membered ring permit enough flexibility to allow the formation of *trans* geometry.

An interesting study of rhodium complexes of 1,3-divinyldisiloxanes and disilazanes has recently been reported in which the nature and geometry of the complex can be altered photochemically[196]. The rhodium atom binds to both vinyl groups of a single disiloxane or disilazane molecule in either a *cis* or *trans* manner, **376** and **377**, or the rhodium bonds to two disiloxane or disilazane molecules, again in a *cis* or *trans* relationship, **378** and **379**, as shown in Scheme 68.

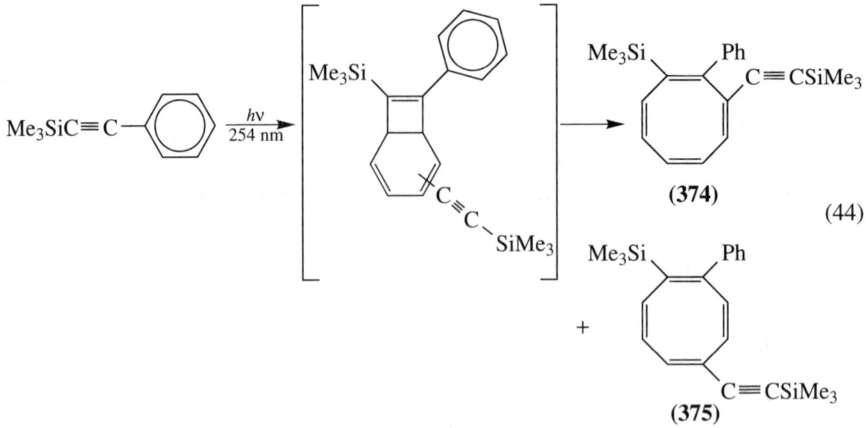

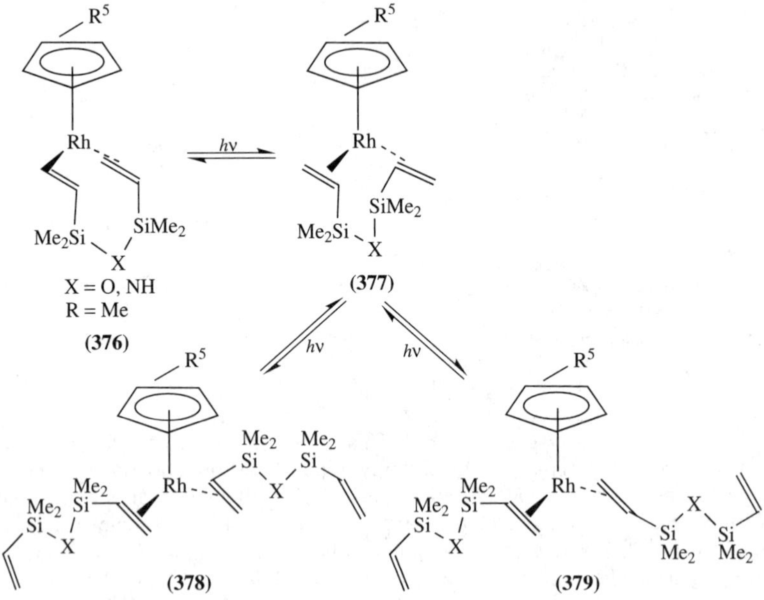

SCHEME 68

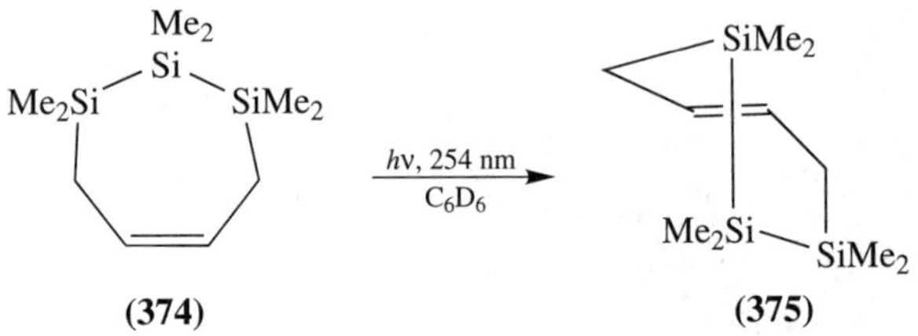

The photoionization of benzylic alcohols leading to carbocations has recently been reported[197]. Taking advantage of the stabilization provided by the β-silicon effect, flash photolysis of **380** in trifluoroethanol gave rise to the fluorenyl cation **381** which was characterized spectroscopically (equation 46).

C. Cophotolyses of Various Organosilicon Compounds with Fullerenes

In the past few years a number of unusual photochemical reactions of fullerenes with a variety of types of organosilicon compounds have been described. It seems most useful to report these studies under a single heading, rather than in the sections corresponding to the particular type of organosilicon compound involved.

Ando and coworkers found that the silylene Dip_2Si: **382** (Dip = 2,6-di-i-propylphenyl), formed on photolysis of the trisilane **382** in the presence of C_{60}, gave an adduct assigned the structure of the silacyclopropane **384** which evidently arose from addition of the silylene across the C=C between two six-membered rings of the fullerene[198]. A segment of the structure of **384** is shown in equation 47.

Photolysis of a mixture of C_{60} with the disilacyclopropane **385** led to ring opening by cleavage of the Si—Si bond, and addition of the silyl ends to the ends of one of the fullerene double bonds of a six-membered ring giving the adduct **386**[199]. When the fullerene was photolyzed with the disilacyclobutane **387** it was suggested that an initial adduct **388** was formed which then rearranged to the final product **389**, obtained in good yield[200]. Finally, when photolyzed with the tetrasilacyclobutanes two types of adduct, **390** and **391**, were obtained[201] (Scheme 69).

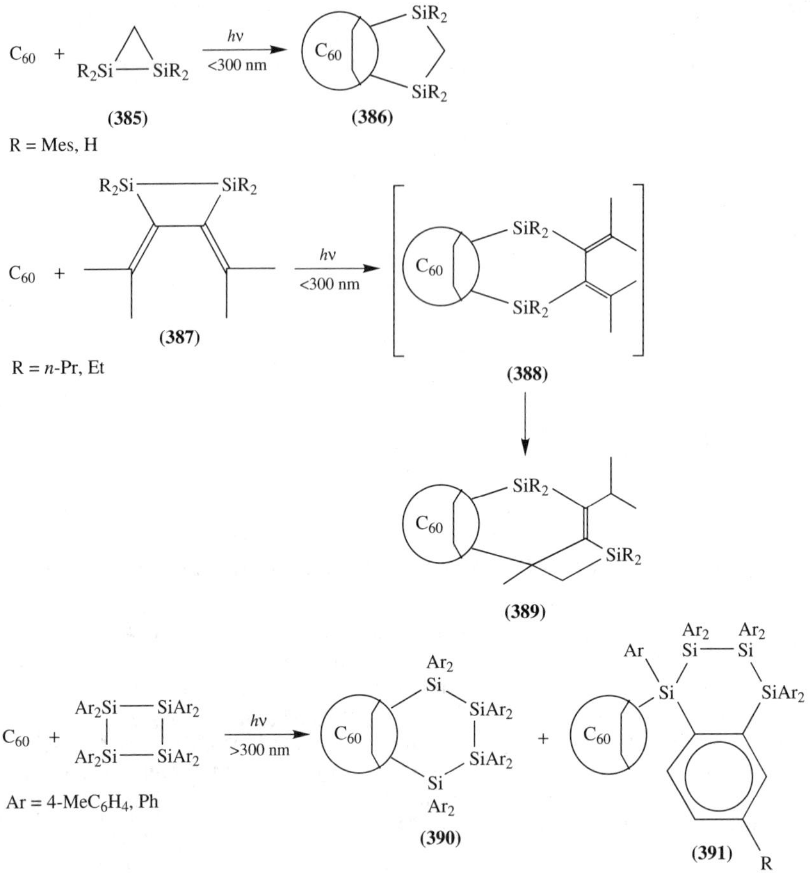

SCHEME 69

More recently, the same research group showed that the same silylene, $Dip_2Si:$, added in a similar manner to the C_{70} fullerene to yield the related silacyclopropane[202].

The dimetallofullerenes $La_2@C_{80}$[203], $La_2@C_{82}$[204] and $Sc_2@C_{84}$[203] reacted similarly with 1,1,2,2-tetramesityl-1,2-disilirane (or other tetraaryldisiliranes) to form the adducts $La_2@C_{80}(Mes_2Si)_2CH_2$, $La_2@C_{82}(Mes_2Si)_2CH_2$ and $Sc_2@C_{84}(Mes_2Si)_2CH_2$, characterized spectroscopically, each of which was believed to contain a 1,3-disilapropylene bridge across the ends of a bond of one of the six-membered rings of the original fullerene.

C_{60} also reacts with silyl ketene acetals when photolyzed with a high-pressure mercury arc. The reaction, illustrated in equation 48, leads to functionalization of the fullerene as an ester, and is said to involve single electron transfer[205].

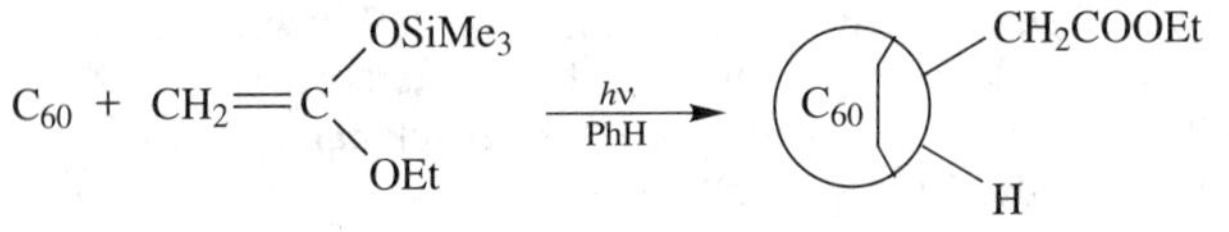

XI. ADDENDUM

Since this chapter was completed at the end of May 1996, several additional papers involving the photolysis of organosilicon compounds have appeared in the literature, or were presented at the XI International Organosilicon Conference held in Montpellier, France from 1 September to 7th September 1996. Some of these new results are described below, using the original subdivisions of the chapter to aid in their proper location.

III. D. Aryldisilanes

Results of the photolysis of both a *meso* and a *racemic* diastereomer of the 1,2-diphenyldisilane **392** indicated that the reaction leading to the silene **393**, followed by its ene reaction with an alkene leading to the adduct **394**, was a diastereospecific process[206]. A concerted mechanism, illustrated in equation 49 for one isomer of the disilane, was proposed to account for the results: such results would not be expected of a process involving silyl radical intermediates, which had been proposed earlier as intermediates in the photolysis of disilanes.

$$(49)$$

(392) (*R,R*)   **(393)** (*E*) (*S,S*)   **(394)** (*R,R*)

Further studies of the photolysis of aryldisilanes in which stable siloxetanes were obtained have been reported by Totyl and Leigh[207].

III. G. Cyclic Disilanes and Related Species

Barton and coworkers have described the interesting rearrangements under photochemical, thermal and Pd-catalyzed conditions of the eight-membered ring **395** which contains both a disilane and alkynyl groups as shown in equation 50[208].

$$(50)$$

(395)

IV. B. Acyclic Polysilanes

Photolyses of some highly silyl-substituted disilanes were found to result in exclusively Si—Si homolysis leading to stable silyl radicals, studied by ESR techniques, as shown in equation 51[209].

$$(Et_nMe_{3-n}Si)_3Si-Si(SiMe_{3-n}Et_n)_3 \longrightarrow (Et_nMe_{3-n}Si)_3Si\cdot \qquad (51)$$

IV. C. 1. Homocyclic Systems

Kira and coworkers recently described the preparation of the first cyclic disilene, **396**[210]. On photolysis at 420 nm 1,2-silyl rearrangement occurred to yield the bicyclic[1.1.0] system, **397** (equation 52), which on standing in the dark reverted to the relatively stabler disilene.

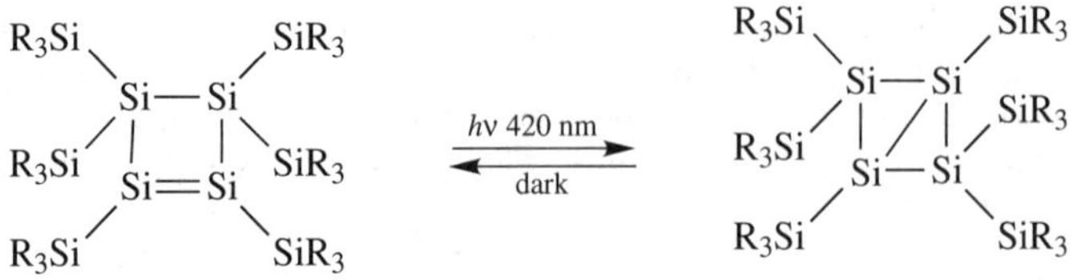

$$ (52) $$

$R_3Si = t\text{-}BuMe_2Si$

(**396**)   (**397**)

The chemistry of octasilacubanes continues to be investigated. Ring-opening halogenation[211] has been described, as has a photochemical oxidation using DMSO as the oxygen source in which an oxygen is inserted into one or two ring Si—Si bonds, as shown in equation 53[212].

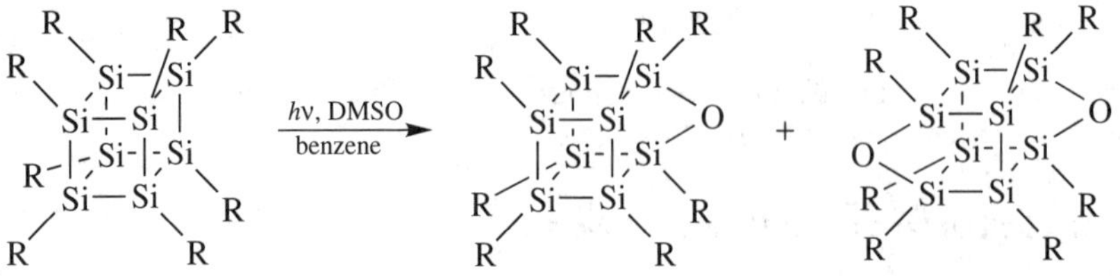

$$ (53) $$

$R = CMe_2CHMe_2$

VII. PHOTOCHEMICAL ROUTES TO SILENES, SILYLENES AND DISILENES

Further reactions of the silylene $t\text{-}Bu_2Si$: produced by photolysis of hexa-t-butylcyclotrisilane with alkenes and cycloalkenes have recently been described[213].

X. A. Photochemically-Induced Electron Transfer Involving Organosilicon Compounds

Tamao and coworkers[214–216] recently demonstrated that several π-conjugated systems containing silole (1-silacyclopenta-2,4-diene) rings are capable of emitting visible light under the influence of electric charge and may be useful in molecular electronics. Structures **398–400** of some of the polymers are given.

(398), X = Br, R = Me

(399)
E = S, NMe

(400)
E = S, NMe, CH=CH, CH=N

XII. SUMMARY

It should be very evident from the above survey that the field of organosilicon chemistry, and in particular the photochemistry of organosilicon compounds, is a very active area of research. The field is of increasing interest not only to chemists whose primary interest lies with the silicon aspect of the subject, but increasingly it is also of interest to organic chemists because of the synthetically novel and useful products of photolysis which can be obtained, and to photochemists where the field presents a wealth of different types of reaction, few of which have been studied mechanistically. It seems certain that there will continue to be great activity in the field of organosilicon chemistry in the near future.

XIII. REFERENCES

1. M. G. Steinmetz, *Chem. Rev.*, **95**, 1527 (1995).
2. A. G. Brook, in *The Chemistry of Organosilicon Compounds* (Eds. S. Patai and Z. Rappoport), Chap 15, Wiley, Chichester, 1989, p. 965.
3. S. T. Reid, *Spec. Period. Rep.: Photochem.*, **22**, 360 (1991); **23**, 350 (1992); **24**, 350 (1993).
4. A. G. Brook and K. M. Baines, *Adv. Organomet. Chem.*, **25**, 1 (1986).
5. G. Raabe and J. Michl, *Chem. Rev.*, **85**, 419 (1985).
6. G. Raabe and J. Michl, in *The Chemistry of Organosilicon Compounds* (Eds. S. Patai and Z. Rappoport), Chap. 17, Wiley, Chischester, 1989, p. 1015.
7. M. Kira and T. Miyazawa, Chap. 22 in this volume.
8. E. W. Colvin, in *Silicon in Organic Synthesis*, Chap. 3, Butterworths, London, 1981, p. 15.
9. R. Walsh, S. Untiedt, M. Stuhlmeier and A. de Meijere, *Chem. Ber.*, **122**, 637 (1989).
10. M. Kira, H. Yoshida and H. Sakurai, *J. Am. Chem. Soc.*, **107**, 7767 (1985).
11. M. Kira, T. Taki and H. Sakurai, *J. Org. Chem.*, **54**, 5647 (1989).
12. A. G. Brook, D. M. MacRae and W. W. Limburg, *J. Am. Chem. Soc.*, **89**, 5493 (1967).
13. A. Padwa, K. E. Krumpe, L. W. Terry and M. W. Wannamaker, *J. Org. Chem.*, **54**, 1635 (1989).
14. P. P. Gaspar, D. Holten, S. Konieczny and J. Y. Corey, *Acc. Chem. Res.*, **20**, 329 (1987).
15. J. Pola, E. A. Volnina and L. E. Gusel'nikov, *J. Organomet. Chem.*, **391**, 275 (1990).
16. J. Pola, D. Cukanova, M. Minarik, A. Lycka and J. Tlaskal, *J. Organomet. Chem.*, **426**, 23 (1992).

17. Z. Bastl, H. Burger, R. Fajgar, D. Pokorna, J. Pola, M. Senzlober, J. Subrt and M. Urbanova, *Appl. Organomet. Chem.*, **10**, 83 (1996).
18. (a) S. Dhanya, A. Kumar, R. K. Vatsa, R. K. Saini, J. P. Mittal and J. Pola, *J. Chem. Soc., Faraday. Trans.*, **92**, 179 (1996).
    (b) H. K. Sharma, F. Cervantes-Lee, L. Párkányi and K. H. Pannell, *Organometallics*, **15**, 429 (1996).
19. H. Sakurai, K. Ebata, K. Kabuto and A. Sekiguchi, *J. Am. Chem. Soc.*, **112**, 1799 (1990).
20. K. Sakamoto, T. Saeki and H. Sakurai, *Chem. Lett.*, 1675 (1993).
21. P. Boudjouk, U. Samaraweera, R. Sooriyakakumaran, J. Chrusciel and K. R. Anderson, *Angew. Chem., Int. Ed. Engl.*, **27**, 1355 (1988).
22. P. Boudjouk, E. Black and R. Kumarathasan, *Organometallics*, **10**, 2095 (1991).
23. H. Saso, W. Ando and K. Ueno, *Tetrahedron*, **45**, 1929 (1989).
24. K. M. Baines and J. A. Cooke, *Organometallics*, **11**, 3487 (1992).
25. (a) M. G. Steinmetz and H. Bai, *Organometallics*, **8**, 1112 (1989).
    (b) W. J. Leigh, C. J. Bradaric, C. Kerst and J.-A. H. Babisch, *Organometallics*, **15**, 2246 (1996).
26. R. T. Conlin, S. Zhang, M. Namavari, K. L. Bobbitt and M. J. Fink, *Organometallics*, **8**, 571 (1989).
27. M. G. Steinmetz, B. S. Udayakumar and M. S. Gordon, *Organometallics*, **8**, 530 (1989).
28. I. N. Jung, D. H. Pae, B. R. Yoo, M. E. Lee and P. R. Jones, *Organometallics*, **8**, 2017 (1989).
29. N. Choi, K. Asano and W. Ando, *Organometallics*, **14**, 3146 (1995).
30. M. G. Steinmetz and Q. Chen, *J. Chem. Soc., Chem. Commun.*, 133 (1995).
31. I. N. Jung, B. R. Yoo, M. E. Lee and P. R. Jones, *Organometallics*, **10**, 2529 (1991).
32. O. M. Nefedov, S. P. Kolesnikov, M. P. Egorov, A. M. Gal'minas and M. B. Ezhova, in *Proc. IXth Int. Symp. Organosilicon Chem.* (Eds. A. R. Bassindale and P. P. Gaspar), Royal. Soc. Chem., Cambridge, UK, 1990, p. 145.
33. A. Sekiguchi, I. Maruki, K. Ebata, C. Kabuto and H. Sakurai, *J. Chem Soc., Chem. Commun.*, 341 (1991).
34. O. M. Nefedov, M. P. Egorov, S. P. Kolesnikov, A. M. Gal'minas, Y. T. Struchkov, M. Y. Antipin and S. V. Sereda *Izv. Akad. Nauk. SSSR, Ser. Khim.*, 1693 (1986); *Chem. Abstr.*, **106**, 196487b (1987).
35. M. G. Steinmetz, K. J. Seguin, B. S. Udayakumar and J. S. Behnke, *J. Am. Chem. Soc.*, **112**, 6601 (1990).
36. M. G. Steinmetz and C. Yu, *Organometallics*, **11**, 2686 (1992).
37. M. G. Steinmetz and C. Yu, *J. Org. Chem.*, **57**, 3107 (1992).
38. H. Hiratsuka, M. Tanaka, T. Okutsu, M. Oba and K. Nishiyama, *J. Chem. Soc., Chem. Commun.*, 215 (1995).
39. K. M. Welsh, J. D. Rich, R. West and J. Michl, *J. Organomet. Chem.*, **325**, 105 (1987).
40. P. Boudjouk, J. R. Roberts, C. M. Golino and L. H. Sommer, *J. Am. Chem. Soc.*, **94**, 7926 (1972).
41. Y. Nakadaira, T. Otsuka and H. Sakurai, *Tetrahedron Lett.*, **22**, 2417 (1981).
42. Y. Nakadaira, T. Otsuka and H. Sakurai, *Tetrahedron Lett.*, **22**, 2421 (1981).
43. M. Ishikawa and M. Kumada, *Adv. Organomet. Chem.*, **19**, 51 (1981).
44. M. Kira, T. Maruyama and H. Sakurai, *J. Am. Chem. Soc.*, **113**, 3986 (1991).
45. R. T. Conlin and K. L. Bobbitt, *Organometallics*, **6**, 1406 (1987).
46. W. J. Leigh, G. W. Sluggett, P. Venneri, M. Ezhova, M. S. K. Dhurjati and R. T. Conlin, *Abstr. XXVIII Organosilicon Symposium*, Gainesville, Fla., March 1995, A20.
47. M. B. Ezhova and R. T. Conlin, *Abstr. XXVII Organosilicon Symposium*, Troy, NY, March 1994, A-8.
48. J. H. Kwon, S. T. Lee, S. C. Shim and M. Hoshino, *J. Org. Chem.*, **59**, 1108 (1994).
49. S. C. Shim and S. T. Lee, *J. Chem. Soc., Perkin Trans. 2*, 1979 (1994).
50. S. T. Lee, E. K. Baek and S. C. Shim, *Organometallics*, **15**, 2182 (1996).
51. M. Ishikawa, T. Fuchikami and M. Kumada, *J. Organomet. Chem.*, **118**, 155 (1976).
52. H. Sakurai, *J. Organomet. Chem.*, **200**, 261 (1980).
53. M. Ishikawa and H. Sakamoto, *J. Organomet. Chem.*, **414**, 1 (1991).
54. M. Ishikawa, H. Sakamoto, F. Kanetani and A. Minato, *Organometallics*, **8**, 2767 (1989).
55. M. Ishikawa, K. Watanabe, H. Sakamoto and A. Kunai, *J. Organomet. Chem.*, **455**, 61 (1993).
56. J. Ohshita, H. Ohsaki, M. Ishikawa, A. Tachibani, Y. Kurosaki, T. Yamabe, T. Tsukihara, K. Takahashi and Y. Kiso, *Organometallics*, **10**, 2685 (1991).

57. J. Ohshita, H. Ohsaki, M. Ishikawa, A. Tochibana, Y. Kurosaki, T. Yamabe and A. Minato, *Organometallics*, **10**, 880 (1991).
58. J. Ohshita, H. Ohsaki and M. Ishikawa, *Organometallics*, **10**, 2695 (1991).
59. G. W. Sluggett and W. J. Leigh, *J. Am. Chem. Soc.*, **114**, 1195 (1992).
60. G. W. Sluggett and W. J. Leigh, *Organometallics*, **11**, 3731 (1992).
61. W. J. Leigh and G. W. Sluggett, *Organometallics*, **13**, 269 (1994).
62. W. J. Leigh and G. W. Sluggett, *J. Am. Chem. Soc.*, **116**, 10468 (1994).
63. O. Ito, K. Hoteiya, A. Watanabe and M. Matsuda, *Bull. Chem. Soc. Jpn.*, **64**, 962 (1991).
64. M. Kira and S. Tikurs, *Chem. Lett.*, 1459 (1994).
65. J. Braddock-Wilking and P. P. Gaspar, *Abstr. IXth International Symposium on Organosilicon Chemistry*, Edinburgh, Scotland, July 1990, B.25.
66. K. Mochida, K. Kimijima, M. Wakasa and H. Hayashi, *Abstr. Xth International Symposium on Organosilicon Chemistry*, Poznan, Poland, Aug. 1993, P-107.
67. K. Mizuno, K. Nakanishi, J.-i. Chosa and Y. Otsuji, *J. Organomet. Chem.*, **473**, 35 (1994).
68. H. Sakamoto and M. Ishikawa, *Organometallics*, **11**, 2580 (1992).
69. K. Takaki, H. Sakamoto, Y. Nishimura, Y. Sugihara and M. Ishikawa, *Organometallics*, **10**, 888 (1991).
70. M. Ishikawa, Y. Nishimura and H. Sakamoto, *Organometallics*, **10**, 2701 (1991).
71. H. K. Sharma and K. H. Pannell, *Chem. Rev.*, **95**, 1351 (1995).
72. H. Tobita, K. Ueno, M. Shimoi and H. Ogino, *J. Am. Chem. Soc.*, **112**, 3415 (1990).
73. H. Tobita, K. Ueno and H. Ogino, *Bull. Chem. Soc. Jpn.*, **61**, 2797 (1988).
74. A. Haynes, M. W. George, M. T. Haward, M. Poliakoff, J. J. Turner, N. M. Boag and M. Green, *J. Am. Chem. Soc.*, **113**, 2011 (1991).
75. K. L. Jones and K. H. Pannell, *J. Am. Chem. Soc.*, **115**, 11336 (1993).
76. C. Hernandez, H. K. Sharma and K. H. Pannell, *J. Organomet. Chem.*, **462**, 259 (1993).
77. K. H. Pannell, L.-J. Wong and J. M. Rozell, *Organometallics*, **8**, 550 (1989).
78. Z. Zheng, R. Sanchez and K. H. Pannell, *Organometallics*, **14**, 2605 (1995).
79. M. Kumada, *J. Organomet. Chem.*, **100**, 127 (1975).
80. H. Sharma and K. H. Pannell, *Organometallics*, **12**, 3979 (1993).
81. H. Sharma and K. H. Pannell, *Organometallics*, **13**, 4946 (1994).
82. M. Kira, K. Sakamoto and H. Sakurai, *J. Am. Chem. Soc.*, **106**, 7469 (1984).
83. A. D. Fanta, D. J. DeYoung, J. Belzner and R. West, *Organometallics*, **10**, 3466 (1991).
84. K. Kabeta, D. R. Powell, J. Hanson and R. West, *Organometallics*, **10**, 827 (1991).
85. H. Suzuki, K. Okabe, S. Uchida, H. Watanabe and M. Goto, *J. Organomet. Chem.*, **509**, 177 (1996).
86. A. J. McKinley, T. Karatsu, G. M. Wallraff, R. D. Miller, R. Sooriyakumaran and J. Michl, *Organometallics*, **7**, 2567 (1988).
87. A. J. McKinley, T. Karatsu, G. M. Wallraff, D. P. Thompson, R. D. Miller and J. Michl, *J. Am. Chem. Soc.*, **113**, 2003 (1991).
88. I. M. T. Davidson, J. Michl and T. Simpson, *Organometallics*, **10**, 842 (1991).
89. B. D. Shepherd, D. R. Powell and R. West, *Organometallics*, **8**, 2664 (1989).
90. G. R. Gillette, G. Noren and R. West, *Organometallics*, **9**, 2925 (1990).
91. R. S. Archibald, Y. Van den Winkel, A. J. Millevolte, J. M. Desper and R. West, *Organometallics*, **11**, 3276 (1992).
92. R. S. Alexander, Y. von den Winkel, D. R. Powell and R. West, *J. Organomet. Chem.*, **446**, 67 (1993).
93. S. Zhang, M. B. Ezhova and R. T. Conlin, *Organometallics*, **14**, 1471 (1995).
94. E. Hengge and R. Janoschek, *Chem. Rev.*, **95**, 1495 (1995).
95. H. Shizuka, K. Murata, Y. Arai, K. Tonokura, H. Tanake, H. Matsumoto, Y. Nagai, G. Gillette and R. West, *J. Chem. Soc., Faraday Trans. 1*, **85**, 2369 (1989).
96. B. J. Helmer and R. West, *Organometallics*, **1**, 1458 (1982).
97. M. Weidenbruch, *Chem. Rev.*, **95**, 1479 (1995).
98. M. Weidenbruch, E. Kroke, H. Marsmann, S. Pohl and W. Saak, *J. Chem. Soc., Chem. Commun.*, 1233 (1994).
99. M. Weidenbruch, H. Piel, A. Lesch, K. Peters and H. G. Von Schnering, *J. Organomet. Chem.*, **454**, 35 (1993).
100. A. Sekiguchi and H. Sakurai, *Adv. Organomet. Chem.*, **37**, 1 (1995)
101. A. Sekiguchi, T. Yatabe, C. Kabuto and H. Sakurai, *J. Am. Chem. Soc.*, **115**, 5853 (1993).

102. H. Suzuki, K. Okabe, R. Kato, N. Sato, Y. Fukuda and H. Watanabe, *J. Chem. Soc., Chem. Commun.*, 1298 (1991).
103. H. Suzuki, K. Okabe, R. Kato, N. Sato, Y. Fukuda, H. Watanabe and M. Goto, *Organometallics*, **12**, 4833 (1993).
104. H. Suzuki, N. Kenmoto, K. Tanaka, H. Watanabe and M. Goto, *Chem. Lett.*, 811 (1995).
105. A. G. Brook and J. M. Duff, *J. Am. Chem. Soc.*, **89**, 454, (1967).
106. A. G. Brook and J. M. Duff, *J. Am. Chem. Soc.*, **91**, 2118 (1969).
107. M. Trommer and W. Sander, *Organometallics*, **15**, 189 (1995).
108. A. G. Brook, A. Baumegger and A. J. Lough, *Organometallics*, **11**, 310 (1992).
109. K. M. Baines, A. G. Brook, R. R. Ford, P. D. Lickiss, A. K. Saxena, W. J. Chatterton, J. Sawyer and B. A. Behnam, *Organometallics*, **8**, 693 (1989).
110. S. Zhang, R. T. Conlin, P. F. McGarry and J. C. Scaiano, *Organometallics*, **11**, 2317 (1992).
111. A. G. Brook, K. Vorspohl, R. R. Ford, M. Hesse and W. J. Chatterton, *Organometallics*, **6**, 2128 (1987).
112. A. G. Brook, A. Baumegger and A. J. Lough, *Organometallics*, **11**, 3088 (1992).
113. A. G. Brook, W. J. Chatterton, J. F. Sawyer, D. W. Hughes and K. Vorspohl, *Organometallics*, **6**, 1246 (1987).
114. A. G. Brook, S. S. Hu, W. J. Chatterton and A. J. Lough, *Organometallics*, **10**, 2752 (1991).
115. A. G. Brook, S. S. Hu, A. K. Saxena and A. J. Lough, *Organometallics*, **10**, 2758 (1991).
116. A. G. Brook, K. Y. Kong, A. K. Saxena and J. F. Sawyer, *Organometallics*, **7**, 2245 (1988).
117. A. G. Brook, A. K. Saxena and J. F. Sawyer, *Organometallics*, **8**, 850 (1989).
118. A. G. Brook, P. Chiu, J. McClenaghnan and A. J. Lough, *Organometallics*, **10**, 3292 (1991).
119. A. G. Brook, W. J. Chatterton and R. Kumarathasan, *Organometallics*, **12**, 3666 (1993).
120. A. G. Brook, R. Kumarathasan and A. J. Lough, *Organometallics*, **13**, 424 (1994).
121. A. G. Brook, R. Kumarathasan and W. J. Chatterton, *Organometallics*, **12**, 4085 (1993).
122. A. G. Brook and A. Baumegger, *J. Organomet. Chem.*, **446**, C9 (1993).
123. P. Lassacher, A. G. Brook and A. J. Lough, *Organometallics*, **14**, 4359 (1995).
124. A. G. Brook, A. Ionkin and A. J. Lough, *Organometallics*, **15**, 1275 (1996).
125. A. Sekiguchi and W. Ando, *Chem. Lett.*, 2025 (1986).
126. A. Sekiguchi and W. Ando, *Organometallics*, **6**, 1857 (1987).
127. A. G. Brook and P. F. Jones, *Can. J. Chem.*, **49**, 1841 (1971).
128. K. Schneider, B. Daucher, A. Fronda and G. Maas, *Chem. Ber.*, **123**, 589 (1990).
129. A. Sekiguchi, T. Sato and W. Ando, *Organometallics*, **6**, 2337 (1987).
130. G. Maas, K. Schneider and W. Ando, *J. Chem. Soc., Chem. Commun.*, 72 (1988).
131. G. Maas, M. Alt, K. Schneider and A. Fronda, *Chem. Ber.*, **124**, 1295 (1991).
132. M. Trommer, W. Sander, C.-H. Ottoson and D. Cremer, *Angew. Chem., Int. Ed. Engl.*, **34**, 929 (1995).
133. M. Trommer, W. Sander and C. Marquand, *Angew. Chem., Int. Ed. Engl.*, **33**, 766 (1994).
134. M. Trommer and W. Sander, *Organometallics*, **15**, 736 (1996).
135. W. Ando, H. Yoshida, K. Kurishima and M. Sugiyama, *J. Am. Chem. Soc.*, **113**, 7790 (1991).
136. W. Ando, M. Sugiyama, T. Suzuki, C. Kato, Y. Arakawa and Y. Kabe, *J. Organomet. Chem.*, **499**, 99 (1995).
137. M. J. Menu, H. König, M. Dartiguenave, Y. Dartiguenave and H. F. Klein, *J. Am. Chem. Soc.*, **112**, 5351 (1990).
138. E. Deydier, M. J. Menu, M. Dartiguenave, Y. Dartiguenave, M. Simard, A. L. Beauchamp, J. C. Brewer and H. B. Gray, *Organometallics*, **15**, 1166 (1996).
139. G. Märkl, W. Schlosser and W. S. Sheldrick, *Tetrahedron Lett.*, **29**, 467 (1988).
140. G. Märkl and W. Schlosser, *Angew. Chem., Int. Ed. Engl.*, **27**, 963 (1988).
141. A. Sekiguchi, H. Tanikawa and W. Ando, *Organometallics*, **4**, 584 (1985).
142. M. Weidenbruch, *Coord. Chem. Rev.*, **130**, 275 (1994).
143. R. Okazaki and R. West, *Adv. Organomet. Chem.*, **39**, 232 (1996).
144. A. G. Brook and M. A. Brook, *Adv. Organomet. Chem.*, **39**, 71 (1996).
145. Y. Huang, M. Sulkes and M. J. Fink, *J. Organomet. Chem.*, **499**, 1 (1995).
146. M. Ishikawa, N. Nakagawa, M. Ishiguro, F. Ohi and M. Kumada, *J. Organomet. Chem.*, **152**, 155 (1978).
147. M. J. Michalczyk, M. J. Fink, D. J. De Young, C. W. Carlson, K. M. Walsh, R. West and J. Michl, *Silicon, Germanium, Tin and Lead Compd.*, **9**, 75 (1986).
148. R. S. Archibald, Y. van den Winkel, D. R. Powell and R. West, *J. Organomet. Chem.*, **446**, 67 (1993).

149. J. Hockemayer, Diploma Thesis, University of Oldenburg, 1990.
150. D. H. Pae, M. Xiao, M. Y. Chiang and P. P. Gaspar, *J. Am. Chem. Soc.*, **113**, 1281 (1991).
151. P. Boudjouk, E. Black and R. Kumarathasan, *Organometallics*, **10**, 2095 (1991).
152. E. Kroke, M. Weidenbruch, W. Saak, S. Pohl and H. Marsmann, *Organometallics*, **14**, 5695 (1995).
153. M. Weidenbruch, H. Piel, A. Lesch, K. Peters and H. G. von Schnering, *J. Organomet. Chem.*, **454**, 35 (1993).
154. K. L. Bobbitt and P. P. Gaspar, *J. Organomet. Chem.*, **499**, 17 (1995).
155. M. J. Fink, D. B. Puranic and M. P. Johnson, *J. Am. Chem. Soc.*, **110**, 1315 (1988).
156. V. N. Khabashesku, V. Balaji, S. E. Boganov, O. M. Nefedov and J. Michl, *J. Am. Chem. Soc.*, **116**, 320 (1994).
157. M. Kira, T. Maruyama and H. Sakurai, *Tetrahedron Lett.*, **33**, 243 (1992).
158. A. Vancik, G. Raabe, M. J. Michalczyk, R. West and J. Michl, *J. Am. Chem. Soc.*, **107**, 4097 (1985).
159. G. E. Miracle, J. L. Ball, D. R. Powell and R. West, *J. Am. Chem. Soc.*, **115**, 11598 (1993).
160. M. Trommer, W. Sander and A. Patyk, *J. Am. Chem. Soc.*, **115**, 11775 (1993).
161. K. M. Welsh, J. Michl and R. West, *J. Am. Chem. Soc.*, **110**, 6689 (1988).
162. G. Maier, H. Pacl, H. P. Reisenauer, A. Meudt and R. Janoschek, *J. Am. Chem. Soc.*, **117**, 12712 (1995).
163. M. J. Fink, D. J. DeYoung and R. West, *J. Am. Chem. Soc.*, **105**, 1070 (1983).
164. M. J. Michalczyk, R. West and J. Michl, *Organometallics*, **4**, 826 (1985).
165. H. B. Yokelson, D. A. Siegel, A. J. Millevolte, J. Maxma and R. West, *Organometallics*, **9**, 1005 (1990).
166. H. Suzuki, N. Tokitoh and R. Okazaki, *Bull. Chem. Soc. Jpn.*, **68**, 2471 (1995).
167. N. Tokitoh, H. Suzuki and R. Okazaki, *Abstr. Xth International Organosilicon Symposium*, Poznan, Poland, 1993, O-62.
168. P. Boudjouk, S. R. Bahr and D. P. Thompson, *Organometallics*, **10**, 778 (1991) *Tetrahedron Lett.*, **33**, 243 (1992).
169. J. G. Radziszewski, D. Littmann, V. Balaji, L. Fabry, G. Gross and J. Michl, *Organometallics*, **12**, 4816 (1993).
170. Mechanisms for many of these reactions will be found in the original article, and also in: M. Kira and T. Miyazaw Chap. 22 in this volume.
171. Y. Nakadaira, S. Kyushin and M. Ohashi, *Abstr. IX International Symposium on Organosilicon Chemistry*, Edinburgh, Scotland, July 1990, 1.12.
172. M. Fagnoni, M. Mella and A. Albini, *Tetrahedron*, **50**, 6401 (1994).
173. K. Nakanishi, K. Mizuno and Y. Otsuji, *Bull. Chem. Soc. Jpn.*, **66**, 2371 (1993)
174. T. Nishiyama, K. Mizuno, Y. Otsuji and H. Inoue, *Tetrahedron*, **51**, 6695 (1995).
175. Y. Kubo, T. Todani, T. Inoue, H. Ando and T. Fujiwara, *Bull. Chem. Soc. Jpn.*, **66**, 541 (1993).
176. S. Fukuzumi, M. Fujita and J. Otera, *J. Chem. Soc., Chem. Commun.*, 1536 (1993).
177. T. Tamai, K. Mizuno, I. Hashida and Y. Otsuji, *Bull. Chem. Soc. Jpn.*, **66**, 3747 (1993).
178. G. Pandey, A. Krishna, K. Girija and M. Karthikeyan, *Tetrahedron Lett.*, **34**, 6631 (1993).
179. K. Mizuno, N. Takahashi, T. Nishiyama and H. Inoue, *Tetrahedron Lett.*, **36**, 7463 (1995).
180. M. G. Steinmetz, C. Yu and L. Li, *J. Am. Chem. Soc.*, **116**, 932 (1994).
181. M. Kako, S. Kakuma, K. Hatakenaka, Y. Nakadaira, M. Yasui and F. Iwasaki, *Tetrahedron Lett.*, **36**, 6293 (1995).
182. E. Meggers, E. Steckhan and S. Blechert, *Angew. Chem., Int. Ed. Engl.*, **34**, 2137 (1995).
183. U. C. Yoon, D. U. Kim, Y. S. Choi, Y.-J. Lee, H. L. Ammon and P. S. Marino, *J. Am. Chem. Soc.*, **117**, 2698 (1995).
184. J. R. Hwu, B. A. Gilbert, L. C. Lin and B. R. Liaw, *J. Chem. Soc., Chem. Commun.*, 161 (1990).
185. H. Suginome, T. Takeda, M. Itoh, Y. Nakayama and K. Kobayashi, *J. Chem. Soc., Perkin Trans. 1*, 49 (1995).
186. M. T. Crimmins and L. E. Guise, *Tetrahedron Lett.*, **35**, 1657 (1994).
187. T. Bach, *Tetrahedron Lett.*, **35**, 5845 (1994).
188. M. C. Pirrung and Y. R. Lee, *J. Org. Chem.*, **58**, 6961 (1993).
189. Z.-Z. Wu and H. W. Morrison, *Photochem. Photobiol.*, **50**, 525 (1989).
190. S. V. Kirpichenko, L. L. Olstikova, E. N. Suslova and M. G. Voronkov, *Tetrahedron Lett.*, **34**, 3889 (1993).
191. K. Nakanishi, K. Mizuno and Y. Otsuji, *J. Chem. Soc., Perkin Trans. 1*, 3362 (1990).

192. K. Nakanishi, K. Mizuno and Y. Otsuji, *J. Chem. Soc., Chem. Commun.*, 90 (1991).
193. S. A. Fleming and S. C. Ward, *Tetrahedron Lett.*, **33**, 1013 (1992).
194. R. S. Archibald, D. Chinnery, A. Fanta and R. West, *Organometallics*, **10**, 3769 (1991).
195. T. Shimizu, K. Shimizu and W. Ando, *J. Am. Chem. Soc.*, **113**, 354 (1991).
196. S. S. D. Brown, S. N. Heaton, M. H. Moore, R. N. Perutz and G. Wilson, *Organometallics*, **15**, 1392 (1996).
197. C. S. Q. Lew, R. A. McClelland, L. J. Johnston and N. P. Schepp, *J. Chem. Soc., Perkin Trans. 2*, 395 (1994).
198. T. Akasaka, W. Ando, K. Kobayashi and S. Nagase, *J. Am. Chem. Soc.*, **115**, 1605 (1993).
199. T. Akasaka, W. Ando, K. Kobayashi and S. Nagase, *J. Am. Chem. Soc.*, **115**, 10366 (1993).
200. T. Kusakawa, Y. Kabe, T. Erata, B. Nestler and W. Ando, *Organometallics*, **13**, 4186 (1994).
201. T. Kusakawa, Y. Kabe and W. Ando, *Organometallics*, **14**, 2142 (1995).
202. T. Akasaka, E. Mitsuhida, W. Ando, K. Kobayashi and S. Nagase, *J. Chem. Soc., Chem. Commun.*, 1529 (1995).
203. T. Akasaka, S. Nagase, K. Kobayashi, T. Suzuki, T. Kato, K. Kikuchi, Y. Achiba, K. Yamamoto, H. Funasaki and T. Takahashi, *Angew. Chem., Int. Ed. Engl.*, **34**, 2139 (1995).
204. T. Akasaka, T. Kato, K. Kobayashi, S. Nagase, K. Yamamoto, H. Funasaki and T. Takahashi, *Nature*, **374**, 600 (1995).
205. K. Mikami and S. Matsumoto, *Synlett.*, 229 (1995).
206. J. Ohshita, H. Niwa, M. Ishikawa, T. Yamabe, T. Yoshii and K. Nakamura, *J. Am. Chem. Soc.*, **118**, 6853 (1996).
207. N. P. Totyl and W. J. Leigh, *Organometallics*, **15**, 2554 (1996).
208. T. J. Barton, Z. Ma and S. Ijadi-Maghsoodi, *Abstr. XI International Symposium on Organosilicon Chemistry*, Montpellier, France, September 1–7, (1996), OB15.
209. S. Kyushin, T. Betsuyaku and H. Matsumoto, *Abstr. XI International Symposium on Organosilicon Chemistry*, Montpellier, France, September 1–7, 1996, PC12.
210. T. Iwamoto, C. Kabuto, and M. Kira, *Abstr. XI International Symposium on Organosilicon Chemistry*, Montpellier, France, September 1–7, 1996, PA99.
211. M. Unno and H. Matsumoto, *Organometallics*, **13**, 4663 (1994); **14**, 4004 (1995).
212. M. Unno, T. Yokota and H. Matsumoto, *Abstr. XI International Syposium on Organosilicon Chemistry*, Montpellier, France, September 1–7, 1996, OD2.
213. E. Kroke, S. Willms, M. Weidenbruch, W. Saak, S. Pohl and H. Marsmann, *Tetrahedron Lett.*, **37**, 3675 (1996).
214. K. Tamao, *Abstr. XI International Symposium on Organosilicon Chemistry*, Montpellier, France, September 1–7, 1996, LA10.
215. K. Tamao, S. Yamaguchi and M. Shiro, *J. Am. Chem. Soc.*, **116**, 11715 (1994).
216. K. Tamao, S. Yamazuchi, Y. Ito, Y. Matsuzaki, T. Yamabe, M. Fukushima and S. More, *Macromolecules*, **28**, 8668 (1995).

CHAPTER **22**

Mechanistic aspects of the photochemistry of organosilicon compounds

MITSUO KIRA

Department of Chemistry, Graduate School of Science, Tohoku University, Aoba-ku, Sendai 980-77, Japan

and

TAKASHI MIYAZAWA

Photodynamics Research Center, The Institute of Physical and Chemical Research, (RIKEN), 19-1399, Koeji, Nagamachi, Aoba-ku, Sendai 980, Japan

I. INTRODUCTION	1311
II. OLIGOSILANES AND POLYSILANES	1312
A. Introduction	1312
B. Excited State Nature of Peralkyloligosilanes	1312
C. Excited State Nature of Peralkylpolysilane High Polymers	1317
D. Mechanistic Aspects of Photochemistry of Peralkyloligosilanes and Polysilanes	1319
III. ARYLPOLYSILANES	1321
A. Photophysics of Aryldisilanes	1321
B. Mechanistic Aspects of Photoreactions of Aryldisilanes	1325
C. Photochemistry of Aryltrisilanes	1328
IV. SILYLENES	1329
V. SUMMARY	1332
VI. ACKNOWLEDGMENT	1332
VII. REFERENCES	1332

I. INTRODUCTION

The photochemistry of organosilicon compounds has been extensively studied, since many types of interesting reactive intermediates such as silylenes, silenes, disilenes and silyl

The chemistry of organic silicon compounds, Vol. 2
Edited by Z. Rappoport and Y. Apeloig © 1998 John Wiley & Sons Ltd

radicals as well as highly strained organosilicon molecules, which are usually very difficult to prepare by other methods, have been created via photochemical reactions. Organosilicon compounds possessing unique photochemical and photophysical properties such as thermochromism, photochromism, photoconductivity, nonlinear optical properties and so on would be expected as modules for future optoelectronic devices. Recent advances in the photochemistry of organosilicon compounds have been thoroughly reviewed by Brook[1,2] and Steinmetz[3], including analysis of the photochemical intermediates and their fates.

A complete mechanism of a photochemical reaction of a molecule should require the knowledge of (1) the electronic structure of the excited states attained by absorption of light, (2) the photophysical events arising from the excited states such as emission, internal conversion, radiationless transition, intersystem crossing, quenching and energy transfer, (3) the photochemical events originating from the excited state such as bond breaking, rearrangement, electron transfer and intermolecular reactions and (4) the dynamics of the photophysical and photochemical events. In this respect, very few studies of the mechanistic aspects of organosilicon photochemistry have been reported so far. The detailed mechanistic photochemistry will be delineated by future generations of organosilicon chemists. In this chapter, only selected topics will be discussed, in cases where the electronic structures of excited states and their responsibility for photoreactions have been investigated.

II. OLIGOSILANES AND POLYSILANES

A. Introduction

Polysilane high polymers possessing fully saturated all-silicon backbone have attracted remarkable attention recently because of their unique optoelectronic properties and their importance in possible applications as photoresists, photoconductors, polymerization initiators, nonlinear optical materials etc. A number of review articles have been published on this topic[4-9]. The studies in this field have stimulated both experimental and theoretical chemists to elaborate on understanding the excited state nature of polysilanes and oligosilanes and of their mechanistic photochemistry.

B. Excited State Nature of Peralkyloligosilanes

Historically, the unique electronic properties of linear peralkyloligosilanes[4] were first reported by Gilman and coworkers[10,11] who reported the electronic spectra of permethyloligosilanes, $Me(SiMe_2)_nMe$ ($n \geq 2$), which exhibited electronic transitions in the ultraviolet region. The properties of open-chain and cyclic oligosilanes have been reviewed[12,13]. The absorption maxima shift to the red on increasing the number of silicon atoms in the chain[10,11,14]. Table 1 records the reported UV absorption maxima for linear permethyloligosilanes.

Whereas the finding by Gilman and coworkers was the first recognition of the so-called σ-conjugation in polysilanes, the origin of the red shift of the absorption maxima on increasing the silicon chain length was the subject of controversial discussions. Thus, Pitt and coworkers ascribed it first to lowering of the LUMO of permethyloligosilanes, which was represented by the linear combination of the unoccupied orbitals localized on individual $-SiMe_2-$ groups[15]. On the other hand, a Russian group[16] and later also Pitt and coworkers[17,18] assigned the electronic transitions to $\sigma \rightarrow \sigma^*$ transitions and treated the transition energies by using the Sandorfy C method[19], which was originally used to describe σ conjugation in alkane chains[20]. Thus, the Hamiltonian is of the Hückel type, with a strongly negative vicinal resonance integral β_{vic} and a more weakly negative geminal resonance integral β_{gem} as shown in Figure 1. This type of σ conjugation results

22. Mechanistic aspects of the photochemistry of organosilicon compounds 1313

TABLE 1. UV absorption maxima of permethyloligosilanes Me(Me$_2$Si)$_n$Me in hydrocarbon solvents at room temperature[a]

n	λ_{max} (ϵ)
2	197 (8500)
3	215 (9000)
4	235 (14700)
5	250 (18400)
6	260 (21100)
7	266.5 (23000)
8	272.5 (38000)
10	279 (42700)
12	285 (43000)[b]
16	293 (68000)[c]
18	291 (44300)[b]
24	293 (45500)[b]

[a]Unless otherwise noted, data were taken from Reference 12.
[b]From Reference 14.
[c]From Reference 45.

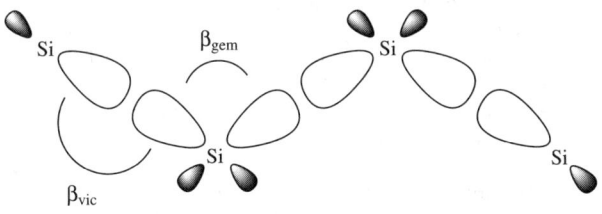

(a) Sandorfy C method

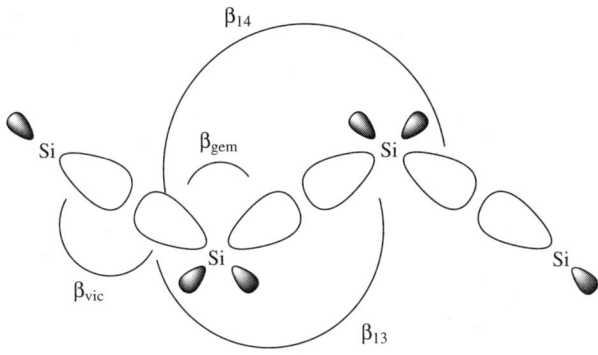

(b) Ladder C method

FIGURE 1. Schematic representation of the interactions between sp^3(Si) hybrid orbitals in linear oligosilanes. (a) Reprinted by permission of The Royal Society of Chemistry, from Reference 33; (b) Reprinted with permission from Reference 17. Copyright 1970 American Chemical Society

in both raising the HOMO levels and lowering the LUMO levels in oligosilanes and explains the red shift of the absorption maxima on increasing the chain length. The existence of the high-lying HOMO in linear and cyclic peralkyloligosilanes is evident from photoelectron[21–23] and charge-transfer[24,25] spectroscopy, and from ESR spectroscopy of the corresponding radical cations[26–28], and that of the low-lying LUMO is evident by ESR of the corresponding radical anions[29–31].

Recently, a deeper understanding of the excited state nature of the oligosilanes has been obtained. In relation to the remarkable thermochromism observed in peralkylpolysilanes (*vide infra*), the dependence of the electronic spectra of peralkyloligosilanes on the conformations of the silicon backbone has been investigated in detail. Plitt and Michl have investigated UV and IR spectra of permethyltetrasilane[32], the shortest peralkylated silicon chain expected to show conformational isomerism in the backbone, in Xe and Ar matrices. The UV spectrum shows a weak peak at 206 nm and a strong peak at 228 nm, the origin of which was first ascribed to the *gauche* and *anti* conformers of permethyltetrasilane, respectively, on the basis of a correlation of the peaks in the IR and UV spectra; irradiation using 206 nm light destroys the *gauche* peaks and 229 nm light destroys the *trans* peaks in the IR spectrum. The simple Sandorfy C model[19] cannot explain the conformational dependence. Michl and coworkers have refined the model and proposed the ladder C model[32,33] where 1,3- and 1,4- orbital interactions are considered and the dihedral angle (θ) dependence is incorporated in the resonance integrals β_{14}, which are -0.40 eV at $\theta = 90°$ and $+0.75$ eV at $\theta = 165°$; $\alpha = -2.63$ eV, $\beta_{vic} = -3.32$ eV, $\beta_{gem} = -0.62$ eV and $\beta_{13} = 0$ eV, which were determined also using nonlinear least-squares optimization of the resonance integrals (Figure 1)[33]. The calculated λ_{max} for *gauche* and *anti* tetrasilane are 209 and 230 nm, respectively, being in good accord with the two observed band maxima for permethyltetrasilane. The calculations reproduce the chain-length dependence of the energies of the intense $\sigma \rightarrow \sigma^*$ transitions of permethylated oligosilanes and their conformational dependence[33]. However, the optimized β_{14} for the *gauche* geometry is unexpectedly more negative than that for the *anti* geometry having better orbital overlap. On the basis of recent *ab initio* MO calculations and more detailed selective irradiation experiments, the explanation has been somewhat modified[34]. Thus, the *ab initio* MO calculations for decamethyltetrasilane have shown an interesting conformational dependence on the ground-state potential surface. Six conformational minima whose energies lie within 1 kcal mol^{-1} of each other and whose dihedral angles are near ± 162 (*anti*), ± 91 (*ortho*) and $\pm 53°$ (*gauche*) have been found. This can be contrasted with the situation in both *n*-butane and *n*-tetrasilane, which have only three minima at 180 (*anti*) and $\pm 60°$ (*gauche*). The IR spectral peaks previously assigned to the *gauche* conformer of permethyltetrasilane should be taken as those of overlapping contributions from two conformers, *gauche* and *ortho* (Figure 2).

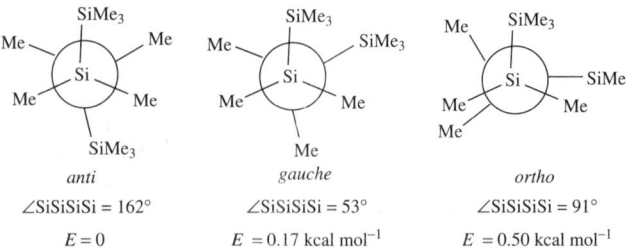

FIGURE 2. Newman projections and energies of the MP2/6-31G* calculated geometries of the *anti*, *gauche* and *ortho* conformers of permethyltetrasilane

The highest occupied and lowest unoccupied molecular orbitals of *anti*-tetrasilane are shown schematically in Figure 3. The calculations predict that one-photon allowed transitions to two *B* states and two-photon allowed transitions to two *A* states for permethyltetrasilane are in the UV region. The two *B* states are of the $\sigma\sigma^*$ and $\sigma\pi^*$ types, the transitions to which are "strongly allowed" and "weakly allowed", respectively. The $\sigma\sigma^*$ state is the lower state at the *anti* geometry (dihedral angle = 180°), while the $\sigma\pi^*$ state at the *cis* geometry (dihedral angle = 0°) is the consequence of an avoided crossing between the two lowest *B* states (Figure 4). Thus, at the *anti* geometry the transition to the 1*B* ($\sigma\sigma^*$) state is intense and the transition to the 2*B* ($\sigma\pi^*$) state is very weak. At the *twisted* (i.e. *gauche* and *ortho*) geometries, symmetry is lower and both transitions into *B* states are predicted to have fairly high intensities due to the mixing of the $\sigma\pi^*$ and

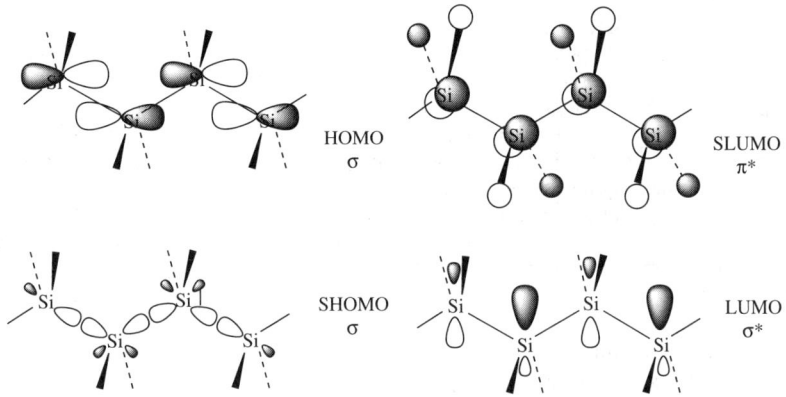

FIGURE 3. Schematic representation of the highest occupied and lowest unoccupied molecular orbitals of tetrasilane. Reproduced by permission of Elsevier Science Ltd from Ref. 35

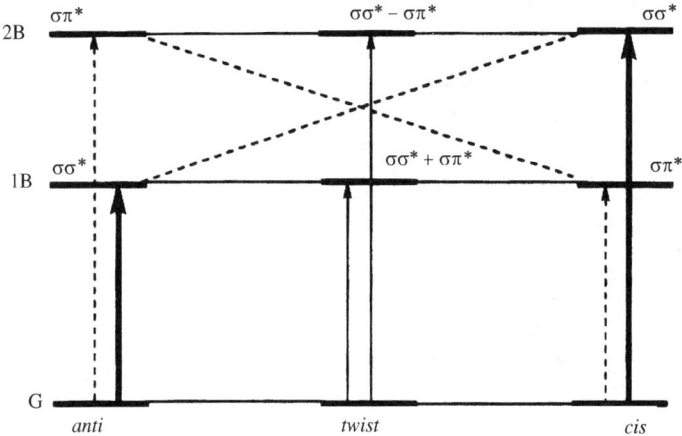

FIGURE 4. Schematic representation of the correlation diagram of *B* states among *anti*, *twist* and *cis* geometries of tetrasilane: ⟶, allowed transition; →, partially allowed transition; ----▸, forbidden transition

$\sigma\sigma^*$ B configurations. Because of the strongly avoided crossing, the energies of both the lower and upper B states are nearly independent of the twist angle, while the intensity is conformation-sensitive. Thus, the weak peak at 206 nm and the strong peak at 228 nm in the UV spectrum of decamethyltetrasilane have now been assigned to intense transitions to the $2B$ ($\sigma\sigma^* - \sigma\pi^*$) states of the *twisted* conformers and to a mixture of an intense transition to the $1B$ ($\sigma\sigma^*$) state of the *anti* conformer and weak transitions to the $1B$ ($\sigma\sigma^* + \sigma\pi^*$) states of the *twisted* conformers, respectively.

Instead of the ladder C model, the ladder H model, where the electrons of the bonds to substituents in addition to those in the backbone are considered, is suggested to be essential for drawing even the simplest picture of σ conjugation[35]. The A states are also of a $\sigma\sigma^*$ and a $\sigma\pi^*$ type, and are calculated to undergo a similar avoided crossing. In view of their very low calculated absorption cross sections, they could be detected only by two-photon absorption spectroscopy.

A number of semiempirical[36-40] and ab initio MO[35,41] calculations have been carried out for singlet excited states of polysilanes and oligosilanes. Ab initio MO calculations have been performed for the excited singlet states of the linear oligosilanes Si_nH_{2n+2}, $n = 2$–5, at SCF-CI levels with several basis sets, in order to establish their nature as a function of chain length[35]. The results suggest that the lowest $\sigma\pi^*$ ($\sigma_{SiSi} \rightarrow \pi^*_{SiH}$) excited state lies below the first $\sigma\sigma^*$ ($\sigma_{SiSi} \rightarrow \sigma^*_{SiSi}$) state in the shortest few oligosilanes, and that the order of the states changes as the number of silicon atoms in the chain increases. Rydberg states are expected to dominate the low energy region of the spectra of oligosilanes shorter than Si_6H_{14}; for longer-chain oligosilanes, all Rydberg excitations will be higher in energy than the lowest valence excitations. The lowest excited state with a large two-photon absorption cross section is well represented by the SHOMO $\rightarrow$ LUMO ($\sigma_{SiSi} \rightarrow \sigma^*_{SiSi}$) configuration and has essentially no doubly excited [HOMO^2 $\rightarrow$ LUMO^2 ($\sigma_{SiSi}^2 \rightarrow \sigma^*_{SiSi}{}^2$) configuration] character. The alkyl substitution at the central silicon of trisilane lowers the $\sigma\sigma^*$ excitation energy, primarily by destabilizing the highest σ_{SiSi} orbital by hyperconjugative interaction with the substituents.

Absorption and emission properties of permethylhexasilane have been investigated in detail[42]. While a hexane solution of Si_6Me_{14} exhibits a strong absorption band at 260 nm at room temperature, the band becomes much narrower ($\Delta\nu = 1400$ cm^{-1}) with a red shift of 640 cm^{-1} in a 3-methylpentane glass at 77 K. Whereas the hexasilane was very weakly fluorescent ($\phi_f < 10^{-4}$) at room temperature in solution, an intense but very broad fluorescence band ($\phi_f = 0.45$, $\Delta\nu = 5350$ cm^{-1}) with a large Stokes shift (9400 cm^{-1}) was observed in a 77 K glass matrix, being independent of the excitation energy. The absorption band is assigned quite reasonably to a $\sigma \rightarrow \sigma^*$ (HOMO–LUMO) excitation. On the other hand, the origin of the fluorescence has been ascribed to the emission from a less strongly allowed and Franck–Condon forbidden excited state lying somewhat lower in energy than the strongly allowed $\sigma\sigma^*$ state that dominates the absorption spectrum, on the basis of the large Stokes shift, the broad fluorescence bandwidth and the temperature effects; the transition from the ground state to the emitting state is masked and hard to see in ordinary absorption spectra. Permethylated tetrasilane and pentasilane exhibit similar broad fluorescence bands with large Stokes shifts as permethylhexasilane in argon matrices. As the Si chain becomes shorter, the emission becomes weaker and its peak red-shifts slightly. Unusually, permethylheptasilane shows a broad fluorescence band with a large Stokes shift in solution at room temperature, but a very slightly Stokes-shifted sharp band at 77 K, similarly to linear long-chain peralkyloligosilanes and polysilanes[43]. In order to explain the anomalous emitting state for short-chain oligosilanes, two models, a self-trapped exciton state, where the excitation is localized in a small fraction of the total segment length, and a $\sigma\pi^*$ state, which is calculated to lie below the lowest $\sigma\sigma^*$ state in the short oligosilane chains, were proposed by Hochstrasser and coworkers[44] and by Sun and Michl[42], respectively. Sun and Michl have excluded the latter assignment and given evidence for the self-trapped exciton model. Michl and coworkers have proposed that the emission is due to a nonvertical excited state with a single greatly stretched SiSi bond[43].

As a model for polysilane high polymers, photophysics of linear permethylhexadecasilane (n-Si$_{16}$Me$_{34}$) which showed similar behavior to linear polysilane polymers have been investigated in detail[45]. At room temperature in dilute hexane solution, the UV spectrum of Si$_{16}$Me$_{34}$ showed a band maximum at 293 nm (ϵ 68,000) with a bandwidth at a half-maximum ($\Delta \nu$) of 6200 cm^{-1}, while the maximum was red shifted and had a narrower bandwidth (λ_{max} 312 nm, $\Delta \nu$ 1400 cm^{-1}) in 3-methylpentane glass at 77 K. A broad Gaussian-like fluorescence band appeared at 327 nm ($\Delta \nu = 1500$ cm^{-1}) at room temperature in solution. The fluorescence quantum yield depends strongly on the excitation energies. Thus, it was about 0.07 at excitation wavelengths (λ_{ex}) shorter than the absorption maximum and 0.63 at the red edge of the absorption spectrum ($\lambda_{ex} = 325$ nm). The fluorescence decay of Si$_{16}$Me$_{34}$ is essentially independent of the excitation energy and the monitored emission frequency ($\tau_F = 179 \pm 20$ ps). The absorption and emission spectra of Si$_{16}$Me$_{34}$ at low temperature solutions and glass are quite different from those in solution at room temperature and depend strongly on the concentration of the sample and also on its thermal history. In dilute 3-methylpentane glass ($c \leqslant 1.6 \times 10^{-7}$ M) at 77 K, two overlapping absorption maxima at 312 and 315 nm are observed but the band is quite narrow. While the fluorescence and the excitation spectra depend strongly on the excitation and the monitored emission energies, respectively, even in the region of $\lambda_{ex} < \lambda_{max}$, a fluorescence spectrum excited at λ_{max} (312 nm) showed two peaks at 320 nm and 315 nm. The combination of absorption, fluorescence and fluorescence excitation spectra suggests the existence of the following three distinct conformers with low energy states in addition to a large number of others: a species responsible for the 315 nm absorption shoulder and the 320 nm emission peak, a species responsible for the fluorescence peak at 315 nm, and a nonfluorescent species responsible for the absorption peak at 312 nm. The ϕ_f value of the dilute sample was ca 0.5, being essentially independent of the excitation wavelengths. Rapidly cooled more concentrated 4.7×10^{-6} M solution of Si$_{16}$Me$_{34}$ showed an additional fluorescence band at 321 nm assignable to the oligosilane dimer or to more highly aggregated clusters. The quantum yield of the cluster emission is about 0.06. Slow cooling of the concentrated Si$_{16}$Me$_{34}$ solution allows the clusters to develop further into microcrystals, which give the same absorption and emission spectra as that of the solid film at 77 K. The phenomena have been understood by a segment model proposed for long-chain polysilane high polymers (Section II.C). Thus, the excitation of Si$_{16}$Me$_{34}$ using $\lambda_{ex} < \lambda_{max}$ generates initially excited short segments, which transfer the energy to the emitting long segments. At low temperatures, they are also generated through adiabatic excited-state conformational changes that extend the length of a particular excited short segment at the expense of its neighbors, presumably producing primarily the all-*trans* conformer, as suggested by the constant ϕ_f, irrespective of the excitation wavelength. The actual value of ϕ_f is 6 times lower for Si$_{16}$Me$_{34}$ than for the model polymer [SiMePr]$_x$ at $\lambda_{ex} < \lambda_{max}$, suggesting that adiabatic generation of long-segment chromophores by conformational change is less efficient than energy transfer along the backbone of a polymer; there are many molecules having only short segments in the hexadecasilane and the excited short segments will undergo dark processes such as intersystem crossing, internal conversion or photochemical reactions.

C. Excited State Nature of Peralkylpolysilane High Polymers

Linear-chain polysilanes have been found to show unusual optical properties[4-9]. At room temperatures in hydrocarbon solution, linear peralkylpolysilanes having more than

50 silicon atoms in a chain, such as $[SiBu_2]_n$, $[SiHex_2]_n$, $[SiOctyl_2]_n$, etc., exhibit an intense absorption band at ca 315 nm and a strong fluorescence peak at ca 340 nm. Upon cooling below $ca - 30\,°C$, the absorption and emission bands become much narrower and shift to about 350 nm, with a very small Stokes shift. The temperature dependence of the absorption spectra is discontinuous; in a very narrow temperature range upon cooling, one absorption band disappears and a new band grows at a longer wavelength. Primarily on the basis of its long-axis polarization, this strong absorption has been assigned to a $\sigma \rightarrow \sigma^*$ transition. It was first proposed that this transformation is due to a coil-to-rod transition[46]. However, rigid-rod-like conformations of macromolecules in solution require either a stiff main-chain structure, particular interaction forces within the chain such as hydrogen bonding, or very strong steric constraints; neither of these requirements is substantially fulfilled for these peralkylpolysilanes. Light scattering experiments[47,48] have shown that aggregation of the polysilane chains is associated with the thermochromism even at concentrations below 10^{-4} M. The remarkable thermochromism would be associated with an aggregation of long, approximately all-*trans* chain segments, intramolecular and/or intermolecular, depending on concentration. Since thermochromism occurs even at low concentrations of polymers, the aggregation should be cooperative with the formation of all-*trans* chain segments.

Using the segment model[5,38,49–51], the photophysical properties of the high-temperature solution form can be understood if the polymer chain behaves as a string of fairly localized, loosely coupled chromophores represented by approximately planar all-*trans* chain segments, separated by one or more stronger (e.g. *gauche*) twists. The $\sigma \rightarrow \sigma^*$ excitation energy of a segment decreases rapidly as a function of its length at first, and reaches a limiting value when the segment contains approximately a dozen silicon atoms. After initial light absorption, the chromophores undergo fluorescence, intersystem crossing and energy transfer from the shorter segments to longer ones, as well as photochemical transformations from both the singlet and triplet excited states. The origin of the inhomogeneous broadening of the room-temperature solution spectra of polysilanes is ascribed to the following three intramolecular causes in addition to the usual solvent effects: a distribution of chain segment lengths, a distribution of deviations from exact planarity within a segment and a distribution of alkyl side-chain conformations. At low temperatures, almost all the segment lengths are presumably so large that a variation in their length has only a very small effect on the excitation energy, so that the first source of inhomogeneous broadening is much less important, the absorption band is much narrower and the optical inhomogeneity is dominated by random disorder within the segments. Whereas a number of band structure calculations have been performed for linear polysilanes[52–58], the interband transition is of limited relevance for the relatively short segments in room-temperature solutions. In these excited segments, the hole and the excited electron are necessarily localized in the same small region of space.

An alternative view of the polysilane structure is depicted using the worm-like model as proposed for poly(diacetylene)s[59], where the linear chain has a large number of small twists without sharp twists playing a special role[60–62]. In this model, a Gaussian distribution of site energies and/or exchange interactions and the coherence of the excitation is terminated by any of the numerous usual random deviations from perfect symmetry.

A two-photon allowed absorption of poly(dihexylsilane) has been observed at 0.9 eV above the one-photon excitation[63–65], in contrast to the two-photon absorption of linear polyenes, which is about 0.5 eV below the intense one-photon absorption. Absorption spectra and the dynamics of other high-lying excited states of the polysilane have been investigated by means of femtosecond time-resolved excited-state absorption spectroscopy[66].

D. Mechanistic Aspects of Photochemistry of Peralkyloligosilanes and Polysilanes

There are two types of basic photoreactions of peralkyloligosilanes and polysilanes:
(1) Homolytic cleavage of a silicon–silicon bond

$$R_3SiSiR'_3 \xrightarrow{h\nu} R_3Si^{\bullet} + {}^{\bullet}SiR'_3 \quad (1)$$

(2) Silylene extrusion

$$R_3SiSiR'_2SiR''_3 \xrightarrow{h\nu} R'_2Si: + R_3SiSiR''_3 \quad (2)$$

A prototypical oligosilane, hexamethyldisilane, reacts upon irradiation of 147 nm light without any trapping reagent to give trimethylsilyl radicals[67]. Recombination and redistribution reactions of the trimethylsilyl radicals give back the starting hexamethyldisilane as well as 1,1-dimethylsilaethene plus trimethylsilane, respectively[68–70] (equation 3).

$$Me_3SiSiMe_3 \xrightarrow{h\nu \,(147\text{ nm})} 2\, Me_3Si^{\bullet} \begin{array}{l} \longrightarrow Me_3SiSiMe_3 \\ \longrightarrow Me_3SiH + Me_2Si{=}CH_2 \end{array} \quad (3)$$

Photolysis of permethyloligosilanes [Me(Me$_2$Si)$_n$Me, $n = 4-8$] in cyclohexane solution with a 254 nm light have been investigated and found to finally give Me$_3$SiSiMe$_2$SiMe$_3$ and Me$_3$SiSiMe$_2$SiMe$_2$H as the sole volatile products[71,72]. These results are well understood when the reactions via two competitive reaction pathways of equations 1 and 2 are taken into account. Photolysis of several branched permethyloligosilanes gives a similar result[71–73]. Photolysis of cyclic oligosilanes like dodecamethylcyclohexasilane[74,75] affords predominantly the corresponding dialkylsilylenes accompanied by ring contraction. Although simple peralkyltrisilanes do not react any more on irradiation with a 254 nm light, irradiation of 1,2,3-trisilacycloheptane derivatives extrudes the central silicon atom as a silylene[76,77].

Much attention has been focused on the detailed mechanism of silylene extrusion from oligosilanes. The following three possible pathways (A-C, equations 4–7) should be first taken into account.

Path A (Stepwise SiSi Bond Cleavages)

$$R_3Si{-}SiR'_2{-}SiR''_3 \xrightarrow{h\nu} R_3Si{-}SiR'^{\bullet}_2 + {}^{\bullet}SiR''_3 \quad (4)$$

$$R_3Si{-}SiR'^{\bullet}_2 \xrightarrow{\Delta} R_3Si^{\bullet} + :SiR'_2 \quad (5)$$

Path B (Concerted Silylene Extrusion)

$$R_3Si{-}SiR'_2{-}SiR''_3 \xrightarrow{h\nu} \left[\begin{array}{c} SiR'_2 \\ \diagup \quad \diagdown \\ R_3Si{-}{-}{-}{-}SiR''_3 \end{array} \right]^{\ddagger} \longrightarrow :SiR'_2 + R_3Si{-}SiR''_3 \quad (6)$$

Path C (Simultaneous Two SiSi Bond Cleavages)

$$R_3Si{-}SiR'_2{-}SiR''_3 \xrightarrow{h\nu} R_3Si^{\bullet} + :SiR'_2 + {}^{\bullet}SiR''_3 \quad (7)$$

Since photon energies are insufficient for simultaneous cleavage of two SiSi bonds to afford two silyl radicals and a silylene, Path C is an unlikely mechanism. The stepwise mechanism (Path A) could also be excluded because thermal α-elimination from pentamethyldisilanyl radical (i.e. equation 5) does not occur at room temperature[78]. The concertedness of the photochemical silylene extrusion from a peralkyltrisilane is corroborated by the high stereoselectivity. Thus, irradiation of E- and Z-1,3-diphenyl-1,2,2,3-tetramethyl-1,2,3-trisilacycloheptane gave, respectively, E- and Z-1,2-diphenyl-1,2-dimethyl-1,2-disilacyclohexane, in a highly stereospecific manner, together with dimethylsilylene[77] (equations 8 and 9).

$$\text{(8)}$$

$$\text{(9)}$$

Several authors have discussed the nature of the excited states responsible for the photochemical silylene extrusion from a trisilane. On the basis of the Woodward–Hoffmann rules, Ramsey concluded that the long-wavelength transition is a $\sigma \rightarrow \sigma^*$ transition and the reaction occurs from the $\sigma\sigma^*$ state[79]. A modified correlation diagram was presented using the energy levels calculated by a pseudopotential method[41]. Bock and coworkers have suggested that Rydberg states can play a role in the photochemical silylene extrusion[80]. The potential energy surface of the S_1 state of 2-methyltrisilane has been studied by using MC-SCF and multireference MP2 methods[81]. These calculations have shown that the avoided crossing first located by Michl and coworkers[42,45] turns out to be points of real crossing when full geometry optimization is performed; two elementary processes, homolysis to produce a silyl radical pair (equation 4; 1st step of Path A) and silylene extrusion (Path B, equation 6) can occur through the $(n-2)$-dimensional conical intersections between the excited S_1 and the ground S_0 states. Fragmentation to two silyl radicals can be achieved in this way and not only through an intersystem crossing between the S_1 and T_1 repulsive states, as suggested by Michl and Balaji[82].

The photochemical behavior of linear peralkylpolysilane high polymers is also understood by invoking the two basic processes, i.e. homolytic cleavage of the SiSi bond and silylene extrusion[83]. In addition, Michl and coworkers[84] have pointed out another pathway, which implies the initial formation of the polysilanylsilylene (equation 10) followed by silylsilylene-to-disilene rearrangement (equation 11). Addition of a silyl radical generated by the SiSi bond cleavage to the disilene give finally a rather stable silyl radical

22. Mechanistic aspects of the photochemistry of organosilicon compounds

(equation 11).

$$\text{(scheme for equations 10 and 11: photolysis of tetrasilane chain yielding silene + silylene, followed by silylene rearrangement to disilene and radical addition)}$$
(10)

(11)

The silylene extrusion should occur from the high polymer, because $Et_3SiSiBu_2H$ appears with no induction period when poly(dibutylsilane) is irradiated in the presence of excess Et_3SiH. Hence the possibility that $R_2Si:$ is produced only from a very short photodegradated silicon chain and not from the high polymer is ruled out[84]. The quantum yield of $Et_3SiSiBu_2H$ decreases as the irradiation wavelength increases and falls to zero above 300 nm, while the polymer absorption band at 315 nm still disappears rapidly and persistent ESR signals are observed[85,86]. Michl and coworkers[84] concluded that silylene production and radical formation occur as two distinct processes, and that the thermal fragmentation of polysilanyl radicals with a sequential loss of a single silylene unit does not occur at room temperature, this being in accord with the result for pentamethyldisilanyl radical[78].

III. ARYLPOLYSILANES

A. Photophysics of Aryldisilanes

In 1964, Sakurai and Kumada[87], Gilman and coworkers[88] and Hague and Prince[89] reported independently that aryl- and vinyl-substituted polysilanes showed appreciable

TABLE 2. ^1L_a absorption maxima of silylbenzenes[87]

Compound	λ_{max} (nm) (ϵ)
PhH	204 (7900)
$PhSiMe_3$	211 (10600)
$PhSiMe_2SiMe_3$	230 (11200)
$PhSiMe_2SiMe_2SiMe_3$	240 (15400)

red shift of the absorption maxima of the corresponding monosilanes (Table 2). Although these and related works[90–98] constitute the first recognition of the conjugation between a silicon–silicon bond and a π system, these conjugating properties were first rationalized in terms of d–π^* interaction in the excited state. Later, the $\sigma - \pi$ conjugation between a silicon–silicon σ orbital and a pπ orbital was taken as the origin of the unique electronic properties, on the basis of the results of the charge-transfer (CT)[99–102] and photoelectron spectroscopic studies of aryldisilanes[103], and the remarkable stereoelectronic effects[104,105] on the absorption spectra. In this model, the high-lying HOMO of an aryl–SiSi system is represented mainly by the linear combination of the σ(SiSi) orbital and the HOMO of the component aromatic π system (Figure 5), while the contribution of the σ(SiSi) orbital in the aryl–SiSi HOMO depends on the HOMO level of the aromatic π system[101] as well as the dihedral angle between an SiSi σ bond and a pπ orbital of the π system[104,105]. The ^1L_a absorption at 230 nm (ϵ ca 10000) of phenylpentamethyldisilane (**1a**) has a strong intramolecular charge-transfer [σ(SiSi) → π^*] nature. In accord with this model, no characteristic band at around 230 nm is observed for the cyclic aryldisilane **2a**, where a SiSi σ bond is in the nodal plane of the π system, while an intense absorption maximum is observed for a cyclic aryldisilane **3a**[104,105].

In 1981, Shizuka and coworkers reported unusual dual fluorescence phenomena of aryldisilanes[106–111]. Thus, in addition to a regular $\pi\pi^*$ (LE, locally excited) emission band from an aromatic π system, a broad band has been observed at a longer wavelength, which is assigned to the emission from a CT excited state ([(2$\pi\pi^*$, 3dπ(Si)] state). In the πd model, the disilanyl group and the aromatic π system are assumed to work as an electron acceptor and an electron donor, respectively. Contrary to the argument by Shizuka and coworkers, Sakurai, Sugiyama and Kira have proposed the intramolecular $\sigma\pi^*$ charge-transfer model[112], in which the disilanyl group and the aromatic π system work as an electron donor and an electron acceptor, respectively, on the basis of a study of substituent and solvent effects on the dual fluorescence of phenylpentamethyldisilane; most distinguishably, the wavelengths of the CT emission band are found to be longer on increasing electron-accepting ability of the substituents.

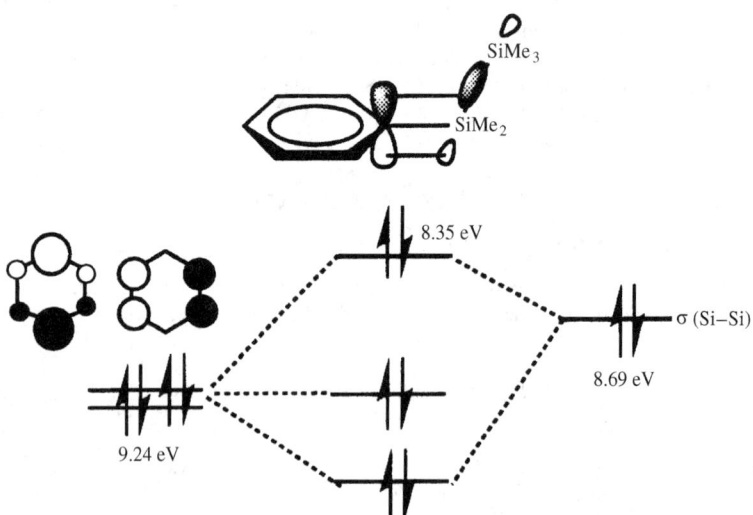

FIGURE 5. Schematic representation of σ(SiSi)–π conjugation. Modified from Reference 103

22. Mechanistic aspects of the photochemistry of organosilicon compounds 1323

$$X-\bigcirc-SiMe_2SiMe_3 \qquad X-\bigcirc\!\!\!\!\!\!\!\!\!\!\!\underset{SiMe_2}{\overset{SiMe_2}{\diagup}}$$

(1) (2)

$$X-\bigcirc\!\!\!\!\!\!\!\!\!\!\!\underset{Me\ SiMe_3}{\overset{Si}{\diagup\!\!\!\diagdown}}$$

(3)

(a) X = H; (b) X = CF$_3$

Dual fluorescence is now a well-documented phenomenon since the discovery by Lippert and coworkers of the dual fluorescence in (p-dimethylamino)benzonitrile (DMABN) in dilute polar solvents[113–115]. Many spectroscopic, thermodynamic and quantum-chemical studies have been devoted to these systems and reviewed recently by Rettig[116a] and by Bhalfacharyya and Chowdury[116b] Grabowski and coworkers have proposed that the charge separation in an $n\pi^*$ intramolecular charge-transfer (ICT) system such as DMABN is attained in a twisted conformation[117], where the two π moieties involved in the charge transfer, the donor and the acceptor, are orbitally decoupled, nearly perpendicular to each other. This led to the terminology of 'twisted intramolecular charge transfer' (TICT) states. However, since CT emission of TICT compounds is not observed in nonpolar solvents, it is obvious that solvation by polar solvents also plays a certain role in the formation of a CT excited state, in addition to internal twisting motion. Such outer contributions for charge separation in $n\pi^*$ ICT systems make the arguments complex and controversial. Since several aryldisilanes exhibit CT emission even in a nonpolar solvent, aryldisilanes may occupy an important position in attempts to understand the intramolecular stabilization of CT excited states. Very recently, (p-cyanophenyl)pentamethyldisilane (**1c**), which bears one of the strongest electron-accepting π systems in the molecule, has been found to show ICT emission even in an isolated jet-cooled condition[118].

$$NC-\bigcirc-SiMe_2SiMe_3$$

(**1c**)

In order to explain the stabilization of the CT excited state even in a nonpolar solvent, the orthogonal intramolecular charge-transfer (OICT) mechanism, which is a conceptually similar term to the TICT mechanism but is used to stress the orbital orthogonality rather than the twisting of the molecular frame, has been proposed. The OICT mechanism for the CT fluorescence of aryldisilanes in nonpolar solvents was supported by the remarkable dependence of the fluorescence spectra on the dihedral angle between a benzene pπ

orbital and an SiSi σ bond in an aryldisilane[105,112]. Thus, 6-(trifluoromethyl)-1,1,2,2,-tetramethyl-1,2-disilaindane (**2b**), where the σ-π overlap is less effective but the CT excited state is well stabilized, showed the typical CT emission, while no intense CT absorption was observed because of forbiddenness of the overlap. On the other hand, 6-(trifluoromethyl)-1-methyl-1-(trimethylsilyl)-1-silaindane (**3b**), which is rigidly held in the perpendicular conformation, showed intense and sharp LE emission but lacked the CT emission. [5-(Trifluoromethyl)-2-methylphenyl]pentamethyldisilane (**1b**) revealed both intense CT absorption and CT fluorescence. The fact that the absorption spectrum of **1b** resembles that of **3b**, while the fluorescence spectrum of **1b** is similar to that of **2b**, is in good agreement with the OICT mechanism for the CT fluorescence of **1b**; the absorption occurs at the preferred conformation **A** and then the OICT state would be formed at the conformation **B** via a dynamic relaxation with internal rotation about the C(aryl)—Si bond (Figure 6). A free jet spectroscopic study on phenylpentamethyldisilane and related compounds also indicated that the most favorable conformation at the ground state is a perpendicular one, which has large overlap between σ (Si—Si) and π orbitals of the aromatic ring[119].

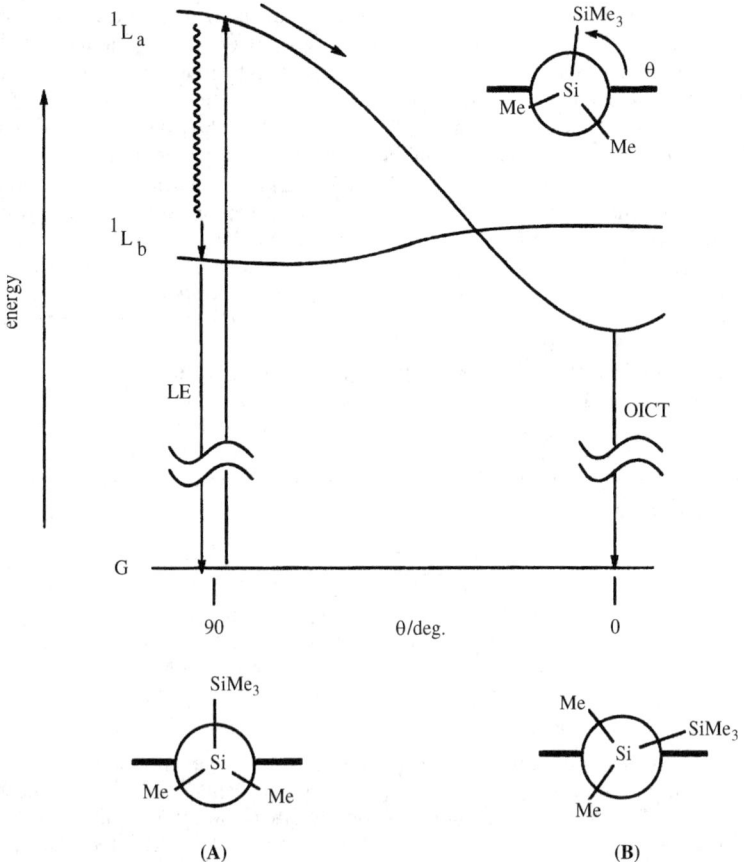

FIGURE 6. Schematic potential surfaces for the OICT mechanism in **1b**. Modified from Reference 105

Contrary to the OICT mechanism, Shizuka and coworkers have concluded that the charge separation does not require internal rotation, on the basis of the very short rise time of the CT band of phenylpentamethyldisilane (<10 ps)[120] in comparison with that of DMABN. However, the rise time may not be a decisive criterion of whether internal rotation is required or not, since the rise time depends not only on the amplitude of the structural change but also on other factors such as the rotational barrier around the pertinent bond[121].

Horn and coworkers have concluded that the $\sigma\pi^*$ charge-transfer model is suitable for explaining the CT fluorescence spectra of (phenylethynyl)pentamethyldisilanes[122].

B. Mechanistic Aspects of Photoreactions of Aryldisilanes

Much attention has been focused on the photochemistry of aryldisilanes[72,123,124], since they generate interesting reactive silicon species such as silylenes, silenes and silyl radicals, as has been shown by chemical trapping. The photolysis of phenyldisilanes in the presence of alcohols gives rise to four types of reaction as shown in Scheme 1: (1) formation of a silaethene via elimination of a hydrosilane (type 1)[125], (2) formation of a silatriene via 1,3-silyl migration (type 2)[126–147], (3) elimination of a silylene (type 3)[148–150] and (4) nucleophilic cleavage of a SiSi bond in the excited state (type 4)[151–153]. The detailed analysis of the steady-state photolysis of (p-trifluoromethylphenyl)pentamethyldisilane **1b**[105] in the presence of alcohols revealed the nature of the excited states responsible for the photoreactions; the 1,3-silyl migration occurs from LE, while the direct alcoholysis of the aryldisilane takes place mainly from the OICT state.

The kinetic profile for the photophysical and photochemical processes of p-(trifluoromethylphenyl)pentamethyldisilane (**1b**) in alcohol–hexane mixtures is represented schematically in Figure 7. Whereas the nature of the excited states of aryldisilanes varies significantly depending on the electronic nature of the aromatic π system, the mechanistic profile shown in Figure 7 would be generally adopted to the photoreactions of aryldisilanes. In contrast to this, Shizuka and coworkers[120] have proposed that 1,3-silyl migration occurs from the πd-type ICT state of phenyldisilanes, on the basis of the time-resolved emission studies of aryldisilanes.

One of the interesting mechanistic problems in photoreactions of phenyldisilanes is whether the 1,3-silyl migration giving a silatriene derivative occurs concertedly, or via the recombination of an intimate silyl radical pair generated by the homolytic cleavage of the SiSi bond in the LE excited state. While Ishikawa and coworkers first suggested a concerted 1,3-migration[133], Sakurai and coworkers have noted that the type 1–type 3 photoreactions of aryldisilanes in nonpolar solvents are all explained by an initially formed silyl radical pair, on the basis of the ESR evidence for the formation of silyl radicals during the photolysis of aryldisilanes[148]. Shizuka and coworkers[120] have shown that the formation of the silatriene from phenylpentamethyldisilane occurs very rapidly, i.e. within ca 30 ps, and supported the concerted mechanism. Quite recently, Ohshita, Ishikawa and coworkers have reported the diastereospecific formation of the corresponding silatriene in the photolysis of meso- and dl-1,2-diethyl-1,2-dimethyl-1,2-diphenyldisilanes (cf **4**) in the presence of olefins to give **5** and carbonyl compounds (equation 12) as more direct evidence for the concerted mechanism[141].

Leigh and Sluggett have demonstrated that the photolysis of 1,1,1-trimethylsilyltriphenylsilane (**6**) gives the rearranged silatriene largely via a singlet excited state of **6** and free silyl radicals are formed via the triplet state[144–146]. Irradiation of **6** and acetone in cyclohexane in the presence of chloroform as a silyl radical trap gives mainly the ene adducts of a silatriene, while a similar irradiation in acetonitrile affords

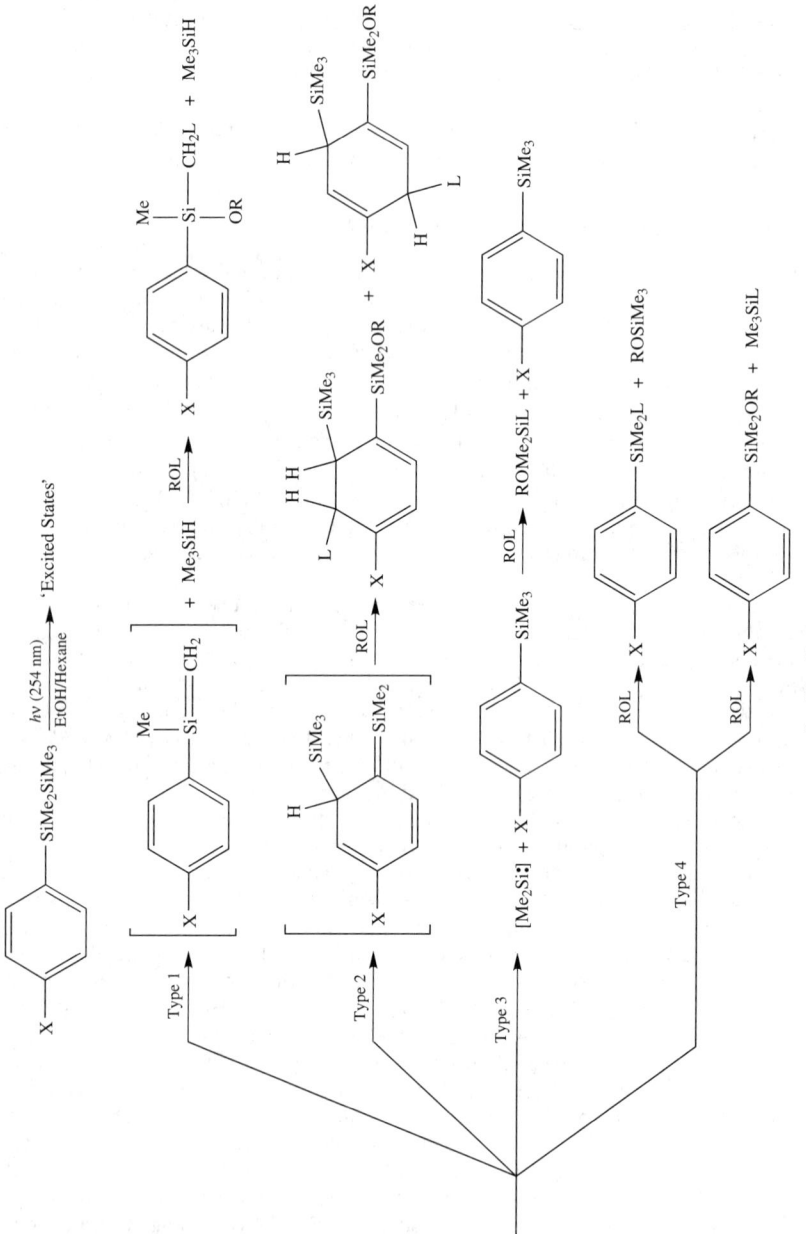

SCHEME 1. Modified from Reference 105

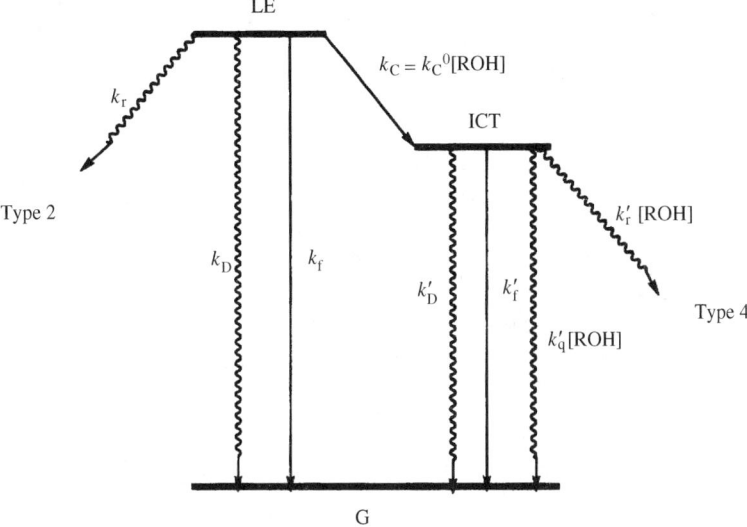

FIGURE 7. Kinetic profile for photophysical and photochemical processes of a phenylpentamethyldisilane in alcohol–hexane mixtures: k_f, k'_f = fluorescence rates; k_D, k'_D = radiationless decay rates; k'_q = rate for CT quenching; k_r, k'_r = reaction rates for 1,3-silyl migration and reaction with alcohol, respectively; k_C = formation rates of ICT state from LE. Modified from Reference 105

triphenylchlorosilane and trimethylchlorosilane as the major products (Scheme 2). Large solvent effects on the relative yields of silatriene- and radical-derived products are ascribed to an intersystem crossing which is enhanced in polar solvents.

(12)

(S,S)- and (R,R)-**4** ⟶ (S,S)- and (R,R)-**5**

meso-**4** ⟶ (S,R)- and (R,S)-**5**

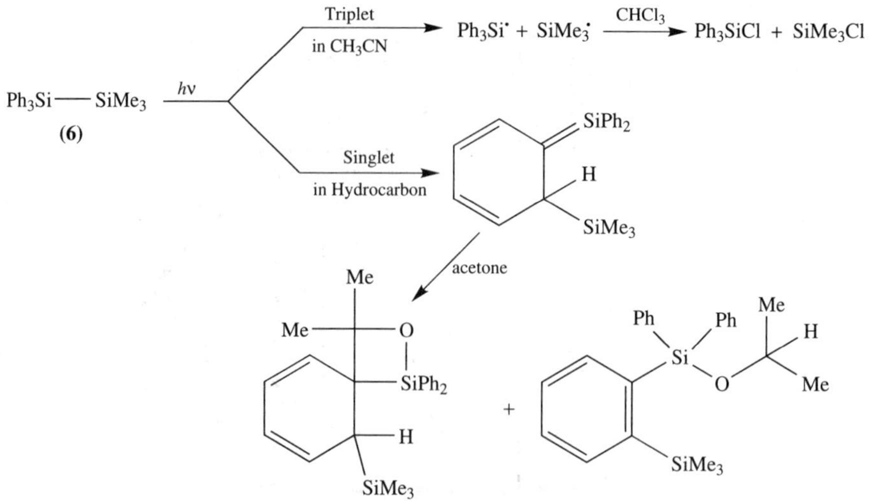

SCHEME 2

In relation to the reactions of CT excited states of aryldisilanes, photochemistry of LE and ICT states of the rigid, *p*-cyano-substituted styryldisilane **7** was investigated by Steinmetz and coworkers[154]. Although no 1,3-silyl shift is observed in this sytem, **7** affords a major product attributable to addition of alcohol across the Si—Si bond in the CT state. The roles of LE and CT states in the formation of additional minor products of silylene extrusion and homolytic Si—Si cleavage have also been elucidated.

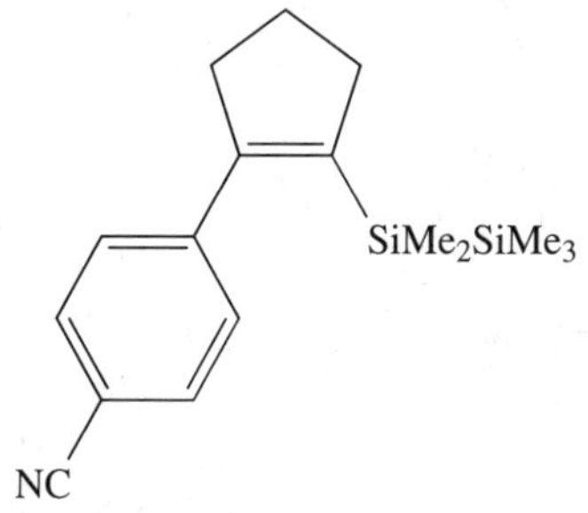

(7)

C. Photochemistry of Aryltrisilanes

Whereas 2,2-diphenylhexamethyltrisilane (**8**) is a well-known photochemical precursor of diphenylsilylene[72,155,156], a 1,3-silyl migration is usually a major side reaction in solution at room temperature. In the presence of excess ethanol, the irradiation of **8** in hexane gives **9** via diphenylsilylene extrusion and **10** (an isomeric mixture) via 1,3-silyl migration in 50 and 37%, respectively (Scheme 3)[157]. Since the product ratio does not depend on the solvent polarity, both reactions, silylene extrusion and 1,3-silyl migration, do not occur via the ICT state but via the nonpolar excited state[158]. However, the excited

22. Mechanistic aspects of the photochemistry of organosilicon compounds   1329

states responsible for these two pathways are suggested to be different from each other by the nonresonant two-photon (NRTP) method, where the highly selective silylene extrusion was observed[157]. Thus, irradiation of a hexane solution of **8** with pulsed 532 nm laser light at room temperature gives **9** and **10** in 80 and 13% yields, respectively, in the presence of ethanol, although **8** shows no absorption at wavelengths longer than 300 nm. The product yields increased in proportion to the square of the laser intensity, being indicative of the simultaneous two-photon nature of the reaction[158].

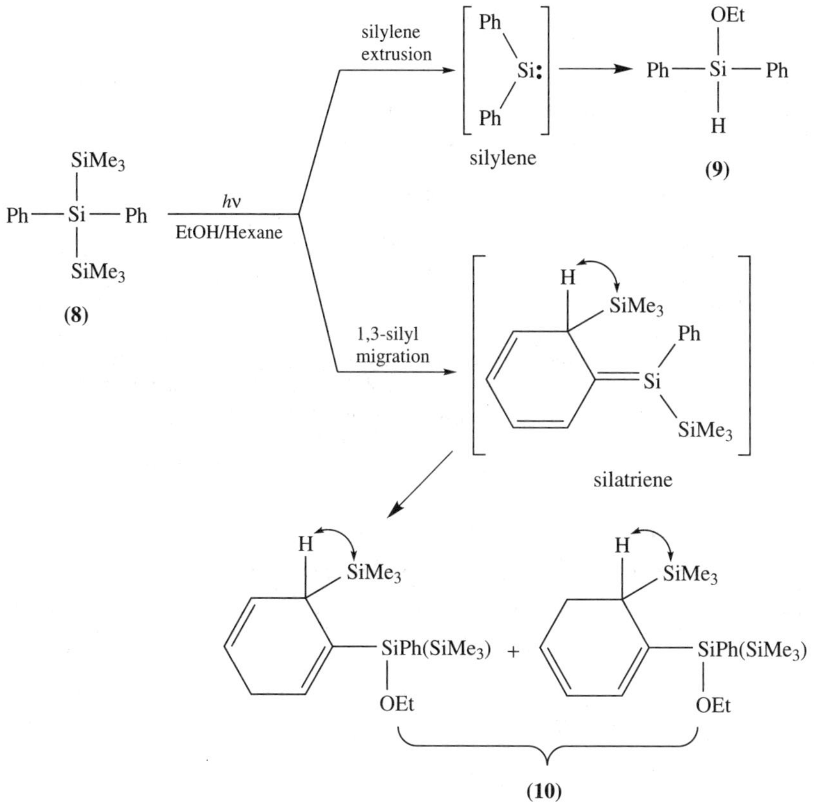

SCHEME 3. Modified from Ref. 157

IV. SILYLENES

The electronic structure of silylenes has been discussed in depth since the first matrix isolation of dimethylsilylene by West and coworkers[159], who observed that dimethylsilylene produced by the irradiation of dodecamethylcyclohexasilane in an argon matrix at 10 K or in a 3-methylpentane glass matrix at 77 K showed a characteristic absorption maximum at 450 nm. All the silylenes investigated so far are known to be ground state singlets and the absorption due to the excitation of a lone-pair electron to the vacant pπ orbital of the silylene (equation 13) is observed in the UV-VIS region[159−170]. The absorption maxima depend strongly on the electronic and steric effects of the substituents. Absorption maxima of various silylenes observed in hydrocarbon matrices at 77 K are shown in Table 3.

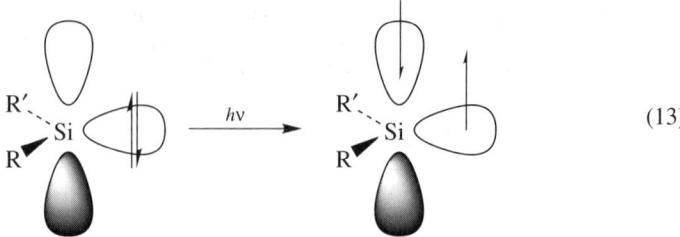

Electronically, the substituent effects can be discussed in terms of their inductive and conjugative effects. Electronegative substituents increase the transition energy (ΔE), as well as the singlet–triplet gap, and thus induce blue shifts. This can be understood in terms of Bent's rule[171–173] by the effect that such substitutents have on the hybridization and energy of the n(Si) orbital. Introduction of electropositive trialkylsilyl and trialkylgermyl substituents on the silylene causes quite large red shifts of n(Si)–3p(Si) absorption; typically, Mes(Me$_3$Si)Si: has its λ_{max} at 760 nm[168–170]. Conjugation between the substituent and the empty 3p(Si) orbital in the ground-state singlet and/or with the half-filled orbitals in the first excited state can have a strong effect on the electronic spectrum of silylenes. Large blue shifts are caused by introduction of n donors such as OR and NR$_2$[162,174], because n donors stabilize the singlet ground state by conjugation with the empty 3p(Si) orbital, while the excited state is actually destabilized by these substituents (Figure 8a). The ethynyl, vinyl and phenyl substituents are all much weaker π donors than OR and NR$_2$, so the stabilization and thus the blue shift resulting from π conjugation are expected to be smaller. However, these unsaturated substituents actually produce red shifts relative

TABLE 3. n(Si)→3p(Si) absorption maxima of substituted silylenes (RR'Si:) in hydrocarbon matrices at 77 K

R	R'	λ_{max} (nm)	References
CH$_3$	CH$_3$	453	159, 163
CH$_3$	Ph	490	163
CH$_3$	Mesityl	495	163
CH$_3$	C≡CSiMe$_3$	473	176
		475	166
		505	166
Mesityl	OPh	400	162
N(Pr-i)$_2$	N(Pr-i)$_2$	335	174
Ph	SiMe$_3$	660	169
Mesityl	SiMe$_3$	760	169

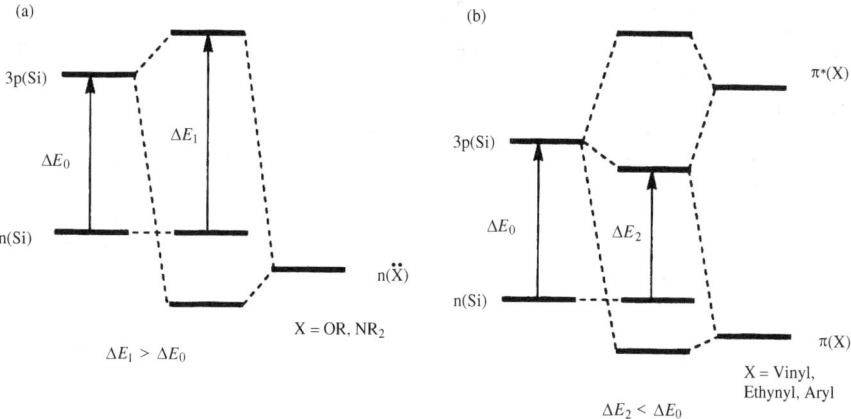

FIGURE 8. Schematic orbital interaction diagram for silylenes having (a) n donors and (b) π groups as substituents. ΔE_0, ΔE_1 and ΔE_2 denote the vertical excitation energies in the parent, an n-donor-substituted and a π-substituted silylenes, respectively. Modified from Reference 176

to Me$_2$Si[163,166,167]. The different behavior of n donors such as OR and a π group such as vinyl is best understood qualitatively as resulting from the presence of a π^* orbital in the unsaturated substituents. According to simple frontier molecular orbital (FMO) theory, the relatively high-lying 3p(Si) orbital can interact more effectively with the low-lying π^* orbital than with the bonding π orbital of the substituents, reducing the n(Si)–3p(Si) energy gap, as displayed schematically in Figure 8b[163,167,175]. Steric effects can affect the apex angle at silicon and thus cause a red shift or a blue shift, depending on whether the angle increases or is reduced.

Substituent effects on the transition energy of silylene have been investigated systematically by *ab initio* MO calculations and good agreement between the calculations and experiment has been attained[175,176].

A number of experimental[177–181] and theoretical studies[182,183] have anticipated formation of silylene–base complexes during reactions of silylenes with oxygen-, sulfur-, nitrogen- and phosphorus-containing compounds. A number of silylene–base complexes have been actually observed in low-temperature glass matrices[167,184–188]. The absorption maxima of silylenes were shifted significantly to the blue by the complexation. The origin of this is ascribed to a weak bonding interaction between the vacant 3p(Si) orbital in a silylene and an n orbital in a base, which raise the LUMO level.

Whereas much effort has been devoted to the search for triplet silylenes in the ground and excited states both experimentally[189–194] and theoretically[195,196], no successful result has been reported so far. Introduction of two bulky trialkylsilyl substituents to a silylene is predicted to lead to a ground state triplet[196].

Irradiation of dimethylsilylene at 450 nm gives the corresponding silaethene, which is reconverted back to dimethylsilylene upon irradiation with 248 nm light (equation 14)[197,198]. The interesting photochemical conversion from silylene to silene in low-temperature matrices has been investigated with polarized light[198]; the 1,2-hydrogen migration has never been discussed on the basis of the excited-state nature of the silylene.

$$\text{(H}_3\text{C)}_2\text{Si:} \xrightleftharpoons[248\text{ nm}]{450\text{ nm}} \text{H(H}_3\text{C)Si}=\text{CH}_2 \qquad (14)$$

V. SUMMARY

Photochemically induced electron-transfer reactions involving electron-donating organosilicon compounds such as oligosilanes, polysilanes, allylic and benzylic silanes have attracted much attention in recent years[2,3]. While these reactions are synthetically interesting, we have not dealt with them in this chapter, because most of the reactions can be understood as due to the behavior of the corresponding radical cations generated by the electron transfer in the environment.

Many interesting photoreactions of organosilicon compounds have been reported without thinking about the nature of the excited states responsible for the photoreactions. Among them, the following reactions are involved: (1) photochemical 1,3-silyl migration in allylic silanes[199,200], (2) photolysis of 1,1-diphenylsilacyclobutane forming diphenylsilaethene[201-203], (3) photochemical rearrangement of disilanyl-iron complexes[204,205], (4) photochemical dissociation of disilene to silylene[206], (5) photochemical isomerization from a tetrasilacyclobutene to a tetrasilabicyclo[1.1.0]butane[207] and so on. Elucidation of the detailed mechanism of these basic organosilicon photoreactions is required not only for the well-balanced development of the basic organosilicon chemistry, but also for future application of organosilicon substances as optoelectronic materials.

VI. ACKNOWLEDGMENT

The authors wish to thank Miss M. Kodaka, Mrs H. Kano and Mr U. Kwon for their assistance in the preparation of the manuscript.

VII. REFERENCES

1. A. G. Brook, in *The Chemistry of Organosilicon Compounds* (Eds. S. Patai and Z. Rappoport) Part 2, Chap. 15, Wiley, Chichester, 1989, p. 965.
2. A. G. Brook, Chap. 21 in this volume.
3. M. G. Steinmetz, *Chem. Rev.*, **95**, 1527 (1995).
4. R. D. Miller and J. Michl, *Chem. Rev.*, **89**, 1359 (1989).
5. J. Michl, J. W. Downing, T. Karatsu, A. J. McKinley, G. Poggi, G. M. Wallraff, R. Sooriyakumaran and R. D. Miller, *Pure Appl. Chem.*, **60**, 959 (1988).
6. J. M. Zeigler, *Synth. Met.*, **28**, C581 (1989).
7. R. West, *J. Organomet. Chem.*, **300**, 327 (1986).
8. R. West, in *The Chemistry of Organosilicon Compounds* (Eds. S. Patai and Z. Rappoport), Part 2, Chap. 19, Wiley, Chichester, 1989, p. 1207.
9. R. West and P. P. Gaspar, Chapter 43 in this volume.
10. H. Gilman, W. H. Atwell and G. L. Schwebke, *Chem. Ind. (London)*, 1063 (1964).
11. H. Gilman, W. H. Atwell and G. L. Schwebke, *J. Organomet. Chem.*, **2**, 369 (1964).
12. M. Kumada and K. Tamao, *Adv. Organomet. Chem.*, **6**, 19 (1968).
13. R. West, in *Comprehensive Organometallic Chemistry* (Eds. G. Wilkinson, F. G. A. Stone and E. W. Abel), Vol. 12, Pergamon, Oxford, 1982, p. 365.
14. W. G. Boberski and A. L. Allred, *J. Organomet. Chem.*, **88**, 65 (1975).
15. C. G. Pitt, L. L. Jones and B. G. Ramsey, *J. Am. Chem. Soc.*, **89**, 5471 (1967).
16. P. P. Shorygin, V. A. Petakhov, O. M. Nefedov, S. P. Kolesnikov and V. I. Shiryaev, *Theor. i Eksperim. Khim. Akad. Nauk Ukr. SSR*, **2**, 190 (1966); *Chem. Abstr.*, **65**, 14660f (1966).
17. C. G. Pitt, M. M. Bursey and P. F. Rogerson, *J. Am. Chem. Soc.*, **92**, 519 (1970).
18. B. G. Ramsey, *Electronic Transitions in Organometalloids*, Academic Press, New York, 1968.
19. C. Sandorfy, *Can. J. Chem.*, **33**, 1337 (1955).
20. K. Fukui, K. Kato and T. Yonezawa, *Bull. Chem. Soc. Jpn.*, **33**, 1197 (1960).
21. H. Bock and W. Ensslin, *Angew. Chem., Int. Ed. Engl.*, **10**, 404 (1971).
22. H. Bock, W. Ensslin, F. Fehér and R. Freund, *J. Am. Chem. Soc.*, **98**, 668 (1976).

23. H. Bock and B. Solouki, in *The Chemistry of Organosilicon Compounds* (Eds. S. Patai and Z. Rappoport), Chap. 9. Vol. 1, Wiley, Chichester, 1989, p. 555.
24. V. F. Traven and R. West, *J. Am. Chem. Soc.*, **95**, 6824 (1973).
25. H. Sakurai, M. Kira and T. Uchida, *J. Am. Chem. Soc.*, **95**, 6826 (1973).
26. H. Bock, W. Kaim, M. Kira and R. West, *J. Am. Chem. Soc.*, **101**, 7667 (1979).
27. T. Shida, H. Kubodera and Y. Egawa, *Chem. Phys. Lett.*, **79**, 179 (1981).
28. J. T. Wang and F. Williams, *J. Chem. Soc., Chem. Commun.*, 666 (1981).
29. R. West and E. Carberry, *Science*, **189**, 179 (1975).
30. E. Carberry, R. West and G. E. Glass, *J. Am. Chem. Soc.*, **91**, 5446 (1969).
31. R. West and E. S. Kean, *J. Organomet. Chem.*, **96**, 323 (1975).
32. H. S. Plitt and J. Michl, *Chem. Phys. Lett.*, **198**, 400 (1992).
33. H. S. Plitt, J. W. Downing, M. K. Raymond, V. Balaji and J. Michl, *J. Chem. Soc., Faraday Trans.*, **90**, 1653 (1994).
34. B. Albinsson, H. Teramae, J. W. Downing and J. Michl, *Chem. Eur. J.*, **2**, 529 (1996).
35. V. Balaji and J. Michl, *Polyhedron*, **10**, 1265 (1991).
36. R. W. Bigelow, *Chem. Phys. Lett.*, **126**, 63 (1986).
37. R. W. Bigelow, *Organometallics*, **5**, 1502 (1986).
38. K. A. Klingensmith, J. W. Downing, R. D. Miller and J. Michl, *J. Am. Chem. Soc.*, **108**, 7438 (1986).
39. Z. G. Soos and G. W. Hayden, *Chem. Phys.*, **143**, 199 (1990).
40. Z. G. Soos, G. W. Hayden and P. C. M. McWilliams, *Polym. Prepr.*, **31**, 286 (1990).
41. E. A. Halevi, G. Winkelhofer, M. Meisl and R. Janoschek, *J. Organomet. Chem.*, **294**, 151 (1985).
42. Y.-P. Sun and J. Michl, *J. Am. Chem. Soc.*, **114**, 8186 (1992).
43. H. S. Plitt, V. Balaji and J. Michl, *Chem. Phys. Lett.*, **213**, 158 (1993).
44. J. R. G. Thorne, S. A. Williams, R. M. Hochstrasser and P. J. Fagan, *Chem. Phys.*, **157**, 401 (1991).
45. Y.-P. Sun, Y. Hamada, L.-M. Huang, J. Maxka, J.-S. Hsiao, R. West and J. Michl, *J. Am. Chem. Soc.*, **114**, 6301 (1992).
46. L. A. Harrah and J. M. Zeigler, *J. Polym. Sci., Polym. Lett. Ed.*, **23**, 209 (1985).
47. P. M. Cotts, R. D. Miller, P. T. Trefonas, III, R. West and G. N. Fickes, *Macromolecules*, **20**, 1046 (1987).
48. P. Shukla, P. M. Cotts, R. D. Miller, T. P. Russell, B. A. Smith, G. M. Wallraff, M. Baier and P. Thiyagarajan, *Macromolecules*, **24**, 5606 (1991).
49. Y.-P. Sun, R. D. Miller, R. Sooriyakumaran and J. Michl, *J. Inorg. Organomet. Polym.*, **1**, 3 (1991).
50. Y.-P. Sun, G. M. Wallraff, R. D. Miller and J. Michl, *J. Photochem. Photobiol. A: Chem.*, **62**, 333 (1991).
51. R. D. Miller, G. M. Wallraff, M. Baier, P. M. Cotts, P. Shukla, T. P. Russell, F. C. De Schryver and D. Declercq, *J. Inorg. Organomet. Polym.*, **1**, 505 (1991).
52. K. Takeda, N. Matsumoto and M. Fukuchi, *Phys. Rev. B*, **30**, 5871 (1984).
53. K. Takeda, M. Fujino, K. Seki and H. Inokuchi, *Phys. Rev. B*, **36**, 8129 (1987).
54. K. Takeda, H. Teramae and N. Matsumoto, *J. Am. Chem. Soc.*, **108**, 8186 (1986).
55. H. Teramae, T. Yamabe and A. Imamura, *Theoret. Chim. Acta (Berl.)*, **64**, 1 (1983).
56. J. W. Mintmire, *Phys. Rev. B*, **39**, 13350 (1989).
57. K. Takeda and N. Matsumoto, *J. Phys. C*, **18**, 6121 (1985).
58. H. Teramae and K. Takeda, *J. Am. Chem. Soc.*, **111**, 1281 (1989).
59. G. Wenz, M. A. Müller, M. Schmidt and G. Wegner, *Macromolecules*, **17**, 837 (1984).
60. A. Tilgner, H. P. Trommsdorf, J. M. Zeigler and R. M. Hochstrasser, *J. Lumin.*, **45**, 373 (1990).
61. A. Tilgner, H. P. Trommsdorf, J. M. Zeigler and R. M. Hochstrasser, *J. Inorg. Organomet. Polym.*, **1**, 343 (1991).
62. A. Tilgner, H. P. Trommsdorf, J. M. Zeigler and R. M. Hochstrasser, *J. Chem. Phys.*, **96**, 781 (1992).
63. J. R. G. Thorne, Y. Ohsako, J. M. Zeigler and R. M. Hochstrasser, *Chem. Phys. Lett.*, **162**, 455 (1989).
64. Y. Moritomo, Y. Tokura, H. Tachibana, Y. Kawabata and R. D. Miller, *Phys. Rev. B*, **43**, 14746 (1991).
65. H. Tachibana, Y. Kawabata, S. Koshihara and Y. Tokura, *Solid State Commun.*, **75**, 5 (1990).

66. J. R. G. Thorne, S. T. Repinec, S. A. Abrash, J. M. Zeigler and R. M. Hochstrasser, *Chem. Phys.*, **146**, 315 (1990).
67. P. Boudjouk and R. D. Koob, *J. Am. Chem. Soc.*, **97**, 6595 (1975).
68. S. K. Tokach and R. D. Koob, *J. Am. Chem. Soc.*, **102**, 376 (1980).
69. B. J. Cornett, K. Y. Choo and P. P. Gaspar, *J. Am. Chem. Soc.*, **102**, 377 (1980).
70. L. Gammie, I. Safarik, O. P. Strausz, R. Roberge and C. Sandorfy, *J. Am. Chem. Soc.*, **102**, 378 (1980).
71. M. Ishikawa, T. Takaoka and M. Kumada, *J. Organomet. Chem.*, **42**, 333 (1972).
72. M. Ishikawa and M. Kumada, *Adv. Organomet. Chem.*, **19**, 51 (1981).
73. I. M. T. Davidson, J. Michl and T. Simpson, *Organometallics*, **10**, 842 (1991).
74. M. Ishikawa and M. Kumada, *J. Chem. Soc., Chem. Commun.*, 612 (1970).
75. M. Ishikawa and M. Kumada, *J. Organomet. Chem.*, **42**, 325 (1972).
76. H. Sakurai, Y. Kobayashi and Y. Nakadaira, *J. Am. Chem. Soc.*, **93**, 5272 (1971).
77. H. Sakurai, Y. Kobayashi and Y. Nakadaira, *J. Am. Chem. Soc.*, **96**, 2656 (1974).
78. J. A. Hawari, D. Griller, W. P. Weber and P. P. Gaspar, *J. Organomet. Chem.*, **326**, 335 (1987).
79. B. G. Ramsey, *J. Organomet. Chem.*, **67**, C67 (1974).
80. H. Bock, K. Wittel, M. Veith and N. Wiberg, *J. Am. Chem. Soc.*, **98**, 109 (1976).
81. A. Venturini, T. Vreven, F. Bernardi, M. Olivucci and M. A. Robb, *Organometallics*, **14**, 4953 (1995).
82. J. Michl and V. Balaji, in *Computational Advances in Organic Chemistry: Molecular Structure and Reactivity* (Eds. C. Ogretir and I. G. Csizmadia), Kluwer Academic Publishers, Dordrecht, 1991, p. 323.
83. P. Trefonas, III, R. West and R. D. Miller, *J. Am. Chem. Soc.*, **107**, 2737 (1985).
84. T. Karatsu, R. D. Miller, R. Sooriyakumaran and J. Michl, *J. Am. Chem. Soc.*, **111**, 1140 (1989).
85. A. J. McKinley, T. Karatsu, G. M. Wallraff, R. D. Miller, R. Sooriyakumaran and J. Michl, *Organometallics*, **7**, 2567 (1988).
86. A. J. McKinley, T. Karatsu, G. M. Wallraff, D. P. Thompson, R. D. Miller and J. Michl, *J. Am. Chem. Soc.*, **113**, 2003 (1991).
87. H. Sakurai and M. Kumada, *Bull. Chem. Soc. Jpn.*, **37**, 1894 (1964).
88. H. Gilman, W. H. Atwell and G. L. Schwebke, *J. Organomet. Chem.* **2**, 369 (1964).
89. D. N. Hague and R. H. Prince, *Chem. Ind. (London)*, 1492 (1964).
90. H. Sakurai, H. Yamamori and M. Kumada, *Bull. Chem. Soc. Jpn.*, **38**, 2024 (1965).
91. H. Sakurai, K. Tominaga and M. Kumada, *Bull. Chem. Soc. Jpn.*, **39**, 1279 (1966).
92. H. Sakurai, H. Yamamori and M. Kumada, *J. Chem. Soc., Chem. Commun.*, 198 (1965).
93. H. Sakurai, M. Ichinose, M. Kira and T. G. Traylor, *Chem. Lett.*, 1383 (1984).
94. H. Gilman and W. H. Atwell, *J. Organomet. Chem.*, **4**, 176 (1965).
95. H. Gilman, W. H. Atwell, P. K. Sen and C. L. Smith, *J. Organomet. Chem.*, **4**, 163 (1965).
96. H. Gilman and P. J. Morris, *J. Organomet. Chem.*, **6**, 102 (1966).
97. C. G. Pitt, *J. Am. Chem. Soc.*, **91**, 6613 (1969).
98. C. G. Pitt, *J. Chem. Soc., Chem. Commun.*, 816 (1971).
99. H. Bock and H. Alt, *J. Am. Chem. Soc.*, **92**, 1569 (1970).
100. C. G. Pitt, R. N. Carey and E. C. Toren, Jr., *J. Am. Chem. Soc.*, **94**, 3806 (1972).
101. H. Sakurai and M. Kira, *J. Am. Chem. Soc.*, **96**, 791 (1974).
102. H. Sakurai and M. Kira, *J. Am. Chem. Soc.*, **97**, 4879 (1975).
103. C. G. Pitt and H. Bock, *J. Chem. Soc., Chem. Commun.*, 28 (1972).
104. H. Sakurai, S. Tasaka and M. Kira, *J. Am. Chem. Soc.*, **94**, 9285 (1972).
105. M. Kira, T. Miyazawa, H. Sugiyama, M. Yamaguchi and H. Sakurai, *J. Am. Chem. Soc.*, **115**, 3116 (1993).
106. H. Shizuka, H. Obuchi, M. Ishikawa and M. Kumada, *J. Chem. Soc., Chem. Commun.*, 405 (1981).
107. H. Shizuka, Y. Sato, M. Ishikawa and M. Kumada, *J. Chem. Soc., Chem. Commun.*, 439 (1982).
108. H. Shizuka, Y. Sato, Y. Ueki, M. Ishikawa and M. Kumada, *J. Chem. Soc., Faraday Trans. 1*, **80**, 341 (1984).
109. H. Shizuka, H. Obuchi, M. Ishikawa and M. Kumada, *J. Chem. Soc., Faraday Trans. 1*, **80**, 383 (1984).
110. H. Hiratsuka, Y. Mori, M. Ishikawa, K. Okazaki and H. Shizuka, *J. Chem. Soc., Faraday Trans. 2*, **81**, 1665 (1984).
111. H. Shizuka, *Pure Appl. Chem.*, **65**, 1635 (1993).

22. Mechanistic aspects of the photochemistry of organosilicon compounds 1335

112. H. Sakurai, H. Sugiyama and M. Kira, *J. Phys. Chem.*, **94**, 1837 (1990).
113. E. Lippert, W. Lüder, F. Moll, W. Nägele, H. Boos, H. Prigge and I. Seibold-Blankenstein, *Angew. Chem.*, **73**, 695 (1961).
114. E. Z. Lippert, Z. *Naturforsch*, **10a**, 541 (1955).
115. E. Lippert, W. Lüder and H. Boos, in *Advances in Molecular Spectroscopy* (Ed. A. Magnani), Pergamon Press, Oxford, 1962, p. 443.
116. (a) W. Rettig, *Angew. Chem., Int. Ed. Engl.*, **25**, 971 (1986).
     (b) K. Bhalfacharyya and M. Chowdhury, *Chem. Rev.*, **93**, 507 (1993).
117. K. Rotkiewicz, K. H. Grellmann and Z. R. Grabowski, *Chem. Phys. Lett.*, **19**, 315 (1973).
118. Y. Tajima, H. Ishikawa, T. Miyazawa, N. Mikami and M. Kira, *J. Am. Chem. Soc.*, **118**, in press (1997).
119. M. Kira, T. Miyazawa, N. Mikami and H. Sakurai, *Organometallics*, **10**, 3793 (1991).
120. H. Shizuka, K. Okazaki, M. Tanaka, M. Ishikawa, M. Sumitani and K. Yoshihara, *Chem. Phys. Lett.*, **113**, 89 (1985).
121. K. B. Eisenthal, in *Ultrashort Laser Pulses* (Ed. W. Kaiser), Springer-Verlag, Berlin, 1988, p. 319 and references cited therein.
122. K. A. Horn, R. B. Grossman, J. R. G. Thorne and A. A. Whitenack, *J. Am. Chem. Soc.*, **111**, 4809 (1989).
123. H. Sakurai, *J. Organomet. Chem.*, **200**, 261 (1980).
124. B. Coleman and M. Jones, Jr., *Rev. Chem. Intermediates*, **4**, 297 (1989).
125. P. Boudjouk, J. R. Roberts, C. M. Golino and L. H. Sommer, *J. Am. Chem. Soc.*, **94**, 7926 (1972).
126. M. Ishikawa, *Pure Appl. Chem.*, **50**, 11 (1978).
127. M. Ishikawa, T. Fuchikami, T. Sugaya and M. Kumada, *J. Am. Chem. Soc.*, **97**, 5923 (1975).
128. M. Ishikawa, T. Fuchikami and M. Kumada, *J. Organomet. Chem.*, **118**, 139 (1976).
129. M. Ishikawa, T. Fuchikami and M. Kumada, *J. Organomet. Chem.*, **118**, 155 (1976).
130. M. Ishikawa, T. Fuchikami and M. Kumada, *Tetrahedron Lett.*, 1299 (1976).
131. M. Ishikawa, T. Fuchikami and M. Kumada, *J. Organomet. Chem.*, **133**, 19 (1977).
132. M. Ishikawa, T. Fuchikami and M. Kumada, *J. Organomet. Chem.*, **162**, 223 (1978).
133. M. Ishikawa, M. Oda, N. Miyoshi, L. Fabry, M. Kumada, T. Yamabe, K. Akagi and K. Fukui, *J. Am. Chem. Soc.*, **101**, 4612 (1979).
134. M. Ishikawa, M. Oda, K. Nishimura and M. Kumada, *Bull. Chem. Soc. Jpn.*, **56**, 2795 (1983).
135. J. Ohshita, H. Ohsaki, M. Ishikawa, A. Tachibana, Y. Kurosaki, T. Yamabe and A. Minato, *Organometallics*, **10**, 880 (1991).
136. K. Takaki, H. Sakamoto, Y. Nishimura, Y. Sugihara and M. Ishikawa, *Organometallics*, **10**, 888 (1991).
137. J. Ohshita, H. Ohsaki, H. Takahashi, M. Ishikawa, A. Tachibana, Y. Kurosaki, T. Yamabe, T. Tsukihara, K. Takahashi and Y. Kiso, *Organometallics*, **10**, 2685 (1991).
138. J. Ohshita, H. Ohsaki and M. Ishikawa, *Organometallics*, **10**, 2695 (1991).
139. M. Ishikawa, Y. Nishimura and H. Sakamoto, *Organometallics*, **10**, 2701 (1991).
140. M. Ishikawa, M. Kikuchi, A. Kunai, T. Takeuchi, T. Tsukihara and M. Kido, *Organometallics*, **12**, 3474 (1993).
141. J. Ohshita, H. Niwa, M. Ishikawa, T. Yamabe, T. Yoshii and K. Nakamura, *J. Am. Chem. Soc.*, **118**, 6853 (1996).
142. G. W. Sluggett and W. J. Leigh, *Organometallics*, **11**, 3731 (1992).
143. G. W. Sluggett and W. J. Leigh, *J. Am. Chem. Soc.*, **114**, 1195 (1992).
144. W. J. Leigh and G. W. Sluggett, *J. Am. Chem. Soc.*, **115**, 7531 (1993).
145. W. J. Leigh and G. W. Sluggett, *Organometallics*, **13**, 269 (1994).
146. G. W. Sluggett and W. J. Leigh, *Organometallics*, **13**, 1005 (1994).
147. J. Braddock-Wilking, M. Y. Chiang and P. P. Gaspar, *Organometallics*, **12**, 197 (1993).
148. H. Sakurai, Y. Nakadaira, M. Kira, H. Sugiyama, K. Yoshida and T. Takiguchi, *J. Organomet. Chem.*, **184**, C36 (1980).
149. M. Kira, K. Sakamoto and H. Sakurai, *J. Am. Chem. Soc.*, **105**, 7469 (1983).
150. H. Sakurai, K. Sakamoto and M. Kira, *Chem. Lett.*, 1213 (1984).
151. H. Sakurai, in *Silicon Chemistry* (Eds. J. Y. Corey, E. Y. Corey and P. P. Gaspar), Chap. 16, Ellis Horwood, Chichester, 1988.
152. H. Okinoshima and W. P. Weber, *J. Organomet. Chem.*, **149**, 279 (1978).
153. S.-S. Hu and W. P. Weber, *J. Organomet. Chem.*, **369**, 155 (1989).

154. M. G. Steinmetz, C. Yu and L. Li, *J. Am. Chem. Soc.*, **116**, 932 (1994).
155. M. Ishikawa and M. Kumada, *J. Organomet. Chem.*, **81**, C3 (1974).
156. M. Ishikawa, K.-I. Nakagawa, M. Ishiguro, F. Ohi and M. Kumada, *J. Organomet. Chem.*, **201**, 151 (1980).
157. M. Kira, T. Miyazawa, S. Koshihara, Y. Segawa and H. Sakurai, *Chem. Lett.*, 3 (1995).
158. T. Miyazawa, H. Sakurai and M. Kira, unpublished results.
159. T. J. Drahnak, J. Michl and R. West, *J. Am. Chem. Soc.*, **101**, 5427 (1979).
160. G. Maier, G. Mihm, H. P. Reisenhauer and D. Littman, *Chem. Ber.*, **117**, 2369 (1984).
161. B. J. Helmer and R. West, *Organometallics*, **1**, 1463 (1982).
162. M. J. Fink, M. J. Michalczyk, K. J. Haller, R. West and J. Michl, *Organometallics*, **3**, 793 (1984).
163. M. J. Michalczyk, M. J. Fink, D. J. De Young, C. W. Carlson, K. M. Welsh, R. West and J. Michl, *Silicon, Germanium, Tin, Lead Compd.*, **9**, 75 (1986).
164. R. West, *Pure Appl. Chem.*, **56**, 163 (1984).
165. R. West, M. J. Fink and J. Michl, *Science*, **214**, 1343 (1981).
166. M. Kira, T. Maruyama and H. Sakurai, *Tetrahedron Lett.*, **33**, 243 (1992).
167. M. Kira, T. Maruyama and H. Sakurai, *Heteroatom Chem.*, **5**, 305 (1994).
168. T. Maruyama, M. Kira and H. Sakurai, XXV Silicon Symposium, Los Angeles, April 1992, paper 72P.
169. M. Kira, T. Maruyama and H. Sakurai, *Chem. Lett.*, 1345 (1993).
170. K. E. Banks, Y. Wang and R. T. Conlin, XXV Silicon Symposium, Los Angeles, April 1992, paper 7.
171. H. A. Bent, *J. Chem. Educ.*, **37**, 616 (1960).
172. H. A. Bent, *J. Chem. Phys.*, **33**, 1258 (1960).
173. H. A. Bent, *Chem. Rev.*, **61**, 275 (1961).
174. S. Tsutsui, K. Sakamoto and M. Kira, to appear.
175. Y. Apeloig, M. Karni, R. West and K. Welsh, *J. Am. Chem. Soc.*, **116**, 9719 (1994).
176. Y. Apeloig and M. Karni, *J. Chem. Soc., Chem. Commun.*, 1018 (1985).
177. D. Seyferth and T. F. O. Lim, *J. Am. Chem. Soc.*, **100**, 7074 (1978).
178. K. P. Steele and W. P. Weber, *J. Am. Chem. Soc.*, **102**, 6095 (1980).
179. T.-Y. Y. Gu and W. P. Weber, *J. Am. Chem. Soc.*, **102**, 1641 (1980).
180. D. Tzeng and W. P. Weber, *J. Am. Chem. Soc.*, **102**, 1451 (1980).
181. A. Chihi and W. P. Weber, *Inorg. Chem.*, **20**, 2822 (1981).
182. K. Raghavachari, J. Chandrasekhar and M. J. Frisch, *J. Am. Chem. Soc.*, **104**, 3779 (1982).
183. K. Raghavachari, J. Chandrasekhar, M. S. Gordon and K. J. Dykema, *J. Am. Chem. Soc.*, **106**, 5853 (1984).
184. G. R. Gillette, G. H. Noren and R. West, *Organometallics*, **6**, 2617 (1987).
185. G. R. Gillette, G. H. Noren and R. West, *Organometallics*, **8**, 487 (1989).
186. M.-A. Pearsall and R. West, *J. Am. Chem. Soc.*, **110**, 7228 (1988).
187. W. Ando, K. Hagiwara and A. Sekiguchi, *Organometallics*, **6**, 2270 (1987).
188. W. Ando, A. Sekiguchi, K. Hagiwara, A. Sakakibara and H. Yoshida, *Organometallics*, **7**, 558 (1988).
189. W. Ando, M. Fujita. H. Yoshida and A. Sekiguchi, *J. Am. Chem. Soc.*, **110**, 3310 (1988).
190. S. Zhang and R. T. Conlin, *J. Am. Chem. Soc.*, **113**, 4272 (1991).
191. S. Zhang, P. E. Wagenseller and R. T. Conlin, *J. Am. Chem. Soc.*, **113**, 4278 (1991).
192. P. Boudjouk, U. Samaraweera, R. Sooriyakumaran, J. Chrusciel and K. R. Anderson, *Angew. Chem., Int. Ed. Engl.*, **27**, 1355 (1988).
193. P. Boudjouk, E. Black and R. Kumarathasan, *Organometallics*, **10**, 2095 (1991).
194. D. H. Pae, M. Xiao, M. Y. Chiang and P. P. Gaspar, *J. Am. Chem. Soc.*, **113**, 1281 (1991).
195. Y. Apeloig, in *The Chemistry of Organosilicon Compounds* (Eds. S. Patai and Z. Rappoport), Vol. 1, Chap. 2. Chichester, 1989, p. 57.
196. R. S. Grev, H. F. Schaefer, III and P. P. Gaspar, *J. Am. Chem. Soc.*, **113**, 5638 (1991).
197. H. P. Reisenauer, G. Mihm and G. Maier, *Angew. Chem., Int. Ed. Engl.*, **21**, 854 (1982).
198. C. A. Arrignton, K. A. Klingensmith, R. West and J. Michl, *J. Am. Chem. Soc.*, **106**, 525 (1984).
199. M. Ishikawa, K.-I. Nakagawa, M. Ishiguro, F. Ohi and M. Kumada, *J. Organomet. Chem.*, **152**, 155 (1978).
200. M. Kira, T. Taki and H. Sakurai, *J. Org. Chem.*, **54**, 5647 (1989).
201. P. Boudjouk and L. H. Sommer, *J. Chem. Soc., Chem. Commun.*, 54 (1973).

202. W. J. Leigh, C. J. Bradaric and G. W. Sluggett, *J. Am. Chem. Soc.*, **115**, 5332 (1993).
203. W. J. Leigh and C. J. Bradaric, *J. Am. Chem. Soc.*, **118**, 8971 (1996).
204. H. Tobita, K. Ueno and H. Ogino, *Bull. Chem. Soc. Jpn.*, **61**, 2797 (1988).
205. H. K. Scharma and K. H. Pannell, *Chem. Rev.*, **95**, 1351 (1995) and references cited therein.
206. H. Suzuki, N. Tokitoh and R. Okazaki, *Bull. Chem. Soc. Jpn.*, **68**, 2471 (1995).
207. M. Kira, T. Iwamoto and C. Kabuto, *J. Am. Chem. Soc.*, **118**, 10303 (1996).

CHAPTER 23

Hypervalent silicon compounds

DANIEL KOST and INNA KALIKHMAN

Department of Chemistry, Ben-Gurion University of the Negev, Beer Sheva 84105, Israel
Fax: +972-7-647-2943;   e-mail: kostd@bgumail.bgu.ac.il

I. INTRODUCTION	1340
II. PENTACOORDINATE ANIONIC SILICON COMPOUNDS	1340
A. Fluorosilicates	1340
1. Monosilicates	1340
2. Fluorosilicates with intramolecular exchange of 'bridged' fluorine	1344
3. Zwitterionic fluorosilicates	1349
B. Pentacoordinate Spirosilicate Anions	1351
1. Spirosilicates	1351
2. Zwitterionic organospirosilicates	1357
3. Polynuclear spirosilicates	1367
C. Hydridosilicates	1370
1. Alkoxyhydridosilicates	1370
2. Gas-phase chemistry of hydridosilicates	1372
3. Computational studies of hydridosilicates	1372
III. PENTACOORDINATE NEUTRAL SILICON COMPLEXES	1373
A. Chelates with Nitrogen–Silicon Coordination	1373
1. Synthesis	1373
2. Structure	1377
a. Crystallographic data	1377
b. ^{29}Si NMR spectroscopy	1380
3. Stereodynamics	1382
4. Intramolecular Lewis-base-stabilized low-valency silanes.	1387
B. Chelates with Oxygen–Silicon Coordination	1390
1. Synthesis	1390
a. Amide-type (O–Si) complexes	1390
b. Amide-type bis-(O–Si) complexes	1392
c. Other (O–Si) complexes	1393
2. Crystal structures	1395
3. ^{29}Si NMR	1397
4. Ligand exchange	1403

The chemistry of organic silicon compounds, Vol. 2
Edited by Z. Rappoport and Y. Apeloig   © 1998 John Wiley & Sons Ltd

	C. Other Neutral Pentacoordinate Silicon Compounds	1406
	1. Sulfur coordination	1406
	2. Fluorine coordination: degenerate fluorine migration in a 'Merry-Go-Round' type mechanism	1407
IV.	CATIONIC PENTACOORDINATE COMPLEXES	1408
V.	HEXACOORDINATE IONIC SILICON COMPOUNDS	1412
	A. Fluorosilicates	1412
	B. Ionic (O−Si) Chelates	1415
	C. Bis-catecholato Complexes with N−Si Coordination	1417
VI.	NEUTRAL HEXACOORDINATE SILICON COMPLEXES	1418
	A. Intramolecular Coordination	1418
	1. Synthetic methods	1418
	2. Structure	1422
	3. Stereodynamics	1424
	B. Intermolecular Coordination	1429
VII.	COMPLEXES WITH HIGHER COORDINATION NUMBERS	1430
VIII.	ADDENDUM	1434
IX.	REFERENCES	1436

I. INTRODUCTION

Hypervalent silicon compounds have generated much interest in recent years. This is evident from the numerous papers published every year on this topic, and from the extensive reviews which have appeared in recent years[1−7]. In this review we have attempted to cover only the most recent (five year) literature. Only for subjects for which we felt that the new work would not be fully appreciated without the previous reports did we include earlier citations, or for those subjects which in our judgment, were not sufficiently covered in earlier reviews.

The most recent comprehensive review focused primarily on the reactivity aspects of hypervalent silicon complexes[6], while the latest review covers only silicon–oxygen coordination[7]. The present chapter focuses on synthesis and structure, silicon-29 NMR spectroscopy, and on the nonrigidity of hypervalent silicon compounds and the resulting kinetic and stereochemical studies.

II. PENTACOORDINATE ANIONIC SILICON COMPOUNDS

A. Fluorosilicates

1. Monosilicates

Penta- and hexacoordinate silicate anions containing four and five fluorine atoms have been known for about 30 years, and were first reviewed by Müller ([RSiF$_4$]$^-$, [RSiF$_5$]$^{2-}$, [R$_2$SiF$_4$]$^{2-}$)[8]. More recently pentacoordinate complexes with fewer fluorine atoms have been reported[9−18] ([R$_2$SiF$_3$]$^-$, as well as the following [R$_3$SiF$_2$]$^-$ silicates: [Ph$_3$SiF$_2$]$^{-11}$, [1-NaphPh$_2$SiF$_2$]$^{-12}$, and [Me$_3$SiF$_2$]$^-$)[15]. The general synthetic route to organofluorosilicates involves the reaction of an organofluorosilane with a fluoride donor, such as alkali metal fluoride, ammonium fluoride or tetraalkyl ammonium fluoride, in protic or aprotic solvents[8−16]. The resulting anionic complexes are usually highly hygroscopic and proved difficult to study in detail. A significant improvement was introduced by Damrauer and coworkers, who prepared the anionic silicates with potassium counter ions in the presence of crown ethers[11]. These products proved to be nonhygroscopic, and hence much easier to study.

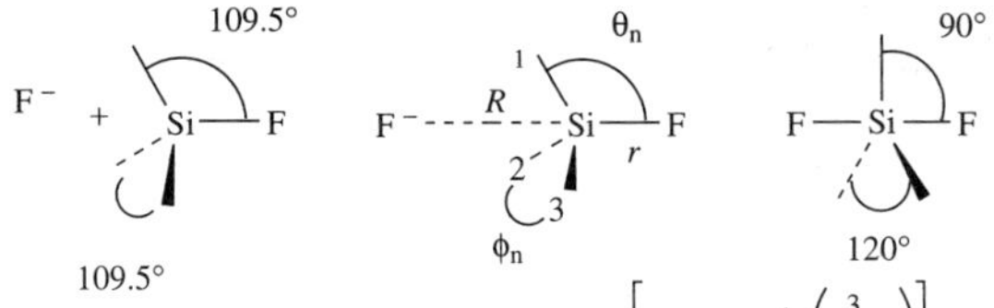

23. Hypervalent silicon compounds

(a) Pentacoordination character (% TBPa) = $\dfrac{\left[109.5° - \dfrac{1}{3}\left(\sum_{n=1}^{3} \theta_n\right)\right]}{(109.5° - 90°)} \times 100$

(b) Pentacoordination character (% TBPa) = $\dfrac{\left[120° - \dfrac{1}{3}\left(\sum_{n=1}^{3} \phi_n\right)\right]}{(120° - 109.5°)} \times 100$

FIGURE 1. Methods for calculating the percent TBP character for fluorosilicates: (a) based on angle θ_n; (b) calculated from angle ϕ_n

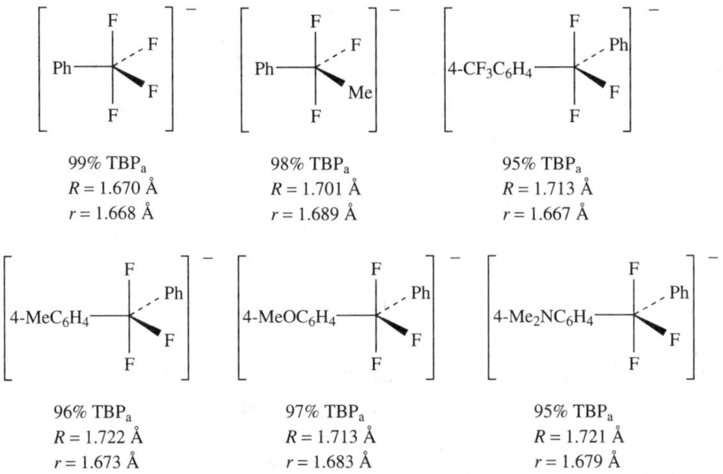

99% TBP$_a$
$R = 1.670$ Å
$r = 1.668$ Å

98% TBP$_a$
$R = 1.701$ Å
$r = 1.689$ Å

95% TBP$_a$
$R = 1.713$ Å
$r = 1.667$ Å

96% TBP$_a$
$R = 1.722$ Å
$r = 1.673$ Å

97% TBP$_a$
$R = 1.713$ Å
$r = 1.683$ Å

95% TBP$_a$
$R = 1.721$ Å
$r = 1.679$ Å

FIGURE 2. Percent TBP character for fluorosilicates where TBP$_a$, R and r are defined in Figure 1 (adapted from Reference 17a)

The geometries of most pentacoordinate fluorosilicates are nearly ideal trigonal bipyramids (TBP) (i.e. the bond angles correspond closely to those of an ideal TBP), with fluorines occupying the apical positions[3,4,10–14,18]. Tamao and coworkers, following an approach developed by Bürgi and Dunitz[19], calculated the TBP character of some fluorosilicates in terms of the correspondence of bond angles to those present in an ideal TBP (Figure 1)[17]. For unconstrained pentacoordinate silicates the TBP percentage was found to be between 95 and 99% (Figure 2). Even in cases of severe steric requirements of the carbon ligands attached to silicon ([Mes$_2$SiF$_3$][K(18-crown-6)][14] and [2,4,6-t-Bu$_3$C$_6$H$_2$SiF$_4$][K(18-crown-6)])[13a], no significant distortion of the basic TBP geometry has been observed. An exception has been reported, in which a fluorine occupies an equatorial position while one of the apical positions is occupied by carbon (**1**)[20]. This is obviously due to the steric strain in the five-membered ring, which can only accommodate the 90° angle between apical and equatorial ligands, rather than the 120° required by two equatorial positions, if a second fluorine were to occupy the apical position.

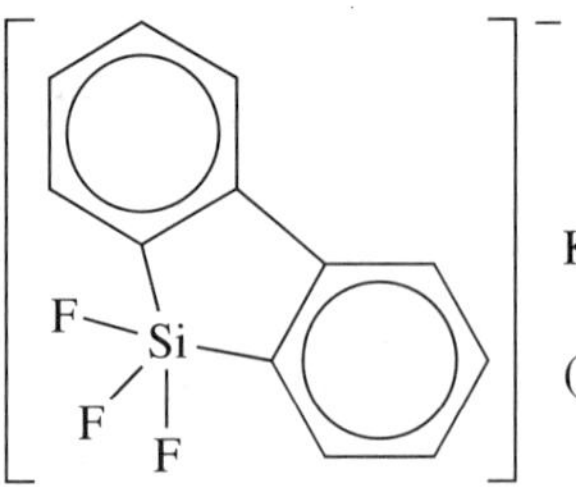

K⁺(18-c-6)

(18-c-6) ≡ (18-crown-6)

(1)

TABLE 1. ^{29}Si NMR spectroscopic data for diorganotrifluorosilicates and organotetrafluorosilicates

Anion	Temperature (°C)	$\delta^{29}Si$ (ppm)	$J(Si-F_{ax})$ (Hz)	$J(Si-F_{eq})$ (Hz)	References
c-(CH$_2$)$_4$SiF$_3$	−68.7	−63.84	255		13b
c-(CH$_2$)$_5$SiF$_3$	−98.2	−76.97	254	213	13b
MePhSiF$_3$	−120	−86.89	250	212	13b
(2-Tol)$_2$SiF$_3$	−92	−91.60	255	218	13b
(Mes)$_2$SiF$_3$	−70	−92.52	262	219	13b
t-BuPhSiF$_3$	−88.2	−94.52	268	222	13b
Ph$_2$SiF$_3$	−107.2	−106.4	252	204	13b
2,2′-BiPhSiF$_3$[a]	−58.2	−91.47	259	220	13b
2,2′-BiBzSiF$_3$[a]	−88.2	−105.9	245	206	13b
4-TolPhSiF$_3$	25	−109.71	237.74		18
(4-MeOC$_6$H$_4$)PhSiF$_3$	25	−110.04	237.82		18
(4-CF$_3$C$_6$H$_4$)PhSiF$_3$	25	−110.44	239.73		18
n-PrSiF$_4$	−78.2	−110.8	225		13a
F$_4$SiCH$_2$CH$_2$SiF$_4$	−98.2	−109.9	227		13a
t-BuSiF$_4$	−58.2	−114.8	237		13a
c-C$_6$H$_{11}$SiF$_4$	−68.7	−114.7	232		13a
BzSiF$_4$	−98.2	−116.4	218		13a
PhSiF$_4$	−58.2	−125.9	210		13a
4-TolSiF$_4$	−62.3	−126.1	209		13a
3-TolSiF$_4$	−58.2	−124.3	214		13a
2-TolSiF$_4$	−58.2	−121.9	216		13a
(4-ClC$_6$H$_4$)SiF$_4$	−58.3	−125.2	213		13a
MesSiF$_4$	21.0	−120.1	219		13a
c-(CH$_2$)$_4$SiF(o-O$_2$C$_6$H$_4$)	25	−52.1	324		20
c-(CH$_2$)$_5$SiF(o-O$_2$C$_6$H$_4$)	25	−68.37	250		20

[a] BiPh = biphenyl; BiBz = bibenzyl

^{29}Si chemical shift has become a powerful diagnostic tool for the silicon coordination number[21]. ^{29}Si chemical shifts and Si−F coupling constants for various fluorosilicates have been collected in Table 1. On average, $\delta(^{29}$Si$)$ for [RSiF$_4$]$^-$ are 51 ppm upfield from the corresponding RSiF$_3$ (average $\delta = -126$ ppm), whereas [R$_2$SiF$_3$]$^-$ (average $\delta = -90$ ppm) are shifted on average 71 ppm upfield from their silane analogs.

Ligand−site exchange is found in all of the fluorosilicates, and monitored primarily by ^{19}F NMR spectroscopy[6,11,13]. The exchange between equatorial and apical positions is very rapid in the case of [RSiF$_4$]$^-$ silicates[13a] and leads to the observation of only one average signal for all fluorine atoms in the ^{19}F NMR spectrum, at temperatures as low as can be practically reached in the solvents used. Steric crowding in [2,4,6-t-Bu$_3$C$_6$H$_2$SiF$_4$][K(18-crown-6)] has been reported to slow down this exchange, to the extent that separate apical and equatorial fluorines can be seen at low temperature (205 K), as well as their eventual coalescence at higher temperature (270 K, $\Delta G^\ddagger = 12.8$ kcal mol^{-1})[11,13a]. The rate of fluorine−site exchange diminishes with the decrease in the number of fluorine atoms. Thus, in trifluorosilicates the isochrony of apical and equatorial fluorine ligands can be removed and the exchange effectively 'frozen', relative to the NMR time scale, with activation barriers ranging between 9−14 kcal mol^{-1}[13b]. By contrast, in difluorosilicates no fluorine−site exchange is observed even at elevated temperatures, as a result of the higher stability of complexes with fluorine atoms in apical positions: replacement of an apical fluorine by a carbon ligand is too costly in terms of energy, and is not observed[11a].

The convenient range of exchange barriers found in trifluorosilicates permitted a more detailed polar substituent effect study[6,18]. A series of diaryltrifluorosilicates with different substituents on one of the phenyl groups (**2**) were prepared, and the free energy barriers for fluorine exchange were measured. The barriers were found to decrease linearly with the electron-withdrawing power of the substituent, as expressed by the σ^+ substituent constant, with a reaction constant $\rho = +2.0$[18]. This result is consistent with the general observation (mentioned above) that the barrier for fluorine interconversion decreases with increasing number of fluorine ligands.

(**2a**) X = 4-CF$_3$; (**2b**) X = 4-Cl; (**2c**) X = H; (**2d**) X = 4-Me; (**2e**) X = 4-OMe; (**2f**) X = 4-NMe$_2$; (**2g**) X = 4-SiF$_2$Ph; (**2h**) X = 3-SiF$_2$Ph

An interesting case of simultaneous exchange of two fluorides has been reported recently[22]. In compounds **2g** and **2h** there is one tetra- and one pentacoordinate silicon. Fluoride transfer from the pentacoordinate silicate to the silane was found to be intermolecular, with a large negative entropy of activation. It was concluded that exchange takes place simultaneously between two complexes, through a cyclophane-like transition state (equation 1).

$$2\left[\underset{PhF_2Si}{}\hspace{-0.5em}\bigcirc\hspace{-0.5em}-SiF_3Ph\right]^- \rightleftharpoons \left[\text{(bridged dimer structure)}\right]^{2-} \quad (1)$$

(2g), (2h)

$$2\left[\underset{PhF_3Si}{}\hspace{-0.5em}\bigcirc\hspace{-0.5em}-SiF_2Ph\right]^-$$

(2g), (2h)

The reactivity of perfluorophenyltetrafluorosilicates ($M^+[C_6F_5SiF_4]^-$, M = K, Cs, Me$_4$N) toward electrophiles has been studied recently[23]. These compounds were found to be highly reactive in the presence of Br$_2$, ICN, IF$_5$, XeF$_2$, NOBF$_4$ which cause immediate C—Si bond cleavage (equation 2).

$$Me_4N^+[C_6F_5SiF_4]^- + Br_2 \xrightarrow{\text{Diglyme}} C_6F_5Br + Me_4N^+Br^- + SiF_4 \quad (2)$$

2. Fluorosilicates with intramolecular exchange of 'bridged' fluorine

An interesting case of fluorine atom bridging between two silicon centers has been reported in fluorosilicates[17]. Compounds **3** and **4**, formally possessing one penta- and one tetracoordinate silicon each, were prepared by reacting KF and 18-crown-6 with the corresponding bis-silanes. The crystal structures show that a central fluorine bridges between the two silicon atoms such that both silicons are essentially pentacoordinate with a distorted TBP geometry. The bridging fluorine and another fluorine on each silicon occupy the apical positions. The bridging fluorine is not symmetrically positioned between the silicons in any of the complexes, not even in **3a**, in which the ligands on the two silicon atoms are equivalent. The authors compared the Si—F bond lengths and calculated the percent TBP in the bridged complexes according to Figure 1. The results given in Table 2 confirm the basic TBP character of the bis silicates, although significant deviations are observed relative to the unconstrained monosilicates.

The bis-silicates **3** and **4** provide an elegant model for the gradual change in geometry from tetrahedral to TBP, associated with the S_N2 reaction on silicon. A linear correlation was found between the calculated percent TBP and the Si—F$_{br}$ (bridging fluorine) and Si—F$_a$ (apical fluorine) bond lengths, as might be expected along the S_N2 reaction coordinate. The change of Si—F$_{br}$ bond length is much greater than that of Si—F$_a$[17].

^{29}Si NMR chemical shifts and Si—F coupling constants were used for a structural analysis of the bis-complexes in solution. At 20 °C all complexes show rapid exchange

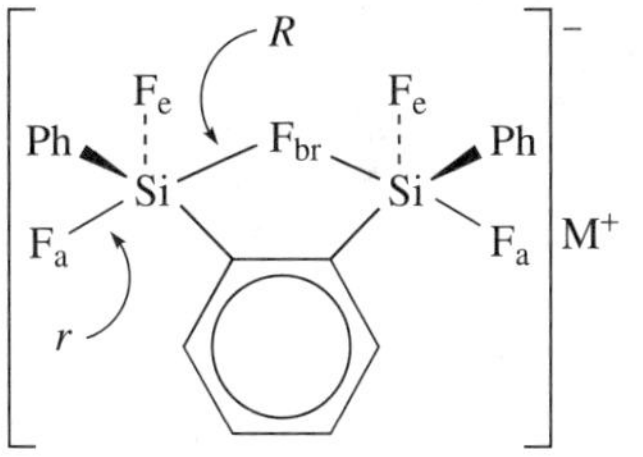

23. Hypervalent silicon compounds

(3a)

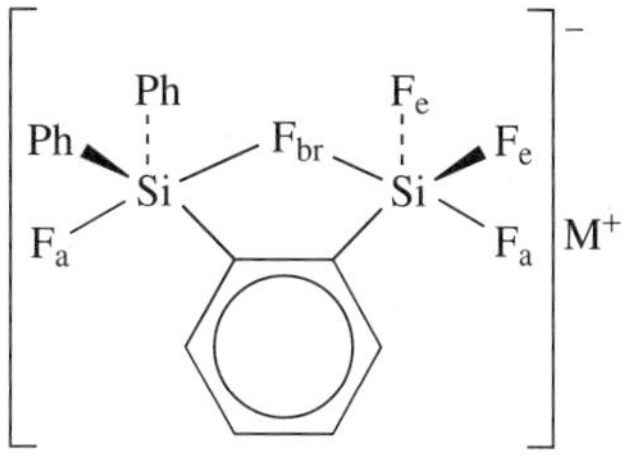

(3b)

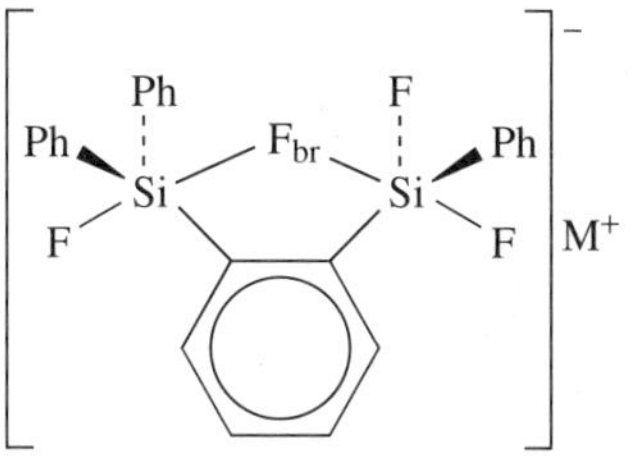

$M^+ = K^+$ (18-c-6)

(4)

TABLE 2. Bond lengths, % TBP and deviations $(\Delta Si)^a$ in **3** and **4**[17a]

	Anion of **3a**		Anion of **3b**		Anion of **4**	
Position	Si2	Si1	Si2	Si1	Si2	Si1
$Si-F_{ap}$ (Å)						
R	2.065	1.898	2.369	1.700	2.090	1.805
r	1.638	1.657	1.639	1.667	1.672	1.669
$Si-F_{eq}$ (Å)	1.601	1.616		1.598		1.624
$Si-C_{eq}$ (Å)	1.871	1.879	1.870		1.860	1.889
			1.878		1.887	
ΔSi (Å)	0.205	0.131	0.314	0.040	0.207	0.078
%TBP	66	78	50	93	68	88

[a]Displacement of the central Si from the plane defined by the three equatorial ligands.

of fluorine sites. Three significant features were noted: (1) Both silicons give rise to one sextet, indicating the equivalence and fast exchange of all fluorines. (2) The ^{29}Si chemical shift for **3a** (−90.03 ppm) is intermediate between similar tetra- (Ph_2SiF_2, −29.00 ppm) and pentacoordinate ($[Ph_2SiF_3]^-$, −109.55 ppm) compounds. The closer resemblance of $\delta(^{29}Si)$ to the pentacoordinate complex is consistent with the average calculated TBP character for **3a** of 66 and 78%, respectively, obtained from the solid state structure. (3) The Si−F coupling constant in **3a** (134.74 Hz) is much smaller than in either tetra- [$^1J(Si-F) = 291$ Hz] or pentacoordinate [$^1J(Si-F) = 238$ Hz] structures. It can be obtained as an average (130 Hz) of two tetracoordinate fluorines (2 × 291 Hz), three pentacoordinate (3 × 238 Hz) and five distant fluorines representing $^4J(Si-F)$ between silicon and the fluorines attached to the neighboring silicon (5 × 0 Hz). This supports rapid exchange between tetra- and pentacoordinate silicons[17].

(a) F_{br} – exchange

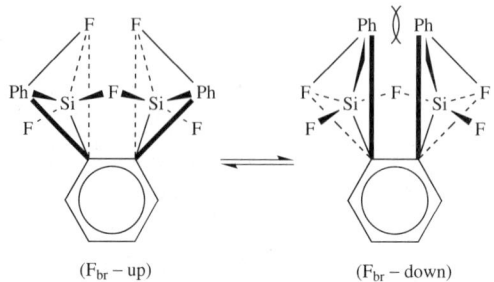

(b) Flipping

(F_{br} – up)   (F_{br} – down)

(c) Si——C rotation

(d) Pseudorotation

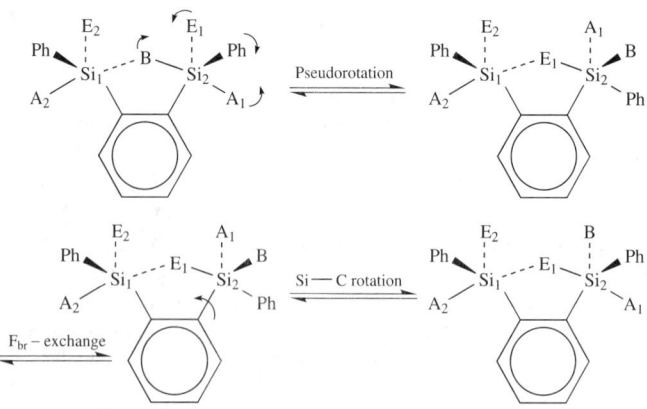

FIGURE 3. Stereodynamics of **3a**: four distinct rate processes were observed[17a]; F_{br}, F_a and F_e are abbreviated B, A and E, respectively

The variable-temperature ^{19}F NMR spectra of **3** and **4** were carefully analyzed and interpreted in terms of four different rate processes (Figure 3). For **3a** the fastest rate process was the exchange between penta- and tetracoordinate silicon, i.e. the interchange of bond lengths between the bridging fluorine and each silicon atom ($F_{br}-Si_1 \leftrightarrows F_{br}-Si_2$, process a in Figure 3). The spectra were consistent with symmetrical silicon complexes down to the lowest temperature measured, $-90\,°C$. While this result does not distinguish a truly symmetrical structure from a rapidly interconverting one, the fact that significantly different $F_{br}-Si_1$ and $F_{br}-Si_2$ bond lengths were found in the crystal structure analysis (Table 2) strongly supports rapid dynamic equilibration in solution.

The second process to become apparent at the NMR time scale as the temperature is raised (Figure 3, process b), is the 'flipping' of the bridging fluorine from one side of the aromatic plane to the other, with 'puckering' of the five-membered heterocycle. This results in interconversion and coalescence of signals due to two diastereomeric conformations, in which the phenyl rings are close or farther away from each other. Further warming of the sample resulted in asymmetric coalescence of signals due to apical and equatorial fluorines, and this was interpreted in terms of fast rotation about the Si—C bonds. Finally, at $-16\,°C$, all the fluorines were completely scrambled, indicating rapid Berry-type pseudorotation. The coalescence temperature for this process falls within the range of known barriers for pseudorotation in $Ph_2SiF_3^-$ and similar silicates[13b,18].

Recently, the intensity of fluoride binding in **3** and **4** was estimated using fluoride-transfer equilibrium reactions[17c]. Surprisingly, no exchange at the NMR time scale was found between the bis-silane $[1,2-(PhSiF_2)_2C_6H_4]$ and **3a**. The equilibrium in this reaction was substantially shifted to the bis-silicate (**3a**) side, and did not permit measurement of K_{eq}. The relative fluoride-binding strength was therefore estimated from the equilibrium reactions shown in Figure 4, for which K_{eq} lies within the measureable range. It is interesting to note that the 'Lewis acidity' toward fluorine of the *ortho*-bis-silanes increases sharply with increasing number of fluorine atoms attached to silicon, i.e. the bridging fluoride is bound more strongly in the more highly fluorinated complexes. The resulting fluoride ion binding constants are shown in Figure 4[17c].

Recently, a case of fluorine exchange among *three* silicon sites has been reported by Corriu and coworkers[24]: when 1,1,3,3,5,5-hexafluoro-1,3,5-trisilacyclohexane was treated with KF[18-crown-6], a fluoride was incorporated into the molecule to produce **5**. The crystal structure of **5** showed that the additional F is closer to one of the Si atoms than to the others. In solution, however, the three silicons were equivalent and gave rise to a sharp singlet at room temperature. The chemical shift for this signal ($\delta^{29}Si = -25.56$ ppm) was intermediate between those for tetracoordinate and purely pentacoordinate analogous compounds. When the temperature was lowered down to 183 K the silicons remained equivalent, however the signal was split to an octet, due to coupling to seven equivalent and hence rapidly exchanging fluorine ligands. Interestingly, the solid state ^{29}Si NMR spectrum of **5** indicated rapid exchange even in the solid, since a broad Si singlet was observed[24].

Another case of bridging was also reported recently in a neutral poly-silicon complex, and is discussed in the section on neutral pentacoordinate complexes (Section III.C.2).

A different class of bis-silicate complexes has also been termed 'bridged'[25], or also 'dinuclear λ^5Si, λ^5Si'-silicon complexes' (see Section II.B.3). These are bis-pentacoordinate silicon anions, connected to each other by a carbon chain or aromatic ring (**6**[13a], **7**[26], **8**[27]). Unlike the previous bridged complexes, in which the bridging atom was directly involved in coordination at silicon, in the present cases the bridges merely connect two silicate anions by covalent bonding.

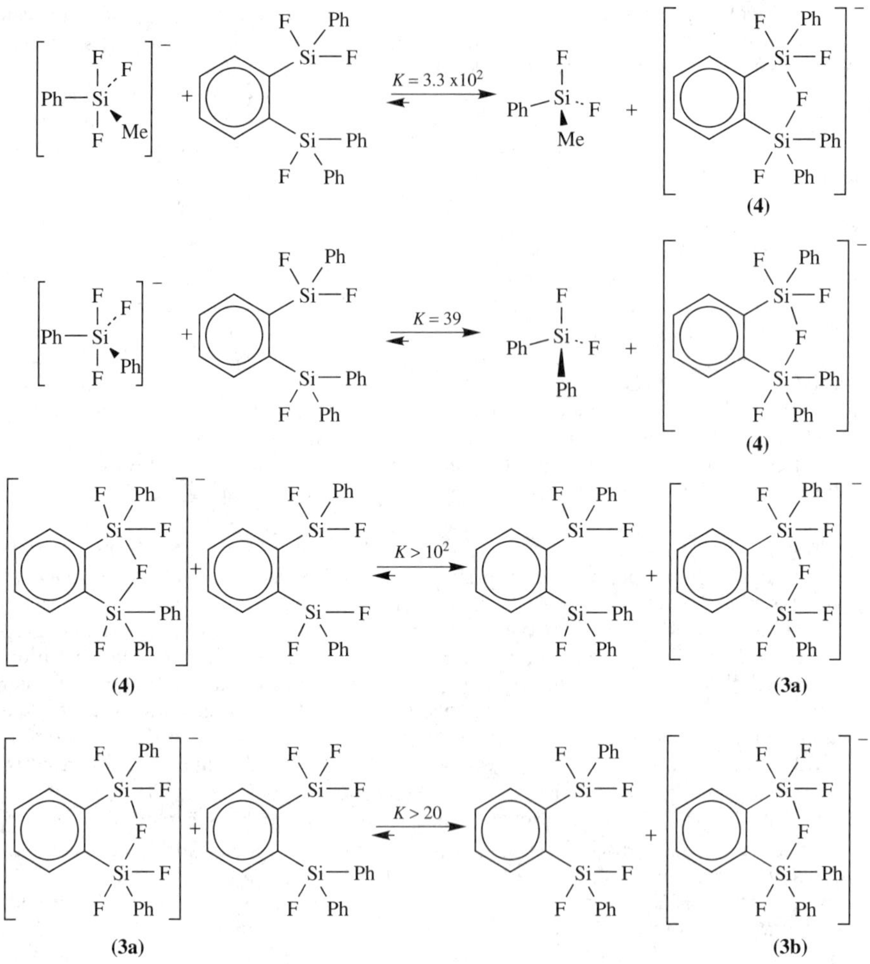

FIGURE 4. Fluoride binding estimated by intermolecular fluoride exchange equilibrium constants[17c]

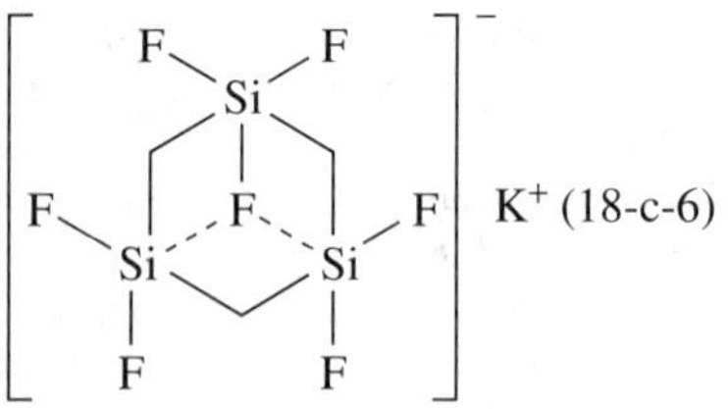

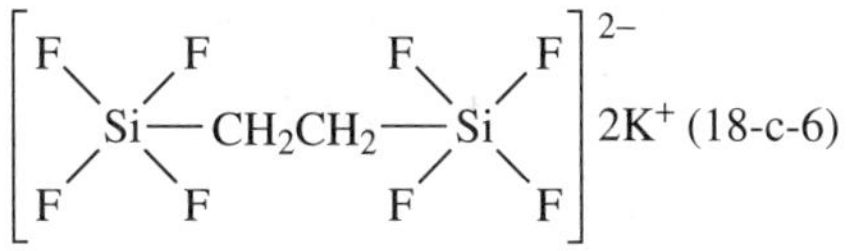

$$\left[\begin{array}{c} F \diagdown \diagup F \\ Si-CH_2CH_2-Si \\ F \diagup \diagdown F \end{array} \begin{array}{c} F \\ F \end{array} \right]^{2-} 2K^+ \text{(18-c-6)}$$

(6)

$$\left[PhF_3Si-\bigcirc-SiF_3Ph \right]^{2-} 2M^+$$

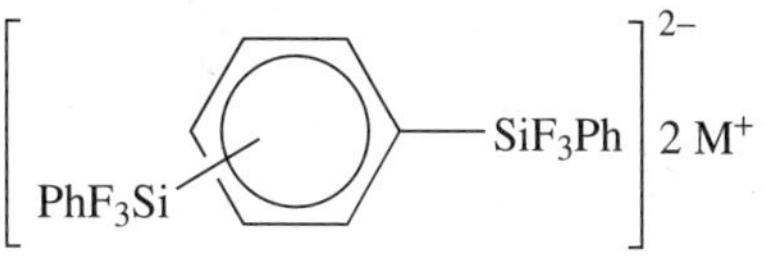

(7a) *para*-isomer; (7b) *meta*-isomer
M = K$^+$ (18-c-6), *n*-Bu$_4$N$^+$

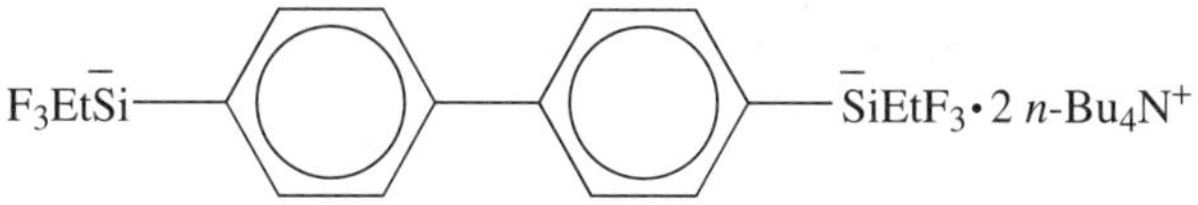

$$F_3Et\bar{S}i-\bigcirc-\bigcirc-\bar{S}iEtF_3 \cdot 2\ n\text{-Bu}_4N^+$$

(8)

3. Zwitterionic fluorosilicates

A new direction in the chemistry of hypervalent silicon compounds has recently been developed by Tacke and coworkers: the chemistry of zwitterionic organosilicates[28−31]. In these compounds the silicon is formally negative, with a cationic nitrogen attached to it by an alkyl chain. One of the groups of products of this general class is the pentacoordinate zwitterionic fluorosilicates[31].

The synthesis is accomplished by HF treatment of (aminoalkyl)polyalkoxysilanes and involves the substitution of alkoxy groups by fluorines (equations 3 and 4)[29].

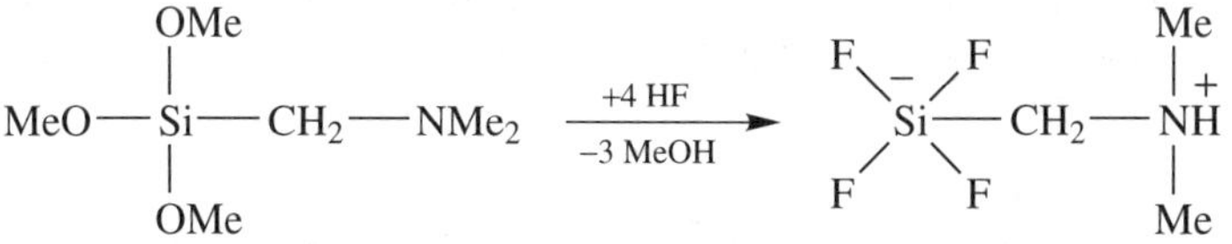

$$\underset{\substack{|\\ OMe}}{\overset{\substack{OMe\\|}}{MeO-Si-CH_2-NMe_2}} \xrightarrow[-3\ MeOH]{+4\ HF} \underset{F \diagup \diagdown F}{\overset{F \diagdown \diagup F}{Si^-}}-CH_2-\overset{\substack{Me\\|\\+\\|\\Me}}{NH} \quad (3)$$

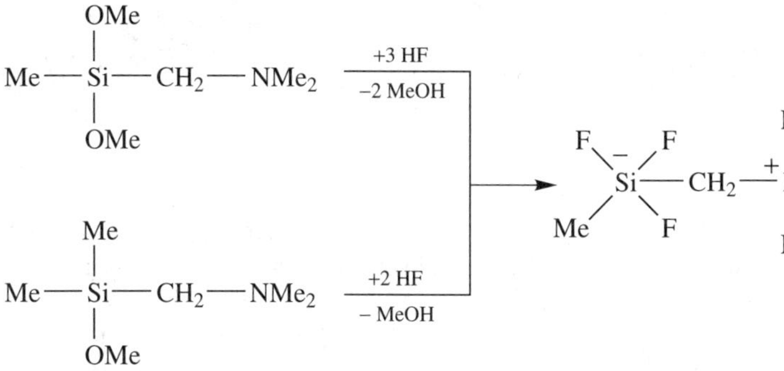

$$\underset{\substack{|\\ OMe}}{\overset{\substack{OMe\\|}}{Me-Si-CH_2-NMe_2}} \xrightarrow[-2\ MeOH]{+3\ HF}$$

$$\underset{Me \diagup \diagdown F}{\overset{F \diagdown \diagup F}{Si^-}}-CH_2-\overset{\substack{Me\\|\\+\\|\\Me}}{NH} \quad (4)$$

$$\underset{\substack{|\\ OMe}}{\overset{\substack{Me\\|}}{Me-Si-CH_2-NMe_2}} \xrightarrow[-MeOH]{+2\ HF}$$

By this method also the zwitterionic silicates 9–15 were obtained

The geometry at silicon in these compounds is TBP, like in anionic and neutral pentacoordinate silicon complexes. A typical crystal structure is shown in Figure 5 for compound 9. This structure apparently also exists in solution (CD$_3$CN), as the ^{29}Si chemical shift for 9 in this solvent (−122.9 ppm) compares well with the solid state CP-MAS shift of −121.0 pm[28,31].

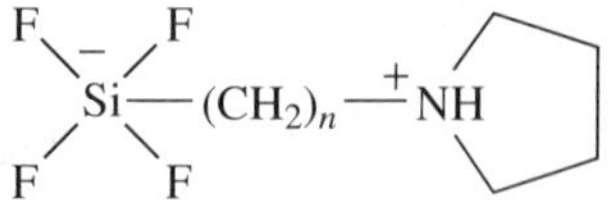

(9) n = 1
(10) n = 2

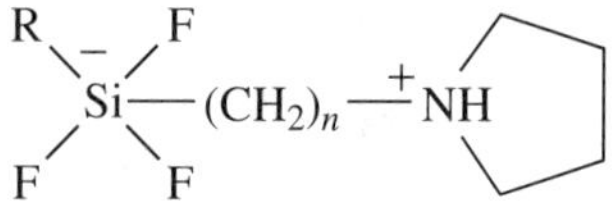

(11) R = Me
(12) R = Ph
(13) R = t-Bu

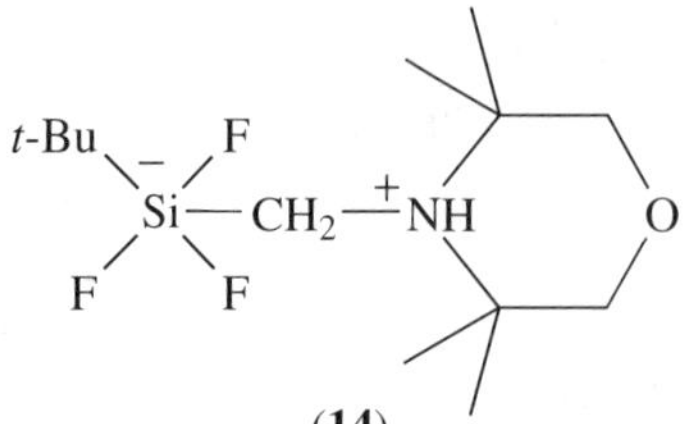

(14)

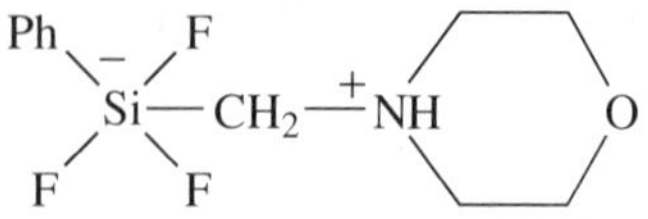

(14a)

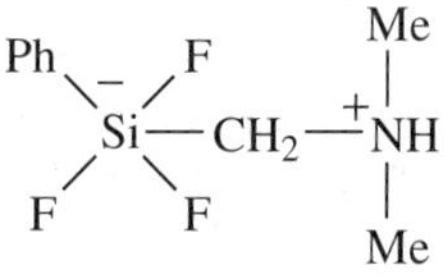

(15)

An interesting reaction found in zwitterionic fluorosilicates is their transformation from acyclic to monocyclic complexes (equation 5), with oxygens replacing fluorine ligands[30a].

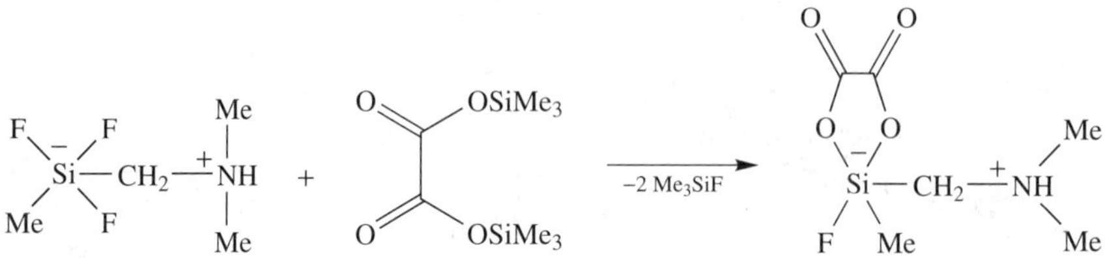

(5)

The zwitterionic trifluorosilicate **14** shows two ligand exchange processes in the ^{19}F NMR spectra[30b]. The first ($\Delta H^{\ddagger} = 9.8$ kcal mol^{-1}) involves interchange of the two axial fluoro ligands, in what must be a rotation of the piperidiniomethyl group around the Si–C bond. The other process ($\Delta H^{\ddagger} = 10.6$ kcal mol^{-1}) averages all three fluoro ligands and was assigned to pseudorotation. The DNMR results were simulated and studied further by high-level *ab initio* SCF-MO calculations of model compounds, $F_4\overset{-}{Si}CH_2\overset{+}{N}H_3$ and $Me\overset{-}{Si}F_3CH_2\overset{+}{N}H_3$. The stationary points (ground and transition states) along the two reaction coordinates: pseudorotation and Si–C bond rotation, were calculated. The calculated barriers for the two processes for the first model were found to be of comparable magnitudes: 5.0 and 5.8 kcal mol^{-1}. For the second (trifluoro) model the calculations resulted in a clear torsional reaction coordinate, with a barrier $\Delta H^{\ddagger}$ of 6.1 kcal mol^{-1}.

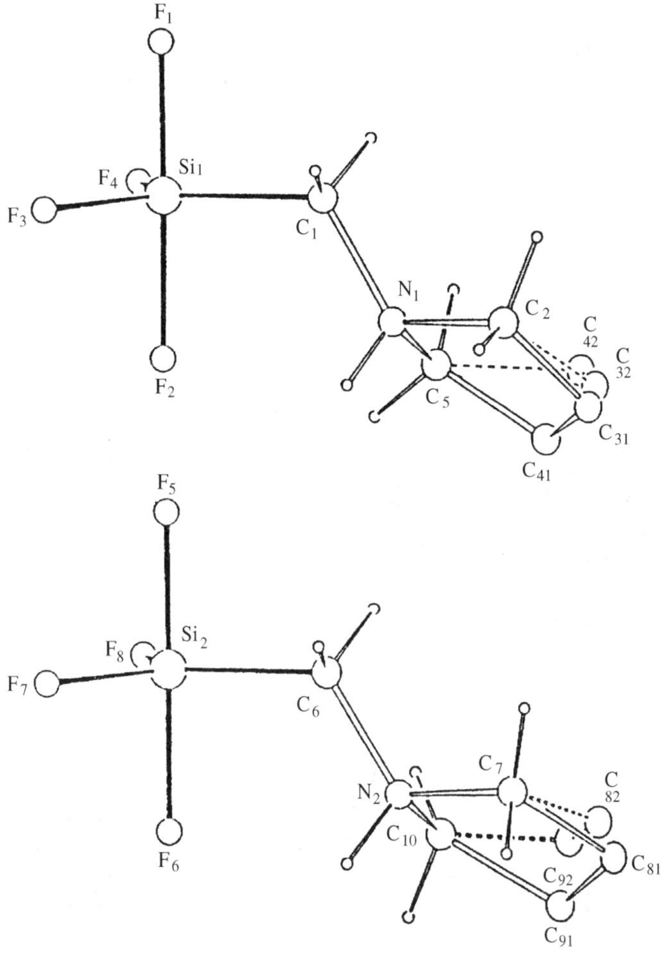

FIGURE 5. X-ray crystallographic structure of **9** (zwitterionic fluorosilicate). Reproduced from Reference 31a by permission of VCH Verlagsgesellschaft

For the exchange of all five ligand atoms bound to silicon two alternative pathways were calculated with six local minima, and comparable activation enthalpies[30b].

B. Pentacoordinate Spirosilicate Anions

1. Spirosilicates

Apart from fluoro ligands at hypervalent silicon complexes, also oxo and aza ligands are known to support penta- and hexacoordination in silicon compounds[1-8]. Among the various pentacoordinate oxo complexes, those with bidentate ligands (1,2-diols, aliphatic or aromatic, as well as α-hydroxycarboxylic acids) are most readily prepared, and are thermodynamically more stable than their acyclic analogs[32]. Two bidentate oxo ligands at the silicon center form a complex with a spiro arrangement around silicon.

Several synthetic pathways for the preparation of spirosilicates have been reported:

(1) From a tri- or tetraalkoxysilane by the reaction with 1,2-diols (catechols and glycols) in the presence of a tertiary amine (for example, equation 6)[32b].

$$RSi(OEt)_3 + 2 \text{ catechol} \xrightarrow[-3EtOH]{Et_3N} [R-Si(O_2C_6H_4)_2]^- Et_3NH^+ \quad (6)$$

(2) Likewise, catechol reacts with a triaminosilane (equation 7)[33].

$$c\text{-}C_6H_{11}SiCl_3 \xrightarrow[-3 Me_2NH_2^+Cl^-]{6 Me_2NH} (Me_2N)_3SiC_6H_{11}\text{-}c \xrightarrow[-2 Me_2NH]{2 \text{ catechol}} [c\text{-}C_6H_{11}Si(O_2C_6H_4)_2]^- Me_2NH_2^+ \quad (7)$$

(3) By the coordination of an anion to a spirosilane, either in the presence of a tertiary amine (equation 8), or with potassium alkoxide in the presence of 18-crown-6 (equation 9)[34].

$$Si(O_2C_2Me_4)_2 + MeOH + Et_3N \longrightarrow [MeO-Si(O_2C_2Me_4)_2]^- Et_3NH^+ \quad (8)$$

$$Si(OCMe_2CMe_2O)_2 + K^+OR^- + 18\text{-}c\text{-}6 \longrightarrow [(OCMe_2CMe_2O)_2SiOR] [K(18\text{-}c\text{-}6)]$$
$$R = Et, i\text{-}Pr \quad (9)$$

(4) The reaction of polyalkoxysilicate anion with pinacol (equation 10). When catechol was used instead of pinacol, a hexacoordinate spirosilicate was formed.

$$[Si(OR)_5] [K(18\text{-}c\text{-}6)] + 2HOCMe_2CMe_2OH \longrightarrow [(Me_4C_2O_2)_2SiOR] [K(18\text{-}c\text{-}6)] + 4ROH$$
$$R = Et, i\text{-}Pr \quad (10)$$

(5) The dilithio derivatives of catechol and hexafluorocumyl alcohol react with organotrichlorosilanes to give spirosilicates (equations 11 and 12)[34,35].

$$RSiCl_3 + 2\ \text{catechol(OH)}_2 \longrightarrow [R-Si(O_2C_6H_4)_2]^- Li^+ + 3LiCl \quad (11)$$

$$[\text{Li-C}_6H_4\text{-C(CF}_3)_2\text{-OLi}] \xrightarrow[-3\text{LiCl}]{Y\text{SiCl}_3} [(C_6H_4(CF_3)_2CO)_2SiY] \xrightarrow[\text{Et}_4N^+Br^-]{-\text{LiBr}} [(C_6H_4(CF_3)_2CO)_2SiY]^- Et_4N^+ \quad (12)$$

$$Y = c\text{-}C_6H_{11}$$

(6) The most intriguing synthetic approach is the conversion of inorganic silica to spirosilicate by the reaction with a 1,2-diol and alkali hydroxide (equation 13)[36].

$$SiO_2 + MOH + \text{HO-diol-OH} \xrightarrow[-3H_2O]{N_2, \Delta} [\text{bis-spirosilicate}]^{2-} 2M^+ \xrightarrow{\text{MeOH}} [\text{Si-OMe}]^- M^+ + [\text{Si-O-CH}_2\text{CH}_2\text{OH}]^- M^+ \quad (13)$$

$$M = Li, Na, K, Cs$$

The structures of spirosilicates were found to range between TBP and square pyramid (SP), depending on the electronegativities of the ligands and the possible formation of hydrogen bonds with the counter ions[4]. Highly electronegative monodentate ligands such as fluoro, or electronegative substituents attached to the bidentate ligand (such as in tetrachlorocatechol), drive the structure toward SP geometry. The same effect results also from hydrogen bonding to the counter ion, such as Et_3NH^+, as opposed to a more TBP-like structure in the presence of K^+(18-crown-6) without hydrogen bonds. The geometries, based on X-ray structure determinations, were analyzed in terms of the percent deviation from TBP toward the SP geometry. Great structural variation was reported, between 9% deviation from ideal TBP to a nearly perfect SP, with a calculated 97% deviation from TBP. This structural variation parallels the progress along the Berry pseudorotation reaction coordinate (in which SP is the assumed transition state), and supports the facility of such a transformation.

More complete discussions with comprehensive compilations of published results were reviewed previously, and are hence omitted from this chapter[3−7].

Anionic five-coordinated silicates have been prepared also with oxygen-containing six- and seven-membered chelate rings **(16–18)**, using suitable aromatic diols, 1,8-dihydroxynaphthalene and 2,2'-dihydroxybiphenyl, respectively[37]. However, spirosilicates

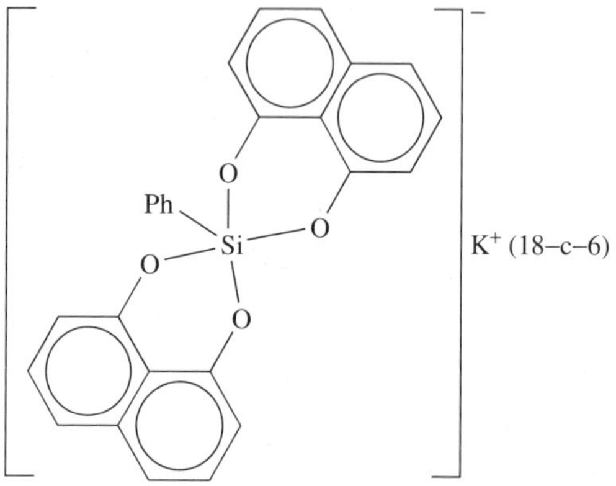

(16) K^+ (18-c-6)

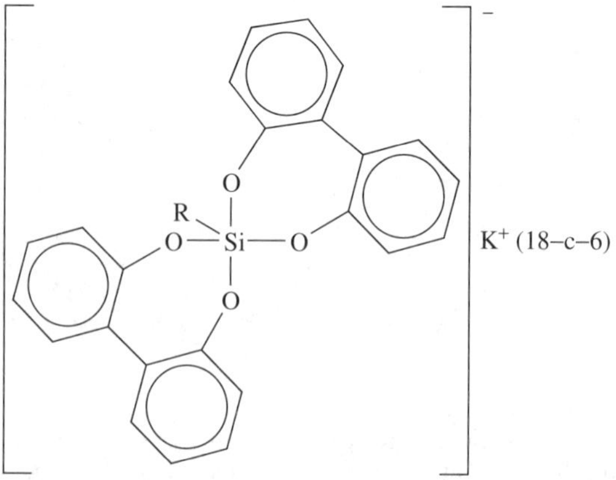

(17) R = OBu-t
(18) R = F

K^+ (18-c-6)

17 and **18**, with the seven-membered chelate ring, were found to be relatively unstable, and slowly decomposed in solution to tetracoordinate silanes and the free or anionic diols[37].

The geometry of **16** around silicon, as revealed by X-ray crystallography, was found similar to that of **19**, the five-membered-ring analog, i.e. both have nearly the same percent distortion from TBP to SP (32.5% and 29.5%, respectively)[4,37].

$$\left[\begin{array}{c}\text{structure}\end{array}\right]^{-} M^{+}$$

(**19**) M = Me$_4$N
(**21**) M = Et$_3$NH
(**22**) M = K(18-c-6)

Interesting structural information about the TBP–SP problem can be obtained from bis-(bidentate ligand)silicates with *unsymmetrically* substituted catechols as ligands[38]. These compounds can, in principle, have five different geometries, shown schematically in Figure 6. However, since single isomers are found in the solution NMR spectra (^{1}H, ^{13}C and ^{29}Si) for the complexes derived from symmetrical catechols, it was concluded that no TBP ⇌ SP isomerism can be seen[38]. In all the unsymmetrical complexes only two isomers could be observed in the different NMR spectra, suggesting that either the SP geometry is the actual structure, or that the two *trans* TBP isomers are rapidly interconverting, presumably through the very close SP transition state or intermediate.

Intermolecular ligand exchange in a solution of two symmetrical complexes **22** and **30** (see Table 3 for structures) produced a mixed complex, with two different catechol ligands, as was evident from the appearance of a new signal in the ^{29}Si and ^{1}H and ^{13}C NMR spectra. When two complexes of unsymmetrical catechols, **20** and **28**, were mixed in solution, two new signals due to the mixed complex were observed. In this case two *trans* and two *cis* isomers are possible in the TBP geometry, and yet only two isomers were observed. Again, this can either be interpreted in terms of an SP structure, with only two possible isomers, or in terms of a large difference in thermodynamic stability of the

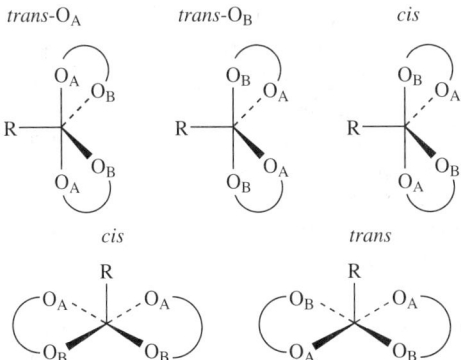

FIGURE 6. Geometrical isomers of pentacoordinate bis(bidentate ligand)silicon complexes

solution isomers preventing the observation of the minor isomers; or it may be interpreted in terms of fast interconversion between the *trans* isomers and between the *cis* isomers, respectively, in a TBP structure[38].

A compilation of ^{29}Si chemical shifts for bis-spirosilicates and related acyclic oxosilicates is given in Table 3. It can be seen that δ^{29}Si is primarily dependent on the monodentate ligand attached directly to silicon, in a manner similar to that found in neutral silanes[40].

No simple correlation can be found between the solid state calculated percent (TBP → SP) deviation and the ^{29}Si chemical shifts in solution. Table 4 demonstrates this analysis for complexes sharing the same ligand environment at silicon, four oxygens and a phenyl group: the crystal structures of four closely related spirosilicates (**19, 21, 49, 50**) were

TABLE 3. ^{29}Si chemical shift in spirosilicates

Number	Compound	$\delta^{29}Si$ (ppm)	References
20	$\{[3,5-(t\text{-Bu})_2-C_6H_2O_2]_2SiPh\}^-$ [K(18-c-6)]$^+$	−88.06, −87.31	38
21	$[(C_6H_4O_2)_2SiPh]^-$[NEt$_3$H]$^+$	−87.53	38
22	$[(C_6H_4O_2)_2SiPh]^-$[K(18-c-6)]$^+$	−87.51	38
23	$[(2,3-C_{10}H_6O_2)_2SiPh]^-$[NBu$_4$]$^+$	−87.14	38
24	$[(2,3-C_{10}H_6O_2)_2SiPh]^-$[NEt$_3$H]$^+$	−87.11	38
25	$[(4-t\text{-Bu}-C_6H_3O_2)_2SiPh]^-$[K(18-c-6)]$^+$	−87.07	38
26	$[(Br_4-C_6O_2)_2SiPh]^-$[K(18-c-6)]$^+$	−86.03	38
27	$[(4\text{-CHO}-C_6H_3O_2)_2SiPh]^-$[K(18-c-6)]$^+$	−85.12	38
28	$[(3\text{-CHO}-C_6H_3O_2)_2SiPh]^-$[K(18-c-6)]$^+$	−85.12, −84.90	38
29	$[(4,5\text{-Cl}_2-C_6H_2O_2)_2SiPh]^-$[K(18-c-6)]$^+$	−84.23	38
30	$[(Cl_4-C_6O_2)_2SiPh]^-$[K(18-c-6)]$^+$	−83.94	38
31	$[(4\text{-NO}_2-C_6H_3O_2)_2SiPh]^-$[NEt$_3$H]$^+$	−83.41	38
32	$[(2,3-C_{10}H_6O_2)_2Si-Pr\text{-}n]^-$[K(18-c-6)]$^+$	−74.83	38
33	$[(C_6H_4O_2)_2SiEt]^-$[K(18-c-6)]$^+$	−74.77	38
34	$[(C_6H_4O_2)_2SiMe]^-$[K(18-c-6)]$^+$	−74.32	38
35	$[(2,3-C_{10}H_6O_2)_2SiEt]^-$[K(18-c-6)]$^+$	−74.13	38
36	$[(4\text{-CHO}-C_6H_3O_2)_2SiMe]^-$[K(18-c-6)]$^+$	−71.45, −71.33	38
37	$[(4\text{-NO}_2-C_6H_3O_2)_2SiMe]^-$[K(18-c-6)]$^+$	−69.41, −69.25	38
38	$[(4\text{-NO}_2-C_6H_3O_2)_2SiEt]^-$[K(18-c-6)]$^+$	−68.88, −68.77	38
39	$[(3,5\text{-}(NO_2)_2-C_6H_2O_2)_2SiMe]^-$[NEt$_3$H]$^+$	−67.19, −65.33	38
40	$[(4,5\text{-}(NO_2)_2-C_6H_2O_2)_2SiMe]^-$[NEt$_3$H]$^+$	−66.81	38
41	$[(2,3-C_{10}H_6O_2)_2SiMe]^-$[NEt$_3$H]$^+$	−73.68	38
16	$[(1,8-C_{10}H_6O_2)_2SiPh]^-$[K(18-c-6)]$^+$	−132.35	37
42	$[(MeO)_4SiPh]^-$[K(18-c-6)]$^+$	−112.4	34
43	$[(EtO)_4SiPh]^-$[K(18-c-6)]$^+$	−117.3	34
44	$[(CF_3CH_2O)_4SiPh]^-$[K(18-c-6)]$^+$	−120.1	34
45	$[(EtO)_5Si]^-$[K(18-c-6)]$^+$	−131.1	34
46	$[(4\text{-Me}-C_6H_3O)_4SiPh]^-$[K(18-c-6)]$^+$	−129.7	34
47	$[(MeO)_3SiPh_2]^-$[K(18-c-6)]$^+$	−100.2	39
48	$[(MeO)_4SiPh]^-$[K(18-c-6)]$^+$	−114.1	39
51g	$[(1,2-C_6H_4C(CF_3)_2O)_2SiCN]^-$[NEt$_4$]$^+$	−91.54	41

TABLE 4. Comparison of solution ^{29}Si chemical shifts and percent deviation (TBP → SP) for tetraoxosilicates

Anion	%(TBP → SP)	Compounds (cation)	References	$\delta^{29}Si$ (ppm)	Compounds (Cation)	References
$(C_6H_4O_2)_2SiPh$	29.5	**19** (Me$_4$N)	42	−87.51	**22** (K,18-c-6)	38
$(C_6H_4O_2)_2SiPh$	59.4	**21** (Et$_3$NH)	43	−87.10	**21** (Et$_3$NH)	40
$(C_6Cl_4O_2)_2SiPh$	89.8	**49** (Et$_4$N)	44	−83.94	**30** (K,18-c-6)	38
$(2,3-C_{10}H_6O_2)_2SiPh$	97.6	**50** (C$_5$H$_6$N)	45	−87.14	**23** (Bu$_4$N)	38
$(2,3-C_{10}H_6O_2)_2SiPh$	97.6	**50** (C$_5$H$_6$N)	45	−87.11	**24** (Et$_3$NH)	38
$(1,8-C_{10}H_6O_2)_2SiPh$	32.5	**16** (K,18-c-6)	37	−132.35	**16** (K,18-c-6)	37

determined, and the corresponding percent deviations from TBP geometries were reported. These differ substantially among the four, while the δ^{29}Si are nearly equal (for the same or analogous complexes differing only in the counter ion), with a relatively small downfield shift in the octachloro complex **30**. This comparison may indicate that δ^{29}Si does not depend on the geometry of the pentacoordinate complex, or, alternatively, one could assume that the substantially different geometries are characteristic of the solid state, and that in solution all of these complexes have essentially the same geometry around silicon, and hence similar chemical shifts.

Conversely, the spirosilicates **16**[37] and **19**[42] (Table 4), having five- and six-membered chelate rings, respectively, are reported to have essentially the same degree of deviation from TBP toward SP structure in the crystal (32.5 and 29.5%, respectively), while their ^{29}Si chemical shifts (**22** is taken as an analog of **19**) are dramatically different (−132.35 and −87.51 ppm, respectively)[37,38,42]. No explanation has been offered for this difference. The ^{29}Si chemical shift in **16** compares favorably with shifts measured in acyclic oxosilicates, **46** and **43**[34,39] (Table 3). It is tempting to conclude that steric strain has a profound effect on δ^{29}Si: the values measured in the strain-free acyclic compounds are nearly equal to that found in **16**, with six-membered chelate rings, while in **22**, with its five-membered rings, the chemical shift is substantially different. However, this leads to the conclusion that six-membered chelate rings are less strained, and hence more stable, than five-membered rings, a result not entirely consistent with known pentacoordinate silicon chemistry.

The percentage TBP → SP character measured in the solid state shows no correlation with the ligand-exchange barriers measured in solution[35b]. The barriers for interchange of CF$_3$ groups in compounds **51** decreased linearly with increasing electron-withdrawing power (σ^* substituent constant) of the ligand Y. On the other hand, the distortion from TBP structure (relative to SP) was determined for two of the complexes, **51d** and **51f**[35b]. These data are presented in Table 5. Clearly, the large difference in activation barriers is not reflected in a parallel difference in the ground state structure[35b].

2. Zwitterionic organospirosilicates

A new class of pentacoordinate zwitterionic silicates (**52–70**) has been developed and reported by Tacke and his coworkers[28–31,46–52] (cf Section II.A.3). These are generally high melting crystalline solids, which are almost insoluble in nonpolar organic solvents, and only slightly soluble in polar solvents.

The general method for the preparation of this class of compounds involves exchange of alkoxy groups in a trialkoxy(aminoalkyl)silane with bidentate ligands, such as catechols

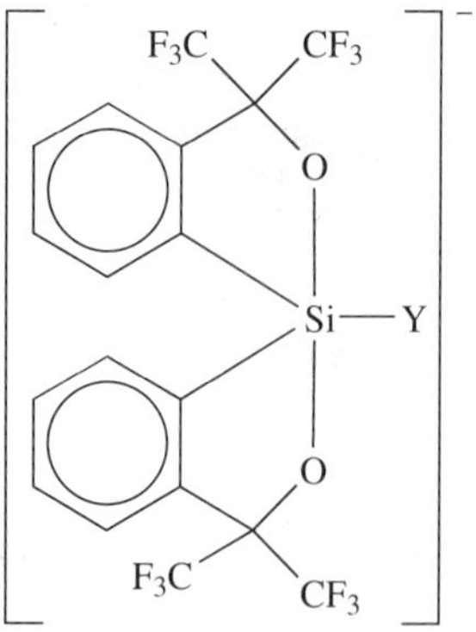

[Structure of spirosilicate 51 shown as a bracketed anion with Si bonded to Y and two bidentate ligands, each being a 2-(hexafluoro-2-hydroxyisopropyl)phenyl group with O–Si bonds]

(**51a**) Y = *n*-Bu
(**51b**) Y = PhMeCH
(**51c**) Y = 4-MeOC$_6$H$_4$
(**51d**) Y = F
(**51e**) Y = 3-CF$_3$C$_6$H$_4$
(**51f**) Y = Ph
(**51g**) Y = CN

TABLE 5. Comparison of ligand-exchange barriers with percent (TBP → SP) deviation in the crystal for spirosilicates **51**[35b]

Compound	Y	% (TBP → SP)	$\Delta G^{\ddagger}$ (kcal mol^{-1})
51d	F	23.9	17.5
51f	Ph	27.4	26.0

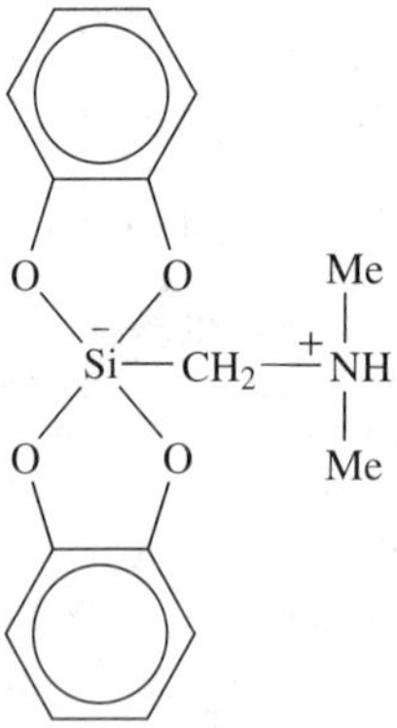

(**52**)[46]

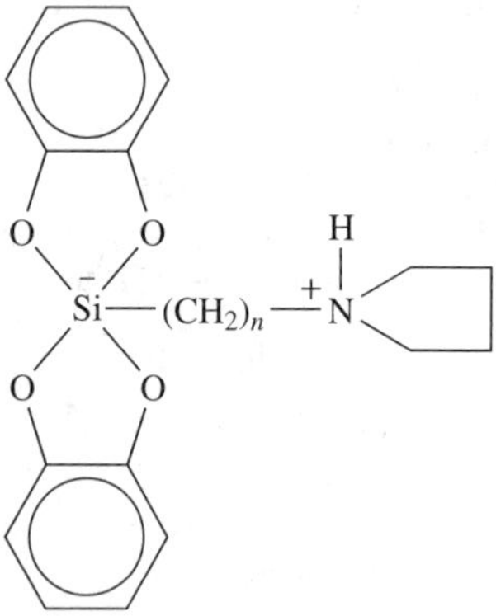

(**53**) $n = 1$[28]
(**54**) $n = 2$[28]
(**55**) $n = 3$[28]

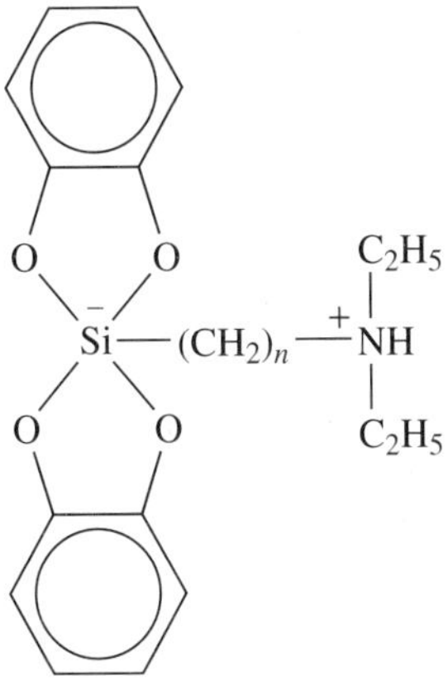

(56)[28]

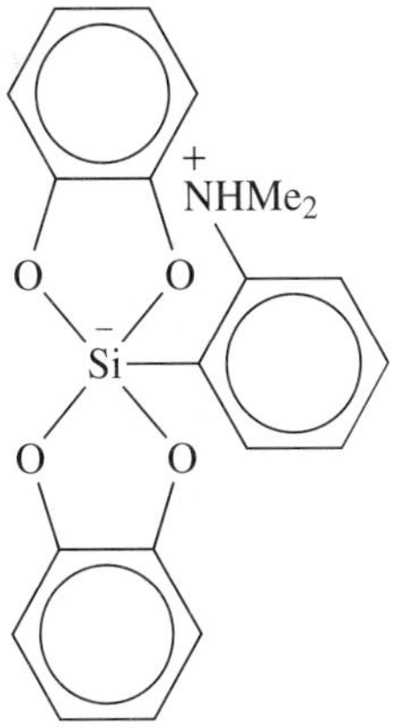

(57)[46]

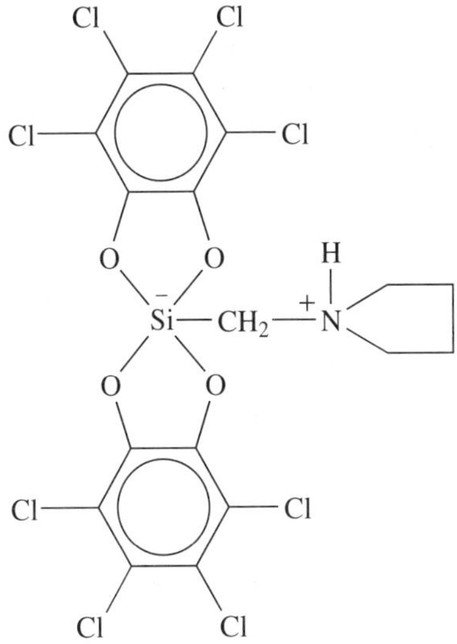

(58)[29]

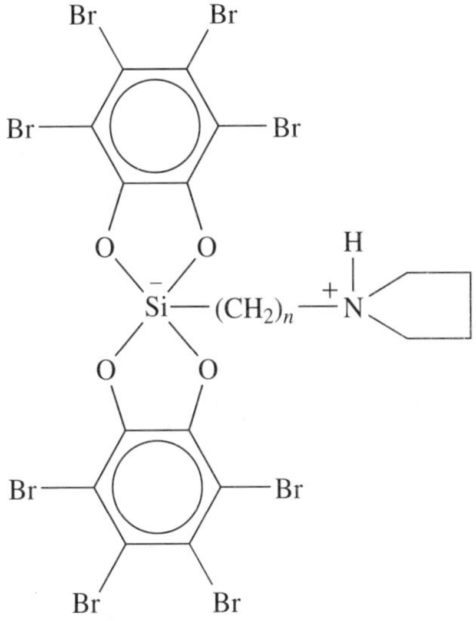

(59) $n = 1$[46]
(60) $n = 2$[47]

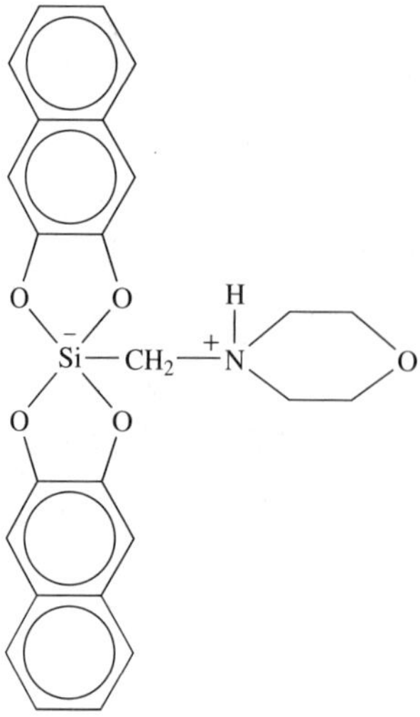

(61)[47]

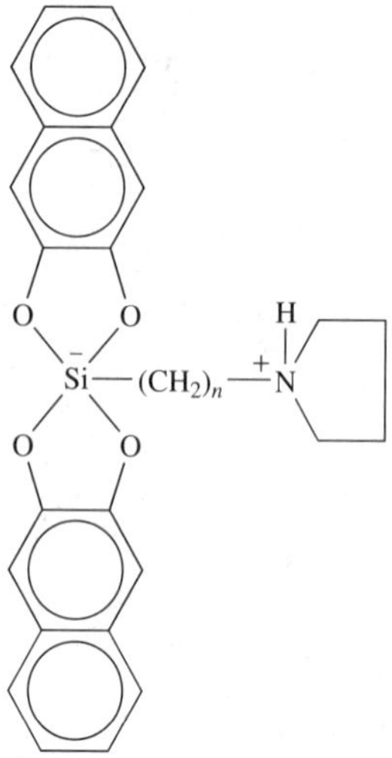

(62) $n = 1$[47]
(63) $n = 2$[46]

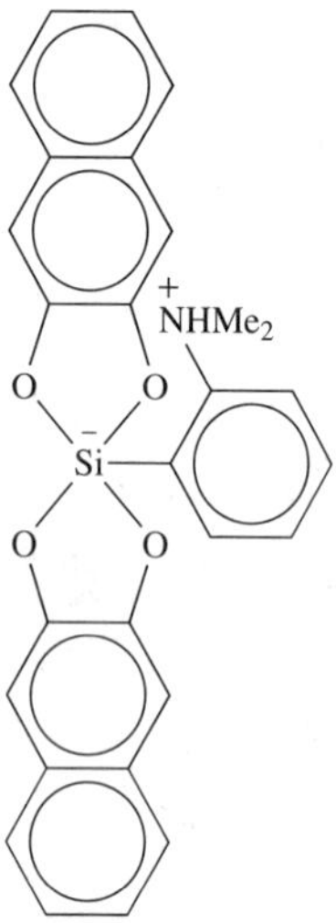

(64)[28]

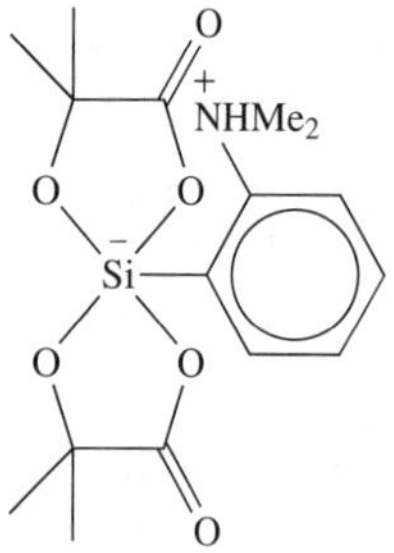

(65) R = H[49]
(66) R = Me[49]
(67) R = Ph[49]

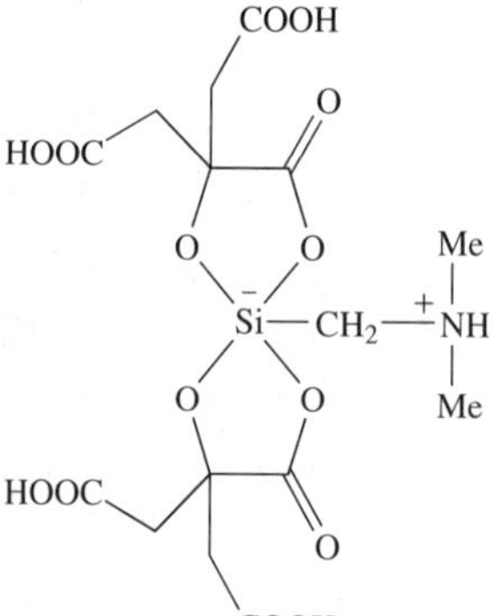

(68)[50]

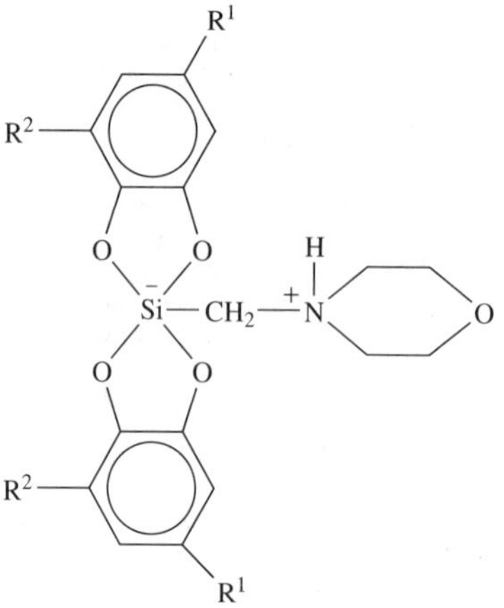

(69) $R^1 = NO_2$, $R_2 = H$[51]
(70) $R^1 = R^2 = t\text{-Bu}$[51]

and similar aromatic diols, as well as glycolic acid and its derivatives, as shown, for example, in equation 14.

Another synthesis, which is less general but of great interest, is the cleavage of Si—C bonds and displacement of alkyl or aryl groups, in addition to alkoxy groups, from alkyldialkoxy(aminoalkyl)silanes or dialkylalkoxy(aminoalkyl)silanes with appropriate 1,2-diols (equation 15)[46]. The cleavage of Si—C bonds even in silanes with only one alkoxy group attached to silicon, seems to indicate a particular stability of these complexes relative to anionic pentacoordinate spirosilicates. The latter are not formed from analogous silanes which do not contain the amino group, since no Si—C cleavage takes place in the absence of the amino group[28].

$$(14)$$

$$(15)$$

In contrast to the syntheses described above, leading to zwitterionic complexes based on the (SiO$_4$C) ligand framework, a direct synthesis of SiO$_5$-based zwitterions has also been reported: simultaneous reaction of one equivalent of 2-(dimethylamino)ethanol and two equivalents of glycolic acid or 2,2-diphenylglycolic acid with tetramethoxysilane afforded the zwitterionic bis-glycolatosilicates **71** and **72** (equation 16)[30].

$$(16)$$

(**71**) R = H
(**72**) R = Ph

Unusual selectivities have sometimes been observed in the displacement reactions leading to zwitterions: equation 17 shows a case of 'inverse selectivity', in which the aminoalkyl group is displaced, rather than the phenyl, leading to an anionic spirosilicate[28].

$$\text{Ph}-\underset{\underset{\text{OMe}}{|}}{\overset{\overset{\text{OMe}}{|}}{\text{Si}}}-(CH_2)_2NEt_2 \quad \xrightarrow[-C_2H_4]{2\ \text{catechol}\atop -2\ \text{MeOH}} \quad [\text{spirosilicate}]^- \ Et_2NH_2^+ \quad (17)$$

The preparation of similar zwitterionic silicates based on glycolic acid was also reported by Erchak and coworkers[53,54].

The crystal structures of several members of the zwitterionic complex family were determined. Like in the anionic spirosilicates, the structures were found to vary continuously between TBP and SP, with Si—O bond lengths essentially equal to those in anionic analogs. More than one crystalline modification was found for some of the zwitterions, and even those (even for the same complex!) were substantially different in molecular geometry: for **60** the percent deviation from TBP → SP was calculated to be 34.9% in the monoclinic crystal, 70.0% in the orthorombic, 86.2% in the crystalline hydrate and 96.3% in another crystal form of the monohydrate[46,47]. These results were interpreted in terms of variation in hydrogen bonds in the different crystal modifications. In the hydrates

TABLE 6. Comparison of solid state vs solution ^{29}Si chemical shifts for several zwitterionic complexes

Compound	δ^{29}Si (ppm) CP-MAS	δ^{29}Si (ppm) DMSO	References
59•MeCN	−84.8	−123.0	46
65	−91.8	−94.4	49
66	−103.2	−103.5	49
67	−102.6		49
62•MeCN	−84.9	−87.0	48
68•H$_2$O	−95.3	−97.3	50
63•MeCN	−81.6	−76.6	46
60a[a]	−73.5	−79.1	47
60b[a]	−79.7		47
60•H$_2$O	−80.8		47
57	−84.8	−85.9	46
61a[b]	−89.1	−85.3	47
61b[b]	−85.1	−85.3	47
61•MeCN	−85.5	−85.3	47
61•Me$_2$CO	−85.3	−85.3	47
61•MeNO$_2$	−83.6	−85.3	47

[a]Two solvent free crystallographic modifications of **60**.
[b]Two solvent free crystallographic modifications of **61**.

hydrogen bonding was intermolecular, and in the nonhydrate crystals the hydrogen bonds were intramolecular, between the NH and one of the chelate ring oxygens. The O··H—N and O—H··O distance requirements for hydrogen bonding promote substantial molecular distortions in the solid state. This suggests that the percent deviation from TBP might be a property characteristic of the crystal form, which may not necessarily reflect the molecular structure in solution.

The ^{29}Si CP-MAS solid state NMR spectra of the crystal modifications of **60** and its hydrate were found to depend on the crystal structure, as shown in Table 6. In fact, the chemical shift difference between the pure (anhydrous) crystalline modifications was used to demonstrate the kinetic transformation in the solid state of one crystal structure to the other (Figure 7)[47].

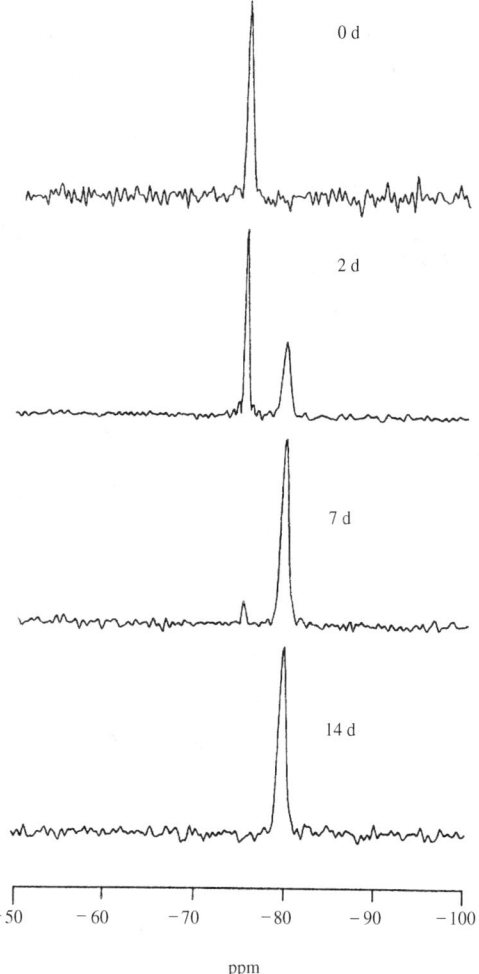

FIGURE 7. CP-MAS ^{29}Si NMR spectra of crystalline **60** at 150 °C showing the transformation of one crystal form to another (d = day). Reproduced from Reference 47 by permission of Hüthig Fachverlage

Table 6 lists comparisons of solid state vs solution ^{29}Si chemical shifts for several other zwitterionic complexes. In general the values obtained in solution agree quite well with those in the solid state, with one notable exception: for **59** dissolved in DMSO-d$_6$ δ^{29}Si is shifted 38.2 ppm upfield relative to the solid state[46].

In the simple symmetrical bis-catecholatosilicates the NMR spectra provided no information on possible exchange phenomena, suggesting that intramolecular ligand exchange, leading to enantiomer interconversion, was rapid on the NMR time scale. Even in cases of unsymmetrical glycolate ligands, such as in **65–67**, the NMR evidence showed a single isomer in solution, indicating fast exchange between the possible isomers[49]. However, when unsymmetrical catecholate ligands were used (**69, 70**), the NMR at room temperature (in DMSO-d$_6$ solution) showed doubling of the signals as a result of slow exchange of diastereomers, with an activation free energy of 16.7 kcal mol^{-1} for **69** and 17.2 kcal mol^{-1} for **70**[51].

To interpret these dynamic NMR observations, high level *ab initio* SCF-MO calculations were carried out for a model anion, **73**[52]. The ground state geometry was confirmed to be an almost ideal TBP (Figure 8, **a**, C_2 symmetry), with two pathways leading from it to

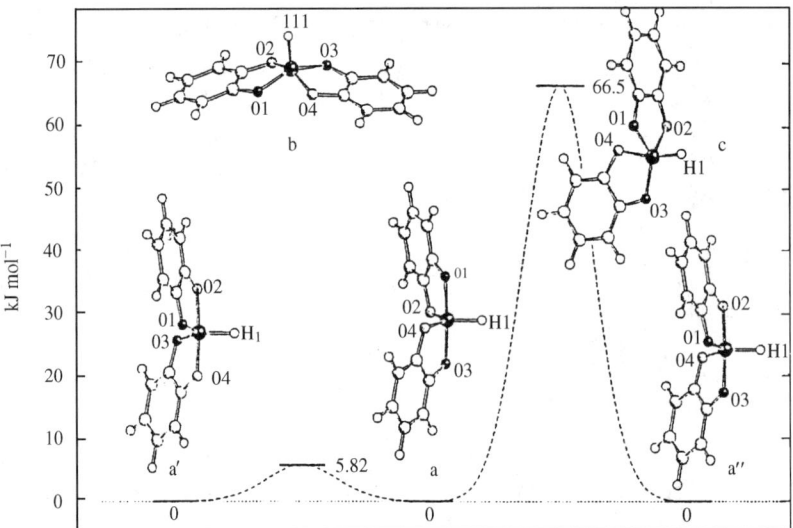

FIGURE 8. Calculated structures (RHF, using an optimized SVP basis set)[52] of **73** and illustration of the different ligand-exchange processes of this anion. **a** = **a'** = **a"** = ground state TBP, **b** = SP transition state, **c** = TBP transition state. Reproduced from Reference 52 by permission of VCH Verlagsgesellschaft

(73)

the diastereomeric TBP geometries, **a'** and **a"** (due to the lack of symmetry of the bidentate ligands). The first exchange mechanism is the Berry pseudorotation using the hydrogen as the pivot atom. This process has a very low activation barrier (1.39 kcal mol^{-1}), and it affects exchange of *both* cyclic ligands such that axial and equatorial positions interchange. Obviously this low barrier is very rapid at the NMR time scale at any temperature studied, and cannot be observed by dynamic NMR. This has also been suggested as the reason for the different geometries observed for various molecules of this family in the solid state: since the potential energy difference between the TBP and SP geometries is so small (the latter being the transition state for pseudorotation, **b** in Figure 8), packing forces associated with the crystallization of each molecule may effect the observed changes.

The second isomerization process in this class of molecules involves two consecutive Berry-type pseudorotations with oxygen as the pivot atoms (Figure 8). This pathway requires a substantially higher activation energy (calculated: 15.9 kcal mol^{-1}), because at the transition state (Figure 8, **c**), one of the bidentate ligands occupies equatorial positions, associated with significant strain. It was therefore concluded that this latter process, which 'rotates' one of the bidentate ligands (i.e. interchanges the equatorial with the axial oxygens for that ligand), is the process actually observed by dynamic NMR spectroscopy[51].

Similar calculations were conducted also for the analogous glycolate complexes, using the hydrido complex **74** as model (Figure 9)[30a]. The calculations confirmed the ground state structure (**I**) established for similar compounds in the solid state. The diastereomeric TBP structures **II** and **III**, as well as the distorted SP structures **IV** and **V**, were calculated

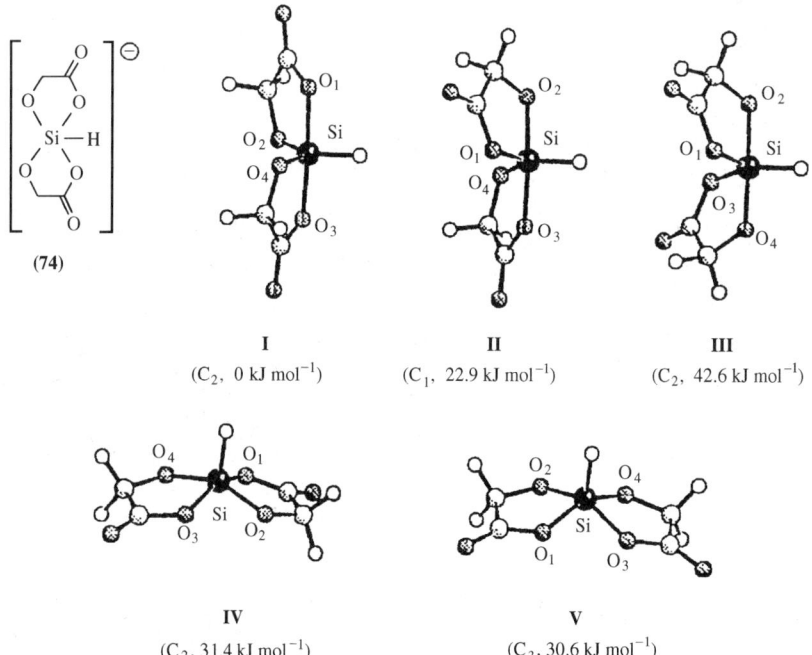

FIGURE 9. Calculated structures and relative energies (RHF, using an optimized SVP basis set, for the isomers **I–V** of the anion **74**. Reproduced from Reference 30a by permission of VCH Verlagsgesellschaft

to be substantially higher in energy than **I**, in agreement with the observation of only one structure in solution, without exchange phenomena[29,49].

Intermolecular ligand exchange has been reported in bis-catecholates as well as in bis-glycolate complexes (equations 18[29] and 19[49], respectively). This was observed after 24 h by the appearance of NMR signals for the mixed complex (**76** and **77** respectively) in the ^{1}H, ^{13}C and ^{29}Si spectra, as well as by FAB MS experiments. A statistical distribution of the three compounds was evident from the 1 : 2 : 1 triplet found in the ^{29}Si spectrum.

3. Polynuclear spirosilicates

A new class of pentacoordinate spirosilicate compounds with two silicon centers connected through a carbon bridge has been reported by several groups[25,55]. The reaction of bridged bis-(trialkoxysilyl)-arylene or -alkylene with catechol in the presence of triethylamine (TEA) yielded arylene- (alkylene) bridged bis(bis-spirocyclic)silicates **78–83**[25] and **84**[55] (equation 20). The crystal structure of **78** was determined, and the geometry was found to conform to a slightly distorted TBP[25].

A bridged bis-spirosilicate with an ethylenedioxy bridge (**85**) (i.e. based on the SiO_5 ligand framework) was prepared directly from inorganic silica and ethylene glycol in the presence of potassium hydroxide (equation 13)[36].

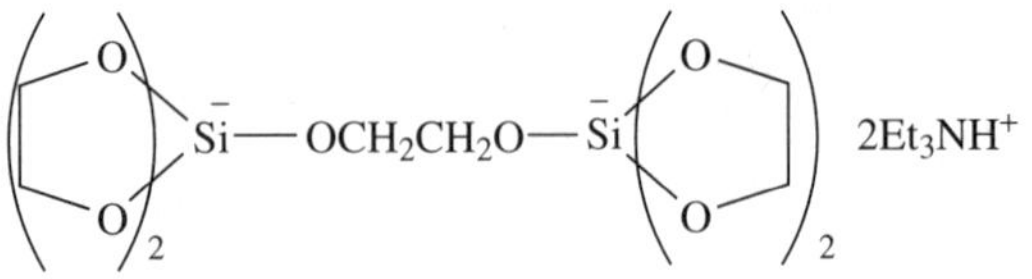

$$\left(\underset{O}{\overset{O}{\diagup}} \overline{Si} - OCH_2CH_2O - \overline{Si} \underset{O}{\overset{O}{\diagdown}} \right)_2 \quad 2Et_3NH^+$$

(**85**)

Erchak and coworkers obtained the bridged bis-silicate **86** based on glycolic acid as the spiro-forming chelate. The synthesis is shown in equation 21[56].

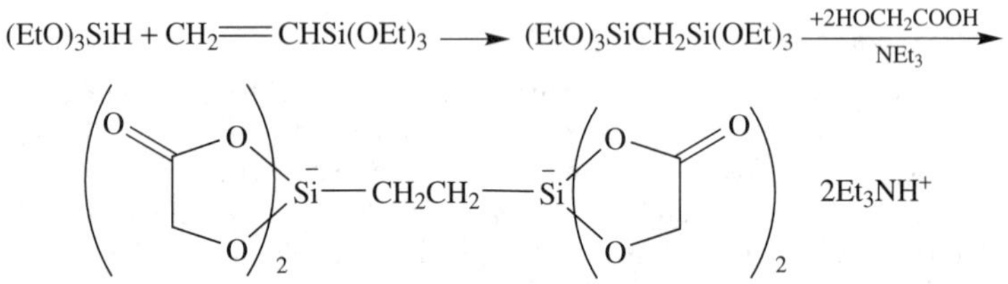

$$(EtO)_3SiH + CH_2\!\!=\!\!CHSi(OEt)_3 \longrightarrow (EtO)_3SiCH_2Si(OEt)_3 \xrightarrow[NEt_3]{+2HOCH_2COOH}$$

$$\left(\underset{O}{\overset{O}{\diagup}} \overline{Si} - CH_2CH_2 - \overline{Si} \underset{O}{\overset{O}{\diagdown}} \right)_2 \quad 2Et_3NH^+ \qquad (21)$$

(**86**)

Dinuclear zwitterionic spirosilicates were also synthesized and characterized by Tacke and coworkers (equations 22 and 23)[30,57]. These differ from previous dinuclear silicates in that they are bis-zwitterionic. The bridge in **87** is through the covalent bonding of a 1,4-(dimethylene)piperazine, whereas in equation 23 a double bridge is formed by two tetradentate ligand molecules, tartaric acid. The crystal structure of **87** proved that the product was the *meso* compound, with opposite configurations at the silicon centers[57a]. When the reaction in equation 23 was run with optically active (R,R)-$(+)$-tartaric acid the first optically active zwitterionic bis-silicate (**88**) was obtained[57b].

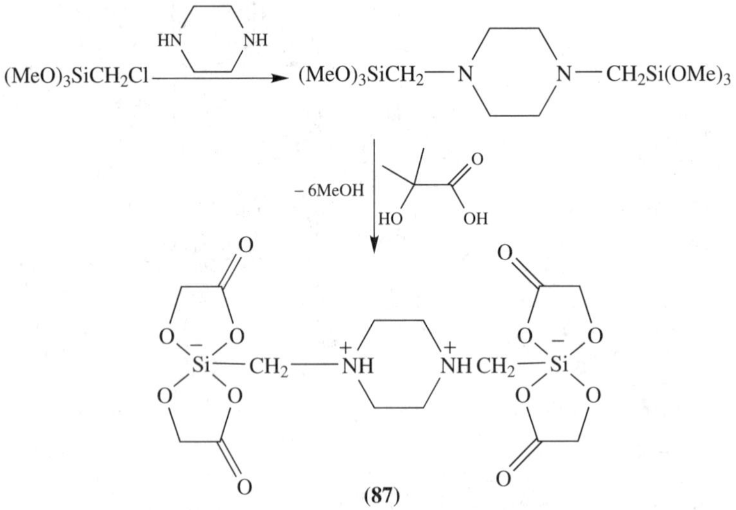

$$(MeO)_3SiCH_2Cl \xrightarrow{HN\diagup\diagdown NH} (MeO)_3SiCH_2 - N\diagup\diagdown N - CH_2Si(OMe)_3$$

$$\downarrow\;-6MeOH\;\;\;HO\diagup\diagdown OH$$

(22)

(**87**)

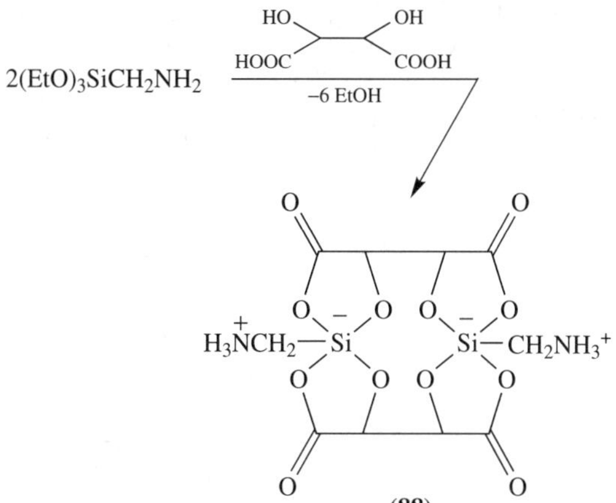

23. Hypervalent silicon compounds

$$2(EtO)_3SiCH_2NH_2 \xrightarrow[-6\ EtOH]{\begin{array}{c}HO\quad\quad OH\\HOOC\quad COOH\end{array}}$$

(88)

(23)

The ^{29}Si chemical shifts for dinuclear pentacoordinate silicates have been compiled in Table 7.

The reaction of 1,2,4,5-tetrahydroxybenzene with phenyltriethoxysilane was reported to yield polysiliconates, with pentacoordinate silicon in a spirocyclic arrangement as part of the polymer chain (equation 24)[58a].

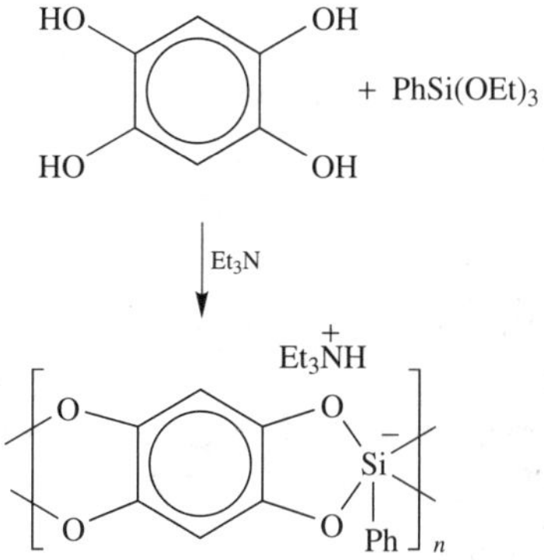

(24)

TABLE 7. ^{29}Si chemical shifts for dinuclear Si,Si′ complexes

Compound	δ^{29}Si (ppm)	References
78	−87.0	25
79	−86.7	25
80	−87.4	25
81	−81.7	25
82	−73.7	25
83	−71.9	25
84	−78.63	55
85	−103.1	36
88	−91.7	58

1370   Daniel Kost and Inna Kalikhman

Similar condensation of phenyltriethoxysilane with a spirocatechol yielded the first macrocyclic tetrasiliconate, a tetraanion containing four pentavalent silicons (equation 24a)[58b]. NMR evidence showed that only one *meso*-stereomer (C_{2h} symmetry) was formed in solution out of the four possible diastereomers. The ^{29}Si chemical shifts of the two unique silicons in DMSO-d_6 solution were −86.3 and −86.9 ppm, respectively, consistent with a pentacoordinate silicate anion.

(24a)

C. Hydridosilicates

1. Alkoxyhydridosilicates

Hydrido- and dihydridosilicates have been prepared and studied by Corriu and coworkers[59−62]. The subject was reviewed in 1993[6], and will only be outlined briefly here for completeness.

Hydridosilicates are readily formed by the reaction of trialkoxysilanes with alkali-metal alkoxides[60] (equation 25). The results are best when potassium alkoxides are used.

$$\text{HSi(OR)}_3 + \text{ROK} \xrightarrow[\text{room temp.}]{\text{THF or DME}} [\text{HSi(OR)}_4]\text{K}$$

$$R = \text{Me, Et, } n\text{-Bu, } i\text{-Pr, Ph} \qquad (25)$$

When trialkoxysilanes are reacted with alkali metal hydrides, mixtures of mono- and dihydrosilicates are obtained (equation 25a)[59,61]. The results of this reaction depend strongly upon several factors: the size of the alkoxy group, the solvent, and in particular whether the hydride is used in the presence of 18-crown-6 or not (Table 8). The Table shows that as the bulk of the alkoxy group increases, the ratio of monohydrido to dihydrido products increases. Use of 18-crown-6 makes the reaction less selective, in the sense that both products are obtained, while without the crown ether quantitative yields

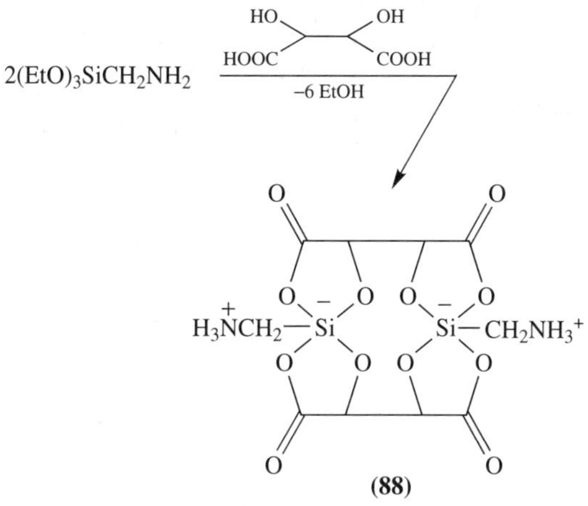

23. Hypervalent silicon compounds

$$2(EtO)_3SiCH_2NH_2 \xrightarrow[-6\,EtOH]{\begin{array}{c}HOOH\\HOOCCOOH\end{array}}$$

(23)

(88)

The ^{29}Si chemical shifts for dinuclear pentacoordinate silicates have been compiled in Table 7.

The reaction of 1,2,4,5-tetrahydroxybenzene with phenyltriethoxysilane was reported to yield polysiliconates, with pentacoordinate silicon in a spirocyclic arrangement as part of the polymer chain (equation 24)[58a].

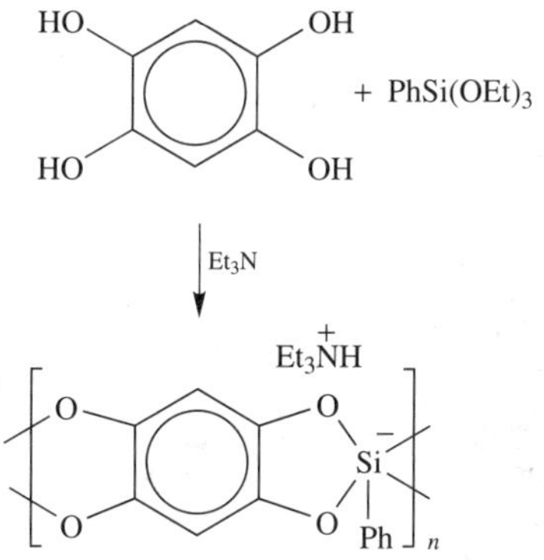

(24)

TABLE 7. ^{29}Si chemical shifts for dinuclear Si,Si' complexes

Compound	δ^{29}Si (ppm)	References
78	−87.0	25
79	−86.7	25
80	−87.4	25
81	−81.7	25
82	−73.7	25
83	−71.9	25
84	−78.63	55
85	−103.1	36
88	−91.7	58

Similar condensation of phenyltriethoxysilane with a spirocatechol yielded the first macrocyclic tetrasiliconate, a tetraanion containing four pentavalent silicons (equation 24a)[58b]. NMR evidence showed that only one *meso*-stereomer (C_{2h} symmetry) was formed in solution out of the four possible diastereomers. The ^{29}Si chemical shifts of the two unique silicons in DMSO-d_6 solution were −86.3 and −86.9 ppm, respectively, consistent with a pentacoordinate silicate anion.

$$\text{(24a)}$$

C. Hydridosilicates

1. Alkoxyhydridosilicates

Hydrido- and dihydridosilicates have been prepared and studied by Corriu and coworkers[59–62]. The subject was reviewed in 1993[6], and will only be outlined briefly here for completeness.

Hydridosilicates are readily formed by the reaction of trialkoxysilanes with alkali-metal alkoxides[60] (equation 25). The results are best when potassium alkoxides are used.

$$\text{HSi(OR)}_3 + \text{ROK} \xrightarrow[\text{room temp.}]{\text{THF or DME}} [\text{HSi(OR)}_4]\text{K}$$

$$R = \text{Me, Et, } n\text{-Bu, } i\text{-Pr, Ph} \tag{25}$$

When trialkoxysilanes are reacted with alkali metal hydrides, mixtures of mono- and dihydrosilicates are obtained (equation 25a)[59,61]. The results of this reaction depend strongly upon several factors: the size of the alkoxy group, the solvent, and in particular whether the hydride is used in the presence of 18-crown-6 or not (Table 8). The Table shows that as the bulk of the alkoxy group increases, the ratio of monohydrido to dihydrido products increases. Use of 18-crown-6 makes the reaction less selective, in the sense that both products are obtained, while without the crown ether quantitative yields of the monohydrido complexes are obtained (for the same alkoxy group).

$$\text{HSi(OR)}_3 \xrightarrow{\text{KH}} \text{K[H}_2\text{Si(OR)}_3\text{]} + \text{K[HSi(OR)}_4\text{]} \quad (25a)$$

^{29}Si chemical shifts in anionic mono- and dihydrosilicates move to higher fields, relative to the analogous hydridosilanes, as is expected for pentacoordinate silicates. Also, the one-bond coupling constant $^1J(^1\text{H}-^{29}\text{Si})$ is reduced relative to the precursor hydrosilanes, as is expected due to the decrease in the s-character at the Si−H bond[59]. The ^{29}Si chemical shifts and $^1J(^1\text{H}-^{29}\text{Si})$ for some of the hydrido and dihydrido silicates are collected in Table 9. Results which disagree with those mentioned above concerning $^1J(^1\text{H}-^{29}\text{Si})$ were communicated recently: for some pentacoordinate complexes the one-bond coupling constant $^1J(^1\text{H}-^{29}\text{Si})$ *increased* relative to the precursor tetracoordinate silane: thus, for MeHSi(OTf)$_2$, $^1J(^1\text{H}-^{29}\text{Si}) = 306$ Hz, while for the silicate anion MeHSi(OTf)$_3^-$, $^1J(^1\text{H}-^{29}\text{Si}) = 351$ Hz[63].

The ^{29}Si and ^{1}H chemical shifts of the dihydridosilicates indicate some dynamic behavior[61]. For the silicates K[H$_2$Si(OR)$_3$] with less bulky alkoxy groups (methoxy, ethoxy) the ^{29}Si signal was a triplet, proving the equivalence of the two hydrogens. This might be the result of either of two reasons: fast exchange between hydrido groups, or the occupation of equivalent, equatorial positions by both hydrogens. In the more bulky alkoxides, R = *i*-propyl and *s*-butyl, in benzene-d$_6$ or toluene-d$_8$ solutions, the Si signal appeared as a doublet of doublets, as a result of the anisochrony (i.e. chemical shift nonequivalence) of the hydrido groups [$J(\text{Si}-\text{H}_1) = 222-227$ Hz, $J(\text{Si}-\text{H}_2) = 192-195$ Hz, hydride exchange barrier at 363 K for the dihydrido-triisopropoxy complex in toluene-d$_8$: 16.3 kcal mol^{-1}]. In solutions of the more strongly solvating THF or DME, as well as in the presence of 18-crown-6, the hydrogens became equivalent

TABLE 8. Effect of reaction conditions on product distribution (equation 25)[61].

Run	HSi(OR)$_3$	Solvent	Reaction time (h)	% K[H$_2$Si(OR)$_3$]	% K[HSi(OR)$_4$]
1	HSi(OMe)$_3$	THF	2	0	100
2	HSi(OEt)$_3$	THF	6	40	60
3	HSi(OEt)$_3$	THF	24	0	100
4	HSi(OEt)$_3$	THF (18-c-6)	26	55	45
5	HSi(OEt)$_3$	DME	2	0	100
6	HSi(OBu-*n*)$_3$	THF	4	50	50
7	HSi(OPr-*i*)$_3$	THF	6	100	0
8	HSi(OPh)$_3$	THF	2	0	100
9	HSi(OPh)$_3$	THF (18-c-6)	2	15	85

TABLE 9. ^{29}Si NMR data for HSi(OR)$_3$ and related hydridosilicates[60]

R	HSi(OR)$_3$		[HSi(OR)$_4$]K		[H$_2$Si(OR)$_3$]K	
	δ (ppm)	$J_{\text{Si}-\text{H}}$ (Hz)	δ (ppm)	$J_{\text{Si}-\text{H}}$ (Hz)	δ (ppm)	$J_{\text{Si}-\text{H}}$ (Hz)
Me	−62.6	290	−82.5	223		
Et	−59.6	285	−86.2	218	−81.8	218
n-Bu	−59.2	286	−86.1	219	−80.8	215
i-Pr	−63.4	285	−90.5	215	−87.1	213
Ph	−71.3	320	−112.6	296		
s-Bu	−62.6	282			−85.6	222, 194

in the i-propoxy silicate. These observations were interpreted as follows: (a) the rate of exchange between hydrido groups decreases with increasing steric bulk of the alkoxy group; (b) the rate also depends on the metal–ion complexation by the solvent and, in poorly solvating solvents, the K^+ is solvated by coordination to the three alkoxy oxygens, forcing one hydrido group to an axial position as depicted in **89**. The results suggest an interplay between the tendency of electronegative ligands to occupy axial positions in TBP complexes, against the steric bulk which dictates a preference for the equatorial position. The available results do not conclusively distinguish between a kinetic or thermodynamic effect in this series of dihydridosilicates: the effect of solvent and steric bulk may either be on the rate of exchange, or alternatively on the equilibrium distribution between equatorial and axial positions.

(**89**)

2. Gas-phase chemistry of hydridosilicates

The formation and properties of alkylhydridosilicates in the gas phase under conditions of flowing afterglow were reported by several authors[64,65] and were recently reviewed[66,67]. Some aspects of this topic are discussed below.

Direct addition of H^- and other nucleophilic anions to primary, secondary and tertiary alkylsilanes formed pentacoordinate silicate anions (equation 26). Interestingly, the parent silicate, SiH_5^-, could not be prepared in this way from silane and hydride[65,68]. SiH_5^- was prepared from alkylhydridosilicates by a hydride transfer reaction (equation 27). The reducing activity of alkylhydridosilicates toward various substrates via hydride transfer in the gas phase was discussed in detail in a previous review[67].

$$RSiH_3 + H^- \longrightarrow RSiH_4^- \qquad (26)$$

$$R = n\text{-pentyl}, n\text{-Bu}, Et$$

$$RSiH_4^- + SiH_4 \longrightarrow SiH_5^- + RSiH_3 \qquad (27)$$

Fluxional behavior was demonstrated for pentacoordinate alkylhydridosilicates in the gas phase, by the reaction of deuterium-labeled alkylsilicate with carbon dioxide[65]. The monodeuterated silicate was first obtained in the flow tube by D^- addition to a neutral alkylsilane, followed by H^-/D^- transfer to CO_2 (equation 28). A statistical distribution of HCO_2^-/DCO_2^- (3 : 1 ratio) indicated complete scrambling of the hydrogens and deuterium in the silicate, presumably by a Berry pseudorotation mechanism.

$$n\text{-}C_5H_{11}SiH_3 \xrightarrow{D^-} n\text{-}C_5H_{11}SiH_3D^- \xrightarrow{CO_2} HCO_2^- + DCO_2^- \qquad (28)$$
$$(3:1)$$

3. Computational studies of hydridosilicates

Numerous high-level theoretical computations of the SiH_5^- anion and SiH_5 neutral species have been reported[69–73]. In contrast to the carbon analog, SiH_5^- is a stable ground state structure with respect to decomposition to SiH_4 and H^-. SiH_5^- has a TBP geometry. The binding energy of the hydride to SiH_4 was calculated to be 16–22 kcal mol^{-1}[69,70]. The silicon anion is stable relative to the neutral species, while in the carbon analogs the opposite is calculated to be true[71a]. However, decomposition of SiH_5^- to SiH_3^- and H_2 is exothermic[65], with a substantial activation barrier which suggests that observation of the pentacoordinate anion should be possible under mild conditions. The question of inversion or retention of configuration at silicon in silanes upon nucleophilic displacement via pentacoordinate species was studied carefully by several groups, and was reviewed by Apeloig[74]. The following brief discussion relates only to recent theoretical calculations relevant to hypervalent hydridosilicates.

It was shown at the MP4/6-31++G** level of theory that the TBP $SiH_3F_2^-$ anion is 9.27 kcal mol^{-1} more stable when both fluorine atoms are at axial positions, relative to the species with one axial and one equatorial fluorines[73]. However, it was pointed out that at higher levels of theory the latter structure may not be a stable species, judging from the calculations of Gordon and coworkers on the SiH_4F^- ion, in which the TBP structure with an equatorial fluorine was not a minimum on the potential energy surface[72a].

Gordon and coworkers have conducted a detailed high-level computational analysis of the Berry pseudorotation mechanism in the series of hydridosilicates SiH_nX_{5-n}, where X = F, Cl, and compared their results with previous studies[71b–71d]. For the tetrahydridosilicates SiH_4F^- and SiH_4Cl^- they report substantially distorted pseudorotation pathways: the TBP structures with equatorial F or Cl are unstable, and the square-pyramidal structures with basal electronegative substituents, presumed to be transition states for the Berry process, were also unstable. For the more highly substituted silicates of this series the behavior was closer to expectation, based on a Berry pseudorotation model. Their major conclusions are as follows[72b]: (1) The higher the substitution at silicon (smaller n), the more closely the potential energy surface resembles an idealized Berry pseudorotation model. (2) The simple electronegativity model proposed by Willhite and Spialter[72c] is generally supported by the high level calculations. (3) In systems of the type SiX_5^-, the Si–X bond lengths follow the trend: Si–X_{axial} > Si–X_{apical} > Si–X_{basal} > Si–$X_{equatorial}$ where the first and last values relate to the TBP and the others to the square-pyramidal structures. (4) The Si–H bonds are consistently longer in the fluoro systems than in the analogous chloro systems. This was attributed to the stronger bonding of F to silicon relative to Cl, causing in turn 'loosening' of the SiH bonds. (5) Most of the structures were found to lie below the X^- dissociation limits.

The possibility of ligand substitution reactions in pentacoordinate silicates $SiH_3F_2^-$ and $SiH_2F_3^-$ via hexacoordinate intermediates was studied by Fujimoto, Arita and Tamao[73]. Attack on each of these silicates by F^- or hydride produced qualitatively similar reaction pathways, leading to stable hexacoordinate intermediates, without significant breaking of the bond between silicon and the leaving group. It was concluded that a nonconcerted displacement mechanism via a hexacoordinate intermediate is likely.

III. PENTACOORDINATE NEUTRAL SILICON COMPLEXES

A. Chelates with Nitrogen–Silicon Coordination

1. Synthesis

The most widely studied compounds possessing N → Si coordination are silatranes, azasilatranes and their analogs[75]. This family of chelates warrants a separate discussion, which can be found elsewhere[75].

Stable N—Si coordinated complexes generally contain a chelate ring with a covalent bond to silicon on one side, and a dative N → Si bond on the other. Only few such bidentate ligand types have been used, and the resulting families of related chelates are depicted in structures **90–100**[76–95].

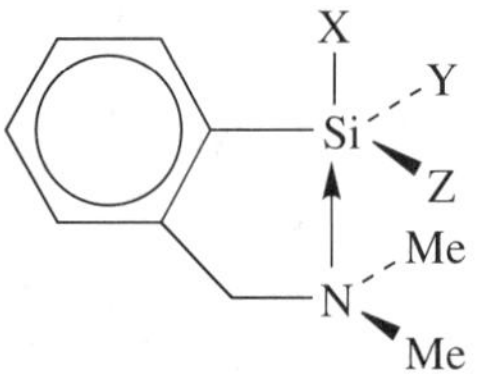

(**90a**) X = Y = F, Z = Me
(**90b**) X = Y = H, Z = 1-Naph
(**90c**) X = Y = Cl, Z = 2-C$_6$H$_4$CH$_2$NMe$_2$
(**90d**) X = Y = Cl, Z = 2-C$_6$H$_4$CH$_2$NH$^+$Me$_2$
(**90e**) X = Cl, Y = H, Z = Me
(**90f**) X = Y = Z = F
(**90g**) X = Y = Z = OEt
(**90h**) X = Me, Y = Z = OEt
(**90i**) X = Y = Z = Me
(**90j**) X = Y = Cl, Z = Ph
(**90k**) X = Y = H, Z = 2-C$_6$H$_4$CH$_2$NMe$_2$
(**90l**) X = Cl, Y = H, Z = 2-C$_6$H$_4$CH$_2$NMe$_2$
(**90m**) X = Y = OEt, Z = 2-C$_6$H$_4$CH$_2$NMe$_2$
(**90n**) X = Y = Z = Cl
(**90o**) X = Y = Z = H
(**90p**) X = Y = Cl, Z = 2–C$_6$H$_4$CH$_2$NMe$_2$
(**90q**) X = Cl, Y = CH=CH$_2$, Z = 2-C$_6$H$_4$CH$_2$NMe$_2$

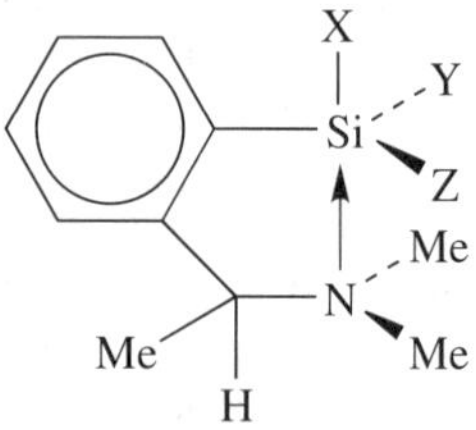

(**91a**) X = F, Y = Ph, Z = Me
(**91b**) X = Y = F, Z = Me
(**91c**) X = Y = Z = F
(**91d**) X = Y = Z = OEt
(**91e**) X = Cl, Y = Z = H

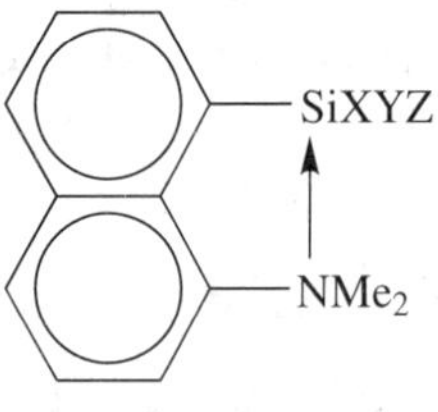

(**92a**) X = Y = Z = F
(**92b**) X, Y = OCMe$_2$CMe$_2$O, Z = H
(**92c**) X = Ph, Y = Z = H
(**92d**) X = Ph, Y = Z = OMe
(**92e**) X = Y = Z = H

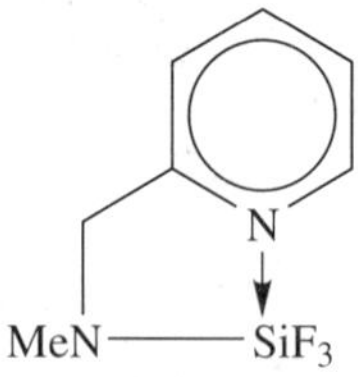

(**93**)

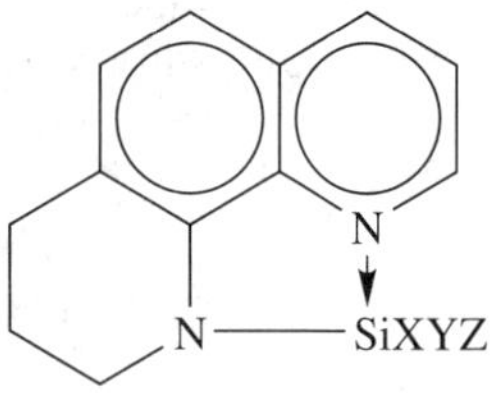

(**94a**) X = Y = Z = Me
(**94b**) X = Cl, Y = Z = Me
(**94c**) X = Y = Cl, Z = Me
(**94d**) X = Y = Z = Cl

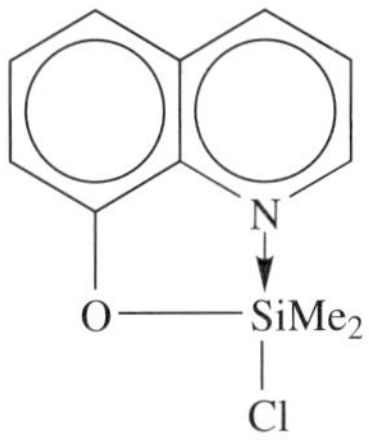

(95)

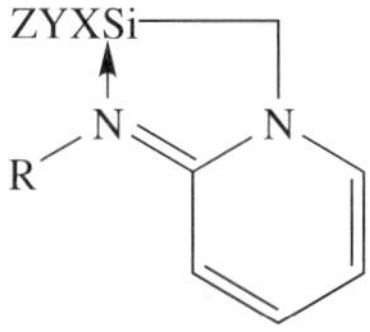

(96a) R = 2-Pyr, X = Cl, Y = Z = Me
(96b) R = 2-Pyr, X = Y = Z = Cl
(96c) R = H, X = Cl, Y = Z = Me

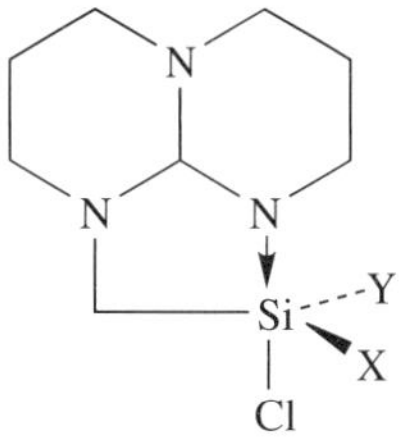

(97a) X = Y = Me
(97b) X = Cl, Y = Me
(97c) X = Y = Cl

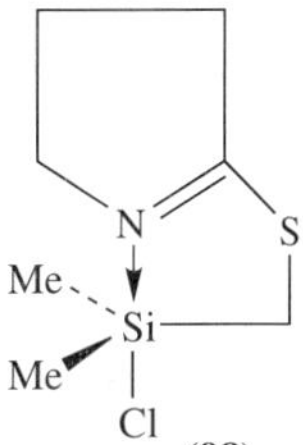

(98)

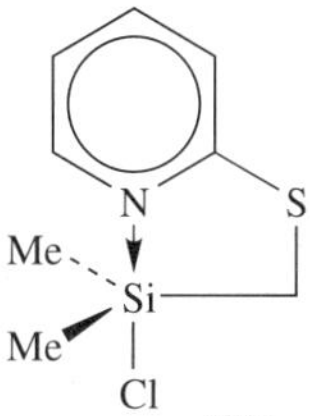

(99)

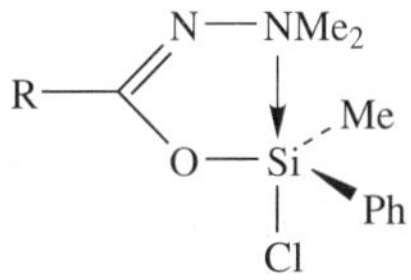

(100a) R = Me
(100b) R = PhCH$_2$
(100c) R = PhMeCH
(100d) R = Ph
(100e) R = 4-MeC$_6$H$_4$
(100f) R = 4-O$_2$NC$_6$H$_4$
(100g) R = 3,5-(NO$_2$)$_2$C$_6$H$_3$
(100h) R = CF$_3$

Neutral pentacoordinate N→Si complexes have been prepared by several methods[6]. The most extensively used method consists of reacting a lithio derivative of an aromatic amine with a tetracoordinated functionalized silane (compounds **90–92**)[76–84] such that a new covalent Si−C bond is formed, and the amino group is now in a geometrically favored position to form a chelate ring by coordinating to silicon. The selective metalation by butyllithium is facilitated by the presence of the neighboring amino group, which directs the proton abstraction to a nearby position (*ortho* position in the case of anilines or benzylamine, 8-position in the case of 1-dimethylaminonaphthalene) presumably through interaction with the lithium ion. Sometimes butyllithium is used to remove a proton from nitrogen (**93,94**)[85–87] or oxygen (**95**)[88], in a similar manner, followed by reaction with the polyhalosilane. Several different substituted silane types were reacted with the lithio amines: di- or trihydridosilanes, di-, tri- or tetraalkoxysilanes[76–78] and sometimes also polychlorosilanes[80–82]. The initial complexes can further be transformed to the desired products by substitution with a variety of reagents[76,84]. A selection of the transformations which were used in one particular case is shown in Figure 10. In addition, the alkoxy complexes can be reduced by LiAlH$_4$ back to the corresponding hydrido complexes. Instead of *N*-chlorosuccinimide (NCLS) used in Figure 10, PCl$_5$ has also been used for the replacement of hydrido by chloro ligands[76]. A variety of transformations leading to differently substituted pentacoordinate complexes has been reported by various groups[76–88].

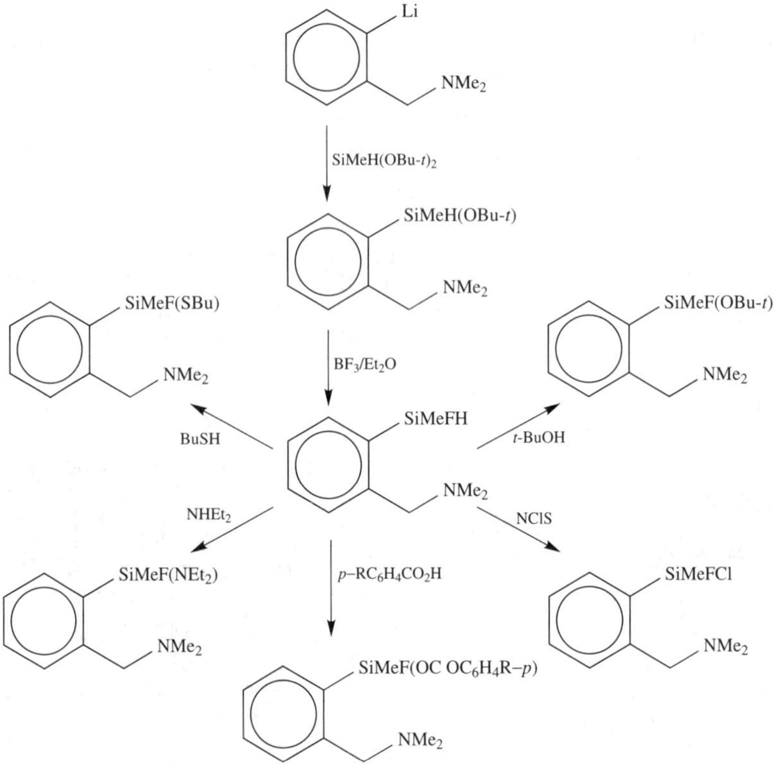

FIGURE 10. Synthetic pathways for some neutral pentacoordinate silicon chelates[76]

A second method for the preparation of neutral pentacoordinate complexes consists of an exchange reaction between chloromethylchlorosilanes (ClCH$_2$SiMe$_n$Cl$_{3-n}$, n = 0–2) and N-trimethylsilylated compounds, as shown for example in equation 29[89]. The compounds of types **96–99** were prepared by this method[89–93]. The same reagent was used extensively for the synthesis of O→Si complexes, and this is discussed in more detail in the corresponding section (III.B.1).

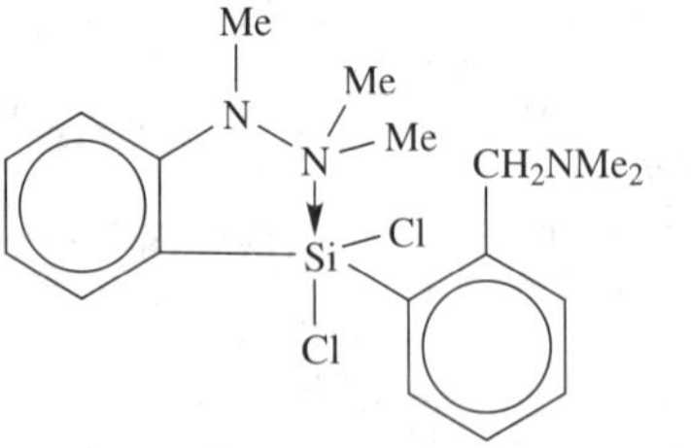

(101)

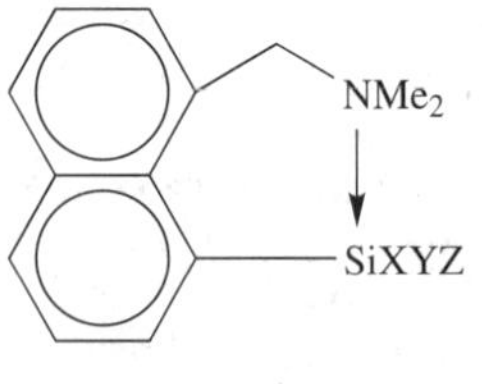

(102a) X = Ph, Y = Me, Z = H
(102b) X = Ph, Y = Z = H
(102c) X = Ph, Y = Z = F

Another route to pentacoordinate N→Si complexes (**100**) consists of the substitution of an O-trimethylsilyl group of O-silylated hydrazides by a halogenated silyl group, capable of coordinating to nitrogen (equation 30)[94,95]. The O-trimethylsilylated precursor molecules are prepared from various hydrazides by reaction with trimethylchlorosilane or other silylation reagents[98].

$$\text{(pyridyl-N=N(SiMe}_3\text{)-pyridyl)} \xrightarrow[-\text{Me}_3\text{SiCl}]{\text{ClCH}_2\text{SiMe}_n\text{Cl}_{3-n} \ (n = 0-2)} \text{(Cl}_{3-n}\text{Me}_n\text{Si chelate)} \quad (29)$$

$$\underset{R}{\overset{\text{Me}_3\text{SiO}}{>}}C=NNMe_2 + PhMeSiCl_2 \xrightarrow{-\text{Me}_3\text{SiCl}} \mathbf{100} \quad (30)$$

2. Structure

a. Crystallographic data. Numerous X-ray crystallographic structures for pentacoordinate silicon complexes have been reported in recent years. All of the reported complexes had essentially TBP structures, with slight distortions. In Table 10 we have collected some of the crystallographic data pertinent to the TBP structures: the bond lengths of silicon to the two apical ligands and, where available, the displacement (Δ) of the silicon atom from the central plane defined by the three equatorial ligands. Several common structural features can be extracted from Table 10 and from the associated literature, as summarized below:

TABLE 10. X-ray data for (N–Si) chelates.

Compound	X	Y	Z	Si–N (Å)	Si–X (Å)	Δ^a (Å)	References
90a	F	F	Me	2.356	1.627	−0.24	83
90b	H	H	$C_{10}H_7$	2.44	1.47		79
90c	Cl	Cl	$PhCH_2NMe_2$	2.291	2.181		81
90d	Cl	Cl	$PhCH_2N^+HMe_2$	2.163	2.210		81
92a	F	F	F	2.318, 2.287	1.612, 1.612		77
92b	$OCMe_2CMe_2O$		H	2.339	1.678	−0.03	78
92c	Ph	H	H	2.584	1.787		99
93	F	F	F	1.974	1.621		85
94a	Me	Me	Me	2.689	1.897		86
94b	Cl	Me	Me	2.028	2.269	−0.135	87
94c	Cl	Cl	Me	2.027	2.207		87
94d	Cl	Cl	Cl	1.984	2.150		87
96a	Cl	Me	Me	1.898	2.598	0.167	89
96b	Cl	Cl	Cl	1.901	2.238	0.01	89
96c	Cl	Me	Me	1.766, 1.777	3.908, 4.111		90
97	Cl	Me	Me	1.852	2.679	0.166	91
98	Cl	Me	Me	1.945	2.423	0.05	85
100d	Cl	Ph	Me	2.264	2.192		87
101	Cl	Cl	$PhCH_2NMe_2$	2.564	2.118		96
102a	Ph	Me	H	2.660	1.910		99

^{a}Deviation of Si atom from equatorial plane toward Cl is negative, toward N is positive.

(1) In general, pentacoordinate silicon complexes have a five-membered chelate cycle. Few exceptions are known with larger (six-membered, **102**)[99] and smaller (four-membered **103**[97a], **104**[97b]) chelate cycles.

(103) **(104)**

(2) The dative bond to nitrogen generally occupies an axial position, with one of the electronegative ligands (F, Cl, O, N) in position *trans* to it. If only one electronegative element is attached to silicon, it always occupies the *trans* position to the dative bond. If more than one electronegative ligand is present, the ligand in the apical position relative to the dative bond is determined by the apicophilicity of the ligand, which follows the order: Cl, OCOR > F ~ SR > OR, NR$_2$ > aryl > alkyl > H[4,77]. The apicophilicity is the aptitude of a ligand to occupy the apical position and support the *trans* coordination[4]. It depends on both the electronegativity of the ligand and the polarizability of its bond to silicon.

(3) An approximate inverse relationship is found between the bond lengths of the opposing axial ligands. For example, the Si−N and Si−Cl axial bonds may be compared in the series of complexes **96c, 96a, 98** and **94b** (Table 10). Along this series the Si−Cl bonds shorten from 3.908 to 2.269 Å, while the coordinative Si−N bonds lengthen from 1.766 to 2.028 Å. This behavior roughly represents a concept of 'constant total bonding' in the two axial bonds to silicon. It agrees well with the hypervalency theory, first developed by Musher[100] and further discussed by Pestunovitch and coworkers[101,102], according to which the axial ligands are bound to the p$_z$ orbital of silicon in a three-center four-electron MO system. This approach is particularly compatible with the observation of a pronounced effect of the two axial ligands on each other and on the overall stability of the hypervalent complex, and the lack of a significant similar effect of the equatorial ligands.

Furthermore, the inverse variation in length of the axial bonds to silicon may be viewed as a representation of different points along a hypothetical reaction coordinate of a bimolecular nucleophilic substitution (S$_N$2) reaction at silicon[4,6,86,103,104]. The progress along this reaction coordinate is reflected in a decrease in the donor–silicon distance, and a parallel increase in the Si–leaving group distance. Another (related) parameter which may serve as a probe for the progress of the nucleophilic substitution reaction is the Δ value shown in Table 10, which is the deviation of the silicon atom from the central (equatorial) plane. Indeed, examination of the Δ values for **96a, 98** and **94b** (Table 10), which have the same equatorial ligands, shows a variation from positive (0.167 Å) through zero (0.05 Å) to negative (−0.135 Å), respectively, as a function of the change in apical ligand opposite the coordination bond to nitrogen. Thus, depending on the position along the reaction coordinate, the silicon geometry has retained or inverted its initial configuration.

(4) An effect of equatorial substituents on the complex strengths has been noted[87,102]: the more electronegative substituents at equatorial positions tend to shorten the axial bonds. This effect is rather small, but apparently significant. For example, compounds **94b, 94c** and **94d** (Table 10) differ only in the number of equatorial chloro groups (0, 1 and 2, respectively); as a result, *both* of the axial bonds decrease in length: Si−Cl, 2.269, 2.207 and 2.150 Å, respectively, and Si−N, 2.028, 2.027 and 1.984 Å, respectively[87].

The equatorial bonds (in an ideal TBP geometry, when $\Delta = 0$) are orthogonal to the three-center MOs associated with the axial ligands, and bond shortening is presumably caused by inductive charge withdrawal from silicon, and a resulting stronger attraction between the latter and the axial ligands.

(5) In complexes where two nitrogen ligands are available for coordination, the more nucleophilic or stronger base will preferentially coordinate to silicon[96]. In **101** (Table 10), a diemthylamino group and a hydrazino nitrogen compete on the coordination site to silicon. The evidence shows that only the hydrazino group is effectively coordinated. However, if both nitrogens are part of hydrazino groups and hence more nucleophilic (as in the symmetrical **105**), *both* are coordinated to silicon and form a hexacoordinate complex[96]. That this is the result of nucleophilic strengths of the ligands, and not from the symmetry, is evident from the symmetric complex **90c**, in which the ligands are dimethylamino groups, and only one nitrogen is coordinated in the solid state and forms a pentacoordinate chelate[80-82].

(105) **(106)**

Another manifestation of the different strength of coordination as a function of the ligand–nitrogen nucleophilicity may be seen in the comparison of complexes **100d** and **106**. Although no crystal structures are available for these compounds, the strength of coordination can be evaluated indirectly by looking at the ^{29}Si chemical shifts: the more upfield shifted silicon resonance represents a more highly coordinated pentacoordinate complex. Indeed, δ^{29}Si was measured for the two complexes, -30.6 and -41.8 ppm, respectively. Thus, the stronger amino base (*iso*-propylidenehydrazide) in **106** coordinates more strongly to silicon than the dimethylhydrazino group in **100d**[105].

(6) Finally, an interesting effect on axial bond lengths was found upon protonation of an adjacent (not coordinated) amino nitrogen[81,82]. Substantial changes in axial bond lengths were observed in **90c** (in the solid state) when the hydrochloride salt was crystallized and its structure determined by X-ray crystallography: the axial Si–Cl distance increased slightly (from 2.18 to 2.22 Å in **90d**), while the dative N–Si bond decreased in length (from 2.29 to 2.16 Å). The interpretation offered by the authors was that protonation of the uncoordinated nitrogen causes some increase in the electronegativity of silicon which, in turn, strengthens the coordination to nitrogen and hence shortens the N→Si bond. Increase of the N→Si coordination is accompanied by a parallel decrease in the Si–Cl bond strength (increasing length), in line with the 'constant total bonding' concept discussed earlier.

b. ^{29}*Si NMR spectroscopy.* Upon coordination of silanes to form pentacoordinate complexes, substantial upfield shifts of the ^{29}Si resonances in the ^{29}Si NMR spectra are observed. However, only in chelates, in which the overall charge on the molecule does not change upon complexation, can an exact contribution of the coordination to the ^{29}Si chemical shift be evaluated. This is accomplished by taking the difference ($\Delta\delta$) between δ^{29}Si(complex) and δ^{29}Si(silane), where both have the same chemical environment around silicon, except that in the complex an additional coordination to a neutral atom is present. $\Delta\delta$ is commonly referred to as the 'coordinative shift', and serves as a measure for the extent of coordination[79,81,88].

TABLE 11. ^{29}Si chemical shifts (δ) and coordinative chemical shifts ($\Delta\delta$) for (N—Si) chelates at 300 K.

Compound	X	Y	Z	δ^{29}Si (ppm)	$\Delta\delta$ (ppm)	References
90f	F	F	F	−102.27	−29.41	79
90a	Me	F	F	−36.13	−27.55	79
90b	H	H	1-Naph	−47.25	−11.62	79
90g	OEt	OEt	OEt	−57.17	0.59	79
90h	Me	OEt	OEt	−18.19	−0.25	79
90i	Me	Me	Me	−4.89	−0.78	79
90j	Cl	Cl	Ph	−27.5	−33.8	81
90c	Cl	Cl	2-C$_6$H$_4$CH$_2$NMe$_2$	−30.1	−36.4	96
90k	H	H	2-C$_6$H$_4$CH$_2$NMe$_2$	−45.0	−11.2	96
90l	Cl	H	2-C$_6$H$_4$CH$_2$NMe$_2$	−54.2	−48.8	81
90m	OEt	OEt	2-C$_6$H$_4$CH$_2$NMe$_2$	−35.6	−1.1	96
90n	Cl	Cl	Cl	−58.2	−57.4	81
90o	H	H	H	−71.5	−11.7	81
90p	Cl	Cl	CH=CH$_2$	−38.7	−42.5	81
90q	Cl	CH=CH$_2$	2-C$_6$H$_4$CH$_2$NMe$_2$	−27.7	−22.4	81
92c	Ph	H	H	−44.1	−8.47	79
92a	F	F	F	−96.0		79
92d	Ph	OMe	OMe	−38.16	−10.56	79
102a	H	Me	Ph	−25.84	−6.03	79
102b	H	H	Ph	−55.52	−19.89	79
94a	Me	Me	Me	−1.5		88
94b	Cl	Me	Me	−53.6	−48.2	88
94c	Cl	Cl	Me	−75.9	−70.9	88
94d	Cl	Cl	Cl	−125.2	−71.2	88
95	Cl	Me	Me	−21.0	−35.1	88
96a	Cl	Me	Me	6.4		89
97a	Cl	Me	Me	17.1		91
97b	Cl	Cl	Me	−19.3		91
97c	Cl	Cl	Cl	−81.9		91
100a	Cl	Ph	Me	−30.6		95
101a	Cl	Cl	2-C$_6$H$_4$CH$_2$NMe$_2$	−31.0	−37.1	96
101b	OEt	OEt	2-C$_6$H$_4$CH$_2$NMe$_2$	−42.1	−7.6	96
101c	H	H	2-C$_6$H$_4$CH$_2$NMe$_2$	−43.8	−10.0	96
103	Cl	Cl	Me	−30.9	−15.5	97

Table 11 lists various N→Si pentacoordinate complexes and their respective ^{29}Si chemical shifts and coordinative shifts. When the latter are near zero, complexation is minimal or nonexistent. Conversely, a large $\Delta\delta$ value represents a substantial effect of coordination, and may be taken to mean a stronger complex. For example, compounds with three alkoxy ligands attached to silicon (**90g**) have very small $\Delta\delta$ values, and are considered barely coordinated. On the other hand, comparison of **90c, 90a, 90b** and **92c** shows a gradual decrease in $\Delta\delta$ (−36.4, −27.6, −11.6 and −8.5 ppm, respectively), with a concomitant increase in coordinative N→Si distances (2.291, 2.356, 2.44 and 2.584 Å, respectively). Thus, in a qualitative manner the two criteria for coordination strength, dative bond length and coordinative shift, seem to agree.

Another interesting observation in connection with ^{29}Si chemical shifts of chelates is their pronounced temperature dependence[1,83,89,91]. In general, at lower temperatures the ^{29}Si signals are further shifted to high field, suggesting stronger complexation. The upfield shift at low temperatures has been interpreted in two ways: (a) a shift of the equilibrium between tetra- and pentacoordination toward the latter, or (b) a shorter dative bond length, in an essentially pentacoordinated system[106–108]. It appears that the available evidence is insufficient to clearly distinguish between these two options, and the question remains open.

In some cases the changes observed in ^{29}Si chemical shifts upon changing the temperature are *opposite* to those described above. For example, in complex **96a** the ^{29}Si chemical shift at 80 °C is −6.2 ppm, while at −80 °C it is shifted *downfield* to +18.4 ppm[89]. This was interpreted by Kummer and coworkers by reference to the equilibrium shown in equation 31: in the general case the equilibrium reaction which responds to changes in temperature is the interconversion I ⇌ II, in which the coordination number increases as the temperature decreases; in the special case of **96a**, however, the opposite is true: the N—Si bond is relatively strong at high temperatures (1.898 Å at RT), and becomes even stronger at low temperature causing ionization of the Si—Cl bond and formation of the tetracoordinate species III[89,91]. The change in temperature is associated in this case with the equilibrium II ⇌ III, and not I ⇌ II.

Nitrogen-coordinated pentacoordinate complexes have been used as stereoselective reducing agents in the preparation of *erythro-(meso)*-1,2-diols from diketones and α-hydroxyketones[109]. The reducing agent was the (1-naphthylamino-8)trihydridosilane **92e**. After formation of the dioxo chelate from the diketone (equation 32), the diol was obtained from the pentacoordinate silicon complex by reduction with LiAlH$_4$. ^{29}Si NMR spectroscopy was used for the product-ratio analysis in this reaction, which was found to yield primarily the *erythro* diols.

(32)

3. Stereodynamics

Ligand exchange phenomena in neutral Si—N complexes differ from those in silicates; not only pseudorotation, as in the latter, but also N—Si cleavage was found to take place in this system[2,5,6,103]. The question of 'regular' (nondissociative) vs 'irregular' (dissociative) mechanism of ligand site exchange was studied extensively[76,77,83]. The two mechanisms can be distinguished by reference to the coordination number: in the 'regular' mechanism the coordination number of silicon remains five throughout the process, i.e. no bond cleavage takes place. In the 'irregular' mechanism the exchange involves reduction in the coordination from five to four at an intermediate stage. This problem was discussed in review articles[5,6].

In many studies the diastereotopic N-methyl groups of a dimethylamino ligand were found to be convenient for monitoring and mechanistic elucidation of the exchange processes[2,5,6,76,77]. However, the stereodynamics of the N-methyl groups alone is insufficient for the exact assignment of the exchange mechanism: for example, in series **92** *both* pseudorotation *and* Si—N cleavage bring about coalescence of the N-methyl groups, and hence the two mechanisms are indistinguishable[77,110]. Coalescence of the silicon ligand signals in these compounds can also lead to mechanistic information, but again it is insufficient by itself. In **92a**, for example, the process rendering the axial fluoro group equivalent to the two equatorial fluorines can be monitored, and should measure the same barrier as the N-methyl coalescence. However, the three fluorines may become equivalent either by Si—N cleavage, followed by rotation about the Si—C and reclosure of the dative bond, or they may equilibrate by pseudorotation.

Exact assignment of either mechanism was possible only by taking advantage of an additional structural feature: (a) If in addition to a chiral silicon atom a second chiral center was present in the molecule (**91a**, or **91b** in which one fluorine is axial and the other equatorial, as well as **100c**). In these cases two diastereomers can be seen in the NMR spectra, in each of which the N-methyl groups are diastereotopic and nonequivalent under conditions of slow exchange[76,95,111]. Two different coalescence phenomena were observed in **91b**, as a result of pseudorotation at silicon and Si—N cleavage[111]. (b) If a chiral neighboring carbon is present, while the ligands at silicon are fluorines or other NMR-active groups and are observed directly, even though silicon may not be chiral. This situation is found, for example, in **91c**: there are no diastereomers in this compound, but pseudorotation can be observed by the equilibration of the axial and equatorial fluoro groups, while Si—N bond cleavage renders the N-methyl groups equivalent[111]. (c) The third structure type allowing, in principle, assignment of pseudorotation and Si—N cleavage is a system containing a chiral silicon and a prochiral group elsewhere, such as in **90e**[76]. Coalescence of the N-methyl groups represents Si—N cleavage, while coalescence of the benzylic methylene protons results from rapid inversion of configuration at the silicon center.

By applying these three methods, Corriu was able to demonstrate the existence of both processes (pseudorotation and Si—N cleavage) in several complexes of type **90-91**, with a chiral or prochiral neighboring carbon[76,77,110,111]. This enabled a discussion of the factors affecting the barriers for Si—N bond cleavage. The barriers ranged between 7.6 (in **91d**) to 16.7 kcal mol^{-1} (in **91e**), as a function of the ligands on silicon, and were taken as measures for the strength of coordination. The following order of ligands was established, corresponding to the relative strength of coordination: Cl, OAc > F, SR > OR, NR$_2$ > aryl > alkyl > H. This order is essentially the same as that of the apicophilicity series[76,77].

Another related effect of ligands on the barrier is due to the number of similar electronegative ligands. In Table 12 are listed the free energy barriers of three complexes differing in the number of fluoro ligands. As expected, the greater total electronegative substitution on silicon effects stronger coordination of nitrogen, and hence results in higher barriers for cleavage of the Si—N coordination bond. This effect parallels the observation discussed earlier (Section III.A.2.a), in which increasing the number of chloro ligands resulted in shorter Si—N distances in the solid state.

TABLE 12. Barriers for pseudorotation and Si—N cleavage in neutral pentacoordinate complexes

Compound	X	Y	Z	$\Delta G^{\ddagger}$ pseudorotation (kcal mol^{-1})	$\Delta G^{\ddagger}$ Si—N cleavage (kcal mol^{-1})	References
91e	F	Me	Ph	>17	10	76
91f	F	F	Me	9.4	11.8	77, 111
91g	F	F	F	13.1	15.8	77, 111
100c	Cl	Ph	Me	18.7	11.4	95

The relationship between complex strengths and the silicon–nitrogen cleavage barriers has recently also been reported in another study[95]. Compounds **100a–100h** were prepared from MePhSiCl$_2$ and trimethylsilylated hydrazides according to equation 30, and their barriers for N-methyl exchange were studied by NMR spectroscopy. In order to unambiguously assign the observed barriers to either pseudorotation or Si–N cleavage, **100c**, possessing two chiral centers, at silicon and carbon, was studied. Both barriers for pseudorotation (18.7 kcal mol^{-1}) and Si–N cleavage (11.4 kcal mol^{-1}) were measured for **100c**, by monitoring the coalescence spectra for the N-methyl groups, and the equilibration of diastereomers caused by epimerization at the silicon center: coalescence of the initial four N-methyl signals to two without concomitant coalescence of signal pairs due to the two diastereomers must result from Si–N cleavage. Conversely, coalescence of signal pairs associated with equilibration of diastereomers is a result of epimerization of the silicon center via pseudorotation.

The single activation barriers determined for each of the rest of the series (**100**) were near or below the lower of the two barriers in **100c**, and it was concluded by analogy that all of the barriers represented Si–N cleavage. A linear correlation was found between the Si–N cleavage barriers and the ^{29}Si chemical shifts of the corresponding complexes (Figure 11). The correlation was interpreted as two manifestations of the coordination strength, as a function of the R substituent: the higher Si–N cleavage barriers represent

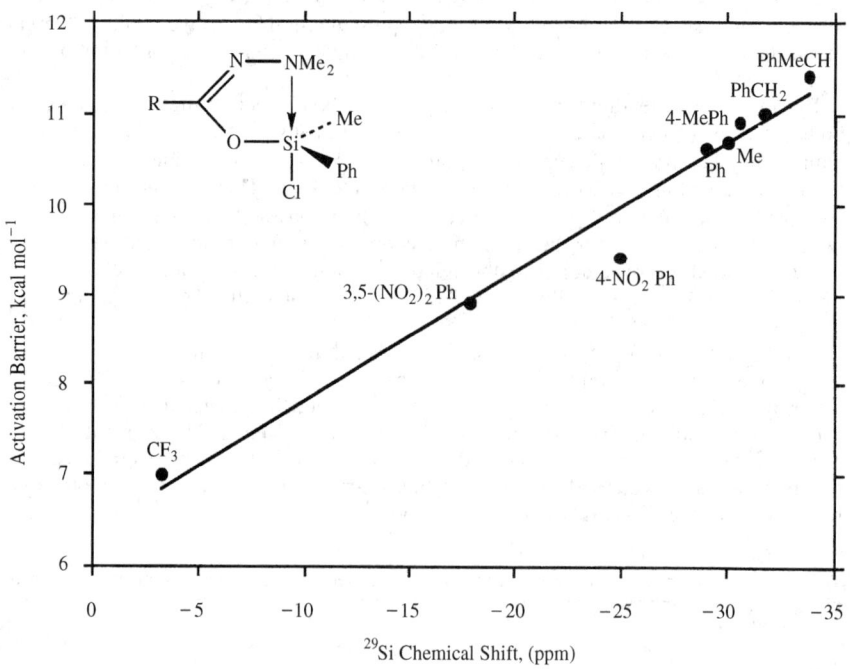

FIGURE 11. Correlation of activation barriers for Si–N cleavage and ^{29}Si chemical shifts (at 300 K) for complexes **100a–100h**[95]

stronger coordination which, in turn, are associated with ^{29}Si chemical shifts more characteristic of pentacoordination, i.e. at higher fields. This conclusion was further supported by a correlation found between ^{29}Si and ^{15}N NMR chemical shifts (taken as 'coordinative shifts': the differences between shifts of penta- and corresponding tetracoordinate silanes, Figure 12)[95].

The parameters affecting the barriers for pseudorotation are more subtle and not completely clear. Corriu has pointed out the existence of two different pseudorotation processes[77]: one which involves only two steps and exchanges between axial and equatorial ligands in LSiRX$_2$-type chelates. The other process is a complete inversion of configuration at silicon, brought about by at least five consecutive pseudorotation steps (Figure 13). This latter process involves a substantially high activation barrier, because it requires that at an intermediate step the chelate be connected in a diequatorial position associated with considerable ring strain (the 90° angle is replaced by 120°). Several barriers of the first type were measured and those ranged (for difluoro or dichloro complexes) between 9 and 12 kcal mol^{-1}[77]. For the trifluoro complex **91c** the barrier was higher, 13.1 kcal mol^{-1}. No explanation was offered for this difference in barriers, despite the conclusion that they represent similar processes: the exchange of axial and equatorial fluorines by the two-step pseudorotation. The barrier for the five-step pseudorotation cannot be measured without a second chiral center(in addition to silicon) in the molecule[76,77].

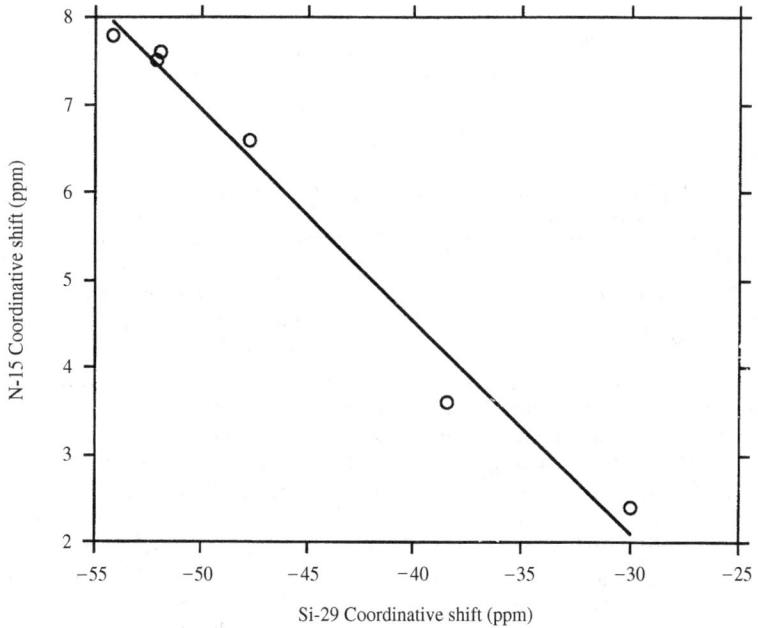

FIGURE 12. ^{15}N NMR coordinative chemical shifts as a function of corresponding ^{29}Si shifts for compounds **100**[95]

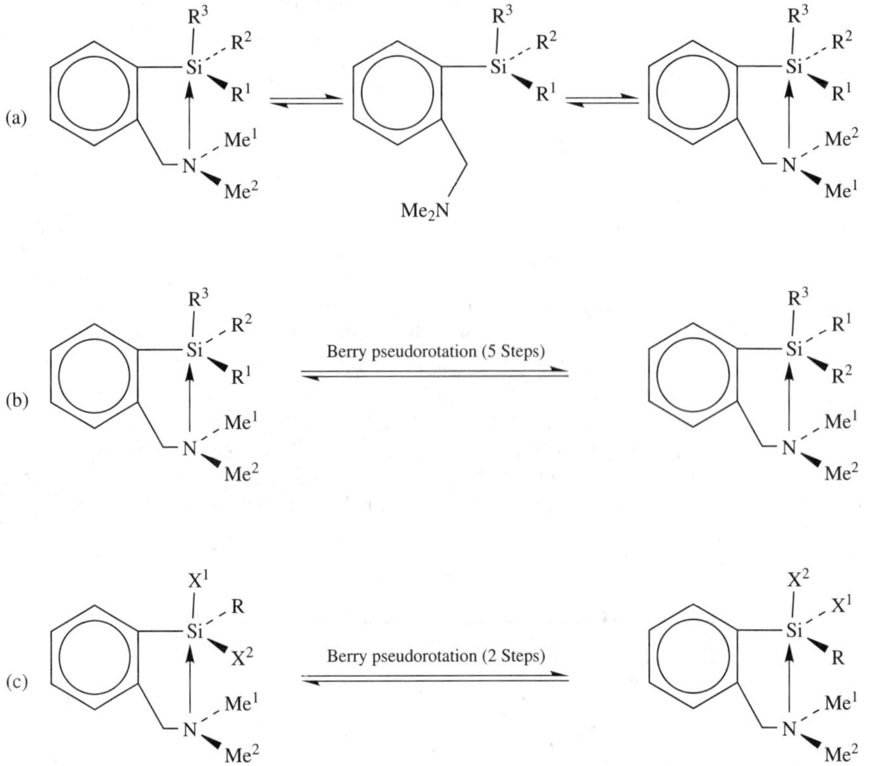

FIGURE 13. Ligand–site exchange mechanisms in **90**: (a) 'irregular' N-methyl exchange via Si–N cleavage; (b) inversion of configuration at silicon through a five-step pseudorotation; (c) exchange of X (fluoro) ligands on silicon without epimerization through a two-step pseudorotation

The barrier for inversion of configuration at silicon was, however, reported recently in a different type of complex, **100c**, in which the chelate cycle is apparently more flexible than in **91a**[76] and allows for the transformation through a 120° angle[95]. The barrier (18.7 kcal mol^{-1}) for pseudorotation (epimerization) was determined from the coalescence of signal pairs of the two diastereomers. Indeed, this barrier is substantially higher than those measured for the two-step pseudorotation, as might have been expected.

An interesting case of ligand exchange through a 'flip-flop' (intramolecular displacement) mechanism was described by Auner and Zybill and coworkers[112,113]. Complex **90c** was shown to be pentacoordinate in the ground state by a crystallographic study (Si–N1 = 2.291 Å while Si–N2 = 4.93 Å)[112], as well as in solution (δ^{29}Si ranged between -30 ppm at room temperature to -54.5 ppm at -70.1 °C)[113]. In solution, the evidence showed equivalence of the two dimethylamino groups and methylene groups, as a result of rapid intramolecular displacement of each other through a hexacoordinate transition state. Two possible transition states of this type could be invoked, **a** and **b** (Figure 14), which *both* have C_2 symmetry and hence retain the configuration at silicon. It was concluded that this process proceeded with retention of configuration at silicon, which caused the two geminal protons in each of the methylene groups to remain diastereotopic.

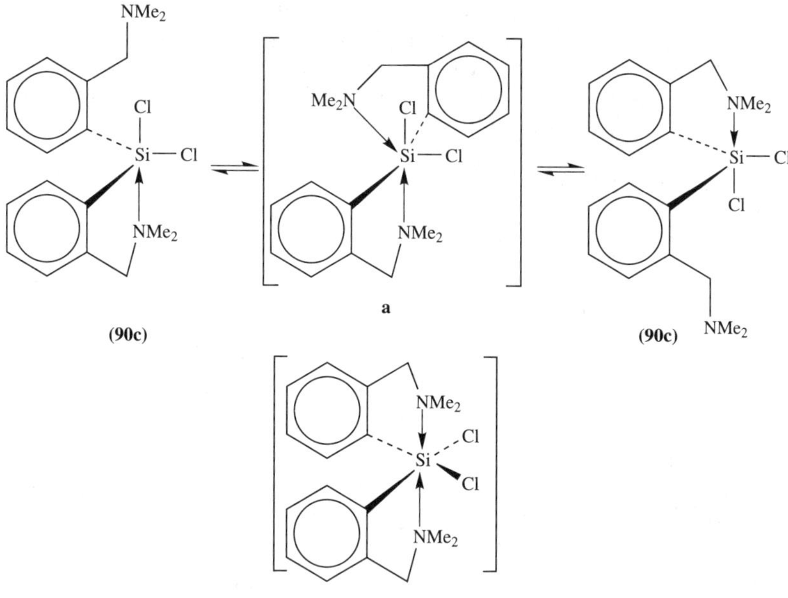

FIGURE 14. Exchange of (dimethylamino)methyl groups in **90c** through hexacoordinate transition states, **a** or **b**[113]

4. Intramolecular Lewis-base-stabilized low-valency silanes

In an attempt to stabilize low-valent silicon compounds, a class of compounds was studied in which N → Si coordination served to stabilize a double bond between silicon and either sulfur[114,115], phosphorus[116], nitrogen[116], oxygen[114,115,117] or a transition metal[113,118–121] (e.g. **107**). In these compounds the silicon is formally pentavalent, though it is coordinated to only four atoms. This topic belongs more appropriately to the chapter on silylenes, and is summarized here briefly for the sake of completeness.

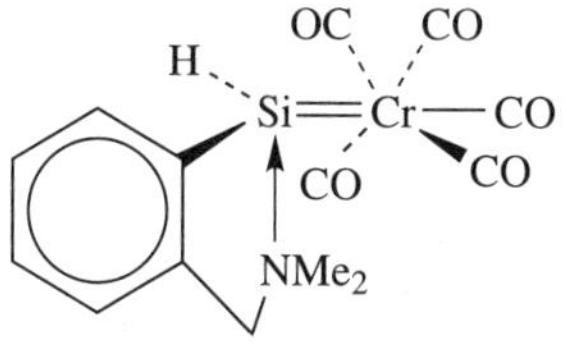

(**107**)

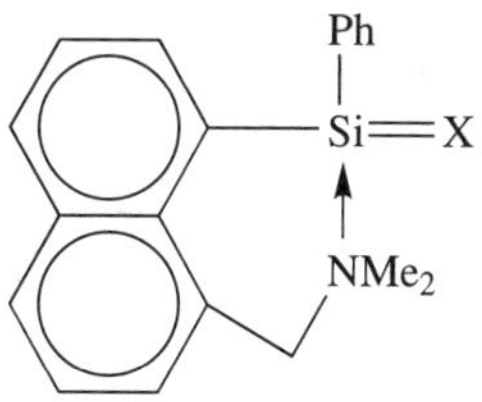

(**108**) X = O
(**109**) X = S
(**110**) X = NBu-t
(**111**) X = PPh
(**112**) X = Fe(CO)$_4$
(**113**) X = Mn(MeCp)(CO)$_2$

Intramolecular base-stabilization of low valency silicon compounds was first reported by Corriu's group in 1988[117]. The reaction of a dihydrido chelate **102b** with CO_2 gave the silyl formate which, upon heating, was shown to yield the transient silanone **108** which spontaneously trimerized by opening the Si=O double bonds (equation 33). Subsequently a stable silanethione (**109**) was isolated from the reaction of CS_2 and **102b**, and an analog containing a 1-naphthyl instead of the phenyl substituent was characterized by X-ray crystallography[114]. The Si–N distance was short (1.96 Å, relative to the common 2.3–2.5 Å) indicating very strong coordination. Several analogous silanethiones with different ligands at silicon were prepared[115].

The reaction of **102c** with the lithium derivative of *t*-butylamine replaced both fluorines and resulted in the silaneimine **110**[116]. A similar reaction with $PhPLi_2$ yielded *syn* and *anti* diastereomers of **111**, observable in the ^{31}P and ^{29}Si NMR spectra[116]. The presence of the stereomers is indicative of a double bond in **111**; however, the ^{29}Si chemical shift (−2.7 and −6.4 ppm for each diastereomer, respectively) is in disagreement with previously reported base free silaphosphenes (150–200 ppm)[122], and suggests that the structure is in fact zwitterionic (with positive charge on nitrogen and negative charge on phosphorus). Likewise, the one-bond couplings ($^1J_{PSi}$ = 9.2 in **102c** and 24.1 Hz in **111**) are in very poor agreement with the previously reported large coupling of 130–150 Hz in base free silaphosphenes[122].

Finally, **102b** reacted readily with $Fe(CO)_5$ to yield the remarkably stable **112**, and with cyclopentadienylmanganesetricarbonyl to yield **113** with double bonds between silicon and the metal[116]. Analogous aryl(dimethylaminomethyl)dihydrosilanes reacted with other transition metal carbonyls in the presence of light and yielded analogous doubly bonded, pentavalent silanes[118–120]. Like in **109** (see above), the Si–N distance in transition-metal silicon complexes was relatively short (1.99 Å in **114**[113] and 1.96 Å in an iron complex[120]) indicating substantial coordination and 'Lewis base stabilization' of the double bonds.

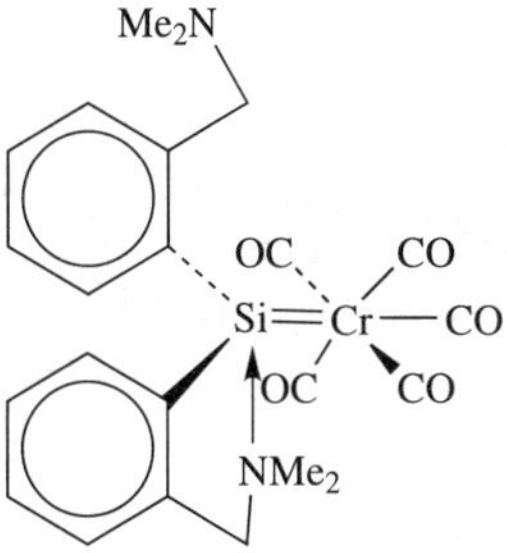

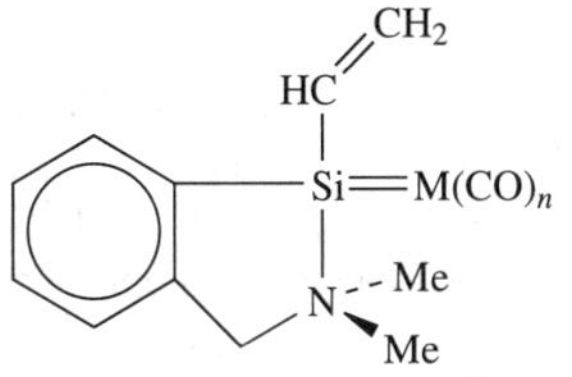

(115a) M = Fe, n = 4
(115b) M = Cr, n = 5

(114)

The analogous complex of chromium (**114**), obtained from **90c** and $Na_2Cr(CO)_5$, showed a 'flip-flop' intramolecular displacement reaction of the two dimethylamino ligands by each other, like **90c** itself[112,113]. This dynamic exchange was observable by the variable-temperature NMR spectra, which gradually changed from a slow exchange (on the NMR time scale) of dimethylamino groups to fast exchange with no discrimination between the two ligand groups, and finally at 58 °C the benzyl-methylene protons also became equivalent. The ^{29}Si chemical shift changed *downfield* with increasing temperature, an observation which was explained as evidence that coordination of *both* dimethylamino groups was no longer significant[113].

The silicon–metal doubly bonded compounds can further react to yield desired derivatives. Recently, vinyl magnesium bromide was reported to react with chloro complexes containing chromium or iron double bonds to silicon, to yield the 1-metalla-2-sila-1,3-diene compounds (**115**, equation 34), which can potentially be used to functionalize silicon polymers in a desired fashion[121].

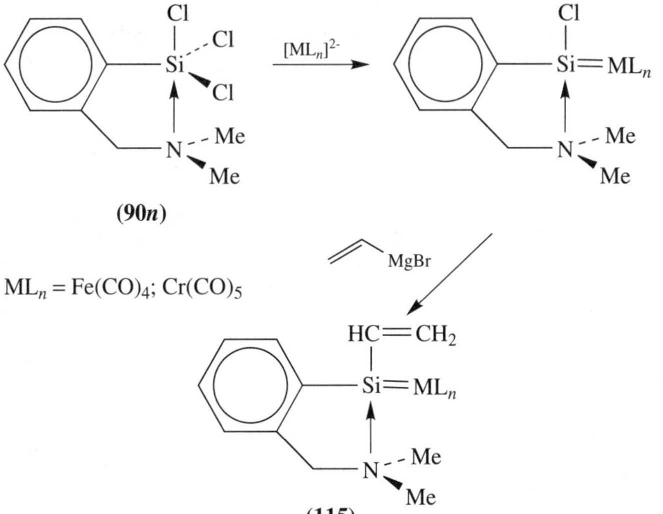

(90n)

$ML_n = Fe(CO)_4; Cr(CO)_5$

(115)

(34)

Attempts to stabilize a sila-ylide (hypercoordinated silylene) by intramolecular Lewis-base coordination did not result in sufficiently stable compounds for actual isolation. Formation of the sila-ylide was demonstrated by trapping with 2,3-dimethylbutadiene[123].

B. Chelates with Oxygen–Silicon Coordination

1. Synthesis

a. Amide-type (O–Si) complexes. The reaction of dimethyl(chloromethyl)chlorosilane (ClCH$_2$SiMe$_2$Cl, **116**) with N- or O-trimethylsilyl-amides[124–127], -lactames[128–130], -ureas[131], -acetylacetamide[132–134], 1,1-dimethyl-2-acylhydrazines[135–137] and 2-oxypyridines[104,138] afforded oxygen–silicon pentacoordinate complexes **117–125**[139].

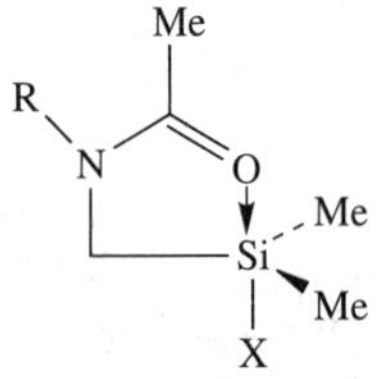

(**117a**) R = CH$_2$SiMe$_2$Cl, X = Cl
(**117b**) R = Me, X = Cl
(**117c**) R = Ph, X = Cl
(**117d**) R = 4-MeOC$_6$H$_4$, X = Cl
(**117e**) R = 4-ClC$_6$H$_4$, X = Cl
(**117f**) R = 4-BrC$_6$H$_4$, X = Cl
(**117g**) R = 4-CF$_3$C$_6$H$_4$, X = Cl
(**117h**) R = 4-NO$_2$C$_6$H$_4$, X = Cl
(**117i**) R = CH$_2$SiMe$_2$F, X = F
(**117j**) R = CH$_2$SiMe$_2$F, X = Cl

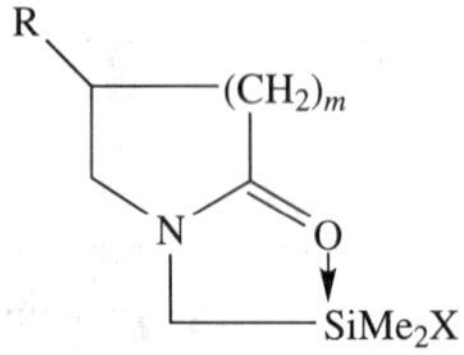

(**118a**) X = F, m = 1, R = H
(**118b**) X = Cl, m = 2, R = H
(**118c**) X = Br, m = 2, R = H
(**118d**) X = I, m = 2, R = H
(**118e**) X = Cl, m = 3, R = H
(**118f**) X = Cl, m = 1, R = Ph
(**118g**) X = F, m = 2, R = H
(**118h**) X = OPh, m = 1, R = H
(**118i**) X = OCOPh, m = 1, R = H
(**118j**) X = OC$_6$F$_5$, m = 3, R = H
(**118k**) X = OSO$_2$CF$_3$, m = 2, R = H
(**118l**) X = OCOCF$_3$, m = 1, R = Ph

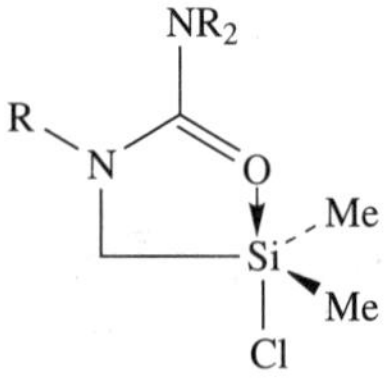

(**119**) R = Alk

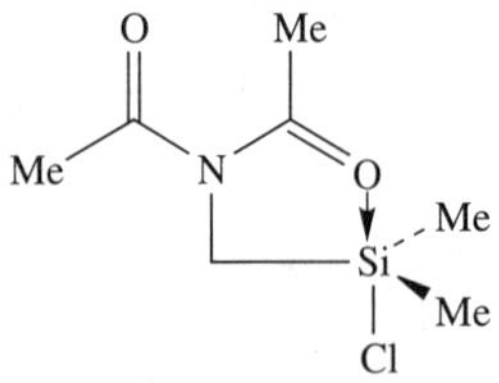

(**120a**) X = Cl
(**120b**) X = Br
(**120c**) X = I
(**120d**) X = OCOMe
(**120e**) X = OCOCF$_3$

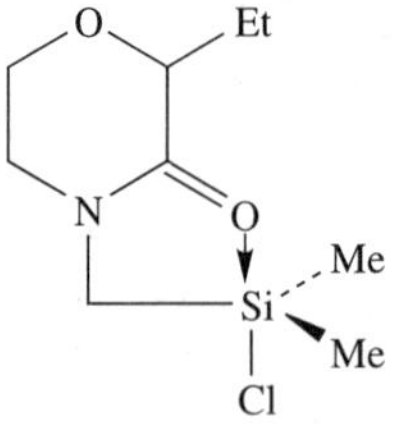

(**121**)

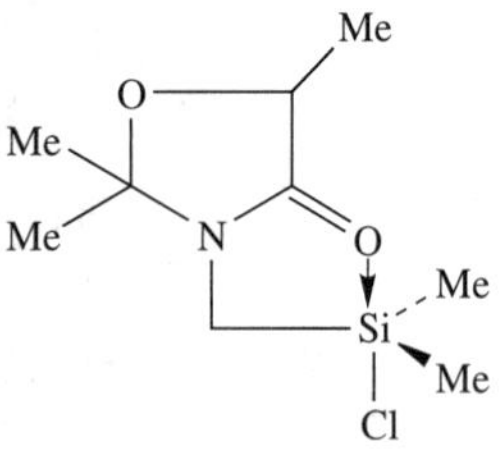

(**122**)

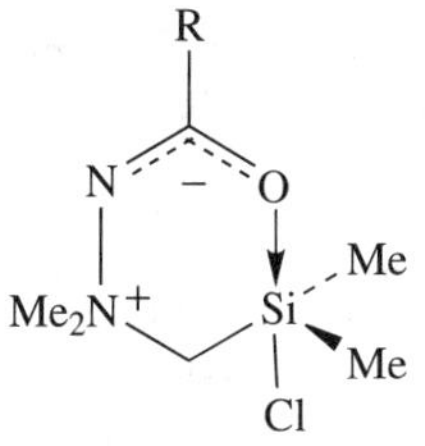

(123a) R = 4-MeOC$_6$H$_4$
(123b) R = CF$_3$
(123c) R = Me
(123d) R = Ph

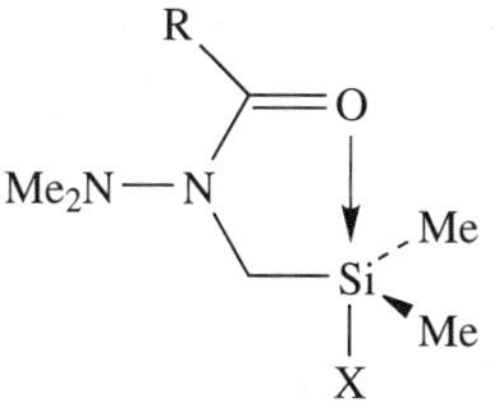

(124a) R = 4-MeOC$_6$H$_4$
(124b) R = CF$_3$
(124c) R = Me
(124d) R = Ph

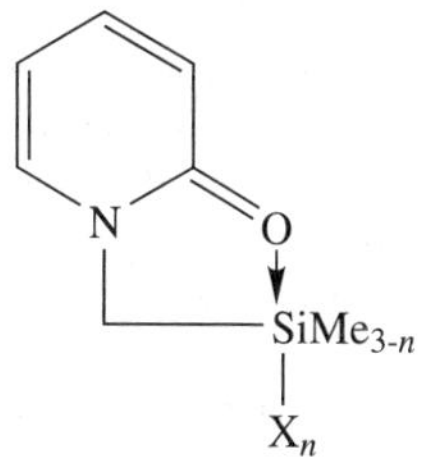

(125a) $n = 1$, X = Cl
(125b) $n = 1$, X = Br
(125c) $n = 2$, X = Cl

The reaction was shown, by NMR monitoring, to proceed in several steps (Figure 15)[140]. Initially, *trans*-silylation takes place resulting in replacement of the Me$_3$Si group by Me$_2$SiCH$_2$Cl (**126**). This intermediate was identified at low temperature (−60 to −80 °C), but was completely converted at −30 to −20 °C by an internal alkylation to **128**, or by a tautomerization to **127**. Under mild conditions (−20 °C) **128** is the main product obtained by kinetic control. At higher temperatures the final O→Si coordinated complex (**117, 118**) is observed, which may result either from a Chapman-like rearrangement[141] of **128**, or by N-alkylation of the O-silylated intermediate **127**, followed by migration of chloride to silicon.

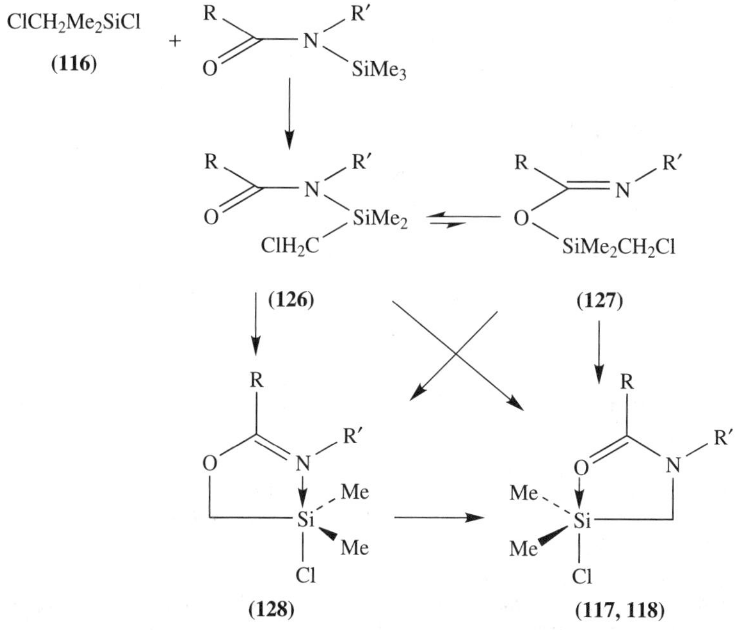

FIGURE 15. Reaction of **116** with trimethylsilylated amides

This reaction pathway appears to be general for a variety of amide-containing complexes; thus, for example, in the case of N,N-dialkyl-N'-(trimethylsilyl)urea the initial chloromethylsilylated intermediate is stable up to 200 °C and is isolated in the reaction together with the final complex **119**[131].

An interesting case of regioselectivity was observed in this synthesis when O-trimethylsilyl-1,1-dimethyl-2-acylhydrazines (**129**) reacted with **116**[135−137]. Depending on the stereochemistry at the double bond of **129** (E or Z), either the six- (**123**) or the five-membered chelate (**124**) was obtained (equations 35 and 36). Both products were stable at ambient temperature. However, heating **123** to its melting point temperature for a short period of time resulted in its irreversible transformation to **124**, in analogy to the Wawsonek rearrangement of ylides[142]. For **124** with R = Ar, CF$_3$ the only known synthetic route for the five-membered chelate is through this rearrangement.

$$\text{(129-Z)} \xrightarrow[-\text{Me}_3\text{SiCl}]{\text{ClCH}_2\text{SiMe}_2\text{Cl}} \longrightarrow \text{(123)} \quad (35)$$

$$\text{(129-E)} \xrightarrow[-\text{Me}_3\text{SiCl}]{\text{ClCH}_2\text{SiMe}_2\text{Cl}} \longrightarrow \text{(124)} \quad (36)$$

Substitution of the chloro ligand on silicon for other groups X was accomplished (for analogous complexes **117**, **118** and **120**) by the appropriate trimethylsilanes, where X = I, Br, F, OCOR, and OR[129,132,139,143]. The fluoro derivative **117i** was also prepared by treatment of the chloro complex (**117a**) with SbF$_3$[144]. The same substitution was also obtained by direct reaction of trimethylsilylated amides or lactams with the appropriate XCH$_2$SiMe$_2$Y[126,127,144].

A single-pot synthesis for complexes **117**, **118** consists of reacting amides with a 1 : 3 mixture of hexamethyldisilazane and **116**[145]. It was suggested that this reaction also proceeds in two consecutive steps as in Figure 15, but without the need to isolate intermediate products.

b. Amide-type bis-(O−Si) complexes. The reaction of **116** with O,O'-bis(trimethylsilyl)-1,2-diacetylhydrazine forms molecules with two (O−Si) chelates having a ClSi(C)$_3$O coordination framework (equation 37)[146,147]. The reaction proceeds in several steps, which could be monitored by multinuclear NMR spectroscopy. The product obtained spontaneously is the tetravalent silane **130**. However, when heat (80 °C) and excess **116** are applied the major product is the bis-complex **131a**. The latter was converted by methanolysis to the dimethoxy compound **131b**, which upon treatment with BF$_3$ was converted to the difluoro (**131c**) complex. From the ^{29}Si chemical shifts of these complexes it was concluded that the O→Si coordination decreased in the order **131a** > **131c** > **131b**. Compounds **131** are chiral due to restricted rotation about the N—N bond and a nonplanar ground state conformation. They are the first examples of substances with two (O—Si) chelated pentacoordinate silicon atoms in one molecule[147].

(37)

Bis-(O—Si) chelates containing two amide groups (**132a** and **133**) have also been prepared by similar reactions from **116** and the cyclic diamides **134** and **135**, respectively (equations 38 and 39)[148,149]. **132a** was converted to the ditriflate complex **132b** by trimethylsilyl triflate. Surprisingly, despite the very similar environments near the two silicon atoms in **133**, the ^{29}Si chemical shifts were substantially different, indicating different O → Si coordination strengths.

c. Other (O—Si) complexes. Other than amide-containing (O—Si) complexes, several complexes have been prepared with other oxygen functionalities[150]. These include triamidophosphate (**136**)[151], (aroyloxymethyl)trifluorosilane (**137**)[152] and sulfoxide (**138**)[153]. **136** was prepared by the reaction of **116** with the lithiated derivative of triamidophosphate (equation 40)[151].

$(Et_2N)_2P(O)NHBu\text{-}t + BuLi \longrightarrow (Et_2N)_2P(O)N(Li)Bu\text{-}t$

The ester-containing **137** were obtained from aroyloxymethyltriethoxysilanes with HF or SF$_4$ (equation 41)[152].

$$4\text{-RC}_6\text{H}_4\text{COOCH}_2\text{Si(OR)}_3 + \text{HF (or SF}_4) \longrightarrow 4\text{-RC}_6\text{H}_4-\underset{\underset{F}{|}}{\overset{\overset{O}{|}}{C}}\underset{F}{\overset{O \rightarrow Si-F}{\diagdown}} \quad (41)$$

(**137a**) R = MeO
(**137b**) R = H
(**137c**) R = NO$_2$

(structure **138**: R–S=O coordinated to SiF$_3$)

(**138a**) R = Me
(**138b**) R = Et
(**138c**) R = PhCH$_2$

The sulfoxide complexes **138a–c** were prepared by oxidation of the corresponding sulfides with hydrogen peroxide[153].

2. Crystal structures

The X-ray crystallographic structures of a number of O–Si complexes were described in the literature and reviewed[139,154,155]. In general, all of these complexes have a near TBP geometry. Most of these complexes have a five-membered chelate ring, with one exception, **123**, having a six-membered ring[105,139]. Some of the crystallographic data are collected in Table 13.

Some interesting features emerge from examination of the data in Table 13:

(a) A general phenomenon common to all of the complexes in Table 13 is the observation that shortening of the dative Si–O distance is accompanied by a parallel relative lengthening of the Si–X distance, where X is the apical ligand.

(b) When compounds containing an acyclic amide function (**117**) are compared with those with a lactam ring (**118b**, six-membered lactam) which otherwise have equivalent ligand frameworks at silicon, the effect of the lactam ring is to somewhat increase the Si–O distance, i.e. the coordination is weaker. In the presence of a lactam, the ring size has a small effect on coordination: for the five-membered lactam **118f** the Si–O distance was determined as 2.050 Å. the six- and seven-membered lactam systems have nearly the same Si–O bond length (**118b**, 1.954 and **118e**, 1.950 Å, respectively), significantly shorter than for **118f**. Introduction of an oxygen atom to the lactam ring has a profound effect on coordination: the Si–O distance in **122** is 2.450 Å, compared to 2.050 Å in **118f**.

(c) The Si–O bond length is dramatically effected by the type of halogen in the apical position: for **118a–118d** the respective change of ligand from F to I is accompanied by a large decrease in Si–O bond length from 2.316 to 1.749 Å[156,157]. The last value is essentially that of a covalent Si–O bond. Clearly the strength of coordination is not promoted by the electronegativity of the halogen (although electronegativity formally increases the acceptor properties of silicon) but by the apicophilicity of the element, which follows the opposite order. A similar trend in Si–O distances is found in series **120** (Table 13)[105,139], although they are slightly longer than in the corresponding **118** analogs, presumably because the second carbonyl reduces the nucleophilicity of the lactam oxygen.

TABLE 13. Crystallographic data for (O−Si) chelates

Compound	m	X	Si−O (Å)	Si−X (Å)	Δ^a (Å)	References
117a		Cl	1.918	2.348	0.029	124
118a	1	F	2.395	1.652	0.286	156
118b	2	Cl	1.954	2.307	0.058	156
118c	2	Br	1.800	3.122	−0.218	156
118d	2	I	1.749	3.734	−0.348	157
118e	3	Cl	1.950	2.315	0.055	156
118f	1	Cl	2.050	2.284		162
118g	2	F	2.316−2.461	1.630−1.665	0.238−0.305	158
118h	1	OPh	2.367	1.711		159
118i	1	OC(O)Ph	2.228	1.778		159
118j	3	OC_6F_5	2.078	1.787		159
118k	2	OSO_2CF_3	1.753	2.785		159
120a		Cl	2.077	2.229	0.131	160
120b		Br	1.978	2.467	0.052	105, 161
120c		I	1.830	3.03	−0.09	105, 161
122		Cl	2.450, 2.425	2.154, 2.148	0.251, 0.251	162
121		Cl	2.021	2.282	0.082	163
123a		Cl	1.788	2.624	−0.178	164
123b		Cl	1.879	2.432	−0.078	165
124d		Cl	1.975	2.294	0.077	105, 161
131a		Cl	2.142	2.223	0.19	105, 161
136		Cl	1.840	2.438	−0.073	151

^{a}Deviation of the silicon atom from the equatorial plane; positive Δ indicates deviation toward X.

(d) When the ligand element in the apical position is oxygen (as in **118h, 118i** in Table 13, in the general complex framework $O-Si(C)_3-O$), the coordination, as represented by the Si−O bond length, is weaker than for Cl, Br or I ligand, in accord with the rules of apicophilicity. When comparison is made between the different oxygen ligands, the dative Si−O distance *decreases* as the electron-withdrawing power of the group attached to oxygen increases **(118h−118j)**, until in **118k**, with the powerful electronegative triflate group, the dative bond is so short that it essentially is a covalent bond, while the triflate group becomes a coordinated anion. The X-ray evidence also shows a similar change in coordination from O→Si−X to O−Si←X in **118c, 118d** and **120c**, in which the apical ligand is a bromo or iodo group. A similar exchange of the coordinating atom was observed for a chloro ligand in a particular series of chelates, **123**, with a $OSi(C)_3Cl$ framework[164,165]. The Si−O and Si−Cl bond lengths (Table 13) clearly show that oxygen is attached to silicon by a nearly covalent bond, while chloride is coordinated. This obviously is a result of the greater donor ability of oxygen in the zwitterionic six-membered hydrazide ring. A second case of reversal of the coordinating O and Cl ligands is found in **136**[151], from the corresponding Si−O and Si−Cl bond lengths in Table 13.

(e) An important structural parameter in Table 13 is Δ, the deviation of the silicon from the plane defined by the three equatorial ligands. This parameter is closely related to the mutual changes in axial bond lengths discussed above, and has been referred to in connection with a model for the S_N2 reaction coordinate[139,156,161,166,167]. The silicon generally deviates towards the more covalently-bonded ligand, and away from the dative ligand. For similar complexes (**118a–118d**), in which only the axial halogen is changed, Δ is positive for the F, near zero for X = Cl, and negative for Br, I. In terms of the reaction coordinate for substitution at silicon, this means that the former (X = F) represents an early step in the reaction, before the tetrahedral silicon configuration has inverted (Figure 16). X = Cl represents a symmetrical, nearly TBP midway state (resembling the transition state for displacement at a central carbon analog) for the reaction. For the heavier halogens the progress of the reaction is more advanced: the oxygen nucleophile has already essentially displaced the halogen, which remains weakly coordinated, and the pyramidal (Walden) inversion of the silicon center is almost complete. This description of the progress along a hypothetical reaction coordinate obtained from a collection of crystal structures follows a Bürgi–Dunitz-type analysis[19,168,169].

From an analysis of the crystal structures of 18 compounds of the type $O-Si(C)_3-Cl$ the sum of Si–Cl and Si–O bond orders was shown to be roughly constant (ranging between 1.172 and 1.230)[167], while the deviations of the Si–Cl and Si–O distances from the standard tetracoordinate values (Δl_{Si-Cl}, Δl_{Si-O}, respectively) related to each other as a hyperbola (i.e. $\Delta l_{Si-Cl} \cdot \Delta l_{Si-O} = $ constant)[139,167]

3. ^{29}Si NMR

Like in the other types of pentacoordinate complexes, the ^{29}Si NMR signal in O–Si chelates is shifted to high field relative to its tetracoordinate analogs (Tables 14 and 15). Therefore the difference between the ^{29}Si chemical shifts of penta- vs. tetracoordinate silanes (the 'coordinative shift') serves as a measure for the extent of coordination (Table 15). It can be seen from the data in Table 14 that in series **117** the silicon signal is shifted downfield as the electron-withdrawing power of the *para*-substituent on the aromatic ring increases, resulting in weaker coordination[170]. This information can also be seen in the ^{14}N chemical shifts, and to a lesser extent in the carbonyl–^{13}C shifts listed in Table 14. Similar trends in the coordinative ^{29}Si shifts as well as in the carbonyl ^{13}C- and ^{17}O-chemical shifts for complexes of type **118** were reported by Pestunovich and coworkers[139].

It is of interest to compare the variation in the ^{29}Si chemical shifts with the corresponding Si–O dative-bond lengths and associated Si–Cl bond lengths in the solid state in compounds sharing the same coordinative center: $OSi(C)_3Cl$ (Table 16). Clearly all three parameters show similar trends in the series from **131a** to **123b**: high field shielding is associated with shorter Si–O and longer Si–Cl distances. The only exception to the general trend is **123a**, in which Si–O appears to be covalent and chloride coordinated, with a corresponding reversal in the deviation of silicon from the central plane of the TBP ($\Delta < 0$, Table 13, cf Section III.B.2).

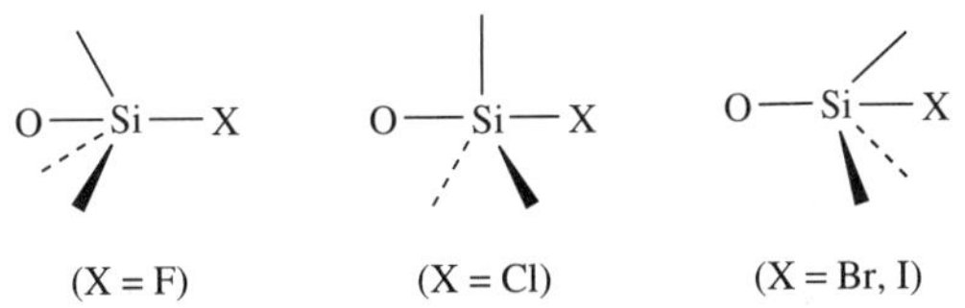

O—Si—X   O—Si—X   O—Si—X

(X = F)   (X = Cl)   (X = Br, I)

FIGURE 16. Simulated S_N2 reaction coordinate: the variation of Δ as a function of X in **118**

TABLE 14. ^{14}N and ^{29}Si NMR chemical shifts for MeCONR(CH$_2$Si(CH$_3$)$_2$Cl) (**117**)[170]

Compound	R	δ^{14}N (ppm)	$\delta^{13}\underline{C}$O (ppm)	δ^{29}Si (ppm)
117b	Me	−252	173.2	−37.6
117c	Ph	−235	173.8	−34.1
117d	4-MeOC$_6$H$_4$	−234	173.6	−34.0
117e	4-ClC$_6$H$_4$	−245	173.3	−30.7
117f	4-BrC$_6$H$_4$	−240	173.2	−31.1
117g	4-CF$_3$C$_6$H$_4$	−250	173.2	−29.0
117h	4-O$_2$NC$_6$H$_4$	−250	173.0	−26.9

TABLE 15. ^{29}Si chemical shifts (δ) and coordinative shifts ($\Delta\delta$) for (O−Si) chelates

Compound	m	R	X	δ^{29}Si (ppm)	$\Delta\delta$ (ppm)	References
118a	1		F	14.2	−9.4	139
118b	2		Cl	−38.5	−62.2	139
118c	2		Br	−22.6	−40.3	139
118e	3		Cl	−35.0	−59.3	139
118g	2		F	−19.6	−44.4	139
118f	1	Ph	Cl	−5.7	−30.0	139
118l	1	Ph	OCOCF$_3$	−14.3		139
120a			Cl	−24.2	−47.7	133
120b			Br	−29.1	−48.0	133
120c			I	−14.9	−15.6	133
123a		MeOPh	Cl	−17.4	−40.4	137
123b		CF$_3$	Cl	−42.9	−65.9	137
123c		Me	Cl	−33.5	−56.5	137
123d		Ph	Cl	−20.8	−43.8	137
124a		MeOPh	Cl	−36.3	−59.3	137
124b		CF$_3$	Cl	11.9	−12.1	137
124c		Me	Cl	−35.9	−58.9	137
124d		Ph	Cl	−34.1	−57.1	137
131a			Cl	−8.8	−31.8	147
131b			MeO	9.1		147
131c			F	4.1		147
133			Cl	−8.9, 13.3		149
137a		MeO	F	−96.3		152
137b		H	F	−94.8		152
137c		NO$_2$	F	−90.9		152
138a		Me	F	−86.4		153
138b		Et	F	−85.7		153
138c		PhCH$_2$	F	−83.6		153
125a	1		Cl	−39.0		171
125b	1		Br	−19.0		171
125c	2		Cl	−52.4		138
125d	3		Cl	−77.7		138

TABLE 16. ^{29}Si chemical shifts and Si—O, Si—Cl bond lengths in complexes with OSiC$_3$Cl ligand framework

Compound	δ^{29}Si (ppm)	References	Si—O (Å)	Si—Cl (Å)	References
131a	−8.8	147	2.142	2.223	105,161
120a	−24.2	133	2.077	2.229	160
124d	−34.1	137	1.975	2.294	105,161
118b	−38.5	139	1.954	2.307	156
118e	−35.0	139	1.950	2.315	156
117a	−42.0	138	1.918	2.348	124
123a	−42.9	137	1.879	2.432	165
123b	−17.4	137	1.788	2.624	164

A similar manifestation of the effects of coordination upon chemical shifts, and the sudden changes in trend associated with reversal of the coordinating ligand (and the change of sign of the measured deviation Δ) can be seen in Figure 17. The figure depicts a reasonable linear correlation between the NMR chemical shifts of silicon and the corresponding Si—O dative bond lengths, as obtained from crystallographic data. As long as oxygen is coordinated to silicon (and the deviation Δ of silicon from the equatorial plane is positive) the correlation holds. For those compounds in which Δ is negative, and hence the halogen is the dative ligand, the points are totally off the linear correlation[105].

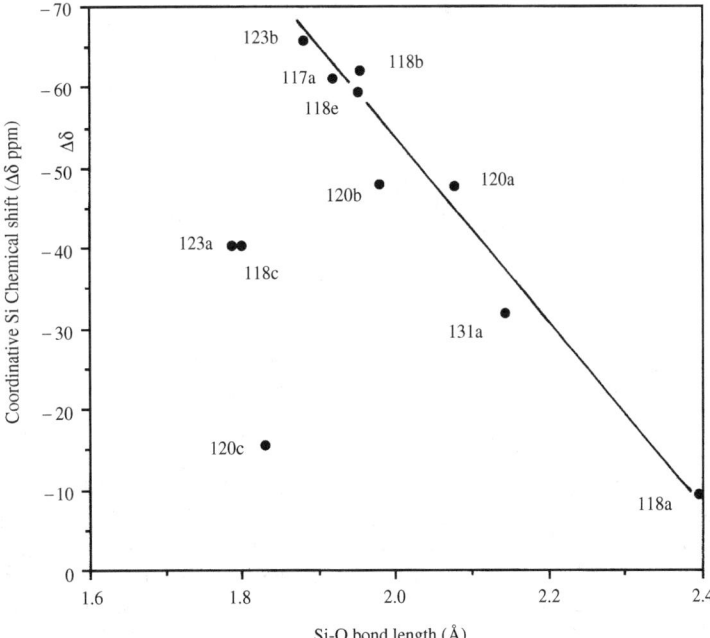

FIGURE 17. Correlation of coordinative ^{29}Si chemical shifts (Δδ) for Si—O pentacoordinate chelates with the Si—O dative bondlength[105]

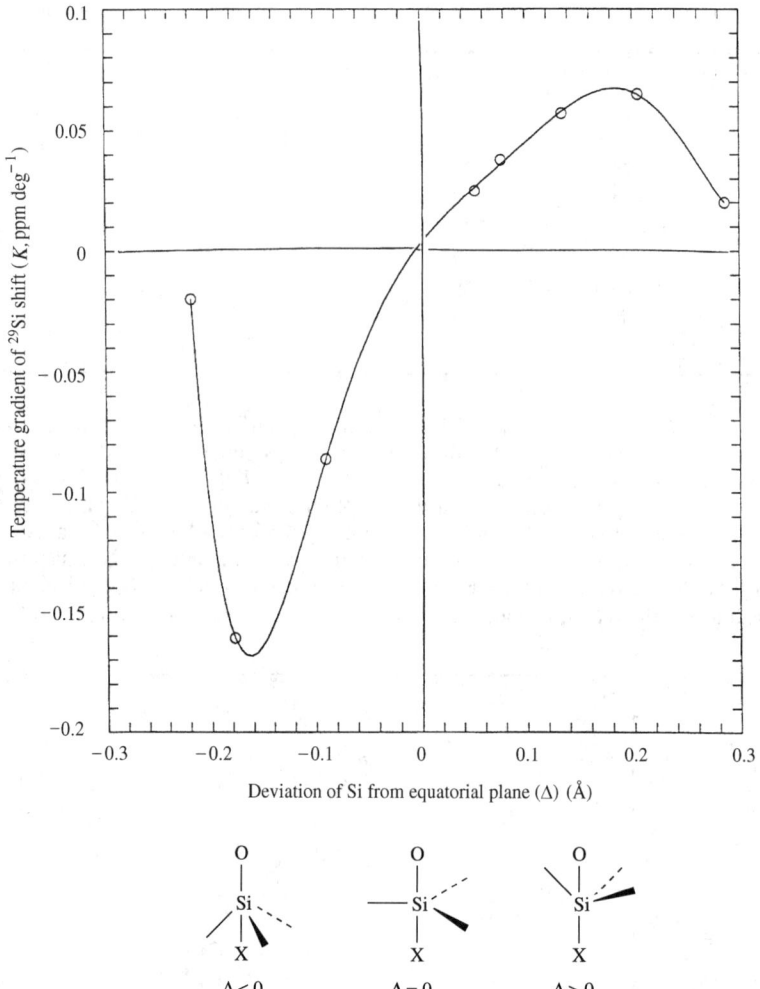

FIGURE 18. The temperature gradient of the ^{29}Si chemical shift in Si−O chelate complexes (in solution) plotted against the deviation (Δ) of silicon from the equatorial plane (in the solid state). The gradient K is expressed as: $K = (\delta^{29}\text{Si}_{T_1} - \delta^{29}\text{Si}_{T_2})/(T_1 - T_2)$

Another criterion for assessing coordination is the temperature dependence of the ^{29}Si chemical shift. In most pentacoordinate complexes the ^{29}Si signal shifts to higher field as the temperature is lowered. This has often been attributed to more intense coordination at lower temperature[139,171]. However, the *extent* to which δ^{29}Si changes with temperature differs among different compounds[137]. It was found that the temperature gradient K of the Si chemical shift is related to the deviation of silicon (Δ parameter) obtained from the crystal structure in a series of compounds sharing similar ligand frameworks at the silicon, $\text{OSi(C)}_3\text{X}$[105,139]. The function describing this relationship is shown in Figure 18[105]. The temperature gradient is positive for as long as Δ is positive, i.e. as long as the coordination is to oxygen. At large deviations from planarity (large positive Δ), the geometry at silicon is essentially tetrahedral, and coordination is negligible. At this region no significant temperature dependence of the chemical shift is found. At intermediate Δ values the gradient goes through a maximum, at which point the chemical shift shows the strongest temperature dependence: lowering the temperature for these molecules is most effective in increasing coordination.

When silicon lies in the equatorial plane ($\Delta = 0$), the coordination is strongest and δ^{29}Si does not significantly depend on the temperature (i.e. $K = 0$)[139]. Negative temperature gradients are found for those complexes in which $\Delta < 0$, i.e. in which halogen is coordinated to silicon[89-91,105,137,171]. The function plotted in Figure 18 provides yet another means of mapping the hypothetical progress along the reaction coordinate of a nucleophilic substitution at silicon, using solution (rather than solid state) data: complexes in which the gradient is positive lie on one side of the reaction coordinate, those with negative gradients are advanced toward the products and chelates with negligible temperature gradient represent the TBP intermediate at the midpoint of the reaction coordinate.

The question of mapping the reaction coordinate by observing changes in chemical shifts has also been addressed by Bassindale[104,138]. By varying the ligands and substituents on a pyridone chelate (**125**), it was possible to follow changes in coordination along the reaction coordinate described by equation 42. This was done by observing the changes in ^{13}C and ^{29}Si NMR chemical shifts as a function of structural modifications in **125**. The ^{13}C shifts were compared against the corresponding shifts of two standard compounds, **139** and **140**, which represent models for the two extreme points along the reaction shown in equation 42. From the relative location of the pyridone-ring ^{13}C chemical shifts of differently substituted chelates between **139** and **140**, the extent of Si—O bond making was assessed (Figure 19). From the variation of ^{29}Si chemical shift between the extremes of tetracoordinate silicon to fully pentacoordinated silicon, it was possible to obtain a map of nucleophilic substitution at silicon for some 25 compounds (Figure 20)[104,138]. The advantage of this method is that it is based completely on solution data, as opposed to the Dunitz–Bürgi[19,168,169] approach which compares crystal data.

(42)

(**125**)

(**139**)

(**140**)

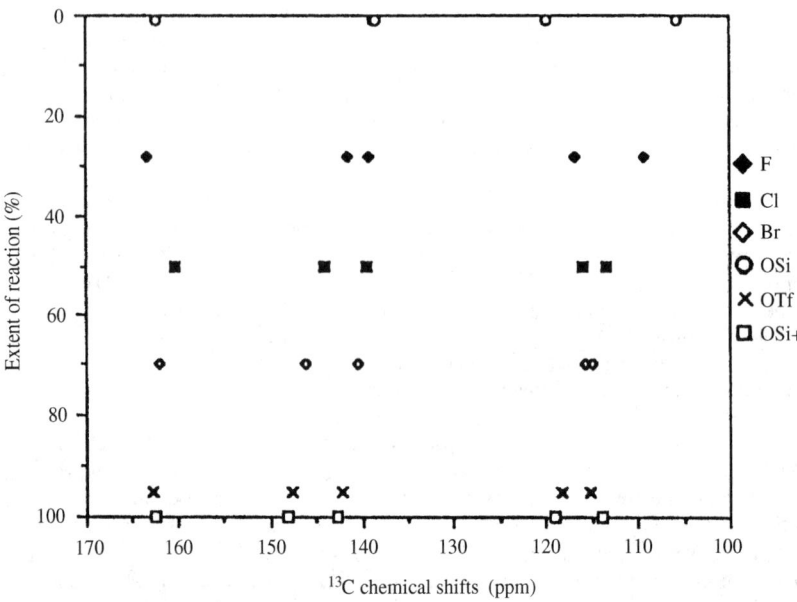

FIGURE 19. Variation of ^{13}C chemical shifts with the extent of Si—O bond making in pyridone complexes **125**[104,138]. Reproduced from Reference 138 by permission of The Royal Society of Chemistry

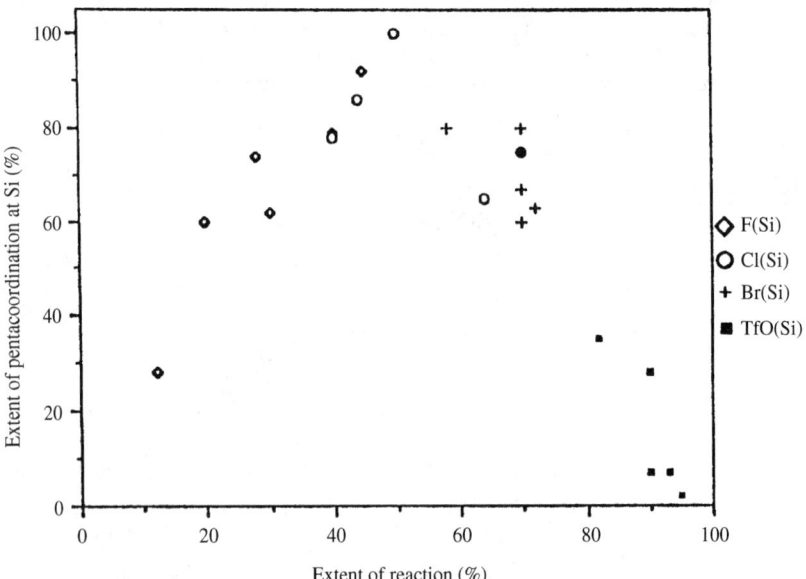

FIGURE 20. A map representing the progress along a reaction coordinate for nucleophilic substitution at silicon in solution, for a series of substituted silylpyridones. Derived from ^{29}Si and ^{13}C chemical shifts[104,138]. Reproduced from Reference 138 by permission of The Royal Society of Chemistry

The above analysis was based on data measured for a collection of silicon complexes. In a recent different approach a gradual change could be observed from O → SiCl to OSi ← Cl coordination in a single compound (**125a**, X = Cl), as the temperature was changed[171]. Figure 21 shows the temperature dependence of the ^{29}Si chemical shift for **125a** over a 180 °C range. Within this range the temperature gradient changes from positive to negative, in accord with a gradual increase of pentacoordinate character when the temperature is first lowered from 100 °C, until a maximum coordination is reached and the gradient is zero at about 10 °C, followed by a decrease in coordination with decreasing temperature after the midpoint has been reached and the coordinating ligand has switched. Figure 21 may also be viewed as a plot of the progress along the S_N2 reaction coordinate[171].

In an approach similar to Bassindale's[138], Kummer and Abdel Halim also probed the variation of ^{29}Si chemical shifts as a function of ^{13}C shifts of the pyridone-ring carbons, but using a single compound (**125b**, X = Br) and modifying the temperature, solvent and concentration[171]. These measurements were also interpreted in terms of various extents of progress along the reaction coordinate of equation 42.

4. Ligand exchange

Unlike in the Si—N complexes, where ligand exchange is conveniently followed by coalescence of the N-methyl groups, in Si—O complexes only the ligands on silicon enable monitoring of the exchange processes. In **118f** and **118l** the presence of a chiral carbon atom in the lactam ring makes the Si—methyl groups diastereotopic. Their coalescence was used to obtain ligand exchange barriers resulting from either pseudorotation or Si—O bond cleavage: 11 and 14 kcal mol^{-1}, respectively[139]. Reference to Table 15 reveals the ^{29}Si chemical shifts for **118f** and **118l**, −5.7 and −14.3 ppm, respectively. In analogy to the linear correlation found between ^{29}Si chemical shifts and Si—N cleavage barriers (cf Figure 11) we tend to conclude that the barriers in **118f** and **118l** are Si—O cleavage barriers.

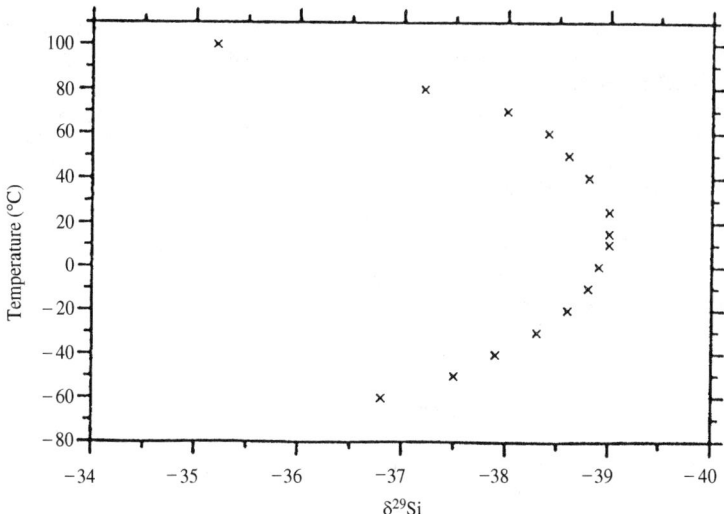

FIGURE 21. Temperature dependence of the ^{29}Si chemical shift for **125a** in CDCl$_3$. Reproduced from Reference 171 by permission of Hüthig Fachverlage

Other cases of diastereotopicity of ligand groups due to the presence of adjacent chirality are found in **138** (chiral sulfoxide) and **141** (chiral carbon). The chirality renders both equatorial fluorine ligands diastereotopic at the slow exchange temperature (*ca* 170 K)[150,153]. Exchange phenomena were measured also in **131**, in which the diastereotopicity of geminal methylene protons and Si—methyl groups results from the N—N chiral axis and the substantial activation barrier for rotation about the N—N bond[147]. Coalescence was measured for the Si—methyls at 70–75 °C, but the nonequivalence of the CH_2 protons was retained. This meant that rotation about the N—N bond was still slow at this temperature, and that exchange of Si—methyls resulted from inversion of configuration at the silicon atoms, either by pseudorotation, or by Si—O cleavage, followed by rotation and reclosure, or by intermolecular exchange. Further evidence indicated that intermolecular exchange was predominant: in **131c**, in which the Si—methyls are coupled to the F-ligand, coalescence of the methyls (two doublets) occurred simultaneously with the collapse of $^{19}F–(Si)–^{13}CH_3$ coupling (singlet). Furthermore, in a mixture of **131a** and **131b** no individual coalescence of Si—methyl groups could be observed, but rather a simultaneous exchange in both compounds, in accord with scrambling of all methyl groups due to intermolecular exchange[147].

(117a) X = Y = Cl
(117i) X = Y = F
(117j) X = Cl, Y = F

(141)

Intermolecular ligand exchange was reported also for **117a**, in which the chloro ligand in the O—Si—Cl fragment is replaced by *N*-methylimidazole more rapidly than the other chlorine, attached to the tetracoordinate silicon[144]. Interestingly, in the difluoro analog **117i** only the fluorine atom attached to the tetracoordinate silicon is replaced, whereas the coordinated fluoro group is retained. Admixture of the two chelates, **117a** and **117i**, resulted in regiospecific exchange to the mixed chelate **117j**. These experiments confirm the greater apicophilicity of chlorine vs. fluorine.

117c was shown to exist in CD_2Cl_2 solution in two different pentacoordinated forms, both by ^{29}Si and ^1H NMR spectroscopy[172]. At low temperature (185 K in the ^1H NMR spectrum) the Si—methyl groups gave two singlets, as did the ^{29}Si signal. This was attributed to the formation of an oligomeric complex: dimer, trimer or larger, in addition to the monomeric chelate, based on the change in relative signal intensities as a function of concentration.

Amide bond rotation barriers were measured in compounds of series **117** in $CDCl_3$ solutions, by measuring the coalescence of two unequal singlets for the Si—methyl groups observed at low temperature[170]. Amide rotation in these compounds (**117c–117h**) is associated with cleavage of the Si—O coordination. The barriers ranged between 13.3–16.9 kcal mol^{-1}, and the population of the open N—C rotamer ranged from 0.5 to 11%, as a function of the *para* substituent on the N-phenyl group. In other solvents (toluene, dichloromethane and THF) no splitting of the Si—methyl signals was observed and the barriers could not be measured[170].

TABLE 17. Bond lengths, ^{29}Si chemical shifts and their temperature gradients [$K = (\delta^{29}\text{Si}_{T_1} - \delta^{29}\text{Si}_{T_2})/(T_1 - T_2)$], and ligand exchange barriers in **120**[133]

Compound	X	Si–O (Å)	δ^{29}Si (ppm)	K (ppm deg^{-1})	δ^{13}CO (ppm)	$\Delta G^{\ddagger}$ (kcal mol^{-1})
120a	Cl	2.077	−24.2	0.07	175.8	8.3
120b	Br	1.978	−29.1	0.04	176.8	9.3
120c	I	1.830	−14.9	−0.11	177.1	10.6
120d	OCOCH$_3$		−11.3	0.11	174.1	7.8
120e	OCOCF$_3$		−30.1	0.04	175.9	9.2

In imide structures **120** the coordination of oxygen to silicon is switched between the two oxygens (equation 43). This was observed for the first time in **120a**, in which both ^{13}C and ^{1}H NMR C-methyl and carbonyl signals split at low temperatures[133]. The barriers for this exchange process depended on the apical ligand X in a manner which roughly follows the order of apicophilicity, and was inversely related to the Si–O distances (Table 17). These observations strongly suggest that the process represents Si–O cleavage rather than rate-determining amide rotation. Interestingly, the correlation of ^{29}Si chemical shifts with activation barriers breaks down for iodine (**120c**). This can be rationalized with reference to the temperature gradient of δ^{29}Si, K, shown in Table 17. For **120c** the temperature gradient is opposite to all other complexes. As discussed earlier (Figure 18, Section III.B.3) this is related to the change from O → Si to I → Si coordination.

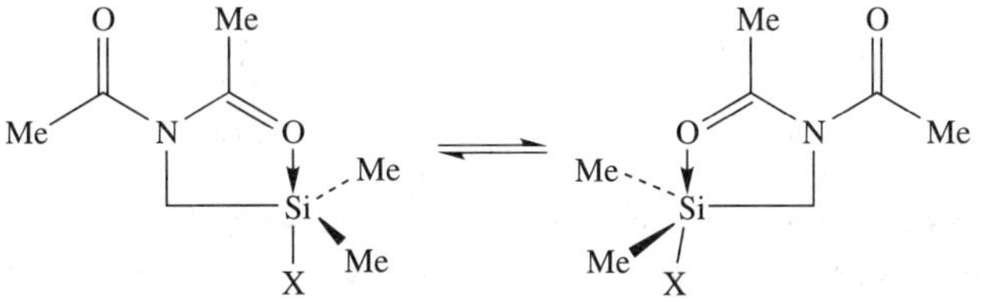

(43)

(**120**)

Similar amide-switching phenomena were reported in other compounds (**142, 143**). X-ray crystallography for **143** showed that only one amide-oxygen was coordinated to silicon (Si–O distance: 1.992 Å)[173]. However, in solution the two amide-moieties were NMR-equivalent and could not be resolved down to −65 °C in toluene-d$_8$ solution[173].

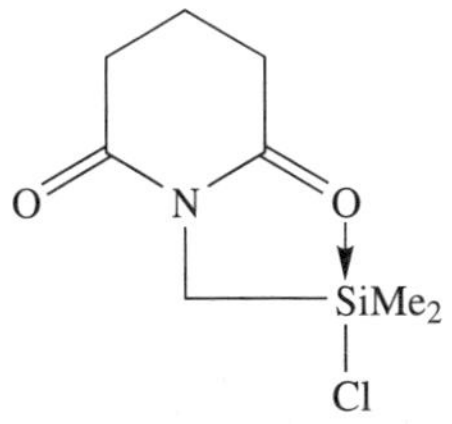

(**142**) 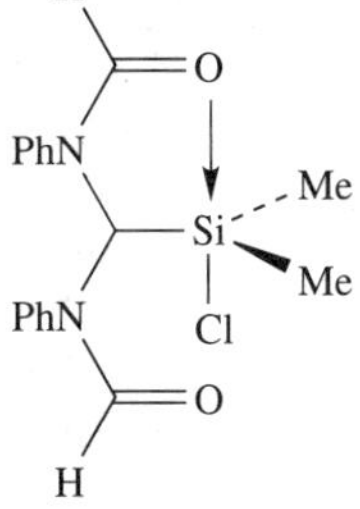   (**143**)

C. Other Neutral Pentacoordinate Silicon Compounds

1. Sulfur coordination

Sulfur to silicon coordination was achieved in a series of eight-membered cyclic silanes, **144** and **145**[174,175]. The complexes were prepared from aromatic diols by reaction with $SiCl_4$ (**144**), Me_2SiCl_2, $MePhSiCl_2$, Ph_2SiCl_2, $Ph(CH_2=CH)SiCl_2$ and $c\text{-}(CH_2)_4SiCl_2$ (**145a–145g**). The silanes were found by crystal analyses to deviate from tetrahedral geometry toward TBP structure, with sulfur coordination to silicon, except in **145b**, where the eight-membered chelate ring is in the chair conformation. Table 18 summarizes some of the crystallographic and NMR data for these compounds. Judging from the change in the ^{29}Si chemical shift of **144** and **145** relative to the tetracoordinate carbon analogs (**146, 147**), the extent of pentacoordination is not very large (ca 7 ppm upfield shift). This conclusion is supported by the calculated TBP character of the complexes, which is as low as 35–54%. The changes measured in the Si–S distances were discussed in terms of conformational changes in the macrocycles[174].

(**144**)

Several analogs of **145** were prepared in which the sulfur atom is oxidized to S=O (**148**)[176a] and to SO_2 (**149**)[176b]. All the sulfinyl compounds (**148**) have the macrocycle in a chair conformation, and hence there is no S–Si coordination[176a]. In the sulfonyl analogs, however, one of the oxygens is coordinated to silicon, as long as the phenyl-*ortho* positions are occupied by *t*-butyl groups[176b].

TABLE 18. Comparison of Si–S bond parameters and ^{29}Si chemical shifts for cyclic silanes[174]

Compound	Si–S (Si–C) (Å)	% TBP[a]	$\delta^{29}Si$ (ppm) (CDCl$_3$)	$\delta^{29}Si$ (ppm) (solid)
144	3.04	50.6		−107.84
	3.11	46.5		−99.35
145	3.292	35.8	−13.22	
	3.280	36.5		
146	3.074	48.6	−13.51	
147	2.978	54.2	−1.62	−6.55
148	(3.418)		−5.62	
149	(3.100)		5.44	5.88

[a]Percent geometrical displacement from a tetrahedron to a trigonal bipyramid.

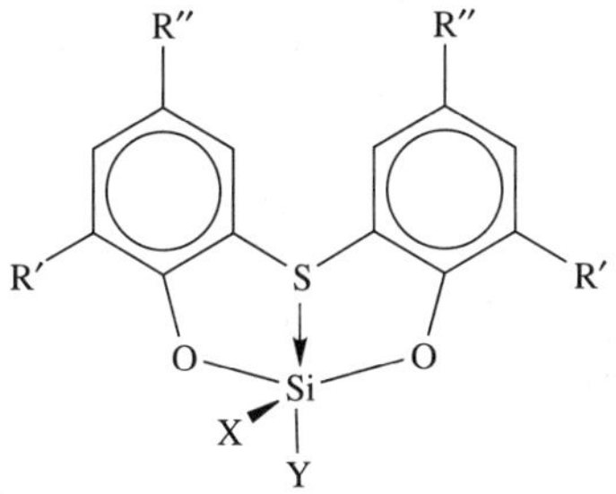

(a) X = Y = R' = R'' = Me
(b) X = Y = Ph, R' = R'' = Me
(c) X = Y = Me, R' = t-Bu, R'' = Me
(d) X = Ph, Y = CH=CH$_2$, R' = t-Bu, R'' = Me
(e) X Y = (CH$_2$)$_4$, R' = R'' = t-Bu
(f) X = Me, Y = Ph, R' = R'' = t-Bu
(g) X = Y = Ph, R' = R'' = t-Bu

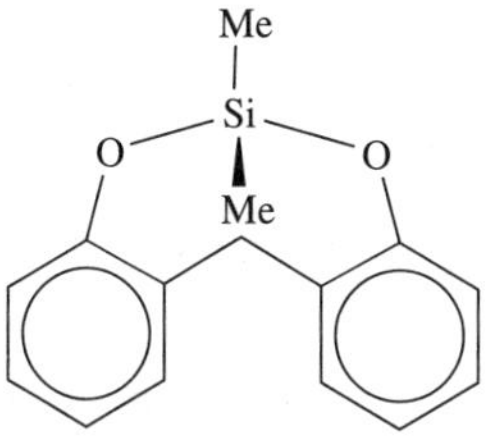

(145)

(146)

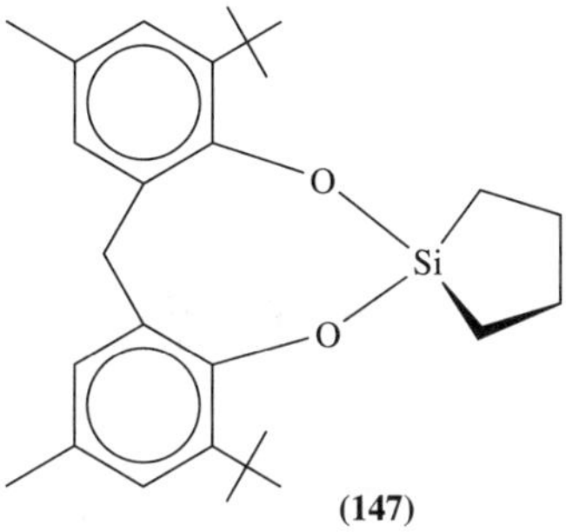

(147)

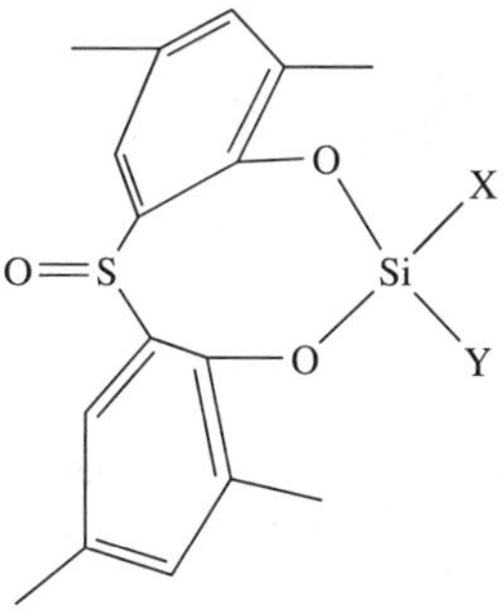

(148)

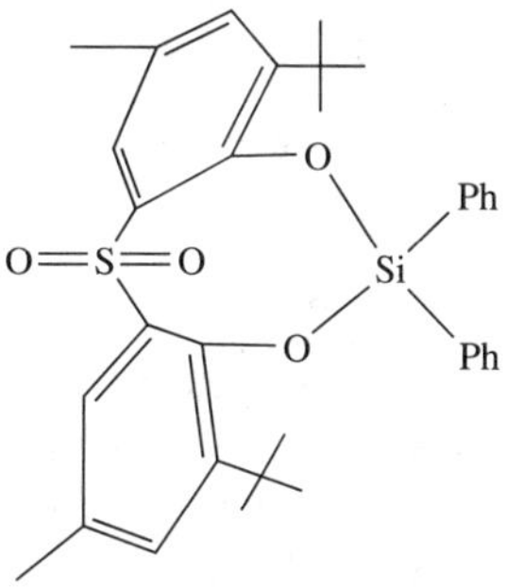

(149)

a, X = Y = Me
b, X = Ph ; Y = Me
c, X = Y = Ph

2. Fluorine coordination: degenerate fluorine migration in a 'Merry-Go-Round' type mechanism

An elegant example of neutral pentacoordinate silicon complexes was reported by Sakurai and coworkers[177]: the hexakis(fluorodimethylsilyl)benzene **(150)** was shown by X-ray crystallography to have nearly TBP silicon centers, arranged in a gear-meshed structure (Figure 22). The Si−F distances (1.68 and 2.39 Å) are not equal, but are typical of SiF bonds (covalent and dative, respectively) in pentacoordinate complexes [i.e. longer than a regular Si−F bond in a silane (1.50 Å) and shorter than the sum of van der Waals radii (2.63 Å), respectively]. All the F−Si−F angles were near 180° (average 176°), as required by a pentacoordinate TBP structure.

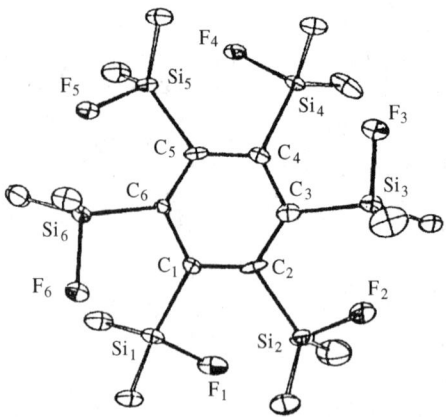

FIGURE 22. An X-ray crystallographic structure of **150**. Reprinted with permission from Reference 177. Copyright 1994 American Chemical Society

(**150**) X = F
(**151**) X = OMe

(**152**)

In solution the bridging fluorines are rapidly transferred between adjacent silicons, such that the ^{29}Si spectrum features a triplet due to two equivalent fluorines ($^{1}J_{Si-F} = 127$ Hz). This triplet did not change to a doublet of doublets at the low-temperature limit in toluene-d_8 solution, and hence 'freezing out' of the fluorine transfer between vicinal silicons could not be achieved. When the temperature was raised, all six fluorines became equivalent, causing the silicon resonance to be a septet ($^{1}J_{Si-F} = 43$ Hz), presumably due to exchange by correlated rotation[178] of the silicon substituents about the Si–benzene bonds. This rotational process was stopped when a molybdenum complex was formed (**152**), presumably as a result of increased steric hindrance for rotation.

It was shown that a D_{6h} symmetric structure for **150** required a 1.88 Å long Si–F bond, and was not possible due to the substantially shorter Si–F bonds (1.68 Å). For this purpose the hexamethoxy analog (**151**) was prepared and studied, and indeed, with the longer (1.88 Å) Si–O bonds, a perfect D_{6h} structure was obtained in a single crystal study[177].

IV. CATIONIC PENTACOORDINATE COMPLEXES

Trivalent silyl cations in the gas phase[179] and coordinated silyl cations in condensed phase were described in recent reviews and papers[180–182].

The first few pentacoordinate cationic silicon complexes (**153–155**) were discussed in an earlier review[2]. The first two were described as trivalent silicenium ions stabilized by *inter*molecular coordination[183,184], while **155** is an intramolecular siliconium ion[21a]. Since then, a few other compounds were reported. Oxidation of the dihydrido-hexacoordinate chelate **156** by the addition of excess iodine produced **157**[185]. The X-ray structure of **157** showed a slightly distorted TBP for the cation, with a well separated (Si–I distance 5.036 Å) I_8^{-2} anion for every two cations. The nitrogens are in the apical positions, and both Si–N distances (2.08 and 2.06 Å) are longer than covalent bonds but well within the coordination distance.

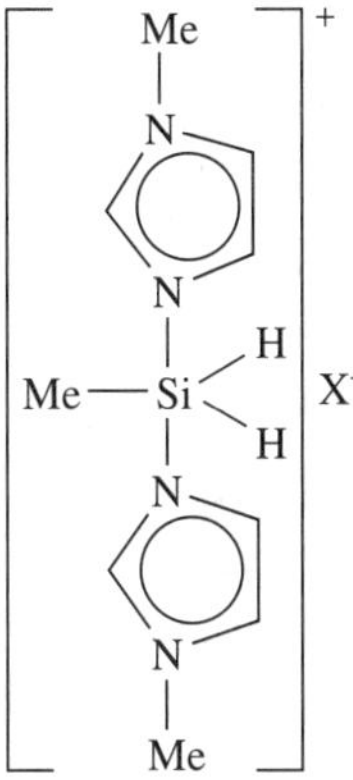

(153) 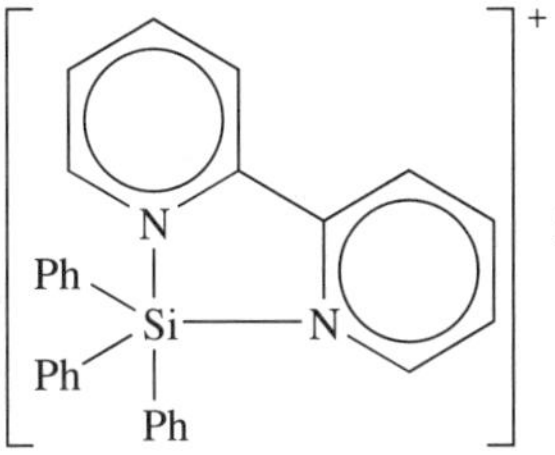 **(154)** 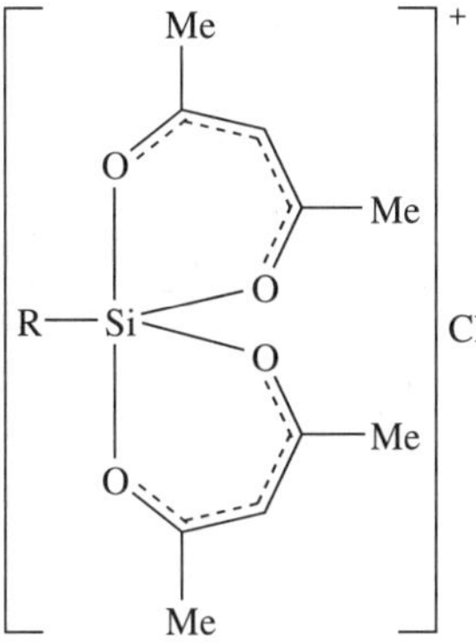 **(155)**

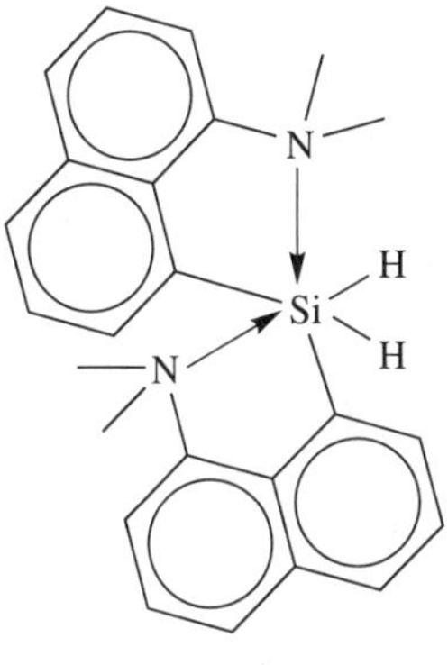

(156) 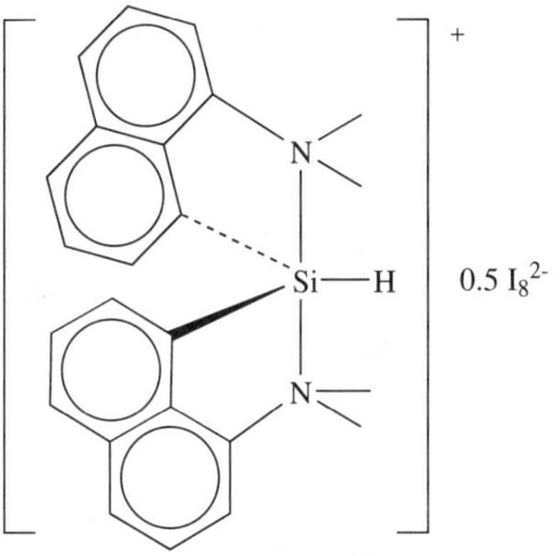 **(157)**

Reaction of the dihydrido complex **158** with a variety of reagents (equation 44) led to abstraction of a hydride and formation of the symmetrically dicoordinated cations **159a–159e**[186a], differing in the counter anion. **159c** was also prepared by a different route: from the 2,6-bis(dimethylaminomethyl)phenyl lithium and phenyldichlorosilane. The ionic character of these complexes was unequivocally demonstrated by conductivity studies. Other cationic complexes (**160, 161**) were similarly prepared from the trihydrido and monohydrido precursors[186a].

The NMR in CDCl$_3$ solution suggested the equivalence of the dimethylamino groups, in analogy to the solid state. This was later disputed in a similar compound, **161b**[187]. In this compound the low temperature ^{1}H and ^{13}C NMR spectra showed splitting of the N-methyl and CH$_2$ signals, suggesting that only one dimethylamino group was coordinated, and that rapid exchange of the two occurred at higher temperatures, i.e. that in solution the silicon was, in fact, tetracoordinate. This complex was reported to be remarkably air-stable, and to dissolve in protic solvents[187].

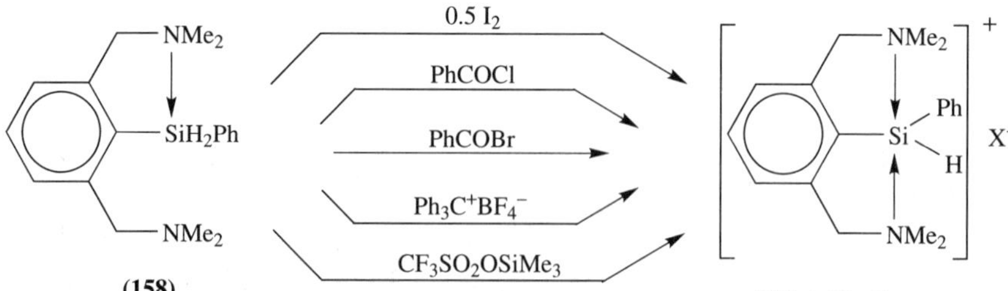

(159a) X = I
(159b) X = Br
(159c) X = Cl   (44)
(159d) X = BF$_4$
(159e) X = CF$_3$SO$_3$

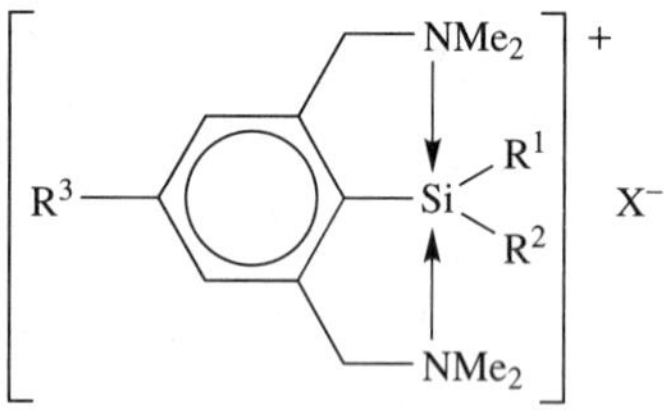

(160a) R^1 = R^2 = R^3 = H, X = I
(160b) R^1 = R^3 = H, R^2 = X = Cl
(161a) R^1 = R^2 = Me, R^3 = H, X = CF$_3$SO$_3$
(161b) R^1 = R^2 = Me, R^3 = t-Bu, X = Cl
(162a) R^1 = Ph; R^2 = Me, R^3 = H; X = Cl
(162b) R^1 = Ph; R^2 = Me, R^3 = H; X = I
(162c) R^1 = Ph; R^2 = Me, R^3 = H, X = [3,5-(CF$_3$)$_2$C$_6$H$_3$]$_4$B
(162d) R^1 = Ph; R^2 = Me, R^3 = H, X = BPh$_4$

A further contribution to this controversy has recently been added by Corriu's group, showing that, depending on the counter anion and solvent, each of the penta and tetra-coordinate structures (**162**) can predominate, and in one particular case (**162d**) both were shown to coexist in nearly equal concentrations[186b].

Two additional cationic complexes were recently reported, **163** and **164**, prepared by the reaction of trimethylsilyl triflate with bis[2-((dimethylamino)methyl)phenyl]silane (**165**) and [2-((dimethylamino)methyl)phenyl]phenylsilane (**166**), respectively[188]. The X-ray crystallographic structure of **163** clearly showed a cationic complex with a distant triflate anion. By contrast, in **164** the anion is directly coordinated to silicon (Si–O distance 1.951 Å), as is the apical dimethylamino-nitrogen (Si–N distance 2.052 Å). The structure is very near a TBP, and may be viewed as a doubly-coordinated silicenium ion, or a tight ion-pair[188].

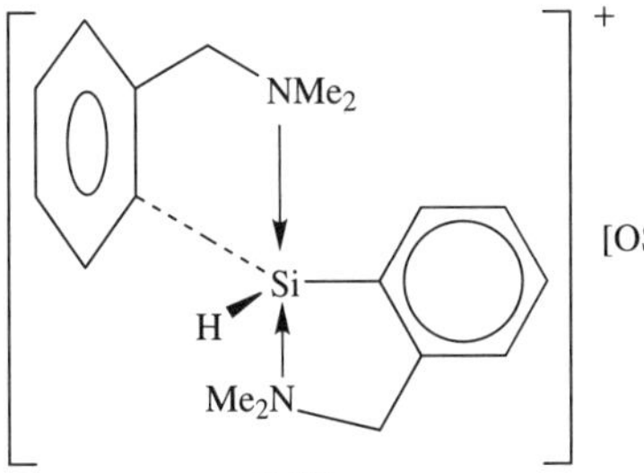

(**163**) 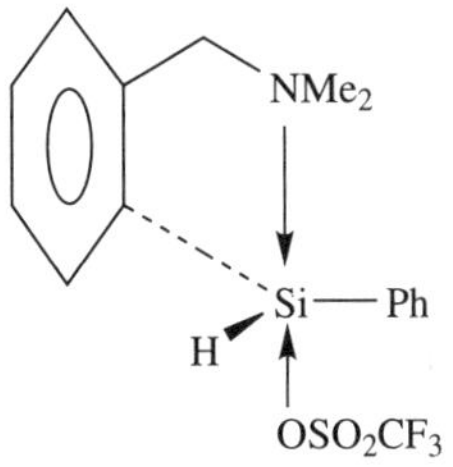   (**164**)

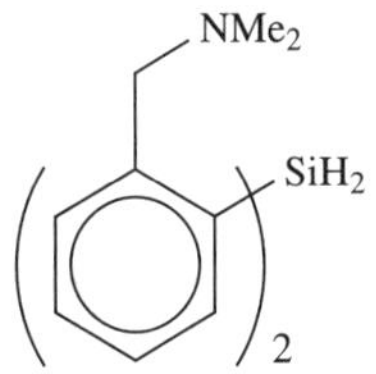

(**165**) 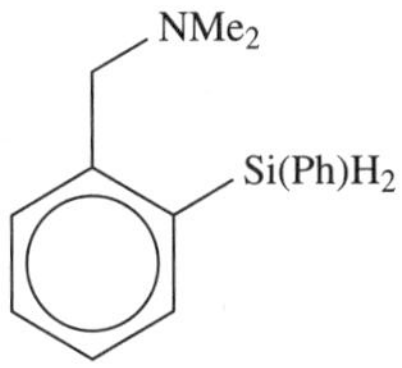   (**166**)

Table 19 presents ^{29}Si chemical shifts and Si–H coupling constants for some of the cationic pentacoordinate silicon complexes and for some of their precursors. No clear trend can be seen for the ^{29}Si chemical shifts of these compounds: some are shifted downfield and others upfield, relative to their pentacoordinated neutral precursors. However, a trend seems to emerge from the one-bond Si–H coupling constants: in the cationic pentacoordinate complexes ^1J(Si–H) is generally greater than for neutral pentacoordinate complexes. This is possibly the result of the greater s-character in the Si–H bonds, in the doubly coordinated (sp^2-hybridized) silicenium cations[185,186].

A formally pentavalent cationic silicenium complex, with intramolecular base-stabilization, is the silylene-iron **167**, with a Fe=Si double bond. The crystal structure of **167** shows that silicon is essentially tetrahedral in this compound, and the solution ^{29}Si chemical shift is $\delta = 118.3$ ppm, compatible with the double bond character rather than with pentacoordination at silicon[189].

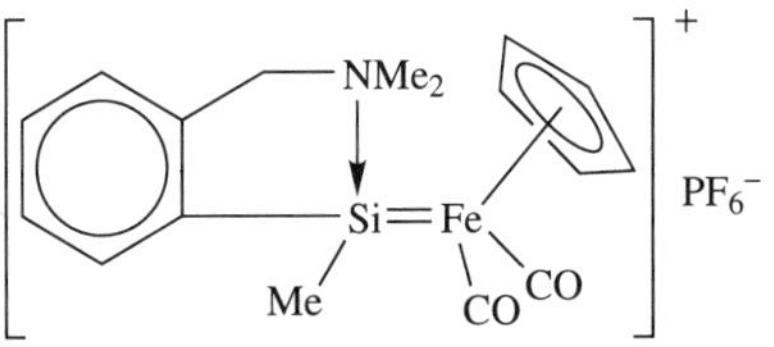

(**167**)

TABLE 19. ^{29}Si NMR data for cationic complexes and their neutral precursors

Complex	Counter anion	δ^{29}Si (ppm)	J(Si−H) (Hz)	References
157	0.5 I$_8^-$	−43.5	290	185
158		−51.5	200	186a
159a	I$^-$	−29.7	280	186a
160a	I$^-$	−46.4	265	186a
160b	Cl$^-$	−40.3	334	186a
161a	CF$_3$SO$_3^-$	−4.1		186a
161b	Cl$^-$	6.57		187
162a	Cl$^-$	−7.0		186b
162b	I$^-$	−8.0		186b
162c	[3,5-(CF$_3$)$_2$C$_6$H$_3$]$_4$B$^-$	−15.0		186b
162d	Ph$_4$B$^-$	−8.5, −14.5		186b
163		−51.6	272	188
164	CF$_3$SO$_3^-$	−56.2	294	188
165		−45.0	216	188
166		−43.3	208	188

V. HEXACOORDINATE IONIC SILICON COMPOUNDS

The main methods for the synthesis of hexacoordinate silicon compounds are similar to those for pentacoordinate complexes and were outlined in a recent review[6]. These methods include: (a) addition of nucleophiles (neutral or anionic) to tetracoordinate silanes; (b) intermolecular or intramolecular coordination to an organosilane; (c) substitution of a bidentate ligand in a tetrafunctional silane. The following discussion focuses mainly on new complexes, reported since the recent reviews[6,7] were published.

A. Fluorosilicates

The first studies of fluorosilicates were reported by Müller[8,32a] and by Kumada and coworkers[190], and later reviewed by Voronkov[9a], Gel'mbol'dt,[9b] Corriu[6] and their coworkers. More recently some interesting hexacoordinate fluorosilicates were reported by Corriu and coworkers[191,192] and by Tacke and Mühleisen[193].

Hexacoordinate fluorosilicates were obtained from neutral pentacoordinate complexes by addition of K[18-crown-6]F[191,192]. In this way compounds **168–173** were prepared from their pentacoordinate precursors. These compounds, in addition to being fluorosilicates, contain in each chelate ring a dative Si−N bond.

An X-ray crystallographic study for **170** revealed a slightly distorted octahedral geometry, as is expected for hexacoordination[191]. The four Si−F bonds in this compound are essentially equal in length, while the Si−N distance (2.213 Å) is shorter than that found in the neutral pentacoordinate precursor (2.318 Å), or in a previously reported neutral hexacoordinate difluorosilane, **174** (2.59, 2.81 Å)[194]. The position of silicon in **170** is 0.16 Å displaced from the mean plane defined by carbon and three fluorine ligands, away from the coordinated nitrogen[191].

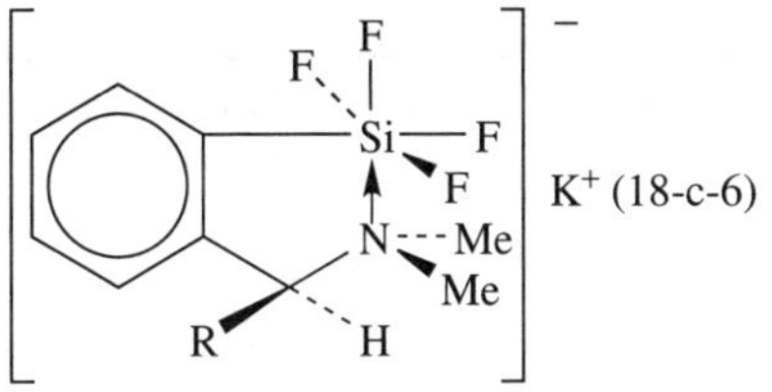

(168) R = H
(169) R = Me

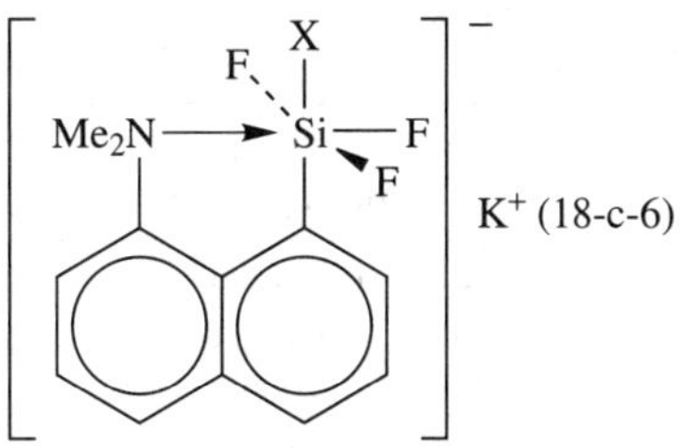

(170) X = F
(171) X = Ph

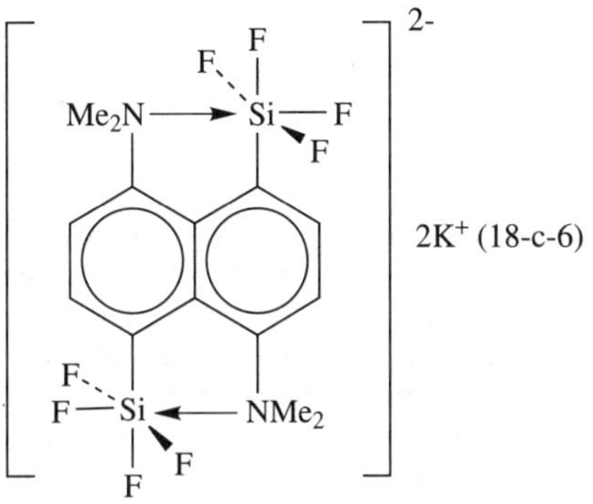

(172)

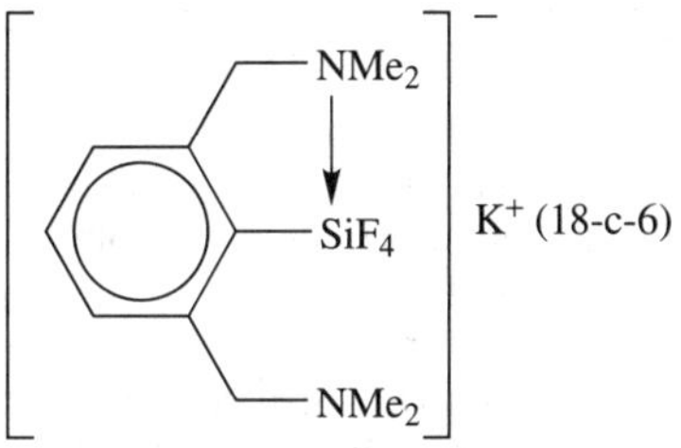

(173)

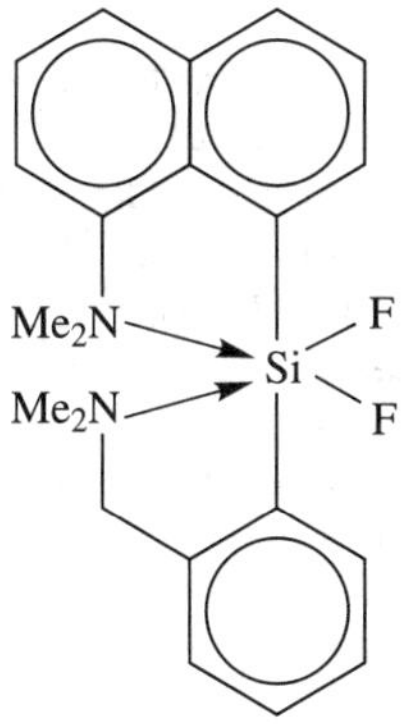

(174)

The ^{29}Si NMR (Table 20) supports a hexacoordinate structure in solution which is similar to that in the solid state. δ^{29}Si for the hexacoordinate fluorosilicates is *ca* 50 ppm shifted upfield relative to the pentacoordinate precursors. The multiplicity of the ^{29}Si resonances indicates the nonequivalence of fluorine ligands at low temperatures (between 25 and −60 °C). These become equivalent as the temperature is raised, such that a symmetrical ^{29}Si multiplet is observed.

TABLE 20. ^{29}Si NMR data and ligand exchange barriers (determined by ^{19}F NMR) for hexacoordinated anionic silicates

Compound	Temp (°C)	δ (ppm) (multiplicity)	J(Si—F)^a (Hz)	$\Delta G_F^{\ddagger}$ (kcal mol^{-1})	References
168	+25	−160.1(quint.)	197.1	11	191
	−60	−161.6 (ddt)	218.4(SiX$_2$),186.7(SiY),147.6(SiZ)		
169	+25	−159.9(quint.)	193.1	11	191
	−40	−161.1(dt)	219.0(SiX$_2$),184.0(SiY),146.3(SiZ)		
170	+25	−156.3(ddt)	223(SiX$_2$),181.2(SiY),149.8(SiZ)	15	191
171	+25	−137.0(dt)	246.2(SiX$_2$),196.7(SiY)	15	191
172	+25	−154.5(ddt)	225.3(SiX$_2$),184.0(SiY),117.6(SiZ)		191
173	+25	−161.0(quint.)	197.1		192
	−90	−162.6(ddt)	217.5(SiX$_2$),191.2(SiY),142.5(SiZ)		

^{a}X, Y and Z refer to non-equivalent fluorine atoms at slow exchange conditions.

Measurement of the temperature dependence of ^{19}F NMR spectra for **169** revealed two distinguishable coalescence phenomena: the averaging of diastereotopic fluorine signals in the ^{19}F spectrum, and the coalescence of the N-methyl groups[191]. The latter process can only result from Si—N cleavage, followed by rotation about the C—N bond and reclosure of the chelate ring. During this process the fluorines can become equivalent. However, the exchange and coalescence of fluorines occurs with a lower activation barrier than the N-methyl exchange (14 kcal mol^{-1}, measured by ^{1}H NMR), and hence it was concluded that fluorine exchange is a regular, nondissociative reaction of the Bailar[195a] or Ray–Dutt (RD)[195b] type. The presence of a chiral group is necessary for the observation and measurement of the two barriers. This was not possible for the other compounds, **168, 170–172**[191].

Complex **173** is unique in that it has a tridentate ligand[192]. It can serve as a model for nucleophilic substitution on hexacoordinate silicon. The NMR provides evidence that only one of the dimethylamino groups is coordinated: both ^{29}Si chemical shift and the ^1J(Si—F) are in accord with those of **168**, in which only one nitrogen ligand is present. However, at room temperature the ^{1}H NMR spectrum of **173** shows a sharp singlet for all 12 N-methyl protons, indicating fast exchange. Also, the fluorine ligands are equivalent at this temperature. At low temperature, the N-methyl singlet splits in two, and the ^{19}F signal splits to three signals, providing evidence for a dynamic equilibration of the two dimethylamino groups via a heptacoordinate C_{2v} transition state, **173a**, in which the fluorines are equivalent.

(173a)

A zwitterionic hexacoordinate pentafluorosilicate (**175**) was recently reported by Tacke[193]. **175** was prepared in the same manner as pentacoordinate zwitterionic complexes (cf Section II.A.3), as shown in equation 45. The crystal structure of **175** showed that one water molecule is present for each molecule, and that the structure is dominated by N−H··F and O−H··F hydrogen bonds. The Si−C bond length (1.943–1.946 Å) is distinctly longer than in the corresponding pentacoordinate zwitterion (**9**), while the Si−F bonds (1.685–1.740 Å) are comparable to those in the pentacoordinate analog. The solid state ^{29}Si NMR of **175** is also compatible with hexacoordination, with δ^{29}Si $= -177.2$ ppm.

$$(EtO)_3SiCH_2Cl \longrightarrow (EtO)_3SiCH_2N\text{-piperazine-}NMe \xrightarrow{HF} \mathbf{(175)} \quad (45)$$

B. Ionic (O−Si) Chelates

The first ionic hexacoordinate complexes, silicon-tris-acetylacetonate cations (**176**), were reported as early as 1903 by Dilthey[196a] and Rosenheim and coworkers[196b]. Subsequently many other β-diketonate complexes were studied[197−199], and the subject was extensively reviewed[7,200] and will not be discussed further here.

A rather similar class of compounds is the group of tris-catecholato anionic complexes, **176a**. These may be obtained by the reactions of catechol and base with either silica (SiO$_2$), Si(OR)$_4$, [SiF$_6$]$^{2-}$2M$^+$ or SiCl$_4$[6,7,201a]. The reaction with silica provides a direct route from an inorganic silicon source to organometallic compounds. This has further been developed to a general and efficient synthesis of organosilanes, using **176a** as intermediate[201b].

(**176**) Cl$^-$·HCl   (**176a**) 2M$^+$

Tris-catecholate complexes were prepared from symmetrically and unsymmetrically substituted catechols (LH$_2$): 4-chlorocatechol, 4,5-dichlorocatechol, 4-nitrocatechol, 3,4-dinitrocatechol and 4,5-dinitrocatechol[202]. All of these complexes are prepared in aqueous solution and are water-stable, down to *ca* pH 4. The ^{1}H NMR signals for the free and complexed catechol moieties are well separated, and enable the determination of formation constants for each of the complexes, according to equation 46, where L is the catechol dianion.

$$3LH_2 + Si(OH)_4 \leftrightarrows SiL_3^{2-} + 4H_2O + 2H^+ \qquad (46)$$

When differently substituted tris-catecholato complexes are mixed in solution they undergo intermolecular ligand exchange, resulting in mixed complexes. Thus the water solution of tris-catecholato and tris-4-nitrocatecholato complexes shows in the ^{29}Si NMR spectrum the signals for all four permutations of tris-ligand complexes. However, solution of the same complexes (as triethylammonium salts) in DMSO did not show signs of ligand exchange after 24 h. As expected, the unsymmetrically substituted catechols give rise to two isomeric complexes: facial and meridional, differing in the relative orientations of

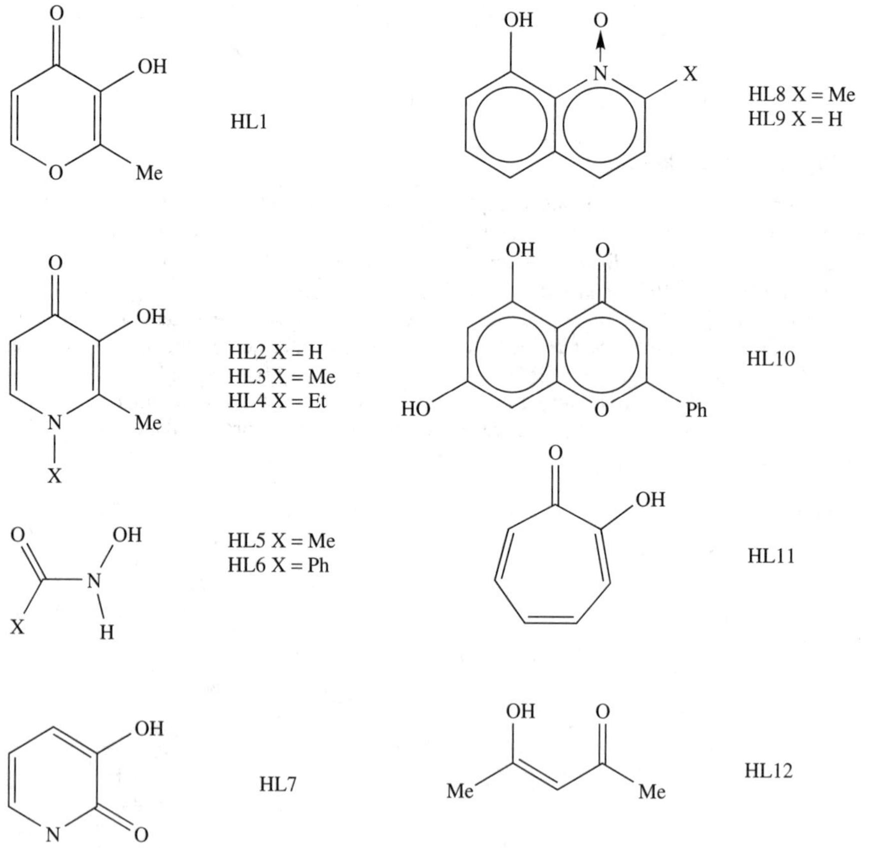

FIGURE 23. Ligand molecules used in hexacoordinate cationic complexes[203]

the nonequivalent oxygens of each catechol unit. The isomers were observed in the ^{1}H NMR spectra[202].

Evans has also reported on the preparation of similar water-soluble hexa-oxosilicon complexes, in which the overall charge is positive[203,204]. Two synthetic methods were used, reacting the bidentate ligands with either SiCl$_4$ or Si(OEt)$_4$. The latter method was reported to be generally preferred. The ligands used are shown in Figure 23[203]. The ^{29}Si chemical shifts of these complexes (Table 21) were analyzed: a substantial effect of chelate-ring size on δ^{29}Si was found (-130 to -150 ppm for the five- vs -190 to -200 ppm for the six-membered chelate rings)[21a,203]. By contrast to this large effect, the nature of the coordinating oxygens, the nature of the counter anion and the overall charge of the complex had little or no effect on δ^{29}Si.

An interesting observation for one group of ligands (L2–L4 in Figure 23) is the fact that they can etch glass at a significant rate, to form the hexacoordinate complex SiL$_3^+$Cl$^-$ from a dilute acid solution[204].

C. Bis-catecholato Complexes with N—Si Coordination

A few ionic bis-catecholato N-coordinated complexes **177–181** have been reported[192,205–207]. Crystallographic studies of **177, 180** and **181** showed all three to have a near-octahedral geometry, with Si—O bond lengths very similar to those in the tris-catecholato complexes, and typical Si—N dative bond lengths of 2.157, 2.085 and 2.173 Å, respectively. The ^{29}Si chemical shifts for **177–181**, which are in line with hexacoordination, are listed in Table 22.

Variable-temperature ^{1}H NMR spectra were measured for **177** and **178** in CD$_2$Cl$_2$ solutions[205]. In the former the N-methyl groups and the methylene protons appeared as sharp singlets down to the lowest attainable temperature, while in **178**, with two chiral centers, signals due to only one diastereomer could be observed, and the initially diastereotopic N-methyls (due to the chiral carbon center) coalesced with an activation barrier of 10.25 kcal mol^{-1}[205]. This is evidence that inversion of configuration at silicon is rapid relative to the NMR time scale at the low temperatures used in the measurements and that the barrier measured for **178** is associated with Si—N bond cleavage.

TABLE 21. ^{29}Si chemical shifts for ionic hexacoordinate complexes (in DMSO)[203]

Complex[a]	δ^{29}Si (ppm)	Complex[a]	δ^{29}Si (ppm)
[Si(L1)$_3$]$^+$[Cl]$^-$	$-134.9, -134.4$	[Si(L6)$_3$]$^+$[CF$_3$SO$_3$]$^-$	-138.7
[Si(L1)$_3$]$^+$[CF$_3$SO$_3$]$^-$	$-134.9, -134.4$	[Si(L7)$_3$]$^+$[CF$_3$SO$_3$]$^-$	-141.6
[Si(L2)$_3$]$^+$[CF$_3$SO$_3$]$^-$	-136.4	[Si(L8)$_3$]$^+$[CF$_3$SO$_3$]$^-$	$-192.3, -192.7$
[Si(L3)$_3$]$^+$[HSO$_4$]$^-$	-135.3	[Si(L8)$_3$]$^+$[Cl]$^-$	$-192.3, -192.7$
[Si(L3)$_3$]$^+$[cam]$^-$	-135.3	[Si(L9)$_3$]$^+$[CF$_3$SO$_3$]$^-$	$-193.4, -193.8$
[Si(L3)$_3$]$^+$[CF$_3$SO$_3$]$^-$	-135.5	Si(L10)$_3$]$^+$[Cl]$^-$	-193.1
[Si(L3)$_3$]$^+$[Cl]$^-$	-135.5	[Si(catecholato)$_3$]$^{2-}$[(Et$_3$NH)$_2$]$^{2+}$	-139.3
[Si(L3)$_3$]$^+$[CF$_3$CO$_2$]$^-$	-135.3	[Si(catecholata)$_3$]$^{2-}$[K$_2$]$^{2+}$	-140.3
[Si(L4)$_3$]$^+$[HSO$_4$]$^-$	-135.4	[Si(L11)$_3$]$^+$[Cl]$^-$	-140.7
[Si(L4)$_3$]$^+$[Cl]$^-$	-135.5	[Si(L11)$_3$]$^+$[SbF$_6$]$^-$	-139.4
[Si(L5)$_3$]$^+$[CF$_3$SO$_3$]$^-$	-137.7	[Si(L12)$_3$]$^+$[ZnCl$_3$]$^-$	-193.7

[a]For the identity of L see Figure 23, [cam]$^-$ — 10-camphorsulfonate anion.

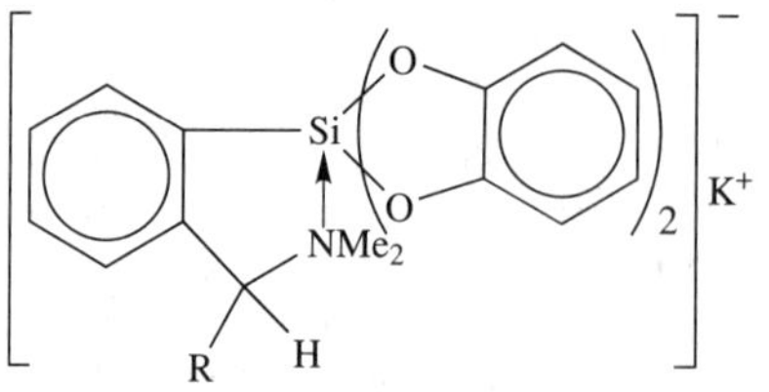

(177) R = H
(178) R = Me

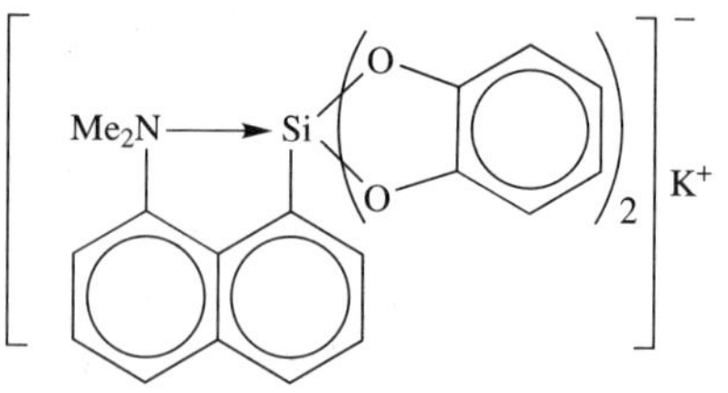

179)

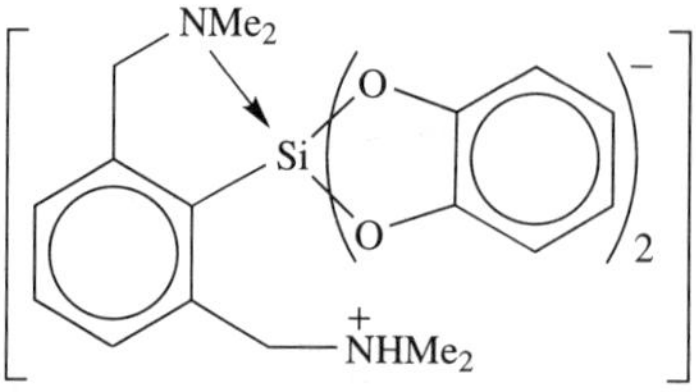

(180)

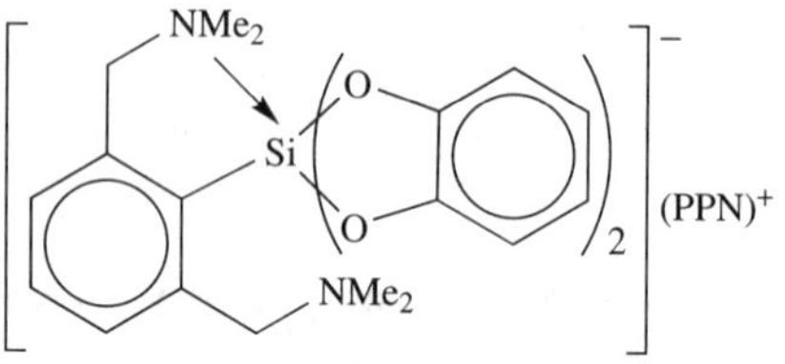

(181)

$PPN^+ = Ph_3P=\overset{+}{N}=PPh_3$

TABLE 22. ^{29}Si chemical shifts for bis-catecholato silicon complexes with N—Si coordination (CD$_2$Cl$_2$)

Compound	δ^{29}Si (ppm)	References
177	−121.1	205
178	−199.88	205
179	−124.5	206
180	−134.9[a]	207
181	−127.2[a]	207

[a] Solid state.

Complex **181**, with its two dimethylamino groups, is octahedral in the solid state with one dative and one loose dimethylamino nitrogens. In solution the NMR evidence shows rapid displacement of one dimethylamino ligand by the other, their NMR resonances being equivalent at 295 K. At 183 K, however, the methyls of one of the NMe$_2$ groups are diastereotopic, while the other group shows a singlet. At this temperature also both methylene groups give rise to AB quartets, as a result of the chirality at silicon[192]. Such internal nucleophilic substitution has been observed previously in pentacoordinate complexes, but it is the first report of dynamic internal substitution in hexacoordinate complexes.

VI. NEUTRAL HEXACOORDINATE SILICON COMPLEXES

A. Intramolecular Coordination

1. Synthetic methods

Relatively few ligand types have been used for the formation of neutral hexacoordinate silicon complexes, resulting in several complex types **182–193**[208–218]. Acetylacetonato (acac) chelates [**182**, (acac)$_2$SiXY] were prepared directly from the reaction of

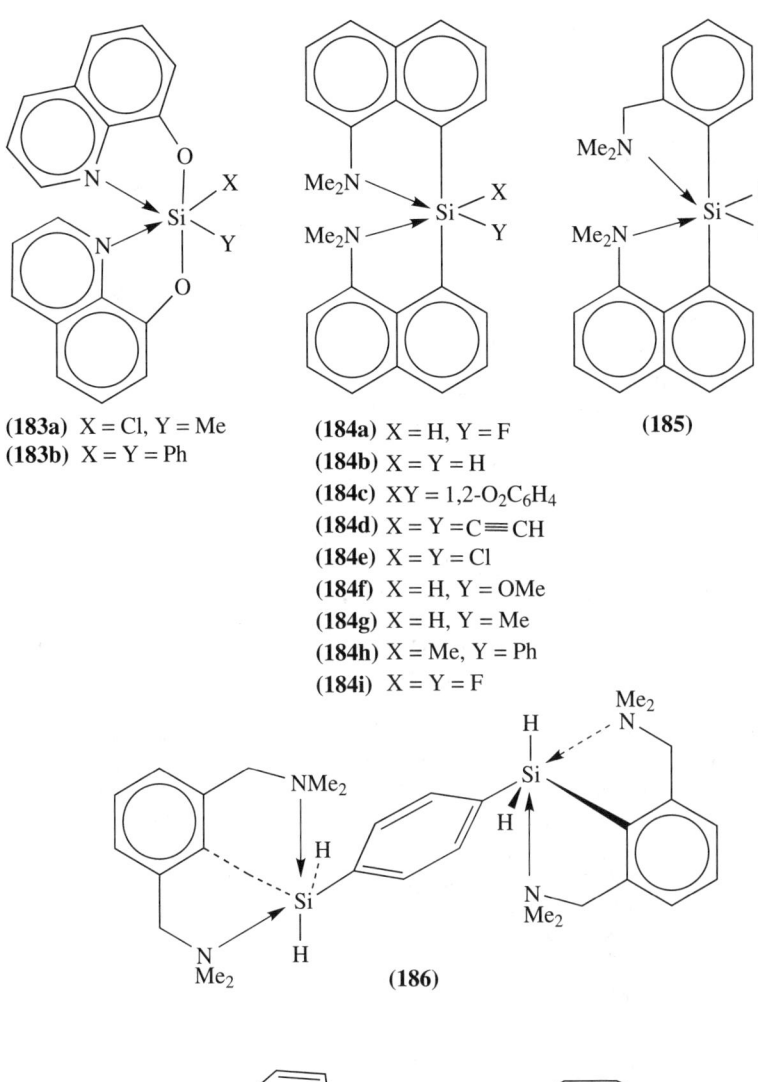

(183a) X = Cl, Y = Me
(183b) X = Y = Ph

(184a) X = H, Y = F
(184b) X = Y = H
(184c) XY = 1,2-O$_2$C$_6$H$_4$
(184d) X = Y = C≡CH
(184e) X = Y = Cl
(184f) X = H, Y = OMe
(184g) X = H, Y = Me
(184h) X = Me, Y = Ph
(184i) X = Y = F

(185)

(186)

(187) R = alkyl, Ar, Cl

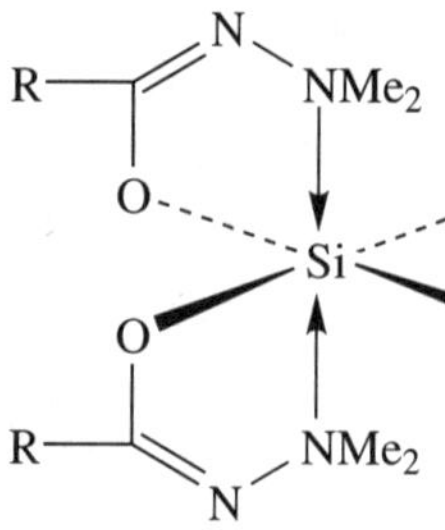

(188a–f) R = Me
(189a–f) R = Ph
(190a–f) R = CF$_3$
(191a–f) R = PhCH$_2$
(192a–f) R = PhMeCH

(a) X = Cl, Y = Me
(b) X = Cl, Y = Ph
(c) X = Cl, Y = H
(d) X = Y = Cl
(e) X = F, Y = Ph
(f) X = Y = F

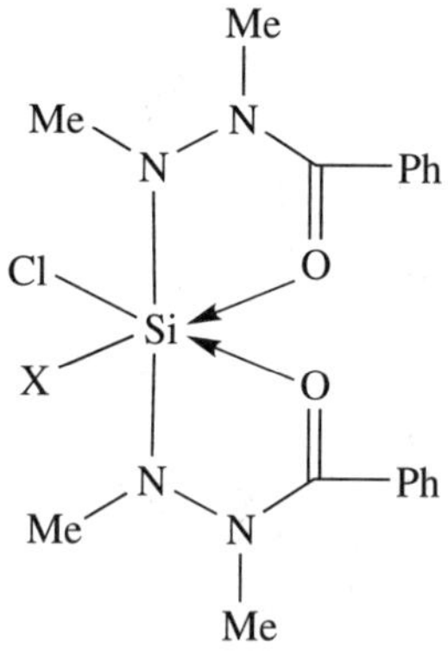

(193a) X = Ph
(193b) X = Cl

acetylacetone with a polyhalosilanes[197,198,200]. These earlier compounds were described in detail and reviewed previously[6,7]. The other types of complexes (**183–193**) were obtained by three main synthetic methods: (a) Reaction of the lithio-derivative of a bidentate ligand with a polyhalo- or polyalkoxy-silane (compounds **183–187**; equation 47[209]). (b) Exchange reaction of a trimethylsilylated derivative of the bidentate ligand with a polyhalosilane (compounds **188–193**; equation 48)[210–212]. (c) Modification of hexacoordinate complexes obtained by (a) or (b): an example of many synthetic transformations of a dihydro complex to other hexacoordinate complexes was presented by Corriu and coworkers, and is shown in Figure 24[213]. Interestingly, chloro ligands are reactive toward nucleophiles, while fluoro ligands are not effected by nucleophiles under similar conditions[206] (Figure 25; cf also work by Bassindale and Borbaruah[144]).

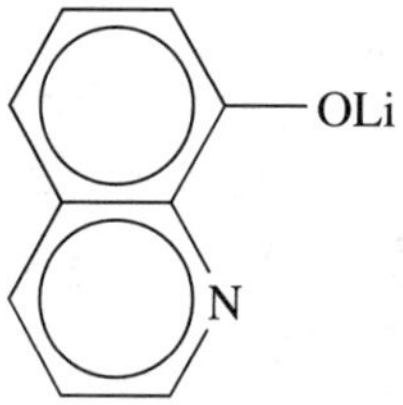

$$\text{quinoline-OLi} + \text{MeSiCl}_3 \longrightarrow \textbf{183a} \tag{47}$$

$$2\text{Me}_2\text{NN=C(Me)OSiMe}_3 \xrightarrow[-2\text{Me}_3\text{SiCl}]{\text{XSiCl}_3} \textbf{188} \tag{48}$$

The synthetic route (c) was recently utilized to prepare unsaturated monomers containing the hexacoordinate silicon unit, followed by polymerization to form novel polymers with hexacoordinate silicon in the polymer chain (equation 49)[214]. The ^{29}Si chemical shift measured in the polymer solution (ca −60 ppm) is very similar to that of the monomer (−63.8 ppm), and is evidence for hexacoordination in the polymer solution.

Method (b), in which ligand exchange between silanes takes place, has been used to obtain both N→Si (equation 48)[210,211] and O→Si complexes (equation 50)[212], depending on the site of trimethylsilylation in the ligand molecule.

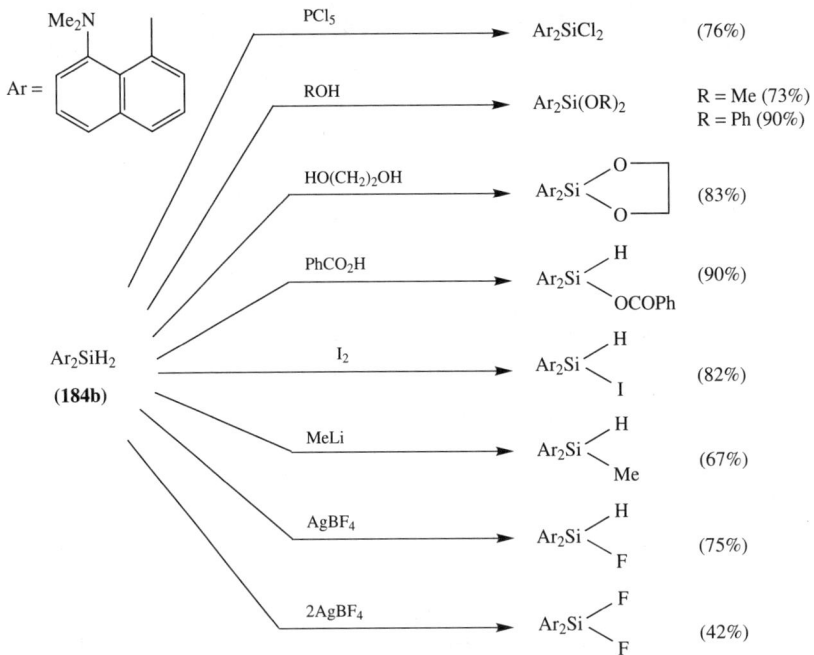

FIGURE 24. Reactivity of **184b**[6,213]

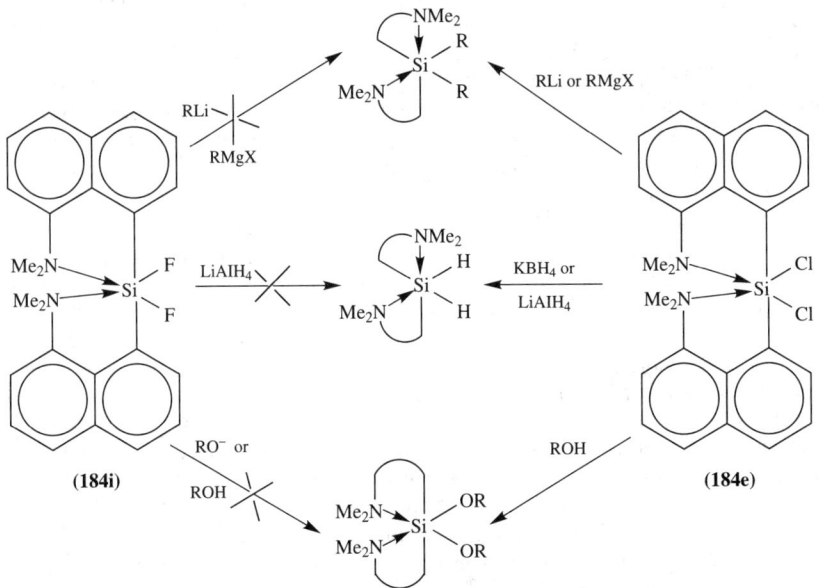

FIGURE 25. Different reactivities of the dichloro (**184e**) vs the difluoro (**184i**) complexes

$$\text{(184f)} \xrightarrow{2\ Me_3SiC\equiv CLi} \text{product with } C\equiv CSiMe_3 \text{ groups} \quad (49)$$

[Scheme showing: (184f) with Me₂N, Me₂N, Cl, Cl ligands on Si between naphthalene units → reacts with 2 Me₃SiC≡CLi → product with Me₂N, Me₂N, C≡CSiMe₃, C≡CSiMe₃ on Si; product labeled, with arrow labeled n-Bu₄NF down to (184d), then Br-Ar-Br arrow to Polymer]

$$2\,Me_3SiN(Me)N(Me)COPh \xrightarrow[-2\,Me_3SiCl]{XSiCl_3} \mathbf{189} \quad (50)$$

186 is an example of a special compound which is both a dinuclear complex and is prepared with a tridentate chelating ligand, in analogy to the synthesis labeled (a)[215].

2. Structure

The X-ray crystallographic studies of neutral and anionic hexacoordinate bischelates show that the pair of monodentate ligands prefer the *cis* orientation with respect to each other[209]. In general, the electronegative ligands occupy *trans* positions relative to the dative ligands. However, exceptions have been reported: **189a** has the two dative N–Si bonds *trans* to each other, despite the presence of the more electronegative O and Cl ligands[216]. Remarkably, hexacoordinate complexes with the same ligand environment, SiO_2N_2CCl (**183**), were found with the oxygens in *trans* position[209].

Geometrically unrestricted hexacoordinate complexes are generally octahedral, as, for example, the anionic SiF_6^{-} [219]. However, intramolecular neutral hexacoordinate complexes contain one or two chelate rings, which may impose strain on the complex and modify the geometry. As a consequence, complexes can have more or less distorted octahedral geometries, depending on the ligands. Thus, while X-ray structure determinations for **183** and **189a**[209,216] (Table 23) show that the complexes have near-regular octahedron geometry, in the more constrained complexes **184–186**[194,206,215] the silicon atom largely maintains its basic tetrahedral geometry, with the two dative ligands loosely coordinated opposite two of the tetrahedron faces. This structure was termed 'bicapped tetrahedron'[194]. The features characteristic of this type of complexes are the substantially longer Si–N distances, ca 2.5–2.8 Å, relative to the octahedral complexes with 2.0–2.1 Å. The Si–N distances in the bicapped tetrahedral complexes are, nevertheless, shorter than the sum of van der Waals radii. Unlike octahedral complexes, in which the ^{29}Si signal is shifted ca 70–80 ppm upfield relative to analogous tetracoordinate complexes[21a,40,211], in the bicapped tetrahedral complexes the ^{29}Si coordinative shift (relative to tetracoordinate analogs) is negligible or very small (Table 24).

TABLE 23. X-ray structural data for neutral hexacoordinate (N—Si) chelate complexes

Complex	X	Y	Si—N (Å)	Si—X or Si—Y, (Å)	Geometry	References
183	Cl	Me	2.014	2.17(Cl)	octahedron	209
			2.016	1.94(C)		
184a	F	H	2.680	1.628(F)	bicapped	194
			2.646	1.550(H)	tetrahedron	
184b	H	H	2.610	1.44	bicapped	194
			2.800	1.54	tetrahedron	
184c	OCH$_2$	OCH$_2$	2.560	1.74	bicapped	206
			2.640	1.72	tetrahedron	
184d	C≡CH	C≡CH	2.836	1.840	bicapped	214
			2.789	1.851	tetrahedron	
185	F	F	2.770	1.617	bicapped	194
			2.594	1.603	tetrahedron	
186	H	Ph	3.008	1.44(H)	bicapped	215
			2.681	1.883(C)	tetrahedron	
189a	Cl	Me	2.036	2.197(Cl)	octahedron	216
			2.015	2.089(C)		

TABLE 24. Comparison of ^{29}Si NMR data for hexa- and tetracoordinate silicon compounds

Compound	X	Y	δ^{29}Si (ppm) (J_{Si-X}, Hz)	Tetracoordinate model	δ^{29}Si (ppm) (J_{Si-X}, Hz)	References
184a	H	F	−37.3 (285.4)			217
184b	H	H	−41.1 (216.4)	(1-Np)$_2$SiH$_2$	−38.7 (197.7)	217
184d	C≡CH	C≡CH	−63.8	Ph$_2$Si(C≡CH)$_2$	−48.2	214
184e	Cl	Cl	−33.6			213
184i	F	F	−61.4 (276.9)	Ph$_2$SiF$_2$	−29.1 (302.7)	213
185a	F	F	−52.8 (273.0)	Ph$_2$SiF$_2$	−29.1 (302.7)	194
186	H	H	−51.3 (194.0)	Ph$_2$SiH$_2$	−33.8	215
188a	Me	Cl	−121.0	MeClSi(OSiMe$_3$)$_2$	−46.2	211
188b	Ph	Cl	−128.8			221
188c	H	Cl	−137.7 (341.8)	HClSi(O—)$_2$	−64.8 (381)	211
188d	Cl	Cl	−147.2			211
188e	Ph	F	−148.6 (272.9)	PhFSi(O—)$_2$		222
188f	F	F	−160.7 (202.0)			213
193a	Me	Cl	−78.9			212
193b	Cl	Cl	−100.1			212

It is interesting to note that in **186** the two N-dimethylamino ligands are in *cis* positions relative to each other, despite the fact that hydrogens, and no electronegative ligands, are in the apical positions, and the likely expectation that the aromatic-ring planarity would prefer the *trans* arrangement[215].

Like ^{29}Si chemical shifts, also the Si−H coupling constants are sensitive to the hybridization at silicon, and not only to the formal coordination number. The increase in coordination number is normally associated with a decrease in ^{29}Si−^{1}H coupling constants (and ^{29}Si−^{19}F coupling as well), due to rehybridization at silicon and the associated decrease in the Si−H (or Si−F) s-character. However, since in the bicapped-tetrahedral complexes there is essentially no change in hybridization at silicon, the coupling constants do not change significantly. To the extent that Si−H coupling constants do change, they have the *opposite* trend, since the weakly coordinated ligands add electron density to the Si−H bonds[217].

Chemical shifts for the hexacoordinate complexes behave differently relative to pentacoordinate complexes. Firstly, no significant temperature dependence of chemical shifts was observed for hexacoordinate complexes[211,215], in contrast to pentacoordinate complexes. Comparison of ^{29}Si chemical shifts for the hexacoordinate octahedral complexes reveals a trend as a function of monodentate ligand electronegativity[211,220]. Thus ^{29}Si resonance in octahedral complexes seems to shift uniformly upfield as the monodentate ligands are more electronegative (−128.8 ppm in **188b** with X = Ph and Y = Cl, and −148.6 ppm in **188e** in which X = Ph and Y = F, i.e. F replaced Cl; likewise we find −137.7 ppm in **188c** with X = H and Y = Cl, and −147.2 ppm in **188d** where X = Y = Cl). This is in contrast to pentacoordinate complexes, in which the chemical shift responds to the *apicophilicity* of the ligands and not to electronegativity[4,77].

3. Stereodynamics

Neutral hexacoordinate silicon complexes can undergo a variety of ligand exchange reactions, which are observable by NMR spectroscopy. These may include intermolecular exchange via an increase in the coordination number (i.e. nucleophilic substitution) or a decrease in coordination of silicon (i.e. dissociation). Alternatively, ligand exchange can be intramolecular and nondissociative, in analogy to pseudorotation in pentacoordinate silicon compounds[211,217]. Such internal ligand permutations have been described as twists of one triangular face of the octahedral skeleton with respect to the opposite face, in what are generally termed Bailar[195a] and Ray−Dutt[195b] twists. The richness of possible exchange mechanisms makes it difficult to determine with certainty in each case which mechanism actually takes place, and considerable work has been published on the stereodynamic behavior of hexacoordinate neutral complexes.

At least three different ligand exchange mechanisms, which operate simultaneously, have been reported in a series of complexes, **188−192**[211,222]. The slowest of the exchange processes is intermolecular, and is effective at the laboratory time scale, i.e. exchange takes place within minutes to several hours. It consists of exchange of the electronegative ligands from a silane to the hexacoordinate complex (equation 51), in a selective manner: Y replaces any X which is positioned to its right in the following priority list: Cl > H > Ph > Me[220]. The exact mechanism and scope of this reaction have not yet been reported.

$$\underset{(188-192)}{\text{Cl}\cdots\underset{\underset{N}{|}}{\overset{\overset{N}{|}}{\underset{X}{\text{Si}}}}\cdots\overset{O}{\underset{O}{}}} + \text{YSiCl}_3 \longrightarrow \text{Cl}\cdots\underset{\underset{N}{|}}{\overset{\overset{N}{|}}{\underset{Y}{\text{Si}}}}\cdots\overset{O}{\underset{O}{}} + \text{XSiCl}_3 \quad (51)$$

In addition, **188–192** undergo two rate processes at the NMR time scale, which have been studied in more detail[211,220]. All of these compounds appear as a single diastereomer in solution, despite the existence, in principle, of six geometrical arrangements for each of the complexes. Since the diastereomer present in solution is chiral, the four N-methyl groups are unique and give rise to four singlets in the NMR spectra at the slow exchange limit temperature. These coalesce, as the temperature is raised, in two distinct steps: first to two singlets, with one activation barrier, and at slightly higher temperature to a singlet, with another barrier. The barriers were independent of concentration, within a tenfold concentration range, indicating intramolecular exchange for both processes. The lower of the two rate processes was shown to involve exchange of the two chelate rings: in **188a**, for example, the initial coalescence of N-methyl signals from four to two was accompanied by a simultaneous exchange of the C-methyl groups, residing on different chelate cycles.

The ligand exchange barriers were found to depend substantially on the solvent. For **189b** the exchange barriers were studied in several solvents (Table 25). The apolar solvents (benzene-d_6, toluene-d_8 and CCl_4) displayed the highest barriers; the π-acceptor solvents (nitrobenzene-d_5, acetone-d_6) had barriers lower by ca 2 kcal mol^{-1}, while the hydrogen-bond donors chloroform-d and dichloromethane-d_2, as well as the strong acceptor nitromethane-d_3, gave the lowest barriers. This suggested that the solvent attaches to the dimethylamino nitrogens and weakens the Si−N coordination.

Additional evidence for the relationship of barriers with the Si−N coordination strength came from the observation of a linear correlation between barriers and the corresponding ^{29}Si chemical shifts (Figure 26). Both the barriers and the ^{29}Si chemical shifts are effected by the strengths of coordination.

These results (solvent and chemical shift dependence of the barriers) suggested that the exchange process observed involves Si−N cleavage. However, evidence for an intramolecular exchange process was recently obtained from ^{15}N NMR spectroscopy[221,222], through direct observation of ^{15}N−(Si)−^{1}H coupling *across the dative bond* [**188c**, ^2J(N−Si−H) = 11.5 Hz; **189c**, ^2J(N−Si−H) = 12.9 Hz; **190c**, ^2J(N−Si−H) = 13.0 Hz]. The persistence of ^2J(N−Si−H) up to 355 K in **188c**–**190c** is evidence that intermolecular ligand exchange does not occur at the time scale at which N-methyl exchange is rapid[221].

TABLE 25. Solvent effect on activation free energies for exchange of NMe groups in **189b**[211]

Solvent	$\Delta G_1^\ddagger$ (kcal mol^{-1})	$\Delta G_2^\ddagger$ (kcal mol^{-1})
CCl_4	16.4	16.9
$C_6D_5CD_3$	15.8	16.4
C_6D_6	16.3	16.7
C_4Cl_6	16.6	16.9
C_5D_5N	13.9	16.1
$C_6D_5NO_2$	13.2	16.2
$(CD_3)_2CO$	13.8	15.0
CD_2Cl_2	10.6	15.0
CD_3NO_2	10.6	16.0

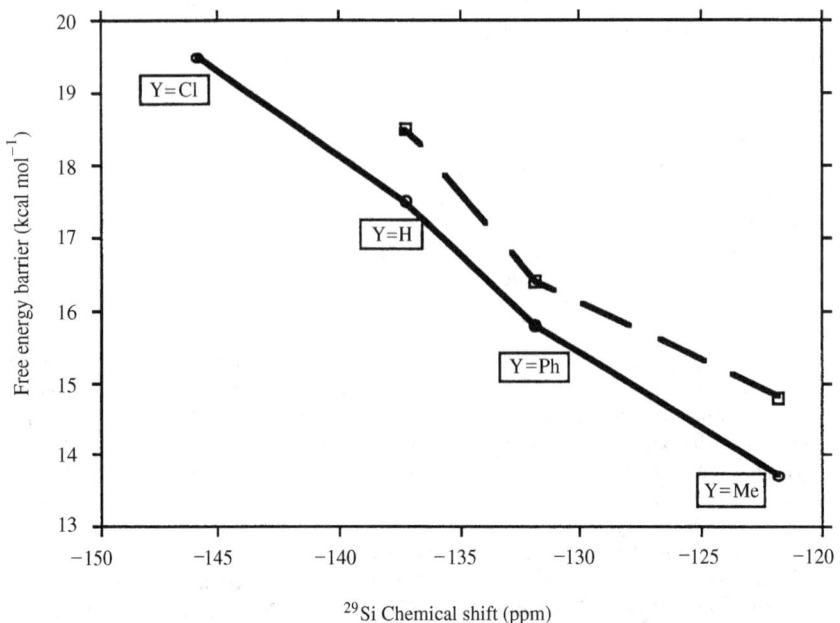

FIGURE 26. Activation free energies for two exchange barriers in **189a–189d** as a function of the ^{29}Si chemical shift[211]

Direct evidence was also obtained that no Si—F dissociation takes place in **188e** during the exchange process of N-methyl groups[222]: The one-bond ^{19}F–^{29}Si coupling constant (273 Hz) persisted essentially unchanged upon heating up to 350 K, at which N-methyl coalescence had occurred. By analogy it was also concluded that in the Si—Cl complexes no dissociation took place, in view of very similar behavior. It was thus shown that both of the observed ligand exchange processes are nondissociative.

Since only one diastereomer was observed in solution, any exchange process must be a topomerization, i.e. an exchange of diastereotopic groups within the original, unchanged diastereomer. Neither of the common exchange mechanisms (Bailar and RD twists) can account for a direct topomerization. However, it was argued that if a series of consecutive Bailar and/or RD twists were to occur and effect an overall topomerization in several steps, it would be difficult to justify two distinct exchange barriers in all of the compounds. The proposed exchange mechanism which seemed to account best for all of the observations in this series (**188–192**) was a 1,2-shift of adjacent ligands via a bicapped-tetrahedron-like transition state or intermediate. The two processes of this kind are the exchange of Cl and Y, and of the two chelate oxygens, labeled (Cl, Y)- and (O, O)-exchange, respectively (Figure 27). The former is a true topomerization, while the second converts the complex to its enantiomer. The arguments put forth in support of this mechanism are: (a) It offers two distinct and yet closely related exchange processes with slightly different activation barriers. (b) The process involves elongation of the dative bonds (up to 32%, according to one analysis)[223], which explains both the solvent dependence and the correlation of barriers with ^{29}Si chemical shifts, without Si—N cleavage. (c) The two processes generate the observed spectral changes and coalescence phenomena: examination of the processes showed that (Cl, Y)-exchange effects equilibration and coalescence of signals of the two chelate rings, while (O, O)-shift exchanges the chelate rings and in addition inverts the configuration at silicon. (d) The reports by Corriu and coworkers of hexacoordinate complexes which have the bicapped-tetrahedron geometry *in the ground state*[194,214] make such a structure a very reasonable intermediate or low-energy transition state.

The distinction between the two 1,2-shift mechanisms (i.e. the assignment of each to the lower and higher energy processes) was achieved by the study of the chiral complex **192b**, made of the optically pure (*R*) ligand[221,222]. The solution of **192b** contains two diastereomers, *RRR* and *RSR*, differing in configuration at the silicon atom. Since the (Cl, Ph)-exchange is a topomerization, and does not effect epimerization at silicon, no interchange of the two diastereomers is involved. On the other hand, the (O, O)-exchange epimerizes the silicon configuration, and constitutes the exchange: *RRR* ⇌ *RSR*. It follows that the former exchange is only *within* each diastereomer, whereas the latter is *between* the diastereomers. A 2D-NOESY-and-Saturation-Transfer experiment at low temperature clearly revealed that the first (low activation) process interchanges signals belonging to the same diastereomer only, i.e. the low-activation process was identified as (Cl, Ph)-exchange[222].

Another series of neutral hexacoordinate complexes for which exchange reactions have been studied is the group of bis-(8-dimethylamino-1-naphthyl)silanes **184**, whose structure in the solid corresponded to a bicapped tetrahedron[194,217]. The nonsymmetrically substituted **184a** and **184f** are chiral and give rise to four N-methyl signals, which coalesce to two signals. The fact that the geminal N-methyl groups remained diastereotopic is evidence that the observed process was a nondissociative (regular) exchange involving epimerization at silicon: Si—N cleavage would have permitted rotation about the N—Ph bond and exchange of geminal methyls, and was ruled out[217]. The intramolecular exchange was attributed to one of the twist mechanisms: Bailar, RD or Springers–Sievers[217]. Table 26 lists the free energies of activation measured in these compounds for N-methyl exchange.

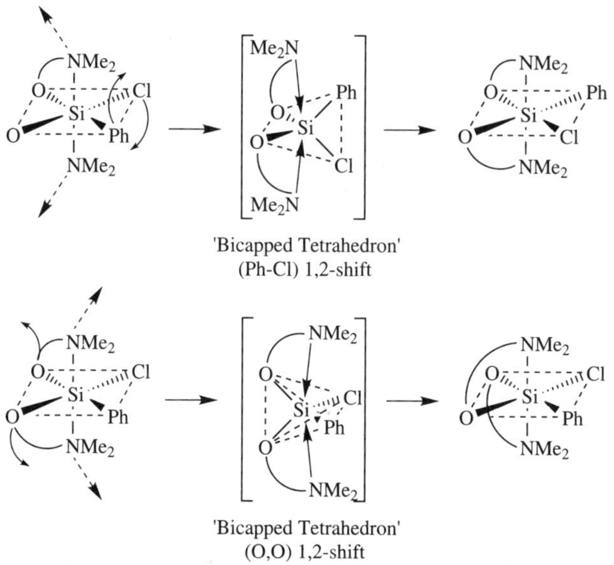

FIGURE 27. 1,2-shift mechanism of topomerization via 'bicapped tetrahedron' intermediate or transition state: upper part, (Ph, Cl)-interchange; lower part, (O, O)-interchange[211]

TABLE 26. Free energies of activation for isomerization of hexacoordinate compounds (at 300 K)[217]

Compound	X	Y	$\Delta G^{\ddagger}$ (kcal mol^{-1})
184a	H	F	14.7
184f	H	OMe	15.2
184g	H	Me	9.3
184h	Me	Ph	12.7

When the monodentate ligands in **184** are equivalent, the molecule has a C_2 symmetry axis and the chelate rings are homotopic and equivalent. As a result the four N-methyl groups give rise to two singlets. In **184c** the coalescence of the N-methyl signals to a singlet was observed, and the barrier for exchange was determined as 20.5 kcal mol^{-1}. It was not possible, however, to determine whether the barrier was due to Si–N dative bond cleavage or to a nondissociative isomerization of ligands around silicon[206].

The DNMR study of **186** revealed, as expected, the equivalence of the two silicon complexes, as well as the equivalence, at room temperature, of all the methyl groups and methylene protons, respectively[215]. When the temperature was lowered, the geminal methyls and the geminal protons became diastereotopic, while the two (dimethylamino)methyl groups attached to each aromatic ring remained equivalent. Over this temperature range the Si–H protons also remained isochronous *(chemical shift equivalent)*. To rationalize this NMR behavior, the authors offered two alternative explanations: either that in solution the predominant conformation was as in the crystal, i.e. with the dimethylamino groups in *cis* positions and the Si–H protons in *cis* positions, with a rapid Bailar-type twist rendering the Si–H and the (dimethylamino)methyl groups equivalent, respectively. Alternatively, the conformation with C_s symmetry depicted in **186a** is predominant in solution, which *a priori* renders the two Si–H and two (dimethylamino)methyl groups equivalent[215]

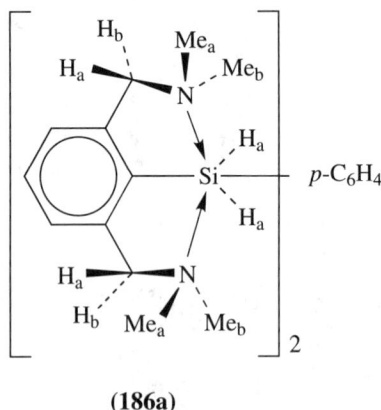

(**186a**)

At lower temperatures the methylene protons and the N-methyl groups decoalesced in two distinct rate processes: $\Delta G^*_{Me} = 8.4$ kcal mol^{-1}; $\Delta G^*_{CH_2} = 11.5$ kcal mol^{-1}. This was interpreted by the authors as Si–N dissociation leading to N-methyl exchange, with a concomitant rotation about the Si-aromatic bond (with a slightly higher barrier) to account for the exchange of geminal methylene protons[215].

B. Intermolecular Coordination

Polyhalosilanes form neutral intermolecular hexacoordinate complexes with donor molecules such as pyridine, triethylamine, 2,2′-bipyridine and 1,10-phenanthroline. This topic has recently been reviewed[6]. It was demonstrated that electronegative substituents on the silicon are essential for the formation of intermolecular complexes. Thus, while $SiCl_4$ and $Cl_2CHSiCl_3$ react with 1,10-phenanthroline and with 2,2′-bipyridine to form hexacoordinate chelates, $MeSiCl_3$ does not react[224,225]. For completion we discuss here a few examples, and compare some of the properties of intermolecular complexes with those of the intramolecular complexes.

An interesting example of an intermolecular complex is the trisilicon complex **194**, in which only the central silicon is coordinated to the bidentate donor molecule[225]. The structure is a regular octahedron, with two tetrahedral termini. The silicon nitrogen bonds are rather short (2.012 and 1.991 Å), and are comparable to those of octahedral intramolecular complexes (Table 23). **194** permits a comparison of Si—Cl bonds in a tetrahedral silicon moiety (2.03 to 2.07 Å) with Si—Cl bonds *trans* to the dative bond in a hexacoordinate silicon (2.39 and 2.21 Å). As expected, the latter are substatntially longer than the regular covalent bonds.

A stereodynamic study in an intermolecular complex, **195a**, was reported by Farnham and Whitney[226]. **195a** undergoes intermolecular exchange, which is observed in the ^{19}F NMR spectrum by warming a mixture of **195a** and **195b**. The equilibrium reaction between **195b**, the donor molecule and **195a** provided the mechanism for both the exchange between diastereomers of **195a**, as well as for enantiomerization (exchange of CF$_3$ groups within **195a**).

A crystallographic structure was reported for an oxygen-coordinated intermolecular complex, [Si(catecholato)$_2$]·2THF **(196)**[201a]. Interestingly, the two catechol ligands in **196** are coplanar (Si−O$_{cat}$ = 1.719, 1.727 Å), while the THF molecules occupy *trans* positions (Si−O$_{THF}$ = 1.930 Å).

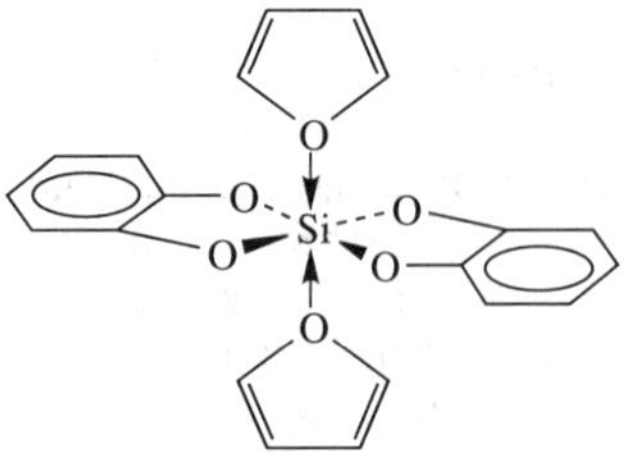

(196)

VII. COMPLEXES WITH HIGHER COORDINATION NUMBERS

A few examples have been reported in which presumably silicon has a coordination number greater than six[227−229]. All of the hepta- and octa-coordinated silicon complexes for which single-crystal X-ray diffraction structures were obtained have a basic tetrahedral geometry, with three or four distantly coordinated ligand groups positioned against faces of the tetrahedron, in analogy to the bicapped tetrahedron structure mentioned earlier (Section VI.A.2).

The first reported synthesis of a heptacoordinate complex of silicon **(197)** was accomplished by the reaction of HSiCl$_3$ with 2-lithio-(dimethylaminomethyl)-benzene[227]. The crystal structure was determined later and confirmed the 'tricapped tetrahedron' geometry[81,82]. Replacement of the hydrido ligand by chloro **(198)** led to the formation of a tetracoordinate silane without any dative bonds, apparently due to the steric bulk associated with the chloro ligand[81].

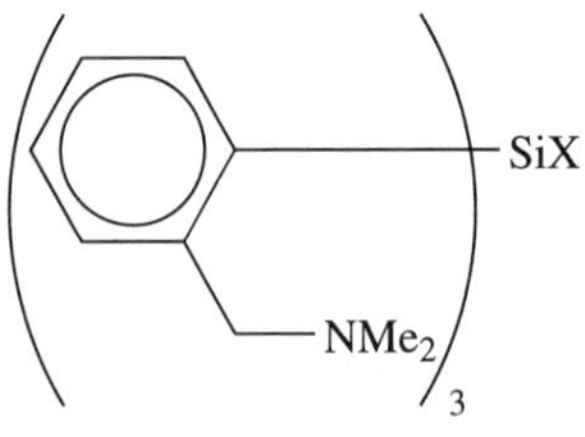

(197) X = H
(198) X = Cl
(199) X = F

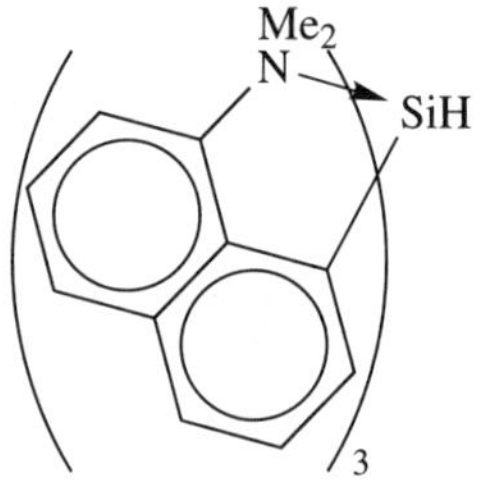

(200)

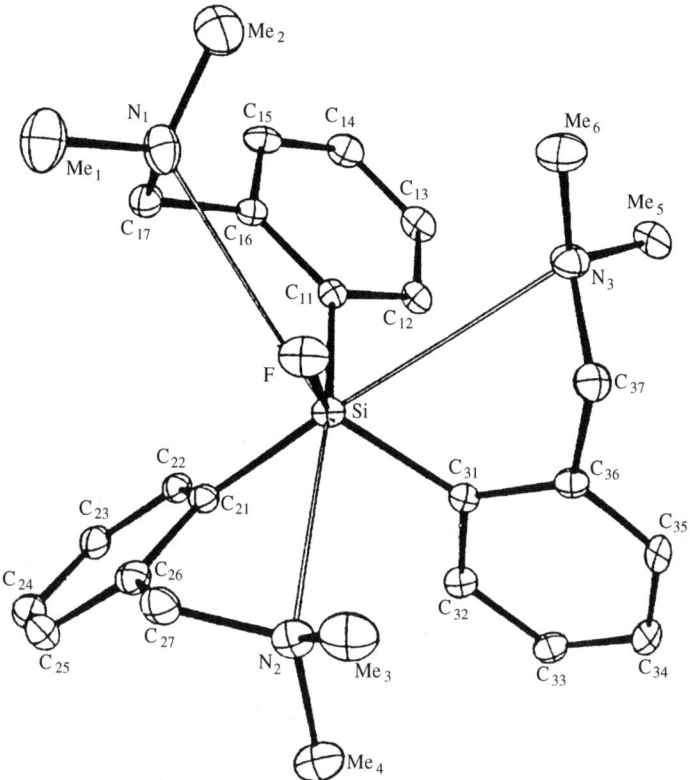

FIGURE 28. Crystallographic structure of the heptacoordinate complex **199**. Reprinted with permission from Reference 228. Copyright 1994 American Chemical Society.

The analogous fluoro compound (**199**) was shown to have heptacoordination[228]. Interestingly, the X-ray crystal structure of **199** (Figure 28) showed that neither of the three nitrogens is coordinated in *trans* position to the Si—F bond, and all three are opposite the Si—C bonds. This was rationalized in terms of the least steric hindrance for approach of the ligand groups from the fluoro side, or, more precisely, from the normals to the triangular faces of the tetrahedron skeleton which have the fluorine in one corner.

A different synthesis of a heptacoordinate silicon complex (**200**) is shown in equation 52. The resulting complex had the same basic tricapped-tetrahedron structure (Figure 29)[228].

Formal octacoordination ('4 + 4')[229] in a silicon complex was recently demonstrated by Corriu's group, who prepared **201** (equation 53). The Si—N distances (Table 27) are long relative to those common in penta- and hexacoordinate compounds, but are in the same range as those found in the heptacoordinate compounds discussed here. Also, the geometry resembles the heptacoordinate complexes, in that it is basically a tetrahedron with dimethylamino donor groups pointing toward the center, in what may be termed a 'tetracapped tetrahedron'.

A similar tetracapped tetrahedral geometry was recently found by X-ray crystallography in bis[2,4,6-tris(trifluoromethyl)phenyl]-fluorosilane (**202**)[230] and -difluorosilane (**203**)[231],

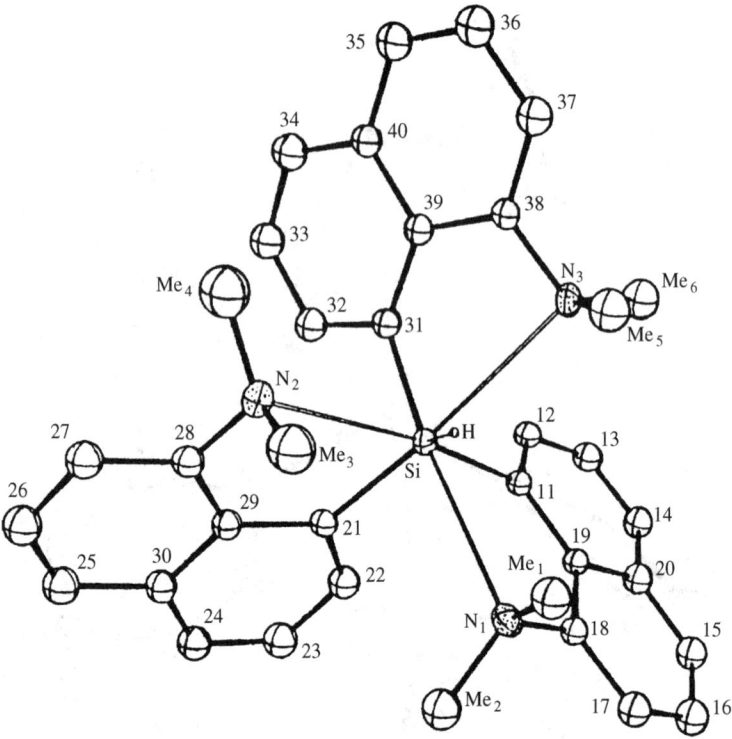

FIGURE 29. X-ray crystallographic structure of **200**. Reprinted with permission from Reference 228, Copyright 1994 American Chemical Society

in which one fluorine atom from each of the *ortho* CF$_3$ groups is coordinated to silicon.

(52)

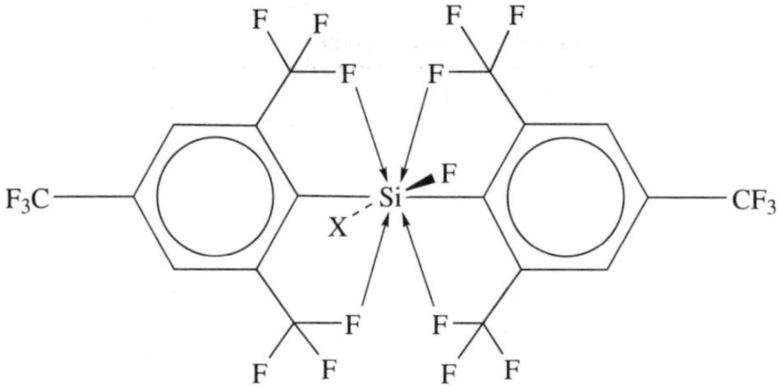

23. Hypervalent silicon compounds

(202) X = H
(203) X = F

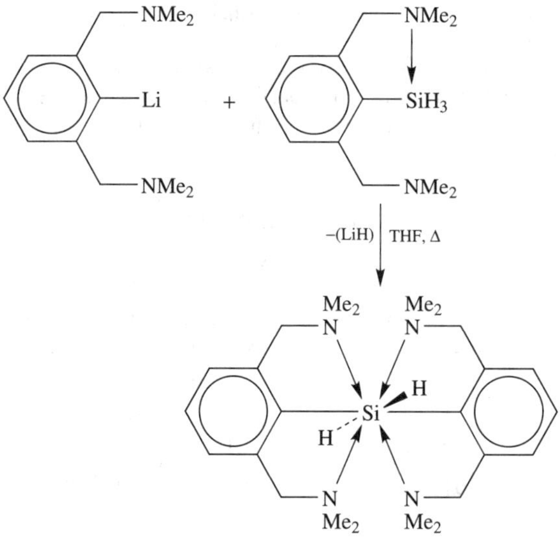

(53)

(201)

Table 27 lists the ^{29}Si chemical shifts for the high coordination complexes. Triphenylsilane serves as a tetracoordinate model for the complexes **197** and **200**, for comparison. It can be seen that the increased coordination in the latter complexes effects a small upfield shift, much smaller than is normally observed upon going from tetra- to penta- (Table 11, Section III.A.2.b) and hexacoordination (Table 24, Section VI.A.2). This is in accord with the crystallographic results, which show that no significant change in hybridization takes place in these compounds, in contrast to TBP and octahedral complexes.

In line with this result, also the one-bond Si—H coupling constants respond differently to hepta- and octacoordination than to TBP and octahedral geometries (Table 27). While in the latter the change in coordination brings about a decrease in the Si—H s-character, with consequent decrease in ^1J(Si—H), in the *higher* coordinated compounds described in this section there is no significant change in s-character, and the only effect is some addition of electron density to the silicon due to donation from the ligands resulting in *increased* ^1J(Si—H).

TABLE 27. X-ray crystallographic and ^{29}Si NMR data for hepta- and octacoordinate silicon complexes and tetravalent model compounds

Compound	Si–N (Å)			δ^{29}Si (ppm) ($J_{\text{Si-H}}$, Hz)	References
197	3.006	3.039	2.997	−34.9(226.4)	81, 82
	2.981	3.009	3.043		
199	3.489	3.004	3.307	−9.86	228
200	2.882	2.876	2.928	−25.86(284.0)	228
201	3.117	2.895		−61.6(213.0)	229
Ph$_3$SiH				−21.1(205.1)	40
Ph$_2$SiH$_2$				−33.8(200.0)	40

The ^{1}H NMR spectrum of **201** showed all four dimethylamino groups to be equivalent, an indication of rapid exchange between them. This presumably occurs by rapid displacement of one group by the other.

Finally, a group of hypercoordinated silicon compounds, the decamethyl silicocenes, in which the formal silicon coordination number is ten, is worthy of mention in connection with this chapter. These analogs of ferrocene have been studied extensively[232], and are described in detail in Jutzi's chapter in this volume.

VIII. ADDENDUM

Since the major parts of this chapter were written in July 1996, a number of papers dealing with hypervalent silicon compounds were published[233–251]. In this late addendum we have tried to cite all the recent relevant publications and give a very brief account of their content. The following citations have been arranged in the order of the most relevant sections in this chapter.

For Section I

Crystal structures of hypervalent silicon complexes were discussed in detail and reviewed by Lukevics and Pudova[233].

For Section II

An equilibrium reaction between a pentaorganosilicate complex (**204**) and its tetracoordinate precursor **205** and methyllithium was observed using ^{29}Si NMR spectroscopy at low temperature[234].

A novel fluoro-bridged complex, pentafluoro-1,4-disilacyclohexane anion (**206**), was reported to undergo unusual intramolecular fluoride ion exchange between two silicon atoms, synchronized with a ring inversion process[235].

A new mass spectrometric technique, triple quadrupole detection, was employed to study 1,1-dimethyl-1-fluorosilacyclobutane anion, a pentacoordinate silicate[236].

A theoretical comparison between the structures and properties of cyclic pentacoordinate anionic pentaoxysilicates and phosphoranes is described[237]. The results are compared with NMR and X-ray crystallographic data.

The preparation of a novel neutral zwitterionic λ^5-spirosilicate and its characterization was reported[238].

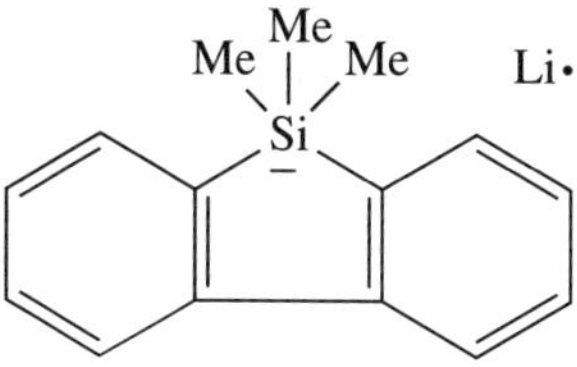

(204)   **(205)**

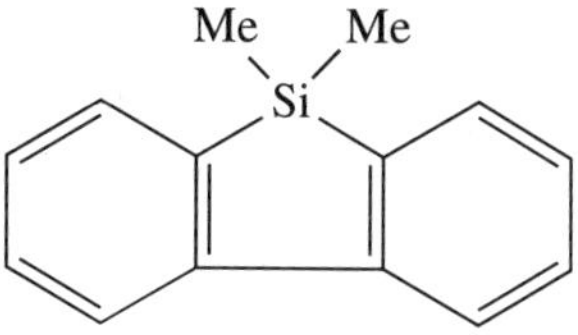

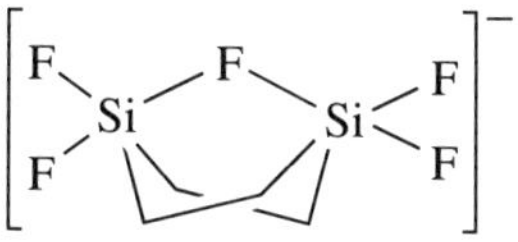

(206)

For Section III

New neutral pentacoordinate silicon complexes with a dative O→Si bond **(207, 208)** have been prepared and studied. The NMR spectral data and crystal structures were reported[239].

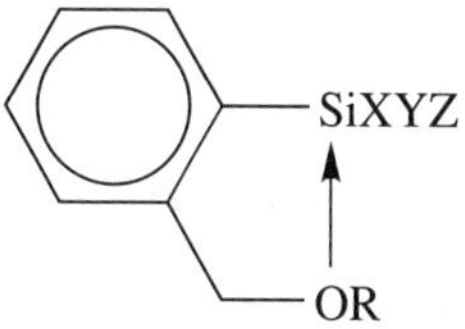

(**207a**) X = Y = Z = F, R = Me
(**207b**) X = Y = Z = F, R = CHMe$_2$
(**208**)   X = H, Y = Ph, Z = OSO$_2$CF$_3$, R = Me

Pentacoordinate O→Si complexes analogous to series **117** with the chiral (R = 1-phenylethyl) ligand were prepared, with various electronegative ligands, (**117k–117n**, X = F, Cl, Br, OCHMe$_2$, respectively). Their stereochemical nonrigidity was studied by variable-temperature ^{29}Si NMR spectroscopy[240].

A study of the solvolytic and thermolytic reactions of a series of heterocyclic pentacoordinate silicon complexes was described[241].

A novel (3,3) sigmatropic rearrangement of a hexacoordinate allyl–silicon complex (neutral tetraoxyspirosilicate) to a pentacoordinate complex was recently described[242]. The allyl group migrates from silicon to the α-carbon of a tropolone ligand[242].

Trapping of an intermediate silylene by reaction with diphenylacetylene to form a pseudo-pentacoordinate silole **(209)** was reported[243]. The silylene was obtained by a Ni(0)-catalyzed degradation of **92f** (X = SiPh$_3$, Y = Me, Z = H)[243].

For Section IV

A full discussion of earlier preliminary results[186] (discussed in this review) has now appeared[244].

A high-level theoretical investigation of the structure of the experimentally studied compound **160a** and some of its analogs is described[245]. The question of solvation of a silicenium cation is addressed and **160a** served as a model[245].

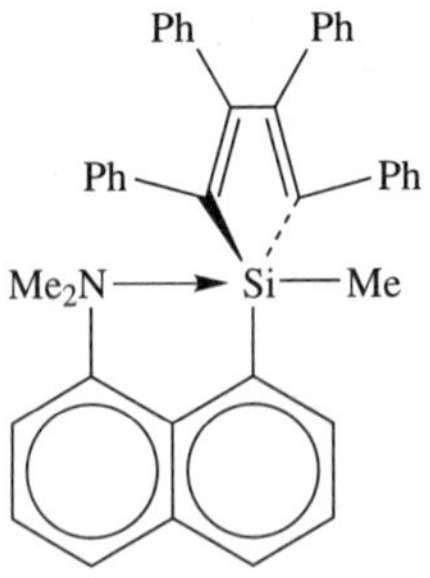

(209)

For Section V

Treatment of tetracoordinated silanes by the NCS$^-$ anion afforded some new pentacoordinate anionic spirobicyclic siliconates [(C$_6$H$_4$O$_2$)$_2$SiNCS]$^-$M$^+$ with various M$^+$ counter cations[246], as well as some hexacoordinate dianions [(C$_6$H$_4$O$_2$)$_2$Si(NCS)$_2$]$^{-2}$.

3-Picoline was found to react in chloroform solution with bis(dichlorosilyl)amine to form two hexacoordinate silicon complexes: the ionic [H$_2$Si(3-MeC$_5$H$_4$N)$_4$]$^{2+}\cdot$2Cl$^-$ and a neutral H$_2$Si(3-MeC$_5$H$_4$N)$_2$Cl$_2$. These complexes are in equilibrium with each other in chloroform solution. The crystal structure of the ionic complex was reported[247].

For Section VI

Karsch and coworkers report the formation of new penta- (**210**) and hexacoordinate (**211**) complexes with novel bidentate phosphorus-donor ligands[248−250]. The phosphorus ligands form four-membered chelate rings (siladiphosphacyclobutane). Interestingly, the crystal structure of **210** indicates a near-square pyramid[248]. The hexacoordinate complexes were found to have the *cis* geometry when the monodentate ligands are methyl groups[250], and *trans* geometry for dichloro complexes[249] (i.e. the monodentate ligands adopt either the *trans* or the *cis* geometry). For **211c** both *cis* and *trans* isomers were obtained, and the former rearranged spontaneously to the latter at room temperature[249].

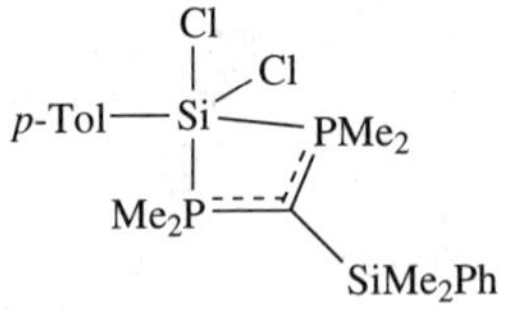

(**210**)

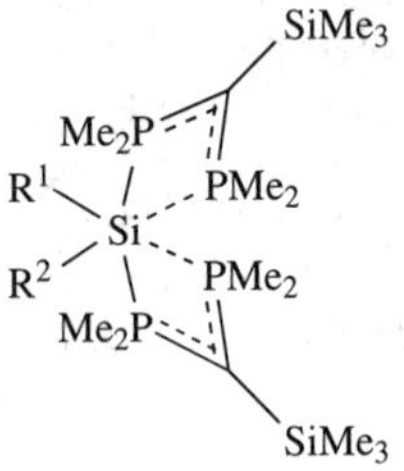

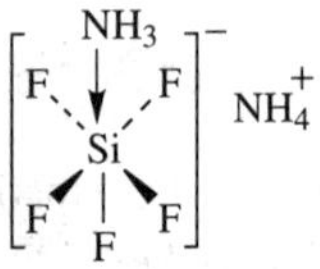

(**211a**) R^{1}R^2 = −(CH$_2$)$_3$
(**211b**) R^1 = R^2 = Me
(**211c**) R^1 = R^2 = Cl

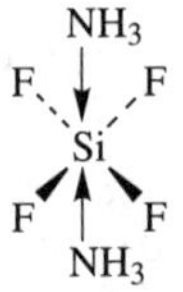

(**212**)   (**213**)

The reaction of silicon powder with NH_4HF_2 at 400 °C in a sealed ampoule afforded single crystals of the hexacoordinate complexes **(212)** and **(213)**. The crystallographic analyses of both compounds showed near-octahedral arrangements, with the NH_3 groups of the latter compound in *trans* positions[251].

IX. REFERENCES

1. S. N. Tandura, M. G. Voronkov and N. V. Alekseev, *Top. Curr. Chem.*, **131**, 99 (1986).
2. R. J. P. Corriu and J. C. Young, in *The Chemistry of Organic Silicon Compounds* (Eds. S. Patai and Z. Rappoport), Wiley, Chichester, 1989, pp. 1241–1288.
3. W. S. Sheldrick, in *The Chemistry of Organic Silicon Compounds* (Eds. S. Patai and Z. Rappoport), Wiley, Chichester, 1989, pp. 227–304.
4. (a) R. R. Holmes, *Chem. Rev.*, **90**, 17 (1990).
   (b) R. R. Holmes, *Chem. Rev.*, **96**, 927 (1996).
5. R. J. P. Corriu, *J. Organomet. Chem.*, **400**, 81 (1990).
6. C. Chuit, R. J. P. Corriu, C. Reye and J. C. Young, *Chem. Rev.*, **93**, 1371 (1993).
7. C. Y. Wong and J. D. Woollins, *Coord. Chem. Rev.*, **130**, 175 (1994).
8. (a) R. Müller, *Organomet. Chem. Rev.*, **1**, 359 (1966).
   (b) R. Müller, *Z. Chem.*, **24**, 41 (1984).
9. Reviews:
   (a) M. G. Voronkov and L. I. Gubanova, *Main Group Metal Chem.*, **10**, 209 (1987).
   (b) V. O. Gel'mbol'dt and A. A. Ennan, *Usp. Khim.*, **58**, 626 (1989); *Chem. Abstr.*, **111**, 174159g (1989).
10. (a) D. Schomburg, *J. Organomet. Chem.*, **221**, 137 (1981).
    (b) D. Schomburg and R. Krebs, *Inorg. Chem.*, **23**, 1378 (1984).
11. (a) R. Damrauer and S. E. Danahey, *Organometallics*, **5**, 1490 (1986).
    (b) R. Damrauer, B. O'Connell, S. E. Danahey and R. Simon, *Organometallics*, **8**, 1167 (1989).
12. J. J. Harland, J. S. Payne, R. O. Day and R. R. Holmes, *Inorg. Chem.*, **26**, 760 (1987).
13. (a) S. E. Johnson, R. O. Day and R. R. Holmes, *Inorg. Chem.*, **28**, 3182 (1989).
    (b) S. E. Johnson, J. S. Payne, R. O. Day, J. M. Holmes and R. R. Holmes, *Inorg. Chem.*, **28**, 3190 (1989).
14. S. E. Johnson, J. A. Deiters, R. O. Day and R. R. Holmes, *J. Am. Chem. Soc.*, **111**, 3250 (1989).
15. F. Scherbaum, B. Huber, G. Muller and H. Schmidbaur, *Angew. Chem., Int. Ed. Engl.*, **27**, 1542 (1988).
16. F. Klanberg and E. L. Muetterties, *Inorg. Chem.*, **7**, 155 (1968).
17. (a) K. Tamao, T. Hayashi and Y. Ito, *Organometallics*, **11**, 2099 (1992).
    (b) K. Tamao, T. Hayashi, Y. Ito and M. Shiro, *J. Am. Chem. Soc.*, **112**, 2422 (1990).
    (c) K. Tamao, T. Hayashi and Y. Ito, *J. Organomet. Chem.*, **506**, 85 (1996).
18. K. Tamao, T. Hayashi, Y. Ito and M. Shiro, *Organometallics*, **11**, 182 (1992).
19. H. B. Bürgi and J. Dunitz, *Acc. Chem. Res.*, **16**, 153 (1983).
20. (a) R. O. Day, C. Sreelatha, J. A. Deiters, S. E. Johnson, J. M. Holmes, L. Howe and R. R. Holmes, *Organometallics*, **10**, 1758 (1991).
    (b) R. O. Day, C. Sreelatha, J. A. Deiters, S. E. Johnson, J. M. Holmes, L. Howe and R. R. Holmes, *Phosphorus, Sulfur Silicon Relat. Elem.*, **100**, 87 (1995).
21. (a) J. A. Cella, J. D. Cargioli and E. A. Williams, *J. Organomet. Chem.*, **186**, 13 (1980).
    (b) V. F. Sidorkin, V. A. Pestunovich and M. G. Voronkov, *Magn. Res. Chem.*, **23**, 491 (1985).
22. M. Kira, T. Hoshi and H. Sakurai, *Chem. Lett.*, 807 (1995).
23. H. J. Frohn and V. V. Bardin, *J. Organomet. Chem.*, **501**, 155 (1995).
24. D. Brondani, F. H. Carre, R. J. P. Corriu, J. J. E. Moreau and M. Wong Chi Man, *Angew. Chem., Int. Ed. Engl.*, **35**, 324 (1996).
25. D. A. Loy, J. H. Small and K. J. Shea, *Organometallics*, **12**, 1484 (1993).
26. M. Kira, T. Hoshi, C. Kabuto and H. Sakurai, *Chem. Lett.*, 1859 (1993).
27. R. O'Dell, *Tetrahedron Lett.*, **36**, 5723 (1995).
28. R. Tacke, J. Becht, A. Lopez-Mras and G. Sperlich, *J. Organomet. Chem.*, **446**, 1 (1993).
29. R. Tacke, J. Becht, O. Dannappel, M. Kropfgans, A. Lopez-Mras, M. Mühleisen and J. Sperlich, in *Progress in Organosilicon Chemistry* (Eds. B. Marciniec and J. Chojnowski), Gordon and Breach, Basel, 1995, pp. 55–68.

30. (a) R. Tacke, O. Dannappel and M. Mühleisen, in *Organosilicon Chemistry II* (Eds. N. Auner and J. Weis), VCH, Weinheim, 1996, pp. 427-446.
    (b) R. Tacke, J. Becht, O. Dannappel, R. Ahlrichs, U. Schneider, W. S. Sheldrick, J. Hahn and F. Kiesgen, *Organometallics*, **15**, 2060 (1996).
31. (a) R. Tacke, J. Becht, G. Mattern and W. F. Kuhs, *Chem. Ber.*, **125**, 2015 (1992).
    (b) R. Tacke, A. Lopez-Mras, J. Becht and W. S. Sheldrick, *Z. Anorg. Allg. Chem.*, **619**, 1012 (1993).
32. For early discussions of this topic see:
    (a) R. Müller and L. Heinrich, *Chem. Ber.*, **94**, 1943 (1961).
    (b) C. L. Frye, *J. Am. Chem. Soc.*, **86**, 3170 (1964).
    (c) C. L. Frye, *J. Am. Chem. Soc.*, **92**, 1205 (1970).
    (d) D. Shomburg, *Z. Naturforsch.*, **37B**, 195 (1982).
33. J. J. Harland, R. O. Day, J. F. Vollano, A. C. Sau and R. R. Holmes *J. Am. Chem. Soc.*, **103**, 5269 (1981).
34. K. C. Kumara Swamy, V. Chandrasekhar, J. J. Harland, J. M. Holmes, R. O. Day and R. R. Holmes, *J. Am. Chem. Soc.*, **112**, 2341 (1990).
35. (a) E. F. Perozzi and J. C. Martin, *J. Am. Chem. Soc.*, **101**, 1591 (1979).
    (b) W. H. Stevenson III, S. Wilson, J. C. Martin and W. B. Farnham, *J. Am. Chem. Soc.*, **107**, 6340 (1985).
36. R. M. Laine, K. Y. Blohowiak, T. R. Robinson, M. L. Hoppe, P. Nardi, J. Kampf and J. Uhm, *Nature*, **353**, 642 (1991).
37. (a) K. C. Kumara Swamy, Channareddy Sreelatha, R. O. Day, J. Holmes and R. R. Holmes, *Inorg. Chem.*, **30**, 3126 (1991).
    (b) K. C. Kumara Swamy, Channareddy Sreelatha, R. O. Day, J. Holmes and R. R. Holmes, *Phosphorus, Sulfur Silicon Relat. Elem.*, **100**, 107 (1995).
38. D. F. Evans, A. M. Z. Slawin, D. J. Williams, C. Y. Wong and J. D. Woollins, *J. Chem. Soc., Dalton Trans.*, 2383 (1992).
39. J. L. Brefort, R. J. P. Corriu, C. Guerin, B. J. L. Henner and W. W. C. Wong Chi Man, *Organometallics*, **9**, 2080 (1990).
40. H. Marsmann, *Nucl. Magn. Reson.*, **17**, 65, (1981).
41. D. A. Dixon, W. R. Hertler, D. B. Chase, W. B. Farnham and F. Davidson, *Inorg. Chem.*, **27**, 4012 (1988).
42. F. P. Boer, J. J. Flynn and J. W. Turley, *J. Am. Chem. Soc.*, **90**, 6973 (1968).
43. R. R. Holmes, R. O. Day, V. Chandrasekhar and J. M. Holmes, *Inorg. Chem.*, **24**, 2009, 2016 (1985).
44. R. R. Holmes, R. O. Day, J. J. Harland, A. C. Sau and J. M. Holmes, *Organometallics*, **3**, 341 (1984).
45. R. R. Holmes, R. O. Day, J. J. Harland and J. M. Holmes, *Organometallics*, **3**, 347 (1984).
46. R. Tacke, A. Lopez-Mras, J. Sperlich, C. Strohmann, W. Kuhs, G. Mattern and A. Sebald, *Chem. Ber.*, **126**, 851 (1993).
47. R. Tacke, M. Mühleisen, A. Lopez-Mras and W. S. Sheldrick, *Z. Anorg. Allg. Chem.*, **621**, 779 (1995).
48. R. Tacke, J. Sperlich, C. Strohmann and G. Mattern, *Chem. Ber.*, **124**, 1491 (1991).
49. R. Tacke, A. Lopez-Mras and P. G. Jones, *Organometallics*, **13**, 1617 (1994).
50. M. Mühleisen and R. Tacke, *Chem. Ber.*, **127**, 1615 (1994).
51. M. Mühleisen and R. Tacke, in *Organosilicon Chemistry II* (Eds. N. Auner and J. Weis), VCH, Weinheim, 1996, pp. 447-452.
52. O. Dannappel and R. Tacke, in *Organosilicon Chemistry II* (Eds. N. Auner and J. Weis), VCH, Weinheim, 1996, pp. 453-458.
53. (a) N. Erchak, A. Kemme and G. Ancens, *Xth International Symposium on Organosilicon Chemistry*, August 15-20, 1993, Poznan, Poland. Abstract, **P - 35**, p. 159.
    (b) N. Erchak, *Xth International Symposium on Organosilicon Chemistry*, August 15-20, 1993, Poznan, Poland. Abstract, **O - 29**, p. 77.
54. (a) N. P. Erchak, Doct. Dissertation Latvian Inst. of Organic Synthesis, Riga 1990.
    (b) N. P. Erchak, G. A. Ancens, A. A. Kemme and E. J. Lukevics, *XI International Symp. on Organosilicon Chemistry*, Montpellier, France 1996, p. PA60.
55. G. Cerveau, C. Chuit, E. Colomer, R. J. P. Corriu and C. Reye, *Organometallics*, **9**, 2415 (1990).
56. N. Erchak, G. Ancens and V. Ryabova, *Xth International Symposium on Organosilicon Chemistry*, August 15-20, 1993, Poznan, Poland. Abstract, **P - 179**, p. 299.

57. (a) M. Mühleisen and R. Tacke., *Organometallics*, **13**, 3740 (1994).
    (b) R. Tacke, M. Mühleisen and P. G. Jones, *Angew. Chem., Int. Ed. Engl.*, **33**, 1186 (1994).
58. (a) K. J. Shea, D. A. Loy and J. H. Small, *Chem. Mater.*, **4**, 255 (1992).
    (b) J. H. Small, D. J. McCord, J. Greaves and K. J. Shea, *J. Am. Chem. Soc.*, **117**, 11588 (1995).
59. B. Becker, R. J. P. Corriu, C. Guerin, B. Henner and Q. Wang, *J. Organomet. Chem.*, **368**, C25 (1989).
60. R. J. P. Corriu, C. Guerin, B. Henner and Q. Wang, *Organometallics*, **10**, 2297 (1991).
61. R. J. P. Corriu, C. Guerin, B. J. L. Henner and Q. Wang, *Organometallics*, **10**, 3574 (1991).
62. R. J. P. Corriu, C. Guerin, B. J. L. Henner and Q. Wang, *J. Organomet. Chem.*, **439**, C1 (1992).
63. A. R. Bassindale and J. Jiang, *J. Organomet. Chem.*, **446**, C3 (1993).
64. R. Damrauer, L. W. Burggrae, L. P. Davis and M. S. Gordon, *J. Am. Chem. Soc.*, **110**, 6601 (1988).
65. D. J. Hajdasz, Y. Ho and R. R. Squires, *J. Am. Chem. Soc.*, **116**, 10751 (1994).
66. C. H. DePuy, R. Damrauer, J. H. Bowie and J. C. Sheldon, *Acc. Chem. Res.*, **20**, 127 (1987).
67. (a) R. Damrauer, *Adv. Silicon Chem.*, **2**, 91 (1993).
    (b) R. Damrauer and J. A. Hankin, *Chem. Rev.*, **95**, 1137 (1995).
68. J. D. Payzant, K. Tanaka, L. D. Betowski and D. K. Bohme, *J. Am. Chem. Soc.*, **98**, 894 (1976).
69. A. E. Reed and P. v. R. Schleyer, *Chem. Phys. Lett.*, **133**, 533 (1987).
70. A. I. Boldyrev and J. Simons, *J. Chem. Phys.*, **99**, 4628 (1993).
71. (a) M. T. Carroll, M. S. Gordon and T. L. Windus, *Inorg. Chem.*, **31**, 825 (1992).
    (b) M. S. Gordon, L. P. Davis, L. W. Burggraf and R. J. Damrauer, *J. Am. Chem. Soc.*, **108**, 7889 (1986).
    (c) L. P. Davis, L. W. Burggraf and M. S. Gordon, *J. Am. Chem. Soc.*, **110**, 3056 (1988).
    (d) M. S. Gordon, T. L. Windus, L. W. Burggraf and L. P. Davis, *J. Am. Chem. Soc.*, **112**, 7167 (1990).
72. (a) T. L. Windus, M. S. Gordon, L. W. Burggraf and L. P. Davis, *J. Am. Chem. Soc.*, **113**, 4356 (1991).
    (b) T. L. Windus, M. S. Gordon, L. P. Davis and L. W. Burggraff, *J. Am. Chem. Soc.*, **116**, 3568 (1994).
    (c) D. L. Willhite and L. Spialter, *J. Am. Chem. Soc.*, **95**, 2100 (1973).
73. H. Fujimoto, N. Arita and K. Tamao, *Organometallics*, **11**, 3035 (1992).
74. Y. Apeloig, in *The Chemistry of Organic Silicon Compounds* (Eds. S. Patai and Z. Rappoport), Wiley, Chichester, 1989, pp. 57–226.
75. (a) J. G. Verkade, *Coord. Chem. Rev.*, **137**, 233 (1994) and references cited therein.
    (b) M. G. Voronkov, *Top. Curr. Chem.*, **84**, 77 (1979).
    (c) M. G. Voronkov, V. P. Baryshok, L. P. Petukhov, V. I. Rachlin, R. G. Mirskov and V. A. Pestunovich, *J. Organomet. Chem.*, **358**, 39 (1988).
    (d) E. Kupce, E. Liepins, A. Lapsina, G. Zelcans and E. Lukevics *J. Organomet. Chem.*, **333**, 1 (1987).
    (e) J. J. H. Edema, R. Libbers, A. Ridder and R. M. Kellogg, *J. Organomet. Chem.*, **464**, 127 (1994).
76. R. J. P. Corriu, A. Kpoton, M. Poirier, G. Royo, A. de Saxce and J. C. Young, *J. Organomet. Chem.*, **395**, 1 (1990).
77. F. Carre, R. J. P. Corriu, A. Kpoton, M. Poirier, G. Royo, J. C. Young and C. Belin, *J. Organomet. Chem.*, **470**, 43 (1994).
78. F. H. Carre, R. J. P. Corriu, G. F. Lanneau and Z. Yu, *Organometallics*, **10**, 1236 (1991).
79. J. Boyer, C. Breliere, F. Carre, R. J. P. Corriu, A. Kpoton, M. Poirier, G. Royo and J. C. Young, *J. Chem. Soc., Dalton Trans.*, 43 (1989).
80. N. Auner, R. Probst, C. Heikenwalder, E. Herdtweck, S. Gamper and G. Müller, *Z. Naturforsch.*, **48b**, 1625 (1993).
81. N. Auner, R. Probst, F. Hahn and E. Herdtweck, *J. Organomet. Chem.*, **459**, 25 (1993).
82. N. Auner, *J. Prakt. Chem.*, **337**, 79 (1995).
83. G. Klebe, *J. Organomet. Chem.*, **332**, 35 (1987).
84. H. Lang, E. Meichel, M. Weinmann and M. Melter in *Organosilicon Chemistry III* (Eds. N. Auner and J. Weis), VCH, Weinheim, in press.
85. G. Klebe, M. Nix and K. Hensen, *Chem. Ber.*, **117**, 797 (1984).
86. G. Klebe, *J. Organomet. Chem.*, **293**, 147 (1985).
87. G. Klebe, J. W. Bats and K. Hensen, *J. Chem. Soc., Dalton Trans.*, 1 (1985).

88. G. Klebe, K. Hensen and J. von Jouanne, *J. Organomet. Chem.*, **258**, 137 (1983).
89. D. Kummer, S. C. Chaudhry, J. Seifert, B. Deppisch and G. Mattern, *J. Organomet. Chem.*, **382**, 345 (1990).
90. D. Kummer, S. H. Abdel Halim, W. Kuhs and G. Mattern, *Z. Anorg. Allg. Chem.*, **614**, 73 (1992).
91. D. Kummer, S. H. Abdel Halim, W. Kuhs and G. Mattern, *J. Organomet. Chem.*, **446**, 51 (1993).
92. (a) M. G. Voronkov, A. I. Albanov, A. E. Pestunovich, V. N. Sergeev, S. V. Pestunovich, I. I. Kandror and Yu. I. Baukov, *Metalloorg. Khim.*, **1**, 1435 (1988); *Chem. Abstr.*, **112**, 77288p (1990).
    (b) A. A. Macharashvili, Yu. E. Ovchinnikov, Yu. T. Struchkov, V. N. Sergeev, S. V. Pestunovich and Yu. I. Baukov, *Izv. Akad. Nauk SSSR, Ser. Khim.*, 189 (1993); *Russ, Chem. Bull.*, **42**, 173 (1993).
93. A. R. Bassindale, S. G. Glynn, J. Jiang, D. J. Parker, R. Turtle, P. G. Taylor and S. S. D. Brown, in *Organosilicon Chemistry II* (Eds. N. Auner and J. Weis), VCH, Weinheim, 1996, pp. 411–426.
94. I. D. Kalikhman, B. A. Gostevskii, O. B. Bannikova, M. G. Voronkov and V. A. Pestunovich, *Metalloorg. Khim.*, **2**, 205 (1989); *Chem. Abstr.*, **112**, 77291j (1990).
95. (a) I. Kalikhman, S. Krivonos, A. Ellern and D. Kost, *Organometallics*, **15**, 5073 (1996).
    (b) I. D. Kalikhman and D. Kost, in *Organosilicon Chemistry III* (Eds. N. Auner and J. Weis), VCH, Weinheim, 1997, pp. 446–451.
96. J. Belzner and D. Schar, in *Organosilicon Chemistry II* (Eds. N. Auner and J. Weis), VCH, Weinheim, 1996, pp. 459–465.
97. (a) T. van den Ancker, B. S. Jolly, M. F. Lappert, C. L. Raston, B. W. Skelton and A. H. White, *J. Chem. Soc., Chem. Commun.*, 1006 (1990).
    (b) H. H. Karsch, F. Bienlein, A. Sladek, M. Heckel and K. Burger, *J. Am. Chem. Soc.*, **117**, 5160 (1995).
98. A. B. Kalinin, E. T. Apasov, S. B. Bugaeva, S. L. Yoffe and B. A. Tartakovskii, *Izv. Akad. Nauk SSSR, Ser. Khim.*, 1413 (1983); *Chem. Abstr.*, **99**, 175871f (1983).
99. C. Breliere, F. Carre, R. J. P. Corriu, M. Poirier and G. Royo, *Organometallics*, **5**, 388 (1986).
100. J. I. Musher, *Angew. Chem., Int. Ed. Engl.*, **8**, 54 (1969).
101. V. F. Sidorkin, V. A. Pestunovich and M. G. Voronkov, *Dokl. Akad. Nauk SSSR*, **235**, 1363 (1977); *Chem. Abstr.*, **87**, 173187Z (1977).
102. V. A. Pestunovich, V. F. Sidorkin and M. G. Voronkov, in *Progress in Organosilicon Chemistry* (Eds. B. Marciniec and J. Chojnowski), Gordon and Breach, Basel, 1995, pp. 69–82.
103. A. R. Bassindale, P. G. Taylor, in *The Chemistry of Organic Silicon Compounds* (Eds. S. Patai and Z. Rappoport), Wiley, Chichester, 1989, pp. 839–892.
104. A. R. Bassindale, in *Progress in Organosilicon Chemistry* (Eds. B. Marciniec and J. Chojnowski), Gordon and Breach, Basel, 1995, pp. 191–208.
105. I. D. Kalikhman, Dissertation, Institute of Organic Chemistry, Sib. Div. Acad. Science SSSR, Irkutsk, 1989.
106. B. J. Helmer, R. West, R. J. P. Corriu, M. Poirier, G. Royo and A. de Saxce, *J. Organomet. Chem.*, **251**, 295 (1983).
107. V. A. Pestunovich, B. Z. Shterenberg, E. T. Lippmaa, M. Ja. Miagi, M. J. Alla, S. N. Tandura, V. P. Baryshok, L. P. Petukhov and M. G. Voronkov, *Dokl. Akad. Nauk SSSR*, **258**, 1410 (1981); *Chem. Abstr.*, **95**, 186219t (1981).
108. V. A. Pestunovich, B. Z. Shterenberg, S. N Tandura, V. P. Baryshok, E. I. Brodskaya, N. G. Komalenkova and M. G. Voronkov, *Dokl. Akad. Nauk SSSR*, **264, 632** (1982); *Chem. Abstr.*, **97**, 144918f (1982).
109. R. J. P. Corriu, G. F. Lanneau and Z. Yu, *Tetrahedron*, **49**, 9019 (1993).
110. R. J. P. Corriu, M. Mazhar, M. Poirier and G. Royo, *J. Organomet. Chem.*, **306**, C5 (1986).
111. R. J. P. Corriu, A. Kpoton, M. Poirier, G. Royo and J. Y. Corey, *J. Organomet. Chem.*, **277**, C25 (1984).
112. R. Probst, C. Leis, S. Gamper, E. Herdtweck, C. Zybill and N. Auner, *Angew. Chem., Int. Ed. Engl.*, **30**, 1132 (1991).
113. H. Handwerker, C. Leis, R. Probst, P. Bissinger, A. Grohmann, P. Kiprof, E. Herdtweck, J. Blumel, N. Auner and C. Zybill, *Organometallics*, **12**, 2162 (1993).
114. P. Arya, J. Boyer, F. Carre, R. Corriu, G. Lanneau, J. Lapasset, M. Perrot and C. Priou, *Angew. Chem., Int. Ed. Engl.*, **28**, 1016 (1989).

115. R. J. P. Corriu, G. F. Lanneau and V. D. Mehta, *J. Organomet. Chem.*, **419**, 9 (1991).
116. R. Corriu, G. Lanneau and C. Priou, *Angew. Chem., Int. Ed. Engl.*, **30**, 1030 (1991).
117. P. Arya, J. Boyer, R. J. P. Corriu, G.F. Lanneau and M. Perrot, *J. Organomet. Chem.*, **346**, C11 (1988).
118. R. J. P. Corriu, B. P. S. Chauhan and G. F. Lanneau, *Organometallics*, **14**, 1646 (1995).
119. R. J. P. Corriu, G. F. Lanneau and B. P. S. Chauhan, *Organometallics*, **12**, 2001 (1993).
120. B. P. S. Chauhan, R. J. P. Corriu, G. F. Lanneau, C. Priou, N. Auner, H. Handwerker and E. Herdtweck, *Organometallics*, **14**, 1657 (1995).
121. (a) M. Weinmann, H. Lang, O. Walter and M. Buchner, in *Organosilicon Chemistry II* (Eds. N. Auner and J. Weis), VCH Weinheim, 1996, pp. 569–574.
    (b) H. Lang, M. Weinmann, W. Frosch, M. Buchner and B. Schiemenz, *Chem. Commun.*, 1299 (1996).
122. C. N. Smit and F. Bickelhaupt, *Organometallics*, **6**, 1156 (1987).
123. (a) K. Tamao, K. Nagata, M. Asahara, A. Kawachi, Y. Ito and M. Shiro, *J. Am. Chem. Soc.*, **117**, 11592 (1995).
    (b) R. J. P. Corriu, G. F. Lanneau, C. Priou, F. Soulairol, N. Auner, R. Probst, R. Conlin and C. Tan, *J. Organomet. Chem.*, **466**, 55 (1994).
124. K. D. Onan, A. T. McPhail, C. H. Yoder and R. W. Hillyard, *J. Chem. Soc., Chem. Commun.*, 209 (1978).
125. R. W. Hillyard, C. M. Ryan and C. H. Yoder, *J. Organomet. Chem.*, **153**, 369 (1978).
126. C. H. Yoder, C. M. Ryan, G. F. Martin and P. S. Ho, *J. Organomet. Chem.*, **190**, 1 (1980).
127. C. H. Yoder, J. A. Gullinane and G. F. Martin, *J. Organomet. Chem.*, **210**, 289 (1981).
128. V. A. Pestunovich, A. I. Albanov, M. F. Larin, M. G. Voronkov, E. P. Kramarova and Yu. I. Baukov, *Izv. Akad. Nauk SSSR, Ser. Khim.*, 2178 (1980); *Chem. Abstr.*, **94**, 29677c (1981).
129. A. I. Albanov, Yu. I. Baukov, M. G. Voronkov, E. P. Kramarova, M. F. Larin and V. A. Pestunovich, *Zh. Obshch. Khim.*, **53**, 246 (1983); *Chem. Abstr.*, **98**, 215656c (1983).
130. Yu. I. Baukov, E. P. Kramarova, A. G. Shipov, G. I. Oleneva, O.B. Artamkina, A. I. Albanov, M. G. Voronkov and V. A. Pestunovich, *Zh. Obshch. Khim.*, **59**, 127 (1989); *Chem. Abstr.*, **112**, 56003b (1990).
131. M. G. Voronkov, A. E. Pestunovich, A. I. Albanov, N. N. Vlasova and V. A. Pestunovich, *Izv. Akad. Nauk SSSR, Ser. Khim.*, 2841 (1989); *Chem. Abstr.*, **113**, 6424q (1990).
132. L. I. Belousova, B. A. Gostevskii, I. D. Kalikhman, O. A. Vyazankina, O. B. Bannikova, N. S. Vyazankin and V. A. Pestunovich, *Zh. Obshch. Khim.*, **58**, 407 (1988); *Chem. Abstr.*, **110**, 114908 x (1989).
133. I. D. Kalikhman, O. B. Bannikova, L. I. Belousova, B. A. Gostevskii, E. Liepinsh, O. A. Vyazankina and N. S. Vyazankin, *Metalloorg. Khim.*, **1**, 683 (1988); *Chem. Abstr.*, **111**, 57840 p (1989).
134. I. D. Kalikhman, O. B. Bannikova, B. A. Gostevskii, O. A. Vyazankina, N. S. Vyazankin and V. A. Pestunovich, *Izv. Akad. Nauk SSSR, Ser. Khim.*, 1688 (1985); *Chem. Abstr.*, **104**, 148977k (1986).
135. I. D. Kalikhman, O. B. Bannikova, L. P. Petuchov, V. A. Pestunovich and M. G. Voronkov, *Dokl. Akad. Nauk SSSR*, **287**, 870 (1986); *Chem. Abstr.*, **106**, 50289d (1987).
136. I. D. Kalikhman, O. B. Bannikova, B. A. Gostevskii, L. I. Volkova, O. A. Vyazankina, N. S. Vyazankin, T. G. Yushmanova, V. A. Lopirev, M. G. Voronkov and V. A. Pestunovich, *Izv. Akad. Nauk SSSR, Ser. Khim.*, 459 (1987); *Chem. Abstr.*, **108**, 94627h (1988).
137. I. D. Kalikhman, V. A. Pestunovich, B. A. Gostevskii, O. B. Bannikova and M. G. Voronkov, *J. Organomet. Chem.*, **338**, 169 (1988).
138. A. R. Bassindale and M. Borbaruah, *J. Chem. Soc., Chem. Commun.*, 1499, 1501 (1991).
139. M. G. Voronkov, V. A. Pestunovich and Yu. I. Baukov, *Organomet. Chem. USSR*, **4**, 593 (1991); *Chem. Abstr.*, **116**, 41503y (1992).
140. I. D. Kalikhman, A. I. Albanov, O. B. Bannikova, L. I. Belousova, M. G. Voronkov, V. A. Pestunovich, A. G. Shipov, E. P. Kramarova and Yu. I. Baukov, *J. Organomet. Chem.*, **361**, 147 (1989).
141. R. Roger and D. G. Nielson, *Chem. Rev.*, **61**, 179 (1961).
142. S. Wawsonek and E. Yeakey, *J. Am. Chem. Soc.*, **82**, 5718 (1960).
143. O. B. Artamkina, E. P. Kramarova, A. G. Shipov, Yu. I. Baukov, A. A. Macharashvili, Yu. E. Ovchinnikov and Yu. T. Struchkov, *Zh. Obshch. Khim.*, **63**, 2289 (1993); *Chem. Abstr.*, **121**, 255881x (1994).

144. A. R. Bassindale and M. Borbaruah, *J. Chem. Soc., Chem. Commun.*, 352 (1993).
145. (a) A. G. Shipov, E. P. Kramarova and Yu. I. Baukov, *Zh. Obshch. Khim.*, **64**, 1220 (1994); *Chem. Abstr.*, **122**, 314629s (1995).
(b) E. P. Kramarova, V. V. Negrebezkii, A. G. Shipov and Yu. I. Baukov, *Zh. Obshch. Khim.*, **64**, 1222 (1994); *Chem. Abstr.*, **122**, 314630k (1995).
146. I. D. Kalikhman, O. B. Bannikova, B. A. Gostevskii, M. G. Voronkov and V. A. Pestunovich, *Izv. Akad. Nauk SSSR, Ser. Khim.*, 492 (1989); *Chem. Abstr.*, **112**, 35947x (1990).
147. I. D. Kalikhman, B. A. Gostevskii, O. B. Bannikova, M. G. Voronkov and V. A. Pestunovich, *J. Organomet. Chem.*, **376**, 249 (1989).
148. A. G. Shipov, O. B. Artamkina, E. P. Kramarova, G. I. Oleneva and Yu. I. Baukov, *Zh. Obshch. Khim.*, **61**, 1914 (1991); *Chem. Abstr.*, **116**, 106355y (1992).
149. O. B. Bannikova, B. A. Gostevskii and I. D. Kalikhman, *Abstracts of VII Conference for Chemistry, Technology and Utility of Organosilicon Compounds*, SSSR, Irkutsk, 1990, p. 59.
150. M. G. Voronkov, V. A. Pestunovich and Yu. L. Frolov, *Advances in Organosilicon Chemistry* (Ed. M.G. Voronkov), Mir Publishers, Moscow, 1985, pp. 54–68.
151. A. I. Albanov, N. M. Kudiakov, V. A. Pestunovich, M. G. Voronkov, A. A. Macharashvili, V. E. Shklover and Yu. T. Struchkov, *Metalloorg. Khim.*, **4**, 1228 (1991); *Chem. Abstr.*, **116**, 83789b (1992).
152. A. I. Albanov, L. I. Gubanova, M. F. Larin, V. A. Pestunovich and M. G. Voronkov, *J. Organomet. Chem.*, **244**, 5 (1983).
153. V. A. Pestunovich, M. F. Larin, M. S. Sorokin, A. I. Albanov and M.G. Voronkov, *J. Organomet. Chem.*, **280**, C17 (1985).
154. V. E. Shklover, Yu. T. Struchkov and M. G. Voronkov, *Russ. Chem. Rev.*, **58**, 211 (1989).
155. M. Yu. Antipin, *Russ. Chem. Rev.*, **59**, 1052 (1990).
156. A. A. Macharashvili, V. E. Shklover, Yu. T. Struchkov, G. I. Oleneva, E. P. Kramarova, A. G. Shipov and Yu. I. Baukov, *J. Chem. Soc., Chem. Commun.*, 683 (1988).
157. A. A. Macharashvili, V. E. Shklover, Yu. T. Struchkov, Yu. I. Baukov, E. P. Kramarova and G. I. Oleneva, *J. Organomet. Chem.*, **327**, 167 (1987).
158. A. A. Macharashvili, V. E. Shklover, N. Yu. Chernikova, M. Yu. Antipin, Yu. T. Struchkov, Yu. I. Baukov, G. I. Oleneva, E. P. Kramarova and A. G. Shipov, *J. Organomet. Chem.*, **359**, 13 (1989).
159. Yu. E. Ovchinnikov, A. A. Macharashvili, Yu. T. Struchkov, A. G. Shipov and Yu. I. Baukov, *J. Struct. Chem.*, **35**, 91 (1994).
160. A. A. Macharashvili, V. E. Shklover, Yu. T. Struchkov, M. G. Voronkov, B. A. Gostevskii, I. D. Kalikhman, O. B. Bannikova and V. A. Pestunovich, *Metalloorg. Khim.*, **1**, 1131 (1988); *Chem. Abstr.*, **110**, 135298 p (1989).
161. A. A. Macharashvili, Dissertation, Institute of Organometallic Chemistry, Acad. Science SSSR, Moscow, 1990.
162. O. B. Artamkina, E. P. Kramarova, A. G. Shipov, Yu. I. Baukov, A. A. Macharashvili, Yu. E. Ovhinnikov and Yu. T. Struchkov, *Zh. Obshch. Khim.*, **64**, 263 (1994); *Chem. Abstr.*, **121**, 255882y (1994).
163. N. A. Orlova, A. G. Shipov, Yu. I. Baukov, A. O. Mozzhukhin, M. Yu. Antipin and Yu. T. Struchkov, *Metalloorg. Khim.*, **5**, 666 (1992); *Chem. Abstr.*, **117**, 251439d (1992).
164. A. A. Macharashvili, V. E. Shklover, Yu. T. Struchkov, B. A. Gostevskii, I. D. Kalikhman, O. B. Bannikova, M. G. Voronkov and V. A. Pestunovich, *J. Organomet. Chem.*, **356**, 23 (1988).
165. A. A. Macharashvili, V. E. Shklover, Yu. T. Struchkov, M. G. Voronkov, B. A. Gostevskii, I. D. Kalikhman, O. B. Bannikova and V. A. Pestunovich, *J. Organomet. Chem.*, **340**, 23 (1988).
166. A. O. Mozzhukhin, Yu. Antipin, Yu. T. Struchkov, A. G. Shipov, E. P. Kramarova and Yu. I. Baukov, *Metalloorg. Khim.*, **5**, 906 (1992); *Chem. Abstr.*, **118**, 102109q (1993).
167. A. O. Mozzhukhin, Yu. Antipin, Yu. T. Struchkov, A. G. Shipov, E. P. Kamarova and Yu. I. Baukov, *Metalloorg. Khim.*, **5**, 917 (1992); *Chem. Abstr.*, **118**, 80993u (1993).
168. H. B. Bürgi, *Angew. Chem., Int. Ed. Engl.*, **14**, 460 (1975).
169. J. D. Dunitz, *X-ray Analysis and Structure of Organic Molecules*, Cornell University Press, Ithaca, 1979.
170. C. H. Yoder, W. D. Smith, B. L. Buckwalter, C. D. Schaeffer, Jr., K. J. Sullivan and M. F. Lehman, *J. Organomet. Chem.*, **492**, 129 (1995).
171. D. Kummer and S. H. Abdel Halim, *Z. Anorg. Allg. Chem.*, **622**, 57 (1996).
172. V. V. Negrebetsky, V. V. Negrebetsky, A. G. Shipov, E. P. Kramarova and Y. I. Baukov, *J. Organomet. Chem.*, **496**, 103 (1995).

173. J. C. Otter, C. L. Adamson, C. H. Yoder and A. L. Rheingold, *Organometallics*, **9**, 1557 (1990).
174. (a) T. K. Prakasha, S. Srinivasan, A. Chandrasekaran, R. O. Day and R. R. Holmes, *J. Am. Chem. Soc.*, **117**, 10003 (1995).
    (b) N. V. Timosheva, T. K. Prakasha, A. Chandrasekaran, R. O. Day and R. R. Holmes, *Inorg. Chem.*, **35**, 3614 (1996).
175. (a) R. O. Day, T. K. Prakasha, R. R. Holmes and H. Eckert, *Phosphorus, Sulfur Silicon Relat. Elem.*, **100**, 211 (1995).
    (b) R. O. Day, T. K. Prakasha, R.R . Holmes and H. Eckert, *Organometallics*, **13**, 1285 (1994).
176. (a) A. Chandrasekaran, R. O. Day and R. R. Holmes, *Organometallics*, **15**, 3182 (1996).
    (b) A. Chandrasekaran, R. O. Day and R. R. Holmes, *Organometallics*, **15**, 3189 (1996).
177. K. Ebata, T. Inada, C. Kabuto and H. Sakurai, *J. Am. Chem. Soc.*, **116**, 3595 (1994).
178. I. I. Schuster, W. Weissensteiner and K. Mislow, *J. Am. Chem. Soc.*, **108**, 6661 (1986).
179. H. Schwarz, in *The Chemistry of Organic Silicon Compounds* (Eds. S. Patai and Z. Rappoport), Wiley, Chichester, 1989, pp. 445–510.
180. J. B. Lambert, L. Kania and S. Zhang, *Chem. Rev.*, **95**, 1191 (1995).
181. J. B. Lambert, S. Zhang and S. M. Ciro, *Organometallics*, **13**, 2430 (1994).
182. G. A. Olah, L. Heiliger, X. Y. Li and G. K. S. Prakash, *J. Am. Chem. Soc.*, **112**, 5991 (1990).
183. A. R. Bassindale and T. Stout, *J. Chem. Soc., Chem. Commun.*, 1387 (1984).
184. J. Y. Corey and R. West, *J. Am. Chem. Soc.*, **85**, 4034 (1963).
185. (a) C. Breliere, F. Carre, R. J. P. Corriu and M. Wong Chi Man, *J. Chem. Soc., Chem. Commun.*, 2333 (1994).
    (b) M. Chauhan, C. Chuit, R. J. P. Corriu and C. Reye, *Tetrahedron Lett.*, **37**, 845 (1996).
186. (a) C. Chuit, R. J. P. Corriu, A. Mehdi and C. Reye, *Angew. Chem., Int. Ed. Engl.*, **32**, 1311 (1993).
    (b) M. Chauhan, C. Chuit, R. J. P. Corriu and C. Reye, *Tetrahedron Lett.*, **37**, 845 (1996).
187. V. A. Benin, J. C. Martin and M. R. Willcott, *Tetrahedron Lett.*, **35**, 2133 (1994).
188. J. Belzner, D. Schär, B. O. Kneisel and R. Herbst-Irmer, *Organometallics*, **14**, 1840 (1995).
189. H. Kobayashi, K. Ueno and H. Ogino, *Organometallics*, **14**, 5490 (1995).
190. K. Tamao, J. Toshida, H. Yamamoto, T. Kakui, H. Yatsumoto, M. Takahashi, A. Kurita, M. Murata and M. Kumada, *Organometallics*, **1**, 355 (1982).
191. C. Breliere, F. Carre, R. J. P. Corriu, W. E. Douglas, M. Poirier, G. Royo and M. Wong Chi Man, *Organometalics*, **11**, 1586 (1992).
192. F. Carre, C. Chuit, R. J. P. Corriu, A. Fanta, A. Mehdi and C. Reye, *Organometallics*, **14**, 194 (1995).
193. R. Tacke and M. Mühleisen, *Angew. Chem., Int. Ed. Engl.*, **33**, 1359 (1994).
194. C. Breliere, F. Carre, R. J. P. Corriu, M. Poirier, G. Royo and J. Zwecker, *Organometallics*, **8**, 1831 (1989).
195. (a) J. C. Bailar, *J. Inorg. Nucl. Chem.*, **8**, 165 (1958).
    (b) P. Ray and N. K. Dutt, *J. Ind. Chem. Soc.*, **20**, 81 (1943).
196. (a) W. Dilthey, *Chem. Ber.*, **36**, 923 (1903).
    (b) A. Rosenheim, W. Loewenstamm and L. Singer, *Chem Ber.*, **36**, 1833 (1903).
197. R. West, *J. Am. Chem. Soc.*, **80**, 3246 (1958).
198. N. Serpone and K. A. Hersh, *J. Organomet. Chem.*, **84**, 177 (1975).
199. T. Shimizutani and Y. Yoshikawa, *Inorg. Chem.*, **30**, 3236 (1991).
200. R. M. Pike, *Coord. Chem. Rev.*, **2**, 163 (1967).
201. (a) F. E. Hahn, M. Keck and K. N. Raymond, *Inorg. Chem.*, **34**, 1402 (1995).
    (b) A. Boudin, G. Cerveau, C. Chuit, R. J. P. Corriu and C. Reye, *Organometallics*, **7**, 1165 (1988).
202. D. F. Evans, J. Parr and E. N. Coker, *Polyhedron*, **9**, 813 (1990).
203. D. F. Evans and C. Y. Wong, *Polyhedron*, **10**, 1131 (1991).
204. D. F. Evans, J. Parr and C. Y. Wong, *Polyhedron*, **11**, 567 (1992).
205. F. Carre, G. Cerveau, C. Chuit, R. J. P. Corriu and C. Reye, *Angew. Chem., Int. Ed. Engl.*, **28**, 489 (1989).
206. F. Carre, G. Cerveau, C. Chuit, R. J. P. Corriu and C. Reye, *New J. Chem.*, **16**, 63 (1992).
207. (a) F. Carre, C. Chuit, R. J. P. Corriu, A. Mehdi and C. Reye, *J. Organomet. Chem.*, **446**, C6 (1993).
    (b) C. Chuit, R. J. P. Corriu, A. Mehdi and C. Reye, *Chem. Eur. J.*, **2**, 342 (1996) [*Angew. Chem., Int. Ed. Engl.*, **35**, 342 (1996)].

208. K. M. Taba and W. V. Dahlhoff, *J. Organomet. Chem.*, **280**, 27 (1985).
209. G. Klebe and D. Tran Qui, *Acta Crystallogr.*, **C40**, 476 (1984).
210. I. D. Kalikhman, B. A. Gostevskii, O. B. Bannikova, M. G. Voronkov and V. A. Pestunovich, *Metalloorg. Khim.*, **2**, 937 (1989); *Chem. Abstr.*, **112**, 118926r (1990).
211. (a) I. Kalikhman and D. Kost, *J. Chem. Soc., Chem. Commun.*, 1253 (1995).
    (b) D. Kost, I. Kalikhman and M. Raban, *J. Am. Chem. Soc.*, **117**, 11512 (1995).
212. I. D. Kalikhman, B. A. Gostevskii, O. B. Bannikova, M. G. Voronkov and V. A. Pestunovich, *Metalloorg. Khim.*, **2**, 704 (1989); *Chem. Abstr.*, **112**, 139113p (1990).
213. C. Breliere, R. J. P. Corriu, G. Royo, W. W. C. Wong Chi Man and J. Zwecker, *Organometallics*, **9**, 2633 (1990).
214. K. Boyer-Elma, F. H. Carre, R. J. P. Corriu and W. E. Douglas, *J. Chem. Soc., Chem. Commun.*, 725 (1995).
215. F. Carre, C. Chuit, R. J. P. Corriu, A. Mehdi and C. Reye, *Organometallics*, **14**, 2754 (1995).
216. A. O. Mozzhukhin, M. Yu. Antipin, Yu. T. Struchkov, B. A. Gostevskii, I. D. Kalikhman, V. A. Pestunovich and M. G. Voronkov, *Metalloorg. Khim.*, **5**, 658 (1992); *Chem. Abstr.*, **117**, 234095w (1992).
217. C. Breliere, R. J. P. Corriu, G. Royo and J. Zwecker, *Organometallics*, **8**, 1834 (1989).
218. (a) K. Tamao, M. Akita, H. Kato and M. Kumada, *J. Organomet. Chem.*, **341**, 165 (1988).
    (b) K. M. Kane, F. R. Lemke and J. L. Petersen, *Inorg. Chem.*, **34**, 4085 (1995).
219. J. A. A. Ketelaar, *Z. Kristallogr.*, **92**, 155 (1935).
220. D. Kost and I. Kalikhman, *Bull. Magn. Reson.*, **17**, 108 (1995).
221. I. Kalikhman, S. Krivonos, D. Stalke, T. Kottke and D. Kost, *Organometallics*, **16**, 3255 (1997).
222. D. Kost, S. Krivonos and I. Kalikhman, in *Organosilicon Chemistry III* (Eds. N. Auner and J. Weis), VCH, Weinheim, 1997, pp. 435–445.
223. A. Rodger and B. F. G. Johnson, *Inorg. Chem.*, **27**, 3061 (1988).
224. D. Kummer, S. C. Chaudhry, T. Debaerdemaeker and U. Thewalt, *Chem. Ber.*, **123**, 945 (1990).
225. D. Kummer, S. C. Chaudhry, W. Depmeier and G. Mattern, *Chem. Ber.*, **123**, 2241 (1990).
226. W. B. Farnham and J. F. Whitney, *J. Am. Chem. Soc.*, **106**, 3992 (1984).
227. C. Breliere, F. Carre, R. J. P. Corriu and G. Royo, *Organometallics*, **7**, 1006 (1988).
228. C. Breliere, F. Carre, R. J. P. Corriu, G. Royo, M. Wong Chi Man and J. Lapasset, *Organometallics*, **13**, 307 (1994).
229. F. Carre, C. Chuit, R. J. P. Corriu, A. Mehdi and C. Reye, *Angew. Chem., Int. Ed. Engl.*, **10**, 1097 (1994).
230. J. Braddock-Wilking, M. Schieser, L. Brammer, J. Huhmann and R. Shaltout, *J. Organomet. Chem.*, **499**, 89 (1995).
231. J.-K. Buijink, M. Noltemeyer and E. T. Edelmann, *J. Fluorine Chem.*, **61**, 51 (1993).
232. (a) P. Jutzi, D. Eikenberg, E. A. Bunte, A. Mohrke, B. Neumann and H. G. Stammler, *Organometallics*, **15**, 1930 (1996).
    (b) P. Jutzi, U. Holtmann, D. Kanne, C. Kruger, R. Blom, R. Gleiter and I. Hyla-Kryspin, *Chem. Ber.*, **122**, 1629 (1989).
233. E. Lukevics and O. A. Pudova, *Khimiya Geterotsiklicheskikh Soedin.*, 1605 (1996). *Chem. Abstr.*, **126**, 157528w (1997).
234. A. H. J. F. de Keijzer, F. J. J. de Kanter, M. Schakel, R. F. Schmitz and G. W. Klumpp, *Angew. Chem., Int. Ed. Engl.*, **35**, 1127 (1996).
235. T. Hoshi, M. Takahashi and M. Kira, *Chem. Lett.*, 683 (1996).
236. R. Damrauer and J. A. Hankin, *J. Organomet. Chem.*, **521**, 93 (1996).
237. J. A. Deiters and R. R. Holmes, *Organometallics*, **15**, 3944 (1996).
238. M. Veith and A. Rammo, *J. Organomet. Chem.*, **521**, 429 (1996).
239. A. Mix, U. H. Berlekamp, H.-G. Stammler, B. Neumann and P. Jutzi, *J. Organomet. Chem.*, **521**, 177 (1996).
240. V. V. Negrebetsky, A. G. Shipov, E. P. Kramarova, V. V. Negrebetsky and Yu. I. Baukov, *J. Organomet. Chem.*, **530**, 1 (1997).
241. D. Kummer, S. H. Abdel Halim and M. F. El-Shahat, *Z. Anorg. Allg. Chem.*, **622**, 1701 (1996).
242. M. Kira, L. Cheng Zhang, C. Kabuto and H. Sakurai, *Organometallics*, **15**, 5335 (1996).
243. K. Tamao, M. Asahara and A. Kawachi, *J. Organomet. Chem.*, **521**, 325 (1996).
244. M. Chauhan, C. Chuit, R. J. P. Corriu, A. Mehdi and C. Reye, *Organometallics*, **15**, 4326 (1996).
245. C.-H. Ottosson and D. Cremer, *Organometallics*, **15**, 5309 (1996).
246. S. P. Narula, R. Shankar and Meenu, *Proc. Indian Acad. Sci., Chem. Sci.*, **108**, 123 (1996).

247. H. Fleischer, K. Hensen and T. Stumpf, *Chem. Ber.*, **129**, 765 (1996).
248. H. H. Karsch, R. Richter and E. Witt, *J. Organomet. Chem.*, **521**, 185 (1996).
249. H. H. Karsch, B. Deubelly, U. Keller, O. Steigelmann, J. Lachmann and G. Muller, *Chem. Ber.*, **129**, 671 (1996).
250. H. H. Karsch, B. Deubelly, U. Keller, F. Bienlein, R. Richter, P. Bissinger, M. Heckel and G. Muller, *Chem. Ber.*, **129**, 759 (1996).
251. C. Plitzko and G. Meyer, *Z. Anorg. Allg. Chem.*, **622**, 1646 (1996).

CHAPTER **24**

Silatranes and their tricyclic analogs

VADIM PESTUNOVICH, SVETLANA KIRPICHENKO and MIKHAIL VORONKOV

Irkutsk Institute of Chemistry, Siberian Branch of the Russian Academy of Sciences, 1 Favorsky St, 664033, Irkutsk, Russia
Fax: +7 3952 356046; e-mail: vadim@irioch.irk.ru

I.	GENERAL INTRODUCTION	1448
II.	SILATRANES	1449
	A. Methods of Synthesis	1449
	1. Introduction	1449
	2. Formation from organyltrialkoxysilanes	1450
	3. Formation from organyltriacetoxysilanes	1456
	4. Formation from organylhalosilanes	1457
	5. Formation from organyltris(dialkylamino)silanes	1458
	6. Formation from organylsilanes	1458
	7. Miscellaneous preparations	1458
	B. Structure and Physical Properties	1460
	1. Molecular structure	1460
	2. Bonding model	1465
	3. Quantum mechanical studies	1466
	4. Thermochemical data	1469
	5. Dipole moments	1469
	6. Photoelectron spectra	1470
	7. Ultraviolet spectra	1471
	8. Vibrational spectra	1471
	9. NMR spectra	1473
	a. ^{1}H NMR spectra	1473
	b. ^{13}C NMR spectra	1474
	c. ^{29}Si NMR spectra	1475
	d. ^{15}N NMR spectra	1476
	e. Other NMR studies	1479

The chemistry of organic silicon compounds, Vol. 2
Edited by Z. Rappoport and Y. Apeloig © 1998 John Wiley & Sons Ltd

    10. Electronic effects of the silatranyl group 1479
    11. Electrochemical studies 1480
    12. Mass spectra 1481
  C. Chemical Properties 1483
    1. Introduction 1483
    2. Hydrolysis 1483
    3. 1-Hydrosilatranes 1486
    4. 1-Halosilatranes 1488
    5. 1-Hydroxysilatrane 1491
    6. 1-Alkoxysilatranes 1491
    7. 1-Alkyl-, 1-aralkyl- and 1-arylsilatranes 1492
    8. 1-Alkenylsilatranes 1494
    9. Carbofunctional silatranes 1500
    10. Silatranyl arene and cyclohexadiene complexes 1503
III. SILATRANE ANALOGS 1505
  A. 3-Homosilatranes 1505
  B. Silatranones 1507
  C. Carbasilatranes 1509
  D. Benzosilatranes 1511
  E. Triazasilatranes 1514
  F. Miscellaneous 1523
IV. ACKNOWLEDGMENTS 1524
V. REFERENCES 1524

I. GENERAL INTRODUCTION

Since the first preparation[1], determination of the unusual structure[1-3] and the sensational discovery of amazing biological activity[2] of silatranes, 2,8,9-trioxa-5-aza-1-silatricyclo[3.3.3.0$^{1.5}$]undecanes (**1**), they became a fascinating class of pentacoordinate organosilicon compounds. Several hundreds of Si-substituted silatranes bearing H, organyl, organoxy, thioorganyl, acyloxy, halogen, pseudohalogen and other groups as substituent X have been synthesized and studied. Numerous C-substituted (mono- and poly-substituted at positions 3, 4, 7 and 10) silatranes and silatrane analogs with partly or completely modified tripodal ligands at the silicon atom are also known. Some of them are shown in Scheme 1, which demonstrates the formal methods of the transformation of silatranes to related compounds by replacing the oxygen atom or a methylene group(s) of the parent compound by another group(s).

    The main features of the silatranes and their analogs are predetermined by their cage structure with a nearly trigonal-bipyramidal (TBP) silicon and an *in*-oriented nitrogen involved in the transannular Si←N bonding. The simplicity of the synthesis and relatively high stability of silatranes and their analogs together with a wide diversity of their substituent X and the tricyclic skeleton character made it possible to use them for detailed studies of some intriguing problems of the chemistry of hypervalent silicon compounds. These are the bonding model, the mutual influence of substituents at the TBP silicon and the structural, spectral and chemical manifestations of the silicon hypervalency as well as modelling of the pathway of the S_N2 substitution at the tetrahedral Si.

    To explain the great interest in silatranes in recent decades one should bear in mind two points. First, for a long time silatranes remained the only accessible class of neutral derivatives of pentacoordinated silicon. Second, the specific physiological action of many silatranes refuted the traditional opinion that organosilicon compounds do not differ much

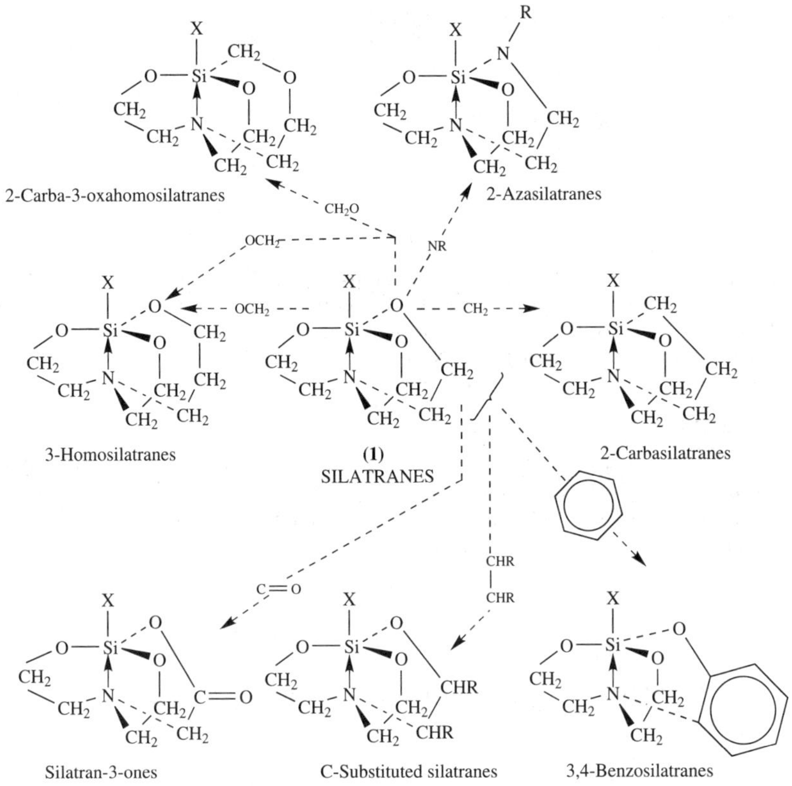

SCHEME 1. Silatranes and some of their analogs.

in biological activity from their organic precursors or analogs. That is why the results of their investigation stimulated an extensive development of the chemistry of hypervalent organosilicon compounds and bioorganosilicon chemistry.

The aim of the present chapter is to give a short review of the present knowledge concerning synthesis, structure and reactivity of silatranes and their analogs. By 1997 several hundred publications on these themes appeared from the research groups of Corriu, Frye, Hencsei, Lukevics, Verkade, Voronkov and others. These publications include several books[4–7] and reviews[2,8–19]. In some recent reviews[20–31] the structure and reactivity of silatranes and their analogs are compared with those of other hypervalent organosilicon compounds and atrane derivatives of nonmetallic main group elements. To avoid superfluous details we shall refer to these earlier publications.

II. SILATRANES

A. Methods of Synthesis

1. Introduction

The most common and convenient methods for the preparation of a variety of Si- and C-substituted silatranes are the reactions of trifunctional silanes with tris(2-hydroxyalkyl)amines (equation 1). Halo, alkoxy, acyloxy and dialkylamino groups can serve as functional substituents at silicon.

$$\text{XSiY}_3 + \begin{matrix} (\text{HOCR}^1\text{R}^2\text{CHR}^3)_n \\ (\text{HOCH}_2\text{CH}_2)_{3-n} \end{matrix} \!\!\!\! \rangle\text{N} \xrightarrow{-3\,\text{HY}} \left[\begin{matrix} \text{O} \!-\! \overset{\overset{\displaystyle X}{|}}{\text{Si}} \!-\! \text{O} \\ \text{R}^1\text{R}^2\text{C} \diagup \!\!\!\! \uparrow \!\!\!\! \diagdown \\ \text{R}^3\text{HC} \diagdown_n \!\!\!\! \text{N} \diagup \end{matrix} \right]_{3-n} \quad (1)$$

It should be emphasized that these reactions involving two trifunctional reagents afford monomeric compounds in high yields instead of polymeric products. There is no question that the driving force of these reactions which generate the tricyclic silatrane skeleton is the formation of the transannular Si←N bond.

2. Formation from organyltrialkoxysilanes

As an organosilicon component for the reaction shown in equation 1, trialkoxysilanes are used predominantly since these are widely available. The transetherification of phenyltriethoxysilane and tetraethoxysilane with triethanolamine [tris(2-hydroxyethyl)amine, THEA] in the presence of the basic catalyst has led to the first silatranes (equation 2)[1,32].

$$\text{XSi(OEt)}_3 + (\text{HOCH}_2\text{CH}_2)_3\text{N} \xrightarrow{-3\,\text{EtOH}} \text{silatrane} \quad (2)$$

(THEA)

X = Ph, EtO

Similar transformations can be used with a wide range of organyltrialkoxysilanes, so that a large number of silatranes can be prepared with yields of up to 95%. Examples include 1-hydrosilatrane[1], 1-alkyl-[1,33], 1-cyclopropyl-[34−36], 1-[(cycloalkyl)alkyl]-[37], 1-alkenyl-[1,38−40], 1-alkynyl-[38,41], 1-aralkyl-[1,33,38] and 1-arylsilatranes[1,33,42−44].

The exchange reactions of organyltrialkoxysilanes with THEA mostly require heating of the components in an appropriate inert solvent (benzene, toluene, xylene, anisole, chloroform, methanol, ethanol etc.) for a long time. However, in some cases the reactions can be carried out at room temperature or, if necessary, with cooling. The transetherification rates and silatrane yields increase in the presence of an alkali metal hydroxide or alkoxide as a basic catalyst.

According to equation 2, disilatranylalkanes can be obtained from THEA and the corresponding bis(triethoxysilyl)alkanes (equation 3)[45]. For example, heating (up to 220 °C) of 1,2-bis(triethoxysilyl)ethane ($n = 2$) with THEA in the presence of KOH leads to the corresponding 1,2-disilatranylethane in 43% yield[45]. A significant yield decrease (to 24%) of this compound is observed when the reaction occurs without catalyst under reflux in xylene even for a longer time (70 h)[46]. Attempts to obtain disilatranylmethane ($n = 1$) failed because of unfavorable steric effects[47]. However, germatranylmethylsilatrane, $N(CH_2CH_2O)_3GeCH_2Si(OCH_2CH_2)_3N$, is formed in moderate yield (30%) under the same conditions[47]. It is of interest that 1-triorganylsilylmethylsilatranes, $R_3SiCH_2Si(OCH_2CH_2)_3N$, are obtained from 1-chloromethylsilatrane in good yields[48].

$$(EtO)_3Si(CH_2)_nSi(OEt)_3 \xrightarrow{THEA} N\rightarrow Si-(CH_2)_n-Si\leftarrow N \quad (3)$$

$$n = 2, 6$$

Starting from heteryltrialkoxysilanes and THEA, 1-heterylsilatranes can be synthesized in high yield (60–90%) (equation 4)[49–51]. As a catalyst, KOH or NaOH are suitable. However, cleavage of the Si—C bond and formation of 1-ethoxysilatrane are observed when 2-furyl- or 5-methyl-2-furyltriethoxysilanes react with THEA in the presence of KOH[50]. Use of H_2PtCl_6 as a catalyst favors transetherification of the two triethoxysilanes as shown by the increased silatrane yields[50].

$$HerylSi(OEt)_3 \xrightarrow{THEA} \text{[heterylsilatrane]}$$

Heteryl	R	Ref.
furyl (R-substituted)	H, Me	50
methylthienyl		50
R—C=N—O (isoxazoline)	Me, Ph	51
O—N=C—R (isoxazoline)	Me, Ph	51

(4)

Use of triethanolamine derivatives allows various substituents to be introduced into the silatrane ring, as illustrated in equations 5–7[52–58].

$$XSi(OEt)_3 + (HOCH_2CH_2)_2NCH_2CH(R)OH \longrightarrow \text{[silatrane]}-R \quad (5)$$

X	R	Yield, %	Ref.
Me	Me		1
Ph	Me		1
Me	Ph	70	52
Ph	Vinyl	75	52
Ph	ClCH$_2$	45	52

$$XSi(OEt)_3 + [HOCR(Me)CH_2]_3N \longrightarrow \text{[silatrane]} \quad (6)$$

X	R	Ref.
Vinyl	H	53
Ph	H	54
4-MeC$_6$H$_4$	H	55
Ph	Me	58

$$XSi(OEt)_3 + \begin{array}{c}[HOCH(CF_3)CH_2]_n \\ (HOCH_2CH_2)_{3-n}\end{array}\!\!N \longrightarrow \left[CF_3\text{-silatrane}\right]_n [\text{silatrane}]_{3-n} \quad (7)$$

X = Me, Ph, ClCH$_2$, MeCHCl, Cl(CH$_2$)$_4$, etc., n = 1–3[57]

Organyltriethoxysilanes react with racemic[59] or optically active triethanolamine derivatives[60,61] to afford 4-substituted 1-organylsilatranes (equation 8).

The reaction of equation 8 can proceed without a catalyst. However, the reaction rates and product yields increase in the presence of sodium methoxide[59]. 4,4-Dimethyl substituted silatranes are formed in a similar way[62].

$$XSi(OR)_3 + [(HOCH_2CH_2)_2NCHR'CH_2OH \longrightarrow$$

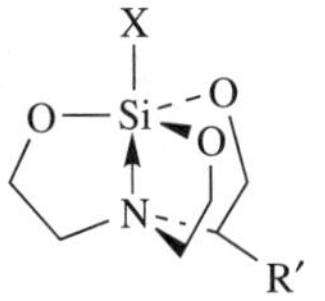

(structure of silatrane with X on Si and R' on carbon)

(8)

X	R	R'	Ref.
Me	Et	Et	59
Vinyl	Et	Et	59
Ph	Et	Et	59
Ph	Me	Me	61
Ph	Me	i-Pr	61
S-(+)-4-ClC$_6$H$_4$	Et	Et	60

Reaction 2 can also be applied to the synthesis of a number of 1-alkoxy-[1,53,61] and 1-aryloxysilatranes[33,63] as illustrated by equations 9[61] and 10[33,63].

$$(MeO)_4Si + (HOCH_2CH_2)_2NCH(R)CH_2OH \longrightarrow MeOSi(OCH_2CH_2)_2(OCH_2CHR)N$$

R = H, Me, i-Pr

(9)

$$4\text{-}YC_6H_4OSi(OEt)_3 \xrightarrow{THEA} 4\text{-}YC_6H_4OSi(OCH_2CH_2)_3N$$

Y = Cl[33,63], Br[33], O$_2$N[63]

(10)

Higher 1-organoxysilatranes can conveniently be obtained by treatment of lower tetraalkoxysilanes with a mixture of THEA and an appropriate hydroxyl-containing compound in high or almost quantitative yields (equation 11)[33,64−66]. An organoxy–alkoxy exchange easily occurs when the formed, low boiling alcohol (MeOH or EtOH) is immediately removed from the reaction mixture by distillation. In some cases, e.g., with bulky alcohols, the use of KOH as a catalyst can be helpful[64].

$$ROH + (R'O)_4Si + THEA \longrightarrow ROSi(OCH_2CH_2)_3N$$

R = Alk, cyclo-Alk, ArCH$_2$[64], Ar[33,65,66]; R' = MeO, EtO

(11)

The classical way shown in equation 2 was used to prepare silatranes having a labile exocyclic S−Si bond (equation 12)[67] or a pseudohalogen function at the silicon (equation 13)[68].

$$RSSi(OMe)_3 + THEA \xrightarrow{THF} RSSi(OCH_2CH_2)_3N$$

R = HCl·H$_2$NCH$_2$CHR' [R' = H (83%), Me (67%)]

(12)

$$SCNSi(OEt)_3 + THEA \xrightarrow{CH_2Cl_2} SCNSi(OCH_2CH_2)_3N$$
84%

(13)

Organyltrialkoxysilanes can be converted to the corresponding silatranes by a one-pot reaction with bis(2-hydroxyalkyl)amines and epoxides[52,69]. According to Frye and coworkers[69] the procedure involves *in situ* formation of trialkanolamine. For instance, treatment of phenyltrimethoxysilane with diethanolamine and styrene oxide affords 1,3-diphenylsilatrane in good yield (equation 14)[69].

$$PhSi(OMe)_3 + (HOCH_2CH_2)_2NH + PhCH\!-\!\!-\!CH_2\ (\text{epoxide}) \longrightarrow$$

[Structure of 1,3-diphenylsilatrane with Ph–Si, three O linkages, N bridgehead, and Ph substituent] (14)

Reaction 2 is a very helpful synthetic route to various substituted carbofunctional silatranes, $Y(CH_2)_n Si(OCHRCH_2)_3N$, where $Y = Hal^{70-77}$, $CF_3^{56,73}$, $HO^{78,79}$; $R'O^{80-82}$, $R'COO^{83,84}$, $HS^{85,86}$, $R'S^{87-89}$, $R'C(O)S^{90}$, SCN^{91}, $R'Se^{92}$, R'_2N ($R = H^{53,93-95}$, $R' =$ Alk or $Ar^{55,96-101}$), $R'CON(R')^{99,102-107}$, $R'CSN(R')^{102,107}$, $(R'O)_2P(O)^{108,109}$, etc., $n = 1-3$. Preparation of several main types of these compounds is illustrated by the examples in equations 15–21.

$$R_FCH_2CH_2Si(OR)_3 \xrightarrow{\text{THEA, R'OM, }\Delta} R_FCH_2CH_2Si(OCH_2CH_2)_3N \quad (15)$$

$R_F = FCH_2, CF_3, R = Me, Et, R' = H, M = K; R' = Et, M = Na^{56}$

$$Me_3SiO(CH_2)_3Si(OEt)_3 \xrightarrow{\text{THEA, MeOH}} HO(CH_2)_3Si(OCH_2CH_2)_3N \quad (16)$$
$$95\%^{78}$$

$$H(CF_2)_nCH_2OCH_2Si(OEt)_3 \xrightarrow{\text{THEA, RT}} H(CF_2)_nCH_2OCH_2Si(OCH_2CH_2)_3N$$
$$n = 2\ (80\%)$$
$$n = 12\ (94\%) \quad (17)$$
$$n = 16\ (96\%)^{82}$$

$$ArZ(CH_2)_nSi(OR)_3 \xrightarrow{\text{THEA, RT}} ArZ(CH_2)_nSi(OCH_2CH_2)_3N \quad (18)$$

$Ar = C_6F_5, n = 3, Z = O\ (100\%)^{81}, Z = S\ (97\%)^{89}; Ar = Ph, Z = Se, n = 1, 2^{92}$

$$MeOOC(CH_2)_nN(Me)CH_2Si(OR)_3 \xrightarrow{\text{THEA}} MeOOC(CH_2)_nN(Me)CH_2Si(OCH_2CH_2)_3N$$
$n = 1, 2^{100}$ (19)

[Lactam structure $(CH_2)_n$–C(=O)–N–CH$_2$Si(OR)$_3$] $\xrightarrow{\text{THEA, KOH, 120 °C, xylene}}$ [Lactam silatrane with CH$_2$Si(OCH$_2$CH$_2$)$_3$N] (20)

$n = 1-3^{106}$

$$(RO)_2P(X)Y(CH_2)_nSi(OR')_3 \xrightarrow{\text{THEA, KOH, xylene}} (RO)_2P(X)Y(CH_2)_nSi(OCH_2CH_2)_3N \quad (21)$$

$X = O, S, Y = NH, n = 3;$
$X = Y = S, n = 1^{108, 109}$

As catalysts in the above reactions, alkali metal hydroxides or alkoxides are used (equations 15, 19–21).

However, in some cases the use of a basic catalyst can result in cleavage of the Si–C bond[59,76]. For example, treatment of dichloromethyltriethoxysilane with trialkanolamine in the presence of sodium methoxide affords the corresponding 1-ethoxysilatrane in 84% yield (equation 22). The target 1-dichloromethyl-4-ethylsilatrane (96%) is formed in the absence of a catalyst[59].

$$Cl_2CHSi(OEt)_3 + (HOCH_2CH_2)_2N(CHEtCH_2OH) \xrightarrow{\text{MeONa}} \text{[1-ethoxy-4-ethylsilatrane]} \quad (22)$$

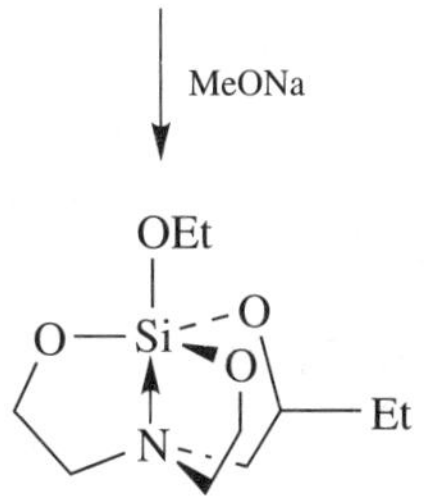

The reaction given in equation 2 cannot be applied for the preparation of 1-(2'-haloethyl)silatranes because of β-elimination leading to cleavage of the Si–C bond (equation 23)[76].

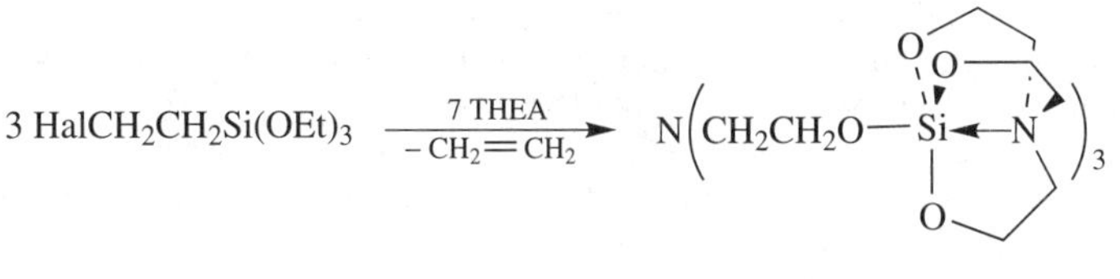

$$3\,HalCH_2CH_2Si(OEt)_3 \xrightarrow[-CH_2=CH_2]{7\,THEA} N(CH_2CH_2O-Si\leftarrow N)_3 \quad (23)$$

$$+$$

$$3(HOCH_2CH_2)_3N \cdot HHal$$

Treatment of chloromethyl- or 3-chloropropyltriethoxysilanes with L-*N*,*N*-bis(2-hydroxyethyl)serine or -threonine in the presence of a catalytic amount of pyridine at 70–80 °C gave a novel class of silatranes, (4*S*)-silatrane-4-carboxylic acids (equation 24)[110,111].

Reaction 24 does not occur with the protonated amine (i.e. with the zwitterionic form of the dihydroxyethylated amino acid) in the absence of pyridine. It seems most likely that the catalytic role of pyridine involves release of the lone-pair electrons of the nitrogen atom, that facilitates the formation of the transannular Si←N bond and, consequently, the silatrane ring[110].

A variety of 1-(heteroarylalkyl)silatranes can be obtained from heteroarylalkyltrialkoxysilanes and THEA according to equation 2[50,112–114]. For instance, 2-(benzoxazolylthiomethyl)trimethoxysilane reacts with THEA at room temperature to give the corresponding silatrane in high (86%) yield. 1-(2′-Benzothiazolylthiomethyl)- and 1-(2′-benzimidazolylthiomethyl)-silatranes are obtained in a similar manner (equation 25)[112].

$Cl(CH_2)_nSi(OEt)_3$ + $(HOCH_2CH_2)_2NCH(COOH)CH(R)OH$

R = H, Me, n = 1,3

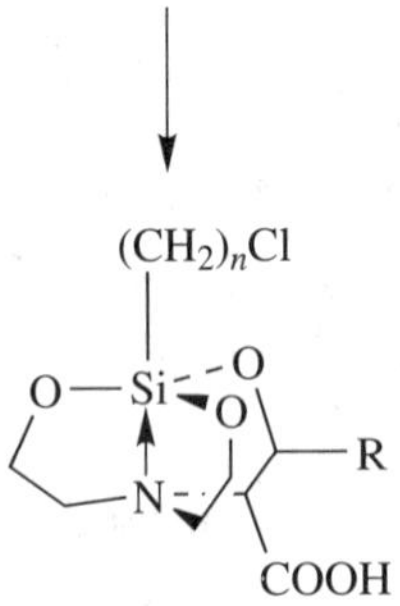

(24)

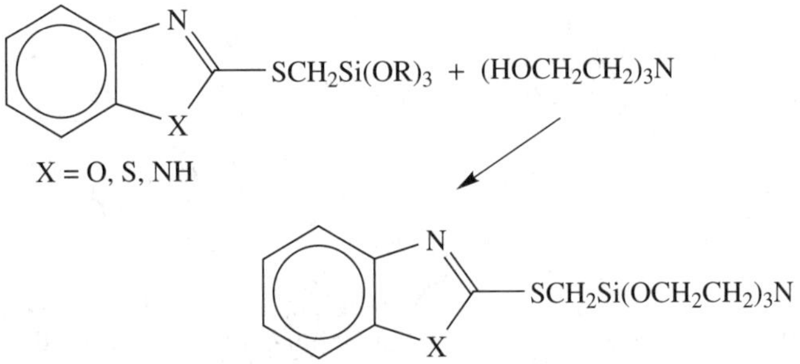

(25)

X = O, S, NH

It should be noted that the stronger (shorter) is the formed transannular Si←N bond, the easier are the above transetherification reactions. For example, 2-$O_2NC_6H_4Si(OEt)_3$ reacts exothermally and rapidly at room temperature, $PhSi(OMe)_3$ requires heating for the reaction to proceed with a reasonable rate and conversion of 4-$Me_2NC_6H_4Si(OMe)_3$ occurs under more drastic conditions[69]. In line with these qualitative observations, the Si←N bond lengths for 1-(2′-nitrophenyl)silatrane and 1-phenylsilatrane were found to be 2.116 Å[115] and 2.193 Å[3], respectively. The reaction rates between organyltrialkoxysilanes $RSi(OEt)_3$ and THEA are enhanced by the electron-withdrawing effect of the substituents in the order: $Cl_2CH > ClCH_2 > Me$[116]. The strength of the Si←N bond decreases in the same sequence[117].

3. Formation from organyltriacetoxysilanes

By using organyltriacetoxysilanes, which are considerably more reactive than organyltrialkoxysilanes, it is possible to obtain the corresponding silatranes by the reaction shown in equation 1 in almost quantitative yields. This method has some advantages, namely mild reaction conditions, no need for a catalyst and convenient isolation of pure products. For example, 1-alkyl- or 1-alkenylsilatranes are formed by treatment of alkyl- or alkenyltriacetoxysilanes with tris(2-hydroxyalkyl)amines in chloroform at 0 °C (equation 26)[37,118,119].

$XSi(OCOCH_3)_3$ + $(HOCHRCH_2)_3N$ ⟶ $XSi(OCHRCH_2)_3N$ + $3CH_3COOH$

(26)

X = Me, Et, n-Pr, n-Bu, n-Oct, Vin, All, etc., R = H[37,118]; X = Vin, R = Me[119]

By comparison, 1-allylsilatrane is prepared in a lower yield (86%) when a mixture of allyltrimethoxysilane and THEA is heated at 75 °C for 4 h in the presence of NaOH[39].

A modified version of this procedure (equation 26) is a one-pot reaction involving conversion of organyltrichlorosilanes into the corresponding organyltriacetoxysilanes without their isolation, followed by treatment with THEA in chloroform at 0 °C (equation 27)[118].

$$MeSiCl_3 + (CH_3CO)_2O + (HOCH_2CH_2)_3N \longrightarrow \underset{70\%}{MeSi(OCH_2CH_2)_3N} \quad (27)$$

4. Formation from organylhalosilanes

A potentially more direct route to 1-organylsilatranes as compared with the conversion of organylalkoxysilanes (or organyltriacetoxysilanes) is the reaction of organyltrichlorosilanes with tris(2-hydroxyalkyl)amines or their hydrochlorides (equation 28)[13,62]. Thus, 1-chloromethyl-3-(2′,2′,2′-trifluoroethyl)silatrane is obtained in 69% yield from chloromethyltrichlorosilane and the corresponding trialkanolamine at 10–15 °C[62].

$$ClCH_2SiCl_3 + [HOCH(CH_2CF_3)CH_2](HOCH_2CH_2)_2N$$

$$\downarrow -3HCl \quad (28)$$

[structure of silatrane with CH$_2$Cl and CH$_2$CF$_3$ substituents]

However, with vinyl- or phenyltrichlorosilanes, the protolytic cleavage of the Si–C bond becomes the main process, leading to the formation of 1-chlorosilatrane or its C-derivatives[62].

This problem can be resolved by the use of silylated tris(2-hydroxyalkyl)amines. For example, 1-vinyl- or 1-phenylsilatranes are prepared from the corresponding organyltrichlorosilanes and tris(2-trimethylsiloxyethyl)amine (equation 29). Pure products are easily isolated in good yield (80 and 83%, respectively)[120].

$$XSiCl_3 + (Me_3SiOCH_2CH_2)_3N \longrightarrow XSi(OCH_2CH_2)_3N + 3Me_3SiCl \quad (29)$$

X = Vin, Ph

Analogously, 1-hydrosilatrane (X = H)[121] and some 1-organylsilatranes, including carbofunctional derivatives [X = ClCH$_2$, Cl(CH$_2$)$_3$, NC(CH$_2$)$_2$], can be synthesized in high yields[120].

With a slight modification of equation 29, 1-allylsilatrane can be obtained in 96% yield (equation 30). Reaction occurs at room temperature in pentane solution[37].

$$H_2C=CHCH_2SiCl_3 + (Et_3SnOCH_2CH_2)_3N$$

$$\downarrow \quad (30)$$

$$H_2C=CHCH_2Si(OCH_2CH_2)_3N + 3Et_3SnCl$$

Treatment of tris(2-trimethylsiloxyethyl)amine with organyltrifluorosilanes in an aprotic solvent (benzene, o-xylene) results in the corresponding 1-organylsilatranes, as shown in equation 31[69,122,123].

$$RCOO(CH_2)_nSiF_3 + (Me_3SiOCH_2CH_2)_3N \xrightarrow[-3\ Me_3SiF]{C_6H_6,\ \Delta} RCOO(CH_2)_nSi(OCH_2CH_2)_3N \quad (31)$$

R = CF$_3$, C$_3$F$_7$, C$_4$F$_9$, $n = 3$[122]; R = Ph, $n = 2$[123]

Tris(2-trimethylsiloxyethyl)amine is also a useful reagent for the preparation of 1-chlorosilatrane (equation 32)[121] including those labelled with ^{15}N[124].

$$SiCl_4 + (Me_3SiOCH_2CH_2)_3N \xrightarrow{hexane,\ -10°C} ClSi(OCH_2CH_2)_3N + 3Me_3SiCl \quad (32)$$
$$95\%$$

Tetrabromosilane reacts with THEA in CHCl$_3$ at $-10\,°C$ to afford 1-bromosilatrane in 31% yield (equation 33)[121].

$$SiBr_4 + (HOCH_2CH_2)_3N \longrightarrow BrSi(OCH_2CH_2)_3N \quad (33)$$

5. Formation from organyltris(dialkylamino)silanes

An alternative synthetic route to silatranes is the reaction of organyltris(dialkylamino)silanes with tris(2-hydroxyalkyl)amines[52,125]. Examples are given in equation 34[52].

$$XSi(NMe_2)_3 + (HOCH_2CH_2)_2NCH_2CH(Ph)OH \xrightarrow{\Delta} XSi(OCH_2CH_2)_2(OCHPhCH_2)N \quad (34)$$

X = Me (55%), Et (35%), Ph (50%), Bz (20%)

6. Formation from organylsilanes

The hydrocondensation reaction of organylsilanes and triethanolamine affords the corresponding 1-organylsilatranes (equation 35). Transition metal compounds can be used as catalysts[126]. However, the method has a limited synthetic utility because of the high cost of the initial silanes.

$$XSiH_3 + (HOCH_2CH_2)_3N \longrightarrow XSi(OCH_2CH_2)_3N + 3H_2 \quad (35)$$

X = Me, n-octyl, Ph, Bz, etc

7. Miscellaneous preparations

The exothermic uncatalyzed reaction of THEA with the tris(O,O-dialkylphosphito)organylsilanes having 3 Si−OP bonds is shown in equation 36[127].

$$XSi[OP(OEt)_2]_3 + (HOCH_2CH_2)_3N \longrightarrow XSi(OCH_2CH_2)_3N \quad (36)$$

X = Me (60%), Ph (63%), ClCH$_2$ (70%)

The cleavage of the Si−O−Si moiety of polyorganylsilsesquioxanes (equation 37), polyorganylsiloxanols (equation 38) or polyorganylhydrosiloxanes (equation 39) by tris(2-hydroxyalkyl)amines results in the corresponding 1-organylsilatranes in 80–90% yields.

The reactions occur in a boiling solvent (toluene, xylene) in the presence of KOH[69,128]. This method can be preferably applied when the initial compounds are more available than organyltrialkoxysilanes.

$$1/n(XSiO_{1.5})_n \xrightarrow{THEA} XSi(OCH_2CH_2)_3N + H_2O \qquad (37)$$

$$1/n[XSiO_{1.5-n}(OH)_{2n}]_m \xrightarrow{THEA} XSi(OCH_2CH_2)_3N + (1.5 + m)H_2O \qquad (38)$$

$$1/n(XSiHO)_n \xrightarrow{THEA} XSi(OCH_2CH_2)_3N + H_2O + H_2 \qquad (39)$$

X = alkyl, alkenyl, aryl

A very elegant reaction is the synthesis of 1-hydrosilatrane and its C-substituted derivatives via readily available boratranes. Aprotic conditions exclude side processes involving the Si—H bond, so that the desired silatranes are formed in high yields (60–95%)[59,129–132]. For example, triethoxysilane reacts with 4-ethylboratrane in xylene in the presence of magnesium propoxide to give 1-hydro-4-ethylsilatrane (92%) (equation 40)[59]. Aluminium alkoxides[129] or aluminum chloride[130–132] are also used as catalysts. The reaction rate decreases sharply in the absence of catalysts. Attempts to prepare 1-(2′-chloroethyl)silatrane by this mild method failed[59].

$$HSi(OEt)_3 + \underset{R=Et}{\text{[boratrane, R=Et]}} \longrightarrow \text{[1-hydro-4-ethylsilatrane]} \qquad (40)$$

R = Et

In some cases, cleavage of a sensitive Si—C bond allows one to use bifunctional diorganylsilanes[133,134] and even tetraorganylsilanes[135] for the preparation of silatranes. Thus, 1-phenylsilatrane is formed in 80% yield from diphenyldimethoxysilane and THEA in the presence of sodium methoxide at 250 °C (equation 41)[133].

$$Ph_2Si(OMe)_2 + (HOCH_2CH_2)_3N \longrightarrow PhSi(OCH_2CH_2)_3N \qquad (41)$$

A versatile approach to 1-heteroarylsilatranes is the cleavage reaction of tetraheteroarylsilanes (equation 42) which are hydrolytically and thermally stable compounds. The latter are more easily available compared with trifunctional heteroarylsilanes. The reaction is effectively catalyzed by CsF[135].

$$X_4Si + (HOCH_2CH_2)_3N \longrightarrow XSi(OCH_2CH_2)_3N$$

X = [furyl], [thienyl] \qquad (42)

In summary, the procedures described in equations 2, 26, 28 and 34 are fairly efficient and versatile for the synthesis of a large number of silatranes. This is clearly illustrated in the preparation of 1-methylsilatrane (equation 43). However, it should be noted that only equation 2 can be considered as a truly general route to this class of compounds.

$$MeSiY_3 + (HOCH_2CH_2)_3N \longrightarrow MeSi(OCH_2CH_2)_3N \qquad (43)$$

Y	Solvent	t (°C)	Cat.	Yield (%)	Ref.
OR	benzene	70	KOH	high	136
OCOCH$_3$	CDCl$_3$	0		95	118
Cl	DMF	140–150		77	120
NEt$_2$	xylene	135		88	125

B. Structure and Physical Properties

1. Molecular structure

Transannular interaction between the silicon and the nitrogen atoms in Si- and C-substituted silatranes is strongly proved by numerous X-ray studies[5–7,13–15,18–22,29]. Table 1 provides structural parameters for a large fraction of the approximately 60 crystallographically determined structures of 1-substituted silatranes which are mainly collected in a recent review[29].

The X-ray data demonstrate that silatranes exist in an *endo* form in which the nitrogen lone pair is oriented toward the silicon. The silicon has a distorted trigonal-bipyramidal (TBP) structure with nearly equatorial disposition of three oxygen atoms and an axial location of the nitrogen and the substituent X. The X−Si←N angle of 174°–179° is close to ideal for the TBP structure value of 180°. The valence O−Si−O and X−Si−O angles vary from 116° to 120° and from 75° to 90°, respectively. The mean length of the equatorial Si−O bonds is about 1.66 Å, compared to 1.65 Å in alkoxysilanes. The Si←N bond distance in crystals of most silatranes is in the range 1.965–2.240 Å and the difference Fourier synthesis shows a high peak of electron density at the region of the transannular bond[154,157]. This bond is considerably shorter than the sum of the van der Waals radii (3.54 Å) but certainly longer than a usual single Si−N bond of 1.7–1.8 Å. The X−Si bond in silatranes is also longer (by 0.02–0.10 Å) than that in related tetrahedral silicon compounds. The mutual weakening of the axial X−Si and Si←N bonds, first noted by Parkanyi and coworkers[158], is very important for understanding the bonding nature in silatranes. Recent investigations show that a visible shortening of the Si←N bond upon π-coordination of transition metals to the phenyl group of 1-phenylsilatrane accompanies a regular elongation of the Si−C bond in complexes[159–161].

As a rule, the Si←N bond length in silatranes decreases on increasing the electron-withdrawing effect of the substituent X (Table 1). The longest Si←N bond (2.89 Å[156]) is found for the silatranylplatinum complex **39**, which should be classified as a 'quasi-silatrane'[27]. The transannular bond of 2.007 Å in 1-isothiocyanatosilatrane **29**[68] is the shortest reported so far in neutral silatranes. In silatranyl oxonium salts **36** the Si←N bond is a record short value, 1.965 Å[150]. However, despite a lower electronegativity of the substituent X, a shorter transannular bond is observed for 1-chlorosilatrane **38** (2.023 Å[155]) than in 1-fluorosilatrane **37** (2.042 Å[154]).

The strengthening of the Si←N bond is accompanied by a reduction in the XSiO angles and an increase in the nitrogen pyramidalization. These are reflected in the corresponding decrease in the deviation of the silicon atom from the plane of the equatorial oxygen atoms (Δ_{Si}) and increase in the deviation of the nitrogen atom from the plane of the neighboring

TABLE 1. Structural parameters for Si-substituted silatranes

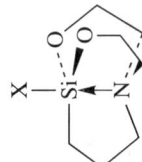

Compound	X	d(N–Si) (Å)	d(X–Si) (Å)	d(Si–O) (Å) (mean)	N–Si–X (°)	N–Si–O (°) (mean)	Δ_{Si} (Å)[a]	Δ_{N} (Å)[b]	Ref.
2	Me	2.175(4)	1.870(6)	1.670(4)	179.4(2)	82.7	0.211	0.379	137
3	Et	2.214(8)	1.881(10)	1.658	178.7(1)	82.0	0.229	0.342	29
4	Cyclopropyl	2.228(9)	1.94(1)	1.664(8)	176.0(7)	81.7	0.24	0.35	34
5	2-Chlorocyclopropyl	2.150(3)	1.877(4)	1.664(3)	187.7(2)	83.2	0.2		35
6	Cyclopropylmethyl	2.149(4)	1.874(5)	1.659(4)	177.2(2)	83.1	0.2		37
7	$H_2N(CH_2)_2NH(CH_2)_3$	2.165(2)	1.875(2)	1.666	179.1(1)	82.7	0.210(1)	0.383(3)	138
8	$HO(CH_2)_3$	2.173(2)	1.869(2)	1.665(1)	179.4(2)	82.7	0.21(1)	0.372(3)	69
9	$HS(CH_2)_3$	2.177(4)	1.872(4)	1.652(4)	178.6(5)	81.7	0.21	0.37	139
10	$Cl(CH_2)_3$	2.181(7)	1.875(8)	1.662	178.2(6)	83.2	0.199	0.368	140
11	2-(1-Silatranyl)ethyl	2.230(5)	1.874(6)	1.650(5)	177.5(2)	81.6	0.24		46
12	$H(O)CCH_2$	2.108(6)	1.914(8)	1.660(5)	179.4(3)	86.0			141
13	MePhSi(H)CH_2[c]	2.208(9)	1.864(10)	1.638(8)			0.222(3)		49
		2.240(9)	1.869(10)	1.642(7)			0.241(3)		
14	MePhSi(OH)CH_2	2.214(9)	1.864(16)	1.655(10)			0.222(3)		49
15	PhCOOCH_2	2.112(7)	1.954(5)	1.661(7)	174.0(1)	78.5	0.16	0.36	142
16	ClCH_2	2.120(8)	1.912(11)	1.676	176.4(4)	84.4	0.163(3)	0.385(8)	143
17	I$^-$Me$_2$S$^+$$CH_2$	2.046(2)	1.930(3)	1.667(2)	179.1(1)	86.0	0.115	0.396	144
18	I$^-$Me$_3$N$^+$$CH_2$	2.080(13)	1.915(15)	1.662(11)	175.7(6)	85.1	0.13	0.41	145
19	Cl_2CH	2.062	1.937		176.4	85.7	0.12	0.38	17
20	CH_2=CH	2.150(6)	1.880(8)	1.664(5)	178.72(25)	83.2	0.20	0.40	146

(continued overleaf)

TABLE 1. (continued)

Compound	X	d(N—Si) (Å)	d(X—Si) (Å)	d(Si—O) (Å) (mean)	N—Si—X (°)	N—Si—O (°) (mean)	Δ_{Si} (Å)[a]	Δ_N (Å)[b]	Ref.
21	Ph(α)	2.193(5)	1.882(6)	1.656(5)	177.90(22)	82.9	0.204	0.34	3
22	Ph(β)	2.156(4)	1.908(5)	1.657(5)	177.04(29)	83.3	0.195	0.379	147
23	Ph(γ)	2.132(4)	1.894(5)	1.656(4)	179.0(2)	83.7	0.183	0.39	148
24	4-MeC$_6$H$_4$	2.171(1)	1.887(1)	1.656	178.8(1)	82.9	0.204	0.366	149
25	3-O$_2$NC$_6$H$_4$	2.116(8)	1.904(9)	1.656(6)	177.4(3)	84.1	0.17	0.39	115
26	2-Furyl	2.112(5)	1.894(6)	1.656(6)	177.2(2)	85.0	0.144	0.387	29
27	3-Furyl	2.130(9)	1.892(11)	1.675	177.7(4)	84.0	0.175	0.384	29
28	2-Thienyl	2.133(9)	1.905(11)	1.666	177.2(4)	84.4	0.162	0.360	29
29	SCN	2.007(3)	1.888(3)	1.639(3)	179.5(1)	87.6(2)			68
30	EtO	2.152	1.658	1.648			0.179		150
31	t-BuO	2.189(4)	1.659(4)	1.650(4)	179.4	83.0	0.201	0.371	151
32	4-MeC$_6$H$_4$O	2.107(2)	1.677(2)	1.649(2)	174.7(2)	84.7	0.154	0.372	152
33	3-ClC$_6$H$_4$O	2.079(2)	1.690(2)	1.656	176.5(2)	85.3	0.14	0.40	42
34	Co$_3$(CO)$_9$CO	2.010(5)	1.677(2)	1.649(2)	178.9(2)	86.2			153
35	EtO··HOOCCF$_3$	2.050	1.710	1.652			0.102		150
36	BF$_4^-$—Me$_2$O$^+$	1.965	1.830	1.642			0.017		150
37	F	2.042(1)	2.042(1)	1.645(2)	179.7(1)	85.9	0.117	0.390	154
38	Cl	2.022(9)	2.022(9)	1.649	179.8(3)	86.7	0.095	0.396	155
39	Cl(PhMe$_2$P)$_2$Pt	2.89(1)	2.292(4)	1.649(9)	176.7	71.2	0.53	0.07	156

[a] Δ_{Si} = the deviation of the Si atom from the plane of the equatorial oxygen atoms.
[b] Δ_N = the deviation of the N atom from the plane of the neighboring carbon atoms.
[c] Two independent molecules in the unit cell.

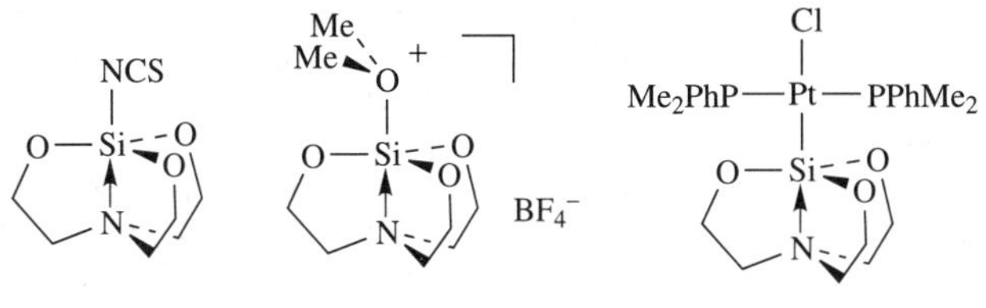

(29)   **(36)**   **(39)**

carbon atoms (Δ_N). Numerous relationships between these geometrical parameters and correlations of the Si←N bond distance with the inductive σ_I and σ^* constants and the electronegativity of the substituent X were found. These regularities were discussed recently in some reviews[17–22]. A general similarity between self-consistent change in the geometry of the silicon polyhedron upon strengthening of the Si←N bond within the silatrane series and the molecular deformation in the course of the intramolecular S_N2 substitution reaction at a tetrahedral silicon atom was found[162,163] by the Bürgi–Dunitz method of structural correlation[164]. In particular, it was demonstrated that with a successive increase in the order of the Si←N bond, the X—Si bond order is reduced[163]. The constancy of the total order of the axial bonds in silatranes is one of the fundamental properties of these and other compounds of a TBP silicon[28,165].

The cage skeleton of most silatranes possesses an approximate threefold (C_3) symmetry and in its five-membered (N—Si) chelate rings the carbon atoms in the α-position to the nitrogen deviate from the plane formed by the other atoms[15,29] (Scheme 2). The partial disordered structure observed for some silatranes may be regarded as a frozen state of the dynamic flapping of the α-C atoms in the crystals[15].

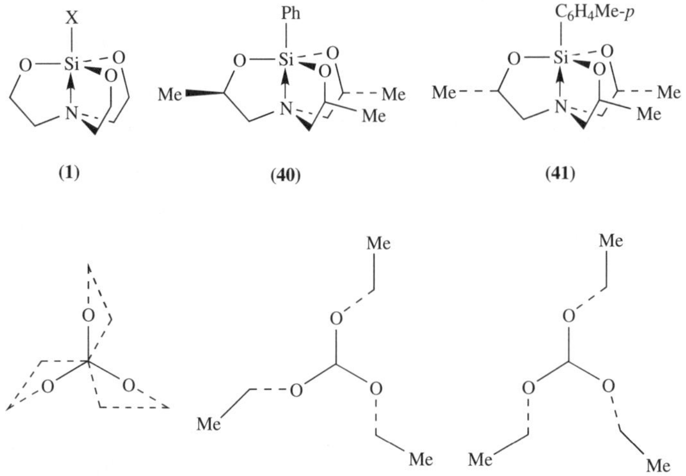

(1)   **(40)**   **(41)**

SCHEME 2. Schematic representation of silatrane moieties of compounds **1**, **40** and **41** as viewed along the X—Si←N axis.

In 3,7,10-trimethylsilatranes the β-carbons are on the flaps of their ring envelopes and the methyl groups occupy a pseudo-equatorial position in the five-membered rings[149,166,167]. Therefore, in contrast with the symmetric isomer **40** of 1-phenyl-3,7,10-trimethylsilatrane[166], the asymmetric isomer **41** of 1-(p-tolyl)-3,7,10-trimethylsilatrane[149] has a skeleton in which the 'flap'-atoms C(3) and C(7) lie on the same side of the ring planes while C(10) is on the opposite side (Scheme 2). The Si–N bond in 3,7,10-trimethylsilatranes seems to be slightly longer in their more strained asymmetric isomer[167].

Both steric interaction between the carbon substituents and the strong electron-withdrawing effect of the carboxylic acid group decrease the donor ability of the nitrogen and hence the strength of the Si←N bond in 1-organyl-3-methylsilatrane-4-carboxylic acids **42**[110,111,168].

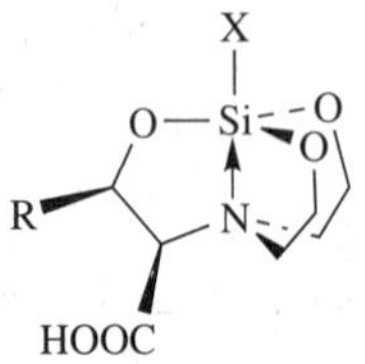

X	R	d(SiN) (Å)	Ref.
Vinyl	H	2.169	110
ClCH$_2$	Me	2.136	111
Cl(CH$_2$)$_3$	H^a	2.176 and 2.208	168
Cl(CH$_2$)$_3$	Me^a	2.198 and 2.244	111

^{a}Two independent molecules.

It should be noted that, unlike many pentacoordinated organosilicates and organosilanes[24,25,165], silatranes are extremely ineffective in participating in a hexacoordinated structure. The silicon atom appears to be pentacoordinate[44,103,104,142] even in silatranes **15**, **43**–**45** bearing a carbofunctional substituent X having a very high (O–Si)- or (N–Si)-chelating ability[24,25,165]. This could be ascribed to a high energetic cost for the cage deformation under the flank side approach of the donor to the silicon.

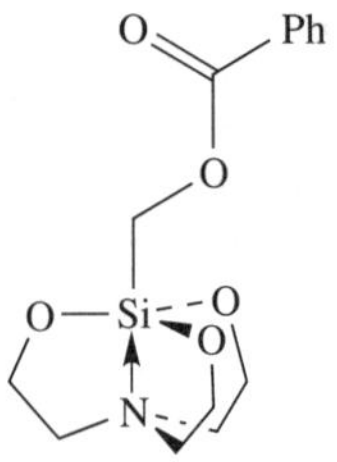

(**15**)[142]

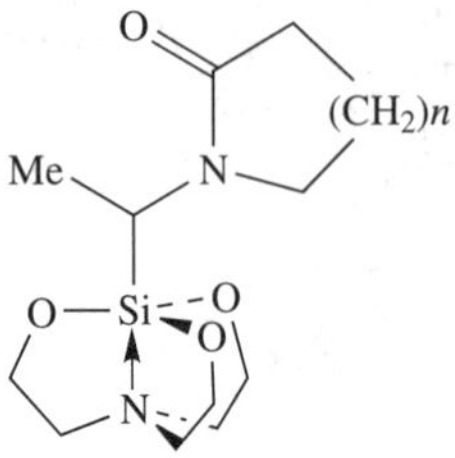

$n = 1$ (**43**)[104]
$n = 2$ (**44**)[103]

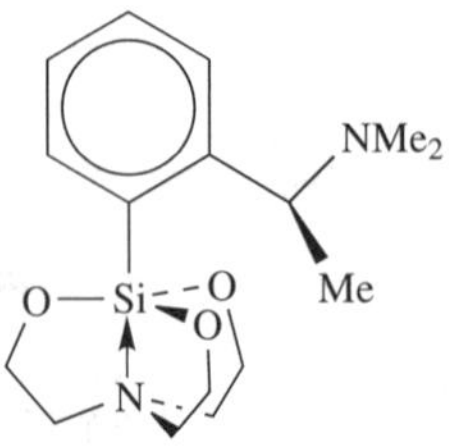

(**45**)[44]

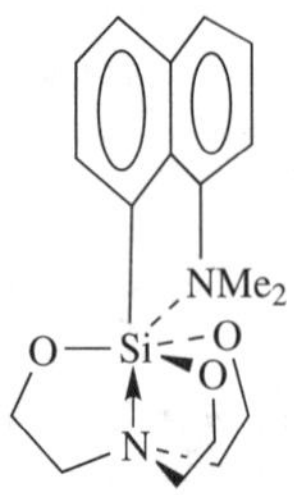

(**46**)[44]

24. Silatranes and their tricyclic analogs 1465

A single hexacoordinated silatrane **46** possessing a very weak additional Si←NMe$_2$ bond (2.952 Å) was obtained recently by Corriu and coworkers[44]. The hexacoordination in **46** is most likely due to the rigid geometry of its 8-dimethylaminonaphthyl group. The reluctance of silicon to become hexacoordinated in the silatrane structure is its special characteristic feature which, as will be shown below (Section II.C), predetermines the low reactivity of silafunctional silatranes toward nucleophiles.

The X-ray data demonstrate that the equatorial oxygen of silatranes bearing acidic OH or COOH groups in the substituent X or the cage can be involved in the intermolecular or intramolecular hydrogen bonding[29]. The formation of a hydrogen bond leads to a visible lengthening of the protonated Si−O bond. Judging by some spectral data (Section II.B.8) and the structures of the hydrogen bonded complex **35**[150] and the oxonium salt **36**[150], the axial oxygen in 1-alkoxysilatranes is more basic than the equatorial oxygens.

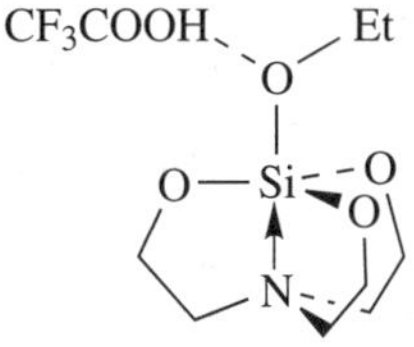

(35)

Mutual weakening of the axial bonds in silatranes predetermines the unprecedented difference between the Si←N bond lengths in the crystal and in the gas phase. Electron diffraction studies of 1-methylsilatrane (**2**)[169] and 1-fluorosilatrane (**37**)[170] found that in the vapors their transannular bond is 0.28 Å longer and the X−Si bonds are somewhat shorter than those in their crystals.

2. Bonding model

The first model of the bonding in silatranes invoked an sp^{3}d hybridization of the silicon atom[2,3,69]. However, this theoretical scheme was of very limited analytical value for understanding the silatrane properties. Moreover, many experimental and quantum chemical data ruled out the valence role of d orbitals in molecules of main group elements[171,172]. In 1977 the Musher model of hypervalency[173] was modified with reference to the Group 14 elements and used for a qualitative analysis of the structural trends in silatranes[12,13,174].

This model implies that the TBP silicon could use its sp^2 orbitals for the bonding with the equatorial oxygen atoms, whereas its p$_z$ orbital could be involved in the interaction with an appropriate orbital of the axial substituent X and a lone electron pair of the donor nitrogen atom. These generate three usual, two-center bonds in the equatorial SiO$_3$ fragment and a hypervalent, three-center four-electron (3c4e) bond in the axial moiety XSiN. The simplest MO diagram of a 3c4e bond may be represented by three molecular orbitals, bonding (b), nonbonding (nb) and antibonding (ab) (Scheme 3).

The nonbonding orbital should have a node in a region of the weakest, Si←N component of a hypervalent X−Si←N bond. This explains a typical distortion of a TBP configuration of the silicon in silatrane and its displacement from the equatorial plane toward the substituent X.

Analysis of the MO diagram of the 3c4e bond shows that the charge transfer from the nitrogen atom to the substituent X should be less than unity. Since the axial and

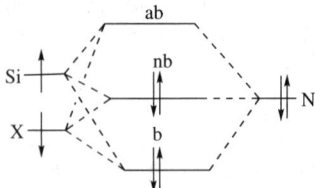

SCHEME 3. The simplest MO diagram of a 3c4e bonding in silatranes

equatorial bonds are practically orthogonal, the proposed model predicts a predominant *trans* effect of the axial substituent X. By use of the second-order perturbation theory, both the X—Si and Si←N components of hypervalent X—Si←N bond were shown to be mutually weakened in comparison with the usual two-center bonds. The corresponding analytical equations describing the energy of the axial bonding predetermine (at the energetic level) a constancy of the total order of the X—Si and Si←N bonds in silatranes. According to the 3c4e model, the degree of the Si←N bonding is determined not only by the electronegativity of the substituent X but also by the strength of the X—Si bond. Displacement of the substituent X by a more electronegative or a better leaving group should increase the energy (decrease the length) of the Si←N component of the X—Si←N bond. In molecules with the same XSiO$_3$N coordination center, a variation in the Si←N bond length (strength) should be accompanied by a change in the opposite direction in the corresponding descriptions of the X—Si bond. Due to a strong influence of the substituent X on the Si←N bond length and a charge transfer in the axial moieties, the values of inductive and resonance effects of the silatranyl group cannot be constant within a series of silatranes. Finally, the 3c4e model predicts that the introduction of a less electronegative substituent in the equatorial sites of silicon TBP should decrease the central atom bonding with both axial substituents.

This model and its consequences appear to be a very constructive and useful concept for a consistent analysis of the physical and chemical properties of silatranes and other organic compounds of TBP silicon[20,24,27,28].

3. Quantum mechanical studies

Despite the extensive experimental evidence for the silicon pentacoordination in silatranes, for a long time there were doubts concerning the true reasons of the *endo* configuration of these molecules (see literature cited elsewhere[12,13]). In particular, it was suggested that this geometry is not realized due to a transannular Si←N bonding, but primarily owing to steric factors typical for medium-ring bicyclic nitrogen compounds[175] which push the nitrogen atom inside the molecular cage. A weak dipole–dipole interaction of the nitrogen lone electron pair with a tetrahedral silicon atom and its neighboring atoms was also considered as a stabilizing factor of the *endo* structure of silatranes.

Calculations of the strain energies of the *endo* and *exo* forms of the 1-methyl-, 1-fluoro- and 1-chlorosilatranes carried out by the methods of molecular mechanics found these assumptions to be unsound[176–180]. Even with an additional consideration of the intramolecular electrostatic interactions[177–179], employment of the usual force field for tetracoordinate silicon led to potential functions of the *endo–exo* isomers (equation 44) with a deeper minimum corresponding to the *exo* form **47**, with *out*-orientation of the nitrogen.

However, it was difficult to exclude a hypothetical *exo* form at least as a minor component in solution or in the gas phase. Since the chemical and physical properties of the *exo* and *endo* forms should differ sharply, their relative thermodynamic and kinetic stability as well as the nature of the bonding in silatranes constitute a major concern in silatrane chemistry.

$$
\begin{array}{c}
\text{(1)} \qquad\qquad \text{(47)}
\end{array}
\tag{44}
$$

The energetic preference of the *endo* form was demonstrated by the first quantum chemical calculations of silatranes performed on 1-hydro- and 1-fluorosilatranes[12,181]. These calculations employed the CNDO/2 semiempirical method with spd basis set and geometries of the *endo* and *exo* forms corresponding to those obtained for 1-methylsilatrane by modified molecular mechanics method. The energy of the Si←N coordinative bonding was found to be about 20–25 kcal mol^{-1} and to be the main contributor to stabilization of the *endo* structure of silatranes.

Subsequent theoretical calculations confirmed the important role of the Si←N interaction in the formation of the silatrane structure. A significant Si←N bond order was determined at the CNDO/2 level using the experimental solid state geometries of silatranes[182–185]. Thus, the Wiberg indexes of this bond in the crystalline α, β and γ modifications of 1-phenylsilatrane were estimated to be 0.35–0.39 and 0.15–0.18 within the spd and sp basis sets, respectively[182]. It was shown that a transannular Si←N bonding decreases the bond order of the Ph−Si bond. Together with a significant bond order of the Si←N bond in 1-methylsilatrane, detected by CNDO/2[186] and MNDO[187] calculations using sp basis set, these data confirmed the validity of the 3c4e bonding model.

The first semiempirical studies of the potential functions of silatranes revealed that their Si←N bond could be very easily deformed by crystal packing and solvation forces[188a]. Direct MINDO/3 and MNDO calculations demonstrated a significant shortening of the equilibrium, gas phase Si←N distance due to the contribution of the solute–solvent interaction energy (within the Onsager reaction field model[188b]) to the total energy of 1-methyl- and 1-fluorosilatranes[188a].

The first *ab initio* calculations on silatranes proved that only a small amount of energy (a few kcal mol^{-1}) is required to decrease the Si←N bond length from the gas-phase to the solid-state value. The single point computations were carried out by using the 3-21G* basis set for two Si←N 'bond-stretch' isomers of 1-methylsilatrane[189], using the solid-state and an approximate gas-phase geometry, and with the 6-31G* basis for 1-hydroxysilatrane with geometries optimized by the AM1 method[190]. According to the Bader electron density analysis, the Si←N bond critical point should still exist even in the gas-phase 1-hydroxysilatrane structure[190].

AM1 and PM3 computation on 1-methyl-[191], 1-fluoro-[180], 1-chloro-[192] and 1-isothiocyanatosilatranes[68] using full geometry optimization confirmed the small energy cost for a shortening of the Si←N bond in silatranes. The potential profiles of these molecules were shown to be very flat in the 2.00–3.60 Å range of the Si←N distance, with two distinct minima related to the *endo* and *exo* forms. Both parametrizations found the *endo* structure to be the more stable in the case of 1-isothiocyanato- and 1-halosilatranes, where the equilibrium Si←N distances are shorter than in 1-methylsilatrane. Both semiempirical methods overestimate the Si←N bond lengths (by 0.11–0.21 Å) with respect to the experimentally observed gas-phase values. Frequency analysis predicts that a separate Si←N vibration frequency should not exist. For 1-methylsilatrane, the Si←N stretching vibration (in the region of 280 cm^{-1}) should be combined with the C–N (at about 450 cm^{-1}), Si–O and Si–C (at about 720 cm^{-1}) vibrations[191] whereas for 1-chlorosilatrane a combined Cl–Si←N vibration should appear at 364 cm^{-1}[192].

In contrast to the semiempirical results, full optimization of the molecular geometry of 1-fluorosilatrane by restricted Hartree–Fock (RHF) calculations using the 3-21G, 3-21G* and 6-31G* basis sets did not find evidence for the existence of an *endo* minimum on the energy hypersurface[193]. Since the *exo* form was never found experimentally, the *ab initio* results support the view of the exclusive existence of silatranes in the *endo* form. As for the equilibrium Si←N distance, the *ab initio* calculations using polarized basis sets overestimated its value (2.556 Å) even more than the AM1 and PM3 methods.

Nearly the same energy profile and Si←N bond length (2.534 Å) of 1-fluorosilatrane were computed by Gordon's group[194] at the RHF/6-31G* level. The use of the extended, 6-31G*EXT basis (two sets of d functions on Si, O, F and N atoms and a diffuse sp shell situated on the same atoms) changed the geometry and decreased the equilibrium Si←N distance by 0.117 Å. However, adding a set of f functions on silicon and all adjacent atoms to this basis set (to produce a 6-31G*EXT+f basis set) led to an apparently large change in the geometry but cancelled 0.08 Å of the shorting achieved in the Si←N distance.

Inclusion of single point MP2 treatment of electron correlation along the SCF 6-31G* potential curve shifted its minimum by about 0.15 Å and brought the computed Si←N bond length (2.29 Å) into extremely close agreement with the experimental gas-phase value[194]. Boggs and coworkers[195] found just the same effect of electron correlation on the equilibrium Si←N bond length of 1-fluorosilatrane by MP2 calculation using the 6-31G* basis set for Si, N, F and O and the 6-31G basis set for C and H at MNDO optimized structures. However, since both latter numerical results[194,195] did not include a full geometry optimization at the MP2 level, they should be viewed with some caution.

Gordon and coworkers performed a quantitative study of the effect of solvation on the molecular geometry of 1-fluorosilatrane[194] using the Onsager reaction field cavity model[188a]. The solute phase Si←N bond length in 1-fluorosilatrane was predicted to be 2.104 Å, a very close value to the known solid state X-ray distance. The energy of solvation and the amount of energy gained by reoptimizing geometry with such a drastic shortening of the Si–N distance were found to be about 10.5 and 4.1 kcal mol^{-1} respectively.

The results of semiempirical and *ab initio* calculations are consistent with the general tendency toward strengthening of the Si←N bond on increasing the electronegativity of the substituent X. However, the experimentally observed shortening of the Si←N bond in 1-chlorosilatrane as compared with that in 1-fluorosilatrane was reproduced successfully only by the PM3 method[180,192,194]. Comparison of the semiempirical results show that the PM3 and MNDO equilibrium geometries of silatranes are generally closer to the experimental data, whereas the AM1 method gives a more reliable picture of charge distribution and orbital energies[193,196].

Calculations of the potential functions governing the Si←N bond stretching motion made it possible to trace regularities in the variation of the atomic charges and stereo-electronic and orbital structure of silatranes with the change in the coordination number of the silicon atom and the degree of the Si←N bonding[180,188,191,192,196]. It was found that the main changes in atomic charges upon variation in the Si←N bond length occur on the O, N and X atoms, rather than on silicon. The values of atomic charges and the signs of their gradients are naturally very sensitive to the computation method.

Previous semiempirical computations of silatranes and some other pentacoordinated organosilicon species found either increase or decrease in the negative silicon charge under pentacoordination, depending on the method and the basis set used[12]. However, the view of a higher positive charge on the pentacoordinate silicon as compared with the tetracoordinate state is widely accepted[184]. Results of *ab initio* 6-31G* calculations on 1-fluorosilatrane[12] demonstrate a definite decrease in the positive charge on silicon, an increase in the negative charge on nitrogen and an elongation of the F−Si and Si−O bonds with a shortening of the Si−N distance. These changes enlarge the values of the dipole moment and the Si←N bond order. The latter value is not negligible in the equilibrium geometry despite the large Si−N distance.

Ab initio study of structural trends[194] led to the conclusion that the dative bonding model rather than the 3c4e model could describe the character of the X−Si←N bonding in silatranes. Within the dative model, which represents a limiting extreme of the basic MO diagram (Scheme 3), the primary interaction includes the formation of bonding and antibonding orbitals between X and Si. The nitrogen lone pair remains essentially unaffected during the molecular formation and the only Si−N binding is provided by a small dative interaction. The model predicts a very weak Si←N bond, a small electron transfer in the axial moiety and an ordinary covalent character of the X−Si bond. In agreement with this prediction, the RHF/6-31G*EXT analysis of the Boys localized orbitals of 1-fluorosilatrane show only slight bonding of the nitrogen lone pair with the silicon atom[194]. Moreover, the RHF/6-31G* calculations found that the X−Si bond in silatranes with X = H, F and Cl is somewhat shorter than in the model trisilanols $XSi(OH)_3$.

However, these arguments call for additional study. The dihedral angles at the Si−O bonds for the model molecules appeared to differ strongly from those in silatranes[197]. The conformation of these trisilanols allows an effective interaction of the oxygen lone pairs with the geminal σ^*-Si−X orbital. This is why the difference in the calculated X−Si bond distances in the molecules compared testifies to a different nature of their anomeric effects[198] rather than to the preference of the dative bonding model for silatranes. Indeed, trisilanols with X = H_3Si, Me, H_2P, HS, H_2N and HO having nearly the same dihedral angles as in silatranes were shown to have a shorter X−Si bond than that in related silatranes[194]. Together with the well known general tendency for the X−Si bond lengthening upon strengthening of the Si←N bond, these results suggest that the bonding in silatranes is closer to the 3c4e picture.

4. Thermochemical data

The enthalpies of combustion and sublimation were determined[199−203] for a number of 1-organyl- and 1-organoxysilatranes. These data made it possible to calculate the enthalpies of formation and atomization of silatranes as well as an energy correction term ($E_{cycl} - E_{Si \leftarrow N}$ = 7.2−11.7 kcal mol^{-1}) combining strain energy and dative bond strength. The E_{Si-N} values were estimated to be 13−22 kcal mol^{-1} on the basis of various experimental spectroscopic and kinetic data[204]. Using the correction terms derived from the thermochemical data, the lower limit of the strain energies of most simple silatranes appears to be close to 15 kcal mol^{-1}, a value known in bicyclo[3.3.3]undecane (manxane)[205]. The thermochemical data indicate also that the enthalpy of formation of the silatranyl group varies only slightly, its value being 186.4 ± 1.7 kcal mol^{-1}[203].

5. Dipole moments

The high polarity of silatranes is one of the most convincing pieces of evidence for the Si←N bonding in silatranes in solution. Their dipole moments (μ) lie in the range

5–11 D, while the influence of the Si and C substituents and the medium on the μ values has been discussed in detail in previous reviews[1,5,6,12,13,28].

By use of the X-ray data and the method of bond moments, the contribution of the Si←N bond to the molecular dipole moment of 1-organylsilatranes was calculated to be 1.5–3.1 D[12,206]. These values correspond to a transfer of approximately 0.2e from the nitrogen to the silicon atom. Use of the known empirical relationship (equation 45) between the heat of formation of the donor–acceptor (DA) bond (ΔH_{DA}) in DA complexes, the dipole moment (μ_{DA}) and the inverse value of the length of the DA bond (d_{DA}) made it possible to estimate $-\Delta H_{Si \leftarrow N}$ values of 5–15 kcal mol^{-1}[207]. Electron-withdrawing substituents increase the dipole moment and energy of the Si←N bond.

$$\Delta H_{DA} = -7.47 \mu_{DA} d_{DA}^{-1} \qquad (45)$$

Within a series of 1-alkylsilatranes, the magnitude of the alkyl group has a negligible effect on the μ value of the molecule[208]. The increase in temperature causes only slight and irregular changes in the dipole moments. This seems to be unexpected, since the IR and NMR data indicate an elongation of the Si←N bond on heating of silatrane solutes (Section II.B.9) and *ab initio* and semiempirical studies predict a decrease in the μ value with the increase in the Si–N interatomic distance.

6. Photoelectron spectra

X-ray photoelectron spectroscopic (ESCA) studies[209–211] found a correlation between the N$_{1s}$ and Si$_{2p}$ binding energies in silatranes, thus providing evidence for a strong Si←N interaction in these molecules. In 1982 an analysis of the Si–K$_\alpha$ chemical shifts in the X-ray fluorescent spectra argued in favor of a higher positive charge on the silicon atom in silatranes than in related Si-substituted triethoxysilanes[212]. Meanwhile, results of previous X-ray fluorescent studies of silatranes suggested that the Si←N bond formation reduced the effective positive charge on the silicon[213,214].

Despite some inconsistencies in the experimental data and their interpretation, the PE spectroscopy studies of silatranes led to some common conclusions[196,215–218]. For most silatranes the IP$_1^v$ is associated with the nitrogen lone pair orbital which involves at least some Si character. The IP$_1^v$ values (9–10 eV) are sensitive to the substituent X and generally increase with strengthening of the Si←N bond. They exceed not only the IP$_1^v$ value of the model compound manxine **48** (7.05 eV)[219], which also has near-planar nitrogen atom, but even the IP$_1^v$ value of triethanolamine (8.7 eV)[215] and tris(2-trimethylsiloxyethyl)amine (7.94 eV)[217], which possess usual, tetrahedral nitrogen. These observations are regarded as a manifestation of the bonding effects in silatranes[215,217].

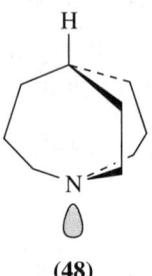

(48)

In silatranes which do not have more easily ionizing groups in the substituent X, the IP_1^v was attributed to the Si←N sigma bond orbital[218] or to a MO localized to a great extent on the axial X−Si←N fragment[196]. The first assignment was based on the *ab initio* STO-3G calculations of silatranes with X = Me, ClCH$_2$, Vinyl, F and Cl using common skeleton geometry based on the solid state structure of 1-methylsilatrane. It was shown, however, that the 3-21G* IP_1^v (assuming the validity of Koopmans' theorem) of the latter compound changes from 10.14 eV for the solid state structure to 9.87 eV for the gas phase structure[189]. The second assignment was based on analysis of the dynamic orbital-correlation diagrams and the results of MNDO and AM1 calculations. It was noted that the increase in IP_1^v on shortening of the Si←N bond is caused by the corresponding increase in the nitrogen atom pyramidality[196], whereas the higher IP_1^v in silatranes relative to those in model compounds can be explained by a probable interaction between the nitrogen lone pair and the antibonding σ^* MOs of the neighboring C−C−O fragments in the former molecules[216,217]. All these results demonstrate that the problems of MO structure and PE spectroscopy of silatranes are still open.

7. Ultraviolet spectra

The ultraviolet absorption spectra of silatranes display only an edge of the n-σ^* band (λ_{max} < 190 nm)[216,220,221]. A hypsochromic shift of this band relative to that observed in the spectra of triethylamine (204 nm), triethanolamine and its tris-TMS derivative (195–196 nm) confirms the Si←N bonding in silatranes. The CNDO/S study of the electronic excited states of 1-methyl- and 1-vinylsilatranes had shown that their UV spectra are determined by two transitions corresponding to charge transfer from the highest occupied molecular orbital, which is localized on the Si←N, Si−C, Si−O and C−O bonds[222]. Despite a high polarity of 1-arylsilatranes (X = Ph, C$_6$F$_5$), the solvent effect on their UV absorption and fluorescence spectra was shown to be only slightly higher than that for the corresponding aryltriethoxysilanes[223]. A silicon pentacoordination does not affect the absorption bands of the aromatic groups in the spectra of 1-aryloxysilatranes[224]. UV absorption spectra of charge-transfer complexes of silatranes [X = Ph, FC$_6$H$_4$CH$_2$[54] and RS(CH$_2$)$_n$[225] where R = Alk, Bz, Allyl and n = 1, 2] and the corresponding trialkoxysilanes with TCNE revealed a higher electron-releasing ability of the silatranyl group. The aromatic substituent and the sulfur were shown to be the electron-donor centers in these complexes.

8. Vibrational spectra

The main features of the IR and Raman spectra of silatranes were reviewed at the beginning of the 1980s[12,13]. There were two hypotheses concerning the manifestation of the Si←N bond vibrations in the spectra. The first presumed that this stretching vibration is in the range 568–590 cm^{-1} since the corresponding absorption band has not been observed in the spectra of the related model compounds, triethanolamine and trialkoxysilanes XSi(OAlk)$_3$. According to the second hypothesis, bands detected in this range should be assigned to the skeleton vibrations of silatranes, whereas the Si←N vibration should be observed in the region of 340–390 cm^{-1}. The main arguments in favor of the latter were obtained from study of the ^{15}N/^{14}N isotopic effect in the IR spectra of silatranes[207].

Since then a number of new results on IR and Raman spectra of silatranes and computation of the force field of their vibrations were published[33,226−232]. Most of them, together with results of frequency analyses based on semiempirical calculations[191,192], show that no pure Si←N vibration frequency exists. Thus, calculations of vibration spectra of 1-hydrosilatrane revealed that at least two completely symmetric vibrations of silatrane molecules should include the Si←N vibration. One of these vibrations is predicted to be at 590 cm^{-1} and to combine with the δ(SiOC)[230] or ν(CCN)[228] vibrations. Another should be displayed in a lower frequency region, combining with δ(CCO)[230] at 450 cm^{-1} or ν(SiO)[228] at 350 cm^{-1}. By normal coordinate analysis for A$_1$-type vibrations of 1-methylsilatrane, the f(Si←N) force constant was estimated to be 0.70 × 10^{-6} cm^{-2} based on Billes' tensor method. A band at 348 cm^{-1} was attributed to the combined vibration of the silatrane cage including 35% ν(Si←N) + 27% ν(C−C) + 20% ν(C−O) + 6% ν(O−C−C)[231]. The calculated ν(Si←N) values were found to be nearly the same for both the gas-phase and solid-state geometries of 1-methylsilatrane. This agrees with an experimentally observed indifference of the corresponding Raman band to melting or dissolving of the crystal, which should be accompanied by lengthening of the Si←N bond[227]. At the same time, in a series of silatranes the Raman ν(Si←N) wavenumbers depend on the substituent X changing from 348 cm^{-1} (X = H) to 398 cm^{-1} (X = F) as shown below[227].

X:	H	Me	H$_2$N(CH$_2$)$_3$	EtO	F
ν(Si←N), cm^{-1}:	348	354	354	366	398

It was shown that introduction of methyl groups in the 3, 7, 10 positions of silatranes shifts the band of skeleton vibration from 560−590 cm^{-1} to 440 cm^{-1} [229].

The bathochromic shift of the ν(Si−H), ν_{as}(Si−C) and ν(Si−Hal) frequencies of 1-hydro-, 1-alkyl- and 1-halosilatranes as compared with those in the IR spectra of triethoxysilanes indicates a weakening of the X−Si bond in silatranes[2,13,69].

A striking sensitivity of silatranes to the physical environment was first detected by the observations of a very strong effect of the phase and solvent on the ν(Si−H) value in the spectra of 1-hydrosilatrane[69,233a] and its 3,7,10-trimethyl and 3,7,10-tris(trifluoromethyl) derivatives[233a]. For example, the ν(Si−H) of solid 1-hydrosilatrane is higher by 83 cm^{-1} than in its solution in CCl$_4$. It was suggested[233a] that a decrease in the ν(Si−H) frequency upon dissolving these compounds in solvent of decreasing polarity is provided by the weakening of the Si←N bond. Subsequent electron diffraction investigation of the gas-phase geometry of 1-methyl-[169] and 1-fluorosilatranes[170] and multinuclear NMR studies of the influence of the physical environment on silatrane structure (see the next section) corroborated this conclusion. Interestingly, a large decrease (of about 80 cm^{-1}) in the ν(Si−H) frequency of C-substituted 1-hydrosilatranes in CCl$_4$ was recently found[233b] under a pressure of 1 kbar.

Spectroscopic studies of hydrogen bond formation between silatranes and such proton donors as phenol and methanol revealed enhanced basicity of silatranes relative to the corresponding organyltriethoxysilanes[150,216,217,221,234]. Analysis of the solvent effects on the spectroscopic basicity of these compounds showed that the oxygen atoms (equatorial in 1-organylsilatranes and axial in 1-alkoxysilatranes) rather than the nitrogen are the proton acceptor centers[216,217,221,234]. A higher basicity of these atoms in silatranes and also in O-trimethylsilylated 2-aminoalkanols was explained by a peculiarity of through-bond interactions in compounds containing the Si−O−C−C−N fragment[216,217]. Recently it was shown, however, that the origin of the observed difference in basicity of silatranes and triethoxysilanes could be due to a difference between the anomeric effects in the molecules[198].

9. NMR spectra

The results of the NMR studies of silatranes gave one of the first unambiguous pieces of evidence for the existence of the transannular Si←N bonding in silatrane solutions[2,12,13]. Moreover, extensive NMR investigations of silatranes revealed a number of reliable spectral indications for the silicon pentacoordination and earlier unknown spectral and structural regularities which appeared to be typical for most organic derivatives of the TBP silicon[20,28].

a. ^{1}H NMR spectra. The NCH$_2$ and OCH$_2$ protons of the three identical cage rings of silatranes give deceptively simple AA′MM′ spectra of two triplets[13,20,235−237]. Chemical equivalence of the geminal protons of both the N- and O-methylene groups is observed even at a very low temperature. This indicates a high flexibility of the unsubstituted silatrane skeleton and a fast rate (on the NMR time scale) of the degenerate transitions between enantiomeric conformers (equation 46). There are two viewpoints concerning the mechanism of epimerization of these conformers. The first suggests a synchronous conversion of all the three rings of the silatrane cage[238], whereas the second proposes their non-correlated successive flips[239].

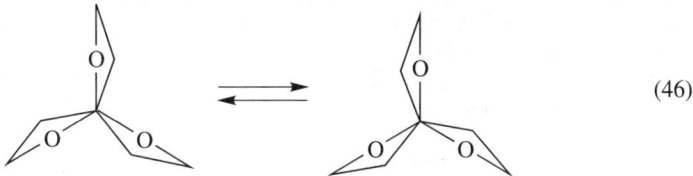

(46)

The ^3J(HH) coupling constants are within the narrow range of 5.75−5.97 Hz and tend to increase on increasing the value of Taft polar substituent constants[236]. This was explained[236] by a decrease in the dihedral angles HCCH on shortening the Si←N bond. Precise determination of ^3J(HH) in the skeleton of some silatranes[237] made it possible to estimate the NCCO torsion angles, which appeared to be comparable to the solid-state angles. This means that in solution the silatrane molecules have a Si←N bond and a conformation similar to that in the solid state.

An increased preference for a *trans* location of the vicinal oxygens in the 2-methoxyethoxy fragment of 1-(2′-methoxyethoxy)silatrane as compared with tetrakis(2-methoxyethoxy)silane was found[237] by analysis of their vicinal ^{1}H−^{1}H coupling constants using a modified Karplus equation. Enhanced electrostatic repulsion between these heteroatoms reflects the electron donation from nitrogen via silicon to the axial ligand in the former compound.

^{1}H NMR studies of 3-methylsilatranes[238], 4-organylsilatranes[61] and diastereomeric 3,7-dimethyl- and 3,7,10-trimethylsilatranes[238,239] revealed conformational rigidity of their substituted five-membered rings and the equatorial position of the C-substituent. It was shown that for most 3,7-dimethyl- and 3,7,10-trimethylsilatranes the diastereomer distributions are nearly statistical, indicating a kinetic control of the reactions leading to the formation of the silatrane cage[238].

The X−Si←N bonding and the corresponding transfer of electron density in silatranes are well reflected in the values of their ^{1}H NMR chemical shifts[13,20,235,236]. Thus, the shielding of the substituent X protons in 1-organylsilatranes is higher than in the model tetracoordinate compounds, organyltriethoxysilanes XSi(OEt)$_3$, whereas the NCH$_2$ protons in silatranes are less shielded than in triethanolamine and its tris-TMS ether. A change in the nature of the substituent X has approximately the same influence on the chemical shifts of both OCH$_2$ and NCH$_2$ groups of most silatranes (Table 2). In onium salts of 1-aminomethyl- and 1-organylthiomethylsilatranes the substituent X strongly affects the OCH$_2$ proton chemical shifts. For 1-alkylsilatranes, their 1'- and 3'-functionally substituted derivatives and 1-alkenylsilatranes, the chemical shift of the cage protons correlate with the lengths of the Si←N bond (in crystals) and the inductive substituent constants $\sigma*$[235]. Within a wider set of silatranes including 1-aryl-, 1-organoxy- and 1-halo derivatives, the δ values are correlated by a dual-substituent parameter equation, using the polar and resonance (σ_I and σ_R^0 or F and R) constants of the substituent X[235].

Due to strong intermolecular dipole–dipole interactions, the ^{1}H NMR chemical shifts of silatranes vary with concentration and solvent[235,240]. For example, the NCH$_2$ protons are more shielded (by 0.8–1.2 ppm) in benzene than in CDCl$_3$ solution. The shifts induced by the aromatic solvent are linearly correlated with the dipole moments of the silatranes[240].

Multiple quantum NMR study found an increase in the $^3J(^{29}$Si$-$O$-$C$-^1$H) coupling constants in the cage moiety of silatranes as compared with those in triethoxysilanes XSi(OEt)$_3$[241]. In contrast, the ^{29}Si$-^1$H coupling across one[242,243], two[241] or three[241] bonds in the axial fragment of 1-hydro-, 1-organyl- and 1-ethoxysilatranes appeared to be lower than those in the model compounds. The $^1J(^{13}$C$-^1$H) coupling constant in 1-methylsilatrane (115.6 Hz) is also lower than in MeSi(OEt)$_3$ (118.9 Hz)[243]. These data reflect the corresponding changes in the bond order and s-character of silicon bonds upon variation of silicon coordination numbers. Very small influence of the substituent X on the $^1J(^{13}$C$-^1$H)[236,244], $^2J(^{15}$N$-^1$H)[245] and $^3J(^{15}$N$-^1$H)[245] couplings in the cage moiety were found for silatranes.

b. ^{13}C NMR spectra. The ^{13}C chemical shifts for the silatrane cage are practically insensitive to the nature of substituent X and are within a narrow range of δ 51–52 ppm for NCH$_2$ and 57–58 ppm for the OCH$_2$ carbons[20,183,236,246,247]. The NCH$_2$ carbons of silatranes are more shielded than those in THEA (δ 57.2), probably due to a strong 1,5-interaction between the hydrogen atoms in the N(CH$_2$)$_3$ silatrane moiety and the nearly ammonium character of its nitrogen[20,236]. Despite the electron transfer via the X$-$Si←N bonding, the α-^{13}C chemical shift of the organyl group of 1-organylsilatranes is

TABLE 2. ^{1}H, ^{15}N and ^{29}Si NMR chemical shifts of silatranes (1) in CDCl$_3$ solutions (sol) and in powders (cryst)[235,255,256,260,263,266]

Compd	X	δ(NCH$_2$)	δ(OCH$_2$)	δ_{Si}(sol)	δ_{Si}(cryst)	δ_N(sol)	δ_N(cryst)
49	H	3.83	2.87	−83.6	−85.8[b]	−354.4	−351.6
2	Me	3.78	2.81	−64.4	−70.8	−359.4	−355.6
16	ClCH$_2$	3.86	2.89	−79.9	−81.9	−354.1	−351.9
19	Cl$_2$CH	3.93	2.96	−87.6	−89.4	−351.8	−349.2
20	Vinyl	3.83	2.85	−81.1	−83.4[b]	−356.8	−353.9[b]
21	Ph	3.91	2.92	−80.2	−82.9	−356.3	−353.6
30	EtO	3.84	2.84	−94.7	−96.0	−353.3	−350.1
37	F	3.91	2.95	−100.5	−101.5	−349.3	−348.3[b]
38	Cl	3.97	3.01	−85.8[a]	−85.2[b]	−348.8	−348.1[b]
50	Br	4.01	3.04	−88.8		−348.0	
51	I	4.04	3.08	−98.6		−346.9	

[a]Datum taken from Reference 236.
[b]Data taken from Reference 254.

24. Silatranes and their tricyclic analogs

downfield relative to that of the corresponding organyltriethoxysilanes[20,183,236,242,246,247]. Perhaps it results from a weakening of the C_α–Si bond in silatranes due to reduction in the average excitation energy of the occupied molecular orbitals of the α-carbon atom. Within Karplus–Pople theory[248], an inverse value of this energy term together with charge characteristics govern the local paramagnetic term of the ^{13}C shielding constant. There is a linear correlation between the α-^{13}C chemical shifts of 1-alkylsilatranes with those of the terminal carbon of related alkanes[236], but the changes in this chemical shift induced by the silatranyl group are not constant and they increase with the strength of the Si←N bond.

A strong electron-releasing effect of the silatranyl group is displayed in the upfield shift of the β- and p-^{13}C signals of the vinyl, ethynyl, and phenyl groups of silatranes relative to those of the parent triethoxysilanes[244,246,248]. By use of the linear correlations between the ^{13}C chemical shifts of the aromatic carbons and the σ_I, σ_R, σ_R^0, σ_R^+ and σ_R^- substituent constants in mono-substituted benzenes, the values of these constants for the silatranyl group were determined from the NMR data for 1-phenylsilatrane[249]. The nature of the substituent R (Me, ClCH$_2$, Vinyl and Ph) at C3 in 1-phenyl-3-substituted silatranes does not affect the ^{13}C chemical shifts of the phenyl group[250].

^{13}C NMR spectroscopy was used for structural studies of silatranyl derivatives of some monosaccharides[251], silatranylmethyl esters of substituted benzoic acids[252], diastereomeric 3,7,10-trimethyl-substituted 1-chloromethylsilatranes[253], several 1-organyl-4-ethylsilatranes[59] and 1-organyl-4,4-dimethylsilatranes[62].

In line with these solution-phase ^{13}C NMR results, the isotropic solid-state ^{13}C NMR chemical shifts of the cage carbons of silatranes appeared to be nearly independent on the Si←N bond length[254].

The absolute values of the $^1J(^{29}\text{Si}-^{13}\text{C})$ coupling constants for silatranes are greater than for the corresponding triethoxysilanes[124,236,242,244]. The $^1J(^{15}\text{N}-^{13}\text{C})$ coupling in silatranes decreases on increasing the electron-withdrawing ability of the substituent X and the polarity of the medium[124].

c. ^{29}Si NMR spectra. Within a series of silatranes, the influence of the substituent X on the solution ^{29}Si NMR chemical shifts is nearly the same as in triethoxysilanes[236,255] (Table 2). Only the enhanced ^{29}Si shielding constants of the former compounds by 11–25 ppm clarifies the pentacoordination state of their silicon atom[236,242,246,255,256]. However, the difference in the ^{29}Si chemical shifts between a silatrane and related triethoxysilane ($\Delta\delta_{\text{Si}}$) is not a direct measure of the Si←N bond strength in a silatrane[255]. Thus, despite a shorter Si←N bond, the absolute value of $\Delta\delta_{\text{Si}}$ for 1-fluorosilatrane is greater than for 1-methylsilatrane[255].

Straightforward interpretation of these peculiarities of the ^{29}Si NMR of silatranes is as yet difficult. It was suggested that a higher positive charge at the silicon in silatranes than in triethoxysilanes, together with the well known chemical shift–effective charge relationship could explain the upfield shift in the ^{29}Si NMR spectra of silatranes[212]. On the other hand, use of the Flygare approach[257] to the calculation of the diamagnetic and paramagnetic molecular terms in the average magnetic shielding of the heavy nuclei revealed a diamagnetic origin of the high-field shift in the silatrane spectra[258]. The calculated difference in the diamagnetic components of the ^{29}Si nuclear shielding constant between silatranes and the related triethoxysilanes appears to be practically insensitive to the nature of the substituent X. The difference in the corresponding paramagnetic terms increases with the strengthening of the Si←N bond in silatranes. These regularities are responsible for the observed trends in the variation of the $\Delta\delta_{\text{Si}}$. It should be added that a recent study of solid-state NMR spectra of silatranes found a very strong influence of the substituent X and the Si←N bonding on the principal elements of the ^{29}Si chemical shift powder patterns of Si-substituted silatranes[254]. Unfortunately, the combination of these effects is too difficult to be easily interpreted.

The ^{29}Si chemical shifts of silatranes depend on whether it is in the crystal or in solution, on the medium and the temperature[243,259,260]. The greatest ^{29}Si shielding is observed in powder silatrane patterns[254,260]. Dissolution of silatranes and decrease in the polarity, polarizability and acidity of the solvent, as well as heating of a solute result in a downfield shift of their ^{29}Si signal[243,259,260]. It indicates once more that the Si←N bond is very deformable and sensitive to the influence of the physical environment[188]. The weaker the Si←N bond in a silatrane crystal, the higher the solvent and temperature effects on the ^{29}Si chemical shift[243]. That is why the latter effects were proposed to be used for a fast estimation of the related strength of the Si←N bond within a series of these compounds[261]. Despite its simplicity, this method proved to be rather reliable and after some modifications it was successfully applied to numerous (N—Si)— and (O—Si)chelate compounds[28,165,262].

3,7,10-Substitution induces small and nonadditive changes in the ^{29}Si chemical shifts of silatranes. High-field shifts of 1–2.5 ppm were observed for the 3-Me, 3-ClCH$_2$, 3-Vinyl and 3-Ph substituted silatranes[57,238,262] but downfield shifts of 2.7 ppm were found for the 3-CF$_3$ derivatives[57]. In the spectra of diastereomeric 3,7-dimethyl- and 3,7,10-trimethylsilatranes, symmetric isomers show more shielded signals than asymmetric isomers[238], due to a stronger transannular bonding in the former isomers.

d. *^{15}N NMR spectra.* The ^{15}N chemical shifts (Table 2) were shown to be a direct measure of the transannular interaction in Si-substituted silatranes. Indeed, their solid-state[254,263] and gas-phase[263] ^{15}N NMR chemical shifts correlate linearly with the Si←N bond lengths determined in the corresponding phase by the X-ray or electron diffraction method (Figure 1). Changes in the isotropic ^{15}N chemical shift are primarily due to changes in the value of the $\delta_\perp$ diagonal element of the ^{15}N chemical shift tensor for powder

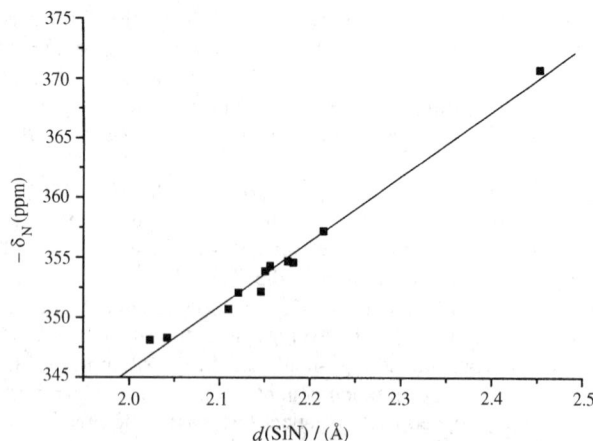

FIGURE 1. Correlation between the ^{15}N chemical shift and the Si←N bond length of powder and gaseous Si-substituted silatranes

patterns[254]. This element, which from local symmetry considerations lies in a plane perpendicular to the Si—N internuclear axis, increases on strengthening the Si←N bond. The observation agrees with the prediction based on Karplus-Pople theory of chemical shifts[248] and the simple 3c4e model of the X—Si← N bonding in silatranes[28,174,254].

Solution-phase ^{15}N NMR chemical shifts which are intermediate between those in the solid state and the gas phase depend on both the substituent X and the solvent. For a series of patterns in the same medium, the ^{15}N chemical shifts of most silatranes vary linearly with the σ^* polar constants of the substituent X[245,264,265]. 1-Halosilatranes are the exception and their ^{15}N shielding constants do not decrease on increasing the $-I$ effect of the halogen but on increasing its atomic number[266]. This is in agreement with the X-ray detection of a shorter Si←N bond in 1-chlorosilatrane[155] than in 1-fluorosilatrane[154].

The solvent effect on the δ_N values of silatrane was shown[243,259,263,267] to be described by equations 47 and 48, which reflect linear solvation energy relationships.

$$\delta_N = A + yY + pP + eE \qquad (47)$$

$$\delta_N = A + s(\pi^* - d\delta) + a\alpha \qquad (48)$$

In these equations the Palm-Koppel[268] or Taft-Kamlet[269] solvatochromic parameters are a measure of the polarity (Y or π^*), the polarizability (P or δ) and the acidity (E or α) of a solvent. The larger these solvent properties, the lower the ^{15}N shielding of the silatrane molecules. Interestingly, in the polar DMSO and in H$_2$O solutions the δ_N and δ_{Si} values are very close to those in the powder patterns[243,260,263]. On the other hand, the δ_N value of -370.6 (± 1.8) predicted for 1-methylsilatrane in the gas phase by equation 48 is the same as that observed experimentally (-370.7)[251]. A change of the solvent and the temperature affects the ^{15}N and ^{29}Si shielding constants in the opposite directions. Linear correlations between the corresponding changes in the δ_N and δ_{Si} values were found for some silatranes[243,259,263]. These indicate the same origin of the solvent effect on the shielding of the donor and acceptor centers in silatranes. It is a change in the Si←N bonding and hence in the corresponding structure reorganization of the silatrane under a variation of physical environments. Solution-phase NMR data imply bond distances which are intermediate between those in the solid state and the gas phase[263]. The amazing softness of the Si←N bond was rationalized by theoretical analysis and semiempirical[188a] and *ab initio* computations[194] of the solvent effect on the silatrane geometry.

According to the Onsager model[188b], the dipole of the neutral spherical cavity solute embedded in a continuum dielectric induces a dipole within the dielectric. The induced dipole, in turns, changes the dipole of the solute and then both dipoles are iterated to self-consistency. The energy of the interaction E_{solv} is represented by equation 49

$$E_{solv} = -\tfrac{1}{2} R \cdot \mu \qquad (49)$$

where the induced dipole R depends on the dielectric constant of the solvent (ϵ), the radius of the cavity containing the solute (a_0), the gas-phase dipole (μ_0) of the solute and its polarizability (α) according to equation 50:

$$R = g\mu, \quad g = 2(\epsilon - 1)/(2\epsilon + 1)a_0^3, \quad \mu = \mu_0(1 + \alpha g) \qquad (50)$$

It is clear that for a given cavity size the energy of solvent-solute interaction will increase on increasing the solvent dielectric constant and the solute dipole. In the case of silatranes, the dipole moment increases when the Si←N bond length decreases and the X—Si bond length increases. Hence, enlarging the dipole under the influence of a solvent will result in a corresponding shortening of the easily compressed Si←N bond. As a result, increasing the solute dipole will increase the interaction with the solvent, providing in turn an additional driving force for a further decrease in the Si−N distance.

Indeed, theoretical analysis (using the harmonic approximation) of the potential function of the deformation of the hypervalent X−Si←N bond of silatranes in solution leads[188a] to equations 51 and 52

$$\Delta d_{SiN} = -2\mu_e(1+\alpha R)^2(\partial\mu_0/\partial d_{SiN} \cdot k_{SiN}^{-1} - \partial\mu_0/\partial d_{XSi} \cdot k_{XSiN} \cdot k_{XSi}^{-1} \cdot k_{SiN}^{-1}) \quad (51)$$

$$\Delta d_{XSi} = -2\mu_e(1+\alpha R)^2(\partial\mu_0/\partial d_{XSi} \cdot k_{XSi}^{-1} - \partial\mu_0/\partial d_{SiN} \cdot k_{XSiN} \cdot k_{XSi}^{-1} \cdot k_{SiN}^{-1}) \quad (52)$$

which reflect the changes in the length of the Si←N and X−Si bonds (Δd_i) in going of these compounds from the gas phase to the solution. Here μ_e is a dipole moment of the isolated molecule at the equilibrium values of the X−Si←N bond and k_i is the force constant of the i-th bond. It could be seen that the absolute Δd_i values should increase on increasing solvent polarity and the sensitivity of the μ_0 to the hypervalent bond deformation. The sign of the Δd_i is determined by the sign of the $\partial\mu_0/\partial d_i$ derivatives. Semiempirical and *ab initio* calculations on silatranes showed that the latter terms are negative for the Si←N bond and positive for the X−Si bonds. Thus, in polar medium, silatranes should have a shorter Si←N bond and a longer X−Si bond than those in a less polar solvent or in the gas phase. The same regularities should be also observed on cooling the solution due to a known increase in the solvent polarity at a lower temperature.

These results enabled one to employ the solution-phase ^{15}N NMR chemical shifts for estimating of the length of the Si←N bond in solute silatrane patterns[28,263,266,270]. This approach made use of equation 53, which was derived from the corresponding solid-state and gas-phase data[263]:

$$d(SiN) = -4.30 - 1.82 \times 10^{-2}\delta_N \quad (53)$$

Table 3 illustrates some examples of such estimations taken from References 28, 263 and 264.

In line with the hypervalent bonding model[174], these data show that 1-iodosilatrane should have one of the shortest Si←N bonds and hence a very weak I−Si bond. This explains why this compound possesses high reactivity as an electrophile (Section II.C).

TABLE 3. Estimated values of the Si←N bond length in solute silatrane patterns

X in **1**	Solvent	$d(SiN)$ (Å)	X in **1**	Solvent	$d(SiN)$ (Å)
Me	(crystal)	2.17[a]	ClCH$_2$	DMSO	2.13
Me	DMSO	2.19	ClCH$_2$	C$_6$H$_5$Cl	2.18
Me	MeNO$_2$	2.21	Cl$_2$CH	DMSO	2.10
Me	CH$_2$Cl$_2$	2.23	Cl$_2$CH	C$_6$H$_5$Cl	2.13
Me	C$_6$H$_5$Cl	2.27	MeO	DMSO	2.11
Me	CCl$_4$	2.31	MeO	C$_6$H$_5$Cl	2.13
Me	(vapour)	2.45[b]	F	CH$_2$Cl$_2$	2.06
H	(crystal)	2.11	Cl	CH$_2$Cl$_2$	2.05
H	DMSO	2.14	Br	CH$_2$Cl$_2$	2.03
H	C$_6$H$_5$Cl	2.19	I	CH$_2$Cl$_2$	2.01

[a] X-ray data[137]
[b] ED data[169].

The $^1J(^{29}Si-^{15}N)$ spin–spin coupling constants in silatranes are 0.2–3.4 Hz and increase with the increased strength of the Si←N bond caused by the substituent X and by solvent effects[124,245,270–272]. However, the nonlinear character of the dependence of $^1J(^{29}Si-^{15}N)$ on the Si←N bond length makes the use of these coupling constants as a measure of a transannular interaction in silatranes too difficult[124,270,272].

e. Other NMR studies. ^{19}F NMR chemical shifts of 1-(*p*-fluorophenyl)silatrane[273] and *m*- and *p*-fluoro-substituted 1-benzylsilatrane[54] were used for estimation of the inductive and resonance effects of silatranyl and silatranylmethyl groups. Observation of a lower $^1J(^{19}F-^{29}Si)$ coupling in 1-fluoro-3,7,10-trimethylsilatrane (131.2 Hz) than in the model FSi(OEt)$_3$ (199.1 Hz) was one of the first pieces of evidence for a reduction in the bond order in axial moieties of silatranes[242]. Very unusual long-range coupling, formally $^8J(^{19}F-^{19}F)$, was found for the asymmetrical diastereomers of 3,7-bis- and 3,7,10-tris(trifluoromethyl)silatranes[57,274]. This through-space coupling reflects a relatively close contact between the corresponding ^{19}F nuclei in these molecules.

The ^{17}O NMR chemical shifts of the equatorial oxygens in silatranes were found to be nearly the same as in related triethoxysilanes[275]. However, the signal of the axial oxygen in 1-ethoxysilatrane is downfield shifted as compared with Si(OEt)$_4$. This shift indicates an electron transfer via a hypervalent X—Si←N bond in the silatranes.

10. Electronic effects of the silatranyl group

The silatranyl group was shown to be a stronger σ-donor than the trialkoxysilyl and even trimethylsilyl groups (Table 4). The Taft inductive constant σ^* of the silatranyl group was derived for the first time from ^{35}Cl NQR frequencies of 1-dichloromethyl- and 1-(1'-chlorovinyl)silatranes by the use of linear relations between $\nu(^{35}Cl)$ and σ_X^* values for the CHCl$_2$X and H$_2$C=CClX derivatives[276]. The corresponding relationships between spectroscopic or potentiometric basicity and the σ_X^* values for a number of mono-substituted acetylenes, organonitriles and N-substituted piperidines were used to estimate the Taft inductive constants of silatranylalkyl and related triethoxysilylalkyl groups[277]. The ^{13}C and ^{19}F NMR chemical shifts of aromatic moieties in the spectra of silatranes with X = Ph and 4-FC$_6$H$_4$CH$_2$ were employed to estimate the values of inductive and resonance constants σ_I and σ_R^0 of silatranyl[249] and silatranylmethyl[54] groups. The σ_R^0 value of the silatranyl group was also determined by the methods of Katritzky from the integral intensities of the corresponding IR bands of 1-vinyl-, 1-ethynyl- and 1-phenylsilatranes[277]. The σ_p^+ value of silatranyl (−0.40) and silatranylmethyl (−0.48) groups were estimated from the frequencies of the charge-transfer band of the corresponding XC$_6$H$_5$·TCNE (TCNE = tetracyanoethylene) complexes which are determined by the ability of the group X to release electrons to the ring[54].

It was assumed that the π-acceptor effect of the silicon influences the $\nu(^{35}Cl)$ values of 1-dichloromethyl- and 1-(1'-chlorovinyl)silatranes and the spectroscopic basicity of 1-ethynylsilatrane. Hence, the value $\sigma^* = -3.49$ derived from ^{13}C NMR data is regarded to be more reliable. This value demonstrates a super electron-releasing effect of the silatranyl group. There is a good correlation between the obtained σ^* values and the IP values of the nitrogen atom of the piperidine fragment in a series of compounds C$_5$H$_{10}$N(CH$_2$)$_n$X with X = Si(OCH$_2$CH$_2$)$_3$N and Si(OEt)$_3$ for $n = 1, 3$[278].

Interestingly, the inductive effect of the N(CH$_2$CH$_2$O)$_3$Si(CH$_2$)$_n$ groups decreases upon increasing the number of methylene groups ($n = 0-3$) fivefold faster than that of the (RO)$_3$Si(CH$_2$)$_n$ group[277]. From our point of view, this means that the σ-donor ability

TABLE 4. Inductive and resonance σ constants of the silatranyl and silatranylalkyl groups

Group	σ^*	σ_I	σ_R^0	Method	Reference
N(CH$_2$CH$_2$O)$_3$Si	−1.2			^{35}Cl NQR	276
N(CH$_2$CH$_2$O)$_3$Si	−3.49^a	−0.56	0.02	^{13}C NMR	249
N(CH$_2$CHRO)$_3$Si	−1.24		0.02−0.10	IR	277
N(CH$_2$CH$_2$O)$_3$SiCH$_2$	−2.24^a	−0.36	−0.21	^{19}F NMR	273
N(CH$_2$CH$_2$O)$_3$SiCH$_2$CH$_2$	−1.48			IR	277
N(CH$_2$CH$_2$O)$_3$SiCH$_2$CH$_2$CH$_2$	−0.32			Potentiometry	277

^{a}Estimated[277] by the relationship $\sigma^* = 6.3\,\sigma_I$.

of the silatranyl moiety within N(CH$_2$CH$_2$O)$_3$Si(CH$_2$)$_n$ groups is not constant and it decreases on increasing n. The reason for this is very clear: the closer the silatranyl group is to the electronegative reaction (indicator) center, the stronger is the SiN bonding and hence the higher the electron-releasing effect of the silatranyl group.

11. Electrochemical studies

Owing to the silicon hypervalency and the cage structure, silatranes are of considerable interest from the standpoint of molecular electrochemistry. However, little work has been devoted to electrochemistry of silatranes[279−281].

The very strong electron-releasing effect of the silatranyl group causes a significant reduction in the redox potential corresponding to the reversible Fe^{II}/Fe^{III} transition of 1-silatranyl-1'-(trimethoxysilyl)ferrocene **52** ($\Delta E_{1/2} = -0.19$ V) and 1,1'-bis(silatranyl)ferrocene **53** ($\Delta E_{1/2} = -0.47$ V) with respect to that of ferrocene ($E_{1/2} = +0.40$ V)[279].

(**52**) R = Si(OMe)$_3$
(**53**) R = Si(OCH$_2$CH$_2$)$_3$N

Electrochemical oxidation of silatranes in carefully dried MeCN on a glassy carbon electrode was shown[280,281] to be a diffusional process which proceeds with one-electron transfer. At room temperature electrooxidation is reversible at scan rates below 100 mV s^{-1} and is partially reversible at higher scan rates (200, 500 mV s^{-1}) but becomes reversible at low temperatures ($< -10\,°$C). It was suggested that the primary products are cation radicals **54** of relatively low stability in which the distance between the silicon and nitrogen atom is higher than in the parent molecules (equation 54).

$$\text{XSi(OCH}_2\text{CH}_2)_3\text{N} - e^- \rightleftharpoons [\text{XSi(OCH}_2\text{CH}_2)_3\text{N}]^{+\bullet} \quad (54)$$

(**54**)

TABLE 5. Electrochemical oxidation peak potentials (E_p, V) of Si-substituted silatranes (**1**)[280,281]

X	E_p	X	E_p	X	E_p
H	1.70	Ph	1.55	3-Thienyl	1.55
Me	1.43	4-ClC$_6$H$_4$	1.60	HC≡C	1.80
Et	1.42	2-Furyl	1.45	MeO	1.53
ClCH$_2$	1.85	3-Furyl	1.60	EtO	1.53
Vinyl	1.52	2-Thienyl	1.60	Cl	no oxidation

Silatranes undergo oxidation at higher potentials than the model compounds containing ordinary nitrogen and silicon atoms. The values of the electrochemical oxidation peak potentials (E_p) of silatranes are in the range 1.42–1.85 V (Table 5) whereas the E_p values of triethanolamine and its tris-O-TMS derivative are 0.90 and 0.92 V, respectively. There are pronounced effects of the substituent X on the oxidation potentials of silatranes and good correlations between their E_p values and the NMR parameters (^{15}N chemical shifts and ^{29}Si–^{15}N spin–spin coupling constants). The values of E_p increase on increasing the strength of the Si←N bond in the silatrane. It is very significant that for 1-chlorosilatrane no oxidation peak is observed in the accessible range. These results indicate the decisive role of the Si←N bonding on the electrooxidation of silatranes.

In wet acetonitrile electrochemical oxidation of silatranes becomes a multielectron, completely irreversible process. Participation of H$_2$O molecules in the electrochemical reaction leads to a product of a total hydrolysis of the silatrane molecule, i.e. to Si(OH)$_4$[281].

12. Mass spectra

The common features of the electron-impact (EI) mass spectra of Si- and C-substituted silatranes have been reviewed[6,13]. As a rule, the intensity of the molecular ion in the spectra is very low and the fragmentation proceeds via a cleavage of the X–Si bond or one of the silatrane rings depending on the nature of the Si and C substituents[6,13,110,282–292].

The EI mass-spectroscopic fragmentation pattern was analyzed recently[285] within the well-known organic mass-spectroscopy concept of the preferential localization of a positive charge. It was assumed that if the silatrane molecules would not have the Si←N bond in the gas phase, the primary cation-radicals M$^{+\bullet}$ (**55**) with the positively charged nitrogen atom should behave as the molecular ions of usual 2-aminoalkoxysilyl compounds and be transformed predominantly to the fragment ions [M – RCHO]$^+$, **A** (Scheme 4, path *i*). Homolytic cleavages of the X–Si and Si–O bonds in the molecular ions (Scheme 4, paths *ii* and *iii*), leading to the formation of the silatranyl cation **B** and bicyclic cation **C**, should be less probable. On the other hand, if the molecular ions M$^{+\bullet}$ were to have a structure **56**, with a pentacoordinate silicon, the formation of ions **A** should become less probable as compared with the formation of cations **B** and **C** (equation 55).

In the spectra of most silatranes the [M – RCHO]$^+$ ions are not observed or their peaks have a very low intensity (<5% of the full electron current), whereas the dependence of the intensity of the silatranyl ion peak on the substituent X differs from that expected for a tetracoordinate silicon species (Table 6). Introduction of methyl or phenyl groups into the silatrane rings increases the probability of elimination of the OCHRCH$_2$ (R = Me, Ph) fragment, whereas introduction of trifluoromethyl groups stabilizes the [M – X]$^+$ ions. Hence, the regularities observed in the fragmentation of silatranes are consistent with the existence of the Si←N bond in silatranes in the gas phase and with the stability of this bond to electron impact[285]. Interestingly, the analysis of the fragmentation products of

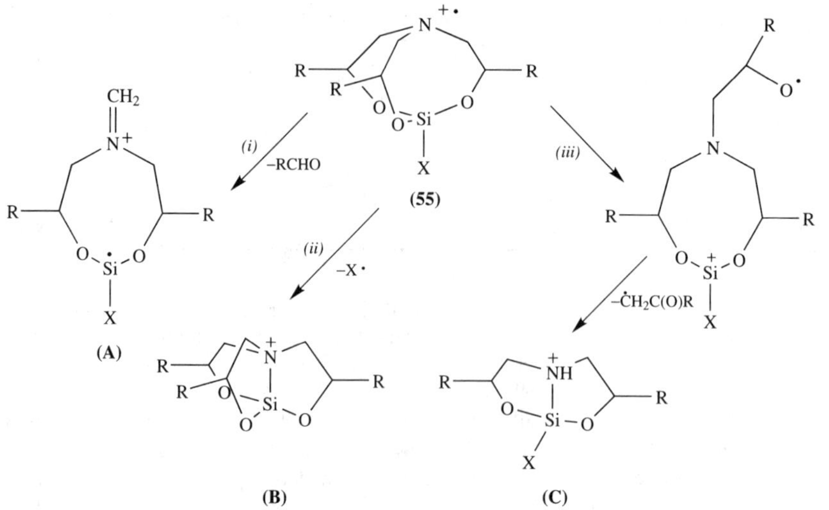

SCHEME 4

germatranes has demonstrated the absence of the Ge←N transannular bond in the gas phase[292].

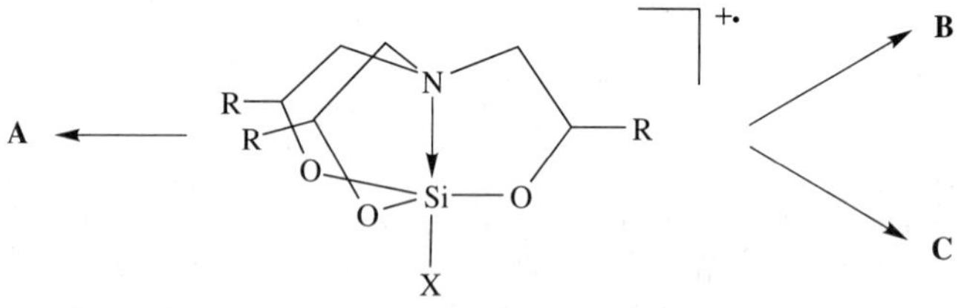

(55)

(56)

TABLE 6. Intensity (% of the total ion current) of the primary fragment ions **A**, **B** and **C** in the mass spectra of Si-substituted silatranes and 3,7,10-trimethylsilatranes[295]

X in **56**	R	A	B	C	X in **56**	R	A	B	C
HC≡C	H	3.8	2.2	11	Cl	Me	13	13	2.7
PhC≡C	H	4.5	4.1	4.1	Br	H	0.6	33	8.2
F	H	2.4	4.6	10	Vinyl	H		73	2.8
F	Me	14	0.6	3.6	Ph	H		80	
EtO	H	1	51	7.9	Ph	Me	2.5	46	6
EtO	Me	5.3	43		H	H	5.6	17	16
t-BuO	H		64	11	Me	H	2.2	19	22
PhO	H		66		Me	Me	6.1	14	5.1
4-O$_2$NC$_6$H$_4$O	H		20		Et	H	3.8	42	3.3
2-O$_2$NC$_6$H$_4$O	H		35		n-Pen	H		86	
3-O$_2$NC$_6$H$_4$O	H		66		ClCH$_2$	H	0.7	56	3
Cl	H	2.1	19	35	ClCH$_2$	Me	1.8	82	

The additional characteristic ions in the EI mass spectra of 1-substituted silatrane-4-carboxylic acids **42** were shown to arise from the loss of a carboxyl group ($[M - COOH]^+$ or, in the case of 3-methyl-substituted compounds, $[M - CO_2]^+$ peaks) and from a ring rupture ($[SiC_4H_{10}O_2N]^+$)[110].

In chemical ionization (CI) mass spectra of 1-phenyl- and 1-heteroarylsilatranes[289,210] (X = 2-furyl, 3-furyl, furfuryl, 2-thienyl and 3-thienyl) employing NH_3 as the carrier gas, the main peak corresponds to the adduct ion $[(M - X)NH_3]^+$.

Due to the Si←N interaction in silatranes, the protonated molecular ions MH^+ are usually less abundant than the molecular ions $M^{+\bullet}$. Further decomposition of both types of ions involves a loss of the substituent X and a subsequent adduct ion formation.

The fast atom bombardment (FAB) mass spectra of 1-organylsilatranes are also characterized by the most intensive $[M - X]^+$ ion peaks[286,288-291]. As a rule, the intensity of their molecular ion peaks is weak, whereas the content of the quasi-molecular MH^+ ions varies from large for 1-alkyl (and 1-alkoxy) derivatives to small for 1-phenyl- and 1-heteroarylsilatranes. An interesting peculiarity of Si- and C-substituted silatranes is the formation of the $[M - H]^+$ ions which are rather rare in FAB spectra[289-291]. In FAB spectra of some silatranes intense peaks of the protonated molecular ion of triethanolamine, $(HOCH_2CH_2)_3NH^+$, were observed. This was ascribed to the alleged reaction of silatrane hydrolysis, caused by the argon atom bombardment[286,288-291].

For silatranes containing alkyl and aryl substituents at the skeleton carbons, additional fragmentation paths were found under FAB. They lead to formation of $[MH - OCHR'CH_2]^+$ and $[MH - OCH_2CR_2]^+$ ions as products of side chain elimination from the MH^+ ions. An α-cleavage with elimination of benzaldehyde molecule from the molecular ions was observed in the spectra of 1-methyl- and 1-benzyl-3,7-diphenylsilatranes[291].

Characteristic features of the FAB spectra of silatranes are also maintained in the spectra of 1-phenyl-2-azasilatrane and 1-phenyl-3,7-dimethyl-10,11-benzosilatrane[291]. However, in the former compound only the nitrogen-containing half cycle is destroyed to form ions $[MH - CH_2NH]^+$ with *m/z* 222 and $[MH - C_2H_2NH]^+$ with *m/z* 208.

C. Chemical Properties

1. Introduction

Both the hypervalent bonding of the silicon and the cage skeleton structure provide some special characteristics to the chemical behavior of silatranes. First, in contrast with four-coordinate analogs and many pentacoordinate organosilicon compounds, silatranes display low activity in nucleophilic substitution reactions at the silicon. The *in*-oriented nitrogen atom appears to be inert toward electrophiles such as MeI[13] and tetracyanoethylene[293]. The strong electron-releasing effect combined with the large bulk of the silatranyl group affect the reactivity of the exocyclic groups of these compounds.

2. Hydrolysis

Owing to the transannular bonding, silatranes are relatively stable to atmospheric moisture, and their hydrolysis and solvolysis are considerably slower than those of the corresponding trialkoxysilanes[13]. Compounds with a relatively short Si←N bond, e.g. 1-fluoro- and 1-chlorosilatranes, can be recrystallized from alcoholic solvents[69], whereas 1-isothiocyanatosilatrane is recovered unchanged from refluxing methanol or ethanol even after 24 h[68]. Some silatranes were found to remain unchanged even in the presence of 0.1 N $HClO_4$ in glacial acetic acid for a long time[69].

Kinetic studies show that hydrolysis of 1-organyl- and 1-alkoxysilatranes in neutral aqueous solutions is a first-order reaction catalyzed by the formed tris(2-hydroxyalkyl)amine[13,294]. As a rule, electron release and steric effects of the substituent X hinder the reaction. However, the hydrolytic stability of 1-methylsilatrane is just below that of 1-chloromethylsilatrane[294]. Successive introduction of methyl groups into the 3, 7 and 10 sites of the silatrane skeleton[13,294] and substitution with ethyl group on C-4[59] retard sharply the hydrolysis rate. It was proposed[294] that nucleophilic attack at silicon by water proceeds via formation of the four-centered intermediate **57** (equation 56).

(57)   (56)

The S_N2-Si mechanism of neutral hydrolysis and the parallel operation of the S_N2-Si and S_N1-Si mechanisms in base-catalyzed hydrolysis of 1-aryloxysilatranes in aqueous solutions and in media of different polarity were proposed[295–297]. The contribution of the S_N1-Si pathway in the latter process was presumed to vary from 0 to 60%, depending on the conditions[297].

The hydrolysis rate of 1-organylsilatranes in acidic medium is much higher than in neutral media[13]. The reaction was found to be of first order in both silatrane and acid[13,298]. For 1-arylsilatranes, electron-releasing substituents in the aromatic ring increase, and electron-withdrawing groups decrease the hydrolysis rate. On the basis of these data and the study of the solvent isotope effect, it was suggested that the rate-determining step involves protonation of the nitrogen with concerted breaking of the transannular bond and a subsequent rapid hydrolytic cleavage of the Si—O bonds[298]. However, a more recent investigation of the acid-catalyzed hydrolysis of 1-aryloxysilatranes assumed that a cleavage of the protonated Si—O bond is the rate-determining stage[295].

Substitution of the cage with the methyl groups increases the solvolytic stability of the silatranes[13,58,298]. The symmetric isomer **40** of 1-phenyl-3,7,10-trimethylsilatrane shows little reaction in acetic acid containing 5% water after 90 h, whereas the half-life of the asymmetric isomer **40a** is about 2–3 h. The half-live of 1-phenyl-3,3,7,7,10,10-hexamethylsilatrane **58** bearing the sterically hindered oxygen atoms is at least one year[58]!

The difference between the solvolysis rate of diastereomers **40** and **40a** made it possible to purify and isolate the former isomer by partial hydrolysis of their mixture[58b]. A similar approach allowed earlier[24] the separation and purification (up to 99.9%) of both isomers of some 1-organyl-3,7,10-trimethylsilatranes. This was performed by partial

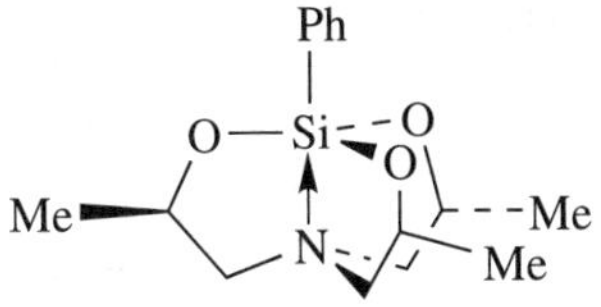

(40) 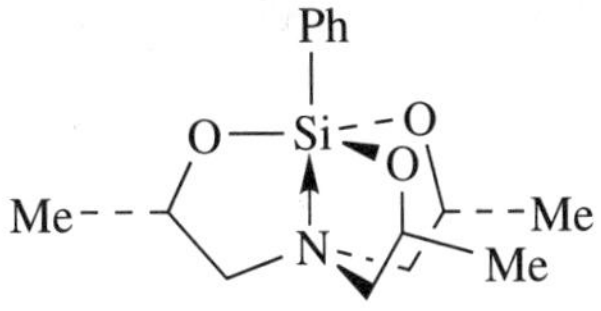   **(40a)**

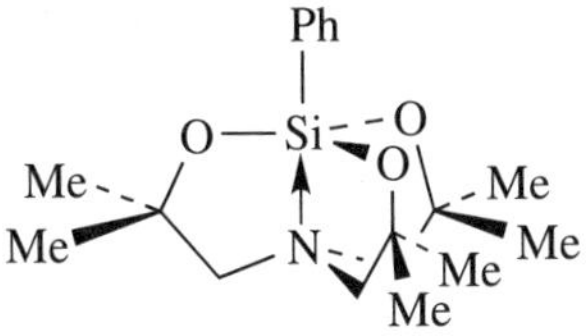

(58)

titrations of their benzene solutions by a saturated solution of dry HCl in C_6H_6 with subsequent isolation of the symmetric and asymmetric isomers from the filtrate and precipitate, respectively.

The problems concerning the ability of the silatrane's nitrogen to be protonated by acid under the solvolysis conditions in organic media are far from being resolved. There are some reports about the preparation of silatrane hydrochlorides, $XSi(OCH_2CH_2)_3N \cdot HCl$, by bubbling dry HCl through a benzene solution of silatranes[62,299] or by exposing solid 1-methylsilatrane to a stream of dry HCl[300]. However, IR spectra show that the products obtained by these methods are not the same. Moreover, multinuclear NMR study proved that two rather than one molecule of HCl are involved in the interaction with silatranes in organic media[301] (equation 57). The exchange between the reagents and the product in the reversible reaction 57 was shown to be fast on the NMR time-scale. Molecular mechanics and semiempirical study, including analysis of the map of the electrostatic potential of silatrane molecules, predicted kinetic and thermodynamic preference for a proton attack on the oxygen[302].

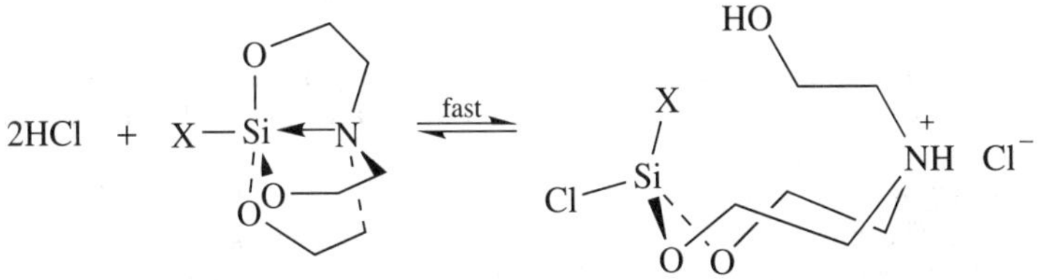

$$2HCl + X-Si \leftarrow N \quad \underset{}{\overset{fast}{\rightleftarrows}} \quad Cl-Si \cdots NH^+ \; Cl^- \quad (57)$$

Taking into account the high electrophilicity of the cage oxygen atoms in silatranes, the weakening of the Si—O bond and the strengthening of the Si←N bond under H-bonding (see Section II.B), these results suggest that protonation of the nitrogen occurs only after the Si—O bond fission. This is corroborated by the extreme resistance of 1-phenyl-3,3,7,7,10,10-hexamethylsilatrane (**58**) to solvolysis[58b].

On the other hand, IR data suggested that some silatrane-4-carboxylic acids might be present in solution as an equilibrium mixture of the acid (**42**) and the salt (**42a**) forms (equation 58). These results, combined with the stability of the protonated equatorial Si—O bond of these acids in their crystals, was regarded as favoring the primary protonation of the axial nitrogen under acidic hydrolysis of silatranes[110,111,168].

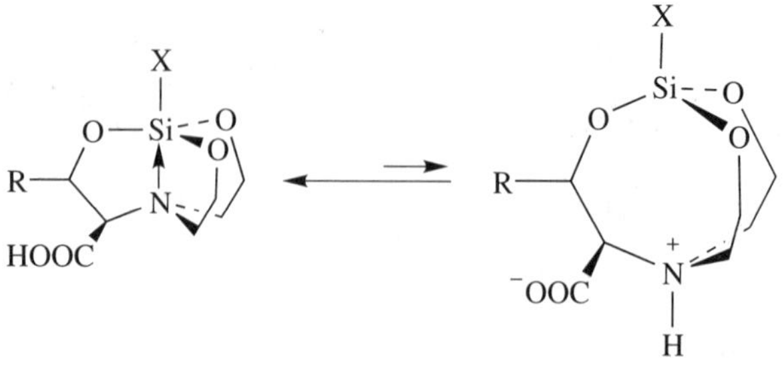

$$
\begin{array}{c}
\text{(42)} \rightleftarrows \text{(42a)}
\end{array}
\tag{58}
$$

3. 1-Hydrosilatranes

1-Hydrosilatrane (**49**) and its 3,7,10-trimethyl derivative (**49a**) can be converted to the corresponding 1-halosilatranes by treatment with halogens (equation 59)[69,129,303], anhydrous HX (X = Cl, Br) (equation 60)[69] or the appropriate *N*-halosuccinimide in a chloroform solution (equation 61)[69]. In reactions 59–61 triethylamine can be used as an acceptor of HX[303]. Attempts to prepare an iodo derivative by the procedure of reaction 61 using *N*-iodosuccinimide were unsuccessful[69].

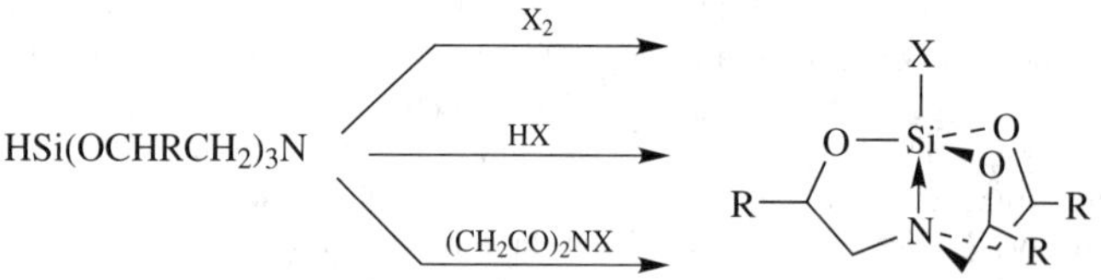

$$
\text{HSi(OCHRCH}_2)_3\text{N} \quad \xrightarrow{\begin{array}{c}X_2\\ \text{HX}\\ (\text{CH}_2\text{CO})_2\text{NX}\end{array}} \quad \text{[silatrane product]}
\tag{59, 60, 61}
$$

(**49**) R = H
(**49a**) R = Me

(**38**) X = Cl, R = H
(**38a**) X = Cl, R = Me
(**50**) X = Br, R = H
(**50a**) X = Br, R = Me

It is remarkable that tetracoordinate silanes such as Et_3SiH are not involved in related reactions. The higher reactivity of the 1-hydrosilatranes arises from the enhanced hydridic character of the Si–H bond in the silatrane[69].

1-Hydrosilatrane (**49**) reacts readily with alcohols and phenols in boiling xylene (equation 62). The process is catalyzed by sodium alkoxides or phenoxides[304]. As the acidity of the phenols decreases, the dehydrocondensation rate increases. An opposite tendency is observed for nucleophilic substitution by alkoxide ions. In this case the steric effect of the bulky alcohol plays a more important role than the electronic effect in governing the reaction rate.

$$
\mathbf{49} + \text{ROH} \longrightarrow \text{ROSi(OCH}_2\text{CH}_2)_3\text{N} + \text{H}_2
\tag{62}
$$

R = Me, *n*-Bu, *i*-Bu, *t*-Bu, Ph, 4-MeC$_6$H$_4$, 4-MeOC$_6$H$_4$, 4-ClC$_6$H$_4$, C$_6$F$_5$, etc

It should be mentioned that the reaction shown in equation 62 does not proceed in the presence of H_2PtCl_6[304]. However, the platinum catalyst is very effective in obtaining alkoxy derivatives from both tetracoordinate triorganosilanes[305,306] and silatrane having an H−Si−CH$_2$ moiety (equation 63)[49].

$$\text{MePh(H)SiCH}_2\text{Si(OCH}_2\text{CH}_2)_3\text{N} \xrightarrow{\text{ROH/H}_2\text{PtCl}_6} \text{MePh(RO)SiCH}_2\text{Si(OCH}_2\text{CH}_2)_3\text{N} \quad (63)$$

R = Me (95%), Et (90%), i-Pr (77%), t-Bu (95%)

Treatment of a twofold excess of 1-hydrosilatrane (**49**) with alkanediols results in the corresponding bis(silatranyloxy)alkanes in a quantitative yield (equation 64)[307].

$$2\text{HSi(OCH}_2\text{CH}_2)_3\text{N} + \text{HOROH} \xrightarrow{-2\text{H}_2} \text{N(CH}_2\text{CH}_2\text{O)}_3\text{SiOROSi(OCH}_2\text{CH}_2)_3\text{N} \quad (64)$$
(**49**)

R = ⎯(CH$_2$)$_n$⎯ (n = 2–6), ⎯CH$_2$CHMe⎯, ⎯CH$_2$CH$_2$OCH$_2$CH$_2$⎯

Reactions of **49** with carboxylic acids take place in the presence of ZnCl$_2$ to give 1-acyloxysilatranes in 40–80% yields (equation 65)[13].

Weakening of the Si−H bond favors radical reactions of **49** with a variety of organohalides and some reactive organic compounds. Thus, polyhalomethanes CH$_2$Cl$_2$, CHCl$_3$ and CHBr$_3$ are reduced by **49** in the presence of organic peroxides (equation 66)[308].

$$\textbf{49} + \text{RCOOH} \longrightarrow \text{RCOOSi(OCH}_2\text{CH}_2)_3\text{N} \quad (65)$$

R = Me, Et, Ph, 3-C$_5$H$_4$N, 2-C$_4$H$_3$O

HSi(OCH$_2$CH$_2$)$_3$N (**49**)

$\xrightarrow{\text{CH}_n\text{X}_{4-n}}$ [silatrane with X] (66)

$\xrightarrow[\text{CH}_2\text{X}_2]{\text{Ph}_3\text{CX}}$ (67)

(**38**) X = Cl
(**50**) X = Br

A similar reaction of **49** with triphenylmethyl halides in the appropriate methylene halide affords the corresponding 1-halosilatranes (equation 67). The reaction rate increases in going from chloro- to bromomethanes (k = 1.56 × 10^{-3} M^{-1} s^{-1} at 24 °C and 2.94 × 10^{-3} M^{-1} s^{-1} at 20 °C for the Cl and Br derivatives, respectively)[308].

A much lower reaction rate of 1-hydrosilatrane compared with triorganylsilanes seems to be due to the steric bulk of the silatranyl group that hinders a side attack at the silicon by the Ph$_3$C$^+$X$^-$ ion pair[308].

According to EPR data, reaction 67 proceeds via a homolytic pathway in nonpolar solvents.

The use of iodo derivatives CH$_2$I$_2$ or CHI$_3$ in reaction 66 results only in polymeric products[308]. However, 1-iodosilatrane (**51**) can be obtained in almost quantitative yield from 1-hydrosilatrane (**49**) and perfluoroalkyl iodides under UV irradiation (equation 68)[309].

$$\text{HSi(OCH}_2\text{CH}_2)_3\text{N} + \text{R}_\text{F}\text{I} \xrightarrow{h\nu} \text{ISi(OCH}_2\text{CH}_2)_3\text{N} + \text{R}_\text{F}\text{H} \quad (68)$$
(**49**) (**51**)

R$_\text{F}$ = CF$_3$, C$_3$F$_7$, C$_6$F$_{13}$

Silatrane **(49)** reduces benzyl bromide, benzoyl chloride, azoxybenzene, nitrobenzene and some carbonyl compounds under rather drastic conditions (140–250 °C); see equations 69–72[131].

$$HSi(OCH_2CH_2)_3N \quad (49)$$

$$\xrightarrow{PhCH_2Br,\ xylene,\ 44\ h} PhCH_3\ (35\%) \quad (69)$$

$$\xrightarrow{PhCOCl,\ xylene,\ 48\ h} PhCHO\ (57\%) \quad (70)$$

$$\xrightarrow{Me_2C=CHC(O)Me,\ (EtOCH_2CH_2)_2O,\ 22\ h} Me_2C=CHCH(OH)Me\ (70\%) \quad (71)$$

$$\xrightarrow{PhN(O)=NPh,\ xylene,\ 62\ h} PhN=NPh\ (28\%) + PhNHNHPh\ (7\%) \quad (72)$$

Both **49** and trialkylsilanes undergo exchange reactions with trimethylhalosilanes in the presence of quinoline to give 1-halosilatranes (equation 73)[16]. It is likely that steric hindrance is responsible for the lack of addition of 1-hydrosilatrane **(49)** to the multiple bond of monosubstituted alkenes and acetylenes in the presence of H_2PtCl_6 or under UV irradiation[129,310].

$$\mathbf{49} + Me_3SiX \longrightarrow XSi(OCH_2CH_2)_3N + Me_3SiH \quad (73)$$

$$\mathbf{(38)}\ X = Cl$$
$$\mathbf{(50)}\ X = Br$$

Silatrane **(49)** reacts with n-butyllithium and with Grignard reagents to give tri-n-butyl or triarylsilane even with a deficiency of the organometallic reagent (equation 74)[311].

$$HSi(OCH_2CH_2)_3N + RM \longrightarrow R_3SiH + R_4Si \quad (74)$$
$$\mathbf{(49)}$$
$$R = n\text{-}Bu,\ Ph$$
$$M = Li,\ MgBr$$

The initial rate-determining nucleophilic attack at the silicon involves cleavage of the silicon–oxygen bond. The reaction selectivity is dependent on the nucleophilic reagent. Thus, a mixture of n-Bu_3SiH (90%) and n-Bu_4Si (10%) is obtained when 3.2 molar equivalents of n-BuLi are used in the reaction, i.e. cleavage of the equatorial Si—O bonds occurs more readily than that of the apical Si—H bond. With PhLi, the prevailing tetrasubstitution reaction results in Ph_4Si (68% yield). At the same time, a slower reaction of **49** with a Grignard reagent leads to the formation of only R_3SiH in high yield (80–95%)[311].

4. 1-Halosilatranes

The halosilatranes are the most interesting members of the silatrane family because of their general unreactivity, except for 1-iodosilatrane. In contrast to chlorotriorganyl- and chlorotrialkoxysilanes, 1-chlorosilatrane **(38)** does not participate in nucleophilic substitution reactions at silicon. The low reactivity of **38** was explained to be a consequence of its unusual geometry. Indeed, a backside attack is forbidden due to its cage structure and a side attack is disfavored since it demands a high energy for deformation of the skeleton upon formation of the hexacoordinate intermediate.

The reaction of **38** with *n*-BuLi followed by reduction with LiAlH$_4$ gave tri-*n*-butylsilane as the major product even when an excess of *n*-BuLi was used (equation 75).

$$\text{ClSi(OCH}_2\text{CH}_2)_3\text{N} + n\text{-BuLi} \longrightarrow n\text{-Bu}_3\text{SiCl} \xrightarrow{\text{LiAlH}_4} n\text{-Bu}_3\text{SiH} \quad (75)$$
$$(\mathbf{38})$$

It was concluded that the initial substitution involved cleavage of an Si—O bond rather than the Cl—Si bond as is usually observed with chloroalkoxysilanes[311]. Since oxygen is more electronegative than chlorine, it is reasonable to suggest that reaction 75 occurs via intermediate **59**.

(**59**)   (**60**)

Weakening of the LiO—Si bond and nucleophilic attack of RCH$_2^-$ on silicon results in generating a six-coordination site. In similar reactions with tetracoordinate ClSi(OR)$_3$ compounds, any such weakening of the LiO—Si bond would be dominated by weakening of the Si—Cl link in a linear 3c4e MO system in **60**[312].

1-Fluorosilatrane (**37**) is formed by treatment of **38** with potassium fluoride in HMPA (equation 76)[16].

$$\text{ClSi(OCH}_2\text{CH}_2)_3\text{N} + \text{KF} \longrightarrow \text{FSi(OCH}_2\text{CH}_2)_3\text{N} \quad (76)$$
$$(\mathbf{38}) \qquad\qquad\qquad (\mathbf{37})$$

1-Bromosilatrane (**50**) can be involved in electrophilic displacement reactions with silver compounds[26,27] to give products **21**, **37**, **61** and **62** (equations 77–80).

BrSi(OCH$_2$CH$_2$)$_3$N (**50**)

- AgBF$_4$
- AgBPh$_4$
- AgClO$_4$, C$_6$H$_6$
- AgX, MeCN ; X = ClO$_4$, BF$_4$

(**37**) X = F	(77)
(**21**) X = Ph	(78)
(**61**) X = OClO$_3$	(79)
(**62**) X = N≡CMe$^+$	(80)

Silatrane **50** reacts with substituted benzoic acids in the presence of pyridine to afford the corresponding 1-acyloxysilatranes (equation 81)[313,314].

$$\mathbf{50} + \text{XC}_6\text{H}_4\text{COOH} \longrightarrow \text{XC}_6\text{H}_4\text{COOSi(OCH}_2\text{CH}_2)_3\text{N} \quad (81)$$
$$\text{X} = \text{4-F, 4-Cl, 4-Br, 4-I, 4-MeO, 4-O}_2\text{N, 3-F, 3-Br, 2-MeO, etc.}$$

If the precursor is 1-bromosilatrane (**50**), the reaction shown in equation 75 occurs in a similar manner[311].

The oxidation reaction of 1-bromosilatrane (**50**) or its 3,7,10-trimethyl derivative (**50a**) and tin dibromide leads to the formation of adduct/solvent complexes **63a,b** (equation 82). Acetonitrile, DMF or methanol are used as the solvent[315].

$$BrSi(OCHRCH_2)_3N + SnBr_2 \xrightarrow{solv.} Br_3SnSi(OCHRCH_2)_3N \cdot (solv)$$
$$(\mathbf{50,50a}) \qquad\qquad\qquad (\mathbf{63,63a})$$
(82)

50/63   R = H
50a/63a R = Me

1-Iodosilatrane (**51**) can behave as a powerful electrophilic reagent with respect to various classes of organic and organometallic compounds. For example, **51** can be involved in cleavage reactions of the C—O or Si—O bonds of a variety of organic and organosilicon compounds to give the corresponding silatrane derivatives as shown by equations 83–88[309,316].

ISi(OCH$_2$CH$_2$)$_3$N (**51**)

- ROR' → ROSi(OCH$_2$CH$_2$)$_3$N + R'I   (83)
- $\overline{CH_2(CH_2)_nO}$, n = 1, 3 → ICH$_2$(CH$_2$)$_n$OSi(OCH$_2$CH$_2$)$_3$N   (84)
- EtOR, R = SiMe$_3$, Si(OCH$_2$CH$_2$)$_3$N → ROSi(OCH$_2$CH$_2$)$_3$N   (85)
- (Me$_3$Si)$_2$O → Me$_3$SiOSi(OCH$_2$CH$_2$)$_3$N + Me$_3$SiI   (86)
- MeCOOEt → MeCOOSi(OCH$_2$CH$_2$)$_3$N + EtI   (87)
- (CF$_3$CO)$_2$O → CF$_3$COOSi(OCH$_2$CH$_2$)$_3$N + CF$_3$C(O)I   (88)

Compared to **51**, the reaction of THF with iodotrimethylsilane requires a more prolonged time and takes place at a higher temperature (60 °C)[317].

1-Iodosilatrane (**51**) readily adds to acetaldehyde to give 1-(1′-iodoethoxy)silatrane (**64**) (equation 89)[19].

$$\mathbf{51} + MeCH{=}O \longrightarrow MeCH(I)OSi(OCH_2CH_2)_3N$$
$$(\mathbf{64})$$
(89)

Compound **51** can be converted to 1-alkoxysilatranes by treatment with alcohols or alkali metal alkoxides[316] (equation 90).

$$\mathbf{51} + ROM \longrightarrow ROSi(OCH_2CH_2)_3N + MI$$
(90)

R = Me, M = H, Na; R = Me$_3$C, M = H, K; R = 1-adamantyl, M = H

It is remarkable that alkyl iodides and trimethylsilanol are formed by treatment of alcohols with iodotrimethylsilane[318].

The conversions described in equations 83–90 can be carried out 'in situ' or 'in statu nascendi', when 1-iodosilatrane (**51**) is prepared from 1-allylsilatrane and C$_3$F$_7$I[319].

The reaction of **51** with alkali thiolates or thiols without a HI acceptor yields silatranes having a S—Si bond (equation 91)[320].

$$\mathbf{51} + RSM \longrightarrow RSSi(OCH_2CH_2)_3N + MI$$
(91)

R = Et, M = H, Na; R = Me$_3$C, Ph, 4-MeC$_6$H$_4$, M = H

Silatrane **51** reacts with acetylenes having an electron-withdrawing substituent to give 1-alkynylsilatranes (equation 92). No reaction takes place with butylacetylene[321].

$$\mathbf{51} + RC\equiv CH \longrightarrow RC\equiv CSi(OCH_2CH_2)_3N + HI \quad (92)$$
$$R = Vinyl, Ph$$

1-Organylsilatranes or 1-chlorosilatrane are obtained from 1-iodosilatrane (**51**) and appropriate organomercury compounds in $CHCl_3$ (equation 93)[322].

$$\mathbf{51} + RR'Hg \longrightarrow RSi(OCH_2CH_2)_3N + R'HgI \quad (93)$$
$$R = R' = Et, Ph; R = Cl, R' = Et$$

This reaction allowed one to prepare the first of the spin-labeled silatranes, (1'-oxy-2',2',6',6'-tetramethyl-3',4'-dehydropiperidinyl-4')silatrane[323].

The Si−N(CS) bond of tetracoordinate silicon compounds is very sensitive towards alcoholysis[324]. By contrast, 1-isothiocyanatosilatrane (**29**) reacts with methanol only in the presence of a strong base such as biguanidine (B) to afford 1-methoxysilatrane (**65**) (equation 94)[68].

$$SCNSi(OCH_2CH_2)_3N + MeOH \xrightarrow{B, CH_2Cl_2, RT} MeOSi(OCH_2CH_2)_3N + B \cdot HNCS \quad (94)$$
$$(\mathbf{29}) \qquad\qquad\qquad\qquad (\mathbf{65})$$

The reactions of **29** with secondary amines such as piperidine and diethylamine and with Lewis acids (including hard, soft and borderline acids in Pearson's sense) afford 1 : 1 adducts[68].

5. 1-Hydroxysilatrane

1-Hydroxysilatrane (**66**) can undergo nucleophilic displacement at silicon. For example, it reacts with acetyl chloride (in the presence of Et_3N) or with acetic acid/acetic anhydride to yield unstable 1-acetoxysilatrane **67** (equation 95)[69]. 1-Trimethylsiloxysilatrane (**68**) was obtained by treatment of **66** with chlorotrimethylsilane (equation 96)[69]. The reaction of **66** with diphenylchlorophosphine in THF led to the substitution product **69** in 37% yield. When treated with sulfur the latter compound was converted to the corresponding thio derivative **70** (equation 97)[325].

(**67**) X = MeCOO	(95)
(**68**) X = Me_3SiO	(96)
(**69**) X = Ph_2PO	
(**70**) X = $Ph_2P(S)O$	(97)

6. 1-Alkoxysilatranes

1-Fluorosilatrane (**37**) or its 3,7,10-trimethyl derivative (**37a**) are formed by the reaction of the corresponding 1-ethoxysilatranes (**30, 30a**) with concentrated HF in *i*-propanol at room temperature[69,154] or with KHF_2 at 200–240 °C[16] (equation 98).

$$EtOSi(OCHRCH_2)_3N \xrightarrow{HF \text{ or } KHF_2} FSi(OCHRCH_2)_3N$$

(30, 30a)   (37, 37a)

(98)

30/37   R = H (25%)

30a/37a   R = Me (73%)

Silatrane (**30**) can be readily converted via alcoholysis or acidolysis reactions to the corresponding organoxy or acyloxy compounds. A variety of higher alcohols[53], glycols[53], phenols[326], alkanolamines[327] and carboxylic acids[69,83] can be used as appropriate hydroxylic reagents (equation 99).

$$EtOSi(OCH_2CH_2)_3N + ROH \longrightarrow ROSi(OCH_2CH_2)_3N$$

(99)

R = Alk, Ar, $Me_2NCH_2CH_2$, R′CO (R′ = Me, Ph), 2-furyl, 2-pyridinyl, 2-thienyl, etc.

Alcoholysis of 1-alkoxysilatranes proceeds more slowly than that of tetraalkoxysilanes due to the bulk of the silatranyl group. The exchange rate falls with the decreasing acidity of the hydroxylic compound, suggesting an initial protonation of the EtO oxygen. Therefore, a Lewis acid catalyst such as $Zn(OAc)_2$ is required for effective silanolysis (reaction 99, R = Ph_3Si)[69].

The high-temperature reaction of **30** with THEA leads to tris-(2-silatranyloxyethyl)amine **71** (equation 100). The latter is a very useful precursor for the preparation of 1-hydroxysilatrane (**66**)[69].

$$30 + (HOCH_2CH_2)_3N \longrightarrow N[CH_2CH_2OSi(OCH_2CH_2)_3N]_3$$

(**71**)

(100)

Silatranes (**72** and **72a**) with an exocyclic S—Si bond are formed by treatment of 1-methoxysilatrane (**65**) with compounds having an HS function (equation 101)[67].

$$MeOSi(OCH_2CH_2)_3N + RSH \xrightarrow{THF \text{ or } PhH, \Delta} RSSi(OCH_2CH_2)_3N + MeOH$$

(**65**)   (**72, 72a**)

R = HCl· H_2NCH_2CHR' (101)

(**72**) R′ = H (83%)

(**72a**) R′ = Me (67%)

7. 1-Alkyl-, 1-aralkyl- and 1-arylsilatranes

1-Alkyl- and 1-aralkylsilatranes are converted to the corresponding tetraorganylsilanes in high yields (70–92%) by treatment with an excess of *n*-BuLi (equation 102)[311].

$$RSi(OCH_2CH_2)_3N + n\text{-BuLi} \longrightarrow n\text{-Bu}_3SiR$$

(102)

R = *n*-Bu, Bz, $Ph(CH_2)_2$

1-Arylsilatranes are reduced by lithium aluminum hydride in ether to the corresponding arylsilanes (equation 103). The reaction does not take place in the case of 1-aralkylsilatranes[311].

$$RSi(OCH_2CH_2)_3N + LiAlH_4 \longrightarrow RSiH_3$$

(103)

R = Ph (32%), 1-Naph (65%), $Ph(CH_2)_2$ (0%)

The apical Si—C bond of 1-organylsilatranes is very sensitive to direct electrophilic attack. As a consequence, alkyl or aryl bromide are formed by treatment of 1-organylsilatranes or the corresponding organyltrialkoxysilanes with N-bromosuccinimide (NBS) (equation 104)[328].

$$RSi(OCH_2CH_2)_3N + NBS \longrightarrow RBr \tag{104}$$

R = Alk, Ph, 4-MeC$_6$H$_4$, 4-ClC$_6$H$_4$, 4-MeOC$_6$H$_4$

The reaction of 1-organylsilatranes with 3-chloroperbenzoic acid (m-ClPBA) in methanol or dichlorometane affords primary alcohols or phenols due to the cleavage of C—Si bond (equation 105)[328].

$$RSi(OCH_2CH_2)_3N \xrightarrow{m\text{-ClPBA}} ROH \tag{105}$$

R = Alk, Ph, 4-MeC$_6$H$_4$, 4-ClC$_6$H$_4$, 4-MeOC$_6$H$_4$

The Si—C bond in 1-organylsilatranes is easily cleaved by bromine or iodine chloride even at −50 °C (equation 106)[329]. This route is observed in CH$_2$Cl$_2$ or CHCl$_3$ as a solvent. By using diethyl ether–bromine, THF–bromine or dioxane–bromine adducts, a mixture of 1-halo- and 1-haloalkoxysilatranes is formed. For example, the reaction of 1-phenylsilatrane (**21**) with dioxane-bromine results in 1-bromosilatrane (**50**) and 1-[2′-(2″-bromoethoxy)ethoxy]silatrane (**73**) in 39% and 12% yield, respectively (equation 107)[329].

Y = Br, I, Z = Cl (106)

(**2**) X = Me, R = H   (**38**) Z = Cl, R = H
(**2a**) X = Me, R = Me   (**38a**) Z = Cl, R = Me
(**20**) X = Vin, R = H   (**50**) Z = Br, R = H
(**21**) X = Ph, R = H

$$PhSi(OCH_2CH_2)_3N \xrightarrow{\text{dioxane}\cdot Br_2} 50 + Br(CH_2)_2O(CH_2)_2OSi(OCH_2CH_2)_3N + PhBr \tag{107}$$

(**21**) (**73**)

Treatment of 1-organylsilatranes with mercury(II) salts in protic or aprotic media gives the corresponding organomercury compound and the substituted silatrane (equation 108)[132,330].

$$XSi(OCH_2CH_2)_3N + HgY_2 \longrightarrow XHgY + YSi(OCH_2CH_2)_3N \tag{108}$$

Y = Cl, I, CH$_3$COO

The relative reaction rates depend on both the steric and electronic effects of the substituent X and decrease in the order Vinyl, Ph, 4-ClC$_6$H$_4$ > Me > Et, n-Pr > ClCH$_2$, Cl$_2$CH, EtO[132]. The parent organyltrialkoxysilanes are essentially inert under similar conditions[331].

Reactions of 1-organylsilatranes with a metal fluoride[332] or heavy metal salts in aqueous solutions containing fluoride ions yield the corresponding organometallic compounds[332–334]. For example, triorganyllead fluorides are formed by treatment of 1-organylsilatranes with a mixture of lead tetraacetate/ammonium fluoride in water (equation 109)[333].

$$3XSi(OCH_2CH_2)_3N + Pb(OAc)_4 + NH_4F + 6H_2O$$
$$\downarrow$$
$$X_3PbF + 3 SiO_2 + 3[HOCH_2CH_2)_3NH]OAc + NH_4OAc \quad (109)$$

X = Me, Vinyl, Ph

8. 1-Alkenylsilatranes

1-Alkenylsilatranes can undergo a variety of useful addition reactions involving carbon-centered or heteroatom-centered species. For instance, 1-vinylsilatrane (**20**) reacts regioselectively with polyhaloalkanes to afford the corresponding terminal adducts (**74**) and (**75**) in high yields of 78–100% (equations 110 and 111)[335,336].

(**74**) X = Cl$_3$CCH$_2$CHY (110)

(**75**) X = R$_F$CH$_2$CHI (111)

R$_F$ = CF$_3$, H(CF$_2$)$_4$, etc.

The reactions occur in the presence of peroxides or under UV irradiation. Under the same conditions, 1-allylsilatrane (**76**) is converted to 1-halosilatrane arising from a thermally unstable adduct (**77a–c**) by a β-cleavage (equation 112)[319]. The reaction provides a simple and successful route to 1-halosilatranes, which are easily isolated in quantitative yields.

$$76 + RX \longrightarrow RCH_2CHXCH_2Si(OCH_2CH_2)_3N \xrightarrow{-RCH_2CH=CH_2} XSi(OCH_2CH_2)_3N$$
$$(\mathbf{77a\text{-}c})$$

(a) R = CCl$_3$, X = Cl
(b) R = CCl$_3$, X = Br   (112)
(c) R = C$_3$F$_7$, X = I

The addition of methyltrichloroacetate to **20** is initiated by benzoyl peroxide or Fe(CO)$_5$ (equation 113). In the latter case, conducting the nucleophile-assisted reaction in an aprotic solvent (HMPA or MeCN) results in a higher yield of the adduct[337].

$$20 + CCl_3COOCH_3 \xrightarrow{(PhCOO)_2} CH_3OOCCCl_2CH_2CHClSi(OCH_2CH_2)_3N \quad (113)$$
$$(\mathbf{78})$$

Silatrane (**20**) reacts with nitrile oxides to afford **79** and **80** in a ratio depending on the method for the *in situ* preparation of the initial oxides (equation 114)[338].

The photochemical thiylation of 1-vinyl- (**20**), 1-allyl- (**76**) and 1-allyl-3,7,10-trimethylsilatranes (**76a**) by alkanethiols gives the terminal adducts in high yield of 90% (equation 115)[339].

$$CH_2=CH(CH_2)_nSi(OCHRCH_2)_3N + R'SH \longrightarrow R'S(CH_2)_{n+2}Si(OCHRCH_2)_3N \quad (115)$$

$$R = H, Me, R' = Alk, n = 0, 1$$

1-Vinyl- and 1-allylsilatranes react at comparable rates, as is evident from the competitive addition of *tert*-butanethiol ($R' = t$-Bu). However, in a similar reaction 1-alkenylsilatranes display higher reactivity compared with the corresponding alkenyltrialkoxysilanes. For example, **76** is about 2.7 times more reactive than allyltrimethoxysilane. It is likely that the increase in the electron density of the C=C bond by the strong $-I$ effect of the silatranyl group facilitates the addition of alkanethiyl radicals[339].

The reaction of 1-allylsilatrane (**76**) with O,O-diethyldithiophosphoric acid in $CHCl_3$ at room temperature gave O,O-diethyl S-(silatran-1-yl)dithiophosphate (**81**) in low yield (15%) instead of the terminal adduct (equation 116)[109].

$$76 + (EtO)_2P(S)SH \xrightarrow{CHCl_3} (EtO)_2P(S)SSi(OCH_2CH_2)_3N \quad (116)$$
$$(81)$$

Addition products are obtained from 1-vinylsilatranes (**20, 20a**) and compounds with a P—H moiety, i.e. dialkyl phosphites (equation 117) and diarylphosphines (equation 118)[340].

(**82**) $X = (R'O)_2POCH_2CH_2$, $R = H$
(**82a**) $X = (R'O)_2POCH_2CH_2$, $R = Me$
(**83**) $X = Ph_2PCH_2CH_2$, $R = H$
(**83a**) $X = Ph_2PCH_2CH_2$, $R = Me$

With **20**, reaction 117 occurs more rapidly in the presence of sodium alkoxide than under UV irradiation. In the latter case, no addition takes place at all when di(*n*-propyl) phosphite is the precursor. Under similar conditions the terminal adduct is produced from **20a** and di(n-propyl) phosphite in high yield (68%) due to the rate acceleration by the methyl ring substituents.

Silatrane (**20**) displays a similar reactivity toward organolithium reagents[311] as the corresponding tetracoordinate organylsilanes[341]. Thus, addition of n-BuLi to the C=C bond of **20** occurs with a simultaneous substitution of the Si—O bonds, while t-BuLi is added to the C=C bond at $-78\,°C$ without attack on the silatrane ring (equation 119).

$$20 \xrightarrow{RLi} RCH_2CH_2Si(OCH_2CH_2)_3N + RCH_2CH_2SiR_3 \quad (119)$$

$$R = n\text{-Bu},\ t\text{-Bu} \qquad R = n\text{-Bu}$$

In contrast to **20**, no addition to the double bond occurs in the case of 1-allylsilatrane (**76**); only the cleavage product of the Si—O bonds, i.e. $CH_2=CHCH_2SiBu\text{-}n_3$, is formed[311].

1-Alkenylsilatranes can add triorganylsilanes or -germanes to give the corresponding β-adducts (equation 120)[342–344]. The reaction is catalyzed by H_2PtCl_6[342,343] or $RhAcac(CO)_2$[344]. The addition of R_3SnH proceeds in a similar manner without any catalyst[344].

$$CH_2=CH(CH_2)_nSi(OCH_2CH_2)_3N + R_3MH \longrightarrow R_3M(CH_2)_{n+2}Si(OCH_2CH_2)_3N$$

$M = Si,\ n = 0,\ R = Et,\ EtO^{344},\ R_3 = Me_m(2\text{-Furyl})_{3-m},\ m = 0\text{–}2^{342,\ 343};$ (120)

$M = Ge,\ n = 0,\ 1,\ R = Et^{344};$

$M = Sn,\ n = 0,\ 1,\ R = Bu^{344}$

Treatment of 1-vinyl- (**20**) or 1-allylsilatrane (**76**) and their 3,7,10-trimethyl derivatives (**20a, 76a**) with $CH_2N_2/Pd(OAc)_2$ in ether affords the corresponding 1-cyclopropyl- and 1-cyclopropylmethylsilatranes (80–100%), respectively (equation 121)[37,345].

(**84**) R = H, n = 0
(**84a**) R = Me, n = 0
(**85**) R = H, n = 1
(**85a**) R = Me, n = 1

Under the same conditions, vinyltriethoxysilane reacts readily with diazomethane to yield the corresponding 1,3-cycloadduct, i.e. 3-triethoxysilyl-1-pyrazoline[345]. In contrast to 1-allylsilatrane, the degree of conversion of allyltriethoxysilane to the corresponding silyl-substituted cyclopropane is rather low (15%)[37].

1-Alkenylsilatranes can react with a variety of electrophiles to give products of substitution or addition. In the latter case, the adducts can undergo subsequent elimination, either thermally or by treatment with a good nucleophile toward silicon, resulting in an overall substitution. Thus, the addition of various derivatives of sulfenyl chlorides to

1-alkenylsilatranes occurs easily even at low temperature ($-50\,^\circ$C) (equations 122)[346,347].

$$CH_2=CH(CH_2)_nSi(OCH_2CH_2)_3N + PhSCl \longrightarrow ClCH_2CH(CH_2)_nSi(OCH_2CH_2)_3N$$
$$|$$
$$SPh \quad \text{(86 and 87)}$$

$$\begin{bmatrix} Cl^- \\ H_2C-CH \quad Si(OCH_2CH_2)_3N \\ \backslash+/ \quad \backslash \quad / \\ S \quad (CH_2)_n \\ | \\ Ph \end{bmatrix}$$

$$\longrightarrow ClSi(OCH_2CH_2)_3N \quad \text{(38)} \quad (122)$$

$$\longrightarrow PhS(CH_2)_nCH=CH_2$$
$$n = 0, 1$$

(88)

However, the thermal stability of the adducts depends on the nature of the sulfenyl chloride and on the structure of the alkenyl group. For instance, 1-chlorosilatrane (**38**) and alkenyl phenyl sulfides are formed by treatment of **20** or **76** with phenylsulfenyl chloride (equation 122). It is likely that the initial adducts (**86** and **87**), which are unstable at $-10\,^\circ$C, undergo elimination reaction via the formation of an episulfonium intermediate (**88**), followed by nucleophilic attack of Cl$^-$ at silicon. In contrast, the adducts from phenylsulfenyl chloride and trialkylalkenylsilanes are thermally stable and decompose by strong bases[348]. The larger electron-donating ability of the silatranyl group compared with the trialkylsilyl group is most probably responsible for these differences in stability.

The adducts from **20** and O,O-dialkylphosphoryl- or dialkylthiophosphorylsulfenyl chlorides are unstable at room temperature. Their transformations proceed in two directions as illustrated by equation 123[346,347].

The reaction course is determined by the heteroatom X. Since the migration of the silatranyl group to the oxygen of the phosphoryl group in the episulfonium intermediate is more preferable than that to sulfur, the a/b ratio is 3 : 2 and 6 : 1 for X = O and S, respectively. Similar sequences are likely to be involved in the reaction with 1-allylsilatranes[346].

A facile reaction between 1-alkenylsilatranes and N,N-dichloroarylsulfonamides in chloroform at room temperature results in almost quantitative yields of the adducts. Depending on the ratio of the starting reagents, the addition may involve one or both chlorine atoms of the dichloroamide (equations 124a and b)[349].

Silatranes **20** and **20a** are readily converted to α,β-epoxy derivatives **89** and **89a** (83–96% yield) by treatment with m-chloroperbenzoic acid in dichloromethane at room temperature (equation 125)[141,350]. Only polymeric products are obtained from 1-allylsilatrane (**76**) under the same conditions[350].

Electrophilic reactions of **20** and **20a** with N-bromosuccinimide in excess of water at room temperature lead to the stable adducts (**90** and **90a**, respectively) in 80% yields (equation 126)[119,351].

Similarly to allylsilanes[352], 1-allylsilatrane (**76**) reacts with carbonyl compounds such as aldehydes, ketones or esters. Thus, treatment of **76** with aldehydes in the presence of TiCl$_4$ followed by hydrolysis afforded γ,δ-unsaturated alcohols (equation 127). Allyl transfer is also catalyzed by other Lewis acids, such as AlCl$_3$ or BF$_3$–OEt$_2$, but the alcohol yields are usually lower[39].

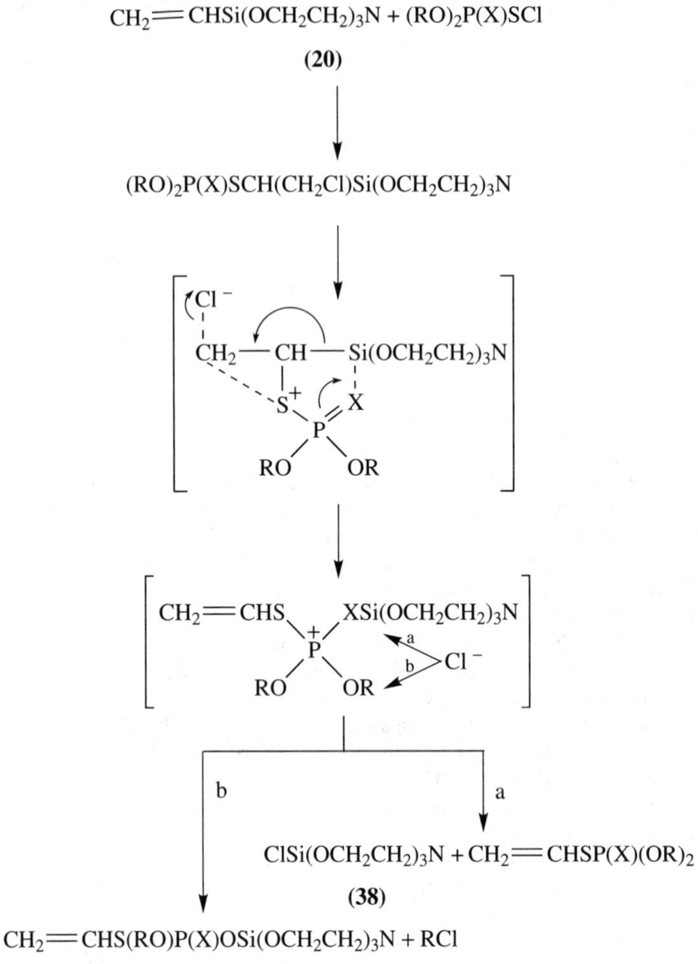

$$CH_2=CHSi(OCH_2CH_2)_3N + (RO)_2P(X)SCl$$
(20)

↓

$$(RO)_2P(X)SCH(CH_2Cl)Si(OCH_2CH_2)_3N$$

↓

$$\left[\begin{array}{c} Cl^- \\ CH_2-CH-Si(OCH_2CH_2)_3N \\ \diagdown S^+ \diagup X \\ P \\ RO \quad OR \end{array}\right]$$

↓

$$\left[\begin{array}{c} CH_2=CHS \quad XSi(OCH_2CH_2)_3N \\ P^+ \\ \diagup \diagdown \quad Cl^- \\ RO \quad OR \end{array}\right]$$ (123)

b ↙ ↘ a

a → $ClSi(OCH_2CH_2)_3N + CH_2=CHSP(X)(OR)_2$
(38)

b ↓
$$CH_2=CHS(RO)P(X)OSi(OCH_2CH_2)_3N + RCl$$

X = O, S

$$20 + ArSO_2NCl_2 \longrightarrow ArSO_2NClCH_2CHClSi(OCH_2CH_2)_3N \quad (124a)$$
$$\searrow ArSO_2N[CH_2CHClSi(OCH_2CH_2)_3N]_2 \quad (124b)$$

Ar = Ph, 4-ClC$_6$H$_4$, 4-MeC$_6$H$_4$

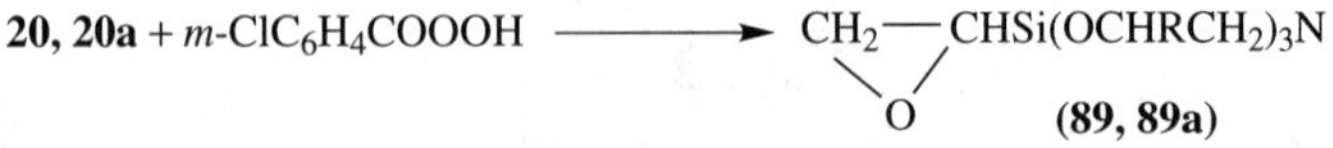

$$\textbf{20, 20a} + m\text{-ClC}_6\text{H}_4\text{COOOH} \longrightarrow \underset{\underset{O}{\diagdown\diagup}}{CH_2-CHSi(OCHRCH_2)_3N}$$
(89, 89a) (125)

20/89   R = H
20a/89a   R = Me

$$\mathbf{20, 20a} \xrightarrow{\text{NBS/H}_2\text{O}} \text{HOCH}_2\text{CHBrSi(OCHRCH}_2)_3\text{N} \quad (126)$$
$$\mathbf{(90, 90a)}$$

20/90   R = H
20a/90a R = Me

$$\mathbf{76} + \text{PhCH}_2\text{CH}_2\text{CHO} \xrightarrow{\text{cat.}} \text{PhCH}_2\text{CH}_2\text{CH(OH)CH}_2\text{CH}=\text{CH}_2 \quad (127)$$

cat. = TiCl$_4$ (96%); AlCl$_3$ (83%);   BF$_3$–OEt$_2$ (52%)

Allyl transfer does not take place with **76** under nucleophilic conditions[39]. By contrast, the reactions of allylsilanes with carbonyl compounds are activated by fluoride ion[353].

Diels–Alder cycloadditions between **20** or **76** and cyclopentadiene or hexachlorocyclopentadiene occur at 170–180 °C. The adduct (**91**) is a mixture of *exo*- and *endo*-isomers in a 3 : 1 ratio (equation 128)[354,355].

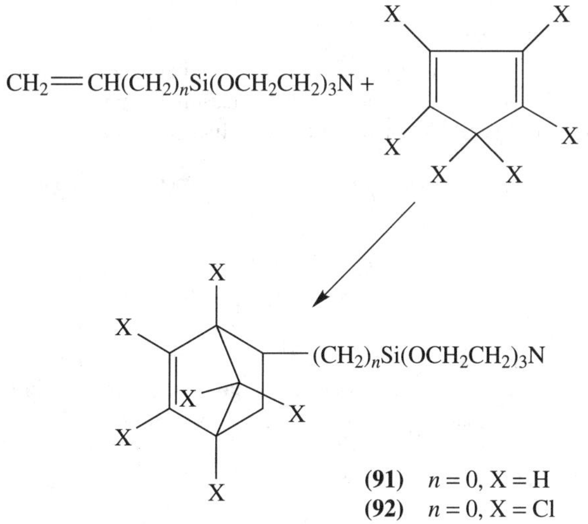

$$\text{CH}_2=\text{CH(CH}_2)_n\text{Si(OCH}_2\text{CH}_2)_3\text{N} + \text{(cyclopentadiene)} \longrightarrow \text{(adduct)} \quad (128)$$

(**91**) n = 0, X = H
(**92**) n = 0, X = Cl
(**92a**) n = 1, X = Cl

Both 1-(butadien-1',3'-yl)silatrane (**93**) and -trialkoxysilane can react by a Diels–Alder-type reaction with tetracyanoethylene (TCNE) or maleic anhydride (MA) to give the corresponding adducts (equations 129 and 130). However, a higher temperature is required for effective conversion of trialkoxysilane[355].

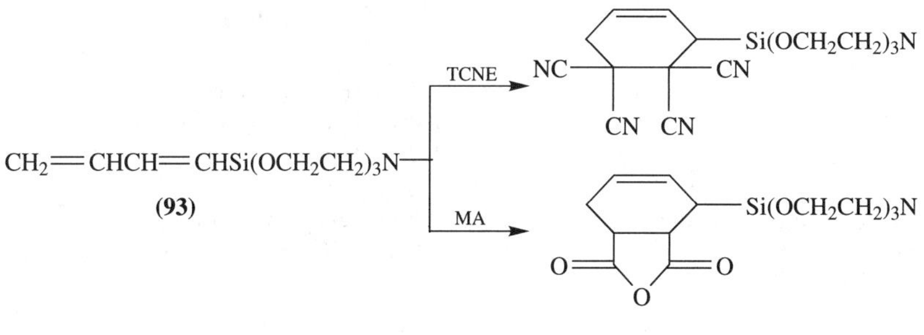

$$\text{CH}_2=\text{CHCH}=\text{CHSi(OCH}_2\text{CH}_2)_3\text{N} \xrightarrow{\text{TCNE}} \text{(TCNE adduct)—Si(OCH}_2\text{CH}_2)_3\text{N} \quad (129)$$

(**93**)

$$\xrightarrow{\text{MA}} \text{(MA adduct)—Si(OCH}_2\text{CH}_2)_3\text{N} \quad (130)$$

9. Carbofunctional silatranes

Generally, carbofunctional silatranes behave in a remarkably similar manner to the corresponding trialkoxysilanes. The reactivity differences may arise from the stereoelectronic influence of the silatranyl group.

1-(Haloalkyl)silatranes can react with a variety of heteronucleophiles. Thus, 1-chloromethylsilatrane (**16**) reacts with sodium or potassium alkoxides to afford 1-alkoxy-2-carba-3-oxahomosilatranes (**94**) instead of 1-alkoxymethylsilatranes. The ring enlargement proceeds via initial cleavage of a Si—O bond by alkoxide ion, followed by intramolecular nucleophilic substitution (equation 131)[356].

$$\text{(16)} + \text{ROM} \longrightarrow \text{(94)} \quad (131)$$

R = Me (54%)
R = Et (40%)
M = K, Na

Cleavage of the Si—C bond to give **96** occurs when 1-iodomethylsilatrane (**95**) is treated with dimethylethanolamine at 150 °C (equation 132)[327]. A related reaction also occurs with tetracoordinated halomethyltriorganylsilanes[357].

$$ICH_2Si(OCH_2CH_2)_3N + Me_2NCH_2CH_2OH \longrightarrow Me_2N(CH_2)_2OSi(OCH_2CH_2)_3N$$

(**95**) (**96**) (132)

1-(Haloalkyl)silatranes react with alkali mercaptides or alkali salts of thiolcarboxylic acids to give **97** (equation 133)[88], with primary or secondary amines to give **98** (equation 134)[99,358] and with potassium diethyldithiophosphate to give **99** (equation 135)[109]. In the latter case no conversion of 1-chloromethyl- or 1-iodomethylsilatrane is observed under the same conditions[109]. The Arbuzov reaction is used to transform 1-(haloalkyl)silatranes into the corresponding dialkyl phosphonate derivatives **100** (equation 136)[359].

$$Y(CH_2)_nSi(OCH_2CH_2)_3N$$

Y = Cl, Br, I
n = 1, 3

Reagents:
- RSM, R′ = Et, Bz, Ac; M = K, Na → (**97**) X = RS (133)
- R_mNH_{3-m}, R = alkyl, cycloalkyl, aryl → (**98**) X = R_mNH_{2-m} (134)
- $(EtO)_2P(S)SK$ → (**99**) X = $(EtO)_2P(S)S$ (135)
- $(R'O)_3P$, R′ = Me, Et, i-Pr → (**100**) X = $(R'O)_2P(O)$ (136)

(**97–100**)

In the reactions in equations 133–136 the chloro derivatives were found to be considerably less reactive than the bromo- and iodoalkylsilatranes. It should be noted that chloro–halogen exchange does not occur when **16** is treated with NaI[72], NaBr, KBr, AlBr$_3$ or PBr$_3$[71]. However, NaI reacts easily with chloromethyltrialkoxysilanes[74].

A variety of onium salts (**101–103**) are formed by reaction of 1-(iodomethyl)silatrane with tertiary amines[358], triorganylphosphines[360] or diorganylchalcogenides (equation 137)[92,361,362].

$$ICH_2Si(OCH_2CH_2)_3N + B \longrightarrow [BCH_2Si(OCH_2CH_2)_3N]I$$

(95)　　　　　　　(101) B = R_3N

　　　　　　　　(102) B = R_3P　　　　　　　　(137)

　　　　　　　　(103) B = R_2Y (Y = S, Te, Se)

The rearrangement of silatranyloxiranes (**89** and **89a**) to silatranylacetaldehydes (**104** and **104a**) occurs at 100–110 °C in the presence of triethyltin bromide as a catalyst (equation 138)[327].

$$\underset{\underset{O}{\diagdown\diagup}}{CH_2-CHSi(OCHRCH_2)_3N} \xrightarrow{Et_3SnBr} H(O)CCH_2Si(OCHRCH_2)_3N$$

89, 89a　　　　　　(**104, 104a**)　　　　　(138)

89/104　R = H

89a / 104　R = Me

2-Silatranylacetaldehyde (**104**) is also formed in a very high yield (98%) by heating 2-silatranyl-2-bromoethanol (**90**) with a small excess of triethylmethoxystannane (equation 139)[119].

$$HOCH_2CHBrSi(OCH_2CH_2)_3N \xrightarrow{Et_3SnOMe} H(O)CCH_2Si(OCH_2CH_2)_3N$$

(**90**)　　　　　　　　　　(**104**)　　　　　　(139)

Under similar conditions, 2-trialkylsilyl-2-bromoethanols are converted to a mixture of three isomers, e.g. trialkylsilyloxiranes (20–30%), trialkylvinyloxysilanes (10%) and trialkylsilylacetaldehydes (60–70%). The latter are unstable and easily rearrange at 180–190 °C to trialkylvinyloxysilanes[363,364]. In contrast, silatranylacetaldehyde **104** does not rearrange at 200 °C after prolonged heating (4 h)[119], probably due to the bulky silatranyl group.

1-(3′-Aminopropyl)silatranes (**105 and 105a**) react with halogen-containing compounds including acyl chlorides (equations 140 and 141)[96,99,365,366] and with aldehydes (equation 142)[366,367] to give the corresponding derivatives (**106–108**).

$$R_2NCH_2CH_2CH_2Si(OCH_2CH_2)_3N \begin{array}{l} \xrightarrow[R=Alk, R'=Alk]{R'Hal} \\ \xrightarrow[R=H, R'=Ar]{R'COCl} \\ \xrightarrow[R=H, R'=Ar]{R'CHO} \end{array}$$

(**105**)　R = H

(**105a**) R = Alk

　　　　　　　　(**106-108**)

(140)
(141)
(142)

(**106**) X = Hal[$R_2R'N(CH_2)_3$]

(**107**) X = R′CO(R)N(CH_2)$_3$

(**108**) X = R′CH=N(CH_2)$_3$

The cyclic adduct **109** is obtained from **105** and divinyl sulfoxide (equation 143)[368,369].

$$\mathbf{105} + (CH_2{=}CH)_2SO \longrightarrow O{=}S\diagup\diagdown NCH_2CH_2CH_2Si(OCH_2CH_2)_3N$$

　　　　　　　　　　　　(**109**)　　　　　　　　(143)

Relative to organic and organosilicon sulfides, 1-(organylthioalkyl)silatranes (**110–112**) show an enhanced tendency for the formation of a variety of onium salts[361]. Due to the very strong $-I$ effect of the silatranyl group, 1-(organothiomethyl)silatranes (**110**) are very reactive toward organic iodides [including iodomethyltrimethylsilane and 1-iodomethylsilatrane] and toward bromides and even toward organic chlorides giving the sulfonium salts **113** (equation 144).

$$RSCH_2Si(OCH_2CH_2)_3N + R'X \longrightarrow [RR'S^+CH_2Si(OCH_2CH_2)_3N]X^- \quad (144)$$
$$\quad\quad (110) \quad\quad\quad\quad\quad\quad\quad\quad\quad\quad (113)$$

R = Me, Et, n-Bu, t-Bu, Bz, 4-MeC$_6$H$_4$

R' = Me, i-Pr, All, HOCH$_2$CH$_2$, PhC(O)CH$_2$, Me$_3$SiCH$_2$, etc.

X = I, Br, Cl

At higher temperature (130–140 °C) the reaction of bulky bis(silatran-1-ylmethyl)sulfide with 1-iodomethylsilatrane results in the formation of $S^+[CH_2Si(OCH_2CH_2)_3N]_3I^-$, having three pentacoordinate silicon atoms.

1-(Organylthioalkyl)silatranes RS(CH$_2$)$_n$Si(OCH$_2$CH$_2$)$_3$N (**111**, $n = 2$; **112**, $n = 3$) react with alkyl iodides more slowly than **110**. The β-sulfonium salts are stable only in nonpolar solvents, whereas they undergo β-cleavage in polar solutions[361].

Hydrogen peroxide (30 or 80%) oxidizes 1-(organylthioalkyl)silatranes (**110** and **111**) in chloroform to the corresponding sulfoxides **114a** and **115a** (at 0–10 °C) or sulfones **114b** and **115b** (at 60 °C) (equation 145)[370].

$$RS(CH_2)_nSi(OCH_2CH_2)_3N + H_2O_2 \xrightarrow{CHCl_3} RS(O)_m(CH_2)_nSi(OCH_2CH_2)_3N \quad (145)$$
$$\quad (110, 111) \quad\quad\quad\quad\quad\quad\quad\quad\quad (114a,b, 115a)$$

110/114a/114b R = Me, Et, Bz, Ph, N(CH$_2$CH$_2$O)$_3$SiCH$_2$, $n = 1$

111/115a R = Me, Et, $n = 2$

(a) $m = 1$, (b) $m = 2$

The reactivity diminishes in going from thiomethyl to thioethyl derivatives. Thus, 1-(phenylthiomethyl)silatrane is oxidized by 80% H$_2$O$_2$ to the corresponding sulfoxide whereas 1-(phenylthioethyl)silatrane is unreactive even under more drastic conditions.

The oxidation of (silatranylmethyl)phenylselenide (**116**) with 30% hydrogen peroxide in ethanol affords the corresponding selenoxide (**117a**) in 91% yield (equation 146). The selenone **117b** is formed by using 70% H$_2$O$_2$[92]. Unstable selenoxide is obtained from (2-silatranylethyl)phenyl selenide under the same conditions[92].

$$PhSeCH_2Si(OCH_2CH_2)_3N + H_2O_2 \longrightarrow PhSe(O)_nCH_2Si(OCH_2CH_2)_3N \quad (146)$$
$$\quad\quad (116) \quad\quad\quad\quad\quad\quad\quad\quad\quad (117a,b)$$

(a) $n = 1$ and (b) $n = 2$

The high nucleophilicity of sulfur in MeSOCH$_2$Si(OCH$_2$CH$_2$)$_3$N favors a more facile reaction with MeI (equation 147)[370] compared with DMSO[371]. Selenoxides can also be involved in the reaction in equation 147[92].

$$RZ(O)CH_2Si(OCH_2CH_2)_3N + MeI \longrightarrow [MeRZ(O)CH_2Si(OCH_2CH_2)_3N]I \quad (147)$$
R = Me, Z = S
R = Ph, Z = Se

Treatment of 1-(mercaptoalkyl)silatranes with organomercury acetate in chloroform affords the corresponding mercury derivatives in 80–85% yield (equation 148). A related reaction with diorganylmercury results in lower product yields[372].

$$HS(CH_2)_nSi(OCH_2CH_2)_3N \begin{array}{c} \xrightarrow{RHgOAc, CHCl_3} RHgS(CH_2)_nSi(OCH_2CH_2)_3N \quad (148) \\ R = Et, n = 2, 3 \\ R = Ph, n = 1, 3 \\ \\ \xrightarrow{HgX_2, X = Cl, OAc} Hg[S(CH_2)_nSi(OCH_2CH_2)_3N]_2 \quad (149) \\ n = 1, 3 \end{array}$$

Bis(silatran-1-ylalkylthio)mercuranes are obtained from 1-(mercaptoalkyl)silatranes with mercury salts (equation 149)[372].

10. Silatranyl arene and cyclohexadiene complexes

1-Phenylsilatranes can act as π-coordinating ligand for transition metals[159–161,330,373–377]. Thus, the reactions of 1-phenylsilatrane with $M(CO)_6$ where M = Cr[159] and W[161] and with $[Mn(CO)_5]ClO_4$[373] afford the corresponding π-coordinated η^6-phenylsilatrane complexes **118**, **119** and **120a** (equations 150a and b).

$$\text{Ph-Si(OCH}_2\text{CH}_2)_3\text{N} \begin{array}{c} \xrightarrow{M(CO)_6, M=Cr, W} (\eta^6\text{-C}_6\text{H}_5)M(CO)_3\text{-Si(OCH}_2\text{CH}_2)_3\text{N} \quad (150a) \\ \textbf{(118)} \ M = Cr \\ \textbf{(119)} \ M = W \\ \\ \xrightarrow{Mn[(CO)_5]ClO_4} [(\eta^6\text{-C}_6\text{H}_5)Mn(CO)_3\text{-Si(OCH}_2\text{CH}_2)_3\text{N}]^+ ClO_4^- \quad (150b) \\ \textbf{(120a)} \end{array}$$

These stable-in-air compounds are versatile precursors for the preparation of a wide range of new complexes. For example, manganese complex **120a** is easily transformed into $[\eta^6$-phenylsilatrane)Mn(CO)$_2$P(OMe)$_3$]ClO$_4$ (**120b**) via a photochemical reaction with P(OMe)$_3$[374].

With NaBH$_4$ and some carbanions, complexes **120a** and **120b** and similar manganese complexes of 4′-substituted 1-arylsilatranes (**120c–g**) undergo a highly regioselective nucleophilic addition reaction (equation 151) yielding mainly one of the isomeric η^5-cyclohexadienylsilatrane complexes (**121–123a–g**)[160,373–375].

Treatment of complex **124** (which was obtained from **120a** via reaction 151) with NOBF$_4$ result in displacement of the carbonyl group to give the product **125** (equation 152)[330].

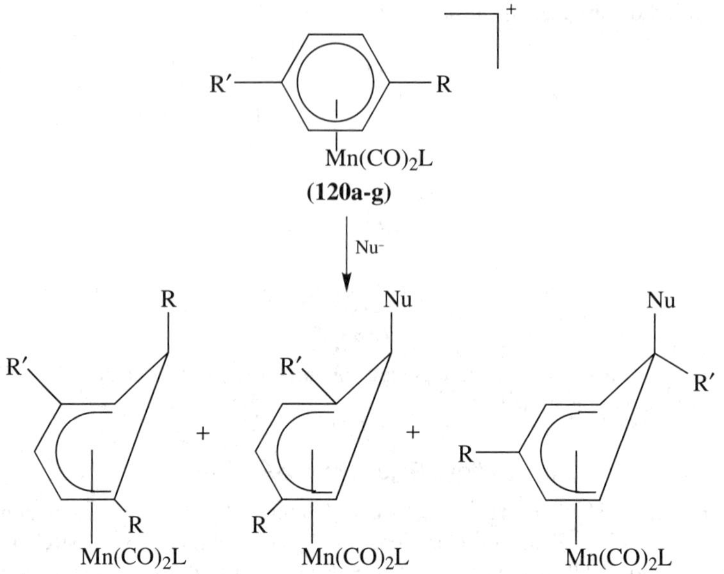

$$R = Si(OCH_2CH_2)_3N$$

(a) R′ = H, L = CO
(b) R′ = H, L = P(OMe)$_3$
(c) R′ = Me, L = CO
(d) R′ = t-Bu, L = CO
(e) R′ = Cl, L = CO
(f) R′ = MeO, L = CO
(g) R′ = H$_2$N, L = CO

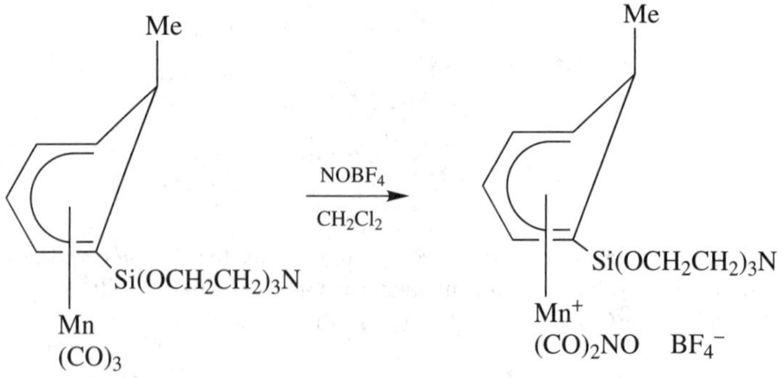

(120a-g) → (121a-g) + (122a-g) + (123a-g) (151)

(124) $\xrightarrow{\text{NOBF}_4,\ \text{CH}_2\text{Cl}_2}$ (125) (152)

The addition reactions of hydride and several kinds of carbon nucleophiles to **125** provide an easy synthetic route to some new silatranyl η^4-cyclohexadienyl complexes, such as **126a** and **126b** (equation 153)[330]. Demetalation of **126a** and **126b** with Me_3NO in refluxing benzene is a very versatile route to 1-(cyclohexadienyl)silatranes **127a** and **127b** (equation 154)[330,376]. Compounds **127a** and **127b** can be transformed to the η^4- and η^5-(silatranylcyclohexadienyl)iron complexes by treatment with $Fe_2(CO)_9$[377].

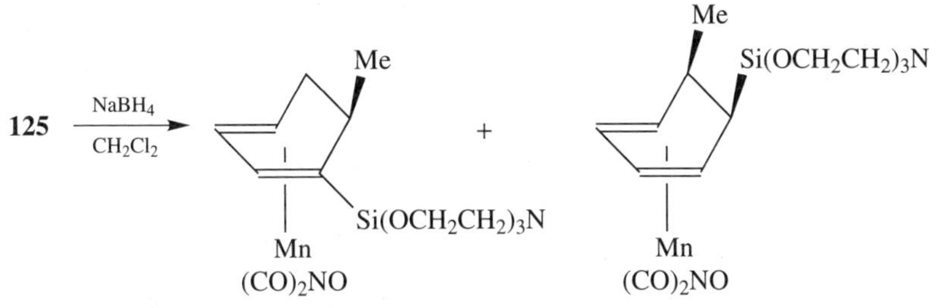
125
NaBH4
CH2Cl2
Me
Si(OCH2CH2)3N
Mn
(CO)2NO
+
Me
Si(OCH2CH2)3N
Mn
(CO)2NO

(153)

(126a) **(126b)**

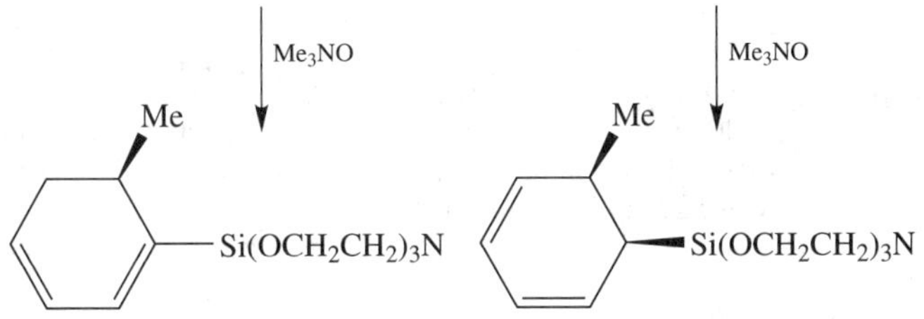
Me3NO
Me3NO
Me
Si(OCH2CH2)3N
Me
Si(OCH2CH2)3N

(154)

(127a) **(127b)**

III. SILATRANE ANALOGS

A. 3-Homosilatranes

The general method for the formation of 1-organyl-2,9,10-trioxa-6-aza-1-silatricyclo[4.3.3.0^{1,6}]dodecanes (1-organyl-3-homosilatranes) having an enlarged ring system is the transetherification reaction of a trifunctional silane with 3-hydroxypropyl-bis(2-hydroxyalkyl)amine in boiling toluene or xylene (equation 155)[326,378]. When Y = OR′, KOH is used as a catalyst. The yields of the desired compounds are excellent (70–90%)[326,378].

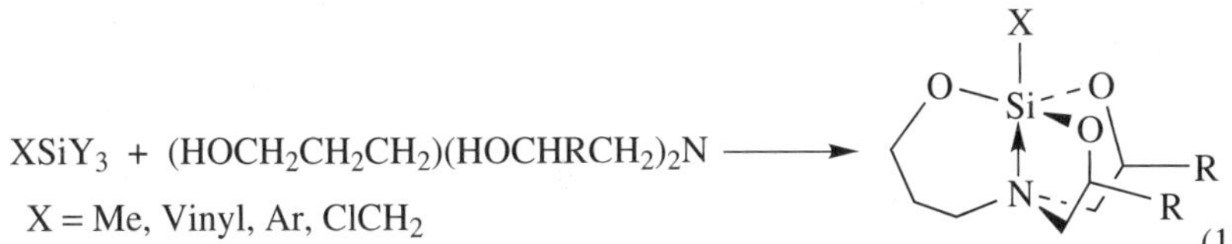
XSiY3 + (HOCH2CH2CH2)(HOCHRCH2)2N
X
O
Si
O
O
N
R
R

X = Me, Vinyl, Ar, $ClCH_2$

Y = OR′, NMe_2

R = H, Me

(155)

Treatment of tetramethoxy- or tetraethoxysilane with phenol and trialkanolamine affords 1-phenoxyhomosilatrane (equation 156)[326].

1-Aryloxyhomosilatranes can also be obtained from 1-alkoxyhomosilatranes and the appropriate phenols (equation 157)[326]. However, reaction 156 proceeds at a higher rate.

$$Si(OR)_4 + PhOH + (HOCH_2CH_2CH_2)(HOCHMeCH_2)_2N$$

R = Me, Et

(156)

[structure: 1-phenoxy-3-homosilatrane with OPh, Si, O, N, Me, Me]

$$ROSi(OCH_2CH_2CH_2)(OCHMeCH_2)_2N \xrightarrow{ArOH} ArOSi(OCH_2CH_2CH_2)(OCHMeCH_2)_2N$$

R = Me, Et; Ar = Ph, 1-Naph

(157)

Attempts to prepare a stable cage structure from organyltrialkoxysilanes and (2-hydroxyethyl)bis(3-hydroxypropyl)-[326] or tris(3-hydroxypropyl)amines[69,326] yielded only macromolecular products. It is not surprising since the strong transannular bonding contributes to the shape of the silatrane cage (see Section II.A.2). Meanwhile, insertion of an additional methylene group into the silatrane framework was shown by some physical methods to weaken the Si←N bond in 3-homosilatranes compared with the parent silatrane analogs[6,13].

The Si←N bond length was measured by X-ray method to be 2.291 Å in 1-chloromethyl-3-homosilatrane[379] and 2.43 and 2.49 Å in two independent molecules of 1-phenyl-7-methyl-3-homosilatrane[6]. Elongation of this bond relative to that observed in the corresponding silatranes is by 0.2–0.3 Å and it may be explained by an increase in the strain energy of the skeleton in 3-homosilatranes: a six-membered chelate ring is obviously more strained than a five-membered ring. The relative weakness of the Si←N bond is accompanied by a small shortening of the apical C–Si bond. Since the silicon electronegativity in 3-homosilatranes and silatranes is nearly the same, these distinctions between their axial X–Si←N moieties agree with the prediction by the 3c4e bonding model[174].

Dipole moment studies[207,378,380] found a lower dipole moment of the Si←N bond in 3-homosilatranes (1–2 D) than in silatranes (1.5–3.1 D). The enthalpy of formation of this bond was estimated to be 3–11 kcal mol^{-1} depending on the substituent X and on the method of the bond dipole calculation[207]. The observation of an absorption band at 570–600 cm^{-1} in the IR spectra suggests a tricyclic structure of 3-homosilatranes in solution[378].

The general trends of the influence of the substituent X on the ^{1}H[239,378], ^{13}C[250], ^{15}N[250] and ^{29}Si[250,261] NMR chemical shifts of 3-homosilatranes are close to those found for silatranes. However, there are some definite spectral indications for a weakening of the Si←N bond in the former molecules[250,261,378].

In the EI mass spectra of 1-organyl-3-homosilatranes[286] the [M − X]$^+$ ion (m/z = 188) is generally the most abundant. The FAB spectra of these compounds contain a very intensive peak of protonated 3-[N,N-bis(2′-hydroxyethyl)]aminopropanol (m/z 164) together with peaks of the very characteristic MH$^+$ and [M − X]$^+$ ions[286]. A decrease in the intensity ratio of the latter peak (m/z 188) to that at m/z 164 upon increasing the total irradiation time indicates that the ring rupture is a FAB-induced reaction.

B. Silatranones

1-Organylsilatran-3-ones (2,8,9-trioxa-5-aza-1-silatricyclo[3.3.3.0$^{1.5}$]undecan-3-ones) can be prepared from organyltrialkoxysilanes and N,N-bis(2-hydroxyethyl)aminoacetic acid in the appropriate boiling solvent (DMF/C$_6$H$_6$, DMSO/C$_6$H$_6$) in the presence of KOH (equation 158). The yields are 30–75%[381].

$$XSi(OMe)_3 + (HOCH_2CH_2)_2NCH_2COOH \longrightarrow \text{[silatranone structure]} \quad (158)$$

X = Alk, Vin, Ar

A slight modification of the synthesis involves the application of silylated amino acids which are more soluble in organic solvents (equation 159). In this case reactions occur rapidly (0.5–1 h) at lower temperatures (40–50 °C, DMF) to give the corresponding silatranones (**128–134**) in good yields. 1-Methylsilatrane-3,7,10-trione (**134**) was isolated as a complex with DMF[382].

$$XSi(OMe)_3 + \begin{matrix}(Me_3SiOCOCH_2)_n \\ (Me_3SiOCH_2CH_2)_{3-n}\end{matrix}\!\!N \longrightarrow \text{[product]} \quad (159)$$

	X	n	Yield (%)
(128)	ClCH$_2$	1	63
(129)	MeO	1	32
(130)	Me	2	69
(131)	Et	2	70
(132)	Ph	2	70
(133)	ClCH$_2$	2	70
(134)	Me	3	83

Organyltriacetoxysilanes can be converted almost quantitatively to 1-organylsilatrane-3,7,10-triones (**134–142**) by heating with aminotriacetic acid at 90–100 °C (equation 160)[383].

$$XSi(OAc)_3 + (HOOCCH_2)_3N \longrightarrow \text{[silatrane-trione structure]} \quad (160)$$

(**134**) X = Me (**139**) X = All
(**135**) X = H (**140**) X = HC≡C
(**136**) X = Et (**141**) X = Ph
(**137**) X = ClCH$_2$ (**142**) X = EtSCH$_2$CH$_2$
(**138**) X = Vin

1-Substituted silatrane-3,7,10-triones **143–147** are formed in high yield (80–95%) from tetrachlorosilane or organyloxytrichlorosilanes and tris(trialkylstannyl) esters of aminotriacetic acid (equation 161). CCl_4, pentane or petroleum ether can be used as a solvent. Tetraethoxy- and tetraacetoxysilanes are ineffective under the same conditions[384].

$$XSiCl_3 + (Me_3SnOOCCH_2)_3N \longrightarrow XSi(OCOCH_2)_3N \qquad (161)$$

	X	Yield (%)
(143)	Cl	96
(144)	MeO	90
(145)	EtO	83
(146)	PhO	95
(147)	MeCOO	72

Monosilatranones are rather stable to hydrolysis and alcoholysis. Thus, N,N-bis(2-hydroxyethyl)aminoacetic acid is formed in only 50% yield when 1-(3′-trifluoromethylphenyl)silatran-3-one) **(148)** is treated with H_2O/THF at room temperature for 7 h. **148** (80%) is recovered from a MeOH/THF mixture upon prolonged (9 h) heating[381].

X-ray studies revealed a shortening of the Si←N bond in 1-organylsilatran-3-ones and 1-methylsilatrane-3,7-dione relative to the corresponding silatranes[385–390] (Scheme 5). This shortening was attributed to a higher effective electronegativity of the silicon which, in accordance with a 3c4e model of silicon bonding[174], should provide (and provides) the strengthening of both components of the hypervalent X–Si←N bond. An elongation of the Si–O(CO) bond is the second peculiarity of silatranones. The influence of the substituent X on the molecular structure of these compounds seems to be similar to that found for silatranes.

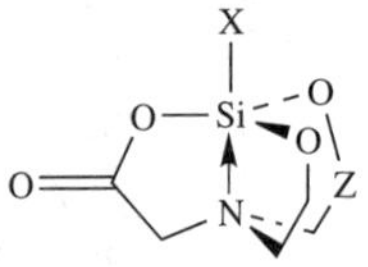

Compound	X	Z	d(SiN) (Å)	Reference
149	Me	CH_2	2.134	385
128	$ClCH_2$	CH_2	2.085	386
150	$Cl(CH_2)_3$	CH_2	2.149	387
151	Ph[a]	CH_2	2.126, 2.111	388
152	4-FC_6H_4	CH_2	2.129	389
148	3-$CF_3C_6H_4$	CH_2	2.106	389
130	Me	C=O	2.146	390

[a]Two independent molecules

SCHEME 5. Bond lengths in 1-organylsilatran-3-ones

The prochirality of the silicon and nitrogen atoms in silatran-3-ones makes the geminal protons of both unsubstituted OCH_2CH_2N fragments anisochronic in the ^{1}H NMR spectra[239,247,381]. The same reason leads to a nonequivalence of the geminal protons of the $OC(O)CH_2N$ groups in silatrane-3,7-diones[247]. Subsequent replacement of the OCH_2 groups in the silatrane skeleton by the $OC(O)$ moieties increases the chemical shifts of all protons[247]. For the silatrane-3,7,10-triones the resonance of the NCH_2 protons was found to be in the range 3.8–4.1 δ (in DMSO) and to be slightly less sensitive to the substituent effect than in the silatrane spectra[383,391]. It was noted that silatrane-3,7,10-triones are able to form 1 : 1 complexes with DMSO and DMF, in which the silicon atom is hexacoordinated[247]. The Si←N bonding in silatranones is displayed in upfield and downfield shifts of their ^{29}Si and ^{15}N NMR signals relative to those observed for model compounds[247].

The values of $^1J(^{29}\text{Si}^{15}\text{N})$ in 1-methyl-substituted silatran-3-one (0.9 Hz) and silatrane-3,7,10-trione (8.2 Hz) indicate an increase in the Si←N bond strength with increase in the number of carbonyl groups in the silatranones[247]. From the relationship between the $^1J(^{29}\text{Si}^{15}\text{N})$ and d(SiN) values in silatranes and their analogs, the lengths of the transannular bond in 1-methyl- (**134**) and 1-chloromethylsilatrane-3,7,10-triones (**137**) were estimated to be 2.05 and 2.01 Å, respectively[124]. The silatranonyl groups were shown to be less electron-releasing than the silatranyl group. Their σ_I constants derived from the ^{13}C NMR spectra of 1-phenylsilatranones are −0.08, 0.04 and 0.04 for mono-, di- and tricarbonyl-substituted groups, respectively[247].

In IR spectra of 1-organylsilatrane-3,7,10-triones (**134–142**) the complex skeleton and Si←N bond vibrations are displayed at 550–570 cm^{-1}[383]. The frequency of the stretching vibration of their carbonyl groups, ν(C=O), is higher than that observed for triacetoxysilanes. A Si←N bonding in 1-hydrosilatrane-3,7,10-trione (**135**) is confirmed by a low-frequency shift of the ν(SiH) band (2234 cm^{-1}) relative to that at 2290 cm^{-1} for HSi(OOCCMe)$_3$[383].

The mass spectra of 1-alkyl- and 1-arylsilatran-3-ones differ from those of silatranes: their most intensive peak is ascribed to $[M - CO_2]^+$ ions, formed perhaps owing to a weakening of the Si—OC(O) bonds[287]. Two pathways, with a primary breaking of the X—Si bond and a splitting of the CO_2, are found for the fragmentation of silatrane-3,7,10-triones[391].

C. Carbasilatranes

2-Carbasilatranes (1-organyl-2,8-dioxa-5-aza-1-silatricyclo[3.3.3.0^{1,5}]undecanes) are derived formally from silatranes by substituting one equatorial oxygen atom with a CH_2 group. 1-Alkyl- and 1-alkoxy-2-carbasilatranes (**153–154**) and their 7,10-dimethyl derivative (**154a**) were obtained from organyl(3-aminopropyl)diethoxysilanes[392,393] or 3-aminopropyltriethoxysilane[69,392] and oxiranes (equation 162).

$$X(\text{EtO})_2\text{SiCH}_2\text{CH}_2\text{CH}_2\text{NH}_2 + \text{RCH}\underset{\text{O}}{-}\text{CH}_2 \xrightarrow{-2\,\text{EtOH}} \quad (162)$$

(**153**) X = Me, R = H
(**154**) X = EtO, R = H
(**154a**) X = EtO, R = Me

An alternative route to 1-organyl-2-carbasilatranes involves the reaction of organyl(3-chloropropyl)dialkoxysilanes with diethanolamine[393–396]. For example, phenyl(3-chloropropyl)dimethoxysilane is converted to **155** (38%) by treatment with diethanolamine in xylene (equation 163). Triethylamine acts as an HCl acceptor[394].

$$\text{Ph(MeO)}_2\text{SiCH}_2\text{CH}_2\text{CH}_2\text{Cl} + (\text{HOCH}_2\text{CH}_2)_2\text{NH} \xrightarrow[-\text{Et}_3\text{NH}^+\text{Cl}^-]{\text{Et}_3\text{N}} \quad (163)$$

(**155**)

Owing to the introduction of the carbon atom into the equatorial site, the effective electronegativity of silicon in 2-carbasilatranes is lower than in silatranes. For this case, the 3c4e bonding model predicts a weakening of both axial X−Si and Si←N bonds of the hypervalent silicon[12,174]. Available X-ray data for 2-carbasilatranes confirmed it as shown below:

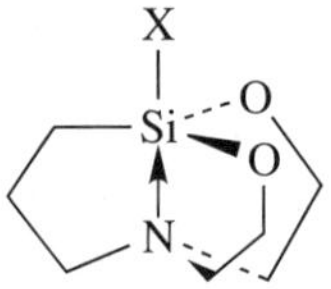

Compound	X	d(SiN) (Å)	d(XSi) (Å)	Δ_{Si} (Å)[b]	Reference
153	Me	2.336(4)	1.898(5)	0.29	397
155	Ph	2.291(1)	1.897(1)	0.27	395
156	4-MeC$_6$H$_4$[a]	2.311(2)	1.892(3)	0.24	166
		2.269(3)	1.905(3)		
157	MeO	2.223(5)	1.672(4)	0.23	396

[a] Two independent molecules.
[b] Δ_{Si} is the deviation of the TBP Si atom from the equatorial plane.

The NMR spectra of Si-substituted 2-carbasilatranes reveal a smaller effect of the substituent X on the ^{1}H chemical shifts of their cage framework relative to that in silatranes[6,12,13]. Diastereotopy of the geminal protons and a high conformational flexibility of the chelate aminoethoxy rings in 1-methyl-2-carbasilatrane (**153**) were detected by ^{1}H NMR study[239]. ^{29}Si NMR study of powder[260] and solute[261] patterns found a high sensitivity of the transannular bond in 2-carbasilatranes to the physical environment. Peculiarities of the solution-phase ^{13}C, ^{15}N and ^{29}Si NMR chemical shifts suggest that the transannular bonding in 2-carbasilatranes is weaker than in the related silatranes and 3-homosilatranes[250,261].

The most intensive peak in the mass spectra of 1-methyl- and 1-phenyl-2-carbasilatranes corresponds to the [M − X]$^+$ ion (m/z = 172) whereas for 1-alkoxy-2-carbasilatranes a cleavage of the skeleton yields the [M − C$_2$H$_5$]$^+$ ions as the major ions[287].

The ability of the silicon atom to extend its coordination number even when it is bonded to three or four carbon atoms was demonstrated by the preparation and structural studies of 2,8,9-tricarbasilatranes **158−160**[398−400].

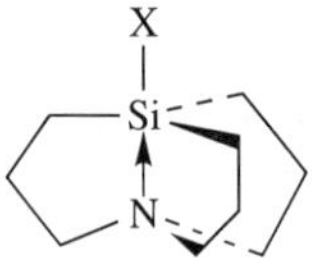

(**158**) X = Me
(**159**) X = Cl
(**160**) X = OSi(CH$_2$CH$_2$CH$_2$)$_3$N

1-Methyl-2,8,9-tricarbasilatrane (**158**) was obtained in 14% yield via the reaction of methyltrichlorosilane with a Grignard reagent (equation 164)[399].

$$\text{MeSiCl}_3 + (\text{ClMgCH}_2\text{CH}_2\text{CH}_2)_3\text{N} \xrightarrow{\text{THF/C}_6\text{H}_6} \textbf{158} \quad (164)$$

The Me−Si bond in **158** was found to be easily cleaved by dimethyldichlorostannane to give 1-chloro-2,8,9-tricarbasilatrane (**159**) in almost quantitative yield (equation 165)[398,399].

$$\textbf{158} + \text{Me}_2\text{SnCl}_2 \longrightarrow \text{ClSi}(\text{CH}_2\text{CH}_2\text{CH}_2)_3\text{N} \quad (165)$$
$$(\textbf{159})$$

The hydrolysis of **159** in benzene in the presence of triethylamine affords the oxygen-bridged bis(tricarbasilatrane) in 89% yield (equation 166)[400].

$$\mathbf{159} + H_2O \xrightarrow{Et_3N} \underset{(\mathbf{160})}{N(CH_2CH_2CH_2)_3SiOSi(CH_2CH_2CH_2)_3N} \quad (166)$$

The typical upfield shift of the ^{29}Si NMR signals of tricarbasilatranes relative to those for the corresponding derivatives of triethylsilane provides strong evidence for a transannular $Si \leftarrow N$ bonding even in the solution state[398–400]. Temperature dependence of the ^1H NMR spectra detects a fast ring flip of their skeleton and a possible strengthening of the $Si \leftarrow N$ bond on cooling of the solutions[398–400].

According to X-ray study, the $Si \leftarrow N$ bond distance (2.477 Å) and the degree of pyramidality of the silicon and nitrogen atoms in compound **160** are rather close to those in silatranes[400]. The steric repulsion between the two cages provides a linear arrangement of the Si—O—Si group. The Si—O bond (1.631 Å) is no longer than in usual siloxanes. This surprising behavior of the apical bond was ascribed to compensation of its expected lengthening by the well-known decrease in Si—O distance on increasing the Si—O—Si angle.

Recent *ab initio* calculations of a wide series of Si-substituted 2,8,9-tricarbasilatranes (X = H, Me, H$_2$N, HO, F, H$_3$Si, H$_2$P, HS and Cl) found a single minimum corresponding to the *endo* form on the potential energy coordinate of the $Si \leftarrow N$ bond deformation[194]. The transannular bond was shown to be very soft and sensitive to the medium. The substituent effect on this bond length and mutual weakness of the axial bonds of the TBP silicon are well described by the 3c4e bonding model.

D. Benzosilatranes

The first monobenzosilatrane, 1-phenylbenzosilatrane (**161**), was synthesized in good yield by treatment of phenyltrimethoxysilane with 2-aminophenol and ethylene oxide (equation 167)[69].

$$PhSi(OMe)_3 + 2\text{-}HOC_6H_4NH_2 + 2H_2C\underset{O}{\overset{}{-\!\!\!-}}CH_2 \longrightarrow \mathbf{(161)} \quad (167)$$

An extension of this method allows one to obtain 1-organyl-3,7-dimethyl-10,11-benzosilatranes from the corresponding organyltrialkoxysilanes (equation 168)[401].

A standard synthesis of 1-organyltribenzosilatranes is the reaction of a trifunctional silane with tris(2-hydroxyphenyl)amine (equation 169)[402].

Tris(2-hydroxyphenyl)amine reacts with organyltrichlorosilanes, e.g. phenyltrichlorosilane, more slowly and at higher temperature than with the acetoxy derivative[402].

The striking contrast between these two reagents was explained by the operation of different mechanisms for the processes. The reaction of tris(2-hydroxyphenyl)amine with acetoxysilane may involve simultaneous nucleophilic attack at the silicon and electrophilic attack upon the leaving acetoxy group via a six-membered transition state. Similar reactions with alkoxy- or chlorosilanes are likely to occur via a disfavored four-membered strained transition state[402].

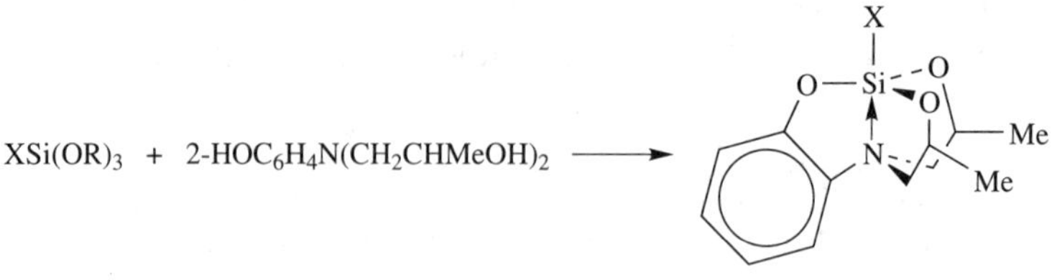

$XSi(OR)_3$ + 2-$HOC_6H_4N(CH_2CHMeOH)_2$ ⟶ [structure] (168)

	X	Yield (%)
(162)	Ph	50
(163)	ClCH$_2$	86
(164)	Cl$_2$CH	42
(165)	CH$_3$CHCl	74
(166)	Cl(CH$_2$)$_3$	73

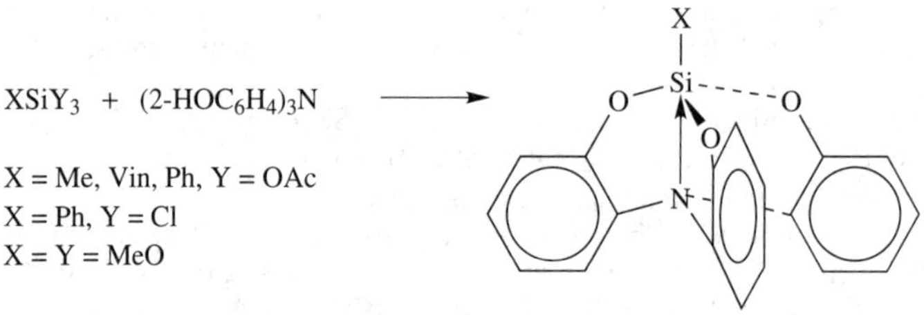

$XSiY_3$ + (2-HOC_6H_4)$_3$N ⟶ [structure] (169)

X = Me, Vin, Ph, Y = OAc
X = Ph, Y = Cl
X = Y = MeO

X	Y	Solvent	T (°C)	Yield (%)
Ph	OAc	CCl$_4$	75	89
Ph	Cl	CCl$_4$	75	0
Ph	Cl	Bu$_2$O	140	74

Nevertheless, more accessible organyltrichlorosilanes are good reagents for the preparation of various organyltribenzosilatranes[403], including *o*-substituted derivatives[404,405]. Tribenzosilatrane **(167)** having four alkyl chains was obtained in 52% yield from *n*-octyltrichlorosilane and tris(2-hydroxy-4-dodecylphenyl)amine by reaction 169[404,405]. This compound was designed for fabricating a new type of material where the overall global shape could facilitate the formation of cubic symmetry mesophases.

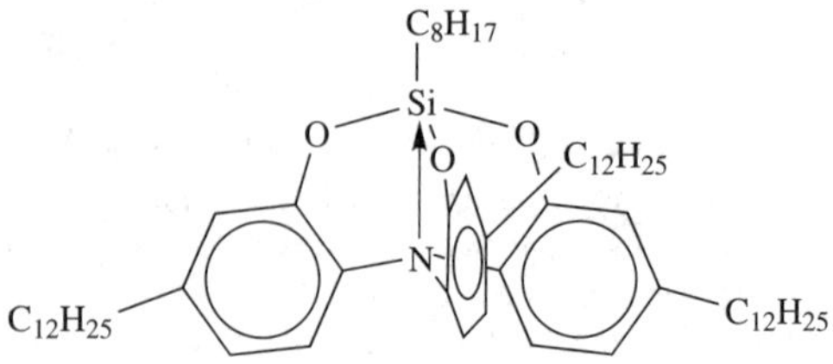

(167)

The reactions of tris(2-hydroxyphenyl)amine with trichloro-, trimethoxy- or triacetoxysilanes in boiling toluene afford the corresponding Si-substituted derivatives instead of 1-hydrotribenzosilatrane (equation 170)[402]. As evident from this reaction, the H—Si bond of 1-hydrotribenzosilatrane is very sensitive to electrophilic attack.

$$HSiX_3 + 3\,(2\text{-}HOC_6H_4)_3N \xrightarrow{-3\,HX} [HSi(OC_6H_4)_3N] \xrightarrow{+HX} XSi(OC_6H_4)_3N \quad (170)$$

$$X = Cl,\ MeO,\ AcO$$

1-Acetoxytribenzosilatrane undergoes hydrolysis to the corresponding disiloxane (**168**) when treated with moist solvents or with H_2O in acetonitrile solution (equation 171)[402].

$$AcOSi(OC_6H_4)_3N + H_2O \longrightarrow N(C_6H_4O)_3SiOSi(OC_6H_4)_3N \quad (171)$$
$$(\mathbf{168})$$

It was suggested that the rate-determining step of the acid-catalyzed hydrolysis of tribenzosilatranes $XSi(OC_6H_4)_3N$ (X = Me, Ph, 4-MeC$_6$H$_4$) involves a rapid protonation of an oxygen atom, followed by cleavage of the Si—O bond[406].

Solvolysis of 1-chloro- or 1-acetoxytribenzosilatranes with methanol or phenol yields the methoxy or phenoxy derivatives (equation 172)[402].

$$XSi(OC_6H_4)_3N + ROH \longrightarrow ROSi(OC_6H_4)_3N \quad (172)$$

$$X = AcO,\ R = Me$$
$$X = Cl,\ R = Ph$$

Unlike the reaction of 1-ethoxysilatrane with CF_3COOH, which results in the hydrogen-bonded adduct, 1-ethoxytribenzosilatrane is converted to disiloxane **168** under the same conditions (equation 173)[27].

$$2\,EtOSi(OC_6H_4)_3N \xrightarrow{2CF_3COOH} \mathbf{168} + 2\,CF_3COOEt + H_2O \quad (173)$$

X-ray studies show that substitution of the ethylene bridge for the *o*-phenylene group in the silatrane framework leads to an essential lengthening of the Si←N bond in Si-substituted 3,7-dimethylbenzosilatranes and tribenzosilatranes compared to that in silatranes[17] (Scheme 6). However, the delayed neutralization of an acidic titrant observed for 1-phenylbenzosilatrane[69] and the very low basicity of the nitrogen in tribenzosilatranes[402] imply a transannular bonding in solution.

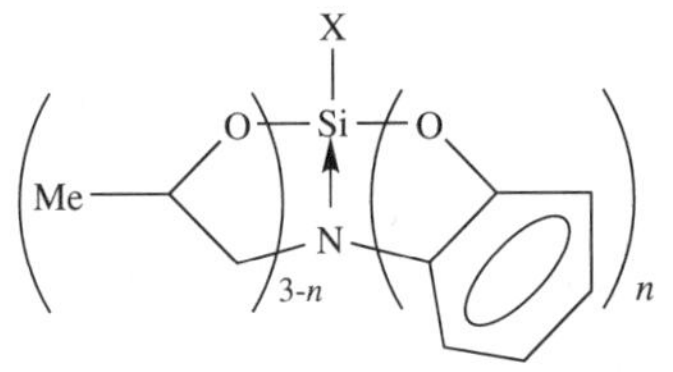

X	n	d(SiN) (Å)	d(XSi) (Å)	Δ_{Si} (Å)[b]	Reference
ClCH$_2$[a]	1	2.177	1.887	0.19	17
		2.185	1.883	0.21	
Ph	1	2.193	1.887	0.21	17
ClCH$_2$	3	2.256	1.864		17
Ph	3	2.344	1.853	0.29	121

[a]Two independent molecules.

[b]Δ_{Si} is the deviation of the TBP Si atom from the equatorial plane

SCHEME 6. Bond lengths in benzosilatranes

^{29}Si NMR chemical shifts of 1-ethoxytribenzosilatrane and the oxygen-bridged bis(tribenzosilatrane) were observed in a region which is typical for five-coordinate compounds[27]. A weaker spectroscopic basicity of the axial oxygen in the former compound as compared with that in related silatrane was found by comparing the O−H stretching frequency of phenol in CCl$_4$ solutions of these compounds[27]. This result is a strong indication of reduced transannular bonding in the tribenzosilatranes.

In the FAB mass spectra of 1-phenyl-3,7-dimethyl-10,11-benzosilatrane, the most intensive peak corresponds to the [M − X]$^+$ ion[291].

E. Triazasilatranes

2,8,9-Triazasilatranes (2,5,8,9-tetraaza-1-silatricyclo[3.3.3.0^{1,5}]undecanes) are very promising analogs of silatranes. Variation of the substituent at the silicon and the equatorial nitrogen atoms provides important changes in the structure and properties of these molecules. Since reactions of triazasilatranes could potentially proceed not only by participation of the silicon atom and its substituent but also by the equatorial nitrogens and their substituents and even by the axial nitrogen as well, their chemistry should be more diverse than the chemistry of silatranes. Many volatile triazasilatranes, primarily 1-aminotriazasilatranes containing five nitrogens in the immediate proximity of silicon, are candidates for the molecular organic chemical vapor deposition (MOCVD) of silicon nitride[26,27].

1-Organyl-2,8,9-triazasilatranes can be obtained from organyltris(dialkylamino)silanes and tris(2-aminoethyl)amines (equation 174)[407,408a]. The use of a solvent[312] or catalytic amounts of Me$_3$SiCl or (NH$_4$)$_2$SO$_4$[409] increases the yields of products **169–178**[312].

$$\text{XSi(NMe}_2)_3 + \text{(HNRCH}_2\text{CH}_2)_3\text{N} \longrightarrow \text{product} \quad (174)$$

	X	R	Yield (%)	Reference
(169)	H	H	72–84	408a, 409
(170)	H	Me	54	409
(171)	H	Me$_3$Si	61	409
(172)	Me	H	∼100	408a
(173)	Et	H	∼100	408a
(174)	Vin	H	∼100	408a
(175)	Ph	H	∼100	408a,b
(176)	EtO	H	84	409
(177)	EtO	Me	58	409
(178)	PhO	H	34	409

This approach is unsuitable for the preparation of 1-chlorotriazasilatrane due to the formation of polymeric mixtures[409]. However, 1-hydrotriazasilatrane (**169**) can be converted into the 1-chloro derivative **179** by treatment with CCl$_4$ (equation 175)[312,409]. Considerable lowering of the yield and increase in the reaction rate are observed in the presence of a catalytic amount of a bis(phosphine)platinum or -palladium dichloride[312,409].

169 or 170 ⟶ [structure] (175)

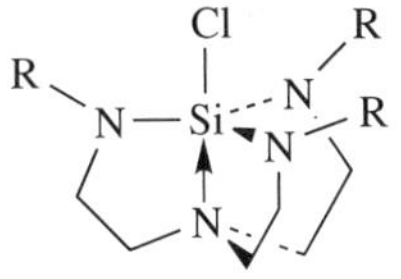

169/179: R = H   (179, 180)
170/180: R = Me

A similar type of transformation of 1-hydro-2,8,9-trimethyl-2,8,9-triazasilatrane (**170**) occurs by using *N*-chlorosuccinimide as a chlorinating agent (equation 175)[312].

Triazasilatranes **179** and **180** react with various nucleophiles such as organometallic reagents (equation 176), metal alkoxides (equation 177) and amides (equations 178 and 179) to give the substitution products **172, 181–184** as well as hydride transfer products **169, 170**. The relative ratios of these products depend on stereoelectronic factors, the nature of the nucleophilic reagents and the reaction conditions[312]. Thus, the reaction of triazasilatrane **180** with *n*-butyllithium affords **181a**, which is the product of substitution, while only 1-hydrotriazasilatrane (**170**) is formed from **180** and *tert*-butyllithium in a hydride transfer process.

179 or 180 + R'Li ⟶ [structure] + [structure]

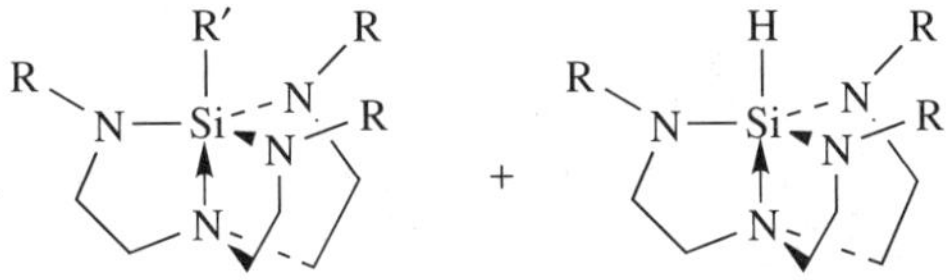

(172, 181–184)   R = H   (169, 170)
(172a, 181a–184a) R = Me

(176)

	R'	Yield (%)		R'	Yield (%)
(172)	Me	26	(172a)	Me	0
(181)	n-Bu	14	(181a)	n-Bu	95
(182)	s-Bu	<10	(182a)	s-Bu	0
(183)	t-Bu	<10	(183a)	t-Bu	0
(184)	Ph	<10	(184a)	Ph	96

179 or 180 + R'OM —THF→ [structure]

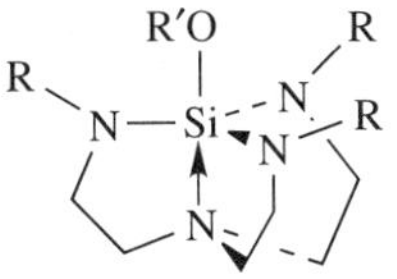

R'	R	M	Yield (%)
Me	H	Na	85
Me	Me	Na	85
Et	Me	Na	81
i-Pr	Me	Li	80
t-Bu	Me	Li	86

(177)

179 or **180**

→ NaNH$_2$ → (**185** or **185a**)   (178)

→ R$_2'$NLi → (**186 – 189**) R = H
(**186a – 189a**) R = Me   (179)

	R'	Yield (%)		R'	Yield (%)
(**186**)	Me	60	(**186a**)	Me	41
(**187**)	Et	19	(**187a**)	Et	25
(**188**)	i-Pr	0	(**188a**)	i-Pr	30
(**189**)	Me$_3$Si	50	(**189a**)	Me$_3$Si	88

It should be noted that similar nucleophilic chloride substitution reactions in the tetra-coordinate compound ClSi(NMe$_2$)$_3$ occur only at higher temperatures over prolonged reaction times[312].

In contrast to **179**, the reaction of **180** with C$_6$F$_5$Li affords not only the expected compound **190**, but also the fluoride transfer product **191** and the tetrafluorobenzene insertion product **192** in 1 : 2 : 1 ratio (equation 180)[312].

180 + C$_6$F$_5$Li ⟶ (**190**) + (**191**) + (**192**)   (180)

Silylation of the NH groups of triazasilatranes is effected by chlorosilanes in a benzene solution in the presence of Et$_3$N as an HCl acceptor[409,410]. The degree of substitution depends on the nature of the triazasilatrane and the chlorosilane (equation 181). In most cases disubstitution is the main process.

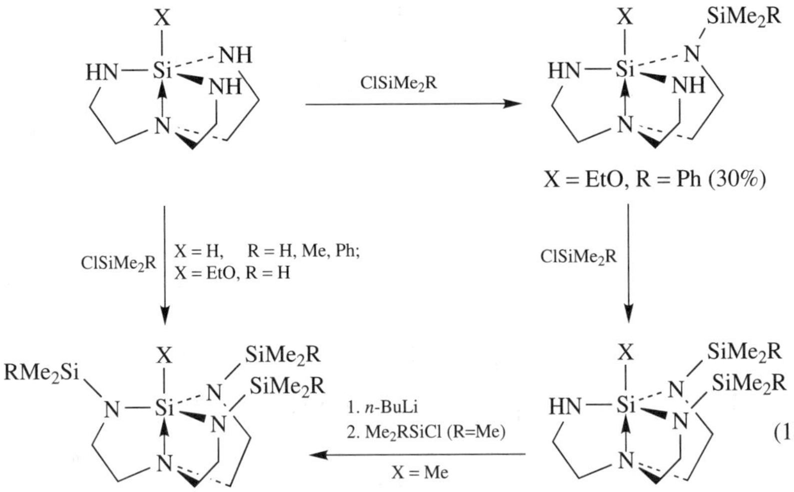

(181)

X	R	Yield (%)
Me	Me	
EtO	Me	74
EtO	Ph	45

The reaction of 1-ethoxytriazasilatrane (**193**) with diphenylchlorophosphine leads to N,N'-bis(diphenylphosphino)triazasilatrane (**195**) via the spectroscopically detected monosubstituted intermediate **194** (equation 182)[411].

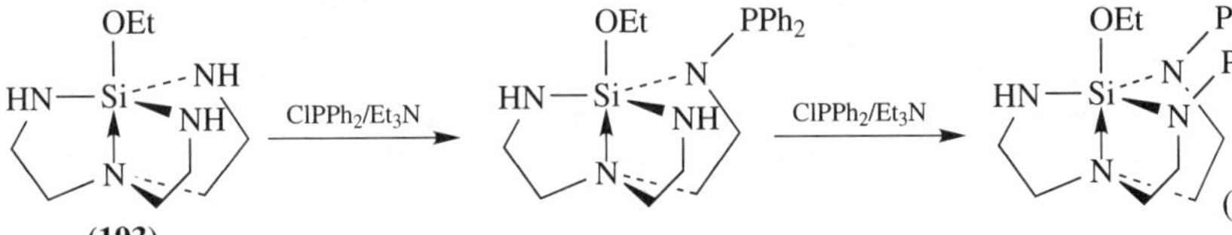

(182)

Triazasilatrane **195** behaves as a bidentate (P,P') or a tridentate (P,P',O) ligand to form chelate complexes with transition metal compounds[411]. On reaction with sulfur or methyl iodide, compound **195** is converted into the corresponding derivatives **196** and **197** (equations 183 and 184)[411].

The greatly increased steric repulsion between the axial and the bulky equatorial groups in trisilylated 1-methyltriazasilatrane (**198**) leads to a significant weakening of the transannular Si←N bond. The Si—N$_{ax}$ distance in **198** is the longest ever recorded in a triazasilatrane (2.775 Å)[410]. It makes the N$_{ax}$ atom sufficiently basic to react with CF$_3$SO$_3$Me, forming the cationic species **199** in which the silicon atom is tetracoordinated (equation 185)[410].

It seemed tempting to involve the N_{eq} atom in triazasilatranes in the quaternization reactions owing to its greater basicity compared with that of O_{eq} in silatranes. However, protonation of 1-hydrotriazasilatrane (169) by CF_3SO_3H or strong electrophiles such as CF_3SO_3Me or $CF_3SO_3SiMe_3$ results in cleavage of the atrane cage[412]. The successful synthesis of 200 and 201 the first stable models of incipient proton-assisted equatorial bond cleavage of five-coordinate intermediates was carried out by the following procedure. Nucleophilic attack of N_{eq} of the triazasilatrane on the silicon of the weakly electrophilic Me_3SiN_3 or Me_3SiSCN affords silylated products 171, 202 and 203 and free acids, HN_3 or HSCN, respectively. The latter acids form with 169 the salts 200 and 201, containing equatorially protonated cations (equation 186)[412,413].

(200) Y = N_3
(201) Y = SCN

(171) $R^1 = R^2 = R^3 = SiMe_3$
(202) $R^1 = H$, $R^2 = R^3 = SiMe_3$
(203) $R^1 = R^2 = H$, $R^3 = SiMe_3$

Presumably, the Si—H bond in the cationic species points toward one of the ammonium protons, thus favoring an easy hydrogen elimination by their thermolysis (equation 187)[413].

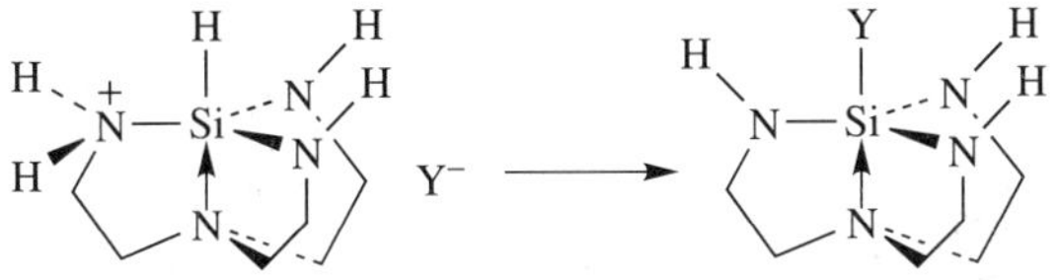

(187)

(200) Y = N$_3$
(201) Y = SCN

(204) Y = N$_3$
(205) Y = SCN

Triazasilatranes are hydrolytically less stable than silatranes[408a,414]. However, it is of interest that 1-aminotriazasilatrane (185) is stable to solvolysis by EtOH in C$_6$D$_6$ at room temperature[312]. Methanolysis of triazasilatranes 169, 172 and 176 at room temperature does not proceed by an initial displacement of an axial substituent but by attack on the tricyclic structure to afford tris(2-aminoethyl)amine and the corresponding trimethoxysilanes (equation 188)[414].

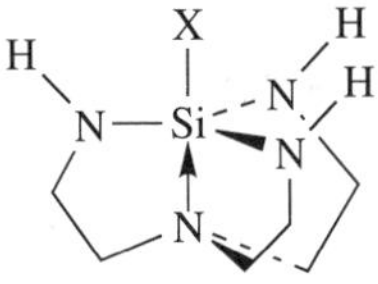

+ 3 MeOH ⟶ XSi(OMe)$_3$ + (H$_2$NCH$_2$CH$_2$)$_3$N

(188)

(169) X = H
(172) X = Me
(176) X = EtO

The rate and pathway of the methanolysis appear to depend strongly on the steric and electronic effects of the equatorial substituents. Because of the steric hindrance at the pentacoordinate silicon induced by N$_{eq}$ substitution in N-silylated triazasilatranes, nucleophiles can attack their four-coordinate silicon atoms (equation 189)[414].

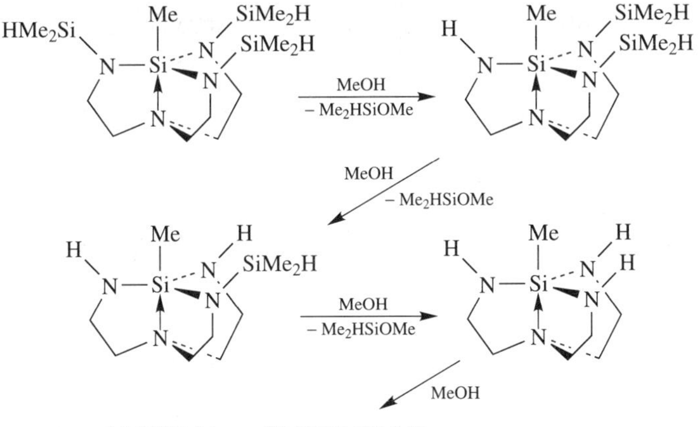

(189)

MeSi(OMe)$_3$ + (H$_2$NCH$_2$CH$_2$)$_3$N

On the other hand, the greatly weakened transannular bond in trisilylated 1-methyltriazasilatrane derivative 198 makes quite comparable solvolytic reactivity of the endocyclic and exocyclic N_{eq}–Si bonds. In this case cleavage of the cage structure can proceed by solvolytic displacement of the exocyclic silyl substituent, so that monocyclic intermediates **206** and **207** were observed in the stepwise opening of the tricyclic structure (equation 190)[414].

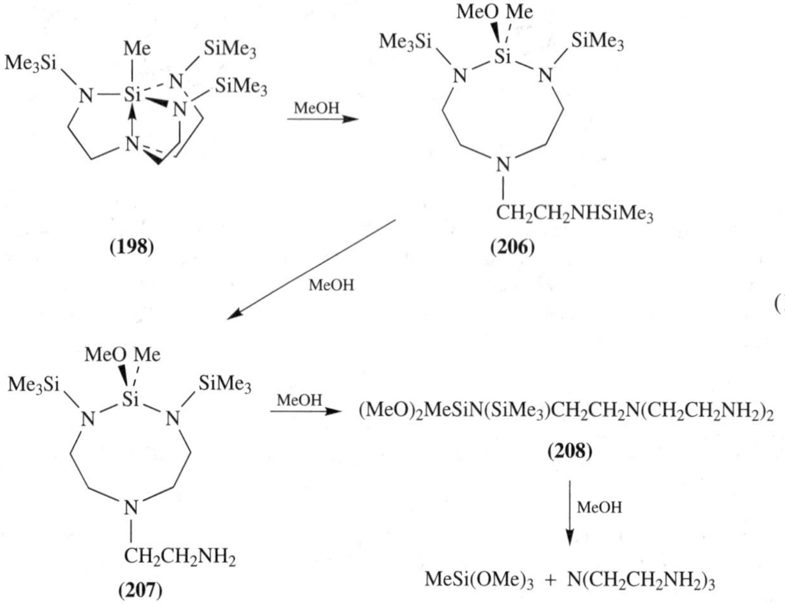

(190)

The reaction of 1-methyltriazasilatrane (**172**) with triethanolamine results in the formation of 1-methylsilatrane (**2**) (equation 191)[415].

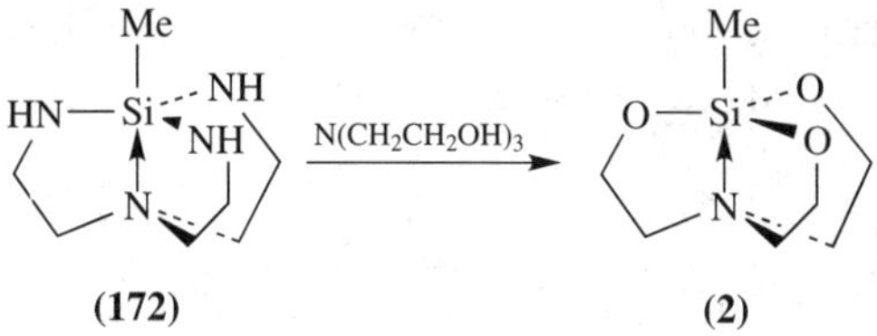

(191)

1-Hydrotriazasilatrane (**169**) easily undergoes dehydrocyclodimerization to give the dehydrocyclodimer **209** in high yield, 77–95% (equation 192). The reaction can occur on heating to 200 °C or at a low temperature (−35 °C) in the presence of $NaNH_2$[416].

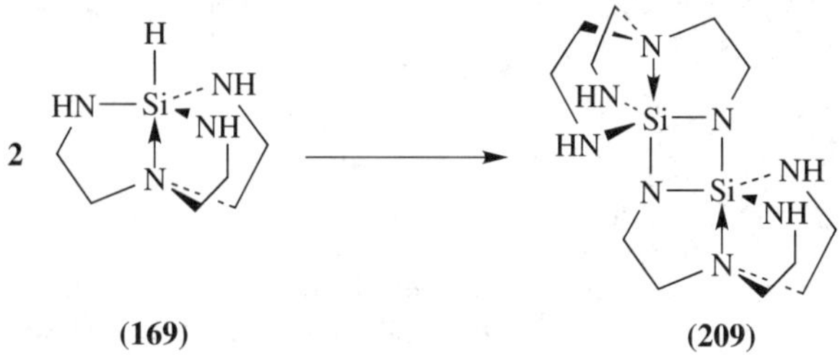

(192)

TABLE 7. ^{29}Si and ^{15}N NMR chemical shifts of triazasilatranes XSi(NHCH$_2$CH$_2$)$_3$N in CDCl$_3$[418]

X	$\delta(^{29}Si)$	$\Delta\delta(^{29}Si)$	$\delta(^{15}NC_3)$	$\delta(^{15}NH)$
H	−82.3	−47.5	−346.8	−350.1
Me	−68.3	−45.0	−354.5	−352.8
Vinyl	−79.2	−46.6	−352.8	−354.0
Ph	−77.2	−44.2	−352.2	−354.1

^{1}H and ^{29}Si NMR studies suggest a stronger Si←N bonding in triazasilatranes bearing the unsubstituted N$_{eq}$ atoms than in silatranes[408a,417]. Similar to other compounds of pentacoordinate silicon, these triazasilatranes are characterized by a higher shielding of the ^{29}Si nuclei than that in the model compounds of tetracoordinate silicon, XSi(NHPr)$_3$. The corresponding ^{29}Si NMR coordination shift $\Delta\delta(^{29}Si)$ is almost twice as large as that in silatranes and is close to −46 ppm (Table 7). As in the spectra of silatranes, the ^{15}N NMR signals for the apical cage nitrogen are shifted downfield on increasing the acceptor ability of the substituent X, i.e. on strengthening the Si←N bond[418].

Reduction of the s character of the axial X−Si bond causes a decrease in the one-bond and two-bond ^{29}Si-^{1}H and ^{29}Si-^{13}C coupling in triazasilatranes relative to the model compounds[417,418]. In the IR spectra the v(SiH) frequency of 1-hydrotriazasilatrane (**169**) is lower than that for HSi(NHAlk)$_3$ (2115 cm^{-1}). The solvent effect on v(SiH) in the IR spectra of **169** is weaker than for 1-hydrosilatrane. The $^1J(^{29}Si^1H)$ and v(SiH) values decrease with increasing solvent polarity as shown below[418].

Solvent	C$_6$D$_{12}$	CDCl$_3$	CD$_3$COCD$_3$	DMSO-d$_6$	Nujol
$^1J(^{29}Si^1H)$ (Hz)	178.4	176.6	175.1	171.6	
v(SiH) (cm^{-1})	1995	1995	1988	1980	1980

Apparently, this is a result of lengthening of the Si−H bond upon strengthening of the Si←N bond in polar media.

As a rule, the ^{29}Si shielding in 2,8,9-triazasilatranes decreases with substitution of the hydrogen on the equatorial NH functions by Me[312], SiR$_3$[414] or PPh$_2$[411] groups. For 1-diethylamino-2,8,9-trimethyltriazasilatrane, the ^{29}Si shielding falls about 38 ppm[312]. In the 2,8,9-tris-TMS derivative of 1-hydrotriazasilatrane the corresponding downfield ^{29}Si chemical shift is about 13 ppm, whereas in 1-ethoxytriazasilatrane bearing three Me$_2$HSi groups at the N$_{eq}$ atoms this shift is 24 ppm[414]. This indicates a strong weakening of the Si←N bond owing to steric crowding by bulky substituents attached to both the silicon and the equatorial nitrogen atoms[312,409,411,414]. However, 2,8,9-trimethyltriazasilatranes, XSi(NMeCH$_2$CH$_2$)$_3$N, with relatively small electronegative groups X (Cl, MeO, H$_2$N and Me) are characterized by upfield shifts relative to their unsubstituted analogs[312]. This phenomenon was explained by stretching of the X−Si bond and hence strengthening of the Si←N bond upon steric congestion.

Investigations carried out by Verkade and coworkers show that the ^{29}Si shielding of most 1-substituted triazasilatranes are sensitive to the phase and temperature[312,409,411,414]. In this respect triazasilatranes resemble silatranes. *Ab initio* calculations predict that the differences between the gas phase and the solution or solid state geometries of triazasilatranes should be rather less than in the case of silatranes[194]. The lower sensitivity of

TABLE 8. Si←N bond distances in triazasilatranes, and their salts

Compound	X	R¹	R²	d(SiN) (Å)	Reference
169	H	H	H	2.109	409
169 (in 210)	H	H	H	2.080	412
200 (in 210)	H	H, H⁺	H	2.087	412
201a	H	H, H⁺	Me₃Si	2.062	413
198	Me	Me₃Si	Me₃Si	2.775	410
175	Ph	H	H	2.132	408b
209	(N)	(Si)	H	2.139	416
177	EtO	Me	Me	2.135	409
196	EtO	H	Ph₂P(S)	2.214	411
191	F	Me	Me	2.034	312

the triazasilatrane structure to the physical environment could be explained by a stronger Si←N bonding and lower polarity of the molecules. The optimized length of the Si←N bond in triazasilatranes was found to be much shorter (by 0.3 Å) than in related silatranes. The dipole moments of triazasilatranes (3–4 D) are considerably lower than those of silatranes[408a].

Available X-ray data demonstrate great changes in the Si←N bond distance, d(SiN), in triazasilatranes on variation in the electronic and steric effects of the substituents at the silicon and equatorial nitrogens (Table 8). The shortest Si←N bond was found to be in 1-fluoro-2,8,9-trimethyltriazasilatrane (**191**), possessing the very electronegative fluorine at the silicon atom. Lengthening of this bond in **196** as compared with **177** and an extremely long Si←N bond distance in **198** reveal a strong destabilizing repulsion between the substituents at Si and N_{eq} on the transannular bonding. A comparison of the d(SiN) values in 1-hydrotriazasilatrane (**169**) and in salt **201a** confirms the expectation from the 3c4e bonding model of strengthening the Si←N bond upon protonation of an equatorial nitrogen. An opposite relation between the corresponding bond lengths in the parent molecule **169** and cation **200** was observed for compound **210**, formed when **169** and **200** were cocrystallized. This could be explained by the differences in the packing of the lattices of **169**, **201a** and **210**.

Data listed in Tables 1 and 7 show that the Si←N bond lengths in triazasilatranes with $R^1 = R^2 = H$ or Me are nearly the same, or just shorter than in the related silatrane. At the same time, the X—Si bonds in the former compounds are definitely longer than in silatranes. For triazasilatranes **175, 177** and **191**, the corresponding elongations are 0.02–0.04 Å. Consequently, it is difficult to judge from these facts in which class of the compared compounds the total energy of the hypervalent (3c4e) X—Si←N bond is higher. The 3c4e bonding model predicts that this bond should be stronger in silatranes, where the silicon atom is surrounded by more electronegative equatorial substituents[12,28,174].

The value of the inductive substituent constant σ_I for the triazasilatranyl group (-0.38) was estimated from ^{13}C NMR data for 1-phenyltriazasilatrane[418] and indicates that the triazasilatranyl group is a stronger electron-releasing substituent compared with Si(NMe$_2$)$_3$ or Si(OEt)$_3$ groups ($\sigma_I = -0.03$[418] and -0.13[249], respectively). However, the donor ability of the triazasilatranyl group is lower than that of the silatranyl group ($\sigma_I = -0.54$[13,20,249]).

Mass spectra measured under EI conditions were shown to reflect the expected order of X—Si energy in triazasilatranes: F—Si > O—Si > N—Si > C—Si[312]. For the 1-alkyl derivatives, the cage cation moiety is generally the base peak.

F. Miscellaneous

There are some kinds of silatrane analogs whose chemistry and structure were not studied enough. Some data concerning their preparation and properties are discussed below.

The reaction of phenyltris(dimethylamino)silane with bis(2-hydroxyethyl)(2-aminoethyl)amine affords 1-phenyl-2-azasilatrane (**210**) in a small yield (equation 193)[291]. The compound is very unstable to moisture.

PhSi(NMe$_2$)$_3$ + (HOCH$_2$CH$_2$)$_2$NCH$_2$CH$_2$NH$_2$ ⟶ (**210**) (193)

Characteristic features of the FAB spectrum of **210** resemble those mentioned for silatranes[291]. However, in **210** only the nitrogen-containing half-cycle is destroyed to form ions [MH − CH$_2$NH]$^+$ with m/z 222 and [MH − C$_2$H$_2$NH]$^+$ with m/z 208.

1-Methyl-2-azasilatran-3-ones (**211a** and **211b**) are prepared from methyltrichlorosilane and silylated derivatives of amides of N,N-bis(2-hydroxyethyl)aminoacetic acid (equation 194)[419].

Me$_3$SiCl$_3$ + (Me$_3$SiOCH$_2$CH$_2$)$_2$N—CH$_2$C(O)NRSiMe$_3$ $\xrightarrow{-3\text{Me}_3\text{SiCl}}$ (194)

(**211a**) R = H (70%)
(**211b**) R = Me (80%)

The reactions of 1,2,2-trifunctional disilanes with triethanolamine, using the 'Dilution principle', result in the formation of compounds **212–214** (equation 195)[420] which are the silyl analogs of 2-carba-3-oxahomosilatranes (**94**).

$$\text{XMeRSiSiMeX}_2 + (\text{HOCH}_2\text{CH}_2)_3\text{N} \longrightarrow \quad (195)$$

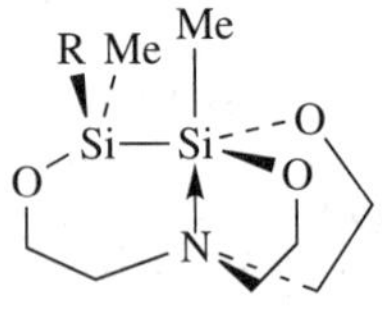

R = X = EtO
R = Me, X = Me$_2$N
R = Ph, X = Me$_2$N

(**212**) R = EtO
(**213**) R = Me
(**214**) R = Ph

The structures of these compounds were determined by ^{1}H NMR spectroscopy and X-ray diffraction. They demonstrate the preference of five- over six-membered chelating rings. The length of the Si←N bond in **213** is 2.768 Å[420].

1-Substituted 2,8,9-trithiasilatranes can be synthesized by the reactions of tris(2-mercaptoethyl)amine with tris(diethylamino)silane derivatives. Examples are given in equation 196[27,421] and equation 197[422].

$$\text{XSi(NEt}_2)_3 + (\text{HSCH}_2\text{CH}_2)_3\text{N} \longrightarrow \quad (196)$$

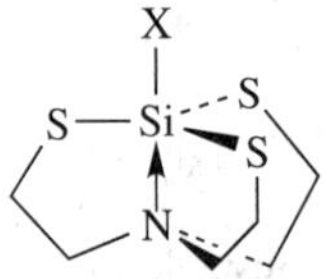

X = H, Me, Ph, EtO

$$\text{Si(NEt}_2)_4 + (\text{HSCH}_2\text{CH}_2)_3\text{N} \xrightarrow{\text{THF}} \text{Et}_2\text{NSi(SCH}_2\text{CH}_2)_3\text{N} \xrightarrow{\text{X-H}} \text{XSi(SCH}_2\text{CH}_2)_3\text{N}$$

$$\downarrow \text{HCl} \quad (197)$$

$$\text{XSi(SCH}_2\text{CH}_2)_3\text{N·HCl}$$

X = HCl·H$_2$NCH$_2$CH$_2$S (55%), 2HCl·H$_2$N(CH$_2$)$_3$NHCH$_2$CH$_2$S (68%)

Pentacoordination of the silicon atom in some 2,8,9-trithiasilatranes was proved by ^{29}Si NMR spectroscopy[27].

IV. ACKNOWLEDGMENTS

The authors wish to thank all our colleagues who have been involved in the chemistry of silatranes and whose names have been cited in the text. We are indebted to Elena Belogolova, Yuri Danilevich, Gennadij Dolgushin, Natal'ya Lazareva, Antonina Pestunovich, Elena Petrova and Nina Semenova for assistance with the preparation of the manuscript. We are also grateful to the RFBR for a grant (96-03-32718).

V. REFERENCES

1. C. L. Frye, G. E. Vogel and J. A. Hall, *J. Am. Chem. Soc.*, **83**, 996 (1961).
2. M. G. Voronkov, *Pure Appl. Chem.*, **13**, 35 (1966).
3. J. W. Turley and F. P. Boer, *J. Am. Chem. Soc.*, **90**, 4026 (1968).

4. (a) M. G. Voronkov, G. I. Zelchan and E. Ya. Lukevit. *Kremnij and Zhizn'*, Zinatne, Riga, 1971.
   (b) M. G. Voronkov, G. I. Zelcian and E. Ya. Lukevit, *Siliciul si Viata*, Editura Stiintifica, Bucuresti, 1974.
   (c) M. G. Voronkov, G. I. Zelchan and E. Lukevitz, *Silicium und Leben*, Akademie-Verlag, Berlin, 1975.
5. M. G. Voronkov and V. M. Dyakov, *Silatrany*, Nauka, Novosibirsk, 1978.
6. I. S. Birgele, A. A. Kemme, E. L. Kupcè, E. E. Liepins, I. B. Mazeika and V. D. Shatz, *Kremnijorganicheskie Proizvodnye aminospirtov* (Ed. E. Lukevics), Zinatne, Riga, 1987.
7. (a) E. Lukevics, O. A. Pudova and R. Sturkovich, *Molekulyarnaya Struktura Kremnijorganicheskikh Soedinenij*, Zinatne, Riga, 1988.
   (b) E. Lukevics, O. A. Pudova and R. Sturkovich, *Molecular Structure of Organosilicon Compounds*, Ellis Horwood, Chichester, 1989.
8. M. G. Voronkov, *Pure Appl. Chem.*, **19**, 399 (1969).
9. M. G. Voronkov, in *XXIVth International Congress of Pure and Applied Chemistry*, Butterworths, London, **4**, 45 (1973).
10. M. G. Voronkov, in *Biochemistry of Silicon and Related Problems* (Eds. G. Bendz and I. Lundqvist), Plenum Press, New York–London, 1978, pp. 395–433.
11. M. G. Voronkov, *Top. Curr. Chem.*, **84**, 77 (1979).
12. V. F. Sidorkin, V. A. Pestunovich and M. G. Voronkov, *Usp. Khim.*, **49**, 789 (1980); *Russ. Chem. Rev. (Engl. Transl.)*, **49**, 414 (1980).
13. M. G. Voronkov, V. M. Dyakov and S. V. Kirpichenko, *J. Organomet. Chem.*, **233**, 1 (1982).
14. P. Hencsei and L. Parkanyi, *Kém. Közl.*, **61**, 319 (1984); *Chem. Abstr.*, **103**, 105019m (1985).
15. P. Hencsei and L. Parkanyi, *Rev. Silicon, Germanium, Tin, Lead Compd.*, **8**, 191 (1985).
16. M. G. Voronkov, V. P. Baryshok, L. P. Petukhov, V. I. Rakhlin, R. G. Mirskov and V. A. Pestunovich, *J. Organomet. Chem.*, **358**, 39 (1988).
17. A. Greenberg and G. Wu, *Struct. Chem.*, **1**, 79 (1990).
18. P. Hencsei, *Struct. Chem.*, **2**, 21 (1991).
19. L. Parkanyi, P. Hencsei and L. Nyulaszi, *J. Mol. Struct.*, **377**, 37 (1996).
20. S. N. Tandura, M. G. Voronkov and N. V. Alekseev, *Top. Curr. Chem.*, **131**, 99 (1986).
21. V. E. Shklover, Yu. T. Struchkov and M. G. Voronkov, *Main Group Metal Chem.*, **11**, 109 (1988).
22. V. E. Shklover, Yu. T. Struchkov and M. G. Voronkov, *Usp. Khim.*, **58**, 353 (1989); *Russ. Chem. Rev. (Engl. Transl.)*, **58**, 211 (1989).
23. R. J. P. Corriu, *J. Organomet. Chem.*, **400**, 81 (1990).
24. M. G. Voronkov, *Izv. Akad. Nauk SSSR, Ser. Khim.*, 2664 (1991); *Bull. Acad. Sci. USSR, Div. Chem. Sci. (Engl. Transl.)*, **40**, 2319 (1991).
25. C. Chuit, R. J. P. Corriu, C. Reye and J. C. Young, *Chem. Rev.*, **93**, 1371 (1993).
26. J. G. Verkade, *Acc. Chem. Res.*, **26**, 483 (1993).
27. J. G. Verkade, *Coord. Chem. Rev.*, **137**, 233 (1994).
28. V. A. Pestunovich, V. F. Sidorkin and M. G. Voronkov, in *Progress in Organosilicon Chemistry* (Eds. B. Marciniec and J. Chojnowski). Gordon and Breach, New York., 1995, pp. 9–82.
29. E. Lukevics and O. A. Pudova, *Khim. Geterotsikl. Soed.*, 1605 (1996); *Chem. Heterocycl. Compd. (Eng. Transl.)*, **32**, 1381 (1996).
30. W. S. Sheldrick, in *The Chemistry of Organic Silicon Compounds*, Part 1 (Eds. S. Patai and Z. Rappoport), Wiley, Chichester, 1989, pp. 227–303.
31. R. J. P. Corriu and J. C. Young, in *The Chemistry of Organic Silicon Compounds*, Part 2 (Eds. S. Patai and Z. Rappoport), Wiley, Chichester, 1989, pp. 1241–1288.
32. A. B. Finestone, US Patent 2953545 (Westinghouse Electric Corp.), 1960; *Chem. Abstr.*, **55**, 4045i (1961).
33. P. Hencsei, L. Bihatsi, I. Kovacz, E. Szalay, E. B. Karsai, A. Szollosy and M. Gal, *Acta Chim. Acad. Sci. Hung.*, **112**, 261 (1983).
34. S. N. Gurkova, A. I. Gusev, N. V. Alekseev and M. A. Ignatenko, *Zh. Strukt. Khim.*, **29** (2), 203 (1988); *J. Struct. Chem. (Engl. Transl.)*, **29**, 345 (1988).
35. S. N. Gurkova, A. I. Gusev, N. V. Alekseev and M. A. Ignatenko, *Metalloorg. Khim.*, **1**, 1251 (1988); *Organomet. Chem. USSR (Engl. Transl.)*, **1**, 684 (1988).
36. K. V. Pavlov, N. A. Viktorov and V. F. Mironov, *Izv. Akad. Nauk, Ser. Khim.*, 756 (1993); *Russ. Chem. Bull. (Engl. Transl.)*, **42**, 723 (1993).
37. G. S. Zaitseva, S. S. Karlov, A. V. Churakov, E. V. Avtomonov, J. Lorberth and D. Hertel, *J. Organomet Chem.*, **523**, 221 (1996).

38. M. G. Voronkov, O. G. Yarosh, L. V. Shchukina, E. O. Tsetlina, S. N. Tandura and I. M. Korotaeva, *Zh. Obshch. Khim.*, **49**, 614 (1979); *Chem. Abstr.*, **91**, 39563q (1979).
39. G. Cereveau, C. Chuit, R. J. P. Corriu and C. Reye, *J. Organomet. Chem.*, **328**, C17 (1987).
40. M. G. Voronkov, O. G. Yarosh, Z. G. Ivanova, V. K. Roman and A. I. Albanov, *Izv. Akad. Nauk. SSSR, Ser. Khim.*, 1403 (1987); *Bull. Acad. Sci. USSR, Div. Chem. Sci. (Engl. Transl.)*, **36**, 1297 (1987).
41. M. G. Voronkov, O. G. Yarosh, L. V. Shchukina and E. E. Kuznetsova, *Izv. Akad. Nauk SSSR, Ser. Khim.*, 2611 (1984); *Bull. Acad. Sci. USSR, Div. Chem. Sci. (Engl. Transl.)*, **33**, 2391 (1984).
42. L. Parkanyi, P. Hencsei and L. Bihatsi, *J. Organomet. Chem.*, **232**, 315 (1982).
43. L. Bihatsi and P. Hencsei, *Magy. Kem. Foly.*, **87**, 137 (1981); *Chem. Abstr.*, **95**, 169078d (1981).
44. F. Carre, G. Cerveau, C. Chuit, R. J. P. Corriu, N. K. Nayyar and C. Reye, *Organometallics*, **9**, 1989 (1990).
45. L. Birkofer and K. Grafen, *J. Organomet. Chem.*, **299**, 143 (1986).
46. Yu. E. Ovchinnikov, T. G. Kovyazina, V. E. Shklover, Yu. T. Struchkov, V. M. Kopylov and M. G. Voronkov, *Dokl. Akad. Nauk SSSR*, **297**, 108 (1987); *Dokl. Chem. (Engl. Transl.)*, **297**, 474 (1987).
47. T. K. Gar, N. Yu. Khromova, V. M. Nosova and V. F. Mironov, *Zh. Obshch. Khim.*, **50**, 1764 (1980); *J. Gen. Chem. USSR (Engl. Transl.)*, **50**, 1433 (1980).
48. (a) V. Gevorgyan, L. Borisova and E. Lukevics, *J. Organomet. Chem.*, **418**, C21 (1991).
(b) V. Gevorgyan, L. Borisova, A. Vjater, J. Popelis, S. Belyakov and E. Lukevics, *J. Organomet. Chem.*, **482**, 73 (1994).
49. E. Lukevics, O. A. Pudova, J. Popelis and N. P. Erchak, *Zh. Obshch. Khim.*, **51**, 369 (1981); *J. Gen. Chem. USSR (Engl. Transl.)*, **51**, 300 (1981).
50. E. Lukevics and M. M. Ignatovich, *Khim. Geterotsikl. Soed.*, 725 (1992); *Chem. Heterocycl. Compd. (Engl. Transl.)*, **28**, 603 (1992).
51. E. Lukevics, V. Dirnens, A. Kemme and J. Popelis, *J. Organomet. Chem.*, **521**, 235 (1996).
52. E. Lukevics, G. Zelchans, T. J. Barton, A. Lapsina and I. Judeika, *Latv. PSR Zina. Akad. Vestis, Khim. Ser.*, 747 (1978); *Chem. Abstr.*, **90**, 137894x (1979).
53. C. M. Samour, US Patent 3118921 (Kendall Co.), 1964; *Chem. Abstr.*, **60**, 10715h (1964).
54. A. Daneshrad, C. Eaborn and D. R. M. Walton, *J. Organomet. Chem.*, **85**, 35 (1975).
55. I. Kovacs, L. Bihatsi and P. Hencsei, *Magy. Kem. Lapja*, **40**, 562 (1985); *Chem. Abstr.*, **106**, 18660g (1987).
56. M. G. Voronkov, V. M. D'yakov, O. N. Florensova, V. P. Baryshok, I. G. Kuznetsov and V. Chvalovsky, *Collect. Czech. Chem. Commun.*, **42**, 480 (1977).
57. M. G. Voronkov, V. P. Baryshok, S. N. Tandura, V. Yu. Vitkovskii, V. M. D'yakov and V. A. Pestunovich, *Zh. Obshch. Khim.*, **48**, 2238 (1978); *Chem. Abstr.*, **90**, 104051s (1979).
58. (a) C. L. Frye and R. D. Streu, *Main Group Metal. Chem.*, **16**, 211 (1993).
(b) C. L. Frye and R. D. Streu, *Main Group Metal. Chem.*, **16**, 217 (1993).
59. V. P. Baryshok, S. N. Tandura, G. A. Kuznetsova and M. G. Voronkov, *Metalloorg. Khim.*, **4**, 1150 (1991); *Organomet. Chem. USSR (Engl. Transl.)*, **4**, 568 (1991).
60. Y. Yang and C. Yin, *Gaodeng Xuexiao Huaxue Xuebo*, **7**, 430 (1986); *Chem. Abstr.*, **107**, 217703u (1987).
61. M. Tasaka, M. Hirotsu, M. Kojima, S. Utsuno and Y. Yoshikawa, *Inorg. Chem.*, **35**, 6981 (1996).
62. M. G. Voronkov, V. P. Baryshok and G. A. Kuznetsova, *Zh. Obshch. Khim.*, **66**, 1943 (1996); *Russ. J. Gen. Chem.*, **66**, 1889 (1996).
63. P. Hencsei and L. Bihatsi, *Period. Polytech., Chem. Eng.*, **26**, 35 (1982); *Chem. Abstr.*, **97**, 38992r (1982).
64. M. G. Voronkov and G. I. Zelchans, *Khim. Geterotsikl. Soed.*, 210 (1965); *Chem. Abstr.*, **63**, 9936b (1965).
65. M. G. Voronkov, I. B. Mazeika and G. I. Zelchans, *Khim. Geterotsikl. Soed.*, 58 (1965); *Chem. Abstr.*, **63**, 5506e (1965).
66. M. G. Voronkov and G. I. Zelchans, *Khim. Geterotsikl. Soed.*, 511 (1966); *Chem. Abstr.*, **66**, 85780q (1967).
67. J. Satge, G. Rima, M. Fatome, H. Sentenac-Roumanou and C. Lion, *Eur. J. Med. Chem.*, **24**, 48 (1989).
68. S. P. Narula, R. Shankar, M. Kumar, R. K. Vhandra and C. Janaik, *Inorg. Chem.*, **36**, 1268 (1997).
69. C. L. Frye, G. A. Vincent and W. A. Finzel, *J. Am. Chem. Soc.*, **93**, 6805 (1971).

70. M. G. Voronkov, V. M. D'yakov and V. P. Baryshok, *Zh. Obshch. Khim.*, **43**, 444 (1973); *Chem. Abstr.*, **79**, 5384u (1973).
71. M. G. Voronkov, V. M. D'yakov and L. I. Gubanova, *Izv. Akad. Nauk SSSR, Ser. Khim.*, 657 (1974); *Chem. Abstr.*, **81**, 13585s (1974).
72. M. G. Voronkov, V. M. D'yakov, Yu. A. Lukina, G. A. Samsonova and N. M. Kudyakov, *Izv. Akad. Nauk SSSR, Ser. Khim.*, 2794 (1974); *Chem. Abstr.*, **82**, 140247d (1975).
73. M. G. Voronkov, V. M. D'yakov and O. N. Florensova, *Zh. Obshch. Khim.*, **45**, 1902 (1975); *Chem. Abstr.*, **83**, 206365v (1975).
74. M. G. Voronkov, V. M. D'yakov, G. A. Samsonova, Yu. A. Lukina and N. M. Kudyakov, *Zh. Obshch. Khim.*, **45**, 2010 (1975); *Chem. Abstr.*, **84**, 17471m (1976).
75. M. G. Voronkov, V. M. D'yakov and V. P. Baryshok, *Zh. Obshch. Khim.*, **45**, 1650 (1975); *Chem. Abstr.*, **83**, 193426p (1975).
76. M. G. Voronkov, V. M. D'yakov and V. P. Baryshok, *Zh. Obshch. Khim.*, **47**, 797 (1977); *Chem. Abstr.*, **87**, 85075x (1977).
77. M. G. Voronkov, N. F. Lazareva, V. P. Baryshok, V. I. Dymchenko and N. A. Nedolya, *Izv. Akad. Nauk SSSR, Ser. Khim.*, 740 (1989); *Bull. Acad. Sci. USSR, Div. Chem. Sci. (Engl. Transl.)*, **38**, 666 (1989).
78. P. Hencsei, L. Parkanyi, V. Fulop, V. P. Baryshok, M. G. Voronkov and G. A. Kuznetsova, *J. Organomet. Chem.*, **346**, 315 (1988).
79. V. P. Baryshok, E. I. Brodskaya, N. F. Lazareva, G. A. Kuznetsova and M. G. Voronkov, *Metalloorg. Khim.*, **5**, 1136 (1992); *Organomet. Chem. USSR (Engl. Transl.)*, **5**, 555 (1992).
80. M. G. Voronkov, V. M. D'yakov, M. S. Sorokin, S. N. Tandura and N. F. Chernov, *Zh. Obshch. Khim.*, **45**, 1901 (1975); *Chem. Abstr.*, **83**, 206363t (1975).
81. M. G. Voronkov, N. F. Chernov, O. N. Florensova, V. P. Baryshok, E. E. Kuznetsova, T. I. Malkova, T. A. Pushechkina and L. I. Tokareva, *Zh. Obshch. Khim.*, **54**, 2017 (1984); *J. Gen. Chem. USSR (Engl. Transl.)*, **54**, 1800 (1984).
82. M. G. Voronkov, N. F. Chernov and E. O. Fedorova, *Zh. Org. Khim.*, **30**, 1263 (1994); *Russ. J. Org. Chem. (Engl. Transl.)*, **30**, 1328 (1994).
83. J. Wang, Q. Xie, R. Liao, J. Li and X. Lin, *Youji Huaxue*, 199 (1987); *Chem. Abstr.*, **108**, 150548x (1988).
84. M. G. Voronkov, V. M. D'yakov and L. I. Gubanova, *Zh. Obshch. Khim.*, **45**, 1905 (1975); *Chem. Abstr.*, **83**, 206369z (1975).
85. K. N. Grundy, J. D. Crabtree and A. E. Johnson, Brit. Patent 1243629 (Fiberglass Ltd.), 1968; South Afr. Patent 6806969, 1970; *Chem. Abstr.*, **73**, 121219u (1970).
86. M. G. Voronkov, M. S. Sorokin, V. M. D'yakov, F. P. Kletsko and N. N. Vlasova, *Zh. Obshch. Khim.*, **45**, 1649 (1975); *Chem. Abstr.*, **83**, 179196g (1975).
87. M. G. Voronkov, M. S. Sorokin and V. M. D'yakov, *Zh. Obshch. Khim.*, **45**, 1904 (1975); *Chem. Abstr.*, **83**, 206368y (1975).
88. M. G. Voronkov, M. S. Sorokin and V. M. D'yakov, *Zh. Obshch. Khim.*, **49**, 605 (1979); *Chem. Abstr.*, **91**, 57093r (1979).
89. N. F. Chernov, S. V. Shilin, O. N. Florensova, E. O. Novikova and M. G. Voronkov, *Zh. Obshch. Khim.*, **62**, 1300 (1992); *J. Gen. Chem. USSR (Engl. Transl.)*, **62**, 1069 (1992).
90. M. G. Voronkov, M. S. Sorokin, F. P. Kletsko, V. M. D'yakov, N. N. Vlasova and S. N. Tandura, *Zh. Obshch. Khim.*, **45**, 1395 (1975); *Chem. Abstr.*, **83**, 131674y (1975).
91. M. G. Voronkov, M. S. Sorokin and V. M. D'yakov, *Zh. Obshch. Khim.*, **45**, 1394 (1975); *Chem. Abstr.*, **83**, 131673h (1975).
92. M. G. Voronkov and M. S. Sorokin, *Zh. Obshch. Khim.*, **56**, 1818 (1986); *J. Gen. Chem. USSR (Engl. Transl.)*, **56**, 1608 (1986).
93. M. G. Voronkov, V. P. Baryshok, N. F. Lazareva, V. V. Saraev, T. I. Vakul'skaya, P. Hencsei and I. Kovacs, *J. Organomet. Chem.*, **368**, 155 (1989).
94. V. D. Sheludyakov and N. S. Fedotov, *Andrianovskie Chteniya*, Abstracts, Moscow, 1995, p. 2.175.
95. J.-M. Lin, L. Fang and W.-T. Huang, *Synth. React. Inorg. Met.- Org. Chem.*, **25**, 1467 (1995).
96. E. J. Lukevics, L. I. Libert and M. G. Voronkov, *Latv. PSR Zinat. Akad. Vestis, Khim. Ser.*, 451 (1972); *Chem. Abstr.*, **77**, 152270r (1972).
97. J. Lukasiak, A. Radecky and Z. Jamrogiewicz, *Rocz. Chem.*, **47**, 1975 (1973); *Chem. Abstr.*, **80**, 83126w (1974).
98. E. Lukevics, R. Ya. Moskovich, E. Liepins and I. S. Yankovskaya, *Zh. Obshch. Khim.*, **46**, 604 (1976); *Chem. Abstr.*, **85**, 21526w (1976).

99. G. Wu, K. Lu and Y. Wu, *Huaxue Togbao, Chemistry*, 10 (1983); *Chem. Abstr.*, **99**, 105333h (1983).
100. N. F. Lazareva, V. P. Baryshok and M. G. Voronkov, *Izv. Akad. Nauk, Ser. Khim.*, 341 (1995); *Russ. Chem. Bull. (Engl. Transl.)*, **44**, 333 (1995).
101. K. Lu and T. Wang, *Huaxue Togbao, Chemistry*, 30 (1995); *Chem. Abstr.*, **124**, 87120q (1996).
102. M. G. Voronkov, A. E. Pestunovich, E. I. Kositsyna, B. Z. Shterenberg, T. A. Pushechkina and N. N. Vlasova, *Z. Chem.*, **23**, 248 (1983).
103. V. E. Shklover, Yu. E. Ovchinnikov, Yu. T. Struchkov, V. M. Kopylov, T. G. Kovyazina and M. G. Voronkov, *Dokl. Akad. Nauk SSSR*, **284**, 131 (1985); *Chem. Abstr.*, **105**, 226714j (1986).
104. Yu. E. Ovchinnikov, V. E. Shklover, Yu. T. Struchkov, V. M. Kopylov, T. G. Kovyazina and M. G. Voronkov, *Zh. Strukt. Khim.*, **27(2)**, 133 (1986); *J. Struct. Chem. (Engl. Transl.)*, **27**, 287 (1986).
105. M. G. Voronkov, V. P. Baryshok, N. F. Lazareva, G. A. Kuznetsova, E. I. Brodskaya, V. V. Belyaeva, A. I. Albanov and L. S. Romanenko, *Metalloorg. Khim.*, **5**, 1323 (1992); *Organomet. Chem. USSR (Engl. Transl.)*, **5**, 648 (1992).
106. E. P. Kramarova, A. G. Shipov and Yu. I. Baukov, *Zh. Obshch. Khim.*, **62**, 2559 (1992); *J. Gen. Chem. USSR (Engl. Transl.)*, **62**, 2113 (1992).
107. M. G. Voronkov, N. N. Vlasova and L. I. Belousova, *Zh. Obshch. Khim.*, **65**, 270 (1995); *Chem. Abstr.*, **123**, 228269p (1995).
108. M. G. Voronkov, N. M. Kudyakov and A. I. Albanov, *Zh. Obshch. Khim.*, **56**, 1094 (1986); *J. Gen. Chem. USSR (Engl. Transl.)*, **56**, 962 (1986).
109. M. G. Voronkov, N. M. Kudyakov and A. I. Albanov, *Izv. Akad. Nauk SSSR, Ser. Khim.*, 1882 (1987); *Bull. Acad. Sci. USSR, Div. Chem. Sci. (Engl. Transl.)*, **36**, 1745 (1987).
110. R. Zhuo, Z.-R. Lu and J. Liao, *J. Organomet. Chem.*, **446**, 107 (1993).
111. Z.-R. Lu, R.-X. Zhuo, L.-R. Shen, X.-D. Zhang and L.-F. Shen, *J. Organomet. Chem.*, **489**, C38 (1995).
112. M. G. Voronkov, N. F. Chernov, O. M. Trofimova and T. N. Aksamentova, *Izv. Akad. Nauk, Ser. Khim.*, 1965 (1993); *Russ. Chem. Bull. (Engl. Transl.)*, **42**, 1883 (1993).
113. M. G. Voronkov, N. F. Chernov, O. M. Trofimova, Yu. E. Ovchinnikov, Yu. T. Struchkov and G. A. Gavrilova, *Izv. Akad. Nauk, Ser. Khim.*, 758 (1993); *Russ. Chem. Bull. (Engl. Transl.)*, **42**, 725 (1993).
114. M. G. Voronkov, V. P. Baryshok, N. F. Lazareva and G. G. Efremova, *Izv. Akad. Nauk, Ser. Khim.*, 384 (1995); *Russ Chem. Bull. (Engl. Transl.)*, **44**, 375 (1995).
115. J. W. Turley and F. P. Boer, *J. Am. Chem. Soc.*, **91**, 4129 (1969).
116. G. Wu and K. Lu, *Youji Huaxue, Org. Chem.*, 109 (1982); *Chem. Abstr.* **97**, 55882x (1982).
117. G. Wu, K. Lu and Y. Wu, in *Fundam. Res. Organomet. Chem., Proc. China–Japan–U.S. Trilateral Semin Organomet. Chem. Peking, June 1980* (Eds. M. Tsutsui, Y. I. Tshui and Y. Huang), van Nostrand Reinhold, New York, 1982, pp. 737–742; *Chem. Abstr.*, **98**, 4601t (1983).
118. M. Nasim, A. K. Saxena, I. P. Pal and L. M. Pande, *Synth. React. Inorg. Met.-Org. Chem.*, **17**, 1003 (1987).
119. M. Nasim, L. I. Livantsova, D. P. Krul'ko, G. S. Zaitseva, J. Lorberth and M. Otto, *J. Organomet. Chem.*, **402**, 313 (1991).
120. M. G. Voronkov, G. A. Kuznetsova and V. P. Baryshok, *Zh. Obshch. Khim.*, **53**, 1682 (1983); *J. Gen. Chem. USSR (Engl. Transl.)*, **53**, 1512 (1983).
121. S. N. Adamovich, V. Yu. Prokopyev, V. I. Rakhlin, R. G. Mirskov and M. G. Voronkov, *Synth. React. Inorg. Met.-Org. Chem.*, **21**, 1261 (1991).
122. M. G. Voronkov, V. M. D'yakov, E. E. Kuznetsova, O. N. Florensova, G. S. Dolgushina, G. V. Kozlova and V. B. Pukhnarevich, *Khim.-Pharm. Zh.*, **18**, 811 (1984); *Pharm. Chem. J. (Engl. Transl.)*, **18**, 467 (1984).
123. M. G. Voronkov, L. I. Gubanova, Yu. L. Frolov, N. F. Chernov, G. A. Gavrilova and N. N. Chipanina, *J. Organomet. Chem.*, **271**, 169 (1984).
124. E. Kupce, E. Liepins, A. Lapsina, I. Urtane, G. Zelchans and E. Lukevics, *J. Organomet. Chem.*, **279**, 343 (1985).
125. E. Lukevics, L. I. Libert and M. G. Voronkov, *Zh. Obshch. Khim.*, **38**, 1838 (1968); *Chem. Abstr.*, **70**, 11742q (1969).
126. H. Sakurai and A. Shirohata, Japan Kokai (Ajinomoto Co., Inc), 7831689, 1978; *Chem. Abstr.*, **89**, 43759j (1978).
127. G. I. Orlov, V. M. D'yakov, T. A. Tandura, E. F. Bugerenko and E. A. Chernyshov, *Zh. Obshch. Khim.*, **56**, 2653 (1986); *J. Gen. Chem. USSR (Engl. Transl.)*, **56**, 2349 (1986).

128. M. G. Voronkov and G. I. Zelchans, *Khim. Geterotsikl. Soed.*, 51 (1965); *Chem. Abstr.*, **63**, 5670d (1965).
129. G. I. Zelchans and M. G. Voronkov, *Khim. Geterotsikl. Soed.*, 371 (1967); *Chem. Abstr.*, **67**, 108112k (1967).
130. S. Cradock, E. A. V. Ebsworth and I. B. Muiry, *J. Chem. Soc., Dalton Trans.*, 25 (1975).
131. M. T. Attar-Bashi, C. Eaborn, J. Vencel and D. R. M. Walton, *J. Organomet. Chem.*, **117**, C87 (1976).
132. J. N. Dirk, J. M. Bellama and N. Ben-Zvi, *J. Organomet. Chem.*, **296**, 315 (1985).
133. I. P. Urtane, G. I. Zelchans, E. E. Liepin'sh, E. L. Kupche and E. Lukevits, *Zh. Obshch. Khim.*, **57**, 1110 (1987); *J. Gen. Chem. USSR (Engl. Transl.)*, **57**, 991 (1987).
134. M. G. Voronkov, V. P. Baryshok, G. A. Kuznetsova, V. Yu. Vitkovskii and A. G. Gorshkov, *Metalloorg. Khim.*, **3**, 181 (1990); *Organomet. Chem. USSR (Engl. Transl.)*, **3**, 99 (1990).
135. V. Gevorgyan, L. Borisova and E. Lukevics, *J. Organomet. Chem.*, **527**, 295 (1996).
136. P. Hencsei, Gy. Zsombok, L. Bihatsi and J. Nagy, *Period. Polytech., Chem. Eng.*, **23**, 185 (1979).
137. L. Párkányi, L. Bihatsi and P. Hencsei, *Cryst. Struct. Commun.*, **7**, 435 (1978).
138. P. Hencsey, L. Párkányi and I. Kovacs, *Khim. Geterotsikl. Soed.*, 1600 (1996); *Chem. Heterocycl. Compd.*, **32**, 1376 (1996).
139. P. Hencsei, I. Kovacs and V. Fülöp, *J. Organomet. Chem.*, **377**, 19 (1989).
140. A. A. Kemme, Ya. Ya. Bleidelis, V. M. D'yakov and M. G. Voronkov, *Izv. Akad. Nauk SSSR, Ser. Khim.*, 2400 (1976); *Chem. Abstr.*, **86**, 71739j (1977).
141. M. Nasim, V. S. Petrosyan, G. S. Zaitseva, J. Lorberth, S. Wocadlo and W. Massa, *J. Organomet. Chem.*, **441**, 27 (1992).
142. E. A. Zelbst, V. E. Shklover, Yu. T. Struchkov, A. A. Kashaev, M. P. Demidov, L. I. Gubanova and M. G. Voronkov, *Dokl. Akad. Nauk SSSR*, **260**, 107 (1981); *Dokl. Phys. Chem. (Engl. Transl.)*, **260**, 403 (1981).
143. A. A. Kemme, Ya. Ya. Bleidelis, V. M. D'yakov and M. G. Voronkov, *Zh. Strukt. Khim.*, **16**, 914 (1975); *Chem. Abstr.*, **84**, 82847c (1976).
144. V. E. Shklover, Yu. T. Struchkov, M. S. Sorokin and M. G. Voronkov, *Dokl. Akad. Nauk SSSR*, **274**, 615 (1984); *Dokl. Phys. Chem. (Engl. Transl.)*, **274**, 36 (1984).
145. M. P. Demidov, V. E. Shklover, Yu. L. Frolov, Yu. A. Lukina, V. M. D'yakov, Yu. T. Struchkov and M. G. Voronkov, *Zh. Strukt. Khim.*, **32(1)**, 177 (1991); *J. Struct. Chem. (Engl. Transl.)*, **32**, 154 (1991).
146. S. Wang and C. Hu, *K'o Hsueh T'ung Pao*, **26**, 603 (1981); *Chem. Abstr.*, **95**, 53095c (1981).
147. L. Párkányi, K. Simon and J. Nagy, *Acta Crystallogr.*, **30B**, 2328 (1974).
148. L. Párkányi, J. Nagy and K. Simon, *J. Organomet. Chem.*, **101**, 11 (1975).
149. L. Párkányi, P. Hencsei, L. Bihátsi, I. Kovacs and A. Szöllösy, *Polyhedron*, **4**, 243 (1985).
150. R. J. Garant, L. M. Daniels, S. K. Das, M. N. Janakiraman, R. A. Jacobson and J. G. Verkade, *J. Am. Chem. Soc.*, **113**, 5728 (1991).
151. A. A. Macharashvili, V. E. Shklover, Yu. T. Struchkov, V. P. Baryshok and M. G. Voronkov, *Dokl. Akad. Nauk SSSR*, **297**, 1123 (1987); *Chem. Abstr.*, **108**, 122284q (1988).
152. I. Kovacs, P. Hencsei and L. Párkányi, *Period. Polytech., Chem. Eng.*, **31**, 155 (1987).
153. M. W. Kim, D. S. Uh, S. Kim and Y. Do, *Inorg. Chem.*, **32**, 5883 (1993).
154. L. Párkányi, P. Hencsei, L. Bihátsi and T. Müller, *J. Organomet. Chem.*, **269**, 1 (1984).
155. A. A. Kemme, Ya. Ya. Bleidelis, V. A. Pestunovich, V. P. Baryshok and M. G. Voronkov, *Dokl. Akad. Nauk SSSR*, **243**, 688 (1978); *Chem. Abstr.*, **90**, 86560a (1979).
156. (a) C. Eaborn, K. J. Odell, A. Pidcock and G. R. Scollary, *J. Chem. Soc., Chem. Commun.*, 317 (1976).
     (b) G. R. Scollary, *Aust. J. Chem.*, **30**, 1007 (1977).
157. Yu. E. Ovchinnikov, Yu. T. Struchkov, N. F. Chernov, O. M. Trofimova and M. G. Voronkov, *Dokl. Akad. Nauk*, **328**, 330 (1993); *Chem. Abstr.*, **119**, 37830x (1993).
158. L. Parkanyi, K. Simon and J. Nagy, *J. Organomet. Chem.*, **101**, 11 (1975).
159. T.-M. Chung, Y. A. Lee, Y. K. Chung and I. N. Jung, *Organometallics*, **9**, 1976 (1990).
160. A.-S. Oh, Y. K. Chung and S. Kim, *Organometallics*, **11**, 1394 (1992).
161. J.-S. Lee, Y. K. Chung, D. Whang and K. Kim, *J. Organomet. Chem.*, **445**, 49 (1993).
162. M. J. Barrow, E. A. V. Ebsworth and M. M. Harding, *J. Chem. Soc., Dalton Trans.*, 1838 (1980).
163. V. A. Pestunovich, V. F. Sidorkin, O. B. Dogaev and M. G. Voronkov, *Dokl. Akad. Nauk SSSR*, **251**, 1140 (1980); *Chem. Abstr.*, **93**, 203827f (1980).
164. H. B. Bürgi and J. D. Dunitz, *Acc. Chem. Res.*, **16**, 153 (1983).

165. M. G. Voronkov, V. A. Pestunovich and Yu. A. Baukov, *Metalloorg. Khim.*, **4**, 1210 (1991); *Organomet. Chem. USSR (Engl. Transl.)*, **4**, 593 (1991).
166. L. Parkanyi, V. Fülöp. P. Hencsei and I. Kovacs. *J. Organomet. Chem.*, **418**, 173 (1991).
167. J. Wang, F. Maio, K. Lu, Y. Wu, G. Wu and S. Dou, *Jiegou Huaxue*, **5**(2), 78 (1986); *Chem. Abstr.*, **106**, 224889s (1987).
168. Z.-R. Lu, R.-X. Zhuo, L.-R. Shen and B.-S. Luo, *Main Group Metal Chem.*, **17**, 377 (1994).
169. Q. Shen and R. I. Hilderbrandt, *J. Mol. Struct.*, **64**, 257, (1980).
170. G. Forgacs, M. Kolonits and I. Hargittai, *Struct. Chem.*, **1**, 245 (1990).
171. Y. Apeloig, in *The Chemistry of Organic Silicon Compounds*, Part 1 (Eds. S. Patai and Z. Rappoport), Wiley, Chichester, 1989, pp. 57–225.
172. E. Magnusson, *J. Am. Chem. Soc.*, **112**, 7940 (1990).
173. J. I. Musher, *Angew. Chem., Int. Ed. Engl.*, **8**, 54 (1969).
174. V. F. Sidorkin, V. A. Pestunovich and M. G. Voronkov, *Dokl. Akad. Nauk SSSR*, **235**, 1363 (1977); *Dokl. Phys. Chem. (Engl. Transl.)*, **235**, 850 (1977).
175. R. W. Alder, *Tetrahedron*, **46**, 683 (1990).
176. M. G. Voronkov, V. V. Keiko, V. F. Sidorkin, V. A. Pestunovich and G. I. Zelchan, *Khim. Geterotsikl. Soed.*, 613 (1974); *Chem. Abstr.*, **81**, 119809h (1974).
177. M. G. Voronkov, V. F. Sidorkin, V. A. Shagun, V. A. Pestunovich and G. I. Zelchan, *Khim. Geterotsikl. Soed.*, 715 (1975); *Chem. Abstr.*, **83**, 96248r (1975).
178. M. G. Voronkov, V. F. Sidorkin, V. A. Shagun, V. A. Pestunovich and G. I. Zelchan, *Khim. Geterotsikl. Soed.*, 1347 (1976); *Chem. Abstr.*, **86**, 89065w (1977).
179. G. K. Balakhchi, V. V. Keiko, V. F. Sidorkin, V. A. Pestunovich and M. G. Voronkov, *Dokl. Akad. Nauk SSSR*, **275**, 393 (1984); *Chem. Abstr.*, **101**, 72816x (1984).
180. G. I. Csonka and P. Hencsei, *J. Organometal. Chem.*, **446**, 99 (1993).
181. V. F. Sidorkin, V. A. Pestunovich, V. A. Shagun and M. G. Voronkov, *Dokl. Akad. Nauk SSSR*, **233**, 386 (1977); *Dokl. Phys. Chem. (Engl. Transl.)*, **233**, 160 (1977).
182. P. Hencsei and G. Csonka, *Acta Chim. Acad. Sci. Hung.*, **106**, 285 (1981).
183. J.-C. Zhu, H.-J. Wu, C. S. Li, G. I. Martin, P.-C. Chen, G.-L. Wu and Z.-G. Lai, *J. Chim. Phys. Phys.-Chim. Biol.*, 407 (1984).
184. Yu. L. Frolov, S. G. Shevchenko and M. G. Voronkov, *J. Organomet. Chem.*, **292**, 159 (1985).
185. A. K. Kozyrev, R. G. Kutlubaev, N. P. Erchak and E. Lukevits, *Khim. Geterotsikl. Soed.*, 1314 (1989); *Chem. Heterocycl. Compd. (Engl. Transl.)*, **25**, 1096 (1989).
186. Zh. E. Grabovskaya, N. M. Klimenko and G. N. Kartsev, *Zh. Strukt. Khim.* **28**(6), 34 (1987); *J. Struct. Chem. (Engl. Transl.)*, **28**, 840 (1987).
187. G. N. Kartsev, N. M. Klimenko, Zh. E. Grabovskaya and G. M. Chaban, *Zh. Strukt. Khim.* **29**(6), 126 (1988); *J. Struct. Chem. (Engl. Transl.)*, **29**, 931 (1988).
188. (a) V. F. Sidorkin, G. K. Balakhchi, M. G. Voronkov and V. A. Pestunovich, *Dokl. Akad. Nauk SSSR*, **296**, 113 (1987); *Dokl. Phys. Chem. (Engl. Transl.)*, **296**, 400 (1987).
 (b) L. Onsager, *J. Am. Chem. Soc.*, **58**, 1486 (1936).
189. A. Greenberg, C. Plant and C. A. Venanzi, *J. Mol. Struct. (Theochem)*, **234**, 291 (1991).
190. M. S. Gordon, M. T. Carroll, J. H. Jensen, L. P. Davis, L. W. Burggraf and R. M. Guidry, *Organometallics*, **10**, 2657 (1991).
191. G. I. Csonka and P. Hencsei, *J. Mol. Struct. (Theochem)*, **283**, 251 (1993).
192. G. I. Csonka and P. Hencsei, *J. Organomet. Chem.*, **454**, 15 (1993).
193. G. I. Csonka and P. Hencsei, *J. Comput. Chem.*, **15**, 385 (1994).
194. M. W. Schmidt, T. L. Windus and M. S. Gordon, *J. Am. Chem. Soc.*, **117**, 7480 (1995).
195. J. E. Boggs, Ch. Peng, V. A. Pestunovich and V. F. Sidorkin, *J. Mol. Struct. (Theochem)*, **357**, 67 (1995).
196. V. F. Sidorkin and G. K. Balakhchi, *Struct. Chem.*, **5**, 187 (1994).
197. M. W. Schmidt, private communication.
198. V. F. Sidorkin, V. A. Shagun and V. A. Pestunovich, *Izv. Russ. Akad. Nauk, Ser. Khim.*, in press.
199. V. A. Klyuchnikov, G. N. Shvetz, M. S. Sorokin and M. G. Voronkov, *Dokl. Akad. Nauk SSSR*, **282**, 1174 (1985); *Chem. Abstr.*, **103**, 167173r (1985).
200. M. G. Voronkov, V. A. Klyuchnikov, T. F. Danilova, A. N. Korchagina, V. P. Baryshok and L. M. Landa, *Izv. Akad. Nauk SSSR, Ser. Khim.*, 1970 (1986); *Bull. Acad. Sci. USSR, Div. Chem. Sci. (Engl. Transl.)*, **35**, 1789 (1986).
201. M. G. Voronkov, V. A. Klyuchnikov, A. N. Korchagina T. F. Danilova, G. N. Shvets, V. P. Baryshok and V. M. D'yakov, *Izv. Akad. Nauk SSSR, Ser. Khim.*, 1976 (1986); *Bull. Acad. Sci. USSR, Div. Chem. Sci. (Engl. Transl.)*, **35** 1795 (1986).

202. M. G. Voronkov, V. P. Baryshok, V. A. Klyuchnikov, A. N. Korchagina and K. L. Pepekin, *J. Organomet. Chem.*, **359**, 169 (1989).
203. M. G. Voronkov, M. S. Sorokin, V. A. Klyuchnikov, G. N. Shvetz and V. I. Pepekin, *J. Organomet. Chem.*, **359**, 301 (1989).
204. E. I. Brodskaya and M. G. Voronkov, *Izv. Akad. Nauk SSSR, Ser. Khim.*, 1694 (1986); *Bull. Acad. Sci. USSR Div. Chem. Sci. (Engl. Transl.)*, **35**, 1545 (1986).
205. W. Parker, W. V. Steele, W. Stirling and I. Watt, *J. Chem. Thermodyn.*, **7**, 795 (1975).
206. P. Hencsei, G. Csonka, G. Zsombok and E. Gergö, *Period. Polytech. Chem. Eng.*, **27**, 263 (1983).
207. I. S. Birgele, I. B. Mazheika, E. E. Liepin'sh and E. Lukevits, *Zh. Obshch. Khim.*, **50**, 882 (1980); *J. Gen. Chem. USSR (Engl. Transl.)*, **50**, 711 (1980).
208. P. Hencsei, L. Ambrus, R. Farkas, L. Morvai, L. Szakacs and M. Gal, *Acta Chim. Acad, Sci. Hung.*, **126**, 145 (1989).
209. R. C. Gray and D. M. Hercules, *Inorg. Chem.*, **16**, 1426 (1977).
210. D. Wang, G. Wu, S. Li and C. Chen, *Fenzi Kexue Yu Huaxue Yanjiu*, **3(3)**, 35 (1983); *Chem. Abstr.*, **100**, 121162w (1984).
211. D. Wang, D. Zhang, K. Lu, Y. Wu and G. Wu, *Sci. Sin., Ser. B*, **26**, 9 (1983).
212. S. G. Shevchenko, V. P. Elin, G. N. Dolenko, V. P. Baryshok, V. P. Feshin, Yu. L. Frolov, L. N. Mazalov and M. G. Voronkov, *Zh. Strukt. Khim.*, **23**, 43 (1982); *J. Struct. Chem. (Engl. Transl.)*, **23**, 360 (1982).
213. A. P. Zemlyanov, A. T. Shuvaev, V. V. Krivitskii and M. G. Voronkov, *Izv. Akad. Nauk SSSR, Ser. Fiz.*, **36**, 255 (1972); *Chem. Abstr.*, **77**, 11885h (1972).
214. T. Shuvaev, A. P. Zemlyanov, Yu. V. Kolodyazhnyi, O. A. Osipov, V. N. Eliseev and M. M. Morgunova, *Zh. Strukt. Khim.*, **15**, 433 (1974); *Chem. Abstr.*, **81**, 63003w (1974).
215. S. Gradock, E. A. V. Ebsworth and J. B. Muire, *J. Chem. Soc., Dalton Trans.*, 25 (1975).
216. M. G. Voronkov, E. I. Brodskaya, V. V. Belyaeva, D. D. Chuvashov, D. D. Toryashinova, A. F. Ermikov and V. P. Baryshok, *J. Organomet. Chem.*, **311**, 9 (1986).
217. E. I. Brodskaya, M. G. Voronkov, D. D. Toryashinova, V. P. Baryshok, G. V. Ratovski, D. D. Chuvashov and V. G. Efremov, *J. Organomet. Chem.*, **336**, 49 (1987).
218. J. B. Peel and D. Wang, *J. Chem. Soc., Dalton Trans.*, 1963 (1988).
219. D. H. Aue, H. M. Webb and M. T. Bowers, *J. Am. Chem. Soc.*, **97**, 4136 (1975).
220. V. A. Petukhov, L. V. Gudovich, G. Zelchans and M. G. Voronkov, *Khim. Geterotsikl. Soed.*, 968 (1969); *Chem. Abstr.*, **72**, 105542t (1970).
221. M. G. Voronkov, E. I. Brodskaya, N. M. Deriglazov, V. P. Baryshok and V. V. Belyaeva, *J. Organomet. Chem.*, **225**, 193 (1982).
222. M. G. Voronkov, D. D. Chuvashov, G. V. Ratovski and E. I. Brodskaya, *Dokl. Akad. Nauk SSSR*, **292**, 384 (1987); *Chem. Abstr.*, **107**, 236819c (1987).
223. M. G. Voronkov, E. I. Brodskaya, V. V. Belyaeva and V. P. Baryshok, *Dokl. Akad. Nauk SSSR*, **261**, 1362 (1981); *Chem. Abstr.*, **96**, 142125a (1982).
224. M. G. Voronkov, Yu. L. Frolov, O. A. Zasyadko and I. S. Emel'yanov, *Dokl. Akad. Nauk SSSR*, **213**, 1315 (1973); *Chem. Abstr.*, **80**, 69917g (1974).
225. M. G. Voronkov, M. S. Sorokin, V. F. Traven', M. I. German and B. I. Stepanov, *Dokl. Akad. Nauk SSSR*, **243**, 926 (1978); *Chem. Abstr.*, **90**, 151155s (1979).
226. M. G. Voronkov, S. G. Shevchenko, E. I. Brodskaya, V. P. Baryshok, P. Reich, D. Kunat and Yu. L. Frolov, *Izv. Sib. Otd. Akad. Nauk SSSR, Ser. Khim.*, 135 (1981); *Chem. Abstr.*, **96**, 34361r (1982).
227. M. Imbenotte, G. Palavit and P. Legrand, *J. Raman Spectrosc.*, **14**, 135 (1983).
228. M. Imbenotte, G. Palavit, P. Legrand, J. P. Huvenne and G. I. Fleury, *J. Mol. Spectrosc.*, **102**, 40 (1983).
229. P. Hencsei, M. Gál and L. Bihátsi, *J. Mol. Struct.*, **114**, 391 (1984).
230. I. S. Ignat'ev, A. N. Lazarev, S. G. Shevchenko and V. P. Baryshok, *Izv. Akad. Nauk SSSR, Ser. Khim.*, 1526 (1986); *Bull. Acad. Sci. USSR, Div. Chem. Sci. (Engl. Transl.)*, **35**, 1375 (1986).
231. P. Hencsei and A. Sebestyen, *Acta Chim. Acad. Sci. Hung.*, **127**, 501 (1990).
232. J.-M. Lin, *Jiegou Huaxue*, **15**, 57 (1995); *Chem. Abstr.*, **124**, 232554r (1996).
233. (a) M. G. Voronkov, E. I. Brodskaya, P. Reich, S. G. Shevchenko, V. P. Baryshok and Yu. L. Frolov, *J. Organomet. Chem.*, **164**, 35 (1979).
    (b) M. G. Voronkov, E. I. Brodskaya, Yu. N. Udodov, Yu. M. Sapozhnikov and V. P. Baryshok, *Dokl. Akad. Nauk SSSR*, **313**, 1153 (1990); *Chem. Abstr.*, **114**, 102175a (1991).
234. M. G. Voronkov, E. I. Brodskaya, V. V. Belyaeva, V. P. Baryshok, M. S. Sorokin and O. G. Yarosh, *Dokl. Akad. Nauk SSSR*, **267**, 654 (1982); *Chem. Abstr.*, **98**, 160752y (1983).

235. S. N. Tandura, V. A. Pestunovich, G. I. Zelchan, V. P. Baryshok, Yu. A. Lukina, M. S. Sorokin and M. G. Voronkov, *Izv. Akad. Nauk SSSR, Ser. Khim.*, 295 (1981); *Bull. Acad. Sci. USSR, Div. Chem. Sci. (Engl. Transl.)*, **30**, 223 (1981).
236. J. M. Bellama, J. D. Nies and N. Ben-Zvi, *Magn. Reson. Chem.*, **24**, 748 (1986).
237. M. H. P. van Genderen and H. M. Buck, *Recl. Trav. Chim. Pays-Bas*, **106**, 449 (1987).
238. S. N. Tandura, V. A. Pestunovich, M. G. Voronkov, G. I. Zelchan, V. P. Baryshok and Yu. A. Lukina, *Dokl. Akad. Nauk SSSR*, **235**, 406 (1977); *Dokl. Phys. Chem. (Engl. Transl.)*, **235**, 688 (1977).
239. K. Kupce, E. Liepins and E. Lukevics, *Khim, Geterotsikl. Soed.*, **129** (1987); *Chem. Abstr.*, **108**, 56176c (1988).
240. E. E. Liepin'sh, I. S. Birgele, E. L. Kupche and E. Lukevits, *Zh. Obshch. Khim.*, **57**, 1723 (1987); *J. Gen. Chem. USSR (Engl. Transl.)*, **58**, 1537 (1987).
241. E. Liepins, I. Birgele, P. Tomsons and E. Lukevics, *Magn. Reson. Chem.*, **23**, 485 (1985).
242. V. A. Pestunovich, S. N. Tandura, M. G. Voronkov, V. P. Baryshok, G. I. Zelchan, V. I. Glukhikh, G. Engelhardt and M. Witanovsky, *Spectrosc. Lett.*, **11**, 339 (1978).
243. V. A. Pestunovich, S. N. Tandura, B. Z. Shterenberg, V. P. Baryshok and M. G. Voronkov, *Dokl. Akad. Nauk SSSR*, **253**, 400 (1980); *Chem. Abstr.*, **94**, 46558g (1981).
244. S. N. Tandura, Yu. A. Strelenko, M. G. Voronkov, N. V. Alekseev and O. G. Yarosh, *Dokl. Akad. Nauk SSSR*, **267**, 397 (1982); *Dokl. Phys. Chem. (Engl. Transl.)*, **267**, 421 (1982).
245. E. Liepins, I. Birgele, G. Zelchans and E. Lukevics. *Zh. Obshch. Khim.*, **49**, 1537 (1979); *Chem. Abstr.*, **91**, 210444c (1979).
246. R. K. Harris, J. Jones and S. Ng, *J. Magn. Reson.*, **30**, 521 (1978).
247. E. Kupce, E. Liepins, A. Lapsina, G. Zelchan and E. Lukevics, *J. Organomet. Chem.*, **251**, 15 (1983).
248. M. Karplus and J. A. Pople, *J. Chem. Phys.*, **38**, 2803 (1963).
249. M. G. Voronkov, V. I. Glukhikh, V. M. D'yakov, V. V. Keiko, G. A. Kuznetsova and O. G. Yarosh, *Dokl. Akad. Nauk SSSR*, **258**, 382 (1981); *Chem. Abstr.*, **95**, 114573f (1981).
250. E. E. Liepin'sh, I. S. Birgele, I. S. Solomennikova, A. F. Lapsinya, G. I. Zelchan and E. Lukevits, *Zh. Obshch. Chem.*, **50**, 2462 (1980); *J. Gen. Chem. USSR (Engl. Transl.)*, **50**, 1989 (1980).
251. J. Schraml, A. M. Krapivin, A. P. Luzin, V. M. Kilesso and V. A. Pestunovich, *Collect. Czech. Chem. Commun.*, **49**, 2897 (1984).
252. M. G. Voronkov, L. I. Gubanova, V. M. D'yakov, V. I. Glukhikh, N. G. Glukhikh and B. Shirchin, *Zh. Obshch. Khim.*, **55**, 1041 (1985); *J. Gen. Chem. USSR (Engl. Transl.)*, **55**, 928 (1985).
253. S. N. Tandura, M. G. Voronkov, A. V. Kisin, N. V. Alekseev, E. E. Shestakov, Z. A. Ovchinnikova and V. P. Baryshok, *Zh. Obshch. Chem.*, **54**, 2012 (1984); *J. Gen. Chem. USSR (Engl. Transl.)*, **54**, 1795 (1984).
254. H. Iwamija and G. E. Maciel, *J. Am. Chem. Soc.*, **115**, 6835 (1993).
255. V. A. Pestunovich, S. N. Tandura, M. G. Voronkov, G. Engelhardt, E. Lippmaa, T. Pehk, V. F. Sidorkin, G. I. Zelchan and V. P. Baryshok, *Dokl. Akad. Nauk SSSR*, **240**, 914 (1978); *Chem. Abstr.*, **89**, 107136p (1978).
256. P. Hencsei and H. C. Marsmann, *Acta Chim. Acad. Sci. Hung.*, **105**, 79 (1980).
257. W. H. Flygare, *Chem. Rev.*, **74**, 653 (1974).
258. V. F. Sidorkin, V. A. Pestunovich and M. G. Voronkov, *Magn. Reson. Chem.*, **23**, 491 (1985).
259. V. A. Pestunovich, S. N. Tandura, B. Z. Shterenberg, V. P. Baryshok and M. G. Voronkov. *Izv. Akad. Nauk SSSR, Ser. Khim.*, 2653 (1978); *Chem. Abstr.*, **90**, 86545r (1979).
260. M. Ya. Myagi, A. V. Samoson, E. T. Lippmaa, V. A. Pestunovich, S. N. Tandura, B. Z. Shterenberg and M. G. Voronkov, *Dokl. Akad. Nauk SSSR*, **252**, 140 (1980); *Chem. Abstr.*, **93**, 238047r (1980).
261. V. A. Pestunovich, B. Z. Shterenberg, S. N. Tandura, G. I. Zelchan, V. P. Baryshok, I. I. Solomennikova, I. P. Urtane, E. Lukevics and M. G. Voronkov, *Izv. Akad. Nauk SSSR, Ser. Khim.*, 467 (1981); *Chem. Abstr.*, **95**, 79412d (1981).
262. D. Kost and I. D. Kalikhman, Chapter 23 in this book.
263. V. A. Pestunovich B. Z. Shterenberg, E. T. Lippmaa, M. Ya. Myagi, M. A. Alla, V. P. Baryshok, L. P. Petukhov and M. G. Voronkov, *Dokl. Akad. Nauk SSSR*, **258**, 1410 (1981); *Dokl. Phys. Chem. (Engl. Transl.)*, **258**, 587 (1981).
264. V. A. Pestunovich, S. N. Tandura, B. Z. Shterenberg, V. P. Baryshok and M. G. Voronkov, *Izv. Akad. Nauk SSSR, Ser. Khim.*, 2159 (1979); *Chem. Abstr.*, **92**, 75393p (1980).
265. J.-C. Zhu, X.-Y. Sun, H.-J. Wu, L.-J. Jiang, B.-Q. Chen and G.-L. Wu, *Acta Chim. Sin. (Engl. Ed.)*, **44** 140 (1986).

266. V. A. Pestunovich, B. Z. Shterenberg, L. P. Petukhov, V. I. Rakhlin, V. P. Baryshok, R. G. Mirskov and M. G. Voronkov, *Izv. Akad. Nauk SSSR, Ser. Khim.*, 1935 (1985); *Bull. Acad. Sci. USSR, Div. Chem. Sci. (Engl. Transl.)*, **34**, 1790 (1985).
267. R. W. Taft and M. J. Kamlet, *Org. Magn. Reson.*, **14**, 485 (1980).
268. I. A. Koppel and V. A. Palm, in *Advances in Linear Free Energy Relationships* (Eds. N. B. Chapman and J. Shorter), Plenum Press, London, 1972, pp. 203–280.
269. J. Kamlet, J. L. M. Abboud and R. W. Taft, *Prog. Phys. Org. Chem.*, **13**, 485 (1981).
270. E. L. Kupche and E. Lukevits, *Khim. Geterotsikl. Soed.*, 701 (1989); *Chem. Heterocycl. Compd. (Engl. Transl.)*, **25**, 586 (1989).
271. V. A. Pestunovich, B. Z. Shterenberg, M. G. Voronkov, M. Ya. Myagi and A. V. Samoson, *Izv. Akad. Nauk SSSR, Ser. Khim.*, 1435 (1982); *Bull. Acad. Sci. USSR, Div. Chem. Sci. (Engl. Transl.)*, **31**, 1284 (1982).
272. E. Kupce and E. Lukevits, *J. Organomet. Chem.*, **358**, 67 (1988).
273. J. Lipowitz, *J. Am. Chem. Soc.*, **94**, 1582 (1972).
274. S. N. Tandura, V. A. Pestunovich, V. I. Glukhikh, V. P. Baryshok and M. G. Voronkov, *Spectrosc. Lett.*, **10**, 163 (1977).
275. E. E. Liepinsh, I. A. Zitsmane, G. I. Zelchan and E. Lukevits, *Zh. Obshch. Khim.*, **53**, 245 (1983); *J. Gen. Chem. USSR (Engl. Transl.)*, **53**, 215 (1983).
276. M. G. Voronkov, V. P. Feshin, V. M. D'yakov, L. S. Romanenko, V. P. Baryshok and M. V. Sigalov, *Dokl. Akad. Nauk SSSR*, **223**, 1133 (1975); *Chem. Abstr.*, **84**, 3940b (1976).
277. M. G. Voronkov, E. I. Brodskaya, V. V. Belyaeva, T. V. Kashik, V. P. Baryshok and O. G. Yarosh, *Zh. Obshch. Khim.*, **56**, 621 (1986); *J. Gen. Chem. USSR (Engl. Transl.)*, **56**, 550 (1986).
278. E. I. Brodskaya, M. G. Voronkov, V. V. Belyaeva, V. P. Baryshok and N. F. Lazareva, *Zh. Obshch. Khim.*, **63**, 2252 (1993); *Russ. J. Gen. Chem. (Engl. Transl.)*, **63**, 1564 (1993).
279. G. Cerveau, C. Chuit, E. Colomer, R. J. P. Corriu and C. Reyé, *Organometallics*, **9**, 2415 (1990).
280. K. A. Broka, V. T. Glezer, Ya. P. Stradyn' and G. I. Zelchan, *Zh. Obshch. Chem.*, **61**, 1374 (1991); *J. Gen. Chem. USSR (Engl. Transl.)*, **61**, 1252 (1991).
281. K. Broka, J. Stradiņš, V. Glezer, G. Zelčans and E. Lukevics, *J. Electroanal. Chem.*, **351**, 199 (1993).
282. T. Müller, P. Hencsei and L. Bihatsi, *Period. Polytech., Chem. Eng.*, **25**, 181 (1981).
283. Yu. W. Fang, S.-W. Hu, G.-L. Wu and M.-S. Xu, *Acta Chim. Sin. (Engl. Ed.)*, **41**, 630 (1983).
284. I. B. Mazheika, A. P. Gaukhman, I. I. Solomennikova, A. F. Lapsinya, G. I. Zelchan and E. Ya. Lukevits, *Zh. Obshch. Khim.*, **54**, 117 (1984); *J. Gen. Chem. USSR (Engl. Transl.)*, **54**, 103 (1984).
285. V. N. Bochkarev, A. E. Chernyshov, V. Yu. Vitkovskii and M. G. Voronkov, *Zh. Obshch. Khim.*, **55**, 1354 (1985); *J. Gen. Chem. USSR (Engl. Transl.)*, **55**, 1209 (1985).
286. L. Yan, W. Chai, G. Wang, G. Wu, K. Lu and Y. Luo, *Org. Mass Spectrom.*, **22**, 279 (1987).
287. P. Hencsei, I. Kovacs, I. Bihatsi, E. B. Karsai, T. Müller and P. Miklos, *Period. Polytech., Chem. Eng.*, **33**, 1 (1989).
288. M. G. Voronkov, V. P. Baryshok, N. F. Lazareva, V. V. Saraev, T. I. Vakul'skaya, P. Hencsei and I. Kovacs, *J. Organomet. Chem.*, **368**, 155 (1989).
289. S. Rozite, I. Mazeika, A. Gaukhman, N. P. Erchak, L. M. Ignatovich and E. Lukevics, *Metalloorg. Khim.*, **2**, 1389 (1989); *Organomet. Chem. USSR (Engl. Transl.)*, **2**, 736 (1989).
290. S. Rozite, I. Mazeika, A. Gaukhman, N. P. Erchak, L. M. Ignatovich and E. Lukevics, *J. Organomet. Chem.*, **384**, 257 (1990).
291. I. Mazeika, S. Grinberga, A. P. Gaukhman, G. I. Zelchan and E. Lukevics, *J. Organomet. Chem.*, **426**, 41 (1992).
292. A. E. Chernyshev and V. N. Bochkarev, *Zh. Obshch. Khim.*, **57**, 154 (1987); *Chem. Abstr.*, **108**, 56230r (1988).
293. V. A. Pestunovich, L. P. Petukhov, T. I. Vakul'skaya, V. P. Baryshok, V. K. Turchaninov, Yu. L. Frolov and M. G. Voronkov, *Izv. Akad. Nauk SSSR, Ser. Khim.*, 1470 (1978); *Chem. Abstr.*, **89**, 89839p (1978).
294. M. G. Voronkov, D.-S. D. Toryashinova, V. P. Baryshok, B. A. Shainyan and E. I. Brodskaya, *Izv. Akad. Nauk SSSR, Ser. Khim.*, 2673 (1984); *Bull. Acad. Nauk USSR, Div. Chem. Sci. (Engl. Transl.)*, **33**, 2447 (1984).
295. J. Lukasiak and Z. Jamrodgiewicz, *Acta Chim. Acad. Sci. Hung.*, **105**, 19 (1980).
296. J. Lukasiak and Z. Jamrodgiewicz, *Acta Chim. Acad. Sci. Hung.*, **115**, 167 (1984).
297. J. Lukasiak and Z. Jamrodgiewicz, *Acta Chim. Acad. Sci. Hung.*, **123**, 99 (1986).

298. A. Daneshrad, C. Eaborn, R. Eidensshink and D. R. M. Walton, *J. Organomet. Chem.*, **90**, 139 (1975).
299. M. G. Voronkov, S. G. Shevchenko, E. I. Brodskaya, Yu. L. Frolov, V. P. Baryshok, N. M. Deriglazov, E. S. Deriglazova and V. M. D'yakov, *Dokl. Akad. Nauk SSSR*, **230**, 627 (1976); *Chem. Abstr.*, **86**, 111751r (1977).
300. M. Imbenotte, G. Palavit and P. Legrand, *J. Raman Spectrosc.*, **15**, 293 (1984).
301. V. A. Pestunovich, L. P. Petukhov, B. Z. Shterenberg and M. G. Voronkov, *Izv. Akad. Nauk SSSR, Ser. Khim.*, 2169 (1981); *Bull. Acad. Nauk Sci., Div. Chem. Sci. (Engl. Transl.)*, **30**, 1784 (1981).
302. L. P. Kuzmenko, M. G. Voronkov, V. V. Keiko and V. A. Pestunovich, *Zh. Obshch. Khim.*, **53**, 105 (1983); *Chem. Abstr.*, **98**, 126217t (1983).
303. M. G. Voronkov, G. I. Zelchan, G. F. Tsybulya and P. G. Volfson, USSR Patent 299510, 1971; *Chem. Abstr.*, **75**, 63951 (1975j).
304. M. G. Voronkov and G. I. Zelchan, *Khim. Geterotsikl. Soed.*, 43 (1969); *Chem. Abstr.*, **71**, 3363x (1969).
305. G. H. Barnes and G. W. Schweitzer, US Patent 2967171 (Dow Corning Corp.), 1961; *Chem. Abstr.*, **55**, 9281e (1961).
306. R. J. Boyer, R. J. P. Corriu, R. Perz and C. Reye, *J. Organomet. Chem.*, **148**, C1 (1978).
307. M. G. Voronkov, G. I. Zelchan, G. F. Tsybulya and L. P. Urtane, *Khim. Geterotsikl. Soed.*, 756 (1975); *Chem. Abstr.*, **83**, 164146s (1975).
308. M. G. Voronkov, L. P. Petukhov, T. I. Vakul'skaya, V. P. Baryshok, S. N. Tandura and V. A. Pestunovich, *Izv. Akad. Nauk SSSR, Ser. Khim.*, 1665 (1979); *Chem. Abstr.*, **91**, 211391v (1979).
309. M. G. Voronkov, L. P. Petukhov, V. I. Rakhlin, V. P. Baryshok, B. Z. Shterenberg, R. G. Mirskov and V. A. Pestunovich, *Izv. Akad. Nauk SSSR, Ser. Khim.*, 2412 (1981); *Bull. Acad. Sci. USSR, Div. Chem. Sci. (Engl. Transl.)*, **30**, 1991 (1981).
310. M. G. Voronkov, V. I. Rakhlin, S. N. Adamovich, L. P. Petukhov and R. G. Mirskov, *Izv. Akad. Nauk SSSR, Ser. Khim.*, 899 (1986); *Bull. Acad. Sci. USSR, Div, Chem. Sci. (Engl. Transl.)*, **35**, 819 (1986).
311. G. Cerveau, C. Chuit, R. J. P. Corriu, N. K. Nayyar and C. Reye, *J. Organomet. Chem.*, **389**, 159 (1990).
312. Y. Wan and J. G. Verkade, *J. Am. Chem. Soc.*, **117**, 141 (1995).
313. J. Wang, Q. Xie, R. Liao, J. Li and X. Lin, *Youji Huaxue*, **4**, 285 (1987); *Chem. Abstr.*, **108**, 94532y (1988).
314. J. Wang, Q. Xie, R. Liao, J. Li and X. Lin, *Chem. J. Chin. Univ.*, **9**, 466 (1988).
315. M. W. Kim, D. S. Un, H. C. Shin, J. Kim and Y. Do, *J. Korean Chem. Soc.*, **38**, 241 (1994); *Chem. Abstr.*, **121**, 49020n (1994).
316. V. A. Pestunovich, L. P. Petukhov, V. I. Rakhlin, B. Z. Shterenberg, R. G. Mirskov and M. G. Voronkov, *Dokl. Akad. Nauk SSSR*, **263**, 904 (1982); *Dokl. Phys. Chem. (Engl. Transl.)*, **263**, 128 (1982).
317. M. G. Voronkov, V. G. Komarov, A. I. Albanov, I. M. Korotaeva and E. I. Dubinskaya, *Izv. Akad. Nauk SSSR, Ser. Khim.*, 1415 (1978); *Chem. Abstr.*, **89**, 109712x (1978).
318. A. H. Schmidt, *Aldrichimica Acta*, **14**, 31 (1981).
319. M. G. Voronkov, L. P. Petukhov, V. I. Rakhlin, B. Z. Shterenberg, S. H. Hangazheev, R. G. Mirskov and V. A. Pestunovich, USSR Patent 1031181, 1985; *Chem. Abstr.*, **104**, 6017y (1986).
320. M. G. Voronkov, V. I. Rakhlin, L. P. Petukhov, R. G. Mirskov, B. Z. Shterenberg and V. A. Pestunovich, *Zh. Obshch. Khim.*, **54**, 2398 (1984); *J. Gen. Chem. USSR (Engl. Transl.)*, **54**, 2144 (1984).
321. M. G. Voronkov, L. P. Petukhov, V. I. Rakhlin, B. Z. Shterenberg, R. G. Mirskov, S. N. Adamovich and V. A. Pestunovich, *Zh. Obshch. Khim.*, **56**, 964 (1986); *J. Gen. Chem. USSR (Engl. Transl.)*, **56**, 849 (1986).
322. M. G. Voronkov, V. A. Pestunovich, L. P. Petukhov and V. I. Rakhlin, *Izv. Akad. Nauk SSSR, Ser. Khim.*, 699 (1983); *Bull. Acad. Sci. USSR, Div. Chem. Sci. (Engl. Transl.)*, **32**, 637 (1983).
323. L. P. Petukhov, E. V. Bakhareva, V. A. Pestunovich, A. B. Shapiro, P. I. Dmitriev and M. G. Voronkov, *Khim.-Pharm. Zh.*, **20**, 979 (1986); *Pharm. Chem. J. (Engl. Transl.)*, **20**, 562 (1986).
324. C. Eaborn, *J. Chem. Soc.*, 3077 (1950).
325. D. S. Uh, Y. Do, J. H. Lee and H. Suh, *Main Group Metal Chem.*, **16**, 131 (1993); *Chem. Abstr.*, **120**, 270557u (1994).

24. Silatranes and their tricyclic analogs

326. E. Lukevics, I. I. Solomennikova and G. Zelchans, *Zh. Obshch. Khim.*, **46**, 134 (1976); *Chem. Abstr.*, **84**, 150696e (1976).
327. M. G. Voronkov, Yu. A. Lukina, V. M. D'yakov and M. V. Sigalov, *Zh. Obshch. Khim.*, **53**, 803 (1983); *J. Gen. Chem. USSR (Engl. Transl.)*, **53**, 703 (1983).
328. A. Hosomi, S. Iijima and H. Sakurai, *Chem. Lett.*, 243 (1981).
329. M. G. Voronkov, V. P. Baryshok and N. F. Lazareva, *Izv. Akad. Nauk, Ser. Khim.*, 2075 (1996); *Russ. Chem. Bull. (Engl. Transl.)*, **45**, 1970 (1996).
330. S. S. Lee, E. Jeong and Y. K. Chung, *J. Organomet. Chem.*, **483**, 115 (1994).
331. M. G. Voronkov, N. F. Chernov and T. A. Dekina, *Dokl. Akad. Nauk SSSR*, **230**, 853 (1976); *Chem. Abstr.*, **86**, 16767s (1977).
332. R. Müller, *Organomet. Chem. Rev.*, **1**, 359 (1966).
333. R. Müller and H. J. Frey, *Z. Anorg. Allg. Chem.*, **368**, 113 (1969).
334. R. Müller and C. Dathe, *J. prakt. Chem.*, **22**, 232 (1963).
335. M. G. Voronkov, L. P. Petukhov, S. N. Adamovich, V. P. Baryshok, V. I. Rakhlin and B. Z. Shterenberg, *Izv. Akad. Nauk SSSR, Ser. Khim.*, 204 (1987); *Bull. Acad. Sci. USSR, Div. Chem. Sci. (Engl. Transl.)*, **36**, 182 (1987).
336. M. G. Voronkov, V. P. Baryshok, E. E. Kuznetsova, I. M. Remez and L. E. Deev, *Khim.-Farm. Zh.*, 44 (1991); *Pharm. Chem. J. (Engl. Transl.)*, **25**, 405 (1991).
337. A. A. Kamyshova, M. Nasim, G. S. Zaitseva and A. B. Terent'ev, *Metalloorg. Khim.*, **4**, 459 (1991); *Organomet. Chem. USSR (Engl. Transl.)*, **4**, 222 (1991).
338. E. Lukevics, V. Dirnens, N. Pokrovska, J. Popelis and A. Kemme, *Main Group Metal Chem.*, **18**, 337 (1995).
339. M. G. Voronkov and M. S. Sorokin, *Zh. Obshch. Khim.*, **54**, 2020 (1984); *J. Gen. Chem. USSR (Engl. Transl.)*, **54**, 1803 (1984).
340. M. G. Voronkov, N. M. Kudyakov and A. I. Albanov, *Zh. Obshch. Khim.*, **49**, 1525 (1979); *Chem. Abstr.*, **91**, 175431x (1979).
341. E. W. Colvin, *Silicon in Organic Synthesis*, Chap. 4, Butterworth, London, 1981, pp. 21–29.
342. E. Lukevics and N. P. Erchak, *Latv. PSR Zinat. Akad. Vestis, Khim. Ser.*, 250 (1975); *Chem. Abstr.*, **83**, 79323c (1975).
343. E. Lukevics and N. P. Erchak, *Zh. Obshch. Khim.*, **47**, 809 (1977); *Chem. Abstr.*, **87**, 53422s (1977).
344. M. G. Voronkov, S. N. Adamovich, N. M. Kudyakov, S. Yu. Khramtsova, V. I. Rakhlin and R. G. Mirskov, *Izv. Akad. Nauk SSSR, Ser. Khim.*, 488 (1986); *Bull. Acad. Sci. USSR, Div. Chem. Sci. (Engl. Transl.)*, **35**, 451 (1986).
345. M. Nasim, V. S. Petrosyan and G. S. Zaitseva, *J. Organomet. Chem.*, **430**, 269 (1992).
346. G. A. Kutyrev, A. A. Kapura, M. S. Sorokin, R. A. Cherkasov, M. G. Voronkov and A. N. Pudovik, *Dokl. Akad. Nauk SSSR*, **275**, 1104 (1984); *Dokl. Phys. Chem. (Engl. Transl.)*, **275**, 141 (1984).
347. G. A. Kutyrev, A. A. Kapura, M. S. Sorokin, R. A. Cherkasov, M. G. Voronkov and A. N. Pudovik, *Zh. Obshch. Khim.*, **55**, 1030 (1985); *J. Gen. Chem. USSR (Engl. Transl.)*, **55**, 918 (1985).
348. F. Cooke, R. Moerek, J. Schwindeman and F. Magnus, *J. Org. Chem.*, **45**, 1046 (1980).
349. S. N. Adamovich, V. Yu. Prokopyev, V. I. Rakhlin, R. G. Mirskov, B. Z. Shterenberg and M. G. Voronkov, *Izv. Akad. Nauk SSSR, Ser. Khim.*, 2839 (1989); *Bull. Acad. Sci. USSR, Div. Chem. Sci. (Engl. Transl.)*, **38**, 2603 (1989).
350. M. G. Voronkov, S. N. Adamovich, V. I. Rakhlin and R. G. Mirskov, *Zh. Obshch. Khim.*, **57**, 1661 (1987); *J. Gen. Chem. USSR (Engl. Transl.)*, **57**, 1481 (1987).
351. M. Nasim, L. I. Livantsova, A. V. Kisin, G. S. Zaitseva and V. S. Petrosyan, *Metalloorg. Khim.*, **3**, 949 (1990); *Organomet. Chem. USSR (Engl. Transl.)*, **3**, 486 (1990).
352. A. Hosomi and H. Sakurai, *Tetrahedron Lett.*, 1295 (1976).
353. T. K. Sarkar and N. H. Andersen, *Tetrahedron Lett.*, 3513 (1978).
354. M. G. Voronkov, S. N. Adamovich, V. I. Rakhlin, R. G. Mirskov and M. V. Sigalov, *Izv. Akad. Nauk SSSR, Ser. Khim.*, 2792 (1984); *Bull. Acad. Sci. USSR, Div. Chem. Sci. (Engl. Transl.)*, **33**, 2558 (1984).
355. M. G. Voronkov, S. N. Adamovich, V. I. Rakhlin and R. G. Mirskov, *Zh. Obshch. Khim.*, **54**, 475 (1984); *J. Gen. Chem. USSR (Engl. Transl.)*, **54**, 425 (1984).
356. M. V. Kilesso, V. I. Kopkov, A. S. Shashkov and B. N. Stepanenko, *Izv. Akad. Nauk SSSR, Ser. Khim.*, 1404 (1986); *Bull. Acad. Sci. USSR, Div. Chem. Sci. (Engl. Transl.)*, **35**, 1276 (1986).
357. J. E. Noll, J. L. Speier and B. F. Daubert, *J. Am. Chem. Soc.*, **73**, 3867 (1951).

358. M. G. Voronkov, Yu. A. Lukina, V. M. D'yakov, Yu. L. Frolov and S. N. Tandura, *Zh. Obshch. Khim.*, **52**, 349 (1982); *J. Gen. Chem. USSR (Engl. Transl.)*, **52**, 301 (1982).
359. M. G. Voronkov, V. M. D'yakov, G. A. Samsonova, N. M. Kudyakov, Yu. A. Lukina and E. K. Vugmeister, *Izv. Akad. Nauk SSSR, Ser. Khim.*, 2059 (1975); *Chem. Abstr.*, **84**, 17475r (1976).
360. V. M. D'yakov, N. I. Liptuga, G. A. Samsonova, Yu. A. Lukina, M. G. Voronkov and A. V. Kirsanov, USSR Patent 572466, 1976; *Chem. Abstr.*, **87**, 167538q (1977).
361. M. G. Voronkov and M. S. Sorokin, *Zh. Obshch. Khim.*, **49**, 2671 (1979); *Chem. Abstr.*, **92**, 164026p (1980).
362. M. G. Voronkov and M. S. Sorokin, *Zh. Obshch. Khim.*, **59**, 590 (1989); *J. Gen. Chem. USSR (Engl. Transl.)*, **59**, 522 (1989).
363. G. S. Zaitseva, A. I. Chernyavskii, Yu. I. Baukov and I. F. Lutsenko, *Zh. Obshch. Khim.*, **46**, 843 (1976); *Chem. Abstr.*, **85**, 46806a (1976).
364. L. Birkofer and W. Quittmann, *Chem. Ber.*, **118**, 2874 (1985).
365. V. E. Udre and E. J. Lukevits, *Khim. Geterotsikl. Soed.*, 493 (1973); *Chem. Heterocycl. Compd. (Engl. Transl.)*, **9**, 454 (1973).
366. E. Lukevics, A. Lapsina, G. Zelchans, A. Dauvarte and A. Zidermane, *Latv. PSR Zinat. Akad. Vestis, Khim. Ser.*, 338 (1978); *Chem. Abstr.*, **89**, 129581g (1978).
367. F. Ye, X. Luo and R. Zhuo, *Huaxue Shiji*, **17**, 186 (1995); *Chem. Abstr.*, **124**, 8886t (1996).
368. M. G. Voronkov, V. M. D'yakov, G. G. Efremova, G. A. Kuznetsova, N. K. Gusarova, S. V. Amosova and B. A. Trofimov, USSR Patent 722913, 1978; *Chem. Abstr.*, **93**, 114696g (1980).
369. M. G. Voronkov, A. E. Pestunovich, N. N. Vlasova, T. V. Kashik, T. I. Nikiforova, A. I. Albanov, Yu. N. Pozhidaev and S. V. Amosova, *Zh. Obshch. Khim.*, **63**, 869 (1993); *Russ. J. Gen. Chem. (Engl. Transl.)*, **63**, 609 (1993).
370. M. G. Voronkov and M. S. Sorokin, *Zh. Obshch. Khim.*, **56**, 1098 (1986); *J. Gen. Chem. USSR (Engl. Transl.)*, **56**, 962 (1986).
371. S. G. Smith and S. Winstein, *Tetrahedron*, **3**, 317 (1958).
372. N. F. Chernov, O. N. Florensova and M. G. Voronkov, *Metalloorg. Khim.*, **2**, 902 (1989); *Organomet. Chem. USSR (Engl. Transl.)*, **2**, 472 (1989).
373. Y. A. Lee, Y. K. Chung, Y. Kim and J. H. Jeong, *Organometallics*, **9**, 2851 (1990).
374. Y. A. Lee, Y. K. Chung, Y. Kim, J. H. Jeong, G. Chung and D. Lee, *Organometallics*, **10**, 3707 (1991).
375. S. S. Lee, J.-S. Lee and Y. K. Chung, *Organometallics*, **12**, 4640 (1993).
376. T.-Y. Lee, Y. K. Kang, Y. K. Chung, R. D. Pike and D. A. Sweigardt, *Inorg. Chim. Acta*, **214**, 125 (1993).
377. S. S. Lee, I. S. Lee and Y. K. Chung, *Organometallics*, **15**, 5428 (1996).
378. E. Lukevics, I. I. Solomennikova, G. I. Zelchans, I. A. Judeika, E. E. Liepins, I. S. Yankovskaya and I. B. Mazeika, *Zh. Obshch. Khim.*, **47**, 105 (1977); *Chem. Abstr.*, **86**, 171535h (1977).
379. A. Kemme, J. Bleidelis, I. Solomennikova, G. Zelchan and E. Lukevics, *J. Chem. Soc., Chem. Commun.*, 1041 (1976).
380. I. S. Jankovska, I. I. Solomennikova, I. Mazeika, G. Zelchans and E. Lukevics, *Latv. PSR Zinat. Akad. Vestis., Khim. Ser.*, 366 (1975); *Chem. Abstr.*, **83**, 96199j (1975).
381. E. Popowski, DDR Patent 106389, 1974; *Chem. Abstr.*, **81**, 169626s (1974); E. Popowski, M. Michalik and H. Kelling, *J. Organomet. Chem.*, **88**, 157 (1975).
382. G. I. Zelchan, A. F. Lapsinya and E. Lukevits, *Zh. Obshch. Khim.*, **53**, 465 (1983); *J. Gen. Chem. USSR (Engl. Transl.)*, **53**, 409 (1983).
383. M. G. Voronkov, S. V. Basenko, R. G. Mirskov, T. D. Toryashinova and E. I. Brodskaya, *Dokl. Akad. Nauk SSSR*, **268**, 102 (1982); *Dokl. Chem. (Engl. Transl.)*, **268**, 1 (1983).
384. M. G. Voronkov, S. V. Basenko, R. G. Mirskov, V. Yu. Vitkovskii and T. D. Toryashinova, *Zh. Obshch. Khim.*, **53**, 944 (1983); *J. Gen. Chem. USSR (Engl. Transl.)*, **53**, 832 (1983).
385. V. Fulop, A. Kalman, P. Hencsei, G. Csonka and I. Kovacs, *Acta Crystallogr.*, **C47**, 720 (1988).
386. J. Dai, J. Zhang, Y. Wu and G. Wu, *Jiegou Huaxue*, **2(3)**, 207 (1983); *Chem. Abstr.*, **101**, 141415e (1984).
387. J. Dai, J. Zhang and Y. Wu, *Jiegou Huaxue*, **2(2)**, 107 (1983); *Chem. Abstr.*, **100**, 94789x (1984).
388. L. Parkanyi, P. Hencsei, G. Csonka and I. Kovacs, *J. Organomet. Chem.*, **329**, 305 (1987).
389. L. Parkanyi, P. Hencsei and E. Popowski, *J. Organomet. Chem.*, **197**, 275 (1980).
390. A. Kemme, J. Bleidelis, A. F. Lapsina, M. Fleisher, G. I. Zelchans and E. Lukevics, *Latv. PSR Zinat. Akad. Vestis, Khim. Ser.*, 242 (1985); *Chem. Abstr.*, **103**, 87961f (1985).

391. M. G. Voronkov, V. Yu. Vitkovskii, S. N. Tandura, N. V. Alekseev, S. V. Basenko and R. G. Mirskov, *Izv. Akad. Nauk SSSR, Ser. Khim.*, 2696 (1982); *Bull. Acad. Sci. USSR, Div. Chem. Sci. (Engl. Transl.)* **31**, 2384 (1982).
392. E. L. Morehouse, US Patent 3032576 (Union Carbide Corp.), 1962; *Chem. Abstr.*, **57**, 9881h (1962).
393. E. Lukevics, L. I. Libert and M. G. Voronkov, *Zh. Obshch. Khim.*, **39**, 1784 (1969); *Chem. Abstr.*, **71**, 124571x (1969).
394. E. Lukevics, L. I. Libert and M. G. Voronkov, USSR Patent 1213588 (1969); *Chem. Abstr.*, **71**, 3465g (1969).
395. P. Hencsei, I. Kovacs and L. Parkanyi, *J. Organomet. Chem.*, **293**, 185 (1985).
396. A. Kemme, J. Bleidelis, G. Zelchans, I. Urtane and E. Lukevics, *Zh. Strukt. Khim.*, **18(2)**, 343 (1977); *J. Struct. Chem. (Engl. Transl.)*, **18**, 268 (1977); *Chem. Abstr.*, **87**, 68452a (1977).
397. F. P. Boer and J. W. Turley, *J. Am. Chem. Soc.*, **91**, 4134 (1969).
398. K. Jurkschat, C. Mugge, J. Schmidt and A. Tzschach, *J. Organomet. Chem.*, **287**, C1 (1985).
399. K. Jurkschat, A. Tzschach, J. Meunier-Piret and M. van Meersche, *J. Organomet. Chem.*, **317**, 145 (1986).
400. A. Tzschach and K. Jurkschat, *Pure Appl. Chem.*, **58**, 639 (1986).
401. M. G. Voronkov, V. P. Baryshok and V. M. D'yakov, *Zh. Obshch. Khim.*, **46**, 1188 (1976); *Chem. Abstr.*, **85**, 46810x (1976).
402. C. L. Frye, G. A. Vincent and G. L. Hauschildt, *J. Am. Chem. Soc.*, **88**, 2727 (1966).
403. M. A. Paz-Sandoval, C. Fernandez-Vincent, G. Uribe and R. Contreras, *Polyhedron*, **7**, 679 (1988).
404. C. Soulie, P. Bassoul and J. Simon, *J. Chem. Soc., Chem. Commun.*, 114 (1993).
405. C. Soulie and J. Simon, *New J. Chem.*, **17**, 267 (1993).
406. R. E. Timms, *J. Chem. Soc. (A)*, 1969 (1971).
407. G. E. Le Grow, US Patent 3576026 (Dow Corning Corp.), 1971; *Chem. Abstr.*, **75**, 37252h (1972).
408. (a) E. Lukevits, G. I. Zelchan, I. I. Solomennikova, E. E. Liepin'sh, I. S. Yankovskaya and I. B. Mazheika, *Zh. Obshch. Khim.*, **47**, 109 (1977); *J. Gen. Chem. USSR (Engl. Transl.)*, **47**, 98 (1977).
    (b) A. A. Macharashvili, V. E. Shklover, Yu. T. Struchkov, A. Lapsina, G. Zelcans and E. Lukevics, *J. Organomet. Chem.*, **349**, 23 (1988).
409. D. Gudat and J. G. Verkade, *Organometallics*, **8**, 2772 (1989).
410. D. Gudat, L. M. Daniels and J. G. Verkade, *J. Am. Chem. Soc.*, **111**, 8520 (1989).
411. D. Gudat, L. M. Daniels and J. G. Verkade, *Organometallics*, **9**, 1464 (1990).
412. J. Woning, L. M. Daniels and J. G. Verkade, *J. Am. Chem. Soc.*, **112**, 4601 (1990).
413. J. Woning and J. G. Verkade, *Organometallics*, **10**, 2259 (1991).
414. D. Gudat and J. G. Verkade, *Organometallics*, **9**, 2172 (1990).
415. J. Pinkas and J. G. Verkade, *Phosphorus, Sulfur, Silicon*, **93–94**, 333 (1994).
416. Y. Wan and J. G. Verkade, *Organometallics*, **15**, 5769 (1996).
417. S. N. Tandura, V. A. Pestunovich, M. G. Voronkov, G. I. Zelchans, I. I. Solomennikova and E. Lukevics, *Khim. Geterotsikl. Soed.*, 1063 (1977); *Chem. Abstr.*, **88**, 5715p (1978).
418. E. Kupce, E. Liepins, A. Lapsina, G. Zelchan and E. Lukevics, *J. Organomet. Chem.*, **333**, 1 (1987).
419. V. A. Pestunovich, N. F. Lazareva, O. B. Kozyreva, L. V. Klyba, G. A. Gavrilova and M. G. Voronkov, *Zh. Obshch. Khim.*, in press
420. J. Grobe, G. Henkel, B. Krebs and N. Voulgarakis, *Z. Naturforsch., B: Anorg. Chem., Org. Chem.*, **39B**, 341 (1984).
421. E. Lukevics, I. I. Solomennikova and G. I. Zelchans, USSR Patent 509049, 1974; *Bull. Izobret.*, No. 9, 217 (1977); *Chem. Abstr.*, **87**, 23483w (1977).
422. F. G. Rima, J. Satge, M. Fatome, J. D. Laval, H. Sentenac-Roumanou, C. Lion and M. Lazraq, *Eur. J. Med. Chem.*, **26**, 291 (1991).

CHAPTER 25

Tris(trimethylsilyl)silane in organic synthesis

C. CHATGILIALOGLU, C. FERRERI* and T. GIMISIS

I.Co.C.E.A., Consiglio Nazionale delle Ricerche, Via P. Gobetti 101, 40129 Bologna, Italy

I. INTRODUCTION	1539
A. General Aspects of Radical Chain Reactions	1540
B. Hydrogen Donor Abilities of Silanes	1541
II. AUTOXIDATION OF TRIS(TRIMETHYLSILYL)SILANE	1542
III. REDUCTIONS	1543
IV. HYDROSILYLATIONS	1552
V. CONSECUTIVE RADICAL REACTIONS	1556
A. Initiation by Atom (or Group) Abstraction	1557
B. Initiation by (TMS)$_3$Si• Radical Addition to Unsaturated Bonds	1572
VI. REFERENCES	1577

I. INTRODUCTION

The synthetic application of free-radical reactions has increased dramatically within the last decade. Nowadays, radical reactions can often be found driving the key steps of multistep chemical synthesis oriented towards the construction of complex natural products[1]. Tributyltin hydride is the most commonly used reagent for the reduction of functional groups and formation of C−C bonds either inter- or intramolecularly (cyclization)[2,3]. However, it is well known that there are several problems associated with organotin compounds[4]. The main drawback consists of the incomplete removal of highly toxic tin by-products from the final material. For this reason, it is inappropriate to extend these reactions to industrial and medicinal chemistry. It is therefore of great importance to find new reagents which are suitable from a toxicological point of view and are analogous to organotin hydrides in their chemistry, so that the knowledge gained in the last thirty years in the field of free radicals can be further applied.

---

*On sabbatical leave. Permanent address: Dipartimento di Chimica Organica e Biologica, Università di Napoli 'Federico II', Via Mezzocannone 16, 80134 Napoli, Italy.

The chemistry of organic silicon compounds, Vol. 2
Edited by Z. Rappoport and Y. Apeloig © 1998 John Wiley & Sons Ltd

Ten years ago, one of us introduced tris(trimethylsilyl)silane, (TMS)$_3$SiH, as an alternative approach[5,6]. (TMS)$_3$SiH has indeed quickly proven to be a valid alternative to tin hydride for the majority of its radical chain reactions, although in some cases the two reagents can complement each other[4,7]. Although the above considerations coupled with the ease of purification and lack of toxicity of (TMS)$_3$SiH and its by-products should have rendered this reducing agent as the reagent of choice in radical reactions, tributyltin hydride still remains the reagent primarily used by synthetic chemists who decide to incorporate one or more radical steps in their synthetic schemes. Attempting to explain this fact by cost arguments is no longer valid, especially when one takes into account the effective loss of valuable products through exhausting purification procedures for the elimination of stubborn tributyltin residues. In our opinion the following two arguments can explain why (TMS)$_3$SiH is still waiting to be fully recognized and widely used as radical-based reducing agent: (a) the fact that Bu$_3$SnH has long been established as a reliable radical reducing agent, supported by a vast literature which highlights its advantages and also clarifies its limitations; (b) more specifically, the fact that the slower hydrogen donation of (TMS)$_3$SiH is considered in many cases a disadvantage for its use in radical chemistry. It is also worth emphasizing that there is a growth of cases where the two reagents behave differently.

It is not our purpose to review the entire chemistry of (TMS)$_3$SiH. Rather, this survey reflects the scientific interest of our group and deals mainly with the recent literature on the application of (TMS)$_3$SiH in organic synthesis. We do not also wish to discuss further the differences between the silicon and tin hydrides, although some salient features of the two reagents will be included. In order to help those readers who are not familiar with radical chemistry, we discuss below some general aspects of radical chain reactions as well as the abilities of group 14 hydrides, particularly silanes, to donate hydrogen atom.

A. General Aspects of Radical Chain Reactions

The majority of radical reactions of interest to synthetic chemists are chain processes[2,3]. Scheme 1 represents the simple reduction of an organic halide by silicon hydride as an example of a chain process. Thus, R'$_3$Si$^{\bullet}$ radicals, generated by some initiation processes, undergo a series of propagation steps generating 'fresh' radicals. The chain reactions are terminated by radical combination or disproportionation. In order to have an efficient chain process, the rate of chain-transfer steps must be higher than that of chain-termination steps. The following observations: (i) the termination rate constants in liquid phase are controlled by diffusion (i.e. 10^{10} M^{-1} s^{-1}), (ii) radical concentrations in chain reactions are about $10^{-7}-10^{-8}$ M (depending upon the reaction conditions) and (iii) the concentration of reagents is generally between 0.05–0.5 M, indicating that the rate constants for the chain-transfer steps must be higher than 10^3 M^{-1} s^{-1}.

The free-radical construction of C−C bonds either inter- or intramolecularly using a hydride as mediator is of great importance in chemical synthesis. The propagation steps for the intermolecular version are shown in Scheme 2. For a successful outcome, it is important (i) that the R'$_3$Si$^{\bullet}$ radical reacts faster with RZ (the precursor of radical R$^{\bullet}$) than with the alkene and (ii) that the alkyl radical reacts faster with alkene (to form the adduct radical) than with the silane. In other words, for a synthetically useful radical chain reaction, the intermediates must be *disciplined*. Therefore, in a synthetic plan one is faced with the task of considering kinetic data or substituent influence on the selectivity of radicals. The reader should note that the hydrogen donation step controls the radical sequence and, often, the concentration of silane provides the variable by which the products distribution can be influenced.

Initiation steps:

$$R'_3SiH \xrightarrow{\text{Radical Initiator}} R'_3Si\bullet$$

Propagation steps:

$$R'_3Si\bullet + RX \longrightarrow R'_3SiX + R\bullet$$
$$R\bullet + R'_3SiH \longrightarrow RH + R'_3Si\bullet$$

Termination steps:

$$\left.\begin{array}{l} 2\,R'_3Si\bullet \longrightarrow \\ R\bullet + R'_3Si\bullet \longrightarrow \\ 2\,R\bullet \longrightarrow \end{array}\right\} \text{no radical products}$$

SCHEME 1

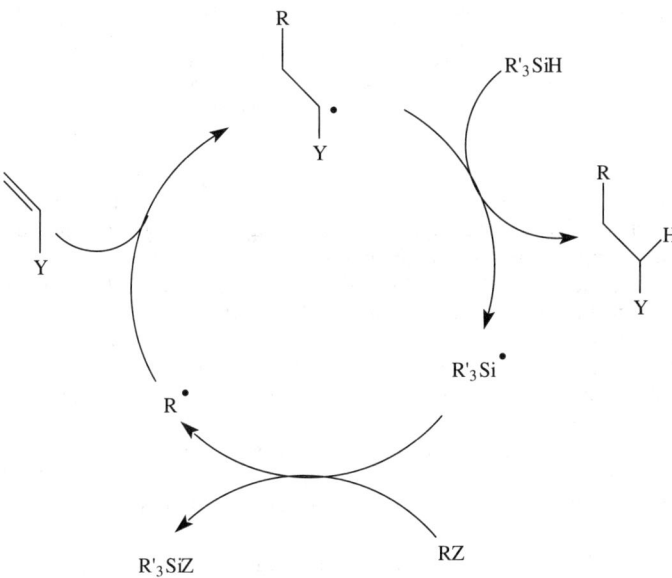

SCHEME 2

B. Hydrogen Donor Abilities of Silanes

We already mentioned that the majority of free-radical reactions applied in synthesis deal with Bu_3SnH. Occasionally, Bu_3GeH has been used as complement to Bu_3SnH. The corresponding silanes, i.e. trialkylsilanes, are not capable of donating hydrogen atom at a sufficient rate to propagate the chain. Rate constants for hydrogen abstraction from

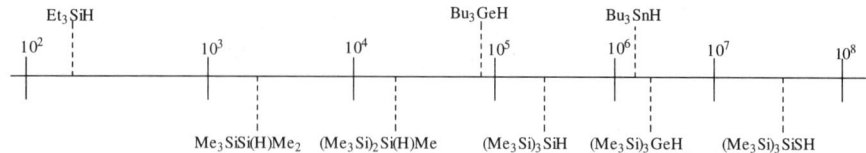

FIGURE 1. Rate constants for hydrogen abstraction (k, M^{-1} s^{-1}) from a variety of group 14 reducing agents by primary alkyl radicals at 25 °C

TABLE 1. Rate constants for the reaction of some radicals with (TMS)$_3$SiH

Radical	$\log A - E_a/\theta^a$	k_{SiH} (M^{-1} s^{-1}) at ca 27 °C	Reference
t-BuO$^{\bullet}$		1.1×10^8	12
RCH$_2^{\bullet}$	$8.9 - 4.5/\theta^b$	3.8×10^5	13
R$_2$CH$^{\bullet}$	$8.3 - 4.3/\theta^b$	1.4×10^5	13
R$_3$C$^{\bullet}$	$7.9 - 3.4/\theta^b$	2.6×10^5	13
Ph$^{\bullet}$		2.8×10^{8c}	13,14
n-C$_7$F$_{15}^{\bullet}$		5.1×10^{7d}	15
RC(O)$^{\bullet}$	$8.2 - 5.4/\theta^{e,f}$	1.8×10^4	16

^{a} Absolute rate expression, i.e. $\log k_{SiH} = \log A - E_a/\theta$, where $\theta = 2.3\ RT$ kcal mol^{-1}
^{b} Depends on literature values for 5-hexenyl-type rearrangements.
^{c} Depends on the rate constant for the cyclization of o-(allyloxy)phenyl radical.
^{d} Depends on the rate constant for the addition of the radical to 1-hexene.
^{e} Similar values are obtained for R being primary, secondary and tertiary.
^{f} Depends on literature Arrhenius parameters for the decarbonylation of propanoyl radical in the gas phase.

a variety of group 14 hydrides by primary alkyl radicals at ambient temperature are reported on a reactivity scale in Figure 1. In the upper part of the diagram the reactivities of trialkyl-tin, -germanium and -silicon hydrides are shown[8-10].

Following the success of (TMS)$_3$SiH, other organosilanes capable of sustaining analogous radical chain reactions have been introduced. In the lower part of Figure 1 the reactivities of a variety of organosilanes towards primary alkyl radicals are reported. In conclusion, a reactivity scale that is useful in synthetic planning has been established. The rate constants cover a range of several orders of magnitude and therefore the hydrogen donating abilities of organosilanes can be modulated by substituents. A recent review by Chatgilialoglu provides in-depth coverage of this class of reactions[11].

Table 1 summarizes the kinetic data available for the reaction of (TMS)$_3$SiH with a variety of radicals. It has been shown that the attacks of both t-BuO$^{\bullet}$ and RCH$_2^{\bullet}$ radicals on the Si–H bond and the methyl groups occur in about 95% and 5% of the cases, respectively[11]. The rate constants for the reaction of primary, secondary and tertiary alkyl radicals with (TMS)$_3$SiH are very similar in the range of temperatures that are useful for chemical transformation in the liquid phase. This is due to compensation of entropic and enthalpic effects through this series of alkyl radicals.

II. AUTOXIDATION OF TRIS(TRIMETHYLSILYL)SILANE

This silane reacts spontaneously at ambient temperature with molecular oxygen to form siloxane as the sole product (equation 1).

25. Tris(trimethylsilyl)silane in organic synthesis

$$(Me_3Si)_3SiH + O_2 \longrightarrow (Me_3SiO)_2Si(H)SiMe_3 \quad (1)$$

Mechanistic studies based on labeling experiments[17] and kinetic studies[18] have indicated a chain reaction to occur in which the silyl radical adds to molecular oxygen to form the corresponding peroxyl species which undergoes a cascade of unimolecular reactions and, finally, abstracts hydrogen to give the observed product and regenerate $(TMS)_3Si^\bullet$ radical. The rate constant for the initiation step is found to be 5.1×10^{-5} $M^{-1} s^{-1}$ at 70 °C.

Two aspects of the autoxidation of $(TMS)_3SiH$ are of interest to synthetic chemists[19]: (i) for radical reactions having long chain length, traces of molecular oxygen can serve to initiate reactions, and therefore no additional radical initiator is needed, and (ii) the oxidized product does not interfere with radical reactions, and therefore the reagent can be used even if partially oxidized, taking into account the purity of the material. From GC analysis, the exact concentration of silane can be deduced.

III. REDUCTIONS

Tris(trimethylsilyl)silane is found to be an efficient reducing agent for a variety of functional groups. In particular, the reduction of halides, chalcogen groups, thiono esters and isocyanides are the most common ones. The efficiency of these reactions is also supported by available kinetic data. The rate constants for the reaction of $(TMS)_3Si^\bullet$ radicals with a variety of organic substrates are collected in Table 2.

Reduction of iodides and bromides is straightforward, and the reactions are complete after a short time. Two examples are given in equations 2 and 3[22,23]. The efficiency depends to a limited extent on the substituent; in alkyl bromides the rate constants decrease along the series benzyl > tertiary alkyl > secondary alkyl > primary alkyl > phenyl (see Table 2). $(TMS)_3SiH$ can be also used as a reagent for driving the reduction of iodides and bromides through a radical mechanism with sodium borohydride, the reductant that is consumed. In equation 4, 1-bromonaphthalene is treated with an excess of $NaBH_4$ (50 equiv) and a small amount of $(TMS)_3SiH$ (0.1 equiv), under photochemical initiation conditions, to give the reduced product in 88% yield[24].

$$\text{substrate} \xrightarrow[50\,°C]{(TMS)_3SiH} \text{product} \quad 98\% \quad (2)$$

$$\text{Br-substrate-Br} \xrightarrow[h\nu(254\,nm),\,r.t.]{(TMS)_3SiH} \text{product} \quad 99\% \quad (3)$$

TABLE 2. Rate constants for the reaction of (TMS)$_3$Si• radicals with some organic substrates at ambient temperature

Halides[a]	k (M^{-1} s^{-1})	Others[c]	k (M^{-1} s^{-1})
CH$_3$(CH$_2$)$_4$Br	2.0 × 10^7	n-C$_{10}$H$_{21}$SPh	< 5 × 10^6
CH$_3$CH$_2$CH(CH$_3$)Br	4.6 × 10^7	n-C$_{10}$H$_{21}$SePh	9.6 × 10^7
(CH$_3$)$_3$CBr	1.2 × 10^8	n-C$_{12}$H$_{25}$C(O)SePh[b,d]	2 × 10^8
PhCH$_2$Br	9.6 × 10^8	c-C$_6$H$_{11}$OC(S)SMe	1.1 × 10^9
C$_6$H$_5$Br	4.6 × 10^6	c-C$_6$H$_{11}$NC	4.7 × 10^7
CH$_3$(CH$_2$)$_5$C(CH$_3$)$_2$Cl	4.0 × 10^5	(CH$_3$)$_3$CNO$_2$	1.2 × 10^7
PhCH$_2$Cl	4.6 × 10^6		
CHCl$_3$	6.8 × 10^6		
CCl$_4$	1.7 × 10^8		
n-C$_{12}$H$_{25}$C(O)Cl[b,d]	7 × 10^5		

[a]From Reference 20. [b]From Reference 16. [c]From Reference 21. [d]At 80 °C.

$$\text{1-bromonaphthalene} \xrightarrow[\substack{(p\text{-MeOC}_6\text{H}_4\text{CO}_2)_2 \\ h\nu(254\text{nm}),\text{r.t.}}]{\text{NaBH}_4/(\text{TMS})_3\text{SiH}} \text{naphthalene} \quad 88\% \tag{4}$$

For tertiary, secondary and primary chlorides the reduction becomes increasingly difficult due to shorter chain lengths. On the other hand, the replacement of a chlorine atom by hydrogen in polychlorinated substrates is much easier. The reduction of the dichloride depicted in equation 5 represented the key step in the total synthesis of dactomelynes[25], natural products isolated from a marine organism. The use of (TMS)$_3$SiH at room temperature allowed one to obtain in high yield and stereoselectively the monochloride ($\beta : \alpha = 13 : 1$), whereas other radical-based reducing systems failed. In this respect, the stereoselectivity in the reduction of *gem*-dichlorides by (TMS)$_3$SiH and Bu$_3$SnH was previously reported[26] and showed that the silane has a stronger preference than tin to transfer a hydrogen atom from the less hindered side of the ring due to the different spatial shapes of the two reagents.

$$\text{dichloride} \xrightarrow[\text{Et}_3\text{B, r.t.}]{(\text{TMS})_3\text{SiH}} \text{monochloride products} \tag{5}$$

98% (13:1)

25. Tris(trimethylsilyl)silane in organic synthesis

(TMS)$_3$SiH does not react spontaneously with acid chlorides, in contrast with Bu$_3$SnH. Under free-radical conditions, the reaction of (TMS)$_3$SiH with acid chlorides, RC(O)Cl, gives the corresponding aldehydes and/or the decarbonylation products depending on the nature of substituent R. The reduction of 1-adamantanecarbonyl chloride is shown in equation 6[27].

$$\text{Adamantyl-C(O)Cl} \xrightarrow[\text{AIBN/80 °C}]{(TMS)_3SiH} \text{Adamantane} \quad (6)$$

90%

Early data from product studies and kinetic measurements established that the reaction of (TMS)$_3$Si• radical with phenyl alkyl sulfides is sluggish[21]. Nevertheless, it was reported soon after that efficient S_H2 cleavage of C–S bonds could be effected by (TMS)$_3$SiH when the resulting carbon-centered radicals were further stabilized by an α-heteroatom[28,29]. Therefore, (TMS)$_3$SiH can induce the efficient radical chain monoreduction of 1,3-dithiolane[28,30], 1,3-dithiane[28], 1,3-oxathiolane[29,30], 1,3-oxathiolanone[29,30] and 1,3-thiazolidine[29,30] derivatives. Three examples are outlined in equation 7. The (TMS)$_3$Si group incorporated in the adducts could be conveniently deprotected with fluoride ions generating a thiolate anion which could be used for further synthetic transformations[31]. The reduction of chiral 1,3-thiazolidine derivatives (as 1 : 1 mixture of diastereoisomers) was used as a model for studying the stereoselectivity in the hydrogen abstraction of α-aminoalkyl radicals[32]. These intermediates abstract hydrogen giving different *anti/syn* ratios, depending on the substituents. An example is given in equation 8.

$$\text{X–S (cyclic)} \xrightarrow[\text{AIBN/80 °C}]{(TMS)_3SiH} \text{X} \cdots \text{S—Si(TMS)}_3 \quad (7)$$

X = S   85%
X = O   84%
X = NC(O)OEt   79%

$$\text{Pr-}i, \text{Ph, S, NH (thiazolidine)} \xrightarrow[\text{AIBN/80 °C}]{(TMS)_3SiH} \text{Pr-}i, \text{Ph, HN, S—Si(TMS)}_3 \quad (8)$$

92% (*anti:syn* = 84:16)

Secondary alkyl selenides are reduced by (TMS)$_3$SiH (equation 9), as expected in view of the affinity of silyl radicals for selenium-containing substrates (Table 2)[21]. Similarly to 1,3-dithiolanes and 1,3-dithianes, 5- and 6-membered selenoacetals can be monoreduced to the corresponding selenides in the presence of $(TMS)_3SiH^{33}$. The silicon hydride approached from the less hindered equatorial position to give *trans/cis* ratios of 30/70 and 25/75 for the 5-membered (equation 10) and 6-membered selenoacetals, respectively, whereas Bu_3SnH transferred a hydride preferentially from the axial position.

$$\text{cyclohexyl-SePh} \xrightarrow{(TMS)_3SiH, AIBN/80\,°C} \text{cyclohexyl} \quad 99\% \tag{9}$$

$$t\text{-Bu-cyclohexyl-(Se-Se selenoacetal)} \xrightarrow{(TMS)_3SiH \mid AIBN/80\,°C} t\text{-Bu-cyclohexyl-Se-CH}_2\text{CH}_2\text{-SeSi(TMS)}_3 \tag{10}$$

trans/cis = 30/70

Phenyl selenoesters have been reported to undergo reduction to the corresponding aldehydes and/or alkanes in the presence of $(TMS)_3SiH$ under free-radical conditions[16]. The decrease of aldehyde formation through the primary, secondary and tertiary substituted series, under the same conditions, indicated that a decarbonylation of acyl radicals takes place. Equation 11 shows an example of a tertiary substituted substrate.

$$\text{cyclohexyl-C(Me)(C(O)SePh)} \xrightarrow{(TMS)_3SiH, h\nu/50\,°C} \text{cyclohexyl-Me} \quad 80\% \tag{11}$$

The removal of the hydroxy group has been achieved from an appropriate selenocarbonate by heating with $(TMS)_3SiH$ and AIBN in benzene. Equation 12 outlines a particular of a multistep synthesis of an alkaloid. The deoxygenation was achieved with 87% efficiency of the two steps[34].

In the above-described displacement reactions (S_H2) of sulfur and selenium containing compounds, the mechanism could either be a synchronous or a stepwise process (Scheme 3). Thus, in the S_H2 stepwise path an intermediate sulfuranyl or selenanyl radical is formed[35a], followed by α-cleavage. Based on competitive studies, a stepwise process was suggested to occur in the reaction of $(TMS)_3Si^•$ radical with *n*-decyl phenyl selenide[21]. On the other hand, *ab initio* calculations favored the synchronous process and

25. Tris(trimethylsilyl)silane in organic synthesis

the same observations were explained in terms of an overall reversible reaction[35b].

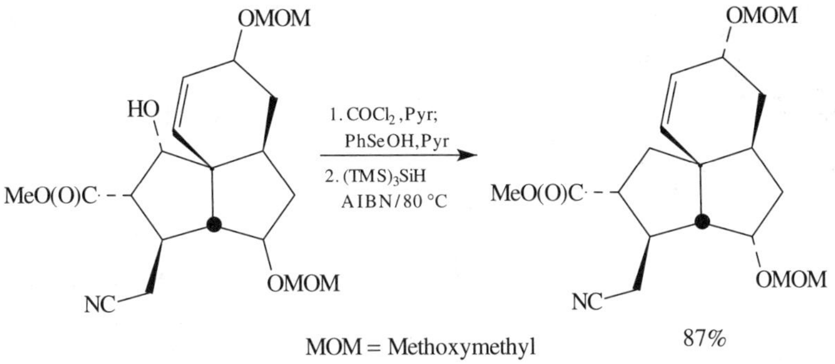

(12)

MOM = Methoxymethyl   87%

$$(TMS)_3Si^\bullet + R\text{—}X\text{—}R' \xrightarrow{S_H2 \text{ synchronous}} (TMS)_3SKR' + R^\bullet$$

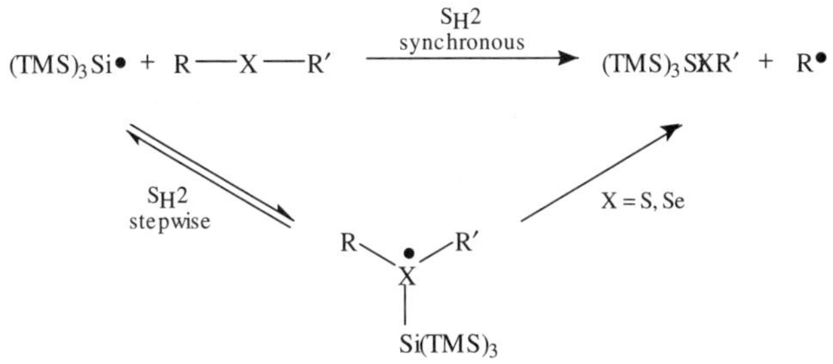

SCHEME 3

Another relevant reductive process which has been largely applied and studied in organic synthesis is the deoxygenation reaction, well known as the Barton–McCombie reaction. The reaction, shown in equation 13, involves thiocarbonyl derivatives, easily obtained from the alcohol, and a radical reducing agent which acts as hydrogen donor, in the presence of initiator and appropriate conditions. The use of silicon hydrides as radical-based reducing agents in the Barton–McCombie reaction and their comparison with Bu_3SnH has recently been reviewed by two of us[7]. Therefore, only examples of $(TMS)_3SiH$ that have appeared in the more recent literature will be discussed.

$$ROH \longrightarrow ROC(S)X \xrightarrow{(TMS)_3SiH} RH \quad (13)$$

An improved homolytic deoxygenation of amino acid derivatives has been reported[36]. The reaction with O-phenyl chlorothionocarbonate/DMAP afforded the phenoxythiocarbonyl derivative of a protected homoserine dimethylphosphonate (95% yield), followed by its treatment with the silane at 80 °C in toluene to give the deoxygenated product. Without purification, acid treatment by TFA and an aqueous extraction gave the final products in 72% yield (equation 14). Using the silane procedure a simplification of the work-up resulted, since the extraction with water was enough to separate the product from silyl and sulfur containing by-products.

The dideoxygenation of 1,6-anhydro-D-glucose with $(TMS)_3SiH$ has been described to yield the desired product in 86% yield (equation 15), whereas other radical-based reducing systems give much poorer yields[37]. Radical dideoxygenation was also useful to prove the structure of a stemodane ring system, achieved by other routes. The dithioimidazolyloxy compound in equation 16 was reduced in quantitative yield under normal experimental conditions[38].

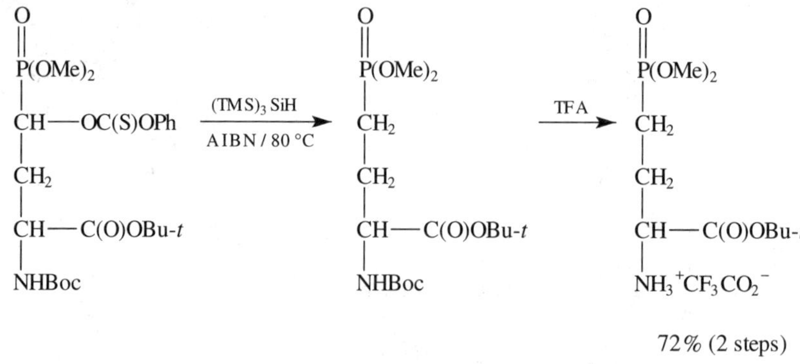

$$\begin{array}{c}\text{O}\\\|\\\text{P(OMe)}_2\\|\\\text{CH—OC(S)OPh}\\|\\\text{CH}_2\\|\\\text{CH—C(O)OBu-}t\\|\\\text{NHBoc}\end{array}\xrightarrow[\text{AIBN/80°C}]{\text{(TMS)}_3\text{SiH}}\begin{array}{c}\text{O}\\\|\\\text{P(OMe)}_2\\|\\\text{CH}_2\\|\\\text{CH}_2\\|\\\text{CH—C(O)OBu-}t\\|\\\text{NHBoc}\end{array}\xrightarrow{\text{TFA}}\begin{array}{c}\text{O}\\\|\\\text{P(OMe)}_2\\|\\\text{CH}_2\\|\\\text{CH}_2\\|\\\text{CH—C(O)OBu-}t\\|\\\text{NH}_3^+\text{CF}_3\text{CO}_2^-\end{array}\quad(14)$$

72% (2 steps)

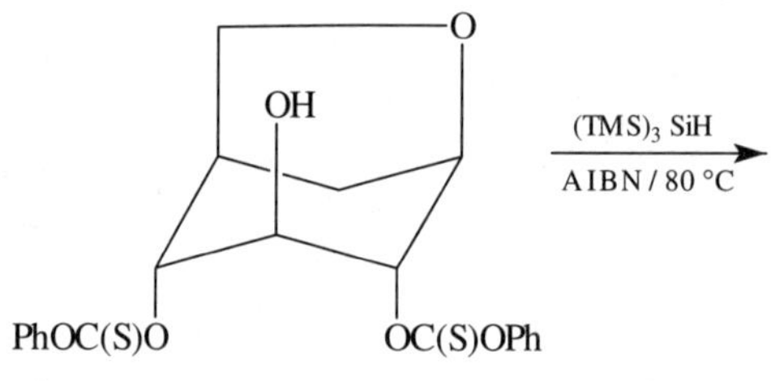

(15) 86%

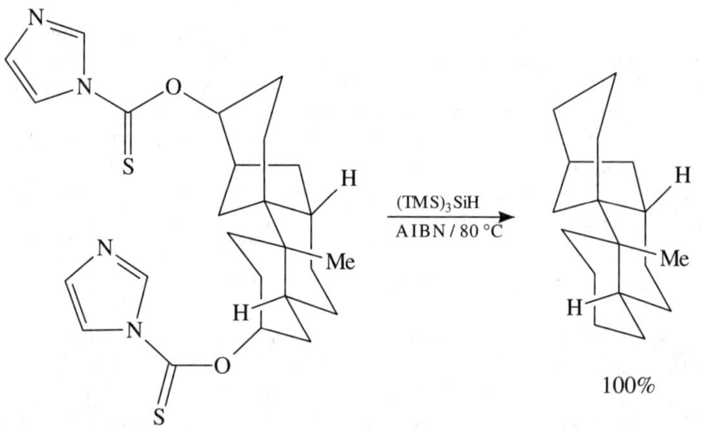

(16) 100%

In order to study the structure–activity relationship of pharmaceutically active derivatives of baccatin III, radical-based deoxygenations were used to generate a systematic defunctionalization of the taxole core. When tris(trimethylsilyl)silane was used as reducing agent of the xanthate functionality (equation 17; yields based on the consumption of the starting material), a cascade of intramolecular rearrangements took place, due to the slower hydrogen donation of silane relative to tin hydrides[39].

It is worth adding that the usual starting derivatives for the deoxygenation of alcohols (i.e. X = SMe, OPh and imidazolyl in equation 13) have recently been extended to include thioxocarbamates (X = HNPh) with the same success[40]. The mechanism of the reduction of thionoesters by (TMS)$_3$SiH is the following: (TMS)$_3$Si$^\bullet$ radical, initially generated by small amounts of AIBN, attacks the thiocarbonyl moiety to form in a reversible manner a radical intermediate that undergoes β-scission to form an alkyl radical. Hydrogen abstraction from the silane gives the alkane and (TMS)$_3$Si$^\bullet$ radical, thus completing the cycle of this chain reaction[7,21].

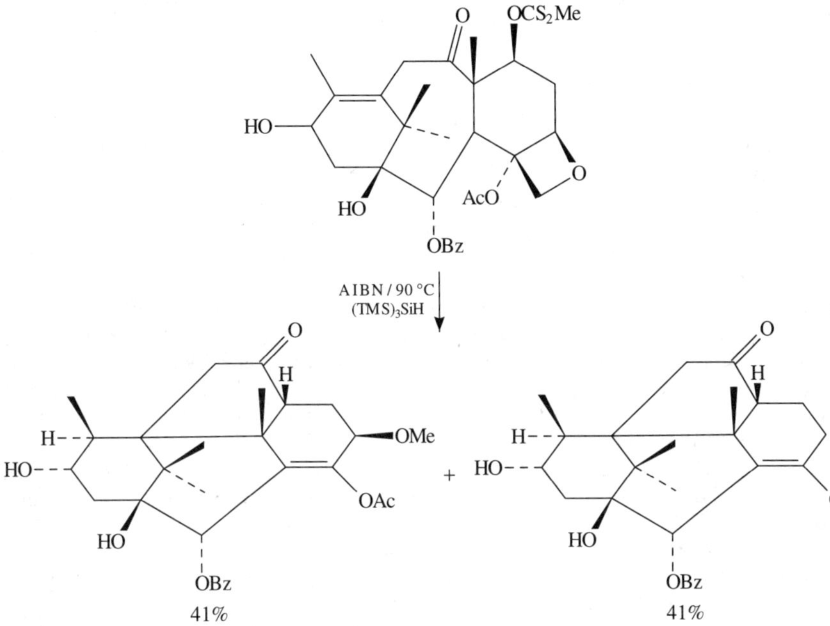

(17)

Isocyanides can be reduced to the corresponding hydrocarbon by (TMS)$_3$SiH. The reaction can be considered as the deamination of primary amines since isocyanides are obtained via formylation of amines and dehydration. The efficiency of the reduction is independent of the nature of the alkyl substituent. That is, primary, secondary and tertiary isocyanides at 80 °C gave the corresponding hydrocarbon in good yields[21]. An example is given in equation 18[4].

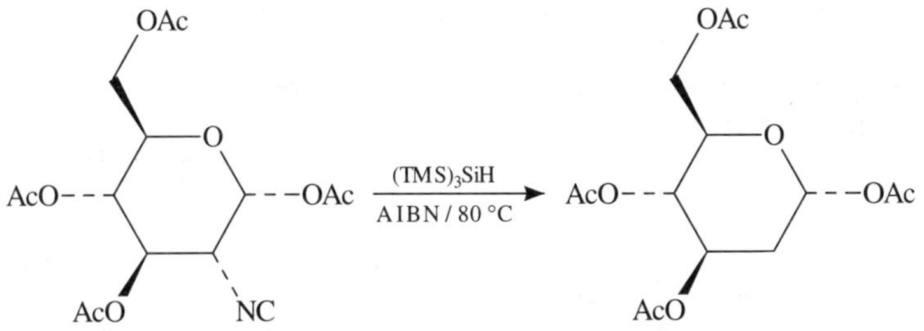

(18)

(TMS)$_3$SiH is not able to reduce tertiary nitroalkanes to the corresponding hydrocarbons whereas tin hydrides are efficient in this process[41]. This 'anomalous' behavior is due to the fact that the nitroxide adducts formed by addition of (TMS)$_3$Si$^\bullet$ radical to the nitro compounds, i.e. RN(O$^\bullet$)OSi(TMS)$_3$, fragment preferentially at the nitrogen–oxygen bond rather than at the carbon–nitrogen bond as in the analogous tin adduct (equation 19)[41a].

1550   C. Chatgilialoglu, C. Ferreri and T. Gimisis

For the reactions of N-nitramines with (TMS)$_3$SiH a similar behavior has also been observed, although in this case the corresponding N-nitrosoamines are obtained in ca 60% yield[41b].

$$R^\bullet + ONOSi(TMS)_3 \longleftarrow \underset{\underset{R}{|}}{\overset{\overset{\bullet}{O}}{\underset{|}{N}}}\underset{O}{\diagdown}Si(TMS)_3 \longrightarrow RNO + \overset{\bullet}{O}Si(TMS)_3$$

(19)

1-Methylquinolinium ion derivatives are reduced regioselectively to 1,4-dihydroquinones by (TMS)$_3$SiH under photochemical conditions[42]. Mechanistic studies demonstrated that the reactions are initiated by photoinduced electron transfer from silane to the singlet excited states of 1-methylquinolinium ion derivatives to give the silane radical cation—quinolinyl radical pairs, followed by hydrogen transfer in the cage to yield 1,4-dihydroquinones and silicenium ion. The one-electron oxidation potential of (TMS)$_3$SiH is 1.30 V[42].

It is worth pointing out that the replacement of a variety of functional groups by a hydrogen described so far is not only an efficient and straightforward process but the work-up is rather simple: the reaction mixtures can be concentrated by evaporation of the solvent and then flash-chromatographed to isolate the products. Furthermore, it has been shown that (TMS)$_3$SiH and its silylated by-products are not toxic; this is very important for pharmaceutical application of the silane reagent, since the biological assays on the final compounds are not affected by the eventually remaining silylated materials[43]. It can be expected that in the future an increasing choice of (TMS)$_3$SiH as reducing agent, especially in the field of natural products synthesis, will take place.

In this respect, some recent reports on the radical reduction of pharmaceutically relevant compounds showed the efficiency and effectiveness of this reagent. Here below we describe some examples in the field of nucleoside chemistry.

The deoxygenation of nucleosides has been well represented in the last few years for at least two main reasons: the radical deoxygenation in C-2' can be used as simple conversion from ribonucleosides to deoxyribonucleosides to be incorporated in DNA oligomers, and many new potent anti-HIV and antiviral drugs having the 2'- or 3'-deoxy as well as 2',3'-dideoxy nucleosides skeleton were recently introduced in therapy. Compared with other routes, the formation of thiocarbonyl derivatives of the alcoholic function on the sugar moiety and its subsequent radical reduction has appeared to provide the easiest access to these substrates, avoiding any other side reactions. 2'-Deoxyapio-β-D-furanosyl nucleosides were prepared by deoxygenation from the corresponding thiocarbonate derivatives (equation 20), and tested against HIV, Herpes simplex and other viruses[44]. Pharmaceutically important 2'-and/or 3'-deoxynucleosides were described in a patent[45] and the key step for the synthesis was achieved by treating the new and economically more convenient (cyanoethylthio) thiocarbonyl derivatives of nucleosides, instead of classical xanthates, with (TMS)$_3$SiH under very mild conditions (equation 21). In the field of nucleosides mimic, β-2'-deoxypseudouridine[46] and β-2'-deoxyzebularine[47] were prepared by radical-based deoxygenation from the corresponding β-pseudouridine and β-zebularine using (TMS)$_3$SiH as reducing agent.

Equations 22 and 23 show two examples of chloride and phenylseleno removal from the sugar moiety under normal radical conditions[48], whereas in equation 24 the removal of the tritylthio protecting group (TrS) on the base moiety has been accomplished in good yield[49]. On the other hand, starting from the bromide in equation 25 a novel type of a β-(acyloxy)alkyl radical rearrangement has been observed, which leads through the generation of a C-1' radical species to the stereoselective preparation of an α-ribonucleoside[48].

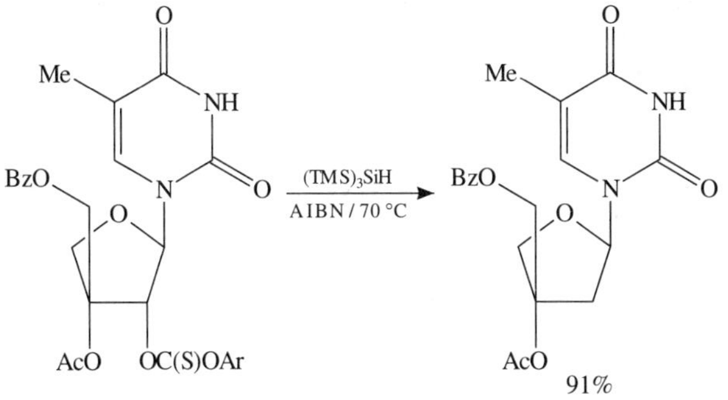

25. Tris(trimethylsilyl)silane in organic synthesis 1551

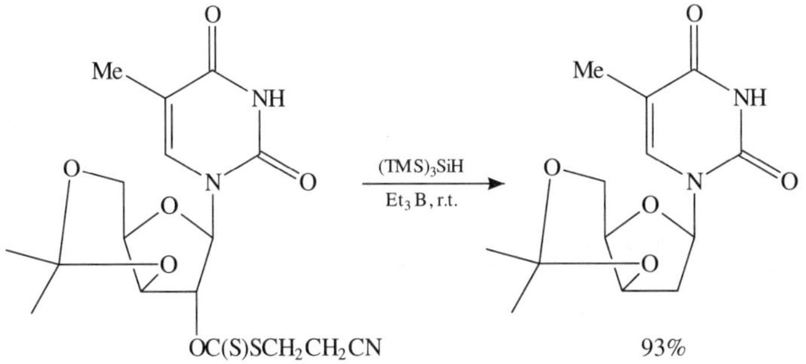

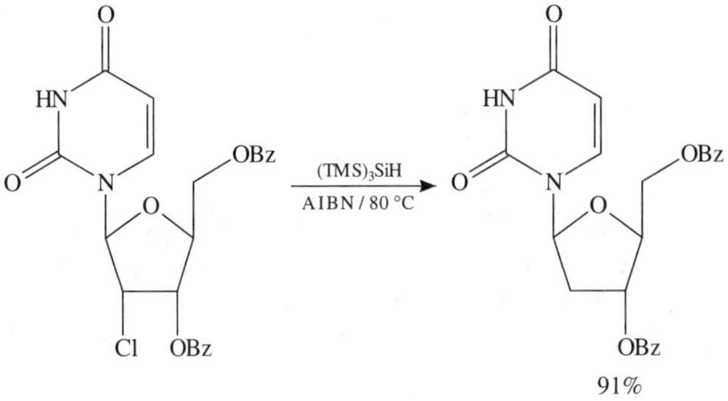

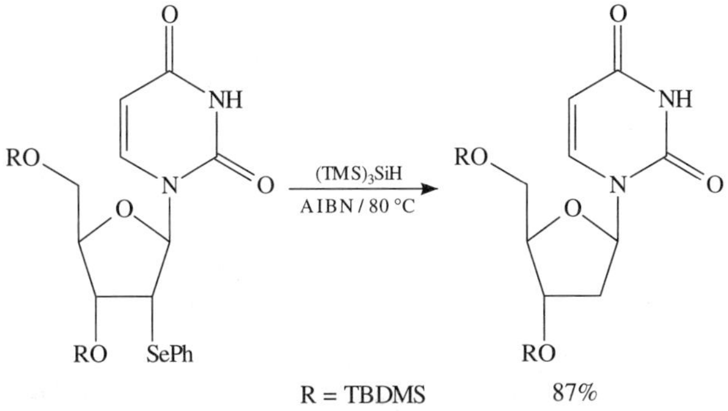

(20)

(21)

(22)

(23)

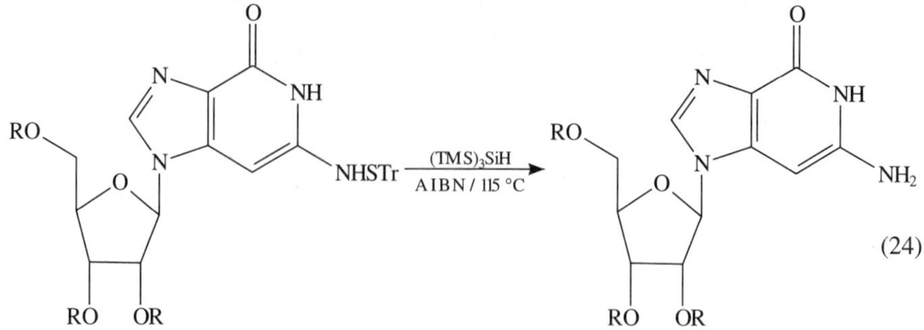

(24)

R = TBDMS   81%

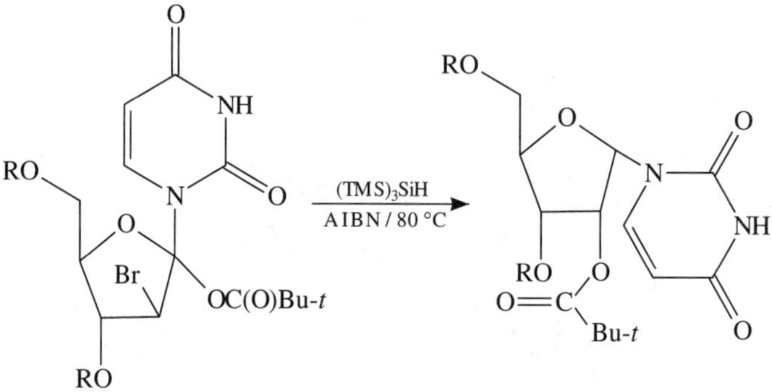

(25)

R = TBDMS   86%

IV. HYDROSILYLATIONS

(TMS)$_3$SiH adds across the C=C and C=O double bonds of a variety of compounds under free-radical conditions. The propagation steps for these hydrosilylation processes are reported in equations 26 and 27. The available rate constants for the reaction of (TMS)$_3$Si• radicals with some ketones and alkenes (equation 26) are collected in Table 3. In the ketone series, the rate constants decrease in the series of quinone > diaryl ketone > dialkyl ketone[50]. On the other hand, the rate constants for the addition of (TMS)$_3$Si• radical to activated alkenes[21] are close to 10^8 M^{-1} s^{-1}.

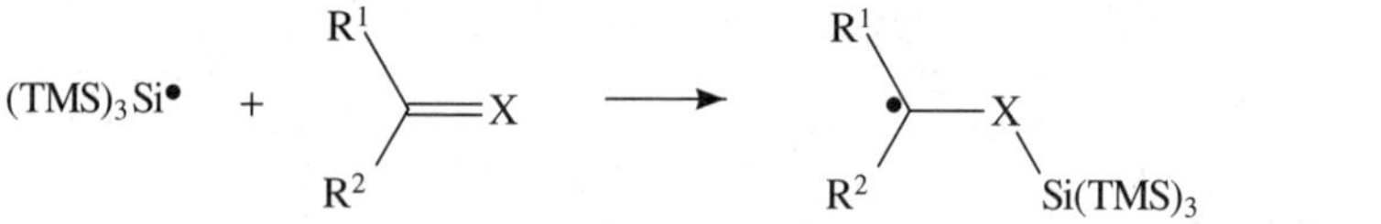

(26)

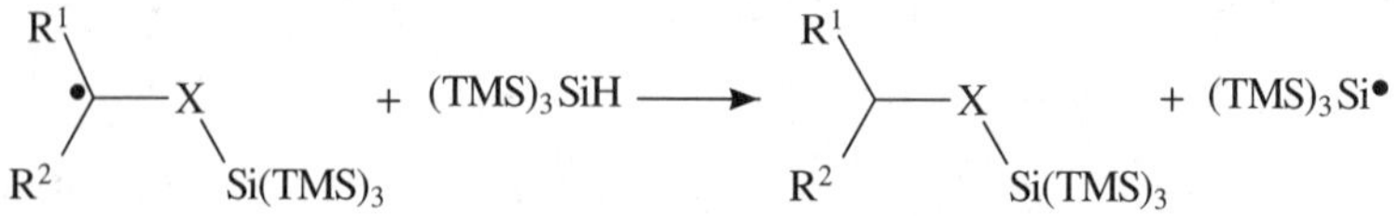

(27)

X = O or CR^{3}R^4

The hydrosilylation of monosubstituted and *gem*-disubstituted olefins (equation 28 and 29) are efficient processes and have been shown to occur with high regioselectivity

TABLE 3. Rate constants at 20 °C for the reaction of $(TMS)_3Si^•$ radicals with some ketones and alkenes

Substrate[a]	k (M^{-1} s^{-1})
Duroquinone[a]	1.0×10^8
Fluorenone[a]	3.8×10^7
Acetone[a]	8.0×10^4
Styrene[b]	5.9×10^7
Acrylonitrile[b]	6.3×10^7
Ethyl acrylate[b]	9.7×10^7

[a]From reference 50. [b]From reference 21

(*anti*-Markovnikov) in the case of both electron-rich and electron-poor olefins[51]. For *cis* or *trans* disubstituted double bonds, hydrosilylation is still an efficient process, although it required slightly longer reaction times (equation 30) and an activating substituent[51]. No hydrosilylation product has been observed with 1,2-dialkyl- and 1,2-diaryl-substituted olefins, due to the reversible addition of $(TMS)_3Si^•$ radical to the double bond[52].

$$\begin{array}{c} \text{CH}_2=\text{CH-OBu} \xrightarrow[\text{AIBN / 80 °C}]{(TMS)_3SiH} (TMS)_3Si-CH_2-CH_2-OBu \\ 92\% \end{array} \qquad (28)$$

$$\begin{array}{c} \text{CH}_2=\text{C(Me)CO}_2\text{Et} \xrightarrow[\text{AIBN / 80 °C}]{(TMS)_3SiH} (TMS)_3Si-CH_2-C(Me)H-CO_2Et \\ 77\% \end{array} \qquad (29)$$

$$\begin{array}{c} \text{MeCH=CH-CN} \xrightarrow[\text{AIBN / 80 °C}]{(TMS)_3SiH} (TMS)_3Si-CH(Me)-CH_2-CN \\ 74\% \end{array} \qquad (30)$$

The above observation on the reversible addition of $(TMS)_3Si^•$ radical to the double bonds is noteworthy from a synthetic point of view. The first consequence is that $(TMS)_3Si^•$ radical is able to isomerize alkenes. That is, this radical adds to (Z)- or (E)-alkenes to form radical **1** or **2**, respectively (equation 31). Interconversion between the two radical adducts by rotation around the carbon–carbon bond, followed by β-scission, can then lead to the formation of either (Z)- or (E)-alkene, depending on the radical-alkene combination. For comparison, we report in Figure 2 the reaction profile for the interconversion of (E)-to (Z)-3-hexen-1-ol and *vice versa* by $(TMS)_3Si^•$ and $Bu_3Sn^•$ radicals, under identical experimental conditions. The choice of either $Bu_3Sn^•$ or $(TMS)_3Si^•$ radical does not influence the isomeric composition of the alkene after completion, i.e. $Z/E = 18/82$, although the equilibration of the two geometrical isomers is reached much faster with tin radicals. Therefore, care must be taken when the isomerization can occur *in situ*, while accomplishing other reactions[52–55]. A postisomerization process can be invoked in several reported cases where the two reagents behave differently (*vide infra*).

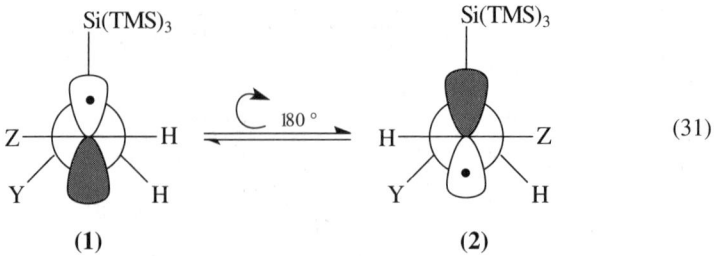

Depending upon the choice of substrates, the hydrosilylation of alkenes can also be highly stereoselective. Two examples are given below. The reaction with methylmaleic anhydride proceeded regiospecifically to the less substituted side, but also diastereoselectively to afford the thermodynamically less stable *cis* isomer. The stereoselectivity decreased by increasing the reaction temperature, indicating the difference in enthalpy of activation for *syn* vs *anti* attack (equation 32). On the other hand, a complete stereocontrol has been achieved in the reaction with the α-chiral olefins (equation 33, R = Me)[56]. The observed stereoselectivity was rationalized in terms of steric and Felkin–Anh

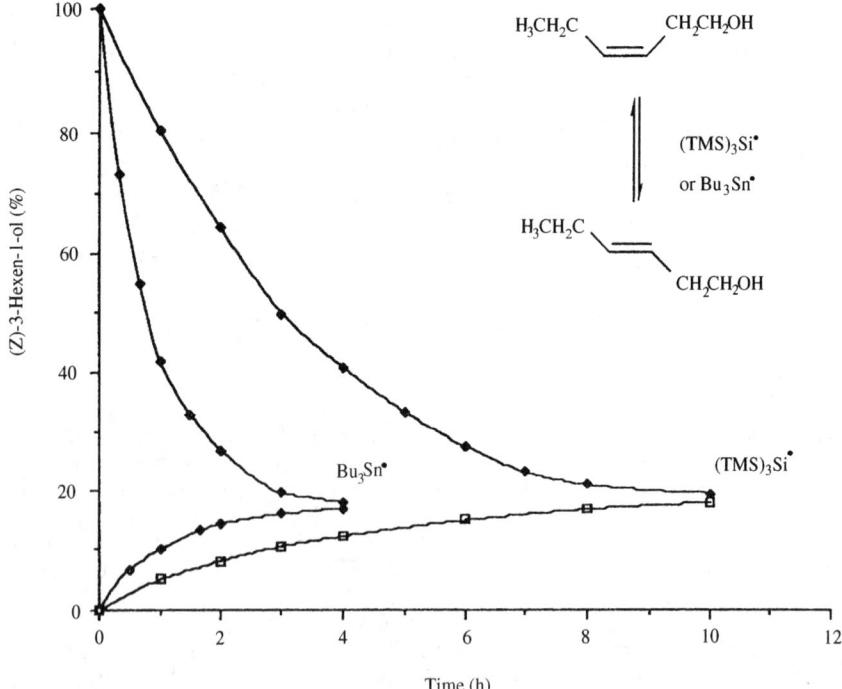

FIGURE 2. Isomerizations of *E*- and *Z*-3-hexen-1-ol. Conditions: 0.5 equiv of hydride and AIBN (5 mol%) at intervals of 2 h in refluxing benzene

stereoelectronic controls.

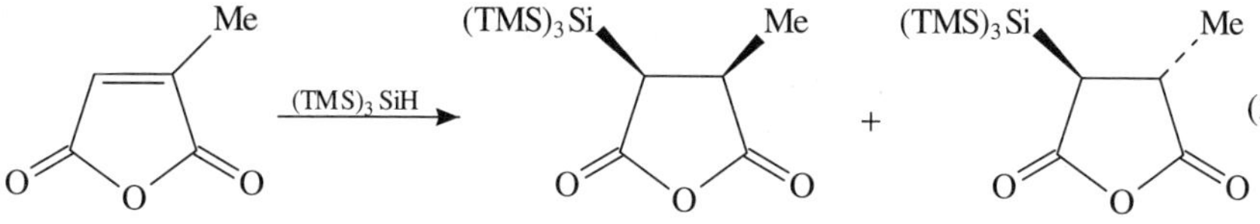

(32)

AIBN / 90 °C   89% (16:1)
Et$_3$B, O$_2$/−15 °C   60% (99:1)

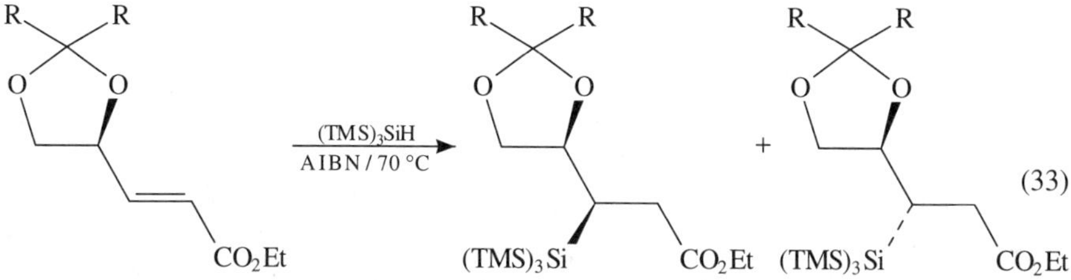

(33)

R = H
R = Me   65% (85:15)
>95% (100:0)

Hydrosilylation of ketones with (TMS)$_3$SiH were also examined[50,57] and resulted in generally good yields. For example, the reactions with cycloalkanones under normal conditions, i.e. AIBN/80 °C, gave the silylated products in high yields (>70%). With 4-*tert*-butylcyclohexanone, the products ratio *trans* : *cis* = 10 : 1 was obtained, indicating that axial attack by the intermediate α-silyloxy radical is favored[57]. By means of the addition of (TMS)$_3$SiH to chiral ketones, the 1,2-stereoinduction in carbon-centered radicals bearing an α-silyloxy substituent has been studied[58]. An example is given in equation 34.

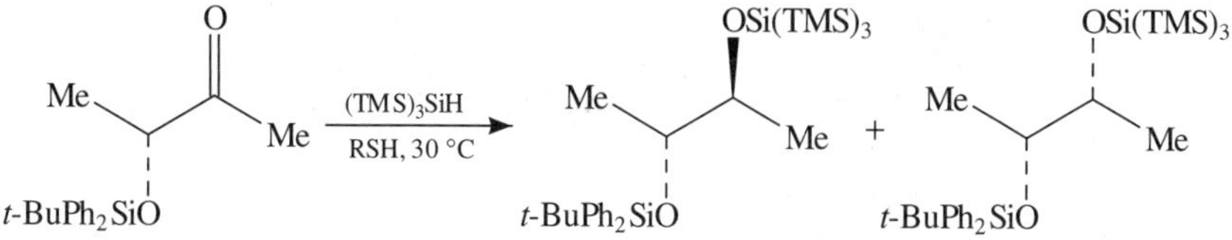

64% (1:4)

(34)

The addition of (TMS)$_3$SiH to a number of mono and disubstituted acetylenes has also been studied[51,59]. The reaction gives (TMS)$_3$Si-substituted alkenes in good yields via a radical chain mechanism where the key intermediate is a vinyl radical. The reaction is highly regioselective (*anti*-Markovnikov) and can also show high *cis* or *trans* stereoselectivity depending on the nature of the substituents at the acetylenic moiety. Two examples are given in equations 35 and 36. Although the (TMS)$_3$Si$^\bullet$ has been proven to isomerize alkenes, the postisomerization of the hydrosilylation adduct could not be observed due to steric hindrance. The (*Z*)–(*E*) interconversion of (TMS)$_3$Si-substituted alkenes was achieved by using Ph$_3$Ge$^\bullet$ radical as an isomerizing agent[59]. The hydrodesilylation and bromodesilylation of adducts were performed in good yields and with retention of stereochemistry (equations 35 and 36)[51,59].

$$n\text{-}C_{10}H_{21}\text{—}C\equiv CH \xrightarrow[\text{Et}_3B, \text{r.t.}]{(TMS)_3SiH}$$

n-$C_{10}H_{21}$ / H Si(TMS)$_3$ / H (alkene)
98% (Z:E > 20:1)

↓ concd. HCl, CH$_3$CN, 82 °C (35)

n-$C_{10}H_{21}$ / H H / H (alkene)
96%

$$Ph\text{—}C\equiv C\text{—}CO_2Et \xrightarrow[\text{AIBN / 80 °C}]{(TMS)_3SiH}$$

Ph / H CO$_2$Et / Si(TMS)$_3$
85%

↓ Br$_2$, CH$_2$Cl$_2$, −78 °C (36)

Ph / H CO$_2$Et / Br
82%

V. CONSECUTIVE RADICAL REACTIONS

The alkyl radical that is created by the initial atom (or group) abstraction by (TMS)$_3$Si$^\bullet$ radical or by addition of (TMS)$_3$Si$^\bullet$ to unsaturated bonds is often designed so that it can undergo a number of consecutive reactions. Care has to be taken in order to ensure that the effective rates of the consecutive radical reactions are higher than the rate of hydrogen abstraction from (TMS)$_3$SiH. Apart from standard synthetic planning based on known rate constants (see Sections I.A and I.B), this is usually effected either by controlling the concentration of the reducing agent present (slow addition by syringe-pump) or, in the case of intermolecular addition reactions, by adding a large excess of the radical acceptor. In both cases, however, tris(trimethylsilyl)silane is advantageous when compared with Bu$_3$SnH due to its lower hydrogen donation ability which provides the intermediate radical species the time to undergo the desired sequential reactions prior to reduction. In fact, in many cases direct addition of (TMS)$_3$SiH often gives results comparable to the syringe-pump addition of Bu$_3$SnH. The purpose of the following paragraphs is to introduce (TMS)$_3$SiH as a reliable radical mediator in the synthetic community, providing also some enlightenment on its advantages and disadvantages based on the updated work published on its use.

A. Initiation by Atom (or Group) Abstraction

The intermolecular C—C bond formation mediated by (TMS)$_3$SiH has been the subject of several investigations. In the initial work, the reaction of cyclohexyl iodide or isocyanide with a variety of alkenes was tested in order to find out the similarities with tin reagents (equations 37 and 38)[60,61]. The propagation steps for these reactions are shown in Scheme 2 in Section I.A. It is worth pointing out that with iodides the reaction yields are comparable with those obtained by the tin method, whereas no C—C bond formation takes place when alkyl isocyanides are used together with Bu$_3$SnH in the presence of alkenes. Similarly, bishomolithocholic acid (**3**) was synthesized from the corresponding iodo derivative (equation 39)[62]. This analog of lithocholic acid with a modification of the side chain was synthesized together with others in order to study their inhibitory effects on glucuronosyl transferase activity in a colon-cancer cell line, thus attempting to identify structural features critical for the enzyme inhibition. The ease of isolation and purification of the product was in accord with the intended purpose of its synthesis, i.e. biological testing. The reactions of β- or γ-phenyl substituted α-methylenebutyrolactones with BuI in the presence of (TMS)$_3$SiH afforded α, β- or α, γ-disubstituted lactones in good yield and with high diastereoselectivity (equation 40)[63]. With β- or γ-alkyl substituted substrates the same reaction was found to be less selective.

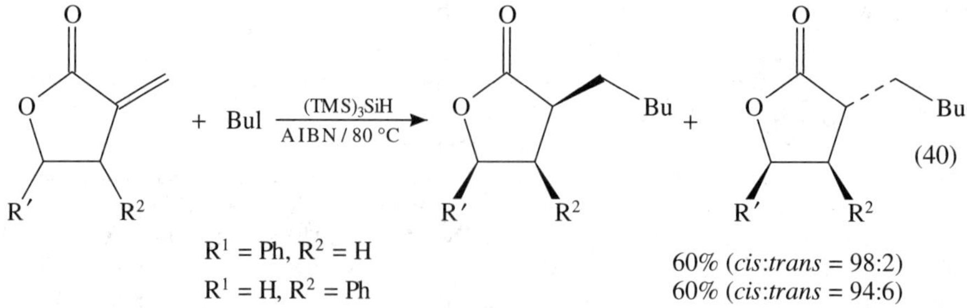

$R^1 = Ph, R^2 = H$
$R^1 = H, R^2 = Ph$

60% (cis:trans = 98:2)
60% (cis:trans = 94:6)

(40)

Radical allylation with 2-functionalized allyl phenyl sulfones has been performed by using $(TMS)_3SiH$ as radical mediator[64]. Yields varied from moderate to good depending on the nature of the starting materials. Two examples are given in equation 41. The reaction proceeded via addition of adamantyl radical to the double bond, giving rise to an intermediate that undergoes β-scission to form $PhSO_2^{\bullet}$ radical which abstracts hydrogen from the silane to regenerate the $(TMS)_3Si^{\bullet}$ radical.

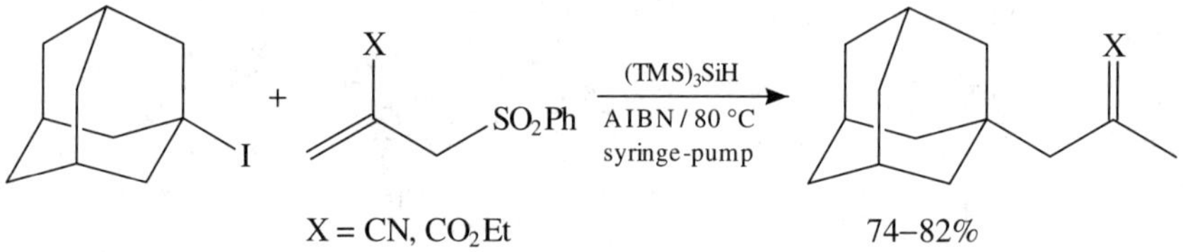

$X = CN, CO_2Et$

74–82%

(41)

$(TMS)_3SiH$ has been used as mediator for the alkylation of heteroaromatic compounds by two independent groups[65,66]. Alkyl bromides or iodides treated with protonated heteroaromatic bases and $(TMS)_3SiH$ either under photochemical (irradiation with 400 W of visible light in CH_2Cl_2) or thermal (in the presence of 1–2 molar amounts of AIBN in benzene) conditions afforded the desired product in moderate to good yields. Two examples are given in equations 42 and 43. The presence of AIBN in at least stoichiometric quantity in this homolytic aromatic substitution reaction ensured rearomatization of the intermediate stabilized cyclohexadienyl-type radical, which can reach a stationary concentration suitable for intercepting an α-cyanoisopropyl radical, thus leading to the substitution product. At the same time, the electrophilicity of the α-cyanoisopropyl radical prevents it from adding to the protonated heteroaromatic base while completing the course of this nonchain process. It is worth pointing out that this method is the first report on the alkylation of heteroaromatic bases under nonoxidative conditions.

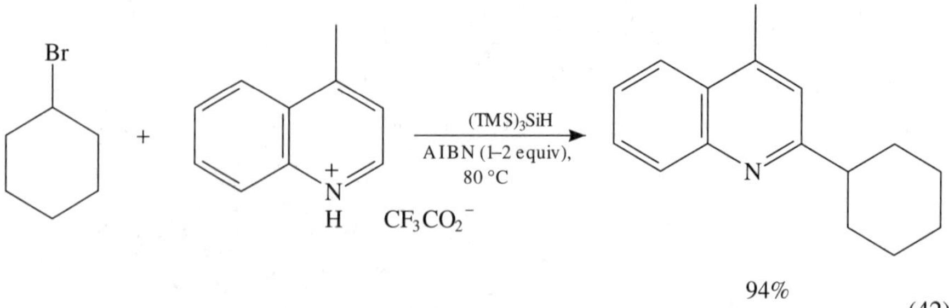

94%

(42)

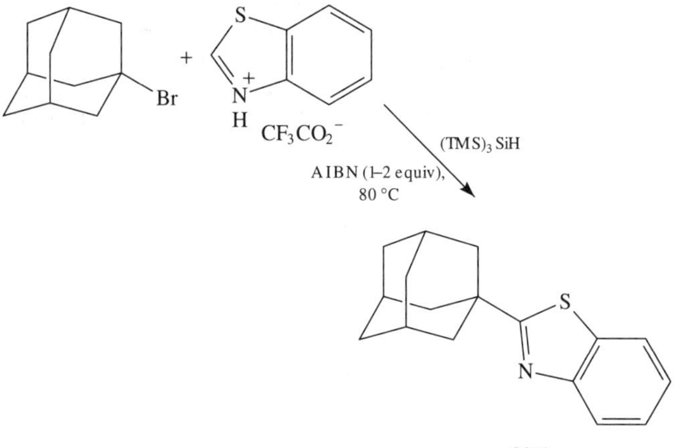

25. Tris(trimethylsilyl)silane in organic synthesis

(43)

83%

The full advantages of using a slow hydrogen donating reagent as $(TMS)_3SiH$ were perhaps best demonstrated in the free-radical carbonylation reactions developed by Ryu and Sonoda where the syringe-pump addition cannot be applied[67]. These radical chain reactions proceeded by formation of an alkyl radical which may be reduced to **4** or add onto carbon monoxide, generating an acyl radical intermediate which, in turn, can either abstract hydrogen from the reducing agent to give the corresponding aldehyde **5** (Scheme 4) or further react with electron-deficient olefins to lead, after reduction, to a formal double alkylation of carbon monoxide (equations 44 and 45)[68]. This three-component coupling reaction requires the generation of four highly disciplined radical species which have specific functions during the chain reaction. Employment of $(TMS)_3SiH$ in place of Bu_3SnH resulted in a multifaceted improvement of the reaction: (i) Since alkyl radicals abstract a hydrogen atom from $(TMS)_3SiH$ approximately ten times slower than from tin hydride (e.g. see Figure 1), $(TMS)_3SiH$ can mediate the carbonylation of free radicals at CO pressures lower than those needed for the tin hydride system (15 atm compared with 50, see Scheme 4). (ii) Similarly, the intermediate acyl radical abstracts hydrogen from $(TMS)_3SiH$ twenty times slower than from tin hydride[16]. This lead to an efficient three-component coupling reaction even in the presence of a stoichiometric amount of

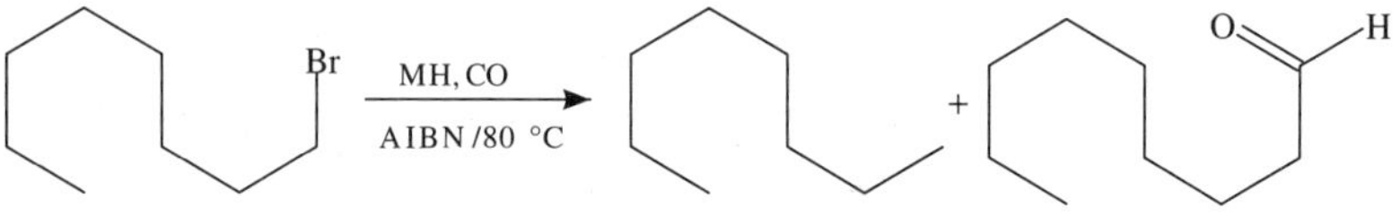

MH	CO (atm)	**4** (%)	**5** (%)
$(TMS)_3SiH$	30	16	80
	15	29	65
Bu_3SnH	50	36	63
	15	49	38

SCHEME 4

the alkene, whereas in the presence of Bu$_3$SnH three to four equivalents of alkene are necessary for efficient coupling. (iii) Finally, since the rate of hydrogen abstraction from tributyltin hydride by vinyl radicals is about 100 times faster than by primary alkyl radicals, vinyl halides cannot participate in free radical carbonylations in the presence of Bu$_3$SnH but they did afford good yields of β'-substituted-α,β-enones in the presence of (TMS)$_3$SiH (equation 45).

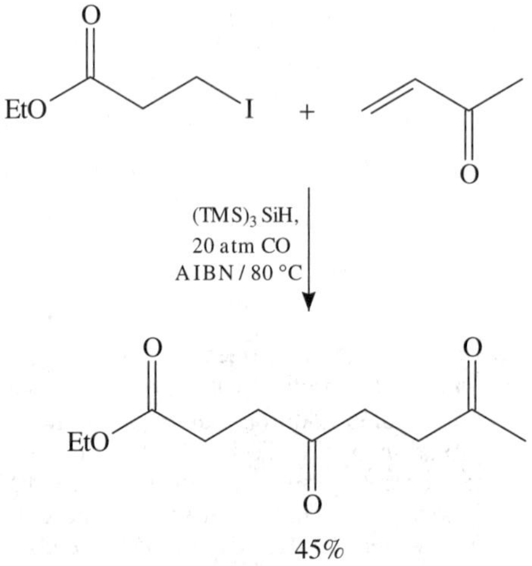

(44)

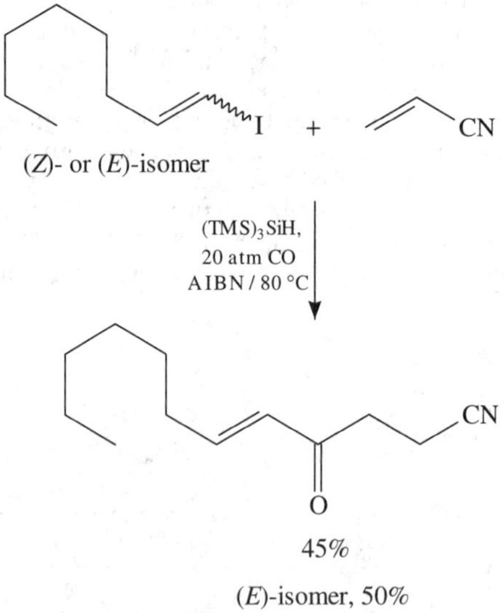

(45)

The field of macrocyclization reactions was recently augmented by the same group[69] with an $n + 1$ radical annulation which results in the incorporation of a CO molecule in the macrocycle. Thus, in the presence of highly diluted (0.005–0.01 M) (TMS)$_3$SiH, ω-iodoacrylates underwent efficient three-step radical chain reaction to generate 10- to 17-membered macrocycles in 28–78% yields, respectively (equation 46). It has also been shown that (TMS)$_3$SiH mediates a 14-*endo-trig* macrocyclization in 55% yield as a key step feature in a concise synthesis of optically active (−)-zearalenone[70].

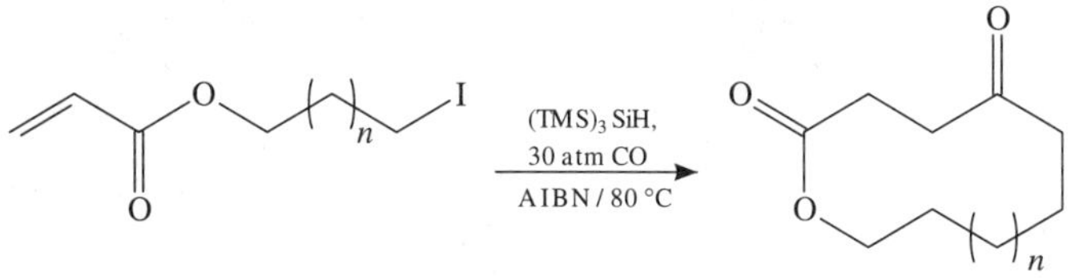

$$n = 1,2,4,6,7,8 \qquad 28–78\%$$

(46)

The intramolecular C−C bond formation (or cyclization) mediated by (TMS)$_3$SiH has been the subject of numerous publications. Scheme 5 presents the simplest and more popular type of 5-hexenyl radical cyclization. Thus, a 50 mmolar solution of 5-hexenyl bromide and silane (or tin hydride) lead to a 24 : 1 (or 6 : 1) ratio of cyclized versus uncyclized products[61]. Therefore, under the same conditions the silane gives higher yields of cyclization than the stannane.

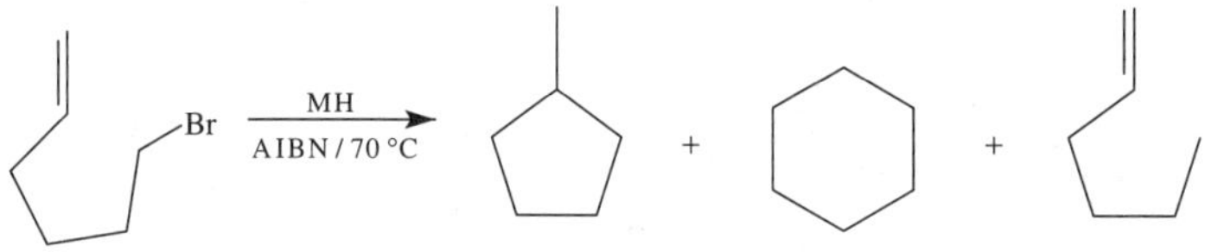

(TMS)$_3$SiH	93%	2%	4%
Bu$_3$SnH	83%	1.5%	15%

SCHEME 5

Tacamonine, an indole alkaloid of the *Iboga* type, isolated from *Tabernaemontana eglandulosa*, the root of which is used to treat snake bites in Zaire, bears structural similarity to the *Hunteria* alkaloids, eburnamonines, which possess vasodilator and hypotensive activities. Its synthesis in racemic and homochiral form was accomplished by incorporating a classic 6-*exo-trig* radical cyclization in the key step of the synthesis (Scheme 6)[71]. The radical precursor **6** was constructed in a 7-step synthesis by starting from racemic or chiral propane-1,3-diol. The radical cyclization of **6** produced the piperidinone in 72% yield as a diastereomeric mixture, which was then transformed into tacamonine.

Pattenden and coworkers have designed and performed a cascade of radical reactions towards the synthesis of angular triquinanes[72]. Irradiation of the refluxed benzene solution containing a 1 : 1 mixture of diastereomers of bromide **7** and (TMS)$_3$SiH gave the corresponding triquinane oxime **8** as a 1 : 1 mixture of α- and β-methyl diastereomers in 38% yield.

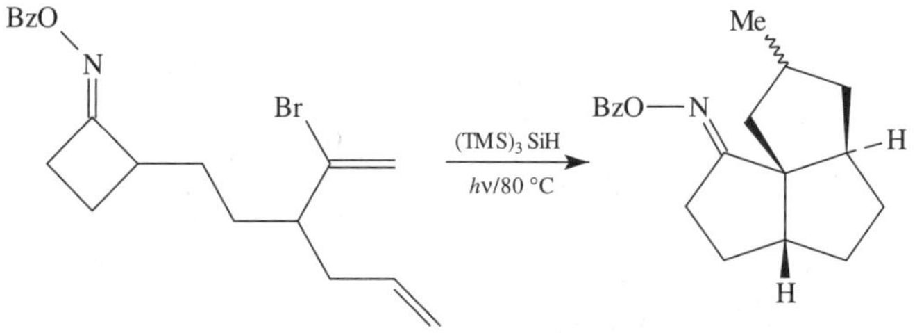

(47)

(7) (8) 38%

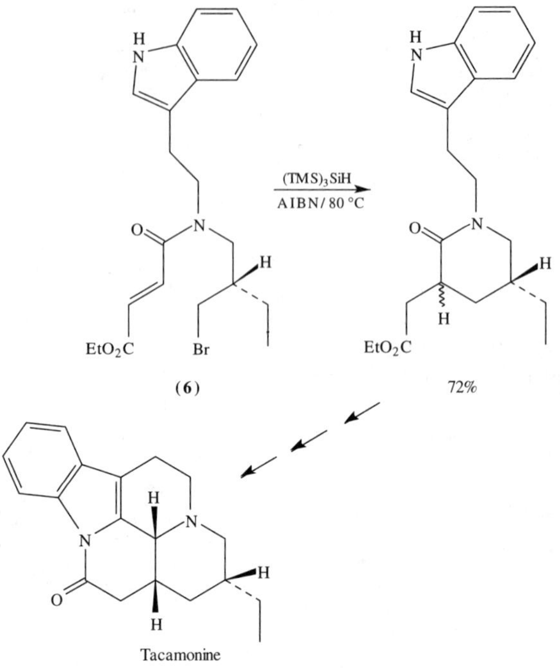

SCHEME 6

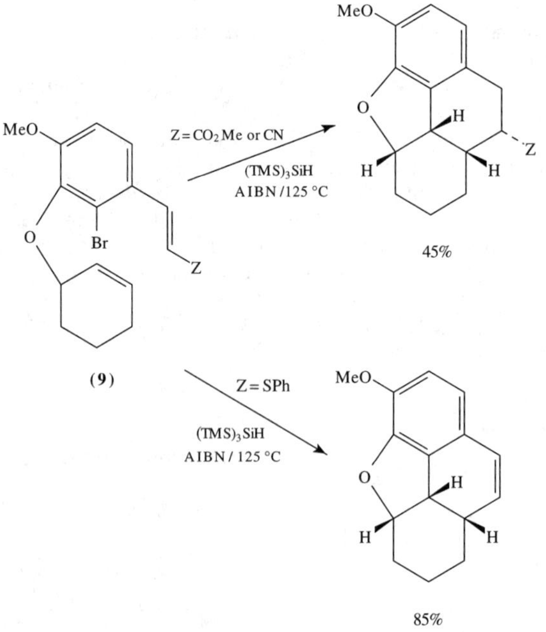

SCHEME 7

A tandem cyclization as key steps in the synthesis of morphine alkaloids has been projected and successfully performed[73]. Three examples are given in Scheme 7. Specifically, the initially formed aryl radical, generated by bromine abstraction from compound **9**, underwent a tandem cyclization to construct the desired carbocyclic skeleton. The product radicals either abstracted hydrogen from silane (Z = CO_2Me or CN) or eliminated thiyl radical (Z = SPh). On the other hand, the planned tandem cyclization starting from the bromide **10**, toward the construction of indole alkaloids skeleton, gave an unexpected result (equation 48)[74a]. Mechanistic considerations for the formation of quinoline have been advanced. In model studies directed towards the synthesis of (±)-gelsemine[74b], 5-*exo-trig*-cyclization of an aryl radical, derived from the vinylogous urethane **11**, onto a methoxymethyl enol ether resulted in partial fragmentation of the intermediate radical species with expulsion of a methoxymethyl radical and generation of the ketone group (equation 49).

It is known that the geometries of the reactants play an important role in the regio- and stereochemical outcome of radical reactions since they are commonly involved in early transition states. Previous attempts to affect rotamer populations during the reaction included, among others, control of temperature and addition of a Lewis acid. It was recently reported[75] that organotin halides, common byproducts of radical reactions, act as Lewis acids and control the course of such reactions. An indicative example of this control is given in Scheme 8; the reaction of oxazolidinone **12** with (TMS)$_3$SiH giving products **13** and/or **14** has been performed in the absence and in the presence of organotin halides. The results demonstrated that higher temperatures and weak Lewis acids were necessary not only for inducing the conformational change from the stable *anti-(s)-Z-***12** to the *syn-(s)-Z-***12** required for the radical cyclization reaction, but also for obtaining high diastereoselectivity.

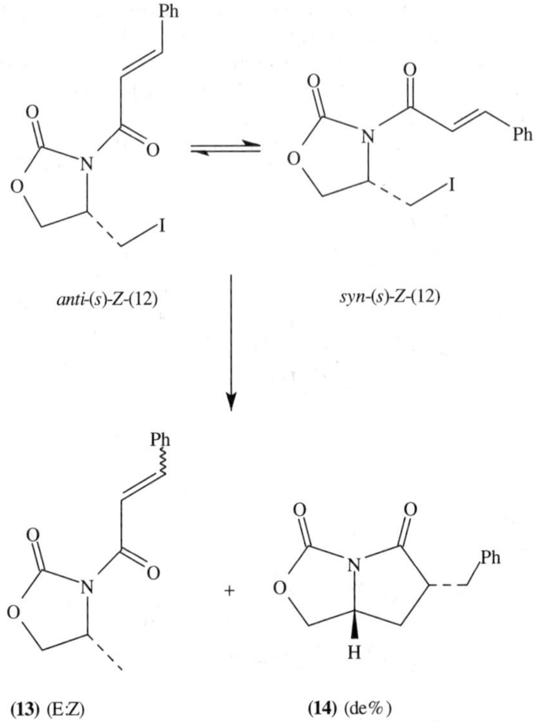

anti-(s)-Z-(**12**)   *syn-(s)-Z-*(**12**)

(**13**) (E:Z)   (**14**) (de%)

(TMS)$_3$SiH, AIBN/80 °C	78 % (100:0)	13 % (87)
(TMS)$_3$SiH, hv/r.t.	72 % (71:29)	0 %
(TMS)$_3$SiH, Bu$_3$SnCl, AIBN/80 °C	0 %	87 % (>97)
(TMS)$_3$SiH, Bu$_3$SnCl, hv/ r.t.	60 % (60:40)	14 % (86)

SCHEME 8

An example in which primary alkylamines are used as alkyl radical precursors for cyclization is reported in equation 50[61], in which the initial transformation to isocyanide and subsequent reaction with (TMS)$_3$SiH gave the desired products in 78% yield after workup.

The intermediate α-heterosubstituted carbon radicals generated by reaction of (TMS)$_3$Si• radical with 1,3-dithiane[28], or *N*-(ethoxycarbonyl)-1,3-thiazolidine[76] derivatives, can participate in consecutive intramolecular C—C bond formation reactions in the presence of proximate 1,2-disubstituted double bonds (equation 51). In the presence of terminal double bonds or in an attempted intermolecular addition of the intermediate radical species to electron-deficient olefins, hydrosilylation of the double bond by (TMS)$_3$SiH competed with the reduction and prevented C—C bond formation.

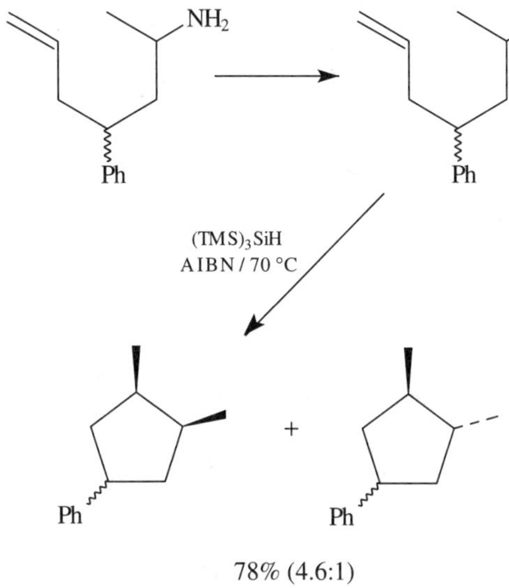

(50)

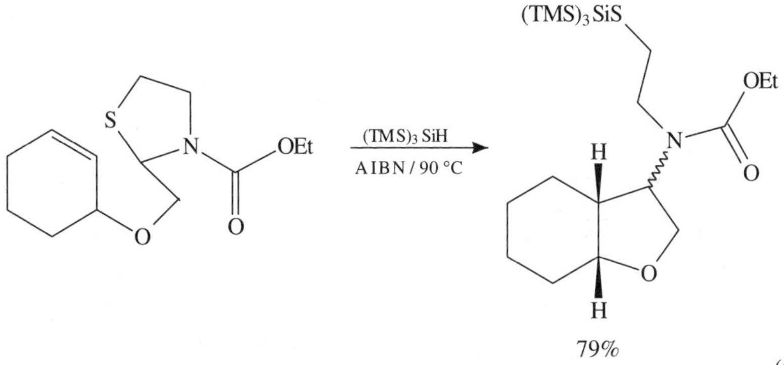

78% (4.6:1)

(51)

79%

Free-radical cyclization of phenyl selenide **15** to indolizidinone **16** represented a key step in the total synthesis of (−)-slaframine (equation 52). The two pairs of diastereomers were first separated and then hydrolyzed to the corresponding alcohols in 76% overall yield[77]. (TMS)$_3$SiH-mediated acyl radical reactions from phenylseleno esters **17** have recently been utilized for the stereoselective synthesis of cyclic ethers[78]. In fact, the experimental conditions reported in equation 53 are particularly good for both improving *cis* diastereoselectivity and suppressing decarbonylation.

The cyclization of secondary alkyl radicals with α, β-alkynyl esters **18** proceeded with high stereoselectivity to give predominantly (Z)-exocyclic alkenes at low temperature upon reaction with (TMS)$_3$SiH (equation 54)[79]. On the other hand, the formation of (E)-exocyclic alkenes predominated with Bu$_3$SnH, the E/Z ratio being 98 : 2 at 80 °C. It has been suggested that the main factor controlling the formation of these products is the ability of (TMS)$_3$Si$^•$ and Bu$_3$Sn$^•$ radicals to isomerize the product alkene. That is, the $(E) \rightleftharpoons (Z)$ isomerization occurred under tin hydride reduction conditions whereas no such transformation was observed with $(TMS)_3SiH$. This class of reactions has also been extended to cyclization of difluoroalkyl radicals[80].

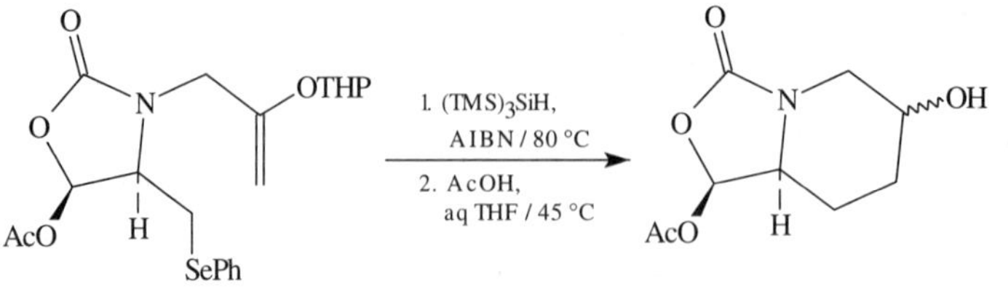

(52)

(15) → (16) 76% (α:β=7:1)

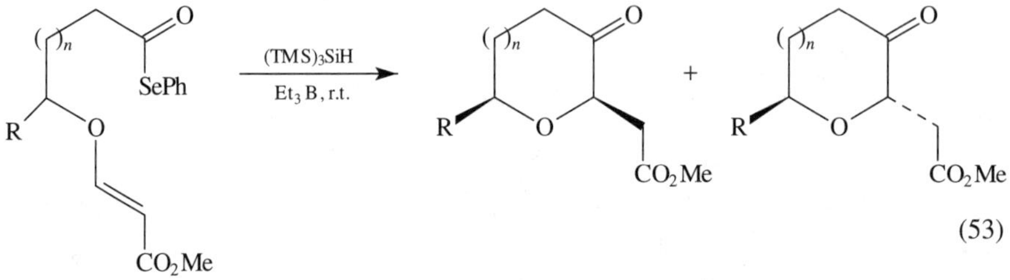

(53)

(17) R = Me, Ph, i-Pr

n = 1	94% (5.7:1)	
n = 2	90% (≥19:1)	

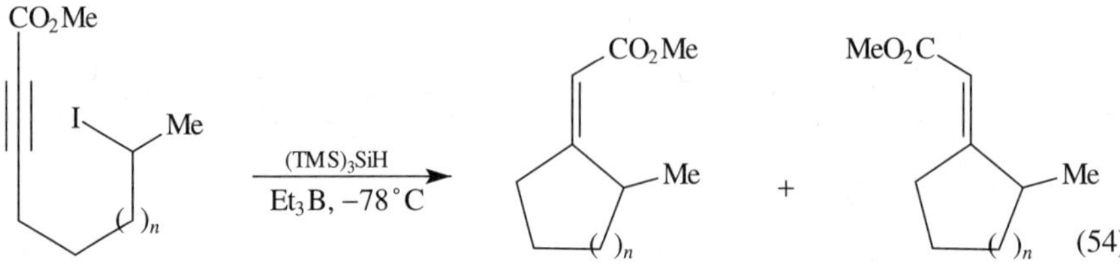

(54)

(18)

n = 1	85% (Z:E = 11:89)
n = 2	82% (Z:E = 9:91)

The radical cyclization of bromomethyldimethylsilyl propargyl ethers has recently been utilized by Malacria and coworkers[81,82] in order to generate precursors for the synthesis of the triquinane framework. In the first of these reports[81], reduction of the bromide **19** in the presence of $(TMS)_3SiH$ and subsequent treatment with MeLi produced the functionalized cyclopentanone precursor **20** as a single diastereomer (Scheme 9). The formation of **20** could be explained by a series of reactions, indicated in Scheme 9, involving a 5-*exo-trig* cyclization of the initial α-silylalkyl radical followed by a [1,5]-radical translocation of the generated σ-type vinyl radical onto the proximal acetal function and a final 5-*exo-trig* process. Stereoselective hydrogen abstraction, dependent on the steric bulkiness of the hydrogen donor, followed by MeLi induced opening of the Si–O bond afforded the final product. The introduction of different substituents on the skeleton, as in compound **21** resulted in a completely different reaction pattern (equation 55)[82].

In this case, the intermediate vinyl radical (cf Scheme 9) underwent a remarkable [1,5]-hydrogen abstraction from the non-activated C—H bond of the proximal isopropyl group. Furthermore, the resulting primary alkyl radical underwent a unique, stereoselective 5-*endo-trig* cyclization onto the adjacent double bond to generate a tertiary radical, which is a precursor of the highly substituted cyclopentanols **22** and **23**. The reaction with Bu_3SnH as radical mediator totally reversed the products ratio obtained in 88% yield, i.e. **22** : **23** = 19 : 81.

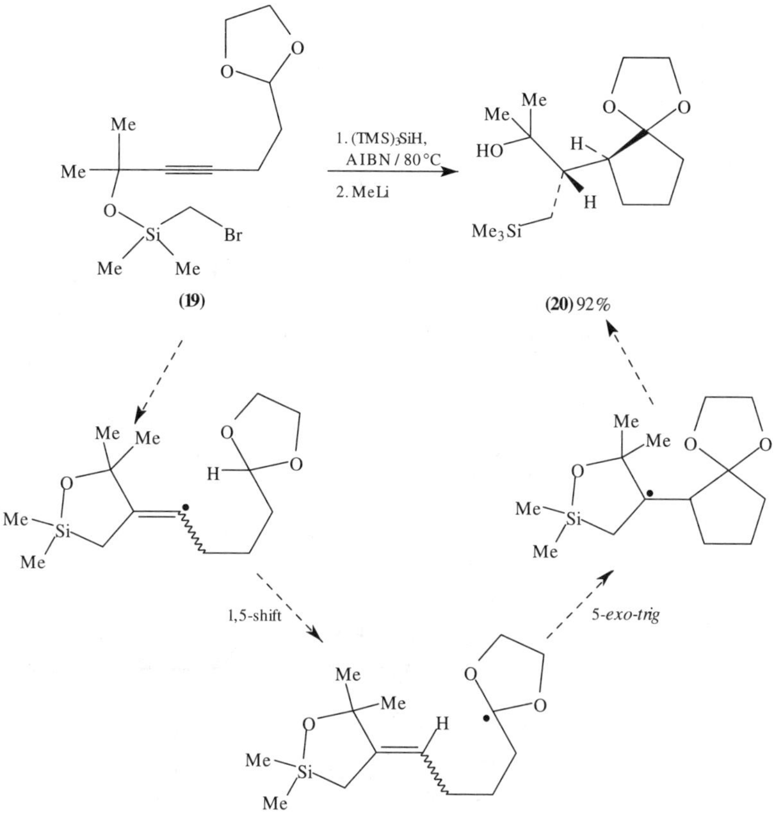

SCHEME 9

Interestingly, another example of a [1,5]-radical translocation coupled with an unusual 5-*endo-trig* radical cyclization was reported in a structurally different system (Scheme 10)[83]. In this case, the α-bromovinyl radical abstracted a hydrogen either from $(TMS)_3SiH$ to give vinyl bromide in 25% yield or from the anomeric position to generate the C-1′ radical which underwent an unusual 5-*endo-trig* cyclization onto the proximal double bond to generate an anomeric mixture of spironucleosides after bromine atom ejection.

The utilization of the azido group as radical acceptor in radical reactions shown in equation 56, has been tested by Kim and coworkers[84,85] using $(TMS)_3SiH$, which is relatively inert towards azides when compared to Bu_3SnH. Therefore, alkyl bromides and thionocarbonates can be used as precursors of radicals, generated by $(TMS)_3SiH$ which

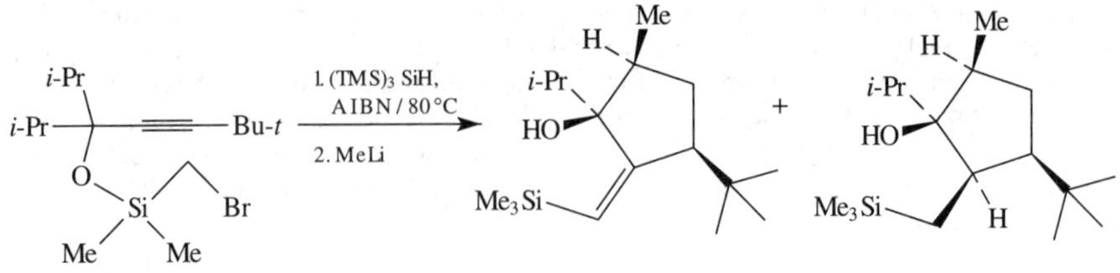

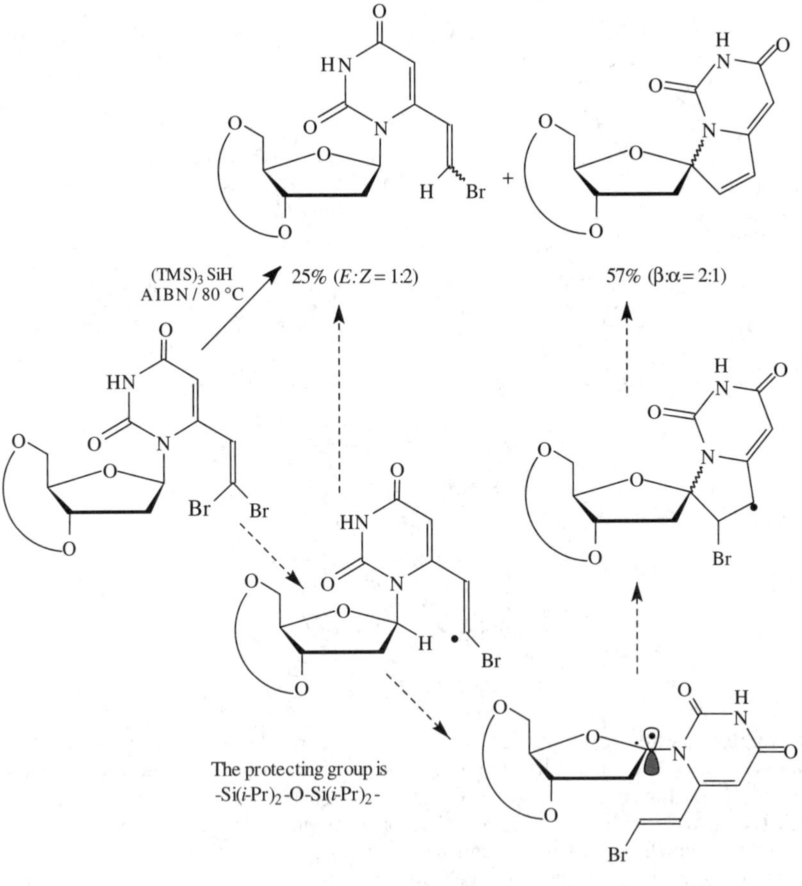

SCHEME 10

in turn can cyclize onto suitably positioned azides to provide nitrogen heterocycles. An example is given in equation 57 where the amine product was tosylated before work-up. For comparison, in the presence of Bu_3SnH only alkyl iodides are reactive enough to be used for the same purpose. An analogous behavior of the azide group was observed by others[86] in a similar cascade of radical reactions.

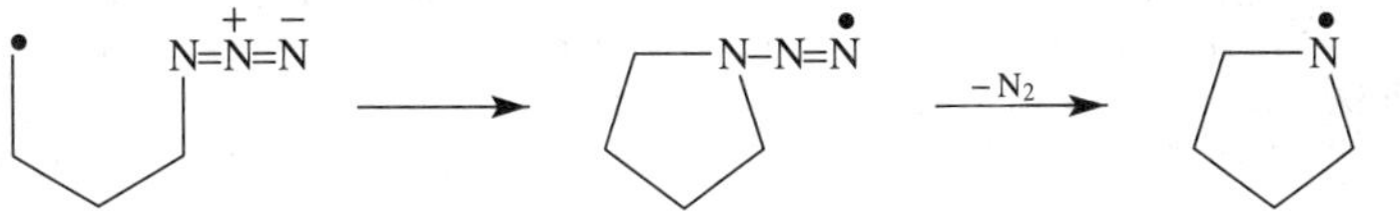

(56)

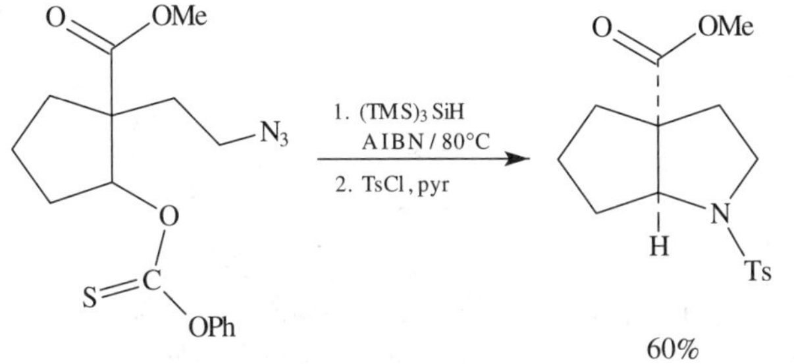

(57)

60%

More recently, the radical cyclization onto azide groups pioneered by Kim was applied in the construction of the *B/E* spirocyclic junction found in the [6.5.6.5] *ABCE* ring system of indole alkaloids such as strychnine, the clinically used anticancer agents vincristine and vinblastine and, in particular, aspidospermine[87]. In model system studies, cyclization of the iodoazide **24**, prepared in 6 steps, in the presence of $(TMS)_3SiH$, produced the N−Si $(TMS)_3$ protected alkaloid **25**, that after washing with dilute acid afforded the amine **26** in 95% yield from **24** (equation 58). The formation of the liable $N-Si(TMS)_3$ bond was considered to arise from the reaction of the product amine **26** with the byproduct $(TMS)_3SiI$.

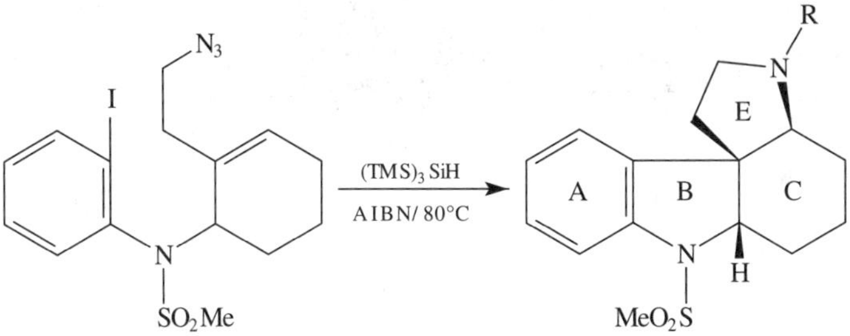

(58)

(24)

(25) $R = Si(TMS)_3$
(26) $R = H, 95\%$

The different H-atom donor abilities of $(TMS)_3SiH$ and Bu_3SnH, which have been encountered many times, can also be useful to solve intriguing radical mechanisms, and an example of this was recently reported[88]. The outcome of the reaction of δ, ε-unsaturated acyl radicals has been the research subject of several groups and can be illustrated as in Scheme 11[89]. Based on the available kinetic data of secondary substituted acyl radicals with the two reagents [i.e. 1.1×10^6 and 7.2×10^4 $M^{-1} s^{-1}$ at 80 °C for Bu_3SnH and $(TMS)_3SiH$, respectively][16], of the decarbonylation of acyl radicals[16], and of different hydrogen donor abilities of $(TMS)_3SiH$ and Bu_3SnH towards alkyl radicals (Section I.B), it was possible to plan and perform kinetic experiments. In particular, the reactions of thionocarbonate **27** with $(TMS)_3SiH$ and of phenylseleno ester **28** with Bu_3SnH were informative on the rate constants of the possible transformations, thus clarifying the overall mechanism (Scheme 12)[88].

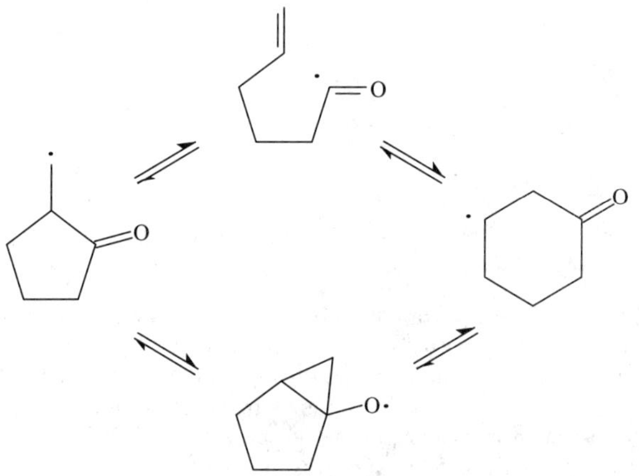

SCHEME 11

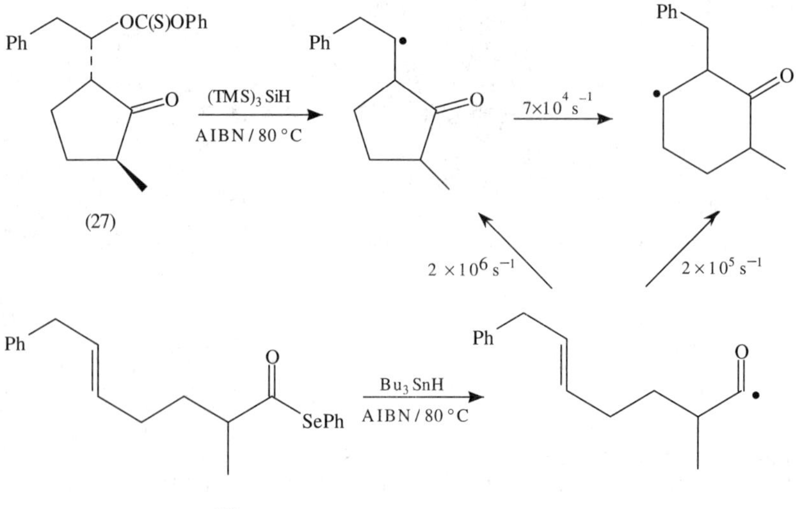

SCHEME 12

The reaction of iodide **29** with (TMS)$_3$SiH under free radical conditions gave the products depicted in equation 59 in quantitative yield[90]. This procedure, which used a homolytic substitution at sulfur, has been proposed as versatile and alternative to phenylseleno esters for the generation of acyl radicals. In fact, the aryl radical formed by iodine abstraction rearranged with expulsion of the acyl radical and concomitant formation of dihydrobenzothiophene. Similarly, internal homolytic substitution of aryl radicals at selenium has been used for the preparation of selenophenes and benzoselenophenes[91]. Scheme 13 illustrates the reaction of iodides **30** with (TMS)$_3$SiH to afford benzoselenophenes in good yield. The presence of (TMS)$_3$SiI after the reduction induced dehydration of the intermediate 3-hydroxybenzoselenophenes, presumably through an intermediate silyl ether.

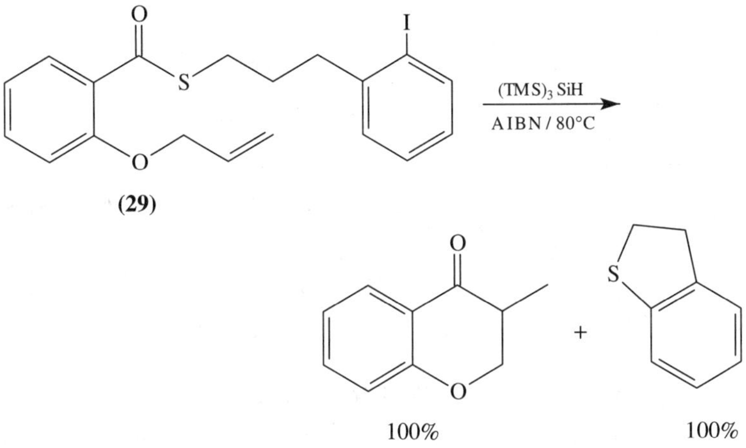

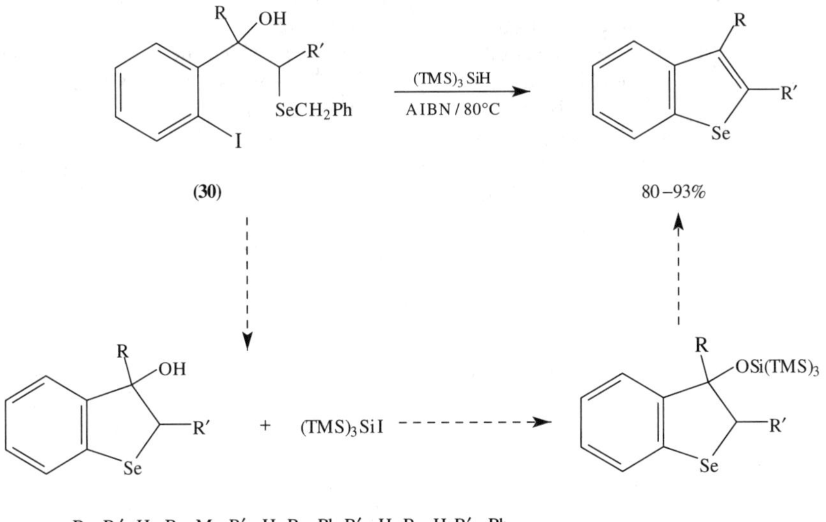

R = R′ = H; R = Me, R′ = H; R = Ph, R′ = H; R = H, R′ = Ph

SCHEME 13

Reflux of bromide **31** in benzene in the presence of small amounts of $(TMS)_3SiH$ and AIBN afforded the silabicycle **32** in 88% yield (equation 60)[92]. The key step for this transformation is the intramolecular homolytic substitution at the central silicon atom which occurred with a rate constant of 2.4×10^5 s^{-1} at 80 °C. The reaction has also been extended to the analogous vinyl bromide (equation 61)[93].

B. Initiation by $(TMS)_3Si^\bullet$ Radical Addition to Unsaturated Bonds

β-Silyl substituted carbon-centered radicals which are produced upon addition of $(TMS)_3Si^\bullet$ to unsaturated bonds can participate in consecutive reactions. A simple example is given in equation 62 where the adduct of $(TMS)_3Si^\bullet$ radical to β-pinene rearranged by opening the four-membered ring prior to H atom transfer[21]. Reactions of unsubstituted and 2-substituted allyl phenyl sulfides with $(TMS)_3SiH$ provided the corresponding allyl tris(trimethylsilyl)silanes in high yields (equation 63)[94]. That is, $(TMS)_3Si^\bullet$ radical adds to the double bond giving rise to a radical intermediate that undergoes β-scission with ejection of thiyl radical. Hydrogen abstraction from the silane completes the cycle of these chain reactions. 2-Functionalized allyl tris(trimethylsilyl)silanes have been employed in the radical-based allylation reactions[95].

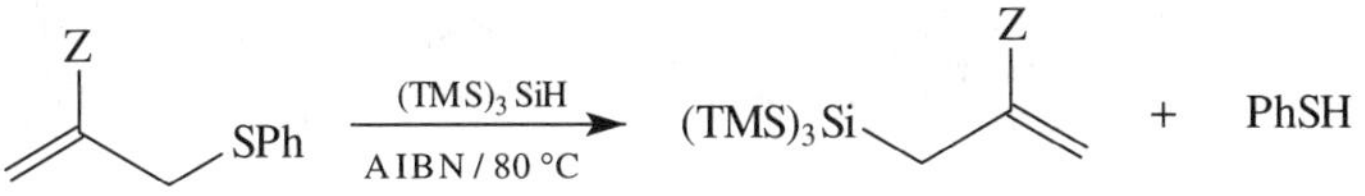

25. Tris(trimethylsilyl)silane in organic synthesis

$$Z = H, Me, Cl, CN, CO_2Et \quad >80\%$$

(63)

The product distribution of the (TMS)$_3$SiH addition to 1,6-dienes depends on the concentration of reducing agent. As an example, diallyl ether reacted with 1.2 equivalents of silane to give 63% yield of **33** in a *cis:trans* ratio of 3 : 1[21]. The same reaction performed by syringe-pump addition of the silicon reagent gave silabicycle **34** in 55% yield together with **33** in 15% yield (equation 64). The reaction mechanism for this reaction is analogous to that described earlier in equation 60[92].

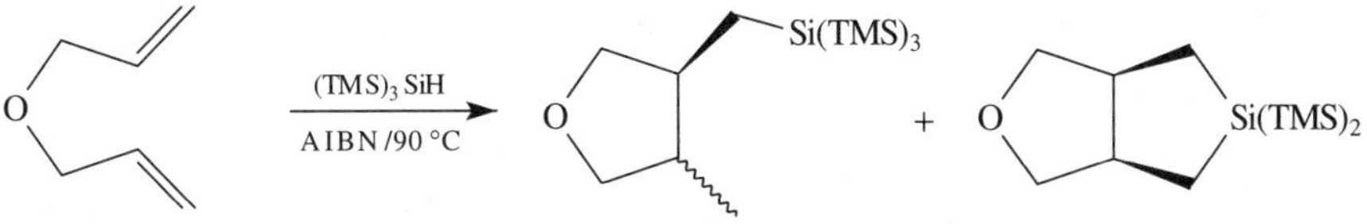

(64)

Reaction of (TMS)$_3$SiH with β-alkenyloxyenones can follow either one or both of two different pathways depending on the substitution at the double bond[96a]. Specifically, in the case of loosely conjugated s-*cis* enone **35**, addition of silyl radical to the terminal double bond was accompanied by a 5-*exo-trig* radical cyclization leading to the diastereomeric cyclic products (equation 65). In the case of nonterminal double bonds such as in **36**, exclusive addition of the (TMS)$_3$Si• radical to the carbonyl was followed by a 5-*exo-trig* cyclization and hydrogen abstraction. In the presence of cyclohexenones or cyclopentenones and terminal double bonds, a combination of both chemistries was observed. Useful bicyclic ring systems are obtained by (TMS)$_3$Si• radical mediated fragmentation of strained alkene precursors[96b]. For example, the ketoalkene **37** reacted with 1.5 equivalents of silane to give 95% of hydrindanone **38** (equation 66).

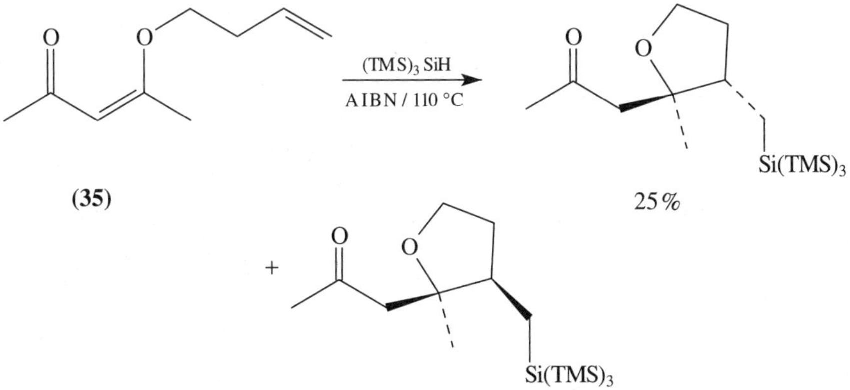

(65)

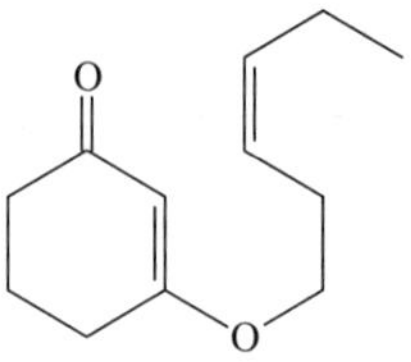

(36)

(66)

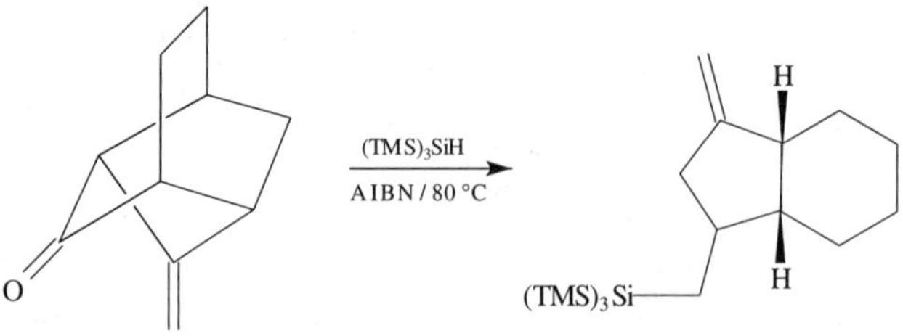

(37) → (TMS)₃SiH, AIBN / 80 °C → (38, 95%)

The α-silyloxy alkyl radical generated by addition of (TMS)$_3$Si• radical to the aldehyde moiety of **39** has been employed in radical cyclization onto β-aminoacrylates (equation 67)[97]; the *trans*-hydroxy ester and the lactone formed in a 2.4 : 1 ratio were the two products. An alternative route to *N*-heterocycles has been developed by Kim and coworkers[84,85] using the azido group as radical acceptor (cf equation 56). Two examples are given in equations 68 and 69. The carbon-centered radical, derived by addition of (TMS)$_3$Si• radical to **40**, cyclized to the azido group to afford the desired product in good yields, independently of the ring size. On the other hand, the radical adduct arising from **41** underwent an opening of the cyclopropyl ring, with formation of **42** as intermediate, prior to the intramolecular addition to the azido group. After tosylation the final product was obtained in 61% yield.

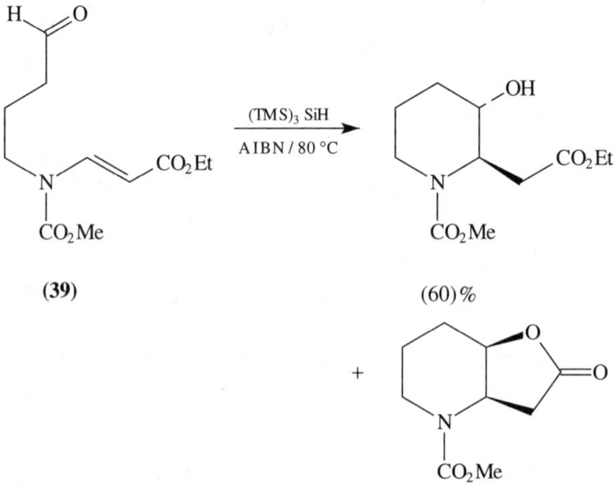

(39) → (TMS)$_3$SiH, AIBN / 80 °C → (60)% + (25)%

(67)

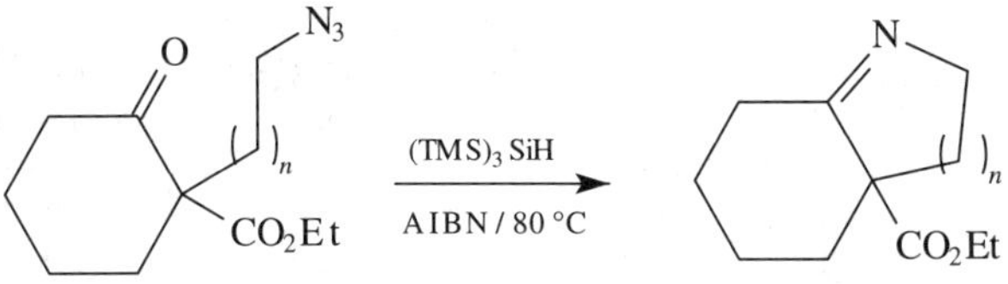

(68)

(40) $n = 1, 2$ 78%

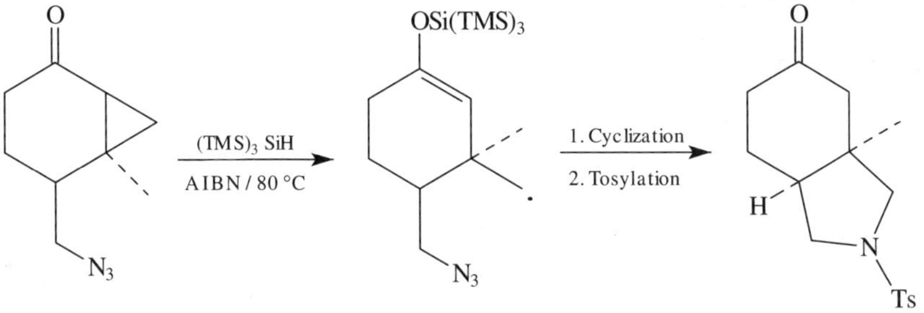

(69)

(41)   **(42)**   61%

The addition of $(TMS)_3SiH$ to the 3-oxo-1,4-diene steroid-type derivative promoted C(9)–C(10) bond cleavage (equation 70)[98]. The removal of the silyl group was achieved by treatment with dilute aqueous hydrochloric acid at room temperature in 86% yield.

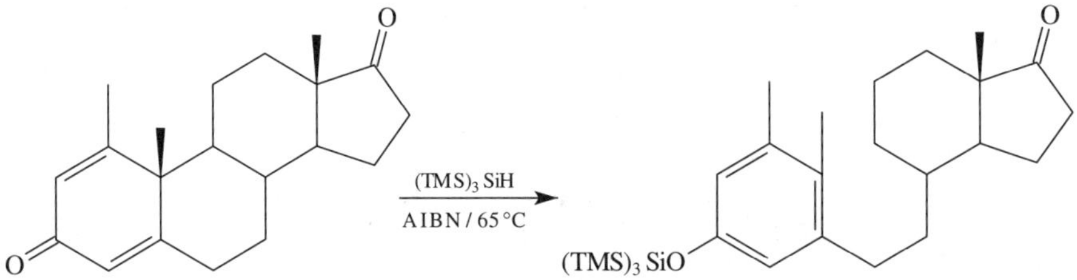

(70)

Pattenden and Schulz have reported that treatment of the acetylene derivative **43** with $(TMS)_3SiH$ leads, in one pot, to the bicyclic compound **44** in 70% yield (equation 71)[99]. The proposed mechanism involves $(TMS)_3Si^{\bullet}$ radical addition to the triple bond to form a vinyl radical followed by a remarkable cascade of radical cyclization–fragmentation–transannulation–ring expansion and termination via ejection of the $(TMS)_3Si^{\bullet}$ radical to afford the bicyclic product.

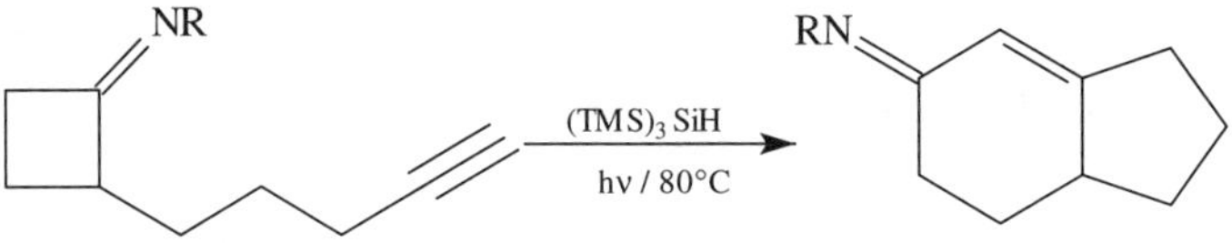

(71)

(43) R = OCH_2Ph **(44)** 70%

Treatment of isothiocyanide derivatives of glycine **45** or **46** with the silane under radical conditions afforded the corresponding pyroglutamates in good yields (equation 72)[100]. The expected products, derived from the cyclization of α-silylthio imidoyl radicals, hydrolyzed spontaneously during chromatography. α-Diazo ketones were found to react with $(TMS)_3SiH$ under free radical conditions to give α-silyl ketones (equation 73)[101]. Mechanistic evidence that the attack of $(TMS)_3Si^\bullet$ takes place at carbon rather than nitrogen, to give a diazenyl radical adduct which decomposes to α-silyl substituted radical and nitrogen, has been obtained. α-Silyl carbonyl compounds have also been obtained in moderate yields from the reaction of α-diazo esters with $(TMS)_3SiH$ in the presence of rhodium(II) complexes as catalysts (equation 74)[102]. This reaction was presumed to be mediated by transition metal carbenoid rather than by radical intermediates.

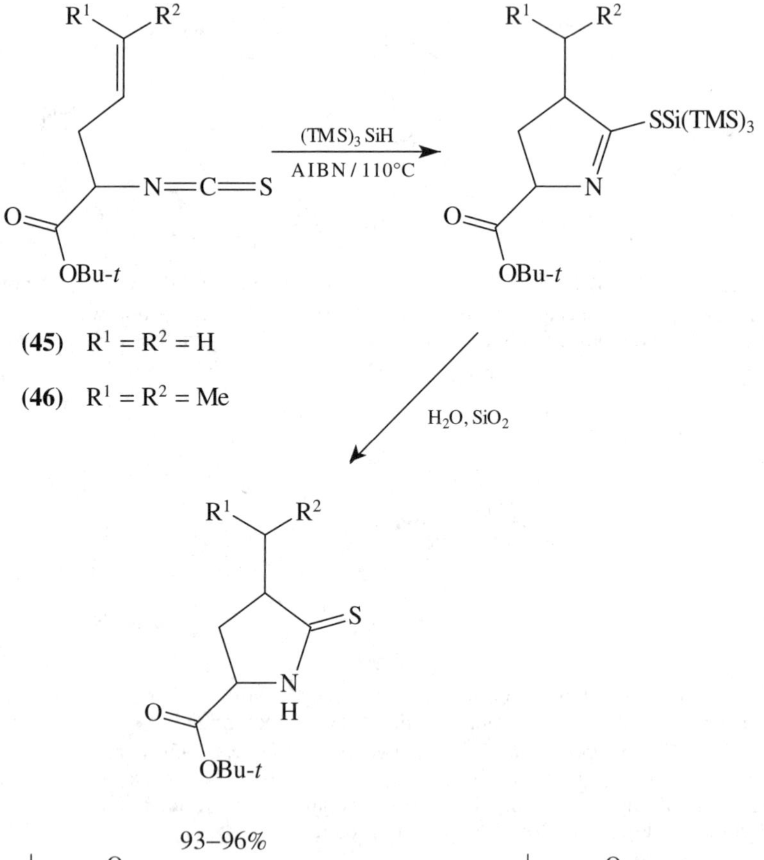

(**45**) $R^1 = R^2 = H$

(**46**) $R^1 = R^2 = Me$

(72)

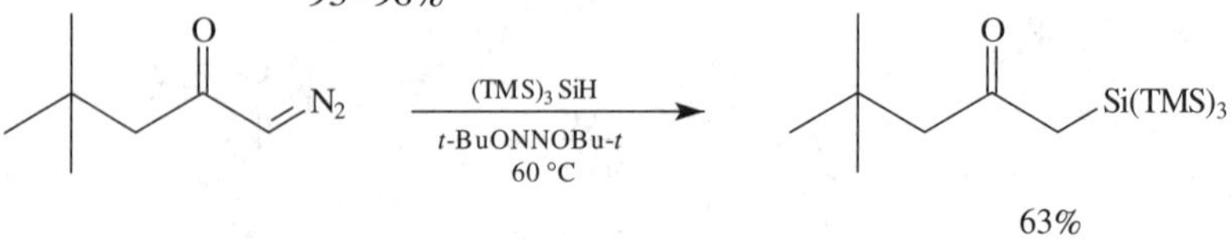

(73)

(74)

55% (34% de)

VI. REFERENCES

1. U. Koert, *Angew. Chem., Int. Ed. Engl.*, **35**, 405 (1996) and references cited therein.
2. For example, see:
   (a) W. B. Motherwell and D. Crich, *Free Radical Chain Reactions in Organic Synthesis*, Academic Press, London, 1992.
   (b) B. Giese, *Radicals in Organic Synthesis: Formation of Carbon–Carbon Bonds*, Pergamon Press, Oxford, 1986.
3. For example, see:
   (a) D. P. Curran, *Synthesis*, 417, 489 (1988).
   (b) C. P. Jasperse, D. P. Curran and T. L. Fevig, *Chem. Rev.*, **91**, 1237 (1991).
   (c) D. P. Curran in *Comprehensive Organic Synthesis* (Eds. B. M. Trost and I. Fleming), Vol. 4, Pergamon Press, Oxford, 1991, pp. 715–831.
4. C. Chatgilialoglu, *Acc. Chem. Res.*, **25**, 188 (1992) and references cited therein.
5. J. M. Kanabus-Kaminska, J. A. Hawari, D. Griller and C. Chatgilialoglu, *J. Am. Chem. Soc.*, **109**, 5267 (1987).
6. C. Chatgilialoglu, D. Griller and M. Lesage, *J. Org. Chem.*, **53**, 3641 (1988).
7. C. Chatgilialoglu and C. Ferreri, *Res. Chem. Intermed.*, **19**, 755 (1993).
8. C. Chatgilialoglu, K. U. Ingold and J. C. Scaiano, *J. Am. Chem. Soc.*, **103**, 7739 (1981).
9. C. Chatgilialoglu and M. Ballestri, *Organometallics*, **14**, 5017 (1995).
10. C. Chatgilialoglu, C. Ferreri and M. Lucarini, *J. Org. Chem.*, **58**, 249 (1993).
11. C. Chatgilialoglu, *Chem. Rev.*, **95**, 1229 (1995).
12. C. Chatgilialoglu and S. Rossini, *Bull. Soc. Chim. Fr.*, 298 (1988).
13. C. Chatgilialoglu, J. Dickhaut and B. Giese, *J. Org. Chem.*, **56**, 6399 (1991).
14. S. J. Garden, D. V. Avila, A. L. J. Beckwith, V. W. Bowry, K. U. Ingold and J. Lusztyk, *J. Org. Chem.*, **61**, 805 (1996).
15. X. X. Rong, H.-Q. Pan, W. R. Dolbier, Jr. and B. E. Smart, *J. Am. Chem. Soc.*, **116**, 4521 (1994).
16. C. Chatgilialoglu, C. Ferreri, M. Lucarini, P. Pedrielli and G. F. Pedulli, *Organometallics*, **14**, 2672 (1995).
17. C. Chatgilialoglu, A. Guarini, A. Guerrini and G. Seconi, *J. Org. Chem.*, **57**, 2208 (1992).
18. C. Chatgilialoglu, V. Timokhin, A. Zaborovskiy and A. Berlin, manuscript in preparation.
19. C. Chatgilialoglu, unpublished results.
20. C. Chatgilialoglu, D. Griller and M. Lesage, *J. Org. Chem.*, **54**, 2492 (1989).
21. M. Ballestri, C. Chatgilialoglu, K. B. Clark, D. Griller, B. Giese and B. Kopping, *J. Org. Chem.*, **56**, 678 (1991).
22. J. A. Robl, *Tetrahedron Lett.*, **35**, 393 (1994).
23. F. Wahl, J. Wörth and H. Prinzbach, *Angew. Chem., Int. Ed. Engl.*, **32**, 1722 (1993).

24. M. Lesage, C. Chatgilialoglu and D. Griller, *Tetrahedron Lett.*, **30**, 2733 (1989).
25. E. Lee, C. M. Park and J. S. Yun, *J. Am. Chem. Soc.*, **117**, 8017 (1995).
26. Y. Apeloig and M. Nakash, *J. Am. Chem. Soc.*, **116**, 10781 (1994).
27. M. Ballestri, C. Chatgilialoglu, N. Cardi and A. Sommazzi, *Tetrahedron Lett.*, **33**, 1787 (1992).
28. P. Arya, C. Samson, M. Lesage and D. Griller, *J. Org. Chem.*, **55**, 6248 (1990).
29. P. Arya, M. Lesage and D. D. M. Wayner, *Tetrahedron Lett.*, **32**, 2853 (1991).
30. M. Lesage and P. Arya, *Synlett*, 237 (1996).
31. P. Arya and D. D. M. Wayner, *Tetrahedron Lett.*, **32**, 6265 (1991).
32. D. P. Curran and S. Sun, *Tetrahedron Lett.*, **34**, 6181 (1993).
33. A. Krief, E. Badaoui and W. Dumont, *Tetrahedron Lett.*, **34**, 8517 (1993).
34. L. A. Paquette, D. Friedrich, E. Pinard, J. P. Williams, D. St. Laurent and B. A. Roden, *J. Am. Chem. Soc.*, **115**, 4377 (1993).
35. (a) C. Chatgilialoglu, in *The Chemistry of Sulphenic Acids and Their Derivatives* (Ed. S. Patai), Wiley, Chichester, 1990, pp. 549–569.
    (b) C. H. Schiesser and B. A. Smart, *Tetrahedron*, **51**, 6051 (1995).
36. J. W. Perich, *Synlett*, 595 (1992).
37. P. Boquel, C. Loustau Cazalet, Y. Chapleur, S. Samreth and F. Bellamy, *Tetrahedron Lett.*, **33**, 1997 (1992).
38. M. Toyota, T. Seishi, M. Yokoyama, K. Fukumoto and C. Kabuto, *Tetrahedron*, **50**, 1093 (1994).
39. S.-H. Chen, S. Huang, Q. Gao, J. Golik and V. Farina, *J. Org. Chem.*, **59**, 1475 (1994).
40. M. Oba and K. Nishiyama, *Tetrahedron*, **50**, 10193 (1994).
41. (a) M. Ballestri, C. Chatgilialoglu, M. Lucarini and G. F. Pedulli, *J. Org. Chem.*, **57**, 948 (1992).
    (b) C. Imrie, *J. Chem. Res. (S)*, 328 (1995).
42. S. Fukuzumi and S. Noura, *J. Chem. Soc., Chem. Commun.*, 287 (1994).
43. D. Schummer and G. Höfle, *Synlett*, 705 (1990).
44. F. Hammerschmidt, E. Öhler, J.-P. Polsterer, E. Zbiral, J. Balzarini and E. DeClercq, *Liebigs Ann.*, 551 (1995).
45. C. K. Chu and Y. Chen, US Patent 5,384,396, (1995); *Chem. Abstr.*, **122**, 265937 (1995).
46. J. R. Grierson, A. F. Shields, M. Zheng, S. M. Kozawa and J. H. Courter, *Nucl. Med. Biol.*, **22**, 671 (1995).
47. J. J. Barchi, Jr., A. Haces, V. E. Marquez and J. J. McCormack, *Nucleosides & Nucleotides*, **11**, 1781 (1992).
48. T. Gimisis, G. Ialongo, M. Zamboni and C. Chatgilialoglu, *Tetrahedron Lett.*, **36**, 6781 (1995).
49. M. Sekine and K. Seio, *J. Chem. Soc., Perkin Trans. 1*, 3087 (1993) and references cited therein.
50. A. Alberti and C. Chatgilialoglu, *Tetrahedron*, **46**, 3963 (1990).
51. B. Kopping, C. Chatgilialoglu, M. Zehnder and B. Giese, *J. Org. Chem.*, **57**, 3994 (1992).
52. C. Chatgilialoglu, M. Ballestri, C. Ferreri and D. Vecchi, *J. Org. Chem.*, **60**, 3826 (1995).
53. D. W. Johnson and A. Poulos, *Tetrahedron Lett.*, **33**, 2045 (1992).
54. C. Ferreri, M. Ballestri and C. Chatgilialoglu, *Tetrahedron Lett.*, **34**, 5147 (1993).
55. G. Emmer and S. Weber-Roth, *Tetrahedron*, **48**, 5861 (1992); corrigenda, *Tetrahedron*, **49**, 291 (1993).
56. W. Smadja, M. Zahouily and M. Malacria, *Tetrahedron Lett.*, **33**, 5511 (1992).
57. K. J. Kulicke and B. Giese, *Synlett*, 91 (1990).
58. (a) B. Giese, W. Damm, J. Dickhaut, F. Wetterich, S. Sun and D. P. Curran, *Tetrahedron Lett.*, **32**, 6097 (1991).
    (b) W. Damm, J. Dickhaut, F. Wetterich and B. Giese, *Tetrahedron Lett.*, **34**, 431 (1993).
    (c) B. Giese, M. Bulliard, J. Dickhaut, R. Halbach, C. Hassler, U. Hoffmann, B. Hinzen and M. Senn, *Synlett*, 116 (1995).
59. K. Miura, K. Oshima and K. Utimoto, *Bull. Chem. Soc. Jpn.*, **66**, 2356 (1993).
60. B. Giese, B. Kopping and C. Chatgilialoglu, *Tetrahedron Lett.*, **30**, 681 (1989).
61. C. Chatgilialoglu, B. Giese and B. Kopping, *Tetrahedron Lett.*, **31**, 6013 (1990).
62. H. Schneider, H. Fiander, K. A. Harrison, M. Watson, G. W. Burton and P. Arya, *Bioorg. Med. Chem. Lett.*, **6**, 637 (1996).
63. H. Urabe, K. Kobayashi and F. Sato, *J. Chem. Soc., Chem. Commun.*, 1043 (1995).
64. C. Chatgilialoglu, A. Alberti, M. Ballestri, D. Macciantelli and D. P. Curran, *Tetrahedron Lett.*, **37**, 6391 (1996).
65. F. Minisci, F. Fontana, G. Pianese and Y. M. Yan, *J. Org. Chem.*, **58**, 4207 (1993).

66. (a) H. Togo, K. Hayashi and M. Yokoyama, *Chem. Lett.*, 641 (1993).
    (b) H. Togo, K. Hayashi and M. Yokoyama, *Bull. Chem. Soc. Jpn.*, **67**, 2522 (1994).
67. For recent reviews on free-radical carbonylation, see:
    (a) I. Ryu, N. Sonoda and D. P. Curran, *Chem. Rev.*, **96**, 177 (1996).
    (b) I. Ryu and N. Sonoda, *Angew. Chem., Int. Ed. Engl.*, **35**, 1050 (1996).
68. I. Ryu, M. Hasegawa, A. Kurihara, A. Ogawa, S. Tsunoi and N. Sonoda, *Synlett*, 143 (1993).
69. I. Ryu, K. Nagahara, H. Yamazaki, S. Tsunoi and N. Sonoda, *Synlett*, 643 (1994).
70. S. A. Hitchcock and G. Pattenden, *J. Chem. Soc., Perkin Trans. 1*, 1323 (1992).
71. (a) M. Ihara, F. Setsu, M. Shohda, N. Taniguchi and K. Fukumoto, *Heterocycles*, **37**, 289 (1994).
    (b) M. Ihara, F. Setsu, M. Shohda, N. Taniguchi, Y. Tokunaga and K. Fukumoto, *J. Org. Chem.*, **59**, 5317 (1994).
72. G. J. Hollinworth, G. Pattenden and D. J. Schulz, *Aust. J. Chem.*, **48**, 381 (1995).
73. K. A. Parker and D. Fokas, *J. Org. Chem.*, **59**, 3927 and 3933 (1994).
74. (a) P. J. Parsons, C. S. Penkett, M. C. Cramp, R. I. West and E. S. Warren, *Tetrahedron*, **52**, 647 (1996).
    (b) D. J. Hart and D. Kuzmich, *J. Chin. Chem. Soc.*, **42**, 873 (1995).
75. M. P. Sibi and J. Ji, *J. Am. Chem. Soc.*, **118**, 3063, (1996).
76. P. Arya and D. D. M. Wayner, *Tetrahedron Lett.*, **32**, 6265 (1991).
77. S. Knapp and F. S. Gibson, *J. Org. Chem.*, **57**, 4802 (1992).
78. (a) P. A. Evans and J. D. Roseman, *J. Org. Chem.*, **61**, 2252 (1996).
    (b) P. A. Evans, J. D. Roseman and L. J. Garber, *J. Org. Chem.*, **61**, 4880 (1996).
79. T. B. Lowinger and L. Weiler, *J. Org. Chem.*, **57**, 6099 (1992).
80. L. A. Buttle and W. B. Motherwell, *Tetrahedron Lett.*, **35**, 3995 (1994).
81. (a) S. Bogen, M. Journet and M. Malacria, *Synlett*, 958 (1994).
    (b) L. Fensterbank, A.-L. Dhimane, E. Lacote, S. Boger and M. Malacria, *Tetrahedron*, **52**, 11405 (1996).
82. S. Bogen and M. Malacria, *J. Am. Chem. Soc.*, **118**, 3992 (1996).
83. T. Gimisis and C. Chatgilialoglu, *J. Org. Chem.*, **61**, 1908 (1996).
84. S. Kim, G. H. Joe and J. Y. Do, *J. Am. Chem. Soc.*, **116**, 5521 (1994).
85. S. Kim, *Pure Appl. Chem.*, **68**, 623 (1996).
86. M. Santagostino and J. D. Kilburn, *Tetrahedron Lett.*, **36**, 1365 (1995).
87. M. Kizil and J. A. Murphy, *J. Chem. Soc., Chem. Commun.*, 1409 (1995).
88. C. Chatgilialoglu, C. Ferreri and A. Sommazzi, *J. Am. Chem. Soc.*, **118**, 7223 (1996).
89. P. Dowd and W. Zhang, *Chem. Rev.*, **93**, 2091 (1993).
90. D. Crich and Q. Yao, *J. Org. Chem.*, **61**, 3566 (1996).
91. J. E. Lyons, C. H. Schiesser and K. Sutej, *J. Org. Chem.*, **58**, 5632 (1993).
92. K. J. Kulicke, C. Chatgilialoglu, B. Kopping and B. Giese, *Helv. Chim. Acta*, **75**, 935 (1992).
93. K. Miura, K. Oshima and K. Utimoto, *Bull. Chem. Soc. Jpn.*, **66**, 2348 (1993).
94. C. Chatgilialoglu, M. Ballestri, D. Vecchi and D. P. Curran, *Tetrahedron, Lett.*, **37**, 6383 (1996).
95. C. Chatgilialoglu, C. Ferreri, M. Ballestri and D. P. Curran, *Tetrahedron Lett.*, **37**, 6387 (1996).
96. (a) J. Cossy and L. Sallé, *Tetrahedron Lett.*, **36**, 7235 (1995).
    (b) C. Dufour, S. Iwasa, A. Fabré and V. H. Rawal, *Tetrahedron Lett.*, **37**, 7867 (1996).
97. E. Lee, T. S. Kang, B. J. Joo, J. S. Tae, K. S. Li and C. K. Chung, *Tetrahedron Lett.*, **36**, 417 (1995).
98. H. Künzer, G. Sauer and R. Wiechert, *Tetrahedron Lett.*, **32**, 7247 (1991).
99. G. Pattenden and D. J. Schulz, *Tetrahedron Lett.*, **34**, 6787 (1993).
100. M. D. Bachi, A. Balanov, N. Bar-Ner, E. Bosch, D. Denenmark and M. Mizhiritskii, *Pure Appl. Chem.*, **65**, 595 (1993).
101. H.-S. Dang and B. P. Roberts, *J. Chem. Soc., Perkin Trans. 1*, 769 (1996).
102. Y. Landais, D. Planchenault and V. Weber, *Tetrahedron Lett.*, **35**, 9549 (1994).

CHAPTER **26**

Recent advances in the direct process[†]

LARRY N. LEWIS

GE Corporate Research & Development Center, Schenectady, NY 12309, USA

I. INTRODUCTION	1582
II. THE EFFECT OF PROMOTERS ON THE DIRECT PROCESS	1582
A. Promoters in the MCS Reaction	1582
B. Use of Promoters to Improve the Yield of Si—H Containing Products	1585
III. THE EFFECT OF SILICON, CATALYST AND PROMOTER MORPHOLOGY ON THE MCS REACTION	1585
A. The Effect of Surface Area	1585
B. The Effect of Silicon Size on the MCS Reaction	1586
C. The Effect of Oxygen in Silicon Metal Used in the Direct Process	1586
D. Preparation of Catalyst, Silicon and Promoters for Use in the MCS Reaction	1586
E. The Effect of Copper Catalyst Preparation on the MCS Reaction	1587
F. The Effects of Intermetallics in the Contact Mass on the MCS Reaction	1587
IV. SURFACE FUNDAMENTALS AND MECHANISTIC STUDIES OF THE MCS REACTION	1588
V. SILICON DIRECT PROCESS REACTIONS WITH REAGENTS OTHER THAN MeCl	1589
A. Other Organic Halides	1589
B. Reactions of Silicon with Alcohols	1590
C. Alternative Methods to Formation of Si—CH_3 Bonds	1591
VI. RECOVERY AND USE OF BY-PRODUCTS FROM THE MCS REACTION	1592
A. High Boiling Residues from the Direct Process	1592
B. Recycle of Solid Waste Products from the MCS Reaction	1593

[†] This review is dedicated to the memory of Alan Ritzer.

The chemistry of organic silicon compounds, Vol. 2
Edited by Z. Rappoport and Y. Apeloig © 1998 John Wiley & Sons Ltd

VII. ACKNOWLEDGMENTS 1594
VIII. REFERENCES 1594

I. INTRODUCTION

The so-called 'Direct Process' is the industrial process for the synthesis of alkyl chlorosilanes. When the direct process produces methyl chlorosilanes the reaction is frequently called the Methyl Chlorosilane Process (MCS) (equation 1)[1]. Prior to the 1940s, alkyl chlorosilanes were primarily prepared via Grignard-like reactions. In 1940 Eugene Rochow of General Electric invented the direct process[2]. Rochow's work, together with work by J. F. Hyde of Corning Glass (later Dow Corning)[3] led to the birth of the industrial silicones industry. Independently, Muller discovered an analog of the MCS reaction but his discovery was not well known in the West until later[4]. The history of the development of the Silicones Industry vis-a-vis the Direct Process is well documented in two excellent books on the subject[5,6].

$$\text{Si} + \text{MeCl} \xrightarrow[\substack{250-330\,^\circ\text{C} \\ \text{fluidized bed}}]{\text{Cu} + \text{promoters}}$$

Compound	%
Me_2SiCl_2(Di)	83–93
$MeSiCl_3$(Tri)	3–10
Me_4Si	0.01–0.5
Me_2SiHCl	0.01–0.5
$MeSiHCl_2$	0.1–1
Me_3SiCl	1–5
High boilers	1–5

(1)

From the 1940s through the 1970s, while the MCS reaction was practiced industrially on a large scale, it was frequently the source of frustrating irreproducibility. The major treatises on the subject of Direct Process failed to mention the importance of promoter elements in the catalyst[7–10]. In addition, the use of substrates other than MeCl was explored only occasionally. Finally, there were few successful attempts to understand the Direct Process on a fundamental level before 1980.

This review attempts to summarize the published literature and information in patents from the 1980s to the present. The effect of promoters will be discussed first because it was the understanding of their function which led to a quasi-second revolution in the area of the Direct Process. The second area discussed will be the effect of silicon morphology on selectivity and yield. The third area of this review will focus on mechanistic studies aimed at understanding the fundamental chemical and physical effects in the Direct Process. The fourth area of review will be the use of substrates other than MeCl in the Direct Process. Finally, a discussion will be presented on the recovery and use of by-products from MCS, previously considered waste products.

II. THE EFFECT OF PROMOTERS ON THE DIRECT PROCESS

A. Promoters in the MCS Reaction

Ward and coworkers at GE described the importance of both tin and zinc in the Direct Process[11,12]. They claimed that their work was the first reported description of the importance of these trace impurities. The Ward promoter work marked a benchmark in Direct Process literature because without controlling the levels of Sn, Zn and other promoters other effects are equivocal. The effect of zinc was known to some extent as early as 1949[4] but this knowledge was not clearly manifest in patents or publications on the subject. Fluidized and stirred bed reactors were used to show that trace quantities of Sn affect the MCS process profoundly and the effects of Sn and Zn were synergistic[11,12]. A catalyst system consisting of Cu, Zn and Sn was discovered which yielded 90% Di with nearly complete Si utilization. The trace elements in silicon used were (ppm): Fe (5600), Al (2700), Ti (850), Mn (200), Ca (160) and Ni (120) and the silicon was ball-milled to give particles with a surface area of 0.5 $m^2 g^{-1}$. Pure CuCl was used and the zinc added also contained (ppm): Pb (1700) and Cd (170). Pure tin and methyl chloride were employed in the experiments. The best rate and Tri/Di were obtained when element ratios were: Cu (5%), Zn (0.5%) and Sn (0.005%). When tin was > 2200 an increase in residue occurred. Further work was done in the GE labs on the effect of tin and zinc[13,14].

Kim and Rethwisch further investigated the affect of tin and zinc on the MCS reaction[15]. Ball milling was used to grind copper and silicon and to impregnate the mixture with tin and zinc. When MeCl was passed over the balled milled mixtures, silanes began to form at 147 °C with a maximum production at 317–347 °C. When the catalyst consisted of only copper and zinc, MeCl decomposed without formation of silanes. However, when MeCl reacted with copper and tin, some Me_xSiCl_{4-x} was observed. Reacting MeCl with SiZnSn gave 64% $MeSiCl_3$. The addition of zinc to the MCS bed decreased the coking rate. Furthermore, surface chlorine was necessary for direct reaction; Zn and Sn promote its formation. Consistent with the synergism reported by Ward and coworkers, zinc was found not to increase the overall rate but increased the selectivity for methylated silanes. Tin increased the selectivity for chlorinated silanes. Tin and zinc together increased the rate of the MCS reaction.

Recently the effects of tin, zinc and antimony were studied in more detail leaving the frustrated researchers to state, 'At present, no justified mechanistic explanation of the effect can be given, because of the total lack of information on nature, number and energetic distribution of the active sites of the catalyst system'.[16] Thus Lieske and coworkers report several important empirical observations[16]. In order to determine the effects of Zn, Sb and Sn, different combinations of silicon and copper catalyst were used: $CuCl_2/Si_{pure}$, $CuCl_2/Si_{tech}$, CuO/Si_{pure} and CuO/Si_{tech}. Zinc promoted the reaction of $CuCl_2/Si_{tech}$, CuO/Si_{pure} and CuO/Si_{tech} but not of $CuCl_2/Si_{pure}$, the latter was active without promoter. CuO/Si_{pure} and CuO/Si_{tech} were nearly inactive without promoters. With $CuCl_2/Si_{pure}$, antimony and tin caused higher activities than zinc but, with CuO/Si_{pure}, addition of Sn and Sb to the reaction did not reach the efficiency obtained with zinc. With $CuCl_2$, antimony promoted Si_{pure} and was more active than Si_{tech}, but with CuO, Si_{tech} was more active than Si_{pure}. Zinc appears to prevent detrimental effects of Si impurities and the authors note that zinc may catalyze formation of CuCl from Cu[17]. Additionally, zinc may help to remove the high oxygen content formed by using CuO[18]. No correlations were found between promoter and amount and size of Cu_3Si, nor was there a correlation between Cu_3Si amount and particle size *and* catalyst properties.

Since the discovery of the importance of some metal impurities on the rate and selectivity of the MCS reaction, several other combinations of promoter elements have appeared in the patent literature. One patent employs tin and/or antimony (they seem to be interchangeable) and zinc in combination with a lanthanide metal[19] (the promoter effects of barium and strontium have also been described[20]). In one case silicon (210 g) was preconditioned by reaction with MeCl (16 L h^{-1}) at 345 °C for 1 h, and then reacted at 315 °C for the duration of the reaction, in the presence of a catalyst consisting of CuCl (16.3 g), a bronze (0.38 g) containing 10% (by wt) tin, $ZnCl_2$ (1.60 g) and K_3LaCl_6 (1.69 g). The yields were: 93.6% Me_2SiCl_2, 2.8% $MeSiCl_3$ and 1.7% Me_3SiCl, with 58% conversion of silicon[19]. The relationship between zinc and antimony is also addressed in a Czech patent[21]. In the reaction of MeCl with Si–Cu substrate, replacing the catalytic mixture

TABLE 1. Percentage of silane products with Zn vs Sb in MCS

Promoter	Me_2SiCl_2	$MeSiCl_3$	Me_3SiCl	$MeHSiCl_2$
Sb	78	18	3	0.5
Zn	25	52	8	7

of $ZnCl_2$ with Sb shifted the product composition in favor of Me_2SiCl_2 as shown in Table 1.

A summary of the various metals used as promoters was prepared by Kanner and Lewis[1] and by Rethwisch and coworkers[22]. Zinc is claimed by Bayer workers as well to improve selectivity of the MCS reaction[23]. Further improvements in the use of zinc as a promoter are described wherein the zinc is in the form of a zinc salt of a weak acid having a pK_a of 3.50–10.25 (e.g. carbonic acid, carboxylic acids)[24].

Alkali metal salts are effective promoters in the direct process. Improved synthesis of Me_2SiCl_2 has been claimed in the presence of copper, tin and zinc and either KCl[25] or CsCl[26]. Alkali earth metals are also effective promoters. Beryllium, magnesium and calcium were claimed to improve the selectivity for Di in the MCS reaction. For example, a mixture of silicon (210 g), CuCl (16.4 g), $CaCl_2$ (1.244 g), $ZnCl_2$ (1.53 g) and bronze (2% Sn, 2.0 g) was pre-activated at 200 °C under nitrogen, whereupon MeCl was fed in at 16–27 L h^{-1} at 330–345 °C for 15.5 h to give 55% conversion of Si and 86.7% selectivity to Me_2SiCl_2[27].

Phosphorus has been added to MCS beds and reported to improve selectivity for Di when tin and zinc are also present. Margaria and others describe the use of phosphorus as a promoter and cite a US patent where the benefits of phosphorus were realized as early as 1954[28]. Dow Corning workers describe the use of various phosphides such as copper phosphide to improve selectivity for Di[29]. Alternatively, phosphorus was introduced into an MCS bed by manufacturing silicon metal from quartz that was naturally high in phosphorus[30,31]. Apparently the use of any phosphorus compound that is not volatile, such as phosphides, leads to improved Di yield. Table 2 shows that use of phosphorus gave improved Di in the MCS reaction where copper, tin and zinc were also present as promoters[32]. Other workers have employed phosphorus as a promoter in volatile or gaseous form[33].

Other main group elements have been used as promoters in MCS beds containing copper, tin and zinc. Phosphorus and indium have been used in the MCS reaction. The reaction of silicon (40 g) with MeCl (2 bar) at 300 °C in the presence of Cu (3.2 g), ZnO (0.05 g), In (0.004 g) and P (0.056 g) gave 1.7% of Me_3SiCl, 0.017% mixture of $MeSiCl_3/Me_2SiCl_2$ and 3.7% of polysilanes[34]. Antimony, just below phosphorus in

TABLE 2. Effect of phosphorus on the MCS reaction

Sample	Analytical data on Si (in ppm)				% Di	Me/Me$_2^a$	Si conv (%)
	Al	Ca	Fe	P			
Conventional Si, no P	4100	5300	3200	0	81.6	0.11	77.8
Conventional Si, 14 ppm P	4500	1900	4900	14	84.6	0.08	80.7
Si refined with P	2170	250	2130	32	83.5	0.09	50
Si refined with P	2520	280	3620	33	84.6	0.08	59.9
Added tricalcium phosphate	2520	330	3560	148	90.5	0.04	82
Added tricalcium phosphate	2760	430	3450	69	88	0.06	79.4

^{a}Me/Me$_2$ = wt% Me_2SiCl_3/wt% Me_2SiCl_2.

group V on the periodic table, also appears to be a promoter as well in the MCS reaction[35,36].

B. Use of Promoters to Improve the Yield of Si—H Containing Products

In 1975 van den Berg and coworkers reported on improved direct synthesis of methyldichlorosilane and dimethylchlorosilane[37]. The key elements of the van den Berg work were the use of hydrogen co-feed with the methyl chloride and the use of promoters. The paper is important because several important patents followed in its wake on improved methods for production of Si—H containing silanes and because it begins to address the use of promoters. The use of hydrogen was shown to improve the yield of $MeHSiCl_2$ (DH) and Me_2ClSiH (MH) vs reactions run without hydrogen. The authors concluded that metal halides, which are promoters for formation of Me_2SiCl_2, are not promoters for formation of DH and MH. Using a contact mass 'without promoter' gave higher per cent of MH and DH than experiments where either Ag, Cd, Zn or Al were added. Unfortunately, precise compositional analysis of the 'un-promoted' experiment was not discussed, leading one to wonder if promoters were present or not.

Both Dow Corning and the old Union Carbide silicones group were active in optimizing conditions for formation of MH and DH. Halm and Zapp of Dow Corning improved the method for making DH[38] by using both hydrogen and HCl as a co-feed with MeCl. By also employing silicon containing Al (0.22%), Ca (0.046%), Fe (0.34%) and by adding brass with tin and CuP alloy, improved yield of DH was obtained. In a control reaction (no HCl/H_2) the crude product contained Di (92.1%), $MeHSiCl_2$ (1.3%), Me_2HSiCl (0.3%) and Tri (4.1%), but with a feed that contained 1% H_2 and 1% HCl the crude product contained Di (76.1%), $MeHSiCl_2$ (12.1%), Me_2HSiCl (1.5%). If HCl is used without hydrogen, lower yield of DH is obtained[39].

The Union Carbide workers report a much higher overall yield of Si—H containing monomers if transition metal promoters are employed[40]. In one case, a high yield of MeH_2SiCl was obtained if a fluidized bed of CuCl, Si and MeCl also contained certain attributes, namely MeH_2SiCl yield improved if the silicon had 0.004–0.02% Ni and 0.002–0.01% Cr (Rh is also a promoter); $CaSi_2$ also helped. Poor results were obtained if the following were present: CuS_x, Cu_xP, Zn, Sn and Cd (last 3 lower selectivity for alkylhalosilanes). A fluidized bed run at 325 °C and which had HCl in the feed gave: Me_2SiHCl (7.4%), $MeSiHCl_2$ (21.5%), Me_3SiCl (2.9%), Tri (12%), Di (55%); total MeSiH 29%. When Ni was present the yield of MH was 16.2% and that of DH was 43.5%, with a total MeSiH of 60%. Earlier work showed the effectiveness of other metal promoters to increase SiH yield[41].

III. THE EFFECT OF SILICON, CATALYST AND PROMOTER MORPHOLOGY ON THE MCS REACTION

A. The Effect of Surface Area

The MCS reaction is a solid/gas reaction, so that it is natural that morphological effects of the solid components of the reaction should be important. Workers at Wacker defined a surface area parameter, QF, which relates the ratio of elongated to round structural forms[42]. MCS promoters controlled by employing Si particles for a given surface area defined the ratio QF as the ratio of the area portion of intermetallic phases at grain boundaries to the area portion of intermetallic phases in primary silicon[43]. These workers found that maximum Di was obtained for silicon particles having QF = 18–60. Furthermore, the rate was effected by QF, so that for CuO, ZnO, and Sn and Si with QF = 29.55, the rate of Di formation was 103.28 mg m^{-2} min^{-1}, but with Si with a QF value of 2.34 the rate was 50.70 mg min^{-1}. The value of QF could be effected by variation of the range of solidification speed from atomization to extremely slow cooling. Atomized silicon has a QF value of 64.

B. The Effect of Silicon Size on the MCS Reaction

Dow Corning workers describe the improvements in the MCS reaction obtained by selecting a particular particle size distribution[44]. They found the best results when the particle size of the silicon was within a range of 1 to 85 microns. Specifically, the most effective silicon was when the particle size of the silicon had a mass distribution characterized by a 10th percentile of 2.1 to 6 microns, a 50th percentile of 12 to 25 microns and a 90th percentile of 35 to 45 microns. Bayer workers specify a silicon particle size distribution of 5–15 microns[45] in one case and 71 to 160 microns in another[46].

C. The Effect of Oxygen in Silicon Metal Used in the Direct Process

Considering the high oxophilicity of silicon, it is not surprising that oxygen has a negative effect on the MCS reaction. Careful research by workers from silicon manufacturer Elkem showed that oxygen in silicon varies from 0.02 to 1 weight percent[47]. There is a native oxide layer of 20 Å on silicon, which for a 50-micron particle contributes about 0.012 weight % oxygen to the total mass of the particle. Oxygen in the silicon is detrimental to its reactivity in the Direct Process. Oxygen is present as a small amount of SiO inclusions and present as oxides of Al_2O_3, CaO, SiO_2 and lesser amounts of MgO, K_2O and Na_2O. Slag in silicon in the Direct Process is detrimental due to the presence of aluminum as an oxide[48]. Also, slag particles accumulate and reduce reactivity because less silicon is available for reaction.

The results reported by Dubrous of silicon manufacturer Pechiney[49] apparently contradict the above work. Dubrous describes a process wherein the crushing and grinding steps of making silicon form an oxide layer and the presence of the oxide layer facilitates the manufacture of Di. When silicon containing Ca (0.08%), Al (0.21%) and Fe (0.39%) was crushed and powdered in the presence of 0.3% silicones under an atmosphere of Ar, it had a 1.2-nm-thick SiO_2 coating, and gave a conversion rate of MeCl in MCS of 88% and MeCl conversion 85%, vs 70 and 85%, respectively, for conventionally coated Si powder.

D. Preparation of Catalyst, Silicon and Promoters for Use in the MCS Reaction

As already discussed, the particle size and surface properties of silicon used in the MCS reaction are important. Several groups have reported improvements in the MCS reaction by different methods of physically combining silicon, copper catalyst and promoters. In Kim and Rethwisch's fundamental studies on activation energies in the MCS reaction (*vide infra*) copper and silicon were ball milled and then tin and zinc promoters were combined by ball milling or by impregnation[50]. Ball milling (specifically energy milling) was also used by SCM workers to prepare copper catalysts (<15 microns) for use in the silicon direct process[51].

More intricate treatment of silicon and copper was reported by Mui[52]. The Si/Cu contact mass was prepared by (1) treating Si particles with Cu catalyst vapor in a mechanical mixer, (2) mixing the Si particles with the Cu catalyst vapor, (3) contacting the vapor with the particles and (4) forming active Cu–Si alloys on the Si particles. The CuCl vapor was formed at 1200 °C and fed to silicon at 320 °C. This author claims that treating the contact mass in this way and by using zinc carbonate and tin oxide there was no induction period, and performance of this contact mass was comparable to that of contact masses with much higher Cu content.

It is possible that the improvements described above are due to pre-formation of silicon copper alloys, as claimed by Elattar[53]. Improved activity of MCS contact mass was obtained by forming a contact mass of particulate Si and Cu (16.9% metallic Cu, 45.5% Cu_2O, 36.9% CuO) by heating the mass in the presence of MeCl gas at 305 °C for a time sufficient to form active spots of Cu–Si alloy on the surface. The nature of these copper–silicon alloys may include those described by some former E. German workers who showed the importance of forming Cu_6Si and $CuSi_2$ among others[54].

E. The Effect of Copper Catalyst Preparation on the MCS Reaction

Several groups have investigated improvements in the MCS reaction wrought by improving the physical and chemical nature of the copper catalyst. SCM workers made an improved catalyst by grinding a charge of Cu(I) particulates with a small proportion of a hydroxide of group IV metals [e.g. Fe(III) oxide monohydrate][55]. The grind charge contained a major proportion of Cu and CuO and was subjected to high energy milling with concomitant crystal lattice distortion. The SCM workers additionally note the importance of copper catalyst size (average 20 microns) tin concentration (400–3000 ppm)[56].

Workers at Wacker employed a more chemical approach to producing and activating their copper catalyst. They added an aqueous solution of $CuCl_2$ to a suspension of iron particles (average particle size 60–70 microns). The catalyst was mixed with Si, $ZnCl_2$ and Zn powder to give a contact mass which was heated to 350 °C and treated with MeCl to give a product mixture containing 80.6% Di[57]. Another chemical process for formation of the copper catalyst was described by workers from Leuna-Werke A.-G where an aqueous $Cu(NO_3)_2$ was added to an aqueous solution of Na_2CO_3 at 40 °C until the resulting suspension was neutral. The product was filtered, washed and treated with oxalic acid and H_2O to bring the solution to a pH of <1. After 1 h stirring, additional oxalic acid was added, followed by stirring and drying at 200 °C to give a product showing 82% selectivity for Di in the MCS reaction[58]. These workers also employed copper salts of mono- and di-carboxylic acids[59] and copper hydroxide/copper carbonate precursors[60]. Published work by GE shows the use of copper oxide precursors for MCS[61].

While most published efforts aim toward maximum production of Di, one set of workers prepared a catalyst which selectively makes Tri. Workers from Wacker prepared a mixture containing SiO_2 powder (75 parts), Fe (75 parts), carbon black (60 parts) and Cu (18 parts) which was pelletized with H_2O, dried at 220 °C and heated at 1100 °C for 17 h[62]. The resulting solid was treated with MeCl at 200–350 °C under Ar to give a mixture containing $MeSiCl_3$ 80%, Me_2SiCl_2 5% and $SiCl_4$ 15 mol%.

F. The Effects of Intermetallics in the Contact Mass on the MCS Reaction

There has been extensive work devoted to correlating the effects of the various recipes described in this section with formation of specific reactive sites and intermetallics on silicon. A quite comprehensive study of this kind was reported by Laroze and others[63]. These authors point out that aluminum is an activator for the overall consumption of silicon in the MCS reaction; however, it does not stabilize Cu_3Si alloy because it increases the rate of alloy consumption by CuCl. Aluminum does not dissolve in the crystal lattice of silicon but is present as an intermetallic, e.g. Si_2Al_2Ca, $Si_8Al_6Fe_4Ca$, $Si_{2.4}FeAl$, Si_2Al_3Fe and $Si_7Al_8Fe_5$. The formation of these silicides and of Si_2FeTi and Si_2Ca is effected by cooling rates and annealing. The reaction of Si_2Al_2Ca with CuCl at 25 °C is violent, giving

Cu_3Si, Cu_5Si and Cu. At lower temperatures the only product is Cu. Intermetallics have an important effect on the direct process. For example, the activating effect of aluminum disappears when added as the quaternary alloy $Fe_4Si_8Al_6Ca$ but an increase in selectivity is noted when the ternary alloy is present. Aluminum added as the ternary alloy $Si_7Al_8Fe_5$, Al_3Si_2Fe and Al_2Si_2Ca favors activity but decreases selectivity.

The formation of intermetallics is clearly critical in the MCS reaction as further shown in recent work by Margaria and coworkers[28]. These inventors describe the harmful effects of the Si_2Al_2Ca ternary phase (must be below 0.3%). The ternary phase was limited during silicon refining by simultaneously regulating the range of Fe, Al and Ca impurities and setting the solidification rate. Silicon with 0.15% ternary phase gave a Di yield of 86.4% while silicon with no ternary phase gave a Di yield of 90.3%.

IV. SURFACE FUNDAMENTALS AND MECHANISTIC STUDIES OF THE MCS REACTION

Eugene Rochow himself wrote about the factors that led to the use of copper catalyst in the direct process. Rochow's knowledge of metallurgy certainly played a part in the discovery of the direct process but serendipity certainly contributed as well[64]! Details of the copper catalyzed MCS reaction at the molecular level are still not well understood over 50 years after its discovery. The most extensive investigations of the surface chemistry occurring during the direct process have been carried out by the Falconer group in Colorado[65-71]. They have used surface analyses to determine the changes that occur on a silicon surface under MCS conditions. Typical experiments were carried out on silicon wafers and characterized before and after exposure to MeCl reaction by X-ray photoelectron spectroscopy (XPS), Auger electron spectroscopy (AES), X-ray diffraction (XRD), scanning electron microscopy (SEM), electron disperssive spectroscopy (EDS) and optical microscopy.

Without zinc, silicon diffusion to the surface is slow under MCS conditions. Furthermore, when only tin is used as a promoter, no Cu is observed at the surface. Table 3 shows the elemental concentration under various conditions. Under MCS reaction conditions, when zinc was present silicon was not depleted from the subsurface, and when zinc was absent the subsurface was depleted in silicon. Zinc causes the rate of silicon diffusion and copper dispersion to increase. Zinc accumulates at grain boundaries and lowers the free energy of Cu_3Si. Tin and zinc appear to work synergistically but tin does not enhance silicon diffusion on its own. Tin does appear to lower the surface energy of silicon/copper.

Further surface studies have addressed the effect of different forms of copper as a catalyst in the direct process[65-69]. Analyzing Si(100) surfaces it was shown that the form of copper is critical for selectivity in the MCS reaction. The most effective combination was 82% Cu and 18% Cu_2O (neither worked very well alone). There was no correlation between the amount of Cu_3Si on the surface and rate or selectivity for the direct process.

TABLE 3. Auger analysis of Cu_3Si after reaction at 277 °C[71]

Promoter	Time (min)	Si^a	Cu^a	C^a	Cl^a	O^a	Sn^a
None	240	8	52	14	24	2	0
Sn	220	61	<0.5	20	12	7	0
Zn	210	23	39	23	11	3	0
Zn,Sn	250	40	15	24	9	9	3

^{a}Atom %.

The orientation of Cu_3Si relative to the Si(100) face did correlate to rate and selectivity; a random orientation gave the best results. One study found that the reaction was inhibited by a SiO_2 layer[65] and a second study found no such inhibition[66]. SEM analysis of several surfaces after reaction with MeCl showed square reaction pits.

Lewis and coworkers have also made significant contributions to the understanding of the MCS reaction via the use of surface studies[72]. XPS and AES analysis of catalytically active surfaces showed that zinc causes a restructuring of the Cu_3Si surface. Additionally, zinc enrichment is enhanced by the addition of $SnCl_4$. Lead is a well known poison for the direct reaction and the Lewis group found that lead suppressed enrichment of the Cu_3Si surface in zinc and silicon.

A 1989 review of the literature by Clarke gave many examples which support the importance of copper–silicon rich phases near the surface during the MCS reaction[73]. Clarke noted that Cu_3Si (eta phase) forms above 880 °C but will form at 350 °C in the presence of chloride ion. Methyl chloride reacts with copper to form copper chloride which then serves as the chloride source needed for formation of the eta phase, thus explaining the shorter induction period obtained using copper chloride vs other copper catalysts. The mechanism of replacement of silicon from the surface is by diffusion of copper into the bulk silicon to reform a copper–silicon rich surface. Iron–silicon phases stabilize the eta phase and metal promoters catalyze chloride transfer, e.g. see equation 2. Silicon also reacts with $ZnCl_2$ and $AlCl_3$. Excess zinc causes unproductive decomposition of MeCl to give methane. Finally, Clarke presented data that ruled out the importance of methyl radicals in the MCS reaction.

$$CdCl_2 + Si \longrightarrow CdCl + SiCl \qquad (2)$$

More recent support has appeared for the importance of the copper–silicon rich phase on the silicon surface in the MCS reaction. Lieske and coworkers[74] showed that redispersion of the eta phase can be an element of the induction period of the MCS reaction and seems to be brought about by the reaction itself. The Cu–Si surface species, perhaps Cu–Si surface compounds or extremely small Cu–Si particles, seem to be of similar importance as X-ray detectable Cu–Si phases.

Finally, Rong and coworkers discuss the roll of surface oxygen on the MCS process[75]. Rong employed a lab-scale stirred bed reactor and then applied XPS to analyze the silicon samples before and after the reaction. The reactivity of silicon depended on the initial thickness of the native oxide on the silicon. After the reaction the surfaces of all of the samples were mostly covered with SiO_2. There was no observed correlation between the surface and bulk O content. XPS analysis showed the presence of Al, Ca and Ti impurities in some samples. Titanium on the surface appeared to increase the reactivity, whereas Ca decreased the selectivity of Di formation. Addition of ZnO to the silicon before CuCl improved reactivity and also decreased the induction period of the reaction. XPS studies of samples prepared in this manner exhibited a lower Zn surface concentration compared to the samples where CuCl, Si and ZnO were mixed together.

V. SILICON DIRECT PROCESS REACTIONS WITH REAGENTS OTHER THAN MeCl

A. Other Organic Halides

There are many examples of the direct reaction of silicon with ethyl chloride, vinyl chloride and chlorobenzene[1]. Vinyl and allylchlorosilanes[76] were first made via a direct process in 1945 as were phenylchlorosilanes[77]. Jung's group has recently extended the direct reaction of silicon with a variety of substrates including allyl chloride[78]. Silicon reacted with allyl chloride in a stirred reactor in the presence of HCl and a copper catalyst at 220–320 °C in a stirred reactor to make (allyl)Cl$_2$SiH as the major product and allylSiCl$_3$ as the minor product. Unlike MeCl reactions cadmium was a good promoter and zinc was an inhibitor. Jung also published (in the Gelest Inc. product catalogue) a review of direct reactions between silicon and a variety of organic halides including methylene chloride, benzyl chlorides, etc. The reaction of α-(chloromethyl)silanes with silicon is shown in equation 3[79]. The reaction of α, α'-dichloro-o-xylene with silicon gives the silicon-containing ring product shown in equation 4[80].

$$R-\underset{\underset{Cl}{|}}{\overset{\overset{R}{|}}{Si}}-CH_2Cl \xrightarrow{Si/Cu} R-\underset{\underset{Cl}{|}}{\overset{\overset{R}{|}}{Si}}-CH_2-\underset{\underset{Cl}{|}}{\overset{\overset{Cl}{|}}{Si}}-CH_2-\underset{\underset{Cl}{|}}{\overset{\overset{R}{|}}{Si}}-R$$

88%

$$+ \quad R-\underset{\underset{Cl}{|}}{\overset{\overset{R}{|}}{Si}}-CH_2-\underset{\underset{Cl}{|}}{\overset{\overset{Cl}{|}}{Si}}-Cl$$

7%

(3)

$$\underset{CH_2Cl}{\underset{|}{\text{o-C}_6H_4}}(CH_2Cl) \xrightarrow{Cu/Si} \text{indane-SiCl}_2 \quad (4)$$

B. Reactions of Silicon with Alcohols

The first attempts at direct reaction of silicon with methanol and ethanol were reported shortly after the methylchlorosilane reaction was discovered[81]. Rochow and Newton continued their interest in the non-halogen direct process and in 1970 described the direct synthesis of (MeO)$_3$SiH and (MeO)$_4$Si[82]. There has been much recent effort in the alcohol-based direct process. Okamoto and coworkers report high silicon conversions and selectivity for (MeO)$_3$SiH by reaction of silicon and methanol in the presence of copper acetate and small amounts of thiophene[83]. Shiozawa and Okumura report similar results in a Japanese patent[84]. Earlier work by the Union Carbide group showed that greater than 80% selectivity for (MeO)$_3$SiH was possible in the copper catalyzed reaction between methanol and silicon[85,86]. Unlike the methyl chloride reaction where tin is a promoter, tin had adverse effects on the reaction between methanol and silicon.

There are several groups working toward large-scale manufacture of alkoxysilanes via a direct process. Workers at Japan's Tonen Corporation report 84% silicon conversion and 96% selectivity for (MeO)$_3$SiH by reacting pulverized silicon, Cu(OMe)$_2$ and methanol[87]. The silicon pulverization was carried out using an alumina vibrator mill under argon to make 150-micron particles. These workers also use silicon essentially free of iron. Mitsubishi workers report that the reaction of silicon with methanol and cuprous chloride in dodecyl benzene at 220 °C gave 69% (MeO)$_3$SiH[88]. The silicon was pretreated by grinding with stainless steel balls at −25 °C. Toa Gosei Chemical Industries in Japan also report a commercial methanol direct process[89]. The (former) Union Carbide silicones group also reported a method for commercialization of a methanol direct process[90]. The Union Carbide group employed a continuous reactor for the synthesis of $(MeO)_3SiH$ and they specified a silicon particle size <420 microns, 0.1 to 2.6 parts (based on silicon) of $Cu(OH)_2$ catalyst and the use of high boiling aromatic solvents. These workers again point out the adverse effects of tin in this reaction.

Some work has appeared discussing the mechanism of the alcohol-based direct process. A kinetic study by Suzuki and coworkers using a fixed bed reactor found that the rate of $(MeO)_3SiH$ formation from methanol, silicon and a copper catalyst was proportional to methanol pressure[91]. Preheating the reactants was found to have an important effect on the kinetics. Further XRD work by Suzuki and Ono linked the methyl chloride process to the methanol process[92]. Cuprous chloride and silicon powder were preheated at 350 °C for 3 h and then reacted with methanol (99 kPa, 270 °C) in a fixed bed reactor for 5 h to give $(MeO)_3SiH$ (98% selectivity, 82% silicon conversion). XRD analysis after reaction showed Cu_3Si formation had occurred. SEM analysis of silicon from the MCS process showed reaction pits. Likewise, Suzuki's group reports formation of pits on silicon after a methanol direct process[93]. The amount of reaction pits increased as silicon conversion increased. Pretreatment of the silicon is important for silicon conversion and selectivity to $(MeO)_3SiH$[94]. When the silicon was pretreated above 350 °C, 65% selectivity for $(MeO)_3SiH$ was obtained and surface analysis such as XRD showed Cu_3Si. When the silicon was pretreated at <280 °C the rate was higher, the selectivity for $(MeO)_3SiH$ increased to >98% and no Cu_3Si was observed.

A key report investigated a variety of substrates in their reaction with silicon in an effort to find evidence for silylene intermediates during the silicon direct process reaction. When silicon, copper and methanol were reacted as described above but in the presence of alkenes, alkyldimethoxysilanes and $(MeO)_3SiH$ were formed[95-97]. The use of allyl propyl ether instead of alkenes gave allyldimethoxysilane, with 38% selectivity. These results and the reaction of silicon with MeCl in the presence of butadiene to give silacyclopent-3-enes indicates intermediate formation of silylenes.

In some cases, the reaction of silicon and methanol has been optimized for formation of $(MeO)_4Si$. As discussed above, thiophene addition favored formation of $(MeO)_3SiH$. Both thiophene and propyl chloride poison copper; copper poisoning seems to favor formation of the trialkoxysilane. High-temperature pretreatment disfavors trialkoxysilane formation; copper is formed on the surface of the silicon during pretreatment at 450 °C[98]. Metallic Cu catalyzes dehydrogenation of alcohols and favors formation of $(RO)_4Si$. Workers from Tonen Corporation reported 50% conversion of silicon to make $(MeO)_4Si$ with 92% selectivity if silicon, methanol and $Cu(OMe)_2$ were pretreated (lower conversion and selectivity without pretreatment) and then reacted at 180 °C and 1 atmosphere[99].

C. Alternative Methods to Formation of Si—CH$_3$ Bonds

Newton and Rochow reported low (1–2%) yields of methylalkoxysilanes during the copper catalyzed reaction of silicon and methanol[82]. In 1988 a group from Lopata R&D Corporation reported a version of the silicon methanol direct reaction wherein they added metal formates such as potassium formate and employed about 100 psi H_2 (out of a total pressure of 900 psi). Under these conditions about 10% $MeSi(OMe)_3$ was produced (<0.1% when potassium formate was not present)[100]. These results represented one of the highest production of silicon-methyl containing products in the absence of halogen. Rochow has stated that the non-halogen based direct process was and continues to be one of the great challenges left for the silicon direct process[64]. In this author's opinion an even greater challenge is a direct process based on SiO_2!

Two groups have made minor breakthroughs in the non-halogen direct process. In 1978 Malek, Speier and coworkers reported the reaction between dimethyl ether and silicon[101]. Prior to their work there were some attempts to employ dimethyl ether as a substrate. Rochow and Zuckerman[102] disputed one claim[103]. Later Newton and Rochow discussed the difficulty of using dimethyl ether in a direct process[82]. Malek's and Speier's work showed that when silicon was reacted with MeOMe in the presence of metal catalysts and MeBr, 64.4% $Me_2Si(OMe)_2$ could be obtained. The silicon (150 g, 98%, 0.44% Fe, 0.26% Al, 0.058% Ca, 0.005% Sn, 0.04% Mn, 0.032% Ti, 0.051% V, 0.002% B, 0.014% Cr, 0.05% Zn and 0.002% Pb) was ground with steel balls to 2.7 microns for 16 h with Cu powder (12 g). One-hundred parts of silicon, 3.3 parts of copper, 0.3 part of iron, 0.43 part of aluminum and 0.13 part of calcium were combined in an autoclave and MeOMe and MeBr were added (mole ratio of Si:MeOMe:MeBr 100:350:1) and then heated for 20 h at 258 °C. While this process was not entirely halogen-free, it was the most successful ether silicon direct process to date.

Since the Malek and Speier work, Lewis and Kanner also reported a dimethyl ether direct process which was a great improvement over Malek's and Speier's because it employed a fluidized bed in place of an autoclave[104,105]. Lewis and Kanner found that silicon (200 g, 98.4%, 0.35% Al, 0.55% Fe), copper cement (5 g, Cu 22%, Cu_2O 50.4% and CuO 18.8%) and $ZnCO_3$ (0.5 g), when preheated in the presence of HCl at 300 °C after N_2 fluidization and then addition of MeCl (0.8 std lit min^{-1}), after 8 h at 325 °C yielded Di (90%) and Tri (1%). If just MeOMe and HCl were added, no silicon products were obtained. If, however, the feed contained MeOMe (0.37 std lit min^{-1}) and MeCl (0.54 std lit min^{-1}), then 8% methylsiloxanes were obtained at atmospheric pressure. Using a fluidized bed and MeBr a product mixture containing Me_3SiOMe, $(Me_3Si)_2O$, $Me_2Si(OMe)_2$, $Me_3Si(OSiMe_2)_nMe$ ($n = 1-4$), $Me_3Si(OSiMe_2)_nSiMe_3$ ($n = 1-4$) and $(Me_2SiO)_n$ ($n = 3-7$) was obtained.

VI. RECOVERY AND USE OF BY-PRODUCTS FROM THE MCS REACTION

A. High Boiling Residues from the Direct Process

The MCS reaction is characterized by production of high boiling fractions and spent silicon solids enriched in various metals. Industry is continually improving their ability to minimize waste due to environmental and economic realities. In the 1980s workers from Bayer described a method to convert a complicated mixture into useful monomers. Their mixture was composed of approximately 14.7% low boilers ($SiMe_4$, 61.3%; Me_2SiHCl, 11.9%), 14.7% high boilers (Me_6Si_2, 18%; Me_5ClSi_2, 18.4%; $Me_2ClSiSiClMe_2$, 11.7%; $MeCl_2SiSiClMe_2$, 3.5%) and 68.6% $MeSiCl_3$. The Bayer mixture was combined with $AlCl_3$ and then heated to give a mixture where 80% of the product had a boiling point less than 80 °C and was composed of Me_2SiCl_2, 58.3%; Me_3SiCl, 5.2%; $MeSiCl_3$, 32.8% and $MeSiHCl_2$, 2.0%[106]. These workers also employed alumina and $SOCl_2$-containing catalysts[107]. An early but excellent discussion of the whole problem of waste recovery reactions from the MCS reaction was published by Calas, Dunogues, Deleris and Duffaut in 1982[108]. The disilane residue from the direct process can be converted into useful materials by disproportionation (equation 5), cleavage by HCl (equation 6), cleavage in the presence of H_2 (equation 7), cleavage of $Cl_3SiSiCl_3$ in the presence of organic halide RCl, such as MeCl or allyl chloride to give $Cl_3SiR + Cl_4Si$ catalyzed by Pd, Pt, Ni, and finally disilane addition to unsaturated molecules like acetylenes and dienes catalyzed by Pd.

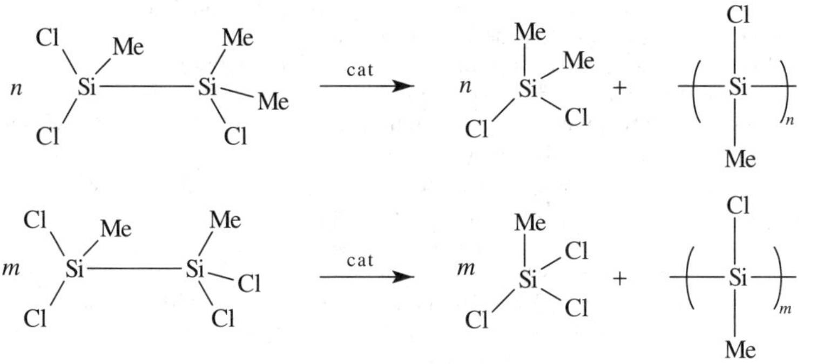

$$\text{Cl}_2\text{MeSi–SiMe}_3 \xrightarrow{\text{cat}} n\ \text{MeSiCl}_2\text{–} + (\text{SiMeCl})_n$$

(schemes shown)

Cat = Bu$_3$N, Me$_4$NCl etc.; most efficient is P(O)(NMe$_2$)$_3$/(HMPA)

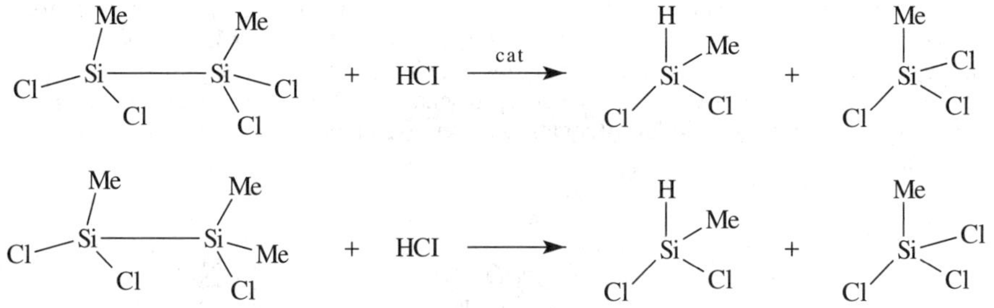

(5)

MeSiCl$_2$–SiMeCl$_2$ + HCl $\xrightarrow{\text{cat}}$ MeSiHCl$_2$ + MeSiCl$_3$

Me$_2$SiCl–SiMeCl$_2$ + HCl $\xrightarrow{\text{cat}}$ MeSiHCl$_2$ + Me$_2$SiCl$_2$

Cat = amines and amides

(6)

Cl$_3$Si–SiCl$_3$ + H$_2$ $\xrightarrow{\text{cat}}$ HSiCl$_3$

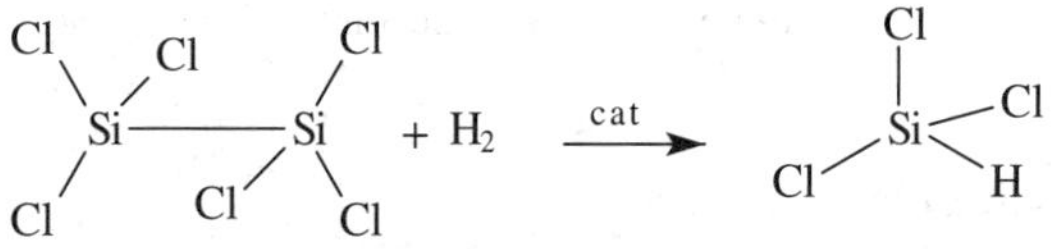

(7)

cat = Ni(0); others include Pd, Ru, etc.

Recently, there has been a lot of activity in the area of recovery of high boilers from the direct process. Dow Corning workers add Me$_3$SiCl to high boilers in the presence of H$_2$ and AlCl$_3$[109] or supported metal catalysts[110,111] to make 91% Me$_y$H$_z$SiCl$_{4-y-z}$, whereas workers from Chemiewerk Nuenchritz employ catalysts derived from alkyl amine-trialkoxysilane on an SiO$_2$ support[112]. Pachaly and Schinabeck from Wacker Chemie convert a mixture composed of 7% Me$_4$Si$_2$Cl$_2$, 46.5% Me$_3$Cl$_3$Si$_2$ and 46.5% Me$_2$Cl$_4$Si$_2$ into a mixture composed of 45% MeCl$_2$SiH, 28% MeSiCl$_3$ and 27% Me$_2$SiCl$_2$ in the presence of Bu$_3$N and HCl[113]. Similar results were obtained by Bokerman and coworkers using supported Pd and Cu catalysts[114].

B. Recycle of Solid Waste Products from the MCS Reaction

In order to maintain a steady-state concentration of bed metals during MCS production, metals are constantly blown out of the bed and replaced by fresh metals. The metal dust is separated from crude chlorosilane in devices called cyclones. Residual bed and cyclone dust represent a serious waste problem because of the high metal content of the waste. Workers from Wacker Chemie have described a method for leaching the metals from the solid waste with water/HCl. The leachate is then recovered and used as a catalyst for further MCS reaction. The precipitated silicon is fed back into the MCS reaction to obtain further silicon utilization and minimize land fill[115]. GE workers Webb, Ritzer and Neely successfully used 'spent contact mass' to catalyze the reaction of silicon with MeCl under MCS conditions to make Di and Tri[116]. Other work by the same researchers has focused on converting the pyrophoric, spent reactor contact mass (obtained from the reaction of metallic Si with MeCl under direct process conditions) into air-stable powder. The spent mass was thermally treated at 1100 °C for 3 min under nitrogen. The treated contact mass could then be used in the MCS reaction, for example it was placed in a stirred bed reactor exposed to flowing MeCl at 300 °C to give Di and Tri[117].

The late Alan Ritzer of GE wrote of the opportunities that environmental regulations present to silicone manufactures[118]. A major strategy of GE Silicones continues to be avoidance of landfilling and incineration and returning key materials back into commerce. Said Ritzer, 'Additional opportunities remain as overall silicon utilization still can be further improved'. Residue cleavage represents one method to create more Di and Tri. Mono and Tri redistribution (equation 8) represents yet another method to improve Di yield.

$$\underset{Me}{\overset{Cl}{\underset{|}{Si}}}\!\!\!\!\!\!\!\!\!\overset{Me}{\underset{Me}{<}} \;\;+\;\; \underset{Cl}{\overset{Me}{\underset{|}{Si}}}\!\!\!\!\!\!\!\!\!\overset{Cl}{\underset{Cl}{<}} \;\;\xrightarrow{\text{cat}}\;\; 2\;\; \underset{Cl}{\overset{Me}{\underset{|}{Si}}}\!\!\!\!\!\!\!\!\!\overset{Me}{\underset{Cl}{<}} \qquad (8)$$

VII. ACKNOWLEDGMENTS

Help in preparing this review in the form of literature search and discussions is due to Caroline Warden and Bill Ward.

VIII. REFERENCES

1. B. Kanner and K. M. Lewis, in *Catalyzed Direct Reactions of Silicon* (Eds. K. M. Lewis and D. G. Rethwisch), Elsevier, Amsterdam, 1993, p. 1.
2. E. G. Rochow and W. Gilliam, *J. Am. Chem. Soc.*, **63**, 798 (1941).
3. J. F. Hyde and R. DeLong, *J. Am. Chem. Soc.*, **63**, 1194 (1941).
4. J. J. Zuckerman, *Adv. Inorg. Chem. Radiochem.*, **6**, 383 (1964) and references cited therein.
5. H. A. Liebhafsky, *Silicones Under the Monogram*, Wiley, New York, 1978.
6. E. L. Warrick, *Forty Years of Firsts*, McGraw-Hill, New York, 1990.
7. W. Noll, *Chemistry and Technology of Silicones*, Academic Press, New York, 1968.
8. A. Petrov, B. F. Mironov, V. A. Ponomarenko and E. A. Chernyshev, *Synthesis of Organosilicon Monomers*, Consultants Bureau, New York, 1976.
9. R. J. H. Voorhoeve, *Methylchlorosilanes: Precursors to Silicones*, Elsevier, New York, 1967.
10. W. Buechner, *Organometallic Reviews, 5th Int. Symp. on Organosilicon Chem.* (Eds. D. Seyferth, A. G. Davies, E. O. Fischer, J. F. Normant and O. A. Keutov), Elsevier, Amsterdam, 1980, p. 409.
11. W. J. Ward, A. Ritzer, K. M. Carroll and J. W. Flock, *J. Catal.*, **100**, 240 (1986).
12. W. J. Ward, A Ritzer and J. W. Flock, U.S. Patent 4,500,724 (1985).
13. W. J. Ward, G. L. Gaines and A. Ritzer, Brit. Pat. Appl. GB 2119808, (1983) **100**, *Chem. Abstr.*, 103630 (1984).
14. W. J. Ward, G. L. Gaines and A. Ritzer, Ger. Offen. DE 3312775 (1983); *Chem. Abstr.*, **100**, 68521 (1984).
15. J. P. Kim and D. G. Rethwisch, *J. Catal.*, **134**, 168 (1992).

16. H. Lieske, U. Kretzshmar and R. Zimmermann in *Silicon Chem. Ind. II, 2nd Int. Cong.* (Eds. H. A. Oye, H. M. Rong, L. Nygaard, G. Schussler and J. Kr. Tuset), Trondheim, Norway, 1994, p. 147.
17. R. A. Turetskaya, K. A. Andrianov, I. V. Trofimova and E. A. Chernyshev, *Usp. Khim.*, **44**, 444 (1975); *Chem. Abstr*, **82**, 154580 (1975).
18. A. F. Hollemann and E. Wiberg, *Lehrbuch der Anorgaischen Chemie*, Walter de Gruyter, Berlin–New York, 1985, p. 1038.
19. J. L. Plagne, G. Godde and R. Cattoz, Eur. Pat. Appl. EP 470020 (1992); *Chem Abstr.*, **116**, 152011 (1992).
20. C. P. Homme, Fr. Demande FR 2577929 (1987); *Chem Abstr.*, **106**, 158883 (1987).
21. J. Mlynar, L. Dvorak and B. Kurka, Czech. CS 207859, (1983); *Chem. Abstr*, **101**, 192187 (1984).
22. L. D. Gasper-Galvin, D. G. Rethwisch, D. M. Sevenich and H. B. Friedrich, in *Catalyzed Direct Reactions of Silicon*, (Eds. K. M. Lewis and D. G. Rethwisch), Elsevier, Amsterdam, 1993, p. 279.
23. K. Feldner, B. Degen, G. Wagner and M. Schulze, Ger. Offen. DE 3823308 (1990); *Chem. Abstr.*, **112**, 179469 (1990).
24. E. Klar and A. Elattar, Eur. Pat. Appl. EP 304144 (1989); *Chem. Abstr.*, **110**, 213073 (1989).
25. C. Prud'Homme and G. Simon, Eur. Pat. 138679 (1985); *Chem. Abstr.*, **103**, 88071 (1989).
26. C. Prud'Homme and G. Simon, Eur. Pat. Appl. EP 138678 (1985); *Chem. Abstr.*, **103**, 88070 (1985).
27. C. Prud'Homme, Eur. Pat. Appl. EP194214 (1986); *Chem. Abstr.*, **105**, 209196 (1986).
28. T. Margaria, B. Degen, E. Licht and M. Schultze, World Patent, WO/95/01303 (1994).
29. R. L. Halm, A. B. Pierce and O. K. Wilding, U.S. Patent 4762940 (1990).
30. R. L. Halm and O. K. Wilding, U.S. Patent 4,946,978 (1990).
31. V. D. Dosaj, R. L. Halm and O. K. Wilding, U.S. Patent 4,898,960 (1990).
32. R. L. Halm and O. K. Wilding, U.S. Patent 5,059,343 (1991).
33. B. Degen, K. Feldner, H. J. Kaiser and M. Schulze, DE 3910665 (1990); U.S. Patent 5,059,706 (1991); *Chem. Abstr.*, **114**, 62348 (1991).
34. B. Degen, K. Felder, E. Licht and G. Wagner, Eur. Pat. Appl. EP 416406 (1991); also US 5,068,385 (1991); *Chem. Abstr.*, **114**, 164496 (1991).
35. W. Walkow, H. Lieske, G. Meier, A. Schenk, K. Dreier, W. Lambrecht, R. Thaetner, K. Ohl and R. Berk, Ger. (East) DD 265407 (1989); *Chem. Abstr.*, **111**, 154103 (1989).
36. G. N. Kozlova, E. S. Starodubtsev, L. M. Khananashvili, L. V. Gasanova and N. I. Tsomaya, *Soobshch. Akad. Nauk Gruz. SSR*, **117**, 329 (1985); *Chem. Abstr.*, **104**, 88657 (1986).
37. M. G. R. T. DeCooker, J. H. N. DeBruyn and P. J. Van Den Berg, *J. Organomet. Chem.*, **99**, 371 (1975).
38. R. L. Halm and R. H. Zapp, U.S. Patent 4,962,220 (1990).
39. R. L. Halm and R. H. Zapp, U.S. Patent 4,966,986 (1990).
40. K. M. Lewis, R. A. Cameron, J. M. Larnerd and B. Kanner, U.S. Patent 4,973,725 (1990).
41. K. M. Lewis, R. A. Cameron, J. M. Larnerd and B. Kanner, Eur. Pat. Appl. EP 348902 (1989); *Chem Abstr.*, **112**, 235609 (1990).
42. B. Pachaly, V. Frey and H. Strausberger, Ger Offen. DE 4303766 (1994); *Chem. Abstr.*, **122**, 58821 (1995).
43. B. Pachaly, in *Silicon Chem. Ind. II, 2nd Int. Cong.*, (Eds. H. A. Oye, H. M. Rong, L. Nygaard, G. Schussler and J. Kr. Tuset), Trondheim, Norway, 1994, p. 55.
44. S. K. Freeburne, R. L. Halm, J. P. Kohane and J. D. Wineland, U.S. Patent 5,312,948 (1994).
45. B. Degen, E. Licht, M. Schulze, G. Wagner and K.-P. Minuth, Eur. Pat. Appl. EP 617039, (1994); *Chem. Abstr.*, **121**, 280876 (1994).
46. K. Feldner, B. Degen, G. Wagner and M. Schulze, Ger. Offen. DE 3841417, (1990); *Chem. Abstr.*, **113**, 132505 (1990).
47. A. G. Forwald, H. M. Rong and G. C. Vogelaar, in *Silicon Chem. Ind. II, 2nd Int. Cong.* (Eds. H. A. Oye, F., H. M. Rong, L. Nygaard, G. Schussler and J. Kr. Tuset), Trondheim, Norway, 1994, p. 107.
48. G. Laroze, in *Silicon Chem. Ind. II, 2nd Int. Cong.* (Eds. H. A. Oye, H. M. Rong, L. Nygaard, G. Schussler and J. Kr. Tuset), Trondheim, Norway, 1994, p. 121.
49. F. Dubrous, Eur. Pat. Appl. EP 494837 (1992); *Chem. Abstr.*, **117**, 133933 (1992).
50. J. P. Kim and D. G. Rethwisch, *J. Catal.*, **134**, 168 (1992).

51. E. Klar, D. H. Hashiguchi and R. J. Dietrich, U.S. Patent 4,504,597 (1985).
52. J. Y. P. Mui, U.S. Patent 5,250,716 (1993).
53. A. A. Elattar, Eur. Pat. Appl. EP 440414 (1991); *Chem. Abstr.*, **115**, 208249 (1991).
54. M. Selenina, H. Lieske, G. Meier and U. Heinze, Ger. (East) DD 250534 (1987); *Chem. Abstr.*, **109**, 212781 (1988).
55. D. Hashiguchi, R. Dietrich and G. Schoepe, U.S. Patent 4,503,165 (1985).
56. D. Hashiguchi, E. Klar and R. Dietrich, U.S. Patent 4,520,130 (1985).
57. W. Streckel, H. Straussberger and B. Pachaly, Ger. Offen. DE 4142432 (1993); *Chem. Abstr.*, **120**, 54691 (1994).
58. R. Thaetner, K. Ohl, P. Birke, W. Koegler, H. Guenschel, H. Lieske, W. Walkow, G. Meier and J. Brumme, Ger. (East) DD 293507 (1991); *Chem. Abstr.*, **116**, 6739 (1992).
59. W. Walkow, H. Lieske, G. Meier, A. Schenk and U. Heinze, Ger. (East) DD 250536 (1987); *Chem. Abstr.*, **109**, 131214 (1988).
60. W. Walkow, H. Lieske, G. Meier and A. Schenk, Ger. (East) DD 250535 (1987); *Chem. Abstr.*, **108**, 204835 (1988).
61. A. Ritzer and H. Lapidot, U.S. Patent 4,450,282 (1984).
62. L. Roesch, G. Kratel and A. Stroh, Ger. Offen. DE3610267 (1987); *Chem. Abstr.*, **108**, 132044 (1988).
63. G. Laroze, J. L. Plagne, G. Weber and B. Gillot, in *Silicon for the Chemical Industry* (Eds. H. A. Øye and H. Rong), Geiranger, Norway, June 1992, p. 151.
64. E. G. Rochow, *Main Group Chemistry News*, **2**, 27 (1995).
65. S. Yilmaz, N. Floquet and J. L. Falconer, in *Silicon Chem. Ind. II, 2nd Int. Cong.* (Eds. H. A. Oye, H. M. Rong, L. Nygaard, G. Schussler and J. Kr. Tuset), Trondheim, Norway, 1994, p. 137.
66. N. Floquet, S. Yilmaz and J. L. Falconer, *J. Catal.*, **148**, 348 (1994).
67. K. A. Magrini, J. L. Falconer and B. E. Kohl, in *Catalyzed Direct Reactions of Silicon* (Eds. K. M. Lewis and D. G. Rethwisch), Elsevier, Amsterdam, 1993, p. 249.
68. T. C. Frank, K. B. Kester and J. L. Falconer, *J. Catal.*, **91**, 44 (1985).
69. J. L. Falconer and S. Yilmaz, in *Silicon Chem. Ind.* (Eds. H. A. Oeye and H. Rong), Norway Inst. Technol., Inst. Inorg. Chem., Trondheim, Norway, 1992, p. 99.
70. T. C. Frank, K. B. Kester and J. L. Falconer, *J. Catal.*, **95** 396 (1985).
71. S. J. Potochnik and J. L. Falconer, *J. Catal.*, **147**, 101 (1994).
72. K. M. Lewis, D. McLeod and B. Kanner, *Stud. Surf. Sci. Catal. 1987*, Vol. 38, Elsevier, Amsterdam, 1988, p. 415.
73. M. P. Clarke, *J. Organomet. Chem.*, **376**, 165 (1989).
74. H. Lieske, H. Fichtner, I. Grohmann, M. Selenina, W. Walkow and R. Zimmermann, in *Silicon Chem. Ind.* (Eds. H. A. Oeye and H. Rong), Norway Inst. Technol., Inst. Inorg. Chem., Trondheim, Norway, 1992, p. 111.
75. G. J. Hutchings, R. W. Joyner, M. R. H. Siddiqui and H. M. Rong, *Silicon Chem. Ind.* (Eds. H. A. Oeye and H. Rong), Norway Inst. Technol., Inst. Inorg. Chem., Trondheim, Norway, 1992, p. 85.
76. D. T. Hurd, *J. Am. Chem. Soc.*, **67**, 1813 (1945).
77. E. G. Rochow and W. F. Gilliam, *J. Am. Chem. Soc.*, **67**, 1772 (1945).
78. S. H. Yeon, B. W. Lee, S.-I. Kim and I. N. Jung, *Organometallics*, **12**, 4887 (1993).
79. I. N. Jung, G. H. Lee, S. H. Yeon and M. Y. Suk, *Bull. Korean Chem. Soc.*, **12**, 445 (1991).
80. I. N. Jung, S. H. Yeon and J. S. Han, *Bull. Korean Chem. Soc.*, **14**, 315 (1993).
81. E. G. Rochow, *J. Am. Chem. Soc.*, **70**, 2170 (1948).
82. W. E. Newton and E. G. Rochow, *Inorg. Chem.*, **9**, 1071 (1970).
83. M. Okamoto, N. Mimura, E. Suzuki and Y. Ono, *Catal. Lett.*, **33**, 421 (1995).
84. M. Shiozawa and Y. Okumura, Jpn. Kokai Tokkyo Koho JP 06306083 (1994); *Chem. Abstr.*, **123**, 199138 (1995).
85. F. D. Mendicino, U.S. Patent 4,727,173 (1988).
86. L. G. Moody, T. E. Childress, R. L. Pitrolo, J. S. Ritscher and R. P. Leichliter, U.S. Patent 4,999,446 (1991).
87. M. Shiozawa and Y. Okumura, Jpn. Kokai Tokkyo Koho JP 06271587 (1994); *Chem. Abstr.*, **122**, 106124 (1995).
88. K. Adachi, H. Kumoyama, Jpn Kokai Tokkyo Koho JP 06293776 (1994); *Chem. Abstr.*, **122**, 106125 (1995).
89. M. Harada and Y. Yoshinori, Ger. Offen DE 4415418 (1994); *Chem. Abstr.*, **122**, 81621 (1995).

90. J. S. Ritscher and T. E. Childress, U.S. Patent 5,084,590 (1992).
91. E. Suzuki, M. Okamoto and Y. Ono, *Chem. Lett.*, 199 (1991).
92. E. Suzuki and Y. Ono, *Chem. Lett.*, 47 (1990).
93. E. Suzuki, T. Kamata and Y. Ono, *Bull. Chem. Soc. Jpn.*, **64**, 3445 (1991).
94. M. Okamoto, M. Osaka, K. Yamamoto, E. Suzuki and Y. Ono, *J. Catal.*, **143**, 64 (1993).
95. Y. Ono, E. Suzuki and M. Okamoto, Jpn. Kokai Tokkyo Koho JP 06263769 (1994); *Chem. Abstr.*, **123**, 83719 (1995).
96. M. Okamoto, N. Watanabe, E. Suzuki and Y. Ono, *J. Organomet. Chem.*, **489**, C12 (1995).
97. Y. Ono, M. Okano, N. Watanabe and E. Suzuki, in *Silicon Chem. Ind. II, 2nd Int. Cong.* (Eds. H. A. Oye, H. M. Rong, L. Nygaard, G. Schussler and J. Kr. Tuset), Trondheim, Norway, 1994, p. 185.
98. M. Okamoto, K. Yamamoto, E. Suzuki and Y. Ono, *J. Catal.*, **147**, 15 (1994).
99. M. Shiozawa and Y. Okumura, Jpn. Kokai Tokkyo Koho JP 06009652 (1994); *Chem. Abstr.*, **121**, 109252 (1994).
100. J. O. Stoffer, J. F. Montle and N. L. D. Somasivi, U.S. Patent 4,778,010 (1988).
101. J. R. Malek, J. L. Speier and A. P. Wright, U.S. Patent 4,088,669 (1978).
102. E. G. Rochow and J. J. Zuckerman, PB 157357 NTIS. U.S. Dept of Commerce (1960).
103. S. Yamada and E. Yasunaga, Jp. Patent 286 (1951); *Chem. Abstr.*, **47**, 3334 (1953).
104. K. M. Lewis and B. Kanner, U.S. Patent 4,593,114 (1986).
105. K. M. Lewis and B. Kanner, Eur. Pat. Appl. EP 175282 (1986); *Chem. Abstr.*, **105**, 60753 (1986).
106. K. Feldner and W. Grape, Ger. Offen. DE 3410644 (1985); *Chem. Abstr.*, **104**, 19670 (1986).
107. K. Feldner and W. Grape, Ger. Offen. DE 3436381 (1986); *Chem. Abstr.*, **105**, 60755 (1986).
108. R. Calas, J. Dunogues, G. Deleris and N. Duffaut, *J. Organomet. Chem.*, **225**, 117 (1982).
109. S. P. Ferguson, R. F. Jarvis, B. M. Naasz, K. K. Oltmanns, G. L. Warrick and D. L. Whitely, U.S. Patent 5,430,168 (1995); *Chem. Abstr.*, **123**, 257007 (1995).
110. K. M. Chadwick, A. K. Dhaul, R. L. Halm and R. L. Johnson, U.S. Patent 5,326,896 (1994); *Chem. Abstr.*, **121**, 206,245 (1994).
111. K. M. Chadwick, A. K. Dhaul, R. L. Halm, R. G. Johnson and R. D. Steinmeyer, U.S. Patent 5,321,147 (1994); *Chem. Abstr.*, **121**, 109261 (1994).
112. J. Albrecht, R. Neumann and W. Geisler, Ger. Offen. DE 4323406 (1995); *Chem. Abstr.*, **122**, 265626 (1995).
113. B. Pachaly and A. Schinabeck, Eur. Pat. Appl. EP574912 (1993); *Chem. Abstr.*, **120**, 191988 (1994).
114. G. N. Bokerman, J. P. Cannady and A. E. Ogilvy, U.S. Patent 5,175,329 (1992); *Chem. Abstr.*, **118**, 191999 (1993).
115. B. Pachaly, H. Straussberger and W. Streckel, in *Silicon Chem. Ind. II, 2nd Int. Cong.* (Eds. H. A. Oye, H. M. Rong, L. Nygaard, G. Schussler and J. Kr. Tuset), Trondheim, Norway, 1994, p. 235.
116. S. W. Webb, A. Ritzer and J. D. Neely, U.S. Patent 5,243,061 (1993).
117. S. W. Webb, A. Ritzer and J. D. Neely, U.S. Patent 5,239,102 (1993).
118. A. Ritzer, in *Silicon Chem. Ind. II, 2nd Int. Cong.* (Eds. H. A. Oye, H. M. Rong, L. Nygaard, G. Schussler and J. Kr. Tuset), Trondheim, Norway, 1994, p. 241.

CHAPTER 27

Acyl silanes

PHILIP C. BULMAN PAGE, MICHAEL J. McKENZIE, SUKHBINDER S. KLAIR and STEPHEN ROSENTHAL

Department of Chemistry, Loughborough University, Loughborough, Leicestershire LE11 3TU, UK
Fax: +44(0)1509 223926; e-mail: p.c.b.page@lboro.ac.uk

I. INTRODUCTION	1600
II. STRUCTURE AND SPECTROSCOPY OF ACYL SILANES	1600
A. Infrared Spectroscopy	1600
B. ^{1}H NMR Spectroscopy	1601
C. ^{13}C NMR Spectroscopy	1602
D. ^{17}O NMR Spectroscopy	1603
E. ^{29}Si NMR Spectroscopy	1603
F. Ultraviolet and Visible Spectroscopy	1603
G. X-ray Diffraction Studies	1605
III. SYNTHESIS OF ACYL SILANES	1606
A. Simple Acyl Silanes	1606
1. Formyl silanes	1606
2. Hydrolysis of acetals	1606
3. Silyl metallic species	1608
4. Palladium-catalysed coupling	1610
5. Oxidation reactions	1611
6. Hydroboration–oxidation of alkynyl silanes	1613
7. Rearrangement of silyloxy carbenes	1614
8. Preparation from enol ethers	1614
9. Silylation of acyl metallic species	1618
10. Non-racemic acyl silanes	1619
B. α-Haloacyl Silanes	1619
C. α-Ketoacyl Silanes	1621
D. α,β-Unsaturated Acyl Silanes	1622
1. Hydroboration–oxidation of enynes	1622
2. Oxidation of allylic carbinols	1622
3. Preparation from enol ethers of acyl silanes	1623
4. Preparation from α,β-alkynyl acyl silanes	1628
5. Miscellaneous methods	1628
E. α-Cyclopropyl and α-Epoxyacyl Silanes	1630

The chemistry of organic silicon compounds, Vol. 2
Edited by Z. Rappoport and Y. Apeloig © 1998 John Wiley & Sons Ltd

IV. REACTIONS OF ACYL SILANES 1631
   A. General Reactivity 1631
      1. Nucleophilic addition 1632
      2. Acyl silanes as acyl anion precursors 1642
      3. Cyclization reactions 1643
      4. Photochemistry 1645
      5. Biotransformations 1649
      6. Miscellaneous 1649
   B. α-Haloacyl Silanes 1651
   C. α-Ketoacyl Silanes 1653
   D. α,β-Unsaturated Acyl Silanes 1653
V. CONCLUSION 1660
VI. REFERENCES 1660

I. INTRODUCTION

Early work on acyl silanes, perhaps spurred by the green or yellow-green colour of simple examples, concentrated on spectroscopic properties, but in addition to their unusual spectroscopic behaviour, acyl silanes exhibit interesting chemistry. In general, they display rather poor stability towards basic conditions and light, but their sensitivity depends to a large degree upon structure and substitution; for example, α,β-unsaturated acyl silanes are relatively stable, while α-ketoacyl silanes are especially sensitive. The last ten to fifteen years have seen a steady increase in the level of investigation into the chemistry of acyl silanes, resulting both in valuable new reactions and in improved methods of synthesis of many types of acyl silanes[1]. This chapter is principally concerned with methods of synthesis of all types of acyl silanes together with the development of their chemistry since the first isolation in 1968. A simple descriptive spectroscopy section is included; readers are referred to publications cited in the text for a more thorough treatment of the structure and bonding aspects of acyl silane chemistry.

II. STRUCTURE AND SPECTROSCOPY OF ACYL SILANES

A. Infrared Spectroscopy

Acyl silanes exhibit interesting and unusual spectroscopic properties, and early work was stimulated by these observations[2-4]. Carbonyl stretching frequencies in their infrared absorption spectra are lower relative to those of simple ketones. For example, benzoyl trimethylsilane and acetyl trimethylsilane have carbonyl stretching frequencies at ca 1620 cm^{-1} and 1645 cm^{-1}, respectively, in their infrared spectra (Table 1, entries 6 and 7). This relative lowering of the carbonyl stretching frequency is usually explained as a consequence of an inductive effect: Polarization of the carbonyl group as a consequence of electron release from the silicon atom towards the carbonyl group weakens the C=O bond[5].

It is perhaps surprising that the position of the carbonyl group absorption of acyl silanes is not significantly altered by changing the type of group attached to the silicon atom, similar absorptions being observed for acyl, trimethyl, triphenyl and tris(p-substituted) phenyl silanes, the last showing a linear correlation of the C=O wavelength with Hammett substituent constants. Further, the position of acyl silane carbonyl absorptions is relatively independent of medium polarity, again unlike the behaviour of simple ketones, where carbonyl absorptions are shifted to lower frequency on increasing the polarity of the medium in which they are measured[6]. The proposed large σ-inductive effect is caused by the low electronegativity of silicon (1.8) relative to carbon (2.5), and by the larger mass

27. Acyl silanes

TABLE 1. Infrared C=O absorption of selected acyl metalloids and related compounds[a]

Entry	Compound	ν (M=C)	ν (M=Si)	ν (M=Ge)	ν (M=Sn)	Reference
1	Ph_3MCOMe		1645	1669	1670	5
2	Ph_3MCOPh	1692	1618	1629	1627	5,7
3	Ph_3MCH_2COPh	1698	1667	1661		5
4	$Ph_3MCH_2CH_2COPh$	1692	1692			5
5	$PhMe_2MCOPh$		1620			5,7
6	Me_3MCOPh	1675	1620	1629		5,7
7	Me_3MCOMe	1710	1645			5,7
8	$Me_3MCO(CH_2)_4Me$	1700	1640			7–9
9	$(Me_3MCOCH_2CH_2)_2$		1640			10
10	Me_3MCOCH_2Ph		1635	1660	1656	11
11	$Me_3MCOCHBr(CH_2)_3Me$		1645,1652			7,12
12	$Me_3MCOCH=CHMe$		1642	1644	1648	11
13	$Me_3MCOCH=CMe_2$		1640	1645	1640	11
14	$Me_3MCOCMe_3$		1636			5
15	$Me_3MCOMMe_3$		1556, 1570			13
16	$Ph_3MCOMPh_3$		1558 + 1592	1616		5
17	$Me_3MCOCOMe$		1713 + 1658			7,14
18	Me_3MCOCO_2Me		1660			15
19	$Me_3MCONEt_2$	1620	1560			16
20	$(Me_3Si)_3MCO_2H$		1630			17

[a]For additional examples, see References 18–26.

of silicon. Such differences in electronegativity, producing significant inductive release of electron density towards the carbonyl group, would be expected to lead to a decrease in the frequency, as is indeed the case[5,6].

The carbonyl groups of acyl silanes and acyl germanes display lower frequency absorptions than those of α-silyl ketones, which in turn absorb at lower frequencies than those of β-silyl ketones (Table 1, entries 2–4) a pattern consistent with inductive release of electrons. Indeed, β-silyl ketone carbonyl groups absorb at similar frequencies to their carbon analogues (Table 1, entry 4). The effect is weaker in the α-silyl ketones, where it must operate through an extra methylene group, and is not apparent at all in the β-silyl ketones, where it must operate through two methylene groups. The strong inductive release in acyl silanes appears to be approximately additive, as shown by the pink-coloured carbonyl bis(trimethylsilane), where the infrared stretching frequency is lowered by approximately 80 cm^{-1} compared with acetyl triphenylsilane (Table 1, entries 14 and 15)[5].

B. ^{1}H NMR Spectroscopy

Generally, protons attached to the α-carbon atoms of acyl silanes are somewhat deshielded relative to their carbon analogues (Table 2). Differences in magnetic anisotropy and electronegativity are no doubt responsible for this effect. It is interesting that α,β-unsaturated acyl silanes appear to be an exception (entry 12).

TABLE 2. ^{1}H chemical shifts (δ_H) at α-carbon atom of selected acyl metalloids

Entry	Compound	M=C	M=Si	M=Ge	M=Sn	Reference
1	Me$_3$MCOC<u>H</u>$_2$Ph		3.77	3.73	3.80	11
2	Me$_3$MCOMe	2.07	2.18,2.20			5,7
3	Ph$_3$MCOMe	2.01	2.30	2.38		5
4	t-BuMe$_2$MCOMe		2.32			18
5	Me$_3$MCOC<u>H</u>$_2$(CH$_2$)$_3$Me	2.33	2.50			7–9
6	Me$_3$MCOC<u>H</u>Et$_2$		2.50			18
7	Et$_3$MCOC<u>H</u>$_2$Me		2.60			19
8	Me$_3$MCOC<u>H</u>PhEt		3.86			18
9	Me$_3$MCOCOMe		2.03			7,14
10	Me$_3$MCOCOC<u>H</u>$_2$Bu-n		2.55			7
11	Me$_3$MCOC<u>H</u>BrBu-n		4.39,4.33			7,12
12	Me$_3$MCOC<u>H</u>=CH$_2$	6.88	6.28,6.38			7,14,20,27
13	Me$_3$MCOC<u>H</u>=CMe$_2$		6.56	6.32	6.40	11

C. ^{13}C NMR Spectroscopy

The ^{13}C NMR signals of carbonyl groups in acyl silanes are dramatically shifted downfield in comparison to the analogous ketones (Table 3)[28,29]. The effect ranges from ca 25 to 100 ppm, and is approximately additive.

^{13}C NMR studies also expose some interesting features of acyl silane bonding. The carbonyl group carbon atom of an alkyl phenyl ketone displays a ^{13}C chemical shift similar

TABLE 3. ^{13}C carbonyl group chemical shifts (δ_c) of selected acyl silanes and related compounds[a]

Entry	Compound	M=Si	M=C	Reference
1	Me$_3$MCOCMe$_3$	249.0	215.1	29
2	Me$_3$MCOMe	244.3,247.6	210.4	29,30
3	Me$_3$MCOPh	233.6,237.5	207.8,209.1	29,30
4	Me$_3$MCOCH=CH$_2$	236.7,237.9		14,20
5	t-BuMe$_2$COC≡CMe	225.7		14
6	Me$_3$MCOCHBr(CH$_2$)$_4$Me	234.8		7
7	Me$_3$MCOSiMe$_3$	318.2,318.8	249.0	13,29
8	Me$_3$<u>M</u>COCO(CH$_2$)$_4$Me	235.1		7
9	Me$_3$<u>M</u>COCOPh	220.4		7
10	Ph$_3$MCOMe	240.1		31
11	(Me$_3$Si)$_3$MCOPh	233.8		31
12	(Me$_3$Si)$_3$MCOCMe$_3$	244.6		31
13	Me$_3$MCOFc[b]	236.8		32
14	Ph$_3$MCOFc[b]	231.3		32
15	PhMe$_2$MCOFc[b]	234.7		32
16	Ph$_2$MeMCOFc[b]	232.9		32

[a]For additional examples see References 10,23 and 33.
[b]Fc = ferrocenyl.

to aliphatic analogues, but the difference between the two corresponding silicon species (e.g. PhCOSiMe$_3$ and MeCOSiMe$_3$) is somewhat more marked. For example, the carbonyl group carbon atom of benzoyl trimethylsilane has a ^{13}C NMR chemical shift *ca* 11–14 ppm upfield of that displayed by acetyl trimethylsilane (Table 3, entries 2 and 3). This effect could be ascribed to participation of resonance structures which could increase the shielding of the carbonyl group carbon atom by reducing the amount of positive charge, so positioning the chemical shift upfield relative to the aliphatic analogue, where such an effect could not operate.

D. ^{17}O NMR Spectroscopy

The oxygen atoms of acyl silanes resonate at higher chemical shifts than do those of the analogous carbon compounds (Table 4). The displacement to higher chemical shifts (i.e. downfield) of ^{17}O and ^{13}C carbonyl group signals in acyl silanes has been attributed to lowering of the HOMO–LUMO gap ΔE^{34}. Partial reversal of this effect is possible by increasing the electron density in the α-carbon atom pπ orbitals (Table 4, entries 1 and 3); even though ΔE is smaller for the aryl derivative, the ^{17}O chemical shift of the alkyl derivative is the larger.

E. ^{29}Si NMR Spectroscopy

The ^{29}Si NMR properties of acyl silanes indicate that the acyl group has a moderate shielding effect upon the silicon atom, similar in magnitude to that observed in vinyltrimethylsilane (Table 5)[28]. As would be expected, the nature of the groups attached to the silicon moiety affects the chemical shift of the silicon atom.

F. Ultraviolet and Visible Spectroscopy

The unusual ultraviolet and visible spectroscopic characteristics of the simpler acyl silanes are perhaps their most immediately obvious feature: they are green in colour. Data for selected acyl silanes and other acyl metalloids are given in Table 6[36–41]. As for simple ketones, acyl silanes show absorptions due to n–π^* and π–π^* transitions. The n–π^* transition in alkyl acyl silanes, however, occurs at *ca* 370 nm, a shift of about 100 nm to longer wavelength compared with the same transition in pinacolone, the analogous carbon compound (ν_{max} 279 nm), corresponding to a *ca* 25 kcal mol^{-1} reduction in the n–π^* transition energy. Comparison of the extinction coefficients of acetyl trimethylsilane and pinacolone ($\epsilon = 126$ and 21, respectively) shows that absorption for the acyl silane derivative is by far the more intense. The λ_{max} values for acyl silanes have been shown to be proportional to the degree of deshielding observed in the ^{17}O NMR spectra, which, in agreement with the Karplus–Pople equation, is *ca* four times larger than is the corresponding effect on the ^{13}C signal of the carbonyl carbon atom[42].

TABLE 4. ^{17}O chemical shifts (δ_o) of selected acyl silanes and related compounds[34]

Entry	Compound	M=Si	M=C
1	Me$_3$MCOMe	692	553
2	Me$_3$MCOCMe$_3$	659	550
3	Me$_3$MCOPh	671	554
4	Me$_3$MCOC$_6$H$_4$Cl-*p*	673	551

TABLE 5. ^{29}Si chemical shifts (δ_{Si}) of selected acyl silanes and related compounds[a]

Entry	Compound	δ_{Si}	Reference
1	Me$_3$SiCOMe	−10.1	30
2	Ph$_3$SiCOMe	−30.4	31
3	(Me$_3$Si)$_4$	−135.5	31
4	Ph$_3$SiCOPh	−28.3	31
5	Et$_3$SiCOPh	−28.3	31
6	Me$_3$SiCOPh	−7.4, −15.1	30,31
7	Me$_3$SiCOSiMe$_3$	−14.4	13
8	Me$_3$SiCH=CH$_2$	−7.6	35
9	(Me$_3$Si)$_3$<u>Si</u>CO−Ad[b]	−78.8	31
10	(Me$_3$<u>Si</u>)$_3$SiCO−Ad[b]	11.5	31
11	Ph(Me$_3$Si)$_2$<u>Si</u>CO−Ad[b]	−44.75	26
12	Ph(Me$_3$<u>Si</u>)$_2$SiCO−Ad[b]	−13.45	26
13	(Me$_3$Si)$_3$<u>Si</u>CO$_2$H	−73.84	17
14	(Me$_3$<u>Si</u>)$_3$SiCO$_2$H	−6.15	17
15	(Me$_3$Si)$_3$<u>Si</u>CO$_2$Me	−74.45	17
16	(Me$_3$<u>Si</u>)$_3$SiCO$_2$Me	−6.34	17
17	Me$_3$SiCOFc[c]	−9.5	32
18	Ph$_3$SiCOFc[c]	−30.9	32
19	PhMe$_2$SiCOFc[c]	−16.5	32
20	Ph$_2$MeSiCOFc[c]	−21.9	32

[a] For additional examples see References 23 and 25.
[b] Ad = 1-adamantyl.
[c] Fc = ferrocenyl.

TABLE 6. UV/Visible absorption of selected acyl metalloids and related compounds[a]

Entry	Compound	n → π* [λ_{max}/nm (ε)]	π → π*	Reference
1	Me$_3$CCOMe	278 (15)		5
2	Me$_3$SiCOMe	372 (126)		5
3	Et$_3$GeCOMe	365 (173)		5
4	Me$_3$SiCO(CH$_2$)$_4$Me	367 (129)		8
5	Me$_3$SiCOCHBr(CH$_2$)$_4$Me	374 (142)		7
6	Ph$_3$CCOPh	329 (299)	251 (11 600)	5
7	Ph$_3$SiCOPh	424 (292)	257 (16 200)	5
8	Ph$_3$GeCOPh	417 (210)	254 (16 200)	5
9	Ph$_3$SnCOPh	435		5
10	Me$_3$SiCOPh	425 (127)	250 (14 500)	5,7
11	Me$_3$GeCOPh	412 (120)	252 (10 700)	5
12	Me$_3$SiCOCH=CHMe	424 (98)	224 (10 300)	7,11
13	Me$_3$GeCOCH=CHMe	416 (109)	258 (11 300)	11
14	Me$_3$SnCOCH=CHMe	432	258	11
15	Me$_3$SiCOC(Me)=CH$_2$	425 (77)	222 (5 900)	20
16	Me$_3$SiCOC≡CMe	420 (170)	227 (7 450)	14
17	Me$_3$SiCOCO$_2$Me	455 (100)	227 (254)	15
18	Me$_3$SiCOCO$_2$Et	455 (97)	230 (388)	15
19	PhMe$_2$SiCOCO$_2$Me	455 (213)	279 (620)	15
20	Me$_3$SiCOCOMe	535 (99)		14
21	Me$_3$SiCOCOPh	518 (117)	275 (6 500)	7
22	Me$_3$SiCONEt$_2$	264 (270)		16

For additional data see References 26 and 33.

Considerable fine structure is often observed in the ultraviolet/visible spectra of acyl silanes, usually consisting of three main bands, sometimes with two additional shoulders at lower wavelengths[5]. This vibrational structure persists in polar solvents, an effect which is not commonly observed in ketones, and is not well understood. In α,β-unsaturated acyl silanes, and in aryl acyl silanes, which can be lime green in colour, the n–π^* transition occurs around 420 nm again shifted to longer wavelength by about 100 nm compared with the analogous carbon compounds. In α-carboxyacyl silanes, which are usually yellow, the transition occurs at around 455 nm, and in the deep crimson α-ketoacyl silanes at ca 520 nm; carbonyl bis(trimethylsilane) is also pink in colour.

The π–π^* transition of the carbonyl group in aryl acyl silanes produces an intense absorption band at around 250–260 nm, and the position and extinction coefficient of this transition are largely independent of the nature of the substituents on the silyl group. As would be expected for conjugated carbonyl group transitions, small red shifts are observed in polar solvents.

A third absorption is usually observed at approximately 185–195 nm. For acyl silanes with aromatic substituents, this may be ascribed to a primary arene band or could arise from a second π–π^* transition, but for other acyl silanes, this transition (at 195 nm, $\epsilon = 4200$ for acetyl trimethylsilane) is presumably the latter. In general, however, the nature of the groups attached to the silicon atom (other than the acyl group) has little effect on the energies of the n–π^* and π–π^* transitions[5].

Several acyl metalloids, including acetyl trimethylsilane and acetyl trimethylgermane, have also been studied by photoelectron spectroscopy[38], and various *ab initio* and semiempirical calculations have been carried out[38,39].

G. X-ray Diffraction Studies

In 1968, Chieh and Trotter reported the single-crystal X-ray analysis of acetyl triphenylsilane[43]. The acetyl group and the three phenyl groups were found to be located tetrahedrally around the silicon atom. The silicon–carbonyl group bond length is considerably elongated, at 1.926 Å; generally, silicon–carbon single bonds fall in the range 1.85–1.90 Å. Trotter ascribes this lengthening of the silicon–carbonyl group bond to a contribution to the structure not only of canonical forms with single C–O bonds, which contribute to the structure of ordinary ketones, but also to a canonical form in which there is no formal bond between the metalloid atom and the acetyl carbon atom. This structure was considered possible as a contributing resonance form because of the considerable differences in electronegativity between silicon and carbon[43,44]. The silicon–phenyl group bond lengths were found to be more typical, at 1.864 Å, and the carbonyl bond length, which one might expect to be abnormally long, reflecting the unusually long wavelength carbonyl group absorption and the enhanced basicity, was found to be 1.21 Å, approximately the same as that found in simple ketones.

More recently, Sharma has completed single-crystal X-ray analyses of four ferrocenyl acyl silanes[32]. In line with the earlier observations, the silicon–carbonyl group bond lengths were significantly elongated, at 1.935 Å. However, they also observed some elongation of the carbonyl bond length. In all four examples, which vary only in the substituents at silicon, the carbonyl bond length was 1.230 Å, somewhat longer than that reported for acetyl ferrocene (1.220 Å)[45], and significantly longer than in the Trotter structure. The Si–C–O bond angle in acetyl triphenylsilane was approximately the theoretical value of 120°, but was compressed, at 115.7°, in the ferrocenyl triphenyl acyl silane.

The structure of acetyl triphenylgermane is very similar to its silicon analogue[44], and shows a lengthening of the germanium–carbonyl group bond length by 0.066 Å over that of the germanium–phenyl group bond. The acetyl triphenylgermane molecule was

III. SYNTHESIS OF ACYL SILANES

A. Simple Acyl Silanes

1. Formyl silanes

Formyl silanes remain rare in the literature, due no doubt to their instability. They are in general thermally stable, but are extremely sensitive to air, spontaneously igniting or decomposing violently upon exposure to atmospheric oxygen. Following several years of speculation[5,46], Ireland and Norbeck achieved strong evidence for the formation of formyl trimethylsilane using a trap with a Wittig reagent following Swern oxidation[47] of trimethylsilyl methanol (*vide infra*, Section III.A.5)[48]. Formyl trimethylsilane was subsequently identified by Tilley using NMR spectroscopy as the product of reaction of the zirconium η^2-sila-acyl complex $(\eta^5\text{-}C_5H_5)_2Zr(\eta^2\text{-}COSiMe_3)Cl$ (**1**) with hydrochloric acid[49], and has recently been prepared by *in situ* reaction of 1,1-dimethylsilene with formaldehyde; the formyl silane rearranged quantitatively to 2,2-dimethyl-2-silapropanal within a few minutes (Scheme 1)[50]. Tilley was later able to prepare, isolate and characterize the first stable formyl silane, formyl tris(trimethylsilyl)silane, from the zirconium η^2-sila-acyl complex (**2**)[51] by treatment with anhydrous hydrogen chloride at $-78\,°C$ in toluene solution. Formyl tris(trimethylsilyl)silane was stable under an inert atmosphere and showed typical carbonyl group reactivity. More recently, the stable greenish-yellow formyl triisopropylsilane was isolated in 91% yield by hydrolysis of the dimethyl acetal, prepared from the dithiane derivative by transacetalization mediated by mercury salts[52a], an approach also used by Silverman in a preparation, from the corresponding dioxolane, of formyl *t*-butyldimethylsilane, which was converted *in situ* into the 2,4-dinitrophenylhydrazone derivative in 70% overall yield[52b]. Katritzky has trapped formyl silanes *in situ* as their 2,4-dinitrophenylhydrazone derivatives by hydrolysis of trialkylsilyl benzotriazolyl carbazolyl methanes in the presence of the hydrazine (Scheme 2)[53]. Formyl silanes prepared *in situ* have been used to produce α,β-acetylenic acyl silanes by lithium acetylide addition and Swern oxidation[54]. Triphenylsilyl bis(cyclopentadienyl) zirconium chloride undergoes carbonyl insertion under carbon monoxide pressure to give an acyl zirconium species which, when treated with anhydrous hydrogen chloride in a benzene matrix at $-196\,°C$ and warmed to room temperature, produces formyl triphenylsilane (*vide infra*, Section III.A.3)[55].

2. Hydrolysis of acetals

The most general approach to the synthesis of acyl silanes is based on hydrolysis of 2-silyl-1,3-dithioacetals, first investigated for 1,3-dithiane derivatives by Brook[56] and Corey[57] in the late 1960s (Scheme 3). Oxathioacetals and protected hemithioacetals have also been used, and some formyl silanes have been isolated by hydrolysis of dioxa-acetals (*vide supra*).

The major disadvantage of the dithiane approach lies in the final deprotection step; the ease and success of hydrolysis of the 2-silyl-1,3-dithiane with retention of the silicon group is highly dependent upon the dithiane 2-substituent and on the size of the groups attached to silicon, production of the corresponding aldehyde often accounting

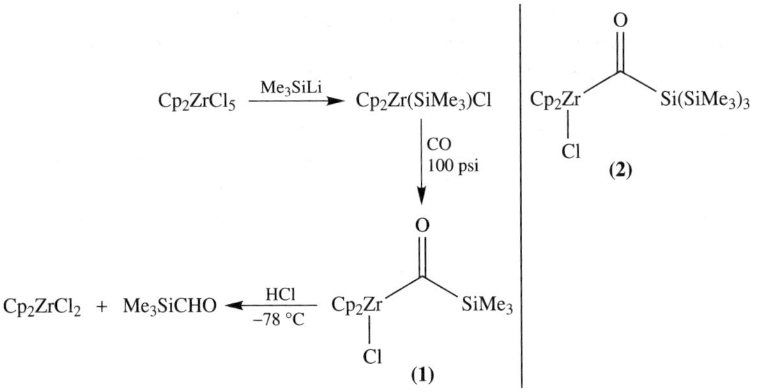

27. Acyl silanes

SCHEME 1

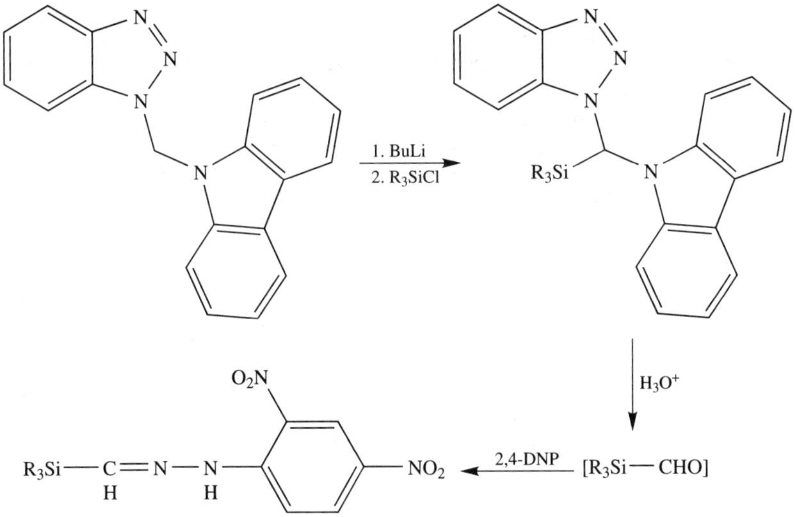

SCHEME 2

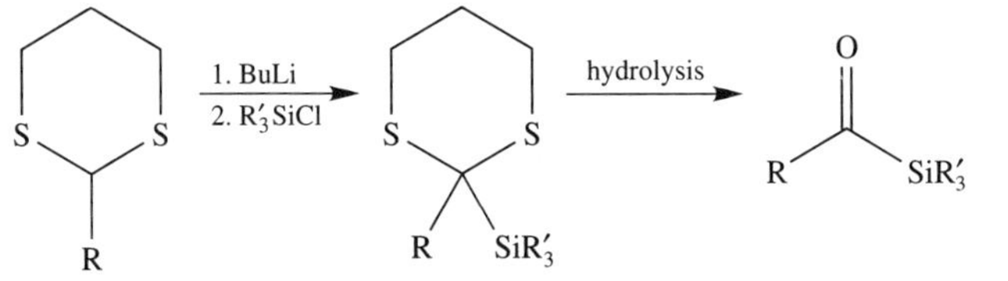

SCHEME 3

for 10–100% of the product mixture. In general, the larger these groups are, the more successful is the final deprotection step. Early methods of hydrolysis involved the use of mercury(II) salts, for example in one synthesis of thienamycin[58], and, although this process has been improved by the use of such reagents as chloramine T hydrate[59–62], this step still remains a significant problem in some cases[7], although anodic oxidation appears to be particularly successful[63]. A number of interesting acyl silanes including chiral[64] and cyclic[60] examples have been prepared using the dithiane method.

A number of more readily hydrolysed acetals have been investigated, including lithio bis(methylthio) methanes[65] and methoxy phenylthio trialkylsilyl methane (3), which acts as an α-silyl acyl anion equivalent[66]. This route (Scheme 4) is successful for a wide range of aliphatic electrophiles and is even successful in conjugate addition to enones[67].

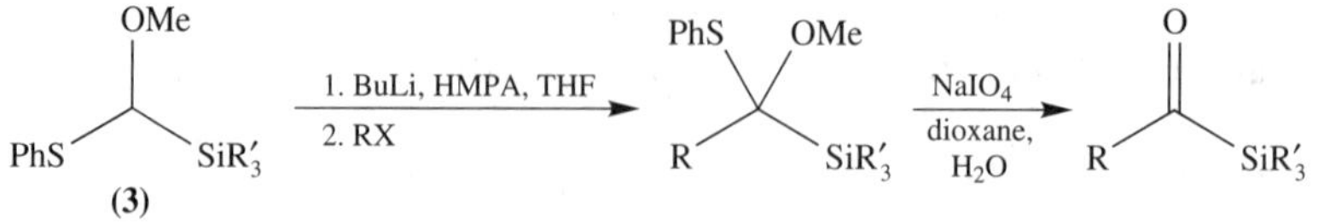

SCHEME 4

O-Trimethylsilyl hemithioacetal (4), an example of a group of acetals which are very readily hydrolysed[68], is an intermediate in the synthesis of carbonyl bis(trimethylsilane). The unstable sulphoxide species (5), prepared as shown in Scheme 5, undergoes a sila-Pummerer rearrangement to give the intermediate hemithioacetal (4)[69], which in turn reacts with the Pummerer intermediate to give a 1 : 1 mixture of carbonyl bis(trimethylsilane) and a dimethyl thioacetal. In a related reaction, the sila-Pummerer rearrangement of bis(trimethylsilyl) phenylseleninyl phenyl methane has been used to prepare benzoyl trimethylsilane in 46% yield[70].

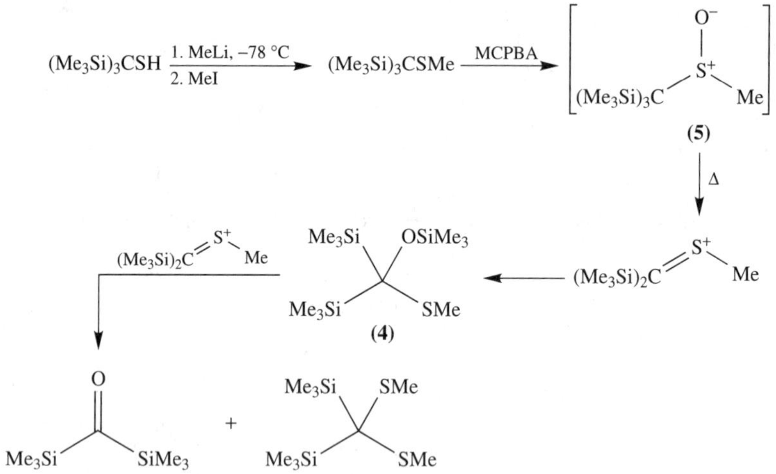

SCHEME 5

3. Silyl metallic species

One of the very earliest syntheses of an acyl silane involved the reaction of benzoyl chloride with triphenylsilyl potassium at low temperature (Scheme 6)[2]. The very low yield of benzoyl triphenylsilane (6%) obtained, and the similar yields of acetyl triphenylsilane obtained from the reactions of silyl lithium reagents with acetyl chloride[71], demonstrate that this is not a useful method for the synthesis of simple acyl silanes. It is, however, successful for the preparation of acyl tris(trimethylsilyl)silanes and some derivatives[24,31,72,73]. Tris(trimethylsilyl) and alkyl bis(trimethylsilyl)silyl lithium reagents react cleanly with acyl chlorides, typically at 0 °C, to give the corresponding acyl silanes in up to ca 85% yields. The tris(trimethylsilyl)silyl ('sisyl') lithium reagent is prepared by deprotonation of tris(trimethylsilyl)silane using an alkyl lithium reagent. The related alkyl bis(trimethylsilyl)silyl lithium species have been prepared by several methods (Scheme 7)[26]. The silyl lithium reagents, which are yellow to orange to brown in colour, can be remarkably stable; for example, phenyl bis(trimethylsilyl)silyl lithium forms a recrystallizable pale yellow solid 'ate' complex with THF with the stoichiometry Ph(Me$_3$Si)$_2$SiLi·3THF. Tris(trimethylsilyl)silyl lithium, which also forms an 'ate' complex with three molecules of THF, reacts with carbon dioxide to give the interesting crystalline tris(trimethylsilyl)silane carboxylic acid after acidic work-up, in 85% yield[17]. The methyl ester and several silyl esters were prepared from this acid by conventional means[28]. Phenyldimethylsilyl lithium reacts rapidly with some methyl esters at −110 °C and with amides at −78 °C to give acyl silanes smoothly. Use of other esters can lead to further addition of silyl lithium species to give disilyl carbinols; these materials may, however, be converted into acyl silanes with loss of one silyl moiety by oxidation with pyridinium dichromate[74].

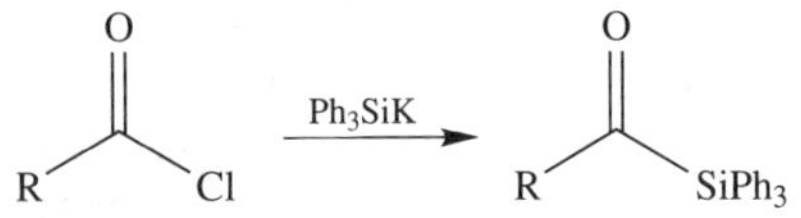

SCHEME 6

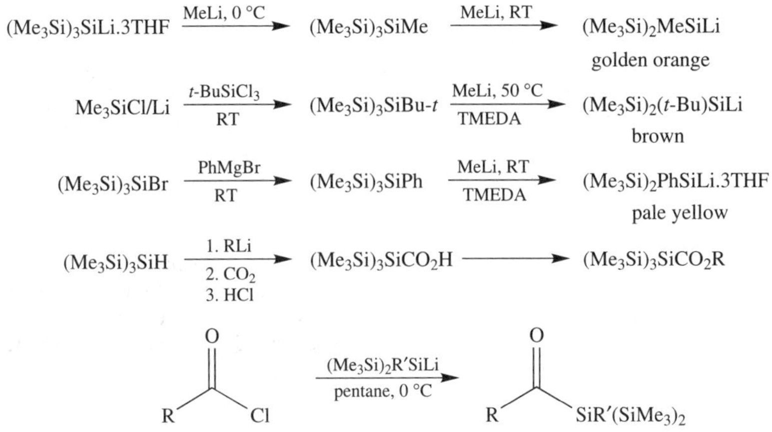

SCHEME 7

Less reactive silyl metal species such as lithium bis(triphenylsilyl) cuprate react with a variety of acyl chlorides to give the corresponding acyl silanes in moderate to good yields[75]. Dilithium bis(trimethylsilyl) cyanocuprate is particularly effective for the preparation of sterically hindered acyl silanes (Scheme 8), and appears to provide a good general preparative method[10]. Other higher-order lithium silyl cuprate species have more recently been shown to be effective in combination with the use of thiolesters as electrophiles[76].

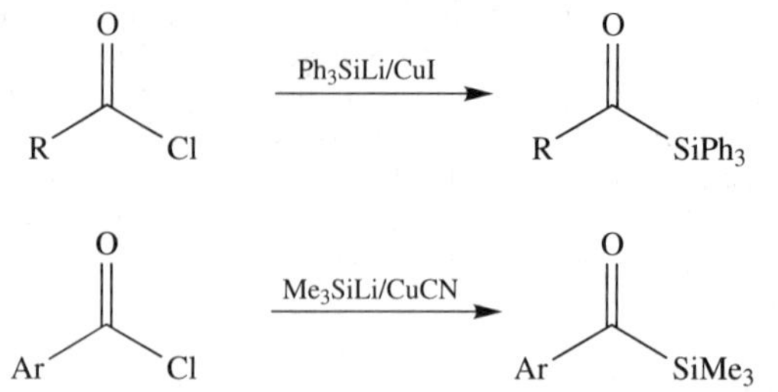

$$R-C(=O)-Cl \xrightarrow{Ph_3SiLi/CuI} R-C(=O)-SiPh_3$$

$$Ar-C(=O)-Cl \xrightarrow{Me_3SiLi/CuCN} Ar-C(=O)-SiMe_3$$

Ar = Ph, 3-MeC$_6$H$_4$, 4-MeOC$_6$H$_4$, 2,4,6-Me$_3$C$_6$H$_2$, 2-Furyl
R = Me, Et, t-Bu

SCHEME 8

The low ionic character of the aluminium–silicon bond has been cleverly utilized to develop a very mild, general and effective synthesis of acyl silanes, successful for aliphatic, aromatic, heteroaromatic, α-alkoxy, α-amino and even α-chiral and α-cyclopropyl acyl silanes[77]. Acyl chlorides are treated with lithium tetrakis(trimethylsilyl)aluminium or lithium methyl tris(trimethylsilyl) aluminium in the presence of copper(I) cyanide as catalyst to give the acyl silanes in excellent yields after work-up[77]. Later improvements include the use of 2-pyridinethiolesters in place of acyl halides, allowing preparation of acyl silanes in just a few minutes in very high yields indeed (Scheme 9)[80], and the use of bis(dimethylphenylsilyl) copper lithium[78] and a dimethylphenylsilyl zinc cuprate species[79,80] as nucleophiles.

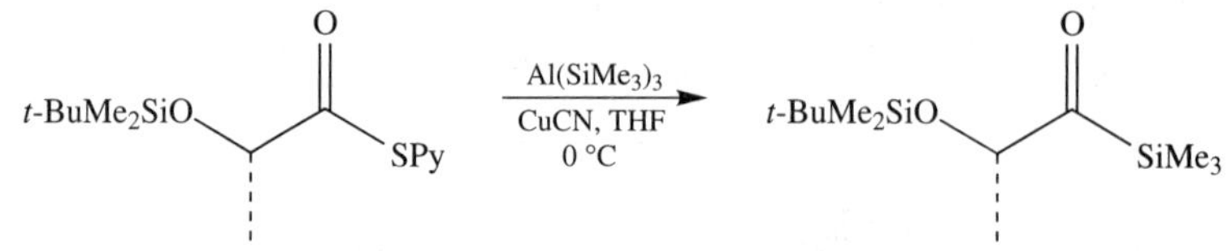

$$t\text{-BuMe}_2\text{SiO}-CH-C(=O)-SPy \xrightarrow[\text{CuCN, THF}]{\text{Al(SiMe}_3)_3} t\text{-BuMe}_2\text{SiO}-CH-C(=O)-SiMe_3$$
$$0\,°C$$

SCHEME 9

Triphenylsilyl zirconium and hafnium derivatives have been prepared from the silyl lithium species. Triphenylsilyl bis(cyclopentadienyl) zirconium chloride undergoes carbonyl insertion under pressure of carbon monoxide (100 psi) to give the corresponding acyl zirconium species which, upon treatment with anhydrous hydrogen chloride in a benzene matrix at $-196\,°C$ and warming to room temperature, gives rise to formyl triphenylsilane[55].

4. Palladium-catalysed coupling

Many aromatic and heteroaromatic acyl silanes have been prepared by transition metal catalysed coupling (e.g. Scheme 10)[23,81]. This is a very successful approach for most aromatic substrates, including furyl, thienyl, pyrryl and electron-deficient aryl acyl silanes, which can otherwise be difficult to prepare.

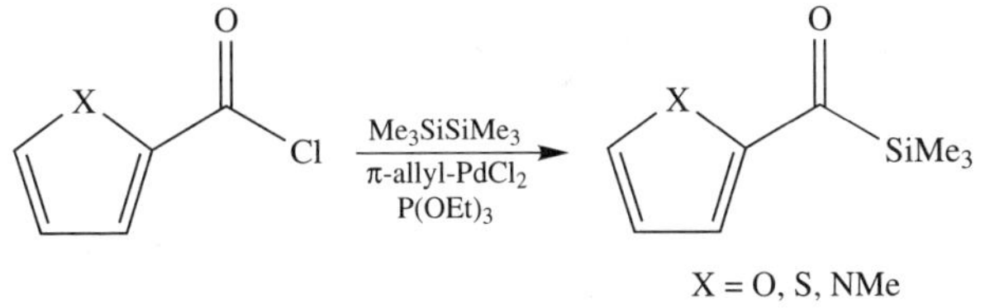

SCHEME 10

Palladium-mediated addition of silyl stannane reagents to alkynyl ethers has been employed for the synthesis of aliphatic acyl silanes in very good yields via the intermediate α-alkoxy-β-stannyl vinyl silanes (enol ethers of acyl silanes)[82]. In a second palladium-catalysed step, the vinyl stannane moiety could be coupled to suitable halides before hydrolysis to the acyl silanes with trifluoroacetic acid (Scheme 11).

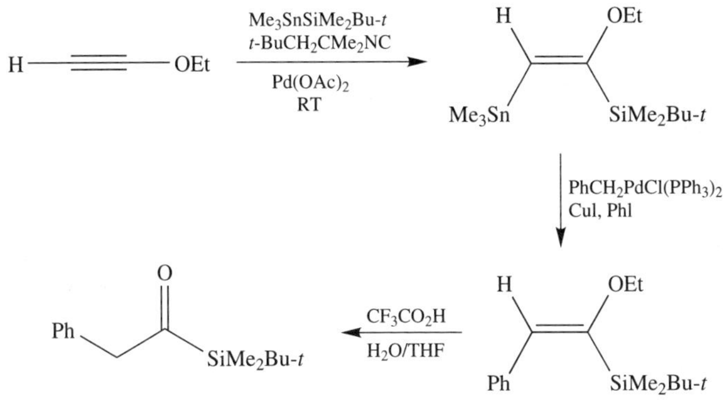

SCHEME 11

5. Oxidation reactions

The oxidation of secondary alcohols to ketones by any of a wide variety of oxidizing agents is a standard reaction, and it might therefore be supposed that the corresponding oxidation of α-hydroxy silanes would be effective for acyl silane preparation[5]. This approach is, however, commonly less than straightforward. A number of oxidizing systems have been examined[83], but the most satisfactory method for oxidation of α-hydroxy silanes to acyl silanes remains the Swern procedure[47,48,54], as used recently to prepare a series of acyl triisopropylsilanes as shown in Scheme 12, where the α-hydroxy silanes were produced by nucleophilic ring cleavage of triisopropylsilyl oxirane[84].

Use of chromic acid-based oxidation reagents often results in silicon–carbon bond cleavage (Scheme 13)[5], although such an oxidation has been used to produce α-cyclopropyl acyl silanes (*vide infra*, Section III.E)[85]. In an interesting development, this handicap of ready silicon–carbon bond cleavage during oxidation of α-hydroxy silanes has been cleverly utilized in an acyl silane synthesis by incorporation of two silicon moieties in the substrate (Scheme 14).

Several benzoyl silanes have been generated in good yields from the corresponding α,α-dibromobenzyl silanes by treatment with the oxidant silver acetate in acetone–ethanol–water mixture[4], or by simple hydrolysis with silica gel[86]. Indeed, the

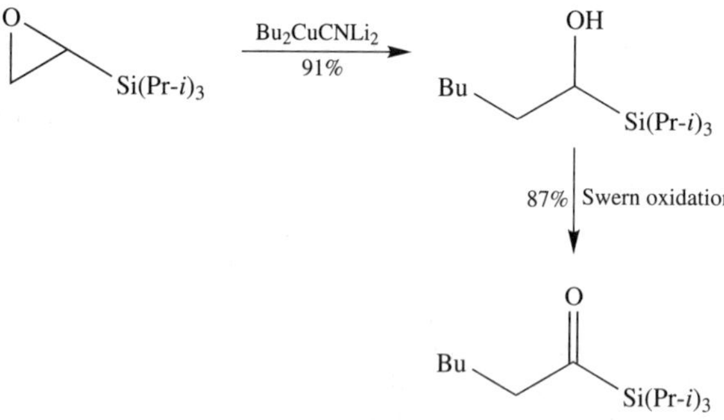

SCHEME 12

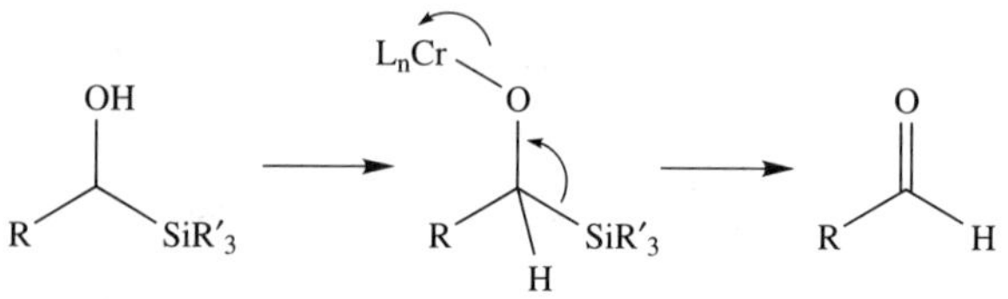

SCHEME 13

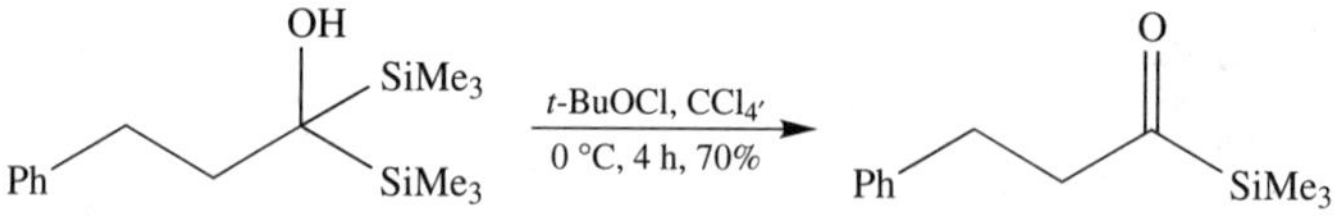

SCHEME 14

very first synthesis of an acyl silane was achieved using this method[4], and Brook has even prepared the diacyl silane **6** in this way[4], but the process is of course restricted to systems where geminal dihalides can easily be procured. Several unusual cyclic acyl silanes such as **7** have also been prepared using this approach[87].

A number of acyl silanes have been isolated from the photosensitized oxygenation of silyl diazo compounds using *meso*-tetraphenylporphine (TPP) as the sensitizer (Scheme 15)[15]. This synthesis is most useful for aromatic acyl silanes and the yellow α-carboxyacyl silanes, but it is not of general applicability, isolation difficulties commonly being encountered in the synthesis of aliphatic and other acyl silanes, resulting in poor overall yields.

A much more general method for acyl silane synthesis involving silyl diazo intermediates is illustrated in Scheme 16[88]. The lithiated derivative of trimethylsilyl diazomethane reacts smoothly with alkyl halides in THF solution to give α-trimethylsilyl diazoalkanes in good yields. Oxidative cleavage of the diazo moiety is effected using 3-chloroperbenzoic acid in benzene solution, to give access to a wide variety of acyl silanes in yields of up to 71%. A phosphate buffer (pH 7.6) is used to prevent side reactions. Aromatic acyl silanes clearly cannot be prepared by this chemistry since an aromatic nucleophilic substitution reaction would be required.

Several aromatic acyl silanes and the fascinating but unstable pink carbonyl bis(trimethylsilane) have been prepared in reasonably good yields by the oxidation of phosphonium ylids (Scheme 17)[13].

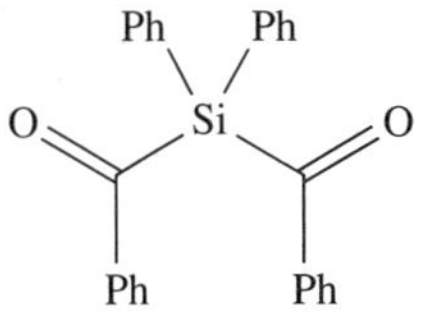

(6) 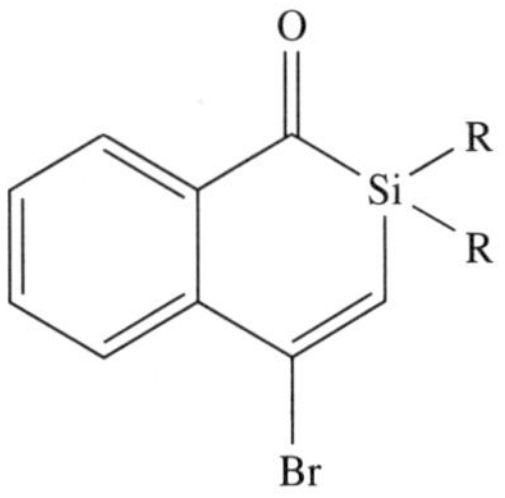 **(7)**

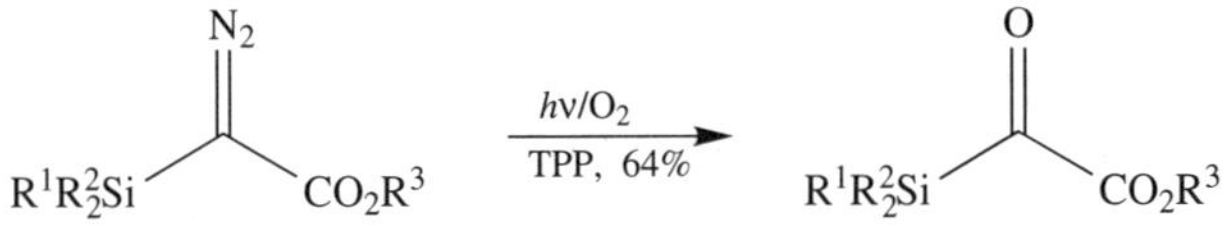

SCHEME 15

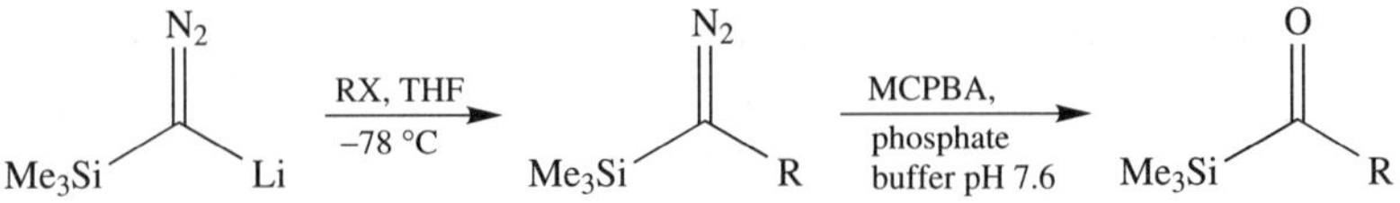

SCHEME 16

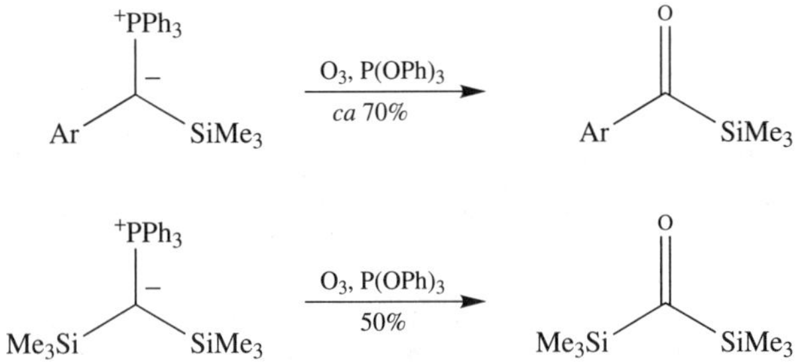

SCHEME 17

6. Hydroboration–oxidation of alkynyl silanes

Hydroboration–oxidation of alkynyl silanes is an excellent general method for synthesis of substituted acyl silanes from readily available starting materials[89,90]. Although the system as originally studied gave only moderate yields, the sequence was later modified by Zweifel and Miller to produce a superior and high-yielding one-pot synthesis of alkyl acyl silanes (Scheme 18), limited only by the reduced yields obtained from very bulky substrates[8]. The use of alkaline peroxide in the oxidation step must be avoided as the acyl silane product is converted into silyl ester by this reagent. It is the opinion of the present authors that this method remains the best general procedure for the preparation of simple aliphatic acyl silanes; aryl acyl silanes obviously cannot be prepared in this way.

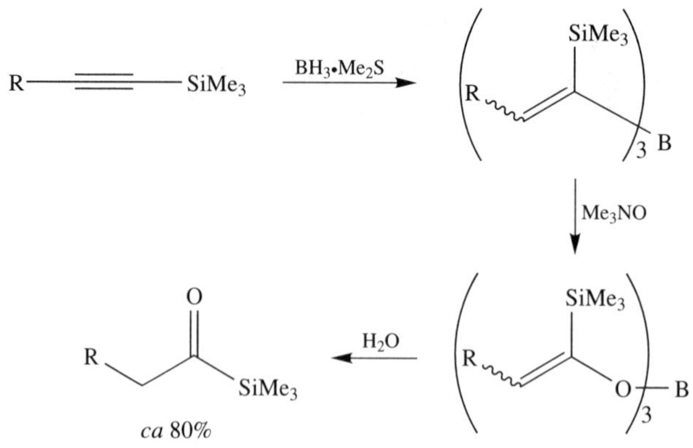

SCHEME 18

7. Rearrangement of silyloxy carbenes

Acyl silanes have been obtained by the pyrolytic rearrangement of silyloxy carbenes derived from α-keto silyl esters (Scheme 19)[91]. The pyrolysis takes place in high yield for aroyl silanes, but is less effective for other substrates.

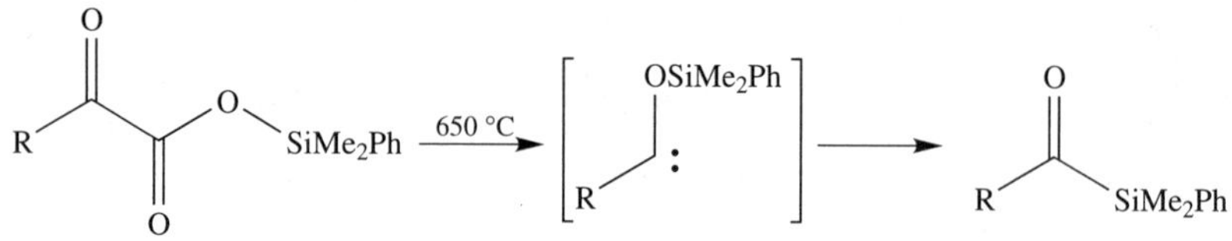

SCHEME 19

8. Preparation from enol ethers

Trialkylsilyl enol ethers of acyl silanes have been prepared using a variety of routes and can be excellent precursors to acyl silanes through simple hydrolysis.

Acyl imidazoles take part in a silyl acyloin reaction to give the corresponding silyl enol ethers in moderate yields. A possible mechanism is outlined in Scheme 20[92]. The silyl enol ethers could be hydrolysed to acyl silanes by treatment with acid.

Lithium alkoxides of bis(trimethylsilyl) carbinols react with benzophenone to produce silyl enol ethers of acyl silanes in good yields[93]. The alcohols were prepared in reasonable yields by hydrolysis of the bis(trimethylsilyl) carbinol silyl ethers[94,95], which in turn were produced from the corresponding esters using another silyl acyloin reaction, which itself, ironically, proceeds through an acyl silane intermediate (Scheme 21)[94].

Picard has reported more direct approaches to acyl silanes and to their silyl enol ethers by reductive silylation of substituted benzoates and of α,β-dihalo-α,β-unsaturated acyl chlorides, respectively, using a similar reagent mixture of trimethylchlorosilane, magnesium and HMPA[96].

A much more general synthesis of these silyl enol ethers, however, is based on the reductive cleavage of the carbon–sulphur bond of the silyl enol ether of a thiolester using sodium metal and chlorotrimethylsilane, once again in a silyl acyloin reaction (Scheme 22)[97,98].

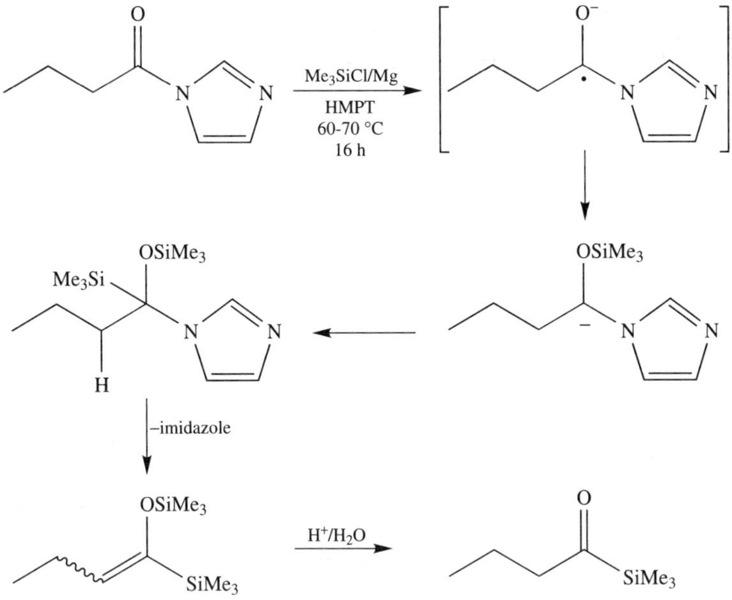

27. Acyl silanes

SCHEME 20

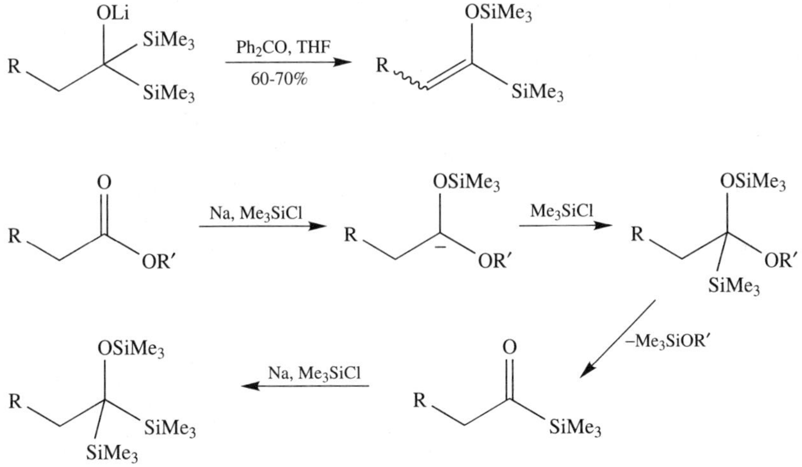

SCHEME 21

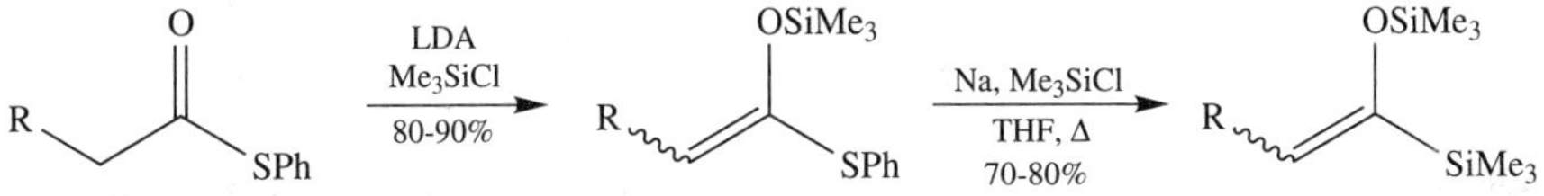

SCHEME 22

A further efficient preparation of these silyl enol ethers proceeds through an intramolecular 1,2-silicon shift in an α-silyl acyl lithium substrate **(8)**, prepared from an α-lithiosilane (Scheme 23)[99]. This method appears very simple to carry out, and produces the silyl enol ethers in good yields with high isomeric purity (usually, E isomer > 90% of mixture).

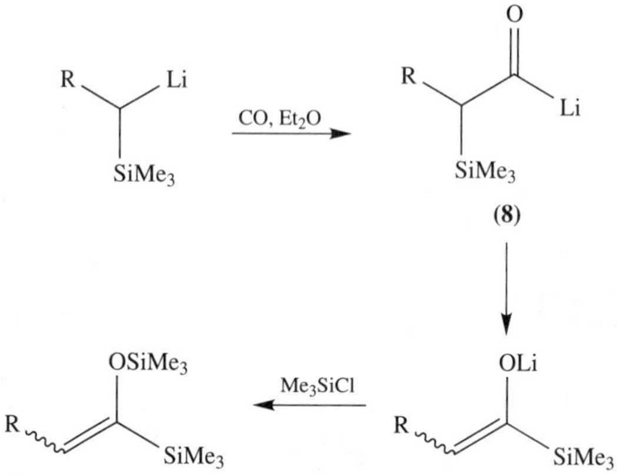

SCHEME 23

In a rather more unusual process, presumably involving tellurium–lithium exchange, acyl tellurides may be converted into silyl enol ethers of acyl silanes by treatment with butyl lithium and trimethylchlorosilane. In this procedure it is the Z isomer which is the predominant product (Scheme 24)[100].

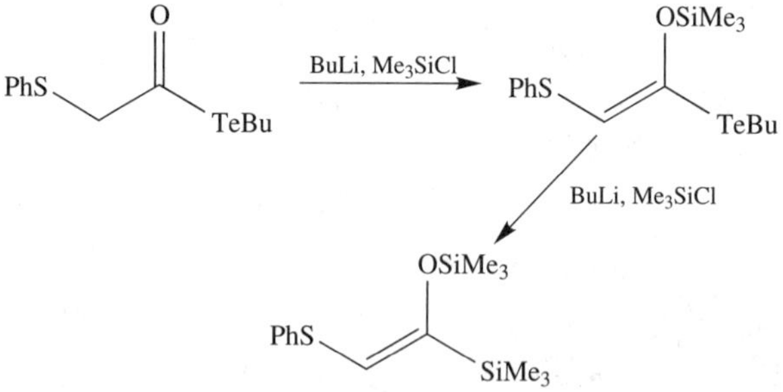

SCHEME 24

Another metal–metal exchange procedure involves tin–lithium exchange in 1-silyloxy vinyl stannanes (silyl enol ethers of acyl stannanes) induced by butyl lithium. The resulting lithio species suffers migration of the silicon moiety from oxygen to carbon to generate the lithium enolates of acyl silanes, which were shown to undergo enolate alkylation and aldol reaction, so producing new, functionalized acyl silanes (Scheme 25)[101].

A number of silyl enol ethers of acyl silanes have been produced from alkenes by subjection to 50 atmospheres of carbon monoxide in the presence of 0.1 equivalents of trialkylsilane and 2 mol% of an iridium catalyst (Scheme 26)[102]. Hydrolysis to the acyl silanes was achieved using hydrochloric acid–acetone.

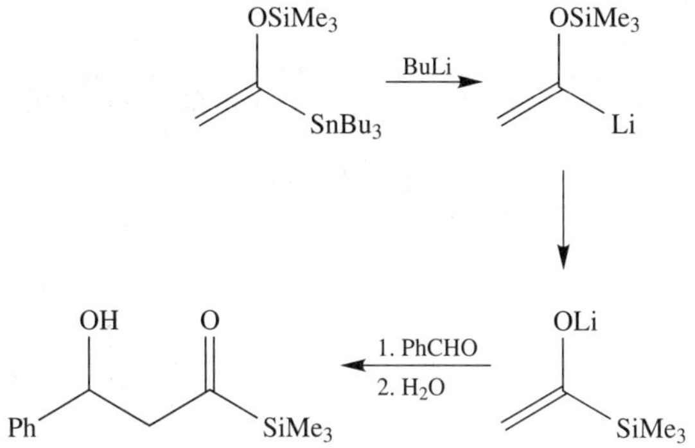

27. Acyl silanes

SCHEME 25

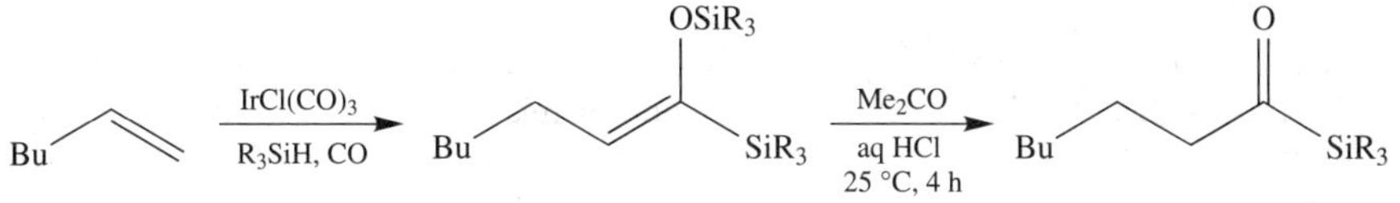

SCHEME 26

Methyl enol ethers of acyl silanes have been prepared in good yield by the silylation of vinyl lithium reagents derived from methyl enol ethers[103]. Indeed, perhaps the simplest preparation of a methyl enol ether of an acyl silane results from addition of α-methoxyvinyl lithium to chlorotrimethylsilane. Acid hydrolysis gave acetyl trimethylsilane in ca 80% yield[104]. A similar reaction has been carried out with phenyl methyl t-butyl chlorosilane. Again, acid hydrolysis gave the acyl silane, which is of course chiral at silicon[64].

In another simple procedure, deprotonation of methoxy bis(trimethylsilyl)methane with butyl lithium and addition of the resulting anion to aldehydes induces Peterson elimination (Scheme 27). The product methyl enol ethers could be hydrolysed to the parent acyl silanes with hydrochloric acid–THF or could be treated with electrophiles such as N-halosuccinimides to give α-haloacyl silanes[105]. Alternatively, treatment with phenyl selenenyl chloride, oxidation at selenium and selenoxide elimination afforded α,β-unsaturated acyl silanes.

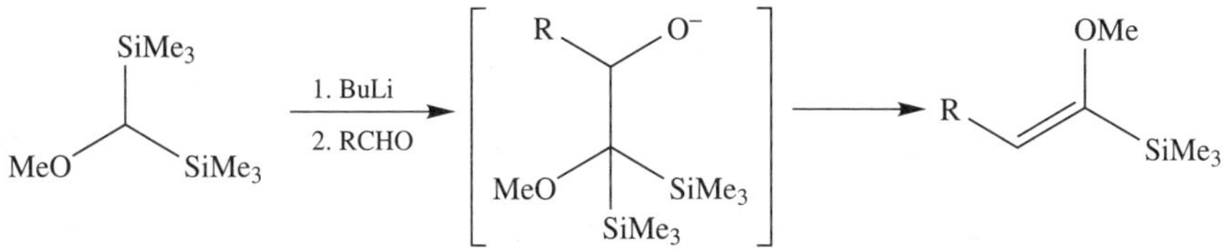

SCHEME 27

Ethyl enol ethers of acyl silanes have been prepared by the palladium-mediated addition of silyl stannanes to alkynyl ethers. Hydrolysis using trifluoroacetic acid gave very high yields of acyl silanes (*vide supra*, Section III.A.4)[82].

9. Silylation of acyl metallic species

Perhaps the most direct method of synthesizing an acyl silane is by reaction of an acyl lithium, prepared by carbonylation of an alkyl lithium at $-110\,^\circ\text{C}$, with a silicon electrophile, illustrated in Scheme 28[106,107]. Although this method is successful for a variety of alkyl acyl silanes in moderate yields, low temperatures must be used, and the method is not suitable for aryl acyl silanes.

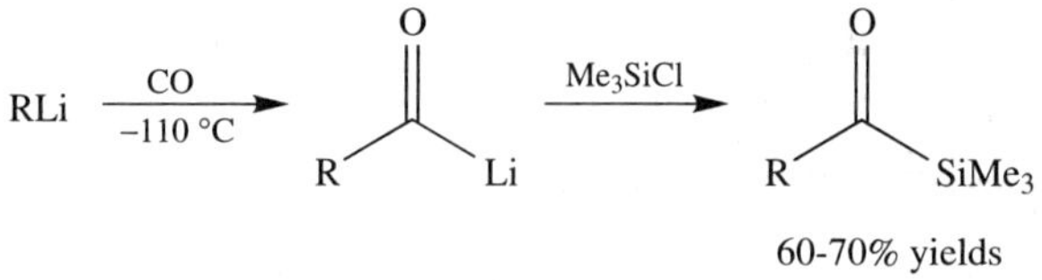

$$RLi \xrightarrow[-110\,^\circ C]{CO} R\underset{Li}{\overset{O}{\|}} \xrightarrow{Me_3SiCl} R\underset{SiMe_3}{\overset{O}{\|}}$$

60-70% yields

SCHEME 28

Several acyl silanes have been prepared by the silylation of metalloaldimines followed by hydrolysis (Scheme 29)[108,109]. One limitation of this scheme is the ready decomposition of the aldimine to give aldehyde in addition to acyl silane in approximately equal amounts.

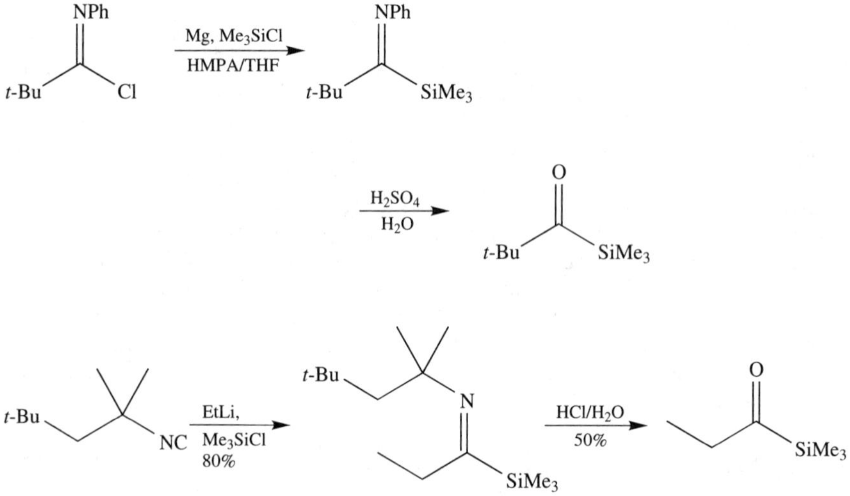

SCHEME 29

N,N-Diethylcarbamoyl trimethylsilane has been prepared by the reaction of bis(trimethylsilyl) sulphide with bis(N,N-diethylcarbamoyl) mercury (Scheme 30)[16]. Silylation of the carbamoyl cuprate reagent derived from a lithium amide, by addition of copper(I) cyanide and subsequent exposure to carbon monoxide (1 atm), is also effective[75,110]. Poor to moderate yields of carbamoyl silanes may be isolated by treatment of lithium silylamides with carbon monoxide and methyl iodide, in a reaction sequence involving a nitrogen to carbon silyl shift in an intramolecular silylation (Scheme 31)[111].

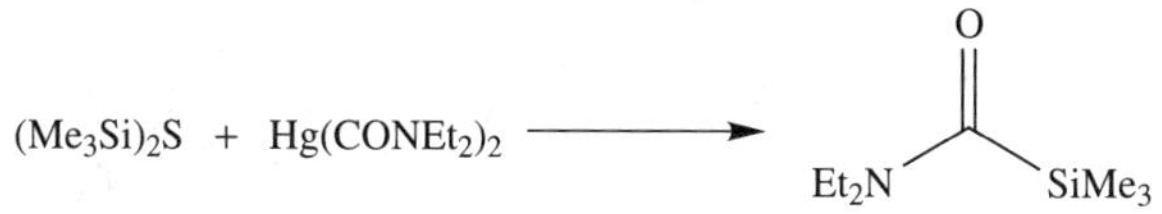

$$(Me_3Si)_2S + Hg(CONEt_2)_2 \longrightarrow Et_2N-C(=O)-SiMe_3$$

SCHEME 30

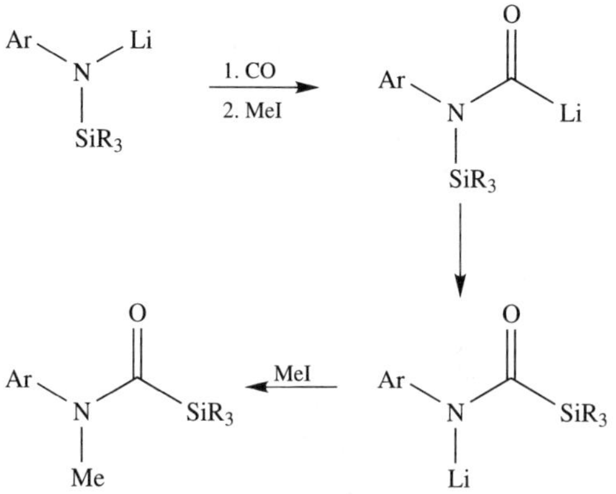

SCHEME 31

10. Non-racemic acyl silanes

A number of acyl trimethyl silanes chiral at the α- or β-carbon atom have been prepared in non-racemic form. Chiral α-alkoxy and α-silyloxy acyl silanes have been generated in very high yields by oxidative rearrangement of enantiomerically pure silyl epoxides, induced by dimethyl sulphoxide and silyl triflates (Scheme 32)[112].

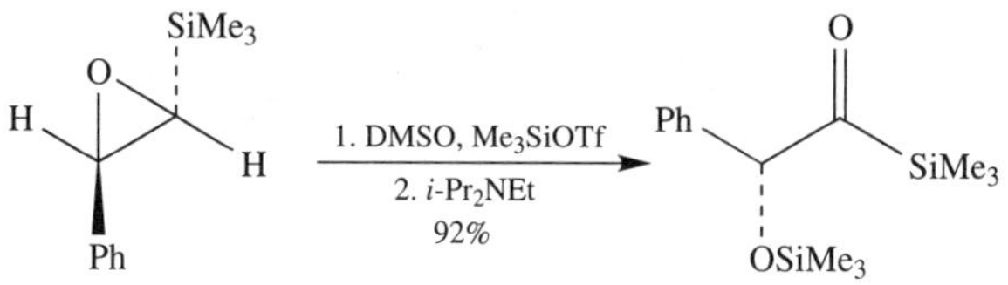

SCHEME 32

Chiral β-amino acyl silanes have been prepared through the addition of 2-lithio-2-trimethylsilyl-1,3-dithiane to enantiomerically pure N-tosylaziridines followed by mercury-mediated thioacetal hydrolysis[113].

Enantioselective Claisen rearrangement of allyl (α-trimethylsilyl)vinyl ethers in the presence of aluminium binaphthol derivatives gives β-chiral γ, δ-unsaturated acyl silanes with good ee (Scheme 33)[114].

B. α-Haloacyl Silanes

The most direct procedure for the synthesis of α-haloacyl silanes is electrophilic halogenation of enolates or enol ethers of acyl silanes. This has been achieved with the silyl enol ethers using bromine at low temperatures, but the reaction suffers from the general

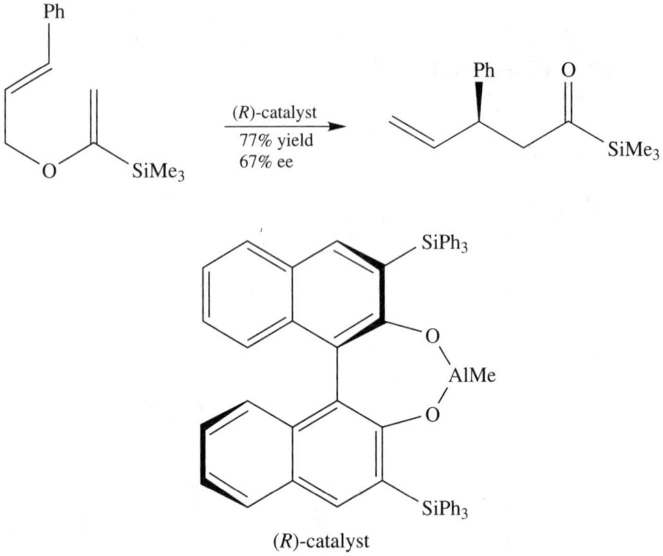

SCHEME 33

sensitivity of both the starting materials and the products[12,115]. Alkyl enol ethers of acyl *t*-butyldimethylsilanes, prepared by deprotonation and silylation of vinyl ethers[103], are also successful substrates[105,116]. α-Iodoacyl silanes were prepared by the same authors by treatment of α-bromoacyl silanes with sodium iodide in acetone.

Bis(trimethylsilyl)carbinols have been reported to react with *N*-bromosuccinimide in carbon tetrachloride to give the corresponding α-haloacyl silanes in moderate yields (Scheme 34)[115].

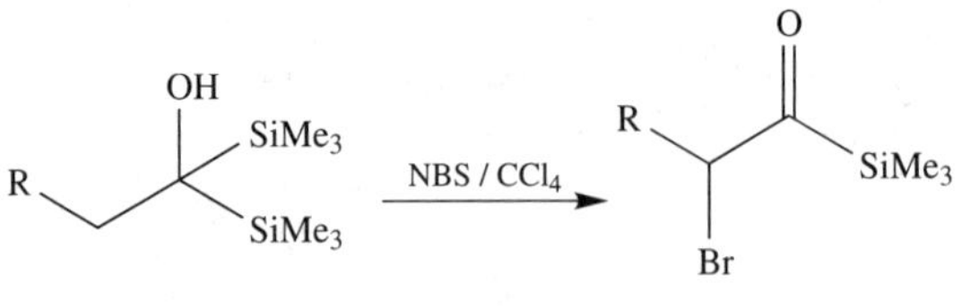

SCHEME 34

Perhaps the best general method to date for preparing α-haloacyl silanes involves bromination of silyl enol borinates (9) at 0 °C, a reaction which proceeds in good yield and involves no sensitive intermediates. This route offers a most convenient one-pot synthesis of α-haloacyl silanes from readily available starting materials, as the intermediate enol borinates are very easily prepared from silyl acetylenes (Scheme 35)[7,117,118].

α-Bromoacyl silanes may be isolated in variable yields by bromination–rearrangement of 1-phenylsulphonyl-1-trimethylsilyl oxiranes induced by magnesium bromide etherate (Scheme 36)[119,120]. The process appears to be rather sensitive towards the reaction conditions used, and can give rise to a number of by-products. In a recently-reported related rearrangement, 1-halo-1-trimethylsilyl oxiranes are converted into α-haloacyl silanes upon exposure to Lewis acids, generally in high yields[121].

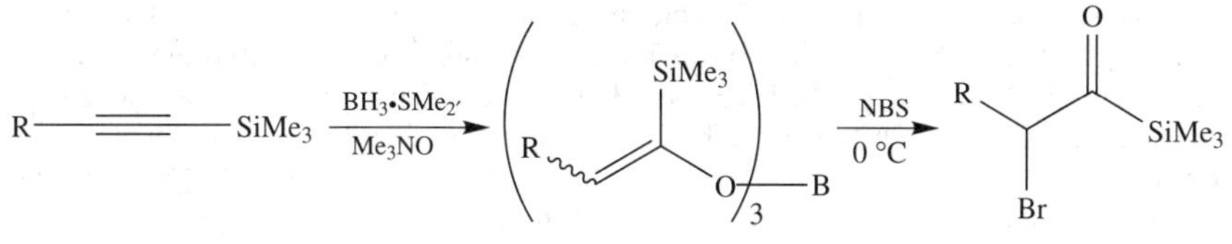

(9)

50-60% overall

SCHEME 35

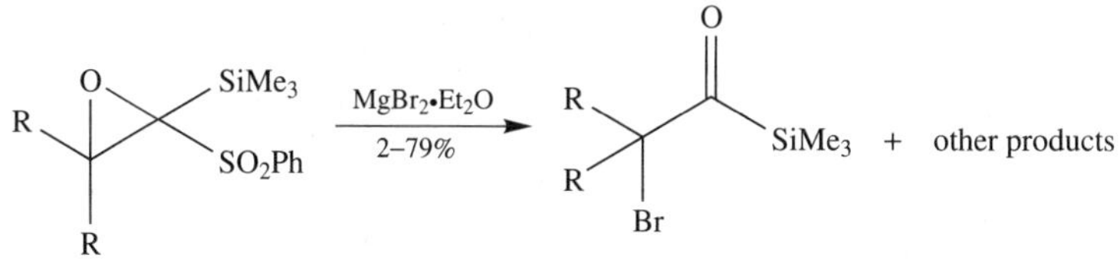

SCHEME 36

The pale yellow crystalline trifluoroacetyl triphenylsilane and a number of other perfluoroalkyl acyl triphenylsilanes have been prepared by simple acylation of triphenylsilyl lithium with perfluoroalkyl anhydrides[122].

C. α-Ketoacyl Silanes

α-Ketoacyl silanes have a deep rich crimson colour and are particularly sensitive to light, often requiring purification by chromatography at −78 °C, preferably in the dark. They were first isolated in 1982 using allene methodology as shown in Scheme 37[14], although the yellow α-carboxyacyl silanes were already known at that time (*vide supra*, Section III.A.4)[15]. Thus, lithiation of allene **(10)** with butyl lithium at −78 °C and subsequent reaction with a chlorosilane, followed by oxidative work-up with *m*-chloroperbenzoic acid, provided the unstable α-ketoacyl silanes in moderate yields, presumably via the epoxides[123].

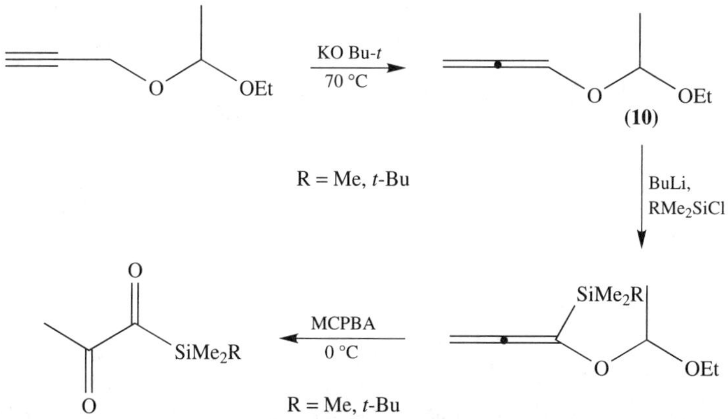

(10)

R = Me, *t*-Bu

R = Me, *t*-Bu

SCHEME 37

A much more general and very simple synthesis requiring a minimum of laboratory manipulation uses a Swern oxidation of the corresponding diols to give the α-ketoacyl silanes directly in useful yields (Scheme 38)[7,124]. Purification in this case was accomplished in the dark, by chromatography at −78 °C or by distillation. α-Ketoacyl silanes appear to be intermediates in the oxidation of silyl acetylenes to α-ketoesters by osmium tetroxide[125], and indeed have been isolated from the oxidation of silyl acetylenes by dimethyl dioxirane[126].

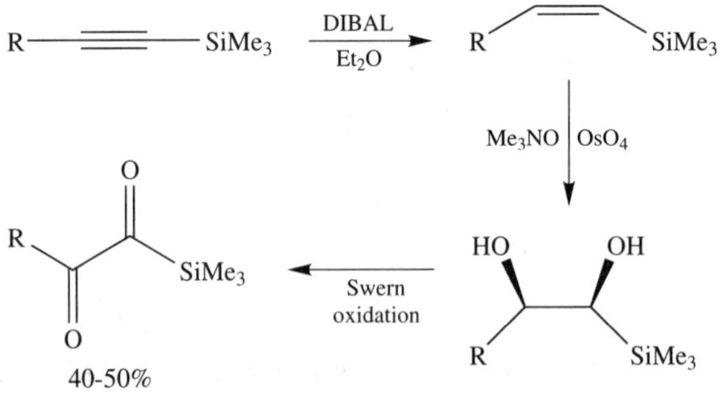

SCHEME 38

D. α,β-Unsaturated Acyl Silanes

α,β-Unsaturated acyl silanes, which are yellow or yellow-green in colour, are generally less sensitive than those types of acyl silane discussed above and have been prepared through a number of different approaches.

1. Hydroboration–oxidation of enynes

α,β-Unsaturated acyl silanes have been prepared by hydroboration methodology, similar to that used in the synthesis of aliphatic acyl silanes (*vide supra*) (Scheme 39)[89]. This synthesis is effective, but suffers from the relative difficulty of synthesis of the necessary functionalized enynes.

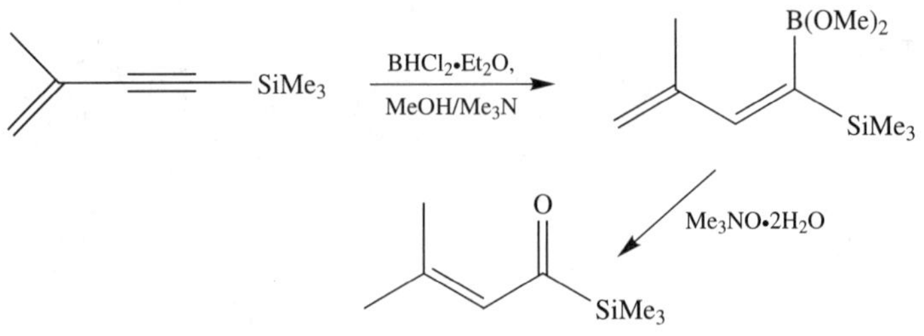

SCHEME 39

2. Oxidation of allylic carbinols

An excellent synthesis of α,β-unsaturated acyl silanes from allyl silyl ethers is shown in Scheme 40[20]. This simple two-step procedure hinges on the Wittig rearrangement[127,128], and is successful on a large scale. The metallation of allyl silyl ethers generates a rapidly interconverting mixture of two organometallic species, **11** and **12**. Although alkylation of this mixture of organometallic derivatives generally proceeds at the C-3 position via **11**, hard electrophiles such as protons react predominantly at the oxygen atom of the alkoxide intermediates (**12**), leading to α-hydroxy allyl silanes (**13**). Swern oxidation leads to α,β-unsaturated acyl silanes.

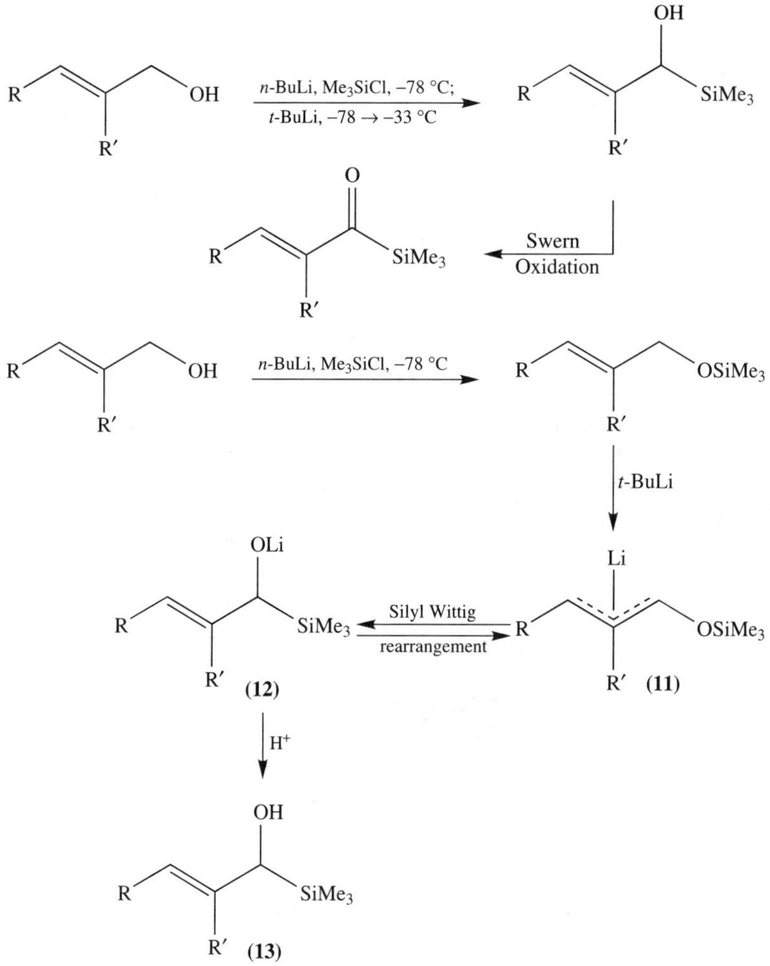

SCHEME 40

3. Preparation from enol ethers of acyl silanes

A number of α,β-unsaturated acyl silanes have been prepared from silyl enol ethers of acyl silanes (Scheme 41)[129]. Addition of phenyl sulphenyl chloride to the silyl enol ether with subsequent elimination of chlorotrimethylsilane gives the α-(phenylthio)acyl silane. Oxidation to the sulphoxide followed by *in situ* elimination of benzenesulphenic acid produces the α,β-unsaturated acyl silane in good yield. A similar sequence has been carried out with the corresponding methyl enol ethers[105].

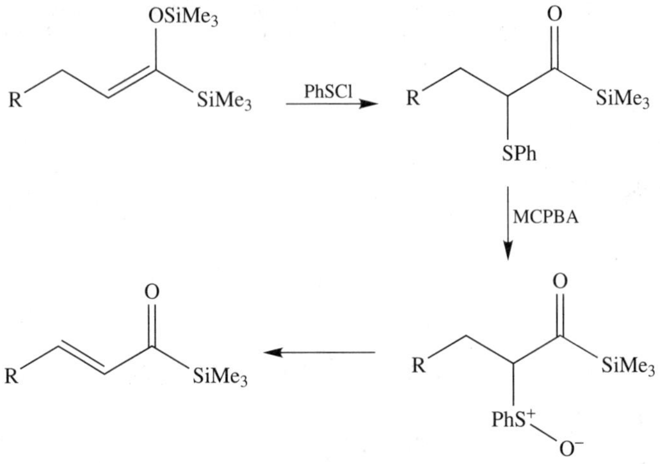

SCHEME 41

One of the simplest methods for preparation of an α,β-unsaturated acyl silane is by hydrolysis of a 1-alkoxy-1-trimethylsilylbutadiene, the conjugated dienol ether of α,β-unsaturated acyl silane, prepared by deprotonation and alkylation of the 1-alkoxydiene (Scheme 42)[11]. This method is generally limited in application to simple substrates, presumably due to the complexity of preparation of more highly functionalized 1-alkoxydienes.

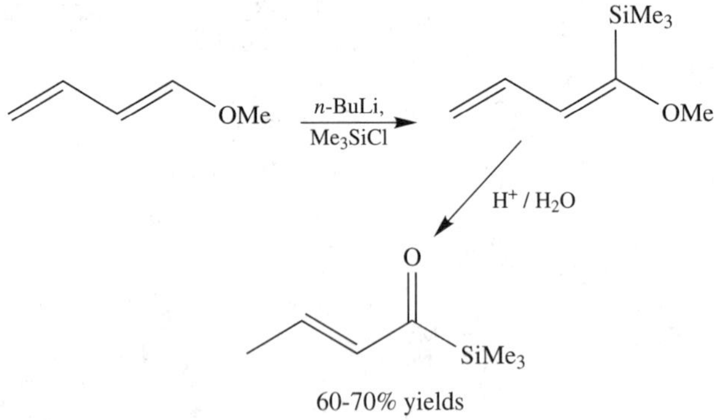

60-70% yields

SCHEME 42

The most versatile synthesis of α,β-unsaturated acyl silanes involves the use of allene methodology, developed by a number of groups[14,22]. Deprotonation and silylation of allenyl ethers followed by hydrolysis gives rise directly to α,β-unsaturated acyl silanes via their enol ethers, 1-alkoxy-1-trimethylsilylallenes (Scheme 43). Indeed, the first example of an α,β-unsaturated acyl silane was prepared by such a route[22a], as was the first example of an allenic acyl silane (from a 1-trimethylsilyl-1-trimethylsilyloxy-1,2,3-alkatriene)[22b].

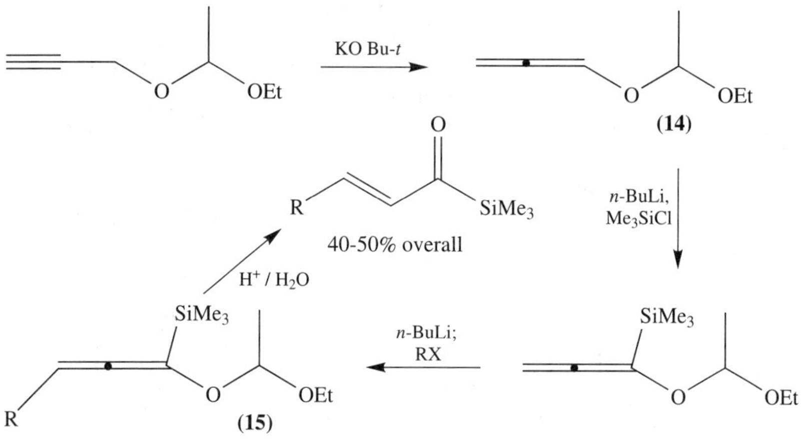

SCHEME 43

Reich uses ethoxyethyl propargyl ether as precursor to the allenyl ether (14), because the relatively large and polar protecting group gives it much better handling characteristics than simpler analogues[14]. The resulting 1-alkoxy-1-trimethylsilylallene may be further deprotonated and functionalized at the C-3 terminus to give the substituted derivatives (15) for hydrolysis to 3-substituted α,β-unsaturated acyl silanes. The 1-trimethylsilyl-1-alkoxyallenes (15) are also excellent precursors to α-halo and α-selenenyl α,β-unsaturated acyl silanes (Scheme 44)[14].

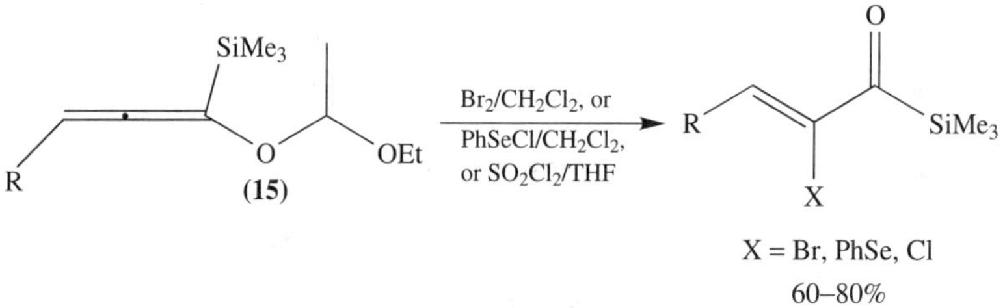

SCHEME 44

In a further development of this approach, the synthesis of α,β-acetylenic acyl silanes has been achieved as shown in Scheme 45[14]. Oxidation of the 3-selenenyl allenyl ethers (16) with m-chloroperbenzoic acid at −78 °C gave the corresponding unstable selenoxides, which underwent in situ [2,3] sigmatropic shift producing acetals (17). Loss of selenenyl ester on work-up gave the α,β-acetylenic acyl silanes in ca 50% yields.

1-Alkoxy-1-trimethylsilylallenes also undergo a Lewis acid-induced rearrangement to give 2-substituted α,β-unsaturated acyl silanes in reasonable yields (Scheme 46)[130]. The related 1-methylthio-1-trialkylsilylallenes undergo Lewis acid-induced aldol and Mukaiyama reactions to produce 2-alkoxyalkylated α,β-unsaturated acyl silanes (Scheme 47)[131].

Lewis acid-mediated addition of (phenylthio)trimethylsilane to acryloyl silane takes place to give 1,3-bis(phenylthio)-1-trimethylsilylprop-1-ene (18). This compound may be deprotonated with t-butyl lithium at the β-position and alkylated to give a range

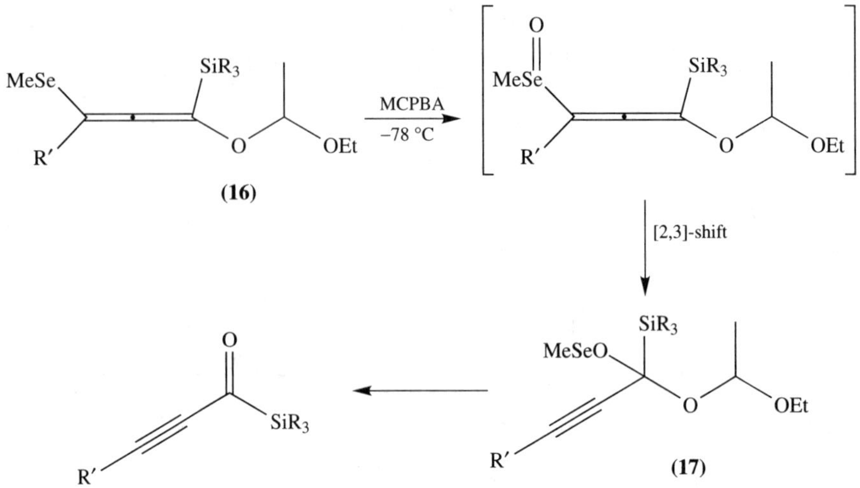

SCHEME 45

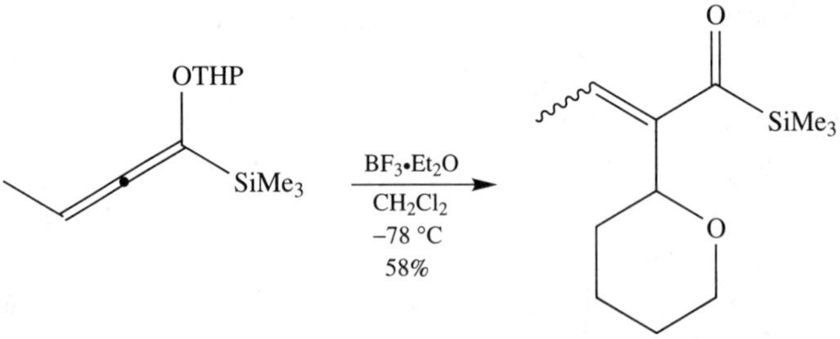

SCHEME 46

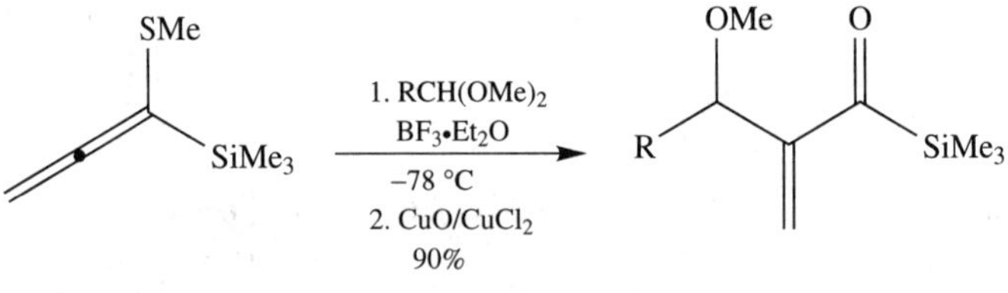

SCHEME 47

of substituted derivatives, which may be converted into β-substituted α,β-unsaturated acyl silanes by hydrolysis–elimination with mercuric chloride in aqueous acetonitrile (Scheme 48) (*vide infra*, Section IV.D)[132]. It should be noted that the β-substituent has here been introduced as an electrophile, complementing other methods of preparation of β-substituted α,β-unsaturated acyl silanes.

α,β-Unsaturated acyl silanes have been prepared in a stereospecific manner from α-trimethylsilylacyl trimethylsilanes by an interesting aldol–Peterson reaction sequence via intermediate **19** (Scheme 49)[133,134].

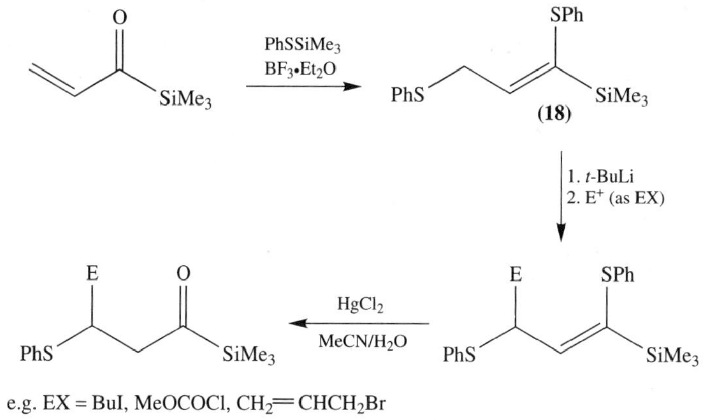

SCHEME 48

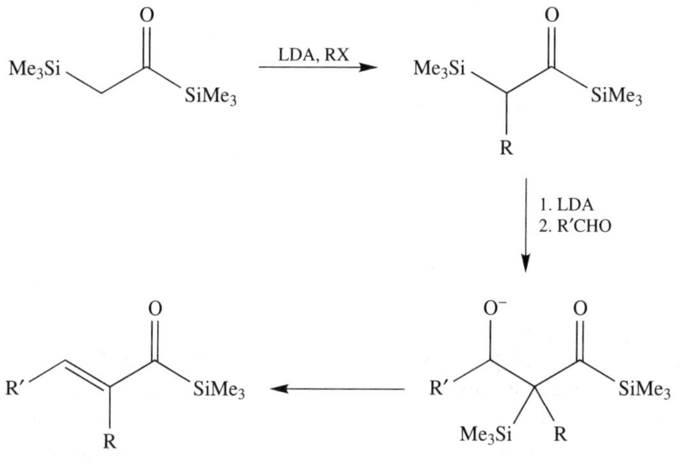

e.g. EX = BuI, MeOCOCl, CH$_2$=CHCH$_2$Br

SCHEME 49

Photochemically-induced addition of bromotrichloromethane to 1-ethoxy-1-trimethylsilylethene, the ethyl enol ether of acetyl trimethylsilane, generates a 1 : 1 adduct which provides 3,3-dichloropropenoyl trimethylsilane (**20**) on solvolysis. Treatment of this material with lithium alkyl cyanocuprates resulted in addition–elimination to give the E-isomers of the 3-substituted α,β-unsaturated acyl silane products (Scheme 50)[135].

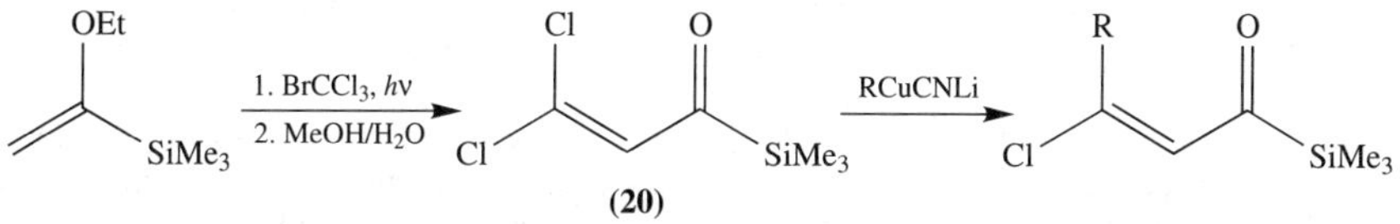

SCHEME 50

4. Preparation from α,β-alkynyl acyl silanes

α,β-Acetylenic acyl triphenylsilanes have been used by the Degl'Innocenti group as precursors of α,β-unsaturated and other acyl silanes[136]. Several reaction types have been utilized. Simple conjugate addition of lithium dialkyl cuprates is successful and proceeds in good to excellent yields[137]. Conjugate addition of heteroatomic nucleophiles is also successful[138], including the addition of a tin nucleophile to give 3-(tributylstannyl)propenoyl triphenylsilane. This last example is interesting because it allows subsequent palladium-catalysed coupling to produce structurally more complex α,β-unsaturated acyl silanes, for example with extended conjugation (Scheme 51)[139]. The β-iodo derivative has been similarly used[140].

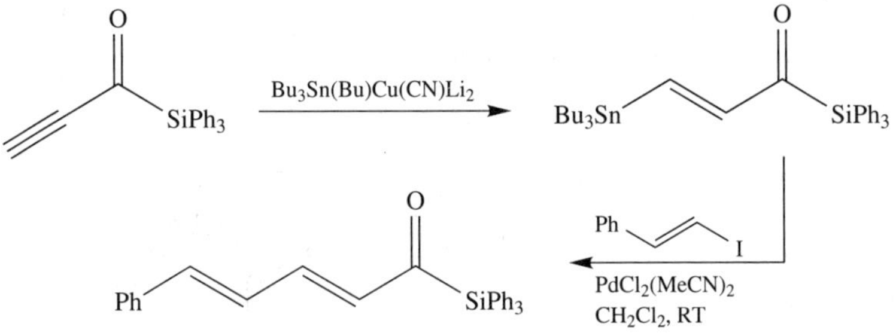

SCHEME 51

5. Miscellaneous methods

A limited number of functionalized acyl silanes have been prepared by the use of 1,3-dithiane methodology as described above[46,141], but this route is unfavourable for the synthesis of α,β-unsaturated acyl silanes[20]. Several β-hydroxy acyl silanes have, however, been produced in high yields using this chemistry; acid-catalysed elimination of water from these compounds does give α,β-unsaturated acyl silanes, and Swern oxidation gives β-ketoacyl silanes (Scheme 52)[142]. Analogous acetylenic 1,3-dioxanes have been used as precursors to α,β-acetylenic acyl silanes[143], as have formyl silanes (vide supra, Section III.A.1)[54].

Nowick and Danheiser have employed the Horner–Emmons reaction of α-phosphonoacyl silanes to prepare α,β-unsaturated acyl silanes in 54–97% yields[116]. The α-phosphonoacetyl silane intermediate (**21**), prepared from α-iodoacetyl t-butyldimethylsilane through the Arbuzov reaction, undergoes enolate alkylation, for example using potassium t-butoxide and methyl iodide; the alkylated products also underwent Horner–Emmons reaction (Scheme 53).

Isomerization of a β-trimethylsilylpropargyl alcohol to give the α,β-unsaturated acyl silane via the intermediate enol form, the 1-hydroxy-1-trimethylsilylallene, occurs upon treatment with tetrabutylammonium per-rhenate and p-toluenesulphonic acid[144]. While several examples of the isomerization reaction are given, only one is able to give an acyl silane product.

In an interesting and unusual reaction, Paquette and Maynard have shown that peracid oxidation of 1,2-disilyl-3,3-dimethylcyclopropene gives rise to an α-silyl-α,β-unsaturated acyl silane, through rearrangement of the intermediate epoxycyclopropane. Further treatment with peracid gave the α-epoxy acyl silane (Scheme 54)[145].

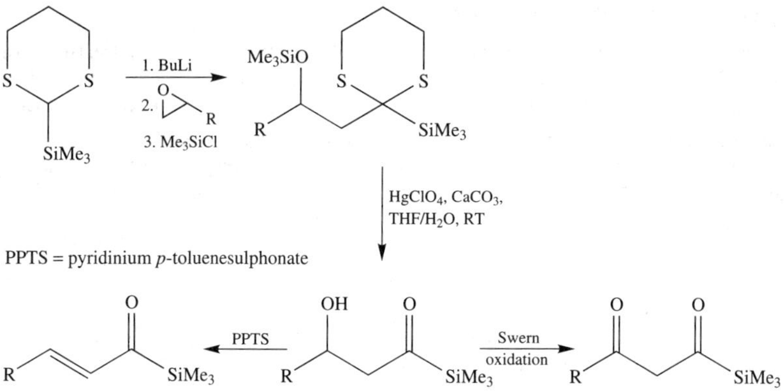

27. Acyl silanes

PPTS = pyridinium *p*-toluenesulphonate

SCHEME 52

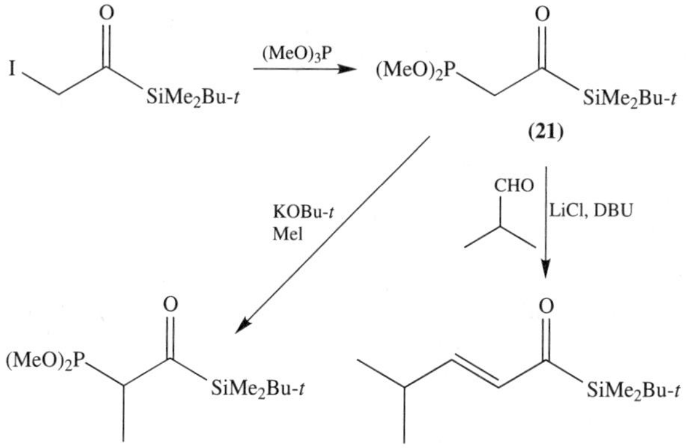

SCHEME 53

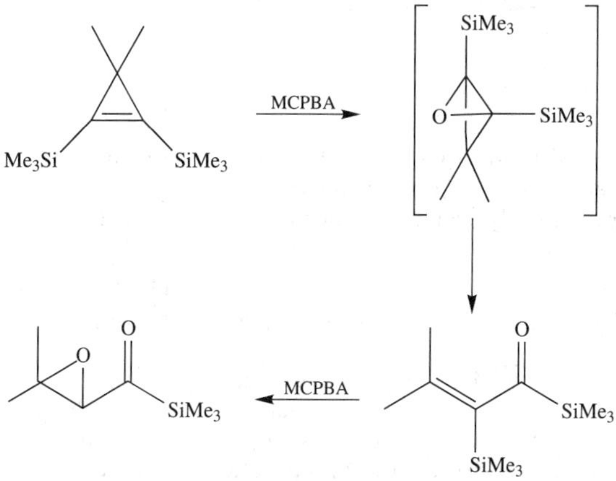

SCHEME 54

E. α-Cyclopropyl and α-Epoxyacyl Silanes

The first α-cyclopropyl acyl silanes to be isolated were generated by treatment of α,β-unsaturated acyl silanes with diazomethane, followed by vapour-phase pyrolysis of the intermediate pyrazoline derivatives (*vide infra*, Section IV.D)[141]. They suffer acid-induced cleavage or rearrangement under more mild conditions than do their carbon analogues[146].

Wittig rearrangement of allyl silyl ethers, followed by Simmons–Smith cyclopropanation and Collins oxidation, produces α-cyclopropyl acyl silanes, e.g. **22**, in 10–85% yields (Scheme 55)[85].

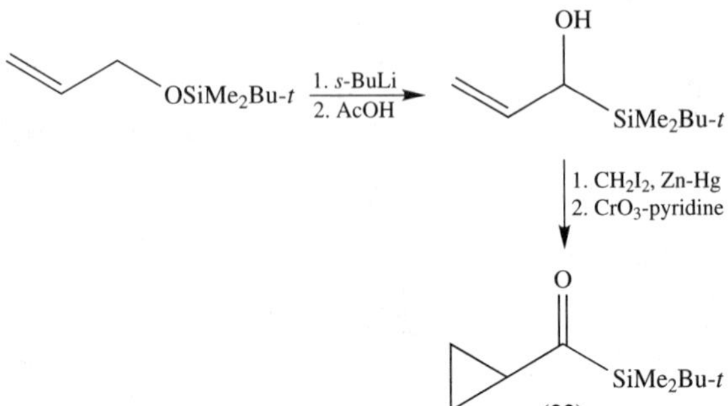

SCHEME 55

Nakajima has shown that α-cyclopropyl acyl silane (**23**) results from reaction of 1-trimethylsilyl cyclopropyl lithium with dichloromethyl methyl ether at low temperature in THF solution, in a reaction said to involve a carbene intermediate and a 1,2-silicon shift (Scheme 56)[147].

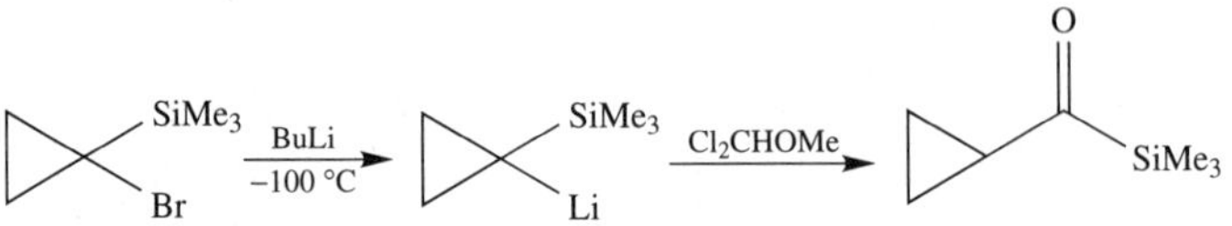

SCHEME 56

One general method for acyl silane synthesis particularly successful for α-cyclopropyl examples (and even an α-cyclobutyl example) involves treatment of acid chlorides with lithium tetrakis(trimethylsilyl) aluminum or lithium methyl tris(trimethylsilyl) aluminium and cuprous cyanide (*vide supra*, Section III.A.3)[77]. For example, cyclopropyl acyl silane (**23**) was obtained in 89% yield by this process. Improved procedures use lithium *t*-butyldimethylsilyl cuprate[78] and a dimethylphenylsilyl zinc cuprate species[79,80] as reagents.

Nowick and Danheiser[148] have explored α-cyclopropyl acyl silane generation from α-haloacyl silanes through McCoy reactions (Scheme 57) and via sulphur ylids (Scheme 58). Ylid species such as **24** were found to be stable in aprotic solvents in the presence of lithium salts, and were used for the cyclopropanation of α,β-unsaturated aldehydes.

α-Epoxy acyl silanes may be prepared by simple epoxidation of α,β-unsaturated acyl silanes[145].

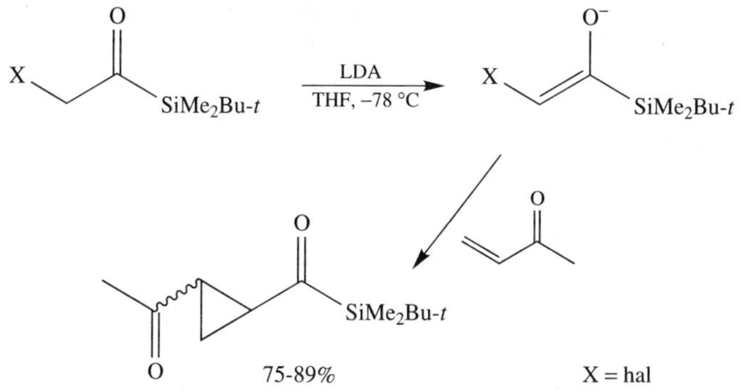

SCHEME 57

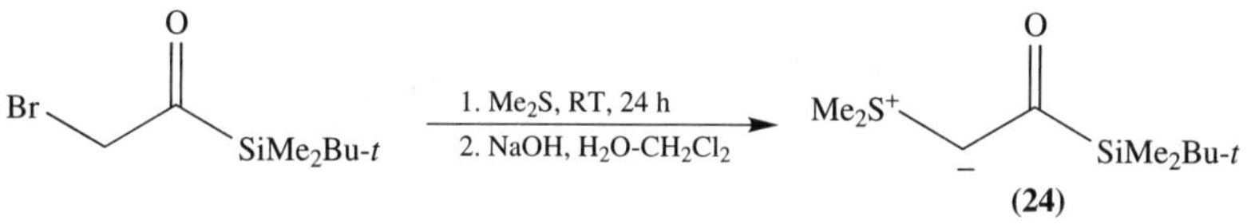

SCHEME 58

IV. REACTIONS OF ACYL SILANES

A. General Reactivity

Acyl silanes, although sensitive to light and to basic media, behave as typical ketones when treated with a wide variety of reagents. Some examples are shown in Scheme 59[5,149−152].

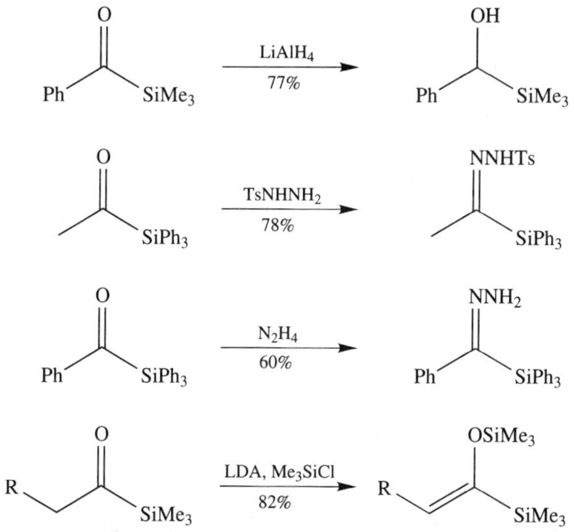

SCHEME 59

Acyl silanes often, however, exhibit abnormal behaviour, for example involving rearrangements leading to silicon–oxygen bond formation, especially when treated with nucleophilic reagents[5,61,149,150].

1. Nucleophilic addition

Acyl silanes are extremely sensitive towards nucleophiles and nucleophilic bases[1,5]; for example, alcoholic solutions of benzoyl triphenylsilane containing a trace of aqueous hydroxide ion rapidly produce triphenylsilanol and benzaldehyde[1,5]. Three reasonable mechanisms may be conceived for this reaction (Scheme 60): S_N2 displacement at the silicon atom (path A); nucleophilic attack at the carbonyl carbon atom followed by Brook rearrangement, initially to give a hemiacetal (path B); and nucleophilic attack at the silicon atom to form a pentacoordinate silicon anionic intermediate, followed by migration of the nucleophile to the carbonyl group and subsequent Brook rearrangement (path C).

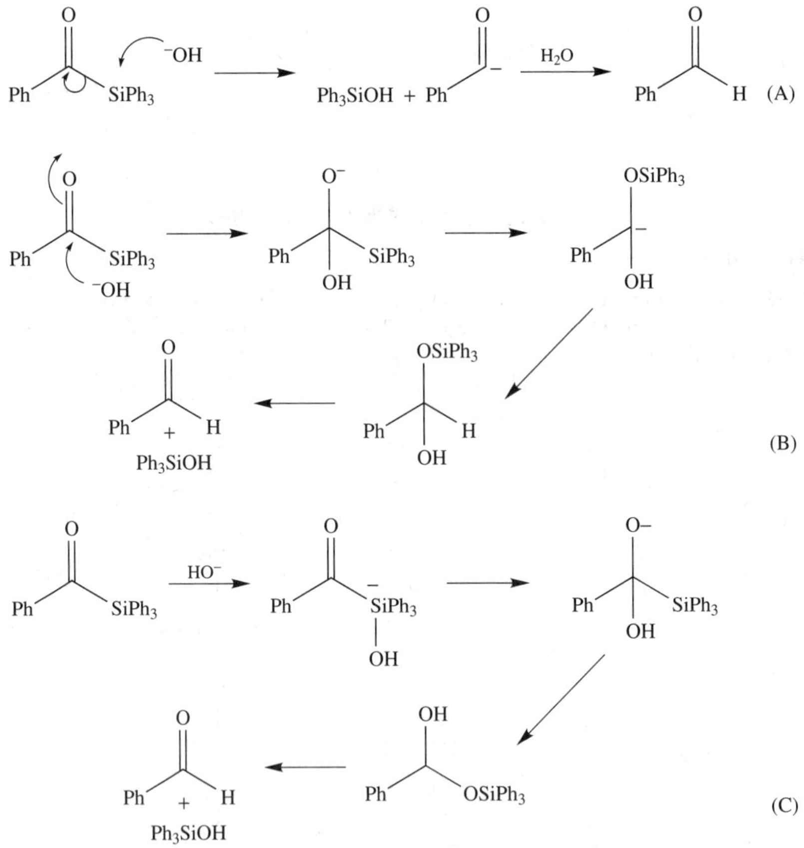

SCHEME 60

The 1,2-migration of a silicon moiety to the oxygen anion of a carbon–oxygen single bond produced by nucleophilic addition to a carbonyl group (Brook rearrangement)[1,5,153,154] is a very common pathway by which acyl silanes react when treated with nucleophiles, the major driving force presumably being formation of the strong Si—O bond. In seminal work, Brook has shown by using various enantiomerically pure acyl silanes, chiral at silicon, that this rearrangement usually occurs with retention of configuration at the silicon atom (*vide infra*)[153]. The stereochemical course of the Brook rearrangement and that of a multitude of other reactions involving nucleophilic additions to silicon atoms can be accounted for if a pentacovalent trigonal bipyramidal silicon intermediate is involved in the substitution process[154b,155]. In the simplest case, inversion of configuration is observed when the intermediate is both formed (**25 → 26**) and undergoes decomposition (**26 → 27**) without pseudorotation taking place (Scheme 61). The base-mediated solvolysis of acyl silanes has been studied in detail by Ricci who, following various kinetic measurements, have suggested that the probable reaction pathway involves direct attack of hydroxide ion at the carbonyl group (Scheme 60, Path B), the rate-determining step being the migration of the trialkylsilyl group from carbon to oxygen[156].

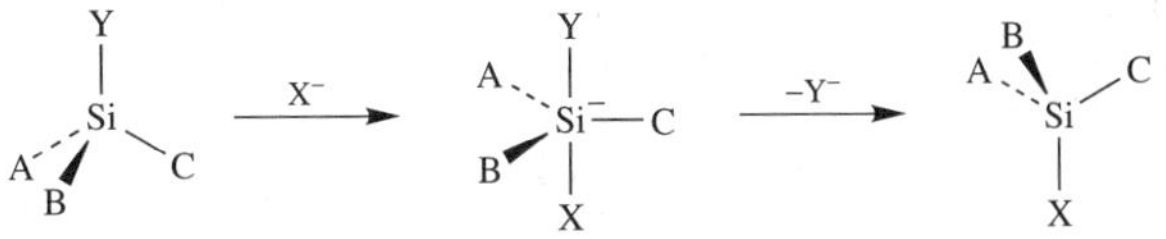

SCHEME 61

The reaction of acyl silanes with alkoxide ions has been studied in great detail (Scheme 62)[157,158]. Again, the reaction pathway may be rationalized by invoking a nucleophilic attack by alkoxide ion at the silicon atom of the ketone, giving a pentacoordinate silicon anionic species which can suffer 1,2-migration of an alkyl group from the silicon atom to the carbonyl carbon atom to give the alkoxide ion. This intermediate then undergoes a Brook rearrangement to give the unsymmetrical dialkoxysilane (**28**) after protonation, which is usually the major product. Other reaction products, such as the alcohol (**29**) and the dialkoxysilane (**30**), arise from a transetherification reaction between the alkoxide ion and the unsymmetrical dialkoxysilane (**28**). A competing reaction, formally corresponding to nucleophilic displacement of the acyl group from the silicon atom, is also observed. This displacement reaction becomes favoured over rearrangement as the polarity of the solvent system increases[5,159].

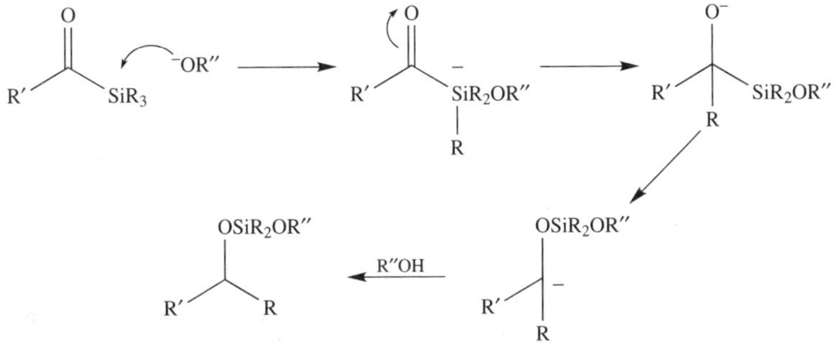

SCHEME 62

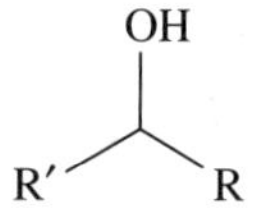

Later elegant work by Brook using *t*-butoxide ion and enantiomerically pure acyl silanes led him to suggest that the cleavage products arise from direct attack of the *t*-butoxide ion at the carbonyl group followed by Brook rearrangement[58]. The evidence for this proposal is outlined in Scheme 63 for (*R*)-acetyl 1-naphthyl phenyl methyl silane (**31**). Should cleavage arise via pentacoordinate anion **32**, without pseudorotation, through attack at silicon, then the (*S*)-*t*-butoxy silane (**33**) would be formed with overall inversion of configuration at silicon relative to starting material. Reduction of **33** would lead, with retention of configuration, to (*R*)-(+)-1-naphthyl phenyl methyl silane (**34**, path B), although experiments have shown that, under the reaction conditions employed by Brook, this reduction is at best very slow. Such rearrangements of alkyl acyl silanes have indeed been observed upon treatment with fluoride ion, giving rearranged secondary alcohols upon desilylation[154b,160].

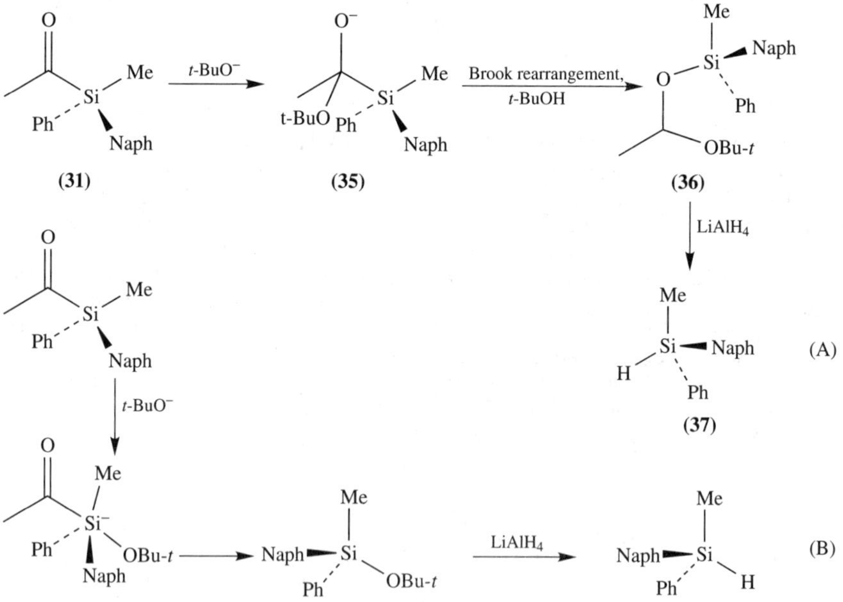

SCHEME 63

Conversely, were nucleophilic attack of the alkoxide ion to occur at the carbonyl group of **31**, then the species formed (**35**) should undergo Brook rearrangement to **36** with retention of configuration at silicon (Path A). Reduction of **36** with lithium aluminium hydride would then produce (*S*)-(−)-1-naphthyl phenyl methyl silane (**37**).

When the reaction was carried out, only the (S)-(−)-silane **37** was isolated from the reduction products of the reaction mixture, leading Brook to suggest that cleavage occurred through nucleophilic attack of the alkoxide ion at the carbonyl group (Path A). No *t*-butoxysilane was detected among the reduction products. Brook does not discuss the possibility of pseudorotation of **32**, which could lead to (R)-*t*-butoxysilane **33** being formed, with overall retention of configuration at the silicon atom, as observed (see Scheme 61). However, the fact that no *t*-butoxysilane was isolated from the reduction products strongly suggests that cleavage does indeed arise through path A. There is in addition the possibility of migration of the *t*-butoxide group from the silicon atom to the carbonyl carbon atom, a sequence which cannot be disproved by these experiments. Such a migration of the *t*-butoxide moiety in **32** seems unlikely, however, as it requires the strong silicon–oxygen bond to be broken[10,75].

In contrast, the carbonyl groups of acyl silanes undergo nucleophilic addition by hydride and by reactive carbon nucleophiles such as Grignard and organolithium reagents to give α-hydroxysilanes. For example, propargyl magnesium, zinc and zinc–copper species add efficiently to aryl and alkyl acyl silanes[161,162], as do allyl magnesium and zinc species, in a reaction used in prostaglandin synthesis[161]. Sila-β-ionone **38**, an intermediate in sila-vitamin A synthesis, has been prepared by addition of an acetylenic Grignard reagent to a cyclic acyl silane[163]. Even Reformatsky reactions are known, although subsequent Brook rearrangement and elimination of trialkylsilanolate, to give alkenes, was observed in some cases (Scheme 64)[164].

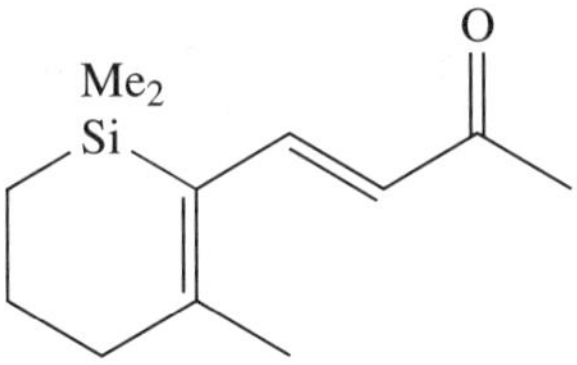

(**38**)

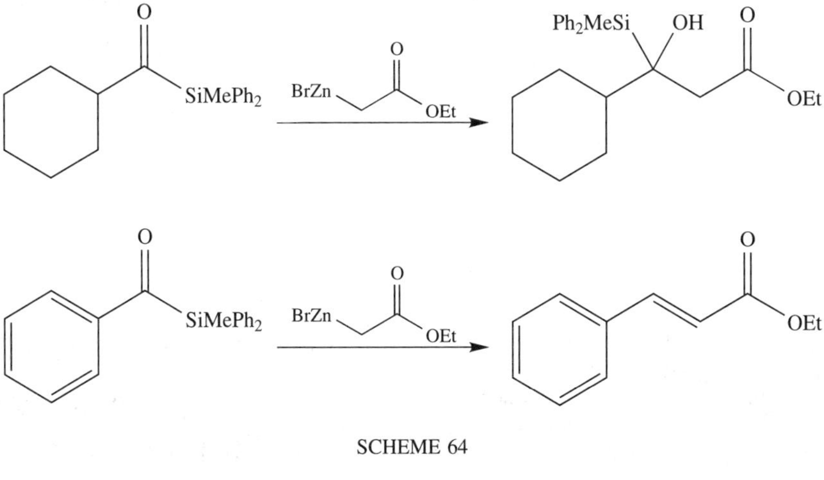

SCHEME 64

Acyl silanes have been used as synthetic equivalents of sterically-hindered aldehydes[165]. Treatment of several aldehydes, including **39** and **40**, with 3-methylpenta-2,4-dienyl lithium gave rise to a mixture of regioisomers. However, when the corresponding acyl silanes **41** and **42** were used, only the conjugated isomers were formed (Scheme 65); selectivity was similar to that observed in the reaction of sterically-hindered ketone **43**. The α-hydroxysilane adducts were readily desilylated with potassium hydride to give the corresponding alcohols, presumably through the Brook rearrangement. Adduct **44** underwent a highly diastereoselective intramolecular Diels–Alder reaction to give alcohol **45** after desilylation without loss of stereochemical integrity (Scheme 66). The authors comment that acyl silanes provide higher overall yields than aldehydes in these reactions as they are less prone to self-condensation and are also superior substrates in that the bulky silyl group may be used for stereocontrol of subsequent reactions.

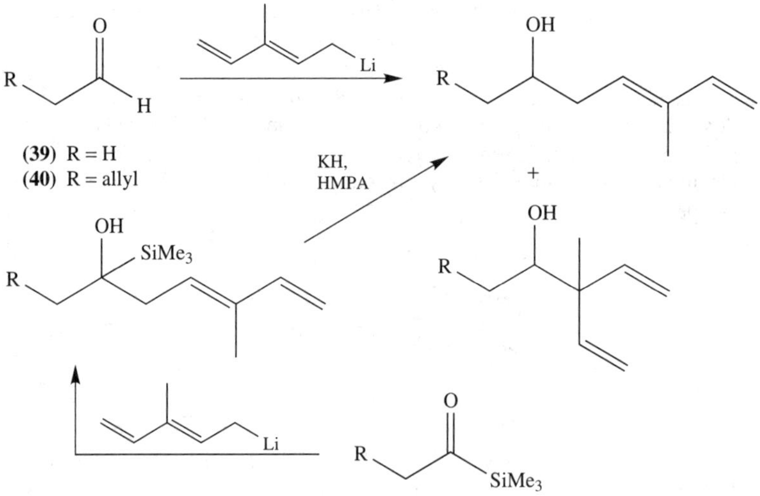

SCHEME 65

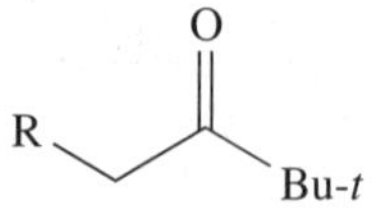

Kuwajima has used acyl silanes as homoenolate equivalents (Scheme 67)[18,166]. Addition of vinyl Grignard reagents give intermediates which subsequently undergo Brook rearrangement to give the homoenolates **46**. Compounds **46** can undergo a further, irreversible, 1,4-silyl group migration, producing the enolates of β-trimethylsilyl ketones **47**, a side-reaction which may be partially suppressed by keeping the temperature low, by using larger alkyl groups attached to the silicon atom, and by using magnesium

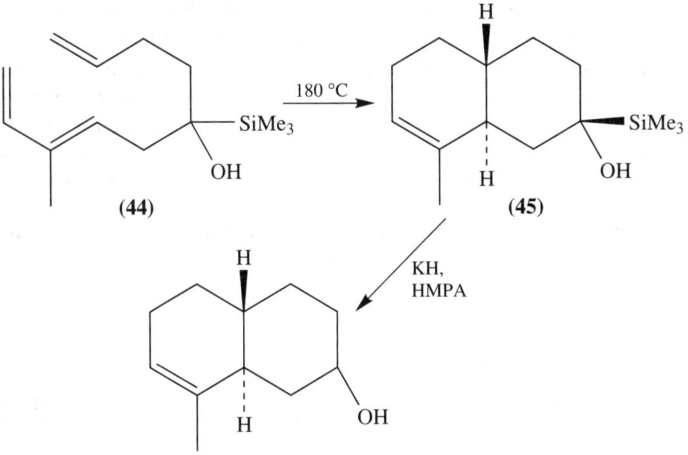

SCHEME 66

instead of lithium enolates. The difference observed between the magnesium and lithium homoenolates may be due to a reduction in the propensity for attack at the silicon atom to form **47**, as a result of the less ionic character of the carbon–magnesium bond. Kuwajima was able to form cuprate reagents from homoenolates **46** by use of copper trimethylsilyl acetylide. The reagents effect various conjugate addition reactions with enones to provide 1,6-dicarbonyl compounds or their equivalents in good overall yields (Scheme 68).

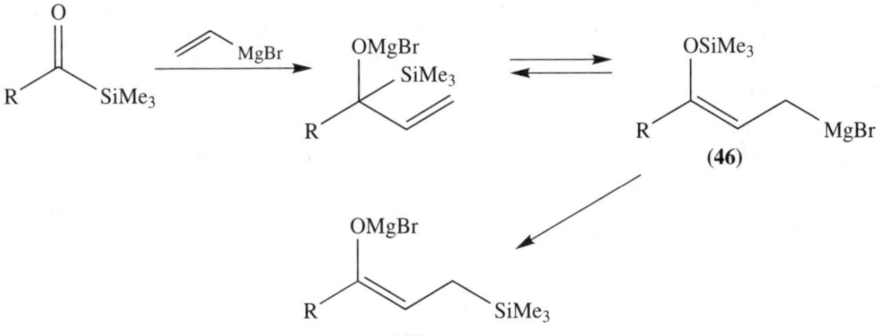

SCHEME 67

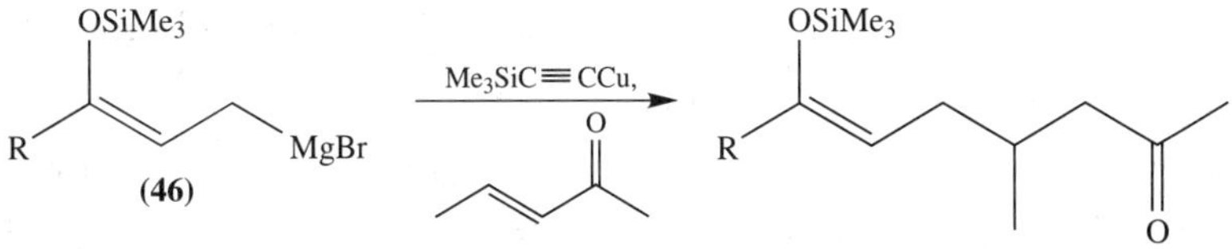

SCHEME 68

The steric bulk of the silyl group in some acyl silanes allows enantioselective reduction of the carbonyl group by asymmetric reducing agents including the Itsuno reagent[167] [a 2 : 1 complex of borane and (S)-2-amino-3-methyl-1,1-diphenylbutan-1-ol][168] and di-isopinocampheyl chloroborane/(+)-α-pinene[169,170], often in better than 95% ee. The resulting α-hydroxy silanes could be converted via their acetates into secondary alcohols of high enantiomeric purity by a thermally-induced stereospecific migration of a substituent from silicon to carbon, followed by oxidative desilylation (Scheme 69)[170,171]. This reduction has been used in a synthesis of (+)-sesbanimide[172]. In a derivative process, treatment of acyl silanes with di-isopinocampheyl allyl borane results in enantioselective formation of allylated α-hydroxy silanes, in some cases in high ee[173].

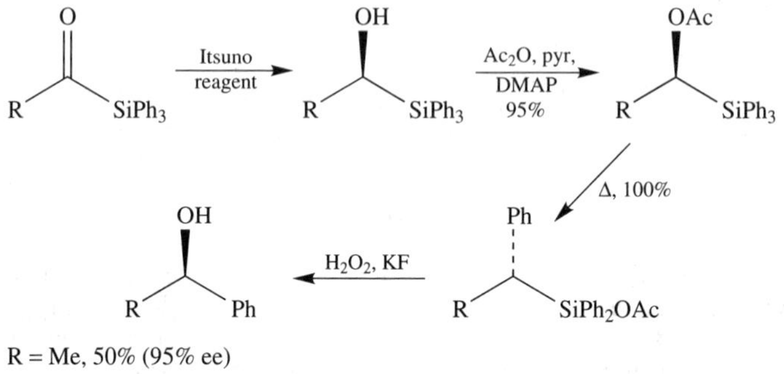

SCHEME 69

Recently, several acyl silanes chiral at silicon have been prepared and shown to undergo diastereoselective addition to the carbonyl groups by hydride, Grignard and organolithium reagents through chelated transition states[174–176]. High diastereoselectivities were observed in some examples, particularly where an alkoxymethyl group was present on the silicon atom (Scheme 70).

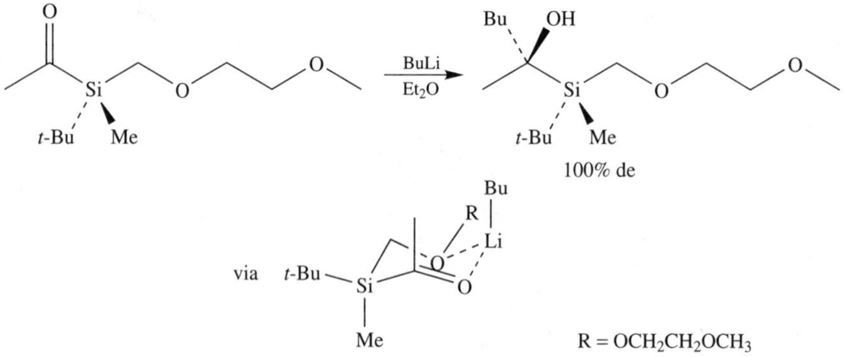

SCHEME 70

Acyl silanes containing chiral centres at the α- and/or β-carbon atoms have also been shown to undergo highly stereoselective addition of organolithium and Grignard reagents, and of various allyl tin and allyl silane reagents in the presence of Lewis acids (Scheme 71). The resulting α-hydroxy silanes, formed in up to 98% yield and with diastereoselectivities up to >100:1, can be protiodesilylated with >99% retention of configuration[177–179].

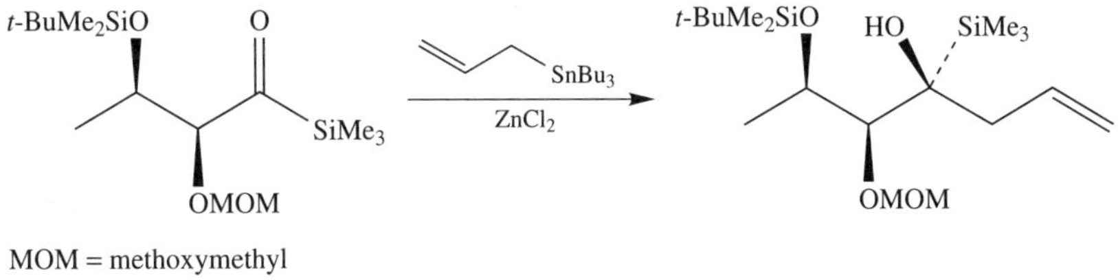

SCHEME 71

MOM = methoxymethyl

Upon treatment with organolithium reagents, α-(phenylthio)acyl silanes give silyl enol ethers with very high *erythro* stereoselectivities (Scheme 72),[180] rationalized by invoking a Felkin–Anh transition state[181]. The alcohols rearrange by a Brook-type migration with concerted expulsion of the phenylthiolate leaving group; because of the stereoelectronic demands of the reaction, the silyl group approaches eclipsing with a hydrogen atom during the reaction pathway leading to carbon to oxygen migration in the major diastereoisomer **48**, but with a benzyl moiety during the reaction pathway leading to the minor, less reactive diastereoisomer **49**.

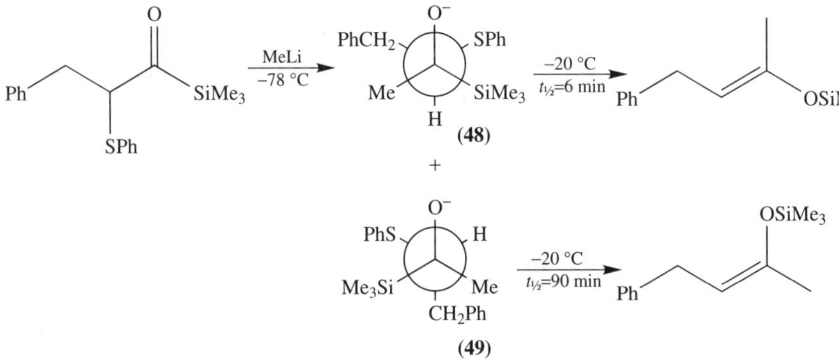

SCHEME 72

Acyl silanes can display disparate behaviour when treated with carbon nucleophiles, even of related types[5,61,149]. For example, when aroyl silanes were treated with a Wittig reagent, none of the expected alkenes was obtained, and the only reaction products isolated were silyl enol ether and triphenylphosphine (Scheme 73)[182,183]. When alkanoyl silanes were treated with Wittig reagents, however, only the normal olefinated vinyl silane products were isolated (Scheme 74)[182–184]. Under soluble lithium salt conditions, Z-vinyl silanes were produced with very high selectivities; the reaction was used to prepare a pheromone component (**50**) of the sweet potato leaf folder moth (Scheme 75)[183].

The pattern of results suggest that, if the substituent at the acyl group is alkyl, and hence relatively carbanion-destabilizing, rearrangement is inhibited relative to the alternative 'normal' Wittig reaction pathway; when the substituent is aromatic, however, and therefore capable of stabilizing incipient carbanion formation as the silicon–carbon bond

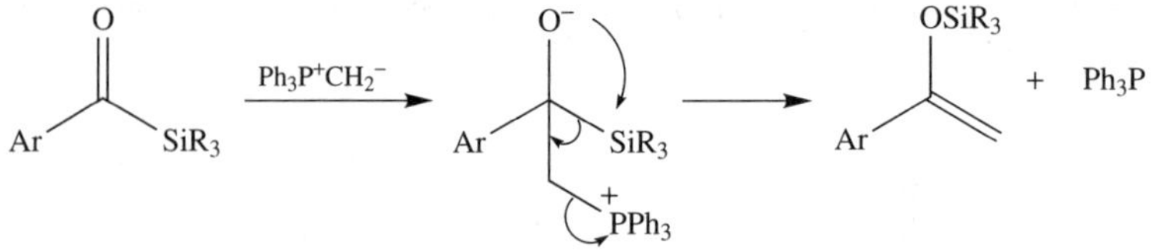

SCHEME 73

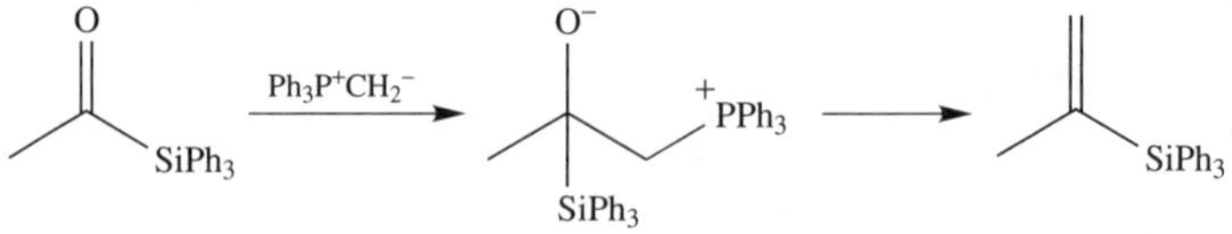

SCHEME 74

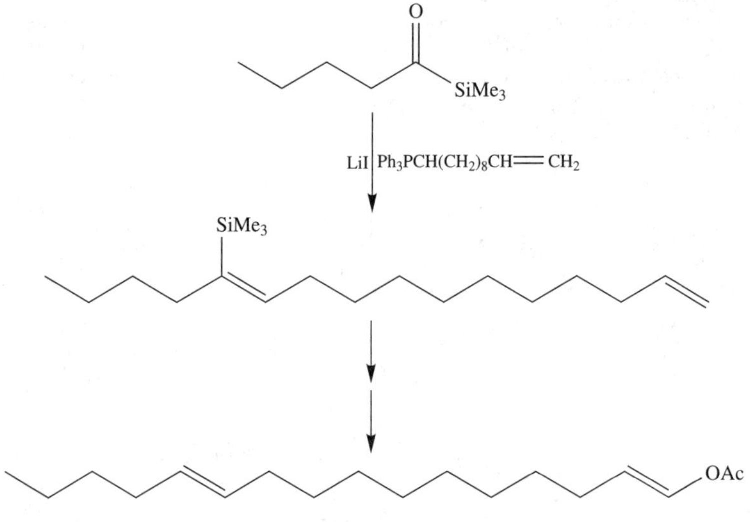

SCHEME 75

cleaves, rearrangement occurs readily, and the silyl ether product is predominant. Diazomethane reacts in a similar manner[5]. An alkyl acyl silane has been shown to react as a normal ketone with dimethyl titanocene, giving the methylenated vinyl silane product in 65% yield[185].

Sulphur ylids also react with acyl silanes by two different, competing pathways, to give either silyl enol ethers, formed under salt-free conditions, or β-ketosilanes, formed in the presence of soluble inorganic salts (Scheme 76)[186].

In an interesting transformation, reaction of benzoyl trimethylsilane with lithium enolates derived from various methyl ketones gives rise to 1,2-cyclopropanediols, predominantly with the *cis* configuration, in good yields (Scheme 77). The reaction, which proceeds through addition, Brook rearrangement and cyclization, is also successful with α,β-unsaturated acyl silanes (*vide infra*, Section IV.D)[187].

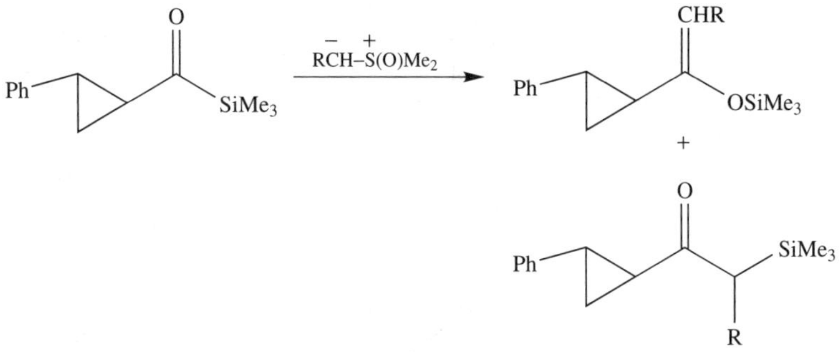

SCHEME 76

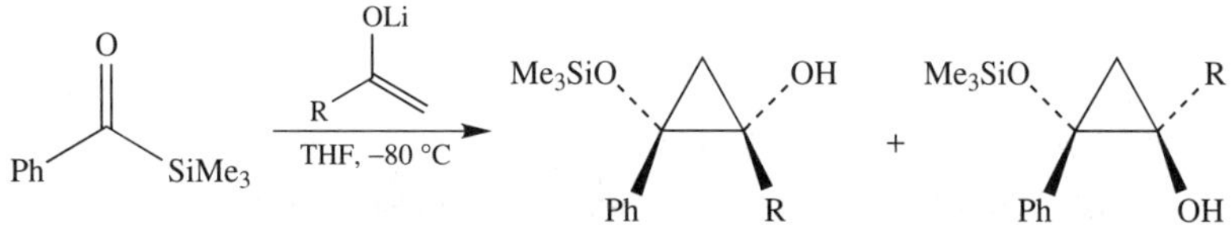

SCHEME 77

Fluoride ion-catalysed addition of trifluoromethyltrimethylsilane to acyl silanes occurs to give 1,1-difluoro-2-trimethylsilyloxyalkenes (silyl enol ethers of difluoromethyl ketones), through nucleophilic addition of trifluoromethyl anion, Brook rearrangement and loss of fluoride. These compounds could be isolated when tetrabutylammonium difluorotriphenylstannate was used as a catalyst; use of tetrabutylammonium fluoride gave the product corresponding to subsequent aldol reaction with the difluoromethyl ketone (Scheme 78)[188].

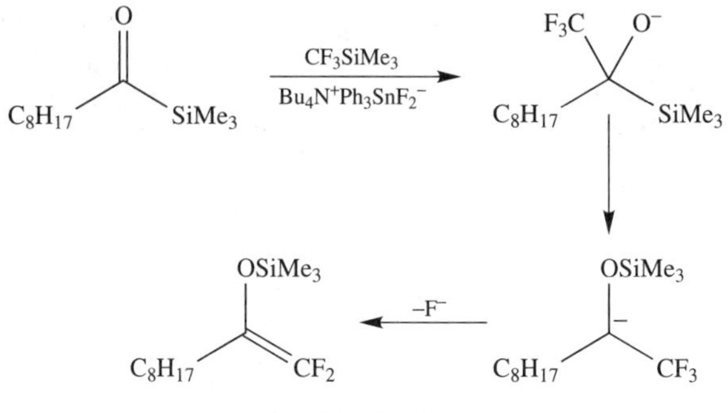

SCHEME 78

A Japanese group have reported an unusual reaction, mediated by lanthanide metals, involving the deoxygenative acylation of diaryl ketones with aryl acyl silanes, to give 1,1-diaryl acetophenones (Scheme 79)[189].

Treatment of α-cyclopropyl acyl silanes with sulphuric or triflic acids results in rearrangements to give cyclobutanones and 2-silyl-4,5-dihydrofurans, respectively, in processes formally involving intramolecular nucleophilic attack at the acyl silane moiety (Scheme 80)[190].

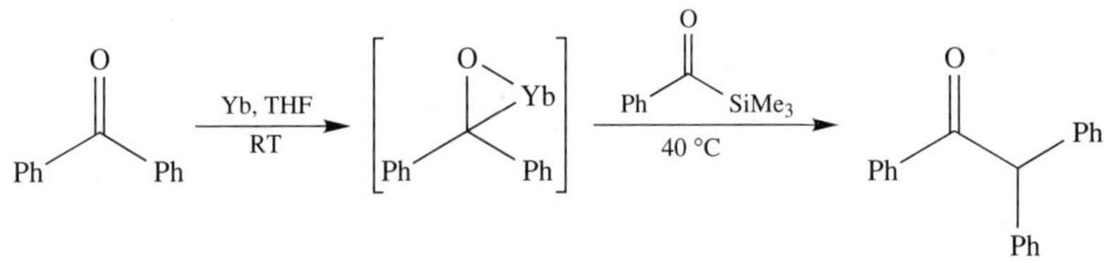

SCHEME 79

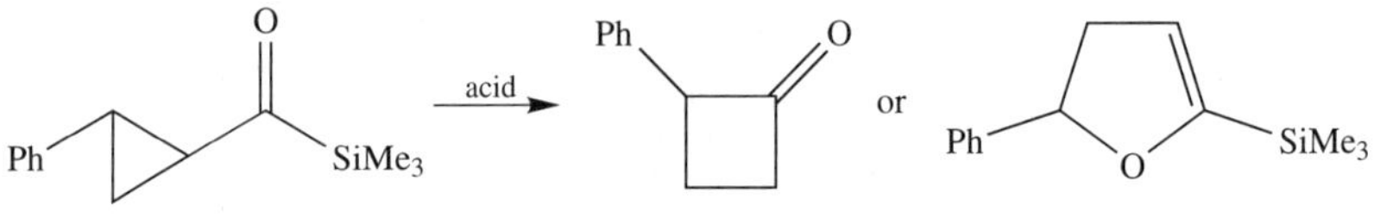

SCHEME 80

2. Acyl silanes as acyl anion precursors

Several aromatic and heterocyclic acyl trimethylsilanes have been used as acyl anion equivalents by treatment with fluoride ion (Scheme 81, path a)[23,133,154b,160,191,192]. Provided that the acyl substituent is electron-withdrawing, and that there are no aryl substituents on the silicon atom, acyl anions can be trapped by various electrophiles in moderate to good yields; indeed, acyl anions and pentacoordinate silicon anionic species have both been detected in gas-phase reactions of acyl silanes with fluoride ion[193].

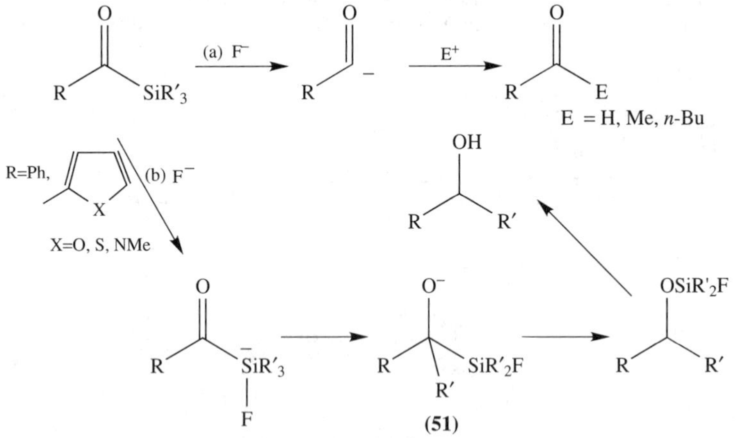

SCHEME 81

An alternative rearrangement pathway may be observed when the silicon atom bears aryl substituents or when simple alkanoyl trialkylsilanes are used. This pathway is similar to one suggested, but not observed, by Brook for reaction of acyl silanes with alkoxide ions[157,158]. Addition of fluoride ion to silicon induces a migration to the carbonyl carbon atom of one of the groups attached to the silicon atom, to give **51**, followed by a Brook-type rearrangement, giving a rearranged alcohol after protic work-up (Scheme 81, path b)[154b,160]. The acyl anion reaction pathway may only be observed for these substrates at higher temperatures in the presence of acid. Both pathways may proceed via a pentacoordinate silicon anionic species as a common intermediate.

Remarkably, carbonyl bis(trimethylsilane) can act as a source of the CO^{2-} dianion in the presence of fluoride ion (Scheme 82)[13].

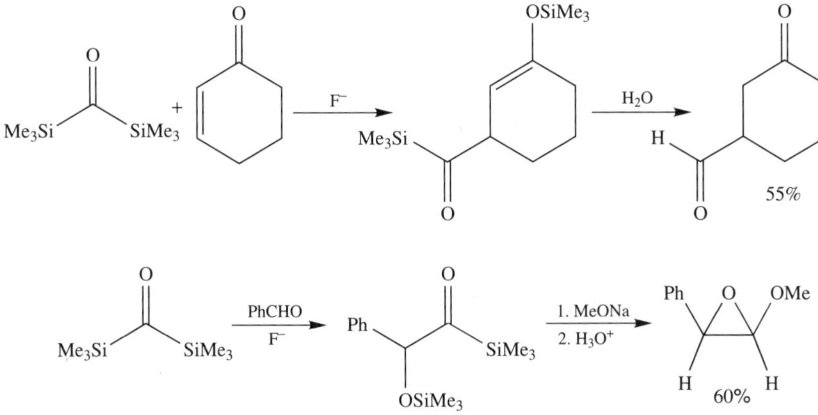

SCHEME 82

3. Cyclization reactions

A number of types of cyclization reaction have been reported in which the reactivity of the acyl silane grouping is a factor in the transformation. These are discussed below.

Molander and Siedem have reported the reaction of 1-methoxy-1,3-bis(trimethylsilyloxy)-buta-1,3-diene, the bis(trimethylsilyl) enol ether of methyl acetoacetate, with 1,4- and 1,5-dicarbonyl species under the influence of trimethylsilyl triflate. Two regioisomeric oxabicyclo[3.2.1]octane products are formed (Scheme 83)[194]. Use of 4- or 5-ketoacyl silanes as the dicarbonyl species can result in a reversal of the sense of regioselectivity.

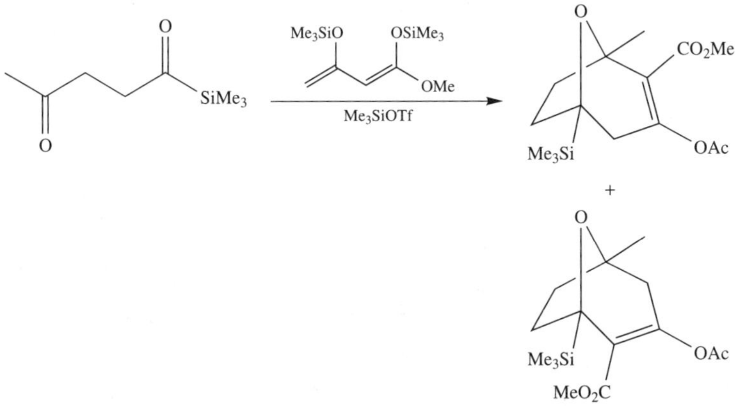

SCHEME 83

Heating 4- and 5-bromoacyl silanes at 100 °C in a polar aprotic solvent induces cyclization through the enol forms to give 2-silyldihydrofurans and 2-silyldihydropyrans, respectively[195]. Similar transformation of 4- to 7-halothioacyl silanes, prepared from the corresponding haloacyl silanes by reaction with hydrogen sulphide, but induced by sodium hydroxide, gave the 2-silylated sulphur heterocycles in excellent yields (Scheme 84)[196]. Intermolecular enolate reactions of acyl silanes are also known (*vide infra*, Section IV.A.6).

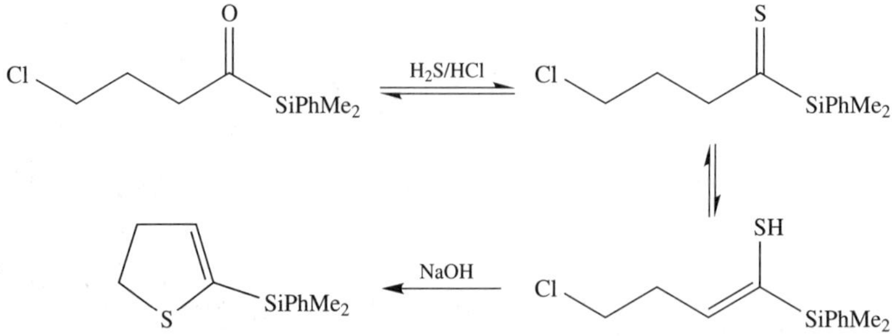

SCHEME 84

5- and 6-haloacyl silanes also undergo radical-mediated cyclization with Brook-type migration of the silyl radical moiety to give the cyclized silyloxy radical intermediates[197], which may in turn undergo further reaction in a tandem process, for example to form spiro or fused bicyclic products (Scheme 85)[198,199]. The acyl silane unit is thus acting as a formal geminal diradical species. An interesting development of this reaction includes a stannyl group at the halo- (or xanthate-) bearing carbon atom, cyclization of an α-stannyl radical then resulting in formation of an α-stannylated silyloxy radical, from which the stannyl unit is lost, providing a regiospecific synthesis of cyclic silyl enol ethers (Scheme 86)[200].

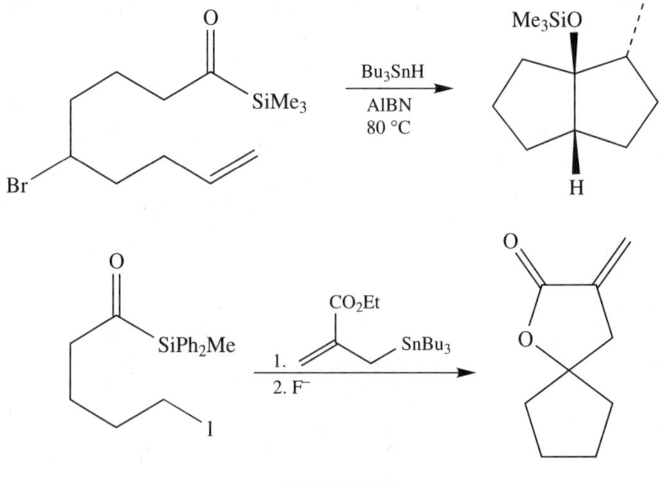

SCHEME 85

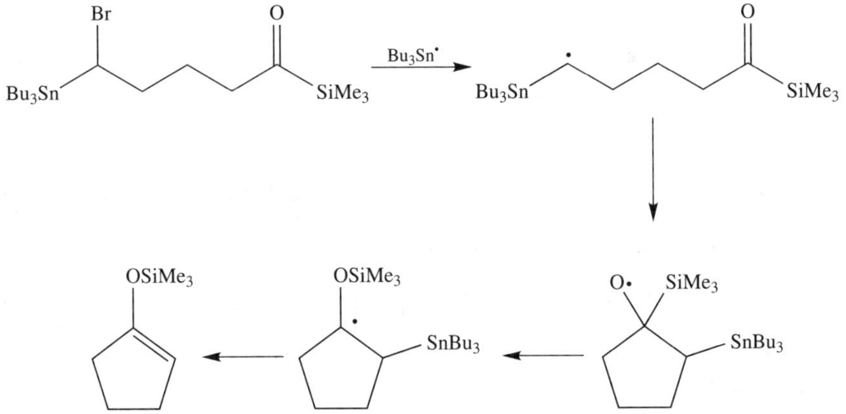

SCHEME 86

Cycloaddition reactions of acyl silanes appear to be rare, but Brook has shown that α-silyloxy bis(trimethylsilyl)silenes (**52**), generated photochemically from acyl tris(trimethylsilyl)silanes (*vide infra*, Section IV.A.4), undergo [2 + 2] and [4 + 2] cycloaddition reactions with ketones, and [4 + 2] cycloaddition reactions with less bulky acyl silanes, as illustrated in Scheme 87[17,24,26,72,73,201]. They do not, however, react with their parent acyl tris(trimethylsilyl)silanes.

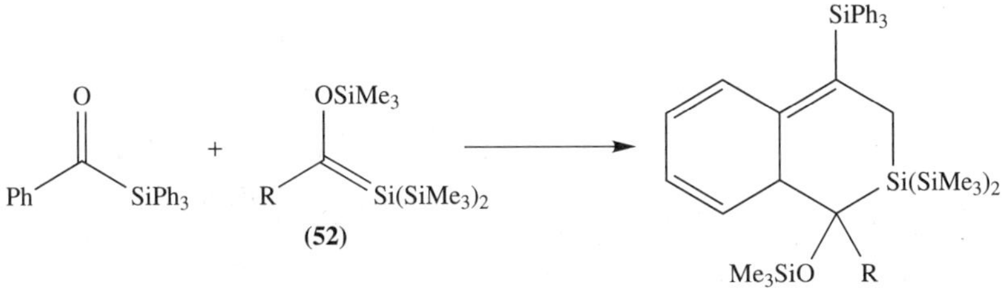

SCHEME 87

4. Photochemistry

Acyl silanes display a range of behaviour upon irradiation, depending upon their structure and the reaction conditions. The interesting photochemistry displayed by acyl silanes has been attributed to the low-energy n → π* carbonyl group transition.

Brook has shown that a wide variety of acyl tris(trimethylsilyl) silanes and acyl alkyl bis(trimethylsilyl)silanes undergo clean 1,3-rearrangements of silyl groups from silicon to oxygen to give α-silyloxy silenes upon irradiation, in an analogue of photoenolization (Scheme 88). Many of the silenes are remarkably stable and even recrystallizable[17,24,26,72,73]. They may undergo head-to-head or head-to-tail [2 + 2] dimerization to give 1,2- or 1,3-disilacyclobutanes, dependent upon the nature of the alkyl groups present[26], and, while α-silyloxy bis(trimethylsilyl)silenes (**52**) do not cycloadd to their parent acyl tris(trimethylsilyl)silanes, they do cycloadd to less bulky acyl silanes (Scheme 87)[202]. The structures of several 1,3-disilacyclobutanes have been determined by X-ray crystallography[201].

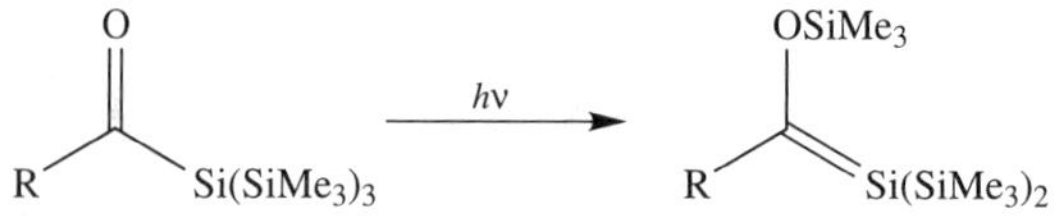

SCHEME 88

Acyl silanes react readily in alcoholic solution upon near-visible irradiation[203]. In the absence of base, the reaction process involves cleavage of the acyl–silicon bond to give a silyl ether and an acetal. Silanol and aldehyde may also be isolated. The proposed mechanism, suggested following experiments involving an enantiomerically pure acyl silane, is shown in Scheme 89[5]. Although also proceeding through a carbene intermediate, quite distinct from this process is the near-quantitative formation of mixed acetal which occurs upon irradiation of an alcoholic solution of an acyl silane in the presence of traces of base (typically pyridine)[5]. In this case, the acetal is formed by the photochemical generation of a silyloxycarbene from the acyl silane, which then inserts into the O–H bond of a solvent molecule (Scheme 90). Dalton has examined the kinetics of the latter reaction and have confirmed that acetal formation occurs exclusively through reaction of alcohol with an intermediate presumed to be the silyloxycarbene, generated from the acyl silane T1 state[204].

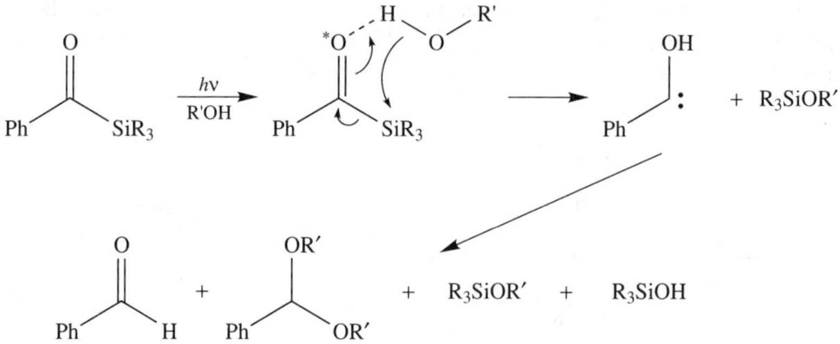

SCHEME 89

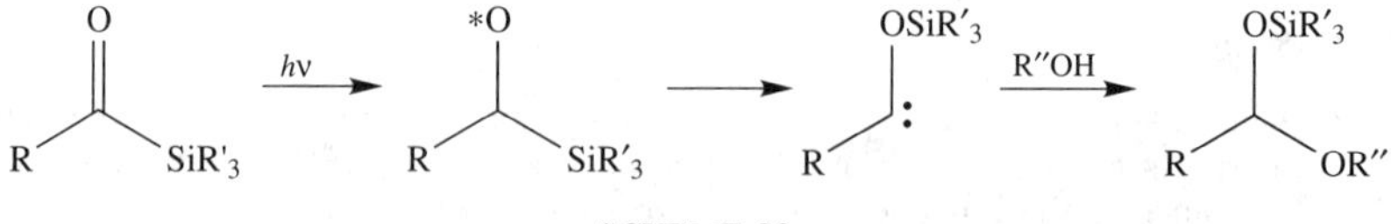

SCHEME 90

Silyloxycarbenes are also formed on heating acyl silanes; intramolecular C–H bond insertion may then occur, as illustrated in Scheme 91[205]. Such an intramolecular insertion reaction of a silyloxycarbene generated from *ortho*-tolyl acyl trimethylsilane has been investigated as a potential route to benzocyclobutenols (Scheme 92)[206]. The unstable benzocyclobutenol silyl ether (53), however, underwent ring opening and further rearrangement as shown, to give the aldehyde 54 in good yield.

2-(Diphenylsila)cyclohexanone, an unusual cyclic acyl silane, has been found to undergo photo-oxidation promoted by ambient light to produce the silicon-containing lactone 55. 1,1-Diphenylsilacyclohexanone was stable in the presence of oxygen in the

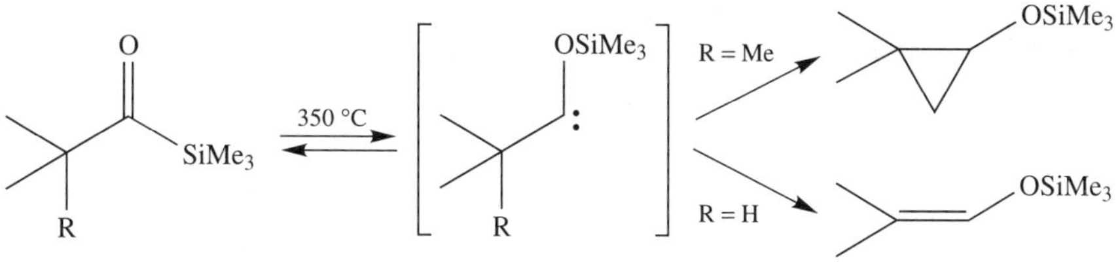

SCHEME 91

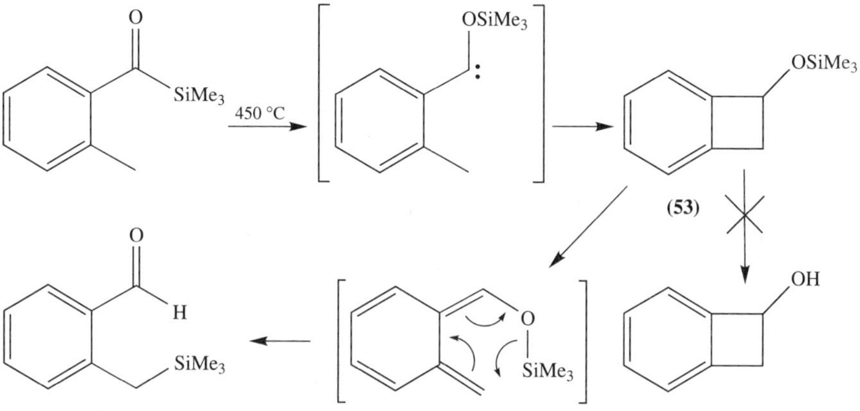

SCHEME 92

absence of light over long periods; subsequent investigations have demonstrated that this photo-oxidation is typical of other alkyl acyl silanes, but that aryl acyl silanes are inert[5]. The lactone **55** was hydrolysed to give a δ-(hydroxysilyl) carboxylic acid (Scheme 93)[5,60].

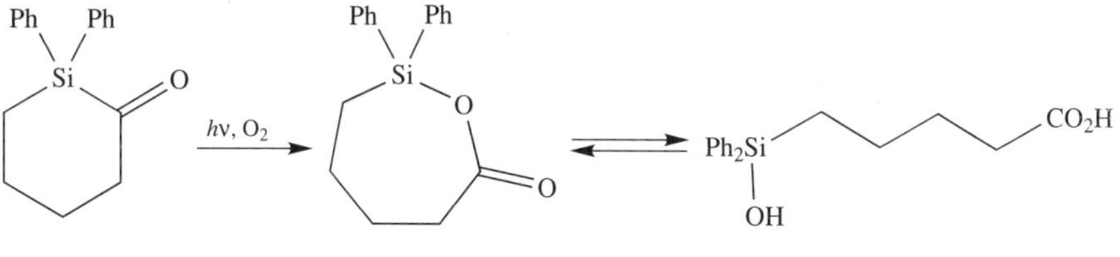

SCHEME 93

The terpene-derived acyl silane **56** undergoes a Norrish type II cleavage reaction as the major pathway upon n → π* excitation, involving hydrogen abstraction and fragmentation to give an acetyl silane and a diene as the major products (Scheme 94)[33]. The corresponding ketone (**57**) behaves in an analogous manner. Acyl silane **56** also displays typical photochemical behaviour, undergoing rearrangement to the silyloxycarbene **58**. Insertion of **58** into the O—H bond of the enol **59** led to compound **60** among others (Scheme 95). The silyloxycarbene **58** therefore reacts through intermolecular insertion into the O−H bond rather than by insertion into a neighbouring C−H bond or intramolecular addition to a carbon–carbon double bond.

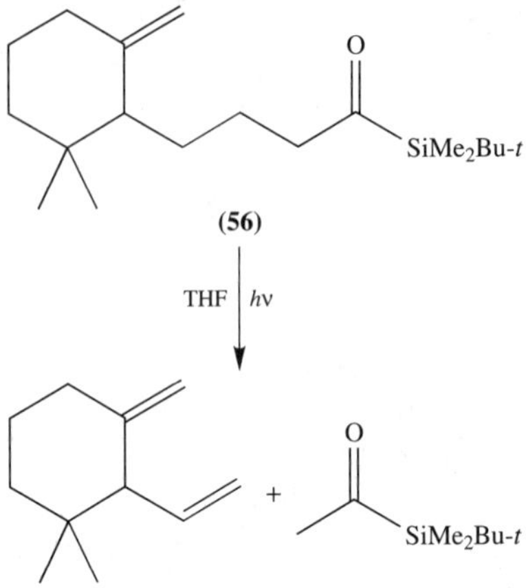

SCHEME 94

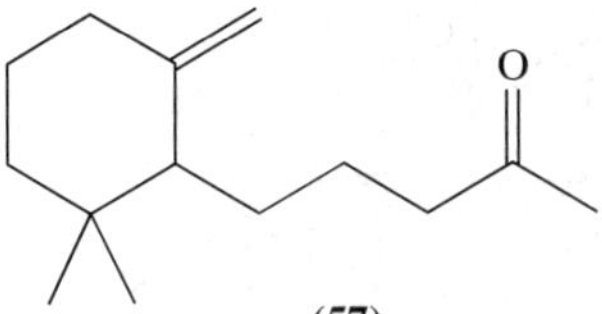

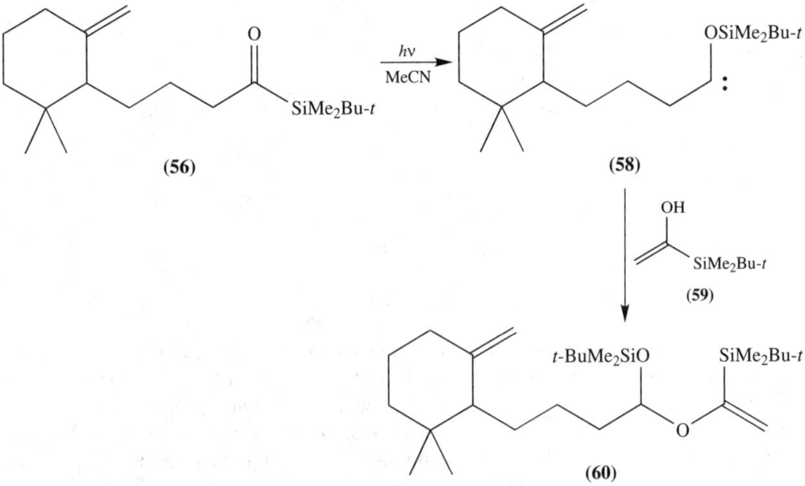

SCHEME 95

In a further unusual process, irradiation of α,β-epoxyacyl silanes in acetonitrile induces cleavage to give a silylketene species and a ketone, with quantitative conversion[207]. The initial step of the mechanism is presumably analogous to the α-(C—O) bond cleavage of α,β-epoxymethyl ketones (Scheme 96).

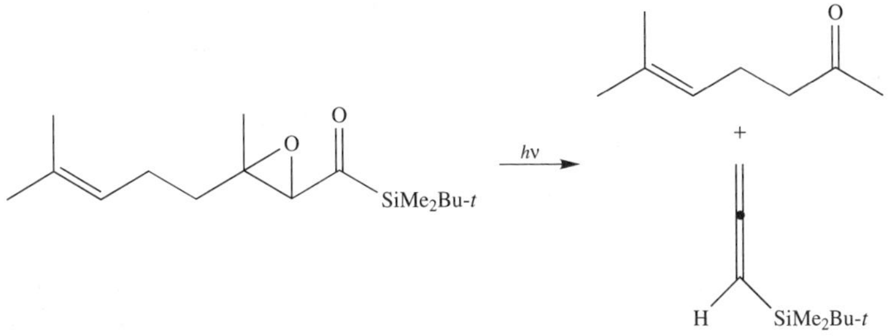

SCHEME 96

5. Biotransformations

A number of reductive biotransformations of acyl silanes into α-hydroxy silanes have appeared recently. Acetyl dimethylphenylsilane is converted into (R)-(1-hydroxyethyl) dimethylphenylsilane by plant cell suspension cultures of Symphytum officinale L. and Ruta graveolens L. in low yields but in 81% ee and 6% ee, respectively[208]. Silanol and disiloxane were observed as by-products. Microbial reduction of racemic acetyl t-butyldimethylsilane has been achieved using Trigonopsis variabilis (DSM 70714) and Corynebacterium dioxydans (ATCC 21766) on a 10- gramme scale with > 96% ee in each case and in up to 78% yield[209a]. Recently, Saccharomyces cerevisiae (DHWS 3), a commercially available form of bakers' yeast, has been shown to reduce acetyl dimethylphenylsilane, again to (R)-(1-hydroxyethyl) dimethylphenylsilane, in 40% yield and >99.5% ee[209b].

6. Miscellaneous

Intramolecular alkylation of the enol forms of 4- and 5-bromoacyl silanes has been observed upon heating at 100 °C in a polar aprotic solvent, inducing cyclization to give 2-silyldihydrofurans and 2-silyldihydropyrans, respectively[195]. Similar transformation of the thiono analogues, 4- to 7-halothioacyl silanes, induced by sodium hydroxide, gave the 2-silylated sulphur heterocycles in excellent yields (vide supra Section IV.A.4)[196].

Enolates or enol ethers derived from acyl silanes have been used by several groups as reagents for stereoselective aldol condensations[19,210,211]. Acyl silanes, treated with LDA followed by aldehydes, give aldol products in up to > 20 : 1 selectivity in favour of the syn products (Scheme 97)[19]; increasing steric bulk around the silicon atom appears to give increased selectivities. Product mixtures were analysed as the carboxylic acids. A similar scheme using acyl silanes chiral at silicon, however, gave an anomalous reaction resulting from aldol reaction as desired followed by Cannizaro-type hydride transfer processes (Scheme 98)[211].

Silyl enol ethers of acyl silanes have been used in Lewis acid-mediated Mukaiyama reactions with acetals. Treatment of the resulting β-alkoxy acyl silanes with tetrabutylammonium hydroxide or tetrabutylammonium fluoride gave the corresponding α,β-unsaturated aldehydes (Scheme 99)[210].

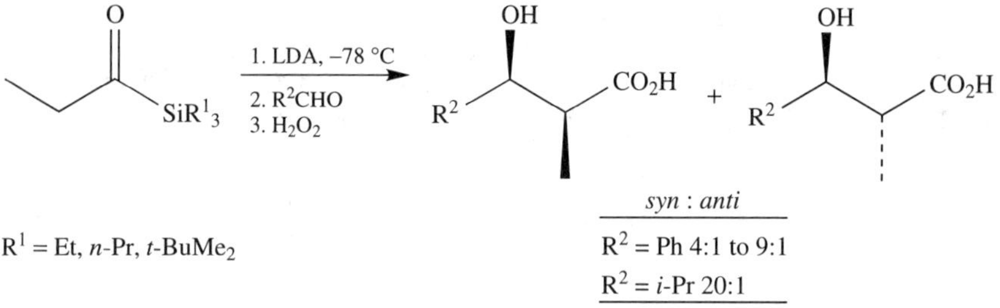

SCHEME 97

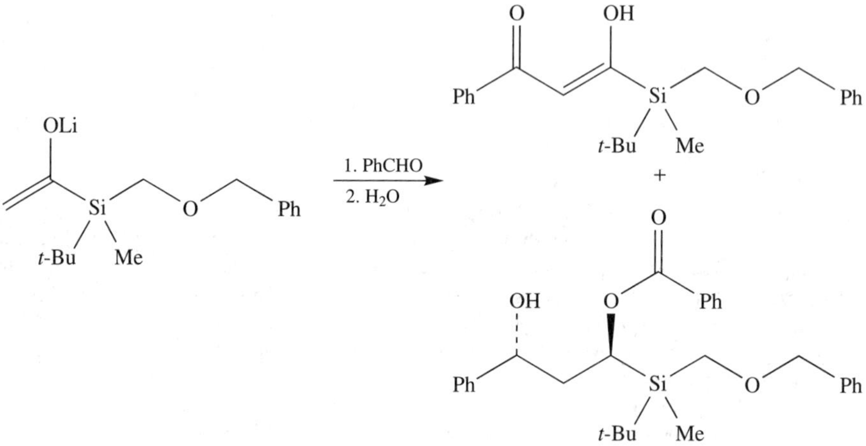

R^1 = Et, n-Pr, t-BuMe$_2$

syn : anti
R^2 = Ph 4:1 to 9:1
R^2 = i-Pr 20:1

SCHEME 98

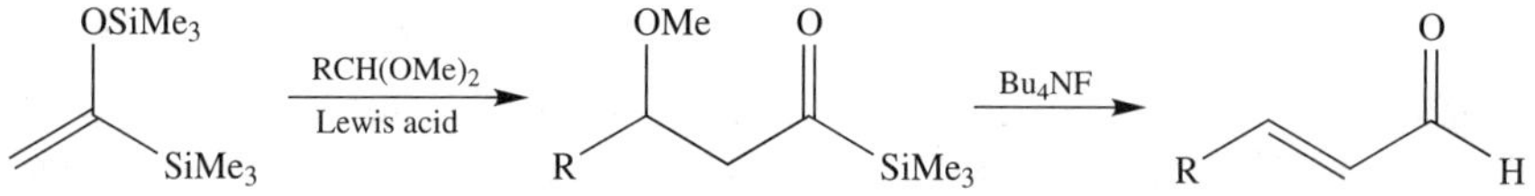

SCHEME 99

Reductive cleavage of the carbon–silicon bond of acyl silanes, including α-alkoxy and α,β-dialkoxy derivatives, to give aldehydes may be accomplished by palladium-catalysed hydrogenolysis at ambient temperature and pressure[212].

The electrochemical oxidation of acyl silanes has been investigated, giving rise to esters and amides when carried out in the presence of alcohols and amines. The oxidation potentials of acyl silanes proved to be much lower than those of the corresponding ketones[213].

Bis(trimethylsilyl) sulphide reacts with acyl silanes in the presence of a cobalt chloride catalyst to afford the corresponding thiocarbonyl derivatives, thioacyl silanes (Scheme 100)[214]. The reaction is mild and proceeds in good yields. It is also applicable to aldehydes. Thioacyl silanes have been prepared from the corresponding haloacyl silanes by reaction with hydrogen sulphide (*vide supra*, Section IV.A.3)[196].

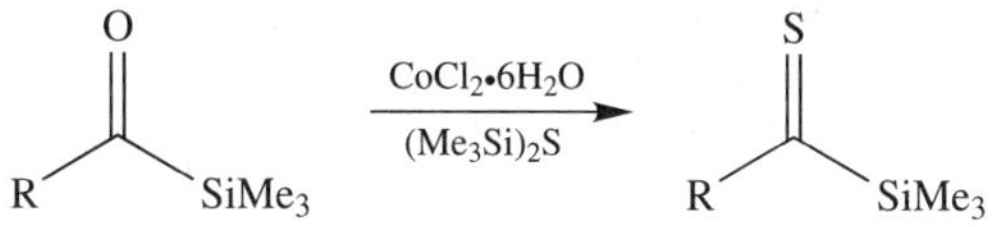

SCHEME 100

Acyl silanes react with ytterbium by two different pathways, depending upon structure. Aroyl silanes undergo reductive coupling to produce diaryl acetylenes, while alkanoyl silanes give the product of carbonyl group reduction (Scheme 101)[215]. Benzoyl silanes react with benzylmanganese pentacarbonyl to give *ortho*-metallated tetracarbonyl manganese complexes, which could be desilylated to give the previously inaccessible benzaldehyde complex[216].

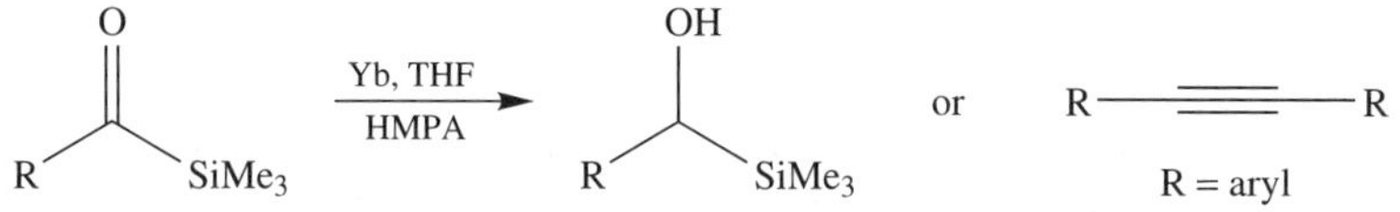

SCHEME 101

B. α-Haloacyl Silanes

α-Chloroacyl silanes react with Grignard reagents by initial addition to the carbonyl group as expected, followed not by Brook rearrangement, but by 1,2-migration of silicon with expulsion of halide ion to afford α-silyl ketones (Scheme 102)[12,151,152]. α-Haloacyl silanes therefore behave as α-trimethylsilyl acylium ion (**61**) equivalents in this reaction. The 1,2-rearrangement may be accelerated compared with the lithium analogue by the more polar character of the oxygen–magnesium bond[217,218].

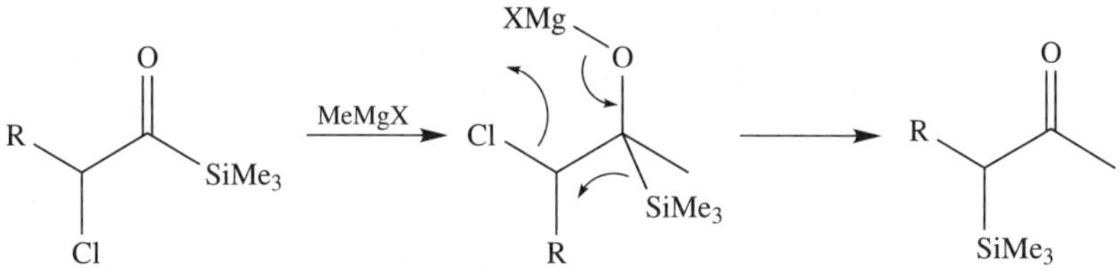

SCHEME 102

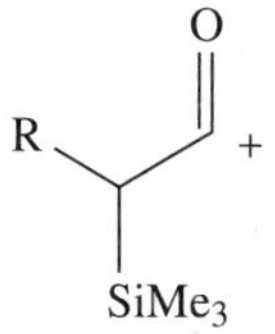

(**61**)

Further addition of the Grignard reagent can take place to give β-hydroxy silanes with high diastereoselectivity (Scheme 103)[12,165]. The hydroxy silanes can of course undergo subsequent stereocontrolled acid or base catalysed elimination of trialkylsilanol to give alkenes.

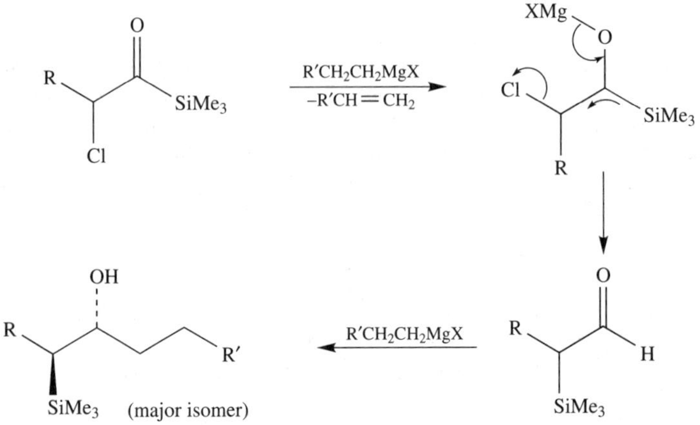

SCHEME 103

Similar reactions of α-haloacyl silanes take place with enolates (Scheme 104)[219].

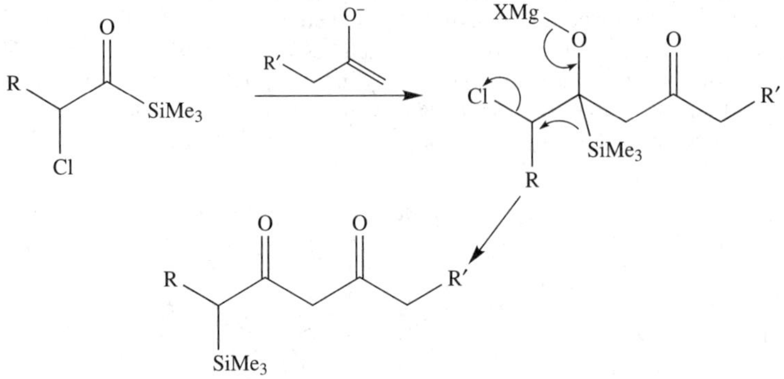

SCHEME 104

It is fascinating that a completely different reaction pathway takes over for trifluoroacetyl silanes. In this case, initial addition of Grignard reagent, in THF solution, to the carbonyl group of the acyl silane is followed by Brook rearrangement, as might be expected from general patterns of acyl silane reactivity. Elimination of fluoride ion then occurs to give 1-silyloxy-2,2-difluoroalkenes (silyl enol ethers of difluoromethyl ketones)[220,221]. The same type of intermediates and subsequent reaction are of course seen in the addition of trifluoromethyl anion to alkanoyl silanes (*vide supra*, Section IV.A.1)[188]. The reaction was applied to a synthesis of a fluorinated brassinosteroid. Similar reaction of vinyl magnesium bromide with trifluoroacetyl triphenylsilane gives rise to 1,1-difluoro-2-triphenylsilyloxybuta-1,3-diene (**62**, Scheme 105) which has been used in a number of [4 + 2] and [2 + 2] cycloadditions[222].

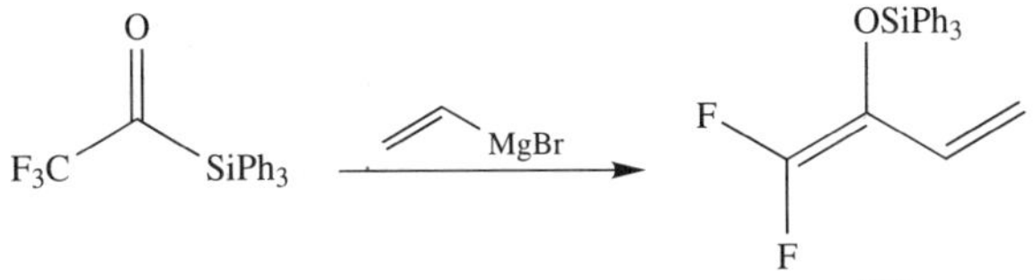

27. Acyl silanes

SCHEME 105

The related Brook rearrangement of α-halo-α,β-unsaturated acyl silanes produces silyloxy allenes (63), from which several sesquiterpenes have been synthesized (Scheme 106)[21]. Silyloxy allenes may also be prepared by the alkylation of silyloxy allenyl lithium reagents; the acyl silane route is, however, less sensitive to solvent effects and other experimental parameters. An outline of the synthesis of dehydrofukinone (64), which elegantly exemplifies this methodology, appears in Scheme 107.

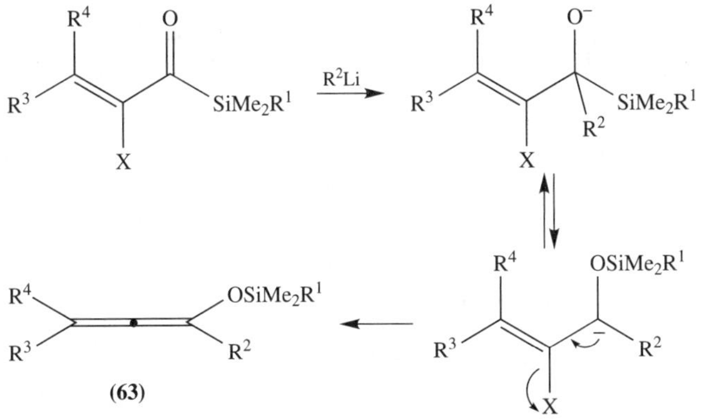

SCHEME 106

α-Phosphonoacyl silanes, which may be prepared from α-iodoacyl silanes through the Arbuzov reaction, undergo enolate alkylation, and are precursors of α,β-unsaturated acyl silanes through the Horner–Emmons reaction (*vide supra*)[116]; α-haloacyl silanes are also precursors of cyclopropyl acyl silanes through conversion into sulphur ylids and through McCoy reactions (*vide supra*)[146].

C. α-Ketoacyl Silanes

α-Ketoacyl silanes are a deep rich red in colour and are especially sensitive to light, often necessitating purification and handling in a darkened room. Very little chemistry of these interesting materials has been reported, although they are implicated as intermediates in the oxidation of trimethylsilyl acetylenes with osmium tetroxide to give α-ketoesters (Scheme 108)[7,125].

D. α,β-Unsaturated Acyl Silanes

A good deal of chemistry of α,β-unsaturated acyl silanes has been published, perhaps reflecting their added stability over other types of acyl silane. Much of this chemistry

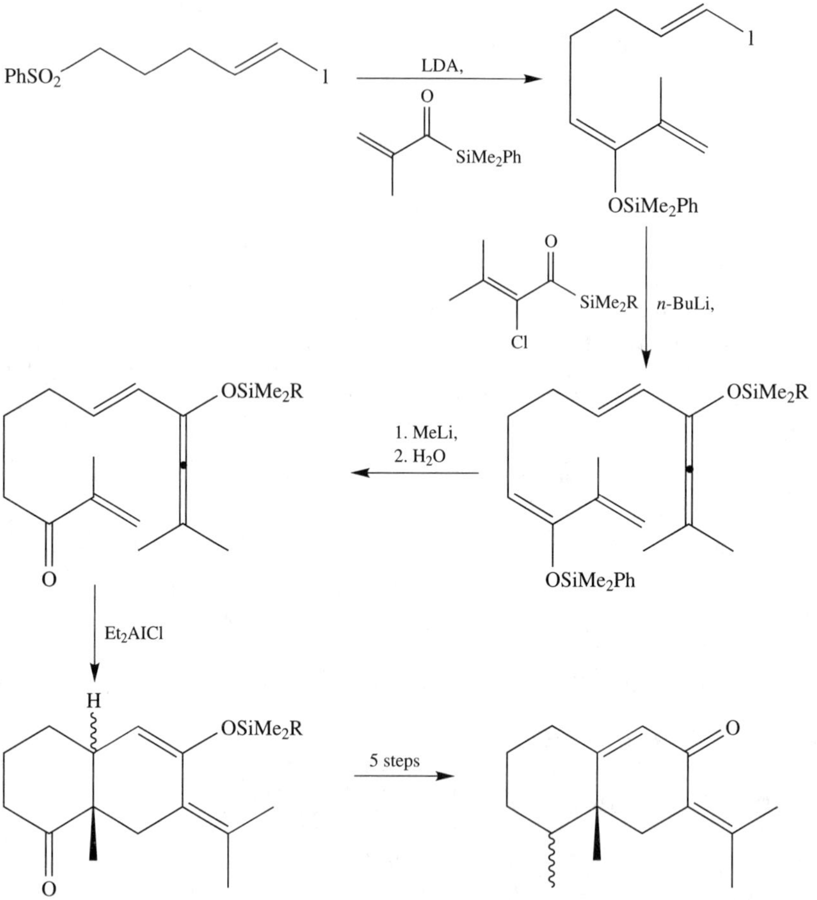

SCHEME 107

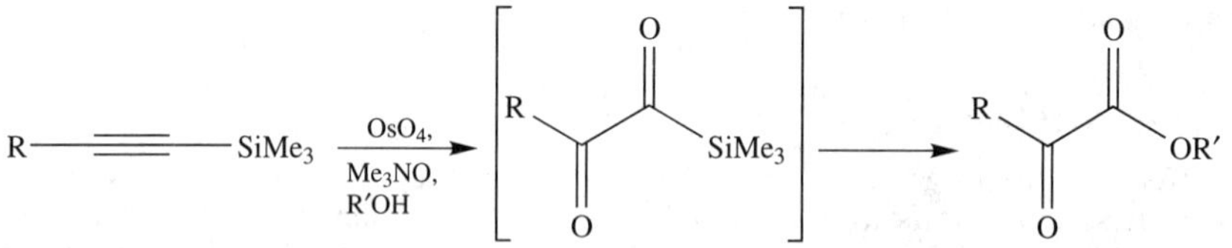

SCHEME 108

involves conjugate addition or cycloaddition involving the carbon–carbon double bond. In one exception, reaction of lithium enolates derived from various methyl ketones with but-2-enoyl *t*-butyldimethylsilane gives rise to *cis* vinyl 1,2-cyclopropanediols through addition to the carbonyl group, Brook rearrangement and cyclization, as well as to the conjugate addition products (Scheme 109). The reaction is also successful with simple aroyl silanes (*vide supra*, Section IV.A.1)[187]. Wittig reactions have also been carried out with the carbonyl group of α,β-unsaturated acyl silanes. Desilylation gives the *E*-alkene, in contrast to the Wittig reaction of the related aldehydes[137].

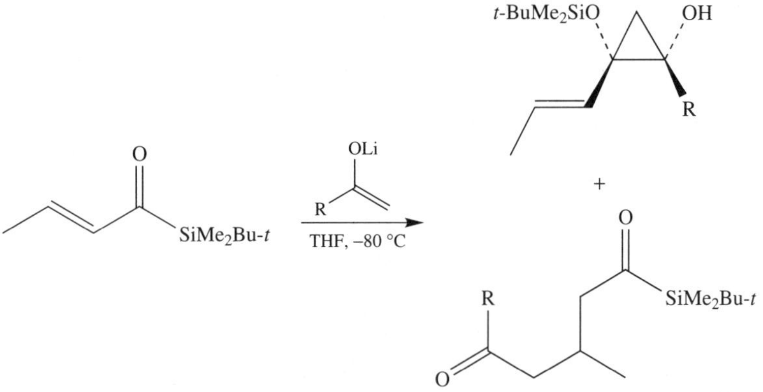

SCHEME 109

α,β-Unsaturated acyl silanes are oxidized to carboxylic acids by alkaline hydrogen peroxide[133], and can therefore be regarded as α,β-unsaturated carboxylic acid equivalents. They have been so used in Lewis acid-mediated conjugate allylation reactions with allyl silane derivatives (Scheme 110)[62,116,141]. The acyl silanes are highly reactive, and the authors suggest that they are much more electrophilic than the corresponding carboxylic acids and esters due to the net destabilizing effect of trialkylsilyl groups on developing α-carbocationic centres in the reaction intermediates[30].

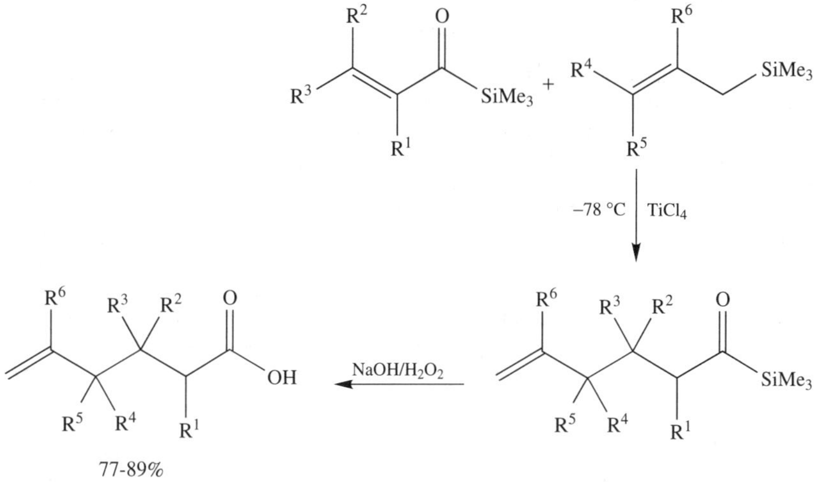

SCHEME 110

The Degl'Innocenti group has published a number of reactions of α,β-acetylenic acyl silanes, mostly involving initial conjugate addition to give α,β-unsaturated acyl silanes with a range of substituents[136]. For example, reaction with dialkyl, diaryl and divinyl lithium cuprates provides β-substituted α,β-unsaturated acyl silanes in high yields[137]. Heteroatomic β-substituents, including β-halo, β-amino, β-thio and β-azido groupings, may be introduced by the use of heterosubstituted trimethylsilanes as reagents[138,140]. Conjugate addition of a tributyltin unit has also been carried out, using a lithium tributylstannyl cuprate species[139].

3-Iodopropenoyl triphenylsilane takes part in palladium-catalysed coupling with alkyl tin derivatives to give β-substituted α,β-unsaturated acyl silanes, normally with the E configuration[140]. In a nicely complementary process, 3-(tributylstannyl)prop-2-enoyl triphenylsilane takes part in palladium-catalysed coupling with vinyl iodides (*vide supra*, Section III.D.4)[139]. Interestingly, the compound is also capable of acting as a nucleophile at the 2-position.

3,3-Dichloropropenoyl trimethylsilane, prepared by photochemically induced addition of bromotrichloromethane to the ethyl enol ether of acetyl trimethylsilane, undergoes completely regioselective addition–elimination upon treatment with lithium cyanocuprate reagents, giving the *E*-3-chloroalk-2-enoyl silane products (*vide supra*, Section III.D.3)[135].

Some β-heteroatom substituted α,β-unsaturated acyl silanes react with methyl ketone enolates in a stepwise stereoselective cyclopentannelation process, formally a [3 + 2] annelation, which may proceed through aldol reaction followed by Brook rearrangement and cyclization (Scheme 111)[223].

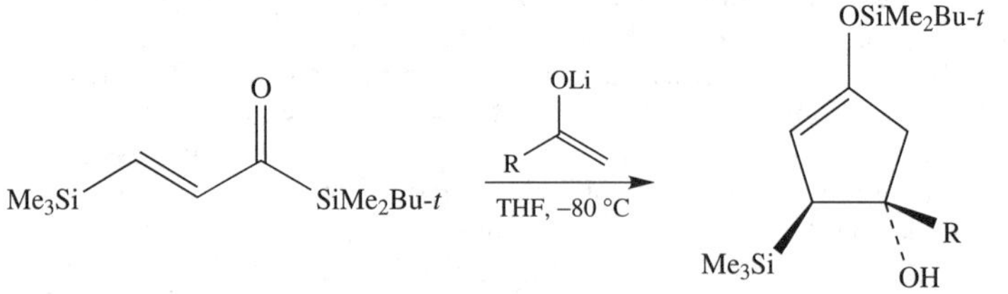

SCHEME 111

α,β-Unsaturated acyl silanes also combine with allenyl silanes in the presence of TiCl$_4$ in [3 + 2] and [3 + 3] annelations to give five- and six-membered carbocycles (Scheme 112)[141]. It is particularly interesting that the course of these reactions may be controlled to produce either five- or six-membered rings as desired by manipulating the trialkylsilyl group, the reaction temperature and the nature of the acyl group. Following regiospecific electrophilic substitution at C-3 of the allenyl silane, cyclization of a rearranged vinyl cationic intermediate (**65**) provides the cyclopentene **66**. If R^1 is alkyl, R$_3$Si is trimethylsilyl, and the reaction is carried out at elevated temperature, cyclopentene **66** can undergo a ring expansion to give the six-membered carbocyclic cation **67**, which undergoes a second 1,2-cationic shift of the trimethylsilyl moiety to produce the cyclohexenone **68**. This transformation can be prevented by employing the less mobile *t*-butyldimethylsilyl acyl silanes, by maintaining the reaction temperature below −78 °C and by minimizing the reaction time.

The first recorded cyclopropyl acyl silane (**69**) was generated by vapour phase pyrolysis of a pyrazoline derived from α,β-unsaturated acyl silane by 1,3-dipolar cycloaddition of diazomethane (*vide supra*, Section III.E)[141]. Exposure of **69** to titanium tetrachloride induced ring expansion to give the cyclobutanone in 75% yield (Scheme 113).

The Diels–Alder reactivity of α,β-acetylenic and α,β-unsaturated acyl silanes is comparable to that of the related methyl ketones, and such reactions have been used to prepare other α,β-unsaturated acyl silanes. For example, the α,β-acetylenic acyl silanes

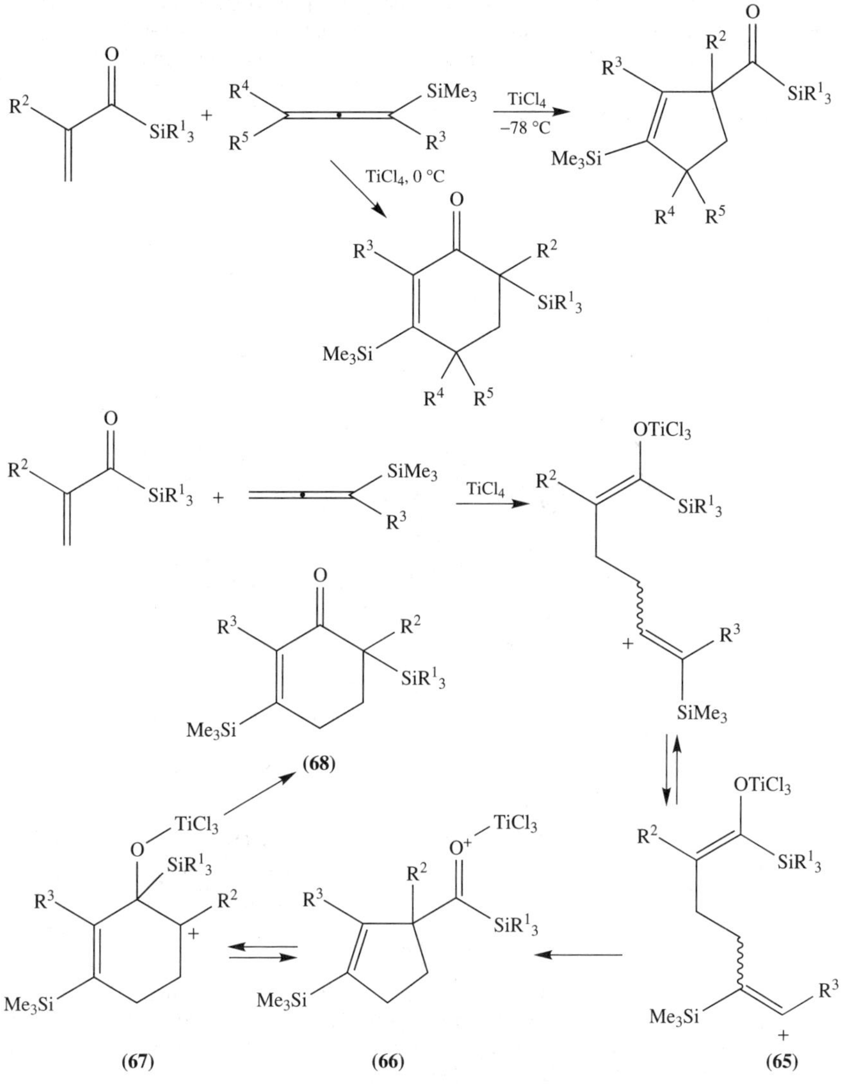

27. Acyl silanes

SCHEME 112

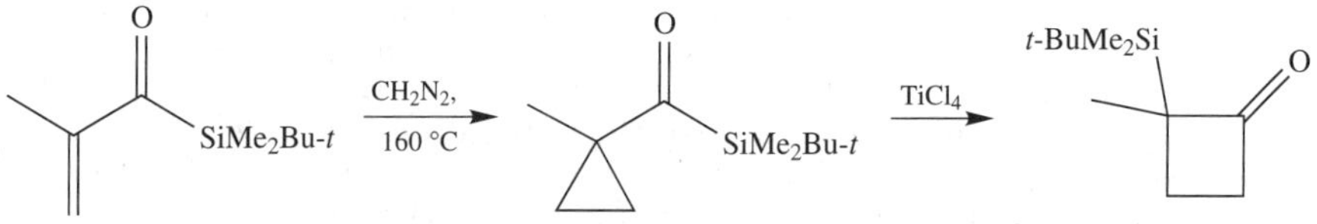

SCHEME 113

70 and **71** react with 2,3-dimethylbuta-1,3-diene and 4-phenyloxazole to give **72** and **73**, respectively, under conditions similar to those appropriate for conventional acetylenic dienophiles (Scheme 114)[14]. Diels–Alder reactions of propenoyl trimethylsilane with thioketones take place at room temperature in ethereal solution to give 6-trimethylsilyl-(4H)-dihydro-1,3-oxathiins **74**, (Scheme 115)[224].

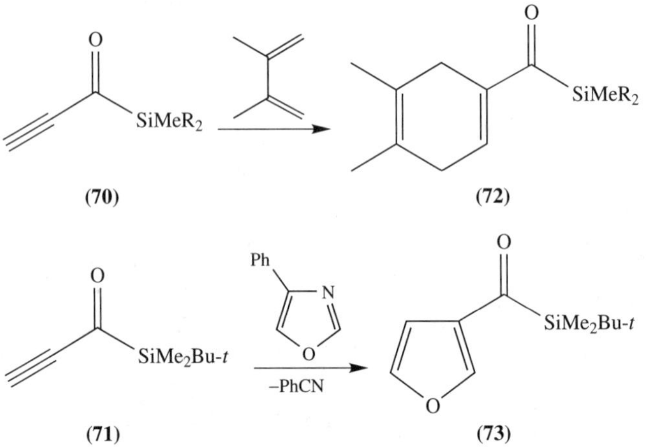

SCHEME 114

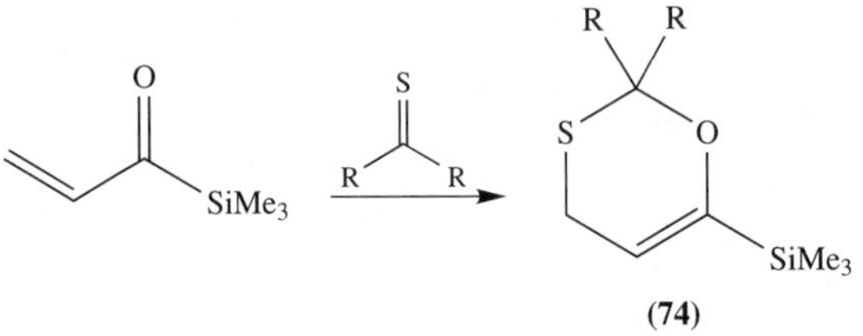

SCHEME 115

The [4 + 2] cycloaddition of α-phenylselenopropenoyl trimethylsilane (**75**) with 2,3-dimethylbuta-1,3-diene is unusual in that a significant portion of product mixture consists of the hetero-Diels–Alder dihydropyran adduct **76**. The phenylselenenyl substituent appears to be responsible for this unusual pattern of reactivity, since propenoyl trimethylsilane gives only the expected regioisomer (**77**, X = H) (Scheme 116)[14]. α-Selenenyl substituted α,β-unsaturated acyl silanes such as **75** were used to prepare a series of substituted dienes in excellent yields through the addition of α-sulphinyl carbanions, Brook rearrangement and expulsion of sulphinate, in a reaction pathway recognisably more typical of acyl silanes (Scheme 117).

Phenylthiotrimethylsilane adds to propenoyl trimethylsilane under the influence of Lewis acid to give 1,3-bis(phenylthio)-1-trimethylsilylprop-1-ene (**18**). This enol thioether may be deprotonated with *t*-butyl lithium and alkylated with any of a large range of electrophiles. Subsequent hydrolysis–elimination with mercuric chloride in aqueous acetonitrile provides β-substituted α,β-unsaturated acyl silanes (*vide supra*, Section III.D.3)[132]. It should be noted that, in this transformation, the β-substituent has

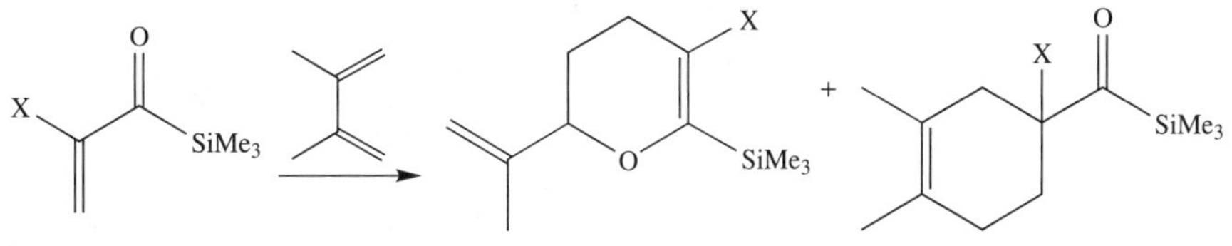

SCHEME 116

(75) X = PhSe

(76)

(77) X = PhSe, 15:85

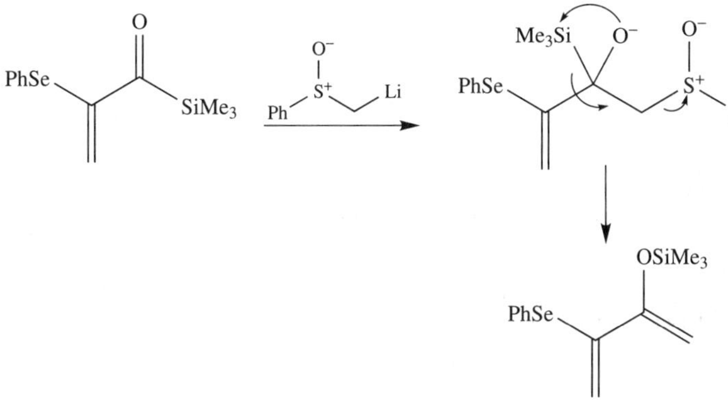

SCHEME 117

been introduced as an electrophile, complementing other methods of preparation of β-substituted α,β-unsaturated acyl silanes.

Acyl tri-isopropylsilanes containing alkene or alkyne functionality undergo smooth hydrozirconation with Schwartz' reagent, in contrast to the corresponding aldehydes, which also suffer addition to the carbonyl group (Scheme 118)[225].

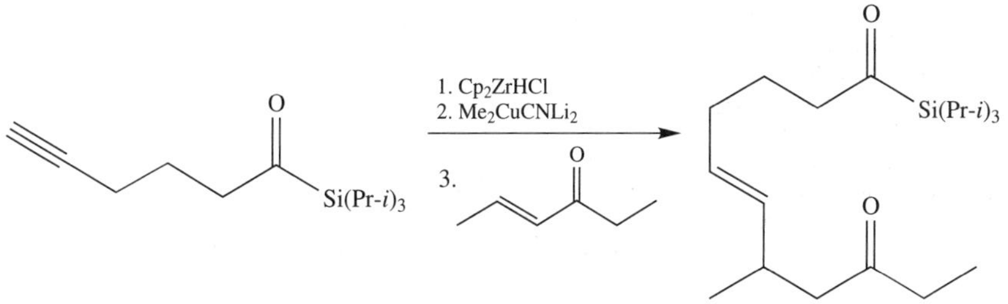

SCHEME 118

Gibson née Thomas, and Tustin have reported the formation of a number of iron carbonyl complexes of α,β-unsaturated acyl silanes[226]. Propenoyl trimethylsilane did not give a stable complex, but the iron tricarbonyl complexes of cinnamoyl silanes were very stable.

V. CONCLUSION

The last few years have seen a continued expansion in acyl silane chemistry, both in synthetic methods and in reactivity. There can now be no doubt that the difficulties encountered in preparing and isolating acyl silanes in the early investigations have been quite surmounted. While acyl silanes have been used merely as alternative carboxylic acid derivatives, they have also become synthetic intermediates in their own right, as some reactions particular to acyl silanes, such as the Brook rearrangement, have been used to great effect as synthetic processes. It can surely be expected that the synthesis of acyl silanes of all types will continue to become more sophisticated, and that, as functionalized acyl silanes become more accessible, these interesting materials will continue to find new uses in organic chemistry for a diverse range of purposes.

VI. REFERENCES

1. A. Ricci and A. Degl'Innocenti, *Synthesis*, 647 (1989); P. C. B. Page, S. S. Klair and S. Rosenthal, *Chem. Soc. Rev.*, **19**, 147 (1990); P. F. Cirillo and J. S. Panek, *Org. Prep. Proced. Int.*, **24**, 555 (1992).
2. A. G. Brook, *J. Am. Chem. Soc.*, **79**, 4373 (1957); A. G. Brook, M. A. Quigley, G. J. D. Peddle, N. V. Schwartz and C. M. Warner, *J. Am. Chem. Soc.*, **82**, 5102 (1960).
3. V. Bazant, V. Chvalovsky and J. Rathousky, *Organosilicon Compounds*, Academic Press, New York, 1965; V. Chvalovsky, *Handbook or Organosilicon Compounds —Advances Since 1961*, Marcel Dekker, New York, 1974.
4. A. G. Brook and R. J. Mauris, *J. Am. Chem. Soc.*, **79**, 971 (1957); A. G. Brook and G. J. D. Peddle, *Can. J. Chem.*, **41**, 2351 (1963).
5. A. G. Brook *Adv. Organomet. Chem.*, **7**, 96 (1968).
6. A. G. Brook, R. Kiviskik and G. E. Le Grow, *Can. J. Chem.*, **43**, 1175 (1965).
7. P. C. B. Page and S. Rosenthal, *Tetrahedron*, **46**, 2573 (1990); P. C. B. Page and S. Rosenthal, unpublished data.
8. G. Zweifel and J. A. Miller, *Synthesis*, 288 (1981).
9. C. Heathcock and J. Lampe, *J. Org. Chem.*, **48**, 4330 (1983).
10. A. Capperucci, A. Degl'Innocenti, C. Faggi, A. Ricci, P. Dembech and G. Seconi, *J. Org. Chem.*, **53**, 3612 (1988).
11. J. A. Soderquist and A. Hassner, *J. Am. Chem. Soc.*, **102**, 1577 (1980); D. I. Gasking and G. H. Whitham, *J. Chem. Soc., Perkin Trans. 1*, 409 (1985).
12. I. Kuwajima, T. Sato, K. Matsumoto and T. Abe, *Bull. Chem. Soc. Jpn.*, **57**, 2167 (1984); T. Sato, T. Abe and I. Kuwajima, *Tetrahedron Lett.*, 259 (1978).
13. A. Ricci, M. Fiorenza, A. Degl'Innocenti, G. Seconi, P. Dembech, K. Witzgall and H. J. Bestmann, *Angew. Chem., Int. Ed. Engl.*, **24**, 1068 (1985).
14. H. J. Reich, M. J. Kelly, R. E. Olson and R. C. Holtan, *Tetrahedron*, **39**, 949 (1983); H. J. Reich and M. J. Kelly, *J. Am. Chem. Soc.*, **104**, 1119 (1982).
15. A. Sekiguchi, Y. Kabe and W. Ando, *Tetrahedron Lett.*, 871 (1979).
16. G. J. D. Peddle and R. W. Walsingham, *J. Chem., Soc., Chem. Commun.*, 462 (1969).
17. A. G. Brook and L. Yau, *J. Organomet. Chem.*, **271**, 9 (1984).
18. I. Kuwajima and J. Enda, *J. Am. Chem. Soc.*, **107**, 5495 (1985).
19. D. Schinzer, *Synthesis*, 179 (1989).
20. R. L. Danheiser, D. M. Fink, K. Okano, Y.-M. Tsai and S. W. Szczepanski, *J. Org. Chem.*, **50**, 5393 (1985).
21. H. J. Reich, E. K. Eisenhart, R. E. Olson and M. J. Kelly, *J. Am. Chem. Soc.*, **108**, 7791 (1986); H. J. Reich and E. K. Eisenhart, *J. Org. Chem.*, **49**, 5282 (1984).
22. (a) R. Mantione and Y. Leroux, *Tetrahedron Lett.*, 591 (1971).
    (b) R. G. Visser, L. Brandsma and H. J. T. Bos, *Tetrahedron Lett.*, **22**, 2827 (1981).
    (c) K. J. H. Kruithof and G. W. Klumpp, *Tetrahedron Lett.*, **23**, 3101 (1982).
    (d) J. C. Clinet and G. Linstrumelle, *Tetrahedron Lett.*, **21**, 3987 (1980).
23. A. Ricci, A. Degl'Innocenti, S. Chimichi, M. Fiorenza, G. Rossini and H. J. Bestmann, *J. Org. Chem.*, **50**, 130 (1985).

24. A. G. Brook, J. W. Harris, J. Lennon and M. El Sheikh, *J. Am. Chem. Soc.*, **101**, 83 (1979); A. G. Brook, S. C. Nyburg, W. F. Reynolds, Y. C. Poon, Y.-M. Chang, J.-S. Lee and J.-P. Picard, *J. Am. Chem. Soc.*, **101**, 6750 (1979); A. G. Brook, S. C. Nyburg, F. Abdesaken, B. Gutekunst, G. Gutekunst, R. K. M. R. Kallury, Y. C. Poon, Y.-M. Chang and W. Wong-Ng, *J. Am. Chem. Soc.*, **104**, 5667 (1982).
25. A. G. Brook, F. Abdesaken and H. Sollradl, *J. Organomet. Chem.*, **299**, 9 (1986).
26. K. M. Baines, A. G. Brook, R. R. Ford, P. D. Lickiss, A. K. Saxena, W. J. Chatterton, J. F. Sawyer and B. A. Benham, *Organometallics*, **8**, 693 (1989).
27. I. Naito, A. Kanishita and T. Yonenitsu, *Bull. Chem. Soc. Jpn.*, **49**, 339 (1976).
28. E. M. Dexheimer, G. L. Buell and C. le Croix, *Spectrosc. Lett.*, **11**, 751 (1978).
29. F. Bernardi, L. Lunazzi, A. Ricci, G. Seconi and G. Tonachini, *Tetrahedron*, **42**, 3607 (1986).
30. G. A. Olah, A. L. Berrier, L. D. Field and G. K. Surya Prakash, *J. Am. Chem. Soc.*, **104**, 1349 (1982).
31. A. G. Brook, F. Abdesaken, G. Gutekunst and N. Plavac, *Organometallics*, **1**, 994 (1982).
32. H. K. Sharma, S. P. Vincenti, R. Vicari, F. Cervantes and K. H. Pannel, *Organometallics*, **9**, 2109 (1990).
33. B. Frei and M. E. Scheller, *Helv. Chim. Acta*, **67**, 1734 (1984).
34. S. Chimichi and C. Mealli, *J. Mol. Struct.*, **271**, 133 (1992).
35. E. Lippmaa, M. Magi, V. Chvalovsky and J. Schraml, *Collect. Czech. Chem. Comm.*, **42**, 318 (1977).
36. R. West, *J. Organomet. Chem.*, **3**, 314 (1965).
37. D. F. Harnish and R. West, *Inorg. Chem.*, **2**, 1082 (1963).
38. B. G. Ramsey, A. G. Brook, A. R. Bassindale and H. Bock, *J. Organomet. Chem.*, **74**, C41 (1974); F. Agolini, S. Klemenko, I. G. Csizmadia and K. Yates, *Spectrochim. Acta*, **24A**, 169 (1968).
39. E. B. Nadler, Z. Rappoport, D. Arad and Y. Apeloig, *J. Am. Chem. Soc.*, **109**, 7873 (1987).
40. L. E. Orgel, in *Volatile Silicon Compounds* (Ed. E. A. V. Ebsworth), Pergamon Press, Oxford, 1963, p. 81.
41. K. Yates and F. Agolini, *Can. J. Chem.*, **44**, 2229 (1966).
42. H. Dahn, P. Péchy and H. J. Bestmann, *J. Chem. Soc., Perkin Trans. 2*, 1497 (1993).
43. P. C. Chieh and J. Trotter, *J. Chem. Soc.*, 1778 (1969).
44. R. W. Harrison and J. Trotter, *J. Chem. Soc.*, 258 (1968).
45. Q. Liu, Y. Hu, J. Huang and F. Li, *Huaxue Xuabao*, **2**, 68 (1986); *Chem. Abstr.*, **104**, 234791b (1986).
46. L. H. Sommer, D. L. Bailey, G. M. Goldberg, C. E. Buck, T. S. Bye, F. J. Evans and F. C. Whitmore, *J. Am. Chem. Soc.*, **76**, 1613 (1954).
47. A. J. Mancuso, D. S. Brownfain and D. Swern, *J. Org. Chem.*, **44**, 4148 (1979).
48. R. E. Ireland and D. W. Norbeck, *J. Org. Chem.*, **50**, 2198 (1985).
49. B. K. Campion, J. Falk and T. D. Tilley, *J. Am. Chem. Soc.*, **109**, 2049 (1987).
50. W. Sander, W. Trommer, C. H. Ottoson and D. Cremer, *Angew. Chem., Int. Ed. Engl.*, **34**, 929 (1995).
51. F. H. Elsner, H.-G. Woo and T. D. Tilley, *J. Am. Chem. Soc.*, **110**, 313 (1988).
52. (a) J. A. Soderquist and E. I. Miranda, *J. Am. Chem. Soc.*, **114**, 10078 (1992). (b) R. B. Silverman, X. Lu and G. M. Banik, *J. Org. Chem.*, **57**, 6617 (1992).
53. A. R. Katritzky, Z. Yang and Q. Hong, *J. Org. Chem.*, **59**, 5097 (1994).
54. R. J. Linderman and Y. Suhr, *J. Org. Chem.*, **53**, 1569 (1988).
55. H. Woo, W. P. Freeman and T. D. Tilley, *Organometallics*, **11**, 2198 (1992).
56. A. G. Brook, J. M. Duff, P. F. Jones and N. R. Davis, *J. Am. Chem. Soc.*, **89**, 4431 (1967).
57. E. J. Corey, D. Seebach and R. Freedman, *J. Am. Chem. Soc.*, **89**, 434 (1967).
58. T. N. Salzmann, R. W. Ratcliffe, B. G. Christensen and F. A. Bouffard, *J. Am. Chem. Soc.*, **102**, 6163 (1980).
59. D. W. Emerson and H. Wynberg, *Tetrahedron Lett.*, 3445 (1971); M. E. Scheller, G. Iwasaki and B. Frei, *Helv. Chim. Acta*, **69**, 1378 (1986).
60. A. G. Brook and H. W. Kucera, *J. Organomet. Chem.*, **87**, 263 (1975).
61. H. J. Reich, J. J. Rusek and R. E. Olson, *J. Am. Chem. Soc.*, **101**, 2225 (1979).
62. R. L. Danheiser and D. M. Fink, *Tetrahedron Lett.*, **26**, 2509 (1985).
63. K. Suda, J. Watanabe and T. Takanami, *Tetrahedron Lett.*, **33**, 1355 (1992).
64. R. Tacke, K. Fritsche, A. Tafel and F. Wuttke, *J. Organomet. Chem.*, **388**, 47 (1990).

65. R. Burstinghaus and D. Seebach, *Chem. Ber.*, **110**, 841 (1977).
66. T. Mandai, M. Yamaguchi, Y. Nakayama, J. Otera and M. Kawada, *Tetrahedron Lett.*, **26**, 2675 (1985).
67. J. Otera, Y. Niibo and H. Nozaki, *Tetrahedron Lett.*, **33**, 3655 (1992).
68. A. Ricci, A. Degl'Innocenti, M. Ancillotti, G. Seconi and P. Dembech, *Tetrahedron Lett.*, **26**, 5985 (1985).
69. D. J. Ager, *Chem. Soc. Rev.*, **11**, 493 (1982).
70. H. J. Reich and S. K. Shah, *J. Org. Chem.*, **42**, 1773 (1977).
71. D. Wittenberg and H. Gilman, *J. Am. Chem. Soc.*, **80**, 4529 (1958).
72. A. G. Brook and K. M. Baines, *Adv. Organomet. Chem.*, **25**, 1 (1986).
73. G. Raabe and J. Michl in *Chem. Rev.*, **85**, 419 (1985); *The Chemistry of Organic Silicon Compounds*, S. Patai and Z. Rappoport, (Eds.), Wiley-Interscience, New York, 1989.
74. I. Fleming and U. Ghosh, *J. Chem. Soc., Perkin Trans. 1*, 257 (1994).
75. N. Duffaut, J. Dunogues, C. Biran, R. Calas and J. Gerval, *J. Organomet. Chem.*, **161**, C23 (1978).
76. A. Degl'Innocenti, *Synlett*, 937 (1993).
77. J. Kang, J. H. Lee, K. S. Kim, J. U. Jeong and C. Pyun, *Tetrahedron Lett.*, **28**, 3261 (1987).
78. M. Nakada, S. Nakamura, S. Kobayashi and M. Ohno, *Tetrahedron Lett.*, **32**, 4929 (1991).
79. B. F. Bonini, F. Busi, R. C. de Laet, G. Mazzanti, J.-W. J. Thuring, P. Zani and B. Zwanenburg, *J. Chem. Soc., Perkin Trans. 1*, 1011 (1993).
80. B. F. Bonini, M. Comes-Franchini, G. Mazzanti, U. Passamonti, A. Ricci and P. Zani, *Synthesis*, 92 (1995).
81. K. Yamamoto, S. Suzuki and J. Tsuji, *Tetrahedron Lett.*, **21**, 1653 (1980).
82. M. Murakami, A. Hideki, N. Takizawa and Y. Ito, *Organometallics*, **12**, 4223 (1993).
83. A. G. Brook and J. Pierce, *J. Organomet. Chem.*, **30**, 2566 (1965).
84. B. H. Lipshutz, C. Lindsley, R. Susfalk and T. Gross, *Tetrahedron Lett.*, **35**, 8999 (1994).
85. M. E. Scheller and B. Frei, *Helv. Chim. Acta*, **69**, 44 (1986).
86. A. Degl'Innocenti, D. R. M. Walton, G. Seconi, G. Pirazzini and A. Ricci, *Tetrahedron Lett.*, **21**, 3927 (1980).
87. R. Corriu and J. Masse, *J. Organomet. Chem.*, **22**, 321 (1970).
88. T. Aoyama and T. Shioiri, *Tetrahedron Lett.*, **27**, 2005 (1986).
89. A. Hassner and J. Soderquist, *J. Organomet. Chem.*, **131**, C1 (1977).
90. G. Zweifel and S. J. Bäcklund, *J. Am. Chem. Soc.*, **99**, 3184 (1977).
91. A. G. Brook, J. W. Harris and A. R. Bassindale, *J. Organomet. Chem.*, **99**, 379 (1975).
92. P. Bourgeois, J. Dunogues, N. Duffaut and P. Lapouyade, *J. Organomet. Chem.*, **80**, C25 (1974).
93. I. Kuwajima, M. Arai and T. Sato, *J. Am. Chem. Soc.*, **99**, 4181 (1977).
94. I. Kuwajima, N. Minami, T. Abe and T. Sato, *Bull. Chem. Soc. Jpn.*, **51**, 2391 (1978).
95. I. Kuwajima, T. Sato, N. Minami and T. Abe, *Tetrahedron Lett.*, 1591 (1976).
96. J.-P. Picard, R. Calas, J. Dunogues and N. Duffaut, *J. Organomet. Chem.*, **26**, 183 (1971); J.-P. Picard, R. Calas, J. Dunogues, N. Duffaut, J. Gerval and P. Lapouyade, *J. Org. Chem.*, **44**, 420 (1979); J.-P. Picard, A. Ekouya, J. Dunogues, N. Duffaut and R. Calas, *J. Organomet. Chem.*, **93**, 51 (1975); P. Bourgeois, *J. Organomet. Chem.*, **76**, C1 (1972).
97. I. Kuwajima, M. Kato and T. Sato, *J. Chem. Soc., Chem. Commun.*, 478 (1978).
98. T. Cohen and J. R. Matz, *J. Am. Chem. Soc.*, **102**, 6900 (1980).
99. S. Murai, I. Ryu, J. Iriguchi and N. Sonoda, *J. Am. Chem. Soc.*, **106**, 2440 (1984).
100. T. Inoue, N. Kambe, I. Ryu and N. Sonoda, *J. Org. Chem.*, **59**, 8209 (1994).
101. J. B. Verlhac, H. Kwon and M. Pereyre, *J. Organomet. Chem.*, **437**, C13 (1992).
102. N. Chatani, S. Ikeda, K. Ohe and S. Murai, *J. Am. Chem. Soc.*, **114**, 9710 (1992).
103. J. A. Soderquist and G. J.-H. Hsu, *Organometallics*, **1**, 830 (1982); E. M. Dexheimer and L. Spialter, *J. Organomet. Chem.*, **107**, 229 (1976).
104. J. A. Soderquist, *Org. Synth.*, **68**, 25 (1990).
105. J. Yoshida, S. Matsunaga, Y. Ishichi, T. Maekawa and S. Isoe, *J. Org. Chem.*, **56**, 1307 (1991).
106. D. Seyferth and R. M. Weinstein, *J. Am. Chem. Soc.*, **104**, 5534 (1982).
107. D. Seyferth and R. C. Hui, *Organometallics*, **3**, 327 (1984).
108. P. Bourgeois, *J. Organomet. Chem.*, **76**, C1 (1974).
109. G. E. Niznik, W. H. Morrison and H. M. Walborsky, *J. Org. Chem.*, **39**, 600 (1974).
110. A. Orita, K. Ohe and S. Murai, *J. Organomet. Chem.*, **474**, 23 (1994).
111. A. Orita, K. Ohe and S. Murai, *Organometallics*, **13**, 1533 (1994).

27. Acyl silanes

112. P. Raubo and J. Wicha, *J. Org. Chem.*, **59**, 4355 (1994).
113. H. M. I. Osborn, J. B. Sweeney and B. Howson, *Synlett*, 675 (1993).
114. K. Maruoko and H. Yamamoto, *Synlett*, 793 (1991); K. Maruoko, H. Banno and H. Yamamoto, *Tetrahedron: Asymmetry*, **2**, 647 (1991).
115. I. Kuwajima, T. Abe and N. Minami, *Chem. Lett.*, 993 (1976).
116. J. S. Nowick and R. L. Danheiser, *J. Org. Chem.*, **54**, 2798 (1989).
117. P. C. B. Page and S. Rosenthal, *Tetrahedron Lett.*, **27**, 5421 (1986).
118. J. Hooz and J. N. Bridson, *Can. J. Chem.*, **50**, 2387 (1972).
119. C. T. Hewkin and R. F. W. Jackson, *Tetrahedron Lett.*, **31**, 1877 (1990).
120. C. T. Hewkin and R. F. W. Jackson, *J. Chem. Soc., Perkin Trans. 1*, 3103 (1991); M. Ashwell, W. Clegg and R. F. W. Jackson, *J. Chem. Soc., Perkin Trans. 1*, 897 (1991).
121. Y. Horiuchi, M. Taniguchi, K. Oshima and K. Utimoto, *Tetrahedron Lett.*, **36**, 5353 (1995).
122. F. Jin, B. Jiang and Y. Xu, *Tetrahedron Lett.*, **33**, 1221 (1992).
123. S. Hoff, L. Brandsma and J. F. Arens, *Recl. Trav. Chim. Pays-Bas*, **87**, 916 (1968).
124. P. C. B. Page and S. Rosenthal, *Tetrahedron Lett.*, **27**, 2527 (1986).
125. P. C. B. Page and S. Rosenthal, *Tetrahedron Lett.*, **27**, 1947 (1986).
126. R. W. Murray and M. Singh, *J. Org. Chem.*, **58**, 5076 (1993).
127. W. C. Still and T. L. Macdonald, *J. Am. Chem. Soc.*, **96**, 5561 (1974).
128. A. Wright and R. West, *J. Am. Chem. Soc.*, **96**, 3214 (1974).
129. N. Minami, T. Abe and I. Kuwajima, *J. Organomet. Chem.*, **145**, C1 (1978).
130. A. Ricci, A. Degl'Innocenti, A. Capperucci, C. Faggi, G. Seconi and L. Favaretto, *Synlett*, 471 (1990).
131. K. Narasaka and T. Shibata, *Bull. Chem. Soc. Jpn.*, **65**, 2825 (1992).
132. A. Degl'Innocenti, P. Uliva, A. Capperucci, G. Reginato, A. Mordini and A. Ricci, *Synlett*, 883 (1992).
133. J. A. Miller and G. Zweifel, *J. Am. Chem. Soc.*, **103**, 6217 (1981).
134. D. J. Ager, *Synthesis*, 384 (1984).
135. R. F. Cunico and C. Zhang, *Tetrahedron Lett.*, **33**, 6751 (1992).
136. A. Degl'Innocenti, A. Ricci, G. Reginato and G. Seconi, *Pure Appl. Chem.*, **64**, 439 (1992).
137. A. Degl'Innocenti, E. Stucchi, A. Capperucci, A. Mordini, G. Reginato and A. Ricci, *Synlett*, 329 (1992).
138. A. Degl'Innocenti, A. Capperucci, G. Reginato, A. Mordini and A. Ricci, *Tetrahedron Lett.*, **33**, 1507 (1992).
139. A. Degl'Innocenti, E. Stucchi, A. Capperucci, A. Mordini, G. Reginato and A. Ricci, *Synlett*, 332 (1992).
140. A. Degl'Innocenti, A. Capperucci, L. Bartoletti, A. Mordini and G. Reginato, *Tetrahedron Lett.*, **35**, 2081 (1994).
141. R. L. Danheiser and D. M. Fink, *Tetrahedron Lett.*, **26**, 2513 (1985).
142. R. Plantier-Royon and C. Portella, *Synlett*, 527 (1994).
143. K. J. H. Kruithof, R. F. Schmitz and G. W. Klumpp, *J. Chem. Soc., Chem. Commun.*, 239 (1983); K. J. H. Kruithof, R. F. Schmitz and G. W. Klumpp, *Tetrahedron*, **39**, 3073 (1983).
144. K. Narasaka, H. Kusama and Y. Hayashi, *Tetrahedron*, **48**, 2059 (1992).
145. L. A. Paquette and G. D. Maynard, *J. Org. Chem.*, **56**, 5480 (1991).
146. T. Nakajima, H. Miyaji, M. Segi and S. Suga, *Chem. Lett.*, 181 (1986).
147. T. Nakajima, H. Miyaji, M. Segi and S. Suga, *Chem. Lett.*, 177 (1986).
148. J. S. Nowick and R. L. Danheiser, *Tetrahedron*, **44**, 4113 (1988).
149. I. Fleming, in *Comprehensive Organic Chemistry*, Vol. 3 (Ed. N. Jones), Pergamon Press, Oxford, 1979, p. 647.
150. P. D. Magnus, T. Sarkar and S. Dujuric, in *Comprehensive Organometallic Chemistry*, Vol. 7, Pergamon Press, Oxford, 1982, p. 631.
151. E. W. Colvin, *Silicon in Organic Synthesis*, Butterworths, London, 1981.
152. E. W. Colvin, *Silicon Reagents in Organic Synthesis*, Academic Press, London, 1988.
153. A. G. Brook, *Acc. Chem. Res.*, **7**, 77 (1974).
154. (a) I. Fleming, in *Comprehensive Organic Chemistry*, Vol. 3 (Ed. N. Jones), Pergamon Press, Oxford, 1979, p. 554.
     (b) P. C. B. Page and S. Rosenthal, *J. Chem. Res. (S)*, **10**, 302 (1990).
155. R. J. P. Corriu and C. Guerin, *Adv. Organomet. Chem.*, **20**, 265 (1982).
156. D. Rietropaolo, A. Ricci, M. Taddei and M. Fiorenza, *J. Organomet. Chem.*, **197**, 7 (1980).

157. A. G. Brook and N. V. Schwartz, *J. Org. Chem.*, **27**, 2311 (1962).
158. A. G. Brook, W. Limburg and T. S. D. Vandersar, *Can. J. Chem.*, **56**, 2758 (1978).
159. E. D. Hughes and C. K. Ingold, *J. Chem. Soc.*, 244 (1935).
160. P. C. B. Page, S. Rosenthal and R. V. Williams, *Tetrahedron Lett.*, **28**, 4455 (1987).
161. A. Yanagisawa, S. Habaue and H. Yamamoto, *Tetrahedron*, **48**, 1969 (1992).
162. M. R. Burns and J. K. Coward, *J. Org. Chem.*, **58**, 529 (1993).
163. R. Munsted and U. Wannagat, *J. Organomet. Chem.*, **322**, 11 (1987); R. Munsted and U. Wannagat, *Monatsh. Chem.*, **116**, 693 (1985).
164. A. Fürstner, G. Kollegger and H. Weidmann, *J. Organomet. Chem.*, **414**, 295 (1991).
165. S. R. Wilson, M. S. Hague and R. N. Misra, *J. Org. Chem.*, **47**, 747 (1982).
166. I. Kuwajima, T. Matsutani and J. Enda, *Tetrahedron Lett.*, **25**, 5307 (1984).
167. J. D. Buynak, J. B. Strickland, T. Hurd and A. Phan, *J. Chem. Soc., Chem. Commun.*, 89 (1989).
168. S. Itsuno, M. Nakano, K. Miyazaki, H. Masuda, K. Ito, A. Hirao and S. Nakahama, *J. Chem. Soc., Perkin Trans. 1*, 2039 (1985).
169. J. A. Soderquist, C. L. Anderson, E. I. Miranda and I. Rivera, *Tetrahedron Lett.*, **31**, 4677 (1990).
170. J. D. Buynak, J. B. Strickland, G. W. Lamb, D. Khasnis, S. Modi, D. Williams and H. Zhang, *J. Org. Chem.*, **56**, 7076 (1991).
171. A. R. Bassindale, A. G. Brook, P. F. Jones and J. M. Lennon, *Can. J. Chem.*, **53**, 332 (1975).
172. P. F. Cirillo and J. S. Panek, *J. Org. Chem.*, **59**, 3055 (1994).
173. J. D. Buynak, J. B. Geng, S. Uang and J. B. Strickland, *Tetrahedron Lett.*, **35**, 985 (1994).
174. B. F. Bonini, S. Masiero, G. Mazzanti and P. Zani, *Tetrahedron Lett.*, **32**, 6801 (1991).
175. A. Chapeaurouge and S. Bienz, *Helv. Chim. Acta*, **76**, 1876 (1993).
176. S. Bienz and A. Chapeaurouge, *Helv. Chim. Acta*, **74**, 1477 (1991).
177. M. Nakada, Y. Urano, K. Susumu and M. Ohno, *J. Am. Chem. Soc.*, **110**, 4826 (1988).
178. M. Nakada, Y. Urano, S. Kobayashi and M. Ohno, *Tetrahedron Lett.*, **35**, 741 (1994).
179. P. F. Cirillo and J. S. Panek, *J. Org. Chem.*, **55**, 6071 (1990).
180. H. J. Reich, R. C. Holtan and S. L. Borkowsky, *J. Org. Chem.*, **52**, 312 (1987).
181. M. Cherest, H. Felkin and N. Prudent, *Tetrahedron Lett.*, 2199 (1968); N. T. Anh and O. Eisenstein, *Nouv. J. Chim.*, **1**, 61 (1977); A. S. Cieplak, *J. Am. Chem. Soc.*, **103**, 4540 (1981).
182. A. G. Brook and S. A. Fieldhouse, *J. Organomet. Chem.*, **10**, 235 (1967).
183. J. A. Soderquist and C. L. Anderson, *Tetrahedron Lett.*, **29**, 2425 (1988); J. A. Soderquist and C. L. Anderson, *Tetrahedron Lett.*, **29**, 2777 (1988).
184. C. L. Anderson, J. A. Soderquist and G. W. Kabalka, *Tetrahedron Lett.*, **33**, 6915 (1992).
185. N. A. Petasis and S.-P. Lu, *Tetrahedron Lett.*, **36**, 2393 (1995).
186. T. Nakajima, M. Segi, F. Sugimoto, R. Hioki, S. Yokota and K. Miyashita, *Tetrahedron*, **49**, 8343 (1993).
187. K. Takeda, J. Nakatani, H. Nakamura, K. Sako, E. Yoshii and K. Yamaguchi, *Synlett*, 841 (1993).
188. T. Brigaud, P. Doussot and C. Portella, *J. Chem. Soc., Chem. Commun.*, 2117 (1994).
189. Y. Taniguchi, A. Nagafuji, Y. Makioka, K. Takaki and Y. Fujiwara, *Tetrahedron Lett.*, **35**, 6897 (1994).
190. T. Nakajima, M. Segi, T. Mituoka, Y. Fukute, M. Honda and K. Naitou, *Tetrahedron Lett.*, **36**, 1667 (1995).
191. C. H. Heathcock and D. Schinzer, *Tetrahedron Lett.*, **22**, 1881 (1981).
192. A. Degl'Innocenti, S. Pike, D. R. M. Walton, G. Seconi, A. Ricci and M. Fiorenza, *J. Chem. Soc., Chem. Commun.*, 1201 (1980).
193. C. H. DePuy, V. M. Bierbaum, R. Damrauer and J. A. Soderquist, *J. Am. Chem. Soc.*, **107**, 3385 (1985).
194. G. A. Molander and C. S. Siedem, *J. Org. Chem.*, **60**, 130 (1995).
195. Y.-M. Tsai, H.-C. Nieh and C.-D. Chemg, *J. Org. Chem.*, **57**, 7010 (1992).
196. B. F. Bonini, M. Comes-Franchini, G. Mazzanti, A. Ricci, L. Rosa-Fauzza and P. Zani, *Tetrahedron Lett.*, **35**, 9227 (1994).
197. J. M. Harris, I. MacInnes, J. C. Walton and B. Maillard, *J. Organomet. Chem.*, **403**, 25 (1991).
198. Y.-M. Tsai, K.-H. Tang and W.-J. Jiaang, *Tetrahedron Lett.*, **34**, 1303 (1993).
199. D. P. Curran, W.-J. Jiaang, M. Palovich and Y.-M. Tsai, *Synlett*, 403 (1993).
200. Y.-M. Tsai and S.-Y. Chang, *J. Chem. Soc., Chem. Commun.*, 981 (1995).
201. K. M. Baines, A. G. Brook, P. D. Lickiss and J. F. Sawyer, *Organometallics*, **8**, 709 (1989).
202. A. G. Brook, R. Kumarathasan and W. Chatterton, *Organometallics*, **12**, 4085 (1993).
203. A. G. Brook and J. M. Duff, *J. Am. Chem. Soc.*, **89**, 454 (1967).

204. R. A. Bourque, P. D. Davis and J. C. Dalton, *J. Am. Chem. Soc.*, **103**, 697 (1981).
205. A. R. Bassindale, A. G. Brook and J. Harris, *J. Organomet. Chem.*, **90**, C6 (1975).
206. C. Shih and J. S. Swenton, *J. Org. Chem.*, **47**, 2668 (1982).
207. M. E. Scheller and B. Frei, *Helv. Chim. Acta*, **75**, 69 (1992).
208. R. Tacke, S. A. Wagner, S. Brakmann, F. Wüttke, U. Eilert, L. Fischer and C. Syldatk, *J. Organomet. Chem.*, **458**, 13 (1993).
209. (a) R. Tacke, S. Brakmann, F. Wüttke, J. Fooladi, C. Syldatk and D. Schomburg, *J. Organomet. Chem.*, **403**, 29 (1991).
    (b) L. Fischer, S. A. Wagner and R. Tacke, *Appl. Microbiol. Biotechnol.*, **42**, 671 (1995).
210. T. Sato, M. Arai and I. Kuwajima, *J. Am. Chem. Soc.*, **99**, 5827 (1977).
211. P. Huber, V. Enev, A. Linden and S. Bienz, *Tetrahedron*, **51**, 3749 (1995).
212. P. F. Cirillo and J. S. Panek, *Tetrahedron Lett.*, **32**, 457 (1991).
213. J. Yoshida, M. Itoh, S. Matsunaga and S. Isoe, *J. Org. Chem.*, **57**, 4877 (1992).
214. A. Ricci, A. Degl'Innocenti, A. Capperucci and G. Reginato, *J. Org. Chem.*, **54**, 19 (1989).
215. Y. Taniguchi, N. Fujii, Y. Makioka, K. Takaki and Y. Fujiwara, *Chem. Lett.*, 1165 (1993).
216. R. C. Cambie, L. C. M. Mui, P. S. Rutledge and P. D. Woodgate, *J. Organomet. Chem.*, **464**, 171 (1994).
217. N. de Kimpe, P. Sulmon and N. Schamp, *Angew. Chem., Int. Ed. Engl.*, **24**, 881 (1985).
218. D. J. Cram and F. A. Abd Elhafez, *J. Am. Chem. Soc.*, **74**, 5828 (1952).
219. I. Kuwajima and K. Matsumoto, *Tetrahedron Lett.*, 4095 (1979).
220. F. Jin, Y. Xu and W. Huang, *J. Chem. Soc., Perkin Trans. 1*, 795 (1993).
221. F. Jin, B. Jiang and Y. Xu, *Tetrahedron Lett.*, **33**, 1221 (1992).
222. F. Jin, Y. Xu and W. Huang, *J. Chem. Soc., Chem. Commun.*, 814 (1993).
223. K. Takeda, M. Fujisawa, T. Makino and E. Yoshii, *J. Am. Chem. Soc.*, **115**, 9351 (1993).
224. B. F. Bonini, S. Masiero, G. Mazzanti and P. Zani, *Tetrahedron Lett.*, **32**, 2971 (1991).
225. B. Lipshutz, C. Lindsley and A. Bhandari, *Tetrahedron Lett.*, **35**, 4669 (1994).
226. S. E. Thomas and G. J. Tustin, *Tetrahedron*, **48**, 7629 (1992).

CHAPTER 28

Recent synthetic applications of organosilicon reagents

ERNEST W. COLVIN

Department of Chemistry, University of Glasgow, Glasgow G12 8QQ, UK

I. INTRODUCTION	1667
II. TRIALKYLSILYL HALIDES	1667
III. TRIALKYLSILYL CYANIDES	1670
IV. TRIALKYLSILYL AZIDES	1672
V. ALKYL SILYL ETHERS	1674
VI. (TRIHALOMETHYL)TRIALKYLSILANES	1675
VII. TRIMETHYLSILYLDIAZOMETHANE	1675
VIII. TRIALKYLSILYL TRIFLUOROMETHANESULPHONATES	1676
IX. THIO- AND SELENOSILANES	1677
X. α,β-EPOXYSILANES	1679
XI. Si–C OXIDATIVE CLEAVAGE	1680
XII. MISCELLANEOUS	1682
XIII. REFERENCES	1682

I. INTRODUCTION

It is the purpose of this chapter to review selectively the advances in the synthetic applications of organosilicon compounds, excluding those organosilanes covered in other chapters. Since the publication of *The Chemistry of Organic Silicon Compounds*, two monographs[1,2] and several reviews[3–5] have been published, as have three conference proceedings[6–8] and a 'symposium-in-print'[9].

II. TRIALKYLSILYL HALIDES

The chemistry of iodotrimethylsilane (TMSI) has been reviewed[10].

The role of chlorotrimethylsilane (TMSCl) in accelerating organocuprate conjugate additions[11] continues to arouse much discussion. It has been proposed that TMSCl acts as a Lewis acid in complexing the enone. However, high level *ab initio* calculations have shown that such complexation is only a weak dipole–dipole interaction, and cannot

The chemistry of organic silicon compounds, Vol. 2
Edited by Z. Rappoport and Y. Apeloig © 1998 John Wiley & Sons Ltd

be responsible for the rate acceleration. The most recent suggestion[12] proposes two fundamentally different modes of action. With the powerful silylating conditions of TMS-Cl/HMPA, direct silylation of the initially formed π-complex **1** can occur. Under relatively weak silylating conditions, i.e. TMSCl alone, the rate of cuprate addition can be enhanced by conversion of the π-complex into a tetravalent copper species **2** which then undergoes reductive elimination to **3**, the transition state for this elimination being stabilized by the complexation which places silicon β to copper (equation 1).

An RCu(LiI)/TMSI combination[13] promotes conjugate addition to α,β-unsaturated ketones, leading directly to TMS enol ethers from a presumed organocopper–enone π-complex, thus ruling out significant amounts of enolate as an intermediate on the reaction pathway (equation 2). However, these results do not allow discrimination between an α-cuprio ketone or a β-silyl copper species.

Activation by TMSCl has also been applied to the allylstannane allylation of aldimines[14], the copper-catalysed conjugate addition of organoaluminium reagents[15] to enones, and in the uncatalysed conjugate addition of organozincs[16] in THF/N-methylpyrrolidone. Its involvement in the conjugate addition of stabilized organolithiums[17] results in yield and selectivity enhancement.

$$\text{(2)}$$

NMR spectroscopic experiments on the role of TMSCl in lithium dialkylamide deprotonations have shown[18] that LDA and TMSCl react at −78 °C to generate enough LiCl to be responsible for the stereoselectivities normally observed, i.e. hindered amide bases are not fully compatible with TMSCl at −78 °C. This does not apply when TBDMSCl or TBDPSCl is used; in such cases, it is probably necessary to add externally prepared LiCl to enhance stereoselectivity.

Motherwell and Roberts' dicarbonyl coupling reaction[19] can be improved by substituting 1,2-bis(chlorodimethylsilyl)ethane for TMSCl. The intermediate organozinc carbenoids can be trapped with alkenes to produce cyclopropanes, as exemplified in the intramolecular case in equation 3.

$$\text{(3)}$$

TMSCl serves as an efficient chlorinating agent[20] in the presence of a catalytic amount of selenium dioxide for a wide variety of alcohols; the reagent system TMSCl/DMSO converts primary and tertiary alcohols into the corresponding chlorides, probably by S_N2 and S_N1 pathways, respectively, although no stereochemical detail was given[21]. Allylic acetates can be converted into allylic bromides using TMSBr with ZnI_2 as catalyst, with the regiochemistry being controlled by the relative stabilities of the product alkenes[22].

The combination of TMSCl and $NaNO_2$ converts anilines efficiently into the corresponding aryl chlorides[23]; use of TMSBr or TMSI gives the corresponding bromides or iodides, respectively. $AlCl_3$ catalysed aromatic nitration[24] can be performed using TMSCl/$NaNO_3$.

TMSCl is essential as a promoter in a low-valent vanadium-catalysed reductive cyclotrimerization[25] of aliphatic aldehydes, producing 1,3-dioxolanes (equation 4). It is also an essential component in the zinc-induced intramolecular imine cross coupling[26] of diarylidene sulphamides; subsequent cleavage provides a route to unsymmetrical 1,2-diaryl-1,2-diaminoethanes (equation 5). It acts as an activating agent in the samarium-promoted cyclopropanation[27] of allylic and α-allenic alcohols.

A synthesis of 1-β-methylcarbapenems has been described[28] in which TMSCl acts both as a Lewis acid in accelerating the Dieckmann-type cyclization and as a trapping agent for the initially liberated thiolate anion (equation 6).

$$RCHO \xrightarrow{Zn, TMSCl, cat. CpV(CO)_4} \text{[dioxolane product]} \quad (4)$$

$$\text{[thiadiazetidine dioxide]} \xrightarrow{Zn, TMSCl} \text{[sulfamide]} \xrightarrow{HBr} \text{[diamine]} \quad (5)$$

$$\text{[azetidinone-COSR]} \xrightarrow[\substack{1.\ NaN(TMS)_2 \\ 2.\ TMSCl \\ 3.\ (PhO)_2P(O)Cl \\ 4.\ TBAF}]{} \text{[carbapenem-SR]} \quad (6)$$

III. TRIALKYLSILYL CYANIDES

The chemistry of TMSCN has been reviewed[29].

α,β-Unsaturated ketones undergo regioselective addition[30] with TMSCN to give 1,4-adducts in the presence of strongly acidic montmorillonite or 1,2-adducts when strongly basic CaO or MgO is employed. Dibutyltin and diphenyltin dichloride are effective catalysts[31] for the addition of TMSCN to aldehydes and unactivated ketones; with an α,β-unsaturated aldehyde, exclusive 1,2-addition was observed.

A key step in a synthesis[32] of 18-O-methyl mycalamide B involved the conjugate addition of TBDMSCN, with TBDMSOTf catalysis, to give a dihydropyranone; this proceeded with very high 1,3-asymmetric induction (equation 7).

$$\text{[dihydropyranone]} \xrightarrow[\text{TBDMSOTf}]{\text{TBDMSCN}} \text{[CN adduct with OTBDMS]} \quad (7)$$

Di-isobutylaluminium hydride reduction[33] of aldehyde-derived O-TBDMS cyanohydrins, followed by hydrolysis with a tartaric acid buffer, gives the corresponding O-protected 2-hydroxyaldehydes. Related methodology has been employed in a diastereoselective synthesis[34] of a potential precursor of taxol®.

Enantioselective TMS cyanohydrin formation with aldehydes can be achieved using a variety of chirally-modified titanium catalytic systems; chiral modifiers include Schiff bases[35], including salen ligands[36] and sulphoximines[37]. A chirally-modified yttrium complex[38] has been employed for the same purpose (equation 8); here, the chiral modifier was a ferrocene-derived 1,3-diketone. This general area has been reviewed recently[39].

$$PhCHO \xrightarrow[\text{cat. } L^*\text{-}Y_5(O)(OPr\text{-}i)_{13}]{\text{TMSCN}} \underset{90\% \text{ ee}}{\overset{\text{OTMS}}{\underset{Ph}{\bigvee}}\text{CN}} \quad (8)$$

There have been further reports[40,41] on the Lewis acid induced opening of epoxides with TMSCN to produce β-trimethylsilyloxy nitriles.

Full details[42] have been published on the conversion of enynes into iminocyclopentenes using a titanium precatalyst in the presence of BuLi and TESCN (equation 9); the resulting iminocyclopentenes can be hydrolysed to cyclopentenones or reduced to allylic silylamines. In a related protocol[43], the tandem insertion of TMSCN and alkenes, alkynes, ketones or isocyanates into zirconacyclo-pentanes or -pentenes leads to cyclopentylamines carrying an α-alkyl, -alkenyl, -1-hydroxyalkyl or -carboxamide substituent, respectively (equation 9).

TMSCN has been employed (equation 10) in an efficient conversion[44] of 1-tetralones into the less readily available 2-tetralones via the derived unsaturated nitriles.

(9)

IV. TRIALKYLSILYL AZIDES

Both samarium iodide and ytterbium triisopropoxide catalyse the ring opening[45] of epoxides by TMS-azide; in the latter case, due to an acidic work-up, the products are isolated as vicinal azido alcohols.

Full details on this ring opening[46] under titanium tetraisopropoxide or aluminium triisopropoxide catalysis have been published. Using chirally modified titanium catalysts, cyclohexene oxide provides[47] trans-2-azidocyclohexanol in up to 63% ee.

More recently, Jacobsen and coworkers have found that epoxides undergo a highly enantioselective ring-opening with TMS-azide when catalysed by (salen)Cr(III) complexes such as (S,S)-**4**[48]. This asymmetric ring opening shows a second-order rate dependence on catalyst concentration[49]. Applications of the process have included kinetic resolution of terminal epoxides[50], an efficient synthesis[51] of (R)-4-(trimethylsilyloxy)cyclopent-2-enone, the dynamic kinetic resolution (equation 11) of epichlorohydrin[52], an enantioselective route[53] to carbocyclic nucleoside analogues and a formal synthesis[54] (equation 12) of the protein kinase inhibitor **5**.

Magnus and coworkers have published full details[55] on the direct α- or β-azido functionalization of triisopropylsilyl (TIPS) enol ethers using an iodosylbenzene–TMS-azide combination (equation 13); the α-pathway, favoured at −78 °C, is an azide radical addition process, whereas the β-pathway, favoured at −15 to −20 °C, involves ionic dehydrogenation. Attempts to extend the β-functionalization to other TMSX derivatives failed.

A novel approach to the synthesis of lactams has been reported[56] which involves reaction of such enol ethers with TMS-azide to give (triisopropylsilyl)azidohydrins, which then undergo a photoinduced Schmidt rearrangement (equation 14).

PPTS = Pyridinium *p*-toluenesulfonate

The reagent system TMS-azide/triflic acid performs efficient amination[57] of arenes, while the combination of TMS-azide and N-bromosuccinimide with Nafion-H® transforms alkenes into β-bromoalkyl azides[58]. On the other hand, the combination of TMS-azide and chromium trioxide converts alkenes into α-azidoketones[59] and aldehydes into acyl azides[60].

Alkyl halides can be converted into alkyl azides with inversion of configuration[61] using TMS-azide in the presence of TBAF.

V. ALKYL SILYL ETHERS

The protection of alcohols as silyl ethers has been reviewed[62], as have the relative stabilities of the different trialkylsilyl groups[63]. Their stability under alcohol oxidation conditions and their oxidative deprotection have been discussed[64]. Methods for selective deprotection of the various silyl ethers have been the subject of an excellent review[65].

The merits of TIPS ether protection have been reviewed[66], with emphasis on those reactions where the incremental structural difference between TIPS and other commonly used trialkylsilyl groups leads to results quite different from those obtained with TMS or TBDMS ethers.

The 'temporary silicon connection', pioneered by Stork,[67a] and based on the protocol of temporarily bringing two reaction partners together by means of an eventually removable silicon atom, often as silyl ethers, has been surveyed[67b]. This concept, i.e. that of converting an intermolecular reaction into its intramolecular equivalent, has proven a very valuable synthetic strategy for various types of reactions.

The selective cleavage of the N-t-butoxycarbonyl group in the presence of TBDMS (t-butyldimethylsilyl) (selective only when the protected alcohol was phenolic) or TBDPS (tert-butyldiphenylsilyl) (completely selective) ethers can be achieved[68] using a saturated solution of HCl in ethyl acetate.

Primary TBDMS ethers can be cleaved selectively[69] in the presence of THP ethers and ketals using ceric ammonium nitrate in methanol. Both phenolic[70] and aliphatic[71] TBDMS ethers undergo cleavage when exposed to catalytic amounts of $PdCl_2(MeCN)_2$; in the latter case, and with longer exposure, this also results in oxidation to the corresponding aldehyde or ketone, if the alcohol was primary or secondary, respectively. This cleavage and oxidation can be performed selectively in the presence of TIPS, TBDPS and benzyl ethers (equation 15).

R = TIPS, TBDPS, PhCH$_2$

(15)

Two closely related methods for the direct transformation of THP ethers into TBDMS ethers, using TBDMSOTf and a tertiary amine base, have been disclosed[72,73]. This reagent system has been reported[74] to perform the same transformation on *t*-butyl and *t*-amyl ethers. In a similar vein[75], *p*-methoxybenzyl ethers are converted directly into the corresponding silyl ethers using TMSOTf and triethylamine.

A new alcohol and phenol protective group, the 1-[(2-trimethylsilyl)ethoxy]ethyl moiety, readily introduced using 2-(trimethylsilyl)ethyl vinyl ether and catalytic pyridinium *p*-toluenesulfonate (PPTS), has been described[76]. Cleavage is achieved under near-neutral conditions using TBAF monohydrate (equation 16).

$$ROH \underset{TBAF \cdot H_2O}{\overset{\diagup\!\!\!\diagdown O \diagdown\!\!\!\diagup TMS, \; PPTS}{\rightleftarrows}} RO\diagdown\!\!\!\diagup O \diagdown\!\!\!\diagup TMS \qquad (16)$$

VI. (TRIHALOMETHYL)TRIALKYLSILANES

Full details[77] have been provided for the optimized preparation of (trifluoromethyl)trimethylsilane (Ruppert's Reagent[78]). It acts as a nucleophilic trifluoromethide equivalent, normally under fluoride ion initiation, reacting readily with a variety of carbonyl electrophiles and related species[79].

It adds (equation 17) to α-amino acid derived oxazolidin-5-ones giving, after acid hydrolysis, protected α-amino trifluoromethyl ketones[80], which are of interest as serine protease inhibitors.

(Trifluoromethyl)tributyl tin, prepared from (trifluoromethyl)trimethylsilane and bis(tributyltin) oxide, has been reported[81] to react with disilyl sulphides to give the corresponding (trifluoromethyl)di- and (trifluoromethyl)trialkylsilanes.

The preparation and reactions of a range of mixed (trihalomethyl)trimethylsilanes have been described[82].

VII. TRIMETHYLSILYLDIAZOMETHANE

The chemistry of TMS-diazomethane has been reviewed[83].

The reaction between lithio TMS-diazomethane and carbonyl compounds, which generates alkylidenecarbenes, has been studied further. Aliphatic ketones give cyclopentene derivatives[84], while aldehydes and aryl alkyl ketones give alkynes[85]. The diazoalkene intermediates from aliphatic ketones can be trapped with diisopropylamine, producing aldehyde enamines and thus a method[86] for the one-carbon homologation of ketones to aldehydes (equation 18).

Lithio TMS-diazomethane reacts with ketenimines to produce 4-TMS-1,2,3-triazoles or 4-amino-3-TMS-pyrazoles[87,88], depending on the keteneimine substitution pattern.

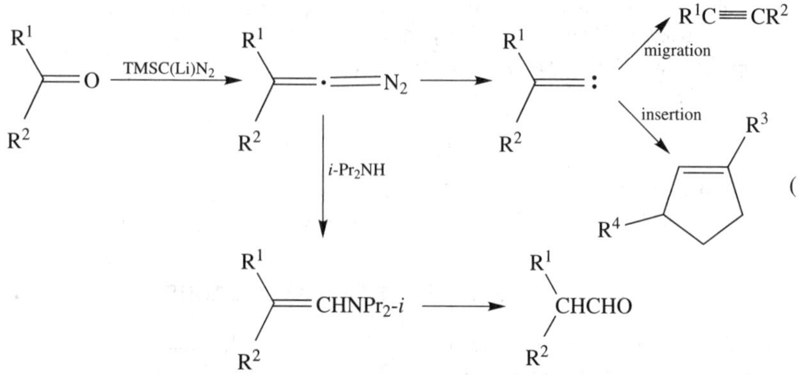

(18)

VIII. TRIALKYLSILYL TRIFLUOROMETHANESULPHONATES

The preparation and utility of trialkylsilyl perfluoroalkanesulphonates have been reviewed[89].

Trimethylsilyl fluorosulphonate, generated by *in situ* reaction of fluorosulphonic acid with either tetramethylsilane or allyltrimethylsilane, has been proposed[90] as a cheaper alternative to TMSOTf.

The combination of TMSOTf, dimethyl sulphide and 1,2-bis(trimethylsilyloxy)ethane provides a method[91] for the selective dioxolanation of ketones in the presence of aldehydes, via intermediate protection of the aldehyde as its silyloxysulphonium salt (equation 19).

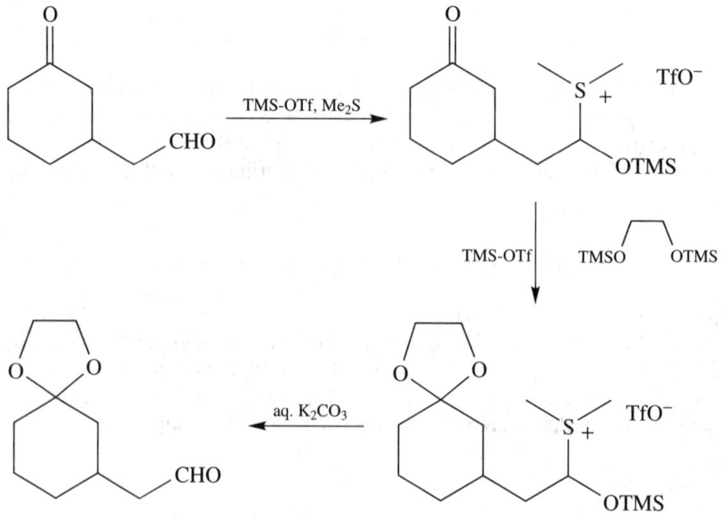

(19)

Dimethoxyamine reacts with simple alkenes in the presence of TMS-OTf to produce N-methoxyaziridines in good yield[92]; cyclic ketones react in a Beckmann-type process to yield N-methoxylactams (equation 20).

$$ (20) $$

The aluminoxy acetal intermediates in ester reduction using DIBALH can be trapped with either TMSOTf[93] or TMS imidazole[94] to give monosilyl acetals (equation 21).

$$ (21) $$

Ring-opening[95] of tetrahydrofuran with TBDMS-Mn(CO)$_5$, generated *in situ* from TBDMSOTf and NaMn(CO)$_5$, proved a key step (equation 22) in the synthesis of spiroketal lactones, precursors of certain insect pheromones.

$$ (22) $$

IX. THIO- AND SELENOSILANES

The preparation and reactions of organosilathianes and other mixed silicon–sulphur compounds have been reviewed[96,97].

Full details[98] have been published on the use of hexamethyldisilathiane in the generation of thioaldehydes and thioketones, which can be trapped *in situ* (equation 23). The corresponding selenium reagent has been employed[99] in the generation of selenoaldehydes.

$$\begin{array}{c} \text{(23)} \end{array}$$

Triphenylsilanethiol, a white crystalline solid, has been recommended[100] for the ring-opening of epoxides to β-hydroxymercaptans or β-dihydroxysulphides, depending on the choice of base (equation 24).

$$\begin{array}{c} \text{(24)} \end{array}$$

Nucleophilic bis-*O*-demethylation of dimethoxybenzenes in one flask is often difficult. It can, however, be achieved by use of sodium TMS thiolate in 1,3-dimethyl-2-imidazolidinone (DMEU) at high temperature in a sealed tube[101]. The same reagent system converts nitriles into primary thioamides at or slightly above room temperature in varying yields[102].

The TMSOTf catalysed reaction of (phenylthio)trimethylsilane with aldehydes to give hemithioacetals has been employed in a stereocontrolled route to polyols, via the 1,3-diol synthon shown (equation 25)[103].

$$\begin{array}{c} \text{(25)} \end{array}$$

TMS isoselenocyanate, Me$_3$SiNCSe, prepared *in situ* from TMSCl and excess KSeCN, shows remarkable chemoselectivity[104] in the transformation of carbonyl compounds into

O-TMS cyanohydrins. Aliphatic aldehydes react smoothly, aromatic and α,β-unsaturated aldehydes react very slowly, with ketones being unreactive.

X. α,β-EPOXYSILANES

The preparation and applications of α,β-epoxysilane α-anions have been described in detail[105], as have the preparation and applications of α,β-epoxysilanes themselves[106].

α,β-Epoxysilanes undergo nucleophilic ring-opening with a regioselectivity normally, but not exclusively, due to α-attack. The question of whether this nucleophilic ring-opening involves prior coordination of the nucleophile with silicon has been addressed further. It has been concluded[107] that whether or not pentacoordinate intermediates are involved, they do not appear to be on the product-determining pathway.

A careful re-investigation[108] of the ring-opening reactions of (triphenylsilyl)ethylene oxide by Grignard reagents with added $MgBr_2$ has shown them to be analogous to that of the trimethylsilyl analogues. The observed regiochemistry can be explained by an initial magnesium halide induced rearrangement to (triphenylsilyl)ethanal followed by reaction with the Grignard reagent (equation 26), and not by direct β-attack.

ϵ-Hydroxy-α,β-epoxysilanes undergo a regioselective intramolecular ring-opening[109]. The regioselectivity is highly dependent[110] on the configuration of the hydroxy epoxysilanes, with the *anti* diastereomers yielding tetrahydropyrans by α-attack and the *syn* stereoisomers leading to tetrahydrofurans by β-attack (equation 27).

α,β-Epoxytriphenylsilanes undergo β-opening[111] when reacted with anions of hindered sulphones. The more usual selective α-opening[112] can be seen in the silica gel catalysed reaction of α,β-epoxytrimethylsilanes with benzenethiol; the products lead to α-hydroxyaldehydes after sila-Pummerer rearrangement and hydrolysis.

Sharpless kinetic resolution of γ-trimethylsilyl allylic alcohols can be highly efficient; in the case shown (equation 28), the epoxyalcohol and the remaining allylic alcohol were both formed in greater than 99% ee. Further synthetic applications of the product chiral epoxysilanes have been reviewed[113]. The preparation of several optically active α,β-epoxysilanes from the corresponding chiral diols has been described[114].

$$\text{(28)}$$

α,β-Epoxydisilanes undergo an acid-catalysed rearrangement[115] to acylsilanes. On the other hand, on treatment with LDA, simple epoxydisilanes rearrange (equation 29) to vinylsilane silanols[116].

$$\text{(29)}$$

α-Halo-α,β-epoxysilanes rearrange in the presence of metal salts such as $ZnCl_2$ to produce α-haloacylsilanes[117]. Chiral β-vinyl-α,β-epoxysilanes undergo a Pd(0)-catalysed rearrangement to α-silyl-β,γ-unsaturated aldehydes[118] in high yield and with high stereoselectivity.

XI. Si—C OXIDATIVE CLEAVAGE

Oxidative cleavage of the carbon–silicon bond has been the subject of a recent and comprehensive review[119], in which emphasis has been placed on the compatibility of the oxidation conditions with various functional groups, with the inclusion of very useful compatibility tables.

A limitation on the use of the phenyldimethylsilyl group as a masked hydroxyl group is that treatment with acidic or electrophilic reagents is required to instal the necessary heteroatom prior to oxidative cleavage. Many of these reagents are incompatible with the presence of an alkene, although Fleming has shown[120] that the 2-methylbut-2-enyl(diphenyl)silyl group can be converted to the alcohol in the presence of certain types

28. Recent synthetic applications of organosilicon reagents

of alkenes (equation 30).

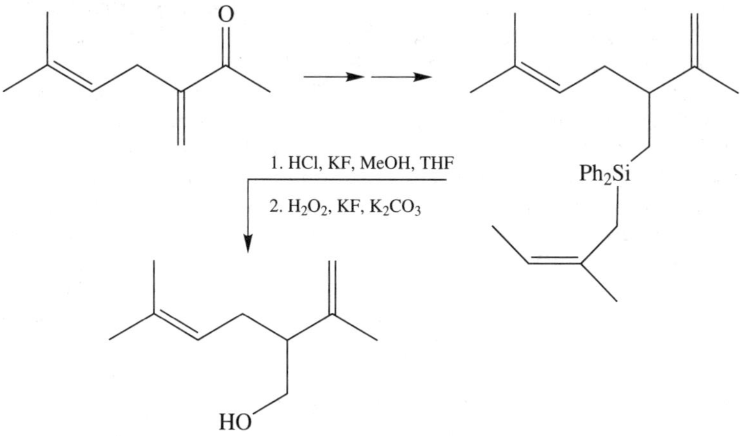

[Structure: starting ketone with isoprenyl group] →
1. HCl, KF, MeOH, THF
2. H$_2$O$_2$, KF, K$_2$CO$_3$

[Ph$_2$Si intermediate structure] (30)

[HO product structure]

Recently, two completely different methods which are compatible with an alkene have been used successfully. The first of these involves[121] prior Birch reduction of the phenyl group to a cyclohexadiene. Subsequent treatment with fluoride ion and basic hydrogen peroxide then completes the overall cleavage (equation 31).

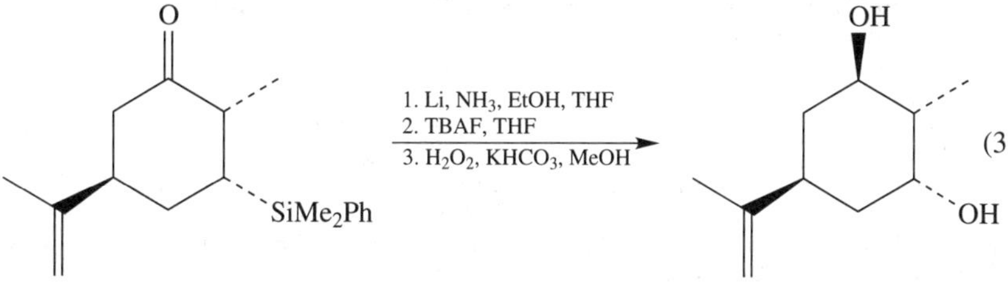

[Cyclohexanone with SiMe$_2$Ph]
1. Li, NH$_3$, EtOH, THF
2. TBAF, THF
3. H$_2$O$_2$, KHCO$_3$, MeOH
→ [diol product] (31)

The second method[122] uses a polar aprotic medium, such as DMF or N-methylpyrrolidone, a strong base, such as CsOH or KH, and t-BuOOH to facilitate nucleophilic attack at silicon; sometimes, the presence of fluoride ion is advantageous. This protocol also cleaves sterically hindered alkoxysilanes. It provided a key step (equation 32) in a total synthesis[123] of (±)-lupinine.

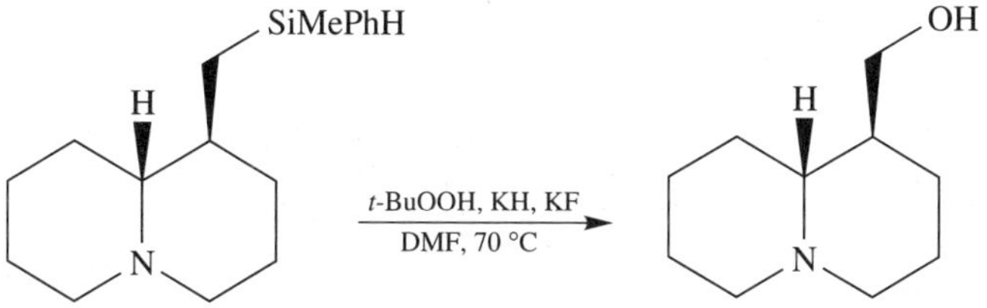

[Quinolizidine with SiMePhH]
t-BuOOH, KH, KF
DMF, 70 °C
→ [Quinolizidine with OH] (32)

The Grignard reagent from (1-chloroethyl)dimethylphenylsilane, prepared in turn by α-methylation of (chloromethyl)dimethylphenylsilane, acts as a hydroxyethylating reagent after oxidative cleavage; its use can be seen in a stereoselective synthesis[124] of lincosamine, the sugar component of the antibiotic lincomycin.

XII. MISCELLANEOUS

The ability of silicon substituents to direct or control certain types of reactions, mainly fragmentation processes, has been reviewed[125].

Use of the urea-hydrogen peroxide complex and N,N'-bis(TMS) urea provides an improved method[126] for the preparation of bis(TMS) peroxide, TMSOOTMS. In the presence of Fe(III)(picolinic acid)$_3$, bis(TMS) peroxide carries out selective oxidation of alkanes to ketones by a non-radical mechanism. The Fe(III)–Fe(IV) manifold is believed to be responsible[127]. On the other hand, using FeCl$_2$ in pyridine, alkyl chlorides are formed through a radical mechanism. Here, the Fe(II)–Fe(IV) manifold has been proposed[128].

Selective inter- and intra-molecular oxidation reactions using α-silyloxyalkyl hydroperoxides have been reviewed[129].

Finally, an efficient method[130] for the synthesis of Mosher's acid has been reported; this involves the addition of TMS trichloroacetate to 1,1,1-trifluoroacetophenone followed by hydrolysis.

XIII. REFERENCES

1. E. W. Colvin, in *Silicon Reagents in Organic Synthesis*, Academic Press, London, 1988; updated and reprinted 1990.
2. S. E. Thomas, in *Organic Synthesis. The Roles of Boron and Silicon*, Oxford University Press, Oxford, 1991.
3. E. W. Colvin, in *Chemistry of the Metal–Carbon Bond*, Vol. 4 (Ed. F. R. Hartley), Wiley, Chichester, 1987, pp. 539–621.
4. E. W. Colvin, in *Comprehensive Organometallic Chemistry II*, Vol. 11 (Ed. A. McKillop), Elsevier, Oxford, 1995, pp. 313–354.
5. M. Wills and E. W. Colvin, in *Comprehensive Organic Functional Group Transformations*, Vol. 2 (Ed. S. V. Ley), Elsevier, Oxford, 1995, pp. 513–547.
6. E. R. Corey, J. Y. Corey and P. P. Gaspar (Eds.) *Silicon Chemistry*, Ellis Horwood, Chichester, 1988.
7. D. Schinzer (Ed.) *Selectivities in Lewis Acid Promoted Reactions*, Kluwer Academic Publishers, Dordrecht, 1989.
8. A. R. Bassindale and P. P. Gaspar (Eds.) *Frontiers of Organosilicon Chemistry*, The Royal Society of Chemistry, Cambridge, 1991.
9. I. Fleming (Ed.) *Organosilicon Chemistry in Organic Synthesis*, in *Tetrahedron*, **44**, 3761 (1988).
10. G. A. Olah, G. K. Surya Prakash and R. Krishnamurti, in *Advances in Silicon Chemistry*, Vol. 1 (Ed. G. L. Larson), Jai Press, London and Greenwich, Connecticut, 1991, pp. 1–64.
11. E. Nakamura, *Synlett*, 539 (1991).
12. S. H. Bertz, G. Miao, B. E. Rossiter and J. P. Snyder, *J. Am. Chem. Soc.*, **117**, 11023 (1995).
13. M. Eriksson, A. Johansson, M. Nilsson and T. Olsson, *J. Am. Chem. Soc.*, **118**, 10904 (1996).
14. D.-K. Wang, L.-X. Dai and X.-L. Hou, *Tetrahedron Lett.*, **36**, 8649 (1995).
15. J. Kabbara, S. Flemming, K. Nickisch, H. Neh and J. Westermann, *Tetrahedron Lett.*, **35**, 8591 (1994).
16. Ch. Kishan Reddy, A. Devasagayaraj and P. Knochel, *Tetrahedron Lett.*, **37**, 4495 (1996).
17. H. Liu and T. Cohen, *Tetrahedron Lett.*, **36**, 8925 (1995).
18. B. H. Lipshutz, M. R. Wood and C. W. Lindsley, *Tetrahedron Lett.*, **36**, 4385 (1995); see also D. C. Harrowven and H. S. Poon, *Tetrahedron Lett.*, **37**, 4281 (1996).
19. W. B. Motherwell and L. R. Roberts, *Tetrahedron Lett.*, **36**, 1121 (1995).
20. J. G. Lee and K. K. Kang, *J. Org. Chem.*, **53**, 3634 (1988).
21. D. C. Snyder, *J. Org. Chem.*, **60**, 2638 (1995).
22. H. H. Seltzman, M. A. Moody and M. K. Begum, *Tetrahedron Lett.*, **33**, 3443 (1992).
23. J. G. Lee and H. T. Cha, *Tetrahedron Lett.*, **33**, 3167 (1992).
24. G. A. Olah, P. Ramaiah, G. Sandford, A. Orlinkov and G. K. Surya Prakash, *Synthesis*, 468 (1994).
25. T. Hirao, T. Hasegawa, Y. Muguruma and I. Ikeda, *J. Org. Chem.*, **61**, 366 (1996).
26. S. V. Pansare and M. G. Malusare, *Tetrahedron Lett.*, **37**, 2859 (1996).

27. M. Lautens and Y. Ren, *J. Org. Chem.*, **61**, 2210 (1996).
28. M. Seki, K. Kondo and T. Iwasaki, *Synlett*, 315 (1995).
29. J. K. Rasmussen, S. M. Heilman and L. Krepski, in *Advances in Silicon Chemistry*, Vol. 1 (Ed. G. L. Larson), Jai Press, London and Greenwich, Connecticut, 1991, pp. 65–187.
30. K. Higuchi, M. Onaka and Y. Izumi, *J. Chem. Soc., Chem. Commun.*, 1035 (1991).
31. J. K. Whitesell and R. Apodaca, *Tetrahedron Lett.*, **37**, 2525 (1996).
32. P. Kocienski, P. Raubo, J. K. Davis, F. T. Boyle, D. E. Davies and A. Richter, *J. Chem. Soc., Perkin Trans. 1*, 1797 (1996).
33. N. Adjé, F. Vogeleisen and D. Uguen, *Tetrahedron Lett.*, **37**, 5893 (1996).
34. B. Muller, F. Delaloge, M. den Hartog, J.-P. Férézou, A. Pancrazi, J. Prunet, J.-Y. Lallemand, A. Neuman and T. Prangé, *Tetrahedron Lett.*, **37**, 3313 (1996).
35. M. Hayashi, T. Inoue, Y. Miyamoto and N. Ogune, *Tetrahedron*, **50**, 4385 (1994) and references cited therein.
36. W. D. Pan, X. M. Feng, L. Z. Gong, W. H. Hu, Z. Li, A. Q. Mi and Y. Z. Jiang, *Synlett*, 337 (1996).
37. C. Bolm and P. Müller, *Tetrahedron Lett.*, **36**, 1625 (1995) and references cited therein.
38. A. Abiko and G. Wang, *J. Org. Chem.*, **61**, 2264 (1996).
39. M. North, *Synlett*, 807 (1993).
40. M. Hayashi, M. Tamura and N. Oguni, *Synlett*, 663 (1992) and references cited therein.
41. P. Van de Weghe and J. Collin, *Tetrahedron Lett.*, **36**, 1649 (1995).
42. F. A. Hicks, S. C. Berk and S. L. Buchwald, *J. Org. Chem.*, **61**, 2713 (1996).
43. G. D. Probert, R. J. Whitby and S. J. Coote, *Tetrahedron Lett.*, **36**, 4113 (1995).
44. D. C. Pryde, S. S. Henry and A. I. Meyers, *Tetrahedron Lett.*, **37**, 3243 (1996).
45. M. Meguro, N. Asao and Y. Yamamoto, *J. Chem. Soc., Chem. Commun.*, 1021 (1995).
46. K. I. Sutowardoyo, M. Emziane, P. Lhoste and D. Sinou, *Tetrahedron*, **47**, 1435 (1991).
47. M. Hayashi, K. Kohmura and N. Oguni, *Synlett*, 774 (1991).
48. L. E. Martínez, J. L. Leighton, D. H. Carsten and E. N. Jacobsen, *J. Am. Chem. Soc.*, **117**, 5897 (1995).
49. K. B. Hansen, J. L. Leighton and E. N. Jacobsen, *J. Am. Chem. Soc.*, **118**, 10924 (1996).
50. J. F. Larrow, S. E. Schaus and E. N. Jacobsen, *J. Am. Chem. Soc.*, **118**, 7420 (1996).
51. J. L. Leighton and E. N. Jacobsen, *J. Org. Chem.*, **61**, 389 (1996).
52. S. E. Schaus and E. N. Jacobsen, *Tetrahedron Lett.*, **37**, 7937 (1996).
53. L. E. Martínez, W. A. Nugent and E. N. Jacobsen, *J. Org. Chem.*, **61**, 7963 (1996).
54. M. H. Wu and E. N. Jacobsen, *Tetrahedron Lett.*, **38**, 1693 (1997).
55. P. Magnus, J. Lacour, P. A. Evans, M. B. Roe and C. Hulme, *J. Am. Chem. Soc.*, **118**, 3406 (1996).
56. P. A. Evans and D. P. Modi, *J. Org. Chem.*, **60**, 6662 (1995).
57. G. A. Olah and T. D. Ernst, *J. Org. Chem.*, **54**, 1203 (1989).
58. G. A. Olah, Q. Wang, X.-Y. Li and G. K. Surya Prakash, *Synlett*, 774 (1991).
59. M. V. R. Reddy, R. Kumareswaran and Y. D. Vankar, *Tetrahedron Lett.*, **36**, 6751 (1995).
60. J. G. Lee and K. H. Kwak, *Tetrahedron Lett.*, **33**, 3165 (1992).
61. M. Ito, K. Koyakumaru, T. Ohta and H. Takaya, *Synthesis*, 376 (1995).
62. T. W. Greene and P. G. M. Wuts, *Protective Groups in Organic Synthesis*, 2nd ed., Wiley, Chichester, 1991.
63. G. Simchen and J. Heberle, *Silylating Agents*, 2nd ed., Fluka Chemie AG, Buchs, Switzerland, 1995.
64. J. Muzart, *Synthesis*, 11 (1993)
65. T. D. Nelson and R. D. Crouch, *Synthesis*, 1031 (1996).
66. C. Rücker, *Chem. Rev.*, **95**, 1009 (1995).
67. (a) G. Stork, T. Y. Chan and G. A. Breault, *J. Am. Chem. Soc.*, **114**, 7578 (1992) and references therein.
    (b) M. Bols and T. Skrydstrup, *Chem. Rev.*, **95**, 1253 (1995).
68. F. Cavelier and C. Enjalbal, *Tetrahedron Lett.*, **37**, 5131 (1996).
69. A. DattaGupta, R. Singh and V. K. Singh, *Synlett*, 69 (1996).
70. N. S. Wilson and B. A. Keay, *Tetrahedron Lett.*, **37**, 153 (1996).
71. N. S. Wilson and B. A. Keay, *J. Org. Chem.*, **61**, 2918 (1996).
72. S. Kim and I. S. Kee, *Tetrahedron Lett.*, **31**, 2899 (1990).
73. T. Oriyama, K. Yatabe, S. Sugawara, Y. Machiguchi and G. Koga, *Synlett*, 523 (1996).

74. X. Franck, B. Figadère and A. Cavé, *Tetrahedron Lett.*, **36**, 711 (1995).
75. T. Oriyama, K. Yatabe and Y. Kawada, *Synlett*, 45 (1995).
76. J. Wu, B. K. Shull and M. Koreeda, *Tetrahedron Lett.*, **37**, 3647 (1996).
77. P. Ramaiah, R. Krishnamurti and G. K. Surya Prakash, *Org. Synth.*, **72**, 232 (1993).
78. I. Ruppert, K. Schlich and W. Volbach, *Tetrahedron Lett.*, **25**, 2195 (1984).
79. G. K. Surya Prakash, in *Synthetic Fluorine Chemistry* (Ed. G. A. Olah, R. D. Chambers and G. K. Surya Prakash), Wiley, New York, 1992.
80. M. W. Walter, R. M. Adlington, J. E. Baldwin, J. Chuhan and C. J. Schofield, *Tetrahedron Lett.*, **36**, 7761 (1995).
81. G. K. Surya Prakash, A. K. Yudin, D. Deffieux and G. A. Olah, *Synlett*, 151 (1996).
82. A. K. Yudin, G. K. Surya Prakash, D. Deffieux, M. Bradley, R. Bau and G. A. Olah, *J. Am. Chem. Soc.*, **119**, 1572 (1997).
83. T. Shioiri and T. Aoyama, in *Advances in the Use of Synthons in Organic Chemistry*, Vol. 1 (Ed. A. Dondoni), Jai Press, London and Greenwich, Connecticut, 1993, p. 51; R. Anderson and S. B. Anderson, in *Advances in Silicon Chemistry*, Vol. 1 (Ed. G. L. Larson), Jai Press, London and Greenwich, Connecticut, 1991, pp. 303–325.
84. S. Ohira, K. Okai and T. Moritani, *J. Chem. Soc., Chem. Commun.*, 721 (1992).
85. K. Miwa, T. Aoyama and T. Shioiri, *Synlett*, 107 (1994).
86. K. Miwa, T. Aoyama and T. Shioiri, *Synlett*, 109 (1994).
87. T. Aoyama, S. Katsuta and T. Shioiri, *Heterocycles*, **28**, 133 (1989).
88. T. Aoyama, T. Nakano, K. Marumo, Y. Uno and T. Shioiri, *Synthesis*, 1163 (1991).
89. G. Simchen in *Advances in Silicon Chemistry*, Vol. 1 (Ed. G. L. Larson), Jai Press, Greenwich, Connecticut, 1991, pp. 189–301.
90. B. H. Lipshutz, J. Burgess-Henry and G. P. Roth, *Tetrahedron Lett.*, **34**, 995 (1993).
91. S. Kim, Y. G. Kim, and D. Kim, *Tetrahedron Lett.*, **33**, 2565 (1992).
92. E. Vedejs and H. Sano, *Tetrahedron Lett.*, **33**, 3261 (1992).
93. K. Shibata, N. Tokitoh and R. Okazaki, *Tetrahedron Lett.*, **34**, 1491 (1993).
94. R. Polt, M. A. Peterson and L. DeYoung, *J. Org. Chem.*, **57**, 5469 (1992).
95. P. DeShong and P. J. Rybczynski, *J. Org. Chem.*, **56**, 3207 (1991).
96. M. D. Mizhiritskii and V. O. Reikhsfel'd, *Russ. Chem. Rev. (Engl. Transl.)*, **57**, 447 (1988).
97. E. Block and M. Aslam, *Tetrahedron*, **44**, 281 (1988).
98. A. Capperucci, A. Degl'Innocenti, A. Ricci, A. Mordini and G. Reginato, *J. Org. Chem.*, **56**, 7323 (1991); A. Degl'Innocenti, A. Capperucci, A. Mordini, G. Reginato, A. Ricci and F. Cerreta, *Tetrahedron Lett.*, **34**, 873 (1993).
99. M. Segi, T. Nakajima, S. Suga, S. Murai, I. Ryu, A. Ogawa and N. Sonoda, *J. Am. Chem. Soc.*, **110**, 1976 (1988); for another silicon-based route to these reactive species, see G. A. Krafft and P. T. Meinke, *J. Am. Chem. Soc.*, **108**, 1314 (1986).
100. J. Brittain and Y. Gareau, *Tetrahedron Lett.*, **34**, 3363 (1993).
101. J. R. Hwu and S.-C. Tsay, *J. Org. Chem.*, **55**, 5987 (1990).
102. P.-Y. Lin, W.-S. Ku and M.-J. Shiao, *Synthesis*, 1219 (1992).
103. S. D. Rychnovsky, *J. Org. Chem.*, **54**, 4982 (1989).
104. K. Sukata, *J. Org. Chem.*, **54**, 2015 (1989).
105. G. A. Molander and K. Mautner, *J. Org. Chem.*, **54**, 4042 (1989); G. A. Molander and K. Mautner, *Pure Appl. Chem.*, **62**, 707 (1990).
106. P. F. Hudrlik and A. M. Hudrlik, in *Advances in Silicon Chemistry*, Vol. 2 (Ed. G. L. Larson), Jai Press, Greenwich, Connecticut, 1993, pp. 1–89.
107. P. F. Hudrlik, D. Ma, R. S. Bhamidipati and A. M. Hudrlik, *J. Org. Chem.*, **61**, 8655 (1996).
108. P. F. Hudrlik, M. E. Ahmed, R. R. Roberts and A. M. Hudrlik, *J. Org. Chem.*, **61**, 4395 (1996).
109. H. Flörke and E. Schaumann, *Synthesis*, 647 (1996).
110. G. Adiwidjaja, H. Flörke, A. Kirschning and E. Schaumann, *Tetrahedron Lett.*, **36**, 8771 (1995).
111. P. Jankowski and J. Wicha, *J. Chem. Soc., Chem. Commun.*, 802 (1992).
112. P. Raubo and J. Wicha, *Synlett*, 25 (1993).
113. F. Sato and Y. Kobayashi, *Synlett*, 849 (1992).
114. A. R. Bassindale, P. G. Taylor and Y. Xu, *Tetrahedron Lett.*, **37**, 555 (1996).
115. D. M. Hodgson and P. J. Comina, *Chem. Commun.*, 755 (1996).
116. D. M. Hodgson and P. J. Comina, *Tetrahedron Lett.*, **37**, 5613 (1996).
117. Y. Horiuchi, M. Taniguchi, K. Oshima and K. Utimoto, *Tetrahedron Lett.*, **36**, 5353 (1995).
118. F. Gilloir and M. Malacria, *Tetrahedron Lett.*, **33**, 3859 (1992).

119. G. R. Jones and Y. Landais, *Tetrahedron*, **52**, 7599 (1996).
120. I. Fleming and S. B. D. Winter, *Tetrahedron Lett.*, **34**, 7287 (1993); I. Fleming and D. Lee, *Tetrahedron Lett.*, **37**, 6929 (1996).
121. D. F. Taber, L. Yet and R. S. Bhamidipati, *Tetrahedron Lett.*, **36**, 351 (1995); D. F. Taber, R. S. Bhamidipati and L. Yet, *J. Org. Chem.*, **60**, 5537 (1995); D. F. Taber, R. S. Bhamidipati and L. Yet, *J. Org. Chem.*, **61**, 1554 (1996); R. Angelaud and Y. Landais, *J. Org. Chem.*, **61**, 5202 (1996).
122. J. H. Smitrovich and K. A. Woerpel, *J. Org. Chem.*, **61**, 6044 (1996).
123. G. A. Molander and P. J. Nichols, *J. Org. Chem.*, **61**, 6040 (1996).
124. F. L. van Delft, M. de Kort, G. A. van der Marel and J. H. van Boom, *J. Org. Chem.*, **61**, 1883 (1996).
125. J. R. Hwu and H. V. Patel, *Synlett*, 989 (1995).
126. W. P. Jackson, *Synlett*, 536 (1990).
127. D. H. R. Barton and B. M. Chabot, *Tetrahedron*, **53**, 487 (1997).
128. D. H. R. Barton and B. M. Chabot, *Tetrahedron*, **53**, 511 (1997).
129. R. Nagata and I. Saito, *Synlett*, 291 (1990).
130. Y. Goldberg and H. Alper, *J. Org. Chem.*, **57**, 3731 (1992).

CHAPTER **29**

Recent advances in the hydrosilylation and related reactions

IWAO OJIMA, ZHAOYANG LI and JIAWANG ZHU

Department of Chemistry, State University of New York at Stony Brook, Stony Brook, New York 11794-3400, USA

I. INTRODUCTION	1688
II. HYDROSILYLATION OF ALKENES	1688
A. Simple and Functionalized Alkenes	1688
1. By Group VIII transition metal catalysts	1688
2. By metallic and colloidal metal catalysts	1693
3. By Group IV metallocenes, lanthanides and other metal catalysts	1697
4. By immobilized catalysts	1701
5. By radical initiators	1703
B. Mechanism of the Alkene Hydrosilylation	1704
C. Intramolecular Hydrosilylation of Alkenes	1710
D. Other Reactions Associated with Alkene Hydrosilylation	1714
III. HYDROSILYLATION OF ALKYNES	1716
A. Simple Alkynes	1717
1. By platinum catalysts	1717
2. By rhodium catalysts	1718
3. By iridium, ruthenium, osmium and other catalysts	1720
4. By radical initiators	1722
B. Mechanism of the Stereoselective Hydrosilylation of 1-Alkynes	1723
C. Intramolecular Hydrosilylation of Alkynes	1725
D. Bis-TMS-ethyne and Bis-TMS-butadiyne	1725
E. Functionalized Alkynes	1727
F. Other Reactions Associated with Alkyne Hydrosilylation	1732
IV. HYDROSILYLATION OF CARBONYL AND RELATED COMPOUNDS	1733
V. ASYMMETRIC HYDROSILYLATION	1743

The chemistry of organic silicon compounds, Vol. 2
Edited by Z. Rappoport and Y. Apeloig © 1998 John Wiley & Sons Ltd

A. Of Prochiral Ketones, Imines and Imine N-Oxides	1743
B. Of Alkenes and 1,3-Dienes	1752
C. Intramolecular Reactions	1756
VI. SYNTHESES OF SILICON-CONTAINING DENDRIMERS AND POLYMERS USING HYDROSILYLATION	1758
A. Silicon-containing Dendrimers	1758
B. Silicon-containing Polymers	1763
C. Modification of Polymers by Hydrosilylation	1768
VII. SILYLCARBONYLATION AND SILYLCARBOCYCLIZATION REACTIONS	1771
A. Silylcarbonylation Reactions	1771
B. Silylcarbocyclization and Related Reactions	1779
VIII. REFERENCES	1785

I. INTRODUCTION

Hydrosilylation, especially of carbon–carbon multiple bonds, is one of the most important reactions in organosilicon chemistry, and thus the reaction has been studied extensively for half a century. Even after five decades of research, it is apparent that interest in this process has been continuing and even increasing in the last several years rather than fading, judging from the number of publications. The reaction is used for the industrial production of organosilicon compounds such as adhesives, binders and coupling agents. In research laboratories, hydrosilylation is a very convenient and efficient method for the syntheses of a variety of organosilicon compounds not only for the study of organosilicon chemistry, but also for silicon-based polymers and dendrimers for new materials. The hydrosilylation of various functional groups catalyzed by transition metal complexes or promoted by radical initiators provides facile routes to various organosilicon reagents and synthetic intermediates in organic syntheses as well as a unique method for the selective reduction of carbon–heteroatom bonds. Catalytic asymmetric hydrosilylation applied to organic syntheses has been substantially advanced[1–3]. New reactions related to hydrosilylation have been discovered and developed such as silylcarbonylation, silylformylation and silylcarbocyclization, which are useful in organic syntheses.

This chapter will summarize the recent advances in hydrosilylation and related reactions catalyzed by transition metal complexes since this subject was reviewed in *The Chemistry of Organic Silicon Compounds* in 1989 that covered the advances till the end of 1986[3].

II. HYDROSILYLATION OF ALKENES

A. Simple and Functionalized Alkenes

1. By Group VIII transition metal catalysts

Platinum and rhodium catalysts have been the most frequently used catalysts among all the metal-based catalysts for the hydrosilylation of alkenes to date. In particular, a variety of rhodium catalysts have been extensively studied[3–6], while the development of other Group VIII transition metal catalysts such as those of palladium and ruthenium continues. In general, the hydrosilylation of an alkene gives the corresponding silylalkanes with varying regioselectivity (equation 1).

$$R^1CH{=}CHR^2 + HSiR_3 \xrightarrow{\text{catalyst}} R^1(R^3{}_3Si)CHCH_2R^2 + R^1CH_2CHR^2(SiR_3) \quad (1)$$

Little attention has been paid to the hydrosilylation reactions using monosilane, SiH_4, for synthetic purposes, mainly because of its limited availability and difficulty in handling. However, SiH_4 has recently emerged as a new raw material in the organosilicon industry in connection with the recent development of polysilicon and amorphous silicon products[7]. Thus, the hydrosilylation of alkenes with SiH_4 may provide a direct and practical route to trihydroalkylsilanes. A preliminary study on the reactions of 1-hexene with SiH_4 catalyzed by Group VIII transition metal and metal complexes at 80 °C and 10 atm for 3 h shows that the reaction gives a mixture of $H_3Si(C_6H_{13}\text{-}n)$ (major) and $H_2Si(C_6H_{13}\text{-}n)_2$ (minor)[7]. The most active catalyst so far examined is $Pt(PPh_3)_4$ which gives $H_3Si(C_6H_{13}\text{-}n)$ in 37.4% yield and $H_2Si(C_6H_{13}\text{-}n)_2$ in 4.7% yield. The relative catalytic activity decreases in the order $Pt(PPh_3)_4 > Rh/C > PtCl_2(PPh_3)_2 > Pt/C \gg NiCl_2(PPh_3)_2$. The reaction of 1,5-hexadiene with SiH_4 using Pt and Rh catalysts affords two major products, $H_3Si(5\text{-hexenyl})SiH_3$ and silacycloheptane in $1:1 - 4:1$ ratio accompanied by a small amount of $H_2Si(5\text{-hexenyl})_2$ in 8–38% total yield[7]. The product ratio is mainly dependent on the catalyst species used.

The reactions of styrene, α-methylstyrene and methyl methacrylate (equation 2) with $HSiEt_3$ and $HSiPh_3$ catalyzed by $PtCl_2$, $PtCl_2(PhCN)_2$ or $PtCl_2(PhCN)_2PR_3$ give the linear hydrosilylation products predominantly[8]. The activity and selectivity of these catalysts are dependent on the nature of the phosphine ligand as well as the Pt/ligand ratio. The use of chelating bidentate diphosphine ligands results in the loss of catalytic activity.

$$\text{CH}_2=\text{C(Me)COOMe} + HSiEt_3 \xrightarrow[\text{CHCl}_3, \text{RT}]{PtCl_2(PhCN)_2} Et_3Si\text{-CH}_2\text{-CH(Me)-COOMe} \quad (2)$$
91% yield

The reactions of enamides (1) and N-vinylureas (2) with $HSiMe_2Ph$ catalyzed by $Rh_2(OAc)_4$ give the corresponding 1-silylalkylamides (3) and 1-silylalkylureas (4) in 38–93% isolated yields with complete regio- and site selectivity (equations 3–5)[9]. The directing effect of the carbonyl oxygen illustrated for the key intermediate 5 is proposed to account for the remarkable regioselectivity. In fact, the reaction of vinyl acetate that can take a similar chelate structure also gives 1-silylethyl acetate 6 as the major product in 61% isolated yield[9]. Other rhodium catalysts, $[Rh(CO)_2Cl]_2$, $HRh(PPh_3)_3$, $Rh(acac)_3$, $[Rh(COD)Cl]_2$ and $RhCl(PPh_3)_3$, are found to be much less active.

$$\text{enamide (1)} + HSiMe_2Ph \xrightarrow[\text{toluene, 110 °C}]{Rh_2(OAc)_4} \text{1-silylalkylamide (3)} \quad (3)$$

$R^1 = CH_2Ph, H, Me; R^2 = Me, NMe_2$

Remarkable effects of non-vicinal, proximate polar groups on the reaction rates are disclosed in the hydrosilylation of substituted bicyclo[2.2.1]hept-5-enes (8) catalyzed by $Pt[(CH_2=CHSiMe_2)_2O]_2$ (Karstedt's catalyst)[10,11] or $PtCl_4$ (equation 6)[12]. *Endo*-2,3-dicarboxylic N-phenylimide **8c** does not react at all and the reaction of *endo*-2,3-dicarboxylic anhydride **8a** is unusually slow. On the contrary, the reaction of *exo*-imide

8d or *exo*-anhydride **8b** is very fast, completing in 30 min or less under the same conditions, while *endo*-ether **8h** reacts normally (equation 6) (Table 1). These results clearly indicate that extremely strong field effects of the electron-deficient imide and anhydride groups are operative in these reactions[12]. Also, the nature of hydrosilane exerts marked influence on the reaction rate, i.e., the reactivity of hydrosilanes is in the order: $HSiMe_2Cl \sim HSiMe_2Ph > HSiMeCl_2 \gg HSiMe_3 \sim HSiCl_3$ [12].

a: $R^1 = R^2 = R^3 = H$; **b:** $R^1 = H, R^2 = Me, R^3 = Et$; **c:** $R^1 = Me, R^2 = H, R^3 = Me$; **d:** $R^1, R^3 = (CH_2)_4, R^2 = Me$

$[Pt] = Pt[(CH_2=CHSiMe_2)O]_2$ (Karstedt's catalyst) or $PtCl_4$

TABLE 1. Hydrosilylation of norbornene derivatives

X,X	Y,Y	Yield (%)	Time (h)
8a CO–O–CO	H,H	95	48
8b H,H	CO–O–CO	99	<0.5
8c CO–N(Ph)–CO	H,H	0	48
8d H,H	CO–N(Ph)–CO	99	0.5
8e CO_2Me,CO_2Me	H,H	99	2
8f H, CO_2Me	H,CO_2Me	99	0.5
8g H,H	CO_2Me,CO_2Me	99	1
8h CH_2OCH_2	H,H	99	<0.5
8i H,H	H,H	99	<0.5

Hydrosilylation of divinyl ether has been applied for the synthesis of silacyclopentane **12** using Speier's catalyst (Scheme 1)[13]. One of the two carbon–carbon double bonds was hydrosilylated first with a dialkyl(ethoxy)silane, giving silylethyl vinyl ether **10** in 53–59% yield, which was reduced with $LiAlH_4$ to hydrosilane **11**. The intramolecular hydrosilylation of **11** affords silacyclopentane **12** in moderate yields (Scheme 1). The reaction with $HSiEt_2(OEt)$ gives **12a** exclusively in 45% yield, while silacyclohexane **13b** is formed as the minor product when $HSiMe_2(OEt)$ is used as the hydrosilane (**12b/13b** = 2.3/1; 50% total yield)[13]. Other intramolecular hydrosilylation reactions useful in organic syntheses will be discussed in the section II.C. (*vide infra*).

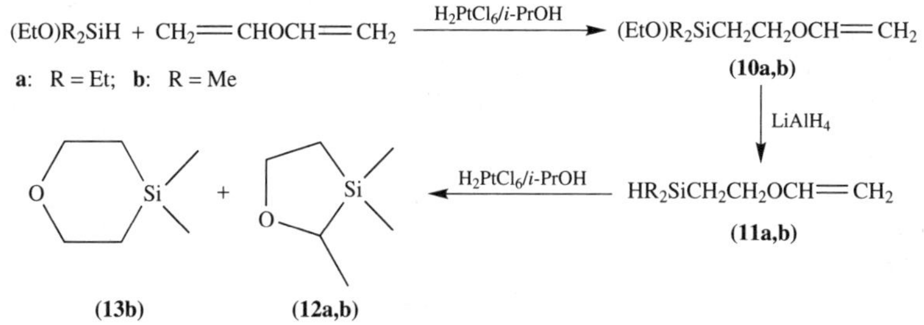
(EtO)R2SiH + CH2=CHOCH=CH2
H2PtCl6/i-PrOH
(EtO)R2SiCH2CH2OCH=CH2
(10a,b)
a: R = Et; b: R = Me
LiAlH4
H2PtCl6/i-PrOH
HR2SiCH2CH2OCH=CH2
(11a,b)
(13b)
(12a,b)

SCHEME 1

The reactions of styrene, 4-chlorostyrene, 4-methoxystyrene, and methyl acrylate with 1,2-bis(dimethylsilyl)benzene (**14**) catalyzed by $Pt(CH_2{=}CH_2)(PPh_3)_2$ give the corresponding "dehydrogenative double silylation" products **15** in good to high yields (equation 7)[14]. When dimethyl maleate is employed, benzo-1,4-disilacyclohexene **16** ($R^1 = R^2 = CO_2Me$) is obtained as the major product (**15/16** = 22/78) (equation 7). In the reactions of 1-alkenes, i.e., ethylene and 1-octene, the formation of monosilylated products is also observed (13–57% yield). On the basis of the fact that no deuterium is incorporated into the products when 1,2-dideuterio-1,2-bis(dimethylsilyl)benzene is used, disilyl-Pt metallacycle **17** is proposed to be the key intermediate of this process (equation 7).

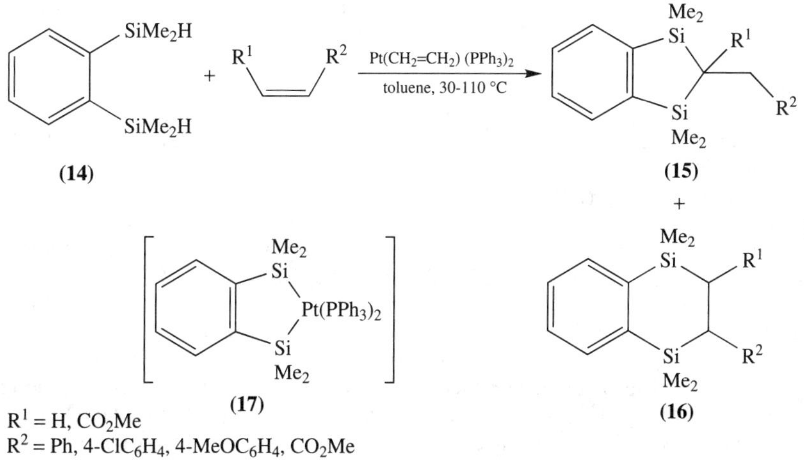
SiMe2H
SiMe2H
(14)
Pt(CH2=CH2) (PPh3)2
toluene, 30-110 °C
(15)
(16)
(17)
Pt(PPh3)2
R1 = H, CO2Me
R2 = Ph, 4-ClC6H4, 4-MeOC6H4, CO2Me

(7)

The reaction of 3,3-dimethyl-1-butene or styrene with 1,2-bis(dimethylsilyl)ethane catalyzed by $RhCl(PPh_3)_3$ gives only linear monosilylation product, $RCH_2CH_2SiMe_2(CH_2)_2 SiMe_2(H)$, in high yield, i.e., there is a stark difference in reactivity of the Si−H moiety between 1,2-bis(dimethylsilyl)ethane and $RCH_2CH_2SiMe_2(CH_2)_2SiMe_2(H)$ toward hydrosilylation[15]. Although clear enhancement of reaction rate is not observed in the reaction of alkenes, marked rate enhancement in comparison with normal trialkylsilanes is exhibited in the reactions of alkynes and ketones (*vide infra*)[15].

The isocyanide complexes of platinum, $PtCl_2(C\equiv N-C_6H_3Me_2-2,6)_2$, show modest catalytic activity for the hydrosilylation of α-methylstyrene with $HSiMe_2Ph$. The corresponding Ni-isocyanide complex shows a low catalytic activity and the Pd-isocyanide complex is inactive[16]. The platinum complex of bidentate 2,2'-bis(isocyano)-1,1'-binaphthyl (DIBN), $Pt_2Cl_4(DIBN)_2$, is more active than those of PPh_3 or the monodentate isocyanide complexes[16]. However, $PtCl_2(C\equiv N-C_6H_3Me_2-2,6)_2$ is an excellent catalyst for the hydrosilylation of 5-hexen-2-one (0.034 mol % catalyst, 25 °C, 3 h), giving 6-silyl-2-hexanone **18** in 98% yield (equation 8)[16]. Under the same reaction conditions, $Pt_2Cl_4(DIBN)_2$ is also highly active (74% yield), but $PtCl_2(PPh_3)_2$ gives only a trace amount of **18**. The isocyanide complexes of rhodium, $RhCl(C\equiv N-C_6H_3Me_2-2,6)_3$ and $[RhCl(DIBN)_2]_n$, are excellent catalysts for this reaction (96–99% yield), whereas $RhCl(PPh_3)_3$ is much less active affording **18** in only 5% yield.

$$\text{CH}_2=\text{CHCH}_2\text{CH}_2\text{C(O)CH}_3 + HSiMe_2Ph \xrightarrow[\text{benzene, 25 °C}]{\text{[Pt] or [Rh]}} PhMe_2Si\text{-CH}_2\text{CH}_2\text{CH}_2\text{CH}_2\text{C(O)CH}_3 \quad (8)$$

(18)

$[Pt] = PtCl_2(CN-C_6H_3Me_2-2,6), Pt_2Cl_4(DIBN)_2$
$[Rh] = RhCl(CN-C_6H_3Me_2-2,6)_3, [RhCl(DIBN)_2]_n$

The reaction of styrene with H_2SiPh_2 catalyzed by $RhCl(PPh_3)_3$ in THF at ambient temperature is shown to give the linear monohydrosilylation product, $Ph_2HSiCH_2CH_2Ph$, in 77–89% isolated yield, accompanied by $Ph_2HSi(CH=CHPh)$ or $PhCH_2CH_3$ (1–7%). The observed linear selectivity as well as product selectivity is substantially better than that of the reaction using $HSiEt_3$ as the hydrosilane[17].

Palladium complexes with bis(diphenylphosphino)methane (dpm), $Pd_2(dpm)_3$, $PdCl_2(dpm)$, $Pd_2Cl_2(\mu\text{-dpm})_2$, and $Pd_2Cl_2(\mu\text{-S})(\mu\text{-dpm})_2$, are found to catalyze the hydrosilylation of 1-hexene, styrene, and vinyltrichlorosilane with $HSiCl_3$ at 120 °C without solvent, giving branched products, i.e., 2-silylhexane, 1-silylethylbenzene, and 1,1-disilylethane, respectively, in moderate to high yields[18]. A similar regioselectivity is observed in the reactions of styrene and 1-hexene with $HSiCl_3$ catalyzed by $PdCl_2(PhCN)_2$ and $Pd(acac)_2$, while opposite regioselectivity predominates when $HSiMeCl_2$ or $HSiEtCl_2$ is employed[19].

Rh-phosphite clusters, $[\{HRh\{P(OR)_3\}_2\}_3$ (R = Me, 2-MeC$_6$H$_4$) and $[RhCl\{P(OR)_3\}_2]_2$ (R = Me, 2-MeC$_6$H$_4$) serve as moderate catalysts effecting the hydrosilylation of 2,3-dimethylbutadiene with $HSiEt_3$ or $HSi(OEt)_3$, yielding a mixture of 1,2- (minor) and 1,4-adduct (major)[20]. The reaction of 1-octene is sluggish with $HSi(OEt)_3$.

With regard to a possible cluster catalysis of $Ru_3(CO)_{12}$, the kinetic study of the hydrosilylation of 1-octene with $HSi(OEt)_3$ in the presence of $Ru_3(CO)_{12}$ at 70 °C in benzene shows that the cluster catalysis is indeed involved, especially during the first 11 min of the reaction[21]. However, the lower nuclearity species becomes predominant as the reaction proceeds. The $Ru_3(CO)_{12}$-catalyzed hydrosilylation of allyl chloride with $HSi(OMe)_3$ gives 3-chloropropyltrimethoxysilane, a useful silane coupling agent, in good yield[22].

2. By metallic and colloidal metal catalysts

Since the intermediacy of metal colloids in Pt-catalyzed hydrosilylation reactions was strongly indicated by Lewis in 1986[23], the use of colloidal metals as catalysts for hydrosilylation has been actively investigated[24-27]. It appears to be recognized to date that the ostensibly homogeneous catalysis of many platinum group metals goes through colloidal intermediates in hydrosilylation reactions. The platinum group metal colloids can be prepared through direct reaction of metal halide salts with excess $HSiMe_2(OEt)$ (equation 9)[28].

$$HSiMe_2(OEt) + MCl_x + D_4^{vi} \longrightarrow H_2 + Si\ products + M_{colloid} \quad (9)$$
(excess)

$M = Pt(no\ D_4^{vi}), Rh, Ru, Ir, Os, Pd$

$D_4^{vi} = [Me(CH_2=CH)Si - O]_4$

Relative catalytic activity of these metal colloids in the hydrosilylation of vinyltrimethylsilane with $HSi(C_6H_{13}\text{-}n)_3$ and $HSi(OEt)_3$ (25 ppm of catalyst at ambient temperature) is determined to be in the order $Pt > Rh > Ru \sim Ir \gg Os$, and Pd-colloid does not show any activity[29]. The catalytic activity depends on the nature of the hydrosilane used, i.e., Ru, Ir, and Os show almost negligible activity with $HSi(C_6H_{13}\text{-}n)_3$, while Ru and Ir display much higher activity, equivalent to Rh, with $HSi(OEt)_3$.

However, Pt-colloid is found to be less active than Rh-colloid when a dihydrosilane or a trihydrosilane is used. Recently, "silahydrocarbon", $Me(n\text{-}C_{10}H_{21})Si(C_8H_{17}\text{-}n)_2$, is attracting research interest as a hydraulic fluid in a wide temperature range, and the double hydrosilylation of 1-octene with $H_2SiMe(n\text{-}C_{10}H_{21})$ is an apparently efficient synthetic route to this compound. The reaction catalyzed by Karstedt's Pt catalyst that is shown to be a precursor of Pt-colloid at 80 °C is very sluggish, giving a mixture of the starting dihydrosilane (35.1%), monohydrosilylation product, $Me(n\text{-}C_{10}H_{21})Si(C_8H_{17}\text{-}n)H$ (58.7%), and $Me(n\text{-}C_{10}H_{21})Si(C_8H_{17}\text{-}n)_2$ (6.2%) after 1 h, while the same reaction catalyzed by Rh-colloid affords $Me(n\text{-}C_{10}H_{21})Si(C_8H_{17}\text{-}n)_2$ in 99.8% yield after 1 h[29]. Accordingly, the relative reaction rate is determined using H_2SiEt_2, $HSiEt_3$, $H_3Si(C_6H_{13}\text{-}n)$, and $HSi(C_6H_{13}\text{-}n)_3$ in the reaction of vinyltrimethylsilane catalyzed by Pt-colloid and Rh-colloid. The results are noteworthy in that the order of the reactivity of hydrosilanes is $H_3SiR > H_2SiR_2 > HSiR_3$ for the Rh-catalyst, whereas the order is $HSiR_3 > H_2SiR_2 > H_3SiR$ for the Pt-catalyst. This rather unusual phenomenon observed for the Pt-catalyst can be ascribed to the poisoning of the active Pt species by H_2SiR_2 and H_3SiR[29].

The attempted syntheses of tetraalkylsilanes through the triple hydrosilylation of monoalkylsilanes with 1-alkenes catalyzed by commercially available platinum catalysts, e.g., Pt/C, $PtCl_2(PPh_3)_2$, $PtCl_2(CH_3CN)_2$, $H_2PtCl_6 \cdot 6H_2O$, and PtO_2 suffer from a low reactivity of monoalkylsilanes[24]. However, oxygen-activated $H_2PtCl_6 \cdot 6H_2O$, and PtO_2 (purging the reaction system by air) catalysts are found to promote the desired triple hydrosilylation of $H_3SiC_6H_{13}\text{-}n$ with 1-octene or 1-decene to give the corresponding tetraalkylsilanes exclusively[24]. The active species in this system is considered to be colloidal Pt(0)[24].

A mechanistic study on the Pt-colloid catalyzed hydrosilylation by Lewis has revealed that molecular oxygen, i.e., dioxygen, serves as the crucial cocatalyst[30]. It is proposed that the catalytically active Pt-colloid, $Pt_x^0(O-O)$ **19** (a yellow color in the reaction mixture), is formed from Karstedt's catalyst by the action of a hydrosilane and molecular oxygen. Then, the active colloid **19** reacts with a hydrosilane to generate the non-classical hydrosilane-Pt complex **20**. Molecular oxygen plays two unique roles in these processes: (1) inhibition of the formation of large Pt-colloid (dark color or black) that has much reduced activity or is inactive, and (2) enhancement of electron deficiency of **20**, thus making it more susceptible to the nucleophilic attack of olefin. The proposed mechanism illustrated in Scheme 2[30] is in accordance with (i) the fact that the reaction stops when the catalytic system is depleted in oxygen, (ii) the dependence of the reaction rate on the electronic nature of substitutents of alkene substrates, i.e., the more electron-rich the alkene is, the faster is the reaction: The reaction rate decreases in the order $H_2C=CH_2$ > $CH_2=CF_2$ > $F_2C=CFCl$ > $F_2C=CF_2$, (iii) $CH_2=CHSiMe_3$ reacts faster than $CH_2=CHSiCl_3$, (iv) $HSi(OEt)_3$ reacts faster than $HSiEt_3$. The substituent effects observed in the reactions of substituted styrenes, 4-X-$C_6H_4CH=CH_2$, with $HSiCl_2Me$, i.e., the rate decreases in the order (X =) Me > H > OMe > Cl, indicate the importance of σ induction effects and the lack of resonance effects of the substituents. The hydrosilane-Pt complex **20** reacts with nucleophiles other than alkenes. Thus, the reaction of **20** with an alcohol or water leads to the generation of molecular hydrogen and an alkoxysilane or silanol via a non-classical complex **22** (Scheme 2). Although the Lewis mechanism does not explain how an alkene reacts with **20** to form another non-classical complex **21** and then the hydrosilylation product, the classical Harrod-Chalk mechanism for a mononuclear Pt catalyst appears to be adopted for this part of the catalytic cycle (Scheme 2)[30].

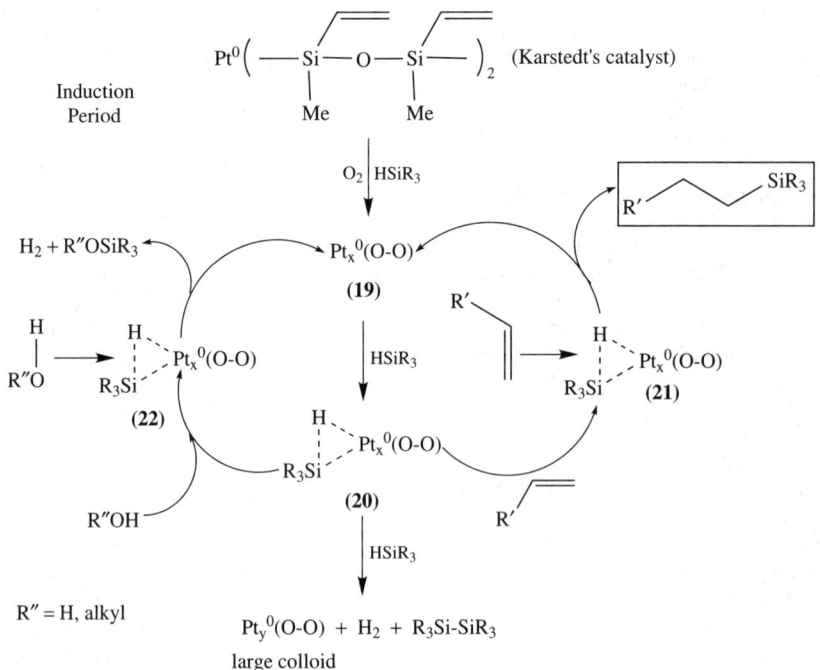

SCHEME 2

As described above, Karstedt's catalyst appears to be an excellent precursor for the generation of catalytically highly active Pt-colloid. Karstedt's catalyst can be prepared by reacting $H_2PtCl_6 \cdot xH_2O$, K_2PtCl_4, $[Pt(CH_2=CH_2)Cl(\mu\text{-}Cl)]_2$, cis-$Pt(\eta^2\text{-}CH_2=CHPh)_2Cl_2$ or $Pt(COD)_2$ with excess $(CH_2=CHSiMe_2)_2O$[10,11,31,32]. However, the mechanism of

Pt(IV) or Pt(II) to Pt(0) reduction and the structure of Karstedt's catalyst needs to be clarified. The X-ray crystallographic structure of the Pt complex isolated from the reaction mixture of $H_2PtCl_6 \cdot xH_2O$ with excess $(CH_2=CHSiMe_2)_2O$ followed by neutralization with $NaHCO_3$ is determined to be $Pt_2[(CH_2=CHSiMe_2)_2O]_3$ (**23**)[32]. The NMR (^{1}H, ^{13}C, ^{29}Si and ^{195}Pt) studies of **23** with $(CH_2=CHSiMe_2)_2O$ and styrene in toluene-d_8 has disclosed equilibrium between **23** and other 16-electron Pt(0) complexes **24** and **25** (Scheme 3)[33].

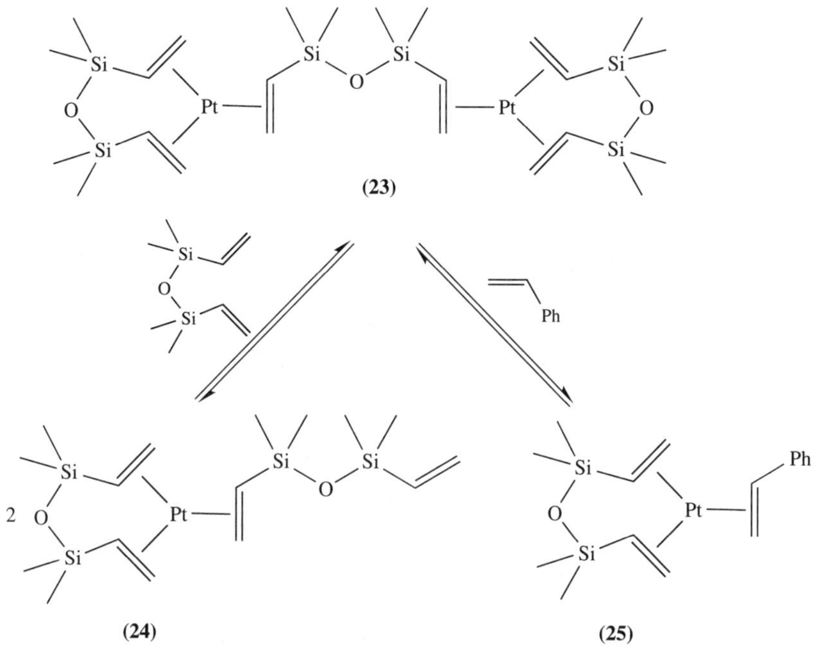

SCHEME 3

On the basis of the multinuclei NMR study and the fact that the reactions of Pt(IV) and Pt(II) chlorides with $(CH_2=CHSiMe_2)_2O$ yield polysiloxanes[31], vinyl chloride[31], 1,3-butadiene[32] and ethene[32], Lappert and coworkers have proposed a plausible mechanism illustrated in Scheme 4[33], which includes a rather unique vinyl–chlorine exchange (**26** → **27**) and reductive elimination of vinyl chloride (**28** → **29**). The homolytic fission of Pt–CH=CH$_2$ bond is also suggested. If a divinyl–Pt complex is formed by double vinyl–chlorine exchange, the observed formation of 1,3-butadiene can be explained as well. This study concludes that 16-electron species such as **24** and **25** are considered to be highly active catalytic species due to the availability of a vacant site for oxidative addition by a hydrosilane[33].

Highly active homogeneous Pt-catalysts are also shown to be generated by photoactivation of Pt(acac)$_2$, Pt(ba)$_2$ (ba = benzoylacetonate) Pt(dbm)$_2$ (dbm = dibenzoylmethanate) and Pt(hfac)$_2$ (hfac = hexafluoroacetylacetonate) in the presence of HSiEt$_3$ and/or CH$_2$=CHSiEt$_3$[34]. These photoactivated Pt catalysts promote the reaction of CH$_2$=CHSiEt$_3$ with HSiEt$_3$ efficiently, giving Et$_3$SiCH$_2$CH$_2$SiEt$_3$ in good to high yield. Once the active catalyst species is generated, the reactions proceed smoothly without

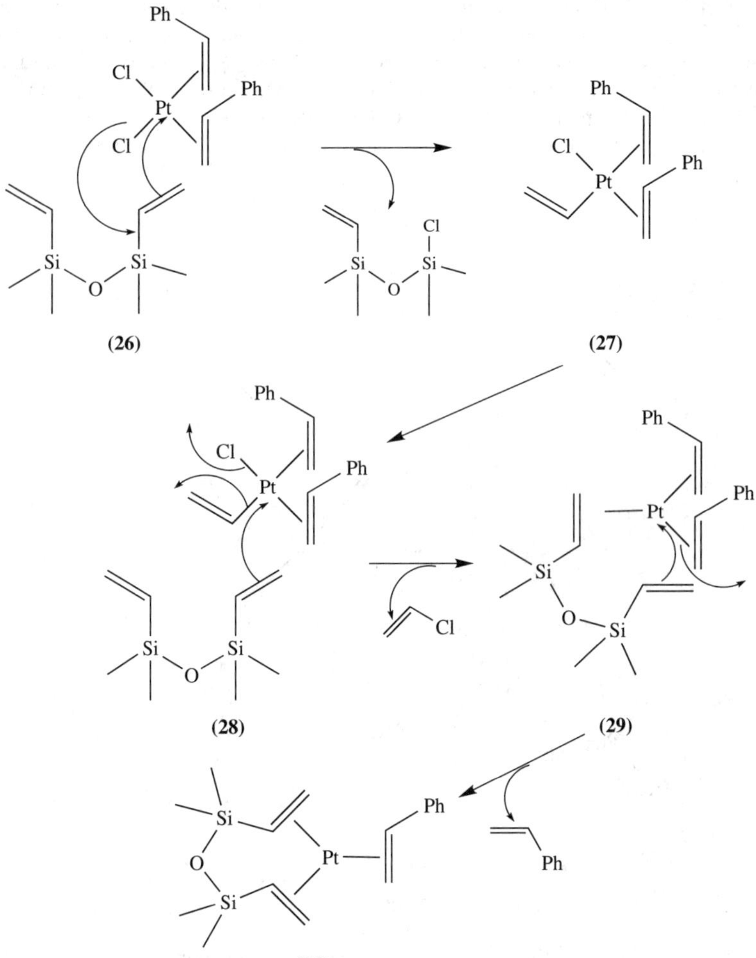

SCHEME 4

irradiation. The presence of oxygen prior to the photoactivation inhibits the generation of active catalyst species, but the addition of oxygen after the photoactivation does not affect the catalytic activity of the active species. In the absence of HSiEt$_3$ or CH$_2$=CHSiEt$_3$, only a less active heterogeneous Pt species is formed either photochemically or thermally. Thus, either HSiEt$_3$ or CH$_2$=CHSiEt$_3$ is essential for the generation of active homogeneous Pt catalyst species.

Highly dispersed metallic platinum catalysts prepared by treating mesitylene-solvated Pt atoms with solid supports such as carbon, graphite and γ-Al$_2$O$_3$ support are found to be active (Pt/hydrosilane = $1.1-4.9 \times 10^{-4}$ g-atom mol^{-1}) for the hydrosilylation of isoprene at 70 °C under neat conditions, giving a mixture of 4-silyl-2-methyl-2-butene (76–83%) and 4-silyl-2-methyl-1-butene (24–17%)[35]. The catalytic activity decreases in the order Pt/C > Pt/graphite $\gg$ Pt/Al$_2$O$_3$[35].

3. By Group IV metallocenes, lanthanides and other metal catalysts

Metallocenes of early-transition metals (Group IV) have recently been recognized as a new class of catalysts for the hydrosilylation of alkenes[36–39]. Metallocenes of Ti, Zr and Hf have been used mainly for the dehydrogenative coupling of hydrosilanes, especially dihydrosilanes to form polysilanes[39–42]. However, Cp_2MMe_2 (M = Ti, Zr) and $Cp_2M(Bu-n)_2$ (M = Ti, Zr, Hf) generated *in situ* from Cp_2MCl_2 and n-BuLi (2 equiv.) are found to be active catalysts for 1-alkenes[36–38]. The reaction of a 1-alkene, e.g. 1-hexene, 1-octene and vinyltrimethylsilane, with H_2SiPh_2 or $H_2SiMePh$ proceeds with complete regioselectivity, giving a 1-silylalkane in excellent yield accompanied by a small amount of an alkane via hydrogenation (equation 10)[37,38]. A small amount of a dehydrogenative silylation product, 1-silyl-1-alkene, is also formed in the reaction of styrene[37]. The reaction of 2-pentene with H_2SiPh_2 gives 1-Ph_2(H)SiC_5H_{11}-n in 85% yield[37]. The result clearly indicates rapid isomerization of the olefinic bond under the reaction conditions. When $HSiPh_3$ is used for styrene hydrosilylation, a 1:1 mixture of $Ph_3SiCH=CHPh$ and PhEt is formed, i.e. no hydrosilylation product is obtained[37]. The relative reactivity of hydrosilanes decreases in the order $H_3SiPh > H_2SiPh_2 \sim H_2SiMePh > H_2SiEt_2 \gg HSiPh_3$[37]. The relative catalytic activity of metallocenes decreases in the order $Cp_2ZrCl_2/2$ n-BuLi $>$ $Cp_2TiCl_2/2$ n-BuLi $>$ $Cp_2ZrMe_2 \gg Cp_2HfCl_2/2$ n-BuLi[37]. Cycloalkenes and substituted alkenes are much less reactive, promoting the dehydrogenative coupling of hydrosilanes[37,38]. These catalysts do not show any activity in the hydrosilylation of phenylacetylene, benzaldehyde and acetophenone[37].

$$RCH=CH_2 + H_2SiRR' \xrightarrow[25\ °C]{[Cp_2M(n-Bu)_2]} R\diagup\!\!\!\diagdown SiHRR' \quad (10)$$

$$+ R\diagup\!\!\!\diagdown \left[+ R\diagup\!\!\!\diagdown SiHRR'\right]$$

(with Cp_2MCl_2, 2 n-BuLi, $-78\ °C$)

Lanthanum (La) and lanthanide metals neodymium (Nd), samarium (Sm) and lutetium (Lu) as well as yttrium (Y) with pentamethylcyclopentadienyl (η^5-Me_5C_5 = Cp*) and a silicon-bridged bis(tetramethylcyclopentadienyl) (Me_4C_5-$SiMe_2$-C_5Me_4 = $Me_2SiCp''_2$) ligand have been proved to be excellent and unique hydrosilylation catalysts which exhibit distinctive features in comparison with traditional Group VIII late-transition metal catalysts[43–45].

The hydrosilylation of vinylarenes including styrene catalyzed by $Cp^*_2LnCH(SiMe_3)_2$ (Ln = La, Nd, Sm, Lu) or $Me_2SiCp''_2SmCH_2(SiMe_3)_2$ with H_3SiPh gives the corresponding 1-silylethylarenes **30** with complete regioselectivity (equation 11)[43] in sharp contrast to the reactions promoted by traditional Group VIII transition metal catalysts which favor the formation of 2-silylethylarenes (*vide supra*). The relative catalytic activity of these complexes in the reaction of styrene decreases in the order $Me_2SiCp_2''SmCH(SiMe_3)_2 \gg Cp^*_2LaCH(SiMe_3)_2 > Cp^*_2NdCH(SiMe_3)_2 > Cp^*_2SmCH(SiMe_3)_2 \gg Cp^*_2LuCH(SiMe_3)_2$[43]. The observed unique regioselectivity can be ascribed to the formation of η^n-benzylic metal species (**31**) through interaction of the electrophilic lanthanide metal center with the arene π-system. The active catalyst species of these complexes are very likely to be the hydride complexes, Cp^*_2LnH and $Me_2SiCp''_2LnH$, that are formed via reductive cleavage of a $Ln-CH(SiMe_3)_2$ bond by the action of H_3SiPh[43].

$$\underset{R}{\overset{Ar}{>}}= + H_3SiPh \xrightarrow[C_6D_6, 23-60\,°C]{[Ln]} \underset{R}{\overset{Ar}{>}}\!\!\underset{SiH_2Ph}{\overset{Me}{<}}$$

(30)

(11)

$[Ln] = Cp^*_2LnCH(SiMe_3)_2$ or $Me_2SiCp''_2SmCH(SiMe_3)_2$

(a) Ar = Ph, R = H;   (b) Ar = 4-MeOC$_6$H$_4$, R = H;   (c) Ar = 4-FC$_6$H$_4$, R = H;
(d) Ar = 2-MeOC$_6$H$_4$, R = H;   (e) Ar = Ph, R = Me;   (f) Ar = Ph, R = Et;
(g) Ar = R = Ph;   (h) Ar = 2-Np, R = H

(31)

The reversed regioselectivity is also observed in the reaction of 1-hexene with H_3SiPh catalyzed by $Me_2Cp''_2SmCH(SiMe_3)_2$, which affords 2-silylhexane as the major product ($\leqslant 76\%$ regioselectivity)[43]. However, the reactions of sterically more demanding 1-alkenes, cyclohexylethene and 2-ethyl-1-butene, with H_3SiPh catalyzed by $Me_2Cp''_2SmCH(SiMe_3)_2$, afford linear products, $PhH_2Si-CH_2CH_2(C_6H_{11}$-$c)$ (96% yield) and $PhH_2Si-CH_2CH(C_2H_5)_2$ (98% yield), exclusively[43]. Exclusive formation of 1-silylalkanes is also observed in the reactions of 1-decene with H_3SiPh, $H_3SiC_6H_{13}$-n and $H_2SiMePh$ catalyzed by Cp^*_2Nd-R [R = $CH(SiMe_3)_2$, H, Cl] at 80°C[44]. The reactions of isoprene, 1,5-cyclohexadienes, and 1,6-heptadiene with H_3SiPh are also effected by $Cp^*_2Nd-CH(SiMe_3)_2$[46]. The reaction of isoprene gives a mixture of 1,4-addition products in which (E)-1-silyl-2-methyl-2-butene is the major (63%) product. The reactions of 1,5- and 1,6-dienes give silylmethylcyclopentane (84% yield) and 1-silylmethyl-2-methylcyclopentane (54%), respectively, as the major products.

Highly regio- and site-selective hydrosilylation of 1-alkenes (equation 12) and nonconjugated alkadienes (Table 2) with H_3SiPh can be achieved using $Cp^*_2YCH(SiMe_3)_2$ as the catalyst in benzene at ambient temperature[45]. It appears that this catalyst is sterically very demanding so that it reacts with the terminal olefinic bond exclusively, delivering the silicon moiety to the terminal carbon. The active catalyst species is most likely to be

29. Recent advances in the hydrosilylation and related reactions

Cp^*_2YH in the same manner as the lanthanide catalysts mentioned above.

(a) $R^1 = n\text{-}C_4H_9$, $R^2 = R^3 = R^4 = R^5 = H$ (95%) (12)
(b) $R^1 = R^2 = H$, $R^3 = R^4 = Me$, $R^5 = H$ (94%)
(c) $R^1 = H$, $R^2 = Me$, $R^3 = R^4 = H$, $R^5 = Me$ (56%)
(d) $R^1 = Cl$, $R^2 = R^3 = R^4 = R^5 = H$ (85%)
(e) $R^1 = OCH_2Ph$, $R^2 = R^3 = R^4 = R^5 = H$ (90%)
(f) $R^1 = OTBDMS$, $R^2 = R^3 = R^4 = R^5 = H$ (83%)

The hydrosilylation of acrylates with chlorohydrosilanes using Group VIII transition metal catalysts has been shown to yield either a mixture of 3-silylpropanoate and 2-silylpropanoate ($Pt^{47,48}$) or 2-silylpropanoate selectively (Ni^{49}, Rh^{50}). However, a binary catalyst system, Cu_2O–TMEDA (TMEDA = N, N, N', N'-tetramethylethylenediamine), is found to promote the exclusive β-addition of $HSiCl_3$, $HSiCl_2Me$ and $HSiCl_2Ph$, giving the corresponding 3-silylpropanoates **32** exclusively in 93–99% yields (equation 13)[51]. A variety of copper salts other than Cu_2O which include $CuCl$, $CuBr$, $CuCl_2$, $CuBr_2$, $CuCN$, CuO, $(CuO)_2 \cdot Cr_2O_3$ can also effect the reaction. It is proposed that TMEDA

TABLE 2. Hydrosilylation of nonconjugated dienes with $PhSiH_3$ catalyzed by $Cp^*_2YCH(SiMe_3)_2$

Substrate	Product	Yield (%)
(CH₂=CHCH₂CH₂CH=CH₂)	(CH₂=CHCH₂CH₂CH₂CH₂SiH₂Ph)	96
(isopropylidene diene)	(isopropylidene–SiH₂Ph)	94
(1,7-octadiene)	PhH_2Si—(octyl)—SiH_2Ph	96
(OTBDMS-substituted diene)	(OTBDMS-substituted, SiH₂Ph)	54
(vinylcyclohexene)	(cyclohexenyl–CH₂CH₂SiH₂Ph)	97

reacts with $HSiCl_2R$ to generate a hydrosilane–amine complex, $[TMEDA-H]^+RCl_2Si^-$, that undergoes conjugate addition to an acrylate, but the role of the copper salt is not clear at present.

$$HSiCl_2R \; + \; \diagup\!\!\!\diagdown COOR' \xrightarrow[TMEDA]{[Cu]} RCl_2Si\diagdown\!\!\!\diagup COOR' \quad (13)$$

(R = Cl, Me, Ph; R' = Me, Et)   **(32)**

The β-addition of chlorohydrosilanes to acrylonitrile, yielding 3-silylpropanenitrile **33**, can be effected by different promoters[3], and $CuCl/NBu_3/TMEDA^{52}$ or a copper salt–isocyanide[53] combination is known to be a good catalyst for a long time. Ultrasound is found to accelerate the reaction catalyzed by Cu_2O–TMEDA at ambient temperature, but the reaction also proceeds effectively under reflux (equation 14)[54]. The relative reactivity of chlorohydrosilanes in this catalyst system decreases in the order $HSiCl_3 >$ $HSiCl_2Ph \sim HSiCl_2Me \gg HSiClPh_2$, and $HSiClMe_2$ does not show any reactivity.

$$HSiCl_2R \; + \; \diagup\!\!\!\diagdown CN \xrightarrow[\text{)))} or reflux]{Cu_2O\;TMEDA} RCl_2Si\diagdown\!\!\!\diagup CN \quad (14)$$

R = Cl, Me, Ph
))) = ultrasound   **(33)**

A manganese(0) carbonyl complex, $Mn_2(CO)_{10}$, is found to be a fairly active catalyst for the reaction of 1-hexene with $HSiEt_3$ or $HSi(OEt)_3$ in toluene or THF at 40 °C[55]. Kinetic study suggests that $HMn(CO)_5$ is the catalytically active species.

The reaction of triethoxyvinylsilane, $CH_2=CHSi(OEt)_3$, with $HSi(OEt)_3$ catalyzed by $Ni(acac)_2$ gives a complex mixture of products, arising from dehydrogenative silylation, hydrogenation, disproportionation, and dimerization besides the normal hydrosilylation product, $(EtO)_3SiCH_2CH_2Si(OEt)_3$[56]. The dimerization product is a mixture of $(EtO)_3Si(CH_2)_4Si(OEt)_3$ and $(EtO)_3SiCH(Me)(CH_2)_2Si(OEt)_3$. A possible mechanism that can accommodate the formation of these products is proposed, which includes key intermediates such as $(EtO)_3Si(CH_2)_2Ni-CH[Si(OEt)_3]CH_2Si(OEt)_3$ and $Ni[(CH_2)_2Si(OEt)_3]^{56}$.

The photocatalytic hydrosilylation of 1,3-dienes in the presence of $Cr(CO)_6$ is known to proceed smoothly at ambient temperature to give (Z)-1,4-addition product(s) exclusively. A detailed follow-up study[57] on the reactions of 1,3-butadiene, 2,3-dimethyl-1,3-butadiene and isoprene with $HSiEt_3$ confirmed the previously reported results using $HSiMe_3$ except for a higher regioselectivity for the reaction of isoprene, i.e. (Z)-$Et_3SiCH_2C(Me)=CHMe/Et_3SiCH_2CH=CMe_2 = 72/28$ (60/40 for $HSiMe_3$). However, the minor product (3% isolated yield; the major product **34** is isolated in 86% yield) in the reaction of 1,3-pentadiene is found to be (E)-product **35**[57] instead of the previously reported (Z)-product (for $HSiMe_3$-reaction[58,59]). The reaction of 2-methyl-1,3-pentadiene affords one regioisomeric product, (Z)-$Et_3SiCH_2C(Me)=CHEt$, exclusively, but in only 14% isolated yield (equation 15)[57]. The reactions of 1,3-cyclohexadiene and 1,4-cyclohexadiene both give a mixture of 3-silyl-1-cyclohexene **36** and 1-silyl-1-cyclohexene **37** (**36/37** = 86/14, 84% yield from 1,3-hexadiene; **36/37** = 88/12, 23% yield from 1,4-hexadiene) (equation 16)[57]. In the latter case, only a low conversion of 1,4-cyclohexadiene is observed, and it is apparent that the isomerization of 1,4-cyclohexadiene to 1,3-cyclohexadiene takes place prior to hydrosilylation. Detailed mechanistic study shows intermediacy of $Cr(CO)_4(\eta^4$-1,3-diene) and $Cr(CO)_3(H)(SiEt_3)(1,3$-diene) complexes, and deuterium labeling experiments unambiguously indicate that a π-allylic silyl chromium species, (η^3-enyl)Cr(CO)$_3$(SiEt$_3$), arising from the insertion of 1,3-diene into the H−Cr bond, is the key intermediate in this catalytic cycle[57].

$$\text{diene} + \text{HSiEt}_3 \xrightarrow[n\text{-hexane, -10 °C}]{\text{Cr(CO)}_6, h\nu} \quad (34) + (35) \quad (25:1) \tag{15}$$

$$\text{cyclohexadiene} \xrightarrow[n\text{-hexane, -10 °C}]{\text{HSiEt}_3, \text{Cr(CO)}_6, h\nu} (36) + (37) \tag{16}$$

4. By immobilized catalysts

The development of Group VIII transition metals immobilized on inorganic or polymer support continues to be an active area in the hydrosilylation reaction, mainly because of easy separation and recycling capability of these catalysts[60−65].

Rhodium(I) complexes immobilized on silica using 3-(3-silylpropyl)-2,4-pentanedionato ligands (38) show good activity in the hydrosilylation of 1-octene with HSi(OEt)$_3$ at 100 °C[60]. The immobilized Rh catalysts are prepared by (i) reaction of (EtO)$_3$Si(CH$_2$)$_3$C(COMe)$_2$Rh(CO)$_2$ with untreated silica (Catalyst A), (ii) reaction of Rh(acac)(CO)$_2$ (acac = acetylacetonato = 2, 4-pentanedionato) with silica modified by [(EtO)$_3$Si(CH$_2$)$_3$C(COMe)$_2$]$^-$ prior to the complexation (Catalyst B), (iii) reaction of [Rh(CO)$_2$Cl]$_2$ with a polycondensate of [(EtO)$_3$Si(CH$_2$)$_3$C(COMe)$_2$]$^-$, Si(OEt)$_4$ and water (Catalyst C) and (iv) sol-gel processing of (EtO)$_3$Si(CH$_2$)$_3$C(COMe)$_2$Rh(CO)$_2$ and Si(OEt)$_4$ (Catalyst D). The Catalysts A and B show ca three times better activity than their homogeneous counterparts, while the Catalyst D exhibits only low activity and the Catalyst C is inactive[60].

(38)

In a manner similar to the four methods mentioned above, rhodium complex catalysts immobilized on silica modified by 2-(MeO)$_3$Si(CH$_2$)$_2$C$_5$H$_4$N and (MeO)$_3$Si(CH$_2$)$_3$OCOCMe=CH$_2$ using [Rh(CO)$_2$Cl]$_2$ as the precursor are prepared[61]. These immobilized pyridine−Rh complexes are shown to be active catalysts in the hydrosilylation of 1-octene, and claimed to be more efficient than their homogeneous counterparts[61].

Polyamide-supported Rh, Pd, Pt and Ru complex catalysts are shown to be active in the hydrosilylation of 2-methyl-1,3-pentadiene and isoprene with $HSiMe_2Ph$ and $HSi(OEt)_3$[63,64]. The catalysts are prepared by immobilizing $[Rh(CO)_2Cl]_2$, $PdCl_2(PhCN)_2$, $PtCl_2(PhCN)_2$ and $RuCl_2(bipy)_2$ on the polyamide supports **39a,b** and **40a,b** that have relatively uniform distribution of micropores with voids from 1.0 to 3.0 nm. The reaction of 2-methyl-1,3-pentadiene catalyzed by these polyamide-supported catalyst ([M]−P) gives a mixture of 1,4-addition product **41** and 1,2-addition product **42** regioselectively, i.e. the silyl group is exclusively delivered to the C-1 of the diene (equation 17). It is noteworthy that $HSiMe_2Ph$ favors 1,4-addition (51–95%) while $HSi(OEt)_3$ does 1,2-addition (75–85%) in the reactions using the Pd, Pt and Ru catalysts. However, the Rh-catalyzed reactions favor 1,4-addition regardless of the hydrosilane used (85–95%). The reactions of isoprene promoted by the Pd, Pt and Ru catalysts follow the same 1,4/1,2 selectivity pattern depending on the hydrosilane used although two regioisomers are formed for both 1,4-addition (**43a,b**) and 1,2-addition (**44a,b**) processes for this substrate (equation 18). In contrast, the Rh-catalyzed reaction gives only (Z)-**43a** as the 1,4-adduct (86–95%) regardless of the hydrosilane used although the 1,2-addition product was a mixture of **44a** and **44b**. These polyamide-immobilized metal catalysts exhibit virtually the same regio- and stereoselectivity as those observed for their homogeneous counterparts. This fact indicates a good match of the dimensions of the key intermediates in the catalytic process to those of the polymer micropores. The Rh complexes with the polyamide supports **40a** and **40b** are found to show highest selectivities among the catalysts examined, and are proved to be extremely stable, keeping the same activity even after six runs[64].

Support **39a**: $n = 2$
Support **39b**: $n = 6$

Support **40a**: $n = 2$
Support **40b**: $n = 6$

[M] = Rh, Pd, Pt, Ru
$SiX_3 = SiMe_2Ph, Si(OEt)_3$

(17)

(41)
and/or

(42)

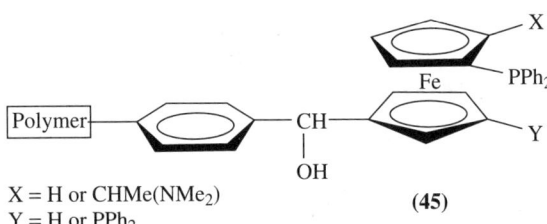

Platinum complexes of ferrocenes attached to cross-linked polystyrene support **45** are shown to be highly active in the hydrosilylation of styrene and 1-hexene with $HSiCl_3$, giving linear products, 2-silylethylbenzene and 1-silylhexane, respectively, in excellent yields with complete regioselectivity[65]. These catalysts can be recycled without appreciable loss of activity. However, the absence of any ligand effects on the selectivity and activity indicates that catalytically active metallic Pt is generated in these systems[65]. The corresponding palladium complexes are also active in the hydrosilylation of styrene with $HSiCl_3$ at 70–90 °C (0.1 mol% of the catalyst for 24–48 h), giving 1-silylethylbenzene in excellent yields with 95–100% regioselectivity[65]. These Pd catalysts possess, however, only a low catalytic activity for the reaction of 1-hexene.

$X = H$ or $CHMe(NMe_2)$
$Y = H$ or PPh_2

(**45**)

5. By radical initiators

The hydrosilylation of alkenes with $HSiCl_3$ and $HSiCl_2Me$ under free-radical conditions using UV irradiation or organic peroxides was studied in the early years since the discovery of the reaction. However, Speier's catalyst and other transition metal catalysts have been almost exclusively used for the reaction afterwards. It has been shown that trialkylsilyl radicals are extremely reactive species for addition to carbon–carbon multiple bonds, but the hydrosilylation of alkenes with trialkylsilanes under free-radical conditions, e.g. $(t\text{-}BuO)_2$ as the initiator, is limited to nonpolymerizable alkenes. Tris(trimethylsilyl)silane, $HSi(SiMe_3)_3$ (TTMSS), however, is found to have a very low Si–H dissociation energy (11 kcal mol^{-1})[66], which is even lower than that of trialkylsilanes. This unique hydrosilane, TTMSS, undergoes smooth addition to simple and functionalized alkenes **46** in the presence of azobis(isobutyronitrile) (AIBN) at 80–90 °C in toluene to give the corresponding hydrosilylation products **47** in high yields (equation 19)[66–68].

The reaction clearly includes carbon-centered radical, which can account for the observed skeletal rearrangement in the reaction of β-pinene (equation 20) and extremely high stereoselectivity in the reactions of methylmaleic anhydride (equation 21) and a methylfumarate (equation 22) using Et_3B/O_2 as the initiator at -15 °C [The reactions using AIBN at 90 °C give the isomer ratios of 16 : 1 (**48a:48b**) and 12 : 1

(49a:49b), respectively].

(a) $R^1 = R^2 = H$, $R^3 = (CH_2)_7CH_3$, Ph, CN, CO_2Me, COMe, OBu, OAc, $P(O)(OEt)_2$ or SPh
(b) $R^1 = H$, $R^2 = Me$, $R^3 = Ph$ or CO_2Me
(c) $R^1 = Me$, $R^2 = CN$ or CO_2Et, $R^3 = H$
(d) $R^1 = R^2 = CO_2Et$, $R^3 = H$

B. Mechanism of Alkene Hydrosilylation

When the hydrosilylation reactions were reviewed in this series and published in 1989[3], the 'Chalk–Harrod mechanism'[69] (Scheme 5) was apparently the most widely accepted mechanism for the alkene hydrosilylation, with minor exceptions of photocatalyzed reactions using $Fe(CO)_5$, $M_3(CO)_{12}$ (M = Fe, Ru, Os) and $R_3SiCo(CO)_4$[59,70,71]. However, extensive mechanistic studies recently conducted have revealed that there are many catalyst systems which do not follow the Chalk–Harrod mechanism.

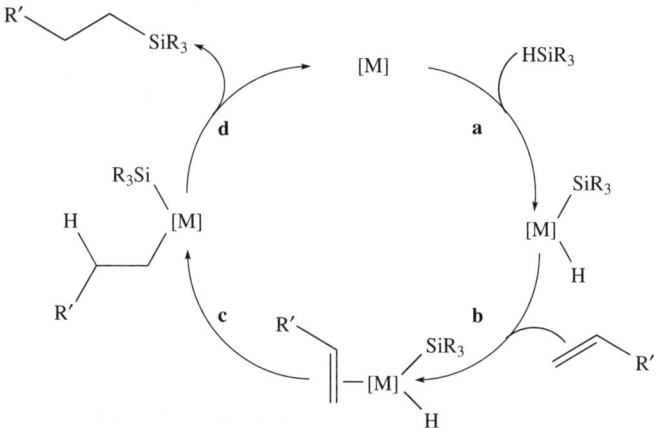

SCHEME 5. Chalk–Harrod mechanism

The hydrosilylation of alkenes often yields silylalkenes through dehydrogenative silylation (*vide infra*) and an alkane by hydrogenation besides normal hydrosilylation product, silylalkane[3]. Since the Chalk–Harrod mechanism cannot accommodate these products, mechanisms including the initial insertion of an alkene to the [M]–Si bond rather than the [M]–H bond (i.e. 'silyl migration pathway'[72]), followed by β-hydride elimination, forming a silylalkene and the subsequent hydrogenation of the alkene by $[M](H)_2$ species generated, were proposed as a part of or as a competing pathway in the hydrosilylation process[59,73–80]. Another important point is the fact that a facile reductive elimination of a silylalkane from an alkyl–[M]–SiR$_3$ species, which is one of the key steps in the Chalk–Harrod mechanism (step **d** in Scheme 5), is not well-established or precedented in stoichiometric reactions.

The first convincing results for the occurrence of the 'silyl migration pathway' were presented by Seitz and Wrighton in the photochemical reaction of $Et_3SiCo(CO)_4$ with ethene[81]. On the basis of FTIR and ^{1}H NMR studies monitoring the reaction, the mechanism illustrated in Scheme 6 is proposed, in which the steps (a)–(c), (f) and (g) are confirmed[81]. For the steps (d) and (e), $MeCo(CO)_4$ is used as a model complex and subjected to the reaction with $HSiMe_3$. The reaction gives CH_4 (detected by ^{1}H NMR) and $Me_3SiCo(CO)_4$ (by FTIR) as the products and the formation of $SiMe_4$ is not observed (equation 23)[81]. Although the formation of a trace amount of $SiMe_4$ cannot be completely ruled out, it is evident that the reductive elimination of alkane is predominant over that of silylalkane (step **e** in Scheme 6), which strongly supports the 'silyl migration pathway'.

$$MeCo(CO)_4 + HSiMe_3 \xrightarrow{-CO} \left[Me-\underset{H}{\overset{SiMe_3}{Co(CO)_3}} \right] \xrightarrow{-CO} Me-H + Me_3SiCo(CO)_4$$

(23)

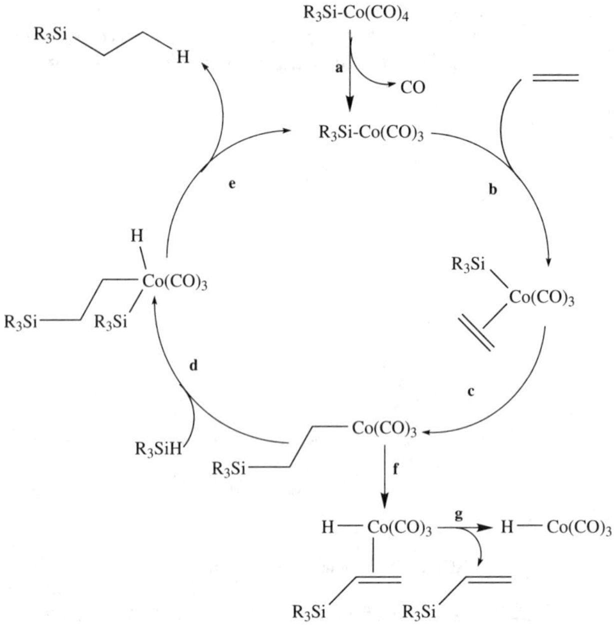

SCHEME 6. Seitz–Wrighton mechanism

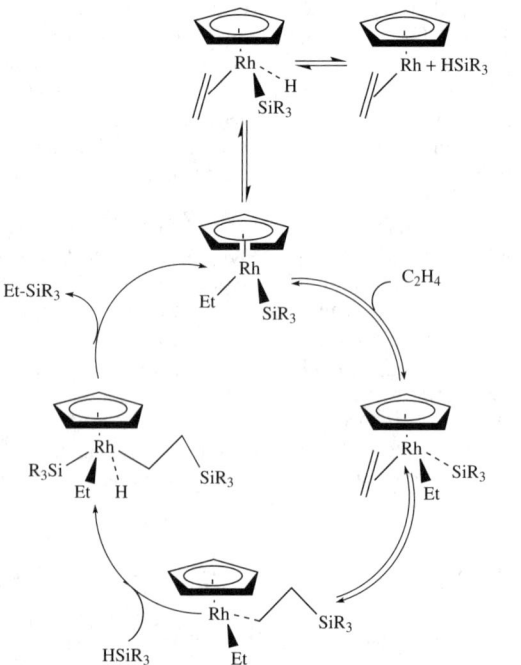

SCHEME 7. Two-silicon cycle by Duckett and Perutz

A detailed mechanistic study on the hydrosilylation of ethene with HSiR$_3$ (R = Et, i-Pr, sec-Bu, Me) catalyzed by (η^5-C$_5$H$_5$)Rh(CH$_2$=CH$_2$)(SiR$_3$)H has demonstrated on the basis of deuterium labeling experiments and a series of control experiments[82] that the Chalk–Harrod mechanism is not operative in this catalyst system. The 'two-silicon cycle' mechanism (another 'silyl migration pathway') including a Rh(V) intermediate is proposed by Duckett and Perutz for this catalysis (Scheme 7)[82]. The integrity of the original ethene (or ethyl) group is proved to be retained. The proposed mechanism is essentially the same as the Seitz–Wrighton mechanism (*vide supra*) if the CpRh–Et moiety is replaced by Co(CO)$_3$ in the catalytic cycle. A possible involvement of Rh(V) species has been proposed by Maitlis and coworkers as well[74,83].

Direct evidence for the 'silyl migration pathway' was presented by Brookhart and Grant in the hydrosilylation of 1-hexene catalyzed by a cobalt(III) complex[72]. The detailed mechanistic study was conducted on the reaction using [Cp*Co{P(OMe)$_3$}CH$_2$CH$_2$–μ–H]$^+$ [BAr$_4$]$^-$ (Ar = 3, 5-CF$_3$C$_6$H$_3$) **(50)** as the catalyst precursor based on the spectroscopic detection of catalyst species in a 'working' catalytic cycle, kinetics, deuterium labeling techniques and a convincing mechanism including the silyl migration step to form the secondary alkyl–Co complex with agostic hydrogen (μ-H) (step **c**) and the unexpected isomerization of **53** to the primary alkyl–Co complex with a μ-H **54** as the 'turnover-limiting step' (step **d**) (Scheme 8)[72]. This unique isomerization process is elucidated by a labeling experiment using DSiEt$_3$, which shows almost exclusive incorporation of deuterium at the C-6 position to yield Et$_3$Si(CH$_2$)$_5$CH$_2$D (>95% based on ^{13}C NMR analysis) (equation 24)[72]. This labeling experiment clearly indicates that the isomerization to **54** takes place prior to the reaction of **53** (or any other possible intermediates involved in

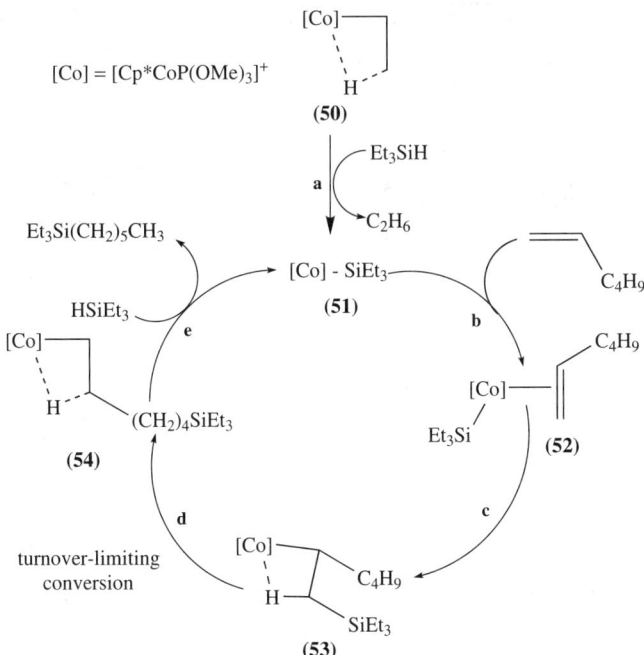

SCHEME 8. Brookhart–Grant mechanism

the isomerization) with HSiEt$_3$ (step **e**). For the final H-shift or reductive elimination step (step **e**), a σ-bond metathesis mechanism either via concerted fashion or via η^2-HSiEt$_3$ coordination to the electrophilic Co(III) metal center is strongly suggested[72] rather than the one including classical oxidative addition of HSiEt$_3$ to form a cationic Co(V) intermediate like the Rh(V) species proposed by Duckett and Perutz[82] (*vide supra*).

$$\text{CH}_2=\text{CHCH}_2\text{CH}_2\text{CH}_3 + \text{DSiEt}_3 \xrightarrow{\text{[Co]}} \text{Et}_3\text{Si-CH}_2\text{CH}_2\text{CH}_2\text{CH}_2\text{CH}_2\text{D} \quad (24)$$

[Co] = [Cp*Co(P(OMe)$_3$)CH$_2$CH$_2$-μ-H]$^+$ (**50**)

In the hydrosilylation of alkenes catalyzed by Group IV metallocene complexes, Cp$_2$MCl$_2$/2BuLi(M = Zr, Hf, Ti)[37,38] (*vide supra*), the 'olefin-first' mechanism including σ-bond metathesis of η^2-alkene-MCp$_2$ and HSiR$_3$ is proposed by Kesti and Waymouth (Scheme 9)[37]. After the formation of β-silylalkyl–M–H species **56** via σ-bond metathesis

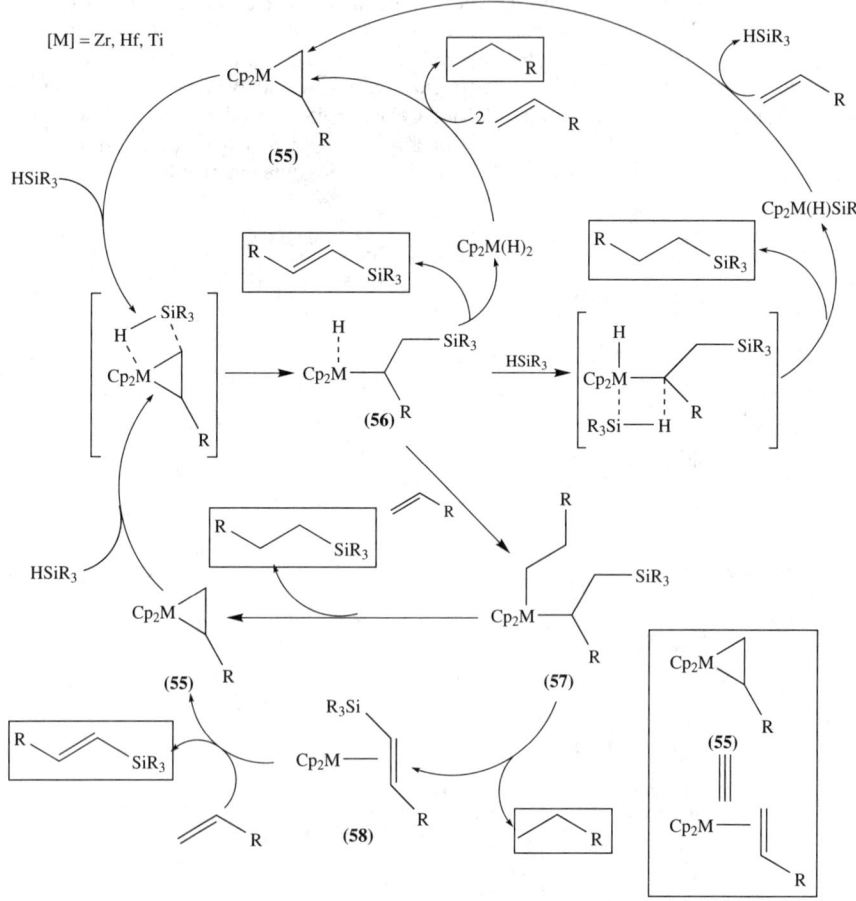

SCHEME 9. Kesti–Waymouth's 'olefin-first' mechanism

of **55**, another σ-bond metathesis of **56** with HSiR$_3$ yields silylalkane (hydrosilylation product) and Cp$_2$M(H)SiR$_3$, while the β-hydride elimination of **56** gives alkenylsilane (dehydrogenative silylation product) and Cp$_2$M(H)$_2$. It appears that the Cp$_2$M(H)SiR$_3$ is not stable[37] and thus an alkene molecule can readily replace the silyl hydride moiety to regenerate η^2-alkene-MCp$_2$ species **55**. Also, Cp$_2$M(H)$_2$ can react with alkene molecules to yield alkane and **55**. In this way, the 'olefin first' catalyst cycle completes. Alternatively, **55** reacts with another alkane molecule to form dialkyl-M species **57**, which undergoes a combination of β-hydride elimination and hydride shift to form either η^2-silylalkene-M complex **58** and alkane (hydrogenation product) or η^2-alkene-M complex **55** and silylalkane (hydrosilylation product) (Scheme 9)[37]. The latter mechanism has close precedents in the Rh and Ru catalyzed reactions proposed by Maitlis and coworkers[74] and Ojima and coworkers[80]. Consequently, the most characteristic feature in the Group IV metallocene catalyzed reaction is σ-bond metathesis processes which are not very common in the Group VIII metal-catalyzed reactions.

The Chalk–Harrod type mechanism is proposed to be operative in the reaction of ethene with HSiMe$_3$, HSiEt$_3$ and HSiPh$_3$ catalyzed by an early–late heterobimetallic complex, Cp$_2$Ta(CH$_2$)$_2$Ir(CO)$_2$[84,85]. The proposed catalytic cycle is very similar to the one shown in Scheme 5, wherein [M] is Cp$_2$Ta(CH$_2$)$_2$Ir(CO), except for the dissociation of CO at the oxidative addition of HSiR$_3$ and the coordination of CO at the reductive elimination of EtSiR$_3$[84]. This catalyst, however, only promotes the hydrosilylation of ethene, and higher alkenes are isomerized[84].

The hydrosilylation of alkenes catalyzed by organolanthanide complexes, Cp*$_2$LnCH(SiMe$_3$)$_2$ (Ln = La, Nd, Sm, Lu) and Me$_2$SiCp″$_2$SmCH(SiMe$_3$)$_2$ (Me$_2$SiCp″$_2$ = Me$_4$C$_5$−SiMe$_2$−C$_5$Me$_4$)[43], or the organoyttrium complex, Cp*$_2$YCH(SiMe$_3$)$_2$[45], is proposed to proceed via the Chalk–Harrod type mechanism except for the inclusion of σ-bond metathesis (steps **b** and **e**) instead of a classical oxidative addition–reductive elimination process (Scheme 10). The organolanthanide complex-catalyzed reactions of

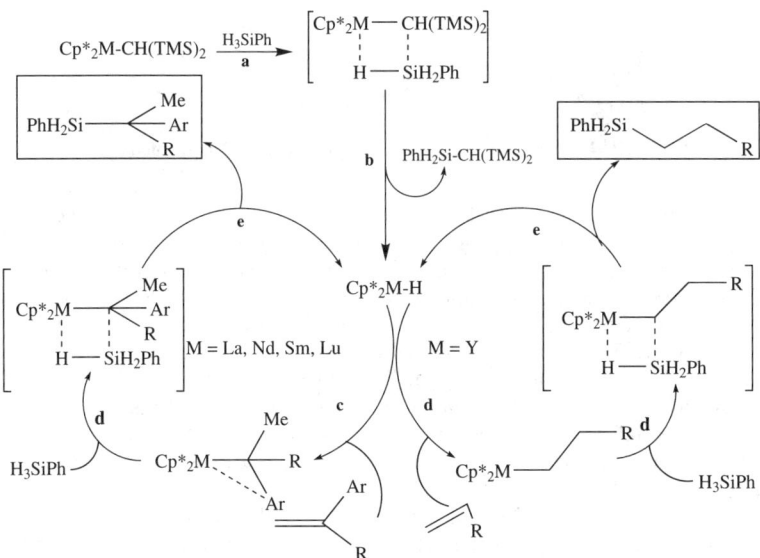

SCHEME 10. Proposed mechanism for organolanthanide or yttrium-catalyzed reaction

vinylarenes with H_3SiPh give the branched hydrosilylation products exclusively via η^n-benzylic metal species **31** (*vide supra*). In the absence of such an arene–Ln interaction, i.e. normal 1-alkenes including 2-alkyl-1-alkenes, these lanthanide catalysts afford 1-silylalkenes except for the reaction of 1-hexene catalyzed by sterically less demanding $Me_2SiCp''_2SmCH(SiMe_3)_2$[43] (*vide supra*). Exclusive formation of 1-silylalkanes with or without functional groups is observed in the $Cp*_2YCH(SiMe_3)_2$-catalyzed reactions[45] (*vide supra*). The observed extremely high terminal selectivity can be ascribed to the substantial steric demands as well as the electrophilic nature of the bulky $Cp*_2M$ moiety in the olefin insertion to the M–H bond (step **c**)[45].

C. Intramolecular Hydrosilylation of Alkenes

Intramolecular hydrosilylations of functionalized alkenes followed by hydrogen peroxide oxidation provide powerful methods for organic syntheses[86–88]. The reactions of allylic *O*-dimethylsilyl ethers **59** promoted by platinum catalysts, e.g. Karstedt's catalyst and $Pt(PPh_3)_2(CH_2=CH_2)$, or rhodium catalysts, e.g. Rh(acac)(COD) and $[RhCl(CH_2=CH_2)_2]_2$, proceed via 5-*endo* cyclization to give oxasilacyclopentanes **60** with a couple of exceptions in which siloxatanes **61** are formed (Scheme 11)[87,89].

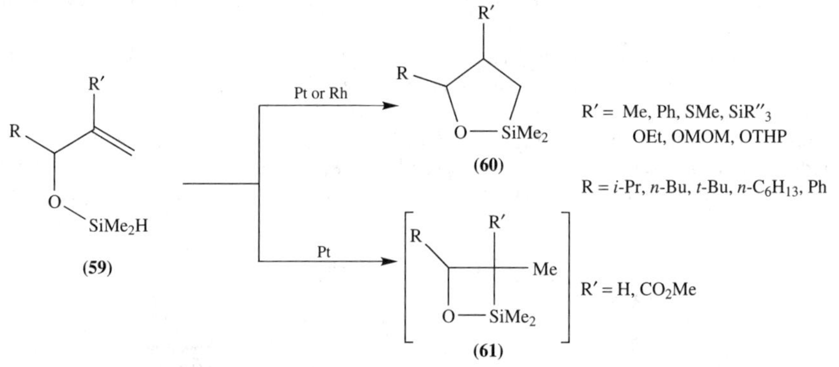

R' = Me, Ph, SMe, SiR''$_3$
OEt, OMOM, OTHP

R = *i*-Pr, *n*-Bu, *t*-Bu, *n*-C_6H_{13}, Ph

R' = H, CO_2Me

SCHEME 11. Intramolecular hydrosilylation of *O*-silylallylic alcohols **59**

When silyloxy enol ethers (**59**: R' = EtO, MOMO, THPO) are used as the substrates, the reactions proceed with excellent *syn*-selectivity (*syn/anti* = 14/1 ∼> 99/1)[87]. The resulting oxasilacyclopentanes **60** are subjected to the Tamao oxidation after the removal of the catalyst to afford the corresponding triol derivatives **62** in good to high overall yields. For example, the reaction of 3-(2-furyl)-3-HMe_2SiO-2-OTHP-1-propene (**59a**: R = 2-furyl; R' = OTHP), prepared from 3-(2-furyl)-2-OTHP-1-propene and $(HMe_2Si)_2NH$, using Karstedt's catalyst, followed by the Tamao oxidation gives the 2-*O*-THP-triol **62a** in 90% isolated yield with >99/1 *syn/anti* ratio (equation 25)[87]. As an application of the process yielding **62**, the stereoselective syntheses of enantiopure pentitols, D-arabinitol and xylitol as their pantaacetates have been achieved in just 4 steps from an enantiopure glyceraldehyde ketal[87].

Remarkable dependence of the regioselectivity on the catalyst species employed is observed in the intramolecular hydrosilylation of allylic *N*-dimethylsilylamines **63**, i.e. the Rh-catalyzed reactions proceed via 5-*endo* cyclization to give 1-aza-2-silacyclopentanes **64** whereas the Pt-catalyzed reactions yield 1-aza-2-silacyclobutanes **65** via 4-*exo* cyclization (Scheme 12)[89,90].

(25)

(i) (HMe$_2$Si)$_2$NH, RT, then *in vacuo*;

(ii) (a) [Pt] = Pt[(CH$_2$=CHSiMe$_2$)$_2$O]$_2$ in toluene, 60°C, (b) EDTA^{2-}.2Na$^+$, hexane;

(iii) 30% H$_2$O$_2$, 15% KOH, MeOH/THF, RT

The 4-*exo* cyclization of open-chain substrates **63** proceeds in *trans*-fashion with moderate to excellent selectivity (*trans/cis* = 77/23 ~> 99/1)[90]. The *trans* selectivity is dependent on the substitution pattern of R^2, R^3 and R^4. The reactions giving *trans*-1-aza-2-silacyclobutanes **65** have been applied to the stereoselective syntheses of *syn*-amino alcohols **66** via the Tamao oxidation as exemplified by the reaction of **63d** (R = HMe$_2$Si), affording **66d** (76% overall yield in 4 steps, *syn/anti* => 99/1) via **65d** (*trans/cis* => 99/1) in equation 26[90]. In the case of 3-*N*-disilylamino-1-cyclohexene **63f** (R = HMe$_2$Si), however, *cis*-1-aza-2-silacyclobutane **65f** is formed exclusively, that is converted to *cis*-2-amino-1-cyclohexanol (**66f**) (equation 27)[90].

Detailed mechanistic study on these intramolecular hydrosilylation of allylic *O*-silyl ethers **59** and allylic *N*-silylamines **63** using deuterium labeling techniques shows that 5-*endo* cyclization giving **60** or **64** proceeds via a Chalk–Harrod type hydrometalation catalytic cycle, while 4-*exo* cyclization process yielding **61** or **65** includes a Seitz–Wrighton type silylmetalation mechanism[89].

In a similar manner, 1-cyclopentene-3-carboxylic acid (**67a**) and *N*-*t*-Boc-(*S*)-3,4-didehydroproline (**67b**) can be transformed to methyl *cis*-2-hydroxycyclopentane-1-carboxylate (**70a**) and methyl (3*R*, 2*S*)-*N*-*t*-Boc-3-hydroxyprolinate (**70b**) in 67% and 76% overall yields (4 steps), respectively (equation 28)[91].

Intramolecular hydrosilylation of *O*-dimethylsilylalkenoate **71**, promoted by Karstedt's catalyst, provides an efficient route to *cis*-4-methyl-5-substituted-γ-lactone **73** (*cis/trans* = 76/24–98/2) in 21–87% yield (equation 29)[92]. The reaction proceeds through 5-*endo* cyclization exclusively. The stereoselectivity of the reaction is highly

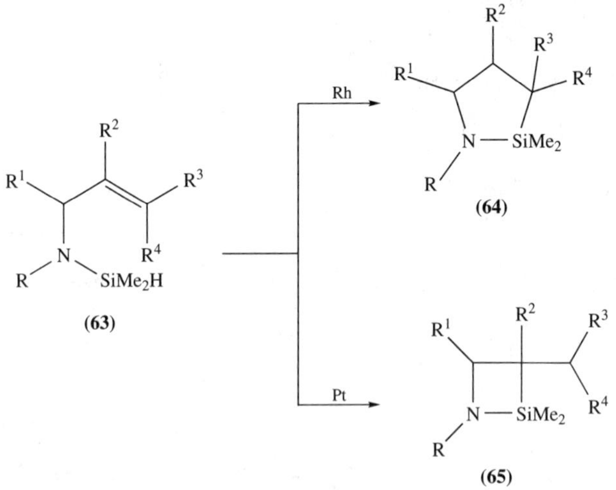

[Rh] = [RhCl(CH$_2$=CH$_2$)$_2$]$_2$ or RhCl(PPh$_3$)$_3$
[Pt] = Pt[(CH$_2$=CHSiMe$_2$)$_2$O]$_2$

R = HMe$_2$Si, t-BuMe$_2$Si, PhCH$_2$
(a) R^1 = R^2 = R^3 = R^4 = H
(b) R^1 = i-Pr, R^2 = R^3 = R^4 = H
(c) R^1 = Ph, R^2 = R^3 = R^4 = H
(d) R^1 = Ph, R^2 = H, R^3 = R^4 = Me
(e) R^1 = Ph, R^2 = Me, R^3 = R^4 = H
(f) R^{1}R^4 = —(CH$_2$)$_3$—, R^2 = R^3 = H

SCHEME 12. Intramolecular hydrosilylation of N-silylallylic amines **63**

dependent on the size of the substituent R, while it is much less sensitive to the substituent on silicon. For bulkier R, stereoselectivity increases, but yield decreases. The A^{1,2} strain can account for the observed *cis*-selectivity[92].

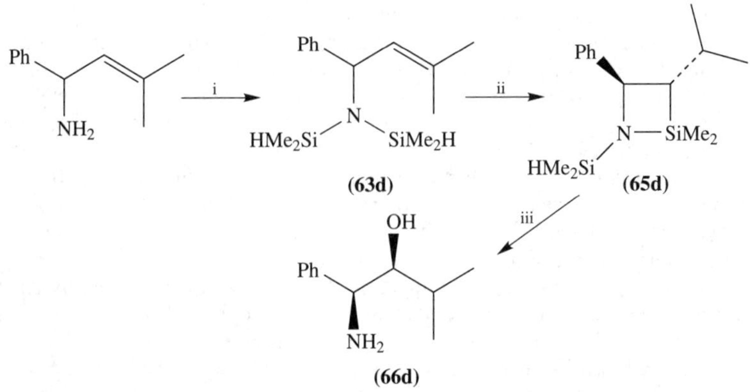

(26)

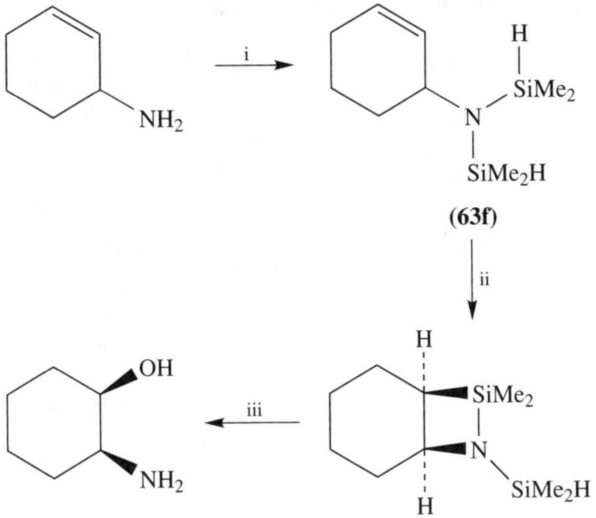

29. Recent advances in the hydrosilylation and related reactions

(27)

(i) *n*–BuLi, then HMe₂SiCl (twice); (ii) (a) [Pt] = Pt[(CH₂=CHSiMe₂)₂O]₂ in toluene, 60 °C

(b) EDTA^{2-}·2Na^{+}, hexane; (iii) 30% H_2O_2, 15% KOH, MeOH/THF, RT

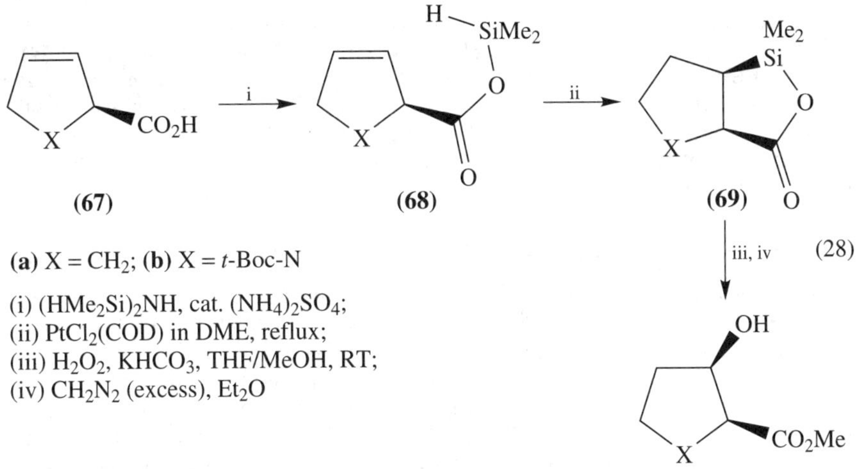

(a) X = CH₂; (b) X = *t*-Boc-N

(i) (HMe₂Si)₂NH, cat. (NH₄)₂SO₄;
(ii) PtCl₂(COD) in DME, reflux;
(iii) H₂O₂, KHCO₃, THF/MeOH, RT;
(iv) CH₂N₂ (excess), Et₂O

(28)

In the reaction of HMe₂SiOCMe₂CH=CH₂ using Karstedt's catalyst in pentane at ambient temperature, an eight-membered ring product **75** is formed (35% yield) besides the normal 5-*endo* cyclization product **74** (35% yield) (equation 30)[93].

R = Me, Et, *i*-Bu, Ph, *i*-Pr, *t*-Bu

A combination of dehydrogenative silylation and the subsequent intramolecular hydrosilylation of allylic alcohols with H_2SiPh_2 and $H_2SiMePh$ catalyzed by Cp_2TiMe_2 or racemic (EBTHI)TiMe$_2$ [EBTHI = ethylene-1,2-bis(η^5-4,5,6,7-tetrahydroindenyl)] gives the desired 5-*endo* cyclization product **60** in up to 75% yield together with a rather complex mixture of redistribution and hydrogenation products[94].

Applications of the intramolecular hydrosilylation to catalytic asymmetric synthesis will be discussed in Section V (*vide infra*).

D. Other Reactions Associated with Alkene Hydrosilylation

Dehydrogenative silylation of alkenes is often observed as a side reaction of hydrosilylation (*vide supra*). However, this reaction becomes a predominant or exclusive process depending on the nature of the catalysts used as well as substrates[3,79,80,95–99].

Exclusive formation of silylstyrenes **76** is achieved when the reactions of styrene and 4-substituted styrenes with $HSiEt_3$ are catalyzed by $Fe_3(CO)_{12}$ or $Fe_2(CO)_9$[100]. Other iron-triad metal carbonyl clusters, $Ru_3(CO)_{12}$ and $Os_3(CO)_{12}$, are also highly active catalysts, but a trace amount of hydrosilylation product **77** is detected in the Ru-catalyzed reactions and the Os-catalyzed reactions are accompanied by 3–12% of **77** (equation 31)[100]. Mononuclear iron carbonyl, $Fe(CO)_5$, is found to be inactive in this reaction[100].

1,5-Hexadiene is also found to undergo extremely selective dehydrogenative silylation in the presence of a catalytic amount of $RhCl(PPh_3)_3$ at the terminal olefin moiety to give (E)-1-silyl-1,5-hexadiene **78** in excellent yields (equation 32)[101]. In order for this reaction to be highly selective, excess 1,5-hexadiene should be used, i.e. a mixture of **78** and the usual hydrosilylation product is obtained when an equal amount or excess hydrosilane is used. The reaction with $HSiEt_2Me$ proceeds with >98% stereoselectivity, giving (E)-**78** in 90% yield, which is accompanied by 9% of an olefin-isomerization product, 1-silyl-1,4-diene **79**. The reaction using $HSi(OEt)_3$ gives **78** in 99% yield (no side product is detected), but stereoselectivity is somewhat decreased ($E/Z = 92/8$). Other Rh-catalysts, e.g. $Rh_6(CO)_{16}$, $RhCl(CO)(PPh_3)_3$, $HRh(CO)(PPh_3)_3$, $RhCl_3 \cdot 3H_2O$ and $Rh_2(OAc)_2$, show good activities in the reaction of 1,5-hexadiene with $HSiEt_2Me$. The results using other 1,5-dienes are exemplified in Table 3[101]. It is noteworthy that only 1,5-dienes yield dehydrogenative silylation products cleanly, i.e. the reactions of 1,4-dienes and 1,6-dienes under the same conditions suffer from the extensive occurrence of hydrosilylation and double-bond isomerization[101].

The formation of allylsilanes **80** is found to take place in the reaction of alkenes with $HSiEt_3$ catalyzed by $Rh_2(pfb)_4$ (pfb = perfluorobutanoate) when the hydrosilane is added to a mixture of an alkene with the catalyst in CH_2Cl_2 (equation 33)[102]. The reactions of simple 1-alkenes, e.g. 1-hexene, 1-octene and 1-decene, give 31–34% of **80**, 3–6% of **81** and 61–66% of **82**, while a highly selective allylsilane formation [91% of **80** ($E/Z => 50/1$), 1% of **81** and 8% of **82**] is observed in the reaction of 3-phenyl-1-propene (R = Ph). 1-Silylalkene **81** (R = n-BuO) is formed as the predominant product (61%) together with **80** (12%) and **82** (27%) when 3-(n-butoxy)-1-propene is employed. Since the reaction should generate molecular hydrogen, substantial hydrogenation of the alkene substrate is also observed during the reaction. It should be noted that usual $Rh_2(pfb)_4$-catalyzed hydrosilylation of these 1-alkenes proceeds with 95–98% selectivity when the

TABLE 3. Dehydrogenative silylation with HSiEt$_2$Me of 1,5-dienes

1,5-Diene	Product	Yield (%) (E/Z at C-1)	Side product (%)
(E/Z = 11/1)	SiEt$_2$Me	85 (E/Z at C$_5$ = 11/1)	7
(cyclohexenyl-vinyl)	SiEt$_2$Me (cyclohexenyl)	90	9
OSiMe$_3$ (diene)	SiEt$_2$Me, OSiMe$_3$	44	trace
	MeEt$_2$Si, OSiMe$_3$	49	

alkene is added to a mixture of the Rh-catalyst and HSiEt$_3$[102].

$$R\text{–CH=CH}_2 + \text{Rh}_2(\text{pfb})_4 \xrightarrow[\text{CH}_2\text{Cl}_2, \text{RT}]{\text{HSiEt}_3} R\text{–CH=CH–SiEt}_3 \quad (80)$$

$$+ \; R\text{–CH=CH–SiEt}_3 \; + \; R\text{–CH}_2\text{–CH}_2\text{–SiEt}_3$$
$$\quad\quad\quad (82) \quad\quad\quad\quad\quad (81)$$

(33)

When a Si–C bond has substantial strain, the addition of hydrosilane to such a Si–C σ-bond takes place. The hydrosilylations of 1,3-disilacyclobutanes, cis-**83** and trans-**83**, catalyzed by Pt(PPh$_3$)$_2$Cl$_2$, proceed stereospecifically to give trisilanes, meso-**84** and dl-**84**, respectively, in quantitative yields (equations 34 and 35)[103].

cis-(**83**) $\xrightarrow[\text{toluene, 70 °C}]{\text{HSiMe}_2\text{Ph} \; \text{Pt(PPh}_3)_2\text{Cl}_2}$ PhMe$_2$Si–Si(Me)(Ph)–CH$_2$–Si(Me)(Ph)–H   meso-(**84**) (34)

trans-(**83**) $\xrightarrow[\text{toluene, 70 °C}]{\text{HSiMe}_2\text{Ph} \; \text{Pt(PPh}_3)_2\text{Cl}_2}$ PhMe$_2$Si–Si(Ph)(Me)–CH$_2$–Si(Me)(Ph)–H   dl-(**84**) (35)

III. HYDROSILYLATION OF ALKYNES

The hydrosilylation of alkynes provides a direct route to a variety of alkenylsilanes that serve as important intermediates for cross-linking silicones as well as versatile reagents

29. Recent advances in the hydrosilylation and related reactions

in organic syntheses. This reaction appears to have been attracting increasing attention among synthetic chemists. The hydrosilylation of alkynes catalyzed by transition metal complexes including clusters proceeds, in general, more smoothly than that of alkenes, and terminal alkynes are much more reactive than internal alkynes[104]. Platinum, especially Speier's catalyst, rhodium and other Group VIII transition metals and their complexes are commonly used as catalysts. The reaction of a 1-alkyne, R−C≡CH, with a hydrosilane, $HSiX_3$, possibly gives three products, i.e. (E)-R−CH=CH−SiX_3, (Z)-R−CH=CH−SiX_3 and R(X_3Si)C=CH_2 (equation 36). Accordingly, the stereoselectivity and the regioselectivity of the reaction are important issues in this reaction.

$$RC\equiv CH + HSiX_3 \xrightarrow{catalyst} (E)\text{-and/or }(Z)\text{-RCH}=CHSiX_3 + R(X_3Si)C=CH_2 \quad (36)$$

A. Simple Alkynes

1. By platinum catalysts

Speier's catalyst (H_2PtCl_6/i-PrOH), the most commonly used catalyst in the hydrosilylation of alkenes (*vide supra*), can be used for the reaction of alkynes[3]. A related potassium salt K_2PtCl_6, however, does not show appreciable catalytic activity. A number of quaternary ammonium, phosphonium and arsonium hexachloroplatinates are shown to be highly active catalysts for the hydrosilylation of phenylacetylene with different hydrosilanes, giving a mixture of (E)-β-silylstyrene (40–95%) and α-silylstyrene (60–65%) in excellent yields under neat conditions (0.1 mol% Pt catalyst) at 80 °C (equation 37)[105]. The complexes with highly lipophilic cations such as $(n\text{-Bu}_4N)_2PtCl_6$, $[(n\text{-C}_8H_{17})_4N]_2PtCl_6$ and $[(n\text{-C}_{18}H_{37})_4N]_2PtCl_6$ are found to be the most active catalysts among the hexachloroplatinates examined. A phase transfer agent, $n\text{-Bu}_4NHSO_4$, and a crown ether, 18-crown-6, are found to activate K_2PtCl_6 by solubilizing $PtCl_6^{2-}$ ion[105]. The results clearly show that the counter cation plays an important role in this catalytic process.

$$PhC\equiv CH + HSiX_3 \xrightarrow[\text{neat, 80 °C}]{[R_4E]_2^{2+}[PtCl_6]^{2-}} \underset{\text{(major)}}{\overset{Ph\quad H}{\underset{H\quad SiX_3}{>=<}}} + \underset{\text{(minor)}}{\overset{Ph\quad H}{\underset{X_3Si\quad H}{>=<}}} \quad (37)$$

E = N, P, As

A quaternary phosphonium hexachloroplatinate bound to the Merrifield resin is also an effective hydrosilylation catalyst which can be used repeatedly without appreciable loss of activity.[105]

Highly dispersed metallic platinum systems prepared by treating mesitylene-solvated Pt atoms with solid supports, carbon, graphite and γ-Al_2O_3 provide highly active catalysts (Pt/hydrosilane = 2.3×10^{-4}–5.8×10^{-5} g-atom mol^{-1}) for the hydrosilylation of 1-hexyne and 2-hexyne at 25–70 °C under neat conditions[35]. Although the regioselectivity achieved by these catalysts is similar to that by H_2PtCl_6, the specific catalytic activity is superior to that of the homogeneous system. The observed high specific activity can be attributed to the high dispersion and small particle size of platinum metal in these systems, i.e. *ca* 20 Å for Pt/Al_2O_3, *ca* 15 Å for Pt/graphite and too small to detect for Pt/C: Pt/C is the most active among the three systems. The catalysts are claimed to be recyclable for several times without loss of activity[35].

2. By rhodium catalysts

The hydrosilylation of 1-alkynes catalyzed by rhodium complexes proceeds predominantly in an *anti*-fashion, giving thermodynamically unfavorable (Z)-alkenylsilanes as the major product (up to 99%)[3,106–108]. For example, the hydrosilylation of 1-hexyne with HSiEt$_3$ catalyzed by RhCl(PPh$_3$), Rh$_4$(CO)$_{12}$, Rh$_2$Co$_2$(CO)$_{12}$ and RhCo$_3$(CO)$_{12}$ in toluene gives (Z)-1-silylhexene **85** as the major product (79–98%), (E)-1-silyl-1-hexene **86** (1–10%) and 2-silyl-1-hexene **87** (1–14%), in which the product ratio depends on the reaction conditions (*vide infra*) (equation 38) (see also Section III.B).

$$n\text{-Bu}-\!\!\equiv\!\!- + \text{HSiEt}_3 \xrightarrow[\text{toluene}]{[\text{Rh}]}$$

(85) (79–98%)

(86) (1–10%)

(87) (1–14%)

[Rh] = RhCl(PPh$_3$)$_3$, Rh$_4$(CO)$_{12}$, Rh$_2$Co$_2$(CO)$_{12}$, RhCo$_3$(CO)$_{12}$

(38)

Other neutral rhodium complex catalysts such as [Rh(COD)Cl]$_2$, HRh(CO)(PPh$_3$)$_3$ and HRh(PPh$_3$)$_4$ also give (Z)-1-silyl-1-hexene **85** as the predominant product when the reactions are run in benzene[109]. However, it has recently been shown that the Z- or E-stereoselectivity of the Rh-catalyzed reactions can be effectively controlled by a proper combination of solvent and ligand[109,110]. For example, the reaction of 1-hexyne with HSiEt$_3$ catalyzed by [Rh(COD)Cl]$_2$ in DMF gives (Z)-1-silyl-1-hexene **85** with 97% selectivity, whereas the same reaction catalyzed by [Rh(COD)Cl]$_2$/2PPh$_3$ in MeCN affords (E)-1-silyl-1-hexene **86** with 97% selectivity (equation 39)[109]. Similar results are obtained for RhCl(PPh$_3$)$_3$-catalyzed reactions. Solvents such as benzene, acetone, THF, CH$_2$Cl$_2$, EtOH, DMF and NEt$_3$ favor the formation of the (Z)-product, while MeCN and n-PrCN promote the (E)-product formation. Triphenylphosphine is necessary to generate the selective catalyst species in MeCN. Based on these results, a cationic rhodium(I) species, [Rh(COD)(PPh$_3$)$_2$]$^+$, is proposed to be the active species that promotes highly selective *cis*-addition of HSiEt$_3$ to a 1-alkyne[109]. In fact, a cationic phosphine-Rh(I) complex, [Rh(COD)$_2$]BF$_4$/2PPh$_3$ is proved to be an excellent catalyst for the formation of (E)-1-triethylsilyl-1-alkenes, regardless of the solvent employed (equation 39)[109].

Although Rh and Rh–Co carbonyl clusters such as Rh$_4$(CO)$_{12}$, Rh$_2$Co$_2$(CO)$_{12}$ and RhCo$_3$(CO)$_{12}$ are highly efficient catalysts promoting the *trans*-addition of hydrosilanes to 1-alkynes[106,107] (*vide supra*), dimeric phosphite–Rh(I) complexes, [RhCl{P(OAr)$_3$}$_2$]$_2$ (Ar = 2-MeC$_6$H$_4$, 2-t-BuC$_6$H$_4$, 2,6-Me$_2$C$_6$H$_3$), effect the *cis*-addition of HSiEt$_3$ to 1-hexyne, yielding (E)-product **86** exclusively in moderate to high yield[20]. The bulkiness of the phosphite ligand is crucial for highly selective *cis*-addition. Thus, the corresponding

P(OMe)$_3$ complex gives a 55 : 45 mixture of **86** and **85**[20].

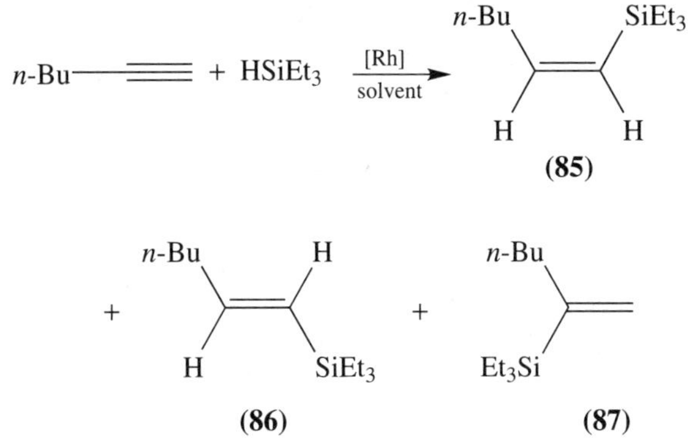

$$n\text{-Bu} \equiv + \text{HSiEt}_3 \xrightarrow[\text{solvent}]{[\text{Rh}]}$$

(85) (Z)-isomer with n-Bu and SiEt$_3$ on same side

+ (86) (E)-isomer + (87) 2-silyl-1-alkene

(39)

[Rh]	Solvent	85	86	87
[Rh(COD)Cl]$_2$	DMF	97%	1%	2%
[Rh(COD)Cl]$_2$	EtOH	94	4	2
[Rh(COD)Cl]$_2$	MeCN	36	33	31
[Rh(COD)Cl]$_2$/2PPh$_3$	MeCN	2	97	1
[Rh(COD)$_2$]BF$_4$/2PPh$_3$	MeCN	2	95	3
[Rh(COD)$_2$]BF$_4$/2PPh$_3$	EtOH	5	95	0
[Rh(COD)$_2$]BF$_4$/2PPh$_3$	Acetone	1	99	0

A dinuclear Rh(II) complex, Rh$_2$(pfb)$_4$ (pfb = perfluorobutanoate), effectively catalyzes the hydrosilylation of 1-alkynes to give (Z)-1-silyl-1-alkenes as the major product when a trialkylsilane is added to the mixture of a 1-alkyne and the catalyst in CH$_2$Cl$_2$[111]. However, the reverse addition, i.e. addition of the 1-alkyne to the mixture of the hydrosilane and the catalyst, affords an allylic silane in place of the silylalkenes in high yield (*vide infra*)[111].

A rhodium(III) complex of Troger's base (TB), TB·2RhCl$_3$ (**88**), prepared by treating RhCl$_3$ with TB is an active catalyst, promoting the hydrosilylation of 1-alkynes, giving a mixture of (Z)-1-silyl-1-alkene (29–95%), (E)-1-silylalkene (5–95%), and 2-silyl-1-alkene (5–20%)[112]. The regio- and stereoselectivity as well as total yield depend on the nature of 1-alkyne and hydrosilane used. The best result is obtained in the reaction of phenylacetylene with HSiCl$_3$ or HSiMe$_2$Ph, giving (Z)-β-silylstyrene (equation 40)[112]. The use of bulky hydrosilanes such as HSiMe$_2$Bu-*t* and HSi(Pr-*i*)$_3$ dramatically changes the selectivity, yielding a mixture of (E)-β-silylstyrene (80–95%) and α-silylstyrene (20–25%)[112].

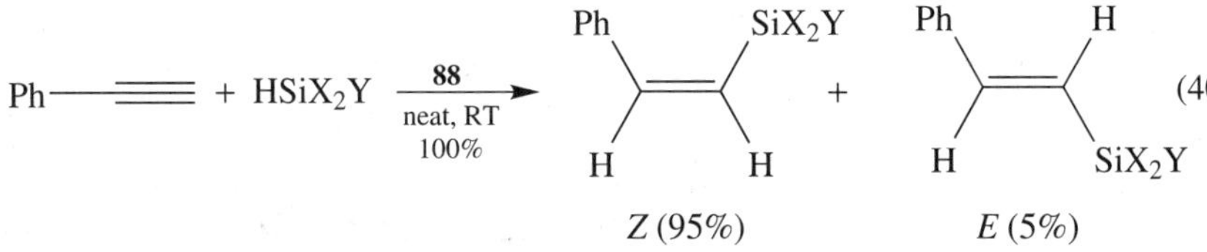

$$\text{Ph} \equiv + \text{HSiX}_2\text{Y} \xrightarrow[\substack{\text{neat, RT}\\100\%}]{88} \underset{Z\,(95\%)}{\text{Ph}\diagup\hspace{-2pt}=\hspace{-2pt}\diagdown\text{SiX}_2\text{Y}} + \underset{E\,(5\%)}{\text{Ph}\diagup\hspace{-2pt}=\hspace{-2pt}\diagdown\text{H}}$$

(40)

(88)

(a) X = Y = Cl
(b) X = Me, Y = Ph

Although Wilkinson's catalyst, RhCl(PPh$_3$)$_3$, usually promotes the *trans*-addition of hydrosilanes to 1-alkynes (*vide supra*), the reactions of 1-alkynes bearing bulky substituents such as *tert*-butyl and cyclohexyl give predominantly (*E*)-1-silyl-1-alkenes via *cis*-addition together with 1-alkynylsilanes via dehydrogenative silylation (*vide infra*)[113]. The use of H$_3$SiPh in the reaction of phenylacetylene also affords (*E*)-PhCH=CHSiH$_2$Ph exclusively[17].

Rhodium complexes with 1,3-bis(di-*tert*-butylphosphino)methane (dtbpm), [(dtbpm)RhCl]$_2$/PPh$_3$ (**89**), (dtbpm)RhSi(OEt)$_3$(PMe$_3$) (**90**) and (dtbpm)RhMe(PMe$_3$) (**91**) are found to be effective catalysts for the hydrosilylation of an internal alkyne, 2-butyne, with HSi(OEt)$_3$ at ambient temperature without solvent to yield (*E*)-2-triethoxysilyl-2-butene with complete stereoselectivity in quantitative yield using a proper concentration of the catalysts, i.e. $\geqslant$0.05 mol% for **89**, >0.4 mol% for **90** and **91**[114]. When the reaction is carried out at lower catalyst concentrations, i.e. 0.1 mol% for **90** or **91**, (Z)-product is formed via *trans*-addition in 7–13% yield.

3. By iridium, ruthenium, osmium and other catalysts

Iridium complexes with a unique oxygen donor ligand, C[P(O)Ph$_2$]$_3$$^-$(triso), such as IrH$_2$(triso)(SiMePh$_2$)$_2$, Ir(triso)(COE)$_2$ (COE = cyclooctene) and Ir(triso)(C$_2$H$_4$)$_2$ are excellent catalysts for the hydrosilylation of 1-alkynes (equation 41)[115,116]. These triso-Ir complex catalysts are extremely chemo-, regio- and stereoselective in that only 1-alkenes, except for *t*-BuC≡CH and HC≡CCO$_2$Me, are hydrosilylated via *trans*-addition, giving (Z)-1-triethylsilyl-1-alkenes with extremely high stereoselectivity (Z/E $\leqslant$ 190/1). Internal alkynes, e.g. 3-hexyne, diphenylacetylene and dimethyl acetylenedicarboxylate, are totally unreactive and the reaction of methyl propynoate only gives trimethyl benzene-1,3,5-tricarboxylate via trimerization in trace amount[115].

Ph—≡≡ + HSiEt$_3$ $\xrightarrow[\text{CH}_2\text{Cl}_2,\ 25\ °C]{\text{[Ir-triso]}}$ [Ph/H C=C SiEt$_3$/H] + [Ph/H C=C H/SiEt$_3$] (41)

(128 : 1)

[Ir-triso] = IrH$_2$(triso)(SiMePh$_2$)$_2$, Ir(triso)(COE)$_2$, Ir(triso)C$_2$H$_4$)$_2$

Other iridium complexes, such as IrH$_2$(SiEt$_3$)(COD)(AsPh$_3$) (COD = 1, 5-cyclooctadiene)[117], [Ir(OMe)(COD)]$_2$/2PPh$_3$ (or AsPh$_3$)[117], [IrH(H$_2$O)(bq)(PPh$_3$)$_2$]SbF$_6$ (bq = 7, 8-benzoquinolinato)[113], [Ir(COD)(η^2-(*i*-Pr)$_2$PCH$_2$CH$_2$NMe$_2$)]BF$_4$, [Ir(TFB)(η^2-(*i*-Pr)$_2$PCH$_2$CH$_2$OMe)]BF$_4$ (TFB = tetrafluorobenzobarrelene)[118], Ir(C≡CPh)(CO)$_2$(PCy$_3$) (Cy = cyclohexyl), Ir(C≡CPh)(TFB)(PCy$_3$)[119], are also found to be active catalysts for the hydrosilylation of phenylacetylene and 1-hexyne, giving a mixture of (Z)- and (*E*)-RCH=CHSiX$_3$, R(X$_3$Si)C=CH$_2$, RC≡CSiX$_3$, and RCH=CH$_2$. The products ratio is highly dependent on the reaction conditions. The reactions catalyzed by [IrH(H$_2$O)(bq)(PPh$_3$)$_2$]SbF$_6$ give (Z)-RCH=CHSiX$_3$ with excellent selectivity comparable to that achieved by using the Ir(triso) complexes[113].

A ruthenium complex with triisopropylphosphine, HRuCl(CO)[P(Pr-i)$_3$]$_2$, is a highly efficient catalyst for the hydrosilylation of phenylacetylene with HSiEt$_3$ (1.0 equiv.) in ClCH$_2$CH$_2$Cl at 60 °C, yielding (Z)-PhCH=CHSiEt$_3$ via *trans*-addition with 100% selectivity in quantitative yield[120]. The corresponding osmium complex, HOsCl(CO)[P(Pr-i)$_3$]$_2$, is an active catalyst as well[121]. The Os-catalyzed reaction using one equivalent of HSiEt$_3$ to the alkene gives (Z)-PhCH=CHSiEt$_3$ with high selectivity ($\leqslant$98%), whereas the (E)-product is formed selectively ($\leqslant$99%) when excess HSiEt$_3$ is used[121]. Ruthenium complexes **92a**, **93a**, and **94a** as well as osmium complexes **92b**, **93b**, and **94b** are also found to be active catalysts for the hydrosilylation of phenylacetylene with HSiEt$_3$ under the same conditions[122]. The reactions catalyzed by the ruthenium complexes give (Z)-PhCH=CHSiEt$_3$ with 91–96% selectivity, while the osmium catalysts are less selective (62–70%)[122].

(92) (93) (94)

(a) M = Ru, (b) M = Os

Nickel(0) complexes with trialkylphosphine or trialkylphosphite ligands with lower linear alkyl groups (C$_1$–C$_4$) can catalyze the hydrosilylation of phenylacetylene with HSiPh$_3$ to afford (E)-PhCH=CHSiPh$_3$ in good yield[123]. The ligand/Ni ratio of 2 as well as the HSiPh$_3$/PhC$\equiv$CH ratio of 2 is essential to promote the hydrosilylation selectively, suppressing the trimerization of phenylacetylene[123].

An organoyttrium complex, Cp*$_2$YCH$_3$(THF) (Cp* = pentamethylcyclopentadienyl), is found to be an excellent catalyst for regioselective hydrosilylation of functionalized unsymmetrical internal alkynes with H$_3$SiPh. The reactions are carried out in cyclohexane at 40–90°C for 24–48 h to give single isomers when one of the alkyl substituents of the acetylene moiety is primary and the other secondary or tertiary[124]. No stereoisomers are formed, i.e. complete *cis*-addition. The regioselectivity drops naturally when both substituents of the acetylene moiety are primary. Nevertheless, good selectivity is observed for 2-alkynes such as 2-nonyne (2-Si/3-Si = 4.1–7.3) and 5-methyl-2-heptyne (2-Si/3-Si = 7.3). Thus, Cp*$_2$YCH$_3$(THF) or its precursor, Cp*$_2$YCH(TMS)$_2$, appears to be the most regioselective catalyst for the internal alkynes to date. Examples are shown in equation 42 and Table 4.

$$R\text{---}\equiv\text{---}R' + H_3SiPh \xrightarrow{Cp^*_2YCH(TMS)_2} \begin{array}{c} R \\ H \end{array}\!\!\!>\!\!=\!\!<\!\!\!\begin{array}{c} R' \\ SiH_2Ph \end{array} \quad (42)$$

(95) (96)

(a) R = cyclohexyl, R' = Me   74%
(b) R = *sec*-Bu, R' = *n*-pentyl   93%

TABLE 4. Selective hydrosilylation of alkynes with H_3SiPh, catalyzed by $Cp^*_2YCH_3(THF)$

Substrate	Product	Yield (%)
t-BuMe₂SiO–C(Me)–C≡C–C₁₀H₂₁-n	t-BuMe₂SiO–C(Me)=C(SiH₂Ph)(H)–C₁₀H₂₁-n (Me above)	89
iPr-C≡C-CH₂CH₂CH₂-OTHP	iPr-C(H)=C(SiH₂Ph)-CH₂CH₂CH₂-OTHP	89
iPr-C≡C-CH₂CH(Me)CH₂CH₂CH=CMe₂	iPr-C(H)=C(SiH₂Ph)-CH₂CH(Me)CH₂CH₂CH=CMe₂	80

4. By radical initiators

Although the hydrosilylation of alkynes under free-radical conditions using UV irradiation or organic peroxides was studied in 1940s and 1950s[4,125,126], the reactions promoted by transition metal catalysts took over the main stream of research and development afterwards[3,4]. As discussed in Section II. A.5, tris(trimethylsilyl)silane, $HSi(SiMe_3)_3$ (TTMSS), is found to undergo smooth addition to simple and functionalized alkenes in the presence of a radical initiator to give the corresponding hydrosilylation products in high yields[66–68] (vide supra). TTMSS also participates in hydrosilylation of alkynes in the presence of AIBN (80–90 °C) or BEt_3/O_2 (0–70 °C). The stereochemistry of the reaction follows usual radical addition to 1-alkynes, i.e. *trans*-addition, yielding the (Z)-product[3]. The Z/E selectivity is dependent on the reaction conditions as well as the steric and electronic nature of the substituents of alkyne. The reactions of 1-alkynes using BEt_3/O_2 as the initiator at lower temperatures (0–25 °C) favor the formation of the (Z)-product **97** ($Z/E = 95/5$–99/1; the Z/E ratio of the same reaction with AIBN at 80–90 °C ranges from 51/49 to 92/8) (equation 43). A mixture of Z- and E-products, **98** and **99**, are formed in the reactions of 2-substituted phenylacetylenes, PhC≡CX, with variable Z/E ratios (45/55–92/8) depending on the conditions and the substituent X (equation 44): The reactions using AIBN at 80–90 °C give either no or a very low selectivity. The reactions of t-BuC≡CH and PhC≡CCO₂Et yield the corresponding (E)-products exclusively, regardless of the reaction conditions.

$$R\text{—}\!\!\equiv\!\!\text{—}H + HSi(SiMe_3)_3 \xrightarrow[\text{toluene} \atop 25\,°C]{BEt_3,\,O_2} \underset{\underset{Z/E = 95/5\text{–}99/1}{(\mathbf{97})}}{\overset{RH}{\underset{HSi(SiMe_3)_3}{\diagup\!\!=\!\!\diagdown}}} \quad (43)$$

$R = n\text{-Bu},\, n\text{-}C_6H_{13},\, Ph,\, CO_2Et$

$$\text{Ph}\equiv\!\!-\!\!\text{X} + \text{HSi(SiMe}_3)_3 \xrightarrow[\text{toluene}]{\text{BEt}_3,\,\text{O}_2}$$

H   X
  \\=/
Ph   Si(SiMe$_3$)$_3$
    (98)

+

Ph   X
  \\=/
H   Si(SiMe$_3$)$_3$
    (99)

(44)

X = CHO, **98/99** = 87/13 (20 °C);
X = CN, **98/99** = 92/8 (0 °C);
X = n-C$_5$H$_{11}$, **98/99** = 67/33 (40 °C)

B. Mechanism of the Stereoselective Hydrosilylation of 1-Alkynes

It has been shown that the stereochemistry of the hydrosilylation of 1-alkynes giving 1-silyl-1-alkenes depends on the catalysts or promoters used. For example, the reactions under radical conditions give the *cis*-product predominantly via *trans*-addition[3,66], while the platinum-catalyzed reactions afford the *trans*-product via exclusive *cis*-addition[3]. In the reactions catalyzed by rhodium complexes, thermodynamically unfavorable *cis*-1-silyl-1-alkenes are formed via apparent *trans*-addition as the major or almost exclusive product. Since the *trans*-addition of HSiEt$_3$ to 1-alkynes catalyzed by RhCl(PPh$_3$)$_3$ was first reported in 1974[108], there have been controversy and dispute on the mechanism of this mysterious *trans*-addition that is very rare in transition-metal-catalyzed addition reactions to alkynes[3]. Recently, iridium[115,116] and ruthenium[120] complexes were also found to give the *cis*-product with extremely high selectivity (*vide supra*).

The mystery of the unusual *trans*-addition has been resolved by essentially the same two mechanisms proposed by Ojima and Crabtree and their coworkers for the rhodium- and iridium-catalyzed reactions, respectively[106,115]. The integrated mechanism is shown in Scheme 13. The most significant feature of the Ojima–Crabtree mechanism is the exclusive insertion of 1-alkyne into the Si–metal bond, i.e. not to the H–metal bond, forming (Z)-silylethenyl-[M] intermediate **101**. Because of severe steric repulsion between the two bulky groups, i.e. R$_3$Si and [M], in **101**, it isomerizes to thermodynamically more favorable (E)-silylethenyl-[M] intermediate **103** via either zwitterionic carbene species **102a** (proposed for the Rh-catalyzed process)[106] or metallacyclopropene species **102b** (proposed for the Ir-catalyzed process)[113,115,116]. Coordination of a hydrosilane molecule to **103** promotes the reductive elimination of (Z)-product **105**, regenerating **100**. Since the reductive elimination is the rate-determining step, (Z)-product **105** is formed as the kinetic product. The isomerization of **101** (intramolecular process) competes with the reductive elimination of **104** (intermolecular bimolecular process). Thus, the concentration of reactants exerts a marked influence on the Z/E selectivity, i.e. low concentrations favor the (Z)-product formation[106]. Also, highly reactive hydrosilanes bearing electron-withdrawing groups, e.g. HSi(OMe)$_3$, HSiMe$_2$Cl and HSiMeCl$_2$, predominantly (65–98%) give (E)-product **104**, whereas HSiEt$_3$, HSiEt$_2$Me and HSiMe$_2$Ph afford (Z)-product **105** as the major product (57–99%)[106]. The bulkiness of the R group has a strong influence on the Z/E selectivity, i.e. when 1-alkynes bearing bulky R groups are used, e.g. c-C$_6$H$_{11}$C≡CH and t-BuC≡CH, the reaction yields (E)-product selectively (89–100%)[113]. It has not been elucidated which of the two species **102a** and **102b** is the intermediate of the unique isomerization of **101** to **103**. A zwitterionic carbene intermediate similar to **102a** was proposed by Brady and Nile[127,106], and the rearrangement of η^1-ethenyl-[M] to η^2-[M] species was shown by Green and coworkers[128,115].

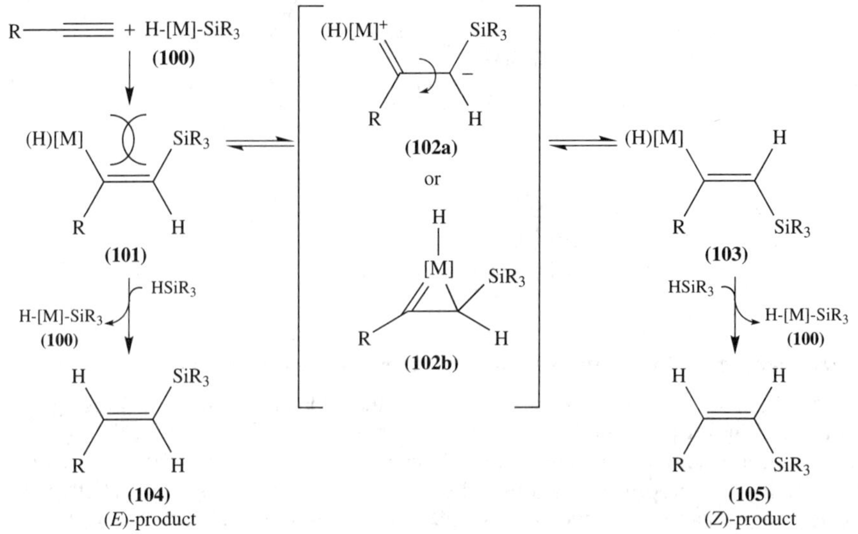

SCHEME 13. The Ojima–Crabtree mechanism for stereoselective hydrosilylation of 1-alkynes

The Ojima–Crabtree mechanism is applicable to the selective formation of (Z)-PhCH=CHSiEt$_3$ (exclusive for Ru; 60–98% for Os) in the reaction of PhC≡CH with HSiEt$_3$ catalyzed by HRuCl(CO)(P(Pr^i$_3$)$_2$[120,122], HOsCl(CO)(P(Pr^i$_3$)$_2$[121], IrH$_2$(SiEt$_3$)(TFB)(PR$_3$) (R = Ph, Cy, i-Pr)[129], [Ir(COD)(η^2–^{i}Pr$_2$PCH$_2$CH$_2$OMe)]BF$_4$[118] and Ir(C≡CPh)L$_2$PCy$_3$ (L = 2 CO or TFB)[119]. (Z)- and (E)-β-ethenyl–Rh complexes **107** have indeed been isolated from the reaction of (dtbpm)Rh[Si(OEt)$_3$](PMe$_3$) **(106)** with dimethyl acetylenedicarboxylate, and their structures identified by ^{31}P NMR analysis (equation 45)[114]. The reaction of 2-butyne with HSi(OEt)$_3$ catalyzed by (dtbpm)Rh[Si(OEt)$_3$](PPMe$_3$) gives (E)-MeCH=C(Me)Si(OEt)$_3$ as the predominant product[114], which is consistent with the previously reported (E)-product formation with reactive HSi(OEt)$_3$ through the path **101** → **104** in Scheme 13.

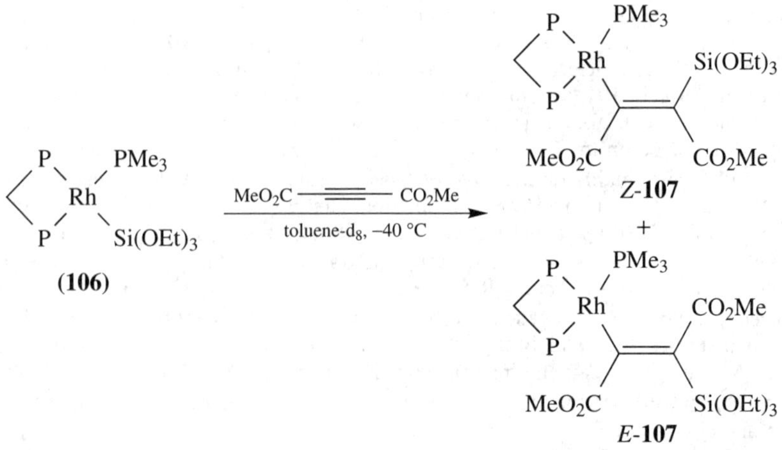

C. Intramolecular Hydrosilylation of Alkynes

The hydrosilylation of disilylalkyne $HMe_2Si[CH_2]_n SiMe_2C{\equiv}CR$ (**108b–e**: R = H or Ph, $n = 2$ or 3) catalyzed by H_2PtCl_6 proceeds via 5-*exo*-dig or 6-*exo*-dig cyclization to give *exo*-cyclic disilylalkene **109** exclusively except for the case of **108c** ($n = 3$, R = H) in which a minor (21%) amount of 6-*endo*-dig cyclization product **110c** is formed (equation 46)[130]. However, no reaction takes place for **108a** ($n = 1$, R = H)[130].

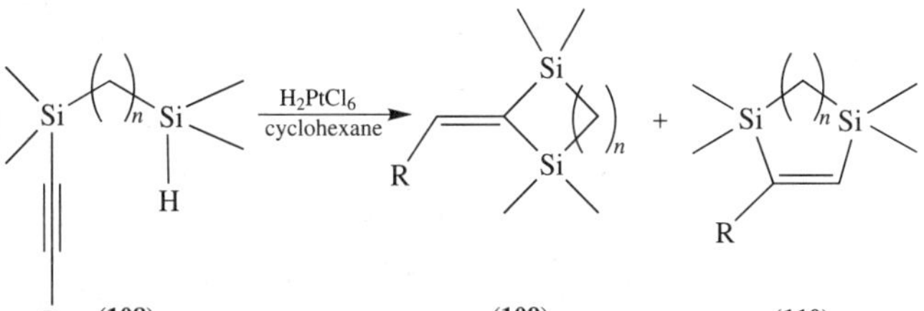

(46)

	Yield (%)	**109**	**110**
(a) $n = 1$, R = H	0	0	0
(b) $n = 2$, R = H	58	100	0
(c) $n = 3$, R = H	71	79	21
(d) $n = 2$, R = Ph	42	100	0
(e) $n = 3$, R = Ph	60	100	0

In a similar manner, the reaction of 5-dimethylsilyl-1-hexyne (**111**) catalyzed by H_2PtCl_6 affords 5-*exo*-dig cyclization product **112** exclusively (equation 47)[130].

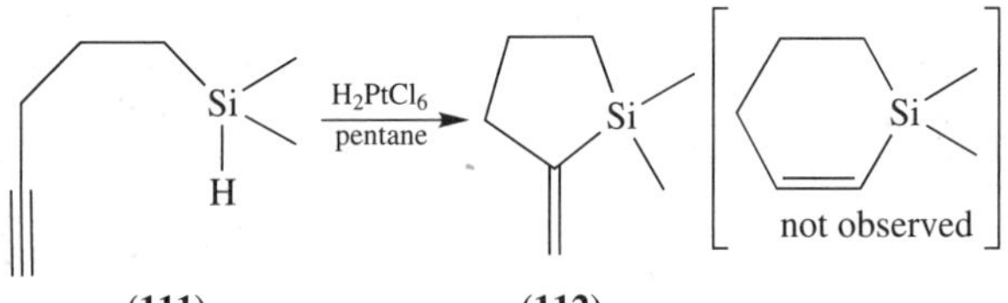

(47)

Intramolecular hydrosilylation of ω-dimethylsiloxyalkynes such as **114**, readily derived from alkynol **113**, provides a convenient method for the regioselective functionalization of internal alkynes in combination with the Tamao oxidation and other transformations. An example of such a process giving **117** via **115** and **116** is illustrated in Scheme 14[87].

D. Bis-TMS-ethyne and Bis-TMS-butadiyne

The reaction of bis(trimethylsilyl)ethyne (**118**) with a high boiling hydrosilane, $HSiMe_2C_6H_{13}$-n, promoted by Karstedt's catalyst affords normal *cis*-addition product **119** in good yield under nitrogen at 90 °C. However, significant amounts of disilylethynes **120** and **121** are unexpectedly formed when the reaction is carried out under dry air and at 130–150 °C (Scheme 15)[131]. The observed scrambling of silyl groups suggests that the hydrosilylation trisilylethene product[119] is not stable under those conditions and undergoes *trans*-elimination of a $HSiMe_3$ forming **120**, followed by the addition of $HSiMe_2C_6H_{13}$-n and elimination of $HSiMe_3$ to yield **121**[132].

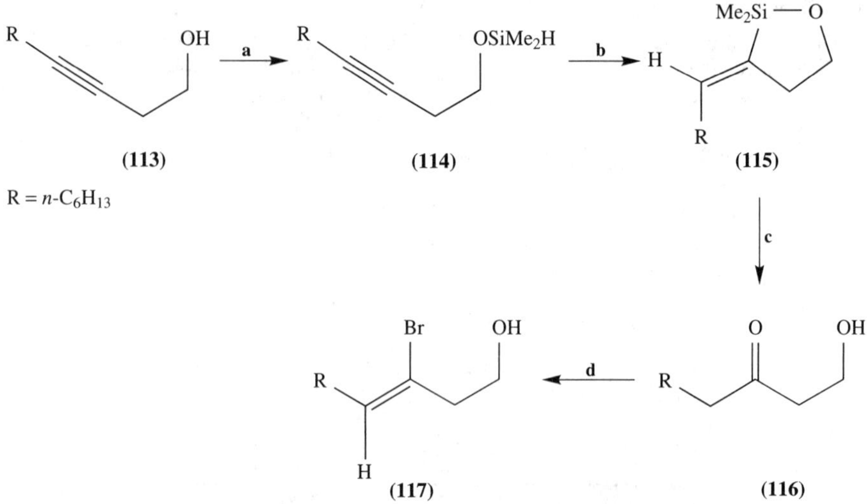

SCHEME 14

a. (HMe$_2$Si)$_2$NH;
b. H$_2$PtCl$_6$·H$_2$O (0.1 mol%) / CH$_2$Cl$_2$ / room temp. / 0.5 h;
c. 30%H$_2$O$_2$ / KF / KHCO$_3$ / MeOH / THF/ 30 °C / overnight
d. (1) Br$_2$ / 0 °C; (2) KHF$_2$ / MeOH / room temp. / overnight.

SCHEME 15

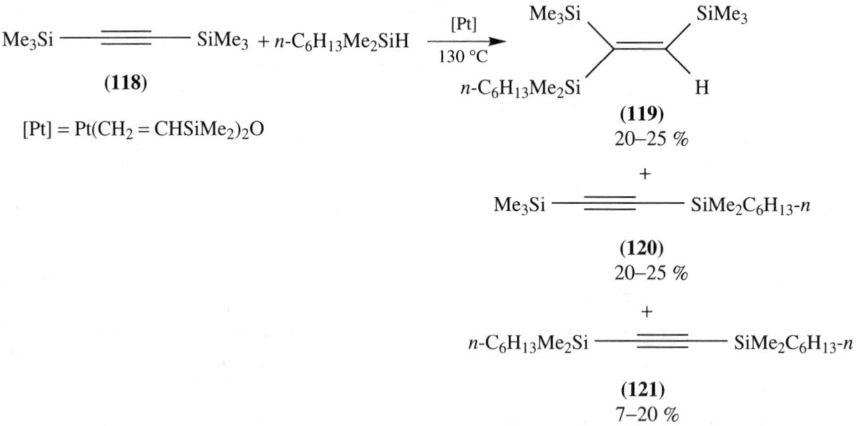

[Pt] = Pt(CH$_2$=CHSiMe$_2$)$_2$O

The hydrosilylation of 1,4-bis(trimethylsilyl)butadiyne (**122**) with HSiMeCl$_2$ in the presence of Speier's catalyst at 80 °C followed by methylation gives 1,1,3,4-tetrakis(trimethylsilyl)-1,3-butadiene (**123**) as the sole product in 54% yield (equation 48)[133]. However, the reactions using HSiMe$_2$Ph at 50–60 °C afford trisilylbutenyne **124a** (70%) and allene **125a** (10%) (equation 49)[133]. The allene **125b** can be obtained exclusively when the reaction is carried out with HSiMe$_3$ using Pt(PPh$_3$)$_4$ as the catalyst at 90 °C for 12 h (**124b**, 1%; **125b**, 94%). When the reaction is stopped at 1 h, butenyne **124b** is formed as the major product (**124b**, 69%; **125b**, 2%). Thus, this is clearly a stepwise reaction process. The RhCl(PPh$_3$)$_3$-catalyzed reactions with different hydrosilanes give **124** and/or **125** with varying ratios depending on the hydrosilane used, e.g. the reaction with HSiMe$_3$ gives **125b** exclusively in 90% yield, while the reaction with H$_2$SiPh$_2$ affords **124c** as the sole product in 49% yield[133]. The reactions with HSiEt$_3$ catalyzed by palladium complexes, PdCl$_2$(PPh$_3$)$_2$ and Pd(PPh$_3$)$_4$, give **124d** selectively in low yields (8–18%)[133].

Me$_3$Si─≡─≡─SiMe$_3$
(**122**)

$\xrightarrow{\begin{array}{c}\text{1. HSiMeCl}_2,\\ \text{H}_2\text{PtCl}_6, i\text{-PrOH}\\ \text{2. MeMgBr}\end{array}}$

H   SiMe$_3$
 \\   /
  C=C
 /   \\
Me$_3$Si   \\
          SiMe$_3$
         /
        C=C
       /   \\
      H   SiMe$_3$
(**123**)   (48)

Me$_3$Si─≡─≡─SiMe$_3$
(**122**)

$\xrightarrow{\text{HSiR}_2\text{R}'\ [M]}$

H   SiR$_2$R'
 \\   /
  C=C
 /   \\
Me$_3$Si   \|\|\|
          SiMe$_3$
(**124**)   (49)

+

Me$_3$Si   SiMe$_3$
      \\   /
       C=C=C
      /   \\
R'R$_2$Si   SiR$_2$R'
(**125**)

[M] = H$_2$PtCl$_6$, Pt(PPh$_3$)$_4$, RhCl(PPh$_3$)$_3$, PdCl$_2$(PPh$_3$)$_2$ or Pd(PPh$_3$)$_4$
(**a**) R = Me, R' = Ph; (**b**) R = R' = Me
(**c**) R = Ph, R' = H; (**d**) R = R' = Et

E. Functionalized Alkynes

The hydrosilylation of propargylic alcohols should provide the most straightforward route to (*E*)-γ-silylallylic alcohols, which are useful building blocks in organic syntheses. However, the reactions promoted by commonly used Speier's catalyst were found to be nonselective, giving an almost 1 : 1 mixture of γ- and β-silylallylic alcohols[134]. Thus, catalyst systems that can achieve high regio- and stereoselectivity are necessary for this reaction to be synthetically useful.

The reaction of propargyl alcohol with HSiMe$_2$Ph or HSiPh$_3$ catalyzed by a phosphine–Pt(0) complex, (*t*-Bu$_3$P)Pt(NBE)$_2$ (NBE = norbornene) or (*t*-Bu$_3$P)Pt(CH$_2$=CH)$_2$Si(OMe)$_2$, gives the desired (*E*)-γ-silylallyl alcohols **126** with excellent regioselectivity (γ/β >19/1) in high yields (equation 50)[135]. No dehydrogenative silylation takes place with the hydroxyl group under the reaction conditions so that no protection is necessary. When a similar Pt(0) catalyst, (Cy$_3$P)Pt(CH$_2$=CH$_2$)$_2$, is employed, the γ/β-regioselectivity drops to 5.3 with HSiMe$_2$Ph, 3.1 with HSiBu$_3$ and 1.2 with HSiPh$_3$[135].

A cationic rhodium complex, [Rh(COD)$_2$]BF$_4$·2PPh$_3$, is a highly efficient catalyst for this reaction, which gives (*E*)-γ-triethylsilylallyl alcohol (**127**) with complete γ-selectivity and >99% *E*-selectivity (equation 51)[136,137].

$$(50)$$

[Pt] = (*t*-Bu$_3$P)Pt(NBE)$_2$ or (*t*-Bu$_3$P)Pt(CH$_2$=CH)$_2$Si(OMe)$_2$
R = Me, Ph

$$(51)$$

α-Substituted γ-silylallylic alcohols can be converted to functionalized chiral allylsilanes through Claisen rearrangement, which are useful intermediates in stereoselective organic syntheses. Accordingly, the highly regio- and stereoselective hydrosilylation of *sec*-propargylic alcohols has a particular synthetic interest for this purpose.

Among the phosphine–Pt(0) complex catalysts mentioned above, (Cy$_3$P)Pt(CH$_2$=CH$_2$)$_2$ exhibits the best results for the reaction of but-1-yn-3-ol with HSiPh$_3$, giving (*E*)-Ph$_3$SiCH=CH-CH(Me)OH with >95% γ-selectivity in 94% yield[135]. In a similar manner, the reactions of other *sec*-propargyl alcohols **128** with HSiMe$_2$Ph catalyzed by (Cy$_3$P)Pt(CH$_2$=CH$_2$)$_2$ give the corresponding (*E*)-γ-triethylsilylallyllic alcohols **129** with >95% γ-selectivity in high yields (equation 52)[135]. The reaction of a *tert*-propargyl alcohol, 3-methylbut-1-yn-3-ol, with HSiMe$_2$Ph catalyzed by (Cy$_3$P)Pt(CH$_2$=CH$_2$)$_2$ afforded (*E*)-PhMe$_2$SiCH=CH-CMe$_2$OH with >95% γ-selectivity in 98% yield[135]. However, the reaction of γ-alkylpropargyl alcohols **131** under the same conditions proceeds with almost no or very low regioselectivity giving a mixture of **132** and **133** (equation 53)[135].

R = *n*-C$_5$H$_{11}$, PhCH$_2$OCH$_2$

(a) R = Me   91%   1.1 : 1
(b) R = *n*-C$_5$H$_{11}$ 99%   2.1 : 1

$$(53)$$

29. Recent advances in the hydrosilylation and related reactions

The use of [Rh(COD)$_2$]BF$_4$·2PPh$_3$ as the catalyst appears to be a better choice for this type of reaction. Thus, the reactions of *sec-* and *tert-*propargylic alcohols **134** with HSiEt$_3$ promoted by this rhodium catalyst system provide the desired (*E*)-γ-silylallyllic alcohols **135** with 100% γ-selectivity and virtually complete stereoselectivity (*E/Z* => 99/1) in excellent yields (equation 54)[136,137].

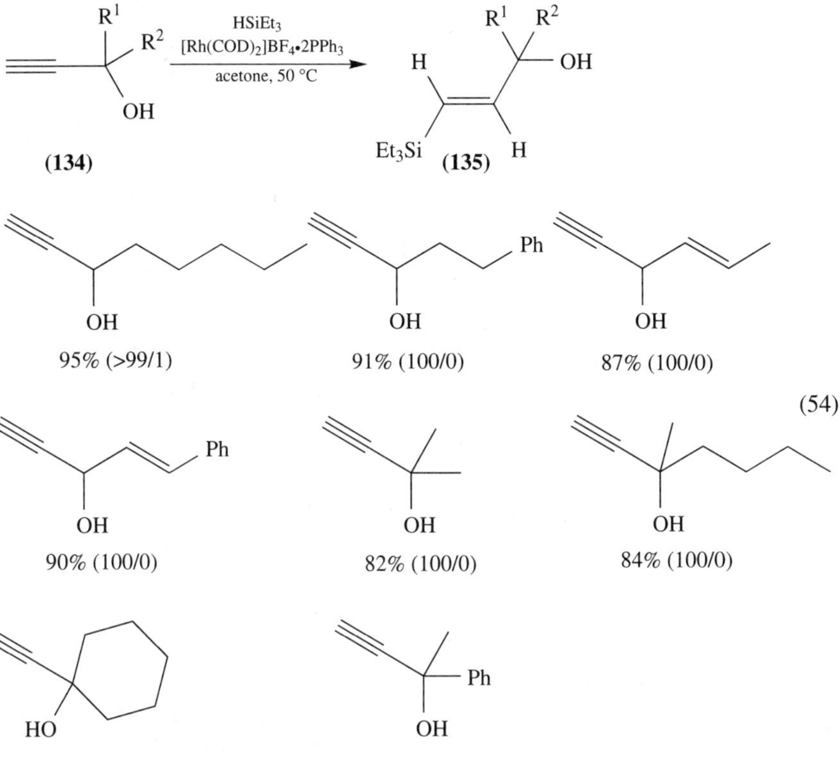

$$\text{(134)} \quad \xrightarrow[\text{acetone, 50 °C}]{\text{HSiEt}_3, \ [\text{Rh(COD)}_2]\text{BF}_4\cdot 2\text{PPh}_3} \quad \text{(135)}$$

95% (>99/1)   91% (100/0)   87% (100/0)

90% (100/0)   82% (100/0)   84% (100/0)

(54)

92% (100/0)   94% (100/0)

The hydrosilylation of 3-butyn-1-ol, i.e. homopropargyl alcohol, with HSiMe$_2$Ph catalyzed by (Cy$_3$P)Pt(CH$_2$=CH$_2$)$_2$ gives a 4.2 : 1 mixture of (*E*)-4-silyl- and (*E*)-3-silyl-3-buten-1-ols[135], whereas the cationic rhodium complex-catalyzed reaction with HSiEt$_3$ affords (*E*)-4-triethylsilyl-3-buten-1-ol (**136**) exclusively in 97% yield (equation 55)[136].

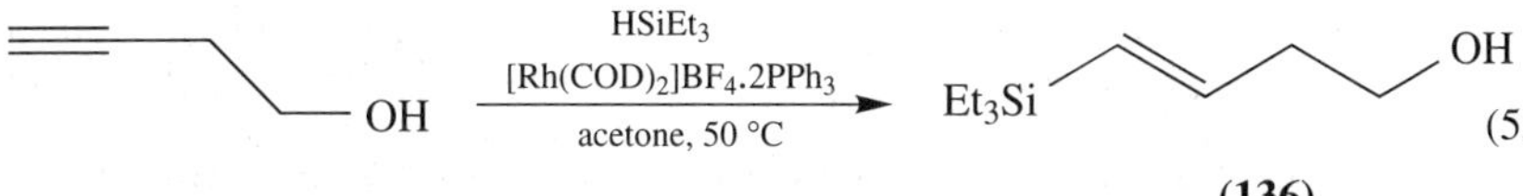

$$\xrightarrow[\text{acetone, 50 °C}]{\text{HSiEt}_3, \ [\text{Rh(COD)}_2]\text{BF}_4\cdot 2\text{PPh}_3} \quad \text{(136)} \quad 97\%$$

(55)

In a similar manner, the cationic Rh-catalyzed reaction of 4-pentynol (**137**) with HSiEt$_3$ gives (*E*)-5-triethylsilyl-4-penten-1-ol (**138**) with excellent selectivity (equation 56)[136].

The tandem hydrosilylation–isomerization process of *sec-*propargyl alcohols **140** provides an easy access to β-silylketones **142** via silylallylic alcohols **141** (equation 57)[136].

The reaction of methyl 3-butynoate (**143**) with HSiMe$_2$Ph catalyzed by (Cy$_3$P)Pt(CH$_2$=CH$_2$)$_2$ gives a mixture of 4-silyl- and 3-silyl-3-butenoates, **144** and **145**, respectively in 71% yield (equation 58)[135].

The reactions of phenylacetylene, 2-(ethynyl)thiophene, 2-(propargyl)thiophene, methyl propynoate, propargyl alcohol, propargyl methyl ether, HC≡C-CH$_2$NR$_2$ [NR$_2$ = NEt$_2$, N(CH$_2$)$_4$, N(CH$_2$)$_5$, N(CH$_2$)$_6$] with HSiMe$_2$(4-XC$_6$H$_4$) (X = H, Me, F, Cl, Br and NMe$_2$) and HSiMe$_2$(2-thienyl) catalyzed by H$_2$PtCl$_6$·6H$_2$O, RhCl(PPh$_3$)$_3$, H$_2$OsCl$_6$·6H$_2$O and Co$_2$(CO)$_8$ give a mixture of *trans*-1-silyl-1-alkene and 2-silyl-1-alkene (50/50–94/6 ratio)[134]. The relative activity of the catalysts in the reaction of *N*-propargylpiperidine with HSiMe$_2$(2-thienyl) decreases in the order H$_2$PtCl$_6$·6H$_2$O > RhCl(PPh$_3$)$_3$ ≫ H$_2$OsCl$_6$·6H$_2$O ≫ Co$_2$(CO)$_8$. The total yield of the hydrosilylation products in the reactions of HC≡C−Y with HSiMe$_2$(2-thienyl) catalyzed by H$_2$PtCl$_6$·6H$_2$O at 50 °C decreases in the order CO$_2$Me ∼ 2-thienylmethyl ≥ HOCH$_2$ > Ph > MeOCH$_2$ >

2-thienyl > R_2NCH_2. The amount of the *trans*-1-silyl-1-alkene decreases in the order R_2NCH_2 > 2-thienylmethyl ⩾ Ph > $MeOCH_2$ > $HOCH_2$ > CO_2Me > 2-thienyl. The effect of the *para*-substituent X of $HSiMe_2(4-XC_6H_4)$ on regioselectivity (1-silyl/2-silyl) is rather minor, i.e. 94/6 (Me_2N)–88/12 (Me, H).

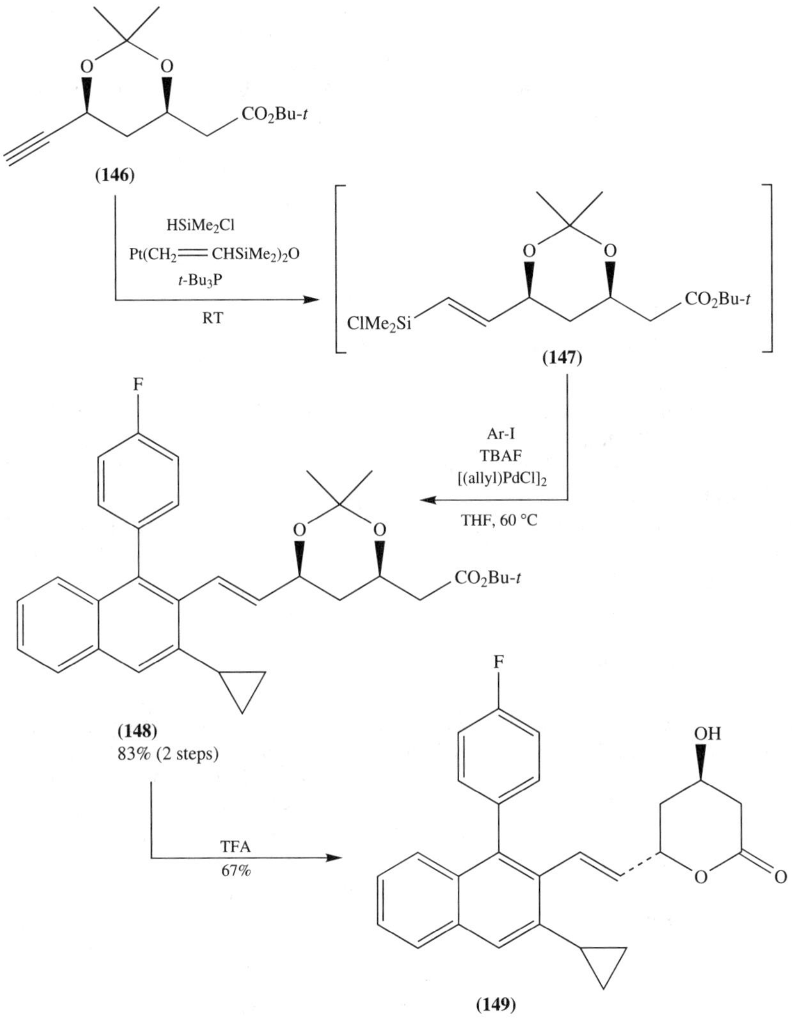

SCHEME 16

The hydrosilylation–cross-coupling process has been applied to the synthesis of the key intermediate to an HMG-CoA reductase inhibitor (Scheme 16)[138,139]. This one-pot process consists of the Pt-catalyzed hydrosilylation of **146**, giving (*E*)-product **147** and its regioisomer, followed by the Pd-catalyzed cross-coupling with 2-cyclopropyl-4-(4-fluorophenyl)quinolin-3-yl iodide (ArI), affording **148** in 83% isolated yield (2 steps). The key intermediate **148** thus obtained is treated with CF_3CO_2H to give the target inhibitor molecule **149**.

F. Other Reactions Associated with Alkyne Hydrosilylation

As described above, $Rh_2(pfb)_4$ and H_2PtCl_6 effectively catalyze the hydrosilylation of 1-alkynes to give (Z)-1-silyl-1-alkenes and (E)-1-silyl-1-alkenes, respectively, as the major products. However, the exclusive formation of allylic silane **150** (68–73% yield; as a mixture of E and Z isomers) is observed when the order of the addition of reactants is reversed, i.e. slowly adding 1-alkyne to the solution of excess hydrosilane (2 equiv.) in CH_2Cl_2 at 25 °C over a period of 1 h (equation 59)[111]. The results imply that two different catalyst species, one for hydrosilylation and the other for the formation of allylic silane, are generated just by changing the order of the addition[111].

$$R^1-C{\equiv}CH + HSiR_3 \xrightarrow[CH_2Cl_2,\ 25\ °C]{Rh_2(pfb)_4\ or\ H_2PtCl_6} R^1\text{-CH=CH-CH}_2\text{-SiR}_3 \quad (59)$$

(slow addn.) (2 equiv.) (**150**)

$HSiR_3$ = $HSiEt_3$, $HSiMe_2Bu$-t

R^1 = n-C_5H_{11}, n-C_3H_7, MeO, $PhCH_2O$

Although iridium complexes such as $IrH_2(SiEt_3)(COD)(AsPh_3)$ (COD = 1,5-cyclo-octadiene)[117], $[Ir(OMe)(COD)]_2/2PPh_3$ (or $AsPh_3$)[117], $[IrH(H_2O)(bq)(PPh_3)_2]SbF_6$[113], $[Ir(COD)(\eta^2\text{-}(i\text{-Pr})_2PCH_2CH_2NMe_2)]BF_4$, $[Ir(TFB)(\eta^2\text{-}(i\text{-Pr})_2PCH_2CH_2OMe)]BF_4$[118], $Ir(C{\equiv}CPh)(CO)_2(PCy_3)$ and $Ir(C{\equiv}CPh)(TFB)(PCy_3)$[119] are active catalysts for the hydrosilylation of 1-alkynes (see Section III.A.3), the dehydrogenative silylation of 1-alkynes is also promoted by these catalysts to give 1-silylalkynes **151** (equation 60)[113,117–119]. This reaction can become the major pathway ($\leqslant$84%) in some cases, especially in the presence of excess 1-alkyne that acts as the hydrogen acceptor.

$$R-C{\equiv}C-H + HSiR_3' \xrightarrow{[Ir]} R-C{\equiv}C-SiR_3' + RHC{=}CHSiR_3' \quad (60)$$

Two different mechanisms have been proposed for this dehydrogenative silylation process. The first mechanism proposed by Oro, Esteruelas and coworkers includes the oxidative addition of 1-alkyne to the Ir–Si bond, followed by the reductive elimination of **151** (equation 61)[117,118]. The proposed mechanism is supported by the identification of $[IrH(C{\equiv}CPh)(\eta^2\text{-}(i\text{-Pr})_2PCH_2CH_2OMe)]BF_4$ in stoichiometric as well as catalytic conditions by $^{31}P\{^1H\}$ NMR analyses[118]. The other mechanism proposed by Jun and Crabtree includes the insertion of 1-alkyne into the Ir–Si bond, followed by isomerization and β-hydride elimination (equation 62)[113], which is consistent with the mechanism proposed for the highly selective formation of (Z)-1-silyl-1-alkenes (see Section III.B)[115].

$$R-C{\equiv}C-H \xrightarrow{[Ir]} R-C{\equiv}C-[Ir]-H$$

$$\downarrow HSiR'_3$$

$$R-C{\equiv}C-\underset{\underset{H}{|}}{\overset{\overset{SiR'_3}{|}}{[Ir]}}-H \longrightarrow R-C{\equiv}C-SiR'_3 \quad (61)$$

$$[Ir](H)_2$$

$$R-C{\equiv}C-H \xrightarrow{\text{[Ir]-SiR}'_3} \cdots \tag{62}$$

IV. HYDROSILYLATION OF CARBONYL AND RELATED COMPOUNDS

Hydrosilylation of various carbonyl compounds, enones and related functional groups catalyzed by Group VIII transition metal complexes, especially phosphine–rhodium complexes, have been extensively studied[1,3], and the reactions continue to serve as useful methods in organic syntheses.

Rhodium carbonyl clusters such as $Rh_4(CO)_{12}$, $Rh_2Co_2(CO)_{12}$ and $RhCo_3(CO)_{12}$ are found to be highly active catalysts for the hydrosilylation of cyclohexenones[140]. The reaction of cyclohexenone with $HSiMe_2Ph$ catalyzed by these Rh and Co–Rh carbonyl clusters proceeds smoothly at ambient temperature to give the 1,4-addition product (cyclohex-1-en-1-yloxy)dimethylphenylsilane exclusively. On the other hand, the reaction with H_2SiPh_2 proceeds at $-35\,^\circ C$ to give the 1,2-addition product (cyclohex-2-en-1-yloxy)diphenylsilane exclusively. Kinetic study shows that the clusters $Rh_4(CO)_{12}$ and $Rh_2Co_2(CO)_{12}$ possess more than one order of magnitude higher catalytic activity than that of $RhCl(PPh_3)_3$. Rhodium phosphite clusters, $[HRh\{P(OR)_3\}_2]_3$ (R = Me, 2-MeC_6H_4) and $[RhCl\{P(OR)_3\}_2]_2$ (R = Me, 2-MeC_6H_4), are shown to be moderately active catalysts in the hydrosilylation of cyclohexenone and mesityl oxide (1,4-addition) with $HSiEt_3$[20].

Stereoselectivity of the hydrosilylation of cyclic ketones catalyzed by $RhCl(PPh_3)_3$ has been shown to depend on the bulkiness as well as electronic nature of the hydrosilane used[3,141–143]. A systematic study of the reaction of 2- and 4-alkylcyclohexanones (alkyl = Me, Pr-*i*, Bu-*t* and Ph) with H_2SiPh_2 catalyzed by $RhCl(PPh_3)_3$ shows that *cis*-1-siloxy-2-alkylcyclohexane is the predominant product in the reactions of 2-alkylcyclohexanone, while *trans*-isomer is formed as the major product in the reaction of 4-alkylcyclohexanone[144]. The results support the previously proposed mechanism for the reaction, involving the insertion of the carbonyl group into the Rh–Si bond[3].

Unusual rate enhancement is observed in the mono-hydrosilylation of ketones with organosilanes bearing two Si–H groups at appropriate distances, **152–156**, catalyzed by $RhCl(PPh_3)_3$[15,145,146]. Other 1,ω-bis(dimethylsilyl)alkanes, $HMe_2SiCH_2SiMe_2H$, $HMe_2Si(CH_2)_4SiMe_2H$, do not show such an enhancement. It is noteworthy that the second hydrosilylation does not have any particular rate enhancement. The reactions of acetone with **152** and **153** are 50 and 120 times faster than that with $HSiMe_2Et$, $HSiMe_2Ph$, $HMe_2SiCH_2SiMe_2H$ or $HMe_2Si(CH_2)_4SiMe_2H$. It should be noted that in the reaction of an α,β-unsaturated ketone, 4-methyl-3-penten-2-one, with **152**, the 1,4-addition product is obtained exclusively although the reaction is as fast as that with H_2SiPh_2 which exclusively gives 1,2-adduct[15]. Mechanistic study has revealed that disilyl–Rh(V) trihydride **157** arising from the double oxidative addition of **152–156** to

$RhCl(PPh_3)_3$ is very likely to be the active catalyst species (Scheme 17)[145]. Although detailed understanding of this unique rate enhancement needs further investigation, the mechanism proposed by Nagashima, Itoh and coworkers[145] has indicated intriguing new possibilities for the mechanism of $RhCl(PPh_3)_3$-catalyzed hydrosilylation of carbonyl compounds and other substrates. (For the proposed mechanisms including Rh(V) species in alkene hydrosilylation, see Section II.B.)

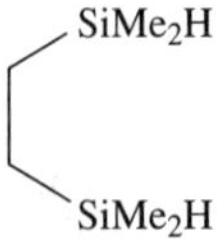

(152)

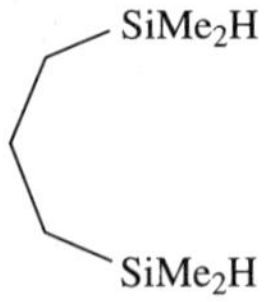

(153)

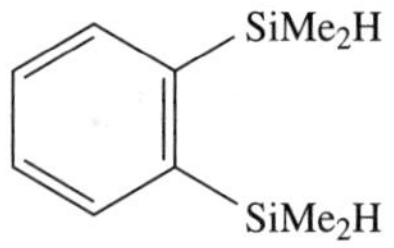

(154)

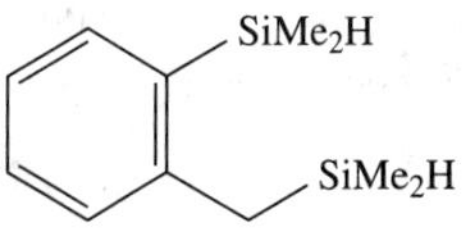

(155)

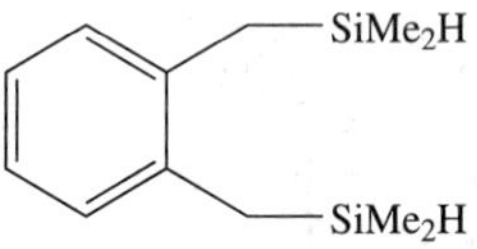

(156)

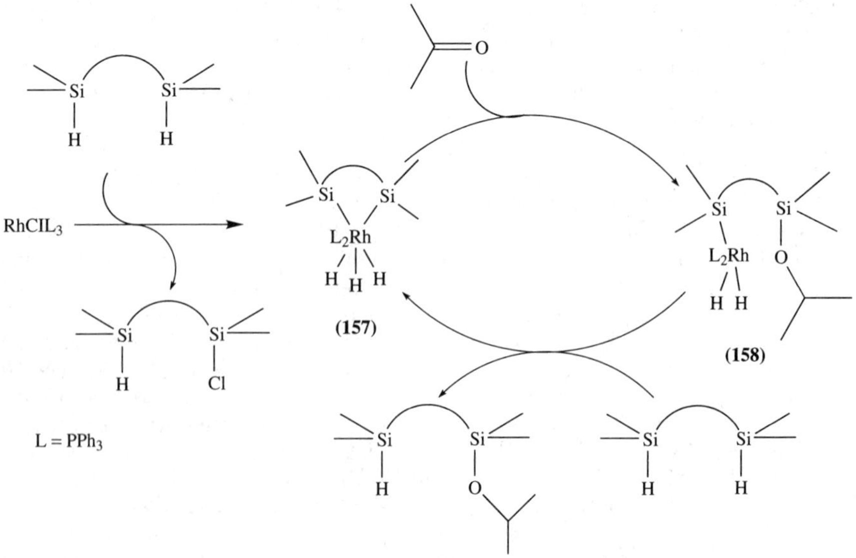

SCHEME 17

The hydrosilylation of cyclopentadienylorganoiron acyl complexes **159** with dihydrosilanes such as H_2SiEt_2, H_2SiPh_2 and $H_2SiPhMe$, catalyzed by $RhCl(PPh_3)_3$, proceeds smoothly at ambient temperature to give α-siloxyalkyliron complexes **160** in 50–91% yields (equation 63)[147]. Dihydrosilane is the silane of choice for the hydrosilylation of **159** since monohydrosilanes do not show reactivity, while trihydrosilanes reduce the acyl group to the corresponding alkyl group[147]. Manganese carbonyl acetyl complexes $(L)(CO)_4Mn-COCH_3$ (**161**, L = CO, PPh_3) are also found to catalyze this reaction efficiently[148].

$$\text{(159)} + H_2SiR'_2 \xrightarrow[\text{0–22 °C}]{\text{RhCl(PPh}_3)_3 \atop \text{THF or benzene}} \text{(160)} \quad (63)$$

R = Me, Et, Pr-*i*, Bu-*i*
SiR'_2 = $SiEt_2$, $SiPh_2$, $SiMePh$

The manganese complex **161** is also found to be an excellent catalyst for the reduction of esters to ethers in high yields using H_3SiPh (equations 64 and 65)[149].

$$R^1\text{COOR}^2 \xrightarrow[\substack{C_6D_6, \text{RT} \\ 61\%-91\%}]{[\text{Mn}] \atop H_3SiPh} R^1CH_2OR^2 \quad (64)$$

$$\gamma\text{-butyrolactone} \xrightarrow[\substack{C_6D_6, \text{RT} \\ 40\%}]{[\text{Mn}] \atop H_3SiPh} \text{tetrahydrofuran} \quad (65)$$

[Mn] = $(Ph_3P)(CO)_4Mn\text{-}COCH_3$ (**161**)

The 1,4-addition of hydrosilanes to α, β-unsaturated esters affords the corresponding silyl ketene acetals that are versatile reagents in organic syntheses. Accordingly, this reaction has been studied extensively using phosphine–rhodium complexes and Speier's catalyst[3,48,50,150,151]. However, the reaction is often accompanied by a small amount of inseparable 1,2-adduct and/or 3,4-adduct[50]. The hydrosilylation of methacrylates **162** with excess $HSiMe_3$ using $RhCl_3 \cdot 6H_2O$ as the catalyst provides an improved route to *O*-TMS-*O*-substituted dimethylketene acetals **163** with excellent product selectivity in high yields (equation 66)[152].

The hydrosilylation of α, β-unsaturated carbonyl compound **164** gives either silyl enol ether **165** via 1,4-addition or *O*-silylallylic alcohol **166** via 1,2-addition, which are desilylated to afford saturated carbonyl compound **167** or allylic alcohol **168** (Scheme 18). Wilkinson's complex, $RhCl(PPh_3)_3$, and Speier's catalyst, $H_2PtCl_6 \cdot 2H_2O/i\text{-PrOH}$, are most commonly used for the 1,4-addition of monohydrosilanes, whereas the combination of $RhCl(PPh_3)_3$ and a dihydrosilane or a trihydrosilane is used exclusively for the 1,2-addition process[3]. The marked dependence of 1,4/1,2-selectivity on the nature of the hydrosilane used was first reported by Ojima and coworkers in 1972 for the $RhCl(PPh_3)_3$-catalyzed reactions[153], and this rather unique dependence of regioselectivity on the nature of hydrosilanes has been proved to be general for phosphine–rhodium complex catalysts and other catalysts[3]. A mechanism that can account for the observed results has been proposed by Ojima and Kogure[154], which includes 1-siloxy-σ-allylic Rh(III)(H) species (for 1,2-addition) as well as 3-siloxy-σ-allylic Rh(III)H species (for 1,4-addition).

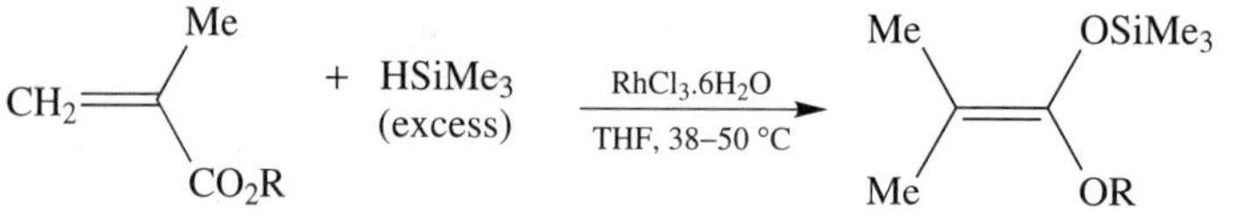

$$CH_2=C(Me)(CO_2R) + HSiMe_3 \text{ (excess)} \xrightarrow[\text{THF, 38–50 °C}]{RhCl_3 \cdot 6H_2O} \text{(163)} \quad (66)$$

(162) **(163)**

R = Me, $CH_2CH_2OSiMe_3$, $(CH_2)_3Si(OMe)_3$, $SiMe_3$, $CH_2CH\!-\!CH_2$ (epoxide)

Scheme:
$R^1C(O)CH=CHR^2$ **(164)** + $HSiXR_2$ —catalyst→ $R^1C(OSiXR_2)=CHCH_2R^2$ **(165)** or $R^1CH(OSiXR_2)CH=CHR^2$ **(166)**

(165) →hydrolysis→ $R^1C(O)CH_2CH_2R^2$ **(167)**

(166) →hydrolysis→ $R^1CH(OH)CH=CHR^2$ **(168)**

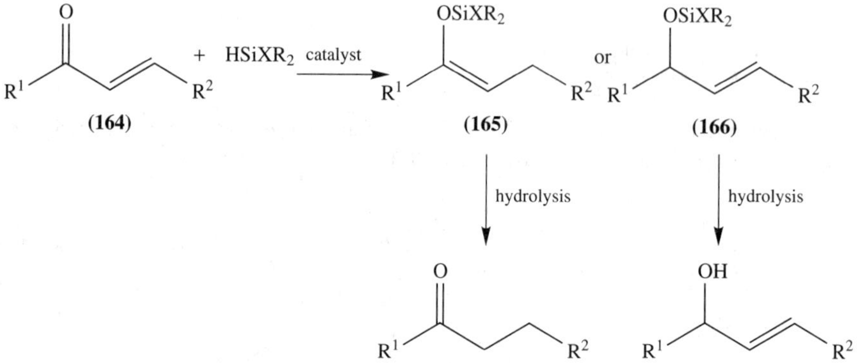

$SiXR_2 = SiEt_3, Si(Pr-n)_3, SiMe_2Ph, SiHEt_2, SiHPh_2$

SCHEME 18

Hydridorhodium(I) complex, $HRh(PPh_3)_4$, is shown to be another efficient catalyst for the 1,2-addition of H_2SiPh_2 to enones giving the corresponding O-diphenylsilylallylic alcohols in high yields[155]. The reaction is highly chemoselective, thus an isolated double bond is unaffected (equation 67)[155]. In the reaction of a diketone **171**, the nonconjugated ketone moiety is selectively reduced (equation 68)[155]. The reactions of α,β-unsaturated ketones, esters and lactones with dimethylphenylsilane and other monohydrosilanes exclusively give 1,4-addition products in high yields (equations 69 and 70)[155].

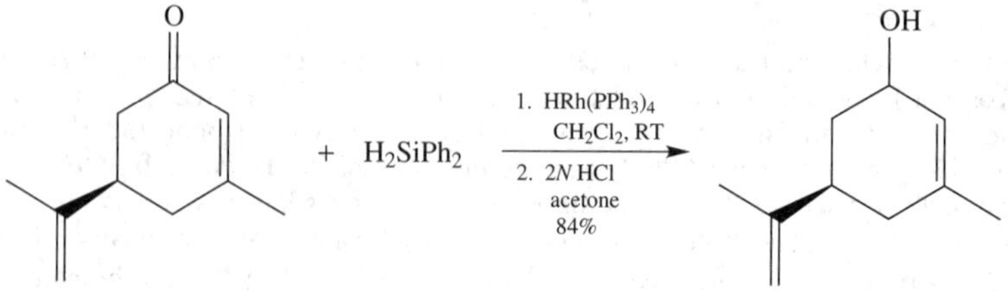

(169) + H_2SiPh_2 $\xrightarrow[\substack{2.\ 2N\ HCl \\ \text{acetone} \\ 84\%}]{\substack{1.\ HRh(PPh_3)_4 \\ CH_2Cl_2,\ RT}}$ **(170)** (syn/anti = 65:35) (67)

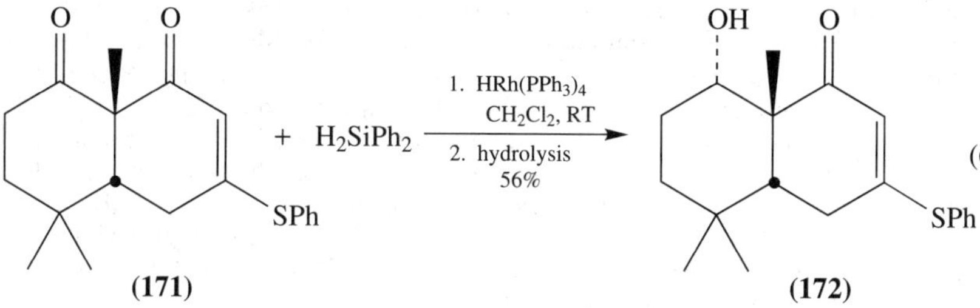

On the basis of the observed kinetic isotope effect ($k_H/k_D = 2$) in the reaction of acetophenone with HSiMe$_2$Ph and DSiMe$_2$Ph catalyzed by HRh(PPh$_3$)$_4$, Zheng and Chan[155] proposed a variation of the Ojima–Kogure mechanism[154], in which a ketone molecule coordinates to silicon rather than rhodium to form alkoxysilyl–Rh(III)H species **(173)** (Scheme 19). This unique mechanism needs further investigation. Nagashima, Itoh and coworkers invoke a possible double oxidative addition of H$_2$SiPh$_2$ to RhCl(PPh$_3$)$_3$ in a manner similar to the one shown in Scheme 17[145].

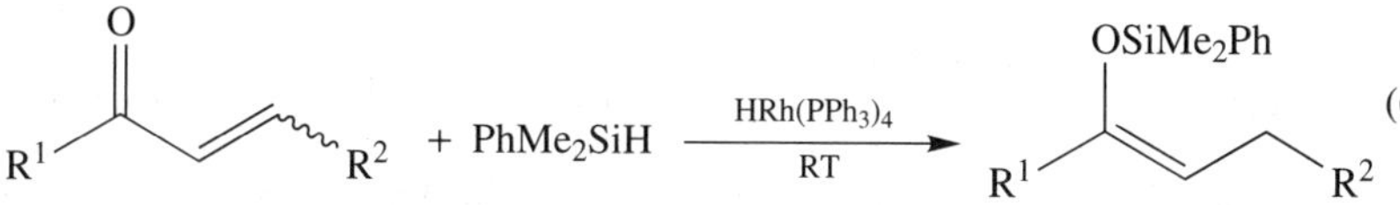

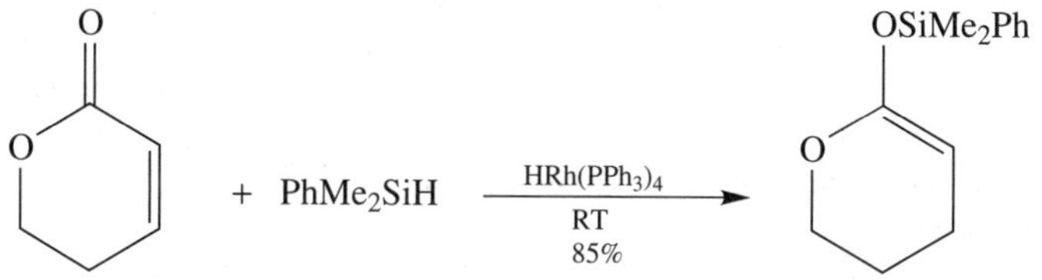

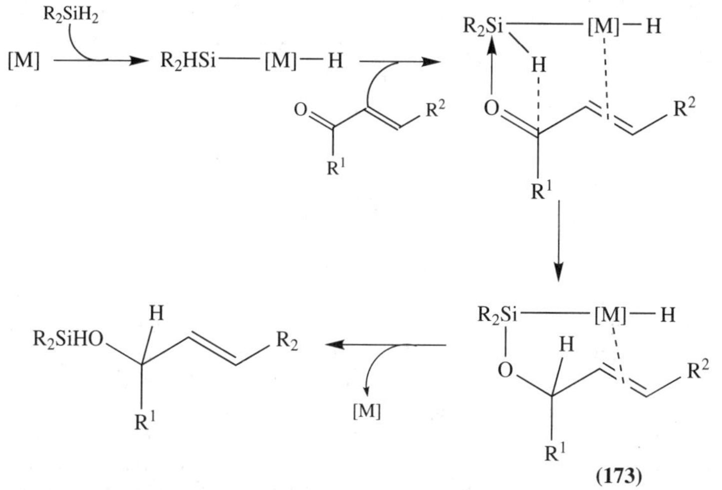

SCHEME 19

The Rh–isocyanide complex, $RhCl(C\equiv N-C_6H_3Me_2\text{-}2,6)_3$, is a highly active catalyst for the hydrosilylation of α,β-unsaturated ketones with $HSiMe_2Ph$, giving 1,4-adducts in nearly quantitative yields[16]. This catalyst is more active than $RhCl(PPh_3)_3$ for the reactions of α,β-unsaturated ketones, acetophenone and 5-hexen-2-one with $HSiMe_2Ph$ under the same conditions, but less active for the reaction of 2-butanone. The corresponding Pt–isocyanide as well as Pd–isocyanide complexes are inactive for the reaction of 2-butanone. However, Pt-isocyanide complexes, $PtCl_2(C\equiv N-C_6H_3Me_2\text{-}2,6)_2$ and $Pt_2Cl_4(DIBN)_2$ (DIBN = 2,2'-isocyano-1,1-binaphthyl), show high activity for the reaction of 5-hexen-2-one. Rhodium complex with DIBN, $[RhCl(DIBN)_2]_n$, is an excellent catalyst (99% yield) for this reaction, while $RhCl(PPh_3)_3$ (5% yield) and $PtCl_2(PPh_3)_2$ (trace) are poor catalysts.

While $Mo(CO)_6$ is only a modest hydrosilylation catalyst[156], oxadiene complexes of molybdenum and tungsten, $M(CO)_2(oxadiene)_2$ (oxadiene = pulegone, pinocarvone and (E)-5-methyl-3-hexen-2-one), exhibit good catalytic activity for the hydrosilylation of α,β-unsaturated ketones and aldehydes with H_3SiPh[157]. Although the 1,4/1,2-selectivity for the reactions of α,β-unsaturated ketones is low to modest at best with one exception for the reaction of pinocarvone catalyzed by $Mo(CO)_2(pinocarvone)_2$ which undergoes 1,4-addition with 92% selectivity, the reactions of α,β-unsaturated aldehydes proceed via 1,2-addition exclusively[157].

Bis(η^5-cyclopentadienyl)diphenyltitanium, Cp_2TiPh_2, is found to be an active catalyst for the hydrosilylation of ketones with H_2SiPh_2, $H_2SiMePh$ at 120 °C[158].

Completely regioselective 1,6-hydrosilylation of dimethyl cis,cis-muconate **174** with $HSiMe_2Ph$, $DSiMe_2Ph$ or $HSiEt_2Me$ is observed using $RhCl(PPh_3)_3$ as the catalyst (equation 71)[159]. However, the reaction of trans,trans-muconate **175** under the same conditions gives only a 3,4-adduct, and that of trans,cis-muconate results in a complicated mixture of adducts[159].

$$\text{MeO-CO-CH=CH-CH=CH-CO-OMe} \ (\mathbf{174}) + HSiR_3 \quad (R = Me_2Ph, MeEt_2)$$

$$\xrightarrow[\text{benzene, 80°C, 100\%}]{RhCl(PPh_3)_3} MeO_2C-CH(H/D)-CH=CH-C(OSiR_3)=CH-OMe \quad (71)$$

Selective reduction of 2,2,4,4-tetramethyl-1,3-cyclobutanedione through hydrosilylation catalyzed by $RhCl(PPh_3)_3$ or $[Rh(COD)Cl]_2$ with various phosphine ligands gives the corresponding hydroxycyclobutanone ($H_2Si(Pr\text{-}i)_2$, $RhCl(PPh_3)_3$, 100%) or dihydroxycyclobutane (H_2SiPh_2, dppp-Rh, 100%, syn/anti = 7/3)[160].

Hydrosilylation of carbonyl compounds catalyzed by fluoride salts such as CsF and KF, originally developed by Corriu and coworkers[161–165], has been modified to homogeneous systems and found applications in organic syntheses as an efficient and selective reduction method.

29. Recent advances in the hydrosilylation and related reactions

Fujita and Hiyama have developed a homogeneous fluoride ion catalyst system consisting of tetrabutylammonium fluoride (TBAF) or tris(diethylamino)sulfonium difluorotrimethylsilicate (TASF) in hexamethylphosphoric triamide (HMPA), which can promote and complete the hydrosilylation of aldehydes and ketones at room temperature or lower temperatures within one hour in most cases (equation 72)[166–169]. The relative reactivity of hydrosilanes decreases in the order: $H_2SiPh_2 \gg HSiMe_2Ph > HSiMePh_2 > HSiPh_3$, and $HSiEt_3$ can be used for the reaction as well[166]. For solvents, HMPA and 1,3-dimethyl-3,4,5,6-tetrahydro-2(1H)-pyrimidinone (DMPU) are the best, while DMF is less effective, THF and CH_2Cl_2 are nonpractical[166].

$$\underset{R^1}{\overset{O}{\|}}\underset{R^2}{} + HSiR_3 \xrightarrow[\text{HMPA, RT}]{\text{TBAF or TASF}} \underset{R^1}{\overset{OSiR_3}{|}}\underset{R^2}{} \qquad (72)$$

On the basis of a significant isotope effect ($k_H/k_D = 1.5$) when using $HSiMe_2Ph$ and $DSiMe_2Ph$, the hydride transfer from hexavalent silicate species $[HSiR_3F(HMPA)]^-$ to the carbonyl functionality is proposed to be the rate-determining step of this reaction[166,170].

The hydrosilane/TBAF in HMPA is a powerful reagent for the *threo*-selective reduction of α-oxy and α-amino ketones **176** at 0 °C or room temperature, giving the corresponding alcohols **177** with excellent diastereoselectivity (*threo/erythro* = 84/16 → 99/1) in high yields (equation 73)[166]. This reagent is also effective for the stereoselective reduction of α-methyl-β-keto amides **178** to *threo*-β-hydroxy amides **179** with excellent selectivity (*threo/erythro* = 98/2 → 99/1) in 86–98% yields as long as R is an aromatic group (equation 74). When R is *tert*-butyl, the *threo/erythro* ratio becomes 91/9, and the reversal of selectivity is observed when R is ethyl (*threo/erythro* = 23/77) or isopropyl (*threo/erythro* = 25/75).

$$\underset{(176)}{\underset{\underset{Me}{|}}{\overset{O}{\underset{R}{\|}}}\!\!\!\!\overset{}{\diagup}\!Z} + HSiR_3 \xrightarrow[\text{2. 1 M HCl or KOH–MeOH}]{\text{1. TBAF/HMPA}} \underset{(177)}{\underset{\underset{Me}{|}}{\overset{OH}{\underset{R}{|}}}\!\!\!\!\overset{}{\diagup}\!Z} \qquad (73)$$

R = Ph, Bu, PhCH=CH; Z = OAc, OBz, OEE, OBu-*t*, OTHP, NMe$_2$

$$\underset{(178)}{\underset{\underset{Me}{|}}{\overset{O\quad O}{\underset{R}{\|}\!\!\diagup\!\!\|}}\!\!X} + HSiR_3 \xrightarrow[\text{2. 1 M HCl}]{\text{1. TBAF/HMPA}} \underset{(179)}{\underset{\underset{Me}{|}}{\overset{OH\quad O}{\underset{R}{|}\!\!\diagup\!\!\|}}\!\!X} \qquad (74)$$

R = Ph, 4-ClC$_6$H$_4$, 4-MeOC$_6$H$_4$, *t*-Bu; X = NEt$_2$, pyrrolidinyl, piperidinyl

When 18-crown-6 is used for the hydrosilane/CsF system, the hydrosilylation of aromatic aldehydes and ketones proceeds smoothly in CH_2Cl_2 under phase-transfer conditions at room temperature[171,172]. The isolation of the products in this HSiMe$_2$Ph/CsF/18-crown-6 system is claimed to be much easier than that in the HSiMe$_2$Ph/TBAF/HMPA system.

Strong acids, e.g. CF$_3$CO$_2$H[173,174] and CF$_3$SO$_3$H,[175,176] as well as Lewis acids, e.g. BF$_3$·OEt$_2$[177,178], are well known to promote the reduction of carbonyl compounds with hydrosilanes. It has recently been shown that the same catalytic mechanism is operating in the Lewis acid (Et$_3$SiClO$_4$)—and Brønsted acid (HClO$_4$)—catalyzed reactions of carbonyl compounds with HSiEt$_3$[179]. The hydrosilane/solid base combination, i.e. HSi(OEt)$_3$/hydroxyapatite [HAp = Ca$_{10}$(PO$_4$)$_6$(OH)$_2$] or CaO, is found to be very effective for the hydrosilylation of a variety of aldehydes and ketones in heptane at 90 °C[176,180]. The reactions of α,β-unsaturated aldehydes and ketones by this combination give the corresponding O-silyl allylic alcohols exclusively via 1,2-addition[176]. The reactions of carbonyl compounds using hydrosilane/solid acid combinations, e.g. HSiEt$_3$/Fe^{3+}-ion exchanged montmorillonite, give deoxygenated products, i.e. hydrocarbons from aldehydes and ketones, or symmetrical ethers (R^{1}R^{2}CH)$_2$O from R^{1}R^{2}C=O[176].

Intramolecular hydrosilylation of β-(hydrosilyloxy)ketone **181**, readily prepared from β-hydroxyketone **180**, catalyzed by SnCl$_4$, MgBr$_2$·OEt$_2$, ZnBr$_2$, ZnCl$_2$, CF$_3$CO$_2$H, TBAF or RhCl(PPh$_3$)$_3$, gives *anti*-**182** as the predominant product. The stereoselectivity depends on the bulkiness of R^1–R^3 as well as the catalyst used. An excellent selectivity (*anti*/*syn* = 120/1) is achieved in the reaction of 2,6-dimethyl-hepta-3,5-dione (R^1–R^3 = *i*-Pr) using SnCl$_4$ as the catalyst at −80 °C, giving the *anti*-diol **183** in 67% overall isolated yield (Scheme 20)[181].

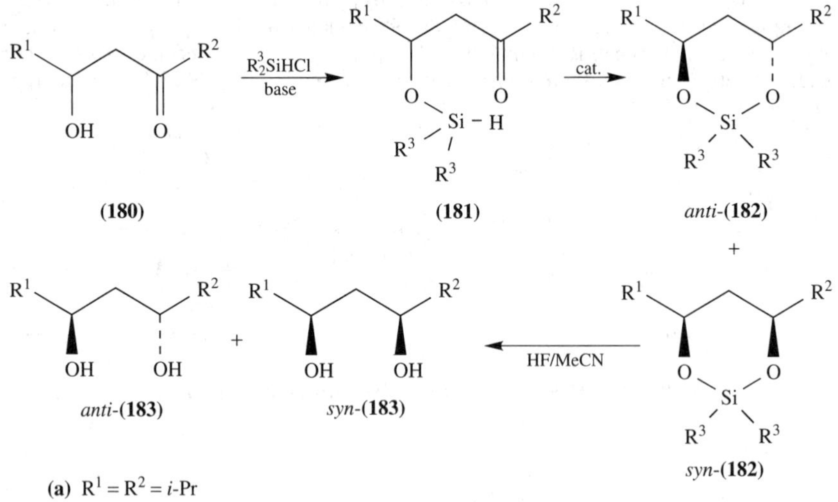

(a) R^1 = R^2 = *i*-Pr
(b) R^1 = R^2 = *n*-Bu
(c) R^1 = *i*-Pr, R^2 = Me
(d) R^1 = Me, R^2 = *i*-Pr

R^3 = *i*-Pr or Me

SCHEME 20

Intramolecular hydrosilylation of β-(hydrosiloxy)alkanoate **184** catalyzed by TBAF in CH$_2$Cl$_2$ at 0 °C, followed by Lewis acid-catalyzed allylation with allyltrimethylsilane in

CH$_2$Cl$_2$, gives *anti*-**186** in high yield with the *anti/syn* ratio up to 23/1 (R^1 = *n*-Bu, R^2 = Me, Lewis acid = TfOH$_2$$^+$B(OTf)$_4$$^-$ at $-50\,°$C) (equation 75)[182,183].

R^1 = *n*-Pr, *n*-Bu, Ph; R^2 = Me, Et

The combination of polymethylhydrosiloxane (PMHS) and catalytic amounts of TBAF (1 mol%) and Cp$_2$Ti(OC$_6$H$_4$Cl-4)$_2$ (2 mol%) provides an efficient and convenient method for the partial reduction of lactones **187** to lactols **189** via hydrosilylation and subsequent hydrolysis of the resulting *O*-silyllactols **188** (equation 76)[184].

While hydrosilylation of imines is known to be effected by rhodium catalysts[3], nickel catalysts prepared *in situ* from $Ni(OAc)_2 \cdot 4H_2O$ and thiosemicarbazones are also found to promote the reactions of *N*-substituted imines with $HSiEt_3$ in dry DMSO at 35 °C, giving the corresponding secondary amines in excellent yields after basic work-up (equation 77)[185].

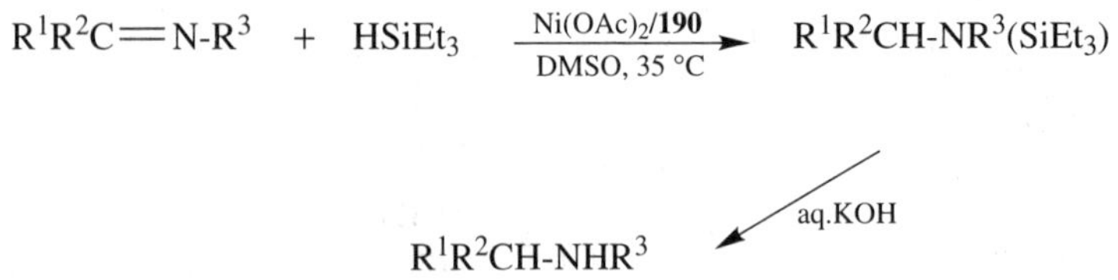

$$R^1R^2C{=}N\text{-}R^3 + HSiEt_3 \xrightarrow[\text{DMSO, 35 °C}]{Ni(OAc)_2/\mathbf{190}} R^1R^2CH\text{-}NR^3(SiEt_3)$$

$$R^1R^2CH\text{-}NHR^3 \xleftarrow{aq. KOH}$$

(77)

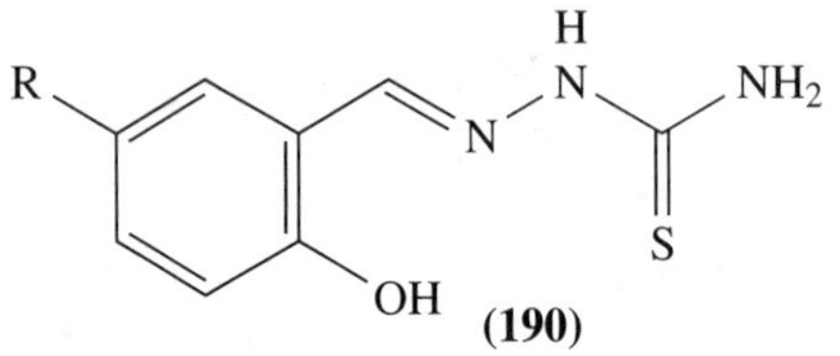

(**190**)   R = H, NaO_3S, *t*-octyl

Cobalt carbonyl-catalyzed hydrosilylation of aromatic nitriles with $HSiMe_3$ proceeds smoothly at 60 °C and ambient pressure of carbon monoxide in toluene to give *N*,*N*-bis-TMS-amines **191** in moderate to excellent yields (equation 78)[186]. Reactions of aliphatic nitriles require 100 °C, except for cyclopropanenitrile, to give **191** in 87–100% yields. The reaction of butanenitrile gives a mixture of **191** (R = *n*-Bu) (44%) and *N*,*N*-bis-TMS-enanime $CH_3CH_2CH{=}CHN(SiMe_3)_2$ (20%). The reactions of α,β-unsaturated nitriles bearing substituent(s) at the α- and/or β-position(s) give *N*,*N*-bis-TMS-enamines **192** (*Z*/*E* mixture) as the major products accompanied by *N*,*N*-bis-TMS-allylic amines **193** and/or saturated *N*,*N*-bis-TMS-amines **194** (equation 79)[186].

$$RC{\equiv}N + HSiMe_3 \xrightarrow[\text{toluene, 60–100 °C}]{Co_2(CO)_8 \\ CO (1 \text{ atm})} RCH_2N(SiMe_3)_2 \quad (\mathbf{191})$$

(78)

(79)

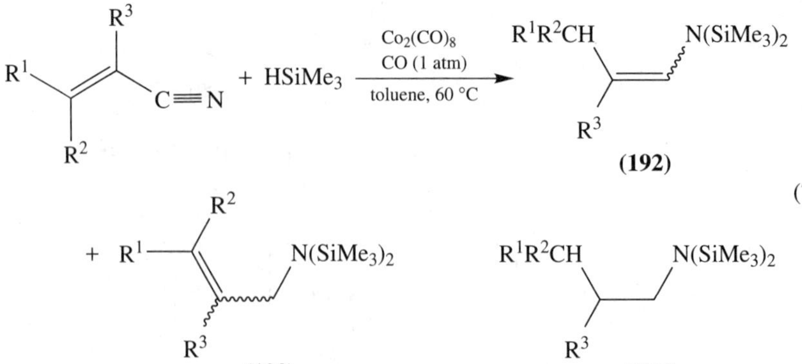

(**192**), (**193**), (**194**)

29. Recent advances in the hydrosilylation and related reactions

V. ASYMMETRIC HYDROSILYLATION

Asymmetric hydrosilylation of prochiral carbonyl compounds, imines, alkenes and 1,3-dienes has been extensively studied and continues to be one of the most important subjects in the hydrosilylation reactions. This topic has been reviewed at each stage of its development as a useful synthetic method based on asymmetric catalytic processes[1,3,187–189]. In the last decade, however, substantial progress has been made in the efficiency of this reaction. Accordingly, this section summarizes the recent advances in this reaction.

A. Of Prochiral Ketones, Imines and Imine N-Oxides

Asymmetric hydrosilylation of prochiral ketones continues to be the most popular reaction to examine the efficacy of new chiral ligands or chiral catalyst systems. This asymmetric catalytic process gives enantiomerically enriched secondary alcohols after facile desilylation of the resulting silyl ethers (equation 80).

$$\underset{R^2}{\overset{R^1}{>}}C=O \ + \ HSiXR_2 \ \xrightarrow{\text{cat.}^*} \ \underset{R^2}{\overset{R^1}{>}}{}^*CHOSiXR_2 \ \xrightarrow[\text{or } F^-]{H^+,\ OH^-} \ \underset{R^2}{\overset{R^1}{>}}{}^*CHOH \quad (80)$$

A variety of newer chiral ligands has been emerging for traditional rhodium catalyst systems[190–223] as well as other transition metal catalyst systems including iridium[209,213], cobalt[224] and titanium complexes[225,226]. Chiral ligands for Rh(I) or Rh(III) catalysts that have been shown to realize high to excellent asymmetric induction ($\geqslant 80\%$ ee) and were not included in *The Chemistry of Organic Silicon Compounds*[3] are summarized in Figure 1. Figure 1 also includes a chiral titanocene catalyst precursor that has exhibited excellent enantioselectivity. Representative results using transition metal catalysts with these chiral ligands as well as the chiral titanocene catalyst precursor are listed in Table 5.

As Figure 1 shows, the most characteristic feature in the development of the newer chiral ligands for Rh-catalysts in the last decade is the emergence of highly efficient bidentate dinitrogen ligands as well as tridentate trinitrogen ligands bearing oxazoline moieties in place of chiral diphosphines and chiral pyridylthiazolidines that were extensively used in 1970s and mid-1980s, respectively[3]. Chiral bis(1-phosphinoethylferrocenyl) (TRAP, **206**)[207] and bis(1-aminoethylferrocenyl)dichalcogenide (**208**)[211] ligands can be pointed out as new developments. Highly effective chiral P–N ligands have also been developed[208,209]. It is worth mentioning that the success of the chiral C_2-symmetrical bis(oxazolinyl)pyridine ligands[190–192,194], pybox **195** and **196**, developed by Nishiyama and coworkers, has stimulated the development of a new series of chiral N–N and P–N ligands. It should also be noted that the chiral Rh-catalyst precursors with pybox[190–192], pymox[190] or bipymox[193] are Rh(III) complexes which are reduced to Rh(I) species by H_2SiPh_2 in the presence of $AgBF_4$.

Another important development is chiral titanocene catalyst using **203** as the catalyst precursor[225]. The (R, R)-(EBTHI)Ti(OR)$_2$ (**203**) is proposed to generate the active catalyst species, '(R, R)-(EBTHI)TiH', upon reaction with n-BuLi (2 equivalents) and polymethylhydrosiloxane (PMHS). This chiral Ti-catalyst system is highly efficient for the reactions of aromatic alkyl ketones achieving $>90\%$ ee in many cases. In sharp contrast to this, only 24% ee is obtained for the reaction of cyclohexyl methyl ketones. However, the reaction of cyclohexen-1-yl methyl ketone achieved 85–90% ee. Thus, it is extremely important for this chiral Ti-catalyst to have a π-system to be effective.

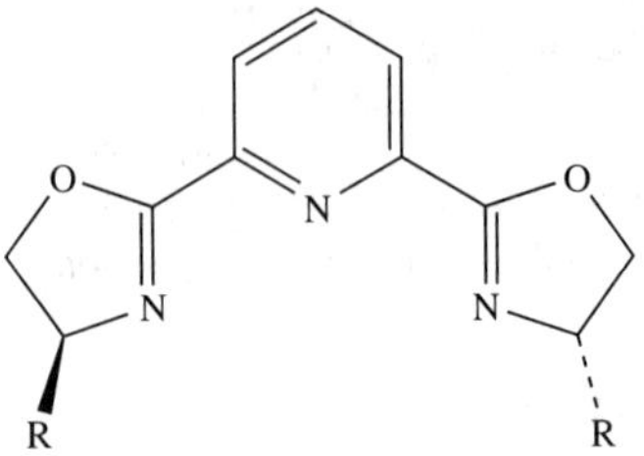

(S,S)-pybox (**195**)

(a) ip-pybox (R = i-Pr)
(b) sb-pybox (R = s-Bu)
(c) tb-pybox (R = t-Bu)

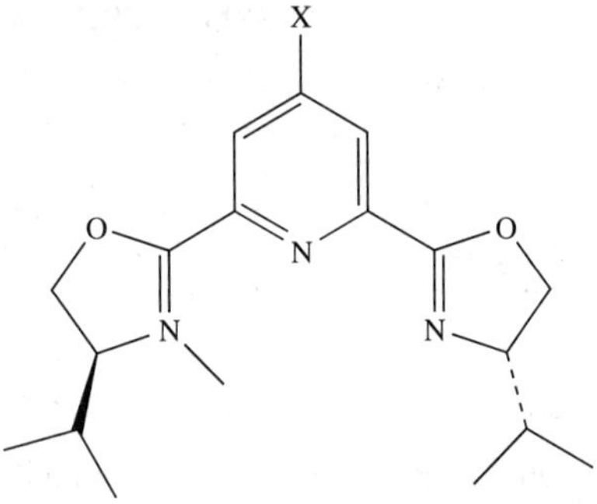

(S,S)-4-X-ip-pybox (**196**)

(a) X = Cl
(b) X = MeO
(c) X = Me$_2$N

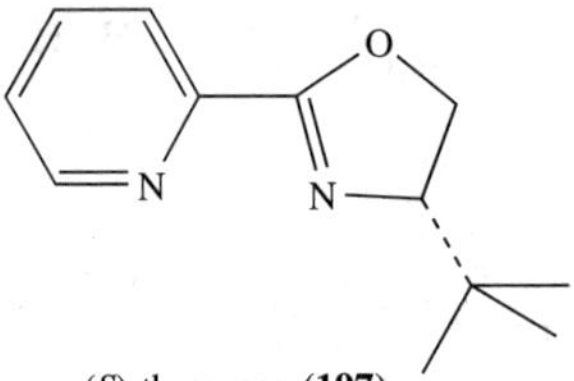

(S)-tb-pymox (**197**)

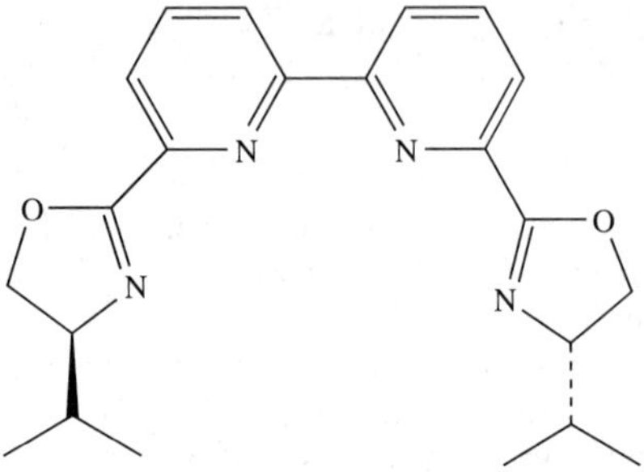

bipymox-(S,S)-Pr-i (**198**)

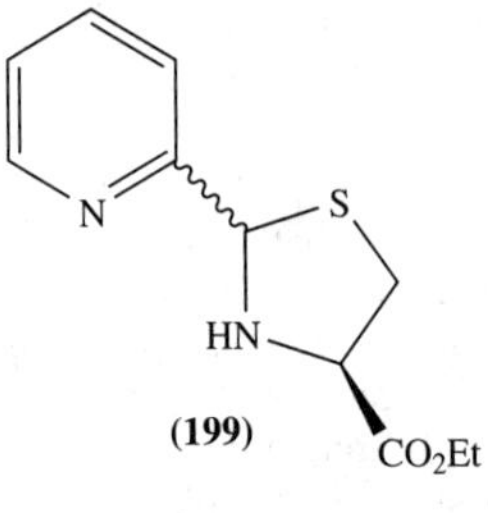

(**199**)

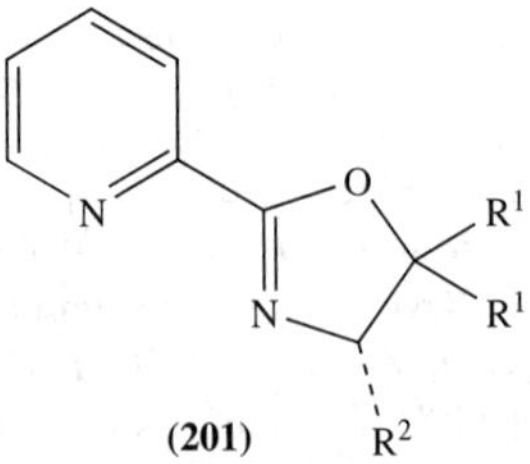

(**200**)

(**201**)

(a) R^1 = H, R^2 = t-Bu
(b) R^1 = Ph, R^2 = CH$_2$Ph
(c) R^1 = Ph, R^2 = i-Pr
(d) R^1 = Ph, R^2 = i-Bu
(e) R^1 = Ph, R^2 = (S)-s-Bu

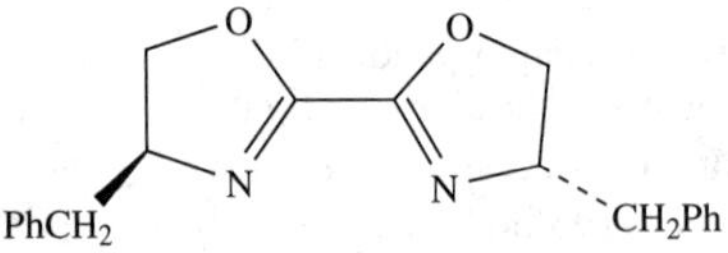

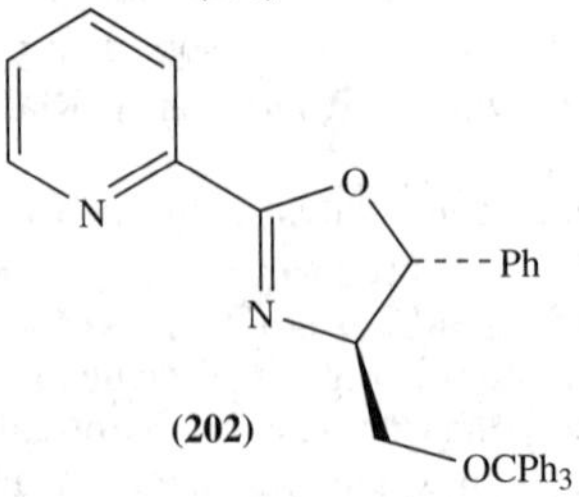

(**202**)

FIGURE 1

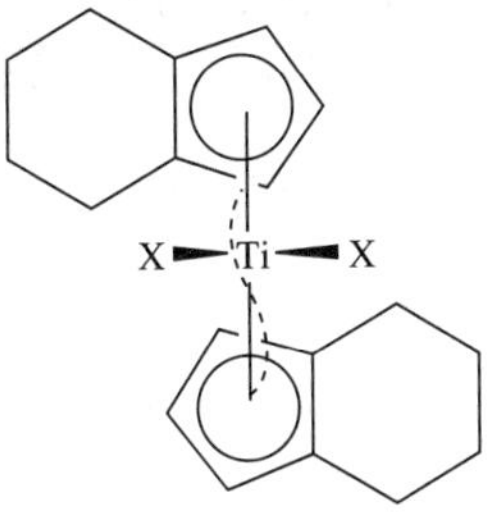

(R,R)-(EBTHI)TiX₂ **(203)**

X₂ = (R,R)-1, 1'-binaphthyl-2, 2'-diolate

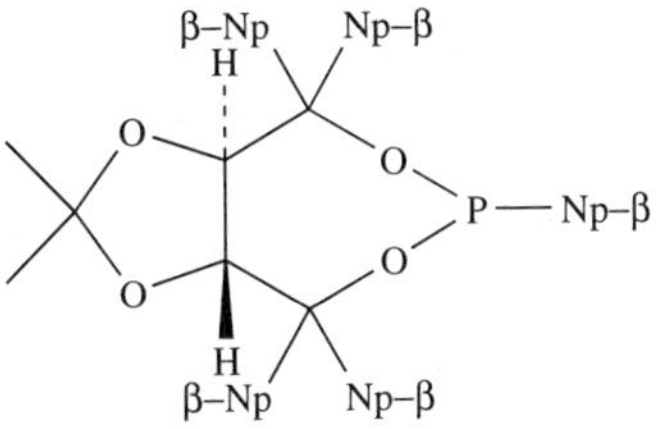

(204)

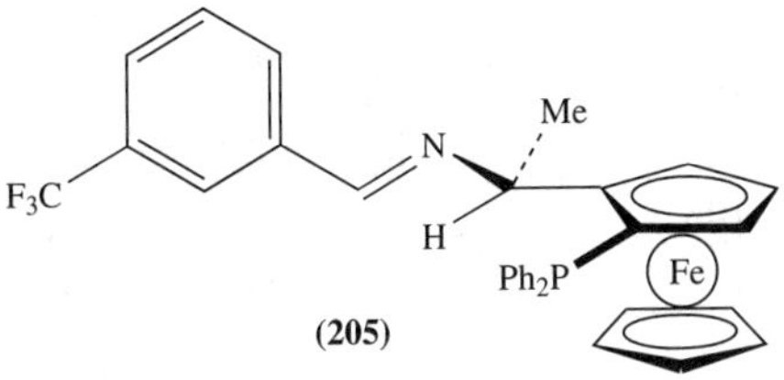

(205)

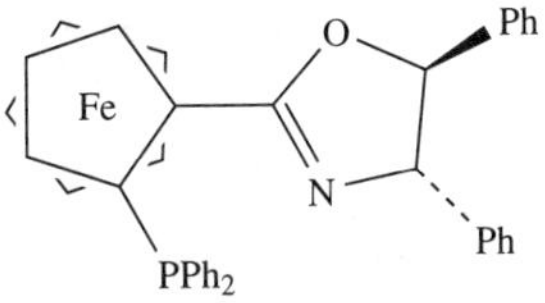

(S,S,S)-DIPOF **(207)**

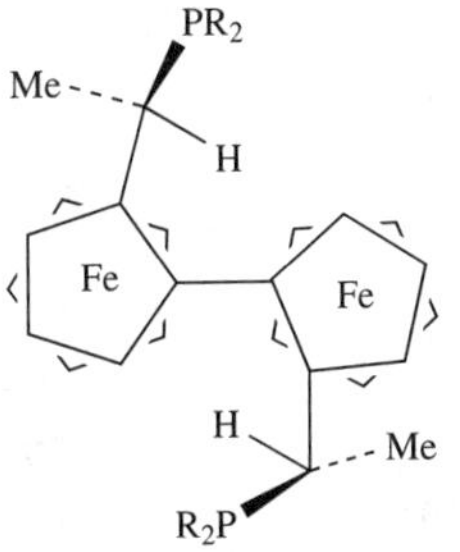

(206)

(a) R = n-Pr; (R,R)-(S,S)-n-PrTRAP
(b) R = n-Bu; (R,R)-(S,S)-n-BuTRAP
(c) R = Et; (R,R)-(S,S)-EtTRAP

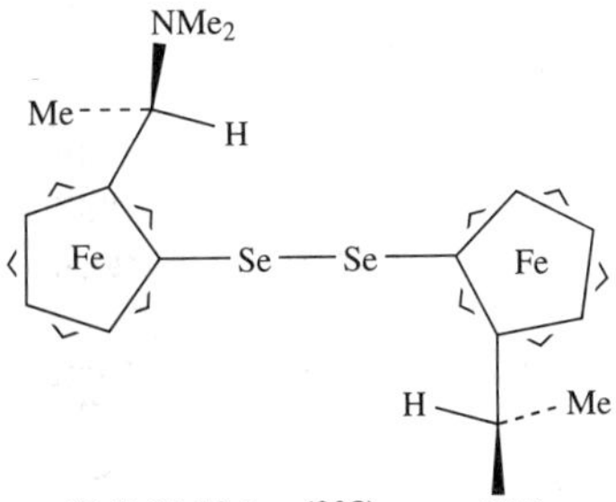

(R,S)-(Fc*Se) **(208)**

FIGURE 1. (*continued*)

As Table 5 indicates, alkyl aryl ketones are the best substrates for this reaction. This has also been the case for the previously developed chiral ligand systems. Nevertheless, there have been substantial improvements in enantioselectivity for the reactions of alkenyl methyl ketones and alkyl methyl ketones, using Rh-catalysts with chiral ligands **195a**[191], **206b**[207], and **207**[209].

A curious reversal of configuration is observed when Ir-catalyst is used instead of Rh-catalyst for the same chiral ligand[209,213]. For example, the asymmetric hydrosilylation followed by desilylation of acetophenone catalyzed by **207**/[Rh(COD)Cl]₂ gives (R)-1-phenylethanol with 91% ee, while the same reaction catalyzed by **207**/[Ir(COD)Cl]₂ yields (S)-alcohol with 96% ee (Scheme 21). The rationale for this remarkable reversal in the direction of asymmetric induction has not been given clearly, but either a change

TABLE 5. Typical results on the asymmetric reduction of prochiral ketones via hydrosilylation

Prochiral ketone	Hydro-silane	Catalyst	Temp. (°C) Solvent	Product	% ee	Reference
PhCOMe	H_2SiPh_2	195a/Rh(III) $AgBF_4$	0 THF	PhMeCHOH (S)	94	191
PhCOMe	H_2SiPh_2	195b/Rh(III) $AgBF_4$	−5 THF	PhMeCHOH (S)	91	191
PhCOMe	H_2SiPh_2	195c/Rh(III) $AgBF_4$	0 THF	PhMeCHOH (S)	83	191
PhCOMe	H_2SiPh_2	196a/Rh(III) $AgBF_4$	−5 THF	PhMeCHOH (S)	94	192
PhCOMe	H_2SiPh_2	196b/Rh(III) $AgBF_4$	10 THF	PhMeCHOH (S)	93	192
PhCOMe	H_2SiPh_2	196c/Rh(III) $AgBF_4$	20 THF	PhMeCHOH (S)	90	192
PhCOMe	H_2SiPh_2	197/Rh(I)	−5 THF	PhMeCHOH (R)	91	192
PhCOMe	H_2SiPh_2	198/Rh(III) $AgBF_4$	5 THF	PhMeCHOH (S)	90	193
PhCOMe	H_2SiPh_2	198/Rh(III) $AgBF_4$	0 → 20 CCl_4	PhMeCHOH (R)	83	195
PhCOMe	H_2SiPh_2	198/Rh(III) $AgBF_4$	0 → 20 CCl_4	PhMeCHOH (R)	89	195
PhCOMe	H_2SiPh_2	198/Rh(III) $AgBF_4$	0 → 20 CCl_4	PhMeCHOH (R)	81	195
PhCOMe	H_2SiPh_2	198/Rh(III) $AgBF_4$	0 → 20 CCl_4	PhMeCHOH (R)	82	195
PhCOMe	H_2SiPh_2	199/Rh(I)	0 → 20 neat	PhMeCHOH (R)	86	203
PhCOMe	H_2SiPh_2	200/Rh(I)	0 CCl_4	PhMeCHOH (R)	84	227
PhCOMe	H_2SiPh_2	201a/Rh(I)	0 → 20 CCl_4	PhMeCHOH (R)	83	201
PhCOMe	$H_2SiPhNp-\alpha$	202/Rh(I)	0 neat	PhMeCHOH (R)	84	223
PhCOMe	PMHS[a]	203 n-BuLi PMHS[a]	room temp. benzene	PhMeCHOH (R)	97	225
PhCOMe	H_2SiPh_2	204/Rh(I)	0 → 20 CCl_4	PhMeCHOH (R)	84	212
PhCOMe	H_2SiPh_2	205/Rh(I)	20 THF	PhMeCHOH (S)	90	208
PhCOMe	H_2SiPh_2	206a/Rh(I)	−40 THF	PhMeCHOH (S)	92	207
PhCOMe	H_2SiPh_2	206b/Rh(I)	−40 THF	PhMeCHOH (S)	92	207

TABLE 5. (continued)

Prochiral ketone	Hydro-silane	Catalyst	Temp. (°C) Solvent	Product	% ee	Reference
PhCOMe	H$_2$SiPh$_2$	207/Rh(I)	25 Et$_2$O	PhMeCHOH (R)	91	209
PhCOMe	H$_2$SiPh$_2$	208/Rh(I)	0 THF	PhMeCHOH (R)	85	211
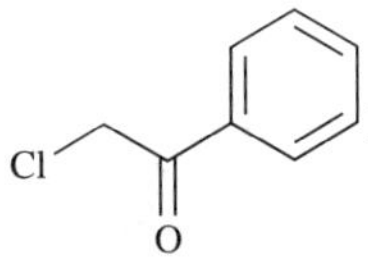 ClCH$_2$COPh	H$_2$SiPh$_2$	208/Rh(I)	0 THF	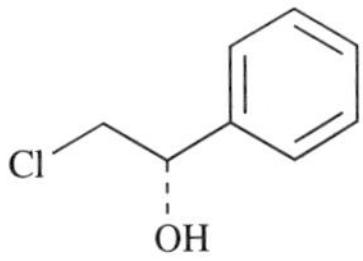 ClCH$_2$CH(OH)Ph	88	210, 211
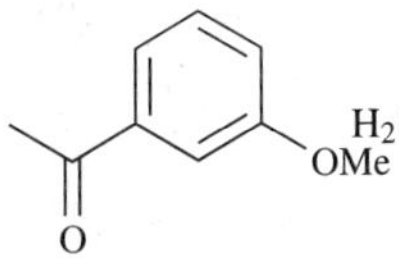 3-MeO-C$_6$H$_4$-COMe	H$_2$SiPh$_2$	200/Rh(I)	0 → 20 neat	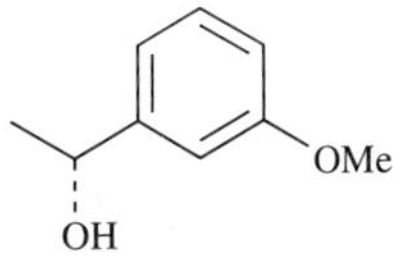 3-MeO-C$_6$H$_4$-CH(OH)Me	93	203
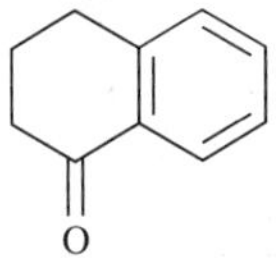 α-tetralone	H$_2$SiPh$_2$	195a/Rh(III) AgBF$_4$	0 THF	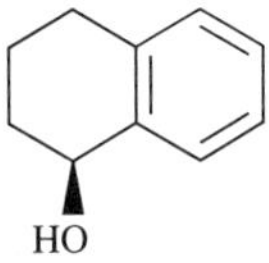 α-tetralol	99	191
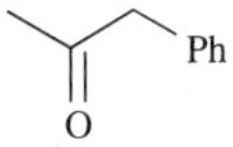 PhCH$_2$COMe	H$_2$SiPh$_2$	196a/Rh(III) AgBF$_4$	−5 THF	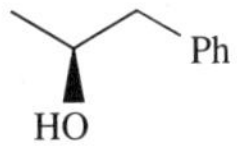 PhCH$_2$CH(OH)Me	80	192
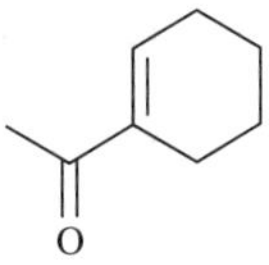 1-acetylcyclohexene	PMHS^a	203/n-BuLi PMHS^a	room temp. benzene	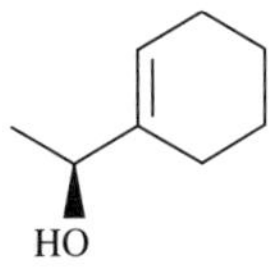 1-(1-hydroxyethyl)cyclohexene	85–90	225
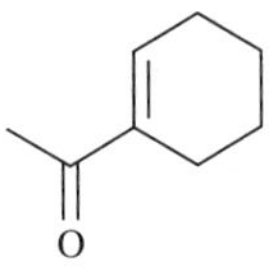 1-acetylcyclohexene	H$_2$SiPh$_2$	206b/Rh(I)	−40 THF	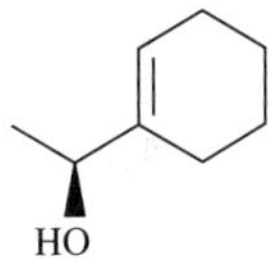 1-(1-hydroxyethyl)cyclohexene	95	207
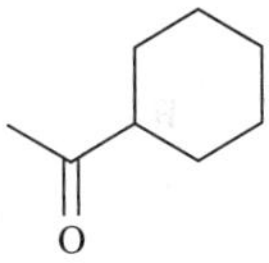 cyclohexyl methyl ketone	H$_2$SiPh$_2$	206b/Rh(I)	−40 THF	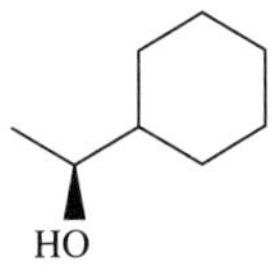 1-cyclohexylethanol	80	207
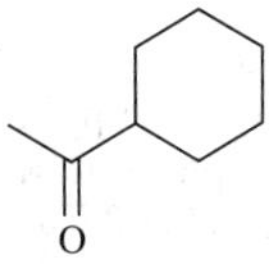 cyclohexyl methyl ketone	H$_2$SiPh$_2$	207/Rh(I)	25 Et$_2$O	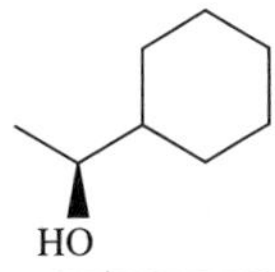 1-cyclohexylethanol	89	209

(continued overleaf)

TABLE 5. (continued)

Prochiral ketone	Hydro-silane	Catalyst	Temp. (°C) Solvent	Product	% ee	Reference
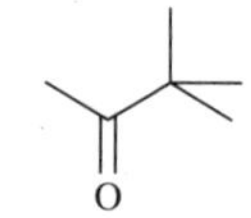 t-Bu-CO-CMe₃ (pinacolone)	H_2SiPh_2	f 207/Rh(I)	25 Et₂O	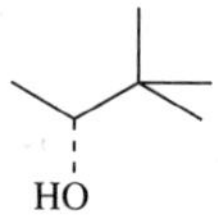 t-Bu-CH(OH)-CMe₃	87	209
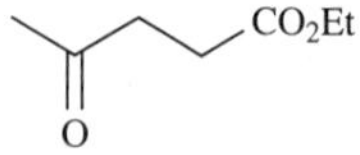 MeCO-CH₂CH₂-CO₂Et	H_2SiPh_2	195a/Rh(III) AgBF₄	0 THF	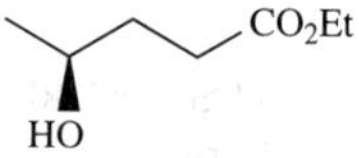 Me-CH(OH)-CH₂CH₂-CO₂Et	95	191
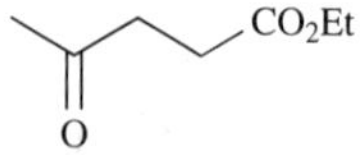 MeCO-CH₂CH₂-CO₂Et	H_2SiPh_2	206c/Rh(I)	room temp. THF	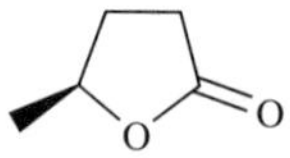 γ-butyrolactone (methyl)	88	206
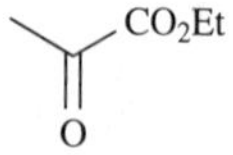 MeCO-CH₂-CO₂Et	H_2SiPh_2	206c/Rh(I)	room temp. THF	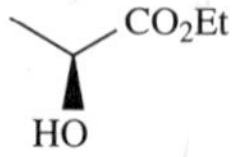 Me-CH(OH)-CH₂-CO₂Et	80	206
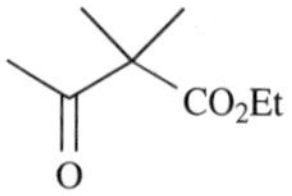 MeCO-CMe₂-CO₂Et	H_2SiPh_2	206c/Rh(I)	room temp. THF	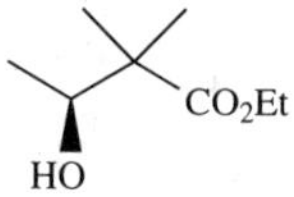 Me-CH(OH)-CMe₂-CO₂Et	93	206

^{a}PMHS = polymethylhydrosiloxane.

in the mechanism or catalyst geometry may account for the phenomenon. An intriguing possibility of the C—H activation of the Me₂N group by Ir(I), but not by Rh(I), is invoked by Faller and Chase[213].

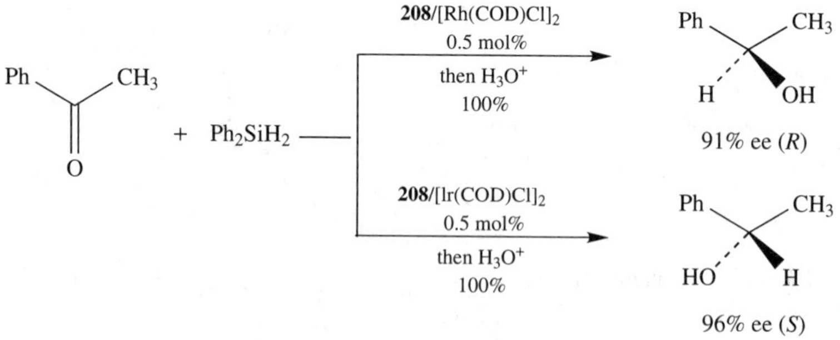

SCHEME 21

Reversal of configuration is also observed when a polystyrene-bound **201**-type pyridyloxazolidine–Rh(I) complex is used as the catalyst for the reaction of acetophenone with H₂SiPh₂ and recycled. This catalyst shows only marginal enantioselectivity (11–27% ee) for (S)-product formation, while it gives (R)-product with 92.5% ee after recycling 9 times, although the chemical yield drops to only 15% under the same reaction conditions. Change in the active catalyst species during recycling is suggested[196].

In comparison with the asymmetric hydrosilylation of prochiral ketones, the reaction of prochiral imines has been more challenging in achieving good enantioselectivity[3] (equation 81). For example, the reaction of PhMeC=NPh with H_2SiPh_2 catalyzed by **208**/[Rh(COD)Cl]$_2$ in ether gives PhMeCH−NHPh with 53% ee but in only 28% yield after 140 h[211]. The reaction of 2-methylquinoxaline with H_2SiPh_2 catalyzed by (+)-DIOP−Rh(H) complex prepared from [Rh(CO)$_2$Cl]$_2$, (+)-DIOP and KBH$_4$ in ethanol gives (S)-2-methyltetrahydroquinoxaline with 19% ee[228]. When the sodium amide of (R)-2,2′-diamino-1,1′-binaphthyl is used as the chiral ligand for this reaction, (R)-product with 36% ee is obtained[228].

$$\begin{array}{c} R^1 \\ \diagdown \\ C=N \\ \diagup \quad \diagdown \\ R^2 \quad R^3 \end{array} + HSiXR_2 \xrightarrow{\text{cat.*}} \begin{array}{c} R^1 \quad SiXR_2 \\ \diagdown \quad \diagup \\ *CH-N \\ \diagup \quad \diagdown \\ R^2 \quad R^3 \end{array} \xrightarrow[\text{or F}^-]{H^+, OH^-} \begin{array}{c} R^1 \\ \diagdown \\ *CH-NHR^3 \\ \diagup \\ R^2 \end{array} \quad (81)$$

A breakthrough for this reaction has just been achieved by Buchwald and coworkers using (S,S)-(EBTHI)TiF$_2$ as a precatalyst[229]. The active catalyst species '(EBTHI)Ti−H' (*vide supra*) is generated by treating (S,S)-(EBTHI)TiF$_2$ with H_3SiPh in the presence of a small amount of methanol and pyrrolidine in THF. This chiral catalyst system can achieve >90% ee for the reactions of various imines. The reaction is proposed to include σ-bond metathesis (see Section IIB) for the cleavage of Ti−N bond with H−SiH$_2$Ph. Representative results are listed in Table 6[229].

TABLE 6. Asymmetric reduction of prochiral imines via hydrosilylation catalyzed by (S,S)-(EBTHI)Ti complex[a] in THF

Prochiral imine	Hydrosilane	Temp. (°C)	Product	% ee
PhMeC=NMe	H_3SiPh	35	(S)-PhMeCH-NHMe	99
3,4-dihydronaphthalen-1(2H)-ylidene NMe	H_3SiPh	room temp.	1,2,3,4-tetrahydronaphthalen-1-yl NHMe	96
Ph−C(=NMe)−(CH$_2$)$_{10}$CH$_3$	H_3SiPh	50	(−) Ph−CH(NHMe)−(CH$_2$)$_{10}$CH$_3$	95
cyclohexyl−C(=NMe)−Me	H_3SiPh	room temp.	cyclohexyl−CH(NHMe)−Me	93
Ph−C=N (pyrroline)	H_3SiPh	room temp.	Ph−CH−NH (pyrrolidine)	99

[a]Catalyst amount = 0.02−1 mol%.

Asymmetric hydrosilylation of prochiral imine N-oxides (nitrones) **(209–211)** catalyzed by Ru$_2$Cl$_4$[(S)-(−)-tolbinap]$_2$(NEt$_3$) with H$_2$SiPh$_2$ gives the corresponding N,N-disubstituted hydroxylamines **(212–214)** with high enantiomeric purity (equations 82–84)[230].

$$(S)\text{-}(-)\text{-tolbinap} \quad \text{with } P(C_6H_4Me\text{-}4)_2 \text{ groups} \tag{82}$$

(209) Me–N$^+$(O$^-$)=C(Me)(C$_6$H$_4$X), X = H, 4-Cl, 4-F, 3-Cl

1. H$_2$SiPh$_2$, (S)-(−)-tolbinap–Ru, dioxane, 0 °C
2. H$_3$O$^+$

→ **(212)** Me–N(OH)–CH(Me)(C$_6$H$_4$X), 83–86% ee (83)

(210) (cyclic nitrone with Ph)

1. H$_2$SiPh$_2$, (S)-(−)-tolbinap–Ru, dioxane, 0 °C
2. H$_3$O$^+$

→ **(213)** 2-phenyl-1-hydroxypyrrolidine, 69% ee (83)

(211) Me–N$^+$(O$^-$)=C(CO$_2$Et)(Ph)

1. H$_2$SiPh$_2$, (S)-(−)-tolbinap–Ru, dioxane, 0 °C
2. H$_3$O$^+$

→ **(214)** Me–N(OH)–CH(CO$_2$Et)(Ph), 91% ee (84)

Asymmetric intramolecular hydrosilylation of α-dimethylsiloxyketones **(216)**, which are prepared from α-hydroxyketones **215**, catalyzed by [(S,S)-R-DuPHOS)Rh(COD)]$^+$CF$_3$SO$_3^-$, **(219)** proceeds smoothly at 20–25 °C to give siladioxolanes **217**. Desilylation of **217** affords 1,2-diols **218** with 65–93% ee in good yields (Scheme 22)[231]. The best result (93% ee) is obtained for the reaction of α-hydroxyacetone using (S, S)-i-Pr-DuPHOS-Rh$^+$ as the catalyst. The same reactions using (S,S)-Chiraphos and (S)-binap give **218** (R′ = Me) with 46 and 20% ee, respectively.

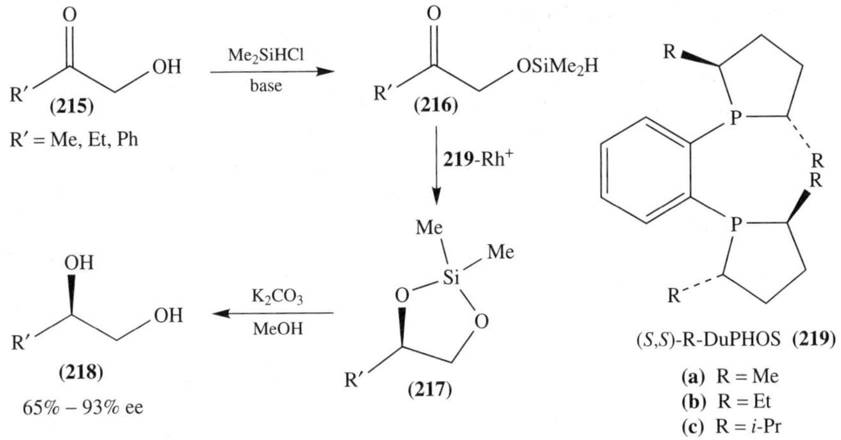

SCHEME 22

Asymmetric synthesis of chiral tertiary hydrosilanes can be realized by means of catalytic asymmetric hydrosilylations of ketones with dihydrosilanes $H_2SiR^1R^2$ ($R^1 \neq R^2$). The asymmetric hydrosilylation of a ketone with such a dihydrosilane gives a silyl ether possessing a chiral silicon center, which is readily converted to the corresponding chiral tertiary hydrosilane by reaction with a Grignard reagent. The reactions of $H_2SiPh(Np-\alpha)$ with 3-pentanone, propiophenone and (−)-menthone using (+)-DIOP–Rh complex as the catalyst, followed by reaction with MeMgBr or EtMgBr, gave HSi*RPh(Np-α) with 46, 36 and 82% ee, respectively[3,232–236]. The same reactions (R = Me) using (R)-Cybinap as the chiral ligand improved the enantiomeric purity of the resulting (R)-HSiMePh(Np-α) (**220**) with up to >99% ee (equation 85)[237]. Chiral ligands, binap, tolbinap (*vide supra*) and *p*-MeObinap are also effective, but the best results are achieved with Cybinap. It should be noted that the absolute configuration of the product obtained with (R)-Cybinap is opposite to those with (R)-binap, tolbinap and *p*-MeObinap[237].

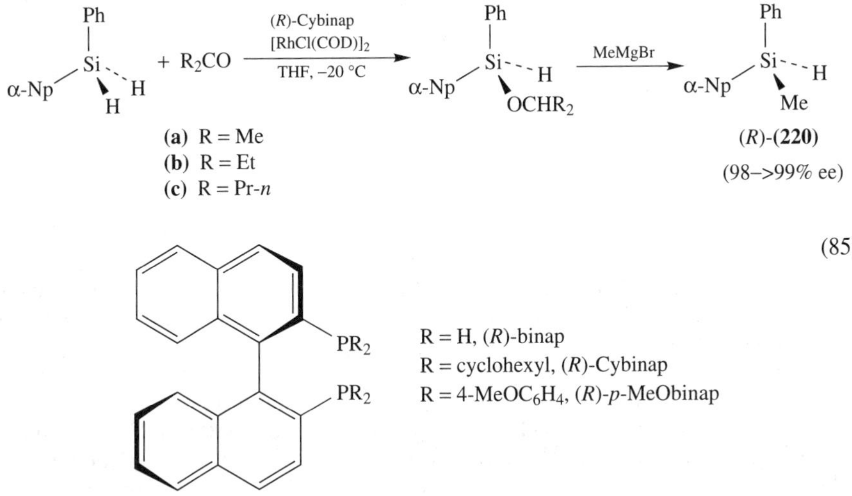

(85)

B. Of Alkenes and 1,3-Dienes

Asymmetric hydrosilylation of prochiral alkenes has been one of the most challenging reactions to achieve high enantioselectivity[3]. A significant breakthrough for this process, however, was brought about by Uozumi and Hayashi in 1991 using a unique chiral monophosphine ligand MOP **(221)**[238–240]. The hydrosilylation of 1-alkenes with HSiCl$_3$ catalyzed by (S)-MOP/[PdCl(η^3-C$_3$H$_5$)]$_2$ gives the corresponding 2-trichlorosilylalkanes **(222)** with 91–97% ee [all R-products except for the case of Ph(CH$_2$)$_2$CH(SiCl$_3$)CH$_3$ (S)] as the predominant product in excellent yields (equation 86) accompanied by a small amount of 1-trichlorosilylalkanes **(222′)**. Since **222** is readily converted to the corresponding secondary alcohols via Tamao oxidation, this process provides an efficient method for the asymmetric synthesis of 2-alkanols **(223)** with high enantiopurity from 1-alkenes (equation 86). The regioselectivity of this reaction using MOP as the chiral ligand is quite unexpected since it is well known that the hydrosilylation of 1-alkenes catalyzed by Pt, Rh and Ni complexes proceeds in an *anti*-Markovnikov manner, yielding 1-silylalkanes selectively (see Section II). The regioselectivity is good to high (**222/222′** = 80/20–94/6) for linear 1-alkenes, but a bulky substituent such as cyclohexyl substantially decreases the selectivity, i.e. **222/222′** = 66/34 for c-C$_6$H$_{11}$CH=CH$_2$, without affecting enantioselectivity (96% ee).

X = MeO (MeO–MOP); OPr-i (i-PrO-MOP); OCH$_2$Ph (BnO–MOP); Et (Et–MOP)

(S)-MOP **(221)**

$$\text{CH}_2=\text{CHR} + \text{HSiCl}_3 \xrightarrow[\text{neat, 40 °C}]{\text{MOP-Pd}} \underset{\textbf{(222)}}{\text{Cl}_3\text{Si-}\overset{*}{\text{CH}}(\text{R})\text{CH}_3} + \underset{\textbf{(222′)}}{\text{Cl}_3\text{Si}\frown\frown\text{R}} \xrightarrow[\text{ii. H}_2\text{O}_2\text{, KF}]{\text{i. EtOH/NEt}_3} \underset{\textbf{(223)}}{\text{HO-}\overset{*}{\text{CH}}(\text{R})\text{CH}_3} \quad (86)$$

R = n-C$_4$H$_9$, n-C$_6$H$_{13}$, n-C$_{10}$H$_{21}$, CH$_2$CH$_2$Ph, c-C$_6$H$_{11}$

Norbornene (**224**, n = 1, X = H), 5,6-bis(*endo*-carbomethoxy)norbornene (**224**, n = 1, X = CO$_2$Me), bicyclo[2.2.2]oct-2-ene (**224**, n = 2, X = H) and norbornadiene (**227**) are excellent substrates for the MOP–Pd-catalyzed asymmetric hydrosilylation, giving *exo*-products exclusively, which are readily transformed to the corresponding *exo*-alcohols (**226, 228b**) with 92–96% ee in excellent yields (equations 87 and 88)[241]. *exo*-2-Trichlorosilylnorbornane (**225**, n = 1, X = H) can also be transformed to *endo*-2-bromonorbornane in 81% yield by reacting with KF and NBS[241]. The reaction of norbornadiene (**227**) with 2.5 equivalents of HSiCl$_3$ gives double hydrosilylation product **229a** with >99% ee and excellent regioselectivity (e.g. 2,5-/2,6- product = 18/1), which is converted to the corresponding diol **229b** and diacetate **229c** (equation 88)[241].

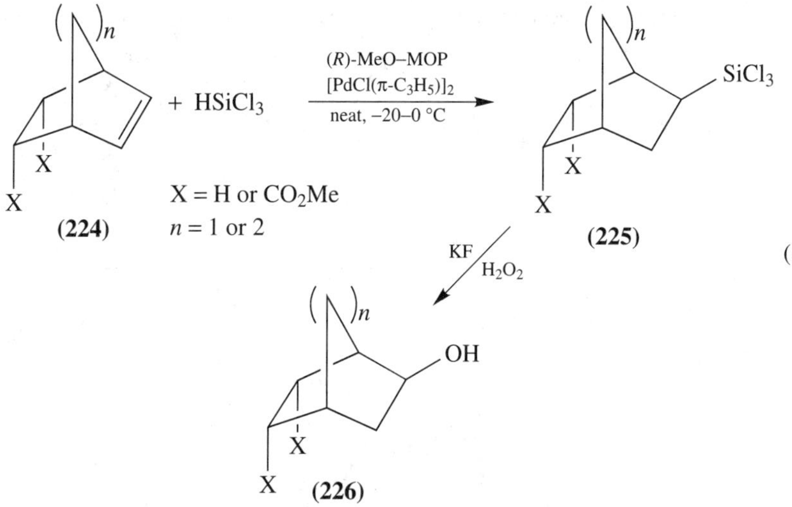

(224) **(225)** **(226)** (87)

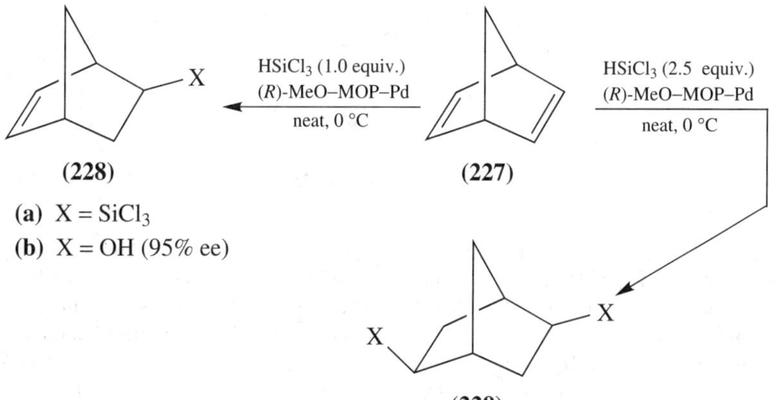

(228)

(a) X = $SiCl_3$
(b) X = OH (95% ee)

(227)

(88)

(229)

(a) X = $SiCl_3$
(b) X = OH
(c) X = OAc (>99% ee)

In a similar manner, the reactions of dihydrofurans **230** and **231** with $HSiCl_3$ catalyzed by (*R*)-MeO-MOP give 3- and 4-trichlorosilyltetrahydrofurans **232** (with *S* absolute configuration) and **233** (whose configuration was not determined) with 95% ee and 82% ee, respectively (equations 89 and 90)[242].

In a similar manner, the reactions of 5,6-disubstituted 7-oxabicyclo[2.2.1]hept-2-enes **(234)** with $HSiCl_3$ catalyzed by (*R*)-MeO–MOP–Pd give the corresponding *exo*-2-trichlorosilyl-7-oxabicyclo[2.2.1]heptanes **(235)** exclusively with excellent enantiopurity (Z = CO_2Me, 20 °C, 95% ee; Z = CH_2OCH_3, −20 °C, 90% ee) (equation 91), which are converted to *exo*-2-hydroxy-7-oxabicyclo[2.2.1]heptanes **(236)** and their esters and carbamates via Tamao oxidation and the subsequent esterification[242]. A trimethylsilyl derivative of **235** (Z = CH$_2$OCH$_3$) is further reacted with MeCOBr to afford 1-bromo-2,3-di(methoxymethyl)-4-acetoxy-5-TMS-cyclohexane stereospecifically in 81% yield[242].

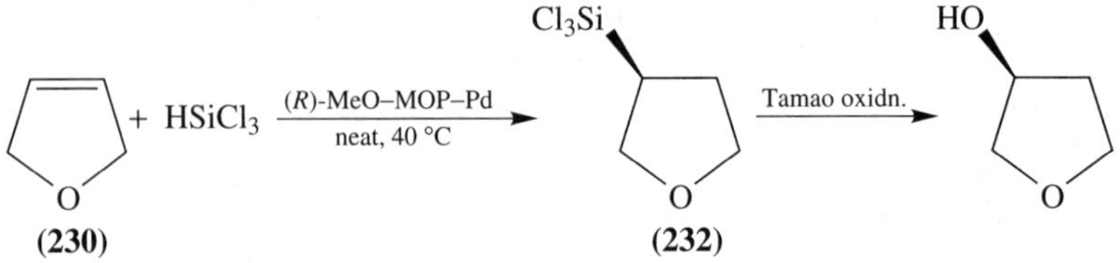

$$(230) + HSiCl_3 \xrightarrow[\text{neat, 40 °C}]{(R)\text{-MeO-MOP-Pd}} (232) \xrightarrow{\text{Tamao oxidn.}} \quad (89)$$

$$(231) + HSiCl_3 \xrightarrow[\text{neat, 40 °C}]{(R)\text{-MeO-MOP-Pd}} (233) \xrightarrow{\text{MeMgBr}} \quad (90)$$

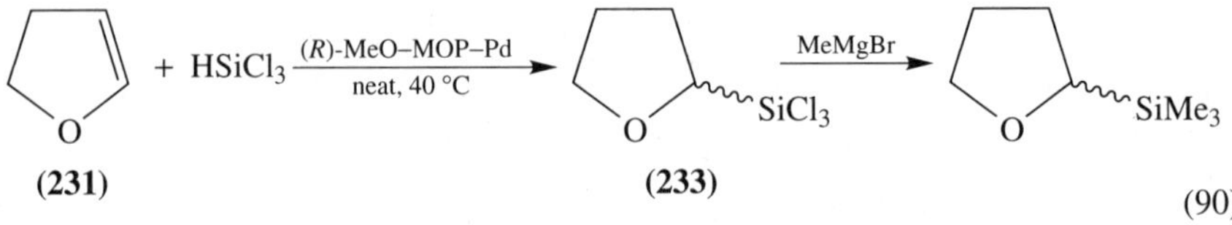

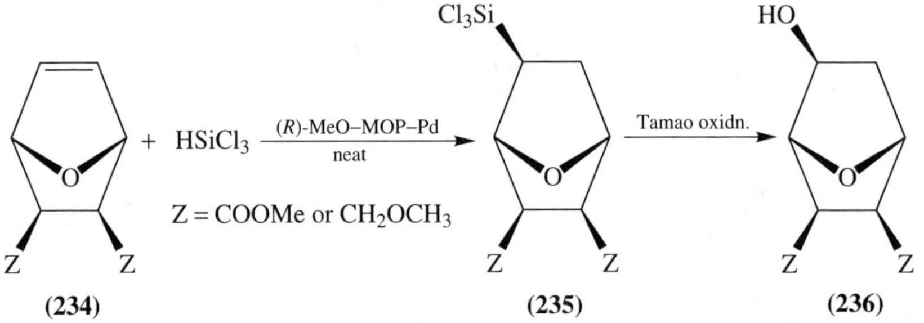

$$(234) + HSiCl_3 \xrightarrow[\text{neat}]{(R)\text{-MeO-MOP-Pd}} (235) \xrightarrow{\text{Tamao oxidn.}} (236) \quad (91)$$

Z = COOMe or CH$_2$OCH$_3$

Although the MOP ligands induce excellent enantioselectivity in the asymmetric hydrosilylation of 1-alkenes, norbornenes and dihydrofurans as described above, the reaction of styrene catalyzed by (*R*)-MeO–MOP–Pd under the same conditions, i.e. neat at 0 °C, gives Ph(Me)CHSiCl$_3$ with only 14% ee[243]. The enantioselectivity can be improved up to 71% ee by using benzene as the solvent at 5 °C[243,244], and the reactions of indene and other 1-arylalkenes catalyzed by MeO–MOP–Rh can achieve 80–85% ee[244]. Other MOP ligands (*vide supra*) or sulfonylaminoalkylphosphines cannot achieve higher enantioselectivity[243,245]. However, a simpler MOP ligand, (*S*)-H–MOP (MOP, X = H, *vide supra*), is found to achieve 89–96% ee in the reactions of styrene and styrene derivatives giving **237** and then **238** (equation 92)[243].

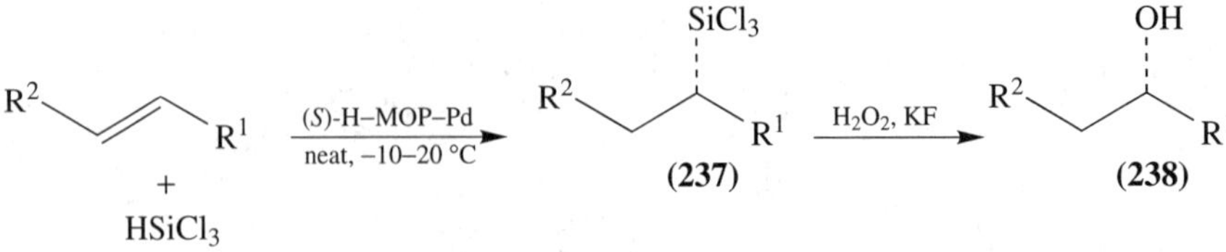

$$R^2\text{-CH=CH-}R^1 + HSiCl_3 \xrightarrow[\text{neat, }-10-20\text{ °C}]{(S)\text{-H-MOP-Pd}} R^2\text{-CH(SiCl}_3)\text{-CH}_2\text{-}R^1 \xrightarrow{H_2O_2, KF} R^2\text{-CH(OH)-CH}_2\text{-}R^1$$

(237) (238)

R^1 = H, 4-MeC$_6$H$_4$, 4-CF$_3$C$_6$H$_4$, 3-ClC$_6$H$_4$ (92)
R^2 = Me, *n*-Bu

Asymmetric hydrosilylation of styrene with $HSiCl_3$ catalyzed by a palladium complex of a chiral ferrocenylphosphine attached to cross-linked polystyrene support at 70 °C gives $PhMeC^*HSiCl_3$ in quantitative yield with only 15.2% ee[65].

Chiral lanthanide catalysts, (*R*)- and (*S*)-$Me_2Si(Me_4Cp)[(-)\text{-menthylCp}]SmCH(SiMe_3)_2$, are effective for the asymmetric hydrosilylation of 2-phenyl-1-butene with H_3SiPh, giving (*R*)- and (*S*)-2-phenyl-2-silylbutanes with 68 and 65% ee, respectively (Scheme 23)[43].

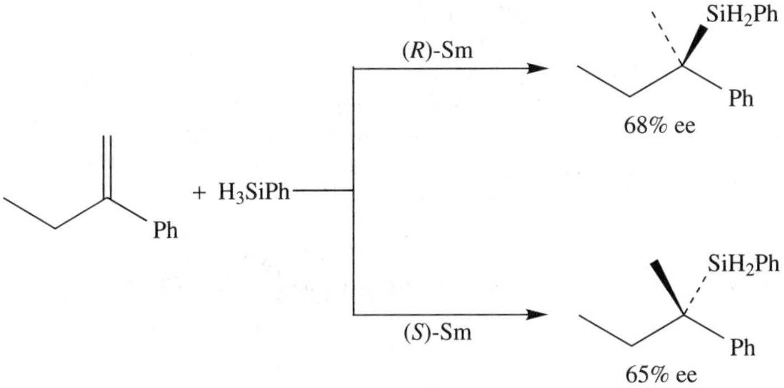

Sm = $Me_2Si(Me_4Cp)[(-)\text{-menthylCp}]SmCH(SiMe_3)_2$

SCHEME 23

Asymmetric hydrosilylation of cyclopentadiene with $HSiCl_3$ catalyzed by a Pd complex of (*S*)-(*R*)-ferrocenylphosphine (**240**, Rf = C_3F_7) proceeds in a 1,4-manner to give 3-silyl-1-cyclopentene (**239**, X = Cl) with up to 60% ee (*R*) (equation 93)[246]. The same reaction using (*S*)-2-(*N*-methanesulfonylamino)-3-methyl-1-diphenylphosphinobutane **(241)** and $HSiMeCl_2$ yields **239** (X = Me) with up to 71% ee (*S*) (equation 93)[245]. In a similar manner, the reaction of 1,3-cyclohexadiene with $HSiF_2Ph$ catalyzed by (*R*)-(*S*)-1-diphenylphosphino-2-(1-acetoxyethyl)ferrocene (PPF–OAc) affords (*S*)-3-silyl-1-cyclohexene **(242)** with 77% ee (equation 94)[247].

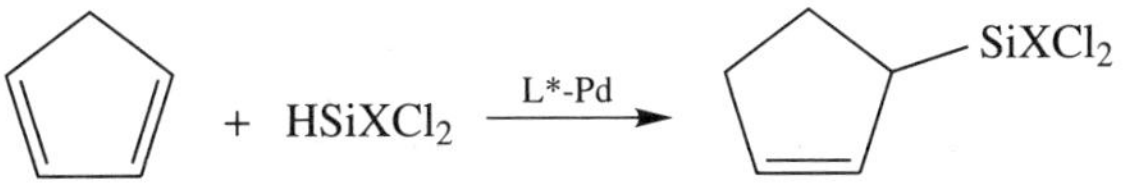

(93)

X = Cl or Me
L* = **240** or **241**

(239)

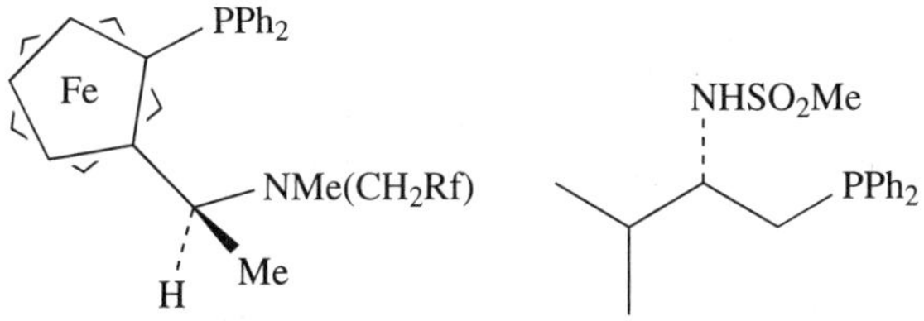

(240) **(241)**

$$\text{cyclohexadiene} + \text{HSiF}_2\text{Ph} \xrightarrow[\text{neat, room temp.}]{\substack{(R)\text{-}(S)\text{-PPFOH} \\ \text{PdCl}_2(\text{PhCN})_2}} \text{product with SiF}_2\text{Ph} \quad (94)$$

(242)

Asymmetric hydrosilylation of linear 1,3-alkadienes (**243**) also proceeds in a 1,4-manner with high regioselectivity to give the corresponding allylic silanes **244** (equation 95). The enantioselectivity of this reaction is 37–66% ee using $HSiCl_3$ and Pd-complexes of chiral ferrocenylphosphines **240** ($Rf = C_3F_7$ or C_8F_{17}) as the catalyst at 30–80 °C[246]. The reactions of 1,3-pentadiene (**243**, R = Me), 1,3-hexadiene (**243**, R = Et) and 1-phenyl-1,3-butadiene (**243**, R = Ph) with $HSiF_2Ph$ catalyzed by (R)-(S)-PPFA–Pd (PPFA = N,N-dimethyl-1-[2-(diphenylphosphino)ferrocenyl]ethylamine) at room temperature give the corresponding (Z)-allylic silanes with 69% ee[247]. The reactions of 1-phenyl-1,3-butadiene (**243**, R = Ph) with phenylhalosilanes, $HSiPh_nX_{3-n}$ (X = F or Cl), catalyzed by Pd-complexes with (R)-OH–MOP (X = OH) at 20 °C yield (S)-(Z)-1-phenyl-1-silyl-2-butene (**244**, R = Ph) with 17–66% ee, while the reactions with (R)-(TBS)–MOP (X = $OSiMe_2Bu$-t) yield (R)-(Z)-1-phenyl-1-silyl-2-butene (**244**, R = Ph) with 9–56% ee[248].

$$R\text{-diene} + HSiX_2Y \xrightarrow{L^*\text{-Pd}} \text{product} \quad (95)$$

(243) **(244)**

R = Me, Et, n-Pr, c-C_6H_{11}, Ph, o-FC_6F_4
X, Y = Ph, F, or Cl (X≠Y)

C. Intramolecular Reactions

Catalytic asymmetric intramolecular hydrosilylation of dialkyl- and diarylsilyl ethers of bis(2-propenyl)methanol (**245**) catalyzed by (R,R)-DIOP–Rh or (R)-binap–Rh complex, followed by Tamao oxidation, gives (2S,3R)-2-methyl-4-pentene-1,3-diol (**247**) with 71–93% ee and excellent *syn* selectivity (*syn/anti* = 95/5– > 99/1) (equation 96)[249]. The enantioselectivity of this reaction depends on the bulkiness of the silyl moiety, i.e. the bulkier the substituent, the higher is the enantiopurity of the product, except for the case of 2-MeC_6H_4: R = Me, 80% ee (binap–Rh); R = Ph, 83% ee (DIOP–Rh); R = 2-MeC_6H_4, 4% ee (DIOP–Rh); R = 3-MeC_6H_4, 87% ee (DIOP–Rh); R = 3,5-$Me_2C_6H_3$, 93% ee (DIOP–Rh). This methodology is successfully applied to the asymmetric synthesis of versatile polyoxygenated synthetic intermediate **249** (equation 97)[249].

Asymmetric intramolecular hydrosilylation of allylic hydrosilyl ethers (**251**) catalyzed by (S)-binap–Rh$^+$ complex gives the corresponding oxasilacyclopentanes (**252**) with moderate to excellent enantiomeric purity (equations 98)[250–252]. The bulkiness of the hydrosilyl group exerts a marked influence on enantioselectivity, i.e. silacyclohexyl (**251**) and silacyclopentyl (**253**, equation 99) give much better selectivity than Ph_2HSi–, i-Pr_2HSi– and Me_2HSi– groups[250]. (S)-Binap is superior to (S,S)-Chiraphos as the chiral ligand for this reaction[250]. An alkenylsilane **255** is also employed as the substrate, affording **256** with 60% ee using (S,S)-Chiraphos as the ligand (equation 100)[250]. Possible mechanisms for asymmetric induction and catalytic cycles are proposed[251,253]. The products of these reactions can be transformed to enantiomerically enriched diols via Tamao oxidation[250].

(245) → (246) [Rh-L*, ClCH$_2$CH$_2$Cl, 30 °C] → (247) [H$_2$O$_2$, KF, KHCO$_3$, MeOH/THF] (96)

Rh-L* = [RhCl(CH$_2$=CH$_2$)]$_2$ + (R,R)-DIOP or (R)-binap

(248) → i. (R)-binap–Rh; ii. Tamao oxid. → (249) → (250) (97)

(251) → (S)-binap–Rh$^+$, acetone, 25 °C → (252) (98)

R^1 = Ph, 3,4-(MeO)$_2$C$_6$H$_3$ or 2-Np; R^2 = R^3 = H: 94–97% ee (R)
R^1 = R^3 = H, R^2 = Ph: 96% ee (R)
R^1 = Ph; R^2 = H; R^3 = Ph or Me: 88–90% ee [(R,R) for R^3 = Me]
R^1 = CO$_2$Me; R^2 = R^3 = H: 78% ee
R^1 = R^2 = H; R^3 = Me: 41% ee
R^1 = R^2 = H; R^3 = t-Bu, PhCO or (MeO)Me$_2$C; 95–96% ee
R^1 = R^2 = H; R^3 = CO$_2$Et, CO$_2$Pr-i, CO$_2$Bu-t or SiMe$_3$: 69–87% ee (CH$_2$Cl$_2$ as solvent)

(253) → (S)-binap–Rh$^+$, acetone, 25 °C → (254) 97% ee (99)

$$\text{(255)} \xrightarrow[\text{acetone, 25 °C}]{(S,S)\text{-Chiraphos-Rh}^+} \text{(256)} \quad 60\% \text{ ee} \tag{100}$$

VI. SYNTHESES OF SILICON-CONTAINING DENDRIMERS AND POLYMERS USING HYDROSILYLATION

A. Silicon-containing Dendrimers

Catalytic hydrosilylation provides a convenient route to silicon-containing dendrimers. Combinations of hydrosilylation with allylation, ethenylation and ethynylation provide powerful protocols for the rapid and efficient syntheses of carbosilane dendrimers. The allylation of $SiCl_4$ with allylmagnesium bromide in ether gives tetraallylsilane (G0). The hydrosilylation of tetraallylsilane (G0) using $HSiCl_3$ and Speier's catalyst or Karstedt's catalyst (*vide supra*) affords the first generation dendrimer **257** (G1-Cl) in quantitative yield (Scheme 24)[254]. The reaction of **257** (G1-Cl) with excess allylmagnesium bromide in ether gives another first generation dendrimer **258** (G1) (Scheme 24). The repetition of the hydrosilylation–allylation cycle then produces second generation (36 allyl end-groups), third generation (108 allyl end-groups), fourth generation (324 allyl end-groups) and fifth generation (972 allyl end-groups) dendrimers in excellent yields[254]. These dendrimers are characterized by ^{1}H, ^{13}C, ^{29}Si NMR, gel-permeation chromatography (GPC) and elemental analyses. In a similar manner, carbosilane dendrimers bearing 64 and 96 allyl end-groups are also synthesized[255,256].

The hydrosilylations of tetraallylsilane with $HSiMe_2Cl$ and $HSiMeCl_2$ in the presence of Karstedt's catalyst (*vide supra*) gives versatile dendrimer units bearing reactive Si−Cl bonds, **259** (G1-Cl) and **260** (G1-Cl$_2$), respectively (Scheme 25)[257]. The bis-allylation of **260** (G1-Cl$_2$) with allyl Grignard reagent followed by hydrosilylation with $HSiMe_2Cl$ using Karstedt's catalyst affords (via **261**) dendrimer **262** that possesses 8 reacting Si−Cl bonds (Scheme 25). The reaction of **262** (G2-Cl) with ferrocenylethylamine in the presence of NEt_3 yields dendrimer **263** (Scheme 26)[257]. In a similar manner, the reaction of **259** (G1-Cl) with ferrocenylethylamine gives a dendrimer bearing four ferrocene units[257]. The reactions of **259** (G1-Cl) and **262** (G2-Cl) with ferrocenyllithium in THF at 0 °C afford the corresponding starburst macromolecules bearing 4 and 8 ferrocene unites, respectively[257]. Each of these compounds shows a single reversible oxidation wave in cyclic voltammetry. Differential pulse voltammetry measurements display only one wave, indicating that oxidation of all ferrocene units takes place at the same potential, i.e. the ferrocene units are essentially noninteracting redox centers. The dendrimers with 8 ferrocene units undergo oxidative precipitation, producing a coated film on the Pt electrode surface when oxidized to the polycation in THF[257].

The hydrosilylation–allylation protocol is applicable to the syntheses of dendritic polyols[258,259]. The hydroboration–oxidation of G0–G3 dendrimers prepared in accordance with the synthetic routes shown in Scheme 24 give the corresponding dendritic polyols, G0-OH–G3-OH, which are characterized by ^{1}H NMR and MALDI-TOF analyses[258]. Further modifications of the dendritic polyols with cholesteryl chloroformate yield unique dendrimers possessing 12, 36 and 108 cholesteryl units, providing the first examples of flexible dendrimers substituted with rigid mesogenic units that are expected to have unusual liquid crystalline properties[259]. The second generation polyol dendrimer **264**

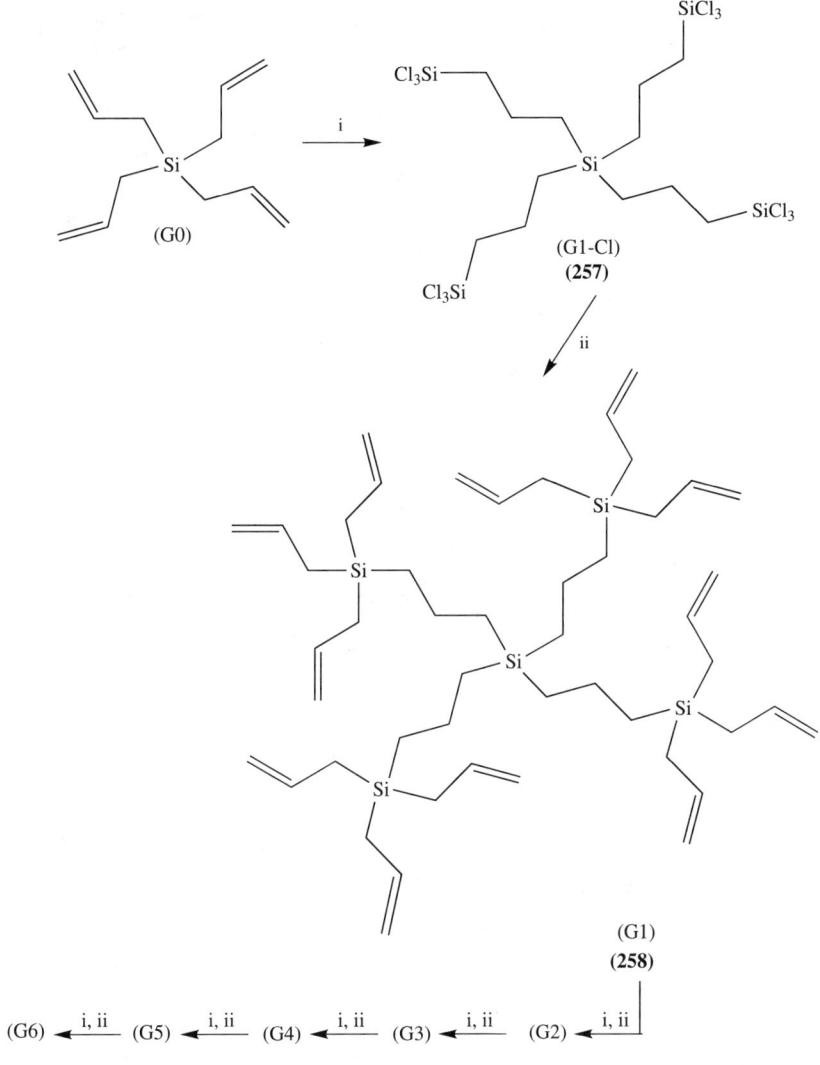

i. HSiCl$_3$, [Pt]; ii. CH$_2$=CHCH$_2$MgBr, Et$_2$O

SCHEME 24

(G2-OH) and cholesteryl dendrimer **265** (G2-DLCP) (DLCP = dendritic liquid crystalline polymer) are shown in Scheme 27 as examples.

In a similar manner, the hydrosilylation-ethenylation cycle starting from tetraethenyl-silane (G0) in THF yields up to fourth generation dendrimers[254,260,261]. This protocol is combined with reduction of the SiCl$_3$ moieties by LiAlH$_4$ to afford starburst dendrimers bearing up to 324 Si−H bonds at its periphery[262]. The structure of the second generation dendrimer **267** (G2-H) is characterized by X-ray crystallography[262]. The synthesis of **267** is shown in Scheme 28 as an example.

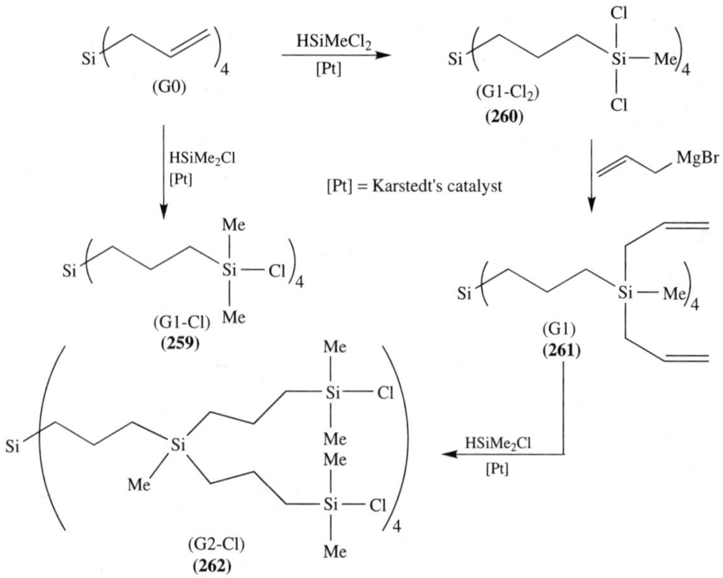

SCHEME 25

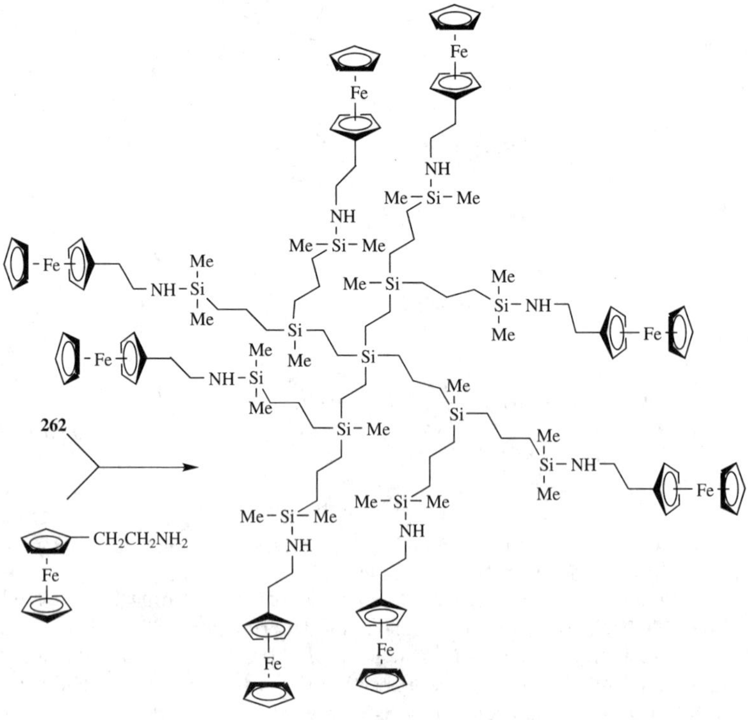

SCHEME 26

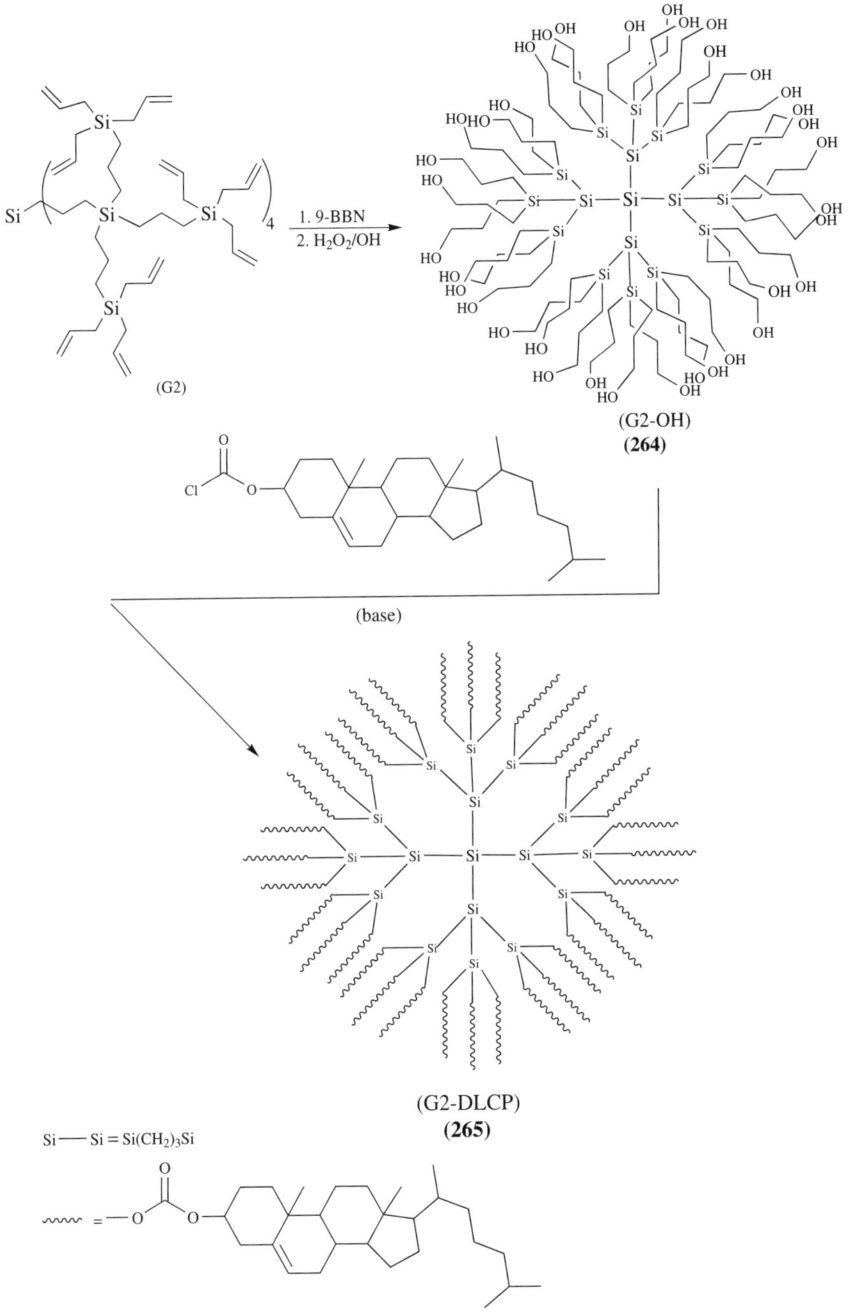

SCHEME 27

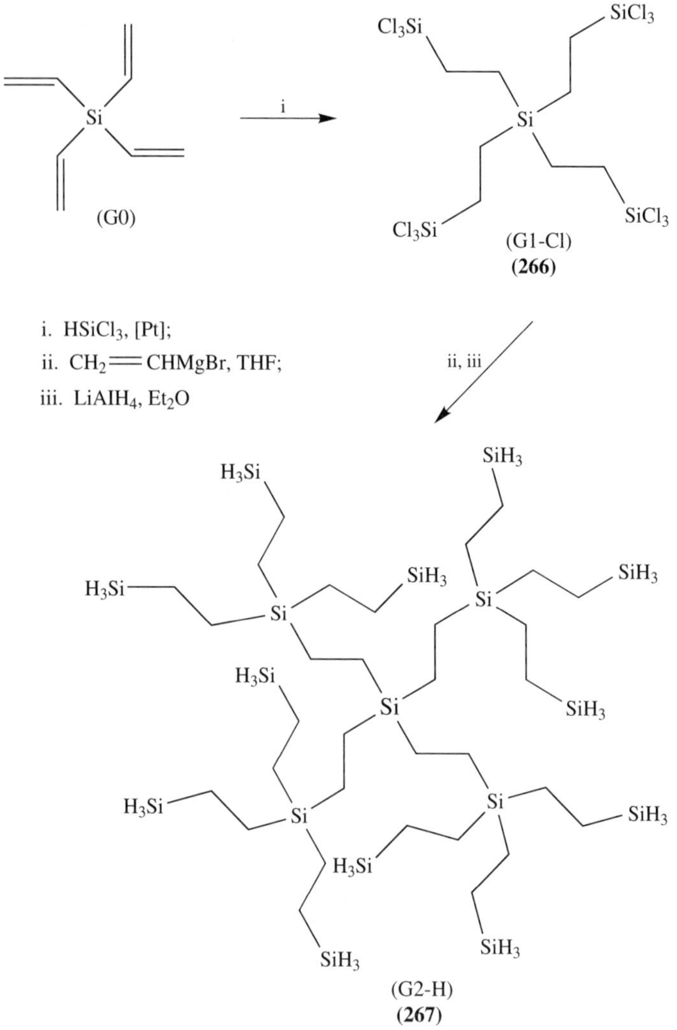

i. HSiCl₃, [Pt];
ii. CH₂=CHMgBr, THF;
iii. LiAlH₄, Et₂O

SCHEME 28

A combination of hydrosilylation–ethynylation and hydrosilylation–ethynylation is successfully applied to the syntheses of carbosilane dendrimers bearing 4, 8 and 12 end-ethynyl groups as well as their $Co_2(CO)_6$ complexes[263]. The synthesis of a second generation dendrimer **268** possessing twelve $Co_2(CO)_6(HC\equiv C-)$ units is shown in Scheme 29 as an example. The structure of **268** is confirmed by X-ray crystallographic analysis[263].

Hyper-branched starburst poly(siloxysilane) polymers are synthesized through hydrosilylation–polymerization of $CH_2=CHSi(OSiMe_2H)_3$ (**269**) and $HSi(OSiMe_2CH=CH_2)_3$ (**270**). The polymer from **269** has Si–H moieties on the surface of the dendrimer, while that arising from **270** possesses vinyl groups on the outer sphere[264].

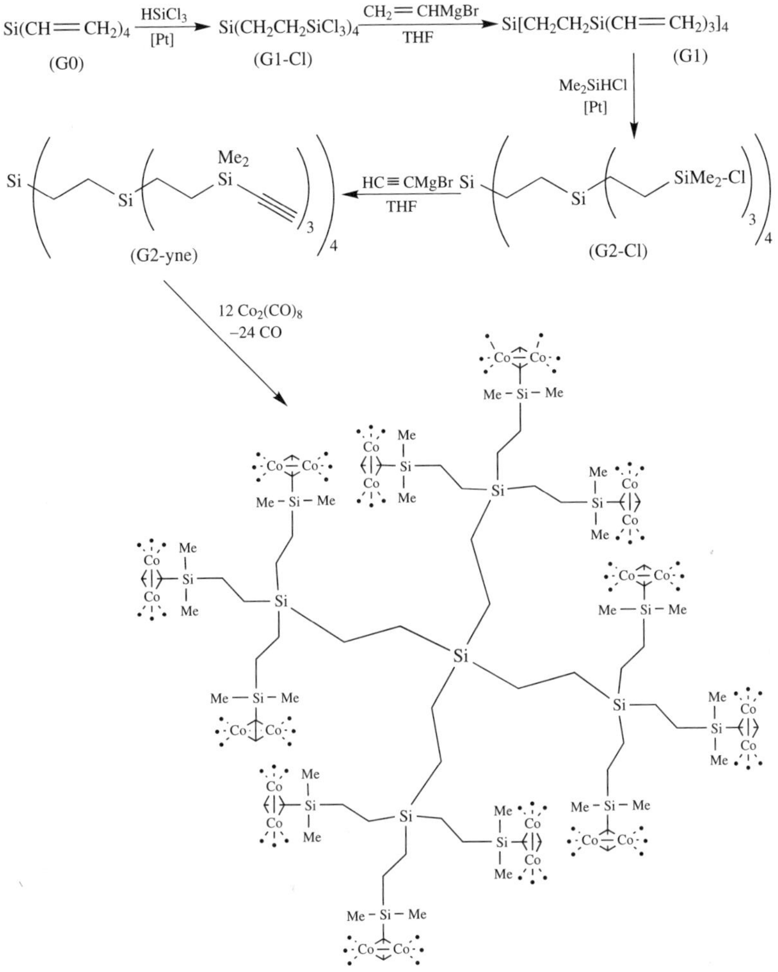

SCHEME 29

Siloxane-based dendritic hindered amine light stabilizers (HALS) such as **271** are synthesized through hydrosilylation of methacrylates containing hindered amines with polyhydrosiloxanes (equation 101)[265].

B. Silicon-containing Polymers

Hydrosilylation of 4-vinylcyclohex-1-ene-1,2-oxide, glycidyl ether, 1,5-hexadiene-5,6-oxide, etc. with $HSiMe_2O(SiMe_2O)nSiMe_2H$ ($n = 0-2$), 1,4-$HSiMe_2-C_6H_4-SiMe_2H$, 1,4-$HSiMe_2-C_6H_4-O-C_6H_4-SiMe_2H$, etc. catalyzed by $RhCl(PPh_3)_3$ gives mono-hydrosilylation products in excellent selectivity[266–268]. The second hydrosilylation of the mono-hydrosilylation products with epoxyalkenes different from those used in the initial reaction provides ambifunctional silicon-containing epoxy monomers[268]. When the second hydrosilylation employs $CH_2=CHSi(OMe)_3$, the reaction gives monomers bearing epoxy and trialkoxysilane moieties[269]. The photopolymerization of these monomers has been studied[269]. In a similar manner, polysiloxanes bearing 1-propenyl ether groups are synthesized through highly chemoselective Pt-hydrosilylation of 1-allyloxy-4-(1-propenyloxy)butane with various hydrosiloxanes, and converted to thin films by photopolymerization[270].

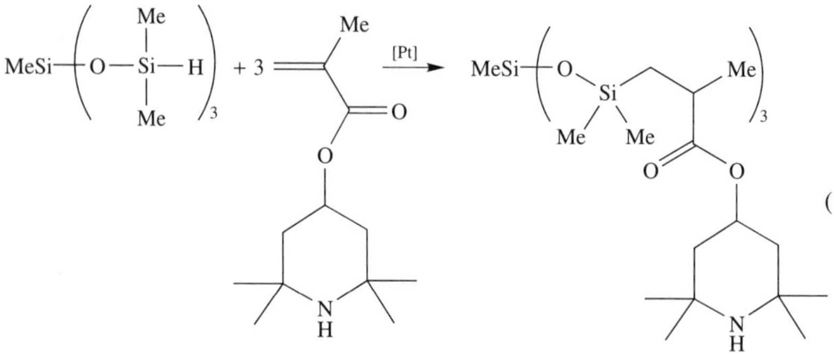

(101)

(271)

The simultaneous hydrosilylation and ring-opening polymerization process provides a powerful method for the construction of a variety of novel polymer networks including comb, block and graft polymers, e.g. **272** (equation 102)[271].

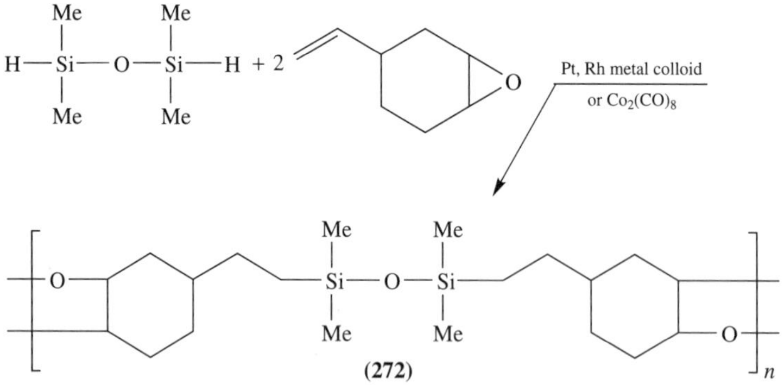

(102)

(272)

Hydrosilylation–polymerization provides a new approach to the syntheses of macromonomers such as poly(arylene-silylethylene) and poly(aralkylene-silylethylene) (**274**) that possess reactive ethenyl and SiH termini for further manipulations through self-polyaddition of **273** promoted by Speier's catalyst (equation 103)[272].

The hydrosilylation–polymerization of 1,3-diethenyltetramethyldisiloxane (**276**) with 1,3-dihydrotetramethyldisiloxane (**275**) proceeds smoothly in the presence of Karstedt's catalyst to give high molecular weight poly[(1,1,3,3-tetramethyldisiloxanyl)ethylene] (**277**) (equation 104)[273,274].

In a similar manner, the hydrosilylation–polymerization of 'diallyl bisphenol A' (**278**) with **275** in the presence of a platinum catalyst affords polymer **279** that possesses a relatively high glass transition temperature for siloxane polymers ($T_g = 28-34\,°C$) and good thermal stability (equation 105)[275].

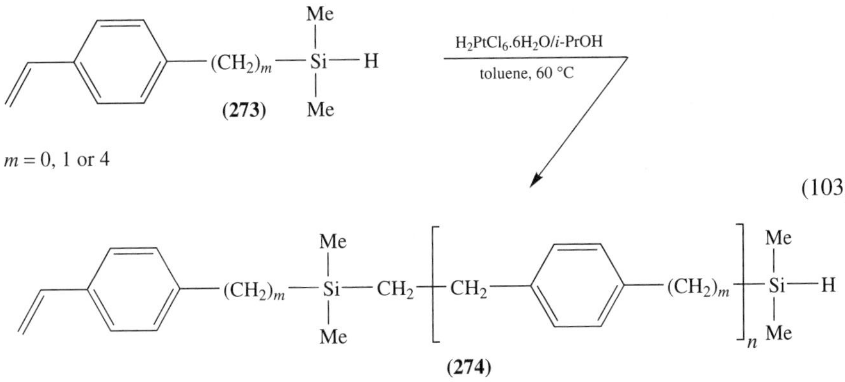

$m = 0, 1$ or 4

(273) → (274) (103)

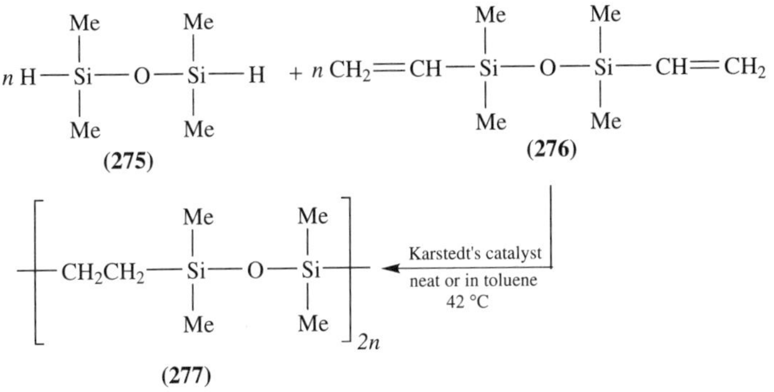

(275) + (276) → (277) (104)

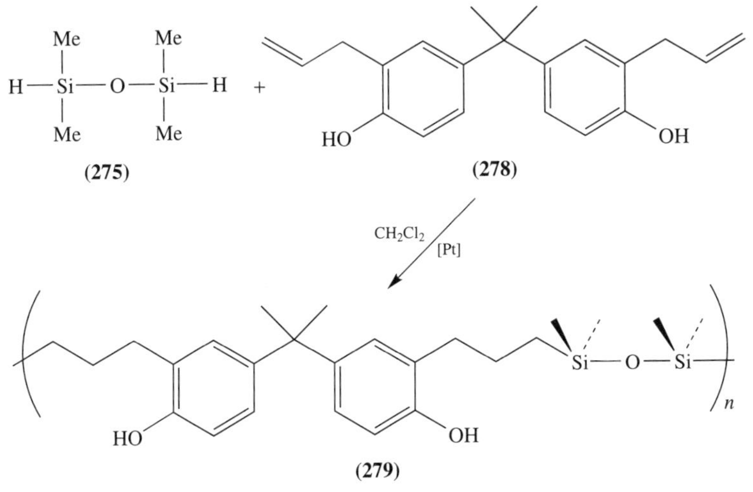

Karstedt's catalyst
neat or in toluene
42 °C

(275) + (278) → (279) (105)

CH$_2$Cl$_2$ / [Pt]

Novel thermoset nonlinear optical (NLO) polymers **283** and **284** are synthesized through the hydrosilylation of NLO chromophores such as **281** and **282** and dicyclopentadiene with tetramethylcyclotetrasiloxane **280** (Scheme 30)[276,277]. Typical NLO chromophores **281a** and **282a** are shown below. The resultant NLO materials exhibit excellent nonlinear optical and thermal properties.

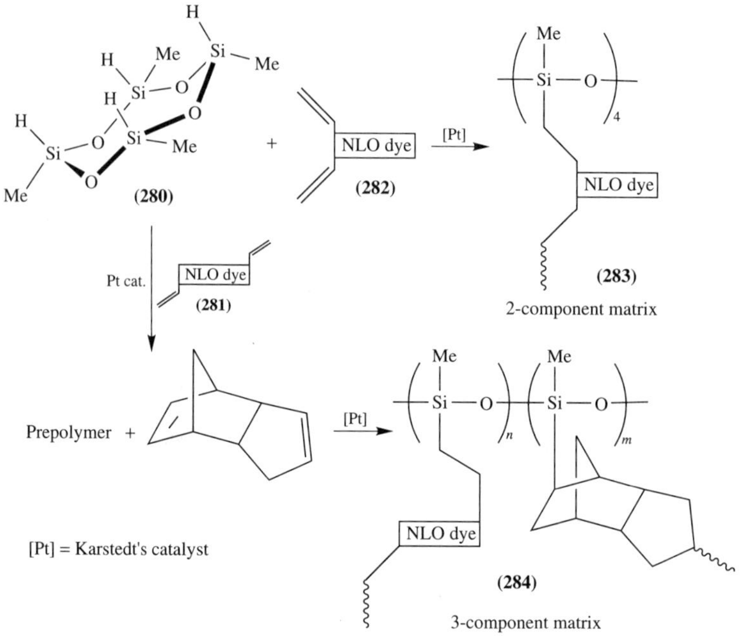

SCHEME 30

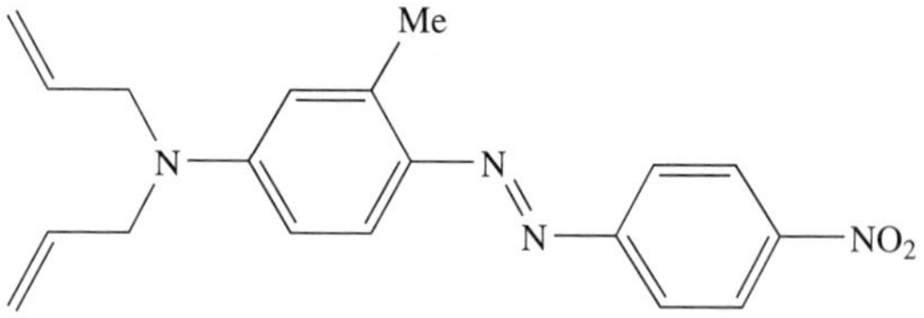

(281a)

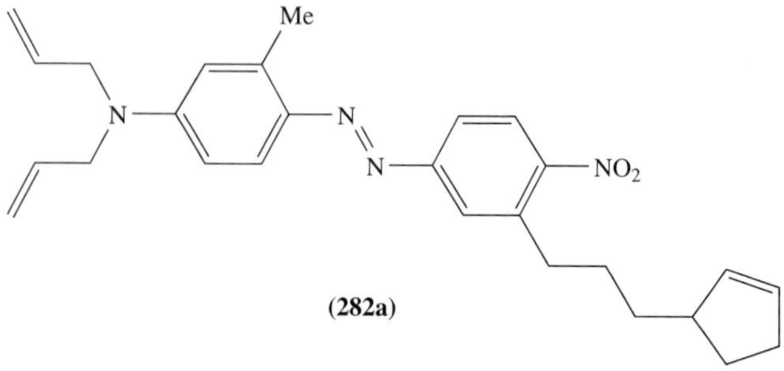

(282a)

Ethynylsilanes **285** undergo self-polymerization via hydrosilylation catalyzed by H_2PtCl_6 in THF to give high molecular weight polymers **286** (equation 106)[278]. The reaction of diethylethynylsilane (**285c**) requires 160 °C to proceed, but yields the highest molecular weight polymer **286c**.

$$H-\underset{\underset{R^2}{|}}{\overset{\overset{R^1}{|}}{Si}}-C\equiv CH \xrightarrow[\text{neat or in THF}]{H_2PtCl_6} \left[\underset{\underset{R^2}{|}}{\overset{\overset{R^1}{|}}{Si}}\diagup\diagdown\right]_n \qquad (106)$$

(**285**) (**286**)

(a) $R^1 = R^2 = Me$; 50 °C; $M_w = 85{,}000$ (neat); 30,380 (THF)

(b) $R^1 = Me$, $R^2 = Ph$; 65 °C; $M_w = 30{,}000$ (neat), 11,400 (THF)

(c) $R^1 = R^2 = Et$; 160 °C; $M_w = 110{,}700$ (neat)

A new functionalized polysiloxane bearing crown ether moieties in the main chain as well as its platinum complex are prepared from 3,19-dihydroxy-1-thia-5,8,11,14,17-pentaoxacycloicosane through ether formation with ω-chloroundecene, followed by hydrosilylation with triethoxysilane, copolymerization with octamethylcyclotetrasiloxane and reaction with potassium chloroplatinate[279].

Polysiloxane-fluoro copolymers with low dielectric constants are prepared through hydrosilylation of 1,3,5-$[CH_2=CHCH_2OC(CF_3)_2]_3C_6H_3$ with polymethylhydrosiloxane catalyzed by Cp_2PtCl_2[280].

Hydrosilylation was also applied to the preparation of thiophene-terminated dimethylsiloxane macromonomer, which is used for the synthesis of thiophene–dimethylsiloxane graft copolymers by oxidative polymerization[281]. Polycarbosilanes have been prepared by hydrosilylation–polymerization of 1,4-bis(ethenylmethylphenylsilyl)benzene with 1,4-bis(methylphenylsilyl)benzene in various ratios[282].

A combination of anionic ring-opening of 1-methyl- or 1-phenyl-1-silacyclobutane (**287**) and the Pt-catalyzed graft hydrosilylation of the resulting carbosilane polymer **288** with functionalized alkenes gives poly(1-silacyclobutane) graft polymers (**289**) possessing functionalized alkyl pendant groups on silicon (equation 107)[283]. The same type of functionalized polycarbosilanes **290** can be synthesized in one step through the simultaneous Pt-catalyzed ring-opening polymerization and hydrosilylation of 1-phenyl-1-silacyclobutane (**287**, R = Ph) with functionalized alkenes (equation 108)[284].

$$\underset{\underset{H}{|}}{\overset{}{\square-Si-R}} \xrightarrow{n\text{-BuLi}} \left(\underset{\underset{H}{|}}{\overset{\overset{R}{|}}{Si}}\diagup\diagdown\right)_n \xrightarrow[\text{[Pt]}]{\diagup\diagdown CN} \left(\underset{\underset{\diagdown CN}{|}}{\overset{\overset{R}{|}}{Si}}\diagup\diagdown\right)_n$$

(**287**) (**288**) (**289**)

R = Me, Ph

(107)

Exclusive polymerization of phenylacetylene takes place, unexpectedly, under hydrosilylation conditions in the presence of a zwitterionic Rh(I) complex, $[Rh(COD)]^+BPh_4^-$, and $HSiEt_3$ to give all-*cis*-poly(phenylacetylene)[285].

A platinum(II) complex-catalyzed hydrosilylation provides a convenient route to polysiloxane macrocycles **291** (equation 109)[286].

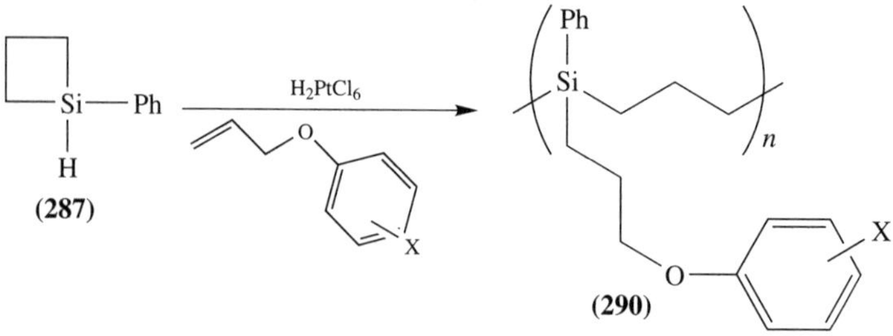

(108)

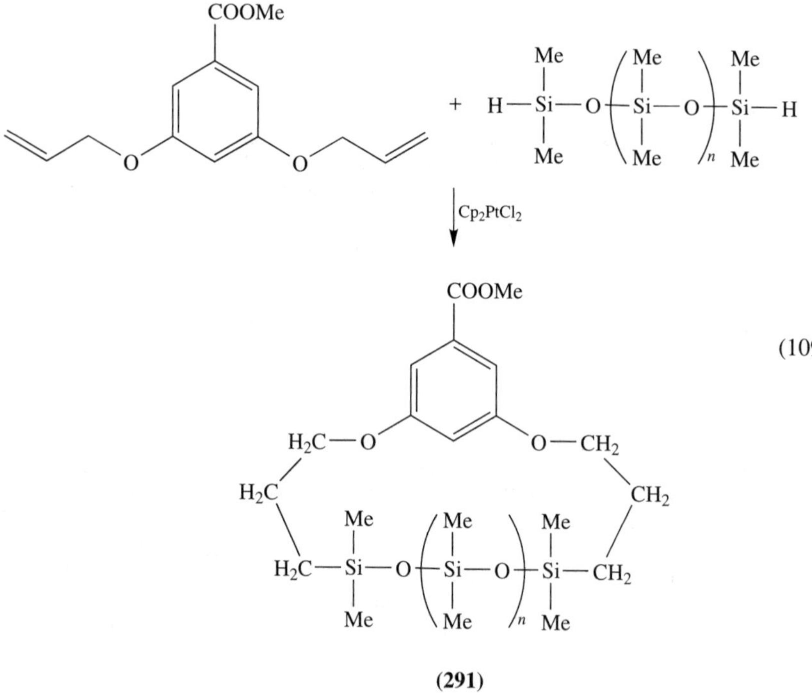

(109)

C. Modification of Polymers by Hydrosilylation

α, ω-Bifunctional fluorine-containing polysiloxanes **293** are prepared through hydrosilylation of 3,3,4,4,5,5,5-heptafluoro-1-pentene with functionalized polysiloxane **292** bearing Si—H moieties catalyzed by H_2PtCl_6 (equation 110)[287]. This modification increased the glass transition temperature by ca 40 °C as compared to the parent polysiloxanes.

Perfluoroalkyl-containing hydrosilanes, $Rf(CH_2)_{n+2}SiMe_2H$, for the syntheses of fluorosilicone-containing polybutadienes, are prepared through the Pt-catalyzed hydrosilylation of $Rf(CH_2)_nCH=CH_2$ with $HSiClMe_2$[288]. In a similar manner, (fluoroalkyl)chlorosilanes, $Rf(CH_2)_nSiR^1R^2Cl$ and $R^1R^2ClSi(CH_2)_2(CF_2)_3(CH_2)_2SiR^1R^2Cl$, and tetra(fluoroalkyl)silanes, $[Rf(CH_2)_3]_3Si(CH_2)_nRf$, are prepared via hydrosilylation of

Rf-containing alkene units[289].

$$\text{X-(CH}_2)_3-\underset{\underset{\text{Me}}{|}}{\overset{\overset{\text{Me}}{|}}{\text{Si}}}-\text{O}\left[\underset{\underset{\text{Me}}{|}}{\overset{\overset{\text{Me}}{|}}{\text{Si}}}-\text{O}\right]_m\left[\underset{\underset{\text{H}}{|}}{\overset{\overset{\text{Me}}{|}}{\text{Si}}}-\text{O}\right]_n\underset{\underset{\text{Me}}{|}}{\overset{\overset{\text{Me}}{|}}{\text{Si}}}-\text{(CH}_2)_3\text{-X}$$

(292)

$\xrightarrow[\text{H}_2\text{PtCl}_6]{\text{CH}_2=\text{CHC}_3\text{F}_7}$ (110)

$$\text{X-(CH}_2)_3-\underset{\underset{\text{Me}}{|}}{\overset{\overset{\text{Me}}{|}}{\text{Si}}}-\text{O}\left[\underset{\underset{\text{Me}}{|}}{\overset{\overset{\text{Me}}{|}}{\text{Si}}}-\text{O}\right]_m\left[\underset{\underset{\text{CH}_2\text{CH}_2\text{C}_3\text{F}_7}{|}}{\overset{\overset{\text{Me}}{|}}{\text{Si}}}-\text{O}\right]_n\underset{\underset{\text{Me}}{|}}{\overset{\overset{\text{Me}}{|}}{\text{Si}}}-\text{(CH}_2)_3\text{-X}$$

(293)

X = CF$_3$ CONH, H$_2$N, HOCH$_2$, HO$_2$C

Chemical modifications of copoly[methylsilylene(alkylene)/1,4-phenylene] bearing Si—H moieties as well as poly(methylhydrosiloxane) are readily performed through Pt-catalyzed graft hydrosilylation with functionalized allyl ethers, 3-cyano-1-propene, and allyl carbamates in a manner similar to the process shown in equation 107 (*vide supra*)[290,291].

Free radical hydrosilylation of poly(phenylsilanes) **(294)** with alkenes, aldehydes or ketones promoted by 2,2′-azo(bisisobutyronitrile) (AIBN) provides functional polysilanes **(295)** possessing a variety of properties (equation 111)[292].

$$(294) \xrightarrow[\text{AIBN, THF}]{X=\hspace{-2pt}\diagdown\hspace{-2pt}^R} (295) \quad (111)$$

X = CH$_2$, O

RCH$_2$X— = C$_5$H$_{11}$O$_9$, C$_6$H$_{13}$, cyclo-C$_6$H$_{10}$O, MeOCO(CH$_2)_3$, HO(CH$_2)_5$, HOCO(CH$_2)_4$, Me$_2$N(CH$_2)_5$

Platinum-catalyzed hydrosilylation is applicable for the crosslinking of copoly(methylsilylene/1,4-phenylene/methylvinylsilylene), yielding a thermoset material[293].

Regioselective hydrosilylation of poly[(dimethylsilylene)but-1,3-diyne] **(296a)** with HSiEt$_3$ using Rh$_6$(CO)$_{16}$ as the catalyst at 40 °C gives poly[2-(triethylsilyl)(dimethylsilylene)but-1-en-3-yne-1,4-diyl][294]. In a similar manner, the reactions of poly[(silylene) but-1,3-diynes] **(296)** with 1,4-bis(methylphenylsilyl)benzene **(297)** catalyzed by Rh_6-$(CO)_{16}$ in benzene or THF yields branched silicon-containing polymers **298** with high molecular weights up to 424,000 in 49–79% yields (Scheme 31)[294,295].

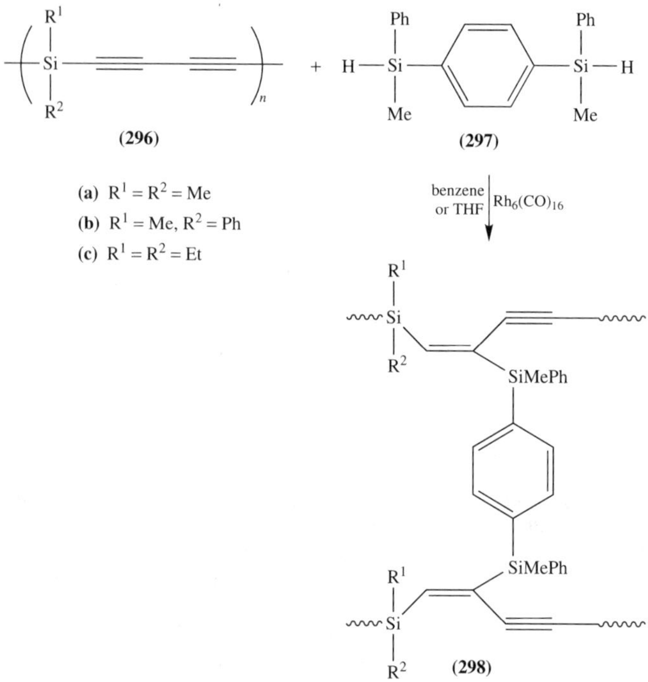

(a) $R^1 = R^2 = Me$
(b) $R^1 = Me, R^2 = Ph$
(c) $R^1 = R^2 = Et$

SCHEME 31

$R' = Me$ or $(EtO)_3Si(CH_2)_3$

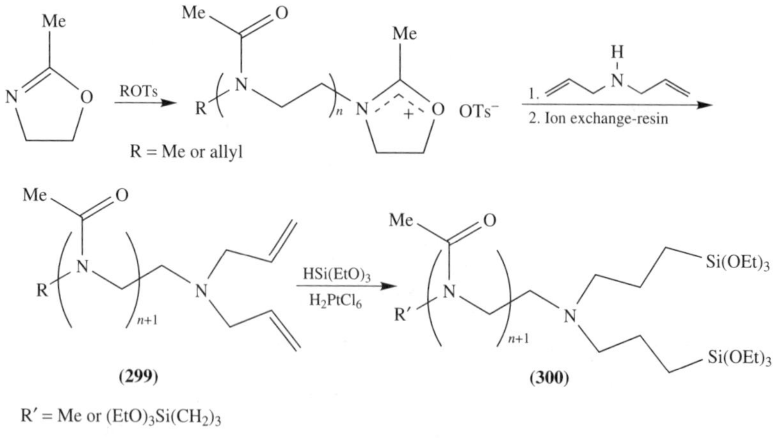

SCHEME 32

Triethoxysilyl-terminated polyoxazolines (**300**) are prepared by the hydrosilylation of allyl-terminated polyoxazolines (**299**), obtained via ring-opening polymerization of 2-methyloxazoline, catalyzed by H_2PtCl_6 (Scheme 32)[296]. Polyoxazolines **300** and related polymers are useful for the syntheses of organic–inorganic polymer hybrid using the 'sol-gel' method.

Hydrosilylations of buckminsterfullerene (C_{60}) with a poly(hydrosiloxane) (**301**) or $H(SiMe_2O)_3SiMe_2H$ in the presence of Karstedt's catalyst (*vide supra*) proceed smoothly at 29 °C to give the corresponding (1 : 1) C_{60}-polysiloxanes (**302**) with molecular weights of 5000 and 4000, respectively (equation 112)[297].

$$Me_3SiO \left(\begin{array}{c} H \\ | \\ Si-O \\ | \\ Me \end{array} \right)_n \left(\begin{array}{c} C_8H_{17}\text{-}n \\ | \\ Si-O \\ | \\ Me \end{array} \right)_{3n} SiMe_3$$

(**301**)

or $H \left(\begin{array}{c} Me \\ | \\ Si-O \\ | \\ Me \end{array} \right)_3 SiMe_2\text{-}H$

$+ \; C_{60} \; \xrightarrow[29\,°C, 1h]{[Pt]} \; C_{60}\text{-polysiloxane}$ (112)

(**302**)

[Pt] = Karstedt's catalyst

Hydrosilylation of polybutadienes bearing olefinic bonds in the backbone or as pendant groups with $HSiMe_xCl_{3-x}$ ($x = 0-2$) in the presence of H_2PtCl_6 provides an efficient method for the syntheses of functional polymers[298]. In a similar manner, high molecular weight ($M_W = 10,000-543,000$) styrene–butadiene block copolymers can be modified by hydrosilylation with pentamethyldisiloxane, heptamethyltrisiloxane and bis(trimethylsiloxy)methylsilane using Karstedt's catalyst[299]. Silicone acrylates are also synthesized through hydrosilylation of polyacryloyloxy functional monomers with copolymers of dimethyl and hydromethylsiloxanes[300]. Platinum-catalyzed hydrosilylation is also applied to the modification of vinyl-functionalized silica with $HSiMe_2$-terminated silicones, eventually leading to the formation of colloidal polymer-grafted silica[301].

VII. SILYLCARBONYLATION AND SILYLCARBOCYCLIZATION REACTIONS

A. Silylcarbonylation Reactions

The silylcarbonylation of alkenes catalyzed by $Co_2(CO)_8$ giving homologous silyl enol ethers was reported by Murai and coworkers in 1977 as a silicon version of hydroformylation of alkenes, and developed as a unique synthetic method[302–304]. Although other catalysts such as $RhCl(PPh_3)_3$, $Ru_3(CO)_{12}$ and $[HRu_3(CO)_{11}]^-$[305] can also effect the reaction, $Co_2(CO)_8$ and its PPh_3 or $P(Bu\text{-}n)_3$ complex are the best catalysts[306]. The cobalt-catalyzed silylcarbonylation of alkenes exhibits strikingly similar features to those of hydroformylations in terms of the effects of CO pressure, phosphine ligands, reactant ratios and reaction temperature on the reaction rate and product distribution[307]. The $Co_2(CO)_8$-catalyzed reactions of oxygen-containing substrates with $HSiR_3$–CO combination have also been found to provide unique and useful organic transformations, e.g. (i) silylformylations of cyclic ethers and aldehydes, (ii) 1,2-siloxyvinylation of THF, (iii) siloxymethylations of aldehydes, esters, cyclobutanones, THF, cyclic ethers[308], glycosyl acetates[309], benzylic esters, cyclic orthoesters[310], acetals and aromatic aldehydes.

The earlier development of these reactions were covered by a couple of reviews[3,306,311]. Recent developments include Rh, Ir and Ru complex catalysts in place of $Co_2(CO)_8$.

Iridium carbonyl complexes, $[Ir(CO)_3Cl]_n$ and $Ir_4(CO)_{12}$, are found to be effective catalysts for the regioselective silylcarbonylation of ethene, norbornene and 1-alkenes with and without functional groups using $HSiEt_2Me$ as the hydrosilane, affording the corresponding silyl ethers of acylsilanes (**303**) as an E/Z mixture in moderate to high yields (equation 113)[312]. Many functional groups such as acetal, ether, nitrile and epoxide are tolerated in this reaction, which makes a sharp contrast to the $Co_2(CO)_8$-catalyzed reactions. Acid hydrolysis of **303** readily provide acylsilanes in quantitative yields (equation 113)[312]. Although the mechanism of this reaction is not fully understood, it is clear that the process does not include an acylsilane as the intermediate based on a control experiment, and thus the intermediacy of a siloxycarbyne complex ($Ir≡C-OSiR_3$) is suggested[312].

$$R = H, n\text{-}Bu, t\text{-}Bu, \text{cyclohexyl}, Ph, n\text{-}BuO, Me_3Si, n\text{-}BuOCH_2, Me_3SiCH_2, Me_3SiOCH_2, (EtO)_2CH, NCCH_2, \text{glycidylmethyl}$$

Regioselective silylcarbonylation of enamines (**305**) with $HSiEt_2Me$ is promoted by $[Rh(CO)_2Cl]_2$, giving α-(siloxymethylene)amines (**306**) as a E/Z mixture ($E/Z = 51/49-84/16$) in modest to high yields, which are readily hydrolyzed to afford α-siloxyketones (**307**) (equation 114)[313]. Enamines of morpholine give much better results than those of diethylamine or benzyl(methyl)amine in this reaction.

Although the $Co_2(CO)_8$-catalyzed ring-opening silylformylation of epoxides gives β-siloxyaldehydes, the products tend to undergo further reactions such as formylation, hydrosilylation and dehydrogenative silylation under the reaction conditions (140 °C and 50 atm of CO)[306,314]. In order to suppress such secondary reactions, the use of excess epoxide as compared to a hydrosilane is necessary, which is an apparent shortcoming of the process. The use of $[RhCl(CO)_2]_2$ and 1-methylpyrazole in place of $Co_2(CO)_8$ is found to circumvent this problem[315,316]. Thus, the reactions of a variety of epoxides with $HSiMe_2Ph$ catalyzed by $[Rh(CO)_2Cl]_2$ in the presence of 1-methylpyrazole (0.4 equivalents to the epoxide) at 50 °C and 50 atm of CO give β-siloxyaldehydes in good yields (equations 115–117)[315]. The reaction is stereospecific, yielding *anti*- and *syn*-products from *cis*- and *trans*-2-butene oxides, respectively (equations 116 and 117).

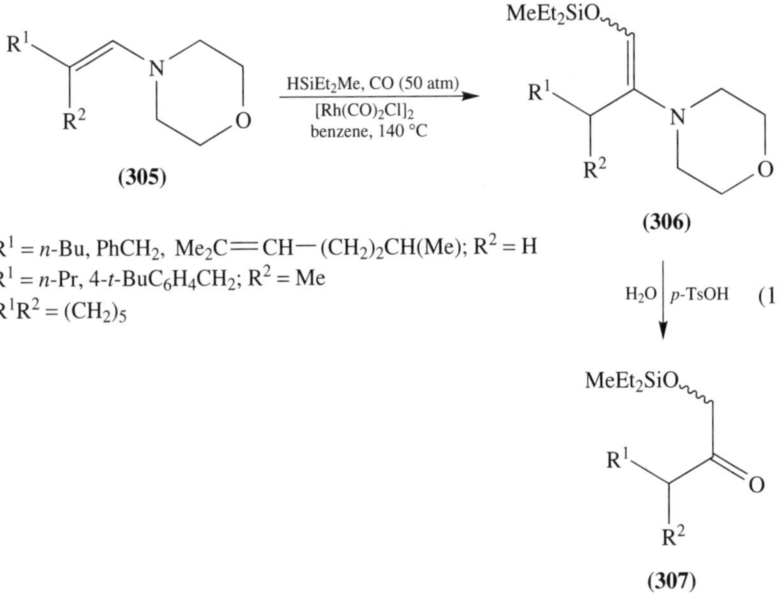

29. Recent advances in the hydrosilylation and related reactions 1773

(305)

$R^1 = n$-Bu, PhCH$_2$, Me$_2$C=CH—(CH$_2$)$_2$CH(Me); R^2 = H
$R^1 = n$-Pr, 4-t-BuC$_6$H$_4$CH$_2$; R^2 = Me
$R^1R^2 = $(CH$_2$)$_5$

(306)

(114)

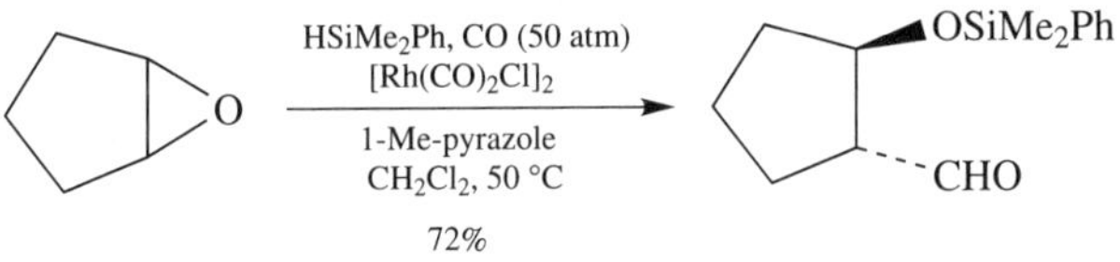

(307)

(115) 72%

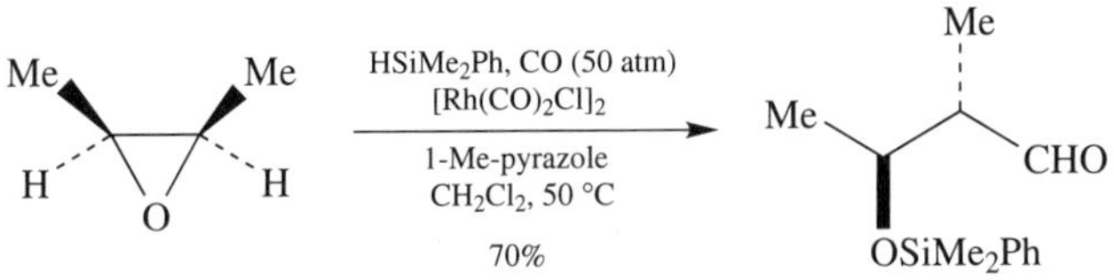

(116) 70%

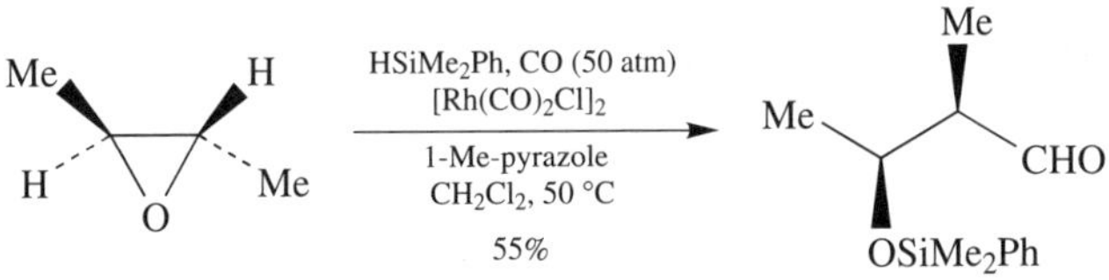

(117) 55%

The Rh-catalyzed reactions of N,N-aminals with HSiEt$_2$Me and CO (50 atm) at 140 °C in benzene give unique siloxymethylation or ring-expansion products, e.g. **308–310** (equations 118–120)[317]. Incorporation of two CO molecules and rearrangement of the dimethylamino group take place in the reaction of N,N,N',N'-tetramethylmethylenediamine (equation 119)[317].

The reaction of N,O-acetal **311** in the presence of NEt$_3$ under the same conditions gives ring-opening siloxymethylation product **312** arising from exclusive C—O cleavage (equation 121)[317]. No reaction takes place with O,O-acetals under these conditions[317].

$$\text{Ph}\overset{\text{NMe}_2}{\underset{\text{NMe}_2}{\diagdown}} \xrightarrow[\substack{\text{benzene, 140 °C} \\ 67\%}]{\text{HSiEt}_2\text{Me, CO (50 atm)} \\ [\text{Rh(CO)}_2\text{Cl}]_2} \text{Ph}\overset{\text{OSiEt}_2\text{Me}}{\underset{\text{NMe}_2}{\diagdown}} \quad (118)$$

(308)

$$\text{Me}_2\text{N}\diagdown\text{NMe}_2 \xrightarrow[\substack{\text{benzene, 140 °C} \\ 55\%}]{\text{HSiEt}_2\text{Me, CO (50 atm)} \\ [\text{Rh(CO)}_2\text{Cl}]_2} \text{Me}_2\text{N}\overset{\text{NMe}_2}{\diagdown}\text{OSiEt}_2\text{Me} \quad (119)$$

(309)

$$\text{Ph}\diagdown\underset{\substack{\text{N} \\ \text{Me}}}{\overset{\text{Me} \\ \text{N}}{\diagdown}} \xrightarrow[\substack{\text{benzene, 140 °C} \\ 51\%}]{\text{HSiEt}_2\text{Me, CO (50 atm)} \\ [\text{Rh(CO)}_2\text{Cl}]_2} \text{Ph}\diagdown\underset{\substack{\text{N} \\ \text{Me}}}{\overset{\text{Me} \\ \text{N}}{\diagdown}} \quad (120)$$

(310)

$$\underset{\text{MeN}}{\overset{n\text{-C}_7\text{H}_{15}}{\diagdown}}\overset{\text{O}}{\diagdown} \xrightarrow[\substack{\text{benzene, 140 °C} \\ 52\%}]{\text{HSiEt}_2\text{Me, CO (50 atm)} \\ [\text{Rh(CO)}_2\text{Cl}]_2/\text{NEt}_3} \underset{\text{MeN}}{\overset{n\text{-C}_7\text{H}_{15}}{\diagdown}}\overset{\text{OSiEt}_2\text{Me}}{\diagdown}\text{OSiEt}_2\text{Me} \quad (121)$$

(311) (312)

Although the silylformylation of alkanals is effected by $\text{Co}_2(\text{CO})_8\text{−PPh}_3$ catalyst at 100 °C and 50 atm of CO, $[\text{Rh(COD)Cl}]_2$ and $[\text{Rh(CO)}_2\text{Cl}]_2$ are found to catalyze the reaction using HSiMe_2Ph as the hydrosilane under much milder conditions, i.e. at room temperature and 17 atm of CO, to give homologous α-siloxyalkanals (313) in 50–90% yields (equation 122)[318,319]. When 1-phenylpropanal is used as a substrate, syn-aldehyde 314 is formed as the predominant product (equation 123)[319]. A double silylformylation of benzaldehyde takes place to yield syn-2,3-disiloxy-3-phenylpropanal (315) when excess hydrosilane is used (equation 124)[319]. The attempted silylformylation of ketones results in the formation of silyl enol ethers, and no reaction takes place with imines[319].

Silylformylation of 1-alkynes giving (Z)-2-formyl-1-silyl-1-alkenes (316) was discovered independently by Matsuda and coworkers[320] and Ojima and coworkers[321–323] using $\text{Rh}_4(\text{CO})_{12}/\text{NEt}_3$ and $\text{Rh}_2\text{Co}_2(\text{CO})_{12}$, respectively, as the catalysts, and first reported by Matsuda in 1989[320] (equation 125). The reactions of 1-alkynes, internal alkynes and functionalized alkynes with HSiMe_2Ph are effected by $\text{Rh}_4(\text{CO})_{12}$ in the presence of NEt_3 at 100 °C and 10–30 atm of CO in benzene[320]. The reaction is extremely regioselective for 1-alkynes, but a mixture of regioisomers is formed for internal alkynes, and the stereoselectivity is highly dependent on the structure of alkyne ($Z/E = 0/100–100/0$). Functional groups such as hydroxyl, olefin, trimethylsilyl and ester are tolerated in this reaction. The reactions of alkynoates, PhC≡CCO$_2$Me and MeC≡CCO$_2$Me, give (Z)-3-formyl-2-silyl-2-propenoates exclusively (equation 126).

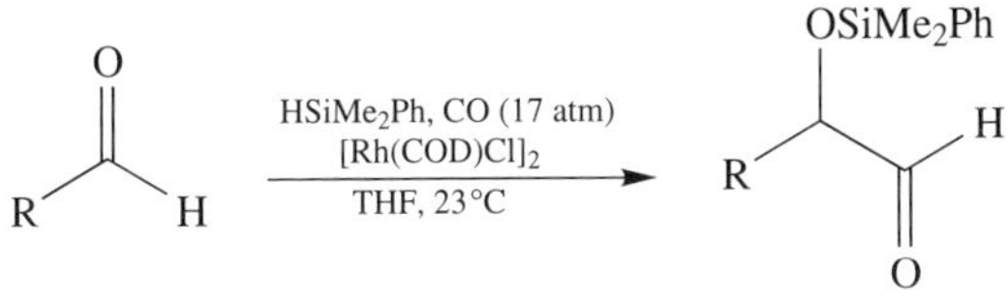

$$\underset{R}{\overset{O}{\underset{H}{\|}}}\quad\xrightarrow[\text{THF, 23°C}]{\text{HSiMe}_2\text{Ph, CO (17 atm)}\atop[\text{Rh(COD)Cl}]_2}\quad R\underset{\underset{O}{\|}}{\overset{\text{OSiMe}_2\text{Ph}}{|}}\text{H}$$

(313)

(122)

R = Ph, 4-BrC$_6$H$_4$, 4-Me$_2$NC$_6$H$_4$, 4-Me$_3$SiOC$_6$H$_4$,
4-Me$_3$SiC≡CC$_6$H$_4$, 4-AcOC$_6$H$_4$, PhCH$_2$, ferrocenyl, 2-furyl, 2-thienyl,
1-methylpyrrol-2-yl, n-Pr, i-Pr,Me$_2$C=CH(CH$_2$)$_2$CH(Me)

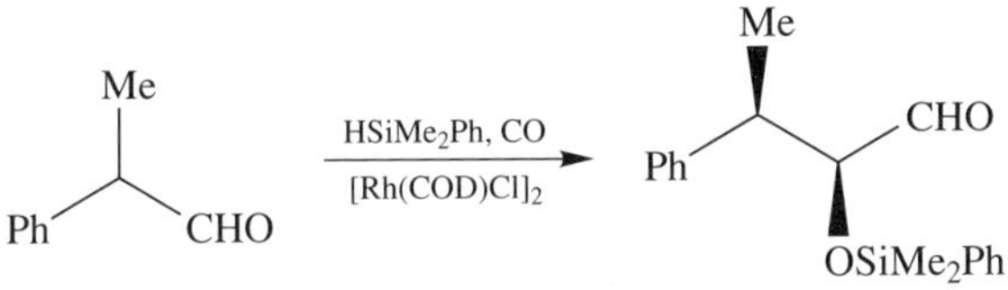

$$\underset{\text{Ph}}{\overset{\text{Me}}{\diagdown}}\text{CHO}\quad\xrightarrow[{[\text{Rh(COD)Cl}]_2}]{\text{HSiMe}_2\text{Ph, CO}}\quad\text{Ph}\underset{\text{OSiMe}_2\text{Ph}}{\overset{\text{Me}}{|}}\text{CHO}$$

(314)

(123)

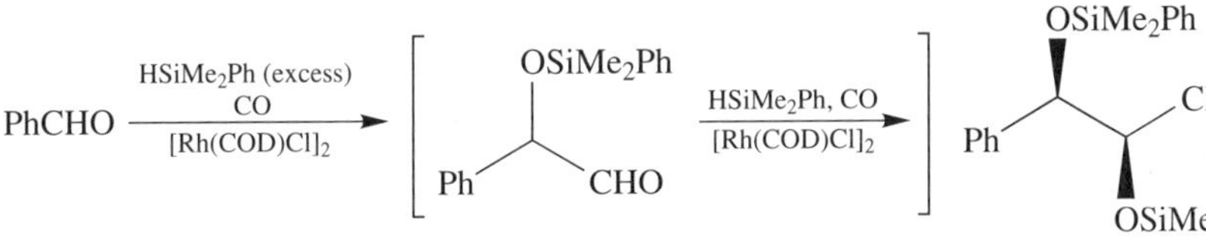

$$\text{PhCHO}\xrightarrow[{[\text{Rh(COD)Cl}]_2}]{\text{HSiMe}_2\text{Ph (excess)}\atop\text{CO}}\left[\text{Ph}\underset{\text{CHO}}{\overset{\text{OSiMe}_2\text{Ph}}{|}}\right]\xrightarrow[{[\text{Rh(COD)Cl}]_2}]{\text{HSiMe}_2\text{Ph, CO}}\text{Ph}\underset{\text{OSiMe}_2\text{Ph}}{\overset{\text{OSiMe}_2\text{Ph}}{|}}\text{CHO}$$

(315)

(124)

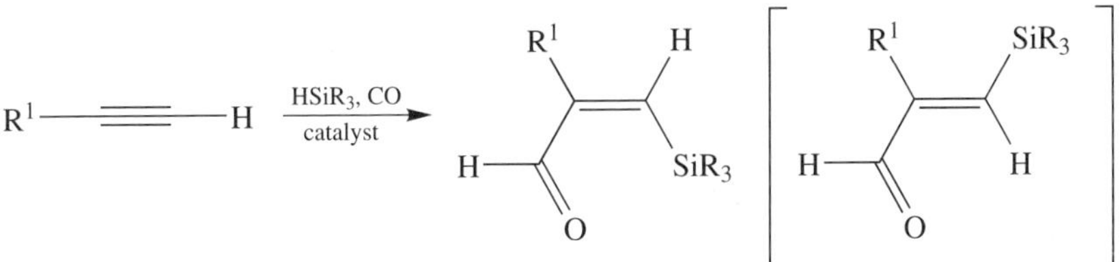

$$\text{R}^1\text{---}\equiv\text{---H}\quad\xrightarrow[\text{catalyst}]{\text{HSiR}_3,\text{CO}}\quad\underset{H\overset{\|}{O}}{\overset{R^1\quad H}{\diagup\!\!\!\diagdown}}\text{SiR}_3\quad\left[\underset{H\overset{\|}{O}}{\overset{R^1\quad\text{SiR}_3}{\diagup\!\!\!\diagdown}}H\right]$$

(E)-(316)   (Z)-(316)

(125)

$$\text{R---}\equiv\text{---CO}_2\text{Me}\xrightarrow[\text{benzene, 100°C}]{\text{HSiMe}_2\text{Ph}\atop\text{CO (50 atm)}\atop\text{Rh}_4(\text{CO})_{12},\text{NEt}_3}\underset{H\overset{\|}{O}}{\overset{R\quad\text{CO}_2\text{Me}}{\diagup\!\!\!\diagdown}}\text{SiMe}_2\text{Ph}$$

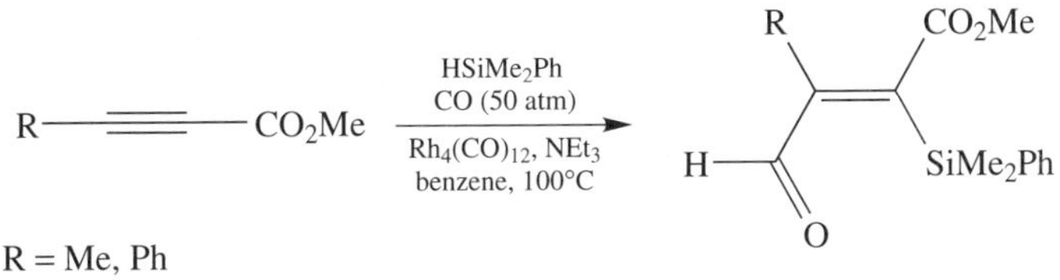

R = Me, Ph

(317)

(126)

The reactions of 1-alkynes with HSiMe$_2$Ph catalyzed by Rh$_2$Co$_2$(CO)$_{12}$ proceed at 25 °C and ambient pressure of CO in toluene, yielding (Z)-2-formyl-1-silyl-1-alkenes

(316) with complete regio- and stereoselectivity[107,323,324]. When HSiMe$_2$Et, HSiEt$_3$ and HSi(OMe)$_3$ are used, substantial amounts of hydrosilylation products are formed. The hydrosilylation can be effectively suppressed by increasing the CO pressure to 10 atm[323,324]. Phenyl-containing hydrosilanes, HSiMe$_2$Ph, HSiMePh$_2$ and HSiPh$_3$, are all very reactive and selective for this reaction[324]. In spite of the increasing bulkiness, HSiMePh$_2$ and HSiPh$_3$ give the silylformylation products with 1-hexyne in 97% and 98% isolated yields, respectively, in the reactions at 25 °C and ambient pressure of CO (equation 125)[324]. The activities of different catalysts (0.2 mol% per Rh metal) in the reaction of 1-hexyne with HSiMe$_2$Ph at ambient temperature and pressure are found to be in the order[324]: Rh(CN-Bu-t)$_4$Co(CO)$_4$ (99%) > Rh(acac)(CO)$_2$ (97%) > Rh$_2$Co$_2$(CO)$_{12}$ (92.5%) > Rh$_4$(CO)$_{12}$ (90%) ≫ RhCl(PPh$_3$)$_3$ (0%)–Co$_2$(CO)$_8$ (0%), i.e. RhCl(PPh$_3$)$_3$ and Co$_2$(CO)$_8$ are inactive. Besides the functional groups listed above, allylic ether and nitrile functionalities are found to be tolerated[324]. Possible mechanisms of silylformylation catalyzed by Rh$_2$Co$_2$(CO)$_{12}$ were proposed, which unveiled the presence of (R$_3$Si)$_2$Rh(CO)$_n$Co(CO)$_4$ (n = 2 or 3) and RhCo(n-BuC≡CH)(CO)$_5$ as key active catalyst species for the reaction, invoking the occurrence of unique homogeneous bimetallic catalysis[107,323].

In addition to these Rh and Rh–Co mixed metal complexes, Rh$_2$(pfb)$_4$ (pfb = perfluorobutanoate) is found to be a highly active catalyst, i.e. the reactions proceed smoothly at 0 °C and ambient pressure of CO in CH$_2$Cl$_2$ with complete regioselectivity and high stereoselectivity (Z/E = 10/1–40/1) (equation 125)[325,326]. When electron-deficient alkynes, HC≡CCOMe and HC≡CCO$_2$Me, are used for the Rh$_2$(pfb)$_4$-catalyzed reaction with HSiEt$_3$ under ambient CO, cyclotrimerization of these alkynes takes place exclusively yielding substituted benzenes. The alkyne cyclotrimerization can be circumvented by using 10 atm of CO, but the yields of the silylformylation products for these alkynes are only 20–24% and almost 1 : 1 mixtures of E- and Z-isomers **318** are formed (equation 127)[326].

$$H\!-\!\!\equiv\!\!-CO_2Me \xrightarrow[\text{Rh}_2(\text{pfb})_4]{\substack{\text{HSiEt}_3 \\ \text{CO (10 atm)} \\ \text{CH}_2\text{Cl}_2,\ \text{RT}}} \underset{(318)}{\text{MeOCO, H, SiEt}_3} \qquad (127)$$

Zwitterionic rhodium(I) complex, Rh$^+$(COD)(η^6-C$_6$H$_5$BPh$_3$)$^-$, is also found to be an efficient catalyst for the silylformylation of 1-alkynes at 40 °C and 40 atm of CO in CH$_2$Cl$_2$ (equation 125) although no reaction occurs with internal alkynes[327]. However, 'silylhydroformylation' takes place when the reaction is carried out under hydroformylation conditions, i.e. in the presence of CO and H$_2$ (CO/H$_2$ = 1/1), to give (E)-2-silylmethyl-2-alkenals (**319**) in 54–92% isolated yields (equation 128). The intermediacy of π-allenyl–Rh species is proposed to account for the formation of **319**[327]. When 4-acetoxy-1-butyne and 4-(p-tosyloxy)-1-butyne are used as the substrates, saturated silylhydroformylation products are obtained[327].

Although the silylformylation of aldehydes is catalyzed by [Rh(COD)Cl]$_2$ or [Rh(CO)$_2$Cl]$_2$, no secondary silylformylation of β-silylenals (**316–318**) takes place, probably due to the electronic nature of the aldehyde functionality conjugated to olefin moiety (*vide supra*). Direct comparison of the reactivity of acetylene and aldehyde functionalities is performed using alkynals[328]. The reactions of 5-hexyn-1-al, 6-heptyn-1-al and 7-octyn-1-al with different hydrosilanes catalyzed by Rh or Rh–Co complexes at

25 °C and 10 atm of CO give the corresponding (Z)-2-(silylmethylene)-1,ω-dialdehydes (**320**) exclusively in good to excellent isolated yields (equation 129), i.e. the reaction takes place at the acetylene moiety with complete chemoselectivity[328]. In the case of 5-hexyn-1-al, the reaction using Rh(CN-Bu-t)$_4$Co(CO)$_4$ or Rh$_2$Co$_2$(CO)$_{12}$ under ambient pressure of CO yields a unique silylcarbocyclization product (*vide infra*).

$$R^1\text{—}CH_2CH_2\text{—}C\equiv CH + HSiR_3 \xrightarrow[CH_2Cl_2,\ 40\ °C]{\substack{CO/H_2\ (1/1)\ (40\ atm) \\ Rh^+(COD)(\eta^6\text{-}C_6H_5BPh_3)^-}} \begin{array}{c} H \quad\quad CHO \\ \diagup=\diagdown \\ R^1 \quad\quad SiR_3 \end{array}$$

(**319**)

(128)

R^1 = Et, Ph, ClCH$_2$
R = PhMe$_2$Si, Et$_3$Si, Ph$_3$Si

$$HC\equiv C(CH_2)_nCHO \xrightarrow[\substack{catalyst \\ toluene\ or\ THF,\ 25\ °C \\ 62-93\%}]{HSiR_3,\ CO\ (10\ atm)} \begin{array}{c} R_3Si \quad\quad CHO \\ \diagup=\diagdown \\ H \quad\quad (CH_2)_nCHO \end{array}$$

n = 3–5

(**320**)

(129)

R$_3$Si = PhMe$_2$Si, Ph$_2$MeSi, Ph$_3$Si, Et$_3$Si, t-BuMe$_2$Si

Catalyst = Rh(acac)(CO)$_2$, Rh(CN-Bu-t)$_4$Co(CO)$_4$, Rh$_2$Co$_2$(CO)$_{12}$,
Rh$_4$(CO)$_{12}$, [Rh(NBD)$_2$]BF$_4$, [Rh(NBD)Cl]$_2$, [Rh(COD)Cl]$_2$

The reactions of α-substituted or α,α-disubstituted propargyl alcohols and propargylamines under silylformylation conditions using Rh$_4$(CO)$_{12}$ and NEt$_3$ or DBU afford the corresponding α-silylmethylene-β-lactones (**321**)[329] and α-silylmethylene-β-lactams (**322**)[330], respectively, in high yields (equation 130). In the absence of base, the reaction tends to give a normal silylformylation product. The use of bulky hydrosilane such as HSiMe$_2$Bu-t favors the cyclization process[329].

$$\begin{array}{c} R^1 \\ \ \ \ |\ \ R^2 \\ \equiv\text{—}C\text{—} \\ \ \ \ | \\ \ \ XH \end{array} + HSiR_3 \xrightarrow[\substack{Rh_4(CO)_{12},\ base \\ benzene,\ 100\ °C}]{CO\ (15-40\ atm)} \begin{array}{c} R_3Si\diagdown\ \ \ \diagup R^1 \\ \quad\ \square\ R^2 \\ O\diagup\ \quad X \end{array}$$

(**321**) X = O
(**322**) X = N-p-tosyl

(130)

R$_3$Si = PhMe$_2$Si, t-BuMe$_2$Si

R^1 = H, Me, i-Bu; R^2 = H, Me; R^1R^2 =—(CH$_2$)$_n$— (n = 4–6)

In a similar manner, the reaction of 1-propargyltetrahydroisoquinoline (**323**) and (R)-2-(3-butynyl)pyrrolidine (**325**) with HSiMe$_2$Ph yields (E)-8-silylmethylene-4,5-benzo-9-indolizidinone (**324**) and (R)-3-(silylmethylene)-2-indolizidinone (**326**) in 65% and 43%

isolated yields, respectively (equations 131 and 132)[331].

(equation 131): Compound (323) + HSiMe$_2$Ph, CO (50 atm), Rh(acac)(CO)$_2$, toluene, 60 °C → Compound (324)

(equation 132): Compound (325) + HSiMe$_2$Ph, CO (1 atm), Rh(acac)(CO)$_2$, toluene, 25 °C → Compound (325)

The reactions of N-alkyl and N,N-dialkylpropargylamines with 2 equivalents of HSiMe$_2$Ph catalyzed by Rh$_4$(CO)$_{12}$ give 2-silylmethyl-2-enals (327) in 40–94% yields via formal silylformylation of allenes generated *in situ* (equation 133)[332].

(equation 133): alkyne with NR^{3}R^4 group + 2 HSiMe$_2$Ph, CO (20 atm), Rh$_4$(CO)$_{12}$, benzene, 100 °C → (327) + [Me$_2$PhSiNR^{3}R^4]

$R^1, R^2 = $ H, Me; $R^3, R^4 = $ Me, CH$_2$Ph

Although the regioselectivity for the silylformylation of 1-alkynes is excellent, that of internal alkynes is low except for 2-alkynoates (*vide supra*). Also, in the reactions of 1-alkynes, the silyl group is always delivered to the terminal position and the formyl to the C-2 position, thus it is impossible to synthesize 3-silyl-2-alkenals, which requires opposite regioselectivity. Intramolecular directed reactions can circumvent these limitations and expand the scope of the silylformylation of alkynes.

The reactions of ω-hydrosilylalkynes (328) catalyzed by Rh$_4$(CO)$_{12}$/NEt$_3$ or Rh$^+$(COD) (η^6-C$_6$H$_5$BPh$_3$)$^-$ give 2-(1-formyl-1-alkylene)-1-silacycloalkanes (329) through 5- or 6-*exo*-trig cyclization in moderate to high yields (equation 134)[333].

In a similar manner, the reactions of ω-hydrodimethylsiloxyalkynes (330) catalyzed by Rh or Rh−Co complexes at 60–70 °C and 10 atm of CO afforded 5- or 6-*exo* cyclization products (331) (equation 135)[334]. When Rh(CN-Bu-i)$_4$Co(CO)$_4$ is used as the catalyst, 331 is formed in 89–99% yield. This reaction is applicable to cyclic systems 332, giving

29. Recent advances in the hydrosilylation and related reactions

333 in quantitative yields (equation 136)[334].

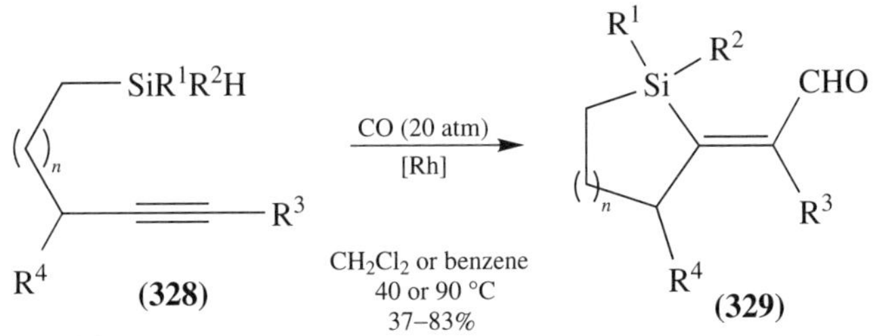

(134)

R^1 = Me, Ph; R^2 = Me, Ph; R^3 = H, Et, n-Bu, Ph; R^4 = H or Me
[Rh] = Rh$^+$(COD)(η^6-C$_6$H$_5$BPh$_3$)$^-$ or Rh$_4$(CO)$_{12}$, Et$_3$N

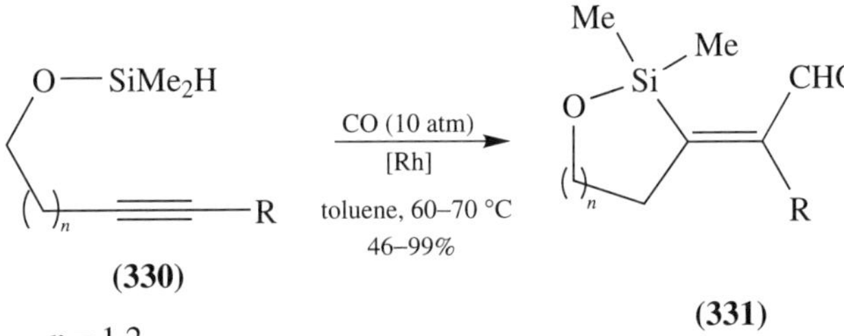

(135)

n = 1,2
R = H, Et, n-Bu

[Rh] = (t-BuNC)$_4$RhCo(CO)$_4$, Rh$_2$Co$_2$(CO)$_{12}$, Rh(acac)(CO)$_2$, Rh$_4$(CO)$_{12}$-Et$_3$N

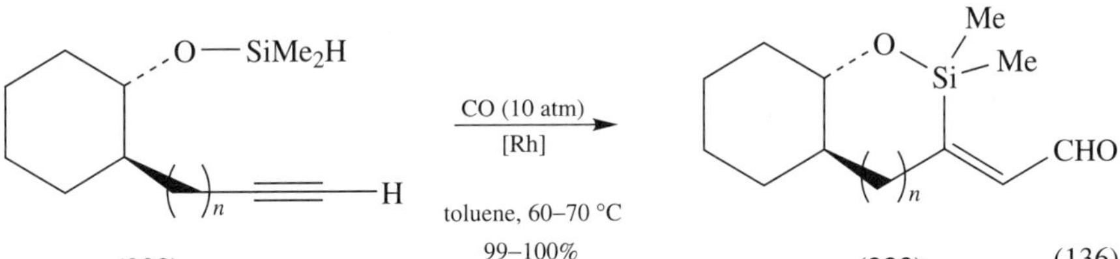

(136)

[Rh] = (t-Bu-NC)$_4$RhCo(CO)$_4$, Rh$_2$Co$_2$(CO)$_{12}$, Rh(acac)(CO)$_2$

Silylformylation is successfully applied to the syntheses of pyrrolizidine alkaloids, ($\pm$)-isoretronecanol and ($\pm$)-trachelanthamidine, from 5-ethynyl-2-pyrrolidinone (**334**) via β-silylenal **335** in combination with amidocarbonylation (Scheme 33)[324,335].

B. Silylcarbocyclization and Related Reactions

In contrast to the reactions catalyzed by Group VIII transition metal complexes (see Section II.A.1), the hydrosilylation of 1,5- or 1,6-dienes with H$_3$SiPh catalyzed by Cp$_2$*NdCH(SiMe$_3$)$_2$ results in the formation of (silylmethyl)cyclopentanes (**336**) via 5-exo carbocyclization (equation 137)[46].

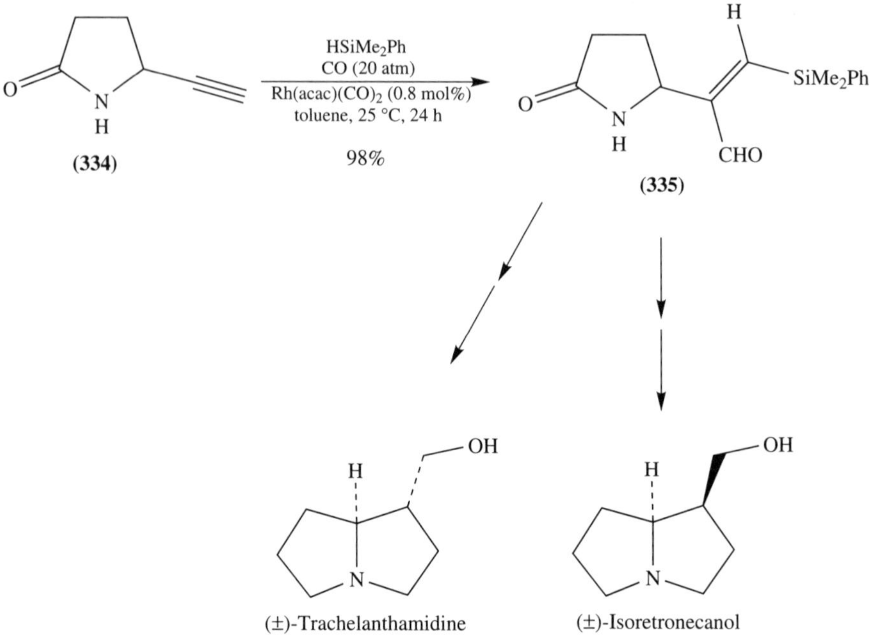

SCHEME 33

Reactions of 1,6-enynes **337** and **340** with $HSiMe_2Ph$ catalyzed by Rh or Rh–Co complexes such as $Rh(acac)(CO)_2$, $Rh_4(CO)_{12}$, $Rh_2Co_2(CO)_{12}$ or Rh(CN-Bu-*t*)$_4$Co(CO)$_4$ give 5-*exo*-trig silylcarbocyclization (SiCaC) to 3-*exo*-silylmethylene-4-methyltetrahydrofuran **(338)** and its pyrrolidine counterpart **(341)**, respectively, in high yields (equations 138 and 139)[336]. The reaction of **337** under CO atmosphere also gives **338** accompanied by a small amount of CO–SiCaC product **339** (9%) (equation 138). The reaction of **340** requires ambient CO atmosphere, i.e. no reaction takes place under N_2[336].

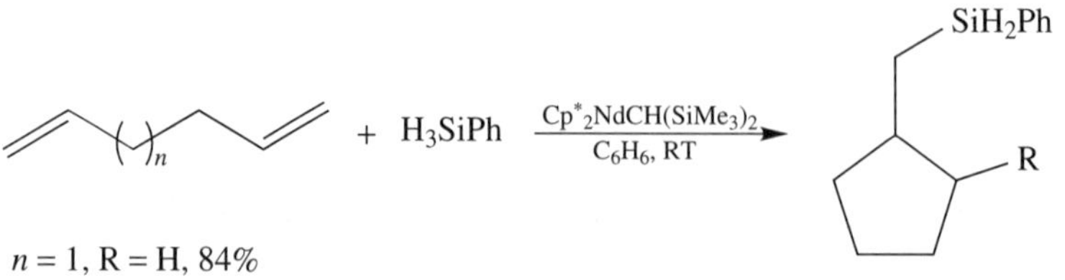

(137)

$n = 1$, R = H, 84%
$n = 2$, R = Me, 54%

(336)

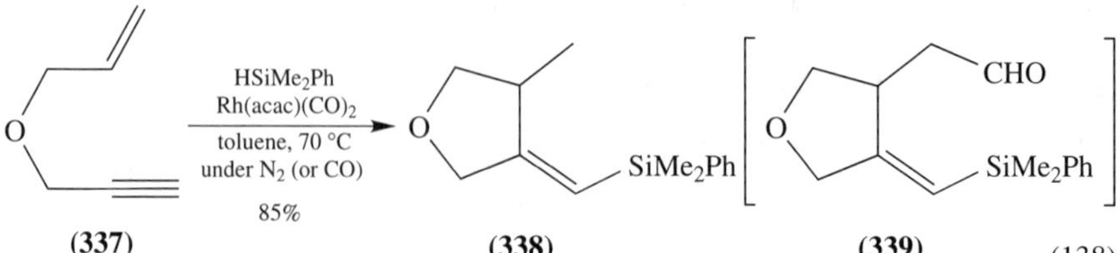

(337) **(338)** **(339)** (138)

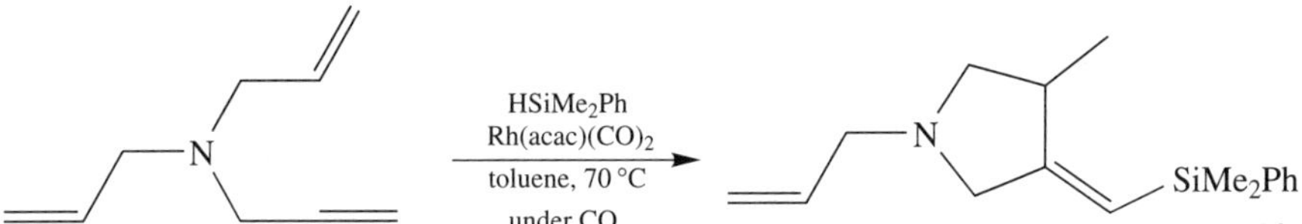

29. Recent advances in the hydrosilylation and related reactions

(340) → (341) (139)

HSiMe₂Ph, Rh(acac)(CO)₂, toluene, 70 °C, under CO

In the reaction of 1,6-enyne **342**, either SiCaC product **343** (98% yield) or CO–SiCaC product **344** (84% yield) can be obtained selectively just by changing reaction conditions, including the addition of P(OPh)₃ to the catalyst (Scheme 34)[337]. The SiCaC reaction also takes place with 5-hexyn-1-al (**345**) using HSiEt₃ under ambient CO atmosphere to give *exo*-(silylmethylene)cyclopentanol (**346**) in high yield (equation 140) through extremely chemoselective silylmetallation of the acetylene moiety[328].

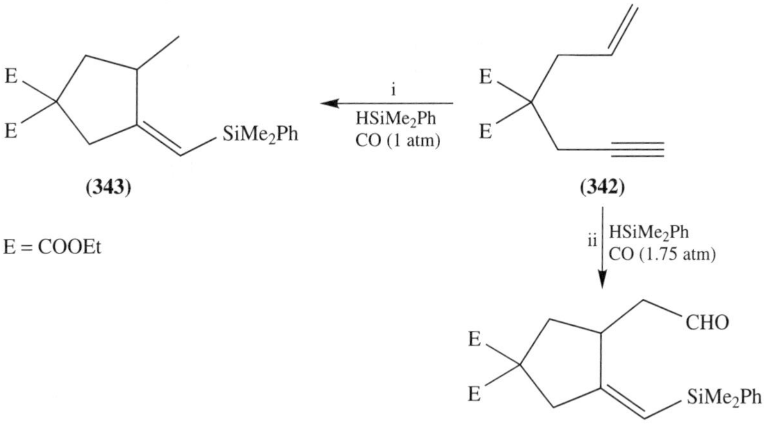

E = COOEt

(i) Rh(acac)(CO)₂, 80 °C, 0.5 M **342** in hexane: **343/344** = 10/1, 98%.
(ii) Rh(acac)(CO)₂, 50 °C, 0.2 M **342** in toluene: **343/344** = 0/100, 84%.

SCHEME 34

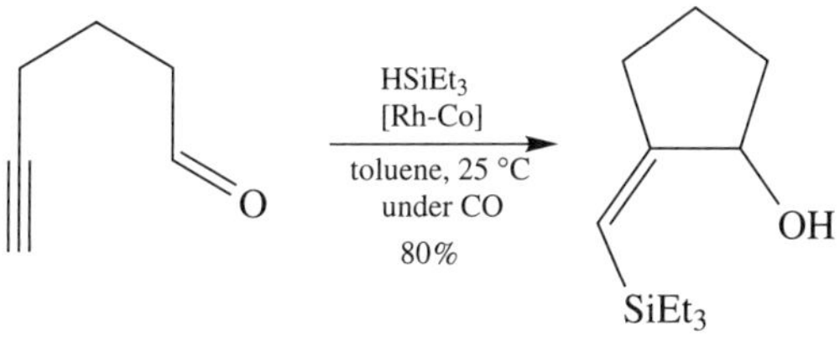

(345) → (346) (140)

HSiEt₃, [Rh-Co], toluene, 25 °C, under CO, 80%

[Rh-Co] = Rh₂Co₂(CO)₁₂ or Rh(CN-Bu-*t*)₄Co(CO)₄

Possible mechanisms of these SiCaC reactions have been proposed, which include β-silylethenyl–metal species (metal = Rh$_n$ or Rh–Co) as a key intermediate, which is trapped by an alkene or an aldehyde moiety[336,337]. The CO−SiCaC reaction further includes CO insertion after the carbocyclization[336,337].

Reactions of 1,6-diynes with HSiMe$_2$Bu-t catalyzed by Rh$_2$Co$_2$(CO)$_{12}$[338], Rh(acac)(CO)$_2$[338] or Rh$_4$(CO)$_{12}$[339] give 2-silylbicyclo[3.3.0]octen-3-ones through silylcarbo*bi*cyclization including the trapping of β-silylethenyl-metal species mentioned above with another acetylene moiety (first carbocyclization) followed by CO insertion and the second carbocyclization.

The reaction of diethyl dipropargylmalonate (**347**), with HSiMe$_2$Bu-t catalyzed by Rh$_2$Co$_2$(CO)$_{12}$ or Rh(acac)(CO)$_2$ at 50 °C and 15–50 atm of CO affords 2-silylbicyclo[3.3.0]oct-Δ^{1,5}-en-3-one (**348**) in excellent yield, which is readily isomerized to bicyclo[3.3.0]oct-1-en-3-one (**349**) (equation 141). The reactions of other 1,6-diynes including benzyldipropargylamine proceed in the same manner, but a mixture of **348**-type and **349**-type products is formed in some cases[338].

$$\text{(141)}$$

The Rh$_4$(CO)$_{12}$-catalyzed reactions of 1,6-diynes at 95 °C and 20 atm of CO in benzene or acetonitrile are less selective than those catalyzed by Rh$_2$Co$_2$(CO)$_{12}$ or Rh(acac)(CO)$_2$, and a different type of silylcarbobicyclization product, azabicyclo[3.3.0]octadienone (**351**), is formed in addition to **349**-type product (**352**) when benzyldipropargylamines (**350**) are used (equation 142)[339]. The formation of a small amount (5%) of the bicyclic pyrrole of **351**-type with N-tosyl group is also observed besides **348**-type (18%) and **349**-type (51%) products in the reaction of N-tosyldipropargylamine[339].

Another type of bicyclo[3.3.0]octadienone formation is observed in the reactions of 4,4-*gem*-disubstituted 1,6-heptadiynes (**353**) with HSiMe$_2$Bu-t catalyzed by Rh(acac)(CO)$_2$ or Rh$_2$Co$_2$(CO)$_{12}$ under forced conditions, i.e. at 120 °C and 50 atm of CO, affording 7,7-disubstituted bicyclo[3.3.0]octa-1,5-dien-3-ones (**354**) in 71–95% isolated yields (equation 143)[340,341].

Cascade silylcarbocyclization of (*E*)- or (*Z*)-dodec-6-ene-1,11-diyne (**355**) with HSiMe$_2$Ph catalyzed by Rh(acac)(CO)$_2$ gives the bis(*exo*-methylenecyclopentyl) (**356**) with complete stereospecificity (equation 144)[342]. The reaction of dodec-1,6,11-triyne (**357**) under the same conditions affords a 3 : 1 mixture of **358** and **359** in 75% yield

29. Recent advances in the hydrosilylation and related reactions 1783

(equation 145)[343,344].

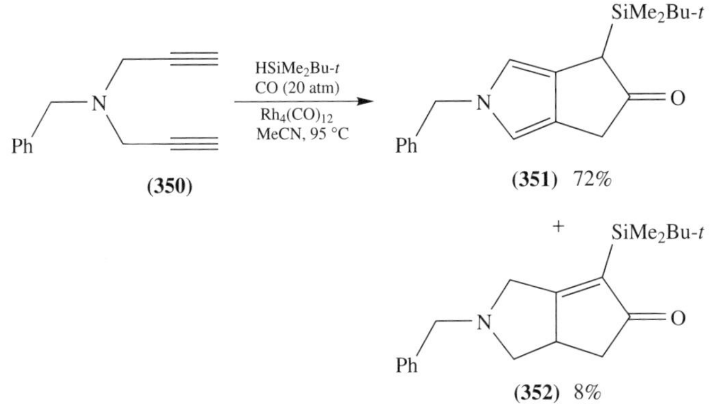

(142)

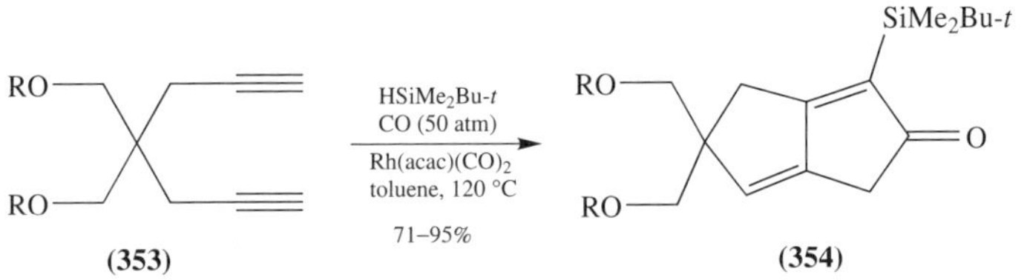

(143)

R = H, t-BuMe$_2$Si, (i-Pr)$_3$Si, Ac

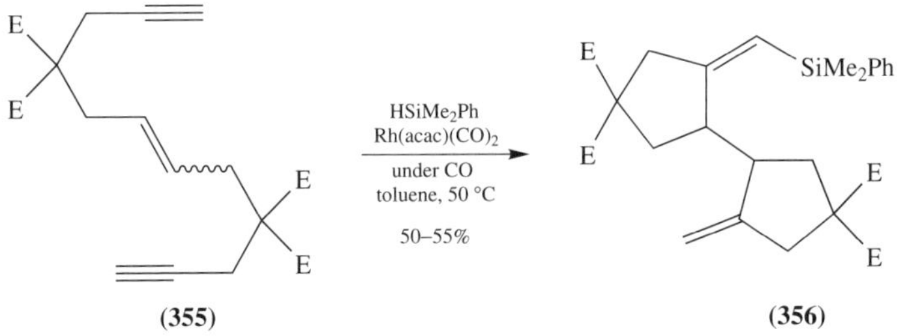

(144)

E = CO$_2$Et

Formal silylcarbocyclization of 1,7-diynes such as **360** and **362** catalyzed by Ni(acac)$_2$/DIBAL (DIBAL = i-Bu$_2$AlH) gives the corresponding vicinal exo-dimethylenecyclohexanes (equations 146 and 147)[345].

Novel carbonylative carbocyclizations of 1,6-diynes promoted by Ru$_3$(CO)$_{12}$/P(hex-c)$_3$ in the presence of HSiMe$_2$Bu-t give bicyclic o-catechol derivatives by incorporating two carbon monoxide molecules as the 1,2-dioxyethenyl moiety (equations 148 and 149)[346]. This reaction is tolerant of functional groups such as ester, ketone, ether and amide. The disilylated product **366** is formed through dehydrogenative silylation of the initially formed mono-silyl product **365** under the reaction conditions.

The proposed catalytic cycle for this unique process includes (i) oxidative addition of a hydrosilane to [Ru–CO] species to form (R$_3$Si)(H)Ru complex **369**, (ii) silyl migration to generate siloxycarbyne–Ru species **370**, (iii) carbyne reaction with CO to form metallacyclopropanone **371**, (iv) hydride shift to generate oxyacetylene–Ru complex **372** and

(v) carbocyclization to form *o*-catechol derivative **373** and regenerate active [Ru−CO] species (Scheme 35)[346].

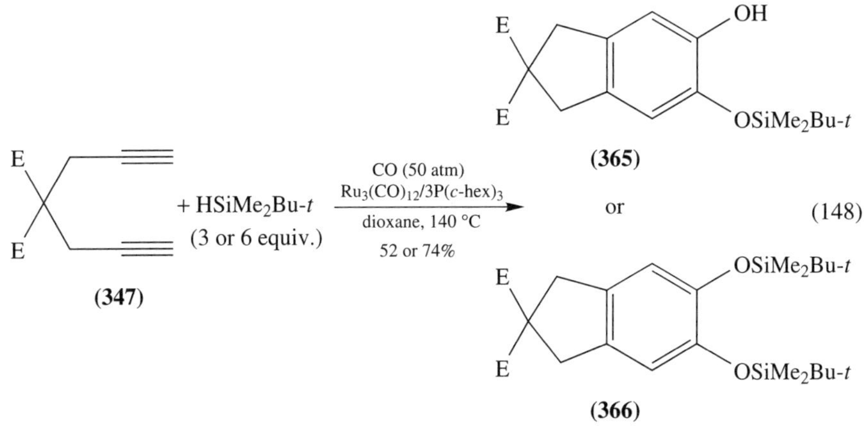

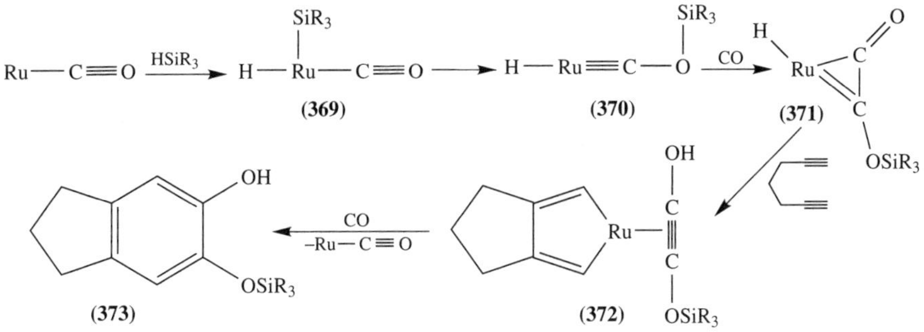

SCHEME 35

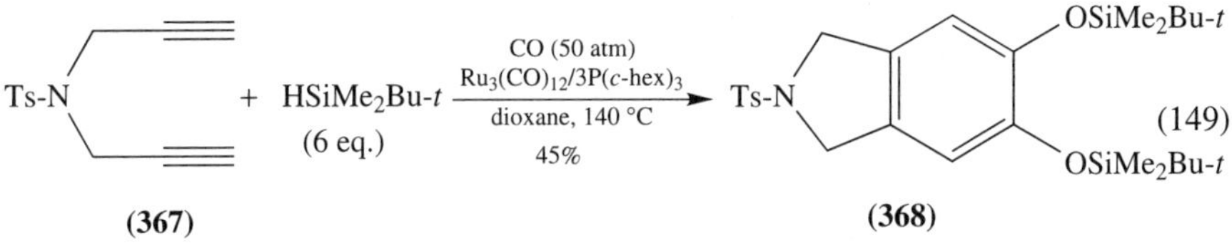

VIII. REFERENCES

1. H. Brunner, H. Nishiyama and K. Itoh, in *Catalytic Asymmetric Synthesis* (Ed. I. Ojima), VCH Publ., New York, 1993, pp. 303–322.
2. R. Noyori, *Asymmetric Catalysis in Organic Synthesis*, Wiley, New York, 1994.
3. I. Ojima, in *The Chemistry of Organic Silicon Compounds* (Eds. S. Patai and Z. Rappoport), Wiley, New York 1989, pp. 1479–1526.
4. J. L. Speier, *Adv. Organomet. Chem.*, **17**, 407 (1979).
5. T. Hiyama and T. Kusumoto, in *Comprehensive Organic Synthesis* (Ed. B. M. Trost), Pergamon Press, Oxford, 1991, p. 763.
6. B. Marcinec and J. Gulinski, *J. Organomet. Chem.*, **446**, 15 (1993).
7. M. Itoh, K. Iwata, R. Takeuchi and M. Kobayashi, *J. Organomet. Chem.*, **420**, C5 (1991).
8. R. Skoda-Foldes, L. Kollar and B. Heil, *J. Organomet. Chem.*, **366**, 275 (1989).

9. T. Murai, T. Oda, F. Kimura, H. Onishi, T. Kanda and S. Kato, *J. Chem. Soc., Chem. Commun.*, 2143 (1994).
10. B. Karstedt, *U.S. Patent*, 3,775,452 (1973); *Chem. Abstr.*, **90**, 16134j (1973).
11. D. N. Willing, *U.S. Patent*, 3,419,593 (1968); *Chem. Abstr.*, **75**, 2123z (1968).
12. V. J. Eddy and J. E. Hallgren, *J. Org. Chem.*, **52**, 1903 (1987).
13. M. G. Voronkov, S. V. Kirpichenko, V. V. Keiko and A. I. Albanov, *J. Organomet. Chem.*, **427**, 289 (1992).
14. M. Tanaka, Y. Uchimaru and H. J. Lautenschlager, *J. Organomet. Chem.*, **428**, 1 (1992).
15. H. Nagashima, K. Tatebe, T. Ishibashi, J. Sakakibara and K. Itoh, *Organometallics*, **8**, 2495 (1989).
16. T. Hagiwara, K. Taya, Y. Yamamoto and H. Yamazaki, *J. Mol. Catal.*, **54**, 165 (1989).
17. J. B. Baruah, K. Osakada and T. Yamamoto, *J. Mol. Catal. A: Chem.*, **101**, 17 (1995).
18. J. Gulinski and B. R. James, *J. Mol. Catal.*, **72**, 167 (1992).
19. L. I. Kopylova, V. B. Pukhnarevich, V. S. Tkach and M. G. Voronkov, *Zh. Obshch. Khim.*, **63**, 1294 (1993); *Chem. Abstr.*, **120**, 134613s (1993).
20. J. M. Chance and T. A. Nile, *J. Mol. Catal.*, **42**, 91 (1987).
21. H. S. Hilal, S. Khalaf and W. Jondi, *J. Organomet. Chem.*, **452**, 167 (1993).
22. M. Tanaka, T. Hayashi and Z.-Y. Mi, *J. Mol. Catal.*, **81**, 207 (1993).
23. L. N. Lewis and N. Lewis, *J. Am. Chem. Soc.*, **108**, 7228 (1986).
24. A. Onopchenko and E. T. Sabourin, *J. Org. Chem.*, **52**, 4118 (1987).
25. K. A. Brown-Wensley, *Organometallics*, **6**, 1590 (1987).
26. W. C. Trogler, *J. Chem. Educ.*, **65**, 294 (1988).
27. A. L. Prignano and W. C. Trogler, *J. Am. Chem. Soc.*, **109**, 3586 (1987).
28. L. N. Lewis and N. Lewis, *Chem. Mater.*, **1**, 106 (1989).
29. L. N. Lewis and R. J. Uriarte, *Organometallics*, **9**, 621 (1990).
30. L. N. Lewis, *J. Am. Chem. Soc.*, **112**, 5998 (1990).
31. G. Chandra, P. Y. Lo, P. B. Hitchcock and M. F. Lappert, *Organometallics*, **6**, 191 (1987).
32. P. B. Hitchcock, M. F. Lappert and N. J. W. Warhurst, *Angew. Chem., Int. Ed. Eng.*, **30**, 438 (1991).
33. M. F. Lappert and F. P. A. Scott, *J. Organomet. Chem.*, **492**, C11 (1995).
34. F. D. Lewis and G. D. Salvi, *Inorg. Chem.*, **34**, 3182 (1995).
35. C. Polizzi, A. M. Caporusso, G. Vitulli, P. Salvadori and M. Pasero, *J. Mol. Catal.*, **91**, 83 (1994).
36. M. R. Kesti, M. Abdulrahman and R. M. Waymouth, *J. Organomet. Chem.*, **417**, C12 (1991).
37. M. R. Kesti and R. M. Waymouth, *Organometallics*, **11**, 1095 (1992).
38. J. Y. Corey and X. H. Zhu, *Organometallics*, **11**, 672 (1992).
39. J. F. Harrod and S. S. Yun, *Organometallics*, **6**, 1381 (1987).
40. J. Y. Corey, X.-H. Zhu, T. C. Bedard and L. D. Lange, *Organometallics*, **10**, 924 (1991).
41. J. F. Harrod in *Organometallic and Inorganic Polymers* (Eds. H. R. Zeldin, H. R. Allcock and K. J. Wynne), American Chemical Society, Washignton, D. C., 1988, p. 89.
42. J. F. Harrod, *Polym. Prepr.*, **28**, 403 (1987).
43. P.-W. Fu, L. Brard, Y. Li and T. B. Marks, *J. Am. Chem. Soc.*, **117**, 7157 (1995).
44. T. Sakakura, H. Lautenschlager and M. Tanaka, *J. Chem. Soc., Chem. Commun.*, 40 (1991).
45. G. A. Molander and M. Julius, *J. Org. Chem.*, **57**, 6347 (1992).
46. S. Onozawa, T. Sakakura and M. Tanaka, *Tetrahedron Lett.*, **35**, 8177 (1994).
47. J. L. Speier, J. A. Webster, and G. H. Barens, *J. Am. Chem. Soc.*, **79**, 974 (1957).
48. E. Yoshii, Y. Kobayashi, T. Koizumi and T. Oribe, *Chem. Pharm. Bull.*, **22**, 2767 (1974).
49. Y. Kiso, M. Kumada, K. Tamao and M. Umeno, *J. Organometal. Chem.*, **50**, 297 (1973).
50. I. Ojima, M. Kumagai and Y. Nagai, *J. Organometal. Chem.*, **111**, 43 (1976).
51. P. Boudjouk, S. Kloos and A. B. Rajkumar, *J. Organomet. Chem.*, **443**, C41 (1993).
52. B. A. Bluestein, *J. Am. Chem. Soc.*, **83**, 1000 (1961).
53. P. Svoboda and J. Hetflejs, *Collect. Czech. Chem. Commun.*, **38**, 3834 (1973).
54. A. B. Rajkumar and P. Boudjouk, *Organometallics*, **8**, 549 (1989).
55. H. S. Hilal, M. Abu-Eid, M. Al-Subu and S. Khalaf, *J. Mol. Catal.*, **39**, 1 (1987).
56. B. Marciniec, H. Maciejewski and J. Mirecki, *J. Organomet. Chem.*, **418**, 61 (1991).
57. W. Abdelqader, D. Chmielewski, F.-W. Grevels, S. Oezkar and N. B. Peynircioglu, *Organometallics*, **15**, 604 (1996).
58. M. A. Wrighton and M. S. Schroeder, *J. Am. Chem. Soc.*, **96**, 6235 (1974).

29. Recent advances in the hydrosilylation and related reactions 1787

59. M. A. Schroeder and M. S. Wrighton, *J. Organometal. Chem.*, **128**, 345 (1977).
60. M. Capka and M. Czakoova, *J. Mol. Catal.*, **74**, 335 (1992).
61. M. Capka, M. Czakoova and U. Schubert, *Appl. Organomet. Chem.*, **7**, 369 (1993).
62. M. Capka, M. Czakoova, J. Hjortkjaer and U. Schubert, *React. Kinet. Catal. Lett.*, **50**, 71 (1993).
63. Z. M. Michalska, B. Ostaszewski, K. Strzelec, R. Kwiatkowski and A. Wlochowicz, *React. Polym.*, **23**, 85 (1994).
64. Z. M. Michalska, B. Ostaszewski and K. Strzelec, *J. Organomet. Chem.*, **496**, 19 (1995).
65. W. R. Cullen and N. F. Han, *J. Organometal. Chem.*, **333**, 269 (1987).
66. B. Kopping, C. Chatgilialoglu, M. Zehnder and B. Giese, *J. Org. Chem.*, **57**, 3994 (1992).
67. C. Chatgilialoglu, *Acc. Chem. Res.*, **25**, 188 (1992).
68. M. Ballestri, C. Chatgilialoglu, K. B. Clark, D. Griller, B. Kopping and B. Giese, *J. Org. Chem.*, **56**, 678 (1991).
69. A. J. Chalk and J. F. Harrod, *J. Am. Chem. Soc.*, **87**, 16 (1965).
70. C. L. Reichel and M. S. Wrighton, *Inorg. Chem.*, **19**, 3858 (1980).
71. C. L. Randolph and M. S. Wrighton, *J. Am. Chem. Soc.*, **108**, 3366 (1986).
72. M. Brookhart and B. E. Grant, *J. Am. Chem. Soc.*, **115**, 2151 (1993).
73. J. C. Mitchener and M. S. Wrighton, *J. Am. Chem. Soc.*, **103**, 975 (1981).
74. A. Millan, M.-J. Fernandez, P. Bentz and P. Maitlis, *J. Mol. Catal.*, **26**, 89 (1984).
75. A. Millan, E. Towns and P. Maitlis, *J. Chem. Soc., Chem. Commun.*, 673 (1981).
76. Y. Seki, K. Takeshita and K. Kawamoto, *J. Organometal. Chem.*, **369**, 117 (1989).
77. Y. Seki, K. Takeshita, K. Kawamoto, S. Murai and N. Sonoda, *J. Org. Chem.*, **52**, 4864 (1987).
78. A. Onopchenko, E. T. Sabourin and D. L. Beach, *J. Org. Chem.*, **49**, 3389 (1984).
79. A. Onopchenko, E. T. Sabourin and D. L. Beach, *J. Org. Chem.*, **48**, 5101 (1983).
80. I. Ojima, T. Fuchikami and M. Yatabe, *J. Organomet. Chem.*, **260**, 335 (1984).
81. F. Seitz and M. S. Wrighton, *Angew. Chem., Int. Ed. Engl.*, **27**, 289 (1988).
82. S. B. Duckett and R. N. Perutz, *Organometallics*, **11**, 90 (1992).
83. J. Ruiz, P. O. Bentz, B. E. Mann, C. M. Spencer, B. F. Taylor and P. M. Maitlis, *J. Chem. Soc., Dalton Trans.*, 2709 (1987).
84. M. J. Hostetler, M. D. Butts and R. G. Bergman, *Organometallics*, **12**, 65 (1993).
85. M. J. Hostetler and R. G. Bergman, *J. Am. Chem. Soc.*, **112**, 8621 (1990).
86. K. Tamao, T. Nakajima, R. Sumiya, H. Arai, N. Higuchi and Y. Ito, *J. Am. Chem. Soc.*, **108**, 6090 (1986).
87. K. Tamao, Y. Nakagawa, H. Arai, N. Higuchi and Y. Ito, *J. Am. Chem. Soc.*, **110**, 3712 (1988).
88. K. Tamao, *J. Syn. Org. Chem. Jpn.*, **46**, 861 (1988).
89. K. Tamao, Y. Nakagawa and Y. Ito, *Organometallics*, **12**, 2297 (1993).
90. K. Tamao, Y. Nakagawa and Y. Ito, *J. Org. Chem.*, **55**, 3438 (1990).
91. M. P. Sibi and J. W. Christensen, *Tetrahedron Lett.*, **36**, 6213 (1995).
92. S. E. Denmark and D. C. Forbes, *Tetrahedron Lett.*, **33**, 5037 (1992).
93. H. K. Chu and C. L. Frye, *J. Organomet. Chem.*, **446**, 183 (1993).
94. S. Xin and J. F. Harrod, *J. Organomet. Chem.*, **499**, 181 (1995).
95. A. N. Nesmeyanov, R. K. Freidlina, E. C. Chukovskaya, R. G. Petrova and A. B. Belyavsky, *Tetrahedron*, **17**, 61 (1962).
96. Y. Seki, K. Takeshita, K. Kawamoto, S. Murai and N. Sonoda, *Angew. Chem., Int. Ed. Engl.*, **19**, 928 (1980).
97. K. Takeshita, Y. Seki, K. Kawamoto, S. Murai and N. Sonoda, *J. Chem. Soc., Chem. Commun.*, 1193 (1983).
98. Y. Seki, K. Takeshita, K. Kawamoto, S. Murai and N. Sonoda, *J. Org. Chem.*, **51**, 3890 (1986).
99. K. Takeshita, Y. Seki, K. Kawamoto, S. Murai, and N. Sonoda, *J. Org. Chem.*, **52**, 4864 (1987).
100. F. Kakiuchi, Y. Tanaka, N. Chatani and S. Murai, *J. Organomet. Chem.*, **456**, 45 (1993).
101. F. Kakiuchi, K. Nogami, N. Chatani, Y. Seki and S. Murai, *Organometallics*, **12**, 4748 (1993).
102. M. P. Doyle, G. A. Devora, A. O. Nefedov and K. G. High, *Organometallics*, **11**, 549 (1992).
103. K. Hayakawa, M. Tachikawa, T. Suzuke, N. Choi and M. Murakami, *Tetrahedron Lett.*, **36**, 3181 (1995).
104. L. N. Lewis, K. G. Sy and P. E. Donahue, *J. Organomet. Chem.*, **427**, 165 (1992).
105. I. G. Iovel, Y. S. Goldberg, M. V. Shymanska and E. Lukevics, *Organometallics*, **6**, 1410 (1987).
106. I. Ojima, N. Clos, R. L. Donovan and P. Ingallina, *Organometallics*, **9**, 3127 (1990).
107. I. Ojima, R. J. Donovan, P. Ingallina and N. Clos, *J. Cluster Sci.*, **3**, 423 (1992).

108. I. Ojima, M. Kumagai and Y. Nagai, *J. Organometal. Chem.*, **66**, C14 (1974).
109. R. Takeuchi and N. Tanouchi, *J. Chem. Soc., Perkin Trans. 1*, 2909 (1994).
110. R. Takeuchi and N. Tanouchi, *J. Chem. Soc., Chem. Commun.*, 1319 (1993).
111. M. P. Doyle, K. G. High, C. L. Nesloney, T. W. J. Layton and J. Lin, *Organometallics*, **10**, 1225 (1991).
112. Y. Goldberg and H. Alper, *Tetrahedron Lett.*, **36**, 369 (1995).
113. C.-H. Jun and R. H. Crabtree, *J. Organomet. Chem.*, **447**, 177 (1993).
114. P. Hofmann, C. Meier, W. Hiller, M. Heckel, J. Riede and M. U. Schmidt, *J. Organomet. Chem.*, **490**, 51 (1995).
115. R. S. Tanke and R. H. Crabtree, *J. Am. Chem. Soc.*, **112**, 7984 (1990).
116. R. S. Tanke and R. H. Crabtree, *J. Chem. Soc., Chem. Commun.*, 1056 (1990).
117. M. J. Fernandez, L. A. Oro and B. R. Mamzano, *J. Mol. Catal.*, **45**, 7 (1988).
118. M. A. Esteruelas, M. Oliván, L. A. Oro and J. I. Tolosa, *J. Organomet. Chem.*, **487**, 143 (1995).
119. M. A. Esteruelas, M. Oliván and L. A. Oro, *Organometallics*, **15**, 814 (1996).
120. M. A. Esteruelas, J. Herrero and L. A. Oro, *Organometallics*, **12**, 2377 (1993).
121. M. A. Esteruelas, L. A. Oro and C. Valero, *Organometallics*, **10**, 462 (1991).
122. M. A. Esteruelas, A. M. López, L. A. Oro and J. I. Tolosa, *J. Mol. Catal. A: Chem.*, **96**, 21 (1995).
123. T. Bartik, G. Nagy, P. Kvintovics and B. Happ, *J. Organomet. Chem.*, **453**, 29 (1993).
124. G. A. Molander and W. H. Retch, *Organometallics*, **14**, 4570 (1995).
125. R. A. Benkeser, *Pure Appl. Chem.*, **13**, 133 (1966).
126. L. H. Sommer, E. W. Pietrusza and F. C. Whitmore, *J. Am. Chem. Soc.*, **69**, 188 (1947).
127. K. A. Brady and T. A. Nile, *J. Organometal. Chem.*, **206**, 299 (1981).
128. S. R. Allen, R. G. Beevor, M. Green, N. C. Norman, A. G. Orpen and I. D. Williams, *J. Chem. Soc., Dalton Trans.*, 435 (1985).
129. M. A. Esteruelas, O. Nurnberg, M. Olivan, L. A. Oro and H. Werner, *Organometallics*, **12**, 3264 (1993).
130. M. G. Steinmetz and B. S. Udayakumar, *J. Organomet. Chem.*, **378**, 1 (1989).
131. T. Suzuki and P. Y. Lo, *J. Organomet. Chem.*, **396**, 299 (1990).
132. T. Suzuki and P. Y. Lo, *J. Organomet. Chem.*, **391**, 19 (1990).
133. T. Kusumoto, K. Ando and T. Hiyama, *Bull. Chem. Soc. Jpn.*, **65**, 1280 (1992).
134. E. Lukevics, R. Y. Sturkovich and O. A. Pudova, *J. Organomet. Chem.*, **292**, 151 (1985).
135. P. J. Murphy, J. L. Spencer and G. Procter, *Tetrahedron Lett.*, **31**, 1051 (1990).
136. R. Takeuchi, S. Nitta and D. Watanabe, *J. Org. Chem.*, **60**, 3045 (1995).
137. R. Takeuchi, S. Nitta and D. Watanabe, *J. Chem. Soc., Chem. Commun.*, 1777 (1994).
138. K. Takahashi, T. Minami, Y. Ohara and T. Hiyama, *Tetrahedron Lett.*, **34**, 8263 (1993).
139. K. Takahashi, T. Minami, Y. Ohara and T. Hiyama, *Bull. Chem. Soc. Jpn.*, **68**, 2649 (1995).
140. I. Ojima, R. J. Donovan and N. Clos, *Organometallics*, **10**, 3790 (1991).
141. M. F. Semmelhack and R. N. Misra, *J. Org. Chem.*, **47**, 2469 (1982).
142. J. Ishiyama, Y. Senda, I. Shinoda and S. Imaizumi, *Bull. Chem. Soc. Jpn.*, **52**, 2353 (1979).
143. I. Ojima, M. Nihonyanagi and Y. Nagai, *Bull. Chem. Soc. Jpn.*, **45**, 3722 (1972).
144. K. Felföldi, I. Kapocsi and M. Bartök, *J. Organomet. Chem.*, **362**, 411 (1989).
145. H. Nagashima, K. Tatebe, T. Ishibashi, A. Nakaoka, J. Sakakibara and K. Itoh, *Organometallics*, **14**, 2868 (1995).
146. H. Nagashima, K. Tatebe and K. Itoh, *J. Chem. Soc., Perkin Trans. 1*, 1707 (1989).
147. E. J. Crawford, P. K. Hanna and A. R. Cutler, *J. Am. Chem. Soc.*, **111**, 6891 (1989).
148. P. K. Hanna, B. T. Gregg and A. R. Cutler, *Organometallics*, **10**, 31 (1991).
149. Z. Mao, B. T. Gregg and A. R. Cutler, *J. Am. Chem. Soc.*, **117**, 10139 (1995).
150. B. Marciniec and J. Gulinski, *J. Organomet. Chem.*, **446**, 15 (1993).
151. J. E. Hill and T. A. Nile, *J. Organometal. Chem.*, **137**, 293 (1977).
152. A. Revis and T. K. Hilty, *J. Org. Chem.*, **55**, 2972 (1990).
153. I. Ojima, T. Kogure and Y. Nagai, *Tetrahedron Lett.*, 5035 (1972).
154. I. Ojima and T. Kogure, *Organometallics*, **1**, 1390 (1982).
155. G. Z. Zheng and T. H. Chan, *Organometallics*, **14**, 70 (1995).
156. E. Keinan and D. Perez, *J. Org. Chem.*, **52**, 2576 (1987).
157. T. Schmidt, *Tetrahedron Lett.*, **35**, 3513 (1994).
158. T. Nakano and Y. Nagai, *Chem. Lett.*, 481 (1988).
159. K. Yamamoto and T. Tabei, *J. Organomet. Chem.*, **427**, 165 (1992).

160. B. Töröek, K. Felföldi, A. Molnár and M. Bartók, *J. Organomet. Chem.*, **460**, 111 (1993).
161. J. Boyer, R. J. P. Corriu, R. Perz and C. Reye, *J. Organometal. Chem.*, **157**, 153 (1978).
162. J. Boyer, R. J. P. Corriu, R. Perz and C. Reye, *J. Organometal. Chem.*, **172**, 142 (1979).
163. C. Chuit, R. J. P. Corriu, R. Perz and C. Reye, *Tetrahedron*, **37**, 2165 (1981).
164. J. Boyer, R. J. P. Corriu, R. Perz, M. Poirier and C. Reye, *Synthesis*, 558 (1981).
165. R. J. P. Corriu, R. Perz and C. Reye, *Tetrahedron*, **39**, 999 (1983).
166. M. Fujita and T. Hiyama, *J. Org. Chem.*, **53**, 5405 (1988).
167. M. Fujita and T. Hiyama, *J. Am. Chem. Soc.*, **107**, 8294 (1985).
168. T. Hiyama, K. Kobayashi and M. Fujita, *Tetrahedron Lett.*, **25**, 4959 (1984).
169. M. Fujita and T. Hiyama, *J. Am. Chem. Soc.*, **106**, 4629 (1984).
170. M. Fujita and T. Hiyama, *Tetrahedron Lett.*, **28**, 2263 (1987).
171. Y. Goldberg, E. Abele, M. V. Shimanskaya and E. Lukevics, *J. Organomet. Chem.*, **372**, C9 (1989).
172. Y. Goldberg, K. Rubina, M. Shymanska and E. Lukevics, *Synth. Commun.*, **20**, 2439 (1990).
173. W. P. Weber, in *Silicon Reagents for Organic Synthesis*, Springer-Verlag, Berlin, 1983, pp. 273–287.
174. M. Fujita and T. Hiyama, *J. Org. Chem.*, **53**, 5415 (1988).
175. G. A. Olah, M. Arvanaghi and L. Ohannesian, *Synthesis*, 770 (1986).
176. M. Onaka, K. Higuchi, H. Nanami and Y. Izumi, *Bull. Chem. Soc. Jpn.*, **66**, 2638 (1993).
177. M. P. Doyle, C. T. West, S. J. Donnelly and C. C. McOsker, *J. Organometal. Chem.*, **117**, 129 (1976).
178. J. L. Fry, M. Orfanopoulos, M. G. Adlington, W. R. Dittman and S. B. Silverman, *J. Org. Chem.*, **43**, 374 (1978).
179. M. Fujita, S. Fukuzumi and J. Otera, *J. Mol. Catal.*, **85**, 143 (1993).
180. Y. Izumi, H. Nanami, K. Higuchi and M. Onaka, *Tetrahedron Lett.*, **32**, 4741 (1991).
181. S. Anwar and A. P. David, *Tetrahedron*, **44**, 3761 (1988).
182. A. P. Davis and S. C. Hegarty, *J. Am. Chem. Soc.*, **114**, 2745 (1992).
183. A. P. Davis and S. C. Hegarty, *J. Am. Chem. Soc.*, **114**, 8753 (1992).
184. X. Verdagnuer, S. C. Berk and S. L. Buchwald, *J. Am. Chem. Soc.*, **117**, 12641 (1995).
185. A. H. Vetter and A. Berkessel, *Synthesis*, 419 (1995).
186. T. Murai, T. Sakane and S. Kato, *J. Org. Chem.*, **55**, 449 (1990).
187. I. Ojima, in *Asymmetric Synthesis*, Vol. 5 (Ed. J. D. Morrison) Academic Press, New York, 1985, pp. 103–146.
188. I. Ojima, K. Yamamoto and M. Kumada, in *Aspects of Homogeneous Catalysis*, Vol. 3 (Ed. R. Ugo,) Reidel, Amsterdam, 1977, pp. 85–228.
189. H. Brunner, *Adv. Chem. Ser.*, **230**, 143 (1992).
190. H. Nishiyama, H. Sakaguchi, T. Nakamura, M. Horihata, M. Kondo and K. Itoh, *Organometallics*, **8**, 846 (1989).
191. H. Nishiyama, M. Kondo, T. Nakamura and K. Itoh, *Organometallics*, **10**, 500 (1991).
192. H. Nishiyama, S. Yamaguchi, M. Kondo and K. Itoh, *J. Org. Chem.*, **57**, 4306 (1992).
193. H. Nishiyama, S. Yamaguchi, S. B. Park and K. Itoh, *Tetrahedron: Asymmetry*, **4**, 143 (1993).
194. H. Nishiyama and K. Itoh, *J. Syn. Org. Chem. Jpn. (Yuki Gosei Kagaku Kyokaishi)*, **53**, 500 (1995).
195. H. Brunner and C. Henrichs, *Tetrahedron: Asymmetry*, **6**, 653 (1995).
196. H. Brunner and P. Brandl, *Z. Naturforsch.*, **47b**, 609 (1992).
197. H. Brunner and P. Brandl, *Tetrahedron: Asymmetry*, **2**, 919 (1991).
198. H. Brunner and C. Huber, *Z. Naturforsch.*, **46b**, 1145 (1991).
199. H. Brunner and P. Brandl, *J. Organomet. Chem.*, **390**, C81 (1990).
200. S. Gladiali, L. Pinna, G. Delogu, E. Graf and H. Brunner, *Tetrahedron: Asymmetry*, **1**, 937 (1990).
201. H. Brunner and U. Obermann, *Chem. Ber.*, **122**, 499 (1989).
202. G. Balavoine, J. C. Clinet and I. Lellouche, *Tetrahedron Lett.*, **30**, 5141 (1989).
203. H. Brunner and A. Kuerzinger, *J. Organomet. Chem.*, **346**, 413 (1988).
204. H. Brunner and H. Fisch, *J. Organomet. Chem.*, **335**, 15 (1987).
205. H. Brunner, U. Obermann and P. Wimmer, *J. Organometal. Chem.*, **316**, C1 (1986).
206. M. Sawamura, R. Kuwano, J. Shirai and Y. Ito, *Synlett*, 347 (1995).
207. M. Sawamura, R. Kuwano and Y. Ito, *Angew. Chem., Int. Ed. Engl.*, **33**, 111 (1994).
208. T. Hayashi, C. Hayashi and Y. Uozumi, *Tetrahedron: Asymmetry*, **6**, 2503 (1995).

209. Y. Nishibayashi, K. Segawa, K. Ohe and S. Uemura, *Organometallics*, **14**, 5486 (1995).
210. Y. Nishibayashi, J. D. Singh, K. Segawa, S. Fukuzawa and S. Uemura, *J. Chem. Soc., Chem. Commun.*, 1375 (1994).
211. Y. Nishibayashi, K. Segawa, J. D. Singh, S. Fuzukawa, K. Ohe and S. Uemura, *Organometallics*, **15**, 370 (1996).
212. J. Sakaki, W. B. Schweizer and D. Seebach, *Helv. Chim. Acta*, **76**, 2654 (1993).
213. J. W. Faller and K. J. Chase, *Organometallics*, **13**, 989 (1994).
214. A. Tillack, M. Michalik, D. Fenske and H. Goesmann, *J. Organomet. Chem.*, **482**, 85 (1994).
215. Y. Goldberg and H. Alper, *Tetrahedron: Asymmetry*, **3**, 1055 (1992).
216. M. E. Wright and S. A. Svejda, *Polyhedron*, **10**, 1061 (1991).
217. M. E. Wright, S. A. Svejda, M. J. Jin and M. A. Peterson, *Organometallics*, **9**, 136 (1990).
218. C. Botteghi, A. Schionato, G. Chelucci, H. Brunner, A. Küerzinger and U. Obermann, *J. Organomet. Chem.*, **370**, 17 (1989).
219. A. Kinting, H. J. Kreuzfeld and H. P. Abicht, *J. Organomet. Chem.*, **370**, 343 (1989).
220. W. R. Cullen and E. B. Wickenheiser, *J. Organomet. Chem.*, **370**, 141 (1989).
221. A. F. M. Mokhlesur Rahman and S. B. Wild, *J. Mol. Catal.*, **39**, 155 (1987).
222. V. A. Pavlov, E. Y. Zhorov, A. A. Voloboev and E. I. Klabunovskii, *J. Mol. Catal.*, **59**, 119 (1990).
223. Y. Vannoorenberghe and G. Buono, *Tetrahedron Lett.*, **29**, 3235 (1988).
224. H. Brunner and K. Amberger, *J. Organomet. Chem.*, **417**, C63 (1991).
225. M. B. Carter, B. Schiøtt, A. Gutiérrez and S. L. Buchwald, *J. Am. Chem. Soc.*, **116**, 11667 (1994).
226. R. L. Halterman, T. M. Ramsey and Z. Chen, *J. Org. Chem.*, **59**, 2642 (1994).
227. G. Helmchen, A. Krotz, K.-T. Ganz and D. Hansen, *Synlett*, 257 (1991).
228. S. Murata, T. Sugimoto and S. Matsuura, *Heterocycles*, **26**, 763 (1987).
229. X. Verdaguer, U. E. W. Lange, M. T. Reding and S. L. Buchwald, *J. Am. Chem. Soc.*, **118**, 6784 (1996).
230. S. Murahashi, S. Watanabe and T. Shiota, *J. Chem. Soc., Chem. Commun.*, 725 (1994).
231. M. J. Burk and E. Feaster, *Tetrahedron Lett.*, **33**, 2099 (1992).
232. R. J. P. Corriu and J. J. E. Moreau, *J. Organometal. Chem.*, **85**, 19 (1975).
233. R. J. P. Corriu and J. J. E. Moreau, *J. Organometal. Chem.*, **64**, C51 (1974).
234. T. Hayashi, K. Yamamoto and M. Kumada, *Tetrahedron Lett.*, 331 (1974).
235. R. J. P. Corriu and J. J. E. Moreau, *J. Organometal. Chem.*, **91**, C27 (1975).
236. R. J. P. Corriu and J. J. E. Moreau, *Nouv. J. Chim.*, **1**, 71 (1977).
237. T. Ohta, M. Ito, A. Tsuneto and H. Takaya, *J. Chem. Soc., Chem. Commun.*, 2525 (1994).
238. Y. Uozumi and T. Hayashi, *J. Am. Chem. Soc.*, **113**, 9887 (1991).
239. T. Hayashi and Y. Uozumi, *Pure Appl. Chem.*, **64**, 1911 (1992).
240. Y. Uozumi, K. Kitayama, T. Hayashi, K. Yanagi and E. Fukuyo, *Bull. Chem. Soc. Jpn.*, **68**, 713 (1995).
241. Y. Uozumi, S. Y. Lee and T. Hayashi, *Tetrahedron Lett.*, **33**, 7185 (1992).
242. Y. Uozumi and T. Hayashi, *Tetrahedron Lett.*, **34**, 2335 (1993).
243. K. Kitayama, Y. Uozumi and T. Hayashi, *J. Chem. Soc., Chem. Commun.*, 1533 (1995).
244. Y. Uozumi, K. Kitayama and T. Hayashi, *Tetrahedron: Asymmetry*, **4**, 2419 (1993).
245. T. Okada, T. Morimoto and K. Achiwa, *Chem. Lett.*, 999 (1990).
246. T. Hayashi, Y. Matsumoto, I. Morikawa and Y. Ito, *Tetrahedron: Asymmetry*, **1**, 151 (1990).
247. H. Ohmura, H. Matsuhashi, M. Yanaka, M. Kuroboshi, T. Hiyama, Y. Hatanaka and K.-I. Goda, *J. Organomet. Chem.*, **499**, 167 (1995).
248. Y. Hatanaka, K. Goda, F. Yamashita and T. Hiyama, *Tetrahedron Lett.*, **35**, 7981 (1994).
249. K. Tamao, T. Tohma, N. Inui, O. Nakayama and Y. Ito, *Tetrahedron Lett.*, **31**, 7333 (1990).
250. S. H. Bergens, P. Noheda, J. Whelan and B. Bosnich, *J. Am. Chem. Soc.*, **114**, 2121 (1992).
251. R. W. Barnhart, X. Wang, P. Noheda, S. H. Bergens, J. Whelan and B. Bosnich, *Tetrahedron*, **50**, 4335 (1994).
252. X. Wang and B. Bosnich, *Organometallics*, **13**, 4131 (1994).
253. S. H. Bergens, P. Noheda, J. Whelan and B. Bosnich, *J. Am. Chem. Soc.*, **114**, 2128 (1992).
254. A. W. van der Made and P. W. N. M. van Leeuwen, *J. Chem. Soc., Chem. Commun.*, 1400 (1992).
255. C. Kim, D.-D. Sung, D.-I. Chung, E. Park and E. Kang, *J. Korean Chem. Soc.*, **39**, 789 (1995).
256. C. Kim, E. Park and E. Kang, *J. Korean Chem. Soc.*, **39**, 799 (1995).

257. B. Alonso, I. Cuadrado, M. Moran and J. Losada, *J. Chem. Soc., Chem. Commun.*, 2575 (1994).
258. K. Lorenz, R. Mülhaupt, H. Frey, U. Rapp and F. J. Mayer-Posner, *Macromolecules*, **28**, 6657 (1995).
259. H. Frey, K. Lorenz and R. Muelhaupt, *Macromol. Symp.*, **102**, 19 (1996).
260. J. Roovers, P. M. Toporowski and L.-L. Zhou, *Polym. Prepr., Am. Chem. Soc., Div. Polym. Chem.*, **33**, 182 (1992).
261. L.-L. Zhou and J. Roovers, *Macromolecules*, **26**, 963 (1993).
262. D. Seyferth, D. Y. Son, A. L. Rheingold and R. L. Ostrander, *Organometallics*, **13**, 2682 (1994).
263. D. Seyferth, T. Kugita, A. L. Rheingold and G. P. A. Yap, *Organometallics*, **14**, 5362 (1995).
264. S. Rubinsztajn, *J. Inorg. Organomet. Polym.*, **4**, 61 (1994).
265. J. Pan, W. W. Y. Lau and C. S. Lee, *J. Polym. Sci., Part A: Polym. Chem.*, **32**, 997 (1994).
266. J. V. Crivello, D. Bi and M. Fan, *J. Polym. Sci., Part A: Polym. Chem.*, **31**, 2563 (1993).
267. J. V. Crivello, D. Bi and M. Fan, *J. Polym. Sci., Part A: Polym. Chem.*, **31**, 2729 (1993).
268. J. V. Crivello and D. Bi, *J. Polym. Sci., Part A: Polym. Chem.*, **31**, 3109 (1993).
269. J. V. Crivello and D. Bi, *J. Polym. Sci., Part A: Polym. Chem.*, **31**, 3121 (1993).
270. J. V. Crivello, B. Yang and W.-G. Kim, *J. Polym. Sci., Part A: Polym. Chem.*, **33**, 2415 (1995).
271. J. V. Crivello and M. Fan, *Macromol. Symp.*, **77**, 413 (1994).
272. S. Itsuno, D. Chao and K. Ito, *J. Polym. Sci., Part A: Polym. Chem.*, **31**, 287 (1993).
273. P. R. Dvornic and V. V. Gerov, *Macromolecules*, **27**, 1068 (1994).
274. P. R. Dvornic, V. V. Gerov and M. N. Govedarica, *Macromolecules*, **27**, 7575 (1994).
275. C. M. Lewis and L. J. Mathias, *Polym. Prepr., Am. Chem. Soc., Div. Polym. Chem.*, **34**, 491 (1993).
276. W. M. Gibbons, R. P. Grasso, M. K. O'Brein, P. J. Shannon and S. T. Sun, *Appl. Phys. Lett.*, **64**, 2628 (1994).
277. W. M. Gibbons, R. P. Grasso, M. K. O'Brien, P. J. Shannon and S. T. Sun, *Macromolecules*, **27**, 771 (1994).
278. Y. Pang, S. Ijadi-Magsoodi and T. J. Barton, *Macromolecules*, **26**, 5671 (1993).
279. Y. Chen, L. Hong and X. Lu, *Youji Huaxue*, **13**, 260 (1993); *Chem. Abstr.*, **119**, 2712286d (1993).
280. H. S. W. Hu, J. R. Griffith, L. J. Buckley and A. W. Snow, *ACS Symp. Ser.*, **614**, 369 (1995).
281. L. M. Pratt, M. Waugaman and I. M. Khan, *Polym. Prepr., Am. Chem. Soc., Div. Polym. Chem.*, **36**, 263 (1995).
282. M. Tsumura, T. Iwahara and T. Hirose, *Polym. J.*, **27**, 1048 (1995).
283. C. X. Liao and W. P. Weber, *Macromolecules*, **26**, 563 (1993).
284. J. Lu and W. P. Weber, *Bull. Soc. Chim. Fr.*, **132**, 255 (1995).
285. Y. Goldberg and H. Alper, *J. Chem. Soc., Chem. Commun.*, 1209 (1994).
286. X. Coqueret and G. Wegner, *Makromol. Chem.*, **193**, 2929 (1992).
287. Y. Chujo and J. E. Mcgrath, *J. Macromol. Sci., Pure Appl. Chem.*, **A32**, 29 (1995).
288. B. Améduri, B. Boutevin, M. Nouiri and M. Talbi, *J. Fluorine Chem.*, **74**, 191 (1995).
289. B. Boutevin, F. Guida-Pietrasanta, A. Ratsimihety and G. Caporiccio, *J. Fluorine Chem.*, **68**, 71 (1994).
290. J. M. Yu, D. Teyssie and S. Boileau, *J. Polym. Sci., Part A: Polym. Chem.*, **31**, 2373 (1993).
291. G. Wang, H. Guo and W. P. Weber, *Polym Prepr., Am. Chem. Soc., Div. Polym. Chem.*, **36**, 523 (1995).
292. Y.-L. Hsiao and R. M. Waymouth, *J. Am. Chem. Soc.*, **116**, 9779 (1994).
293. M. W. Chen, C. X. Liao and W. P. Weber, *J. Inorg. Organomet. Polym.*, **3**, 241 (1993).
294. M. Ishikawa, E. Toyoda, T. Horio and A. Kunai, *Organometallics*, **13**, 26 (1994).
295. A. Kundai, E. Toyoda, I. Nagamoto, T. Horio and M. Ishidawa, *Organometallics*, **15**, 75 (1996).
296. Y. Chujo, E. Khara, S. Kure and T. Saegusa, *Macromolecules*, **26**, 5681 (1993).
297. R. West, M. Miller, H. Takahashi, T. Gunji and K. Oka, *Polym. Prepr., Am. Chem. Soc., Div. Polym. Chem.*, **34**, 227 (1993).
298. A. Iraqi, S. Seth, C. A. Vincent, D. J. Cole-Hamilton, M. D. Watkinson, I. M. Graham and D. Jeffrey, *J. Mater. Chem.*, **2**, 1057 (1992).
299. A. H. Gabor, E. A. Lehner, G. Mao, L. A. Schneggenburger and C. K. Ober, *Polym. Prepr., Am. Chem. Soc., Div. Polym. Chem.*, **33**, 136 (1992).
300. B. J. Kokko, *J. Appl. Polym. Sci.*, **47**, 1309 (1993).
301. H. A. Ketelson, M. A. Brook and R. H. Pelton, *Polym. Adv. Technol.*, **6**, 335 (1995).
302. Y. Seki, A. Hidaka, S. Murai and N. Sonoda, *Angew. Chem., Int. Ed. Engl.*, **16**, 174 (1977).
303. Y. Seki, S. Murai, A. Hidaka and N. Sonoda, *Angew. Chem., Int. Ed. Engl.*, **16**, 881 (1977).

304. Y. Seki, A. Hidaka, S. Makino, S. Murai and N. Sonoda, *J. Organomet. Chem.*, **140**, 361 (1977).
305. G. Süss-Fink and J. Reiner, *J. Mol. Catal.*, **16**, 231 (1982).
306. S. Murai and N. Sonoda, *Angew. Chem., Int. Ed. Engl.*, **18**, 837 (1979).
307. Y. Seki and K. Kawamoto *J. Organomet. Chem.*, **403**, 73 (1991).
308. T. Murai, E. Yasui, S. Kato, Y. Hatayama, S. Suzuki, Y. Yamasaki, N. Sonoda, H. Kurosawa, Y. Kawasaki and S. Murai, *J. Am. Chem. Soc.*, **111**, 7938 (1989).
309. N. Chatani, T. Ikeda, T. Sano, N. Sonoda, H. Kurosawa, Y. Kawasaki and S. Murai, *J. Org. Chem.*, **53**, 3387 (1988).
310. N. Chatani, Y. Kajikawa, H. Nishimura and S. Murai, *Organometallics*, **10**, 21 (1991).
311. S. Murai and N. Sonoda, *J. Mol. Catal.*, **41**, 197 (1987).
312. N. Chatani, S. Ikeda, K. Ohe and S. Murai, *J. Am. Chem. Soc.*, **114**, 3710 (1992).
313. S. Ikeda, N. Chatani, Y. Kajikawa, K. Ohe and S. Murai, *J. Org. Chem.*, **57**, 2 (1992).
314. Y. Seki, S. Murai, I. Yamamoto and N. Sonoda, *Angew. Chem., Int. Ed. Engl.*, **16**, 789 (1977).
315. Y. Fukumoto, N. Chatani and S. Murai, *J. Org. Chem.*, **58**, 4187 (1993).
316. N. Chatani and S. Murai, *Synlett*, 415 (1996).
317. S. Ikeda, N. Chatani and S. Murai, *Organometallics*, **11**, 3494 (1992).
318. M. E. Wright and B. B. Cochran, *J. Am. Chem. Soc.*, **115**, 2059 (1993).
319. M. E. Wright and B. B. Cochran, *Organometallics*, **15**, 317 (1996).
320. I. Matsuda, A. Ogiso, S. Sato and Y. Izumi, *J. Am. Chem. Soc.*, **111**, 2332 (1989).
321. I. Ojima, *22nd Organosilicon Symposium, April 7-8, Philadelphia, PA, U.S.A., Abstracts*, Plenary 7 (1989).
322. P. Ingallina, N. Clos and I. Ojima, *23rd Organosilicon Symposium, April 20-21, Midland, MI, U.S.A., Abstracts*, G2 (1990).
323. I. Ojima, P. Ingallina, R. J. Donovan and N. Clos, *Organometallics*, **10**, 38 (1991).
324. I. Ojima, R. J. Donovan, M. Eguchi, W. R. Shay, P. Ingallina, A. Korda and Q. Zeng, *Tetrahedron*, **49**, 5431 (1993).
325. M. P. Doyle and M. S. Shanklin, *Organometallics*, **12**, 11 (1993).
326. M. P. Doyle and M. S. Shanklin, *Organometallics*, **13**, 1081 (1994).
327. J.-Q. Zhou, and H. Alper, *Organometallics*, **13**, 1586 (1994).
328. I. Ojima, M. Tzamarioudaki and C.-Y. Tsai, *J. Am. Chem. Soc.*, **116**, 3643 (1994).
329. I. Matsuda, A. Ogiso and S. Sato, *J. Am. Chem. Soc.*, **112**, 6120 (1990).
330. I. Matsuda, J. Sakakibara and H. Nagashima, *Tetrahedron Lett.*, **32**, 7431 (1991).
331. I. Ojima, D. Machnik, R. J. Donovan and O. Mneimne, *Inorg. Chim. Acta*, **251**, 299 (1996).
332. I. Matsuda, J. Sakakibara, H. Inoue and H. Nagashima, *Tetrahedron Lett.*, **33**, 5799 (1992).
333. F. Monteil, I. Matsuda and H. Alper, *J. Am. Chem. Soc.*, **117**, 4419 (1995).
334. I. Ojima, E. Vidal, M. Tzamariondaki and I. Matsuda, *J. Am. Chem. Soc.*, **117**, 6797 (1995).
335. M. Eguchi, Q. Zeng, A. Korda and I. Ojima, *Tetrahedron Lett.*, **34**, 915 (1993).
336. I. Ojima, R. J. Donovan and W. R. Shay, *J. Am. Chem. Soc.*, **114**, 6580 (1992).
337. I. Ojima, M. Tzamarioudaki, Z. Li and R. J. Donovan, *Chem. Rev.*, **96**, 635 (1996).
338. I. Ojima, D. A. Fracchiolla, R. J. Donovan and P. Banerji, *J. Org. Chem.*, **59**, 7594 (1994).
339. I. Matsuda, H. Ishibashi and N. Li, *Tetrahedron Lett.*, **36**, 241 (1995).
340. I. Ojima, D. A. F. Kass and J. Zhu, *Organometallics*, **15**, 5191 (1996).
341. D. A. Fracchiolla, C. S. Takeuchi, and I. Ojima, *Eighth IUPAC Symposium on Organometallic Chemistry directed toward Organic Syntheses (OMCOS), August 6-10, Santa Barbara, U.S.A., Abstracts*, 248 (1995).
342. I. Ojima, J. V. McCullagh and W. R. Shay, *J. Organometal. Chem.*, **521**, 421 (1996).
343. J. V. McCullagh and I. Ojima, *212th American Chemical Society National Meeing, Orlando, Florida, August 25-29, 1996; Abstracts*, ORGN 0013 (1996).
344. J. V. McCullagh and I. Ojima, *10th International Symposium on Homogeneous Catalysis, August 11-16, Princeton, New Jersey, U.S.A., Abstracts*, PP-B43 (1996).
345. K. Tamao, K. Kobayashi and Y. Ito, *J. Am. Chem. Soc.*, **111**, 6478 (1989).
346. N. Chatani, Y. Fukumoto, T. Ida and S. Murai, *J. Am. Chem. Soc.*, **115**, 11614 (1993).